Advanced Concepts in Particle and Field Theory

Uniting the usually distinct areas of particle physics and quantum field theory, gravity and general relativity, this expansive and comprehensive textbook of fundamental and theoretical physics describes the quest to consolidate the basic building blocks of nature, by journeying through contemporary discoveries in the field, and analyzing elementary particles and their interactions.

Designed for advanced undergraduates and graduate students and abounding in worked examples and detailed derivations, as well as including historical anecdotes and philosophical and methodological perspectives, this textbook provides students with a unified understanding of all matter at the fundamental level. Topics range from gauge principles, particle decay and scattering cross-sections, the Higgs mechanism and mass generation, to spacetime geometries and supersymmetry. By combining historically separate areas of study and presenting them in a logically consistent manner, students will appreciate the underlying similarities and conceptual connections to be made in these fields.

This title, first published in 2016, has been reissued as an Open Access publication on Cambridge Core.

Tristan Hübsch is a Professor of Physics at Howard University, Washington DC, where he specializes in elementary particle physics, field theory and strings.

Advanced Concepts in Particle and Field Theory

TRISTAN HÜBSCH
Howard University, Washington DC, USA

CAMBRIDGE
UNIVERSITY PRESS

CAMBRIDGE
UNIVERSITY PRESS

Shaftesbury Road, Cambridge CB2 8EA, United Kingdom
One Liberty Plaza, 20th Floor, New York, NY 10006, USA
477 Williamstown Road, Port Melbourne, VIC 3207, Australia
314–321, 3rd Floor, Plot 3, Splendor Forum, Jasola District Centre, New Delhi – 110025, India
103 Penang Road, #05–06/07, Visioncrest Commercial, Singapore 238467

Cambridge University Press is part of Cambridge University Press & Assessment, a department of the University of Cambridge.

We share the University's mission to contribute to society through the pursuit of education, learning and research at the highest international levels of excellence

www.cambridge.org
Information on this title: www.cambridge.org/9781009291521

DOI: 10.1017/9781009291507

First published 2016
Reissued as OA 2022

A catalogue record for this publication is available from the British Library.

ISBN 978-1-009-29152-1 Hardback
ISBN 978-1-009-29151-4 Paperback

To the clouds that are fuzzy,
to the brooks that babble,
and to curiosity.

Contents

Preface

I think we may yet be able to [understand atoms].
But in the process we may have to learn what
the word "understanding" really means.
— Niels Bohr, cited by W. Heisenberg [267, p.41]

PHYSICS MAY BE DEFINED AS THE DISCIPLINE OF UNDERSTANDING NATURE. This definition is about as good as any other I can think of, although – or perhaps exactly because – much of the material in the following chapters is required even just to more precisely describe what it is we are to understand under *discipline*, *understanding* and *Nature*. That is, what is the nature of disciplining our understanding of something of which we ourselves are a part: Nature.

True to the meaning of the Greek original ($\varphi\acute{\upsilon}\sigma\iota\varsigma$), physics is indeed concerned with all aspects of Nature. Molecular phenomena are the objects of study in both chemistry and physics, which disciplines are separate but tightly related through quantum physics [477]. The science with which we study phenomena of continental proportions is called geology (but *areo*logy on Mars), whereas (planetary, stellar, galactic, cosmic) events that are at least a few orders of magnitude larger are labeled as astrophysics. Living things and events are the object of study in biology, but life itself and its characteristics quite probably derive from quantum physics [477]. Extending this point of view, phenomena of thought and feeling (commonly labeled as "psychology") may well be shown to be caused and determined by definite physical processes in the brain, so that social phenomena may be regarded as the "psychology of large ensembles of people," just as thermodynamics is the "mechanics of large ensembles of particles."[1]

Of course, a mere reduction of all phenomena to a common denominator achieves very little other than irking those who would rather keep up the appearance of separateness or those who insist on "irreducible wholeness." Hoping that this has nudged the Reader to think along (or against) such sweepingly unifying avenues of human understanding of Nature, let us turn to the real focus of this tome: to the *fundamental* physics of elementary particles.

Subject This book represents an attempt of a compact but comprehensive review of some of the key questions in contemporary *fundamental* physics, traditionally called both *elementary particle physics* and *high energy physics*. The correlation between these concepts is not at all accidental: The voyage towards an idealized but also pragmatically useful *fundamental* understanding of Nature really does lead through the world of ever smaller objects, the study of which requires ever larger energies in a complementary way.

The concept of "elementary particles" is in this sense a Democritean *ideal*, but it is also an evolving idea: On one hand, we follow this twenty-five-century hypothesis that the World around us may be understood as a complex system, ultimately consisting of certain basic and indivisible

[1] This paragraph was evidently meant provocatively; other fields of study do not *reduce* to physics, but emerge from it, and are "caused and determined" by it. Similarly, the babbling of a brook and the fluffiness of a cloud "derives" from hydrodynamics, but additional ideas from acoustics, nonlinear dynamics, turbulence, chaos, etc., are indispensable in a fuller (and still incomplete) understanding of these phenomena.

objects – *elementary particles*. On the other hand, the past two centuries of the history of science warn us that concrete things (and ideas) in Nature, which we at times identify as *elementary*, not infrequently later turn out to be themselves composed of *more elementary* things (and ideas). In this sense, the list of elementary particles was very short in the first third of the twentieth century. Everything in Nature was understood to consist of either the elementary particles (matter) the electron e^-, the proton p^+, the neutron n^0 and (hypothetically) the neutrino ν_e – or a form of their interaction, which could also be represented in terms of exchanging elementary particles such as the photon γ. Soon enough, however, hundreds of new particles were discovered. Already their unrelentingly growing number vanquished all hope that all these particles could *really* be elementary. Indeed, even the proton and the neutron were soon shown to be consistently describable as composite systems; they both consist of more elementary *quarks*.

To date, no experiment indicates revoking "elementariness" from quarks (u, d, c, s, t, b) and leptons $(e^-, \nu_e, \mu^-, \nu_\mu, \tau^-, \nu_\tau)$; see Table 2.3 on p. 67. Similarly, the electroweak, strong nuclear and gravitational fundamental interactions exhaustively describe all the known interactions of these particles. The model that includes these particles and their interactions is then rightfully called the *Standard Model*. As understood today, this model also requires the existence of the so-called Higgs particle, which has only recently been experimentally confirmed [25, 109]; see also [493, 494, 475]. Besides, the intricate structure and symmetries of the Standard Model also indicate a possible *more fundamental* description of physics.

Inspired tourist guides (see, for example, Refs. [329, 469, 162, 183, 184, 551, 404, 405, 585, 166, 267, 161, 34, 553, 164, 119, 163, 231, 263, 456, 232, 449, 233, 389, 505, 93, 234, 94, 27] but also a critique of superstring theory [489, 490, 577] and a recent response [145]), very recent lecture notes [525, 384, 539, 448, 427], textbooks (such as [407, 35, 64, 63, 306, 48, 106, 45, 218, 257, 238, 580, 241, 239, 307, 249, 240, 221, 554, 555, 159, 504, 422, 423, 538, 484, 250, 116, 588, 355, 243, 589, 7, 590] and worked out problem collections [107, 341, 446], among others) certainly provide excellent sources. In addition, internet sources such as Wikipedia are ever better organized and increasingly more complete – web-pages may be and are constantly corrected, amended and extended. No book can possibly compete with *that*. Instead, the aim of this book is fourfold:

1. a review of our subject matter and its central ideas, sharpened and re-focused by the benefit of hindsight,
2. a presentation of the structure of the theoretical description of fundamental physics and its origins within experimental results,
3. an indication of *some* recent additions in this structure, to some less traveled avenues, some shortcuts, some detours, and even some traps,
4. a general overview for novices as well as the more relaxed but valiant Readers.

This book makes it possible to present the Reader with the facts of the (fundamental) physics of elementary particles and their organization. This is accompanied by my view of the unifying philosophical *woof* that permeates not only this subject, but also the contemporary understanding of fundamental physics and science in general: Nature is one and can only so be understood. Since our goal is the description of the *fundamental* basics of understanding Nature, a discussion of this philosophical woof is an unavoidable part of the journey.

In turn, the intent of this review is to present the main factors in the challenging process of fully grokking[2] Nature. This intent stems from the Democritean idea that substance is finitely divisible, and that it has ultimately indivisible parts – elementary particles. This then provides

[2] To *fully grok* the meaning of the verb *to grok*, the Reader is kindly referred to Ref. [266], the title of which aptly summarizes the feelings of most sincere Students of the (fundamental) physics of elementary particles.

the *warp* (to the philosophical woof from the previous paragraph) of the fabric of contemporary theoretical fundamental physics. By the end of the twentieth century it became evident that these elementary objects cannot be the "material points" used in classical physics, and we are led to fundamental *strings*. Our discussion therefore must also include the questions: what are strings, where does "stringiness" manifest, why strings and not points or something else, and how are strings woven into our incessantly and asymptotically improved understanding of Nature?

Aperitif The gauge principle and its consequences constitute our contemporary description of all fundamental interactions, and form the third strand – *weft* – in the triply woven fabric of our current understanding of Nature [☞ lexicon entry on p. 508, in Appendix B.1]. Gauge theories of the commutative and non-commutative (Yang–Mills) type, the corresponding conservation laws and interactions are the subject matter of Chapters 5 and 6, but are also the quoin of the Standard Model from the very description of the subject matter. Formal similarities between the gauge theories (models) of Yang–Mills type and Einstein's general theory of relativity are exhibited in Chapter 9. This clearly implies that this (gauge) principle unifies *all* symmetries, *all* conserved quantities and conservation laws of the Noether–Gauss–Ampère type, and *all* known fundamental interactions. It also gives them all a geometrical description [☞ Chapter 11].

Similarly, *quantumness* is also an indubitable feature of Nature. Students are well acquainted with this, although mostly within the non-relativistic formalism. However, the study of *quantum* and *relativistic* gauge theories discovered the phenomenon of *anomalies* as well as the unquestionable necessity of canceling these indicators of inconsistency [☞ especially Section 7.2.3]. By including finally also the only known universal mechanism for stabilizing the vacuum – supersymmetry – we arrive at the complete picture displayed in Table P.1, the business card of understanding Nature as presented herein.

Table P.1 A telegraphic summary of the characteristics of our description of Nature; see Section 11.2

	Characteristic	Universal property	Unifies/describes
	Quantumness	Stabilizes atoms	Waves and particles
Gauge principle	Special relativity	Links symmetries, conservation laws, forces/interactions and geometry	Spacetime, energy–momentum
	General relativity		Acceleration–gravitation, mass–inertia
	Relativity of phases (of wave-functions)		(Electro-magneto) + weak, and strong interactions
	Supersymmetry[a]	Stabilizes vacuum	Bosons and fermions

[a] Supersymmetry is the only characteristic listed here that is not yet experimentally verified, but is the only (known!) universal characteristic the consequences of which include vacuum stabilization.

Organization This extended textbook is written for courses such as *Elementary Particle Physics* and *High Energy Physics*, and for Students near their undergraduate to graduate education transition. It aims to remind us that Nature is one; that the various courses the Student has so far mastered are only perforce separated parts of a whole, the reintegration of which into a coherent single vision remains with the Student.

The structure of this book largely follows that of the two-week block-course *Elementary Particle Physics*, as I have been teaching it annually since 2009 at the Department of Physics, University of Novi Sad. This was extended into a regular two-semester course taught in 2011/12 at the Department of Physics and Astronomy, Howard University. That plan started with D. Griffiths's textbook [243], but was iteratively and repeatedly modified, in response to student questions but also rooted in my own learning. A detailed map of all sources and their influences on this book is thus impossible; the unavoidably limited list of references and citations will, I hope, provide a reasonable collection of starters.

The book starts with an introductory chapter, numbered 1, where I summarize my philosophical and formal motivations for the study of the (*fundamental*) physics of elementary particles. Chapter 2 gives a historical review of the developments in this field of physics and so presents a rationale for the final structure, since the latter half of the twentieth century called the Standard Model. The technically (read: mathematically and predictively) detailed description of this subject begins with Chapter 3 and gradually introduces the elements of the Standard Model, through Chapters 4–7. Chapters 8, 9 and 10 give a basic introduction to the contemporary developments in this field and the research beyond the Standard Model. This leads towards a "unified theory of everything" for which the current favorite is described in Chapter 11. That chapter also summarizes the physical and philosophical sense of this subject, and the birth of a new subject in studying Nature: the study of complex systems and their emergent characteristics, complementing the Democritean idea. The appendices summarize various technical results and data that are useful in reading the main part of the book – and *working* through it.

The presentation and organization of the subject matter has a few formatting elements intended to help with the reading: Other than the main body, the book has a lexicon of less familiar terms and concepts in Appendix B, an index of main terms at the end of the book, as well as indicated digressions, conclusions and worked examples scattered throughout. The digressions (boxed) contain detailed computations and derivations that are not mandatory for following the main narrative. The impatient Reader is welcome to skip them on first reading. Similarly, the worked examples and comments (also boxed) serve to additionally illustrate and discuss the main narrative, and provide the derivations of results that are used later, but the mastering of which is not necessary for following through the main narrative. The in-line questions [✎ *so labeled]* prompt the Reader to pause, think through the presented argument and verify it. Frequent references and explicit citation of earlier results, conclusions, examples, etc., will hopefully help the Reader to find their way in the unavoidably multiply and nonlinearly connected presentation, and to find the information sought.

Research in contemporary fundamental physics is technically extraordinarily demanding: On one hand, the historical development and the very nature of *fundamental physics* indicate a synthesis of ideas and methods from many diverse areas in physics. On the other, one uses methods and results from many areas of mathematics such as the theory of groups and algebraic structures, differential geometry, topology, homological algebra, etc. A complete review of these areas is impossible within the confines of any one book, and the Reader is directed to the indicated references as well as earlier courses. The more ambitious Readers are directed to the textbooks [18, 457, 508, 62, 536, 287, 210, 565, 258, 581, 201, 256, 80, 260, 333, 447], with the *ominous warning*: it is impossible to first learn "all the necessary mathematics," and then turn to understanding physics. The mathematical language is best learned *en route*, as needed. In this, we are frequently limited to citing the needed results, presenting concrete examples and perhaps the motivation or basic idea behind the so-borrowed methods and techniques. In spite of this and for the sake of a minimal notion of completeness, this book contains more than enough material for a standard course, and choosing the route through the book is left to the instructor. In this, the diagram of dependencies in Figure P.1 should be useful.

Finally, most sections end with a list of problems. The serious Student is expected to work through these problems and solve them as completely as possible, first using only the material

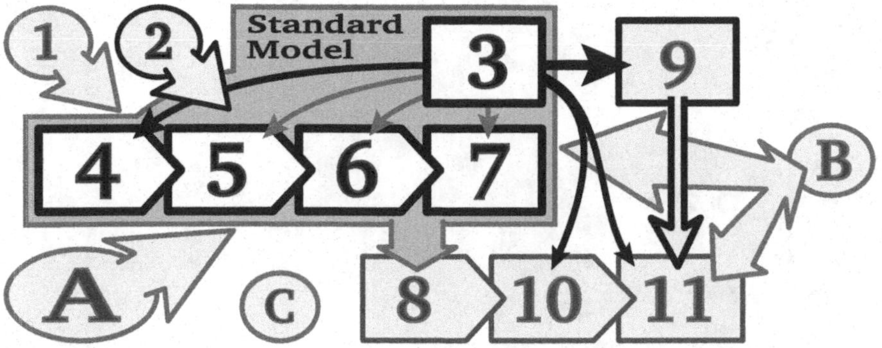

Figure P.1 The arrows indicate the dependencies between chapters and appendices, implying the recommended sequence of reading. The appearance of the boxes and chapter numbers indicates the relative significance of the chapters. The package-framed boxes indicate the minimal content, (**2**→)**3**→**4**→**5**→**6**→**7** (skipping the technically demanding sections), for a one-semester course "elementary particle physics."

presented herein, then comparing with the cited literature – and certainly before looking up a solution on the internet.

— ❦ —

I am grateful, first of all, to the Department of Physics at the University of Novi Sad, Serbia, where I have been lecturing annually on elementary particle physics since May 2009, and especially to my friend and colleague, Professor Miroslav Vesković (Department of Physics, University of Novi Sad), for making this possible. In turn, I am grateful to the Department of Physics and Astronomy at Howard University, Washington DC, for presenting the opportunity to translate this course a quarter of the Earth's circumference westward, and present this interactively modified version to the Washington DC metropolitan area students. This book would of course not exist were it not for my three decades of research in this field, and I am grateful to all my collaborators and colleagues for their uncountable corrections and comments, which have in so many ways shaped my understanding. Even just listing their names is prohibitive and I cannot but resort to a simple collective "thank you, all." I should like to thank all the Students and Colleagues who have attended my lectures and contributed to the evolution of the course – not the least of which by proofreading. In particular, Shawn Eastmond, Tehani Finch, Philip Kurian, Henry Lovelace, Sidi Maiga and Branislav Nikolić have contributed diligently to the current version. I can only hope for such a continued evolution of both the course and this book. Special thanks go to Prof. Darko Kapor (Department of Physics, University of Novi Sad), whose astute, critical and exacting reading of the first drafts provided the invaluable and inexorable impetus for persevering through the project and eventually completing it. Finally, I should also like to thank the staff at Cambridge University Press for their help in finalizing the project, and especially Ms. Patterson for her constructive proofreading.

All the remaining errors[3] are, however, entirely and solely mine.

Tristan Hübsch

[3] Errare divine est, aliter nos non simus.

Part I

Preliminaries

1

The nature of observing Nature

1.1 Fundamental physics as a natural science

The ultimate aim of this course is to present the contemporary attempt to perceive the (fundamental) structure and nature of Nature. First, however, we must examine the (methodo-)logical framework at the foundation of this aim.

1.1.1 Not infrequently, things are not as they seem

Although an erudite historian will certainly and readily cite earlier quotations of the thought expressed in the title of this section, I should like to introduce this *leitmotif* as a Copernican legacy. The readiness to abandon the "obvious," "generally accepted" and "common sense" for unusual insights – those we can actually check – is certainly an essential element. This motif permeates the development of our understanding of Nature, and reappears in its contemporary form as *duality* [☞ Section 11.4].

Of course, not just any unusual insight will do: a *luni*centric or an *iovi*centric system, for example, would offer no advantage over the *geo*centric cosmological system. Most significantly, heliocentricity simplifies both the conceptual structure and the practical application of the planetary system, and makes it more uniform. Although still assuming circular orbits and so in need of corrections,[1] Copernicus' model is *essentially* simpler; maybe this could be regarded as a variant of Ockham's principle.

This idea is not yet Newton's universal law of gravity, but already contains its germ, its unifying motif: all planets follow the same type of regular motion and only appear to wander randomly (as their original Greek name implies). Also, the ultimate test of this model is easily identifiable: the positions and the motions of the planets determined by (the simpler) computations within the heliocentric system agree with astronomical observations.

Examples of this leitmotif begin at such a simple level that they are rarely noticed:

1. The shadow of an object is often distorted and many times larger than the object itself. Nevertheless, only very little children are afraid of the shadow of a wolf or a monster, however aptly conjured by the artists in a puppet theater.

[1] Only after Kepler's *ad hoc* postulate of elliptical orbits (which Newton explained *a posteriori*) did heliocentricity achieve its really convincing technical simplicity and precision.

2. Viewed from a large plateau (without mountains on the horizon), the Earth does look flat. Yet, Eratosthenes (*c.* 276–195 BC) not only proved that the Earth was round, but even computed its size (to about 10–15% of the modern-day value!). This computation was based on the length of the summer solstice noon shadows in Syene (a.k.a. Aswan) and in Alexandria, the distance between these cities, and using geometry that is two millennia later regarded as elementary. In time, Eratosthenes' results and reasoning became "politically inconvenient," were suppressed and forgotten for some sixteen centuries, and were rediscovered in the West only centuries later, in the Renaissance. Although by now few people doubt that the Earth is round, when (if?) humankind expands into Space, the once obvious flatness of the Earth will become unthinkable; just as once its roundness was.

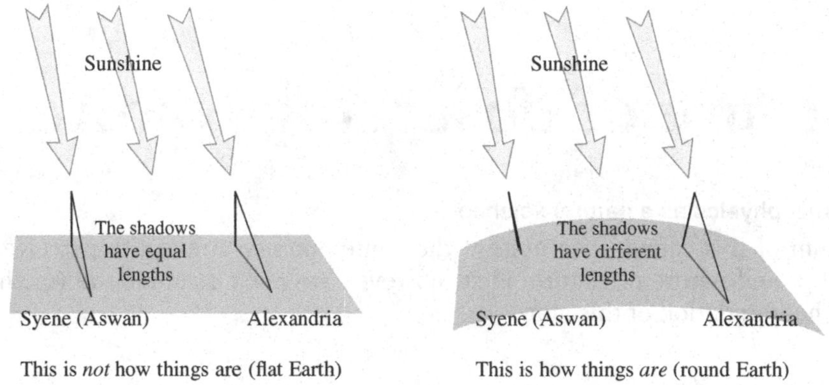

Figure 1.1 Eratosthenes' analysis which, by means of measuring angles and distances, gives (depending on the precise value of ancient units he would have used) the size of our planet Earth to about 16% at worst and 2% at best! (The shadows in the illustration are exaggerated.)

3. Everyday experience convinces us: the Sun and the Moon revolve around the Earth. This was indeed known to the ancient Greek science, as reported in Claudius Ptolemy's (*c.* AD 90–165) *Almagest*. This suppressed the teachings of Aristarchus (*c.* 310–230 BC), who not only advocated the heliocentric system, but also estimated that the Sun is about 20 times further away from Earth than the Moon and about 20 times bigger.[2] It took sixteen centuries for the West to rediscover this.
4. To the naked eye, our blood seems homogeneous and continuous. So it was believed to be until 1683, when the Royal Society published the first detailed pictures of red blood cells, as seen through a microscope and drawn by Antoni van Leeuwenhoek (1632–1723). In 1932, Ernst August Friedrich Ruska (1906–88) designed the first electronic microscope, the modern versions of which permit us to see – in the most direct way possible – individual molecules and even atoms, of which all matter around and within us is composed.

This last insight (quite literally!) is due to technical development, and it fully convinces us of the finite divisibility of *things* around us. Seemingly continuous things: fluids, air, metals... in fact consist of an enormous number of teensy particles! Whence stems the conviction that there exist "elementary particles" – the smallest building blocks of which everything else consists. Although

[2] The 20-fold error in Aristarchus' result stems from insufficient precision in angular measurements of the time; his reasoning and geometry were essentially correct. Also, the ratio of the diameters of the Sun and the Moon indeed does equal the corresponding ratio of their average distances from the Earth, but is ≈ 400, not 20.

this idea is fantastically successful in explaining Nature and even predicting its behavior, it behooves us to keep in mind that the "particulate nature" of Nature mirrors our gradually improving understanding of Nature, and that this insight is subject to verification and periodic audits.

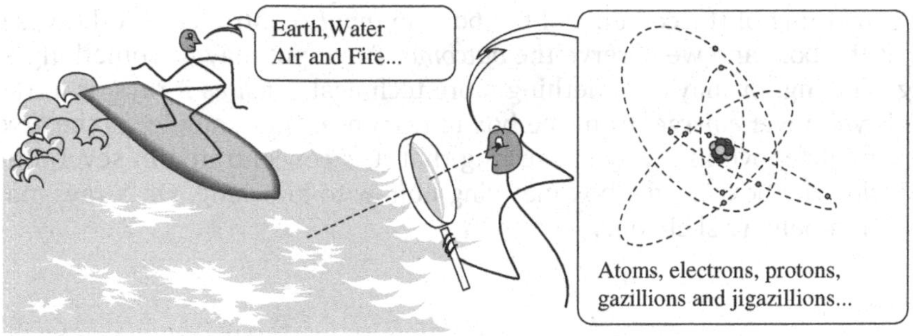

Figure 1.2 What at humanly characteristic scales seems smooth, homogeneous and continuous, may well look completely different under sufficiently closer scrutiny.

The Reader will certainly have no difficulty extending this list with many other and possibly more interesting and amusing examples, evidencing our basic leitmotif. Standard human perception, so well adapted to our daily routine, does not serve us well when concerning scales and proportions that are not as commonplace. From the typical, everyday vantage point and at characteristic human scales, planetary and stellar events appear warped. We must apply our (patiently educated and disciplined) mind to correct this picture. Indeed, once so educated, the Sun in the sky never again seems the same! In our mind's eye, we can actually *see* the Earth upon which we stand, as it rotates around the star we call the Sun. Similarly, once educated about the blood cells, our mind's eye has no difficulty *seeing* the erythrocytes as they stream through the blood plasma in our veins, and the leukocytes as they attack the blood-borne bacterial invaders.

Yesterday's unbelievable and ridiculed "nonsense" (that diseases are caused by germs too tiny to be visible was indeed widely ridiculed) may well turn out to become an evident truth of today – and such realizations turn out well remembered. So-called "evident" truths must not be exempt from verification just because they are considered evident: not infrequently, "evident" is simply that which is familiar and what are we used to. Not yet having doubted something is no guarantee of its truth.

However, we must then inquire which claims should we doubt and how do we establish the truth of any particular claim if everything is to be doubted? Following Descartes' rationale, *everything* that may be doubted without self-contradiction should be doubted. However, physicists are usually more pragmatic than that.[3] With a nod to the principle "if it ain't broke, don't fix it," physics models and theories are doubted and re-examined when they start predicting things that *are not*, or fail to predict the things that *are*... And, predictions are derived from a model as much as technically and practically possible.

In fact, it is our *duty* to "churn out" everything one possibly can from every scientific model. This is both for the sake of economy (the predictions of a model are its "products") and in order to establish if the model is in as full an agreement with Nature as it is possible to determine at any given time.

[3] ... and even without the persnickety conclusion that Descartes' motto *cogito, ergo sum* leads into solipsism, or recalling Hume's demonstration just how destructive such infinitely regressive doubting may be...

1.1.2 The black box: a template of learning

To formalize our approach, let us picture the scrutinized system as a *black box*, representing the lack of knowledge about its contents. What follows may then be regarded as the three pillars of (exact, natural) science.

I. To learn something of the contents of the box, an *input* (controlled or otherwise known) is directed at the box, and we observe the *outcome*. The input may be something as simple as knocking, shaking, or maybe something more technical, such as X-rays or ultrasound. The outcome is whatever *emerges* from the box in response. For example, as the box is shaken, its weight might move in a way suggesting that it is concentrated in several distinct subsystems inside the box. Or, the box may ring hollow to knocking. Or, X-rays may show the image of Thumbelina's skeleton...

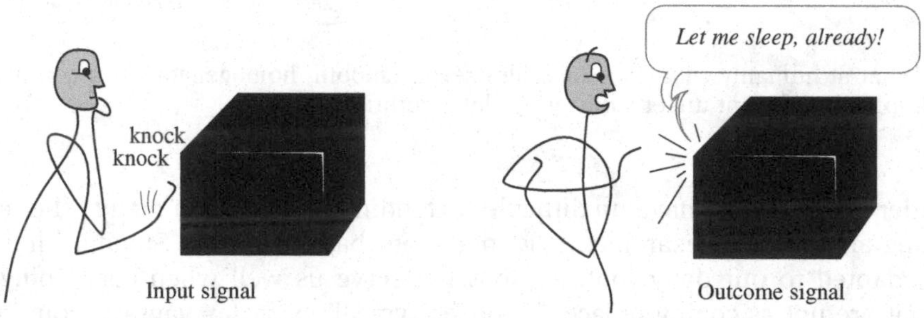

Figure 1.3 The black box experiment template.

II. Using the information about the box in the form of a "response to the input," where both input and outcome are adequately *quantified*, we develop a *mathematical model* that faithfully reproduces all received outcome signals as a response to the corresponding input signals. Needless to say, both input and outcome signals must be measured, and will therefore be known only up to measurement errors. This defines the resolution/precision/tolerance of the model. Of course, a resolution of the mathematical model cannot be guaranteed to be better than this; and this must then be understood as the resolution of the model as a whole.

III. This mathematical model is then used to derive the consequences of the conceptual model: One computes the response of the system (as represented by our model, in the role of the black box) to new, as yet untested input signals. These responses then need to be tested, if and when that becomes possible.

Herein then lies the clue as to "what and when to doubt" and "how to test truthfulness." Physics (and, more generally, scientific) models must be re-verified, wherein one or more of the "ingredients" are doubted and perhaps even replaced, if the model does not reproduce and correctly correspond to Nature to within the resolution of the model [☞ also Comment 10.5 on p. 388]. This shows that:

Conclusion 1.1 *Exact science always errs, but is exact about how much.*

Comment 1.1 *"Physics students learn this very quickly, through a shock, when they proudly obtain the required results of the first lab exercise, and the teaching assistants quiz them about the errors at least as much as about the obtained results." D. Kapor*

This three-step process, "observe–model[4]–predict," repeats iteratively and infinitely, in counterpoint to the above-cited leitmotif, and guaranteed by Gödel's incompleteness theorem,[5] since the research subject is sufficiently complex and is not easily exhaustible (unlike, e.g., "tic-tac-toe," which is exhaustible as a game of strategy) [211]; see the lexicon entry in Appendix B.1, as well as Appendix B.3. When the model is constructed, the predictions of the model are derived and checked experimentally, as well as possible in practice. As human ingenuity incessantly improves the technology, and new techniques and methods (both experimental and theoretical) are being continually developed, new predictions are being continually derived and checked with an increasing precision. Sooner or later, these new checks (both experimental and theoretical) indicate the shortcomings and uncover statements that can be neither proven nor disproven within the given theoretical system.[6] If such a new statement can be experimentally checked as true or false, the model needs to be extended so as to include this new fact about Nature. When the so-extended model successfully reproduces all (known) "new" facts, additional predictions of the now extended model are derived and checked, and these typically indicate further directions of extension and improvement, upon which yet more additional predictions may be derived, and so on.

Comment 1.2 *To illustrate, the phenomena we now label as* ***electrodynamics*** *are describable by equations that are easily written down within the theoretical system of classical mechanics of particles and fields, but can be neither proven nor disproven within this system. The Maxwell equations (5.72) and the electrodynamics laws that they represent, provide* ***new axioms*** *to the theoretical system of the classical mechanics of particles and fields applied to charged particles and electromagnetic fields. In turn, Section 5.1 shows these equations to follow from the gauge principle, which therefore is the one (overarching)* ***new axiom****; see also Appendix B.3.*

1.1.3 Philosophers are not scientists

A second glance at this framework of thought reveals something extraordinary! The scientific models[7] described here, and systems of such models forming theories and theoretical systems, are improved and extended, but not *literally falsified*, i.e., proven to be unconditionally false! (For the most part, it is rather our mental imagery and philosophical "underpinnings" of the scientific model that are taken too seriously, and may have to be abandoned as false.) Radical revisions of course do occur in scientific research – and not so infrequently – but that does not *falsify* established models and theories, only perhaps an unwarranted trust that those models and theories would be exact and absolute truths. Properly understood within their qualifications, models and theories of fundamental physics have not been falsified throughout the past three centuries, but have been and continue to be refined, extended and often united.

Reasons for this are found in comparing scientific models with earlier efforts and doctrines. Scientific models unify the inspiration of (experimental) induction with the rigor, self-consistency and persistence of (rigorous mathematical and logical) deduction.

---

[4] In this context, the verb "to model" encompasses the creation of the mathematical model that describes the scrutinized phenomenon, and that can be summarized into an applicable formula. Whence stems the *law* for the system wherein the phenomenon is observed, and "to model" then includes "to introduce as a law of Nature." However, this is not an absolute and inviolable law by decree, but one that is subject to verifications in comparisons with Nature, and adaptations to this one and ultimate arbiter.

[5] ... barring the dismal logical possibility of the scientific spirit dying out or becoming exterminated...

[6] These are essentially undecideable statements; see the lexicon entry on Gödel's incompleteness theorem, in Appendix B.1, and Appendix B.3 in particular.

[7] A *scientific model* includes the mathematical model together with its concrete interpretation: formulae, algorithms, programs, together with their physical meaning, i.e., a dictionary between the symbols of the mathematical model and the corresponding quantities in Nature. In this sense, a "model" then also implies a "law" – in the sense of Newton's, Ampère's or Gauss's law, not in the sense of a decree of some legislative body. The notion of "natural law" is thus integrally woven into the scientific modeling of Nature, far from it having been abandoned, as sometimes opined [533].

This complementary combination of *quantitative measurements* and *mathematical modeling* is often attributed to the revolution in the philosophical approach in studying Nature, and is most often linked to Galileo and Newton. However, Eratosthenes' and Aristarchus' above-cited planetological results were clearly based on this same combination of methods. This idea is therefore over two millennia old. Suppressed through most of the past two millennia, this same combination of measurements and mathematics was methodically and consistently revived by Galileo, Newton and their followers. With the development of mathematics – and especially of calculus, invented for that purpose by Newton, Leibniz and contemporaries[8] – physics engaged into *warp* drive (the superluminal propulsion from the sci-fi series *Star Trek*).

Roughly, measurements translate quantities describing observed natural phenomena into corresponding quantities in a mathematical model. This model is then used as a faithful (as best as known) representative and replacement of the natural phenomenon. It is also a persistently rigorous tool for deductive predictions about that natural phenomenon. Those predictions are then checked in turn, the model adapted, corrected and improved, if and when the predictions turn out to differ from what is observed in Nature.

Thus, Einstein's theory of relativity does not *falsify* Newton's mechanics but *extends* it: When all relative speeds in a system are much less than the speed of light in vacuum, relativistic corrections to Newton's mechanics are negligible and Newton's mechanics yields a perfectly usable model of reality. If some of the relative speeds increase, the corresponding corrections become relevant, Newton's mechanics is no longer a good enough approximation (the errors, about which physics always must be precise, become unacceptably large), and we must use the relativistic formulae. In turn, Einstein's relativistic physics cannot be claimed to be absolutely true/exact either, but merely that it is more accurate than Newton's. After all, we already know that quantum physics may well force us to revise the structure (and perhaps even the nature) of spacetime itself when approaching Planck-length scales. Science can only make qualified statements, the "truth" of which will always depend on precision (resolution) – and which continues to improve in ways that no one can foresee.

Insisting on the iteration of this precision-sensitive "observe–model–predict" cycle immediately discards "theories" such as the one about *phlogiston*, the supposed intangible substance of heat. That "theory" neither explained nor predicted quantitative data, and may be called a "theory" only in common, non-technical parlance. A similar fate befell the so-called "plum pudding" model of the atom, which explained and predicted very little (and incorrectly), and which its Author humbly called a "model" worth exploring, and mercilessly abandoning if found faulty; which it was – both faulty and abandoned.

It is absolutely crucial that what we intend to call a *scientific* theory must be subject to verification through comparison with Nature, at least in principle. This implies that a theory must be *quantitative*, i.e., a theory must explain and predict experimental data, which can be checked. Quantitative predictions may be as simple as "yes/no" results; whether one predicts a single bit of information or an entire googolplex[9] of them – predictions must contain *new* information.

A word of warning: "subject to testing" does not mean that we can simply call up the local lab, order some results, and expect a twenty-minute delivery. Nor does it mean that even a planetary budget could fund the required experiment (not that there will be a planetary budget any time soon). Nor does it mean that anyone has even the faintest hint of an idea for a concrete experiment, even with a pan-galactic budget and a post-*Star Trek* technology. However, the theory must be

[8] It has recently been discovered that Archimedes knew about the concepts of limit and the principle of exhaustion [382], but that this knowledge has been neglected and forgotten for the better part of two millennia.

[9] *Googol* (which must not be confused with *Google*) is the number 10^{100}; *googolplex* is the number $10^{10^{100}}$. For comparison, there are only about $\mathbf{N} := 10^{80} \ll 10^{100}$ particles in the universe, but the number of all their k-fold relations is immensely larger than googol, $\sum_k \binom{\mathbf{N}}{k} = 2^{\mathbf{N}} \ggg 10^{100}$, and the number of all relationships between those relationships (as a second-order estimate of complexity) is much larger than googolplex, $2^{2^{\mathbf{N}}} \ggg 10^{10^{100}}$.

"subject to testing, in principle": thought experiments may be envisioned rigorously, and their execution is obstructed by neither political economy nor practical "minutiae" such as magnetizing a mountain-size apparatus. Of course, the models that may be tested may be either demonstrated as *tentatively* established,[10] or discarded if they can be shown to disagree with Nature.

It cannot be over-emphasized (see, however, also Digression 1.1 below, as well as Sections 8.3.1 and 11.1.4 and Appendix B.3):

Conclusion 1.2 *Models that can (in principle) be refuted are scientific.*

Interestingly, a verb (in Chinese) is, by definition, a word that can be negated [578]. However, the correct application of this criterion, so simply stated, supposes a detailed understanding of the structure of scientific systems, to which we return in Section 8.3.

Digression 1.1 The principle of Conclusion 1.2 reminds us of the principle of falsifiability, popularized by Karl Popper [443, 444]. Intending to describe the historical process of the evolution of science, he concluded that experiments about atoms *falsify* classical physics, which is then *substituted* by quantum physics since that successfully describes atoms. So understood, the *principle of falsifiability* harbors at least two equivocations: (1) the naive version equates it with the related "testability" and presupposes direct and unequivocal experimental testing, and (2) equivocation in categories. Both equivocations are dangerous to the socio-political status of science. Also, the tacit assumption that all statements of a model are necessarily either confirmable or falsifiable, which simply need not be the case [☞ the lexicon entry on Gödel's incompleteness theorem, in Appendix B.1, and also Appendix B.3].

The first equivocation is based on a restriction of physics as a science to a "directly empirical" science, whereby a theory that we cannot experimentally test is being denied its "scientificity." However, there exist (in the scientific and the sci-fi literature and media) effects that contradict no known science, but for the experimental testing of which [☞ also Refs. [171, 505]]:

1. the resources are too expensive (e.g., a synchrotron around the Earth or around the Sun and Proxima Centauri, not to dream of a tokamak from here to Andromeda),
2. the requisite procedures are prohibited by moral or ethical reasons (e.g., cloning, bionic, and certain educational, behavioral and nutritional experimentation),
3. the resources require an as yet unknown technology (e.g., painting the ceiling of a room with neutronium would cancel gravity in the room – if "only" we knew how to produce neutronium paint and how to paint the ceiling without it caving in),
4. a new concept and/or methodology is needed (e.g., for a direct measurement of an *upper* limit of the proton's lifetime).

It is already intuitively clear that not one of these obstructions for experimental testing should take away from the "scientificity" of a theory. And, even simpler, it is clear that experiments with stars, positions of the constellations and the development of our own universe cannot be performed at will, nor is setting up an experimental control group

---

[10] Being forever subject to future and additional testing, "established" can in this context only ever be understood as *tentative*; this is a "small" detail that is rarely stated explicitly, but must always be understood.

possible even in principle! Nevertheless, it is just as clear that astrophysics, astronomy and cosmology are no less "scientific" for this.

— ❦ —

The second possible equivocation is more subtle, and so also more dangerous. Also, it has at least two aspects. On one hand, there is the danger of confusing the category to which a certain *theoretical structure* belongs. For example, "classical physics" is not a particular model with particular predictions that may be experimentally tested, but a scientific system of assumptions (axioms) and procedures of derivation; this then may be applied to concrete phenomena, such as a pendulum, a bob on a spring, or the atom. The incorrectness of any one concrete model – as in the case of the classical model of the atom (see however Footnote *11* on p. 310 as well as example B.2) – may imply an error in the *application* of classical mechanics or in classical mechanics *itself*, or perhaps even elsewhere in the underlying complete chain of reasoning. We must explore precisely which of the assumptions lead to the observed disagreement with Nature. In fact, the application itself may turn out to harbor an error for various reasons, from a minor technicality to a fundamental inappropriateness. That, after all, is the usual advisory about all proofs by contradiction. However, it would evidently be silly to deny the "scientificity" of classical physics as a whole because of its inability to model the atom.

On the other hand, the very idea that a scientific theory *falsifies* another is a dangerous equivocation. Both in common parlance and in legal practice, the verb "to falsify" implies that the statement being falsified is being shown to be a falsehood. This, in turn, implies the tacit expectation of a binary true/false value. However, it is – or should be – very well known that the relation between quantum and classical physics is *continuous* and depends on the context and "resolution." For any process under scrutiny, we must compute the ratio of $\hbar$ with all characteristic actions and all other commensurate physical quantities.[11] If each of these ratios is *sufficiently smaller* than 1, the numerical errors in the results computed using classical physics are *negligible*. It is evident that "sufficiently small" here implies a finite and an a-priori established tolerance. Therefore, the answer to questions such as "is classical physics applicable even to a single particular event?" essentially depends on at least one *continuous* parameter, and the answer cannot possibly be an unconditional "yes/no." Classical physics is therefore extended/generalized and not *falsified* by quantum physics. The situations with relativity, field theory, and even superstring theory are analogous.

Generally, physicists understand that quantum physics does not simply *falsify* classical physics, but *extends* it into a domain where classical physics is *not sufficiently precise*. Unfortunately, philosophers of science are not physicists. This pragmatic approach should be compared with a similar vantage point of philosophers of natural sciences such as Thomas S. Kuhn [323], where one needs to know that Kuhn obtained his BS (1943), MS (1946) and PhD (1949) degrees in *physics* at Harvard, where he lectured on history of physics 1948–56. However, Kuhn opines that theories (and paradigms) are chosen by the group of researchers that is more successful than others, and assigns this choice a degree of socio-politically pliable subjectivity. This seems all too alien to most physicists I know,

---

[11] In elementary particle physics one uses so-called natural units, based on the natural constants $\hbar$ and c, whereupon these are not written explicitly, and formally one says that "$\hbar = 1 = c$." This practice may well be used in any complete unit system: once in agreement to use SI units, "length of 10" may only mean "10 m," "force of 5" may only mean "5 N=5 kg m/s^2," etc. However, as the purpose of this book is to *introduce* the Reader into this practice, factors of $\hbar$ and c are herein written explicitly, but in gray ink.

and which I myself (perhaps all too naively) cannot accept for physics itself, nor any other science, but only and at most for the admittedly capricious socio-historical process of development in a particular subfield.

Suffice it here then to just assert without a historically and statistically justified argument – as a manifesto, if need be: Theories and theoretical systems in physics are chosen by Nature itself, through our long and patient communication with it (in the sense of the caricature in Figure 1.3 on p. 6). Albeit extremely challenging and difficult at times, this is always well worth the effort and ardor.

Of course, it is logically impossible for a science to be exact without being quantitative. That is, "exactness" must be accepted as the requirement that it must be possible to develop a system of questions that can be answered by precise yes/no answers. Subsequently, these answers (easily written as a binary number) may quantitatively characterize the events to be modeled – and to be predicted. If these yes/no answers follow a statistical (probabilistic) distribution, this is only a *technical* complication and does not take away from the "exactness" in this sense.[12] This is always true of all branches of physics. While statistical physics and quantum mechanics are probabilistic, this only complicates the techniques and dictates the style of research. In fact, many fine (mathematical) techniques of statistics specify precisely which questions are meaningful to ask, which are meaningless, and which among the former ones have a definitive answer, which "only" a probabilistic one. For example, the temperature (as the average kinetic energy) of a fluid may be predicted precisely, but the kinetic energy of a single molecule is subject to fluctuations; the distribution of these fluctuations is predictable precisely, but not their individual values in practice, owing to the too large number of contributing factors, such as the repeated collisions with 10^{26}'s of molecules. The kinetic energy of a single molecule may in practice then only be known probabilistically – even if it were possible to mark and follow a single molecule without disturbing and changing it.

From this point of view, physics and science in general may be accused of being pragmatic, which they indeed are to a considerable degree. However, it is pragmatic physics and science that brought us Moon rocks, pictures of the surface of Jupiter's satellites and of distant galaxies and nebulae, and which can find extrasolar planets; that produce artificial heart valves that the human immune system accepts; that can provide early signs of hurricanes, cyclones and tsunamis so as to warn the endangered population. Unfortunately, ethically and morally wrong, and just plain uninformed application of science may also lead to our planet radioactively glowing in the dark of the universe, or "only" to lose all ice and heat up to a point where life as we know it is no longer possible. Through this feedback, science also affects our thinking, our opinions and convictions, and so influences almost everything else, thereby being far more than "just pragma."

The foregoing also uncovers the price to pay: although physics is about Nature, it describes Nature indirectly, by way of the models that are sufficiently (and ever more) precise.

Example 1.1 The statement "in Rutherford's planetary model, the atom consists of a nucleus at the center and the electrons orbiting around it" does not mean that atoms literally exist in such simple tinkertoy form. More precisely, one means that the mathematical model developed from this *mental caricature* nevertheless reproduces the so far addressed real observations with sufficient accuracy.

[12] Let's recall: exact science always errs, but knows precisely how much [☞ Conclusion 1.1 on p. 6].

In fact, the stability of atoms requires that Rutherford's planetary model is amended by additional "quantization" rules, which in turn lead to Bohr's model and the "old quantum mechanics." Subsequent observations caused a further development of this and subsequently developed models, leading through "quantum mechanics" to "quantum field theory," and even to "quantum theory of (super)strings." In this development, each stage completely contains the previous. Of course, it must be admitted that the current favorite for the *fundamental theory* – (super)string theory – is (by far) not confirmed experimentally as a theory of Nature: It has not even been shown that (super)string theory really can reproduce all the details of the "real World" as known so far, but it is the first one for which no indication to the contrary is currently known.

To be fair, (super)string theory is not one concrete theory but a theoretical system, just as classical mechanics is not limited to the description of a concrete physical system, but is a systematic approach to describing a class of concrete physical systems. The surprising abundance of unexplored possibilities in the theoretical system of (super)strings and the fact that no contraindication has been found, together provide hope that amongst the (super)string models an optimal candidate for the so-called Theory of Everything [☞ Chapter 11.5] can be found – with the requisite warning: this will continue to require a lot of hard work.

1.1.4 Scientific predictions: useful and inevitable

We are already acquainted with the three-step, observe–model–predict, as well as the logical inevitability of repeating this iterative cycle infinitely while developing, testing and asymptotically improving scientific models. Indeed, we may regard it both as a curse and as a boon that the idea of an *ultimate and complete* Theory of Everything is a vanishing point: a theory to which all scientific endeavors aim, asymptotically [☞ Section 11.5].

Apart from this asymptotic (un)reachability, Theory of Everything is also a misnomer, since it refers only to fundamental interactions in Nature. However, neither does a pile of rocks (and other building blocks) make a palace, nor does a few pounds of protein, lipids, some fat and calcium make a Schrödinger's cat. Even if the "ultimate" theory of all fundamental interactions were known, a hazy road would remain to lead from there, through atoms and molecules, to... us, our ambitious thoughts, and beyond.

Not only is there much room for "filling in the details" even if we stick to this 1-dimensional arrangement by size, but very often tiny portions in this all-encompassing size scale produce "pockets of knowledge" of fantastic and baffling complexity, a characteristic perhaps not unlike fractals. Suffice it here just to mention that the complexity of collective phenomena (behaviors beyond the "thermodynamic" average, such as eddies, tornadoes, the shapes and the dynamics of clouds, market crashes...) has only recently been subject to serious scientific thought. Also, life as we know it – and so biology – occupies merely a few orders of magnitude, roughly between 10^{-6} and

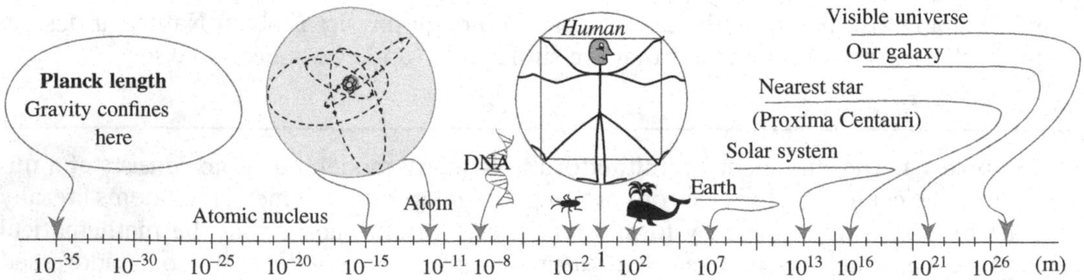

Figure 1.4 A logarithmic scale of sizes, from the Planck length, where everything looks like a black hole (from within which no information can be extracted), to the largest distances, from which the light only now reaches us (and from beyond which information has not yet reached us).

10^2 m; chemistry occupies an even smaller niche around 10^{-9} m. Their complexity and richness are, however, obvious to every Student who has taken exams in these subjects.

Heeding the adage "when eating an elephant, take a bite at a time," physics *analyzes* natural phenomena (systems), identifying their sub-processes (sub-systems). These are usually more easily grasped and understood, upon which it however remains to (re-)integrate them. Along this journey, certain characteristics of the whole are recognized simply as a sum of these parts, while others are identified as intrinsically "collective" – unexplained by the characteristics of the integral parts and inextricably rooted in the complexity of the whole rather than the nature of the constituents. Whereas the analysis of the "parts" says little about the collective phenomena, it certainly permits a better specification and discussion of the properties that are *not* collective, thereby leaving a clearer path towards this complementary front in understanding Nature.

Following this "glory road" of scientific discovery, it is important to realize that:

> ***Claim 1.1*** *Whatever follows logically from the assumed axioms/postulates of a model is a prediction of the model.*

That is, if a model reproduces perfectly the original input/output data, and produces a number of testable predictions even just one of which turns out to disagree with Nature, there is something wrong about the model. It may happen that its minor modification will both fix the glitch and retain the model's fidelity otherwise; if so, this modification must be incorporated as an integral element of the (revised) model, subject of course to any further test that can be conjured. If such a revision or extension cannot be found – off with its head.[13]

All predictions are derived as unavoidable consequences of the given assumptions, and are ensured by the rigor of mathematical deduction. Those very consequences and predictions are sometimes precisely the goal of developing the model; at other times, those are byproducts or afterthoughts. Once in a while, however, they are quite spectacularly unexpected discoveries:

> *The Heitler–London bond is a unique, singular feature of the [quantum] theory, not invented for the purpose of explaining the chemical bond. It comes in quite by itself, in a highly interesting and puzzling manner, being forced upon us by entirely different considerations.* Erwin Schrödinger [477]

The Heitler–London bond is one of the basic "ingredients" in modern chemistry, and we may rightly understand chemistry as based upon quantum statistical physics. Similarly spectacular was Dirac's prediction of the anti-electron (positron), and with it – as a logical consequence – an antiparticle for each other type of matter particle, as well as Pauli's prediction of the particle that Fermi named "neutrino," and which was confirmed experimentally only two decades later!

1.2 Measurement units and dimensional analysis

An exhaustive and detailed description of most real-life physical phenomena is often complex and requires technically demanding computations. However, satisfactory estimates may often be obtained by comparison with a similar and well-known case, by means of "dimensional analysis."

The next few examples in Section 1.2.1 illustrate this practice, using straightforward similarity in the form of simple proportions (scaling) of physical dimensions. Sections 1.2.3 through 1.2.5 show that dimensional analysis and a few other general properties of the quantities that interest

[13] Fortunately, and unlike their human inventors, theories can be resurrected, and this happens on occasion. The glitch that had formerly killed off a model may turn out to be "repairable" at a later time, when a better understanding of the model and requisite techniques and methods of analysis are attained [☞ Section 11.1, for example]. In turn, it can (and does) also happen that the experiments were carried out or analyzed in error, and this is revealed only much later. The corrected analysis may well turn out to agree with what was formerly thought of as a glitch of the model.

us importantly limit the possible answers, and may well at times even fully determine the form of those answers.

1.2.1 Lilliputians

In Jonathan Swift's *Gulliver's Travels*, Lemuel Gulliver meets the people of Lilliput, who are identical to humans, but smaller: their average height is 45 mm. In other words, Lilliputians are about $\lambda = 40$ times smaller copies of ordinary humans.

1. **How much weight (in units of their own weight) can a Lilliputian lift?**
   (a) The weight of the burden is determined by the force available for lifting it, and force is proportional to the area of the muscle cross-section. As the Lilliputian is built just like a normal human, the cross-section of his muscle is $\lambda^2 = 1{,}600$ times smaller.
   (b) The weight of the Lilliputian himself is $mg = \rho V g$, where g is the gravitational *constant*, ρ the density which equals that of ordinary humans, and V is the volume. Since the Lilliputian is 40 times shorter, his volume is $\lambda^3 = 64{,}000$ times smaller than in ordinary humans, so that his weight is also 64,000 times smaller.

   It follows that a Lilliputian can lift a $1/1{,}600$ of the burden an ordinary human can, which is however $\lambda^{-2}/\lambda^{-3} = \lambda = 40$ times more – in units of his own (64,000 times smaller) weight – than an ordinary human. *Proportionally*, Lilliputians are $\lambda = 40$ times stronger than ordinary humans: if an ordinary human can lift his own weight, a Lilliputian can lift 40 times that much!
2. **How fast does the heart of a Lilliputian beat?**
   The frequency with which the heart pumps blood is determined by the quantity of blood it moves in a single beat, and by the quantity of blood needed to circulate in a unit of time. In warm-blooded beings, one of the main functions of blood circulation is to maintain the temperature and life in tissues by carrying oxygen. (When the circulation fails, tissues die and cool.)
   (a) The volume of a Lilliputian's heart is $\lambda^3 = 64{,}000$ times smaller than the volume in an ordinary human. The same holds for the volume of blood that the heart pumps in a single beat.
   (b) The body cools through the surface of the skin, which is $\lambda^2 = 1{,}600$ times smaller in a Lilliputian than in an ordinary human.

   It follows that a Lilliputian's heart must beat $\lambda^{-2}/\lambda^{-3} = \lambda = 40$ times faster than that in an ordinary human, so it would achieve the circulation of λ^{-2} times smaller volume with its (λ^{-3} times smaller) "pumping units." That is about $40\times70 = 2{,}800$ times per minute, or about 46.67 times every second. That is the tone of 46.67 Hz frequency, very near the second "F-sharp" from left, on a standard piano: A Lilliputian's heart thus *hums* – a little deeper than the humming of an AC/DC adapter. The skin of most small warm-blooded animals is covered by fur to reduce the heat loss, among other things, also so that their heart will not have to beat so fast.
3. **How high can a Lilliputian jump?**
   The jump-height is determined by the energy available for lifting the body: Energy E lifts a body of mass m to a height of $h = E/mg$.
   (a) The mass of a Lilliputian is $\lambda^3 = 64{,}000$ times smaller than the mass of an ordinary human.
   (b) The energy available for the jump stems from the work invested by the muscle force F that contracted $\triangle L$: $E = F\triangle L$. The muscle force is proportional to the area of its cross-section, which is $\lambda^2 = 1{,}600$ times smaller in a Lilliputian than in an ordinary

human. The change in the muscle length is $\lambda = 40$ times shorter, making the energy available for the jump $(\lambda^2)(\lambda) = 64{,}000$ times smaller.

(c) Since both the energy and the mass in a Lilliputian are $\lambda^3 = 64{,}000$ times smaller than in an ordinary human, and the gravitational constant is a *constant*, it follows that the height of the jump is also a constant: a Lilliputian can jump just as high as an ordinary human. In units of his own height, however, an ordinary Lilliputian can jump $\lambda = 40$ times as high as an ordinary human.

— ❦ —

Only the physical proportions of the human body were considered in these examples – height and width, in correlation with the basic function of some of the body parts and the survival of the whole. Rightly, the above examples may seem like a pastime and are indeed too naive for a complete account [☞ Exercise 1.2.3, then e.g. Ref. [561], to begin with], but it should be clear that they indicate some of the basic principles behind the fact that there are no insects as big as storks (or horses), nor can warm-blooded animals be as small as an ant, nor can land animals grow (akin to King Kong) to the size of the largest whales.[14]

Let's turn, however, to more "concrete" applications, and with a more detailed application of physical "dimensions," i.e., units. Students of natural sciences are familiar with unit systems based on specifying the units for some three "basic" physical quantities; by a conventional standard, these are mass (M), length (L) and time (T). Table 1.1 on p. 24 gives numerical data for the unit systems that we use. Suffice it to say, every physical quantity may be measured in units that are of the form $M^\alpha L^\beta T^\gamma$, for some (α, β, γ) [☞ also Table C.4 on p. 527]. In the next examples, our goal is to determine the triple (α, β, γ) for the desired physical quantities and thereby, to a large extent, the physical quantities themselves. A more general treatment of this "dimensional analysis" may be found in the book [208], and the books [244, 415, 416] abound with examples from everyday life where a little critical and mathematical analysis leads to sometimes unexpected results.

1.2.2 Characteristic scales

Take, for instance, a pendulum of length ℓ and mass m. Without writing down and solving the equations of motion, we can estimate the frequency as follows:

1. Neglecting dissipative forces, the frequency may depend only on the physical properties of the pendulum, length ℓ and mass m, and the gravitational constant g.
2. Using dimensional analysis, we have

$$[\ell] = L, \quad [m] = M, \quad [g] = \frac{L}{T^2}, \quad \text{we need} \quad [\nu] = \frac{1}{T}. \tag{1.1}$$

Assuming that the frequency ν is an analytic function of ℓ, m and g, we seek solutions of the form $\nu \propto k^\alpha m^\beta g^\gamma$, and find

$$\frac{1}{T} = [\nu] = [\ell]^\alpha [m]^\beta [g]^\gamma = L^\alpha M^\beta \left(\frac{L}{T^2}\right)^\gamma = \frac{L^{\alpha+\gamma} M^\beta}{T^{2\gamma}}. \tag{1.2}$$

[14] A mature blue whale (*Balaenoptera musculus*) can reach 30 m in length. Fossils indicate that some of the prehistoric, swamp-dwelling animals could reach the length of ~60 m (*Amphicoelias fragilimus*). However, the build of such animals was mostly horizontal, with a long and massive neck and tail, considerably different from modern warm-blooded land animals; *T. Rex*, built akin to a kangaroo, was no longer than about 13 m.

This implies the system of equations

$$\left.\begin{aligned} \alpha+\gamma&=0,\\ \beta&=0,\\ 2\gamma&=1, \end{aligned}\right\} \quad\Rightarrow\quad \left\{\begin{aligned} \alpha&=-\tfrac{1}{2},\\ \beta&=\ \ 0,\\ \gamma&=\ \ \tfrac{1}{2}. \end{aligned}\right. \tag{1.3}$$

Note that this result indicates that the frequency of oscillations is independent of the mass of the swinging bead, and specifies

$$\nu \propto \sqrt{\frac{g}{\ell}}. \tag{1.4}$$

Up to a numerical factor that depends on the definition of "frequency" (ν *vs.* $\omega = 2\pi\nu$), the expression (1.4) is in fact the exact formula, even for large oscillations (while the oscillation angle remains within $\frac{\pi}{2}$ to either side of the equilibrium point)! The result (1.4) merely acquires an overall $O(1)$ multiplicative numerical correction, growing monotonously from 1 to about 1.18 as a function of the amplitude. The fact that the frequency of (small) oscillations does not depend on the mass of the pendulum ($\beta = 0$) may come as a surprise, but is easily verified by simple experiments.

— ❦ —

A similar, but in some ways more interesting, example is presented by a bead of mass m, oscillating at the end of an elastic spring, well described as producing a linear restoring force, $F = -kx$, when stretched or compressed a length x. Now consider having this spring with the bead hanging vertically, and we proceed as before:

1. The frequency may depend only on the physical properties of the hanging spring, (k, m), and the gravitational constant g, where we again neglect dissipative forces.
2. To use dimensional analysis, we must first determine the dimensions (physical units) of the spring constant k. To this end, we may use the restoring force law, $F = -kx$, knowing that a force must have units in agreement with Newton's 2nd law:

$$[F] = [m\,a] = M\cdot\frac{L}{T^2}, \quad [F] = [-k\,m] \quad \text{so} \quad [k] = \frac{[F]}{[x]} = \frac{M}{T^2}. \tag{1.5}$$

3. To use dimensional analysis, we again list

$$[k] = \frac{M}{T^2}, \quad [m] = M, \quad [g] = \frac{L}{T^2}, \quad \text{we need} \quad [\nu] = \frac{1}{T}, \tag{1.6}$$

and again seek solutions in the form $\nu \propto k^\alpha m^\beta g^\gamma$:

$$\frac{1}{T} = [\nu] = [k]^\alpha\,[m]^\beta\,[g]^\gamma = \Big(\frac{M}{T^2}\Big)^\alpha M^\beta \Big(\frac{L}{T^2}\Big)^\gamma = \frac{L^\gamma\,M^{\alpha+\beta}}{T^{2\alpha+2\gamma}}. \tag{1.7}$$

This implies the system of equations

$$\left.\begin{aligned} \gamma&=0,\\ \alpha+\beta&=0,\\ 2\alpha+2\gamma&=1, \end{aligned}\right\} \quad\Rightarrow\quad \left\{\begin{aligned} \alpha&=\ \ \tfrac{1}{2},\\ \beta&=-\tfrac{1}{2},\\ \gamma&=\ \ 0. \end{aligned}\right. \tag{1.8}$$

This result clearly indicates that the frequency of oscillation of the bead on the spring is independent of the gravitational acceleration ($\gamma = 0$). This implies that the bead on the spring oscillates with the same frequency regardless of the orientation of the spring in the gravitational field – and even in the absence of any gravitational field! The result (1.8) implies

$$\nu \propto \sqrt{\frac{k}{m}}, \tag{1.9}$$

which is, up to the numerical factor 2π, again the exact formula and holds even for large oscillations as long as the spring "constant" is an analytic function of the displacement x.

Comment 1.3 *Note that no combination of ℓ, m, g and of k, m, g, respectively, is dimensionless. [✐ Verify.] With the uniqueness of the solutions (1.3) and (1.8), this implies the exactness of the results (1.4) and (1.9), respectively. Compare this with the situation described in Section 1.2.4.*

Comment 1.4 *Mathematically, the results (1.3) and (1.8) are very similar. Physically, however, the facts that the result (1.4) does not depend on the mass of the pendulum nor the result (1.9) on the gravitational acceleration both imply a much wider applicability of these results than initially conceived – but in two different ways. The first lets us freely swap the bobs of a fixed-length pendulum near the surface of the Earth, while the latter lets us predict the oscillations of a spring in Skylab, on the surface of the Moon or Mars, or anywhere else where the gravitational acceleration is approximately constant!*

This realization – that a given mathematical model may be far more widely applicable than originally intended – tends to be extremely important in practice.

1.2.3 Larmor's formula

Larmor's formula for the energy per unit time (therefore, power) that an electric charge q loses during deceleration $\vec{a}$ is

$$P = \frac{2}{3}\frac{q^2\,a^2}{c^3} \;(\text{cgs}), \qquad P = \frac{q^2\,a^2}{6\pi\epsilon_0\,c^3} = \frac{2}{3}\frac{1}{4\pi\epsilon_0}\frac{q^2\,a^2}{c^3} \;(\text{SI}), \qquad \text{where} \quad a = |\vec{a}|. \tag{1.10}$$

Dimensional analysis We are interested in *power* – energy loss per unit time, for which the units are

$$[P] = \frac{[\Delta E]}{[\Delta t]} = \frac{M\,L^2/T^2}{T} = \frac{M\,L^2}{T^3}. \tag{1.11}$$

Energy that changes (decreases) in the process of electromagnetic radiation is certainly of electromagnetic origin, and for electrostatic (Coulomb potential) energy V_C the units are

$$\frac{M\,L^2}{T^2} = [V_C] = \Big[\frac{1}{4\pi\epsilon_0}\frac{q_1\,q_2}{r}\Big], \;\Rightarrow\; \Big[\frac{q}{\sqrt{4\pi\epsilon_0}}\Big] = \frac{M^{1/2}\,L^{3/2}}{T}. \tag{1.12}$$

The electric charge may thus be expressed in "mechanical" units, $\sqrt{\mathrm{kg\,m^3}}/\mathrm{s}$. The so rescaled quantity $q' := \frac{q}{\sqrt{4\pi\epsilon_0}}$ is sometimes called the *rationalized* electric charge [☞ [407, 29, 339] for an application in quantum mechanics].

The power lost through electromagnetic radiation through deceleration must depend on the acceleration, $\vec{a}$. As this is a vector and power is a scalar, the power may depend only on the magnitude of the acceleration, $|\vec{a}|^\beta = a^\beta$. Other than this, the power may only depend on the *universal constant*, c (speed of light in vacuum) – and, of course, on the electric charge[15]:

$$[q^\alpha\,a^\beta\,c^\gamma] = \Big(\frac{M\,L^3}{T^2}\Big)^{\alpha/2}\Big(\frac{L}{T^2}\Big)^{\beta}\Big(\frac{L}{T}\Big)^{\gamma} = \frac{M^{\alpha/2}\,L^{3\alpha/2+\beta+\gamma}}{T^{\alpha+2\beta+\gamma}} \stackrel{!}{=} \frac{M\,L^2}{T^3}. \tag{1.13}$$

[15] The symbol "$\stackrel{!}{=}$" denotes an equality that is *required* to hold [☞ Tables C.7 and C.8 on p. 529].

Comparing, it follows that

$$\left.\begin{aligned} \tfrac{1}{2}\alpha \qquad\qquad &= 1,\\ \tfrac{3}{2}\alpha + \ \beta + \gamma &= 2,\\ \alpha + 2\beta + \gamma &= 3, \end{aligned}\right\} \quad \Rightarrow \quad \left\{\begin{aligned} \alpha &= \ \ 2,\\ \beta &= \ \ 2,\\ \gamma &= -3, \end{aligned}\right. \tag{1.14}$$

so that

$$P \propto \frac{q^2\, a^2}{c^3}. \tag{1.15}$$

The numerical factor $\frac{2}{3}$ in Larmor's formula (1.10) cannot be determined by dimensional analysis, whereas the presence or absence of the $\frac{1}{4\pi\epsilon_0}$ factor is determined by the choice of units – SI or cgs, for example.

1.2.4 Perturbations of stationary states in quantum mechanics

Assume a 1-dimensional (non-relativistic) quantum system specified with the Hamiltonian[16] H_0, for which the stationary solutions are known:

$$H_0|n\rangle = E_n^{(0)}|n\rangle, \qquad U_0|n\rangle = e^{-i\omega_n t}|n\rangle, \qquad \omega_n := E_n^{(0)}/\hbar, \tag{1.16a}$$

$$H_0\Psi_n(x,0) = E_n^{(0)}\Psi_n(x,0), \qquad \Psi_n(x,t) = e^{-i\omega_n t}\,\Psi_n(x,0). \tag{1.16b}$$

In addition, using the Gram–Schmidt orthogonalization procedure, we may always arrange the space of solutions so that

$$\mathscr{H} := \Big\{ |n\rangle \ : \ H_0|n\rangle = E_n^{(0)}|n\rangle,\ \langle n|m\rangle = \delta_{n,m},\ \sum_n |n\rangle\langle n| = \mathbb{1} \Big\} \tag{1.16c}$$

is the Hilbert space of states of the system. The notation must be understood symbolically: Here, n represents any system of numbers: one or more, discrete and/or continuous, including also hybrid value-sets. The latter is the case with the familiar hydrogen atom, where the symbol n in the results (1.16) stands for the familiar collection of "quantum numbers" (n, ℓ, m, m_s). For bound states with negative energies, all four of these numbers vary over discrete values; for scattering states with positive energies, one of those numbers [✎ *which one?*] varies over continuous values, while the other three remain quantized.

Consider now a similar quantum system, differing from (1.16a)–(1.16c) by a *perturbation* Hamiltonian, $H' = H - H_0$. To begin with, let H' be independent of time and let H' – as an operator! – be *small*. That is, the effect of the change $H_0 \to H$ on the energies of stationary states and on the states themselves is, we assume, small. More precisely, we assume that these changes are analytic, i.e., may be expanded in a power series, which has been named after Brook Taylor since 1715. We then have that

$$E_n^{(1)} = \langle n|H'|n\rangle; \tag{1.17}$$

$$|n\rangle^{(1)} = -\sum_{m\neq n}\!\!\!\!\!\!\!\int\ \frac{\langle m|H'|n\rangle}{E_m^{(0)} - E_n^{(0)}}\,|m\rangle; \tag{1.18}$$

$$E_n^{(2)} = -\sum_{m\neq n}\!\!\!\!\!\!\!\int\ \frac{\big|\langle m|H'|n\rangle\big|^2}{E_m^{(0)} - E_n^{(0)}}; \qquad \text{etc.} \tag{1.19}$$

[16] After the Irish mathematician William Rowan Hamilton.

A "shoestring" explanation Perturbation corrections of kth order must be proportional to the kth power of the perturbation operator H' – were it not for H', there would be no corrections. Thus:

1. $E_n^{(1)}$ is a real quantity that must be proportional to the first power of H'. Other than this, $E_n^{(1)}$ may depend only on the nth state, and so must be the expectation value of H' in the nth state, as given in equation (1.17). Besides, however $|n\rangle$ and $\langle n|$ may be normalized, $\langle n|m\rangle$ must be dimensionless; thus $\langle m|A|n\rangle$ must have the same dimensions (physical units) and physical character (scalar, vector... time dependence...) as does A, so that the dimensions (physical units) in equation (1.17) also agree.
2. The first correction of the $|n\rangle$ state cannot be proportional to that same state, as an addition of such a correction to a state would change the norm:

$$\big\||n\rangle\big\|^2 = 1 \quad \to \quad \big\||n\rangle + \epsilon|n\rangle\big\|^2 = \langle n|n\rangle + 2\epsilon\langle n|n\rangle + \epsilon^2\langle n|n\rangle \\ = 1 + 2\epsilon + O(\epsilon^2) \not\approx 1. \tag{1.20}$$

Whence "$m \neq n$" in the sum/integral (1.18), and

$$\big\||n\rangle\big\|^2 = 1 \quad \to \quad \big\||n\rangle + \epsilon|m\rangle\big\|^2 = \langle n|n\rangle + \epsilon\big(\langle m|n\rangle + \langle n|m\rangle\big) + \epsilon^2\langle m|m\rangle \\ = 1 + O(\epsilon^2) \approx 1, \tag{1.21}$$

since $\langle m|n\rangle = 0$ for $m \neq n$. Furthermore, since $|m\rangle$ form a complete basis (cf. Sturm–Liouville theorem for eigen-problems of Hermitian operators), $|n\rangle^{(1)}$ must be expandable in $|m\rangle$'s, as in equation (1.18). Comparing the left- and the right-hand side, the coefficients in the sum must be proportional to a matrix element of H'. Since $|n\rangle$ is on the left-hand side of the equation, it must also occur on the right-hand side, so that a linear change in the basis $|n\rangle$ would change both sides of the equation equally.

It follows that the coefficient in the right-hand sum must depend linearly on $\langle m|H'|n\rangle$, which is the amplitude of probability that H' will change $|n\rangle \to |m\rangle$. As that matrix element has the dimensions of energy as does H', it must be divided by some energy – whence "$E_m^{(0)} - E_n^{(0)}$" in the denominator, which is the energy of the transition $|n\rangle \to |m\rangle$ described by the matrix element $\langle m|H'|n\rangle$.
3. The result (1.19) follows on applying first H' and then $\langle n|$ to equation (1.18).

Assume now that the perturbation Hamiltonian depends on time, $H' = H'(t)$. The probability amplitude for the transition from the initial (i) into the final ($f \neq i$) state[17] is then

$$a_{fi}^{(1)}(T) = \frac{1}{i\hbar}\int_{t_0}^{T>t_0} \mathrm{d}\tau\, \langle f|H'(\tau)|i\rangle\, e^{i\omega_{fi}\tau}, \qquad \omega_{fi} := |E_f - E_i|/\hbar, \tag{1.22}$$

to first order in perturbation theory. This result may be – up to the $1/i$ factor – explained by the same type of "shoestring" arguments and dimensional analysis as the above results for stationary perturbative corrections.

Note that for transitions with very large differences in energies, the frequency is very large, the $e^{i\omega_{fi}\tau}$ factor oscillates very fast, and this causes an effective cancellation in the integral. By contrast, for transitions with very small energy difference, the frequency is very small, and the $e^{i\omega_{fi}\tau}$ factor a priori does not diminish the contribution to the integral. A very similar behavior in integration occurs in the Feynman–Hibbs method of quantization [☞ Procedure 11.1 on p. 416, and Ref. [165]].

[17] More precisely, this is the probability amplitude that if the system was prepared in the initial system i at time t_0, at a later time $T > t_0$ it may be detected in the state $f \neq i$.

1.2.5 And, caution!

Consider now a hydrogen-type atom. The binding energy of such an atom must depend on the (reduced) electron mass, m_e, the electron charge, $-e$, and the charge of the nucleus, $+Ze$. The Coulomb force, which holds the atom together, is proportional to the product of charges, for which the relation (1.12) holds. Notice that the atomic number Z always accompanies e^2, going with the one factor of e that stems from the electric charge of the nucleus, which has Z protons. It then follows that the combination $(m_e)^\alpha(Ze^2)^\beta$ has units of $M^{\alpha+\beta}L^{3\beta}T^{-2\beta}$ and there is no choice of α,β for $(m_e)^\alpha(Ze^2)^\beta$ to have the dimensions (units) of energy, $\frac{ML^2}{T^2}$. For a formula that specifies the energy of a hydrogen-type atom, we need at least one more characteristic quantity, the dimensions (units) of which differ from those of all monomials $(m_e)^\alpha(Ze^2)^\beta$.

Such a characteristic quantity may well be provided by the natural constant c; its units indeed differ from those of $(m_e)^\alpha(Ze^2)^\beta$. More importantly, however, although the electron and the proton may be moving non-relativistically within the atom, they are connected by the electromagnetic field, which certainly does propagate relativistically. We thus seek a Coulomb solution to

$$[E_C] = \frac{M\,L^2}{T^2} = [(m_e)^x][(Ze^2)^y][c^z] = M^x\left(\frac{M\,L^3}{T^2}\right)^y\left(\frac{L}{T}\right)^z = \frac{M^{x+y}\,L^{3y+z}}{T^{2y+z}}, \tag{1.23}$$

i.e.,

$$\left.\begin{aligned} x+y\quad &= 1,\\ 3y+z &= 2,\\ 2y+z &= 2, \end{aligned}\right\} \Rightarrow \left\{\begin{aligned} x &= 1,\\ y &= 0,\\ z &= 2, \end{aligned}\right. \tag{1.24}$$

and obtain

$$E_C \propto m_e c^2 \approx 0.511\,\text{MeV}. \tag{1.25}$$

This is, of course, incorrect: $E_C = 0.511\,\text{MeV} \gg |E_1| = 13.6\,\text{eV}$. Besides, this estimate for E_C turns out to be independent of the electric charge, which is just plain wrong for the binding energy of the atom: were it not for the electric charge, there would be no atom as a bound state. Moreover, it is impossible to construct a dimensionless quantity

$$[(m_e)^x][(Ze^2)^y][c^z] = M^x\left(\frac{M\,L^3}{T^2}\right)^y\left(\frac{L}{T}\right)^z = \frac{M^{x+y}\,L^{3y+z}}{T^{2y+z}} \overset{!}{=} \frac{M^0\,L^0}{T^0}, \tag{1.26}$$

i.e.,

$$\left.\begin{aligned} x+y\quad &= 0,\\ 3y+z &= 0,\\ 2y+z &= 0, \end{aligned}\right\} \Rightarrow \left\{\begin{aligned} x &= 0,\\ y &= 0,\\ z &= 0. \end{aligned}\right. \tag{1.27}$$

The binding energy of a hydrogen-type atom therefore *must* depend on a *fourth* characteristic quantity! Only then will it be possible to construct a dimensionless monomial from these four characteristic quantities. A suitable power of this dimensionless monomial could then rescale $E_C = 0.511\,\text{MeV} \to |E_1| = 13.6\,\text{eV}$.

Bohr's postulate introduces just such a "new" characteristic quantity: $\hbar$, a unit of angular momentum, with physical dimensions $[\hbar] = \frac{ML^2}{T}$. With this new quantity, we have

$$\begin{aligned} [E_H] = \frac{M\,L^2}{T^2} &= [(m_e)^x][(Ze^2)^y][c^z][\hbar^w] = M^x\left(\frac{M\,L^3}{T^2}\right)^y\left(\frac{L}{T}\right)^z\left(\frac{M\,L^2}{T}\right)^w\\ &= \frac{M^{x+y+w}\,L^{3y+z+2w}}{T^{2y+z+w}}, \end{aligned} \tag{1.28}$$

and it follows that

$$\left.\begin{aligned} x+\ y\quad\ + \ w &= 1,\\ 3y+z+2w &= 2,\\ 2y+z+\ w &= 2, \end{aligned}\right\} \Rightarrow \left\{\begin{aligned} x &= 1,\\ y &= -w,\\ z &= 2+w, \end{aligned}\right. \tag{1.29}$$

or

$$E_{\mathrm{H}} \propto m_e\,(Ze^2)^{-w}\,c^{2+w}\,\hbar^{w} = \left(\frac{\hbar c}{Ze^2}\right)^{w} m_e c^2. \tag{1.30}$$

Owing to the fact (1.12), the quantity raised to the wth power is dimensionless. Leaving Z to vary with the atom, we evaluate $\frac{(4\pi\epsilon_0)\hbar c}{e^2} \approx 137.036$ and see that the value $w = -2$ would indeed provide the desired $E_C = 0.511\,\mathrm{MeV} \to |E_1| = 13.6\,\mathrm{eV}$ rescaling.

Indeed, equation (1.30) is remarkably close to Bohr's formula, which may be written as

$$E_n = -2\alpha_e^2\,(m_e c^2)\,\frac{Z^2}{(2n)^2}, \qquad \alpha_e := \frac{e^2}{(4\pi\epsilon_0)\hbar c} \approx \frac{1}{137.036}. \tag{1.31}$$

The so-called "fine structure constant" (coupling parameter) α_e is indeed the dimensionless monomial predicted by equation (1.30). To be precise, dimensional analysis can predict only that

$$E_n \propto f(\alpha_e; n)\,(m_e c^2), \tag{1.32}$$

with no information about the *arbitrary* dimensionless function $f(\alpha_e; n)$.

It is also known that the "fine structure" corrections to the energy levels depend on relativistic corrections and the spin–orbital interaction – neither of which introduces a new physical quantity:

$$\Delta E_{\mathrm{fs}} = -\alpha_e^4\,(m_e c^2)\,\frac{1}{(2n)^2}\left(\tfrac{2n}{j+\frac{1}{2}} - \tfrac{3}{2}\right), \qquad j := \ell \pm \tfrac{1}{2} \quad \to \text{degeneracy}. \tag{1.33}$$

This degeneracy refers to the spectrum (the collection of all eigenvalues) of the Hamiltonian, as there exist two states with the same energy for every ℓ, having $j = \ell \pm \frac{1}{2}$. Since the rest energy of the electron is $m_e c^2$, the sequence

$$|m_e c^2| \;:\; |E_n| \;:\; |\Delta E_{\mathrm{fs}}| \quad = \quad \alpha_e^0 \;:\; \alpha_e^2 \;:\; \alpha_e^4 \tag{1.34}$$

suggests that the hydrogen atom energy is an analytic function of the formal variable "α_e^2" and not of α_e:

$$E_n(\alpha_e) = m_e c^2 \sum_{k=0}^{\infty} C_k\,\alpha_e^{2k}, \tag{1.35}$$

$$C_0 = 1, \quad C_1 = -\tfrac{1}{2n^2}, \quad C_2 = -\tfrac{1}{4n^2}\left(\tfrac{2n}{j+\frac{1}{2}} - \tfrac{3}{2}\right), \quad \text{etc.} \tag{1.36}$$

Indeed, Sommerfeld's relativistic formula[18] [407] from 1915:

$$E_{nk} = \frac{m_e c^2}{\sqrt{1 + \left(\frac{\alpha_e}{n-k+\sqrt{k^2-(Z\alpha_e)^2}}\right)^2}}, \qquad k = 1, 2, \ldots, n, \quad n = 1, 2, \ldots \tag{1.37}$$

gives an excellent description of the bound states of hydrogen-type atoms and supposes elliptic orbits for the electron, where n, k quantify the size and ellipticity of the classical orbit. It is easy to see that Sommerfeld's expression (1.37) depends analytically on α_e and that the Taylor expansion only has even powers.

---

[18] Sommerfeld's derivation assumes that $k = \ell + 1$ measures the electron's angular momentum. Subsequent derivations [122, 121, 216] based on Dirac's relativistic theory of the electron obtain the same final result, but with $k = j + \frac{1}{2}$, where $j = \ell \pm \frac{1}{2}$ owing to the electron's spin, which explains the residual and observed two-fold degeneracy of the bound states.

Digression 1.2 However, the conclusion that the energy is an analytic function of α_e^2 is not *completely* true of the hydrogen-type atoms: There exists the so-called Lamb shift, for which

$$\triangle E_{\text{Lamb}} \approx \alpha_e^5\,(m_e c^2)\,\frac{1}{(2n)^2}\,\frac{1}{n}\Big(E_L(n,\ell) \pm \frac{1}{\pi(j+\frac{1}{2})(\ell+\frac{1}{2})}\Big) \tag{1.38a}$$

is an adequate approximation, with $|E_L(n,\ell)| < 0.05$ and where an odd power of α_e appears manifestly. Also, this "hyperfine" structure of the energy levels is further complicated by contributions from the interactions with the nucleus. These contributions then depend also on the proton mass m_p, through its magnetic moment:

$$\vec{\mu}_p := \gamma_p \frac{e}{m_p c}\vec{S}_p \quad \text{as compared to} \quad \vec{\mu}_e := \frac{e}{m_e c}\vec{S}_e; \quad \gamma_p = 2.7928. \tag{1.38b}$$

The ratio $(m_e/m_p) \approx 1/2{,}000$ provides a new dimensionless constant, whereby the energy formula complicates additionally:

$$\triangle E_{\text{hfs}} = \Big(\frac{m_e}{m_p}\Big)\,\alpha_e^4\,(m_e c^2)\,\frac{4\gamma_p}{2n^3}\,\frac{\pm 1}{(f+\frac{1}{2})(\ell+\frac{1}{2})} + \cdots, \tag{1.38c}$$

where $f(f+1)\hbar^2$ is the eigenvalue of the operator $(\vec{L} + \vec{S}_e + \vec{S}_p)^2$.

Comment 1.5 *Expanding on the result (1.30), a general property of hydrogen-type atoms is worth noting: The binding energy of hydrogen-type atoms must depend on* **four** *characteristic constants of the system:*

1. *the reduced mass*[19] *of a sub-system that is bound to the other – here,* m_e*;*
2. *the interaction coefficient – here, the product of charges,* (Ze^2)*;*
3. *the speed with which the interaction travels between the sub-systems – here,* c*, for the electromagnetic interaction;*
4. *the unit (quantum) of the interaction action (since the classical atom is unstable, and is stabilized by angular momentum quantization) – here,* $\hbar$*.*

The existence of more than three characteristic constants of the system $(m_e, e, c, \hbar)$ *guarantees the existence of a dimensionless characteristic constant* $\alpha_e = \frac{e^2}{(4\pi\epsilon_0)\hbar c}$*, since the system (1.29) consists of only three linear equations in four unknowns. The existence of the dimensionless* α_e *then permits a formula such as (1.32), which we may expand:*

$$E_n = E_{n,0} + \alpha_e\,E_{n,1} + \alpha_e^2\,E_{n,2} + \cdots \tag{1.39a}$$

and notice that for the binding energy of a hydrogen-type atom:

$$E_{n,0} = 0, \qquad E_{n,1} = 0, \qquad E_{n,2} \neq 0. \tag{1.39b}$$

Only the coefficients of the second (and fourth, then fifth...) order in the α_e*-expansion differ from zero!*

[19] By "mass" of a particle, we always mean the relativistic-invariant quantity, which is also (needlessly) called the "rest mass" [☞ Section 3.1.3, and especially the result (3.36)].

Definition 1.1 *Bound-state systems may be roughly classified as:*

$$\left.\begin{array}{r}\textbf{weakly bound}\\ \textbf{strongly bound}\\ \textbf{very strongly bound}\end{array}\right\} \text{ if } \left(\frac{\text{binding energy}}{\text{rest energy}}\right) \begin{cases} <1, & (1.40a)\\ \approx 1, & (1.40b)\\ >1. & (1.40c)\end{cases}$$

Conclusion 1.3 *Since $\alpha_e \approx \frac{1}{137} < 1$, the result (1.39b) implies that the hydrogen atom is a weakly bound system. Indeed, the ratio $13.6\,\text{eV}/m_e c^2 = \alpha^2/2 \approx 2.67 \times 10^{-5}$.*

1.2.6 Exercises for Section 1.2

✎ **1.2.1** *Taking into account that both the load-bearing ability of bones and the muscle force is proportional to the cross-section area, and the height of a Lilliputian is $\lambda = 40$ times smaller than the height of an ordinary human, estimate:*

1. *the width of the legs and torso in a Lilliputian body for which the strain in the bones and muscles are about the same as in ordinary humans;*
2. *the weight of a typical Lilliputian;*
3. *the ensuing corrections in the previous estimates of weight-bearing, heartbeat and height of jump.*

✎ **1.2.2** *In his subsequent travels, Lemuel Gulliver found himself in Brobdingrag,[20] where the population is about $\Lambda = 40$ times taller than ordinary humans. Estimate:*

1. *How much weight (in units of their own weight) can a Brobdingragian lift?*
2. *How fast is a Brobdingragian's heartbeat?*
3. *How high can a Brobdingragian jump?*
4. *If a Brobdingragian is $\Lambda = 40$ times taller than an ordinary human, estimate:*
   (a) *the width of the legs and torso in a Brobdingragian body for which the strain in the bones and muscles are about the same as in ordinary humans;*
   (b) *the weight of a typical Brobdingragian;*
   (c) *the ensuing corrections in the previous estimates of weight-bearing, heartbeat and height of jump.*

✎ **1.2.3** *If humankind ever colonizes Mars, one would expect that the native generations will in time adapt to the four times weaker gravity. (Suppose that the breathing equipment is of negligible weight.) Estimate the changes in the height : width ratio in the body of a fully adapted* Homo Aresiensis, *and from there the other characteristics mentioned in the previous questions.*

✎ **1.2.4** *Estimate the lifetime of the hydrogen atom caused by electron bremsstrahlung, using the Larmor formula to estimate the radiation energy loss. The atom may be regarded as collapsed when the electron "falls" into the nucleus, i.e., when the radius of the electron's orbit reduces from $\sim 10^{-10}$ m to about $\sim 10^{-15}$ m.*

✎ **1.2.5** *Prove the statement made in Comment 1.3 on p. 17.*

---

[20] Complete editions of Jonathan Swift's novel contain a note supposedly added after the first printing wherein the fictitious Lemuel Gulliver explains to his cousin Sympson (who mediated the publishing of the novel) that the printers erroneously printed the name of the land as "Brobdingnag."

✎ ***1.2.6*** *Considering the discussion around the equations (1.16) in the first paragraph of Section 1.2.4, identify which of the quantum numbers (n, ℓ, m) in the Bohr model of the hydrogen atom becomes continuous for scattering states and which must remain discrete. Prove this by re-examining the familiar wave-function (4.8c) upon changing $(E < 0) \to (E > 0)$; the discussion in Appendix A.3 should be helpful for the complementary part of the question.*

✎ ***1.2.7*** *Compute what exactly changes in the formulae (1.31) and (1.33):*

1. *if the electron in a hydrogen atom is replaced by a muon: $m_\mu \approx 207\, m_e$;*
2. *if the electron in a hydrogen atom is replaced by an antiproton: $m_p \approx 1{,}836\, m_e$;*
3. *if the proton in a hydrogen atom is replaced by a positron, $e^+ : m_{e^+} = m_e$.*

✎ ***1.2.8*** *Would the formulae (1.31) and (1.33) hold for a hypothetical $Z > 137$ atom? Why? (Hint: consider the consequences of the relations (1.40a)–(1.40c).)*

1.3 The quantum nature of Nature and limits of information

Nature is both quantum and relativistic; the constants $\hbar$ and c are *universal*. Also, Newton's law of gravity – *extended* by Einstein's general theory of relativity – is also universal, and so also is Newton's constant, G_N. Its units are

$$F_G = G_N\,\frac{m_1\,m_2}{r^2}, \quad \Rightarrow \quad [G_N] = \frac{[F_G][r^2]}{[m^2]} = \frac{M\,L\,L^2}{T^2\,M^2} = \frac{L^3}{T^2\,M}. \tag{1.41}$$

Table 1.1 Natural (Planck) units and their SI equivalent values

Name	Expression	SI equivalent	Practical equivalent
Length	$\ell_P = \sqrt{\frac{\hbar G_N}{c^3}}$	$1.616\,25\times10^{-35}$ m	
Mass	$M_P = \sqrt{\frac{\hbar c}{G_N}}$	$2.176\,44\times10^{-8}$ kg	$1.220\,86\times10^{19}$ GeV/c^2
Time	$t_P = \sqrt{\frac{\hbar G_N}{c^5}}$	$5.391\,24\times10^{-44}$ s	
El. charge[a]	$q_P = \sqrt{4\pi\epsilon_0\,\hbar c}$	$1.875\,55\times10^{-18}$ C	$e/\sqrt{\alpha_e} \approx 11.706\,2\,e$
Temperature	$T_P = \frac{1}{k_B} M_P c^2$	$1.416\,79\times10^{32}$ K	

[a] $\alpha_e \approx 1/137.035\,999\,679$ in low-energy scattering experiments, but grows to about 1/127 near ~200 GeV energies [☞ Section 5.3.3].

From this, we define the Planck, i.e., *natural* units, listed in Table 1.1. A comparison with SI equivalents makes it clear that the natural units are in no way reasonable when describing everyday events of typical human proportions: it would be hilariously ridiculous to try buying milk in units of Planck volume (1 gal $= 8.964 \times 10^{98}\, \ell_P^{\,3}$), hot dogs in units of Planck mass (16 oz $= 2.084 \times 10^7\, M_P$), or measure the time to the recess bell in units of Planck time (45 min $= 5.008 \times 10^{46}\, t_P$). However, natural units do indicate certain limiting values, and this is worth exploring when considering ever smaller systems.

In fact, the natural units in Table 1.1 are not very convenient even for typical contemporary elementary particle physics: the electron mass is $m_e = 4.185\,45 \times 10^{-23}\, M_P$! Therefore, one frequently uses units such as "MeV/c^2," so $m_e = 0.510\,999$ MeV/c^2. In this system of "particle physics" units, we formally state $\hbar = 1 = c$ – that is, we use the unit system where $\hbar$ and c are two of the basic three units, *and then do not write them*. All physical quantities can then be expressed as various powers of one particular unit of measurement, for which the usual choice is energy. For this, one typically uses the "eV" unit, with the usual SI prefixes. Table 1.2 lists some relations useful in typical calculations.

Table 1.2 Some typical physical quantities, expressed in "particle physics" and SI units

Quantity	Particle physics	SI equivalent
Energy	$x\,\text{MeV}$	$= x \times 1.602\,18 \times 10^{-13}\,\text{J}$
Mass	$x\,\text{MeV}/c^2$	$= x \times 1.782\,66 \times 10^{-30}\,\text{kg}$
Length	$x\,\hbar c/\text{MeV}$	$= x \times 1.973\,27 \times 10^{-13}\,\text{m}$
Time	$x\,\hbar/\text{MeV}$	$= x \times 6.582\,12 \times 10^{-22}\,\text{s}$

1.3.1 *Smaller, and smaller, and . . .*

To a great extent, the division and analysis of phenomena, processes and systems happens just as it does in the most obvious application of the black box paradigm, e.g., in microscopy. Light hits the object under scrutiny (figuratively, the black box) and the *reflected* light is guided through a system of lenses and/or mirrors to form a magnified image for the observer to see. The difference between the so-reflected light and that which would have arrived at the observer's eye had the observed object not interfered is the image of the object contrasted with its background. As an amplification of our natural eye, a microscope is used to (quite literally) see into the structure of various material objects. In this, it is worth noting the important limitation. Standard optical microscopes cannot resolve structures finer than 10^{-6} m, regardless of the precision and perfection of the optical elements: lenses, mirrors, etc. The reason for this is the wave nature of visible light, with wavelengths in the range of about 380–760 nm. When considering an object that is smaller than that, light diffracts around it. The image is so fuzzy that no detail smaller than $\sim$380 nm can be discerned.

In perfect analogy, the sounds that humans normally hear easily circumnavigate objects of sizes smaller than about 17 mm. It is therefore humanly impossible to *hear* a marble that stands between us and the sound source. Humanly audible sounds have wavelengths within the 17 mm–17 m range, and all but the shortest wavelengths (which only a small number of people can hear well and which are also typically masked by sounds of longer wavelengths) easily circumnavigate objects of typical hand-held sizes. We say that the *resolution* is of the order of magnitude of the wavelength, understanding that only objects larger than the wavelength of the probing wave may be successfully *resolved*.

The alert Reader will, however, recall that *ultrasound* can be used to image objects of human size and smaller – and is routinely used to make a *sonogram* of, e.g., a fetus inside the womb. As a higher frequency corresponds to a shorter wavelength, the resolution of ultrasound is better, i.e., ultrasound may be used to image finer details than one can do with humanly audible sounds. Recall that the humanly visible light is but a tiny portion of the spectrum of electromagnetic waves. In full analogy, electromagnetic waves of a frequency higher than those in visible light (and so of shorter wavelengths) should produce finer resolution in appropriately constructed microscopes. Indeed, there are many types of electromagnetic waves with wavelengths that are shorter than those in visible light (ultraviolet light, X-rays, etc.), which could be used to construct stronger microscopes. In practice, however, the construction of such microscopes is hampered by the fact that very few if any materials can be used for lenses: ordinary optical lenses do not refract X-rays, the wavelengths of which are much smaller than those in visible light.

A solution is presented by the quantum nature of Nature: Material particles, such as electrons, also can behave like waves, and the basic relationship is that the wavelength of the probing beam

Table 1.3 Some "landmark" objects and events, and their characteristic sizes and the corresponding characteristic energies. Compare with Figure 1.4 on p. 12; 1 eV $\approx 1.6\times10^{-19}$ J.

Objects, events	Size	Energy (in eV)	
Crystalline lattice spacing	$\sim10^{-10}$ m	$\sim10^{3}$	($\sim$1 keV)
Typical size of atoms[a]	$\sim10^{-10}$ m	$\sim10^{3}$	($\sim$10 eV[a])
Typical size of atomic nuclei	$\sim10^{-15}$ m	$\sim10^{8}$	($\sim$100 MeV)
Proton radius	$\sim10^{-16}$ m	$\sim10^{9}$	($\sim$1 GeV)
Range of weak nuclear interaction	$\sim10^{-18}$ m	$\sim10^{11}$	($\sim$100 GeV)
So-called "Grand unification"	$\sim10^{-31}$ m	$\sim10^{24}$	($\sim10^{15}$ GeV)
Quantum gravity, strings	$\sim10^{-35}$ m	$\sim10^{28}$	($\sim10^{19}$ GeV)

[a] The Bohr radius is $\alpha_e^{-1} \approx 137$ times smaller than the naive estimate $\sim\hbar c/E_C$ [☞ Section 2.4].

is *inversely proportional* to the energy of the probe. (Even a single electron can exhibit wave-like behavior, so that by a "beam" we mean herein one or arbitrarily many particles, as the case may be.) Table 1.3 lists a few objects and events in Nature, together with their characteristic size and corresponding energy; that is, the listed energies provide a minimum that a probe must have to resolve the details of the given size. Thus, to any "probe" (beam, ray, test-particle, radiation, etc.) with energy less than about 10 keV, typical atoms appear to be indivisible, structureless and featureless, point-like objects. Of course, a probe with (much) less energy would not even "see" an atom, but instead only the (much) larger structure comprised of atoms. To "see" the structure of the atom, one needs a probe with more than about 10 keV energy (per particle). This principle – that seeing ever smaller structures requires ever bigger energies – is the reason for the dual name of the game: "elementary particle physics" is rightfully also called "high energy physics."

With increased energy, the probability that the probe will change the scrutinized object (or at least some of its characteristics) also grows. What is observed is then not the exclusive property of the scrutinized object, but of the interacting object–probe system. This non-negligibility of the probe and its interaction with the scrutinized object is of essential importance and is a basic fact of quantum theory – and especially of atomic and sub-atomic systems. In this sense, testing and observing of a system irreversibly changes this system. This is sometimes expressed by saying that quantum observation – and so all empirical knowledge – is achieved with active participation of the observer. This causes an *essential* indeterminacy in all kinds of empirical exploration, and therefore also in all empirical knowledge. This quality is expressed in Heisenberg's "indeterminacy principle," which may be regarded as one of the fundamental principles of quantum theory.

The principle of indeterminacy is very precisely stated starting with quantities defined in classical (pre-quantum) theory. Again, quantum theory does not falsify but rather *extends* classical theory. To every degree of freedom and its corresponding variable (coordinate) that is used in the description of a physical system, classical theory corresponds a precisely defined *conjugate momentum*: let q and p denote such a pair. The indeterminacy relation then reads:

$$\triangle q \, \triangle p \;\geqslant\; \tfrac{1}{2}\hbar\,, \tag{1.42}$$

where $\triangle q$ and $\triangle p$ are the indeterminacies in observation and measurement of q and of p, respectively. Thus, if the position of a particle is measured to, say, 1.00×10^{-15} m precision, its (linear) momentum in the same direction cannot be measured better than to 0.525×10^{-19} kg m/s. The errors caused by apparatus imperfections are typically bigger than this, but there do exist measurements where this essential indeterminacy is detectable. To repeat: exact science always errs, but it knows precisely how much [☞ Conclusion 1.1 on p. 6]. Furthermore, the measurement of another quantity q', which is independent of q and p,[21] does not affect the measurements of q and p. That is, once we have measured q' with arbitrary resolution of the measuring instrument, the precision of the simultaneous measurements[22] of *either q or p* is limited only by the precision of the measuring instrument. For the general and precise statement, see Digression 2.7 on p. 73.

1.3.2 Breaking up is hard to do

The careful Reader may have questioned the persistent use of "structure" instead of "divisibility." The latter term is often taken to be a synonym of "being a composite," i.e., that the object "has/shows structure." This tacitly implies that a system that shows structure is in fact composite, and furthermore that it may somehow be divided into its constituent parts. Unfortunately, this is only a prejudice, borne out by everyday experiences: once broken, an egg can no longer be put back together whole; it may be possible to glue a broken glass goblet together with superglue, but the cracks remain, however fine.

In turn, something unusual happens when we divide something as teensy as an atom. Imagine ionizing a hydrogen atom, separating its nucleus (a single, positively charged proton) from its negatively charged electron. This may be accomplished, for example, by applying a sufficiently strong electrostatic field (with $\gtrsim$13.6 eV potential energy). That proton and that electron can thus be moved away from each other arbitrarily many light years; at least in a *thought experiment* such as this, the rest of the universe may be ignored. Leave them so separated for some time, and... the electrostatic force will reunite them! Owing to the unbounded distances to which electrostatic forces reach, the electron and the proton that once formed an atom are never truly separated; their mutual interaction (via the electrostatic field) remains present through the whole "separation" experiment, so that this "separation" is quite fictitious.

This brings up another question. The forces that held the parts of the glass goblet together before it was broken in fact also reach to infinite distances. So, how is it that these forces do not reunite the parts of the broken goblet (however long the Reader is prepared to wait)? The answer is not only in the *distance* of the action, but also in the dependence of the force intensity on the distance. The intensity of the electrostatic force decays with the distance as $1/r^2$, while the intensity of "molecular forces" decays much faster. Imagine now testing the action of such a force at a distance of r, and assume, for simplicity, that the force field is spherically symmetric. That is, we observe the same action at every point of a radius-r sphere centered at the source of the force. As the surface area of the sphere grows as r^2, the *flux* (the product of the surface area and the electrostatic field) through the whole surface of the sphere remains unchanged. Through the same sphere, however, molecular forces that decay as $\sim 1/r^6$ (or faster) produce a flux that decays as $\sim 1/r^4$ (or faster) and so quickly fades at ever larger distances from the source.

Conclusion 1.4 *Molecular forces are said to be* ***localized*** *and have* ***finite range*** *(although the force need not in fact vanish at arbitrarily large distance). Coulomb-like forces that obey the "inverse square law" are said to have* ***infinite range****.*

So, the intensity of both molecular and Coulomb-like forces decays with distance. Distant particles interact weakly, whereas those near each other interact strongly. Thus, low-energy probes are deflected gently from their initial direction, while high-energy probes (with a small wavelength and fine resolution) are sometimes deflected at a large angle – as much as 180°! Precisely this correlation of the angular distribution and probe energy is the "hallmark" of Rutherford's experiments that confirmed the existence of a positively charged nucleus within the atom.

[21] The technical requirement is that q' *commutes* with both q and p; see Digression 2.7 on p. 73.

[22] In the context of quantum physics, "simultaneous measurements" do not mean "at the same time" – most often, this is trivially impossible. Instead, it implies successive measurement of two quantities, which is independent of the order in which the measurements are done.

However, this is not so in collision experiments that are essentially the same as Rutherford's, but where the probes have >100 MeV energy [☞ Table 1.3 on p. 26]: These interactions differ significantly from Coulomb-like forces, and may be ascribed to so-called strong nuclear interactions. At distances where the action of these strong nuclear forces may be measured, their intensity *stagnates* with the distance between the colliding centers – as if the connection between those could be represented (modeled) by a string [☞ Chapter 11]! By itself, this may not seem unusual, but some of its consequences definitely are; see Chapter 6.

To stretch a string, one must invest work, and this increases the potential energy of the stretched string. At a certain point, determined by the string elasticity, it simply breaks. Analogously, two particles (so-called quarks) bound by the strong nuclear interaction may be separated to ever larger distances only by incessant investment of ever more energy. This could be done arbitrarily long, and the two quarks could be separated arbitrarily far from each other, except that the invested work sooner or later becomes sufficient to create a particle–antiparticle pair. Each one of these newly created particles then binds with one of the "old" ones, so that the attempt to separate two quarks to more than about 10^{-15} m fails. Instead of having separated one quark from the other, the quark we were trying to move becomes bound with the newly created antiquark, and the other "old" quark is joined by the newly minted quark replacing the old one. These quark–antiquark pairs form two new systems (so-called mesons) that really can be separated arbitrarily far, but the original quarks remain confined within these newly minted mesons [☞ Figure 1.5].

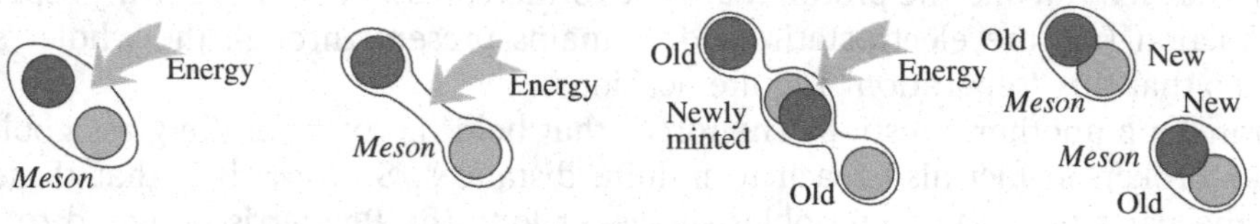

Figure 1.5 Inseparability of quarks and antiquarks in spite of investing ever more energy.

Thus, quarks (to most precise experimental verification and theoretical prediction) cannot be extracted arbitrarily far from one another, and remain "confined" – either in the original system, or in a newly minted system, joined with (anti)quarks created by investing ever more energy.

However, as long as the distance between the quarks is less than about 10^{-15} m, their binding energy is sufficiently small and they move effectively freely. Thus, the concept of "divisibility" (as it is usually understood) is definitely not synonymous with the concept of "compositeness," and those two notions must be clearly distinguished:

1. In all experiments performed to date, the electron behaves as a point-like particle, i.e., it shows no structure.
2. The proton shows structure (three quarks) through the complexity of the angular dependence in scattering, i.e., through deviations from Rutherford's formula – and does so when the collision energy surpasses a well-defined threshold; however, the quarks cannot be extracted arbitrarily far without creating new quark–antiquark pairs [☞ Figure 1.5].
3. Atomic nuclei show structure in collisions: they may be broken into smaller nuclei and/or their constituents, protons and neutrons (jointly called *nucleons*); the resulting parts may be permanently separated, i.e., the restoring forces have a limited, finite range.
4. Atoms show structure in collisions: they may be ionized by extracting one or more electrons; the restoring force between the positive ion and the extracted electron, however, has an infinite range, and the separation is not permanent: if left alone, the atom recombines.

5. Molecules show structure in collisions: they may be broken (dissociated) into smaller molecules and/or atoms; the resulting parts may be permanently separated, i.e., the restoring forces have a finite range.

A remark is in order. Strong nuclear interactions that bind quarks into a proton are related to but not the same as the forces between the proton and the neutron (within atomic nuclei). The latter forces are called "residual," just as some of the molecular forces between otherwise neutral atoms stem from electromagnetic interactions between somewhat separated constituent parts of these atoms (electrons and nuclei). However, even these residual forces are much stronger than the electromagnetic ones, as they overpower the Coulomb repulsion between positive protons within the nucleus. The weak nuclear interaction also has the characteristics of a string-like restoring force, but its characteristic short range and weakness stem from being mediated by the massive particles $W^{\pm}$ and Z^0. In turn, both the strong nuclear force and the electromagnetic force are mediated by massless particles, called gluons and photons, respectively.

1.3.3 ... and smallest: limits of information

It would seem that the (spatial) resolution of measuring devices could, at least in principle, be made arbitrarily fine, but this is not the case. A glance at Figure 1.4 on p. 12 shows that something unusual should happen if the spatial resolution were to improve to the point of detecting details smaller than about 10^{-35} m, around the Planck length; see Ref. [518] and Ref. [99] for a recent discussion. Recall that, for detecting ever smaller details, the probe must have an ever larger energy, since the de Broglie wavelength of the probe is

$$\lambda_p = \frac{2\pi\hbar}{p_p} = \begin{cases} \frac{2\pi\hbar}{\sqrt{2m_p T_p}}, & \text{(non-relativistic)} \\ \frac{2\pi\hbar c}{T_p}, & \text{(relativistic)} \end{cases} \tag{1.43}$$

where

$$T_p = \begin{cases} \frac{p_p^2}{2m_p}, & \text{(non-relativistic)} \\ \sqrt{p_p^2 c^2 + m_p^2 c^4} - m_p c^2, & \text{(relativistic)} \end{cases} \tag{1.44}$$

is the kinetic energy of the probe of mass m_p [☞ Section 3.1.3].

During interaction with the "target," the probe and the target temporarily form a combined system. The total mass of this combined system is greater than or equal to the sum of the target mass and the probe *mass-equivalent* of the total energy of the probe: $m_{t+p} = m_t + (m_p + T_p/c^2)$. Thus, as the kinetic energy of the probe grows, so does the total mass of the temporary target+probe system during the interaction.

Now, the situation becomes interesting owing to gravitational effects, and the fact that the gravitational field of the system grows (linearly) with the *total* mass of the system. Since the gravitational field grows, so does the "separation speed," v_{sep}.[23] Furthermore, the gravitational field is not constant but grows unboundedly, as $1/r$ when $r \to 0$, so that the separation speed is much larger near the gravitating center than further away from it. If the target and the probe are both smaller than the distance between them, at the Schwarzschild distance between their centers,

$$r_S = \frac{2\,G_N\,m_{t+p}}{c^2}, \qquad \text{so that} \qquad v_{\text{sep}} := \sqrt{\frac{2G_N\,m_{t+p}}{r}} \xrightarrow{r \to r_S} c. \tag{1.45}$$

[23] This speed is often called the "escape velocity," as it pertains to the successful launching of a projectile with a speed that suffices for the projectile to escape from the gravitational field of a (much larger) planet. The principle is, however, perfectly general, and applies equally to the separation of two objects after their collision.

That is, the separation speed $v_{\rm sep}$ becomes equal to the speed of light in vacuum, and no probe can separate from the target it has hit. This result holds (both the target and the probe are "sufficiently small") if neither the target nor the probe has any structure larger than the distance r_S as given in equation (1.45).

Alternatively, recall that a probe may be treated as a wave with the de Broglie wavelength $\lambda = 2\pi\hbar/p$, owing to the quantum nature of Nature. This wavelength is smallest for ultra-relativistic probes with $p_p \approx E_p/c$, so $\lambda_p \lesssim 2\pi\hbar c/E_p$. When the de Broglie wavelength becomes as small as the Schwarzschild radius, i.e., when the probe energy grows so that its resolution equals the Schwarzschild radius, the target "swallows" the probe and it cannot extract any information from within a sphere of radius r_S: the target+probe system now looks like a black hole:

$$\lambda_p \sim r_S \quad \Rightarrow \quad \frac{2\pi\hbar c}{E_p} \sim \frac{2\,G_N\,(m_t c^2 + E_p)}{c^4}, \tag{1.46a}$$

$$\Rightarrow \quad E_p \sim \tfrac{1}{2}\big[\sqrt{4\pi M_P^2 + m_t^2} - m_t\big]\,c^2, \tag{1.46b}$$

$$\xrightarrow{m_t\to 0} \quad E_p \sim \sqrt{\pi}\,M_P\,c^2, \tag{1.46c}$$

where we used that $M_P = \sqrt{\hbar c/G_N}$ [☞ Table 1.1 on p. 24]. The formal limit notation $m_t \to 0$ may also be obtained using the leading term in the expansion of the equation (1.46a) when $m_t c^2 \ll E_p$, or as the leading term in the expansion of the result (1.46b) when $m_t \ll M_P$. Indeed, for an ultra-relativistic probe, $E_p/c^2 \gg m_t, m_p$, and $p_p \approx E_p/c$. (In the non-relativistic limiting case, when $E_p/c^2 \ll m_t, m_p$ and $p_p \approx \sqrt{2m_p E_p}$, the probe has insufficient resolution to approach the target and test its structure.)

Of course, this argument extrapolates over many orders of magnitude in distances, and is based on current understanding of gravity and quantum mechanics. However, note that the *qualitative* part of the argument relies on the facts:

1. the minimal size of resolved details *decreases* (the resolution improves) with increasing probe energy,
2. the distance where the "separation speed" becomes inaccessibly big *increases* with the total mass of the source of gravity,
3. the mass (source of gravity) and the rest energy (measure of the ability to do work *not* owing to motion) are proportional to each other.

This already implies that there exists a minimal object (system) size or distance in Nature. If we can furthermore rely on the quantitative details of the argument, the minimal resolvable distance of about $\ell_P \sim 10^{-35}$ m follows unambiguously.

If Nature consists of elementary *particles* (which, by definition, have no constituents), then they must seem like miniature black holes. Their *event horizon*[24] must form a closed surface of which no detail smaller than $\sim 10^{-35}$ m can be resolved. This suggests that massless elementary particles would have to look like minimal, $\sim 10^{-35}$ m, spherically symmetric black holes. Massive particles could have a bigger horizon and a more complicated shape, but again the resolvable details must be bigger than about 10^{-35} m [☞ Digression 9.5 on p. 340].

Conclusion 1.5 *The unknowability of the "inside" of the event horizon of elementary particles indicates that there is no sense in regarding them as ideal point-like objects. Willy-nilly, elementary particles are extended in space.*

[24] The term "event horizon" denotes the border (bordering surface) in space that fully encloses a black hole, and from within which nothing can come out because of confiningly strong gravity. It is *believed* that all naturally occurring black holes are so wrapped by an event horizon; see Comment 9.4 on p. 337.

This conflict between (1) the extrapolated results of the general theory of relativity (which describes gravity) for *point-like* elementary particles, and (2) quantum mechanics (whereby the points cannot really be smaller than about 10^{-35} m in diameter) is of course a prediction of such a combined model, and a result of the model itself. To avoid the conflict, we must leave behind some aspects of this model of gravitating point-like quantum elementary particles, but retain enough of it so as to continue reproducing its experimentally verified properties, for energies $\lesssim 10^2$ GeV, i.e., for distances $\gtrsim 10^{-18}$ m.

There is another (also intuitive and not formally rigorous) argument that indicates the incompatibility of the general theory of relativity and the quantum theory of point-like particles: Heisenberg's principle of indeterminacy implies that the position and the linear momentum in the same direction cannot both be determined with infinite precision. On the other hand, in the general theory of relativity, the presence of matter curves spacetime and so defines a class of coordinate systems: a massive point-like particle curves spacetime in which the position of this particle is determined with perfect precision. Furthermore, this particle is at perfect rest in this class of coordinate systems, so that both the position and the linear momentum are determined with infinite precision. This is in direct and unavoidable contradiction with quantum theory.

— ❦ —

In turn, Conclusion 1.5 has a very important consequence: The variables with which we represent physical objects depend on the variables with which we represent spacetime and the abstract space of various other properties of this physical object. Thereby, for example, the wave-function of an electron is a function of spacetime coordinates and also of numbers that determine its mass, charge, spin, chirality... Recall that every function is simply a rule that assigns to every value of its arguments – a point in the domain space – a value of its own. This value is represented as a point in the target space, i.e., the range of the function. Since spacetime coordinates of any physical system cannot be specified *within* the event horizon of that system but only up to the *surface* of this horizon, it follows that the domain spaces of the functions with which we describe physical systems are not volumes that permeate the open sets of spacetime, but the *surfaces* that enclose such volumes. This insight may be argued to engender the *holography principle*, which was introduced into the fundamental physics of elementary particles by Gerardus 't Hooft and which was in the 1990s gradually built into the (super)string theory and its M-theoretical extension [☞ Chapter 11] (first by Leonard Susskind, and then in his collaboration with Tom Banks, Willy Fischler, and Stephen Shenker).

1.3.4 Unification: the smaller, the more similar

Technically, the incompatibility between the general theory of relativity and quantum theory for point-like elementary particles introduces unavoidable divergences: When computing physically measurable quantities, prediction results are obtained in the form of hopelessly divergent (indefinite) mathematical expressions. By contrast, in the amalgamation of the *special* theory of relativity and quantum theory, in "relativistic field theory," formally divergent results may be removed through the process of "renormalization." The quantum theories of electromagnetic, weak, and strong nuclear interactions (together with all known matter) really do include the special theory of relativity and form a logically consistent structure.[25] Developed akin to quantum electrodynamics (the predictions of which are confirmed to an amazing 12 significant figures [293, 1]), the quantum field theory of electroweak interaction describes the observed electromagnetic and weak

[25] In fact, the existence of the "top" quark (which was only recently convincingly confirmed in experiments) was predicted using so-called anomaly cancellations. That is, without the "top" quark, the Standard Model of elementary particles and interactions would be self-contradictory!

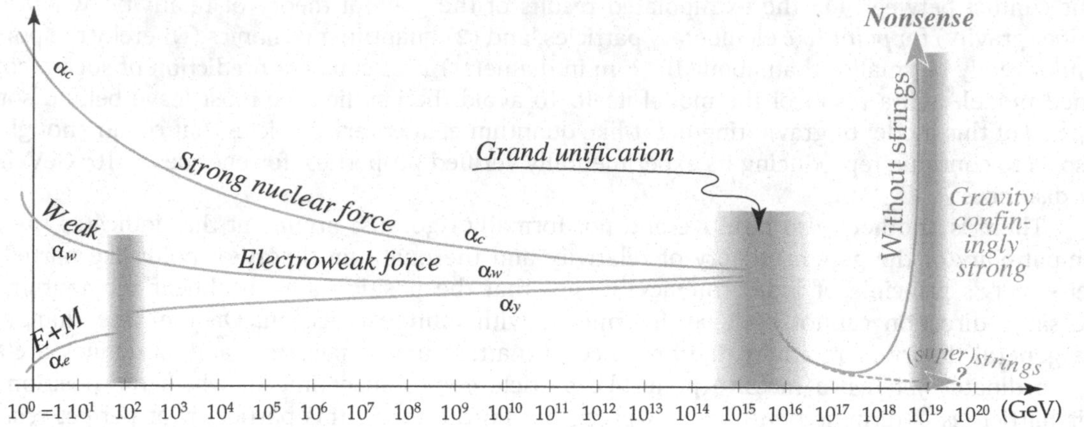

Figure 1.6 A logarithmic plot of coupling parameters, from 1 GeV (≈ proton's rest energy, $\approx 10^5$ times bigger than the hydrogen atom ionization energy), to 10^{19} GeV, where gravity becomes confiningly strong and point-like theories become nonsense, while string theories "pass" through a phase transition of sorts, albeit insufficiently known so far.

nuclear interactions as well as their unification very precisely. Both the theoretical and the experimental precision in both the electroweak and the strong interaction model are considerably more humble, but the results do agree. Several models of quantum field theory have been developed that describe the unification of the electroweak and the strong interaction [☞ Figure 1.6], but only further experiments, starting with the LHC facility at CERN, can decide which of these unifying models – if any – describes the "real World."

In all of these cases, the transition regimes (the shaded areas in Figure 1.6) where a unification happens indicate a phase transition of sorts, in the sense that the qualitative properties of the theory are drastically different on one and the other "side" of the transition region. While these may well be related to (World-scale) bulk-material phase transitions that have presumably happened in the early universe, the subject of particle physics probes the related physics phenomena in individual particle collision experiments performed at high energies – where no actual bulk-material phase transition occurs. For example, below about 10^2 GeV, there is a clear distinction between electromagnetic and weak nuclear processes, each of which can happen without the other. At energies above about 10^2 GeV, however, these processes mix inseparably. This situation is very similar to the fact that electric and magnetic phenomena are well distinguished in stationary systems and often occur one without the other, but become inseparably united and involve electromagnetic waves when the electric charges move and create non-stationary currents.

In the case of electro-magnetic unification, whether or not electric and magnetic fields are distinguishable depends on the speeds, taken in ratio with the speed of light in vacuum ($c \approx 299{,}792$ km/s). For example, it is well known that an electrical current (flow of electric charges) creates a magnetic field around it. The speed of the individual charged particles is typically small as compared to the speed of light in vacuum. However, the speed of *momentum transfer* within that current is very close to the speed of light in vacuum. In turn, the electromagnetic field itself adapts to the motion of electric charges – of course – at the speed of light. When the electric charges either do not move at all or form stationary currents, the ratio of the speed of changes in this current and the speed of light in vacuum (which is the reference parameter here) is zero, and the electric and the magnetic fields are well distinguishable. When the electric charges move so they do not form stationary currents, this ratio is (near or equal to) one, and the electromagnetic field can no longer be separated into an independent electric and an independent magnetic field. Furthermore, non-stationary currents cause a variable magnetic field, the changes in which create an additional electric field, the changes in which modify the magnetic field, etc. This feedback between the electric and the magnetic field for non-stationary currents produces the new phenomenon: electromagnetic waves, which carry away some of the energy carried by the current.

For the electro-weak unification ("electro-magneto-weak" would be more accurate, but is too much of a mouthful), the unifying parameter is the ratio of energies of the processes as compared to the $W^{\pm}$ and Z^0 particle masses. (Just a few years after their discovery at CERN, these particles were routinely observed and studied; their masses are close to 10^2 GeV/c^2.) By contrast, the mass of the particles of light is zero; the total energy of light is entirely of kinetic nature and there is no coordinate system in which light is at rest. Clearly, when an experiment is conducted at energies much below about 10^2 GeV, real $W^{\pm}$ and Z^0 particles cannot be produced and cannot contribute to the processes we study. Weak nuclear processes then happen by exchanging *virtual*[26] $W^{\pm}$ and Z^0 particles [☞ Definition 3.5 on p. 104]. This happens owing to the Heisenberg principle of indeterminacy and the fact (Pauli, 1933 [29, p. 334]) that the *energy* and the *characteristic duration of time* of every process also satisfy the indeterminacy relations:

$$\triangle E\,\triangle\tau \;\geqslant\; \tfrac{1}{2}\hbar\,. \tag{1.47}$$

Roughly, during the time $\triangle\tau \leqslant \frac{\hbar}{2\triangle E}$ for $\triangle E \sim 10^2$ GeV, a $W^{\pm}$ or a Z^0 particle may be freely produced – if it also decays within this time. The necessity that *two* consecutive processes (creation and decay) must happen in such a short time decreases the probability of the *joint* process mediated by an intermediate $W^{\pm}$- or Z^0-boson, and this permits an unambiguous identification of the said process as weak nuclear, rather than electromagnetic. However, when the energies in the experiment become much bigger than 10^2 GeV, *real* $W^{\pm}$ and Z^0 particles are produced with the same probability as the electromagnetic waves (photons). Owing to charge conservation, W^+- and W^--radiation do not mix with the others, but the Z^0-radiation and electromagnetic radiation do mix, inextricably, and form new kinds of phenomena – very similar to the unification of (variable and mutually inducing) electric and magnetic fields [☞ comparative Table 8.1 on p. 299].

A similar novel phenomenon is expected around $\gtrsim 10^{15}$ GeV, where the electroweak and the strong interactions tend towards having the same strength. The entire graph in Figure 1.6 on p. 32, is, however, based on experimental data at currently available energies, $\lesssim 10^2$ GeV, and so is necessarily an extrapolation. The assumption that neither new phenomena nor new particles will be found between $\sim 10^2$ GeV and $\sim 10^{15}$ GeV is often called the "grand desert hypothesis." This follows from Ockham's principle, whereby novelties are introduced only if necessary. The subsequent ideas and arguments rely on this at least in part, and must be re-examined as soon as there is compelling evidence that the "grand desert" turns out to be populated. Several so-called Grand-Unified Theory (GUT) models have been developed attempting to sort through the possible phenomena that could occur in this region, but in ways that leave the physics below $\sim 10^2$ GeV unchanged, in agreement with the experiments performed so far. Such (and all other) models are expected to predict at least some event that could soon be experimentally verified (or refuted).

— ❦ —

A few quick and perhaps overdue remarks: Just *what* exactly do these weak and strong nuclear interactions in fact *do*? Besides esoteric phenomena of particle physics, these interactions are in fact responsible for our own existence! Electromagnetic radiation – and in particular light – is what brings the energy from the Sun to the Earth and makes life as we know it possible. The fundamental process that produces the immense energy of our Sun is *nuclear fusion*, in which the nuclei of

[26] In distinction from real particles, the virtual ones – by definition – cannot be directly observed.

deuterium and *tritium* (two heavier isotopes of hydrogen) fuse into helium and release a neutron and energy. The reason that there is surplus energy is due to the details of strong nuclear interactions. Finally, note that nuclei of pure hydrogen would not fuse; instead, deuterium and tritium are needed. The hydrogen nucleus is a single proton; deuterium and tritium nuclei consist of a proton and respectively one and two neutrons – all held together by strong nuclear forces. These much needed neutrons are all being produced in weak nuclear interactions such as the inverse β-decay, $p^+ + e^- \to n^0 + \nu$. These can – and indeed do – occur within stars [☞ equations (7.118)], and also occurred in the young universe, long before the stars were formed. In addition to providing the required fuel for (strong nuclear) fusion, weak nuclear interactions also moderate this process, thereby preventing our Sun from burning out in one brilliant explosion.

> **Conclusion 1.6** *Thus, by making the Sun burn in the first place, by making it burn at a steady pace that we are familiar with, and by bringing its energy to the Earth, the strong nuclear, the weak nuclear, and the electromagnetic interactions, respectively, bring about the conditions on the Earth that sustain our life and our asking about it. Finally, the fourth fundamental interaction – gravity – keeps the Earth from flying asunder and also keeps it in a stable orbit near the Sun. Were it not for these four interactions, dear Reader, you would not exist to read this book.*

1.3.5 A shift in understanding

The relativity of Nature prevents us from thinking of space and time as two disparate "things," and forces us to join them into a single, undivided spacetime. The concept of simultaneity is recognized to be relative, which then disperses the so-called paradoxes of twins/clocks, of the ladder and the barn, of the ruler and the hole in the table, etc.

The quantumness of Nature disillusions us from thinking of "things" around us as unchanged, and clearly separable from their environment, and forces us to think of them as determined by imposed circumstances. Thus, the electron may behave *both* as a point-like particle and as a wave – and is in fact neither, but "something" that in appropriate circumstances *may look* like a particle or like a wave. Similarly, instead of talking about entangled states of two separate objects in EPR (Einstein–Podolsky–Rosen)-type experiments, it would be wiser to talk about a single, undivided state of a single system, which in certain circumstances may be interpreted as two spatially separated sub-systems.

Besides, the quantumness of Nature indicates the importance of the Hilbert space as a very real space in which processes occur – although neither the Hilbert space nor the unfolding of events in it can ever be seen by unaided human eye or mind. By contrast, although the very spacetime itself is just as "invisible," we do see the unfolding of events in spacetime. This makes thinking about events unfolding in spacetime intuitively easier, while thinking of events unfolding in the Hilbert space appears to be much less natural and very counter-intuitive. That makes the quantumness of Nature baffling.

The combination of (special-)relativistic and quantum physics is then *doubly* baffling and counter-intuitive, but is no less rigorous as a scientific discipline than the familiar and intuitive classical mechanics. In fact, the most precise agreement between theoretical and experimental physics occurs exactly in the realm of quantum field theory. For some of the characteristic and observable quantities such as the fine structure constant and the so-called magnetic moment anomaly, the comparison of various measurements and theoretical computations in quantum electrodynamics agrees with experimental data to an all-time record of *12 significant figures* [1, 293]!

The fundamental physics and its description of Nature – which must include both quantum physics and general (not only special) relativity – are then even more baffling as compared to commonplace experiences. Onward then, into this mutliply baffling journey!

1.3.6 Exercises for Section 1.3

✎ **1.3.1** *Express the value of Newton's gravitational constant,* $G_N = 6.6742 \times 10^{-11}\,\frac{m^2}{kg\,s}$*, in "particle units,"* $\hbar^x c^y (\text{MeV})^z$*, for some* (x, y, z).

✎ **1.3.2** *Using the definitions in Table 1.1 on p. 24, compute the value in Planck units (suitable powers of* ℓ_P*,* M_P *and* t_P*) of the SI units: (a) 1 m/s (speed), (b) 1 m/s² (acceleration), (c) 1 N s (linear momentum), (d) 1 J s (angular momentum), (e) 1 J (work and energy), and (f) 1 W (power).*

✎ **1.3.3** *Using the data in Table 1.2 on p. 25, compute the* MeV ↔ SI *conversion factors for (a) speed, (b) angular speed, (c) acceleration, (d) angular acceleration, (e) linear momentum, (f) angular momentum, (g) work, and (h) power.*

✎ **1.3.4** *Using the definitions in Table 1.1 on p. 24, compute the value in Planck units (suitable powers of* ℓ_P*,* M_P *and* t_P*) of the SI units: (a) 1 C (charge), (b) 1 N/C (electric field), (c) 1 T (magnetic field), (d) 1 A (electric current), (e) 1 V (voltage), and (f) 1 K (temperature).*

✎ **1.3.5** *Using Table 1.2 on p. 25, and the result (1.12),*[27] *compute the* MeV ↔ SI *conversion factors for (a) charge, (b) electric field (from* $\vec{F}_C = q\vec{E}$*), (c) magnetic field (*$\vec{F}_M = q\vec{v} \times \vec{B}$*), (d) electric current (*$I := dq/dt$*), (e) voltage, a.k.a. potential (from* $P = VI$*), and (f) temperature (= average kinetic energy/*k_B*).*

✎ **1.3.6** *Using that* $M_P = \sqrt{\hbar c / G_N}$*, verify that the leading term in the expansion of the result (1.46b) when* $m_t \ll M_P$ *agrees with the solution of the leading term in the expansion of the equation (1.46a) when* $m_t c^2 \ll E_p$*, i.e., with the result (1.46c).*

✎ **1.3.7** *Considering that* $M_P = 2.18 \times 10^{-8}$ *kg does not seem very large in everyday terms, obtain the leading term in the expansion of the result (1.46b) when* $m_t \gg M_P$*, and compute the corresponding range of values for* E_p*. How feasible is it to provide an elementary particle probe with the lowest such energy?*

[27] Recall that the speed of light $c = 1/\sqrt{\epsilon_0 \mu_0}$, and that the Boltzmann constant $k_B = 1.38 \times 10^{-23}$ J/K is a conversion factor between units of energy and units of temperature, which statistical physics defines as average translational kinetic energy of molecules in some large ensemble, while other forms of energy of the molecules (vibrational, rotational, binding...) contribute to the so-called internal energy.

2

Fundamental physics: elementary particles and processes

This chapter serves to familiarize the Student with the physics of elementary particles, where new concepts are introduced in their historical context and without a precise, technical definition. The subsequent chapters will clarify these concepts with more details, examples and applications.

2.1 The subject matter

The task of elementary particle physics is explaining of what and how the World is fundamentally made. Amazingly, and almost exactly in a Democritean sense, substance (tangible matter) comprises tiny particles, and our task includes a coherent classification:

1. both a systematic inventory of these "elementary particles,"
2. and an understanding of the "elementary processes" between them, i.e., their "fundamental interactions."

In principle, these fundamental interactions determine how collections of otherwise independent elementary particles bind into ever larger structures, to macroscopical and even astronomical proportions. However, all except the teeniest in this hierarchy of structures are outside the scope of this subject.

One must *actively* keep in mind that the seemingly homogeneous and continuous substance consists of only a few types of particles, amongst which each one occupies but a tiny volume, and between which most of the space is practically empty. Less than a *trillionth* of the volume of any given substance is occupied by the particles forming the substance. Corresponding to their tininess, these particles come in fantastic numbers, and these countless copies are all identical. Not only "practically equal," but two particles of the same kind really cannot possibly be distinguished from one another: any one of the 10^{29-30} electrons in our body is identical and exchangeable with any of the other electrons. It is absolutely impossible to distinguish one electron from another, except by the state in which an electron is and by its interactions with the rest of the considered system. The same holds for protons and neutrons.

We will see that elementary particles are determined by their types of interaction with other particles. The seeming void between particles is in fact filled with interaction *fields*: in ordinary substances, this is the electromagnetic field. In this sense, it is incorrect to focus exclusively on particles, however elementary. However, the whole image of particles is just a picturesque *model*, a caricature; we must recall the quantum nature of Nature, and the particle–wave duality. Elementary particles must be thought of as neither particles nor waves, but as objects that in certain circumstances appear as very tiny and precisely localized particles, and in other circumstances look like dispersed, continuous waves. Similarly, the electromagnetic field is an object that in certain circumstances behaves as a continuous and spatially very much extended wave, but in other circumstances it looks like a well-localized elementary particle, the photon. Thus, there is no conceptual difference between the objects that we most often encounter as either particles, waves or fields; in field theory, all objects are represented by fields, the quanta of which may be particles.

— ❦ —

We manipulate macroscopic objects directly. Thus, Coulomb could experiment with mica spheres suspended by silk thread, and Cavendish with lead spheres suspended on a torsion swing. Particle physicists, however, cannot catch an electron with tweezers or string up a few protons on a thread. Experiments in elementary particle physics are thus reduced to studying (**1**) collisions and scattering,[1] (**2**) decays, and (**3**) bound states of elementary particles. The laws and rules of interactions are then reconstructed from the results of such studies.

As is well known, if any relative speed of any two sub-systems is comparable with the speed of light in vacuum, one must use relativistic mechanics. Also, if the Hamilton action[2] for a given process is comparable with $\hbar$, one must use quantum mechanics. In our case, we have to use physics that is both relativistic and quantum, i.e., quantum field theory. On the other hand, in an introduction such as this, we do not have to introduce the whole mathematical–technical apparatus of field theory, but rely as much as possible on picturesque models and analysis that is *conceptually* not much more demanding than the usual mathematical apparatus of non-relativistic quantum mechanics.

Some of the characteristics of elementary particle physics are essentially relativistic effects, while other properties stem from quantumness. For example, the 4-vector of energy and momentum (henceforth, "*4-momentum*") is always conserved in so-called real states – i.e., states that may be observed and measured, but mass is not. On the other hand, in *virtual states* the 4-momentum conservation laws need not hold; see Section 2.4.2. Also, Nature's relativity permits the existence of particles with identically vanishing mass: The particle of the electromagnetic field, the photon, makes no sense in non-relativistic physics; see p. 94. Moreover, the combination of relativity and quantum theory leads to results that can be obtained in neither the theory of relativity nor quantum mechanics. The existence of antiparticles, the proof of Pauli's exclusion principle (1925) and the so-called spin-statistics theorem and the so-called "CPT-theorem" all stem from the *combination* of quantum and relativistic, and all relativistic quantum models must include them.

By about 1978, the so-called Standard Model had taken form in elementary particle physics; it encompasses all phenomena involving the elementary particles and their interactions as known to date, and is in full agreement with the experimental data observed in the last three decades; see [307, 221, 422, 159, 423, 538, 250, 243] and also [458] for a first-person account of the 1960/70s excitements from an experimentalist's vantage point. Our main goal is to acquaint ourselves with that Standard Model and the basic principles of its structure, such as the gauge

[1] While the terms "collision" and "scattering" are often used interchangeably, the former will here tend to refer to the physical event of colliding or its bringing about, while the latter will tend to refer to the process and its results, often focusing on the individual particles involved, and often being inelastic.

[2] This is indeed *Hamilton's principal function*, the time-integral of the Lagrangian, familiar to Students from classical physics, where the integrand determines the physical system, and the boundary data and limits of integration specify the process considered.

principle, which is at the foundation of all fundamental interactions and links symmetries[3] with conservation laws via Noether's theorem.

That is, here we are interested in the *theoretical* physics of elementary particles and their elementary processes, via fundamental interactions. However, we must first, even if briefly, turn to the experimental aspects – to know what it is that we have set out to explore and describe theoretically.

2.2 Elementary particles: detection and predisposition in experiments

2.2.1 Production

Most instrumentation used in experimental elementary particle physics is familiar from the literature in nuclear physics, so only a brief review is given here.

Producing electrons for laboratory use is almost trivial: Metals, when heated or irradiated with UV light, emit electrons, which are then easy to "catch" and direct with electric and magnetic fields. Protons – when needed, say, in a beam – may be produced by ionizing hydrogen. Since protons are charged, they can be directed with electric and magnetic fields, just as the electrons. On the other hand, since the electron mass is $\sim 1,836$ times smaller than the proton mass, the electrons may be neglected for many experimental purposes, so that a tank full of hydrogen is in practice a tank full of protons. Of course, many more particles were discovered in the past century, and these particles stem, mainly, from three sources:

Cosmic rays and their interaction with the atmosphere. It is not possible to identify the particular process at the source of any particular cosmic ray, nor do we know all the types of processes that create them; we do know, however, that particles of even very high energy incessantly bombard the Earth and collide with the atomic nuclei of atmospheric gases. Particles resulting from these collisions further collide with atomic nuclei of atmospheric gases, in cascading collisions. Clearly, this way of producing elementary particles is completely uncontrolled and subject to happenstance. However, it *is* a source of extremely high-energy particles, which we otherwise cannot produce in the lab. Besides, this source is also completely free.

Nuclear reactors and sources Atomic nuclei in radioactive materials spontaneously decay and in such processes not infrequently emit neutrons, α-particles (helium atomic nuclei), β-particles (electrons or positrons, depending on the source) and γ-particles (photons). Also, irradiating materials with so-called synchrotron (electromagnetic) radiation frequently either directly produces new particles, or makes those materials radioactive.

Particle accelerators and colliders The basic idea is to direct and accelerate previously produced particles along well-established paths, and then either bombard a target with these particles, or direct two such beams at one another.[4] Let us mention here Van de Graaff's machine, Cockcroft and Walton's linear accelerator, Lawrence's cyclotron, and finally Wildröe and Touschek's betatron. In this last device, oppositely charged particles are accelerated in opposite directions and within nearly identical and approximately circular paths. These then collide at the intersections of these paths; this is the basic idea in contemporary colliders. Since particles are directed and accelerated using electric and magnetic fields, this clearly applies only to charged particles. In contemporary practice, particles are mostly directed along circular paths – following huge circular tunnels of miles-long radii. Thus, particles that "missed" in one "turn" get to collide again in the next one. As we will show in Section 3.2.2, the energy

[3] Symmetries have the mathematical structure of groups, so the mathematical subject of group theory turns out to be very useful in studying and using this structure. Appendix A provides a telegraphical review of the most needed results from this mathematical subject.

[4] Apparati for colliding beams of particles against each other are called colliders.

available in such collisions is "used" for creating new particles. Creation of lighter particles of course requires less energy, so that lighter particles are produced and discovered more easily, and the heavier ones are harder – or have not yet been created/discovered.

2.2.2 Nomenclature

In 1899, after having become known for his "investigations into the disintegration of elements and the chemistry of radioactive substances" (for which he would be awarded the Nobel Prize in Chemistry in 1908), Ernest Rutherford classified the radioactivity emitted by natural samples as α- and β-rays, distinguishing them by their penetrating power. Four years later, he found that radium, discovered in 1900 by Paul Villard, emitted a type of radiation that surpassed both α- and β-rays in penetrating power and named it γ-rays. By this time, he had (1) explaned that radioactivity in natural samples is caused by spontaneous disintegration of the atoms of the sample, (2) identified the exponential decay law and its application to use the constant decay rate as a "clock," and (3) identified the particles of α-rays as probably fully ionized helium atoms, i.e., helium nuclei. His classification was merely refined over the years and is still in use:

α-rays and particles are helium nuclei and consist of two protons and two neutrons.

β-rays and particles may be either electrons (also known as *cathode rays*), or positrons, depending on the process that created them. For example, the negatively charged cathode has a surplus of electrons, which a strong electric field may be able to free from the cathode and direct as a cathode ray. In nuclear processes, the so-called weak interaction can produce both electrons (in the β-decay of the neutron) and positrons (in the β-decay of the proton[5]), which are then emitted from the source material.

Table 2.1 The names of various bands of electromagnetic radiation

Name	Frequencies	Wavelength	Energy
γ-rays	> 30 EHz	< 10 pm	> 124 keV
Hard X-rays	3 – 30 EHz	10 – 100 pm	12.4 – 124 keV
Medium X-rays	0.3 – 3 EHz	0.1 – 1 nm	1.24 – 12.4 keV
Soft X-rays	30 – 300 PHz	1 – 10 nm	0.124 – 1.24 keV
Ultraviolet rays	0.79 – 30 PHz	10 – 380 nm	3.27 – 124 eV
Visible light	400 – 790 THz	380 – 750 nm	1.65 – 3.27 eV
Near infrared	30 – 400 THz	0.75 – 10 μm	0.124 – 1.65 meV
Medium infrared	3 – 30 THz	10 – 100 μm	12.4 – 124 meV
Far infrared	0.3 – 3 THz	0.1 – 1 mm	1.24 – 12.4 meV
Radio and micro-waves	< 0.3 THz	> 1 mm	< 1.24 meV

Standard prefixes: $\mathrm{E} = 10^{18}$, $\mathrm{P} = 10^{15}$, $\mathrm{T} = 10^{12}$, $\mathrm{k} = 10^{3}$, $\mathrm{m} = 10^{-3}$, $\mu = 10^{-6}$, $\mathrm{n} = 10^{-9}$, $\mathrm{p} = 10^{-12}$.

γ-rays and particles (photons) are the known quanta of electromagnetic radiation. More precisely, Table 2.1 shows the division of the electromagnetic radiation spectrum and names the various bands. Traditionally, "γ-rays" meant electromagnetic radiation that stems from spontaneous nuclear γ-decay, while "X- or Röntgen-rays" referred to electromagnetic radiation produced artificially. Initially, γ-rays were meant to denote electromagnetic radiation with energies higher than in X-rays, but contemporary accelerators produce X-rays with energies far above the energy of typical γ-rays, and these two bands overlap. By tradition and for simplicity,

[5] A free proton cannot decay into a positron and a neutron because of energy conservation. [✎ *Verify.*]

high-energy electromagnetic radiation is called γ-radiation regardless of its origin. In fact, "γ" is used as a symbol for the photon, the particle (quantum) of electromagnetic radiation in general and regardless of its frequency (energy) and wavelength.

Independently of this nomenclature, particles in relativistic physics are also classified depending on their mass and possible speeds of propagation, as compared with photons:

Tardion is a particle that in vacuum moves slower than light and has a real and positive mass; all known matter (and antimatter!) consists of tardions; see below.

Luxon is a particle that moves in vacuum at the speed of light in vacuum, c, and has no mass. All particles mediating gauge interactions that correspond to unbroken gauge symmetries are luxons.

Tachyon is a particle that moves in vacuum faster than light, and has an imaginary mass. The emergence of tachyons indicates that the ground state of the system (i.e., vacuum) is not stable [☞ Digression 7.1 on p. 261].

This classification refers to the Lorentz-invariant mass, defined precisely in Section 3.1.3.

2.2.3 Detection

Particle detection relies on the interaction between the particle and its environment, so that all detectors are more or less selective.

Geiger counter detects ionizing radiation, usually β- and γ-radiation, but there exist models that can also detect α-particles. It consists of a tube filled with an inert gas (usually helium, neon or argon, with halogen additives) that becomes conducting when radiation (partially) ionizes it. The spark created between the electrodes when the gas becomes conducting because of ionization provides a signal that is amplified by a cascading array of electrodes. The so-produced current is usually shown on a galvanometer, a pilot-lamp or by a speaker – hence the characteristic crackling. As the density of the gas in the tube is relatively small, very high-energy particles pass through without detectable interaction.

Neutrons are neutral and by themselves do not trigger the Geiger counter; however, they can be detected indirectly, by using boron trifluoride and a moderator that slows neutrons and in the process creates (charged and easily detectable) α-particles.

Scintillation counter consists of a transparent crystal and a fluorescent material that reacts to ionizing radiation; a sensitive photo-multiplier is used to amplify the signal.

Cherenkov counter is based on the fact that there exist materials through which some particles can travel faster than light, albeit slower than the speed of light in vacuum. Such particles also interact with this material and emit electromagnetic radiation, which then forms a cone-shaped shock wave with the opening angle $\theta_c = \arccos(c/nv)$, where v is the speed of the particles and n is the refraction index of this material. This effect was discovered by Pavel Cherenkov, while Ilja Frank and Igor Tamm explained it theoretically, for which the three of them shared the 1958 Nobel Prize.

Cloud chamber also known as the Wilson chamber, after Charles T. R. Wilson, is filled with super-cooled vapor (water or alcohol), in which the passing particle engenders condensation. This forms a sequence of condensed droplets that faithfully trace the particle's passage.

Bubble chamber (invented by Donald Glaser) is filled with superheated liquid (usually liquid hydrogen at $-253\,^\circ$C, propane or some other appropriate liquid), in which the passing particle engenders evaporation. As in the (Wilson) cloud chamber, this forms a sequence of evaporated bubbles that faithfully trace the particle's passage.

Spark chamber (invented by Shuji Fukui and Sigenori Miyamoto) is woven through with wires that are kept at various voltages, in very short (~10 ns) impulses. When a particle passes between two wires and ionizes the gas between the wires, the gas (specially chosen for this purpose, usually an argon–methane mixture) becomes conducting and a spark jumps between the wires; the brevity of the voltage pulses prevents an avalanche of sparks. The spatial distribution and time sequence of the sparks faithfully trace the particle's passage and also depict its speed.

Proportional chamber is Georges Charpak's modification of the spark chamber that enables measurement of the quantity of ionized gas, which is proportional (whence the name) to the kinetic energy of the particle that produced the ionization. Charpak received the 1992 Nobel Prize for this invention.

Photographic emulsion contains molecules of silver halide, which react with passing charged particles. When the film is processed afterwards, the trace in the emulsion faithfully depicts the particles' paths. Clearly, this method captures faithfully only two-dimensional events that happen to be coplanar with the emulsion.

2.2.4 Predisposition in experiments

Experiments in elementary particle physics grew from the relatively humble confines of individuals' labs, such as Rutherford's in Manchester at the turn of the twentieth century, into humongous multi-national installations such as CERN. This evolution of experimental physics has side-effects that in many ways limit the scope, variety and intellectual freedom of the experiments and so limit the advancement of elementary particle physics, and even physics in general.

Even in principle, experiments are performed to test one concrete hypothesis or another, and hypotheses are of course limited by the imagination of the physicists who design the experiments. This creates a selection effect in experimental science: On one hand, we can derive/compute a result from a given concrete theoretical model and then design an experiment to test this result. Or, we can re-test some earlier result, but to a greater precision than was possible up to that point. Not infrequently, the crucial improvements in such experiments require an ingenuity and inspiration that is rightfully impressive and even awesome. However, the fact remains that such experiments – for the most part – "only" test existing/known theory.

On the other hand, experiments may also be designed to test hypotheses that are inspired by science fantasy or even child's play, which start with "what if ... " questions, where the ellipses represents a hypothesis not limited by any concrete result from any concrete theoretical model. One would expect *such* experiments to have a much larger chance of discovering wholly unexpected effects and phenomena. To list but a few examples:

1. Thales (of Miletus, *c*. 620–625 BC!) noted that, after being rubbed against fur, amber attracts particles of dust.
2. The more systematic experiments of Alessandro Volta (1745–1827) showed that frog legs were induced to twitch by poking them with wires of *certain* metals, although the legs were cut off from the frogs and so were evidently dead.
3. Hans Christian Ørsted (1777–1851) observed that the magnetic compass needle changes its direction when brought near a wire through which a current is passing.
4. Henri Becquerel (1852–1908) discovered that certain (radioactive) materials affect photographic material via means wholly invisible to the human eye.

Many discoveries have occurred exactly through such "unbridled" and even entirely *accidental* activity. Had it not been for such spontaneously freewheeling experiments, who knows if electrodynamics would ever have developed into the theoretical model (and its ubiquitous practical applications!) that we know today? It is certain, however, that the history of physics would have turned out very differently.

Returning to elementary particle physics: evidently, such "unbridled" experiments were possible to design up to the mid-twentieth century, and there did indeed occur some completely accidental discoveries. However, in a situation where new discoveries – such as are expected of the LHC experiment at CERN – require budgets, logistics and political agreements of dozens of countries and over several years and even decades, for "unbridled" and "accidental" experimenting there is no chance. Even when these big experiments are very carefully and diligently designed, the socio-political climate can cancel them – or worse, can nix them half-way through (as happened with the Superconducting Super Collider, in 1993) – because of reasons that are not directly related to physics and research.

> ***Conclusion 2.1*** *This very nonlinear feedback between experimental physics, finance and the socio-political climate amplifies the selection effect, and produces an ever stronger and limiting influence – a predisposition – on the types of experiments that we can at least hope to perform. It is easily possible that this is one of the most important factors in the evident slow-down in experimental physics discoveries during the last quarter of the twentieth century and the first decade of the third millennium.*

As it is hard to believe that the financial and socio-political aspects will radically improve, experimental elementary particle physics must adapt if it is to survive; hopefully, by means of some radically new and clever (financially and socio-politically less limited) methodology [☞ also Digression 1.1 on p. 9].

2.3 A historical inventory of the fundamental ingredients of the World

The Democritean[6] idea of the smallest, indivisible constituents of the World was revived as the idea that there exists a smallest quantity of every chemical substance that retains the chemical properties unchanged. Carefully following the proportions in which chemical substances interact, chemists have established the existence of molecules and atoms, and even estimated their size, $\sim 10^{-10}$ m. The tininess of the molecules and atoms reciprocally implies their enormous number in macroscopically "normal" quantities of such substances. For example,

$$\text{there are } \approx 6.69 \times 10^{26} \text{ molecules in a (2 dl) glass of water.} \tag{2.1}$$

Molecules were known to consist of atoms, and – within the methods and techniques of nineteenth century chemistry – atoms are really indivisible: ἄτομος is ancient Greek for "uncuttable."

2.3.1 The electron

In 1897, Joseph J. Thomson showed that atoms are not indivisible, and that a much smaller *electron* may be extracted from them: He showed that the cathode rays may be bent using electric and magnetic fields. The deflection in the electric field depends only on the electric charge of the particles that form the cathode ray, and the deflection in the magnetic field depends both on

[6] D. Griffiths [243] refers to Democritus as a *meta*physicist, and rightly so: many ancient Greek philosophers' teachings have reached us as "armchair philosophy" and with no reference to experimental data in support. We thus carefully ascribe the *scientific* idea of atoms to the nineteenth century chemists, but the *inspiration* for that worldview to the *philosophy* of Democritus and Leucippus. It is worth noting that these Democritean ideas have spread mostly by way of the Latin epic *De rerum natura* [Titus Lucretius Carus, first century BC], which has been fully preserved till today. In six books, it presents the naturalist philosophy of the ancient Greek philosopher Epicurus, according to which the World consists of atoms that move in otherwise empty space.

the electric charge and on the particle speed. Thus, by manipulating the electric and the magnetic fields through which he let the cathode rays pass, Thomson determined that the ratio of the electric charge to the mass of the particles that form the cathode ray is several thousand times larger than the same ratio for any then known ion. It follows that cathode rays consist of particles of which either the electric charge is several thousand times larger, or the mass is several thousand times smaller than those of the then known ions.

Digression 2.1 If a particle enters with the speed v, in the direction of the positive x-axis, into a region with a constant electric field $\vec{E} = E_0\,\hat{e}_y$ and a constant magnetic field $\vec{B} = B_0\,\hat{e}_z$, it is affected by the Lorentz force

$$\vec{F} = q\,\vec{E} + q\,\vec{v}\times\vec{B} = q\,E_0\,\hat{e}_y + q\,v\,B_0\,(\hat{e}_x\times\hat{e}_z) = q\left(E_0 - v\,B_0\right)\hat{e}_y. \tag{2.2a}$$

If the particle does not deflect from its straight path, it follows that the total force vanishes, from which it follows that

$$v = \frac{E_0}{B_0}. \tag{2.2b}$$

If we now switch the electric field off, leaving the particle to follow a circular path of radius R, it follows that the magnetic (Lorentz) force provides the centripetal acceleration, so

$$q\,v\,B_0 = m\,\frac{v^2}{R} \qquad \Rightarrow \qquad \frac{q}{m} = \frac{v}{B_0\,R} = \frac{E_0}{B_0^{\,2}R}. \tag{2.2c}$$

Between the two possible interpretations, the one stating that cathode rays consist of particles smaller than atoms seemed much more reasonable to Thomson. He referred to these particles as *corpuscles* and to their electric charge as the *electron*, but the latter name became universally accepted for the particles themselves.

Digression 2.2 Ironically, Walter Kaufmann (Berlin, Germany) had performed the same experiments as J. J. Thomson – at about the same time and more precisely! However, he did not leap to the same conclusion as Thomson. Adhering to the philosophical (epistemological) doctrine of *positivism*,[7] he could/would not conceive of the explanation in Democritean atomistic terms, explaining the cathode ray as a beam of particles too little to observe [553].

2.3.2 The proton

Since atoms were known to be electrically neutral, it followed that atoms consist of electrons and a positive part that is thousands of times more massive than electrons. The simplest supposition was that this positive part fills the volume of each atom and that the electrons are embedded within this positively charged mass; this was J. J. Thomson's so-called plum pudding model.

[7] Roughly, positivism restricts science to only address directly observable phenomena and discuss them only in terms of directly observable quantities.

In contrast, Ernest Rutherford[8] had, with his students Hans Geiger and Ernest Marsden, performed an epoch-making experiment in 1909: Bombarding a thin golden foil with α-particles, it was shown that atoms of gold (and so also of all other elements) are mostly empty. In 1911, Rutherford derived his classical formula, which is easy to rewrite into a contemporary form:

$$\frac{d\sigma}{d\Omega}=\left(\frac{e^2/4\pi\epsilon_0}{2m_\alpha v_0{}^2}\right)^2\frac{1}{\sin^4(\theta/2)}\xrightarrow{\frac{e^2}{4\pi\epsilon_0}\to\alpha_e\hbar c}\left(\frac{\alpha_e\hbar c}{2m_\alpha v_0{}^2}\right)^2\frac{1}{\sin^4(\theta/2)}, \tag{2.3}$$

where $\alpha_e=\dfrac{e^2}{(4\pi\epsilon_0)\hbar c}$ is the fine structure constant, m_α the α-particle mass, v_0 their speed of approach to the foil, and θ the angle of deflection from the original direction of motion. The ratio $\dfrac{d\sigma}{d\Omega}$ is called the differential cross-section, and gives the probability distribution as a function of the probe's deflection angle θ. Besides, α-particles are positively charged, and owing to the Coulomb repulsion can approach the positively electrically charged portion of the atom only to a distance b, which is determined from equating the energies:

$$\tfrac{1}{2}m_\alpha\,v^2=\frac{1}{4\pi\epsilon_0}\frac{q_{\text{Au}}\,q_\alpha}{b}\qquad\Rightarrow\qquad b=\frac{1}{4\pi\epsilon_0}\frac{2q_{\text{Au}}\,q_\alpha}{m_\alpha v^2}, \tag{2.4}$$

where q_{Au} and q_α are the electric charges of the positive part of the gold atom and the α-particles. Direct measurements show that the minimal value for b was around 2.7×10^{-14} m, which is three to four orders of magnitude smaller than the size of the gold atom.

These initially established characteristics of the α-particle scattering pattern clearly "mapped" the Coulomb repulsive force field of the positively charged parts of the gold atom. Further experiments and more detailed analysis of this α-particle scattering pattern during the next decade managed to obtain indications of non-Coulomb scattering and so establish that the positively charged part of the gold atom is localized within a radius of only about 3.5×10^{-15} m.

This gave rise to the so-called planetary model of the atom, in which all atoms have a positively charged nucleus, of a radius $\sim10^{-15}$ m, around which electrons revolve in orbits of radii $\sim10^{-10}$ m. The nucleus of the simplest atom, hydrogen, was named *proton* by Rutherford.

Having obtained his PhD in May 1911 and spent six months at Cambridge working with J. J. Thomson, Niels Bohr came to the University of Manchester in May 1912 to work with Rutherford. By 1913, he had postulated that the electron's angular momentum in these orbits is limited to integral multiples of a constant, $\hbar$. With this ad hoc quantization, Bohr successfully computed not only the binding energies of the electron in the hydrogen atom, but also derived a general formula for the wavelengths of the photons emitted in transitions, and which was in full agreement with the observed series in the line spectra named after Balmer, Lyman, Paschen, etc.

Digression 2.3 In the last decades of the twentieth century it became clear that the integrality of the angular momentum in atoms of the hydrogen type – Bohr's ad hoc postulate – is also the essential characteristic that stabilizes atomic orbitals, and in fact a concrete manifestation of a general principle. The modern understanding of theoretical physics uses the language of geometry and topology; topological invariants are, roughly,

[8] Rutherford was J. J. Thomson's graduate student at Cambridge, but these atomic nuclei experiments were designed and conducted (a decade after Thomson's) at the University of Manchester, where Rutherford was chair of physics from 1907 – after eight years at McGill University in Canada during which he worked on radioactive materials and for which he was awarded the 1908 Nobel Prize in Chemistry. A decade later, in 1919, Rutherford succeeded Thomson as the director of the Cavendish Laboratory at Cambridge.

quantities that depend on integers in a critical way, so that invariance follows from the impossibility of a continuous change of these integral characteristic quantities.

In retrospect then, with the benefit of hindsight, we may conclude that Bohr's postulate must be accepted because it stabilizes the atomic orbitals. On the other hand, it may in turn be explained by (reduced to) the de Broglie particle–wave duality. *[✎ Why?]*

By 1917, Rutherford's experiments had shown that atoms can split in collisions, and by 1932, Rutherford and his students John Cockroft and Ernest Walton had developed experimental techniques to split some atoms in a fully controlled fashion.

The "natural" assumption that atomic nuclei then consist of just the right number of protons to neutralize the total charge of the electrons that are in orbit around that particular nucleus, however, did not find support in experimental data: Even in the late nineteenth century, it was known that the next atom by mass, the helium atom, has two electrons, but a mass that is not twice but *fourfold* larger than the hydrogen atom. Lithium is the next element, with three electrons, but the mass of its atom is *six* or *seven* times bigger than that of the hydrogen atom. This exemplifies the tendency of atoms to be two or more times heavier than the product of their atomic number in the periodic table and the mass of the hydrogen atom.

2.3.3 The neutron

The disproportionally larger masses of atomic nuclei were explained in 1932, when James Chadwick (Rutherford's student) discovered that atomic nuclei contain another type of particle, besides protons. As that other kind of particle is neutral, he called them *neutrons*. Being the building blocks of nuclei, protons and neutrons are collectively called *nucleons*.

Also, the existence of neutrons helped explain the mismatch in many heavier nuclei: it was known that the nuclear masses are generally (for the first few rows in the periodic table) about twice as large as the electric charge of the nucleus. This induced the supposition that nuclei are composed of twice as many protons as necessary to cancel the charge of the orbiting electrons, and that the surplus of positive charge in the nuclei is neutralized by additional electrons that are confined to the nucleus. However, it was also known that protons and electrons have an intrinsic angular momentum (spin) of $\frac{1}{2}\hbar$. In many nuclei the total sum of angular momenta from both the protons and the electrons – both those in the $\sim 10^{-10}$ m orbits and those (hypothetical ones) inside the nuclei – did not agree with experimental data on angular momenta.

For example, nitrogen-14 would in this model have 14 protons and 7 electrons in the nucleus, and 7 electrons in orbit, a total of 28 spin-$\frac{1}{2}$ particles. The total angular momentum of nitrogen-14 would then have to be an integral multiple of $\hbar$. By contrast, the measured value of the total angular momentum of nitrogen-14 atoms is always a half-integral multiple of $\hbar$. The discovery of the neutron made it clear that the nitrogen-14 nucleus consists of 7 protons and 7 neutrons, which guarantees the total angular momentum of the nucleus to be an integral multiple of $\hbar$ (and half-integral for the whole atom), in agreement with all observations.

Conclusion 2.2 *In hindsight, the 1932 theory of elementary particles now seems fantastically simple: The World consisted of* ***electrons***, ***protons*** *and* ***neutrons***, *it was already known that the first two of these interacted by exchanging* ***photons***, *and it "only" remained to figure out how these particles form the bigger structures: atoms, molecules, etc. Of course, that turned out to be a very naive point of view, and as we will soon see, one that is rather far from the full picture.*

2.3.4 The photon

Newton's original idea that light consists of *corpuscles* did not stand the test of experiments of his day: By the twentieth century, the wave nature of light was fully obvious. This implied the same wave-like nature for all the different types of electromagnetic radiation, in the spectrum of which the visible light occupies but a tiny region.

True to our leitmotif from the introductory Section 1.1.1, the twentieth century began with a systematic demolition of this image: At the turn of the nineteenth into the twentieth century, Max Planck studied the problem of the so-called ultraviolet catastrophe. Classical statistical physics had up to that point in time perfectly explained all known thermodynamic processes. However, its application to electromagnetic waves emitted by all hot objects, such as a piece of ember or a star, produced nonsense – that the total power of the emitted radiation is *infinite*, as well as that the radiation intensity grows unboundedly with growing frequency. Experimental data with no exception show that the intensity does grow with frequency, but only up to a particular maximum (proportional to the temperature of the hot object) and decays thereafter. Also, it is patently obvious that the total power emitted by a hot object must be finite.

In 1900, Planck showed that the ultraviolet catastrophe may be avoided and the experimentally known spectra may be explained theoretically – if we assume that hot objects emit radiation in "packets," the *action* (roughly, energy times the duration of time) of which is an integral multiple of a constant, h – soon enough called the Planck constant. No other viable resolution of this problem has been found since.

In 1905, Albert Einstein showed that the photoelectric effect unambiguously indicates that electrons also absorb electromagnetic radiation in the same "packets," with *identically(!)* the same constant h. From this, Einstein concluded that not only is electromagnetic radiation both emitted and absorbed in such "packets," but that it also *exists* only in the form of such packets. This revolutionary idea encountered enormous resistance: Even a decade later and its quantal emission and absorption notwithstanding, most physicists agreed that electromagnetic radiation was nevertheless of wave-like nature.

In 1910, Peter Debye proved that Planck's result follows from supposing that the Fourier modes of the electromagnetic field have energies that are integral multiples of the $h\nu$ product. Nevertheless, only 15 years after that (in 1925, by which time the quantum nature of the electromagnetic radiation was largely accepted) had Max Born, Werner Heisenberg and Pascual Jordan correctly interpreted the electromagnetic field Fourier mode of energy $nh\nu$ as n particles with energy $h\nu$ each.

In the interim, Arthur H. Compton showed in 1923 that Einstein's claim of the quantum nature of electromagnetic radiation is the *only* known one that successfully explains the scattering of visible light and "soft" X-rays on free electrons. Compton showed that the analysis of the scattering as a collision of particles of light with electrons – by using the 4-momentum conservation laws – gives the formula that today bears his name:

$$\lambda' = \lambda + \lambda_c(1 - \cos\theta), \qquad \lambda_c = \frac{h}{m\,c} = \frac{2\pi\hbar}{m\,c}, \tag{2.5}$$

where λ_c is the so-called Compton wavelength for a particle of mass m. Compton's original analysis was meant for electrons, $m = m_e$, but it is obvious that the formula applies to the scattering of electromagnetic radiation on any free charged particle; see Exercise 2.4.2. The classical wave analysis of the scattering of electromagnetic radiation from charged particles, the so-called Thomson scattering, predicts a change in the radiation wavelength for sufficient radiation intensity. However, when the intensity is too small, Thomson's effect vanishes. By contrast, Compton's effect gives the correct change in the wavelength of the scattered photon regardless of the radiation intensity, and even for a single photon of energy $E_\gamma = 2\pi\hbar\nu = 2\pi\hbar c/\lambda$.

The name *photon* itself was given to the particles of light by the chemist Gilbert Lewis, as late as 1926, together with his proposal that photons can be neither destroyed nor created; the details of Lewis's proposal proved not to be what Nature has to offer, but the name stuck. Niels Bohr, Hendrik Kramers and John Slater as late as 1924 tried to "save" the non-corpuscular nature of the electromagnetic field, by proposing the so-called "BKS model" [70], which required that:

1. energy and momentum are conserved only in the average, but not in processes where a single charged elementary particle absorbs or emits electromagnetic radiation;
2. causality should not hold in such elementary processes.

The BKS model was swiftly given a "decent burial" [406], but inspired Werner Heisenberg in 1925 to co-develop with Max Born and Pascual Jordan the so-called "matrix mechanics" [547]. Also, Einstein received the Nobel Prize only in 1921 "for his services to theoretical physics, and especially for his discovery of the law of the photoelectric effect" – and not for the quantal understanding of light. It thus took over a quarter of a century from Planck's original hypothesis for physicists to accept the quantum nature of light.

2.3.5 *Duality and locality*

In 1924, de Broglie defended his doctoral dissertation with the fundamental idea of the *duality* between particles and waves (for which he received the 1929 Nobel Prize), whereby *every* particle of momentum $\vec{p}$ may be represented by a wave of

$$\text{wavelength} \quad \lambda = \frac{h}{|\vec{p}|}, \qquad \text{i.e., wave-vector} \quad \vec{k} = \frac{h}{\vec{p}^2}\vec{p}, \tag{2.6}$$

and vice versa. Thus, by about 1926, accepting the particle-like nature of light no longer implied abandoning the wave-like nature, but accepting a more general view whereby the objects we call electron, proton, neutron, photon, etc., under certain circumstances behave as particles, but as waves under other circumstances. We will see later, combining quantum theory and the special theory of relativity in field theory, all these objects may be unambiguously represented by appropriate *fields*, of which "particles" and "waves" are certain limiting forms.

— ❦ —

The quantum nature of electromagnetic radiation was at first very hard to accept, and even then mostly owing to the practical ease in explaining Compton's effect. It turns out, however, that such an understanding of electromagnetic radiation has a very deep consequence regarding the essential understanding of the fundamental electromagnetic interaction – and following this template, later, also the other interactions.

The classical understanding of the interaction between two charged bodies relies on the idea that each charged particle creates around itself an electric field. Then, on any other charged particle probing this field there acts a force equal to the product of the probing charge and the electric field being probed. This electric field then simply equals the Coulomb force per unit probing charge. By its definition, the electric field is thus a crutch for predicting the Coulomb force – which then *acts at a distance*. The role of the magnetic field is, in this classical understanding of the electromagnetic interaction, conceptually identical, and their sum gives the full Lorentz force.

One might argue that this action at a distance does not literally contradict the special theory of relativity, according to which all information propagates locally, from a point to another in a continuous space and at most at the speed of light in vacuum. That is, the establishing and all changes in this classical electromagnetic field travel at the speed of light in vacuum or slower than that when traversing a substance that interacts with the electromagnetic field. However, once established, the Coulomb field *instantaneously* produces the force upon a test charge, regardless of how far the test charge is from the source of the Coulomb field. It is this assumed instantaneity that does contradict the fundamental idea of locality, which is in turn woven into the special theory of relativity. Thus, *classical* field theory in fact does incorporate a conceptual contradiction.

The quantum nature of light replaces the concept of an everywhere present electromagnetic field (that instantaneously produces a force at a distance) with the concept wherein charged particles constantly emit and re-absorb photons, and this continual exchange of photons between two charged particles mediates the electromagnetic interaction between the two charges. The elementary interaction here is a charged particle emitting or absorbing a photon. Then, (**1**) a charged particle at some point emits a photon, (**2**) which travels at the speed of light in vacuum to another charged particle, (**3**) which then absorbs it. This photon has thereby mediated the interaction between the two charged particles. As this mediating photon cannot possibly be observed without changing the interaction it mediates, it is a *virtual particle*. The 4-momentum conservation laws cannot be applied to it, since neither energy nor momentum can be observed and measured so as to check. Therefore, there may well be an infinite number of emitted and absorbed virtual mediating photons, and only their combined effect is observed as the effective interaction between the two charged particles. The electromagnetic interaction between two charged particles thus occurs not via an instantaneous action at a distance, but at the speed of light and via the exchange of photons that *mediate* the electromagnetic interaction.

However, it should be clear that this is not a simple kinematic exchange. For example, two ice skaters (to limit the relevance of friction) throwing snowballs at each other certainly exert an interaction mediated by the snowballs. However, this purely kinematic method cannot describe an attractive force: electromagnetic forces may well also be attractive, and even more complicated than that when the charges move with respect to each other.[9]

It will turn out that understanding interactions as *mediated* is generally applicable and in fact a fundamental idea, which will lead to a unified understanding of all fundamental interactions [☞ Chapters 5–7 and 9 for details].

With this in mind, and for ease of locution, we will speak of elementary *particles*, but implicitly understand the de Broglie duality with waves, as well as the essentially more fundamental but technically more demanding representation by fields, the quanta (smallest packets) of which are the elementary particles.

2.3.6 Mesons

Elementary particle physics as described so far had no answer to the obvious question: What keeps the nucleons (protons and neutrons) within the atomic nucleus? It is clear that this force cannot be of electromagnetic origin – neutrons are neutral, and protons repel each other. What's more, this *strong nuclear force* must be stronger than the electromagnetic one, so as to overpower the Coulomb repulsion between protons – at least as long as they are within the atomic nucleus, i.e., at distances no larger than $\sim 10^{-15}$ m. However, at distances much larger than $\sim 10^{-15}$ m, this nuclear force must become negligible even between neutrons, and certainly as compared with the Coulomb repulsion between protons. Since the Coulomb force decays uniformly as $\sim 1/r^2$, it follows that at distances larger than $\sim 10^{-15}$ the strong nuclear force must decay suddenly, much faster than $\sim 1/r^2$.

In 1932, Werner Heisenberg proposed the formalism of *isospin*,[10] to explain the significant similarity between protons and neutrons: their masses differ only by 0.14%, they both have spin $\frac{1}{2}$,

[9] For example, the force with which a small magnetic dipole acts upon an approaching charged particle is orthogonal to the direction between the dipole and the charged particle in motion.

[10] The name itself was bestowed by Eugene Wigner, in 1937.

and the strong interaction within the atomic nuclei does not differentiate between protons and neutrons. Heisenberg thus assigned [☞ Section 4.3 for details]

$$p^+ \mapsto |\tfrac{1}{2},+\tfrac{1}{2}\rangle \qquad \text{and} \qquad n^0 \mapsto |\tfrac{1}{2},-\tfrac{1}{2}\rangle, \tag{2.7}$$

with formal operators $\vec{I}^2, I_3$ defined after those of angular momentum [☞ Appendix A], so that

$$\vec{I}^2|\tfrac{1}{2},\pm\tfrac{1}{2}\rangle = \tfrac{1}{2}(\tfrac{1}{2}+1)\,|\tfrac{1}{2},\pm\tfrac{1}{2}\rangle \qquad \text{and} \qquad I_3|\tfrac{1}{2},\pm\tfrac{1}{2}\rangle = \pm\tfrac{1}{2}\,|\tfrac{1}{2},\pm\tfrac{1}{2}\rangle. \tag{2.8}$$

This model implies that the *exchange* of isospin occurs by exchanging the isospin states

$$\begin{aligned} |1,+1\rangle: &\quad |1,+1\rangle|\tfrac{1}{2},-\tfrac{1}{2}\rangle \to |\tfrac{1}{2},+\tfrac{1}{2}\rangle,\\ |1,0\rangle: &\quad |1,0\rangle|\tfrac{1}{2},\pm\tfrac{1}{2}\rangle \to |\tfrac{1}{2},\pm\tfrac{1}{2}\rangle,\\ |1,-1\rangle: &\quad |1,-1\rangle|\tfrac{1}{2},+\tfrac{1}{2}\rangle \to |\tfrac{1}{2},-\tfrac{1}{2}\rangle, \end{aligned} \tag{2.9}$$

so that by absorbing the state $|1,+1\rangle$, the state $|\tfrac{1}{2},-\tfrac{1}{2}\rangle = |n^0\rangle$ becomes $|\tfrac{1}{2},+\tfrac{1}{2}\rangle = |p^+\rangle$, etc. Overall, in all such processes isospin is conserved.

In 1934, Hideki Yukawa proposed the potential

$$V(r) = -g^2\frac{e^{-r/r_Y}}{r}, \tag{2.10}$$

where the coefficient g^2 is analogous to the product of two electric charges in Coulomb's law, and r_Y is the effective range of the force, for which experiments produce a value of about $r_Y \sim 10^{-15}$ m. Unlike the Coulomb potential that is established by photons – particles with no mass,[11] Yukawa's potential is mediated by particles with mass

$$m_\pi \sim \frac{\hbar}{r_Y c}, \qquad \text{so that} \qquad m_\pi \sim 150\text{–}200\,\text{MeV}/c^2, \tag{2.11}$$

and this mass is responsible for the exponential decay of the potential (2.10). Yukawa's proposal thus predicted a new particle, the *pion*, with a mass (2.11) that is between the electron mass ($m_e \approx 0.511\,\text{MeV}/c^2$) and the proton mass ($m_p \approx 938\,\text{MeV}/c^2$). Combining with Heisenberg's isospin proposal, Yukawa's theory predicts three pions:

$$\pi^+ \leftrightarrow |1,+1\rangle, \quad \pi^0 \leftrightarrow |1,0\rangle, \quad \pi^- \leftrightarrow |1,-1\rangle, \tag{2.12}$$

so that the relations (2.9) correspond to the processes

$$\begin{aligned} |1,+1\rangle|\tfrac{1}{2},-\tfrac{1}{2}\rangle \to |\tfrac{1}{2},+\tfrac{1}{2}\rangle, &\qquad \pi^+ + n^0 \to p^+,\\ |1,0\rangle|\tfrac{1}{2},\pm\tfrac{1}{2}\rangle \to |\tfrac{1}{2},\pm\tfrac{1}{2}\rangle, &\qquad \pi^0 + (p^+ \text{ or } n^0) \to (p^+ \text{ or } n^0),\\ |1,-1\rangle|\tfrac{1}{2},+\tfrac{1}{2}\rangle \to |\tfrac{1}{2},-\tfrac{1}{2}\rangle, &\qquad \pi^- + p^+ \to n^0. \end{aligned} \tag{2.13}$$

What is unusual as compared with electromagnetic interactions: unlike the photon, which is single and electrically neutral, Yukawa's proposal included *three* pions, with charges $-1, 0, +1$, and they would span the 3-dimensional (nontrivial!) representation of the non-abelian (non-commutative) isospin group $SU(2)_I$. This proposal is the forerunner of so-called non-abelian theories of gauge symmetry, which would be introduced two decades later by Chen-Ning Yang and Robert L. Mills, and independently also by Ronald Shaw in his PhD dissertation under Abdus Salam [473].[12]

[11] For a particle of mass m, we have that $E^2 = \vec{p}^2c^2 + m^2c^4$; for photons $m_\gamma = 0$, so that $E_\gamma = |\vec{p}_\gamma|c$.

[12] Ironically, Ernst Stückelberg had independently come up with a proposal very similar to Yukawa's, but Pauli's critique discouraged him from developing his idea. On the other hand, Pauli himself worked independently on a non-abelian generalization of electromagnetism, but was discouraged by his own critical views regarding difficulties in applications to weak interactions – which were resolved only much later by means of the Higgs mechanism – so that he never published on this topic [538].

In 1937, two groups of researchers (C. D. Anderson and S. H. Neddermeyer on the West Coast and J. C. Street and E. C. Stevenson on the East Coast of the USA) independently verified the existence of particles with a mass of the order of magnitude (2.11) when analyzing cosmic ray processes. Later and more precise measurement showed that the particles observed in cosmic rays have a mass very close to 100 MeV/c^2 – which is less than Yukawa's result (2.11), and soon enough different measurements started showing differing results. World War II interrupted these studies, but in 1946, it was shown in Rome, Italy, that the particles of mass $\sim$ 106 MeV/c^2 discovered in cosmic rays interact very weakly with atomic nuclei – completely contrary to the particles in Yukawa's proposal that were supposed to be the very mediators of the strong nuclear force!

In 1947, Robert Marshak and Hans Bethe proposed, and Cecil Powell (in collaboration with C. M. G. Lattes, H. Muirhead and G. P. Ochialini) also verified experimentally that cosmic rays actually involved two types of particles:

1. $\mu^{\pm}$-particles with a mass of $m_\mu \approx 106$ MeV/c^2, which interact very weakly with atomic nuclei and behave like a ~206 times more massive copy of the electron;
2. π-particles with masses of 135 MeV/c^2 (π^0) and 140 MeV/c^2 ($\pi^{\pm}$), which interact strongly with atomic nuclei, and do fit Yukawa's prediction of 12 years earlier.

Later measurements and the quark model would show that the pions cannot be identified with the mediators of the strong interaction, although they do interact by means of the strong interaction, both amongst each other, and also with protons and neutrons; see the discussion leading to equation (6.77).

2.3.7 Antiparticles

Non-relativistic quantum mechanics was completed in only three years, 1923–6: it was conceptually clear that the Schrödinger equation could be adapted to every quantum system, whereupon it "only" remained to solve the differential equation subject to appropriate boundary conditions. Relativistic quantum theory, however, was a tougher nut to crack

In 1927, Paul Dirac discovered the equation that bears his name, and which was supposed to describe free electrons. However, that differential equation has a solution of energy $E = -\sqrt{\vec{p}^2c^2 + m^2c^4}$ for every solution of energy $E = \sqrt{\vec{p}^2c^2 + m^2c^4}$. To avoid the preposterous possibility that the electron interminably loses energy as it successively falls into states of ever lower *negative energy*, Dirac initially proposed that all (infinitely many!) negative-energy states are filled, so that Pauli's principle prevents free electrons from falling into any of those filled states. This "sea" of infinitely many negative-energy electrons (one in each negative-energy state) is totally uniform, so that only the individual electrons with positive energy may be observed.

Furthermore, if any one of these negative-energy electrons were to acquire enough energy to become a positive-energy electron, in its former place in the infinite "Dirac sea" there would remain a "hole" with positive electric charge. Dirac initially hoped that these positively charged holes might be identified with protons, but Hermann Weyl soon showed that the inertial mass of these "holes" is equal to the mass of the free electrons, and so about 1,836 times too small for protons. The other problem with Dirac's idea of an infinite "sea" of electrons with arbitrarily negative energies is that the universe would have an infinitely negative total electric charge, and an infinitely large total mass. Ernst Stückelberg, and then Richard Feynman a little later and in more detail, re-interpreted the theory by introducing the concept of antiparticles (although Feynman attributes this invention to Dirac [166]): according to this by now standard understanding, an antiparticle is a particle traveling backwards in time (both with positive eneries), so that there is no infinitely deep and infinitely charged (for charged particles) Dirac sea [☞ Chapter 3.3].

Digression 2.4 The following is a simplified discussion from Ref. [166]: In all versions of the Fourier transformation there is a factor $e^{-i(E/\hbar)t}$. Changing the sign of energy E is here equivalent to flipping the direction of the passage of time t. Also, the Fourier transform $\widetilde{f}(\omega) := \frac{1}{\sqrt{2\pi}} \int \mathrm{d}t\, e^{i\omega t} f(t)$ has the key property that if $\widetilde{f}(\omega) \neq 0$ for only $\omega := (E/\hbar) \geqslant 0$, then $f(t)$ does not vanish in any continuous interval of time. It then follows that for a process that happens as a successive occurrence of two sub-processes (which are not observed/measured separately and independently but where all exchanged particles have non-negative energy), the time sequence of the two sub-processes depends on the (relativistic!) choice of observer's coordinate system and is not Lorentz-invariant. In turn, it then follows that:

1. antiparticles with positive energy are ordinary particles with positive energy traveling backwards in time;
2. the operations of *parity* (P), time-reversal (T) and charge conjugation (C) satisfy the relation $PT = C$ [☞ Section 4.2.3];
3. a particle–antiparticle pair may be created from vacuum and may (re-)annihilate;
4. for probability conservation:
   (a) two fermions must not be in the same quantum state, so the creation and the re-annihilation of a fermion–antifermion pair contributes to the amplitude of probability negatively,
   (b) two bosons can be in the same quantum state, and that increases the amplitude of probability of the (sub-)process.

See the diagrams (3.82) as well as Feynman's very intuitive and yet sufficiently detailed explanation in [166], where most of Feynman's "half" is dedicated to this connection.

In 1931, Anderson experimentally verified the existence of the *positron* and so also Dirac's theory. However, this implied that all other particles also must have their antiparticles,[13] of the same mass and of the opposite electric charge. Indeed, the antiproton was experimentally verified in 1955, and the antineutron the very next year.

Standard notation for antiparticles is the symbol of the particle with an over-bar: a proton is denoted by p, an antiproton by $\overline{p}$; n denotes a neutron, $\overline{n}$ an antineutron. However, charged leptons [☞ Table 2.3 on p. 67] are customarily denoted by means of the positive charge in the superscript: e^- denotes an electron while e^+ is a positron; μ^- denotes a muon while μ^+ is an anti-muon; τ^- is a tau lepton while τ^+ is the anti-tau lepton. Some of the neutral particles are their own antiparticles, such as the photon, γ, the neutral pion, π^0, and the weak nuclear interaction mediator, Z^0.

2.3.8 *Crossing symmetry and detailed balance*

There exists a general principle called ***crossing symmetry***, according to which for every process

$$A + B \to C + D, \tag{2.14a}$$

where A, B, C, D are particles partaking in the process, the following processes are also possible:

$$A \to \overline{B} + C + D, \tag{2.14b}$$

[13] Strictly, this prediction of antiparticles pertains only to particles of spin $\frac{1}{2}\hbar$, to which Dirac's equation applies.

$$A+\overline{C}\to\overline{B}+D, \tag{2.14c}$$

$$\overline{C}+\overline{D}\to\overline{A}+\overline{B},\qquad \text{etc.}, \tag{2.14d}$$

provided that the ***kinematic*** (4-momentum) conservation laws permit them. In addition, the principle of ***detailed balance*** further predicts the existence of the reverse processes:

$$C+D\to A+B,\qquad \text{etc.}, \tag{2.14e}$$

again, provided that the kinematic conservation laws permit them. We will see later that the computations of the probability for these processes consist of two stages. The first is identical for all of the listed processes (2.14), while the second depends on the kinematics and the 4-momentum conservation laws and may well be drastically different.

Comment 2.1 *Note the difference between the "crossing symmetry" and the "principle of detailed balance" in their present use: The process (2.14d), obtained from (2.14a) using the crossing symmetry, certainly differs from the process (2.14e), which was obtained applying the principle of detailed balance. In this sense, the crossing symmetry permits "moving" any one particle amongst the results of a process into its antiparticle at the input of the process. On the other hand, the application of the principle of detailed balance is equivalent to reversing the direction of time: compare the process (2.14a) with (2.14e) [☞ Section 4.2.2].*

One of the direct applications of the crossing symmetry is the relationship between Compton scattering

$$e^-+\gamma\to e^-+\gamma \tag{2.15}$$

and electron–positron annihilation

$$e^-+e^+\to 2\gamma, \tag{2.16}$$

which shows that the electron–positron annihilation produces *two* photons. That it cannot produce a single photon follows from 4-momentum conservation; see Exercise 3.2.7.

2.3.9 Neutrinos

Lisa Meitner and Otto Hahn had shown in 1911 that the β-decay of atomic nuclei seems to violate the energy conservation law.[14] In decays

$${}^{A}_{Z}\mathrm{X}\xrightarrow{\ \beta\ }{}^{\ \ A}_{Z+1}\mathrm{Y}+e^-,\qquad \text{such as}\qquad {}^{40}_{19}\mathrm{K}\xrightarrow{\ \beta\ }{}^{40}_{20}\mathrm{Ca}+e^-, \tag{2.17}$$

the electron's total relativistic energy (and also the magnitude of the linear momentum) is completely determined by the 4-momentum conservation law [☞ Digression 3.8 on p. 96]:

$$E_e=\Big(\frac{m_X^2+m_e^2-m_Y^2}{2\,m_X}\Big)c^2\;=\;c\sqrt{m_e^2c^2+\vec{p}_e^{\,2}}, \tag{2.18}$$

$$|\vec{p}_e|=c\frac{\sqrt{\big[(m_X+m_e)^2-m_Y^2\big]\big[(m_X-m_e)^2-m_Y^2\big]}}{2\,m_X}. \tag{2.19}$$

Measurements of all energies and linear momenta in β-decays of the type (2.17) showed that the value (2.18) is only the *maximal* value, and that the electron energy varies from case to case. Niels

[14] The neutron was first discovered in 1932, so that one did not know about its role in β-decays of atomic nuclei.

Bohr thus proposed that the 4-momentum conservation laws do not hold in processes involving such small particles.[15]

Opposing Bohr, Wolfgang Pauli proposed in 1930 that – so as to preserve the energy conservation law – β-decay (2.17) actually produces a third particle, with a very small mass and no electric charge, and so difficult to observe. However, this third particle carries away some of the energy, so that the measured values of the electron energy vary from case to case. Pauli proposed the name *neutron*, but was "scooped" for this name by Chadwick, who already used it (in published form by 1932) for the neutral particles that are a little heavier than the protons and which make up about half of the atomic nuclei of most elements.

In 1931, Enrico Fermi named Pauli's particle the *neutrino*[16] and by 1934 had published his theory of β-decay, based on the neutron decay

$$n \xrightarrow{\beta} p + e^- + \overline{\nu}_e, \tag{2.20}$$

which was very successful in describing the experimental observations and measurements. Only later was it established that Pauli's particle was in fact the antineutrino and of the electron species, as correctly denoted in the process (2.20).

2.3.10 Leptons

During 1937–47, particles of about 100–150 MeV/c^2 mass were found in photographs of cosmic ray induced processes, and these were at first identified with Yukawa's mediators of the strong interaction. However, in 1947, Cecil Powell showed that these photographs involve two very different kinds of particles: In a characteristic cascade decay (frequently found on the same photograph) we have

$$\pi^- \longrightarrow \mu^- + \overline{\nu}_\mu, \tag{2.21a}$$

$$\searrow e^- + \overline{\nu}_e + \nu_\mu. \tag{2.21b}$$

Using the derivation (3.44)–(3.49) below, the first decay (2.21a) shows that precisely one particle is not recorded in the photograph (because it is not charged), as the energy of the recorded muon is fixed. Again using the derivation (3.44)–(3.49) below, the second decay (2.21b) produces two invisible particles, because the energy of the visible electron varies. The invisible particles produced in the processes (2.21a) and (2.21b) have here been correctly denoted as $\overline{\nu}_\mu$ and $\overline{\nu}_e + \nu_\mu$, respectively, although in the analyses before 1962 [☞ Table 2.4 on p. 69] one did not know about the difference between the electron- and the muon-neutrino, nor which one in these processes is an *anti*neutrino.

In the first two decades of Pauli's proposal, theoretical proofs that neutrinos must exist abounded. However, no experimental verification was known.

In 1956, Frederick Reines and Clyde Cowan published the results of one of the first big "waiting experiments": a huge tank of water with detector-studded walls, where they waited to observe the so-called inverse β-decay, the process

$$\overline{\nu}_e + p^+ \longrightarrow n^0 + e^+, \tag{2.22}$$

guaranteed to exist by the crossing symmetry. By clever and detailed analysis of a large number of measurements, Reines and Cowan managed to provide an unambiguous experimental proof of the

---

[15] Supposedly [243], Bohr also vigorously fought Einstein's proposal that electromagnetic radiation exists in quanta; he opposed Dirac's electron theory and Pauli's neutrino proposal, ridiculed Yukawa's π-meson theory, and discouraged Feynman from his diagrammatic approach to quantum electrodynamics; he also advised Heisenberg against publishing his uncertainty relation [119].

[16] In Italian, "neutrino" is the diminutive of "neutron," i.e., the "little neutron."

neutrino's existence. Additionally, their analysis showed that antineutrinos interact with ordinary matter extraordinarily feebly: By contemporary estimates (which are by now independently confirmed) the antineutrino flux through their detector was $\sim 5 \times 10^{17}$ antineutrinos per second per meter squared (obtained from the Los Alamos reactor), yet only a handful of type (2.22) reactions were registered per hour.

Thus, in 1956 – a quarter of a century after his original proposal – Pauli's insistence on preserving the conservation laws triumphed! As we will see later, conservation laws are directly related to symmetries – which is the content of Amalie Emmy Noether's theorem. Reliance on symmetries and conservation laws was thereby irrevocably infused into the understanding of Nature – which is highly ironic, given Pauli's denigrating attitude towards group theory [☞ p. 150], which however governs the structure of symmetries. We will see that the success of relativistic physics also may be understood from the point of view of symmetries, although this was definitely not evident at the time.

Applying crossing symmetry to the processes (2.20) and (2.22), we know that the process

$$\nu_e + n^0 \longrightarrow p^+ + e^- \tag{2.23}$$

must also occur. To check if the neutrino is its own antiparticle, Raymond Davis, Jr. and Donald S. Harmer then looked for signals of the analogous reaction

$$\overline{\nu}_e + n^0 \longrightarrow p^+ + e^-, \qquad \text{(this does not occur!)} \tag{2.24}$$

and found no signal although the set-up and analysis was analogous to that of the earlier Reines–Cowan experiment. From the fact that the process (2.22) does occur while (2.24) does not, it follows that a neutrino is distinguishable from an antineutrino, and that we may associate with them opposite values of a conserved quantity.

In fact, in 1953, Konopinski and Mahmoud [319] proposed such a conserved quantity. With a small adaptation of their original proposal, we may call this conserved quantity the *lepton number*, so that

$$L = +1: \quad e^-, \nu_e; \quad \mu^-, \nu_\mu; \quad \tau^-, \nu_\tau; \tag{2.25a}$$

$$L = -1: \quad e^+, \overline{\nu}_e; \quad \mu^+, \overline{\nu}_\mu; \quad \tau^+, \overline{\nu}_\tau; \tag{2.25b}$$

$$L = 0: \quad \text{all other particles.} \tag{2.25c}$$

Of course, in 1953 one knew nothing of the existence of the $\tau^\pm$-leptons and tau-neutrinos, and even the existence of muon-neutrinos (as distinct from electron-neutrinos) was not clear. Nevertheless, using the values (2.25), the reactions (2.20)–(2.23) are permitted, whereas (2.24) is forbidden by the lepton number conservation law.

Conservation of the lepton number – as defined by the values (2.25) – is a law that may be used much as electric charge conservation, although the values (2.25) are ascribed to particles so as to explain the occurrence of processes like (2.20)–(2.23) and the absence of processes like (2.24). Just as electric charge conservation is related to a certain continuous $U(1)$ symmetry [☞ Chapter 5.1–5.3], so lepton number conservation has its "own" symmetry, which will be important when discussing the Lagrangian for the theory of lepton interactions.

Finally, none of the so far mentioned conservation laws prevents the potential decay

$$\mu^- \overset{?}{\longrightarrow} e^- + 2\gamma, \tag{2.26}$$

but this process was never observed. Experience with quantum physics of atomic transitions and reactions and also the increasing number of processes with elementary particles indicates the rule:

Conclusion 2.3 *(Murray Gell-Mann's "totalitarian principle," cited from Ref. [567])*
Everything not forbidden is compulsory.

Thus, the absence of the decay (2.26) requires an explanation[17] in the form of a proposal that leptons $e^{\pm}$, ν_e and $\overline{\nu}_e$ have a separately conserved number, as do $\mu^{\pm}$, ν_μ and $\overline{\nu}_\mu$, and also the later discovered $\tau^{\pm}$, ν_τ and $\overline{\nu}_\tau$.

The absence of the process (2.26) is thus seen as the manifestation of the ban imposed by the separate muon- and electron-number conservation laws. These laws in turn do permit the decays (2.21a)–(2.21b): the net sum of each of the three separate lepton numbers on the "before" side of either of those processes equals the sum of those same lepton numbers on the "after" side. The same holds for all other listed processes, including Fermi's formula (2.20) for β-decay. This illustrates the basic principle that laws of Nature must have no exception and must hold universally.[18]

This analysis, following the proposal by Konopinski and Mahmoud [319], formally introduced a separate conservation law for the electron-, muon- and tau-lepton numbers, in addition to the conservation laws of electric charge, 4-momentum (i.e., energy and 3-dimensional linear momentum), and of the angular momentum.

— ❦ —

Finally, it is worth noting that the name *lepton* stems from the Greek adjective λεπτός (light), because of the relatively small electron mass, as compared to that of the proton and the neutron. For the latter two, the collective name is *baryon*, from the Greek adjective βαρύς (heavy). Particles with masses between $m_e \approx 0.511$ MeV/c^2 and $m_p \approx 938$ MeV/c^2 were thus named *mesons*, from the Greek word μέσος (middle). However, the discovery of the muon and the verification that it is identical to the electron – except for being ~206 times heavier – suggested that the original and naive nomenclature had to change. By the mid-twentieth century, these names were re-purposed according to the interaction type, as shown in Table 2.2. The fact that the lepton and the baryon numbers are conserved in all processes, whereas the number of mesons is not, is a feature of Nature that slowly became ever clearer, through the analysis of an ever larger number of processes. The name *hadron* stems from the Greek word ἁδρός (thick, bulky).

Table 2.2 Defining collections of elementary particles according to their interactions

	Group	Nuclear interactions	Spin	Number
	Leptons	Only weak	Half-integral	Conserved
Hadrons	**Mesons**	Both strong and weak	Integral	Not conserved
	Baryons	Both strong and weak	Half-integral	Conserved[a]

[a]Baryon number (albeit not under that name) conservation proposed in 1938 by Ernst Stückelberg, to explain the absence of the $p^+ \to e^+ + \pi^0$ proton decay.

2.3.11 Strange particles

By mid-1947, there was perfect experimental proof of the existence of the electron, the proton and the neutron, of which practically all substance around us is composed. Yukawa's π-meson

[17] Candidates for a law of Nature are not proven but disproven by exceptions.

[18] One often says that for small speeds non-relativistic physics holds and that for speeds that are near the speed of light in vacuum relativistic physics holds. Literally taken, this is false: what is true is that relativistic physics holds always, but that for small enough speeds the non-relativistic approximations *suffice in practice*. That is, the difference between particular concrete results of relativistic computations and their non-relativistic approximations cannot be experimentally detected.

was also experimentally detected, so that there existed a real chance for a theoretical description of the strong nuclear interactions to be developed so as to adequately reproduce the experimental facts about atomic nuclei. Fermi's theory of β-decay adequately described all known effects of weak nuclear interaction. The antiparticle of the electron that Dirac predicted was also detected experimentally and there was no doubt that, upon appropriate development of experimental devices, all other antiparticles would be experimentally produced. The existence of the neutrino had experimentally still not been verified, but at least ever more theorists agreed that it did have to exist.

Thus, only the existence of the muon presented a capricious puzzle of Nature: this about 206 times heavier copy of the electron was completely unexpected and unexplained.

— ❦ —

In December of 1947, George D. Rochester and Clifford C. Butler opened Pandora's box: They published the results of their analysis of photographs of cosmic rays in the (Wilson) cloud chamber, from which it followed that there existed a neutral particle with a mass of about half the neutron mass, and which decays

$$K^0 \longrightarrow \pi^+ + \pi^-. \tag{2.27}$$

In 1949, Cecil Powell published the experimental discovery of a new charged particle that decays

$$K^+ \longrightarrow \pi^+ + \pi^+ + \pi^-. \tag{2.28}$$

In 1950, Carl D. Anderson's group at CalTech discovered a new neutral particle that decays

$$\Lambda^0 \longrightarrow p^+ + \pi^-. \tag{2.29}$$

Only by 1956 did it transpire that K^0 and K^+ are closely related, just as if they were heavier copies of π^0 and π^+, but even in these early years it was clear that these new particles were rather unusual: $K^0, K^\pm, \Lambda^0$ – and soon several more were discovered – were all produced in very fast ($\sim 10^{-23}$ s) reactions, but their half-life (and lifetime) was relatively long: $\sim 10^{-10}$–10^{-8} s.

It soon turned out that these new particles were created in pairs, so Abraham Pais proposed the concept of "associated production." In 1953, Kazuhiko Nishijima and Tadao Nakano transformed this proposal into a concept of the "eta charge," and in 1965, Murray Gell-Mann independently introduced the "strangeness" charge. Under that name, the idea was finally adopted: When "strange" particles are created, they are created in "strange–antistrange" pairs, which indicates a strangeness conservation law. Thereafter, the decay of a "strange" ($S_0 \neq 0$) particle into a collection of particles the total strangeness of which is $\sum_i S_i \neq S_0$ would be forbidden by this strangeness conservation law, and so could happen only via an interaction that violates this law explicitly. This then establishes that strange particles are created by one (strong) interaction, and decay by another (weak). In addition, this proposal also contained the so-called GNN formula:

$$Q = I_3 + \tfrac{1}{2}(B + S), \tag{2.30}$$

after the initials of Gell-Mann, Nishijima and Nakano. Here, Q denotes the electric charge, I_3 is the isospin [☞ discussion around relations (2.7)–(2.12)], B the baryon number and S denotes the *strangeness*. The fact that these quantum numbers may be consistently assigned to a growing number of particles (both mesons and hyperons[19]) while satisfying the relation (2.30) indicates a regularity that needed an explanation.

[19] The word "hyperon" was initially used for particles heavier than the neutron; nowadays, it is used for strange baryons with neither charm, nor beauty, nor truth [☞ table in display (2.44a)].

2.3.12 The eightfold way

In the early 1930s, the list of elementary particles was short and really simple: Substance consists of electrons, protons and neutrons. These particles interact via electromagnetic forces, mediated by photons, weak nuclear forces formulated by Fermi as a so-called contact interaction (with no mediator), and strong nuclear forces for which Yukawa's theory with pions as mediators seemed a good candidate. The fourth fundamental interaction, gravitation, had by then been described by Einstein's general theory of relativity.

However, by 1960, this list of elementary particles was joined by so many new particles that a systematization became necessary, somewhat akin to Mendeleev's periodic table of elements. Murray Gell-Mann noticed that a 2-dimensional plot of the first eight pseudo-scalar[20] mesons:

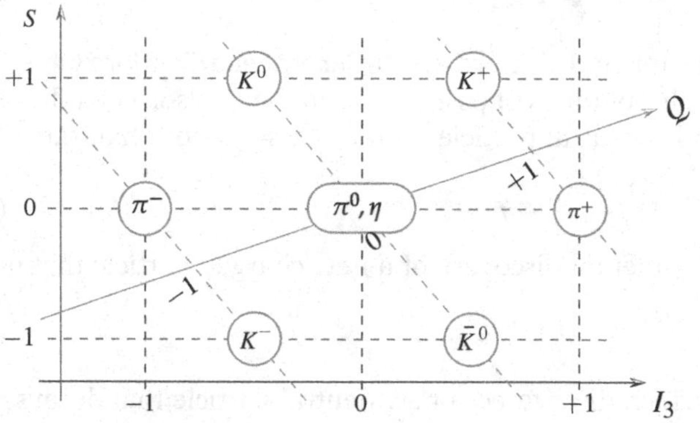

$$\begin{aligned} m_{\pi^0} &= 135.0\,\text{MeV}/c^2\\ m_{\pi^\pm} &= 139.6\,\text{MeV}/c^2\\ m_\eta &= 547.9\,\text{MeV}/c^2\\ m_{K^\pm} &= 493.7\,\text{MeV}/c^2\\ m_{K^0} &= 497.6\,\text{MeV}/c^2\\ &\text{spin} = 0, \end{aligned} \tag{2.31}$$

looks very similar to the analogous plot of baryons:

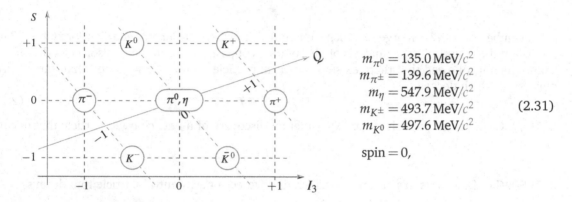

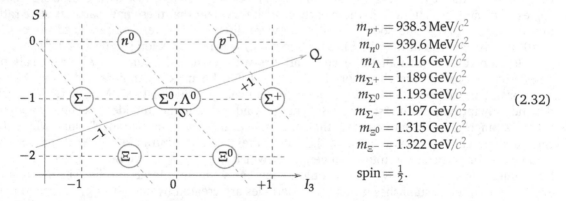

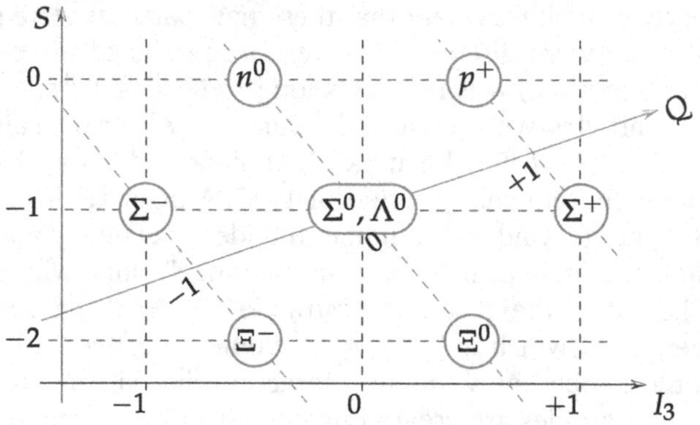

$$\begin{aligned} m_{p^+} &= 938.3\,\text{MeV}/c^2\\ m_{n^0} &= 939.6\,\text{MeV}/c^2\\ m_\Lambda &= 1.116\,\text{GeV}/c^2\\ m_{\Sigma^+} &= 1.189\,\text{GeV}/c^2\\ m_{\Sigma^0} &= 1.193\,\text{GeV}/c^2\\ m_{\Sigma^-} &= 1.197\,\text{GeV}/c^2\\ m_{\Xi^0} &= 1.315\,\text{GeV}/c^2\\ m_{\Xi^-} &= 1.322\,\text{GeV}/c^2\\ &\text{spin} = \tfrac{1}{2}. \end{aligned} \tag{2.32}$$

Besides, Gell-Mann had in 1961, and Susumu Okubo independently in 1962, noticed that the masses of the particles in these diagrams satisfy (up to a small percentage error) the relations

$$2\Big[\Big(\frac{m_{K^-}+m_{\bar{K}^0}}{2}\Big)^2+\Big(\frac{m_{K^0}+m_{K^+}}{2}\Big)^2\Big] \approx 3m_\eta^{\,2}+\Big(\frac{m_{\pi^-}+m_{\pi^0}+m_{\pi^+}}{3}\Big)^2, \tag{2.33}$$

$$2\Big[\frac{m_p+m_n}{2}+\frac{m_{\Xi^-}+m_{\Xi^0}}{2}\Big] \approx 3m_\Lambda+\frac{m_{\Sigma^-}+m_{\Sigma^0}+m_{\Sigma^+}}{3}. \tag{2.34}$$

[20] A scalar function has spin 0, and does not change under rotations; the prefix *pseudo* then indicates that unlike real scalars that are invariant also with respect to the $\vec{r} \to -\vec{r}$ reflection, pseudo-scalars change their sign.

The next collection of baryons forms a somewhat different figure:

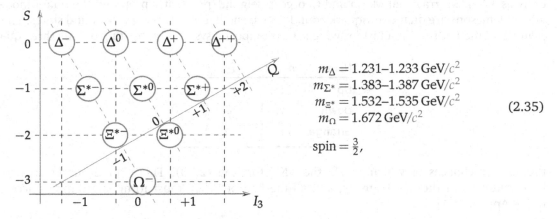

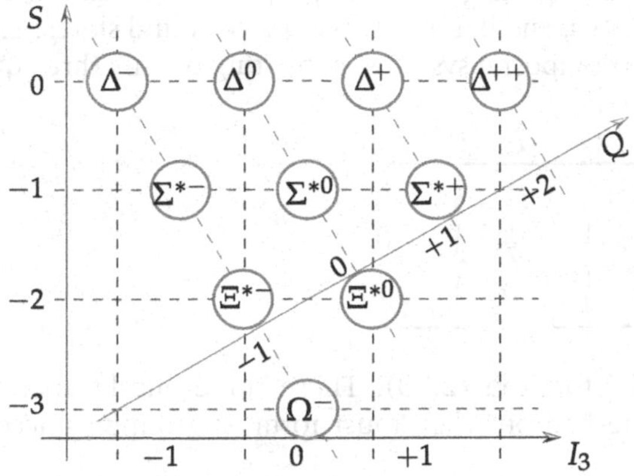

$$m_\Delta = 1.231\text{–}1.233\,\text{GeV}/c^2$$
$$m_{\Sigma^*} = 1.383\text{–}1.387\,\text{GeV}/c^2$$
$$m_{\Xi^*} = 1.532\text{–}1.535\,\text{GeV}/c^2 \qquad (2.35)$$
$$m_\Omega = 1.672\,\text{GeV}/c^2$$
$$\text{spin} = \tfrac{3}{2},$$

and where Ω^- has not yet been discovered. Following the approximate relation

$$\frac{9m_\Lambda + m_{\Sigma^-} + m_{\Sigma^0} + m_{\Sigma^+}}{12} - \frac{m_p + m_n}{2} \approx \frac{m_{\Xi^-} + m_{\Xi^0}}{2} - \frac{9m_\Lambda + m_{\Sigma^-} + m_{\Sigma^0} + m_{\Sigma^+}}{12}, \qquad (2.36)$$

between the masses of the baryons in the octet (2.32), i.e., that their average masses grow uniformly with the increasing of the absolute value of strangeness,[21] Gell-Mann postulated the relation

$$m_\Delta - m_{\Sigma^*} = m_{\Sigma^*} - m_{\Xi^*} = m_{\Xi^*} - m_\Omega, \qquad (2.37)$$

where m_Δ, m_{Σ^*} and m_{Ξ^*} are the average masses of the particles in the upper three rows in the plot (2.35). This predicted the existence of the Ω^- particle with a mass of about 1.70 GeV/c^2, up to a small percentage. In 1964, this particle was experimentally detected, with a mass that differs only 0.6 % from the value predicted by Gell-Mann's relation (2.37). The Reader may have asked why, e.g., four Δ-particles are not in the same collection with the proton, the neutron, and the Λ- and Σ-particles, since the masses of these baryons are closer to the mass of the Δ-particle than to the mass of the Ξ-particle. As it was done in those early 1960s, the answer is pragmatic: because the values of the spin indicated in the plots (2.32) and (2.35), and the formulae (2.34), (2.36) and (2.37) gather particles into collections as shown and not otherwise; [☞ Section 4.4].

This proved the classifying scheme that Gell-Mann called the "eightfold way" (alluding to the "noble eightfold way" of Buddhism). Even though in the 1960s this was not entirely clear, these results crucially use $SU(3)$ symmetry – because of which the plots (2.31), (2.32) and (2.35) are nowadays usually drawn with the S-axis at an angle of 120° with respect to the Q-axis, as is customary for the root system of the $SU(3)$ group [581, 105, 256, 447].

It should be noted that the meson collections, such as (2.31), contain both particles and antiparticles, in diametrically opposite positions: π^- is anti-π^+, K^- is anti-K^+, etc. On the other hand, the baryon groups (2.32) and (2.35) contain only particles, and their antiparticles form identical collections but with charges that all have opposite signs.

2.3.13 Quarks

The true meaning of Gell-Mann's "eightfold way," however, is the systematic use of group theory in particle classification, as was independently proposed by Yuval Ne'eman about the same time. Whereas Mendeleev's periodic table (1869) waited many years for a final explanation (1925) by

[21] The sign and an additive constant in the definition of the quantum number of strangeness are for all present purposes completely arbitrary.

way of quantum mechanics and Pauli's exclusion principle, the "eightfold way" was explained as early as 1964: Murray Gell-Mann and George Zweig independently proposed the quark model,[22] whereby mesons are quark–antiquark bound states, and baryons three-quark bound states. This explained all the hadrons in a fast growing list as composite systems, consisting of only three quarks:

Name	$q: Q$	I_3	B	S
Up	$u: +\frac{2}{3}$	$+\frac{1}{2}$	$\frac{1}{3}$	0
Down	$d: -\frac{1}{3}$	$-\frac{1}{2}$	$\frac{1}{3}$	0
Strange	$s: -\frac{1}{3}$	0	$\frac{1}{3}$	-1

(2.38)

the various charges of which satisfy the GNN formula (2.30). These three quarks span the **3** representation of the $SU(3)$ group, and all the hadrons then must form as groupings according to [☞ Appendix A]

$$\text{mesons} = (q\bar{q}) \; : \qquad \mathbf{3} \otimes \mathbf{3}^* = \mathbf{1} \oplus \mathbf{8}, \tag{2.39}$$

$$\text{baryons} = (qqq) : \quad \mathbf{3} \otimes \mathbf{3} \otimes \mathbf{3} = \mathbf{1} \oplus \mathbf{8} \oplus \mathbf{8} \oplus \mathbf{10}, \tag{2.40}$$

which predicts that mesons must form *singlets* (separate states) or *octets* (collections of eight), whereas baryons must form *singlets* or *octets* or *decuplets*; see Example A.6. This is consistent with experiments: for each collection of particles there exist formulae of the forms (2.33)–(2.34) and (2.36)–(2.37). The combinatorial details of the quark model (and why there are in fact no singlet baryons) are the subject matter of Chapter 4.

Together with the success in classifying hadrons, the quark model also had two important problems:

1. no experiment could produce a free quark;
2. in some baryons it seemed that the existence of the three-quark state violated Pauli's exclusion principle.

The impossibility of extracting *free* quarks does not imply that they were impossible to verify experimentally: One merely needs an adequate probe to "see" the composite structure of a hadron without freeing its constituent building blocks, just as Rutherford bombarded gold atoms with α-particles and "saw" the atomic nuclei without (even partially) ionizing the atoms. In the late 1960s such *deep inelastic* collisions were performed at SLAC (*Stanford Linear Accelerator Center*), bombarding protons with electrons, and later in the 1970s at CERN,[23] bombarding protons with neutrinos, and then also with protons. Much as in Rutherford's experiment, the probes pass right through the protons – in growing number at growing energies and with little deflection, while a small number deflect at a large angle (even to 180°). However, the details of the scattering indicate that the proton is very well represented as a composite system consisting of *three* particles, in agreement with the quark model. Also, while we *can* ionize atoms and so fully liberate their nuclei, quarks cannot be extracted free from the hadrons. Repeated failures to do so created an undercurrent of mistrust in the quark model, whereby most experimentalists rather used Feynman's term, *partons*, for the scattering centers within the hadrons.

[22] Interestingly, Gell-Mann nevertheless advocated that the quarks are not necessarily "real," concrete particles in the usual sense with well-localized and detectable position [☞ Digression 4.2 on p. 151]. In this respect, he drastically differed from Richard Feynman (also at CalTech). Feynman advocated for real particle constituents, but called them "partons" to avoid Gell-Mann's quarks. A decade later, Gell-Mann's quarks (together with gluons) proved to be Feynman's partons. On the other hand, Zweig used the term "ace" (presumably alluding to "ace up the sleeve") for quarks, but that never caught on [592, 119].

[23] In 1954, CERN was established as *Conseil Europeen pour la Recherche Nucleaire*, the provisional council that established the laboratory *Organisation Europeenne pour la Recherche Nucleaire*, keeping however the acronym.

On the other hand, in 1964, Oscar W. Greenberg noticed that the quark model seems to violate Pauli's exclusion principle: certain spin-$\frac{3}{2}$ states such as $\Delta^- = (d\,d\,d)$, $\Delta^{++} = (u\,u\,u)$ and the celebrated $\Omega^- = (s\,s\,s)$ should be forbidden owing to Pauli's exclusion principle. In all three of these cases, experiments indicate that the three *otherwise identical* quarks are in the *S*-state, with parallel spins, and so in the very same quantum state – contradicting Pauli's exclusion principle. Greenberg thus proposed [229] that quarks satisfy *para-fermionic* (anti-)commutation rules: Formally, while bosonic creation and annihilation operators commute and the analogous fermionic operators anticommute,

$$[\,b_i\,,\,b_j^{\,\dagger}\,] = \delta_{ij}, \qquad [\,b_i\,,\,b_j\,] = 0 = [b_i^{\,\dagger},b_j^{\,\dagger}], \qquad \text{bosons,} \tag{2.41a}$$

$$\{\,f_i\,,\,f_j^{\,\dagger}\,\} = \delta_{ij}, \qquad \{\,f_i\,,\,f_j\,\} = 0 = \{f_i^{\,\dagger},f_j^{\,\dagger}\}, \qquad \text{fermions,} \tag{2.41b}$$

para-fermion creation and annihilation operators satisfy the hybrid relations:

$$\left.\begin{aligned} \{\,\tilde{f}_{i,\alpha}\,,\,\tilde{f}^{\dagger}_{j,\alpha}\,\} &= \delta_{ij}, & \{\,\tilde{f}_{i,\alpha}\,,\,\tilde{f}_{j,\alpha}\,\} &= 0 = \{\,\tilde{f}^{\dagger}_{i,\alpha}\,,\,\tilde{f}^{\dagger}_{j,\alpha}\,\}, & & (2.41c)\\ [\,\tilde{f}_{i,\alpha}\,,\,\tilde{f}^{\dagger}_{j,\beta}\,] &= \delta_{ij}, & [\,\tilde{f}_{i,\alpha}\,,\,\tilde{f}_{j,\beta}\,] &= 0 = [\,\tilde{f}^{\dagger}_{i,\alpha}\,,\,\tilde{f}^{\dagger}_{j,\alpha}\,], & \alpha \neq \beta, & \quad (2.41d) \end{aligned}\right\}\ \text{para-fermions,}$$

where $\alpha,\beta = 1,2,3$. In January 1965, Boris V. Struminsky proposed in a paper presented at Dubna (Moscow region, Russia) an "additional" quantum number to resolve this problem, and continued working on this with his mentor Nikolay Bogolyubov, and collaborator Albert Tavchelidze. In May 1965, Tavchelidze presented this idea at ICTP, in Trieste (Italy), without his collaborators' knowlege. Six months later, Moo-Young Han and Yoichiro Nambu independently proposed a model where a new degree of freedom, α,β in (2.41c)–(2.41d) is a new kind of "charge" and has its own interaction with eight new "photons," $g_\alpha{}^\beta$ where $g_\alpha{}^\alpha = 0$. This is what is called *color* today, and which differentiates the quarks in the hadrons $\Delta^-,\Delta^{++},\Omega^-$ so Pauli's exclusion principle would not forbid their existence. In their model, quarks had integer, but color-dependent electric charges [☞ Digression 5.14 on p. 214]. The final version of the formalism of *color* as a charge for strong interaction and which is independent from the electric charge was completed by William Bardeen, Harald Fritzsch and Murray Gell-Mann, in 1974.

On the third hand, in 1964, Sheldon L. Glashow and James D. Bjorken had proposed the existence of a fourth quark, dubbed *charm*, and with it an extension of the classification $SU(3)$ group into $SU(4)$. In 1970, Glashow, John Iliopoulos and Luciano Maiani provided a theoretical proof that the fourth quark must exist based on the absence of weak decays that would change strangeness:

$$\frac{K^+ \to \pi^+ + \overline{\nu} + \nu}{K^+ \to \pi^0 + \mu^+ + \mu^-} = \frac{(u\bar{s}) \xrightarrow{Z^0} (u\bar{d}) + \overline{\nu} + \nu}{(u\bar{s}) \xrightarrow{W^+} (u\bar{u}) + \mu^+ + \mu^-} < 10^{-5}. \tag{2.42}$$

Besides, the appearance of the so-called Adler–Bell–Jackiw (ABJ) $U(1)$ *anomaly* [☞ Section 7.2.3, and the lexicon entries about anomaly and canonical quantization] indicates an inconsistency in the gauge theory of weak interaction: the quark model with only three quarks exhibits a symmetry that is broken by quantum effects, but this breaking cancels through contributions from the fourth quark. That is the first application of a detailed quantum analysis of symmetries; this was later developed into a very powerful theoretical method and gave rise to certain exact (non-perturbative) results in field theory. The c-quark was experimentally discovered four years later. However, a more detailed analysis requires details of both gauge theories and of left–right asymmetric weak interactions, and we will return to this topic in Sections 7.2.2 and 7.2.3.

Finally, in 1973, Makoto Kobayashi and Toshihide Maskawa showed that the so-called *indirect* CP-violation – which James W. Cronin and Val L. Fitch observed back in 1964 (and for which they received the Nobel Prize as late as 1980) in the unevenness of the $K^0 \leftrightarrow \bar{K}^0$ transmutation – can happen only if there exist at least six quarks. The direct *CP*-violation was observed in the 1990s, in agreement with the Kobayashi–Maskawa proposal. By that time the fifth, *b*-quark had been experimentally produced, while the sixth, *t*-quark was produced five years later, in 1995. The species of quarks

$$u \text{ (up)}, \quad d \text{ (down)}, \quad s \text{ (strange)}, \quad c \text{ (charm)}, \quad b \text{ (bottom)}, \quad t \text{ (top)}, \tag{2.43}$$

are dubbed *flavors*, so that the symmetry group that arises from approximate identification of these quarks: $SU(3)$, $SU(4)$, $SU(5)$ and $SU(6)$ is typically labeled by the subscript "f." These $SU(n)_f$ approximate "flavor" symmetries are less and less practical for larger and larger n. The quark masses are more and more different, and the symmetries are less and less precise:

(2.44a)

Name	q	**Mass*** (MeV/c^2)	Q	I_3	B	S	C	B'	T	Y
Up	u :	1.5–3.3	$+\frac{2}{3}$	$+\frac{1}{2}$	$\frac{1}{3}$	0	0	0	0	$+\frac{1}{3}$
Down	d :	3.5–6.0	$-\frac{1}{3}$	$-\frac{1}{2}$	$\frac{1}{3}$	0	0	0	0	$+\frac{1}{3}$
Strange	s :	$105\{^{+25}_{-35}$	$-\frac{1}{3}$	0	$\frac{1}{3}$	-1	0	0	0	$-\frac{2}{3}$
Charm	c :	$1{,}270\{^{+70}_{-110}$	$+\frac{2}{3}$	0	$\frac{1}{3}$	0	$+1$	0	0	$+\frac{4}{3}$
Bottom	b :	$4{,}200\{^{+170}_{-70}$	$-\frac{1}{3}$	0	$\frac{1}{3}$	0	0	-1	0	$-\frac{2}{3}$
Top	t :	$171{,}300\{^{+1{,}100}_{-1{,}200}$	$+\frac{2}{3}$	0	$\frac{1}{3}$	0	0	0	$+1$	$+\frac{4}{3}$

* Inertial mass without the binding energy, which depends on the hadron

$$Q = I_3 + \tfrac{1}{2}\underbrace{(Baryon + Strange + Charm + B'eauty + Truth)}_{=Y,\ \text{so-called (strong) hypercharge [☞ Section 7.2.1]}} \tag{2.44b}$$

During 1964–74, feelings about the quark model were rather mixed. While experimentalists rightfully decried the fact that quarks seemed to be impossible to extract free – for which there was no theoretical explanation – even the quantum number of color, introduced to reconcile the quark model with Pauli's exclusion principle, seemed more of a mnemonic crutch than a real physical property. Just as the quarks seemed impossible to extract, so was the color impossible to detect directly: The three quarks in a baryon each have a different color, red–blue–yellow, so that baryons are "colorless." The quark and the antiquark in a meson have opposite colors (red–green, blue–orange, or yellow–purple), so that mesons are also "colorless." As a classifying system, this rule perfectly predicted hadronic states [☞ Chapter 4]. However, the skeptical physicists could not escape the impression that the color formalism was "invented" so as to "explain" the otherwise unexplained fact: the formalism gave no *reason* for hadrons to be "colorless" [☞ Section 6.1].

Finally, note that the quark model revised the elementary particle image: all substance consists of quarks (u, d, s, c, b, t) and leptons $(e^-, \nu_e, \mu^-, \nu_\mu, \tau^-, \nu_\tau)$, which returns simplicity into the list of elementary constituents of the World: by 1974, the number of hadrons had approached a hundred and no one could possibly consider them elementary. By contrast, the list of (then known) quarks and leptons was short and even fairly "symmetric."

In spite of this *theoretical* and *aesthetic* attractiveness, the event that was crucial in winning the confidence of most physicists in the quark model was the *experimental* detection of the so-called J/ψ particle, towards the end of 1974. During the summer of 1974, C. C. Samuel Ting's research group in Brookhaven discovered a 3.0969 GeV/c^2-mass particle (a little over three times the proton mass), and with a half-life measured to be around 10^{-20} s – which is some 1,000 times longer than the typical hadron half-life! Ting insisted on careful checking of this astounding result, so that the discovery remained a secret till November of that year, when Burton Richter's research group at SLAC discovered the same particle, and the two research groups published their results back-to-back.

Over the next few months, it became clear that the J/ψ particle was a $(c\bar{c})$ bound state, i.e., a meson,[24] which, being akin to positronium, is called "charmonium." The stability of this system follows from the combination of the so-called OZI rule (Susumu Okubo, George Zweig and Jugoro Iizuka) and the fact (discovered soon) that the pair of lightest mesons with a single c- or a single $\bar{c}$-quark (the so-called D, i.e., $\bar{D}$ meson, respectively) is heavier than the J/ψ particle. Thus, a J/ψ does not have enough mass to decay into a D–$\bar{D}$ pair, and the decays that require the c–$\bar{c}$ annihilation are slowed down by the OZI rule.[25] When the J/ψ particle was discovered, it became obvious that its existence fits so perfectly in the quark model that all doubt in the model vanished: As should be clear from the foregoing, and also Table 2.4 on p. 69, the fourth quark was already predicted, both from aesthetic (Bjorken and Glashow) and also technical (what was only later understood as most stringent and rigorous) reasons of anomaly cancellation (Glashow, Iliopoulos and Maiani).

In 1975, a new lepton, τ^-, was discovered, supporting the prediction (Kobayashi and Maskawa, two years earlier) of another lepton and quark pair, so as to explain *CP*-violation. Just as with the J/ψ particle, a $(b\bar{b})$ bound state was discovered in 1977 and dubbed Υ. The sixth, t-quark was finally detected in 1995, and the τ-neutrino, ν_τ, in 2000. Thus, by the dawn of the third millennium, the list of elementary particles (2.44a) was experimentally confirmed.

Digression 2.5 It should be noted that the charming story with charmonium, and even with *bottomonium* (i.e., with the $(b\bar{b})$-state Υ) will probably not be repeated with *toponium*: The t-quark itself has a 174.2 GeV/c^2 mass, so that according to Ref. [215] and contemporary data [293] the $(t\bar{t})$-state would have to have a mass of about 344.4 GeV/c^2, and a standard deviation of $\sigma \approx$ 193.6 MeV/c^2. Thus, even when the experiments reach energies over about 344.4 GeV, the large absolute value of the standard deviation implies that the toponium will decay much faster than the J/ψ and the Υ [☞ Section 2.4], and it may well turn out that it will behave practically as a virtual particle, which by definition cannot be detected directly.

2.3.14 Nuclear force intermediaries

The introduction of *color*, the new quantum number assigned to quarks but not to leptons, not only explained the existence of spin-$\frac{3}{2}$ baryons such as $\Delta^-, \Delta^{++}, \Omega^-$ in terms of three-quark S-state bound states, but also brought about the ultimate explanation of the strong nuclear interaction. The exchange of color between quarks is mediated by gluons, which thus mediate the strong interaction; the details of this mechanism will be examined in Chapters 5–6. Suffice it to say, the theoretical basis for the so-called Yang–Mills theory of non-abelian (non-commutative) gauge symmetry had been introduced back in 1954–5: Chen-Ning Yang and Robert L. Mills and, independently, Ronald Shaw in his PhD dissertation under Abdus Salam, showed how the electromagnetic

[24] By 1974, it became evident that the old motivation for the nomenclature, whereby *mesons* had masses between those of the electron and the proton, was no longer practical. Instead, the quark model convention was adopted, wherein every three-quark bound state was called a *baryon*, and every quark–antiquark bound state was called a *meson*. Hypothetical states of other composition, such as (qq), $(qq\bar{q})$, $(qqqq)$, etc., were referred to as exotic particles.

[25] A decade later, the OZI rule was derived from quantum chromodynamics, but in 1974 it was still only a phenomenological *rule*.

interaction with $U(1)$ symmetry group [☞ Sections 5.1–5.3] can be generalized into a gauge interaction with a non-abelian (non-commutative) symmetry [☞ Section 6.1, as well as Appendix A]. However, the crucial (and in fact complementary) qualities of the quark model,

1. that quarks cannot be extracted free from hadrons (so-called confinement), and
2. that the closer quarks are to each other, the weaker they interact (so-called asymptotic freedom, experimentally observed as early as 1967)

are not obvious consequences of the model. This latter quality was discovered by David Gross with his student Frank Wilczek, and independently by David Politzer in 1973,[26] almost two decades after the discovery of non-abelian gauge theory itself. Although the theoretical proof of the first quality (confinement) is still not rigorously complete☝, the proof of asymptotic freedom as its complementary quality caused an overnight universal acceptance of quantum chromodynamics (QCD) as *the* theory of strong nuclear interactions. In addition, numerical computations in so-called "lattice QCD" (where the infinite and continuous spacetime is approximated by a finite-size lattice with a nonzero lattice spacing) and "Monte Carlo simulations" soon showed that QCD correctly reproduces many of the ratios of hadron masses as well as many other parameters of so-called hadron spectroscopy, so that the crucial unsolved problem☝ is "only" to compute the absolute value of a characteristic mass unit for hadrons. It is worth noting that the theoretical discovery of asymptotic freedom and the experimental discovery of the J/ψ particle practically coincided. Without a doubt, that tandem advancement was decisive in the sudden turn of tide in accepting the quark model together with quantum chromodynamics, and this combination of events is sometimes referred to as the "November Revolution" of 1974.

— ❦ —

Meanwhile, for weak nuclear interactions, Enrico Fermi formulated the so-called 4-fermion contact interaction in 1931–4, which within the quark model stated is as

$$d \longrightarrow u + e^- + \overline{\nu}_e, \tag{2.45a}$$

together with all possible related processes obtained via crossing symmetry and the principle of detailed balance (2.14), such as

$$d + \overline{u} \to e^- + \overline{\nu}_e, \qquad d + e^+ \to u + \overline{\nu}_e, \qquad u + e^- \to d + \nu_e, \qquad \text{etc.} \tag{2.45b}$$

Since the range of the weak interactions is small ($\sim 10^{-15}$ m), Fermi's approximation, where the weak interaction happens in a point, is quite satisfactory up to energies of about $\hbar c/(10^{-15}\,\text{m}) \sim 200\,\text{MeV}$. However, within two decades after Fermi's theory, such energies had been surpassed in accelerators and a better theory was needed. The contact interaction (2.45b) is analogous to describing a scattering such as of a positron and a proton, $e^+ + p^+ \to e^+ + p^+$, by neglecting the repulsive interaction field and pretending that the two like-charged particles actually touch during the collision. Akin to electromagnetic interactions, these processes may be described also by introducing *intermediaries*:

[26] There is a sad story tied to this discovery [366]: David Gross mentored two students, Frank Wilczek and William E. Caswell. Gross asked Wilczek to compute the so-called β-function for the $SU(n)$ gauge theory to first order in perturbation theory and Caswell to second. When Wilczek discovered that the β-function had the opposite sign from the abelian case, Gross realized the fantastic importance of the result – the theoretical proof that non-abelian gauge theory guarantees asymptotic freedom – and published this immediately with Wilczek. This is one of the most cited papers in the second half of the twentieth century; Gross and Wilczek shared with Politzer, the 2004 Nobel Prize. Caswell's contribution, which confirmed Wilczek's and correctly showed that second-order perturbations do not spoil the newly discovered asymptotic freedom was hardly noticed. William E. Caswell died in the Pentagon crash of the American Airlines Flight 77, on September 11, 2001.

$$
\begin{aligned}
d &\to u + W^- \quad\quad\quad\;\; \to u + (e^- + \overline{\nu}_e), && \text{(2.46a)}\\
d + \overline{u} &\to W^- \quad\quad\quad\quad\quad\;\; \to e^- + \overline{\nu}_e, && \text{(2.46b)}\\
d + e^+ &\to (u + W^-) + e^+ \to u + \overline{\nu}_e, && \text{(2.46c)}\\
&\searrow d + (W^+ + \overline{\nu}_e) \;\to u + \overline{\nu}_e, && \text{(2.46d)}\\
u + e^- &\to (d + W^+) + e^- \to d + \nu_e, && \text{(2.46e)}\\
&\searrow u + (W^- + \nu_e) \;\;\to u + e^-, && \text{(2.46f)}
\end{aligned}
$$

and so on. These examples make it clear that we have postulated the intermediaries $W^\pm$ for the weak interaction. Their elementary processes,

$$d + \overline{u} \leftrightarrow W^-, \qquad \overline{d} + u \leftrightarrow W^+, \tag{2.47}$$

$$e^- + \overline{\nu}_e \leftrightarrow W^-, \quad e^+ + \nu_e \leftrightarrow W^+, \tag{2.48}$$

as well as all related processes obtained using the crossing symmetry and the principle of detailed balance (2.14), and by replacing the u, d quarks with the heavier pair c, s (and also t, b), as well as by replacing the e^-, ν_e lepton pair with μ^-, ν_μ (and also τ^-, ν_τ) might seem amply sufficient to describe all known examples of weak interaction. However, that would be false: From the observed decay

$$K^0 = (d\,\overline{s}) \to \pi^+ + \pi^- = (u\overline{d}) + (d\,\overline{u}) \tag{2.49}$$

it follows that the weak elementary processes

$$s + \overline{u} \to W^-, \quad \overline{s} + u \to W^+ \tag{2.50}$$

must also exist, which then also explains the decay

$$\Lambda^0 = (u\,d\,s) \to (u\,d\,W^-\,u) \to (u\,u\,d) + W^- \to (u\,u\,d) + (d\,\overline{u}) = p^+ + \pi^-, \tag{2.51}$$

and so on. Comparing the elementary processes (2.47) and (2.50), it follows that the weak decays “mix” the d- and the s-quark and so violate the conservation of strangeness. That is, the true eigenstates of the Hamiltonian terms responsible for weak interactions[27] are not the $u, d, s, \ldots$ quarks, but the combinations

$$|u\rangle, \quad |d_w\rangle := \cos(\theta_c)|d\rangle + \sin(\theta_c)|s\rangle, \quad |s_w\rangle := \cos(\theta_c)|s\rangle - \sin(\theta_c)|d\rangle, \quad \ldots \tag{2.52}$$

This was proposed in 1963 – before the discovery of the c-quark – by Nicola Cabibbo and after whom the angle θ_c was named. In 1973, Kobayashi and Maskawa generalized this parametrization by proposing

$$\begin{bmatrix}|d_w\rangle\\|s_w\rangle\\|b_w\rangle\end{bmatrix} := \begin{bmatrix}V_{ud} & V_{us} & V_{ub}\\ V_{cd} & V_{cs} & V_{cb}\\ V_{td} & V_{ts} & V_{tb}\end{bmatrix}\begin{bmatrix}|d\rangle\\|s\rangle\\|b\rangle\end{bmatrix}, \tag{2.53a}$$

$$= \begin{bmatrix} c_{12}c_{13} & s_{12}c_{13} & s_{13}\,e^{-i\delta_{13}}\\ -s_{12}c_{23} - c_{12}s_{23}s_{13}\,e^{i\delta_{13}} & c_{12}c_{23} - s_{12}s_{23}s_{13}\,e^{i\delta_{13}} & s_{23}c_{13}\\ s_{12}s_{23} - c_{12}c_{23}s_{13}\,e^{i\delta_{13}} & -c_{12}s_{23} - s_{12}c_{23}s_{13}\,e^{i\delta_{13}} & c_{23}c_{13}\end{bmatrix}\begin{bmatrix}|d\rangle\\|s\rangle\\|b\rangle\end{bmatrix}, \tag{2.53b}$$

$$\text{where} \quad c_{ij} := \cos(\theta_{ij}), \quad s_{ij} := \sin(\theta_{ij}), \quad i, j = 1, 2, 3 = d, s, b,$$

[27] The Student is expected to remember how one computes with Hamiltonians of the form $H = H_0 + H'$ in quantum mechanics, where the eigenstates and eigenvalues for H_0 are known, where H' is treated as a perturbation, and where the eigenstates of H_0 need not be the eigenstates of H', i.e., $[H_0, H'] \neq 0$.

where in the second row the now standard parametrization is given in terms of Euler angles in the (d, s, b)-space. The general form and parametrization of non-Hermitian matrices were known back in 1939 [151], from which it follows that a non-Hermitian $n \times n$ matrix has $\binom{n-1}{2}$ complex phases. It follows that one needs at least three quarks of $-\frac{1}{3}$ electric charge for the Cabibbo–Kobayashi–Maskawa (CKM) matrix to be able to contain one non-removable complex phase, which then can parametrize *CP*-violation, as observed back in 1964. The elements of the matrix $\mathbb{V}$ (2.53) are denoted to indicate their application:

$$\text{Probability}(d + W^+ \to u) = \left|\langle u|W^+|d\rangle\right|^2 \propto |V_{ud}|^2, \tag{2.54a}$$

$$\text{Probability}(s + W^- \to u) = \left|\langle u|W^+|s\rangle\right|^2 \propto |V_{us}|^2, \tag{2.54b}$$

and so on. Present-day observed values are $\theta_{12} = \theta_{ds} = (13.04 \pm 0.05)^\circ$, $\theta_{13} = \theta_{db} = (0.201 \pm 0.011)^\circ$, $\theta_{23} = \theta_{sb} = (2.38 \pm 0.06)^\circ$, and $\delta_{13} = \delta_{db} = (1.20 \pm 0.08)^\circ$, giving

$$\begin{bmatrix} |V_{ud}| & |V_{us}| & |V_{ub}| \\ |V_{cd}| & |V_{cs}| & |V_{cb}| \\ |V_{td}| & |V_{ts}| & |V_{tb}| \end{bmatrix} \approx \begin{bmatrix} 0.974 & 0.226 & 0.004 \\ 0.226 & 0.973 & 0.041 \\ 0.009 & 0.041 & 0.999 \end{bmatrix}. \tag{2.55}$$

To estimate the $W^\pm$ particle mass, we need an estimate for the range of weak nuclear forces, and such a direct estimate does not exist. It is known, however, that the weak nuclear interaction does occur within the atomic nucleus in the form of the β-decay, but it is not known how close the particles must be for the scattering to also include weak interactions. For example, in antineutrino–proton scattering, the contribution of the weak interaction would be seen also as the inelastic collision (2.22), which the quark model represents as the consequence of two alternative collisions:

$$\overline{\nu}_e + p^+ = \overline{\nu}_e + (u\,u\,d) \begin{matrix} \nearrow (e^+ + W^-) + (u\,u\,d) \searrow \\ \searrow \overline{\nu}_e + (W^+ + (u\,d\,d)) \nearrow \end{matrix} e^+ + (u\,d\,d) = e^+ + n^0. \tag{2.56}$$

The range of the weak nuclear interaction mediated by $W^\pm$ is thus probably not larger than the diameter of the nucleus where the process takes place, and may well be (much) smaller than the proton and neutron diameter. Taking $R < 10^{-15}$ m for the range produces $m_W > \frac{\hbar}{Rc} \sim 200\,\text{MeV}/c^2$, which is a *lower limit*, and very weak as an estimate: It was known by the late 1940s that no appropriate particle of such a mass exists. As the experiment energies grew, it was expected that the scatterings would begin to show traces of the intermediary bosons $W^\pm$, but such data would be obtained only in January of 1983 (and the Z^0 particle by mid-1983), for which Carlo Rubbia and Simon van der Meer were to receive the 1984 Nobel Prize.

By about 1958, the possibility was noted that there might exist weak neutral processes of the type

$$q + \overline{q'} \leftrightarrow Z^0, \quad \text{and} \quad \ell + \overline{\ell} \leftrightarrow Z^0, \tag{2.57}$$

where q and q' are any two different quarks of the same electric charge – revealing the "mixing" parametrized by the CKM matrix (2.53). Such processes would also produce a correction of the electromagnetic interactions based on the elementary processes

$$q + \overline{q} \leftrightarrow \gamma \quad \text{and} \quad \ell + \overline{\ell} \leftrightarrow \gamma, \tag{2.58}$$

without the CKM mixing and which, of course, do not include the neutrinos as they are electrically neutral. In the processes where q and q' are not the same quark, such as

$$s \to Z^0 + d, \tag{2.59}$$

the quark "flavor" changes, and such hypothetical processes were dubbed "flavor-changing neutral currents" (FCNC). The experimental detection of such processes was crucial for confirming the Glashow–Weinberg–Salam model of electroweak interaction unification based on the $SU(2) \times U(1)$ symmetry group, and the refutation of the competing model based on the $SO(3)$ symmetry group, which was proposed by Sheldon Glashow,[28] developing the idea of his mentor, Julian Schwinger, and with later collaboration by Howard Georgi. Based on this crucial confirmation of their weak interaction model, Glashow, Salam and Weinberg were awarded the 1979 Nobel Prize – five years before the direct detection of $W^{\pm}$ and Z^0 bosons!

The point is that the proof of existence of the FCNC processes sufficed to establish the existence of a Z^0 virtual particle, which then exists only during a time shorter than that estimated by Heisenberg's indeterminacy relations $\Delta t < \frac{\hbar}{2 m_Z c^2}$, and which was therefore not yet directly detected at the time. Such processes can happen (and were detected for the first time in 1973, at CERN) even when the total energy of the collision is not enough to create a "real" Z^0 particle, which then could have been detected directly. It took several years to "improve the statistics" of the results, i.e., to remove the "noise" of the much stronger electromagnetic interaction: In collisions of two particle beams, most processes occur via the much stronger electromagnetic interaction. The probability of identifying a "true" individual process for analyzing the weak interaction in the sea of electromagnetic processes is then very small and requires ingenious technique and methodology in detection as well as an enormous investment in the form of patience.

2.3.15 The Standard Model

By the mid-1980s, the universally accepted list of elementary particles was as given in Table 2.3, and presents the so-called *Standard Model* in its most succinct form. The subsequent chapters will discuss the details of this model (and there are plenty!), but let us note here that the listed 12 spin-$\frac{1}{2}$ particles also have their antiparticles, and that every quark, in addition, has the additional degree of freedom called color, with three distinct values. Thus some Authors [243] count 12 leptons and 36 quarks in Table 2.3. Of course, since these are spin-$\frac{1}{2}$ particles, one should also count the fact that each one of these 48 particles has spin projections $\pm\frac{1}{2}$, which may be regarded as a *doubling* in counting "particles." Similarly, the photon is usually regarded as a single particle, but one must know that every photon has two possible polarizations, which according to this logic should be counted as *two photons*. For nuclear interaction mediators, this number is bigger: For weak interactions there are three intermediary bosons: $W^{\pm}, Z^0$ and each has *three* polarizations,[29] which then gives $3 \times 3 = 9$ particles. Section 6.1 will show that there are 8 gluons and that

Table 2.3 The content of the Standard Model of elementary particle physics; see equation (2.44a)

Substance (spin-$\frac{1}{2}$ fermions)				
Gen.	**Leptons**		**Quarks**	
1.	ν_e	e^-	u	d
2.	ν_μ	μ^-	c	s
3.	ν_τ	τ^-	t	b

	Interactions	(bosons)
γ	electromagnetic interaction	(spin 1)
$W^{\pm}, Z^0$	weak nuclear interaction	(spin 1)
gluons	strong nuclear interaction	(spin 1)
$\delta g_{\mu\nu}$	gravitation	(spin 2)

Higgs boson: gives mass to the particles with which it interacts (spin 0)

[28] Yes, it is the same Sheldon Lee Glashow, a coauthor of both competing proposals [☞ Footnote *12* on p. 276].

[29] Since the $W^{\pm}, Z^0$ particle mass is not zero, these move with a speed smaller than the speed of light, and have a longitudinal polarization. This is unlike with photons that move at the speed of light so the amplitude of the longitudinal polarization reduces to zero owing to FitzGerald–Lorentz contraction.

they are massless and so have two polarizations each: that adds $8 \times 2 = 16$ particles. Table 2.3 includes gravity, although it is, strictly speaking, not part of the Standard Model; the gravitational field quanta are represented by fluctuations of the metric, and so by a rank-2 tensor. However, those fluctuations propagate at the speed of light and have no mass, and so again have only two polarizations [☞ Chapter 9].

Finally, an integral part of the Standard Model is also the Higgs scalar, which has now been confirmed experimentally [293], the detection of which was one of the original goals of the LHC (Large Hadron Collider) at CERN. Chapter 7 will show that a single real, scalar (spin-0) elementary Higgs particle is predicted, which must be its own antiparticle. While the photon γ, the weak intermediaries $W^{\pm}$ and Z^0, the gluons and even the gravitons are mediating quanta of fundamental interactions, the one real Higgs particle, which has been detected, is a *remnant*: Chapter 7 will show that the Higgs field has *four* real degrees of freedom, three of which are Goldstone modes for spontaneously broken $SU(2)_L$ symmetry. The practical role of the Higgs field is to mediate in giving masses to particles, including the mediating gauge bosons $W^{\pm}, Z^0$ – as if the Higgs field were to slow the particles with which it interacts, reminiscent of the effect of viscosity in materials.

The Goldstone modes in the Higgs field cannot be detected as separate particles, but they can be identified as the additional longitudinal polarizations of the $W^{\pm}$ and Z^0 gauge bosons: In the phase without symmetry breaking there are two complex degrees of freedom of Higgs particles (which may be counted as 2×2 real particles) and three massless gauge bosons of the weak interaction – and so with two polarizations each tallying up to 3×2 real particles; together, that's 10 real particles. In the phase with broken symmetry there is only one real Higgs particle and three massive gauge bosons (three polarizations each), tallying up to 3×3 real particles; together, that's again 10 real particles.

The final sum in this detailed counting is

$$\begin{aligned} \text{fermions} &= (3\times2\times2\times2)+(3\times2\times2\times2)\times3 &&= 96,\\ \text{bosons} &= 1\times2+3\times3+8\times2+(1\times2)+1 &&= 30,\\ \hline & &&= 126. \end{aligned} \tag{2.60}$$

In some ways, this *is* the correct counting – and we will see subsequently in what sense one needs to distinguish all these degrees of freedom, as there are physical observables that depend on this level of detail. However, I should like to hope it is clear to the Reader that the complaint "a system of 126 particles does not look elementary" is not fair: It is crucial that these 126 degrees of freedom are systematically presented in the simple Table 2.3. Finally, the particles listed in that table fully explain the by now many hundreds of experimentally detected mesons and baryons, and so all experimentally detected forms of substance (atoms, molecules, etc.), while they themselves show no sign of compositeness or structure. Therefore, the so-called Standard Model with the business card in Table 2.3, and described in more detail in subsequent chapters, fully satisfies the goal of our original search.

Table 2.4 lists a telegraphic review of the most prominent elementary particle physics milestone discoveries.

— ❦ —

Motivated mostly by the economy of symmetries in Table 2.3 on p. 67, and a little also by the large number of degrees of freedom (2.60), as early as 1974, there were classification systems that in various ways represent at least some of the 126 particles (2.60) as bound states of even more elementary constituents. Generally speaking, in these proposals quarks (and sometimes also the leptons, and/or the mediating bosons) are bound states of *preons*. Different preon models suppose different dynamics, and then also different combinatorial rules for preons, all with the aim to faithfully reproduce the contents of Table 2.3, and the counting (2.60). However, except for economy

Table 2.4 A timeline of significant discoveries in elementary particle physics

Year	Particle	Discovered
1895	X-rays	Wilhelm C. Röntgen (X-rays were later identified as photons)
1897	e^-	Joseph J. Thomson
1899	α-particle	Ernest Rutherford
1900	γ-rays	Paul Villard (γ-rays were later identified as photons)
1911	Atomic nucleus	Hans Geiger and Ernest Marsden, under Ernest Rutherford
1919	p^+	Ernest Rutherford
1932	n^0	James Chadwick
1932	e^+	Carl D. Anderson (predicted by Paul A. M. Dirac, 1927)
1937	μ^-	Seth H. Neddermeyer and Carl D. Anderson, Jabez C. Street and Edward C. Stevenson (erroneously identified as pion until 1947)
1947	$\pi^\pm, \pi^0$	Cecil Powell (predicted by Hideki Yukawa, 1935)
1947	K^0	George D. Rochester and Clifford C. Butler
1949	$K^\pm$	Cecil Powell
1947–1953	$\Lambda^0, \Sigma^\pm, \Sigma^0$	Several research groups
1955	$\overline{p}^-, \overline{n}^0$	Owen Chamberlain, Emilio Segrè, Clyde Wiegand and Thomas Ypsilantis
1956	ν (directly)	Frederick Reines and Clyde Cowan (predicted by Wolfgang Pauli, 1931)
1962	$\nu_\mu \neq \nu_e$	Leon M. Lederman, Melvin Schwartz and Jack Steinberger
1969	Partons, and u, d, s quarks	So-called deep inelastic collisions, SLAC (predicted by Murray Gell-Mann and George Zweig, 1963)
1974	J/ψ ($[c\text{-}\bar{c}]$-state)	Burton Richter and C. C. Samuel Ting, proof of the c-quark existence (predicted by James D. Bjorken and Sheldon L. Glashow in 1964, and Glashow, John Iliopoulos and Luciano Maiani in 1970)
1975	τ-lepton	Martin Perl and collaborators, SLAC
1977	Υ (upsilon) ($[b\text{-}\bar{b}]$-state)	Leon Lederman and collaborators, Fermilab (b-quark predicted by Makoto Kobayashi and Toshihide Masakawa in 1973)
1979	Gluon	$e^- - e^+$ collisions in PETRA experiment at DESY
1983	$W^\pm, Z^0$	Carlo Rubbia, Simon van der Meer, CERN UA-1 collaboration (predicted by Sheldon L. Glashow in 1963, Abdus Salam and Steven Weinberg in 1967)
1995	t-quark	Tevatron, Fermilab (predicted by Makoto Kobayashi and Toshihide Masakawa in 1973)
2000	ν_τ	DONUT collaboration, Fermilab
2012	Higgs	ATLAS and CMS collaborations, LHC at CERN – pending interaction details [25, 109] ☝

in classification and "purely aesthetic" advantages, there exists no *experimental* reason for preon models. In all experiments thus far (at distances $\gtrsim \hbar c/E$, where $E \approx 250\,\text{GeV}/c^2$ is the maximal collision energy), quarks, leptons and gauge bosons behave as ideal point-like (elementary) particles. That is, they show no internal structure [☞ Conclusion 1.5 on p. 30]. Of course, the absence of a proof of a structure within quarks and leptons is not a proof of the absence of such a structure.

2.4 Lessons

During the twentieth century the quantumness and relativity of Nature became universally accepted basic ideas of fundamental physics. The third idea that had similarly taken root in our understanding of Nature is the fundamental role of symmetry. On one hand, this links symmetries and conserved quantities: in the form of Amalie Emmy Noether's theorem in classical physics, rather directly in quantum mechanics, and via so-called Ward–Takahashi identities for gauge theories in quantum field theory. On the other hand, this also links symmetries and interactions, in the form of the gauge principle. Especially in the second half of the twentieth century, symmetries and the algebraic structure of *groups* that those symmetries form so focused the research in fundamental physics that some philosophers of natural sciences [533] acquired the impression that the concept of *law* had been abandoned for symmetry principles. However, the point of view of those who actually do such research is that symmetry groups provide the much needed cohesive (algebraic) structure both to various conservation laws and to the modes of interaction. Symmetries have thus become an integral part of the contemporary understanding of laws of Nature.[30] This difference in understanding the link between symmetries and the laws of Nature reminds us of the comments in Digression 1.1 on p. 9.

The link between symmetry and interaction in the form of gauge theory will be explored in detail in subsequent chapters. Here, we review the conservation laws: We reconsider the logic and rules of their application, provide some cautionary remarks about that application, and list the conservation laws as they are used in elementary particle physics.

2.4.1 *The logic and rules of application*

The growing majority of particles that are studied in elementary particle physics – the hundreds of mesons and baryons – actually are not elementary, but are bound states of quarks and/or antiquarks. All hadrons (except the proton) as well as the $\mu^{\pm}$- and $\tau^{\pm}$-leptons decay, and rather fast [293]: A free n^0 in 15 min, $\mu^{\pm}$ decays in 2.2×10^{-6} s, π^+ in 2.6×10^{-8} s, π^0 in 8.4×10^{-17} s, the J/ψ-particle in 7.05×10^{-21} s, and most of the hadrons decay in a time of about 10^{-23} s! From all those hundreds of particles, only the electron and the proton are stable in the traditional sense – the decay of not one of these has ever been observed.

During 10^{-23} s, light passes only about 3×10^{-15} m in vacuum, which is the order of magnitude of the diameter of the atomic nucleus. It is clear that such short decay lifetimes cannot be measured with a stop-watch. Instead, we use the indeterminacy relation, $\Delta E\,\Delta\tau \geqslant \frac{1}{2}\hbar$, so that $\Delta E = (\Delta m)c^2$ implies

$$\overline{\tau} := \frac{\hbar}{(\Delta m)c^2}, \qquad t_{1/2} := \ln(2)\,\overline{\tau} = \frac{\ln(2)\,\hbar}{(\Delta m)c^2}. \tag{2.61}$$

From the experimentally obtained distribution of the values of the particle mass, one computes the standard deviation of mass and uses it as Δm in equation (2.61). The so-computed average duration time of the particle (state), $\overline{\tau}$, is used as the particle lifetime. Then, $t_{1/2}$ is its half-life: $N(t) = N(0)\,e^{-t/\overline{\tau}} = N(0)\,2^{-t/t_{1/2}}$ is the number of a certain type of particles at the time $t > 0$ in a sample where the number at the time $t = 0$ was $N(0)$.

Also, many hadrons (and leptons too!) can decay in several distinct ways; a few examples for illustration purposes are [293]:

Decay results		Decay results			
$\Lambda^0 \to p^+ + \pi^-$	$(35.8\pm0.5)\%$,	$\to n^0 + \pi^0$	$(63.9\pm0.5)\%$,	etc.	
$K^+ \to \mu^+ + \nu_\mu$	$(63.44\pm0.14)\%$,	$\to \pi^+ + \pi^0$	$(20.9\pm0.12)\%$,		(2.62)
$\to 2\pi^+ + \pi^-$	$(5.590\pm0.031)\%$,	$\to \pi^+ + e^+ + \nu_e$	$(4.98\pm0.07)\%$,	etc.	
$\tau^- \to \mu^- + \nu_\tau + \overline{\nu}_\mu$	$(17.36\pm0.05)\%$,	$\to$ all else:	$(82.64\pm0.05)\%$.		

[30] Roughly, all *laws of Nature*, as they are understood here, contain both conservation laws and interaction laws.

One of the tasks of elementary particle physics is the computation of the relative probabilities of decay, as well as the total lifetimes of various particles. This includes both the elementary particles [☞ Table 2.3 on p. 67], such as the τ-lepton, and also the bound states of these elementary particles. This second group, much larger and growing, consists of hadrons (mesons and baryons) as well as the so far only hypothetical and experimentally unverified bound states that consist purely of gluons,[31] and possibly also the so-called exotic hadrons – quark bound states that are neither mesons, $(q\bar{q})$, nor baryons, (qqq); for example, so-called *dibaryons* are hypothetical bound states of six quarks.

> ***Conclusion 2.4*** *The primary focus of the so-called "elementary particle physics" is on the elementary particles as identified in the Standard Model [☞ Table 2.3 on p. 67]. However, this then covers also the dynamics of these elementary particles, and so also their bound states: all mesons and all baryons [☞ Table 2.5 and Section 11.2].*

By the feature indicated in this conclusion, high energy physics currently differs from all other disciplines in physics: The domains of study of several closely related physics disciplines are sketched in Table 2.5. Unlike all other disciplines in this table, "elementary particle physics" (also known as "high energy physics") studies (at least) *two* levels of elementarity.

Table 2.5 The domains of several physics disciplines of "small" systems and objects

Discipline	Domain of study
Molecular physics	Molecules (chemically bound states of atoms)
Atomic physics	Atoms (electromagnetic bound states of a nucleus and electrons)
Nuclear physics	Atomic nuclei (bound states of protons and neutrons)
Elementary particle physics a.k.a. **High energy physics**	Elementary particles [☞ Table 2.3 on p. 67]; Bound states of these (mesons and baryons)

Finally, by the end of the twentieth century, high energy physics had also brought on an essential shift in the understanding of the Democritean idea of "elementary *particles*": The hierarchy

1. molecules consist of atoms,
2. atoms consist of electrons and a nucleus,
3. nuclei consist of nucleons (protons and neutrons),
4. nucleons (and all other hadrons) consist of quarks,

experimentally stops here, for now. It is reasonable to expect that contemporary "high energy physics" will soon effectively split into "hadron physics" and "fundamental physics," although their respective domains do not yet seem to be sufficiently differentiated.

According to (super)string theory, this hierarchical halt is also conceptually significant: In that theoretical system, the fundamental objects are not any new (smaller, constituent) *particles*, but (super)strings; the particles of the Standard Model, as given in Table 2.3 on p. 67, are not bound states of these more elementary (super)strings, but are their *modes of vibration*. This is an essential shift in understanding: quarks, leptons, gauge and Higgs bosons are not at all bound states consisting of other, more elementary things! Rather, the same string contains amongst its vibrations (its Fourier modes) all the elementary particles of the Standard Model (and indefinitely more) *simultaneously* [☞ Chapter 11].

[31] Unlike the chargeless photons in abelian (commutative) gauge theory of quantum electrodynamics, the mediators of non-abelian (non-commutative) interactions in quantum chromodynamics (the gluons) themselves have color charge and so can form bound states, so-called *glueballs* [☞ Chapter 6.2].

Returning then to elementary *particle* physics, consider the correlation between the mass of a fundamental interaction mediator and the range of that interaction. In 1931, Yukawa reasoned that the total energy is $E = c\sqrt{m^2c^2 + \vec{p}^{\,2}} \geqslant mc^2$ for a mediating particle of mass m. To produce such a particle during the interaction, at least mc^2 energy is needed. Heisenberg's indeterminacy relations permit "borrowing" that much energy for no longer than $\sim \frac{\hbar}{E} \leqslant \frac{\hbar}{mc^2}$, during which this mediating particle may traverse a distance no larger than [✐ *why?*]

$$R \sim \frac{\hbar}{mc}. \tag{2.63}$$

Numerical factors such as $\frac{1}{2}$ in $\triangle E \triangle \tau \geqslant \frac{1}{2}\hbar$ were neglected here as the estimate (2.63) is a rough *upper limit*. Using the fact that the strong interaction must have a range that is *at least* comparable to the size of the atomic nucleus – so as to keep the nucleons in the bound state – Yukawa estimated the mass of the mediators for the strong nuclear interaction as $m \sim \frac{\hbar}{Rc} \approx 200\,\text{MeV}/c^2$.

Digression 2.6 Warning! Using the same reasoning to estimate the range of the electromagnetic interaction from the size of the atom, $a_0 = 5.291\,772\,108 \times 10^{-10}$ m, implies that the photon mass is $\sim$4 keV$/c^2$ – which is wrong. Of course, we know that the range of the electromagnetic interaction is infinite, which agrees with the relation (2.63) and the fact that the photon mass is zero. The error in the first estimate stems from the fact that the binding energy of the hydrogen atom (1.31) is less than the rest energy of the electron by the dimensionless factor $\alpha^2 \sim 5.33 \times 10^{-5}$; bound states where the binding energy is a few orders of magnitude smaller than the rest energy are called "weakly bound" [☞ relations (1.40)]. Heisenberg's relations are *inequalities*, and thus can only provide a *lower limit* for the mass of the mediating particle from the size of the bound state, and usefully so only for "strongly bound" states!

In the case of weak interactions, the mass estimate for the mediating bosons, $W^{\pm}, Z^0$, is additionally hampered by the fact that there do not exist states bound by the weak nuclear interaction. Since the β-decay evidently does happen within the atomic nucleus, we know that the range may well be of the order of the atomic nucleus diameter, but this does not permit estimating either limit for the range: The range may be smaller and the β-decay occurs at distances much smaller than the nucleus diameter. Or, it may be much larger than the nucleus – the involved particles are confined within the nucleus anyway, by the strong interactions.

Complementary to the range, we may compare the times required for a decay to happen. Strong interaction decays typically happen within 10^{-23} s, while electromagnetic decays occur within 10^{-16} s – ten million times slower. Weak interactions, however, vary: decays may be as fast as 10^{-13} s (for the τ-lepton) to as long as 881.5 s (for the free neutron) [293] – which is a spectrum of 16 orders of magnitude! If the decay results in photons, it happened by means of the electromagnetic interaction; similarly, the appearance of a neutrino in the decay results is the "hallmark" of weak interactions. For decays where the result contains neither a photon nor a neutrino, the type of the interaction is harder to determine, so the duration of the decay is a useful indicator.

While it is known that all particles except e^-, p^+, ν_e and γ decay, the decay patterns of all particles are very regular. A systematic analysis of their decays, and also of their inelastic collisions and scatterings then implies:

Conclusion 2.5 *All particles decay into lighter particles, and in all manners that are permitted by conservation laws. (The electron, for example, does not decay as there are no lighter particles into which to decay.)*

This conclusion is related to Conclusion 2.3 on p. 56. This logic led to a successful application of the $SU(3)_f$ and later also of the $SU(4)_f$ approximate symmetries, as well as several conservation laws that became an integral part of the Standard Model. The subsequent review of these laws will have to be amended later, when the technical details become familiar.

2.4.2 Strict conservation laws

By "strict conservation laws" we understand those laws that hold for all interactions and in all situations. For each of these laws, the Standard Model exhibits an explicit symmetry, linked with that conservation law by way of Emmy Noether's theorem. The book [383] is dedicated to various forms and applications (not only in physics!) of this important theorem.

4-momentum, angular momentum, parity

One of the most important lessons from the historical review in Section 2.3 is the reliance on conservation laws of the 4-momentum: both the (relativistic) total energy and the vector of linear momentum are strictly conserved quantities in so-called real states – i.e., in states where these quantities can be observed and measured. On the other hand, in *virtual* states neither the total energy nor the momentum can be measured, and the conservation laws are not applicable. Thus, it is not that the energy and/or momentum conservation law is violated in virtual states, but rather there is neither measurable energy nor measurable momentum for which to apply the law.

The conservation of the 4-momentum is the consequence of Noether's theorem for the spacetime translation symmetry, for a system's real states.

Digression 2.7 Heisenberg's indeterminacy relations are not infrequently cited as the assertion that physical quantities that can be simultaneously measured must correspond to operators that commute. That, in fact, is not quite true, because it neglects the essential dependence in quantum mechanics on the state, or a class of states, in which the considered system is prepared.

A very general proof[32] of this statement follows from considering two Hermitian operators, A and B, which define

$$C := -i\big[A, B\big], \qquad C^\dagger = C. \tag{2.64a}$$

Defining $A_0 := A - \langle A\rangle$ and $B_0 := B - \langle B\rangle$, we have

$$\big[A_0, B_0\big] := \big[(A - \langle A\rangle), (B - \langle B\rangle)\big] = iC, \tag{2.64b}$$

so that

$$0 \leqslant \langle |A_0 - i\omega B_0|^2\rangle = \langle A_0^2\rangle - i\omega\langle [A_0, B_0]\rangle + \omega^2\langle A_0^2\rangle \tag{2.64c}$$

$$= \Delta_A^2 + \omega\langle C\rangle + \omega^2\Delta_B^2. \tag{2.64d}$$

The right-hand side expression is minimized by $\min(\omega) = -\langle C\rangle/2\Delta_B^2$, producing

$$\Delta_A^2\,\Delta_B^2 \geqslant \tfrac{1}{4}\langle C\rangle^2, \qquad \text{that is,} \qquad \Delta_A\,\Delta_B \geqslant \tfrac{1}{2}\,|\langle C\rangle| = \tfrac{1}{2}\,|\langle [A, B]\rangle|, \tag{2.64e}$$

which are Heisenberg's indeterminacy relation for the physical quantities represented by the operators A, B. This manifestly depends on the state in which the indicated expectation values $\langle C\rangle$ and $\langle [A, B]\rangle$ are computed – and may well be zero although $[A, B] \neq 0$.

---

[32] This follows the variational derivation in Refs. [295, 97], refining the original derivation by Robertson [460] and Schrödinger [476]; see also Ref. [242].

Example 2.1 In the best known example, we have $A = p_x$ and $B = x$, and $C = \hbar\mathbb{1}$ is a constant, and the *non-vanishing* of the right-hand side of equation (2.64e) is in fact state independent. However, for the case of the angular momentum,

$$[J_x, J_y] = i\hbar\, J_z \qquad \Rightarrow \qquad \Delta_{J_x}\Delta_{J_y} \geqslant \tfrac{1}{2}\big|\langle J_z\rangle\big|. \tag{2.65}$$

For states where $\langle J_z\rangle = 0$ we have that $\Delta_{J_x}\Delta_{J_y} \geqslant 0$. Thus, although the operators J_x and J_y do not commute, the product of their indeterminacies may well vanish in states of the system where $\langle J_z\rangle = 0$. In those quantum states, Heisenberg's indeterminacy principle does not preclude the simultaneous measurement of J_x and J_y although they do not commute as operators.

Just like 4-momentum, angular momentum is strictly conserved in real states. The conservation law is a consequence of Noether's theorem for rotation symmetry.

Parity, P, is the operation that changes the sign of all Cartesian spatial coordinates; in spherical coordinates, this is the $(r,\theta,\phi) \to (r,\pi-\theta,\phi+\pi)$ transformation. With respect to this, so-called polar vectors (position, velocity, acceleration, electric field, etc.) all change sign. In turn, so-called axial vectors (angular momentum, torque, magnetic field, etc.) do not change sign. Scalar functions of position (temperature, atmospheric pressure, density, etc.) are invariant and do not change sign, while *pseudo-scalar* functions of position (e.g., the volume element $\mathrm{d}^3\vec{r}$) change sign. Tsung Dao Lee and Chen Ning Yang discovered that parity is strictly conserved in all electromagnetic and strong processes, but that by 1956 parity conservation had not been verified in weak interactions. Thus, they proposed several direct experimental tests. During 1956–7, Madam Chien-Shiung Wu found, with her collaborators, clear indication of P-violation in the ${}^{60}_{27}$Co β-decay, which was immediately confirmed by R. L. Garwin, L. Lederman and R. Weinrich by means of precise measurements of cascading decays (2.21). It later turned up that R. T. Cox, C. G. McIlwraith and B. Kurrelmeyer had published experimental results back in 1928 [118, 245] about double scattering β-particles ($e^{\pm}$), which indicated P-violation, but those 28 years earlier no one was willing to consider that as an explanation.

Electric charge

The Maxwell equations (5.72) straightforwardly produce the so-called equation of continuity:

$$\begin{aligned}
\vec{\nabla}\cdot\vec{E} &= \frac{1}{4\pi\epsilon_0}4\pi\,\rho_e \;\Rightarrow & \frac{\partial(\vec{\nabla}\cdot\vec{E})}{\partial t} &= \frac{1}{4\pi\epsilon_0}4\pi\frac{\partial\rho_e}{\partial t},\\
\vec{\nabla}\times(c\vec{B}) - \frac{1}{c}\frac{\partial\vec{E}}{\partial t} &= \frac{1}{4\pi\epsilon_0}\frac{4\pi}{c}\vec{j}_e \;\Rightarrow\; \vec{\nabla}\cdot(\vec{\nabla}\times(c^2\vec{B})) - & \frac{\partial(\vec{\nabla}\cdot\vec{E})}{\partial t} &= \frac{1}{4\pi\epsilon_0}4\pi\vec{\nabla}\cdot\vec{j}_e,\\
\hline
&& \Rightarrow\quad 0 &= \frac{\partial\rho_e}{\partial t} + \vec{\nabla}\cdot\vec{j}_e,
\end{aligned} \tag{2.66}$$

since $\vec{\nabla}\cdot(\vec{\nabla}\times\vec{X}) \equiv 0$ for all $\vec{X}$. It follows, integrating

$$\frac{\partial\rho_e}{\partial t} = -\vec{\nabla}\cdot\vec{j}_e \qquad \Rightarrow \qquad \frac{\mathrm{d}Q_{e,V}}{\mathrm{d}t} = -\oint_{\partial V}\mathrm{d}^2\vec{r}\cdot\vec{j}_e, \tag{2.67}$$

so that the change of the total amount of electric charge $Q_{e,V}$ contained within a volume V equals the flux of the electric current through the (surface) boundary of that volume. The conservation law of electric charges is thus an exact and *inevitable* consequence of the fundamental laws of electromagnetism.

A violation of this law in any process would then indicate a contradiction with the Maxwell equations, and so also with electrodynamics as a whole. In the weak interactions, such as

$$\overline{\nu}_e + p^+ \begin{array}{c} \nearrow \\ \searrow \end{array} \begin{array}{c} e^+ + W^- + p^+ \\ \overline{\nu}_e + W^+ + n^0 \end{array} \begin{array}{c} \searrow \\ \nearrow \end{array} e^+ + n^0, \tag{2.56}$$

the electric charge of individual particles transfers: the neutral antineutrino becomes the positive positron and the positive proton becomes the neutral neutron. However, the *total* electric charge remains conserved in any arbitrarily small volume that contains the interacting particles. In the alternative intermediate processes, we see that the $W^\pm$ also carry electric charge, so that the electric charge is conserved in each of the individual elementary processes:

$$\overline{\nu}_e \to e^+ + W^-, \qquad u \to W^+ + d, \qquad W^- + u \to d, \qquad W^+ + \overline{\nu}_e \to e^+. \tag{2.68}$$

Digression 2.8 The very fact that the weak nuclear interaction mediators carry also electric charge indicates that the weak nuclear interaction is not fully independent of the electromagnetic one. Chapter 7.2 will more precisely examine this link. However, one of the *consequences* of this link has already emerged, in the generalized GNN formula (2.44b).

Color

Chapter 6.1 will show that the *color* in quantum chromodynamics generalizes the electric charge: whereas the electric charge is a scalar quantity, color is a 3-component quantity, i.e., a 3-vector in an abstract 3-dimensional space, just as the 3-vectors of position, velocity and force are vectors in the "real" space in which we ourselves move.

The fundamental differential equations of chromodynamics are the corresponding generalization of the Maxwell equations, and so also follow a corresponding conservation law of *color* as a charge. During elementary chromodynamical processes, quarks change their color, but this change is carried by gluons (the strong nuclear interaction mediators), so that color is conserved in every process. Since all detectable (real) particles are *colorless*, the color conservation law and the corresponding global symmetry are somewhat trivial. However, the gauge (local, i.e., spacetime variable) color symmetry is the reason for the existence of the strong nuclear interaction; see Chapter 6.

Lepton numbers

Unlike electric charge and color, which are conserved charges of gauge symmetries and which thus correspond to electromagnetic and strong nuclear interactions, the lepton numbers are strictly conserved, but are not the conserved charges of a gauge symmetry and do not correspond to any interaction. For example, in the decay

$$\mu^- \to e^- + \overline{\nu}_e + \nu_\mu, \tag{2.69}$$

the muon lepton number ($L_\mu = +1$) is carried by μ^- and ν_μ, and we have $L_\mu = +1$ input, $L_\mu = +1$ output. The electron lepton number is carried by e^- and ν_e: $L_e(e^-) = L_e(\nu_e) = +1$, so that $L_e(\overline{\nu}_e) = -1$. Here we have $L_e = 0$ input, $L_e = +1 + (-1) = 0$ output. Following the original proposal [319], a systematic analysis of all so far observed processes (except for neutrino mixing; see Section 7.3.2) indicates a strict conservation of all three lepton numbers, L_e, L_μ, L_τ, as defined in the table:

	ν_e, e^-	$\overline{\nu}_e, e^+$	ν_μ, μ^-	$\overline{\nu}_\mu, \mu^+$	ν_τ, τ^-	$\overline{\nu}_\tau, \tau^+$
$L_e =$	$+1$	-1	0	0	0	0
$L_\mu =$	0	0	$+1$	-1	0	0
$L_\tau =$	0	0	0	0	$+1$	-1

(2.70)

An analogous conservation of quark numbers, *separately* for the (u,d), (c,s), and (t,b) pairs does not exist, because of the CKM mixing (2.53) of so-called "lower" quarks, d, s, b. The question of lepton mixing, i.e., neutrino mixing, will be addressed in Section 7.3.2; let us note here merely that this possibility was proposed back in 1962 [353], although there was no strong experimental indication until recently that such a mixing really happens [369, 370].

In this sense is the reason for the existence of the (approximate) conservation law of three separate lepton numbers and the absence of a conservation law of three separate quark numbers a phenomenological and not a fundamental law – and an open question☝!

Baryon/quark number

The quark model redefined the baryon number simply as the triple of the quark number, where antiquarks have negative quark number. In the Standard Model, that definition remains, and also explains the absence of a meson conservation number: since mesons are $(q\overline{q})$ bound states, their quark number is zero. Since quarks cannot be extracted, it remains a convention to count baryons, and quarks have $\frac{1}{3}$ of the baryon number.

The baryon number conservation law is also strict – in that it holds in all processes. However, just as the (separate) lepton number conservation laws, this too is a phenomenological and not a fundamental law.

2.4.3 *Approximate conservation laws*

Besides strict conservation laws, there also exist approximate conservation laws, which are nevertheless useful precisely because of their approximate validity, whereby they help in estimates and computations.

Flavor

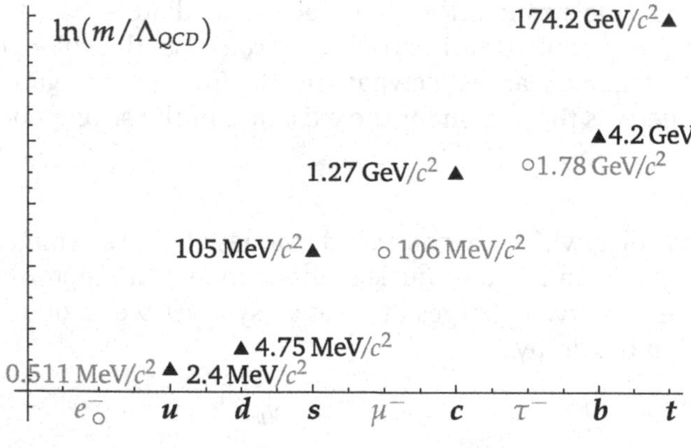

Figure 2.1 Quark (▲) and charged lepton (○) masses plotted on a logarithmic scale.

Table (2.44a) shows that the differences between the consecutive quark masses grow with these masses, as seen on the plot 2.1. In experiments done at the average energy of $\Lambda_{QCD} = 200\,\text{MeV}/c^2$ per process and with the experimental error at about 10% (so about $20\,\text{MeV}/c^2$), it is not possible to distinguish u- and d-quarks purely by their masses; within experimental error, their masses are the same. On the other hand, there is enough energy to produce an s-quark, which indeed can be distinguished from the u- and the d-quarks purely by its mass: $105 \pm 20\,\mathrm{MeV}/c^2$ cannot be confused for m_u, m_d even when identified within the $\pm 20\,\mathrm{MeV}/c^2$ experimental error.

Nevertheless, Gell-Mann proposed to:

1. First consider m_u, m_d, m_s as sufficiently near in masses to be distinguished; this indicated an $SU(3)_f$ symmetry.
2. Then take into account the difference between m_s *vs.* $m_u \approx m_d$; this breaks the symmetry $SU(3)_f \to SU(2)_{u,d}$.

This strategy led to his classification system "eightfold way," the plots such as (2.31), (2.32) and (2.35), the phenomenological formulae (2.33)–(2.34), and also to the discovery of the Ω^- baryon. Thereby, Gell-Mann introduced and established the use of symmetry – even if only approximate.

Within the Standard Model, such classifying schemes are, based on the "flavor" $SU(n)_f$ symmetry, very clearly phenomenological schemes. The conservation laws of individual "flavors" or groups of "flavors" are also only approximate rules, broken by the CKM mixing (2.53)–(2.55).

Digression 2.9 This induced the idea that approximate symmetries are (perhaps, sometimes) "only" broken symmetries, prompting us to uncover the reason and mechanism of breaking. In this sense is the origin of quark masses, and lepton masses too, as well as the CKM matrix, one of the basic questions to which the Standard Model has no answer☝. The quest for this origin is one of the basic motivations for most proposals that go "beyond" the Standard Model. This includes various electroweak and strong interaction unification models, and in these models at least some of the unexplained characteristics of the Standard Model are supposed to be derived and "predicted."

The OZI rule

There is a very general regularity in decays: The speed and probability of a decay,

$$X \longrightarrow Y_1 + Y_2 + \cdots + Y_k, \tag{2.71}$$

both grow with the change in mass, $\Delta m := (m_X - \sum_i m_{Y_i})$. Thus, between two decays that occur by means of the same kind of interaction, the one for which Δm is larger happens more often. Deviations from this regularity require an explanation.

In the 1960s, Susumu Okubo [392], George Zweig [593] and Jugoro Iizuka [289] independently discovered a significant correlation: decays that require the full annihilation of all "incoming" quarks and antiquarks are delayed (i.e., the probability of such decays is diminished) as compared to decays of the same system where at least some of the incoming quarks or antiquarks pass through into the decay result. During the 1960s, his correlation so successfully "explained" the delayed decays that it acquired the nickname the "OZI rule." For example, the probability of the $\phi \to \pi^+ + \pi^- + \pi^0$ decay is diminished as compared to the probability of the $\phi \to K^+ + K^-,\ K^0 + \overline{K}^0$ decay; experiments show that ϕ decays over 83% of the time into kaons and not into pions, although

$$\Delta m(\phi \to 2K) \approx 32.1\ \mathrm{MeV}/c^2, \qquad \text{while} \qquad \Delta m(\phi \to 3\pi) \approx 605\ \mathrm{MeV}/c^2. \tag{2.72}$$

Analogously, the J/ψ particle was supposed to decay predominantly into the pair of mesons $D^+ = (c\bar{d})$ and $D^- = (d\bar{c})$. However, the total mass of the $D^+ + D^-$ meson pair is *larger* than the mass of the J/ψ meson, so that the decay $J/\psi \to D^+ + D^-$ is *kinematically forbidden* [☞ Section 3.2], and only the decays into charmless hadrons remain possible, for all of which the probability is diminished by the OZI rule. This then is the combination of reasons that induces J/ψ to have unusually long ($\sim 10^{-20}$ s) lifetime, which is some 1,000 times longer than most other hadrons.

Symmetries and models

Symmetries have now been mentioned and even used several times, relying on the intuitive understanding of their nature and physical meaning. However, to be more precise – and especially with the discussion of (mathematical) models started in Section 1.1.2 in mind – recall that these mathematical models serve to faithfully reproduce all characteristics of the considered system. The model is therefore automatically identified with the physical system, object or quantity that the model represents [☞ Section A.1.3].

For example, strictly speaking, the 3-dimensional vector $\vec{B}$ is the abstract mathematical construct used as a model for the magnetic field of a concrete magnet, in a concrete point of space and in a concrete moment in time. The union of continuously many vectors $\vec{B}$ in space around the particular magnet in the same moment in time provides the abstract mathematical construct (it is impossible to measure continuously many points) that is automatically identified with the concrete magnetic field of that magnet. The abstract mathematical property of this union of vectors $\vec{B}$ – that it does not change if the whole union is rotated about the axis of the magnet itself – is automatically ascribed to the concrete magnetic field of the concrete magnet.

By the same token, we have more generally:

> ***Conclusion 2.6*** *Symmetries and other significant properties of the abstract mathematical model are automatically ascribed to the concrete physical system, object or quantity that the model faithfully represents.*

We then say that the (concrete, physical) magnetic field of the magnet has axial symmetry, even though this symmetry – strictly speaking – is a property of the mathematical model of this magnetic field. It is therefore of the essence that models do not introduce unnecessary (fictitious) degrees of freedom, concepts and properties.

> ***Conclusion 2.7*** *Ideally, the mathematical models of physical systems, objects and quantities must be* ***optimal****: minimal in the number and complexity of structure of intermediary and auxiliary means, (self-)consistent and faithful in representing the physical system, object and quantity for the description of which it is used.*

On one hand, this requirement of optimality reminds us that physics does not describe Nature directly, but *through* mathematical models that are continually improved. On the other hand, this requirement is a variation of Ockham's principle, which crucially limits the possibilities at our disposal when improving the existing models or creating new ones.

This practice largely determines the development of physics.

2.4.4 Exercises for Section 2.4

✎ ***2.4.1*** *A particle for which the relation (2.1) determines the speed and the ratio of electric charge by mass enters under the same conditions into a region where, however, now the* magnetic field is turned off. Compute the distance and direction of deflection in the (y, z)-plane from the x-axis, when the particle has traversed the length ℓ along the positive x-axis, and show that this deflection again depends on the ratio of charge by mass.

That is, show that successive measurements on the **same** cathode ray, with either one, or the other, or both fields $\vec{E} = E_0\,\hat{e}_y$ and $\vec{B} = B_0\,\hat{e}_z$ cannot determine independently both the electric charge and the mass of the particles that make up the cathode ray.

✎ **2.4.2** A photon of energy $E_\gamma = h\nu$ and linear momentum $\vec{p}_\gamma = \frac{1}{c}E_\gamma\,\hat{e}_x$ collides with an electron at rest. Upon the collision, the photon continues in the direction $\cos(\phi)\,\hat{e}_x + \sin(\phi)\,\hat{e}_y$ with the energy E'_γ and linear momentum $\vec{p}'_\gamma$, while the electron recoils in the direction $\cos(\theta)\,\hat{e}_x - \sin(\theta)\,\hat{e}_y$ with the linear momentum $\vec{p}_e$. Show that the linear momentum and energy conservation laws (a) produce the result (2.5), and (b) forbid that a free electron simply absorbs a photon.

✎ **2.4.3** Show that the energy and linear momentum conservation forbids **all** processes that may be obtained from (2.15)–(2.16) when either of the two photons is deleted and the remaining particles are rearranged using the crossing symmetry and the principle of detailed balance.

Part II

The Standard Model

3

Physics in spacetime

High-energy collisions, scatterings and most decays are, for the most part, relativistic quantum processes. It is therefore imperative to recall relativistic kinematics and the basic rules of tensor calculus; see Appendix B.2 for a more complete introduction. However, this chapter neither replaces nor competes with complete treatments of the special theory of relativity such as the illustrative but perfectly detailed text [512] or the *first and original* text [566]. Instead, the purpose of this chapter is to provide an introduction and the results that will be useful in following the subsequent material.

3.1 The Lorentz transformations and tensors

When describing physical processes, one necessarily uses a mathematical apparatus such as a (reference[1]) coordinate system – equipped with a specific and appropriate collection of coordinates. The choice of any one such coordinate system is arbitrary and should not affect the characteristics of the natural laws.

The basic idea (oft cited as one of the two postulates) of Einstein's theory of *relativity* is that the change in the choice of the *coordinate system* and corresponding *coordinates* used to describe spacetime must not change the meaning of natural laws – and so must not have any measurable, i.e., observable consequences.

Digression 3.1 Chapter 5 will show that the so-called gauge principle is simply the generalization of this relativity to the spaces of so-called internal degrees of freedom (also a type of coordinates), such as the phase of the complex wave-function of any charged particle.

In the *special theory of relativity*, this idea of relativity is limited to so-called *inertial* (coordinate) systems, which are usually defined as coordinate systems that differ from each other

[1] The term "coordinate system" is used instead of "reference system" to remind the Reader that its choice necessarily includes a choice of a particular system of four variables, and the specification of the metric tensor in the space of those variables; [☞ Chapter 9 for the general case, and here the relations (3.15)–(3.19)].

only in moving one with respect to the other with a constant relative velocity.[2] However, against all intentions, this "definition" does not exclude, for example, a pair of coordinate systems that co-rotate about the same axis with the same angular velocity, but move with a constant relative velocity along the co-rotating axis. Intuitively, the actual intention was to define a class of coordinate systems that move with a constant velocity with respect to a system...*at rest!* This shows that our intuition was infiltrated by the Newtonian idea of absolute space and time; it would be inconsistent to define the relativistic inertial frames using Newtonian ideas.

A definition that relies on a relative property between two members of the class being defined must imply that at least one member of this class can be unambiguously identified, so that other members of the class would be defined by comparison with this chosen reference. However, the very *essence* of the theory of relativity is that no such singled-out reference system can exist, which makes such a relative definition essentially insufficient.

The practical property of all inertial coordinate systems in non-relativistic physics is that Newton's first law holds in them. As it was our intention anyway for this law to hold, following Griffiths [243], we adopt it as a definition:

Definition 3.1 *A coordinate system is* ***inertial*** *if Newton's first law, i.e., the law of inertia, holds in it: every body moves at a constant velocity in a straight line if and only if the sum total of all forces that act upon it vanishes.*

Comment 3.1 *It is not hard to show that Definition 3.1 implies that the relative velocity between any two inertial systems is constant. Thus, Definition 3.1 implies the usually assumed property of inertial coordinate systems, which were meant to be selected. It also excludes the non-inertial systems such as the above-mentioned co-axially translating co-rotating systems, which are known to be accelerated.*

Comment 3.2 *Evidently, for a specified coordinate system, we must know what is a "straight line" and what qualifies as a "constant velocity." As the first notion is purely geometrical, and the second requires differential calculus in the specified coordinate system, Definition 3.1 presupposes this level of mathematical knowledge of the specified coordinate system. However, this requirement is logically acceptable and even to be expected. Also, this definition of an inertial system depends on a presupposed familiarity with the concept of force; Chapter 9 will show that amongst* ***all*** *coordinate systems the concept of force may be exchanged for the concept of curvature of the coordinate system. In contradistinction then, all inertial coordinate systems are also* ***flat****, i.e., have no curvature. To be pedantic, one must also require a "trivial topology," i.e., no globally nontrivial features such as multiple connectedness.*

Comment 3.3 *Finally, note that Definition 3.1 implies the testing of certain numerical values: deviation from a straight line, the magnitude and direction of the vector of acceleration. Since every measurement is subject to error, both criteria are subject to the limitations of real, physical measurements. Definition 3.1 is therefore a truly* ***physical*** *one. For example: in nearly all experiments, the "laboratory system" is considered to be inertial, although it is not really so. The laboratory is on the surface of the planet Earth, the gravitational field of which bends the trajectories of objects and accelerates them. Also, the Earth rotates about its axis, so that there are also Coriolis-type forces. Furthermore, the Earth is in the*

[2] It should be kept in mind that the special theory of relativity [179, 55, 69, to name but a few textbooks] is only the linear (*flat* spacetime) approximation to the general theory of relativity [508, 62, 367, 548, 66, 96]. It is possible to extend the use of special relativity so as to include relatively accelerated systems where the incurred nonlinear effects may be consistently neglected. The Reader is expected to have used the formalism of the special theory of relativity at the level of standard texts in electrodynamics [296].

gravitational field of the Moon and the Sun; the Earth rotates about the Sun; together with the solar system, they also rotate about the galactic center, etc. For all practical purposes – and to the precision needed in most experiments – these effects are either negligible or can be accounted for by computation. Only if all these (both conceptual and computational) corrections are negligible may the "laboratory system" be regarded as inertial – to within the stated tolerance. The same applies to all other practical applications of Definition 3.1.

Typically cited as the second postulate is the statement that the speed of propagation of light in vacuum, c, is constant. In the "particle system of units" that we adopt herein, c and $\hbar$ are used as basic units, and both quantities are automatically regarded as universal *constants* of Nature.

3.1.1 Space and time mixing

The next step is the realization that *space* and *time* in relativistic physics are not independently specifiable quantities. The transition from one inertial coordinate system S into another inertial coordinate system S', one that moves with the constant velocity $\vec{v}$ with respect to S, is achieved by means of the so-called Lorentz *boosts*[3] [☞ Exercise 3.1.1]:

$$\vec{r}\,' = \vec{r} + (\gamma-1)(\hat{v}\cdot\vec{r})\,\hat{v} - \gamma\vec{v}t, \qquad \vec{r} = \vec{r}\,' + (\gamma-1)(\hat{v}\cdot\vec{r}\,')\,\hat{v} + \gamma\vec{v}t', \tag{3.1a}$$

$$t' = \gamma\Big(t - \frac{\vec{v}\cdot\vec{r}}{c^2}\Big), \qquad t = \gamma\Big(t' + \frac{\vec{v}\cdot\vec{r}\,'}{c^2}\Big), \tag{3.1b}$$

$$\gamma := \Big(1 - \frac{\vec{v}^{\,2}}{c^2}\Big)^{-\frac{1}{2}}, \qquad \hat{v} := \frac{\vec{v}}{\sqrt{\vec{v}^{\,2}}}. \tag{3.1c}$$

The inverse transformation (in the right-hand column) is formally identical to the original (in the left-hand column), only with a flipped sign of the relative velocity between the two inertial systems S and S'. Also, note that the formulae for the corresponding Galilean transformation in non-relativistic physics emerge in the formal limit $c \to +\infty$, where $\gamma \to 1$.

It is essential to understand that all relativistic effects stem from boosts (3.1) – which after all are the novelty of Lorentz transformations. For a swift motivation for Lorentz symmetries with the benefit of hindsight of a transpired century, see Digression 8.1 on p. 295. Suffice it to say, Lorentz transformations are the correct symmetry of the Maxwell equations, and therefore also of any matter system that interacts with the electromagnetic field.

Relativity of simultaneity If two events A and B are simultaneous in system S so $t_A = t_B$, they need not be simultaneous in system S':

$$t_i' = \gamma\Big(t_i - \frac{\vec{v}\cdot\vec{r}_i}{c^2}\Big), \quad i = A, B, \qquad \Rightarrow \qquad t_A' - t_B' = \gamma\,\frac{\vec{v}\cdot(\vec{r}_B - \vec{r}_A)}{c^2}, \tag{3.2}$$

which vanishes only if $\vec{v}$ is orthogonal to the difference vector (the extent) $\vec{r}_B - \vec{r}_A$, but not otherwise.

[3] Herein, "boost" denotes the mathematical change of coordinates from one inertial system into another, and which moves with a *constant* velocity $\vec{v}$ with respect to the former. The physical process implementing this change would of course require acceleration, to which the special theory of relativity is explicitly *not* applicable. By "Lorentz transformation," some earlier texts [326] mean only boosts, which leads to contradictory-sounding statements such as "Lorentz transformations do not form a group": indeed, boosts alone do not form a group, as their combination may also be a simple rotation. To avoid this nonsense, by "Lorentz transformation" I mean an arbitrary element of the so-called Lorentz group, which contains both rotations and boosts; see Appendix A.5.

Relativity of distance/extent Although a tad trivial, notice that by the length of an object (as measured in an inertial coordinate system S) we mean the extent between the positions of the end-points (A and B) of that object, $L = |\Delta\vec{r}| = |\vec{r}_B - \vec{r}_A|$, as measured *simultaneously*. Since simultaneity is not absolute – see equation (3.2) – neither can we expect length to be.

Consider the two positions $\vec{r}_A$ and $\vec{r}_B$ in the system S, spanning the extent $\Delta\vec{r} := (\vec{r}_B - \vec{r}_A)$. Using equation (3.1a) in the inertial system S', this extent measures

$$\Delta\vec{r}\,' = \Delta\vec{r} + (\gamma-1)(\hat{v}\cdot\Delta\vec{r})\hat{v} - \gamma\,\vec{v}(t_B - t_A). \tag{3.3}$$

If the two positions $\vec{r}_A$ and $\vec{r}_B$ have been established simultaneously in the system S (such as the case of measuring the extent between end-points, i.e., the length of an object), then $t_B - t_A = 0$, and we have that

$$\Delta\vec{r}\,' = \Delta\vec{r} + (\gamma-1)(\hat{v}\cdot\Delta\vec{r})\hat{v} = \Delta\vec{r}_\perp + \gamma\Delta\vec{r}_\parallel, \tag{3.4}$$

where the special cases are

$$\Delta\vec{r}\,'_\parallel = \gamma\,\Delta\vec{r}_\parallel, \qquad \Delta\vec{r}_\parallel := (\hat{v}\cdot\Delta\vec{r})\,\hat{v}, \tag{3.5a}$$

$$\Delta\vec{r}\,'_\perp = \Delta\vec{r}_\perp, \qquad \Delta\vec{r}_\perp := \vec{r} - (\hat{v}\cdot\Delta\vec{r})\,\hat{v}. \tag{3.5b}$$

Formula (3.5a) is the well-known FitzGerald–Lorentz contraction: For an object (and its system S') that moves lengthwise with velocity $\vec{v}$ with respect to the system S, the measurement of the length of the object in the latter system is $L = \Delta\vec{r}_\parallel = L'/\gamma = \gamma^{-1}\Delta\vec{r}\,'_\parallel$. Since $\gamma \geqslant 1$, it follows that $L \leqslant L'$. In turn, formula (3.5b) shows that there is no FitzGerald–Lorentz contraction in directions perpendicular to the relative velocity of the two coordinate systems.

Relativity of the duration of time Consider two moments of time t'_A and t'_B in the inertial system S', which moves with velocity $\vec{v}$ with respect to the inertial system S. Using equation (3.1a) then gives

$$t_B - t_A = \gamma(t'_B - t'_A) + \gamma\frac{\vec{v}\cdot(\vec{r}\,'_B - \vec{r}\,'_A)}{c^2}. \tag{3.6}$$

If the two moments of time t'_A and t'_B have been measured in the same place in system S' (such as the case when the duration of a localized process is observed within the system S'), then $\vec{r}\,'_A = \vec{r}\,'_B$, and $\Delta t' := t'_B - t'_A$ is the duration of time in this "moving" system S'. Then

$$\Delta t = \gamma\,\Delta t', \tag{3.7}$$

is the well-known time dilation formula: $\Delta t \geqslant \Delta t'$. The S-measurement of the duration of time between the events A and B is longer than measured in system S', where A and B are in the same place.

For elementary particle physics, this effect is priceless: in any system with respect to which they move, particles "live" longer than as measured in the system where they are at rest. Thus, a muon created in the higher layers of the Earth's atmosphere nevertheless arrives at the Earth's surface, although its lifetime is only 2.197 µs in its rest-frame. Equivalently, from the muon's point of view, the trip through the Earth's atmosphere is, owing to the FitzGerald–Lorentz contraction, shorter and allows the muon to arrive at the Earth's surface within its lifetime of only 2.197 µs. This also explains how particles with lifetimes of only $\sim 10^{-23}$ s are nevertheless observable.

Addition of velocities For an object that moves with respect to an inertial system S so that it traverses the extent $\Delta\vec{r}$ during the duration of time Δt, the (average) velocity is $\vec{u} = \Delta\vec{r}/\Delta t$. In the inertial system S', which moves with the constant velocity $\vec{v}$ with respect to S, for that same object one measures the velocity $\vec{u}\,' = \Delta\vec{r}\,'/\Delta t'$, so that

$$\vec{u} := \frac{\Delta\vec{r}}{\Delta t} = \frac{\Delta\vec{r}\,' + (\gamma-1)(\hat{v}\cdot\Delta\vec{r}\,')\hat{v} + \gamma\vec{v}\,\Delta t'}{\gamma(\Delta t' + \frac{(\vec{v}\cdot\Delta\vec{r}\,')}{c^2})} = \frac{\gamma^{-1}\vec{u}\,' + (1-\gamma^{-1})(\hat{v}\cdot\vec{u}\,')\hat{v} + \vec{v}}{(1+\frac{(\vec{v}\cdot\vec{u}\,')}{c^2})}$$
$$= \frac{\vec{u}\,'_{\parallel} + \vec{v}}{(1+\frac{(\vec{v}\cdot\vec{u}\,')}{c^2})} + \frac{\vec{u}\,'_{\perp}}{\gamma(1+\frac{(\vec{v}\cdot\vec{u}\,')}{c^2})}, \quad \text{where} \quad \vec{u}\,'_{\parallel} = (\vec{u}\,'\cdot\hat{v})\hat{v}, \quad \vec{u}\,'_{\perp}\cdot\hat{v} = 0. \tag{3.8}$$

The first term is the familiar formula for relativistic addition of collinear velocities, and the second term provides the lesser-known $\hat{v}$-orthogonal contribution to the velocity $\vec{u}\,'$. Notice that the bigger the velocity $\vec{v}$, the bigger the factor γ, and the lesser the contribution from the second (orthogonal) term. It induces an element of rotation – which is a consequence of the algebraic fact that two Lorentz boosts generate a rotation (A.103); see Appendix A.5 for more details.

— ❦ —

As they will be useful, consider the following approximations:

$$\gamma = \frac{1}{\sqrt{1-\beta^2}} \approx 1 + \frac{1}{2}\beta^2 + \frac{3}{8}\beta^4 + \frac{5}{16}\beta^6 + O(\beta^8), \qquad \beta := \frac{v}{c} \ll 1; \tag{3.9a}$$

$$\text{or} \qquad \approx \frac{1}{\sqrt{2\epsilon}}\Big[1 + \frac{1}{4}\epsilon + \frac{3}{32}\epsilon^2 + \frac{5}{128}\epsilon^3 + O(\epsilon^4)\Big], \qquad \epsilon := \Big(1 - \frac{|\vec{v}|}{c}\Big) \ll 1; \tag{3.9b}$$

$$\gamma^{-1} = \sqrt{1-\beta^2} \approx 1 - \frac{1}{2}\beta^2 - \frac{1}{8}\beta^4 - \frac{1}{16}\beta^6 + O(\beta^8); \tag{3.9c}$$

$$\text{or} \qquad \approx \sqrt{2\epsilon}\Big[1 - \frac{1}{4}\epsilon - \frac{1}{32}\epsilon^2 - \frac{1}{128}\epsilon^3 + O(\epsilon^4)\Big]. \tag{3.9d}$$

The expansions (3.9a) and (3.9c) are appropriate approximations for small (non-relativistic, $v \ll c$) velocities, while the expansions (3.9b) and (3.9d) are convenient for large (ultra-relativistic, $v \approx c$) velocities.

3.1.2 Spacetime and the index notation

Since the 3-vector $\vec{r}$ (spatial position) and the moment of time t were in the previous section shown to not be independently specifiable quantities, introduce the 4-vector spacetime[4]

$$\mathrm{x} := \sum_{\mu=0}^{3} x^{\mu}\,\hat{\mathrm{e}}_{\mu}, \qquad \text{where} \qquad x^0 = ct, \;\; \vec{r} = \sum_{i=1}^{3} x^i \hat{\mathrm{e}}_i, \tag{3.10}$$

and where $\hat{\mathrm{e}}_1, \hat{\mathrm{e}}_2, \hat{\mathrm{e}}_3$ are usual unit vectors in some (e.g., Cartesian) inertial coordinate system, and $\hat{\mathrm{e}}_0$ is the additional, fourth unit vector in the direction of time. From now on, we will adopt the strict Einstein convention, whereby summation is implied over any pair of indices precisely if one is a superscript and the other a subscript; thus, $\sum$-symbols are no longer written except for emphasis. Also, Greek indices range over values $0, 1, 2, 3$, while Latin indices are restricted to range over $1, 2, 3$.

[4] It is customary in the literature to denote 4-vectors by a Latin letter without any index or arrow – just like scalars. *Usually*, the context clarifies which of the two is meant; however, without an explicit note, this convention leaves it unclear if a particular "a" denotes a scalar or a 4-vector. Since the purpose of this book is to *introduce* the Reader to the material, "upright" Latin letters will be used for 4-vectors: *herein*, "a, b, c, …" denote 4-vectors, while "$a, b, c, \dots$" are scalars.

Digression 3.2 Note the difference in transformations:

$$\left.\begin{aligned} \mathrm{d}x^\mu &= \Big(\frac{\partial x^\mu}{\partial y^\nu}\Big)\,\mathrm{d}y^\nu, \\ \frac{\partial}{\partial x^\mu} &= \Big(\frac{\partial y^\nu}{\partial x^\mu}\Big)\frac{\partial}{\partial y^\nu}, \end{aligned}\right\} \textit{mutually reciprocal transformations} \qquad \begin{aligned}(3.11a)\\(3.11b)\end{aligned}$$

when changing coordinates $x^\mu \to y^\mu$. Taking a cue from the transformations (3.11a)–(3.11b), any 4-vectors the components of which transform:

$$A^\mu(\mathrm{x}) = \Big(\frac{\partial x^\mu}{\partial y^\nu}\Big)\,A^\nu(\mathrm{y}) \qquad \text{are called } \textbf{contravariant}; \tag{3.11c}$$

$$B_\mu(\mathrm{x}) = \Big(\frac{\partial y^\nu}{\partial x^\mu}\Big)\,B_\nu(\mathrm{y}) \qquad \text{are called } \textbf{covariant}. \tag{3.11d}$$

Digression 3.3 Note that the respectively reciprocal transformations automatically imply that combinations such as

$$A^\mu(\mathrm{x})\,B_\mu(\mathrm{x}), \quad A^\mu(\mathrm{x})\frac{\partial}{\partial x^\mu}, \quad B_\mu(\mathrm{x})\,\mathrm{d}x^\mu, \quad \text{etc.} \tag{3.12a}$$

are invariant with respect to coordinate transformations. Therefore, sums such as

$$\mathrm{A}(\mathrm{x}) := A^\mu(\mathrm{x})\,\hat{\mathrm{e}}_\mu \quad \text{and} \quad \mathrm{B}(\mathrm{x}) := B_\mu(\mathrm{x})\,\hat{\mathrm{e}}^\mu \tag{3.12b}$$

specify the vectors $\mathrm{A}(\mathrm{x})$ and $\mathrm{B}(\mathrm{x})$ *invariantly*. That is, no matter which coordinate system we select, the components $A^\mu(\mathrm{x})$ and $B_\mu(\mathrm{x})$ will transform oppositely from the basis vectors $\hat{\mathrm{e}}^\mu$ and $\hat{\mathrm{e}}_\mu$, respectively, leaving the expressions (3.12b) invariant; see Comment B.1 on p. 512.

Mathematical literature favors this invariant notation, but we will follow the physics notation, using components specified with respect to an implicitly chosen coordinate system, as done in Digression 3.2. Furthermore, a quick comparison of equations (3.12a) and (3.12b) shows that $\frac{\partial}{\partial x^\mu}$ and $\mathrm{d}x^\mu$, being natural vector quantities in any coordinate system, may well serve as explicit choices of basis vectors $\hat{\mathrm{e}}_\mu$ and $\hat{\mathrm{e}}^\mu$, respectively.

In this 4-vector notation, the general Lorentz transformations may be compactly written as

$$y^\mu = \mathrm{L}^\mu{}_\nu\,x^\nu \quad\Leftrightarrow\quad \mathrm{y} = \mathbf{L}\,\mathrm{x} \quad\Leftrightarrow\quad \begin{bmatrix} y^0\\ y^1\\ y^2\\ y^3 \end{bmatrix} = \begin{bmatrix} \mathrm{L}^0{}_0 & \mathrm{L}^0{}_1 & \mathrm{L}^0{}_2 & \mathrm{L}^0{}_3\\ \mathrm{L}^1{}_0 & \mathrm{L}^1{}_1 & \mathrm{L}^1{}_2 & \mathrm{L}^1{}_3\\ \mathrm{L}^2{}_0 & \mathrm{L}^2{}_1 & \mathrm{L}^2{}_2 & \mathrm{L}^2{}_3\\ \mathrm{L}^3{}_0 & \mathrm{L}^3{}_1 & \mathrm{L}^3{}_2 & \mathrm{L}^3{}_3 \end{bmatrix} \begin{bmatrix} x^0\\ x^1\\ x^2\\ x^3 \end{bmatrix}. \tag{3.13a}$$

Comparing equation (3.13) with (3.1), rewriting as the analogous system of equations, all $4 \times 4 = 16$ matrix elements $\mathrm{L}^\mu{}_\nu$ for concrete boosts may be identified:

$$\mathbf{L} = \begin{bmatrix} \gamma & -\gamma\frac{v_x}{c} & -\gamma\frac{v_y}{c} & -\gamma\frac{v_z}{c}\\ -\gamma\frac{v_x}{c} & 1+(\gamma-1)\frac{v_x^2}{\vec{v}^{\,2}} & (\gamma-1)\frac{v_x v_y}{\vec{v}^{\,2}} & (\gamma-1)\frac{v_x v_z}{\vec{v}^{\,2}}\\ -\gamma\frac{v_y}{c} & (\gamma-1)\frac{v_y v_x}{\vec{v}^{\,2}} & 1+(\gamma-1)\frac{v_y^2}{\vec{v}^{\,2}} & (\gamma-1)\frac{v_y v_z}{\vec{v}^{\,2}}\\ -\gamma\frac{v_z}{c} & (\gamma-1)\frac{v_z v_x}{\vec{v}^{\,2}} & (\gamma-1)\frac{v_z v_z}{\vec{v}^{\,2}} & 1+(\gamma-1)\frac{v_z^2}{\vec{v}^{\,2}} \end{bmatrix}. \tag{3.13b}$$

In the general case, Lorentz transformations also include the familiar rotations in addition to the boosts (3.13) and are represented by constant (independent of spacetime coordinates) 4×4 matrices of unit determinant:

$$\frac{\partial \mathrm{L}^{\mu}{}_{\nu}}{\partial x^{\rho}} = 0, \quad (\mu, \nu, \rho = 0,1,2,3), \qquad \det(\mathbf{L}) = 1. \tag{3.14}$$

Digression 3.4 By comparison, the transformation (3.13) is seen to be the special case of the general case (3.11a), when the matrix $\frac{\partial x^{\mu}}{\partial y^{\nu}} = \mathrm{L}^{\mu}{}_{\nu}$ satisfies the additional conditions (3.14), turning the coordinate change $x^{\mu} \to y^{\nu}$ linear ($y^{\nu} = \mathrm{L}^{\mu}{}_{\nu}\, x^{\nu} + C^{\mu}$) and homogeneous ($C^{\mu} = 0$).

Now, just as the rotation group $SO(3)$ leaves the Euclidean length invariant, general Lorentz transformations leave the quantity

$$(c\,\tau)^2 := c^2 t^2 - \vec{r}^{\,2} \;=\; c^2 t^2 - \big[(x^1)^2 + (x^2)^2 + (x^3)^2\big] \tag{3.15}$$

invariant [☞ Appendix A.1.4]. Since c is constant, the quantity τ is also Lorentz-invariant and is called the "proper time." The name stems from the fact that, in any particular inertial system, for any two separate moments in time in the same place we have $\triangle t = t_B - t_A \neq 0$ and $\triangle \vec{r} = \vec{r}_B - \vec{r}_A = 0$, so that

$$\triangle\tau^2 := (t_B - t_A)^2 - c^{-2}\big[\underbrace{(x_B^1 - x_A^1)^2 - (x_B^2 - x_A^2)^2 - (x_B^3 - x_A^3)^2}_{=0}\big] = (t_B - t_A)^2. \tag{3.16}$$

Note that time dilation (3.7) implies that the proper time for any process is always the shortest; in any other inertial system, the duration of that process can only be longer than the proper times or equal to it. Indeed, since $\triangle\tau$ is invariant, in any inertial system where $\triangle\vec{r} \neq 0$, and the events A and B do not happen in the same point in space, $\triangle t$ must be bigger so that $(\triangle t)^2 - c^{-2}(\triangle\vec{r})^2$ remains constant, i.e., invariant with respect to the transformation from that inertial system into the inertial (rest-)system where $\triangle\vec{r} = 0$.

The invariant quantity (3.15) may be more compactly written as[5]

$$c^2\,\tau^2 = \mathrm{x}^2 = \mathrm{x}\cdot\mathrm{x} := x^{\mu}\,\eta_{\mu\nu}\,x^{\nu}. \tag{3.17}$$

An operation "x·y" denotes the (Lorentzian) scalar product of 4-vectors:

Definition 3.2 *For 4-vectors* x *and* y*, the invariant (scalar) product is*

$$\mathrm{x}\cdot\mathrm{y} = x^{\mu}\,\eta_{\mu\nu}\,y^{\nu}. \tag{3.18}$$

The quantity $\mathrm{x}^2 := \mathrm{x}\cdot\mathrm{x}$ *is, simply, the "4-vector* x *square." The matrix*

$$\boldsymbol{\eta} \;=\; [\eta_{\mu\nu}] = \begin{bmatrix} 1 & 0 & 0 & 0 \\ 0 & -1 & 0 & 0 \\ 0 & 0 & -1 & 0 \\ 0 & 0 & 0 & -1 \end{bmatrix} \tag{3.19}$$

*is the **metric tensor** – the **metric** – of the empty (flat) spacetime. The number of positive and negative eigenvalues in the matrix* $[\eta_{\mu\nu}]$ *is called the **signature**, and spacetime and its metric are said to have signature* $(1,3)$.

[5] The scalar product of two n-vectors a and b is denoted "$a{\cdot}b$"; the Reader must understand from the context if this denotes the Euclidean, Lorentzian or some other scalar product. Following this tradition, note that the notation herein is unambiguous, as Euclidean 3-vectors are indicated by an over-arrow and Lorentzian 4-vectors are denoted by "upright" Latin letters. Therefore, $\vec{a}\cdot\vec{b}$ is the Euclidean scalar product, while a·b is Lorentzian.

With this definition, it is possible to verify that the 4×4 matrices representing the general Lorentz transformations satisfy the $\boldsymbol{\eta}$-orthogonality condition

$$\mathbf{L}^T\boldsymbol{\eta} = \boldsymbol{\eta}\mathbf{L}^{-1}, \quad \text{i.e.,} \quad \mathbf{L}^T\boldsymbol{\eta}\mathbf{L} = \boldsymbol{\eta}, \quad \text{or} \quad \mathrm{L}^{\rho}{}_{\mu}\eta_{\rho\sigma}\mathrm{L}^{\sigma}{}_{\nu} = \eta_{\mu\nu}. \tag{3.20}$$

This generalizes the orthogonality relation $\boldsymbol{R}^T\mathbb{1}\boldsymbol{R} = \mathbb{1}$, satisfied by the usual rotation matrices, where the identity matrix serves as the metric for the Euclidean invariant scalar product, $\vec{r}\cdot\vec{r} = x^i\delta_{ij}x^j$. Just as the rotation group is denoted $SO(3)$, the Lorentz group is then denoted $SO(1,3)$ – reminding us that the signature of the metric $\boldsymbol{\eta}$ used to define the Lorentz-invariant scalar product (3.18) is (1,3) [☞ Appendix A.5].

Also,

Definition 3.3 *A 4-vector* v *in spacetime with the metric tensor* $\eta_{\mu\nu}$ *is called*

$$\textbf{time-like (temporal)}, \text{ if } \quad \mathrm{v}^2 > 0, \tag{3.21a}$$

$$\textbf{space-like (spatial)}, \text{ if } \quad \mathrm{v}^2 < 0, \tag{3.21b}$$

$$\textbf{light-like (null)}, \text{ if } \quad \mathrm{v}^2 = 0. \tag{3.21c}$$

It should be fairly straightforward that the replacement $t \to (it)$ changes the sign of η_{00}, the signature into $(0,4)$, and the boosts (3.13). The qualitative nature of this change is easiest to spot in the special case when the coordinate system is chosen so that $\vec{v} \to v\,\hat{\mathrm{e}}_1$ in the relation (3.13):

$$[\mathrm{L}^{\mu}{}_{\nu}] = \begin{bmatrix} \gamma & -\gamma\frac{v}{c} & 0 & 0 \\ -\gamma\frac{v}{c} & \gamma & 0 & 0 \\ 0 & 0 & 1 & 0 \\ 0 & 0 & 0 & 1 \end{bmatrix} = \underbrace{\begin{bmatrix} \cosh(\phi) & -\sinh(\phi) & 0 & 0 \\ -\sinh(\phi) & \cosh(\phi) & 0 & 0 \\ 0 & 0 & 1 & 0 \\ 0 & 0 & 0 & 1 \end{bmatrix}}_{\text{hyperbolic "rotation"}}, \tag{3.22}$$

where we defined the formal variable $\phi := \cosh^{-1}(\gamma)$, so that $v = c\tanh(\phi)$ and $\frac{v}{c}\gamma = \sinh(\phi)$. [✎ *Verify.*] Upon the replacement $t \to it$:

$$[\mathrm{L}^{\mu}{}_{\nu}] \xrightarrow[\varphi=-i\phi]{t\to it} \begin{bmatrix} \cos(\varphi) & -\sin(\varphi) & 0 & 0 \\ \sin(\varphi) & \cos(\varphi) & 0 & 0 \\ 0 & 0 & 1 & 0 \\ 0 & 0 & 0 & 1 \end{bmatrix}, \tag{3.23}$$

so that Lorentz boosts in the x^1-direction become

$$\begin{bmatrix} (ict') \\ x'^1 \\ x'^2 \\ x'^3 \end{bmatrix} = \begin{bmatrix} \cos(\varphi) & -\sin(\varphi) & 0 & 0 \\ \sin(\varphi) & \cos(\varphi) & 0 & 0 \\ 0 & 0 & 1 & 0 \\ 0 & 0 & 0 & 1 \end{bmatrix} \begin{bmatrix} (ict) \\ x^1 \\ x^2 \\ x^3 \end{bmatrix} \tag{3.24}$$

rotations in the $((ict), x^1)$-plane in the so-called Wick-rotated spacetime $((ict), x^1, x^2, x^3)$. Although Henry Poincaré was the first to notice that the complex transformation

$$(ct, x^1, x^2, x^3) \to ((ict), x^1, x^2, x^3) \tag{3.25}$$

turns the group $SO(1,3)$ of Lorentz transformations[6] into the group of rotations $SO(4)$, this was first used by Hermann Minkowski to restate the Maxwell equations and the special theory

[6] We will see later that the Lorentz group is actually $Spin(1,3)$, the double covering of the $SO(1,3)$ group, for spinors to be describable by single-valued spacetime functions [☞ discussion around the relations (5.45)–(5.48)].

of relativity into the 4-dimensional notation. This result solidified the physical irreducibility of 4-dimensional *spacetime*, which is why it is often referred to as "Minkowski space." To emphasize the mixed signature of space+time, the term "spacetime" will be used throughout.

Digression 3.5 Following the example of Digression 3.2 on p. 88, the 4-vector with components x^μ is Lorentz-contravariant, whereby the vector with components $x_\nu := (x^\mu \eta_{\mu\nu})$ is Lorentz-*covariant*, as the quantity $x^\mu\,\eta_{\mu\nu}\,x^\nu$ is defined to be Lorentz-invariant:

$$x^\mu \to \tilde{x}^\mu \qquad = \mathrm{L}^\mu{}_\nu\, x^\nu, \quad \text{contravariant 4-vector;} \tag{3.26a}$$

$$\Rightarrow \quad x_\mu := (x^\nu \eta_{\nu\mu}) \to (\tilde{x}^\nu \tilde{\eta}_{\nu\mu}) = \tilde{x}_\mu = \mathrm{L}^{-1}{}_\rho{}^\nu x_\nu, \quad \text{covariant 4-vector.} \tag{3.26b}$$

Here, $\mathrm{L}^{-1}{}_\mu{}^\nu = [\mathbf{L}^{-1}]_\mu{}^\nu = \frac{\partial x^\nu}{\partial \tilde{x}^\mu}$ are the components of the matrix-inverse of the matrix of Lorentz transformations $\mathrm{L}^\mu{}_\nu = [\mathbf{L}]^\mu{}_\nu = \frac{\partial \tilde{x}^\mu}{\partial x^\nu}$. We then compute, respectively, in the new and in the old coordinates:

$$\tilde{x}_\mu = \tilde{\eta}_{\mu\nu}\tilde{x}^\nu = \tilde{\eta}_{\mu\nu}\mathrm{L}^\nu{}_\sigma x^\sigma, \quad = \mathrm{L}^{-1}{}_\mu{}^\rho x_\rho = \mathrm{L}^{-1}{}_\mu{}^\rho \eta_{\rho\sigma} x^\sigma. \tag{3.26c}$$

This implies that

$$\tilde{\eta}_{\mu\nu}\mathrm{L}^\nu{}_\sigma = \mathrm{L}^{-1}{}_\mu{}^\rho \eta_{\rho\sigma}, \quad \text{i.e.,} \quad \tilde{\eta}_{\mu\nu} = \mathrm{L}^{-1}{}_\mu{}^\rho \eta_{\rho\sigma} \mathrm{L}^{-1}{}_\nu{}^\sigma, \quad \text{i.e.,} \quad \tilde{\boldsymbol{\eta}} = [\mathbf{L}^{-1}]^T \boldsymbol{\eta}\, \mathbf{L}^{-1}. \tag{3.26d}$$

The metric components $\eta_{\mu\nu} = [\boldsymbol{\eta}]_{\mu\nu}$ thus form a twice covariant tensor. However, as $\boldsymbol{\eta}$ and $\tilde{\boldsymbol{\eta}}$ are numerically the same matrix (in the x- and the $\tilde{\mathrm{x}}$-coordinate system, respectively), they are Lorentz-invariants, i.e., remain unchanged under Lorentz transformations. Then

$$\mathrm{x}{\cdot}\mathrm{x} \to \tilde{\mathrm{x}}{\cdot}\tilde{\mathrm{x}} = \tilde{x}^\mu \tilde{\eta}_{\mu\nu}\tilde{x}^\nu = (\mathrm{L}^\mu{}_\rho x^\rho)\tilde{\eta}_{\mu\nu}(\mathrm{L}^\nu{}_\sigma x^\sigma) \overset{(3.26\text{d})}{=} x^\rho\, \mathrm{L}^\mu{}_\rho\, \mathrm{L}^{-1}{}_\mu{}^\nu\, \eta_{\nu\sigma} x^\sigma \tag{3.26e}$$

$$= x^\rho \delta^\nu_\rho \eta_{\nu\sigma} x^\sigma = x^\rho \eta_{\rho\sigma} x^\sigma = \mathrm{x}{\cdot}\mathrm{x}. \tag{3.26f}$$

Note that the result (3.26d) implies

$$\tilde{\boldsymbol{\eta}} = [\mathbf{L}^{-1}]^T \boldsymbol{\eta}\, \mathbf{L}^{-1} \qquad \Leftrightarrow \qquad \boldsymbol{\eta} = [\mathbf{L}]^T \tilde{\boldsymbol{\eta}}\, \mathbf{L}, \tag{3.26g}$$

so that the Lorentz transformation matrices are η-orthogonal; see equation (3.20). This provides the desired spacetime (Lorentzian) generalization of the more familiar (Euclidean) definition of orthogonal matrices $\mathbb{O}^T\mathbb{1}\mathbb{O} = \mathbb{1}$ by replacing $\mathbb{1} \to \boldsymbol{\eta}$; see Appendix A.5.

Further details on tensor calculus and with arbitrary coordinate systems may be found in Chapter 9 and many books; see Refs. [508, 62, 367, 548, 66, 96], to begin with.

The symbol $\eta^{\mu\nu}$ denotes (the components of) the matrix-inverse to $\eta_{\mu\nu}$, so that

$$\eta^{\mu\nu}\,\eta_{\nu\rho} = \eta_{\rho\nu}\,\eta^{\nu\mu} = \delta^\mu_\rho, \qquad \text{so} \qquad x_\mu := \eta_{\mu\nu}\,x^\nu, \quad x^\mu = \eta^{\mu\nu}\,x_\nu. \tag{3.27}$$

Note that $(x^\mu) = (ct, x^1, x^2, x^3)$ and $(x_\mu) = (ct, x_1, x_2, x_3) = (\eta_{\mu\nu}x^\nu) = (ct, -x^1, -x^2, -x^3)$: the value of the covariant spatial components of a 4-vector have the opposite sign from the values of the contravariant spatial components of the 4-vector.

3.1.3 *Mass, energy and linear momentum*

The Hamilton action of a free particle is chosen to be proportional to the length of the "worldline," so Hamilton's least action principle would minimize this length. In turn, the worldline can be parametrized by the proper time τ of the same particle:

$$S = -\int_{\tau_A}^{\tau_B} \mathrm{d}(c\tau)\,\alpha \overset{(3.7)}{=} -\int_{t_A}^{t_B} \mathrm{d}t\,\frac{\alpha c}{\gamma}, \tag{3.28}$$

where α is some positive constant specific for the considered particle, and the sign is negative so that the resting particle would constitute the *minimum*[7] of the function S, in agreement with Hamilton's least action principle. The expression (3.28) implies that the Lagrangian[8] of a free particle

$$L = -\alpha c\sqrt{1-\frac{v^2}{c^2}} \approx -\alpha c + \frac{1}{2}\alpha c\frac{v^2}{c^2} + \alpha c\,O\Big(\frac{v^4}{c^4}\Big), \tag{3.29}$$

where we used the non-relativistic expansion (3.9c). Since the initial constant, $-\alpha c$, is irrelevant for dynamics, comparing the v^2-term with the one in the non-relativistic expression $L_{\rm NR} = \frac{1}{2}mv^2$ fixes $\alpha = mc$, and the relativistic Lagrangian of a free particle is determined to be

$$L = -mc^2\gamma^{-1} = -mc^2\sqrt{1-\frac{\vec{v}^{\,2}}{c^2}} = -mc^2\sqrt{1-\frac{1}{c^2}|\dot{\vec{r}}|^2}. \tag{3.30}$$

Relativistic momentum and energy From equation (3.30) and using the *canonical definition*, we have

$$\vec{p} := \frac{\partial L}{\partial \dot{\vec{r}}} = \frac{\partial L}{\partial \vec{v}} = m\gamma\vec{v} \overset{(3.9a)}{\approx} m\vec{v} + \cdots, \tag{3.31}$$

where we dropped the terms that are at least $O(\frac{v^2}{c^2})$ smaller than $m\vec{v}$, and this canonical definition indeed agrees with the usual non-relativistic definitions, for velocities sufficiently smaller than c. Also, the Hamiltonian, i.e., the energy of a free particle, is, by the *canonical definition* ($H = p_i\dot{q}^i - L$),

$$E := \vec{p}\cdot\dot{\vec{r}} - L = m\gamma\,\vec{v}\cdot\vec{v} + mc^2\gamma^{-1} = m\gamma c^2, \tag{3.32a}$$

$$\overset{(3.9a)}{\approx} \underbrace{mc^2}_{\text{rest energy}} + \underbrace{\tfrac{1}{2}m\vec{v}^{\,2}}_{\substack{\text{non-relativ.}\\ \text{kin. energy}}} + \underbrace{\tfrac{1}{2}m\vec{v}^{\,2}\Big[\tfrac{3}{4}\tfrac{\vec{v}^{\,2}}{c^2} + \tfrac{5}{8}\tfrac{\vec{v}^{\,4}}{c^4} + \cdots\Big]}_{\text{relativistic corrections}}. \tag{3.32b}$$

Recall that the energy, by its definition, is a measure of the ability to do work. From the result (3.32a), it follows that a free particle has the ability to do work not only by virtue of its motion (the kinetic energy), but also owing to simply having a nonzero mass! Indeed, the expression (3.32a) clearly expresses energy as a function of velocity, one that does not vanish in the rest-frame of a particle, in which it is of course at rest:

$$E_0 := E\big|_{\vec{v}=0} = mc^2, \qquad \text{rest energy.} \tag{3.33}$$

The discovery contained in the relation (3.33) is Einstein's best known formula. This is the ideal place to cite Professor Okun's warning [393], that the relation (3.33) – and not "$E = mc^2$" – is the *real* Einstein formula [☞ Exercise 3.1.2].

Of course, the kinetic energy of a particle is then[✎ *why?*]

$$T := E - E_0 = m(\gamma-1)c^2 \approx \underbrace{\tfrac{1}{2}m\vec{v}^{\,2}}_{\substack{\text{non-relativ.}\\ \text{kin. energy}}} + \underbrace{\tfrac{1}{2}m\vec{v}^{\,2}\Big[\tfrac{3}{4}\tfrac{\vec{v}^{\,2}}{c^2} + \tfrac{5}{8}\tfrac{\vec{v}^{\,4}}{c^4} + \cdots\Big]}_{\text{relativistic corrections}}. \tag{3.34}$$

[7] The time between the events A and B is maximal in the system where A and B are in the same place, hence the worldline from A to B is entirely along the time coordinate. In all other systems, the worldline from A to B also extends partially in the spatial directions, and the time $t_B - t_A$ is shorter [☞ time dilation (3.7)].

[8] The term "Lagrangian" and its derivatives honor the French mathematician Joseph Louis Lagrange.

The energy–momentum 4-vector On par with the spacetime 4-vector $\mathrm{x} = ((ct), x^1, x^2, x^3)$, we define also the 4-momentum [☞ Digressions 3.6 and 3.7]:

$$\mathrm{p} = (p_\mu) := (-E/c, \vec{p}) \;=\; (-m\gamma c, m\gamma\vec{v}). \tag{3.35}$$

From this, we have that

$$\mathrm{p}^2 := p_\mu\, \eta^{\mu\nu}\, p_\nu = E^2/c^2 - \vec{p}^{\,2} = m^2\gamma^2c^2 - \vec{p}^{\,2} = m^2\gamma^2c^2\Big(1 - \frac{v^2}{c^2}\Big) = m^2c^2. \tag{3.36}$$

As the left-hand side quantity is evidently Lorentz invariant [☞ Exercise 3.1.3], so then is the mass m. Just as proper time is the Lorentz-invariant magnitude of the position 4-vector $\mathrm{x} = (ct, \vec{r})$, (the c-multiple of) mass is the Lorentz-invariant magnitude of the 4-momentum $\mathrm{p} = (-E/c, \vec{p})$. A very useful formula follows from equation (3.36):

$$E^2 \;=\; \vec{p}^{\,2}c^2 \;+\; m^2c^4. \tag{3.37}$$

Rewriting this as $(mc^2)^2 = E^2 - (c\vec{p})^2$ exhibits the direct parallel with equation (3.15). In turn, the 4-momentum is indeed a covariant 4-vector, as defined in equation (3.26b), and its components transform under Lorentz transformations as $p'_\mu = \mathrm{L}^{-1}{}_\rho{}^\nu\, p_\nu$.

Digression 3.6 To justify the definition (3.35) – the covariance and signs of the components (3.35) – it is simplest to rely on quantum mechanics, where in coordinate representation the components of the *operator* of 4-momentum p become $p_\mu = \frac{\hbar}{i}\frac{\partial}{\partial x^\mu}$:

$$p_0 = \frac{\hbar}{i}\frac{\partial}{\partial x^0} = \frac{\hbar}{i}\frac{\partial}{\partial(ct)} = -\frac{1}{c}\,i\hbar\frac{\partial}{\partial t} = -\frac{1}{c}H, \quad \text{but} \quad \vec{p} = +\frac{\hbar}{i}\vec{\nabla}. \tag{3.38}$$

The peculiar negative sign in the identification of $p_0 = -\frac{1}{c}H$ owes to the *standard* identifications in non-relativistic quantum mechanics, $H = i\hbar\frac{\partial}{\partial t}$ *vs.* $\vec{p} = \frac{\hbar}{i}\vec{\nabla}$, and to insisting that the non-relativistic energy operator of a system should be the limit of the relativistic one, *with the same sign.*

Digression 3.7 The same conclusion may also be derived classically, i.e., non-quantum mechanically. Note first that the components of the canonical linear momentum 3-vector are naturally covariant. This is seen from the explicitly written definition (3.31):

$$p_i := \frac{\partial L}{\partial v^i}, \quad \text{where} \quad v^i := \frac{\partial x^i}{\partial t}, \qquad i = 1,2,3. \tag{3.39a}$$

To extend this canonical definition to the relativistic 4-vector, use the earlier defined (3.10) 4-vector $(x^\mu) = ((x^0 := ct), x^1, x^2, x^3)$, so that

$$(v^\mu) := \frac{\partial x^\mu}{\partial t} = (c, \dot{x}^1, \dot{x}^2, \dot{x}^3). \tag{3.39b}$$

In turn, the Hamilton action (3.28)–(3.30) may be rewritten as

$$S = -\int_{t_A}^{t_B} \mathrm{d}t\; mc^2\sqrt{1 - \frac{\vec{v}^{\,2}}{c^2}} = \int_{x^0_A}^{x^0_B} \mathrm{d}x^0\; L_0, \qquad L_0 := (L/c) = -m\sqrt{c^2 - \vec{v}^{\,2}}, \tag{3.39c}$$

where we note that $[L_0] = \frac{ML}{T}$ has the physical dimensions of linear momentum and not those of energy as does mc^2/γ. From this we have (3.39b):

$$v^0 := \frac{\partial x^0}{\partial t} = \frac{\partial(ct)}{\partial t} = c, \tag{3.39d}$$

as well as that (x^1, x^2, x^3) *depend* on t and so also on $x^0 = ct$:

$$p_\mu := \frac{\partial L_0}{\partial \frac{\partial x^\mu}{\partial x^0}} = \frac{\partial L_0}{\frac{1}{c}\partial \dot{x}^\mu} = c\,\frac{\partial L_0}{\partial v^\mu} \quad \Rightarrow \quad \begin{cases} p_0 := c\,\dfrac{\partial\big(-m\sqrt{c^2-\vec{v}^2}\big)}{\partial c} = -m\gamma c = -E/c, \\ p_i := c\,\dfrac{\partial\big(-m\sqrt{c^2-\vec{v}^2}\big)}{\partial v^i} = m\gamma\,\delta_{ij}\,v^j, \end{cases} \tag{3.39e}$$

which reproduces equation (3.35).

By the way, the expression $S = \int \mathrm{d}x^0\, L_0$ of course does not seem to be Lorentz-invariant, since the coordinate x^0 is singled out. However, the spacetime Lagrangian L_0 may be expressed as a spatial integral of the Lagrangian density:

$$L_0 = \int_V \mathrm{d}^3\vec{r}\,\mathscr{L}, \qquad \text{such that} \qquad S = -\int_{(t_A,V)}^{(t_B,V)} \mathrm{d}^4x\,\mathscr{L}, \tag{3.40}$$

where $\mathscr{L}$ is a scalar density: with respect to coordinate change $x^\mu \to y^\mu$, we have that $\mathrm{d}^4x \to |\frac{\partial x}{\partial y}|\mathrm{d}^4y$, where $|\frac{\partial x}{\partial y}|$ is the determinant of the matrix of partial derivatives $\frac{\partial x^\mu}{\partial y^\nu}$. For the Hamilton action to be independent of any (invertible) choice and/or change of coordinates, it must be that $\mathscr{L}(x) \to |\frac{\partial x}{\partial y}|^{-1}\mathscr{L}(y)$, which is the defining property of scalar densities of weight -1 [☞ Section B.2].

Massless particle In non-relativistic physics, a particle with no mass is nonsense: for such a particle both the linear momentum and the kinetic energy would also have to vanish. Then, its response to the action of a force could not be computed by Newton's laws, since the formula $a = \frac{1}{m}F$ would imply that any finite force would cause its infinitely large acceleration. On the other hand, the *relativistic* formulae are self-consistent. Indeed, from the relation (3.36), it follows that

$$m = 0 \quad \Leftrightarrow \quad E^2 = \vec{p}^{\,2}c^2 \quad \Leftrightarrow \quad E = c|\vec{p}|, \tag{3.41}$$

which, when combined with results (3.31) and (3.32a), gives

$$\gamma mc^2 = \gamma mc|\vec{v}| \qquad \Rightarrow \qquad |\vec{v}| = c. \tag{3.42}$$

That is, a massless particle must move at the speed of light. So far, only the photons provide a manifest and directly observable example.

3.1.4 Exercises for Section 3.1

✎ ***3.1.1*** *Simplify the relations (3.1) for the oft-cited case* $\vec{v} = v\,\hat{\mathbf{e}}_z$.

✎ ***3.1.2*** *Without consulting Ref. [393], prove that the equality "*$E = mc^2$*" is nonsense, contradicting the provided definitions and the physical meaning of energy* E *and mass* m*.*

✎ ***3.1.3*** *Prove that the quantities* $\mathrm{p}^2 := p_\mu\,\eta^{\mu\nu}\,p_\nu$ *and* $\mathrm{x}\cdot\mathrm{p} = x^\mu p_\mu$ *are Lorentz-invariant.*

✎ ***3.1.4*** *Verify the transformations (3.22)–(3.23)–(3.24).*

3.2 Relativistic kinematics: limitations and consequences

The essential reason for defining the 4-momentum (3.31) with (3.32a) is the fact that this 4-vector physical quantity is conserved [☞ Footnote *15* on p. 54] and transforms akin to (3.13). Because of the typical application, we will consider collisions and decays.

Using the definitions (3.32a), (3.34) and (3.35), for collisions we have:

1. The sum of relativistic 4-momenta is strictly conserved.
2. The sum of relativistic kinetic energies:
   (a) is conserved in *elastic collisions*;
   (b) grows in "exo-energetic" (fissile or explosive) processes;
   (c) is diminished in "endo-energetic" (fusing, implosive or sticking) processes.

Since the mass equals $(E - T)/c^2$, it is conserved only in elastic collisions. In explosive/fissile collisions, the total mass is diminished, which supports the impression that part of the mass was "converted" into kinetic energy; in implosive/fusing/sticking processes, the total mass grows, as if part of the kinetic energy was "converted" into mass. One must keep in mind that the total mass of a composite system equals (up to the coefficient of proportionality, c^2) the rest energy, which includes various "internal forms of energy," as these are usually called in non-relativistic physics. Thus, e.g., the total mass of a hydrogen atom equals $(m_p + m_e)c^2 + E_b$, where E_b is the binding energy of the hydrogen atom in the particular state, in the first approximation given by Bohr's formula (1.31).

Example 3.1 Two equal snowballs of mass m fly with the same speed $|\vec{v}_i| = \beta c$, $0 < \beta \leqslant 1$, towards each other, then collide and fuse into one large snowball. For what speed of the colliding snowballs will the resulting snowball have a mass of $M = 3m$ (so that "$m + m \to 3m$")?

Solution Given that $\vec{p}_1 = -\vec{p}_2$, conservation of the linear momentum 3-vector gives that $\vec{p}_{1+2} = \vec{p}_1 + \vec{p}_2 = 0$. That is, the resulting snowball is at rest (which should be obvious). Conservation of p_0 now gives $E_A + E_B = E_{A+B}$, i.e.,

$$2m\gamma c^2 = Mc^2 \quad \Rightarrow \quad M = \frac{2m}{\sqrt{1-\beta^2}} > 2m, \quad \text{since} \quad \beta > 0. \tag{3.43}$$

Inserting $M = 3m$, solve the equation (3.43) for $\beta = \frac{v_i}{c}$ to obtain $|v_i| = \frac{\sqrt{5}}{3}c \approx 74.54\%\, c$.

—❦—

Part of the analysis of this process, the one that relies exclusively on applications of the 4-momentum conservation law is usually referred to as "kinematics." Sometimes, that term also implies the application of the conservation law of angular momentum. For the remainder of this chapter, angular momentum considerations are omitted, and a few simple processes are studied "kinematically" as a user's guide for application in general.

3.2.1 Decays

Two-particle decays

The simplest decay is of the form $A \to B + C$. Label the 4-momenta:

$$A \to B + C, \qquad \mathrm{p}_A = (-m_A c, \vec{0}), \quad \mathrm{p}_B = (-E_B/c, \vec{p}_B), \quad \mathrm{p}_C = (-E_C/c, \vec{p}_C), \tag{3.44}$$

where we used the fact that, before the decay, particle A (with $m_A \neq 0$) defines an inertial system where it is at rest, so that its total relativistic energy reduces to rest energy, $E_A = m_A c^2$. The 4-momentum conservation law provides

$$\mathbf{p}_A = \mathbf{p}_B + \mathbf{p}_C, \qquad \text{or} \qquad \mathbf{p}_B = \mathbf{p}_A - \mathbf{p}_C, \tag{3.45}$$

which includes the usual, 3-momentum conservation:

$$\vec{p}_B = 0 - \vec{p}_C. \tag{3.46}$$

Squaring relation (3.45) for the 4-momenta produces[9]

$$\begin{aligned} \mathbf{p}_B^2 &= (\mathbf{p}_A - \mathbf{p}_C)^2 = \mathbf{p}_A^2 + \mathbf{p}_C^2 - 2\,\mathbf{p}_A \cdot \mathbf{p}_C, \\ \| \quad & \qquad\qquad\qquad\qquad \| \\ m_B^2\,c^2 \quad & \qquad m_A^2\,c^2 + m_C^2\,c^2 - 2\frac{E_A}{c}\frac{E_C}{c} = m_A^2\,c^2 + m_C^2\,c^2 - 2\,m_A\,E_C. \end{aligned} \tag{3.47}$$

From this, it follows that

$$E_C = \Big(\frac{m_A^2 + m_C^2 - m_B^2}{2m_A}\Big)c^2. \tag{3.48}$$

The magnitude of the linear momentum 3-vector is now determined from the relation (3.37), $E_C = c\sqrt{m_C^2 c^2 + \vec{p}_C^{\,2}}$:

$$\begin{aligned} |\vec{p}_C| &= \sqrt{\frac{E_C^2}{c^2} - m_C^2\,c^2} = c\sqrt{\Big(\frac{m_A^2 + m_C^2 - m_B^2}{2m_A}\Big)^2 - m_C^2} \\ &= c\,\frac{\sqrt{(m_A + m_B + m_C)(m_A - m_B + m_C)(m_A + m_B - m_C)(m_A - m_B - m_C)}}{2m_A} \\ &= c\,\frac{\sqrt{m_A^4 + m_B^4 + m_C^4 - 2m_A^2 m_B^2 - 2m_A^2 m_C^2 - 2m_B^2 m_C^2}}{2m_A}. \end{aligned} \tag{3.49}$$

From the relation (3.46) it follows that $|\vec{p}_B| = |\vec{p}_C|$, which also follows from the $B \leftrightarrow C$ symmetry of the formula (3.49). The analogous derivation gives $E_B = \big(\frac{m_A^2 + m_B^2 - m_C^2}{2m_A}\big)c^2$. Note that both E_B and E_C are fully determined by decay kinematics. This was crucial in the discussion on p. 54, and induced Bohr to question the validity of the energy conservation law, and Pauli to predict the neutrino in order to save the energy conservation law. On the other hand, besides the relation (3.46) amongst the magnitudes, there is nothing to determine the direction of $\hat{p}_B = -\hat{p}_C$, which thus remains arbitrary. This implies that, in a large ensemble of $A \to B + C$ decays, the angular distribution of the direction of $\hat{p}_B = -\hat{p}_C$ is expected to be uniform.

Digression 3.8 The same result is obtained starting with equation (3.45), written in Cartesian components, say,

$$E_A = E_B + E_C, \tag{3.50a}$$

$$\vec{0} = \vec{p}_A = \vec{p}_B + \vec{p}_C. \tag{3.50b}$$

[9] The final result, of course, may just as well be obtained by combining the separately stated conservation laws of the linear momentum 3-vector and the relativistic energy. However, by squaring directly the 4-vector equality (3.45), the result (3.48) is obtained faster, because of the simplifying circumstance that three of the components of p_A vanish.

From the equation (3.50b), it follows that $\vec{p}_C = -\vec{p}_B =: \vec{p}$, and in equation (3.50a), express all three energies in terms of the linear momenta and masses using the general relation (3.37):

$$m_A c^2 = c\sqrt{m_B^2 c^2 + \vec{p}^{\,2}} + c\sqrt{m_C^2 c^2 + \vec{p}^{\,2}}. \tag{3.50c}$$

From here, by squaring, rearranging terms to isolate the square-root, then by squaring again, we obtain [✎ verify]

$$\left[m_A^4 + (m_B^2 - m_C^2)^2\right]c^2 = 2m_A^2\left[(m_B^2 + m_C^2)c^2 + 2\vec{p}^{\,2}\right]. \tag{3.50d}$$

Solving this for $|\vec{p}\,|$ one re-derives the result (3.49). [✎ Verify.]

Digression 3.9 On the other hand, if we express (in the equation (3.50a)) E_A and one of E_B, E_C in terms of linear momenta and masses using the general relation (3.37), we obtain, e.g.,

$$m_A c^2 = c\sqrt{m_B^2 c^2 + \vec{p}^{\,2}} + E_C, \qquad \text{i.e.,} \qquad m_A c^2 - E_C = c\sqrt{m_B^2 c^2 + \vec{p}^{\,2}}, \tag{3.51a}$$

the square of which gives, after a little simplifying [✎ verify],

$$E_C^2 - 2m_A c^2\, E_C + \left[(m_A^2 - m_B^2)c^4 - \vec{p}^{\,2}c^2\right] = 0. \tag{3.51b}$$

After inserting the previous result (3.49) and simplifying, the solutions of this quadratic equations are [✎ verify]

$$E_C^{(\pm)} = \left[m_A \pm \frac{m_A^2 + m_B^2 - m_C^2}{2m_A}\right]c^2, \tag{3.51c}$$

where $E_C^{(-)}$ equals the result (3.48).

That the solution $E_C^{(+)}$ is not physical is quickest to see from the special case when $m_B = m_C = 0$, as is the case in the decay $\pi^0 \to 2\gamma$. For this case,

$$E_B^{(+)} = E_C^{(+)} = \frac{3}{2}m_A c^2, \quad \text{and so} \quad m_A c^2 = E_A \overset{(3.50a)}{=} E_B^{(+)} + E_C^{(+)} = 3m_A c^2, \tag{3.51d}$$

which is clearly a contradiction. This leaves $E_C^{(-)}$ in the result (3.51c) as the only consistent solution for the energy of the product in a two-particle decay, confirming the result (3.48).

The technical advantage in using the square of a suitably chosen form of the 4-momentum conservation equation (3.47) is fully understood only through filling in the derivation steps that had been omitted here (mostly, in rearranging and simplifying). The diligent Student is therefore highly recommended to complete these alternate computations.

Many-particle decays

The analysis of a decay of a particle into more than two "fragments" is of course more complicated. However, the starting point is again the 4-momentum conservation, which may be written in any of the following forms:

$$\mathrm{p} = \sum_i \mathrm{p}_i, \qquad \mathrm{p}_i = \mathrm{p} - \sum_{j\neq i} \mathrm{p}_j, \qquad \mathrm{p} - \mathrm{p}_i = \sum_{j\neq i} \mathrm{p}_j, \quad \forall i, \tag{3.52a}$$

$$\mathrm{p}_i + \mathrm{p}_j = \mathrm{p} - \sum_{k\neq i,j} \mathrm{p}_k, \qquad \mathrm{p} - \mathrm{p}_i - \mathrm{p}_j = \sum_{k\neq i,j} \mathrm{p}_k, \quad \forall i,j, \quad \text{etc.} \tag{3.52b}$$

Squaring the 4-vector equations (3.52) in the rest-frame of the decaying particle, where

$$\mathrm{p} = (-E/c, \vec{0}), \qquad \text{so that} \qquad \mathrm{p}^2 = m^2c^2 = E^2/c^2, \tag{3.53}$$

we respectively obtain the equations:

$$\tfrac{1}{2}(m^2 - \sum_i m_i{}^2)c^4 = \sum_{j>i} (E_iE_j - |\vec{p}_i||\vec{p}_j|c^2\cos(\phi_{ij})), \tag{3.54a}$$

$$\tfrac{1}{2}(m^2 - m_i^2 + \sum_{j\neq i} m_j{}^2)c^4 = mc^2 \sum_{j\neq i} E_j - \sum_{\substack{j<k\\ j,k\neq i}} (E_jE_k - |\vec{p}_j||\vec{p}_k|c^2\cos(\phi_{jk})), \quad \forall i, \tag{3.54b}$$

$$\tfrac{1}{2}(m^2 + m_i^2 - \sum_{j\neq i} m_j{}^2)c^4 = mc^2E_i + \sum_{\substack{j<k\\ j,k\neq i}} (E_jE_k - |\vec{p}_j||\vec{p}_k|c^2\cos(\phi_{jk})), \quad \forall i, \tag{3.54c}$$

$$\begin{aligned}\tfrac{1}{2}(m^2 - m_i^2 - m_j^2 + \sum_{k\neq i,j} m_k{}^2)c^4 &= mc^2 \sum_{k\neq i,j} E_k + (E_iE_j - |\vec{p}_i||\vec{p}_j|c^2\cos(\phi_{ij}))\\ &\quad - \sum_{\substack{k<\ell\\ k,\ell\neq i,j}} (E_kE_\ell - |\vec{p}_k||\vec{p}_\ell|c^2\cos(\phi_{k\ell})), \quad \forall i,j,\end{aligned} \tag{3.54d}$$

$$\begin{aligned}\tfrac{1}{2}(m^2 + m_i^2 + m_j^2 - \sum_{k\neq i,j} m_k{}^2)c^4 &= mc^2(E_i + E_j)\\ &\quad + \sum_{\substack{k<\ell\\ k,\ell\neq i,j}} (E_kE_\ell - |\vec{p}_k||\vec{p}_\ell|c^2\cos(\phi_{k\ell})), \quad \forall i,j,\end{aligned} \tag{3.54e}$$

etc.

$$\text{where} \quad E_i^2 = m_i^2c^4 + |\vec{p}_i|^2c^2, \qquad \text{but} \quad E = E_0 = mc^2, \tag{3.54f}$$

using the particular consequences of the general relation (3.37) and also that

$$\mathrm{p}_i\cdot\mathrm{p}_j = p_{i\mu}\eta^{\mu\nu}p_{j\nu} = \left(-\frac{E_i}{c}\right)\left(-\frac{E_j}{c}\right) - \vec{p}_i\cdot\vec{p}_j = \frac{E_iE_j}{c^2} - |\vec{p}_i||\vec{p}_j|\cos(\phi_{ij}). \tag{3.55}$$

The combinatorially growing system (3.52)–(3.54) contains more equations than unknowns, which is convenient, as we can select a subset of the equations (3.52)–(3.54) that provides the simplest way to solve for the desired quantities. (Since the relations (3.52) are all just variants of the same equation, it is clear that the system (3.54) cannot be over-determined.)

3.2.2 Scattering

Besides decays, in elementary particle physics one most often considers the scattering of two particles. The 4-momentum conservation here has the general form

$$\mathrm{p}_1 + \mathrm{p}_2 = \sum_{i>2} \mathrm{p}_i, \tag{3.56}$$

which may, of course, be rewritten in several different ways, just like equations (3.52) are different forms of $\mathrm{p} = \sum_i \mathrm{p}_i$ for a decay. Also, collisions may be analyzed either[10]

[10] The term "CM system" stands for "center-of-momentum system" and is defined by the property that the total linear momentum 3-vector vanishes in it, clearly adapting equations (3.53).

$$\text{CM system, before:}\quad \mathrm{p}_1+\mathrm{p}_2=\Big(-\frac{E_1}{c}-\frac{E_2}{c},\vec{0}\Big),\qquad \text{as}\quad \vec{p}_1+\vec{p}_2=0, \tag{3.57}$$

$$\|$$

$$\text{CM system, after:}\quad \sum_{i>2}\mathrm{p}_i=\Big(-\sum_{i>2}\frac{E_i}{c},\vec{0}\Big),\qquad \text{as}\quad \sum_{i>2}\vec{p}_i=0, \tag{3.58}$$

or in the target system (choosing, say, target = "2," so that $\mathrm{p}_2=(-m_2c,\vec{0})$):

$$\text{target system, before:}\quad \mathrm{p}'_1+\mathrm{p}'_2=\Big(-\frac{E'_1}{c}-m_2c,\vec{p}'_1\Big),\qquad \text{as}\quad \vec{p}'_2=\vec{0}, \tag{3.59}$$

$$\|$$

$$\text{target system, after:}\quad \sum_{i>2}\mathrm{p}'_i=\sum_{i>2}\Big(-\frac{E'_i}{c},\vec{p}_i\Big),\qquad \text{where}\quad \sum_{i>2}\vec{p}'_i=\vec{p}'_1. \tag{3.60}$$

Here, the vertical equality between (3.57)–(3.58) and (3.59)–(3.60) respectively is, of course, the statement of the 4-momentum conservation law.

What's more, by using only Lorentz-invariant expressions (such as squares of 4-vectors), we may combine *both* systems! That is, the 4-vectors in the "vertical" equation (3.57)–(3.58) and the 4-vector on the left-hand side of (3.59)–(3.60) are, of course, not equal, $\mathrm{p}_1+\mathrm{p}_2\neq\mathrm{p}'_1+\mathrm{p}'_2$. However, the squares of these 4-vectors *are* equal – as is the square of any 4-vector – and this provides the continued equality:

$$(\mathrm{p}_1+\mathrm{p}_2)^2=\Big(\sum_{i>2}\mathrm{p}_i\Big)^2=(\mathrm{p}'_1+\mathrm{p}'_2)^2=\Big(\sum_{i>2}\mathrm{p}'_i\Big)^2=\dots, \tag{3.61}$$

where "..." denotes similar equalities for the square of the same 4-momentum specified in any other coordinate system that we may choose for its need or convenience.

For two-particle collisions, $A+B\to C+D$, one defines:

***Definition 3.4** Mandelstam's Lorentz-invariant variables:*

$$s:=-(\mathrm{p}_A+\mathrm{p}_B)^2c^2,\qquad t:=-(\mathrm{p}_A-\mathrm{p}_C)^2c^2,\qquad u:=-(\mathrm{p}_A-\mathrm{p}_D)^2c^2. \tag{3.62}$$

These variables are often used in computations as they are Lorentz-invariant; keep in mind, however, that the 4-momentum conservation law and the relation (3.36) produce the linear relation

$$-(s+t+u)=\Big[3\mathrm{p}_A^2+\mathrm{p}_B^2+\mathrm{p}_C^2+\mathrm{p}_D^2+2\mathrm{p}_A\cdot\underbrace{(\mathrm{p}_B-\mathrm{p}_C-\mathrm{p}_D)}_{-\mathrm{p}_A}\Big]c^2=\sum_{i=A}^{D}\mathrm{p}_i^2c^2=\sum_{i=A}^{D}m_i^2c^4. \tag{3.63}$$

In turn, albeit not Lorentz-invariant, (lab-frame) energies and angles $\phi_{ij}:=\arccos(\hat{p}_i\cdot\hat{p}_j)$ are more convenient variables for comparison with experiments.

Fusing collisions

Generalizing Example 3.1 on p. 95, consider the collision of two particles that fuse into a single one, with a specified mass m_C. This process is evidently a time-reversed version of the two-particle decay, so that the computation (3.44)–(3.8) may be used by adapting the notation. However, in this case, instead of the inertial system (3.44), where the end-product is at rest, select the inertial system where the particle B ("target") is at rest:

$$A+B\to C,\qquad \mathrm{p}_A=(-E_A/c,\vec{p}_A),\quad \mathrm{p}_B=(-m_Bc,\vec{0}),\quad \mathrm{p}_C=(-E_C/c,\vec{p}_C). \tag{3.64}$$

Conservation of 4-momentum gives

$$\mathrm{p}_C = \mathrm{p}_A + \mathrm{p}_B, \quad \text{i.e.,} \quad \Big(-\frac{E_C}{c}\,,\,\vec{p}_C\Big) = \Big(-\frac{E_A}{c} - m_B c\,,\,\vec{p}_A\Big), \tag{3.65}$$

from which it follows that $\vec{p}_C = \vec{p}_A =: \vec{p}$, as well as that $E_C = E_A + m_B c^2$. Squaring the 4-momentum version of equation (3.65) produces, straightforwardly,

$$\mathrm{p}_C{}^2 = \mathrm{p}_A{}^2 + \mathrm{p}_B{}^2 + 2\mathrm{p}_A{\cdot}\mathrm{p}_B, \tag{3.66}$$

$$m_C^2 c^2 = m_A^2 c^2 + m_B^2 c^2 + 2E_A m_B \quad\Rightarrow\quad E_A = \frac{m_C^2 - (m_A^2 + m_B^2)}{2m_B}\,c^2. \tag{3.67}$$

The same relation can, of course, also be obtained using the conservation of energy and 3-momentum, and the diligent Reader is invited to do so, then compare the relative ease of this computation. *[✎ Do it.]* Since $E = mc^2 + T$, we have the condition

$$T_A = \frac{m_C^2 - (m_A + m_B)^2}{2m_B}\,c^2. \quad \textit{[✎ Verify.]} \tag{3.68}$$

That is, the "probe" A must have the precisely specified kinetic energy (3.68) for it to fuse with the target; for any other value of T_A, the 4-momentum conservation law strictly forbids the fusing. The process $A + B \to C$ is said to be *kinematically forbidden* except when the relation (3.68) is satisfied.

Process threshold

Following a worked-out example from Ref. [243], we can determine the "threshold" of the reaction (3.56), i.e., the minimal kinetic energy with which the probe "1" must collide with the target "2" for the particles in the product of the process (3.56) to be created. For this minimal energy, the particles in the product of the process (3.56) will have no kinetic energy, and we have that

$$\mathrm{p}_i\big|_{\text{threshold}} = (-m_i c\,,\,\vec{0}), \qquad i > 2, \quad \text{CM system.} \tag{3.69}$$

On the other hand, before the collision, we have equation (3.59). Using the equality of the second and the third term in equation (3.61), we have for the special "threshold" (minimal energy) case

$$\begin{aligned}
\min\Big[(\mathrm{p}_1' + \mathrm{p}_2')^2\Big] &= \Big(\sum_{i>2} \mathrm{p}_i\big|_{\text{threshold}}\Big)^2,\\
\min\Big[\mathrm{p}_1{}^2 + \mathrm{p}_2{}^2 + 2\mathrm{p}_1'{\cdot}\mathrm{p}_2'\Big] &= \Big(\sum_{i>2} (m_i c\,,\,\vec{0})\Big)^2,\\
\min\Big[m_1{}^2 c^2 + m_2{}^2 c^2 + 2E_1'\,m_2\Big] &= \Big(\sum_{i>2} m_i c\Big)^2,\\
(m_1{}^2 + m_2{}^2)c^2 + 2\min(E_1')\,m_2 &= \sum_{i,j>2} m_i\,m_j\,c^2.
\end{aligned} \tag{3.70}$$

It follows that the occurrence of the process (3.56) requires

$$E_1' \geqslant \frac{1}{2m_2}\Big[\sum_{i,j>2} m_i m_j - (m_1^2 + m_2^2)\Big]\,c^2, \tag{3.71}$$

and thus

$$T_1' \geqslant \frac{1}{2m_2}\sum_{i,j>2} m_i m_j\,c^2 - \frac{(m_1 + m_2)^2}{2m_2}\,c^2. \quad \textit{[✎ Verify.]} \tag{3.72}$$

Thus, e.g., for the process $X + X \to 3X + \overline{X}$, for any particle X, the first X-particle must collide with the second, stationary X-particle with at least $6\, m_X\, c^2$ of kinetic energy.

This threshold is larger than the naive expectation, whereby the kinetic energy would need to be "only" sufficient to produce the $(3X{+}\overline{X}) - (X{+}X) = X{+}\overline{X}$ particles, i.e., $2m_X c^2$. The reason for this is the inefficiency of a moving probe collision with a stationary target: Before the collision, the total linear momentum of the incoming probe-X and stationary target-X system is not zero, and must equal the total linear momentum of the $3X{+}\overline{X}$ system of particles after the collision. Since the linear momentum of the out-coming $3X{+}\overline{X}$ particles differs from zero, so does the total kinetic energy, which increases the process threshold.

Head-on collisions

In the CM system, where $\vec{p}_1 = -\vec{p}_2 = \vec{p}$, so that $E_1 = E_2 =: E$ if $m_1 = m_2$, we have

$$\min\left[(\mathrm{p}_1 + \mathrm{p}_2)^2\right] = \Big(\sum_{i>2} \mathrm{p}_i\big|_{\text{threshold}}\Big)^2,$$

$$\Big(-\frac{2\min(E)}{c}, \vec{0}\Big)^2 = \Big(\sum_{i>2} m_i c, \vec{0}\Big)^2 \quad \Rightarrow \quad \min(E) = \tfrac{1}{2}\sum_{i>2} m_i c^2. \tag{3.73}$$

Since both particles have the same minimal energy (as they are identical) before the collision, and $T = E - mc^2$, we have that

$$\min\big(\sum T_X\big) = \Big(\sum_{i>2} m_i - 2m_X\Big)c^2 \overset{2X\to 3X+\overline{X}}{=} (4-2)m_X c^2 = 2m_X c^2, \tag{3.74}$$

exactly as expected naively. Thus, for the $X + X \to 3X + \overline{X}$ process, head-on collisions of two particles that move with the same speed towards each other (as observed in the CM system) are three times as efficient as colliding a probe-X with a stationary target-X. For the head-on collision, the apparatus must provide only $2m_X c^2$ of energy ($m_X c^2$ per X-particle before the collision) to create the $3X + \overline{X}$ system, while colliding a moving X-particle with a stationary X-particle requires providing the moving X-particle $6m_X c^2$ of energy.

The difference of $4m_X c^2$ in energy threshold for a probe colliding with a stationary target ensures that the resulting $3X + \overline{X}$ particle system has the kinetic energy required by the conservation of linear momentum, and which is the extra "price" in the kinetic energy of the probe before the collision. No such extra energy is needed in head-on collisions, where *all* of the kinetic energy is thus available to produce new particles – providing the basic rationale for building colliders.

The relative kinetic energy

The other aspect of the efficacy of head-on collisions is the *relative* kinetic energy: The previous section showed that if two X-particles move in the lab coordinate system one against another with the kinetic energy $m_X c^2$ each, in the inertial system of one of the two X-particles (wherein it itself is at rest), the other particle moves with a kinetic energy of $6m_X c^2$. More generally, use the equality of the first and third term in equation (3.61):

$$(\mathrm{p}_1 + \mathrm{p}_2)^2 = (\mathrm{p}_1' + \mathrm{p}_2')^2,$$

$$\Big(-\frac{E_1 + E_2}{c}, \vec{0}\Big)^2 = \Big(-\frac{E_1'}{c} - m_2 c, \vec{p}_1\Big)^2.$$

If $m_1 = m_2 = m$, then from $\vec{p}_1 = -\vec{p}_2$ it follows that $E_1 = E_2 =: E$; also, write $E_1' = E'$. Using the results of the previous computations, we arrive at

$$4E^2 = 2mc^2(E' + mc^2), \tag{3.75}$$

or

$$T' = 4T\Big(1+\frac{T}{2mc^2}\Big), \quad [\text{✎ verify}] \tag{3.76}$$

where the second term is the fast-growing relativistic correction:

T/mc^2:	1	2	5	10	20	50	100	$\cdots$
T'/mc^2:	6	16	70	240	880	5,200	20,400	$\cdots$

(3.77)

When two particles of the same mass collide head-on with a kinetic energy of $100\,mc^2$ each, i.e., total $200\,mc^2$, the collision has the same effect as if a particle at rest was hit by another with a kinetic energy of $20{,}400\,mc^2$ – which is 102 *times* more!

3.2.3 Lessons

This is an excellent place to highlight the differences between conserved and invariant quantities:

1. Energy is *conserved but not Lorentz-invariant*:
   (a) The total energy of each particle at any point in time (before, during, after) in a process equals this same quantity at all other points in time.
   (b) Energy (its $-\frac{1}{c}$-multiple) is the 0th component of a 4-vector, and cannot be Lorentz-invariant: It changes – it mixes with the components of $\vec{p}$ – when the observer changes from one inertial coordinate system to another.
2. Mass is *Lorentz-invariant but not conserved*:
   (a) Its value does not change when the observer changes from one inertial coordinate system to another.
   (b) Mass is not conserved, as should be obvious from Example 3.1 on p. 95.

Note that *Lorentz-invariant* means "unchanged under transforming amongst inertial coordinate systems," i.e., with respect to Lorentz transformations, while *conserved* means "unchanged during any isolated process, as *time* passes." That is, the very definition of "conservation" implies a preferred choice of *time*, which cannot possibly be a notion invariant with respect to Lorentz transformations of coordinates.

The relativistic 4-momentum of a particle is *conserved but not Lorentz-invariant* – just like its 0th component, the relativistic energy, as well as its remaining 3 components known also as the "relativistic 3-momentum."

3.2.4 Exercises for Section 3.2

✎ **3.2.1** *Using Bohr's formula (1.31), compute the relative difference $\frac{m(3p)-m(1s)}{m(1s)}$ between the hydrogen atom mass when it is in a 3p state (where $n=3$, $\ell=1$, while $|m_\ell|\leqslant\ell$ and $m_s=\pm\frac{1}{2}$ are arbitrary) and when it is in a ground, 1s state (where $n=1$, $\ell=0=m_\ell$, while $m_s=\pm\frac{1}{2}$ is arbitrary).*

✎ **3.2.2** *Compute the relative contribution of the correction (1.33) to the relative difference between the masses computed in Exercise 3.2.1.*

✎ **3.2.3** *If a particle of mass M at rest decays into two particles of equal masses, $m_1=m_2=m$, compute the speed with which the particles leave the decay locus. Compute the relative speed with which the resulting particles move away from each other.*

✎ **3.2.4** *If a particle of mass M at rest decays into two particles of different masses, $m_1 > m_2$, compute the difference between their kinetic energies as a function of only the masses M, m_1, m_2, including the special case when $m_2 = 0$.*

✎ **3.2.5** *Show that the system of equations (3.54) reproduces all derived and stated results for the special case of a two-particle decay.*

✎ **3.2.6** *Show that for the case of a two-particle decay, the results (3.48)–(3.49) and analogously for $E_B, |\vec{p}_B|$ together with equation (3.54a) produce $\phi_{BC} = 180°$, in agreement with the result obtained using the linear momentum conservation.*

✎ **3.2.7** *Show that a free electron can neither absorb nor emit a single photon, i.e., that the simple processes $\gamma + e^- \to e^-$ as well as $e^- \to e^- + \gamma$ are kinematically forbidden [☞ Chapter 3.3 for explanation].*

✎ **3.2.8** *Reconsider the fusing collision computations (3.64)–(3.68) and assume that the probe A flies into the target B with total energy E_A and fuses with it. Compute the 4-momentum and mass of the resulting fused object C.*

3.3 Feynman's diagrams and calculus

In the analysis of Section 3.2, as well as in the corresponding exercises and especially in Exercise 3.2.7, it is tacitly assumed that all particles in the analyzed processes can be observed directly, i.e., that all kinematic parameters (mass, energy, linear momentum, angular momentum, etc.) can be measured.

However, that is not always the case.

Recall Conclusion 2.3 on p. 56. Of course, this has to do with the consequence of Heisenberg's indeterminacy principle [☞ Digression 2.7 on p. 73]. That is, for kinematics, this involves the specific relations (1.42) and (1.47):

$$\triangle p_0 \, \triangle x^0 = \triangle E \, \triangle \tau \geqslant \tfrac{1}{2}\hbar, \qquad \triangle p_i \, \triangle x^i \geqslant \tfrac{1}{2}\hbar, \quad i = 1,2,3. \tag{3.78}$$

The indeterminacy principle permits the two-step process

$$\Big(e^- + \gamma \xrightarrow{1+2} e^- + \gamma\Big) = \Big(e^- + \gamma \xrightarrow{1} (e^{*-}) \xrightarrow{2} e^- + \gamma\Big), \tag{3.79}$$

even if the process "1" and the process "2" were kinematically forbidden – by themselves. Indeed, if the intermediate, "excited" electron, e^{*-}, exists only during a time shorter than

$$\Delta\tau \sim \frac{\hbar}{2(E_\gamma + m_e c^2)}, \tag{3.80}$$

that is, if the time that elapses between process "1" and process "2" is shorter than the one given by the indeterminacy relation (3.80), then the particle e^{*-} cannot possibly be observed directly: It is then possible neither to measure its kinematic parameters, nor to check the 4-momentum conservation.

Thus, the 4-momentum conservation law is neither violated nor *broken*[11]; Heisenberg's indeterminacy relations (3.78) have to do with a fundamental natural limitation of measuring. That is:

[11] It is important to differentiate between these terms. "Violation" typically refers to a particular case, event or process in which a rule, law or symmetry is not satisfied, while "breaking" applies to all cases, events and processes in a particular phase of the system. Also, in the present context, "breaking" usually refers to symmetry breaking, and in cases of gauge symmetry it also refers to the breaking of the corresponding continuity equation and charge conservation, by extension.

Conclusion 3.1 *The 4-momentum conservation law is strict, applies to all processes, and down to the measurement resolution (precision, tolerance) dictated by Heisenberg's indeterminacy principle.*

To effectively differentiate the precision of the application of the 4-momentum conservation law, we define:

Definition 3.5 *States of a (system, object, particle, etc.) that cannot be directly observed owing to Heisenberg's indeterminacy* **principle** *are called* **virtual**. *Processes that relate real incoming and real outgoing states are real; all others are virtual.*

Comment 3.4 *The processes (3.79) labeled "1" and "2" are virtual, but the process "1+2" is real. [✎ Why?] A virtual particle is also said to be "off-shell," i.e., off the mass shell, which is the hyperboloid* $\mathrm{p}^2 = p_\mu p^\mu = m^2c^2$ *in the 4-momentum space. That is, the 4-momentum of a particle "on-shell" satisfies the relations (3.36)–(3.37), whereas that of a particle "off shell," is not so restricted; to this end, I write* $\mathrm{p}^2 \mathrel{\rlap{\,\,|}=} m^2c^2$ *– in distinction from "*$\mathrm{p}^2 \neq m^2c^2$*," which means that* p^2 **must not** *equal* m^2c^2*.* [☞ Tables C.7 on p. 529 and C.8 on p. 529.]

3.3.1 Diagrams

Processes between particles are naturally represented graphically, by so-called Feynman diagrams.[12] It is important to understand that these diagrams must not be taken as a literal rendition of a process in the "real" space, but as a schematic tool, the primary task of which is to help in the estimation and computation regarding physical processes that they represent. For example, the Feynman diagrams

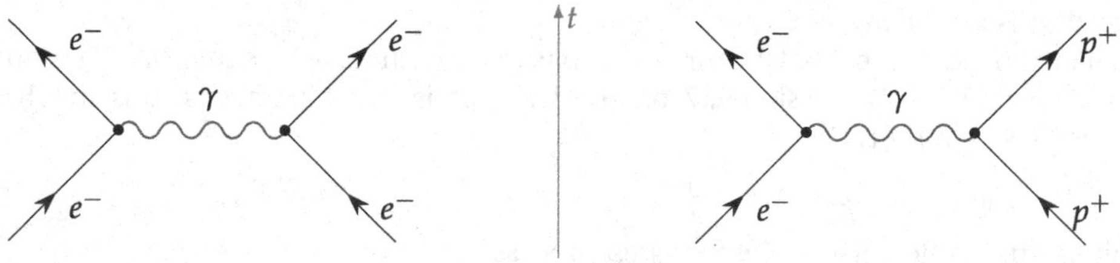

(3.81)

look identical although the left-hand diagram depicts the repulsive effect of (the Coulomb force due to) the exchange of one photon between two electrons, and the right-hand one depicts the attractive effect of (the Coulomb force due to) the exchange of one photon between an electron and a proton.

Except when noted differently, all Feynman diagrams herein are, by convention, drawn with time passing predominantly upward and the lines of simultaneity being oriented predominantly left–right. The tilt (angle with respect to the chosen time axis) of these lines depends on the choice of the observer,[13] which changes the interpretation of the diagram:

[12] The graphical representation of interactions is very intuitive and clear. Feynman certainly did not come up with this idea first, but he did contribute to their popularity as he worked out the technical details that make those diagrams into a useful computational tool. Ernst Stückelberg was the first to use the idea for the individual processes, before Feynman, but had no actual drawings; Freeman Dyson was the first to rigorously establish the link between these diagrams and the well-known perturbative computations. Feynman linked these diagrams to the so-called path integrals, which became a standard only years later.

[13] To be precise, the tilt of all lines changes depending on the observer. However, the tilt of virtual lines – which represent particles that are unobservable in principle and so do not satisfy any classical equation of motion – may change radically, and represent the motion of a massive, light-like, or even tachyonic particle. In distinction, the wave-functions of real particles satisfy their classical equations of motion, and so have the same character for all observers: either massive or light-like. (Or tachyonic – should they ever be experimentally detected [☞ Digression 7.1 on p. 261].)

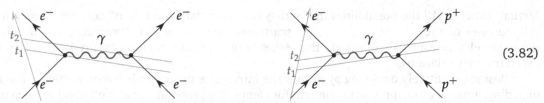

(3.82)

According to the interpretation on the left-hand side, the left-hand particle first emits a photon, which the right-hand particle then absorbs; according to the interpretation on the right-hand side, the right-hand particle emits a photon first, which the left-hand particle then absorbs. Thus we simply speak of an "exchanged" photon, and a diagram such as (3.82) is identified as a schematic representation of this process, and not as a literal, real depiction of the process in spacetime.

The exchanged photon must be virtual (after all, it is by definition never directly observed!), since the individual processes (the left-hand half and the right-hand one)

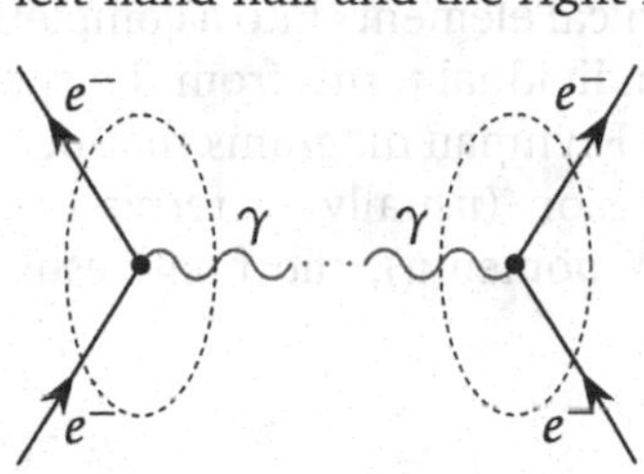

(3.83)

would be kinematically forbidden [☞ Exercise 3.2.7] – while the whole process (3.83) and those in (3.81) are real. However, this implies that processes such as either one of (3.81) must be understood as one of the contributions to the process that may be depicted as

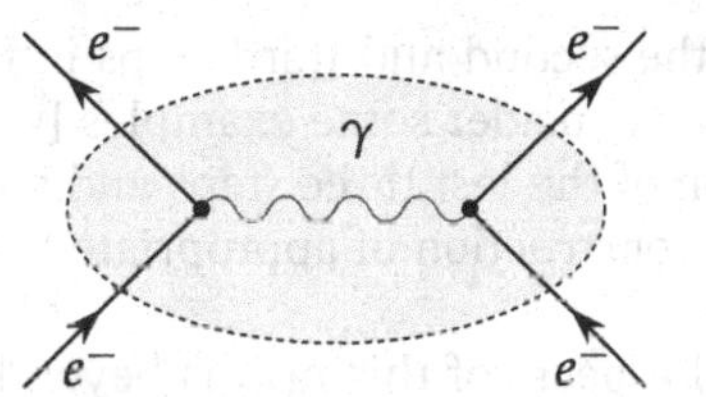

(3.84)

where the schematic region in the shaded ellipse is the *Heisenberg zone*; particles and processes that are entirely within this region can be neither observed nor measured directly as a matter of (Heisenberg's indeterminacy) principle. On the other hand, that also means that within the shaded region of indeterminacy, *all* possible sub-processes may well occur, and in fact do occur [☞ Conclusion 2.3 on p. 56]. It remains to determine the hierarchy of their contributions to the physical quantity being computed for the considered physical process (specified by the particles *outside* the Heisenberg zone of indeterminacy!):

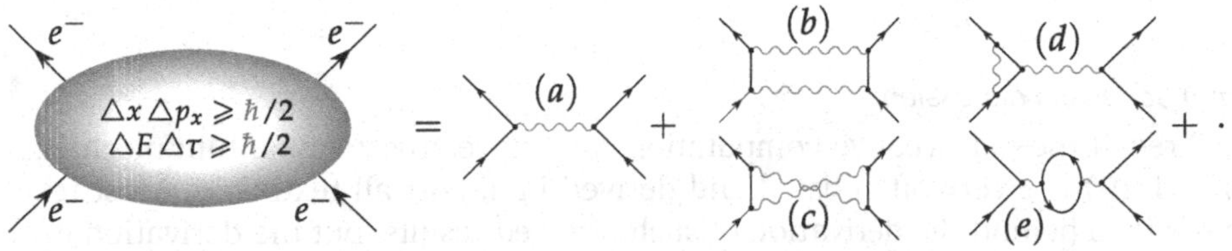

(3.85)

In classical physics, it makes perfect sense to ask: "In a concrete $e^- + e^- \to e^- + e^-$ scattering, which *one* of the processes happened, *either* (*a*) *or* (*b*) *or* (*c*) *or* (*d*) *or* (*e*) *or*…?" In quantum physics, this question makes no sense: *As a matter of principle,* not one of the processes shown in the expansion on the right-hand side of the equality (3.85) can possibly be singled out as the "actual" process. *All* the possibilities that satisfy the "boundary conditions" contribute, as (virtual) *sub-processes* of the $e^- + e^- \rightarrow e^- + e^-$ scattering. In this context, "boundary conditions" are the data reliably established outside the Heisenberg zone, the region obscured by Heisenberg's indeterminacy relations.

Besides intuitively depicting by graphs the interactive processes between particles, the Feynman diagrams are also a precise instrument for computing probabilities as well as other measurable parameters of the considered process. The goal of every application of Feynman diagrams is the establishment of precise 1–1 correspondences between:

1. the fundamental theory that designs the considered process, usually in terms of a specified Lagrangian density,
2. individual Feynman diagram elements as the graphical representation of individual terms from the specified Lagrangian density,
3. the rules of linking these graphical elements into a complete diagram, as a graphical depiction of the computation with the individual terms from the specified Lagrangian density,
4. the rules of listing all possible Feynman diagrams that need to be included in a computation,
5. the final mathematical expression (usually, in terms of an algebraic sum of various multiple integrals over various 4-momenta), the final result of which is the desired physical quantity,

and finally,

6. the *computation* (or, more often, an *estimate*) of the value of the mathematical expression depicted by the Feynman diagram.

Here, we skip the derivations of the second and third steps in this listing; that would be the task of a field theory course. Instead, we consider some examples [☞ Chapters 5–7] from the Standard Model, to illustrate the application of the last three steps and will only heuristically motivate their relationship to the first step – the construction of appropriate Lagrangians, which however we will discuss at length.

A complete discussion of all aspects of this task is beyond the scope of an *introductory* text such as this. Reference [305] describes the early history of Feynman diagrams and the reasons for the variety of "styles" and conventions in their application; see, e.g., Refs. [61, 474, 537], the field-theory texts [64, 63, 48, 257, 307, 221, 159, 422, 423, 538, 250, 389, 243, 45, 580, 238, 241, 239, 240], as well as those specializing in path-integral methods [459, 165, 123, 277]. However, since the Feynman diagram technique is quite widespread – even in topics well outside elementary particle physics [☞ e.g., Refs. [357, 316]] – we first turn to non-relativistic quantum mechanics, where the well-known perturbative computations are also representable graphically.

3.3.2 Quantum-mechanical digression

As a "warm-up," recall the perturbative computations in non-relativistic quantum mechanics: the relations (1.17)–(1.19) are very often listed and derived in almost all textbooks. Most textbooks also give the basic idea behind the derivation of such oft-used results, but the derivation itself and the results are hardly ever given for corrections of higher order. However, adopting the standard derivation, we write

$$H\,|n\rangle = E_n\,|n\rangle, \qquad \text{where} \qquad H := H_0 + \lambda\,H' \tag{3.86}$$

is the "true" Hamiltonian, given as a sum of a "known" Hamiltonian H_0 and a "perturbation" H', and where λ serves to consistently count the order of perturbation. Suppose that for the "known" system (designated by the Hamiltonian H_0) the complete system of orthonormalized solutions is known:

$$H_0|n;0\rangle = E_n^{(0)}|n;0\rangle, \qquad \begin{cases} \langle n;0|n';0\rangle &= \delta_{n,n'}, \\ \sum_n |n;0\rangle\langle n;0| &= \mathbb{1}, \end{cases} \tag{3.87}$$

and the solutions of equation (3.86) may be found in the analytic form

$$E_n = \sum_{k=0}^{\infty} \lambda^k E_n^{(k)}, \qquad |n\rangle = \sum_{k=0}^{\infty} \lambda^k |n;k\rangle, \tag{3.88}$$

with the normalizations

$$\langle m;k|n;k\rangle = \delta_{mn}, \ \forall m,n, \qquad \text{and} \qquad \langle n;k|n;\ell\rangle = \delta_{k\ell}, \ \forall k,\ell. \tag{3.89}$$

The energy $E_n^{(k)}$ is the kth order perturbative correction to the original, unperturbed energy $E_n^{(0)}$, and $|n;k\rangle$ is the kth order perturbative correction to the original, unperturbed state $|n;0\rangle$. The treatment of the general situation with (partial) continuous and/or degenerate spectrum is only technically more complicated,[14] and so will not be discussed here.

Introducing the definition

$$\widehat{\Pi}_n^\alpha := \sum_{m\neq n} \frac{|m;0\rangle\langle m;0|}{(E_n^{(0)}-E_m^{(0)})^\alpha}, \qquad \text{so} \qquad \widehat{\Pi}_n^\alpha \widehat{\Pi}_n^\beta = \widehat{\Pi}_n^{\alpha+\beta}, \tag{3.90}$$

where the superscript in $\widehat{\Pi}_n^\alpha$ really behaves like an exponent, the standard recursive formulae[15] for the kth correction to the state and energy are

$$|n;k\rangle = \widehat{\Pi}_n^1 H'|n;k-1\rangle - \sum_{i=1}^{k-1} E_n^{(i)} \widehat{\Pi}_n^1 |n;k-i\rangle, \qquad k>0, \tag{3.91a}$$

$$E_n^{(k)} = \langle n;0|H'|n;k-1\rangle. \tag{3.91b}$$

The first several iterations of these recursive formulae are:

$$E_n^{(1)} = \langle n;0|H'|n;0\rangle, \tag{3.92a}$$

$$|n;1\rangle = \widehat{\Pi}_n^1 H'|n;0\rangle, \tag{3.92b}$$

$$E_n^{(2)} = \langle n;0|H' \widehat{\Pi}_n^1 H'|n;0\rangle, \tag{3.92c}$$

$$\begin{aligned} |n;2\rangle &= \widehat{\Pi}_n^1 (H' - E_n^{(1)})|n;1\rangle \\ &= \widehat{\Pi}_n^1 H' \widehat{\Pi}_n^1 H'|n;0\rangle - \widehat{\Pi}_n^1 \widehat{\Pi}_n^1 H'|n;0\rangle\langle n;0|H'|n;0\rangle \\ &= \big[\widehat{\Pi}_n^1 H' \widehat{\Pi}_n^1 - \widehat{\Pi}_n^2 H'|n;0\rangle\langle n;0|\big] H'|n;0\rangle, \end{aligned} \tag{3.92d}$$

---

[14] The basis of states $|n;k\rangle$ must be redefined so as to eliminate the meaningless terms such as $\frac{\langle m;k|H'|n;k\rangle}{E_m^{(0)}-E_n^{(0)}} \sim \frac{1}{0}$ for $m \neq n$ – which is always possible, by (at least a partial) diagonalization of the perturbation matrix $\langle m;k|H'|n;k\rangle$.

[15] Most quantum mechanics texts list only the results for $E_n^{(1)}$, $|n;1\rangle$ and $E_n^{(2)}$; for a more complete treatment, see e.g., Ref. [362, pp. 685–695].

$$
\begin{aligned}
E_n^{(3)} &= \langle n;0|H'|n;2\rangle \\
&= \langle n;0|H'\big[\,\widehat{\Pi}_n^1\,H'\,\widehat{\Pi}_n^1 - \widehat{\Pi}_n^2\,H'|n;0\rangle\langle n;0|\big]\,H'|n;0\rangle \\
&= \langle n;0|H'\,\widehat{\Pi}_n^1\,H'\,\widehat{\Pi}_n^1\,H'|n;0\rangle - \langle n;0|H'\,\widehat{\Pi}_n^2\,H'|n;0\rangle\langle n;0|H'|n;0\rangle
\end{aligned}
\tag{3.92e}
$$

$$
\begin{aligned}
|n;3\rangle &= \widehat{\Pi}_n^1\,\big((H'-E_n^{(1)})|n;2\rangle - E_n^{(2)}|n;1\rangle\big) \\
&= \widehat{\Pi}_n^1\,H'|n;2\rangle - \widehat{\Pi}_n^1\,|n;2\rangle\langle n;0|H'|n;0\rangle - \widehat{\Pi}_n^1\,|n;1\rangle\langle n;0|H'\,\widehat{\Pi}_n^1\,H'|n;0\rangle \\
&= \widehat{\Pi}_n^1\,H'\,\widehat{\Pi}_n^1\,H'\,\widehat{\Pi}_n^1\,H'|n;0\rangle - \widehat{\Pi}_n^1\,H'\,\widehat{\Pi}_n^2\,H'|n;0\rangle\langle n;0|H'|n;0\rangle \\
&\quad - \widehat{\Pi}_n^2\,H'\,\widehat{\Pi}_n^1\,H'|n;0\rangle\langle n;0|H'|n;0\rangle - \widehat{\Pi}_n^2\,H'|n;0\rangle\langle n;0|H'\,\widehat{\Pi}_n^1\,H'|n;0\rangle \\
&\quad + \widehat{\Pi}_n^3\,H'|n;0\rangle\langle n;0|H'|n;0\rangle^2,
\end{aligned}
\tag{3.92f}
$$

and so on. The expressions after (3.92c) indeed become increasingly more and more tedious, and very quickly. However, the particular ordering in these expressions uncovers a simple algorithm for finding all subtractions, in the form of "excisions" from the original expression:

$$
\begin{aligned}
|n;3\rangle = {}& \widehat{\Pi}_n^1\,H'\,\widehat{\Pi}_n^1\,H'\,\widehat{\Pi}_n^1\,H'|n;0\rangle && \leftarrow \text{original expression} \\
& - \widehat{\Pi}_n^1\,\underline{[H']}\,\widehat{\Pi}_n^1\,H'\,\widehat{\Pi}_n^1\,H'|n;0\rangle && \leftarrow \text{1st excision} \\
& - \widehat{\Pi}_n^1\,H'\,\widehat{\Pi}_n^1\,\underline{[H']}\,\widehat{\Pi}_n^1\,H'|n;0\rangle && \leftarrow \text{2nd excision} \\
& - \widehat{\Pi}_n^1\,H'\,\widehat{\Pi}_n^1\,H'\,\widehat{\Pi}_n^1\,\underline{[H']}|n;0\rangle && \vdots \\
& - \widehat{\Pi}_n^1\,\underline{[H']}\,\widehat{\Pi}_n^1\,\underline{[H']}\,\widehat{\Pi}_n^1\,H'|n;0\rangle \\
& - \widehat{\Pi}_n^1\,\underline{[H'\,\widehat{\Pi}_n^1\,H']}\,\widehat{\Pi}_n^1\,H'|n;0\rangle \\
& - \widehat{\Pi}_n^1\,H'\,\widehat{\Pi}_n^1\,\underline{[H'\,\widehat{\Pi}_n^1\,H']}|n;0\rangle
\end{aligned}
\tag{3.93}
$$

where the $\underline{[\cdots]}$-bracketed factors in the original expression are successively "excised." For example,

$$
\widehat{\Pi}_n^1\,\underline{[H']}\,\widehat{\Pi}_n^1\,H'\,\widehat{\Pi}_n^1\,H'|n;0\rangle = \widehat{\Pi}_n^1\,\widehat{\Pi}_n^1\,H'\,\widehat{\Pi}_n^1\,H'|n;0\rangle\,\langle n;0|H'|n;0\rangle,
\tag{3.94a}
$$

$$
\widehat{\Pi}_n^1\,\underline{[H'\,\widehat{\Pi}_n^1\,H']}\,\widehat{\Pi}_n^1\,H'|n;0\rangle = \widehat{\Pi}_n^1\,\widehat{\Pi}_n^1\,H'|n;0\rangle\,\langle n;0|H'\,\widehat{\Pi}_n^1\,H'|n;0\rangle, \quad \text{etc.}
\tag{3.94b}
$$

The right-most "excisions" in (3.93) vanish:

$$
\widehat{\Pi}_n^1\,H'\,\widehat{\Pi}_n^1\,H'\,\widehat{\Pi}_n^1\,\underline{[H']}|n;0\rangle = \widehat{\Pi}_n^1\,H'\,\widehat{\Pi}_n^1\,H'\,\underbrace{\widehat{\Pi}_n^1\,|n;0\rangle}_{=0}\langle n;0|H'|n;0\rangle,
\tag{3.95a}
$$

$$
\widehat{\Pi}_n^1\,H'\,\widehat{\Pi}_n^1\,\underline{[H'\,\widehat{\Pi}_n^1\,H']}|n;0\rangle = \widehat{\Pi}_n^1\,H'\,\underbrace{\widehat{\Pi}_n^1\,|n;0\rangle}_{=0}\langle n;0|H'\,\widehat{\Pi}_n^1\,H'|n;0\rangle,
\tag{3.95b}
$$

owing to the fact that the normalization (3.89) guarantees

$$
\widehat{\Pi}_n^\alpha\,|n;0\rangle = \sum_{m\neq n}\frac{|m;0\rangle\langle m;0|}{(E_n^{(0)}-E_m^{(0)})^\alpha}|n;0\rangle = \sum_{m\neq n}\frac{1}{(E_n^{(0)}-E_m^{(0)})^\alpha}|m;0\rangle\,\underbrace{\langle m;0|n;0\rangle}_{=0\ (\because\ m\neq n)}.
\tag{3.96}
$$

Since only factors of the form $(H'\,\widehat{\Pi}_n^\alpha\cdots\widehat{\Pi}_n^\beta\,H')$ have a non-vanishing expectation value in the original, "known" state $|n;0\rangle$, only such factors may be "excised." The relations (3.91) may then be written as

$$
|n;k\rangle = (\widehat{\Pi}_n^1\,H')^k|n;0\rangle - \text{all "excisions"}, \qquad k \geqslant 0,
\tag{3.97a}
$$

$$E_n^{(k)} = \langle n;0|H'(\widehat{\Pi}_n^1 H')^{k-1}|n;0\rangle - \text{all "excisions".} \qquad k \geqslant 1. \tag{3.97b}$$

Using this "excising" notation, e.g., the expression (3.92e) becomes

$$\begin{aligned} E_n^{(3)} &= \langle n;0|H'\,\widehat{\Pi}_n^1\,H'\,\widehat{\Pi}_n^1\,H'|n;0\rangle - \langle n;0|H'\,\widehat{\Pi}_n^1\,\underline{[H']}\,\widehat{\Pi}_n^1\,H'|n;0\rangle \\ &= \langle n;0|H'\,\widehat{\Pi}_n^1\,H'\,\widehat{\Pi}_n^1\,H'|n;0\rangle - \langle n;0|H'\,\widehat{\Pi}_n^2\,H'|n;0\rangle\langle n;0|H'|n;0\rangle. \end{aligned} \tag{3.98}$$

The diligent Student is expected to verify *[✎ do it]* that the formulae (3.97a)–(3.97b) reproduce at least the above results (3.92d)–(3.92f).

Digression 3.10 It is not hard to see that the expression (3.98) has no other non-vanishing "excisions." Take, for instance, the candidate

$$\begin{aligned} \langle n;0|\underline{[H']}\,\widehat{\Pi}_n^1\,H'\,\widehat{\Pi}_n^1\,H'|n;0\rangle &:= \langle n;0|\widehat{\Pi}_n^1\,H'\,\widehat{\Pi}_n^1\,H'|n;0\rangle\,\langle n;0|H'|n;0\rangle \\ &= \underbrace{\langle n;0|\widehat{\Pi}_n^1}_{=0}\,H'\,\widehat{\Pi}_n^1\,H'|n;0\rangle\,\langle n;0|H'|n;0\rangle = 0. \end{aligned} \tag{3.99}$$

The results (3.91) may be depicted graphically, drawing

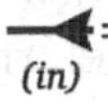 $= |n;0\rangle$, (in) 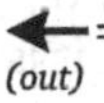 $= \langle n;0|$, (out) 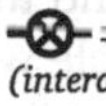 $= H'$, (interaction) 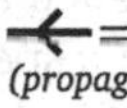 $= \widehat{\Pi}_n^1$, (propagator)  $= \widehat{\Pi}_n^2$, (2nd order propagator) etc. (3.100)

Then we have

$\langle n;0|H'|n;0\rangle \overset{(3.92a)}{\longmapsto} E_n^{(1)} =$ 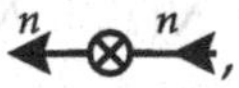

, (3.101a)

$\widehat{\Pi}_n^1\,H'|n;0\rangle \overset{(3.92b)}{\longmapsto} |n;1\rangle =$ 

, (3.101b)

$\langle n;0|H'\,\widehat{\Pi}_n^1\,H'|n;0\rangle \overset{(3.92c)}{\longmapsto} E_n^{(2)} =$ 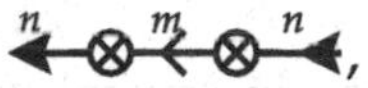

, (3.101c)

$\widehat{\Pi}_n^1\,H'\,\widehat{\Pi}_n^1\,H'|n;0\rangle - \widehat{\Pi}_n^1\,\widehat{\Pi}_n^1\,H'|n;0\rangle\langle n;0|H'|n;0\rangle$

$\overset{(3.92d)}{\longmapsto} |n;2\rangle =$ 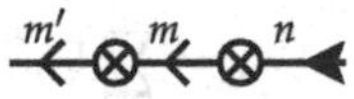

$-$ 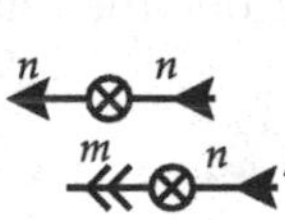

, (3.101d)

$\langle n;0|H'\,\widehat{\Pi}_n^1\,H'\,\widehat{\Pi}_n^1\,H'|n;0\rangle - \langle n;0|H'\,\widehat{\Pi}_n^2\,H'|n;0\rangle\langle n;0|H'|n;0\rangle$

$\overset{(3.92e)}{\longmapsto} E_n^{(3)} =$ 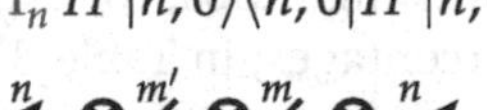

$-$ 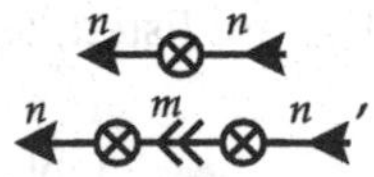

, (3.101e)

and so on, where the subtractions are shown as stacks of diagrams, and represent a product of the corresponding factors. The "excising" algorithm (3.92d) may be graphically depicted also as

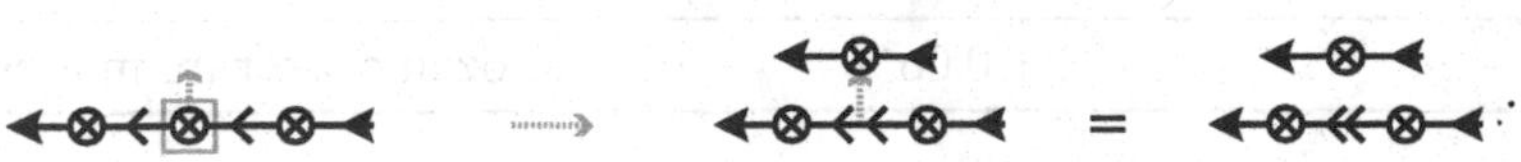 (3.102)

Thus, the whole stationary perturbation theory in non-relativistic quantum mechanics exemplified by (3.101a)–(3.101e) may be written unambiguously and precisely using the graphical symbols (3.100). Similarly, the whole perturbation theory in field theory may be faithfully written in terms of Feynman diagrams.

The detached portions in these "excision" diagrams, such as $\leftarrow\!\otimes\!\leftarrow$ in (3.101d)–(3.102), may well be thought of as the quantum-mechanical analogue of "vacuum diagrams" in field theory: These diagrams begin and end at the same state in the Hilbert space, $|n;0\rangle$; these being stationary states, they do not change in time; finally, fixing $|n;0\rangle$ to be the ground state would indeed refer to the "vacuum."

3.3.3 *Decays, scattering and calculations*

In elementary particle physics, one studies decays, collisions/scatterings, and bound states of these elementary particles. The analysis of bound states uses very successfully the non-relativistic quantum mechanics in Schrödinger's picture, with perturbatively added relativistic corrections [☞ Section 4.1]. On the other hand, decays and collisions/scatterings typically require relativistic analysis. Our goal here will be to estimate the lifetime for the particle A decaying as $A \to B + C$, and the differential as well as the total cross-section (probability) of the $A + B \to C \to A + B$ scattering. The relativistic computations using Feynman diagrams are convenient for this, and we follow the standard approach, adopting Griffiths's conventions [243].

Decays and the half-life

Particles (and even composite systems such as atoms and atomic nuclei) decay probabilistically: It is not possible to specify precisely when a specific particle will decay, but it is possible to determine the average lifetime τ, i.e., half-life, $t_{1/2} = \ln(2)\tau$, where

$$N(t) = N(0)\, e^{-t/\tau} = N(0)\,\left(\tfrac{1}{2}\right)^{t/t_{1/2}} \tag{3.103}$$

is the number of certain particles at time $t > 0$ within a sample where there existed $N(0)$ particles at time $t = 0$. The decay rate (a.k.a. the decay constant) is defined as

$$\Gamma := \frac{1}{\tau} = \frac{\ln(2)}{t_{1/2}}. \tag{3.104}$$

Most particles decay in several ways; in 99.80% cases, π^0 decays into two photons, but in 1.20% cases into an $e^- + e^+$-pair. Other particles have many more "modes" of decay: Each particular decay mode then has a corresponding decay rate Γ_i, and of course

$$\Gamma_{\text{tot}} = \sum_i \Gamma_i, \qquad \tau = \frac{1}{\Gamma_{\text{tot}}}. \tag{3.105}$$

The ratios $\Gamma_i/\Gamma_{\text{tot}}$ are called branching ratios; for the five most significant decay modes of the K^+ particle, $\Gamma_i/\Gamma_{\text{tot}}$ are listed, as percentages, in Table 3.1.

Table 3.1 The significant decay modes of the K^+ meson

$\mu^- + \bar{\nu}_\mu$	$63.44 \pm 0.14\,\%$	$\pi^+ + \pi^0$	$20.92 \pm 0.12\,\%$
$\pi^+ + \pi^+ + \pi^-$	$5.590 \pm 0.031\,\%$	$\pi^0 + e^+ + \nu_e$	$4.98 \pm 0.07\,\%$
$\pi^0 + \mu^+ + \nu_\mu$	$3.32 \pm 0.06\,\%$	plus a dozen or so rare modes	

Scattering and the effective cross-section

The "effective cross-section" is used to design scattering of one particle on another. Conceptually, this is the literal geometrical description of the target, as seen by the incoming probe. For an archer, the probability of a hit is proportional to the cross-section of the target: that's why swordsmen turn sideways, for the opponent to "see" a smaller cross-section, so as to diminish the probability of a stab.

This literal geometric figure is really faithful [☞ Example 3.2] in the case of "hard" targets, as in the case of a collision of a pool ball, a marble, a cannonball, etc. Such objects have a "binary" interaction: they either collide or they miss. That is, there exists a very well determined *critical* distance, d_c, between the centers of such objects. Should the objects pass by each other so that the distance between them is always bigger than d_c, they do not interact at all. For two regular spheres, d_c equals the sum of their radii.

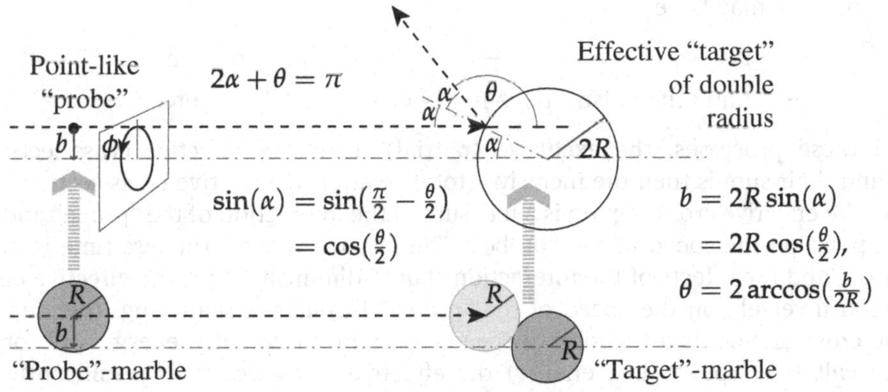

Figure 3.1 The collision of two marbles.

Example 3.2 The classical collision of "hard" marbles of radius R may be analyzed geometrically, as shown in Figure 3.1, where the left-hand marble is replaced by a material point, and the radius of the right-hand marble is doubled. The left-hand marble plays the role of a "probe," and the right-hand one that of the "target." If the orthogonal distance b from the target center is changed a little, $b \to b + \mathrm{d}b$, the scattering angle θ also changes, $\theta \to \theta + \mathrm{d}\theta$. As the collision geometry has axial symmetry, the same result holds if the "probe" approaches the target from any other angle ϕ, so that the "probe" passes through the "surface" element $\mathrm{d}\sigma = |\mathrm{d}b\, b\, \mathrm{d}\phi|$. The out-coming space-angle is then $\mathrm{d}\Omega = |\sin\theta\, \mathrm{d}\theta\, \mathrm{d}\phi|$, so that the ratio

$$\frac{\mathrm{d}\sigma}{\mathrm{d}\Omega} = \left|\frac{b}{\sin\theta}\left(\frac{\mathrm{d}b}{\mathrm{d}\theta}\right)\right| = \left|\frac{2R\cos(\frac{\theta}{2})}{\sin\theta}\Big((2R)\big[-\tfrac{1}{2}\sin(\tfrac{\theta}{2})\big]\Big)\right| = R^2. \tag{3.106}$$

That produces the total effective cross-section

$$\sigma = \int \mathrm{d}\sigma = \int \mathrm{d}\Omega\, \frac{\mathrm{d}\sigma}{\mathrm{d}\Omega} = 4\pi R^2 = \pi(2R)^2, \tag{3.107}$$

which is the cross-section of a circle of radius $2R$: Every "probe" the center of which passes through this effective circle of double radius will collide with the "target," all other probes miss.

The "hard" target *models* evidently cannot hold for scattering of two charged particles, since the electromagnetic interaction extends infinitely far, and the two charged particles always interact, regardless of the smallest distance between them. Of course, the *intensity*, i.e., the force of interaction, diminishes with the square of the distance. However, there is no regime in which the interaction completely vanishes. In distinction from the previous, "hard" targets, such targets are then called "soft."

Molecular forces, which decay as $\sim 1/r^n$ where $n > 2$, as well as forces of Yukawa type (which decay as $\sim e^{-r/r_0}/r^2$) evidently behave between the two limiting cases. The *effective cross-section* is then a measure of the mutual "hardness" of the target and the probe.

Besides, the collision probability may also depend on the nature of the probe as well as the target, of the interaction, and even the number and type of out-coming particles. Indeed, the elastic collision $e^- + p^+ \to e^- + p^+$ is relatively simple at sufficiently small energies. However, at growing energy collisions, we may have

$$e^- + p^+ \to e^- + p^+ + \gamma, \qquad e^- + p^+ \to e^- + p^+ + \pi^0, \qquad e^- + p^+ \to e^- + n^0 + \pi^+,$$
$$\text{and then also} \quad e^- + p^+ \to \nu_e + \Lambda^0 + K^0, \qquad \text{etc.} \tag{3.108}$$

For each of these processes, the exclusive (partial) scattering effective cross-section may be computed, and their sum is then the inclusive (total) scattering effective cross-section.

Finally, the effective cross-section is a measure of the interaction of the "probe" and the target, and must depend on the speed of the "probe": The faster it moves, the less time is available for the interaction, and the effects of the interaction should diminish. Thus, the effective cross-section should depend inversely on the speed of the "probe." In realistic scattering, this dependence of the effective cross-section as a function of speed – or, more frequently, energy – is not so simple: near certain values of speed (i.e., energy) the effective cross-section is significantly amplified. Because of the similarity with the amplification of alternating current when its frequency is near a natural frequency of the circuit, this effect is also called "resonance." In such resonant collisions, the collision energy is just right for the "probe" and the "target" to produce a virtual intermediate state that decays before it could be detected directly [☞ equations (3.67) and (3.48)], and this is the most frequent way of (indirect) observation of new particles.

Example 3.2 on p. 111 shows that the physical meaning of the effective cross-section coincides with the naive measure of interaction – the cross-section of the effective target of doubled radius. In the general case, instead of a point-like "probe" one uses a beam of "probes," of luminosity L, defined as the number of point-like "probes" in unit time and unit area. Thus, we have that

$$\mathrm{d}N = L\,\mathrm{d}\sigma \quad \Rightarrow \quad \frac{\mathrm{d}N}{\mathrm{d}\Omega} = L\,\frac{\mathrm{d}\sigma}{\mathrm{d}\Omega}, \qquad \text{i.e.,} \qquad \frac{\mathrm{d}\sigma}{\mathrm{d}\Omega} = \frac{1}{L}\frac{\mathrm{d}N}{\mathrm{d}\Omega}. \tag{3.109}$$

This shows that the differential cross-section may be understood as the number of point-like "probes" that reach the detector in the interval of space angles $[\Omega, \Omega + \mathrm{d}\Omega]$, per unit luminosity. The first of these relations (3.109) gives the number of scattered probe-particles expected to be observed in the detectors placed in the interval $[\Omega, \Omega + \mathrm{d}\Omega]$, if the total luminosity of the beam of probes is L; that is the theoretical (computed) result that may be compared with experimental results directly.

For dimensional analysis, and to check the results, note the following relation between Γ and σ: For a decay of a two-particle bound state, Γ must be proportional to the effective cross-section, σ, of the collision of the two particles within the bound state, to the relative speed of these particles, as well as the value of the probability distribution in the place where the particles meet:

1. If the collision effective cross-section vanishes, there is no direct interaction between them, and there can be no decay of their bound state.

2. If the relative speed of the two particles vanishes, they will never meet, nor interact.
3. If the probability of the two particles to be in the same place vanishes, the direct interaction cannot happen, nor can the decay.

Dimensional analysis in fact even fixes the *linear* dependence on v, σ and $|\Psi(\vec{0},t)|^2$:

> **Conclusion 3.2** *The physical units for the decay rate are evidently T^{-1}, and for the effective cross-section (both total and differential) they are L^2. It follows that (see Exercise 3.3.2):*
>
> $$\Gamma \propto \sigma v |\Psi(\vec{0},t)|^2. \tag{3.110}$$

Fermi's golden rule

The basic idea of the so-called Fermi's golden rule is that the computation of a physical quantity such as a decay rate or a scattering effective cross-section, both total (inclusive) and exclusive (partial) may be written (up to conventional numerical factors) as a product of two factors:

1. the modulus-squared of the so-called "matrix element," i.e., "amplitude" of the process,
2. the sum/integral over the "phase space" – i.e., over aspects of the process that are *not* being measured/observed, and so do not specify the process.

This approach gives the formulae, cited here from Ref. [243] without derivation:

$$A \to C_1 + C_2 + \cdots \qquad \text{decay}: \tag{3.111}$$

$$\mathrm{d}\Gamma = |\mathfrak{M}|^2 \frac{S}{2\hbar m_A} (2\pi)^4 \delta^4(\mathrm{p}_A - \Sigma_i \mathrm{p}_i) \prod_j \frac{c\,\mathrm{d}^3\vec{p}_j}{2(2\pi)^3 E_j}, \tag{3.112}$$

where S is the product of "statistical factors," one $(k!)^{-1}$ factor for every group of k identical particles amongst the decay products.

$$A + B \to C_1 + C_2 + \cdots \qquad \text{collision/scattering}: \tag{3.113}$$

$$\mathrm{d}\sigma = |\mathfrak{M}|^2 \frac{\hbar^2 S}{4\sqrt{(\mathrm{p}_A \cdot \mathrm{p}_B)^2 - (m_A m_B c^2)^2}} (2\pi)^4 \delta^4(\mathrm{p}_A + \mathrm{p}_B - \Sigma_i \mathrm{p}_i) \prod_j \frac{c\,\mathrm{d}^3\vec{p}_j}{2(2\pi)^3 E_j}, \tag{3.114}$$

where, in both results, the energy of the jth particle amongst the process products is a function of the linear momentum:

$$E_j \equiv E_j(\vec{p}_j) = c\sqrt{m_j^2 c^2 + \vec{p}_j^{\,2}}, \tag{3.115}$$

since all particles in these processes are real, i.e., they can be observed directly in detectors, and so are "on-shell," i.e., on the $E^2 = m^2c^4 + \vec{p}^{\,2}c^2$ hyperboloid. In all these formulae, the indices i, j count the process products $(C_1, C_2, \dots)$, not the components of the linear momenta.

Example 3.3 Consider the two-particle decay $A \to C_1 + C_2$, where the products have masses m_1 and m_2, respectively, and where the linear momenta of the products are not measured, and so must be integrated over. Adapting equation (3.112), we have

$$\Gamma = \frac{S}{2\hbar m_A} \int |\mathfrak{M}|^2 (2\pi)^4 \delta^4(\mathrm{p}_A - \mathrm{p}_1 - \mathrm{p}_2) \frac{c\,\mathrm{d}^3\vec{p}_1}{2(2\pi)^3 E_1(\vec{p}_1)} \frac{c\,\mathrm{d}^3\vec{p}_2}{2(2\pi)^3 E_2(\vec{p}_2)}. \tag{3.116}$$

Translating first into the rest-frame of particle A, we have $\mathrm{p}_A = (m_A c, \vec{0})$. The 4-dimensional δ-function factorizes: $\delta^4(\mathrm{p}_A - \mathrm{p}_1 - \mathrm{p}_2) = \delta(m_A c - E_1/c - E_2/c)\delta^3(-\vec{p}_1 - \vec{p}_2)$. Using the 3-dimensional factor that imposes $\vec{p}_2 = -\vec{p}_1$ and cancels the $\mathrm{d}^3\vec{p}_2$-integration, we have

$$\Gamma = \frac{S}{2(4\pi)^2\hbar m_A} \int \mathrm{d}^3\vec{p}_1\, |\mathfrak{M}|^2\, \frac{\delta\Big(m_A c - \sqrt{m_1^2c^2 + \vec{p}_1^{\,2}} - \sqrt{m_2^2c^2 + (-\vec{p}_1)^2}\Big)}{\sqrt{m_1^2c^2 + \vec{p}_1^{\,2}}\sqrt{m_2^2c^2 + (-\vec{p}_1)^2}}. \tag{3.117}$$

On one hand, we know that $|\mathfrak{M}|$ is a Lorentz-invariant (scalar) function of the vectors $\vec{p}_1$ and $\vec{p}_2 = -\vec{p}_1$. So $\mathfrak{M} = \mathfrak{M}(\vec{p}_1, \vec{p}_2) = \mathfrak{M}(\vec{p}_1, -\vec{p}_1) = \mathfrak{M}(\vec{p}_1)$ may depend on the direction of $\vec{p}_1$ only if it also depends on some other (reference) vector quantity in the resulting particles, such as their spin: Then $\mathfrak{M}$ may depend also on the scalars $\vec{p}_i \cdot \vec{S}_j$ and $\vec{S}_i \cdot \vec{S}_j$, where $\vec{S}_i$ is the (operator of) spin of the ith particle, and $i, j = 1, 2$. In turn, if we may assume that the decay process does not depend on any such additional vector quantities, every scalar function of the vector $\vec{p}_1$ must in fact depend only on the modulus $\rho := |\vec{p}_1|$. It is therefore convenient to use spherical coordinates for the $\mathrm{d}^3\vec{p}_1$-integration. Angular integration gives a factor 4π, and we remain with

$$\Gamma = \frac{S}{8\pi\hbar m_A} \int_0^\infty \frac{\rho^2 \mathrm{d}\rho\, |\mathfrak{M}|^2}{\sqrt{m_1^2c^2 + \rho^2}\sqrt{m_2^2c^2 + \rho^2}}\, \delta\Big(m_A c - \sqrt{m_1^2c^2 + \rho^2} - \sqrt{m_2^2c^2 + \rho^2}\Big). \tag{3.118}$$

To simplify the integral, introduce

$$\mathcal{E} = c\Big(\sqrt{m_1^2c^2 + \rho^2} + \sqrt{m_2^2c^2 + \rho^2}\Big), \tag{3.119}$$

from which it follows that

$$\frac{\mathrm{d}\mathcal{E}}{\mathcal{E}} = \frac{\rho(\mathcal{E})\, \mathrm{d}\rho}{\sqrt{m_1^2c^2 + \rho^2}\sqrt{m_2^2c^2 + \rho^2}} \quad \text{so} \quad \frac{\rho^2 \mathrm{d}\rho}{\sqrt{m_1^2c^2 + \rho^2}\sqrt{m_2^2c^2 + \rho^2}} = \rho(\mathcal{E})\frac{\mathrm{d}\mathcal{E}}{\mathcal{E}}, \tag{3.120}$$

where we intentionally leave $\rho = \rho(\mathcal{E})$ as is, and have

$$\Gamma = \frac{S}{8\pi\hbar m_A} \int_{(m_1+m_2)c^2}^\infty \frac{\mathrm{d}\mathcal{E}}{\mathcal{E}}\, |\mathfrak{M}|^2\, \rho(\mathcal{E})\, \delta(m_A c - \mathcal{E}/c). \tag{3.121}$$

Since $\delta(m_A c - \mathcal{E}/c) = c\,\delta(\mathcal{E} - m_A c^2)$, we finally use the δ-function to cancel the $\mathrm{d}\mathcal{E}$-integral:

$$\Gamma = \begin{cases} \dfrac{S\rho_0}{8\pi\hbar m_A^2 c}\, |\mathfrak{M}(\rho_0)|^2, & \text{if} \quad m_A > m_1 + m_2; \\ 0, & \text{if} \quad m_A \leqslant m_1 + m_2, \end{cases} \tag{3.122}$$

where $\rho_0 = |\vec{p}_1|_0$ solves the relation (3.119) with $\mathcal{E} = m_A c^2$:

$$\rho_0 = |\vec{p}_1|_0 = \frac{c}{2m_A}\sqrt{m_A^4 + m_1^4 + m_2^4 - 2m_A^2\, m_1^2 - 2m_A^2\, m_2^2 - 2m_1^2\, m_2^2}, \quad [\text{✎ verify}] \tag{3.123}$$

and satisfies the linear momentum conservation law. It is useful to list a few simplifications: When the two products have the same mass but are not the same particle, $S = 1$ and

$$\Gamma = \frac{\sqrt{1-(2m/m_A)^2}}{16\pi\hbar m_A}\left|\mathfrak{M}\left(\tfrac{c}{2}\sqrt{m_A^2-(2m)^2}\right)\right|^2. \tag{3.124}$$

If, furthermore, $m_1 = 0 = m_2$ but the resulting particles are still distinct (e.g., a neutrino and a photon, or two *different* neutrinos and where $m \approx 0$ is a pretty good *approximation* for neutrinos), we have

$$\Gamma = \frac{1}{16\pi\hbar m_A}\left|\mathfrak{M}\left(\tfrac{1}{2}m_A c\right)\right|^2. \tag{3.125}$$

Finally, if the products really are identical particles, $S = \frac{1}{2}$, and the decay rate is one half of the previously listed results (3.122), and (3.124)–(3.125).

Example 3.4 Consider the inelastic scattering $A + B \to C_1 + C_2$, where the particles have masses m_A, m_B, m_1 and m_2, respectively, and where the products' linear momenta are not measured, and so must be integrated over. The expression (3.114) must be integrated over $d^3\vec{p}_1 d^3\vec{p}_2$, and the procedure is similar to that in Example 3.3. However, this time note that $\mathfrak{M}$ in principle depends on all four linear momenta, $\vec{p}_A, \vec{p}_B, \vec{p}_1$ and $\vec{p}_2$. Since $\mathfrak{M}$ is a scalar function, it may depend only on the scalar quantities constructed from these four 3-vectors.

However, if these 3-vectors are expressed in the coordinate system where $\vec{p}_A + \vec{p}_B = \vec{0}$, it follows that $\vec{p}_A = -\vec{p}_B = \vec{p}_i$ (initial) and $\vec{p}_1 = -\vec{p}_2 = \vec{p}_f$ (final). Scalar functions of these two 3-vectors can only depend on $|\vec{p}_i|$, $|\vec{p}_f|$ and $\vec{p}_i \cdot \vec{p}_f = |\vec{p}_i||\vec{p}_f| \cos\vartheta$, where ϑ is the angle between the initial and the final linear momentum, $\vec{p}_A$ and $\vec{p}_1$.

Since the initial linear momentum $\vec{p}_A$ is known, amongst the integration variables in the integral of the expression (3.114), $\mathfrak{M}$ may depend only on $|\vec{p}_1|$ and ϑ. Choosing the spherical coordinate system where $\hat{e}_z \| \vec{p}_A$, we have

$$d^3\vec{p}_1 = \rho^2 d\rho \sin(\vartheta) d\vartheta\, d\varphi = \rho^2 d\rho\, d\Omega. \tag{3.126}$$

Repeating the simplification of the integral just as done in Example 3.3 and using the result (3.163), we obtain

$$\frac{d\sigma}{d\Omega} = \left(\frac{\hbar c}{8\pi}\right)^2 \frac{S\,|\mathfrak{M}|^2}{(E_A + E_B)^2} \frac{|\vec{p}_f|}{|\vec{p}_i|}. \tag{3.127}$$

To compute the final expression for the total effective cross-section, σ, by angular integration over $d\Omega = \sin(\vartheta) d\vartheta\, d\varphi$, the angular dependence of $\mathfrak{M}$ on ϑ, φ must be known.

3.3.4 A simple toy-model example

In this section we consider the Feynman calculus in a very simple toy-model. In Chapters 5 and 6, this procedure will be applied to concrete and realistic processes in the Standard Model. This toy-model is "borrowed" from Ref. [243], where it is attributed to Max Dresden; ultimately of course, the number of very simple but nontrivial models is very limited.

— ❦ —

There are only three types of particles in this model, A, B, C, with $m_A > m_B + m_C$, and such that there exists only one elementary process:

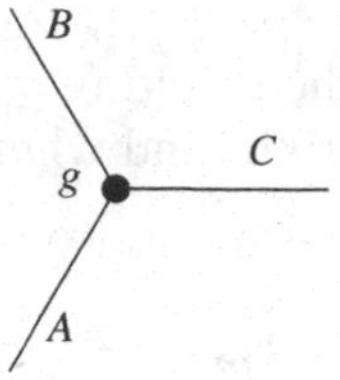

(3.128)

We assume that the constant (charge/strength) of interaction g is sufficiently small to serve as a perturbation parameter, at least formally. The computation of any physical quantity is thus organized as a power series in g, and we compute all contributions of order g^n, ranging from the lowest possible value of $n \geqslant 0$, towards increasingly higher values of n.

Procedure 3.1 *The contribution to the matrix element (amplitude) $\mathfrak{M}$ corresponding to a Feynman diagram in the ABC-model is computed following the algorithm:*

1. ***Notation:*** *Denote the incoming and outgoing 4-momenta by $\mathrm{p}_1, \mathrm{p}_2, \ldots$ and the "internal" 4-momenta (assigned to lines that connect two vertices within the graph) $\mathrm{q}_1, \mathrm{q}_2, \ldots$ Orient each line, selecting the positive sense of the corresponding 4-momentum.*
2. ***Vertices:*** *Assign to each vertex the factor $-ig$.*
3. ***Lines:*** *Assign to the jth internal line the factor $\frac{i}{\mathrm{q}_j^2 - m_j^2 c^2}$, the so-called* **propagator**. *As this depicts a virtual particle,[16] $\mathrm{q}_j^2 \mathrel{+\!\!\!\!=} m_j^2 c^2$.*
4. ***4-momentum conservation:*** *Assign to each vertex the factor $(2\pi)^4 \delta^4(\sum_j \mathrm{k}_j)$, where k_j $(-\mathrm{k}_j)$ are the 4-momenta entering (leaving) the vertex.*
5. ***4-momentum integration:*** *Assign to the jth internal line the $\int \frac{\mathrm{d}^4 \mathrm{q}_j}{(2\pi)^4}$-integral.*
6. ***Reading off the amplitude:*** *The above procedure produces*

$$(-i\,\mathfrak{M})\;(2\pi)^4 \delta^4\big(\sum_j \mathrm{p}_j\big), \tag{3.129}$$

where the $(2\pi)^4 \delta^4(\sum_j \mathrm{p}_j)$ represents the 4-momentum conservation law, and from where the amplitude (matrix element) $\mathfrak{M}$ is read off.

The $A \to B + C$ decay

The lowest order contribution is of the order g^1:

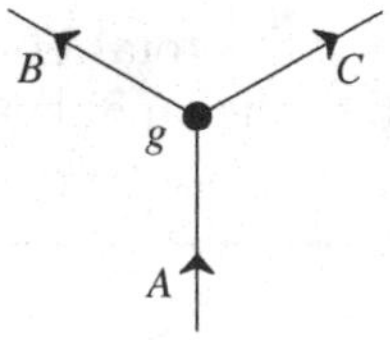

(3.130)

[16] For virtual particles, it is not that q^2 is required to not equal m^2c^2 (i.e., $\mathrm{q}^2 \neq m^2c^2$), but rather that q^2 is not required to equal m^2c^2. In distinction from "does not equal," the relation "not required to equal" will herein be denoted by the non-standard symbol "$\mathrel{+\!\!\!\!=}$" [☞ Tables C.7 on p. 529 and C.8 on p. 529].

The time axis is directed vertically, upward. The next contributions are of the order g^3:

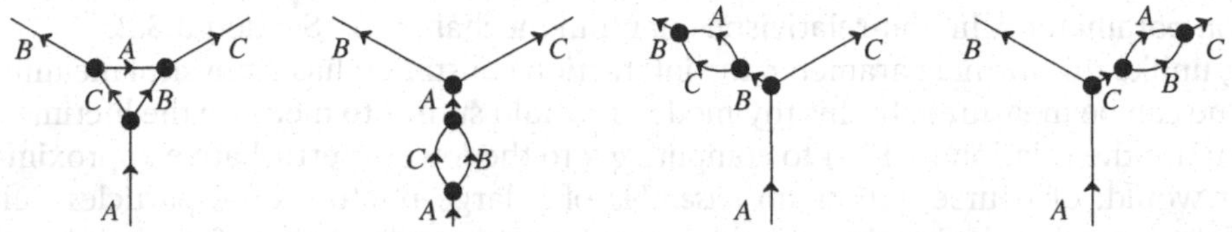

(3.131)

and so on. The lowest order contribution (3.130) is depicted by a tree-graph (with no closed loop). The subsequent contributions (3.131) all have precisely one closed loop and are of the order g^3; there are no contributions of even order g^{2k}. However, starting with the next (g^5) order, a novelty appears, which can be seen by comparing the following two graphs:

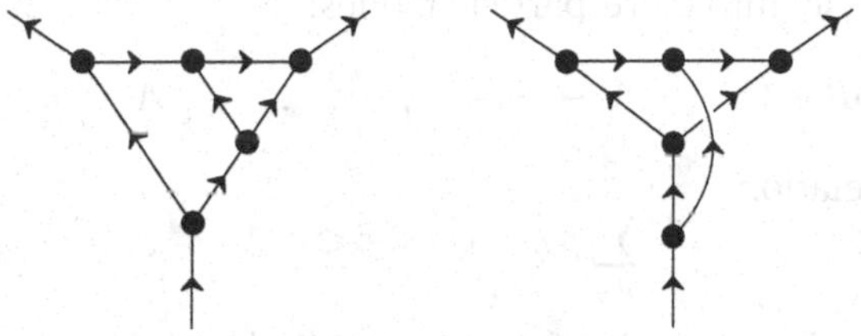

(3.132)

The left-hand graph is planar, but the right-hand graph is not. This property of planarity may be used for a finer classification of graphs, and proves to be very useful in computations for the strong nuclear interaction [511]; for a recent review, see Ref. [349]. Also, only *connected* diagrams contribute: diagrams such as any one of the above but with a disconnected component (e.g., ) added do not contribute; this recalls the "excisions" in Section 3.3.2, the contribution of which had to be subtracted in non-relativistic stationary state perturbation theory.

Return to the contribution of the lowest order (3.130), where there are no internal lines. Procedure 3.1 reduces to:

1. Let the "external" 4-momenta be p_A, p_B and p_C.
2. Assign $-ig$ to the vertex.
3. There are no internal lines, and so no propagators either.
4. Assign $(2\pi)^4\delta^4(\mathrm{p}_A - \mathrm{p}_B - \mathrm{p}_C)$ to the vertex.
5. There are no internal lines, and so no integration either.
6. We've obtained

$$-ig\,(2\pi)^4\delta^4(\mathrm{p}_A - \mathrm{p}_B - \mathrm{p}_C) = (-i\,\mathfrak{M})\,(2\pi)^4\delta^4(\mathrm{p}_A - \mathrm{p}_B - \mathrm{p}_C), \tag{3.133}$$

from where $\mathfrak{M}$ is read off – here, to order g^1. Thus, $\mathfrak{M}^{(1)} = g$.

Inserting this result into the expression (3.122) we obtain

$$\Gamma^{(1)} = \frac{g^2|\vec{p}_B|_0}{8\pi\hbar m_A^2 c}, \qquad \text{so that} \qquad \tau^{(1)} = \frac{1}{\Gamma^{(1)}} = \frac{8\pi\hbar m_A^2 c}{g^2|\vec{p}_B|_0}, \tag{3.134}$$

where

$$|\vec{p}_B|_0 = \frac{c}{2m_A}\sqrt{m_A^4 + m_B^4 + m_C^4 - 2m_A^2 m_B^2 - 2m_A^2 m_C^2 - 2m_B^2 m_C^2} = |\vec{p}_C|_0. \tag{3.135}$$

The result $\mathfrak{M}^{(1)} = g$, and so also (3.134), is analogous to the result (3.92a): $\mathfrak{M}^{(1)} = g$ is the first-order result in $\mathfrak{M}$ expanded in a power series over g, as is $E_n^{(1)}$ the first-order result in a power-series expansion of the energy over λ. In this sense, the constant of interaction g serves as the perturbation parameter, and its numerical value must be sufficiently "small" so that such a power series would make sense,[17] so that the interaction parameter g has the same formal role as the perturbation parameter λ in non-relativistic quantum mechanics, in Section 3.3.2.

However, unlike this formal parameter, the interaction constant g has a physical meaning and its physical value can be measured. In this toy-model, it would suffice to measure the lifetime of the A-particle, then use the relation (3.134) to compute g – to the lowest perturbative approximation. An experiment would, of course, follow an *ensemble* of a large number of A-particles, and the diminishing of their number during time would determine the average value of the half-life $t^{(1)}_{1/2} = \tau^{(1)}\ln(2)$.

Note, however, that there would occur an additional "correction": Processes (3.130), (3.131), (3.132), and others of the same order in g (and then also the higher-order ones) are contributions only to the exclusive (partial) decay $A \to B + C$. If the mass m_A is sufficiently bigger than m_B, m_C, the A-particle may also decay into more-particle modes:

$$A \to 3B + C, \qquad A \to B + 3C, \qquad \dots \qquad A \to pB + qC, \tag{3.136a}$$

which are limited by the relation

$$\sum_{p,q} p\, m_B + q\, m_C < m_A, \tag{3.136b}$$

as well as the nature of the decay graphs, from which it follows, e.g.,

$$(p,q) \neq (1,2), (2,2), (2,3), \dots \tag{3.136c}$$

The A+A→ B+B scattering

The constant g may also be measured – in a thought experiment since this is a toy-model – more directly, by measuring the intensity of the interaction during scattering. Griffiths [243] analyzes the inelastic decay $A + A \to B + B$, where "incoming" A-particles are assigned the 4-momenta p_1 and p_2, and the outgoing B-particles p_3 and p_4.

By definition and using the relation (3.37), we have that $\frac{E_i}{c} = \sqrt{m_i^2c^2 + \vec{p}_i^{\,2}}$ for $i = 1,2,3,4$. In the CM system, where

$$\Big(-\frac{E_1}{c}, \vec{p}_1\Big) + \Big(-\frac{E_2}{c}, \vec{p}_2\Big) = (p_0, \vec{0}) = \Big(-\frac{E_3}{c}, \vec{p}_3\Big) + \Big(-\frac{E_4}{c}, \vec{p}_4\Big), \tag{3.137}$$

the total linear momentum vanishes so $\vec{p}_1 = -\vec{p}_2$ and $\vec{p}_3 = -\vec{p}_4$. Denote $\theta := \measuredangle(\vec{p}_1, \vec{p}_3) = \measuredangle(\vec{p}_2, \vec{p}_4)$, so that $\measuredangle(\vec{p}_1, \vec{p}_4) = \measuredangle(\vec{p}_2, \vec{p}_3) = (\pi - \theta)$. We also have that

$$E_1 = c\sqrt{m_A^2c^2 + \vec{p}_1^{\,2}} = c\sqrt{m_A^2c^2 + (-\vec{p}_1)^2} = E_2, \tag{3.138}$$

$$E_3 = c\sqrt{m_B^2c^2 + \vec{p}_3^{\,2}} = c\sqrt{m_B^2c^2 + (-\vec{p}_3)^2} = E_4. \tag{3.139}$$

From conservation of energy, i.e., the energy component (3.137), it follows that

$$E := E_1 = E_2 = E_3 = E_4. \tag{3.140}$$

[17] It would be ideal if this power series would converge. Within field theory, in practice – if this can be determined at all – one mostly obtains asymptotic or even formally divergent sums, for which one must independently establish if the sum may be unambiguously assigned a particular value [☞ [259] for "summability"] for the given value of the constant g as its argument. Not infrequently, one only knows that the first several orders of perturbative computations are ever smaller "corrections," but with no formal proof about the nature of the whole infinite series. On the other hand, practical computations in quantum electrodynamics show ***unprecedented precision***: both perturbative computations and experimental measurements are found to agree with a relative error $< O(10^{-10})$ [293], better than anywhere else in natural sciences!

Equating the squares of these energies, we obtain

$$m_A^2c^2 - \vec{p}_1^{\,2} = m_A^2c^2 - \vec{p}_2^{\,2} = m_B^2c^2 - \vec{p}_3^{\,2} = m_B^2c^2 - \vec{p}_4^{\,2}. \tag{3.141}$$

In the limiting case of this toy-model, when $m_A = m_B = m$, but $m_C = 0$, the relation (3.141) gives

$$|\vec{p}| := |\vec{p}_1| = |\vec{p}_2| = |\vec{p}_3| = |\vec{p}_4|. \tag{3.142}$$

Scattering In this case, following Procedure 3.1 we have:

1. Denote the "incoming" 4-momenta by p_1, p_2, and the "outgoing" ones by p_3, p_4:

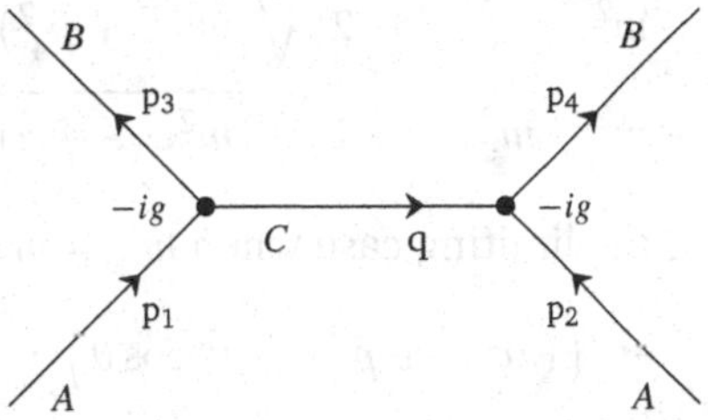

(3.143)

2. Assign to both vertices a factor of $-ig$.
3. Assign to the internal line the 4-momentum q, and the propagator $\frac{i}{q^2 - m_C^2c^2}$.
4. Assign to the vertices the factors $(2\pi)^4\delta^4(p_1 - q - p_3)$ and $(2\pi)^4\delta^4(p_2 + q - p_4)$.
5. Integrate over $\frac{d^4q}{(2\pi)^4}$.
6. We have thus obtained:

$$\begin{aligned}
-i\,\mathfrak{M}\,(2\pi)^4\delta^4(p_1+p_2-p_3-p_4)&\\
&= \int \frac{d^4q}{(2\pi)^4}\,(-ig)^2 \frac{i}{q^2 - m_C^2c^2}\,(2\pi)^4\delta^4(p_1-q-p_3)\,(2\pi)^4\delta^4(p_2+q-p_4)\\
&= -ig^2(2\pi)^4 \int \frac{d^4q}{q^2 - m_C^2c^2}\,\delta^4(p_1-q-p_3)\,\delta^4(p_2+q-p_4)\\
&= -i\,\frac{g^2}{(p_4-p_2)^2 - m_C^2c^2}\,(2\pi)^4\delta^4(p_1+p_2-p_3-p_4).
\end{aligned} \tag{3.144}$$

However, this is not the only Feynman diagram that produces a g^2 contribution; holding the positions of the outgoing lines and their assigned 4-momenta, it is clear that a "topologically" distinct diagram is obtained by swapping the vertices to which the outgoing lines connect:

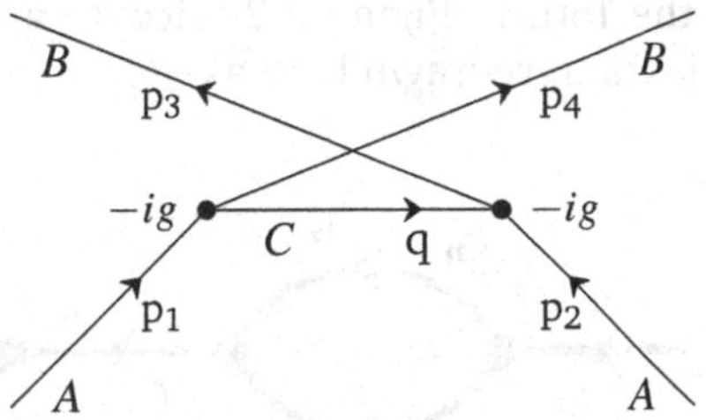

(3.145)

and which clearly produces a contribution of the same form (3.144), however, with the exchange $p_3 \leftrightarrow p_4$. As there are no other Feynman diagrams, the amplitude $\mathfrak{M}$ is read off from the sum of these two contributions:

$$\mathfrak{M} = \frac{g^2}{(p_4-p_2)^2 - m_C^2c^2} + \frac{g^2}{(p_3-p_2)^2 - m_C^2c^2}. \tag{3.146}$$

This amplitude is to be substituted into the expression (3.127):

$$\frac{d\sigma}{d\Omega} = \left(\frac{\hbar c}{8\pi}\right)^2 \frac{S\,|\mathfrak{M}|^2}{(E_1+E_2)^2}\frac{|\vec{p}_3|}{|\vec{p}_1|}, \tag{3.147}$$

where $S = \frac{1}{2}$, we used the result (3.163), and chose to use a spherical coordinate system in which the angle θ equals the angle $\measuredangle(\vec{p}_1, \vec{p}_3)$.

The denominators in the ratios (3.146) are

$$\begin{aligned}(\mathrm{p}_4 - \mathrm{p}_2)^2 - m_C^2c^2 &= (m_A^2 + m_B^2 - m_C^2)c^2 - 2\Big(\frac{E_4}{c}\frac{E_2}{c} - \vec{p}_4{\cdot}\vec{p}_2\Big)\\ &= (m_A^2 + m_B^2 - m_C^2)c^2 - 2\Big(\sqrt{(m_B^2c^2 + \vec{p}_4^{\,2})(m_A^2c^2 + \vec{p}_2^{\,2})} - \vec{p}_4{\cdot}\vec{p}_2\Big),\\ (\mathrm{p}_3 - \mathrm{p}_2)^2 - m_C^2c^2 &= (m_A^2 + m_B^2 - m_C^2)c^2 - 2\Big(\sqrt{(m_B^2c^2 + \vec{p}_3^{\,2})(m_A^2c^2 + \vec{p}_2^{\,2})} - \vec{p}_3{\cdot}\vec{p}_2\Big),\end{aligned}$$

which significantly simplifies in the limiting case when $m_A = m_B = m$ and $m_C = 0$:

$$(\mathrm{p}_4 - \mathrm{p}_2)^2 - m_C^2c^2 = 2m^2c^2 - 2\Big((m^2c^2 + \vec{p}^{\,2}) - \vec{p}^{\,2}\cos\theta\Big) = -2\vec{p}^{\,2}(1-\cos\theta), \tag{3.148}$$

$$(\mathrm{p}_3 - \mathrm{p}_2)^2 - m_C^2c^2 = 2m^2c^2 - 2\Big((m^2c^2 + \vec{p}^{\,2}) - \vec{p}^{\,2}\cos(\pi{-}\theta)\Big) = -2\vec{p}^{\,2}(1+\cos\theta). \tag{3.149}$$

Thus, in the limiting case $m_A = m_B = m$, and $m_C = 0$:

$$\mathfrak{M} = -\frac{g^2}{\vec{p}^{\,2}\sin^2\theta}, \qquad \text{so} \qquad \left(\frac{d\sigma}{d\Omega}\right) = \frac{1}{2}\left(\frac{\hbar c}{16\pi}\frac{g^2}{E\,\vec{p}^{\,2}\sin^2\theta}\right)^2, \tag{3.150}$$

which may be used to measure – in a thought experiment for this toy-model – the interaction constant g by measuring the differential effective cross-section as a function of the deflection angle $\theta := \measuredangle(\vec{p}_1, \vec{p}_2)$, and the energy and linear momentum of the "incoming" particles, $\vec{p} = \vec{p}_1$ and $E = E_1$, respectively.

Of course, the diagrams (3.143) and (3.145) and so also the result (3.150) are all only the contributions of the lowest order. In the next, $O(g^4)$, order, we have the diagrams listed in Figure 3.2.

The number of Feynman diagrams shown in Figure 3.2[18] indicates the volume of the task in computing physical quantities, such as the differential effective cross-section, order by order in perturbation theory. It is fairly obvious that the task of computing even just the first few order contributions (in the expansion organized into growing powers of the interaction constant) to a physical quantity is already a very demanding exercise, so that discussions and analyses of the convergence of the whole perturbative sum must limit to general properties.

Besides, diagrams such as the 15th in Figure 3.2 uncover a new property: divergences and renormalization. Consider this diagram, redrawn here as

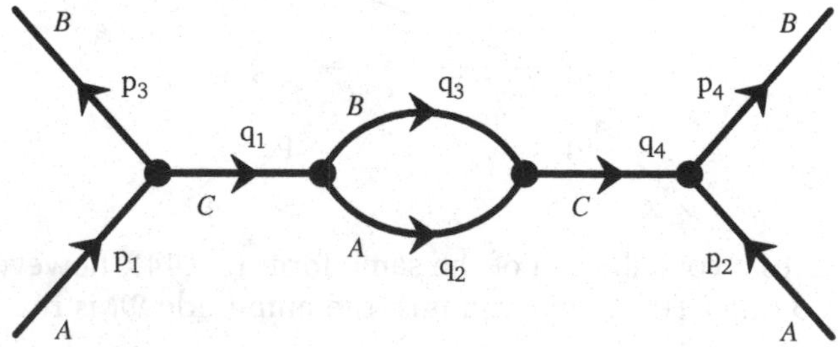

(3.151)

[18] The counting given in Ref. [243, 1st edn.] was imprecise: one-third of the diagrams counted there are either impossible in the *A-B-C* toy-model (diagrams 4 and 7, in Figure 3.2), or are counted twice (diagrams 11, 12 and 14). However, this is seen only when the lines are assigned particles and one verifies all vertices to be of the form (3.128).

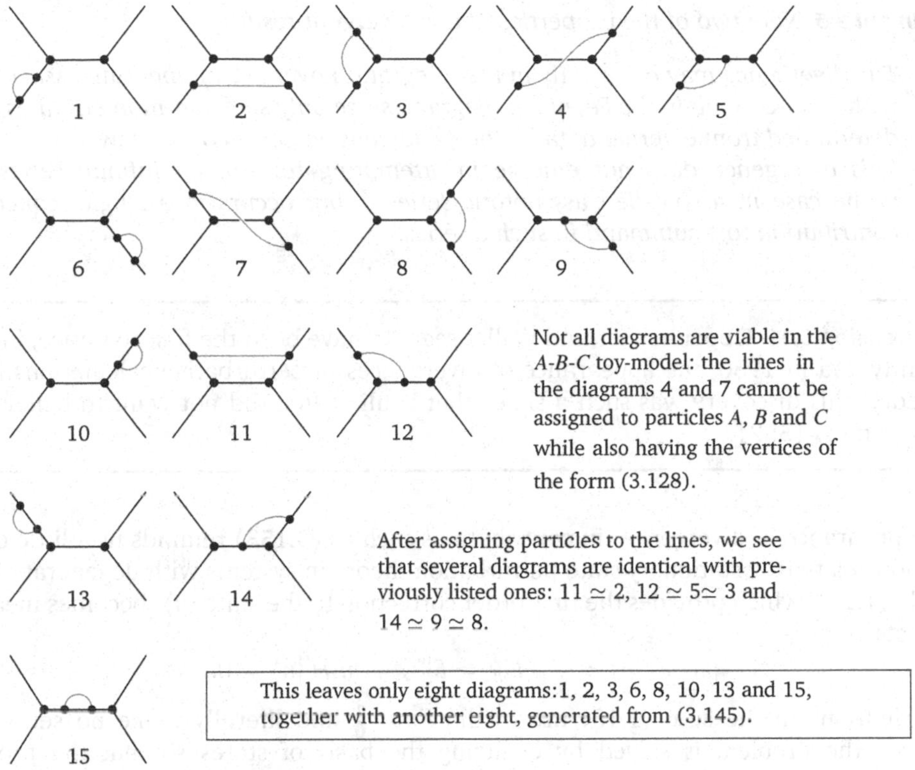
1
2
3
4
5
6
7
8
9
10
11
12
Not all diagrams are viable in the A-B-C toy-model: the lines in the diagrams 4 and 7 cannot be assigned to particles A, B and C while also having the vertices of the form (3.128).
13
14
After assigning particles to the lines, we see that several diagrams are identical with previously listed ones: 11 ≃ 2, 12 ≃ 5≃ 3 and 14 ≃ 9 ≃ 8.
15
This leaves only eight diagrams:1, 2, 3, 6, 8, 10, 13 and 15, together with another eight, generated from (3.145).

Figure 3.2 Fifteen possible $O(g^4)$ diagrams, before assigning particles to lines. All diagrams are generated from the $O(g^2)$ diagram (3.143), by adding one internal line. When assigning particles to lines, maintain the "external conditions": two incoming A-type particles (initial, lower in the diagrams) and two outgoing B-type particles (final, upper in the diagrams). These conditions reduce the total number to eight.

Following the Procedure 3.1, we obtain

$$(-ig)^4 \int \frac{i\,\mathrm{d}^4\mathrm{q}_1}{\mathrm{q}_1^2 - m_C^2c^2}\frac{i\,\mathrm{d}^4\mathrm{q}_2}{\mathrm{q}_2^2 - m_A^2c^2}\frac{i\,\mathrm{d}^4\mathrm{q}_3}{\mathrm{q}_3^2 - m_B^2c^2}\frac{i\,\mathrm{d}^4\mathrm{q}_4}{\mathrm{q}_4^2 - m_C^2c^2}$$
$$\times\,\delta^4(\mathrm{p}_1 - \mathrm{p}_3 - \mathrm{q}_1)\,\delta^4(\mathrm{q}_1 - \mathrm{q}_2 - \mathrm{q}_3)\,\delta^4(\mathrm{q}_2 + \mathrm{q}_3 - \mathrm{q}_4)\,\delta^4(\mathrm{p}_2 + \mathrm{q}_4 - \mathrm{p}_4), \tag{3.152}$$

where the first δ-function cancels the $\mathrm{d}^4\mathrm{q}_1$-integral and replaces $\mathrm{q}_1 \to (\mathrm{p}_1 - \mathrm{p}_3)$, while the last δ-function cancels the $\mathrm{d}^4\mathrm{q}_4$-integral and replaces $\mathrm{q}_4 \to (\mathrm{p}_4 - \mathrm{p}_2)$. Thereafter, the second δ-function cancels the $\mathrm{d}^4\mathrm{q}_2$-integral and replaces $\mathrm{q}_2 \to (\mathrm{q}_1 - \mathrm{q}_3) = (\mathrm{p}_1 - \mathrm{p}_3 - \mathrm{q}_3)$ and turns the remaining, third δ-function into the expected factor $\delta^4(\mathrm{p}_1 + \mathrm{p}_2 - \mathrm{p}_3 - \mathrm{p}_4)$. This then leaves

$$\mathfrak{M} = \frac{(g/2\pi)^4}{[(\mathrm{p}_1 - \mathrm{p}_3)^2 - m_C^2c^2]^2} \int \frac{\mathrm{d}^4\mathrm{q}_3}{[(\mathrm{p}_1 - \mathrm{p}_3 - \mathrm{q}_3)^2 - m_A^2c^2](\mathrm{q}_3^2 - m_B c^2)}. \tag{3.153a}$$

This $\mathrm{d}^4\mathrm{q}_3$-integral necessarily diverges: in 4-dimensional spherical coordinates, we have that $\mathrm{d}^4\mathrm{q}_3 = \rho^3\mathrm{d}\rho\,\mathrm{d}\Omega_{(3)}$, so the "radial" integral becomes, near the upper (infinite) limit:

$$\sim \int^{\infty} \frac{\rho^3\mathrm{d}\rho}{\rho^4} = \lim_{R\to\infty} \int^{R} \frac{\mathrm{d}\rho}{\rho} = \lim_{R\to\infty} \ln(R), \tag{3.153b}$$

which diverges logarithmically.

***Comment 3.5** Note two of the properties of this divergent result:*

1. *The divergence may occur only in the integration over a 4-momentum associated with a closed loop in the Feynman diagram, since only such a 4-momentum is not determined from external data by the 4-momentum conservation law.*
2. *This divergence does not emerge in attempting to sum an infinite series as is the case in a so-called asymptotic series,[19] but occurs in a single, concrete contribution to a summand in such a series.*

Digression 3.11 Oppenheimer and Waller seem to have been the first to notice, independently and in 1930, the appearance of divergences in perturbative calculations in field theory; this discovery was such a shock that Pauli at first did not want to believe in its correctness [552].

The appearance of divergences in contributions such as (3.153) reminds us a little of the situation in non-relativistic stationary state perturbation theory in systems with degeneration. There, the formula (1.18), which provides the first-order correction to the state $|n\rangle$, becomes meaningless if there exists

$$|m\rangle \neq |n\rangle \ : \qquad E_m^{(0)} = E_n^{(0)}, \quad \langle m|H'|n\rangle \neq 0. \tag{3.154}$$

Such contributions in the sum (1.18) are of the form $\frac{1}{0}$ and literally make no sense. In that simpler case, the problem is solved by changing the basis of states so that the problematic combinations (3.154) and terms of the type $\frac{1}{0}$ in the sum (1.18) no longer occur.

Digression 3.12 To eliminate the offending situation (3.154), one defines

$$|m'\rangle = c_{mm}|m\rangle + c_{mn}|n\rangle, \quad |n'\rangle = c_{nm}|m\rangle + c_{nn}|n\rangle, \tag{3.155a}$$

and requires that $\langle m'|H'|n'\rangle = 0$. This implies that H_0 and H' have been simultaneously diagonalized over the $\{|m\rangle, |n\rangle\} \in \mathscr{H}$ subspace of the Hilbert space. In turn, this implies that

$$[H_0, H'] = 0 \quad \text{over} \quad \{|m\rangle, |n\rangle\} \in \mathscr{H}. \tag{3.155b}$$

In this sense the "$\frac{1}{0}$-divergence" is "removed." It is also clear from the structure of the sums in the results (1.18), (1.19), and also the entire algorithm given in Section 3.3.2, that the re-diagonalization of the basis $|n\rangle$ that removes the $\frac{1}{0}$-divergences from the sum (1.18) to the first perturbative order also removes all divergences of this type in the whole perturbative procedure.

However, divergences of the form (3.153) are harder to "remove," and their treatment has halted the first physically significant field theory – quantum electrodynamics – for almost two decades. The work of many physicists on this problem culminated in independent and equivalent methods by Richard Feynman, Julian Schwinger and Shin-Ichiro Tomonaga, which were systematized by Freeman Dyson.

[19] H. Poincaré defined the series $\sum_k c_k x^k$ where $\lim_{x\to\infty} x^k R_k(x) = 0$ for any fixed k, but $\lim_{k\to\infty} x^k R_k(x) = \infty$ to be *asymptotic* (semi-convergent). Here $R_k(x) := [f(x) - \sum_{i=0}^{k} c_i x^{-i}]$, and $f(x)$ asymptotically agrees with the sum for large $x \sim \infty$.

Digression 3.13 The systematic procedure and *idea* of renormalization stems from an older and unrelated idea: In 1902, Max Abraham proposed the model [4] in which the electron was a sphere of finite radius on the surface of which the electron charge is uniformly distributed. In 1904, Hendrik Lorentz [346] developed this idea, so that the model is now called the Abraham–Lorentz model. In this model, the work required to assemble the electron charge by bringing it from infinity into the Gaussian sphere of radius r_e contributes to the mass of the electron. If this is the only contribution, that work may be equated with the (electromagnetic) rest energy of the electron:

$$m_{\rm em}c^2 = \tfrac{1}{2}\int \mathrm{d}^3\vec{r}\,\vec{E}^2 = 2\pi\int_{r_e}^{\infty} r^2\mathrm{d}r\left(\frac{e}{4\pi\epsilon_0\,r^2}\right)^2 = \frac{1}{2}\frac{e^2}{4\pi\epsilon_0}\frac{1}{r_e}, \tag{3.156a}$$

if the electron charge is distributed uniformly on the surface of the sphere of radius r_e. If the charge is distributed uniformly throughout the entire sphere, the factor $\frac{1}{2}$ is replaced by $\frac{3}{5}$. Equating $m_{\rm em}$ with the measured electron mass (as if the electron mass stems entirely from the electric field that this electron produces, and the charged shell has no mass of its own) and neglecting the numerical factor $\frac{1}{2}-\frac{3}{5}$, we obtain the classical electron radius:

$$r_e = \frac{e^2}{4\pi\epsilon_0}\frac{1}{m_e\,c^2} = \alpha_e\frac{\hbar}{m_e c} = 2.817\,940\,325\times 10^{-15}\,\mathrm{m}. \tag{3.156b}$$

Because of the dependence $m_{\rm em} \propto r_e^{-1}$, it follows that the electron cannot be ideally point-like: if it were, its mass would be infinitely large. Since the total effective mass must also include (realistically, a non-vanishing) mass of the spherical shell, the "true" electron radius may differ from the result (3.156b). However, for the radius to be *smaller* than r_e, it would be necessary for the mass of the spherical shell to be *negative*. In the limiting case of the point-like electron, this infinitely small shell would have to have an infinitely negative mass, which is an evidently meaningless value.

However, from these considerations about the Abraham–Lorentz model of the electron stems the idea that the measured values of a physical quantity may consist of several contributions, which – in the limiting case – may each diverge, as long as their sum (which is what is compared with the experimental data!) is a finite quantity.

The basic idea, schematically, is that for each parameter there exists

$$m_{\rm physical} = m_{\rm bare} + \delta m, \qquad g_{\rm physical} = g_{\rm bare} + \delta g, \qquad \text{etc.} \tag{3.157}$$

where the "bare" version of the parameter is the one showing up in the classical Lagrangian theory, and the "quantum correction" δm often diverges. However, $m_{\rm bare}$ is defined so as to also diverge,[20] and precisely so that the physical value of the parameter remains finite and comparable with the experiments. Besides, the systematic procedure of renormalization guarantees that the so-defined finite part and divergent part of the result may be consistently separated order by order in perturbation theory for quantum electrodynamics, which then serves as a template for all other existing field theory models.

[20] One of the methods of "canceling" divergences requires that the integrals are computed in finite limits, $\pm\Lambda$, so that one can isolate the portion of the contributions that are independent of Λ in the sums of the form (3.157). In the $\Lambda\to\infty$ limit, that Λ-independent portion represents the desired physical quantity. There exist several other methods for isolating the "finite part" from divergent integrals, but there is no general formal proof that the finite result does not depend on the method of its isolation.

Most nontrivial models in field theory are in practice *defined* by means of the perturbative computations, including some variant of the renormalization procedure. Certain results in a growing class of models can be computed by non-perturbative means – or by using an essentially different perturbative method where individual terms of "lower order" may represent a sum of a large number of (contributions from) standard Feynman diagrams. A formal and rigorous proof of finiteness of all possible observables is not known in general, and in this sense field theory is in general not formally rigorously defined, nor does one know if field theory in general – or even a certain concrete model, such as quantum electrodynamics – is formally self-consistent![21] Nevertheless, perturbative and other concrete results offer enough useful data to compare with experiments, which suffices for a pragmatic acceptance of this theoretical system – all the more so, since (1) the various renormalization prescriptions invariably produce final results agreeing for observables, and (2) no contradiction has been detected in anomaly-free theoretical models [☞ Section 7.2.3].

The appearance and the conspicuous cancellation of divergent contributions in perturbative computations of evidently measurable (and finite!) physical quantities is still cited as the cause for a principled disagreement with the entire renormalization procedure [☞ e.g., Ref. [29]]. However, the number of living physicists who openly oppose this procedure is decreasing.[22] The Reader with a piqued interest in the subject should turn to the texts on quantum field theory [63, 48, 441, 459, 154, 474, 249, 240, 425, 554, 555, 484, 588, 496, 446, 589, 316, 7, 586, 277, 590], texts on renormalization itself [113, 212], and research articles, such as [44, 431, 343, 146, 253].

3.3.5 Exercises for Section 3.3

✎ **3.3.1** *In the special case when $\vec{v} = v\,\hat{\mathbf{e}}_z$, show that the transformations (3.1) acquire the well-known form:*

$$x' = x, \quad y' = y, \quad z' = \gamma(z - v\,t), \qquad x = x', \quad y = y' \quad z = \gamma(z' + vt'), \tag{3.158a}$$

$$t' = \gamma\Big(t - \frac{v\,z}{c^2}\Big), \qquad\qquad t = \gamma\Big(t' + \frac{v\,z'}{c^2}\Big), \tag{3.158b}$$

with the usual $\gamma = (1 - \vec{v}^2/c^2)^{-1/2}$.

✎ **3.3.2** *Using that* $[\Gamma] = T^{-1}$, $[\sigma] = L^2$, $[v] = LT^{-1}$ *and* $\big[|\Psi|^2\big] = L^{-3}$, *prove equation (3.110) and Conclusion 3.2.*

✎ **3.3.3** *For the elastic collision $A + B \to A' + B'$, in a system where B is originally at rest (and is the target), derive*

$$\frac{\mathrm{d}\sigma}{\mathrm{d}\Omega} \approx S\Big(\frac{\hbar}{8\pi}\Big)^2 \frac{|\mathfrak{M}|^2}{m_B} \frac{\vec{p}_{A'}^{\,2}}{\big|\,|\vec{p}_A|\big(|\vec{p}_{A'}|(E_A + m_B c^2) - |\vec{p}_A| E_{A'} \cos\theta\big)\big|}. \tag{3.159}$$

Here A and A' denote the incoming and outgoing, but otherwise identical particles, just as do B and B'.

✎ **3.3.4** *Show that the result of the previous problem simplifies when $(m_A/m_B) \ll 1$:*

$$\frac{\mathrm{d}\sigma}{\mathrm{d}\Omega} \approx S\Big(\frac{\hbar\,E_{A'}}{8\pi\,E_A}\Big)^2 \frac{|\mathfrak{M}|^2}{m_B^2}. \tag{3.160}$$

---

[21] In view of Gödel's incompleteness theorem, a formally rigorous proof of self-consistency of field theory may turn out to be a pipe dream, since theoretical axiomatic systems that are sufficiently strong (e.g., include standard arithmetics) turn out to also be incapable of proving their own consistency [☞ Appendix B.3].

[22] To paraphrase Max Planck [428, pp. 33–34], new scientific truths do not convince their opponents, they outlive them.

✎ ***3.3.5*** *For the elastic collision in Exercise 3.3.3 but in the case when the recoil of the target after the collision may be neglected since* $m_B c^2 \gg E_A$*, derive*

$$\frac{d\sigma}{d\Omega} \approx \left(\frac{\hbar}{8\pi m_B c}\right)^2 |\mathfrak{M}|^2. \tag{3.161}$$

✎ ***3.3.6*** *For the inelastic collision* $A + B \to C_1 + C_2$*, in a system where* B *(the target) is originally at rest, and* $(m_{C_i}/m_A) \ll 1$ *and* $(m_{C_i}/m_B) \ll 1$*, derive*

$$\frac{d\sigma}{d\Omega} \approx \left(\frac{\hbar}{8\pi}\right)^2 \frac{S\,|\mathfrak{M}|^2}{m_B(E_A + m_B c^2 - |\vec{p}_A| c\,\cos\theta)}\,\frac{|\vec{p}_{C_1}|}{|\vec{p}_A|}, \tag{3.162}$$

where θ *is the angle between* $\vec{p}_1$ *and* $\vec{p}_3$*.*

✎ ***3.3.7*** *Why is the cascade decay* $A \to B + C \to 2B + A$ *in the toy-model of Section 3.3.4 forbidden, but* $A \to B + C \to 2B + A \to 3B + C$ *may be allowed? What is the condition for the latter process to be viable?*

✎ ***3.3.8*** *Using only Feynman diagrams to analyze the possible decay modes of particle* A*, show that* p *and* q *in relation (3.136) must both be odd integers.*

✎ ***3.3.9*** *For a head-on collision of two particles of masses* m_1 *and* m_2*, we have* $\mathrm{p}_1 = (E_1/c, \vec{p})$ *and* $\mathrm{p}_2 = (E_2/c, -\vec{p})$ *in the CM system. Show that*

$$c\sqrt{(\mathrm{p}_1\cdot\mathrm{p}_2)^2 - (m_1 m_2 c^2)^2} = (E_1 + E_2)|\vec{p}|. \tag{3.163}$$

✎ ***3.3.10*** *Prove that, in the A-B-C toy-model and with* $g < 1$*, the elastic collisions are* $O(g^4)$ *times less probable than the inelastic collisions such as (3.143).*

✎ ***3.3.11*** *Prove that equation (3.155b) is satisfied for all of the finitely many degenerate states (3.154), so that the standard procedure described in Digression 3.12 on p. 122 is always possible.*

The quark model: combinatorics and groups

The quark model was initially introduced to explain the emerging plethora of hadrons as bound states of quarks, and this is how we start with its study.

4.1 Bound states

Hadrons – mesons and baryons – are bound states of (anti-)quarks. Mesons are quark–antiquark bound states, held together by the strong nuclear force for which the mediating quanta (particles) are the *gluons*. Baryons consist of three quarks, bound by gluons. It is then clear that the three-particle bound states – baryons – are more complicated than mesons. Furthermore, even the description of mesons as bound states is hard in the case of the "light" quarks. For these one needs a relativistic description of bound states, which is considerably more complicated than the non-relativistic one, and needs more development. One can reliably discuss, using the methods of non-relativistic quantum theory, only the mesons consisting of the "heavier" quarks: the t-, b- and c-quarks and with less precision also the bound states with the s-quark.

The bound states may be analyzed in this way, as a non-relativistic system, if the mass of all constituents is sufficiently bigger than the binding energy. For example, the binding energy of the hydrogen atom (13.6 eV) is 2.66×10^{-5} times that of the electron rest energy, and 1.45×10^{-8} times the proton rest energy, so that the non-relativistic analysis of the hydrogen atom is very accurate. This analysis may then be adapted to many hadronic systems, and we first recall some of the important results about the hydrogen atom.

Unlike the hydrogen atom and the similar positronium, the bound states of quarks and antiquarks will additionally require combinatorial and group-theoretical results since, in addition to the electric charge and spin, quarks also have a "flavor" (u, d, c, s, b, t) and a "color" (red, yellow, blue). With this in mind, the Reader is referred to the group-theoretical results collected in Appendix A to begin with, and the literature [565, 258, 581, 256, 80, 260, 333, 447] for more complete explanations, proofs and detailed theory, which also offer a more complete and pedagogical organization.

4.1.1 The non-relativistic hydrogen atom without spin

Together with the linear harmonic oscillator, the hydrogen atom is the most frequently discussed system in all books on quantum mechanics [362, 363, 471, 328, 480, 134, 391, 407, 360, 472, 29, 339, 242, 3, 110, 324, for example] [☞ also Section 1.2.5]. It is well known that in this two-particle system one can separate the dynamics of the atom as a whole and the relative motion of the electron in the CM-system. Since the proton (nucleus) mass is 1,836.15 times larger than the electron mass, the so-called reduced mass, $\frac{m_e\,m_p}{m_e+m_p} \approx m_e$ differs from the electron mass only by a $\frac{1}{1{,}837.15}$ fraction, which may usually be neglected. Besides, the coordinate origin of the CM-system is very close to the proton location,[1] and one approximates the electron as moving within the electrostatic field of the stationary proton.

The Schrödinger equation is

$$i\hbar\,\frac{\partial}{\partial t}\Psi(\vec{r},t) = H\Psi(\vec{r},t), \qquad H = \Big[-\frac{\hbar^2}{2m_e}\vec{\nabla}^2 + V(r)\Big], \tag{4.1}$$

where H is the non-relativistic Hamiltonian and $V(r)$ is a central potential, which for the hydrogen atom is the Coulomb potential. This may be solved in spherical coordinates, looking for solutions in the form of so-called stationary states:

$$\Psi_{n,\ell,m}(\vec{r},t) = e^{-i\omega_{n,\ell,m}t}\,R_{n,\ell}(r)\,Y_\ell^m(\theta,\phi), \qquad \omega_{n,\ell,m} := \frac{E_{n,\ell,m}}{\hbar}, \tag{4.2}$$

since we know that

$$\vec{\nabla}^2(\cdots) = \frac{1}{r^2}\Big(\frac{\partial}{\partial r}r^2\frac{\partial}{\partial r}\cdots\Big) - \frac{1}{r^2}L^2(\cdots) = \frac{1}{r}\Big(\frac{\partial^2}{\partial r^2}(r\cdots)\Big) - \frac{1}{r^2}L^2(\cdots), \tag{4.3}$$

$$L^2 Y_\ell^m(\theta,\phi) = \ell(\ell{+}1)\,Y_\ell^m(\theta,\phi). \tag{4.4}$$

The $Y_\ell^m(\theta,\phi)$ are the spherical harmonics, the eigenfunctions of the angular part of the Laplacian:

$$L^2(\cdots) := -\frac{1}{\sin\theta}\frac{\partial}{\partial\theta}\Big(\sin\theta\frac{\partial}{\partial\theta}\cdots\Big) - \frac{1}{\sin^2\theta}\Big(\frac{\partial^2}{\partial\phi^2}\cdots\Big), \tag{4.5}$$

which is the square of the operator of so-called "dimensionless angular momentum," i.e., of the angular momentum divided by $\hbar^2$. The radial part of $\vec{\nabla}^2$ and the identity

$$\frac{1}{r^2}\Big(\frac{\partial}{\partial r}r^2\frac{\partial}{\partial r}\cdots\Big) \equiv \frac{1}{r}\Big(\frac{\partial^2}{\partial r^2}r\cdots\Big) \tag{4.6}$$

suggest the substitution $R_{n,\ell}(r) = \frac{u_{n,\ell}(r)}{r}$, whereby the radial differential equation becomes

$$-\frac{\hbar^2}{2m_e}\frac{d^2u_{n,\ell}}{dr^2} + \Big[V(r) + \frac{\hbar^2}{2m_e}\frac{\ell(\ell{+}1)}{r^2}\Big]u_{n,\ell} = E_{n,\ell}\,u_{n,\ell}, \tag{4.7}$$

which is effectively a one-dimensional problem, with $r \in [0,\infty)$ and where the effective potential is the sum of the "actual" potential and the "centrifugal barrier," $\frac{\hbar^2}{2m_e}\frac{\ell(\ell+1)}{r^2}$. For the hydrogen atom, we have the Coulomb potential,

$$V(r) = -\frac{1}{4\pi\epsilon_0}\frac{e^2}{r} = -\frac{\alpha_e\hbar c}{r}, \qquad \alpha_e := \frac{e^2}{4\pi\epsilon_0\hbar c}, \tag{4.8a}$$

[1] To be precise, the coordinate origin of the CM-system is at $1/1{,}837.15 = 5.443\,21\times10^{-4}$ of the distance between the proton center and the electron center, i.e., of the Bohr radius, i.e., $2.880\,42\times10^{-14}$ m from the proton center. Rutherford's experiment [☞ p. 45] showed that the atom nucleus must be smaller than about 2.7×10^{-14} m – which we certainly expect for the simplest, hydrogen nucleus, with a single proton. Thus, the coordinate origin of the CM-system is just outside the proton that forms the atom nucleus, and more precise analyses must take into account the complementary motion of the proton in the CM-system.

for which the solutions are well known:

$$E_n = -\frac{1}{2}\alpha_e^2\, m_e c^2\, \frac{1}{n^2}, \qquad n = 1,2,3,\dots \tag{4.8b}$$

$$\Psi_{n,\ell,m}(\vec{r},t) = \sqrt{\Big(\frac{2}{n\,a_0}\Big)^3 \frac{(n-\ell-1)!}{2n[(n+1)!]^3}}\; e^{-r/(na_0)} \Big(\frac{2r}{na_0}\Big)^{\ell} L^{2\ell+1}_{n-\ell-1}\Big(\frac{2r}{na_0}\Big)\, Y^m_\ell(\theta,\phi), \tag{4.8c}$$

where

$$L^q_{k-q}(x) := (-1)^q \frac{\mathrm{d}^q}{\mathrm{d}x^q}\Big[e^x \frac{\mathrm{d}^k}{\mathrm{d}x^k}(e^{-x}x^k)\Big] \qquad \text{are the Laguerre polynomials,} \tag{4.8d}$$

$$a_0 := \frac{4\pi\epsilon_0\hbar^2}{m_e e^2} = 0.529\times 10^{-10}\,\mathrm{m} \qquad \text{is the Bohr radius.} \tag{4.8e}$$

Recall that the complex phase – and so also the sign – of the wave-functions $\Psi_{n,\ell,m}(\vec{r},t)$ is not measurable [☞ Chapter 5], so different Authors may use different sign conventions in these definitions (4.8c)–(4.8d) for convenience in some particular computations.

A discussion of this solution for the hydrogen atom may be found in every quantum mechanics textbook, and it is well known that Bohr's spectrum of the hydrogen atom (4.8b) is degenerate: Since the energy depends only on the principal quantum number n, states with different (permitted) values of the quantum numbers ℓ, m (and spin, s and m_s) have the same energy. Since [☞ Appendix A.3]

$$n = 1,2,3,\dots, \qquad \ell = 0,1,2,\dots(n-1), \qquad |m| \leqslant \ell,\ m \in \mathbb{Z}, \qquad s = \pm\tfrac{1}{2}, \tag{4.9}$$

it follows that the number of states with the same energy equals

$$\sum_{\ell=0}^{n-1}\sum_{m=-\ell}^{\ell} 2 = 2\sum_{\ell=0}^{n-1}(2\ell+1) = 2n^2, \tag{4.10}$$

where the factor 2 stems from two possible values of spin. Since the potential is central, i.e., it depends only on the distance between the center of the Coulomb field and the electron that moves in that field, the system manifestly has rotational symmetry. In 3-dimensional space, rotation transformations form the *Spin*(3) group. This symmetry would explain the independence of the energy from the quantum number m (quantifying the direction of the angular momentum) and spin, but not the independence from ℓ, which quantifies the *intensity* of the angular momentum.[2]

Indeed, the hydrogen atom – and more generally, the Coulomb, i.e., the Kepler problem – has another symmetry, generated by the components of the so-called Laplace–Runge–Lenz vector:

$$\text{for } V(r) = -\frac{\varkappa}{r}, \qquad \vec{A} := \vec{p}\times\vec{L} - m_e\frac{\varkappa}{r}\vec{r}. \tag{4.11}$$

It may be shown that the Cartesian components of the Laplace–Runge–Lentz vector commute with the Hamiltonian H in equation (4.1). In turn, the dimensionless operators

$$L_i := \frac{1}{\hbar}(\vec{r}\times\vec{p})_i = -i\,\varepsilon_{ij}{}^k\, x^j \frac{\partial}{\partial x^k} \tag{4.12}$$

[2] The continuous group of rotations is generated by operators of the dimensionless angular momentum (4.12): Each rotation may be represented as the result of the action of the operators $R(\vec{\varphi}) := \exp\{\varphi^i L_i\}$, which change the direction of the atom. Since rotations are *symmetries*, it follows that the result of a rotation is not measurable, and the energy of the atom cannot depend on its direction. However, neither of these operators changes ℓ: $R(\vec{\varphi})Y^m_\ell = \sum_\mu c^m_\mu Y^\mu_\ell$. Therefore, the rotation symmetry does not explain the fact that states of the hydrogen atom with different ℓ nevertheless have the same energy (4.8b).

satisfy the relations

$$[L_j, L_k] = i\varepsilon_{jk}{}^{l} L_l, \qquad [L_j, A_k] = -i\varepsilon_{jk}{}^{l} A_l, \qquad [A_j, A_k] = \pm i\varepsilon_{jk}{}^{l} L_l, \ \text{for} \ \left\{{E<0, \atop E>0;}\right. \tag{4.13}$$

where the operators

$$A_j = \frac{1}{\sqrt{2m_e H}} \left[\frac{\hbar}{2i} \varepsilon_j{}^{kl} \left(\frac{\partial}{\partial x^k} L_l + L_l \frac{\partial}{\partial x^k} \right) - \frac{m_e\,\varkappa}{\hbar}\, \hat{\mathbf{e}}_j \right] \tag{4.14}$$

are the components of the quantum dimensionless Laplace–Runge–Lenz vector, normalized by the energy of the stationary state upon which the operators A_j act. The structure of the symmetry group generated by the operators L_j and A_j depends on the choice of the sign in the third of the commutator relations (4.13). When acting on bound states (for which $E < 0$), the commutator relations (4.13) specify the continuous group $Spin(4)$; [☞ Section A.5]. When acting on $e^- + p^+ \to e^- + p^+$ scattering states (for which $E > 0$), L_j and A_j generate the group $Spin(1,3)$. The operators A_i change ℓ through its full range $\ell \in [0, n-1]$ while the operators L_j change m through its full range $m \in [-\ell, \ell]$, both in unit increments. This extended $Spin(4)$, i.e., $Spin(1,3)$, symmetry of the hydrogen atom implies that the energy, as obtained by Bohr's formula (4.8b), does not depend on ℓ, m and m_s.

Thus, the $Spin(4)$ symmetry fully explains the number and classification of hydrogen atom bound states. The lesson from this simple and very well known system is that symmetries may well be of great use in listing and classifying the possible states – very similar to the situation in Section 2.3.12.

Of course, as we know from the discussion in standard quantum mechanics textbooks, Bohr's formula (4.8b) is not the end of the story, and the value for energy acquires "corrections" owing to several different physical phenomena which we briefly review in the subsequent sections. It is well known that these corrections split the degeneracy, and so "break" the approximate $Spin(4)$ symmetry of the hydrogen atom to $Spin(3) \subset Spin(4)$. The quark model uses this correlation in reverse, and deduces some of the details of the quark dynamics from the hierarchy of approximate symmetries.

4.1.2 Relativistic corrections

The approach in Section 4.1.1 may easily be amended using stationary-state perturbation theory, and this "corrects" the energy values (4.8b). One of these corrections stems from the fact that the non-relativistic physics is of course only an approximation, and that the relativistic kinetic energy is

$$\begin{aligned} T_{\text{rel}} &= m_e c^2 \left[\sqrt{1 + (\vec{p}/m_e c)^2} - 1 \right] = m_e c^2 \sum_{k=1}^{\infty} \binom{\frac{1}{2}}{k} \left(\frac{\vec{p}^2}{m_e^2 c^2} \right)^k, \\ &\approx \frac{\vec{p}^2}{2m_e} - \frac{(\vec{p}^2)^2}{8m_e^3 c^2} + \frac{(\vec{p}^2)^3}{16 m_e^5 c^4} - \cdots . \end{aligned} \tag{4.15}$$

The first and second relativistic corrections to the Hamiltonian are then represented by the operators

$$H'_{\text{rel}} := -\frac{\hbar^4}{8m_e^3 c^2} (\vec{\nabla}^2)^2, \qquad H''_{\text{rel}} := +\frac{\hbar^6}{16 m_e^5 c^4} (\vec{\nabla}^2)^3, \tag{4.16}$$

and the first-order perturbative correction of the energy is

$$E_n^{(1,r_1)} = \langle n | H'_{\text{rel}} | n \rangle = \int \mathrm{d}^3\vec{r}\; \Psi^*_{n,\ell,m}(\vec{r})\, H'_{\text{rel}}\, \Psi_{n,\ell,m}(\vec{r}). \tag{4.17}$$

This is calculable by simple substitution of the wave-functions, the application of $H'_{\rm rel}$ on $\Psi_{n,\ell,m}(\vec{r})$, and computation of the ensuing integral. However, it is faster to use that

$$\frac{\hbar^2}{2m_e}\vec{\nabla}^2 = V(r) - H, \tag{4.18}$$

so that

$$H'_{\rm rel} = -\frac{1}{2m_e c^2}\big[V(r) - H\big]^2 = -\frac{1}{2m_e c^2}\big[V^2 - HV - VH + H^2\big], \tag{4.19}$$

$$\begin{aligned} H''_{\rm rel} &= +\frac{1}{2m_e^2 c^4}\big[V(r) - H\big]^3 \\ &= -\frac{1}{2m_e^2 c^4}\big[V^3 - VHV - V^2H + VH^2 - HV^2 + H^2V + HVH - H^3\big], \end{aligned} \tag{4.20}$$

the matrix elements of which are easier to compute, since $H^\dagger = H$ acts equally on $|n,\ell,m,m_s\rangle$ as well as on $\langle n,\ell,m,m_s|$, producing its eigenvalue, E_n, given by the relation (4.8b).

For second-order corrections, one ought to compute both the second-order perturbation correction stemming from $H'_{\rm rel}$ as well as the first-order perturbation correction stemming from $H''_{\rm rel}$. For the first of these two contributions, we must re-diagonalize the basis $|n,\ell,m,m_s\rangle$ to avoid the $\frac{1}{0}$-divergences in the formula (1.19), whereas the second contribution requires a little more attention for the term $\langle VHV\rangle$ owing to the fact that $(\vec{\nabla}^2\frac{1}{r}) = -4\pi\,\delta(r)$. However, this pointillist contribution is limited to the cases $\ell = 0 = m$, *[✎ why?]* which are not hard to compute separately. In the general case, we'll need the results [407, 471, 242, 472, 29, 328, 362, 363, 360, 3, for example]:

$$\langle r^2\rangle = n^4 a_0^2\Big[1 + \tfrac{3}{2}\big(1 - \tfrac{\ell(\ell+1)-\frac{1}{3}}{n^2}\big)\Big], \tag{4.21a}$$

$$\langle r\rangle = n^2 a_0\Big[1 + \tfrac{1}{2}\big(1 - \tfrac{\ell(\ell+1)}{n^2}\big)\Big], \tag{4.21b}$$

$$\langle r^{-1}\rangle = \frac{1}{n^2\, a_0}, \tag{4.21c}$$

$$\langle r^{-2}\rangle = \frac{1}{(\ell+\frac{1}{2})n^3\, a_0^2}, \tag{4.21d}$$

$$\langle r^{-3}\rangle = \frac{1}{\ell(\ell+\frac{1}{2})(\ell+1)n^3\, a_0^3}. \tag{4.21e}$$

The first-order perturbative energy correction stemming from $H'_{\rm rel}$ is

$$\begin{aligned} E_n^{(1,r_1)} &= \langle n,\ell,m,m_s|H'_{\rm rel}|n,\ell,m,m_s\rangle = -\frac{1}{2m_e c^2}\Big[\langle V^2\rangle - 2E_n^{(0)}\langle V\rangle + (E_n^{(0)})^2\Big] \\ &= -\frac{1}{2}\alpha_e^4\, m_e c^2\,\frac{1}{4n^4}\Big[\frac{4n}{(\ell+\frac{1}{2})} - 3\Big]. \end{aligned} \tag{4.22}$$

4.1.3 Magnetic corrections

Besides their electric charges, the electron and the proton both also have an intrinsic (dipole) magnetic field: $\vec{\mu}_e$ and $\vec{\mu}_p$, respectively. Since the electron and the proton move one with respect to the other, the motion of the electron produces a current that, by the Biot–Savart law, creates a magnetic field proportional to the angular momentum of the electron about the proton, $\vec{B} \propto \vec{L}$, and this magnetic field interacts with the intrinsic magnetic dipole of the proton. Of course, it would be nonsense saying that this same magnetic field, caused by the motion of the electron, also interacts with the intrinsic magnetic dipole field of the electron: In its own coordinate system, the electron of course does not move, and so produces neither an electric current nor a magnetic field.

However, in the electron's rest-frame it is the proton that moves. This then produces a current and a corresponding magnetic field $\vec{B}'$, which interacts with the intrinsic magnetic dipole of the electron. To relate $\vec{B}'$ and $\vec{B}$, one must transform the vector of this "rotating" magnetic field from the electron's coordinate system into the proton's. Since the electron's coordinate system rotates about the proton, one must iterate this transformation from moment to infinitesimally adjacent moment, approximated by successive infinitesimal Lorentz boosts. The resulting effect is called Thomas precession and provides the relation $\vec{B}' = \frac{1}{2}\vec{B}$ [296].

With two intrinsic magnetic dipoles $\vec{\mu}_e, \vec{\mu}_p$, and the "orbital" magnetic field $\vec{B}$, there then exist three additions to the hydrogen atom Hamiltonian:

$$H_{S_eO} = -\vec{\mu}_e \cdot (\tfrac{1}{2}\vec{B}), \qquad H_{S_pO} = -\vec{\mu}_p \cdot \vec{B},$$
$$H_{S_eS_p} = -\frac{\mu_0}{4\pi}\Big[\Big(3(\vec{\mu}_e\cdot\hat{r})(\vec{\mu}_p\cdot\hat{r}) - \vec{\mu}_e\cdot\vec{\mu}_p\Big)\frac{1}{r^3} + \frac{8\pi}{3}\vec{\mu}_e\cdot\vec{\mu}_p\,\delta^3(\vec{r})\Big], \tag{4.23}$$

taking the dipole–dipole interaction term from standard texts such as Ref. [296].

Digression 4.1 One of the original motivations for the Abraham–Lorentz model of the electron was also the attempt to explain – with classical physics – the origin of the electron's intrinsic dipole moment. In this model, the electron was supposed to be a teeny electrically charged sphere. If that sphere rotated, the charge distribution on the sphere would also rotate and so produce a circular current, which would in turn produce a magnetic field by the Biot–Savart law. This is the source of the idea that the electron rotates about its own axis, has *spin* (= intrinsic angular momentum), and that its intrinsic magnetic dipole moment is a consequence of this rotation and proportional to this spin. For a classical rotating electric charge q for which the charge and mass (m) distribution coincide, the magnetic dipole is proportional to the angular momentum:

$$\vec{\mu} = \frac{q}{2m}\vec{L}, \tag{4.24a}$$

and $\mu_e := e/2m_e$ is called the Bohr magneton (for the electron).

In fact, this identification is completely backwards: It is the electron's magnetic moment that may be measured and so has a *real* physical meaning; the rotation of the electron about its own axis – *spin* – is a fictitious quantity, *defined* through the relation (4.24a) in terms of the intrinsic magnetic moment. This backwards-engineered explanation stems from G. E. Uhlenbeck and S. A. Goudsmit, who measured the magnetic dipole moment of the electron in 1925, then concluded that this magnetic moment stems from a rotation of the electron about its own axis [528]; all along, they assumed the electron to be represented as an electrically charged sphere, following the Abraham–Lorentz model [☞ Digression 3.13 on p. 123].

Besides, the operators L_j that generate rotations close the $Spin(3) \cong SU(2)$ algebra, which has two classes of representations: tensors and spinors [☞ Digression A.2 on p. 465]. It is easy to show that 360°-rotations around any axis map tensor functions into themselves, but spinors into their negative multiple. Because of this property, physically observable quantities cannot be spinors. Since all real functions over the phase space are observables in classical physics, it follows that there is no room for spinors in classical physics. In quantum physics, however, wave-functions (and abstract state vectors in the

Hilbert space) are not directly observable, and so can be spinorial representations. In this sense, half-integral spin is an exclusively quantum-mechanical phenomenon.

It then also follows that the classical relation (4.24a), based on the fictive rotation of a fictive sphere in the Abraham–Lorentz model of the electron need not hold for the electron, which *is* a spin-$\frac{1}{2}$ particle.[3] Indeed, Dirac's relativistic theory of the electron provides for the electron's magnetic moment a result that is twice as large as the classical value, which is further corrected by quantum field theory effects:

$$\vec{\mu}_e = 2\Big[1 + \underbrace{\frac{\alpha_e}{2\pi} + \cdots}_{\text{quantum field theory}}\Big]\frac{(-e)}{2m_e}\,\vec{S}. \tag{4.24b}$$

Of course, if ever the electron turns out to show a structure, this relation will have to be revisited, just as the proton's magnetic moment is today determined using the fact that it is composed of three quarks, as well as a variable number of gluons (which hold those three quarks in the bound state) and virtual quark–antiquark pairs.

For our purposes, write [☞ relation (4.24b)]

$$\vec{\mu}_e = -g_e\mu_B\vec{S}_e, \qquad \mu_B := \frac{e}{2m_e}, \qquad g_e = 2.002\,319\,304\,361\,1(46) \approx 2, \tag{4.25}$$

$$\vec{\mu}_p = +g_p\mu_N\vec{S}_p, \qquad \mu_N := \frac{e}{2m_p}, \qquad g_p = 2.7928, \tag{4.26}$$

where μ_B and μ_N are, respectively, Bohr's (electron) and nucleon magnetic moments. Note that the electron "g-factor," g_e, is measured to a precision of 12 significant figures, and is in full agreement with the result of quantum electrodynamics [293]. The value $\frac{1}{2}(g_e-2)$ is also referred to as the "anomalous magnetic moment," in the sense that g_e deviates from the "bare" value of 2 in the Dirac theory of the electron; this should not be confused with the (quantum) anomalies mentioned elsewhere in this book [☞ Section 7.2.3].

Inserting the expressions for the magnetic dipole moments, the three additions (4.23) to the hydrogen atom energy become, after a little algebra (see Refs. [362, 363, 407, 471, 328, 242, 472, 29, 360, 3] for example),

$$H_{S_eO} = -\Big(\frac{g_e(-e)}{2m_e}\hbar\vec{S}_e\Big)\cdot\Big(\frac{1}{2}\frac{e}{4\pi\epsilon_0 m_e r^3}\hbar\vec{L}\Big) \approx \frac{e^2}{4\pi\epsilon_0}\frac{\hbar^2}{2m_e^2c^2}\frac{1}{r^3}\,\vec{L}\cdot\vec{S}_e, \tag{4.27a}$$

$$H_{S_pO} = \frac{g_p\,e^2}{4\pi\epsilon_0}\frac{\hbar^2}{m_e m_p c^2}\frac{1}{r^3}\,\vec{L}\cdot\vec{S}_p, \tag{4.27b}$$

$$H_{S_eS_p} \approx \frac{g_p\,e^2}{4\pi\epsilon_0}\frac{\hbar^2}{m_e m_p c^2}\Big[\Big(3(\vec{S}_e\cdot\hat{r})(\vec{S}_p\cdot\hat{r}) - \vec{S}_e\cdot\vec{S}_p\Big)\frac{1}{r^3} + \frac{8\pi}{3}\vec{S}_e\cdot\vec{S}_p\,\delta^3(\vec{r})\Big], \tag{4.27c}$$

where "≈" indicates the use of the approximation (4.25). Comparing the constant pre-factors (which all have the same units), one expects the latter two contributions to be of the same order of magnitude, and about $m_p/2g_pm_e \approx 329$ times smaller than the first, so the first perturbation dominates.

[3] The adjective "spin-j" simply specifies that the particle (or wave, or field) has a characteristic orientation of sorts, so that its representative, its wave-function, transforms as one of the representations of the angular momentum algebra, $|j,m\rangle$ with $|m| \leqslant j$; the special case $j=0$ denotes rotation invariance [☞ Appendix A.3]. Conceptually, a "spin-j field" is simply a generalization of the electric and magnetic spin-1 (vector) fields, and "spin-j particles" are the quanta of spin-j fields.

Furthermore, one expects that $\langle\frac{1}{r^3}\rangle \sim (a_0)^{-3}$ and $\langle L\cdot\vec{S}_e\rangle = O(1)$, so that the first of these three contributions is of the order

$$\langle H_{S_eO}\rangle = \frac{e^2}{4\pi\epsilon_0}\frac{\hbar^2}{2m_e^2c^2}\left\langle\frac{1}{r^3}\vec{L}\cdot\vec{S}_e\right\rangle \sim \frac{\alpha_e\hbar^3}{2m_e^2c}\cdot\frac{1}{a_0^3} = \frac{\alpha_e^4 m_e c^2}{2}. \tag{4.28}$$

This result is of the same order as the relativistic correction (4.22), although its origins are completely different; cf. equation (4.16). Using the result (4.21e), we have

$$E_n^{(1,SO)} = \alpha_e^4 m_e c^2\frac{j(j{+}1) - \ell(\ell{+}1) - \frac{3}{4}}{4n^3\ell(\ell{+}\frac{1}{2})(\ell{+}1)} = \alpha_e^4 m_e c^2\frac{1}{4n^3}\begin{cases}\frac{1}{(\ell+\frac{1}{2})(\ell+1)},\\ -\frac{1}{\ell(\ell+\frac{1}{2})},\end{cases} \tag{4.29}$$

where we used the relation

$$\vec{J} := \vec{L} + \vec{S} \qquad \Rightarrow \qquad \vec{L}\cdot\vec{S} = \tfrac{1}{2}\big[J^2 - L^2 - S^2\big]. \tag{4.30}$$

The corrections (4.29) and (4.22) are indeed very similar, and add up:

$$E_n^{\rm fs} = E_n^{(1,r_1)} + E_n^{(1,SO)} = -\alpha_e^4\, m_e c^2\frac{1}{4n^4}\left[\frac{2n}{(j+\frac{1}{2})} - \frac{3}{2}\right], \qquad \begin{cases} j = \ell + \frac{1}{2},\\ j = \ell - \frac{1}{2};\end{cases} \tag{4.31}$$

providing for the so-called fine structure of the hydrogen atom spectrum.

— ❦ —

It remains to compare the contributions:[4]

$$E_n^{(1,r_2)} = \langle H''_{\rm rel}\rangle \sim \frac{1}{m_e^2c^4}\left\langle\left(\frac{e^2}{4\pi\epsilon_0 r}\right)^3\right\rangle \sim \frac{1}{m_e^2c^4}\frac{(\alpha_e\hbar c)^3}{n^3 a_0^3} \sim \frac{\alpha_e^6\, m_e c^2}{n^3}; \tag{4.32}$$

$$E_n^{(2,r_1)} = \sum_{n'\cdots\neq n\cdots}\frac{|\langle n',\cdots|H'_{\rm rel}|n,\cdots\rangle|^2}{E_n^{(0)} - E_m^{(0)}} \sim \frac{|E_n^{(1,r_1)}|^2}{|E_n^{(0)}|} \sim \frac{(\alpha_e^4 m_e c^2/n^3)^2}{\alpha_e^2 m_e c^2/n^2} \sim \frac{\alpha_e^6 m_e c^2}{n^4}; \tag{4.33}$$

$$E_n^{(1,S_pO)} = \langle H_{S_pO}\rangle = \frac{g_p\, e^2}{4\pi\epsilon_0}\frac{\hbar^2}{m_e m_p c^2}\left\langle\frac{1}{r^3}\vec{L}\cdot\vec{S}_p\right\rangle \sim \frac{g_p\,\alpha_e\hbar^3}{m_e m_p c}\cdot\frac{1}{n^3a_0^3} \sim g_p\left(\frac{m_e}{m_p}\right)\frac{\alpha_e^4 m_e c^2}{n^3}; \tag{4.34}$$

$$E_n^{(1,S_eS_p)} = \langle H_{S_eS_p}\rangle \sim \frac{g_p\, e^2}{4\pi\epsilon_0}\frac{\hbar^2}{m_e m_p c^2}\left\langle\vec{S}_e\cdot\vec{S}_p\frac{1}{r^3}\right\rangle \sim g_p\left(\frac{m_e}{m_p}\right)\frac{\alpha_e^4 m_e c^2}{n^3}. \tag{4.35}$$

It is not hard to see that

$$E_n^{(1,r_2)} : E_n^{(2,r_1)} : E_n^{(1,S_pO)} : E_n^{(1,S_eS_p)} \approx n\alpha_e^2 : \alpha_e^2 : g_p\left(\frac{m_e}{m_p}\right) : g_p\left(\frac{m_e}{m_p}\right). \tag{4.36}$$

Since $\alpha_e^2 \approx 5.33\times 10^{-5}$ and $g_p(\frac{m_e}{m_p}) \approx 1.52\times 10^{-3}$, the last two contributions are about 28 times larger than the first two. Therefore, we neglect the first two of these contributions in comparison with the latter two, the sum of which gives

$$E_n^{\rm hfs} = E_n^{(1,S_eS_p)} + E_n^{(1,S_pO)} = \left(\frac{m_e}{m_p}\right)\alpha_e^4\, m_e c^2\,\frac{g_p}{2n^3}\frac{\pm 1}{(f+\frac{1}{2})(\ell+\frac{1}{2})}, \qquad \begin{cases} f = j + \frac{1}{2},\\ f = j - \frac{1}{2};\end{cases} \tag{4.37}$$

and similarly provide for the so-called *hyperfine* structure of the hydrogen atom spectrum. This last result introduces the so-called f-spin: $\vec{F} := \vec{J} + \vec{S}_p = \vec{L} + \vec{S}_e + \vec{S}_p$, as a vector sum of all three angular momenta.

[4] From Equation (4.27c), it is actually the last, δ-function part that contributes to the result (4.35) and produces the "21 cm hydrogen line," well-known in microwave radio astronomy [407].

When $\ell = 0$, the electron and proton spins are either parallel or antiparallel, giving the so-called triplet and singlet states: Denote $\vec{Z} := \vec{S}_e + \vec{S}_p$, so that the eigenvalue of Z^2 equals $z(z+1)$. When the electron and proton spins are parallel, $z = 1$ and $m_z = \pm 1, 0$; for antiparallel spins $z = 0$ and $m_z = 0$. In the result (4.37), $\ell = 0$, $j = s_e$ and $f = z$, so the numerical factor becomes

$$\frac{\pm 1}{(f+\frac{1}{2})(\ell+\frac{1}{2})} = \begin{cases} +\frac{4}{3}, \\ -4; \end{cases} \qquad \begin{cases} z = 1 \ \text{(triplet)}, \\ z = 0 \ \text{(singlet)}. \end{cases} \tag{4.38}$$

Owing to this split in the energies, a transition is possible (for the same n) between these states, emitting a photon of energy equal to this difference in energies, and a wavelength of 21.0807 cm (for $n = 1$). This result (to first perturbative order) differs less than 1% from the precisely measured wavelength of 21.106 114 054 13 cm, very well known in microwave astronomy.

4.1.4 The Lamb shift

The corrections in the previous Sections 4.1.2–4.1.3 were computed with standard methods of non-relativistic quantum mechanics, in the approach that might be called *semi-quantum*, since the particles (electron and proton) receive a quantum treatment, while the binding (electromagnetic) field is treated classically.

There exist, however, measurable consequences of electromagnetic field quantization, for the computation of which field theory is needed. Consider here only qualitatively the following three Feynman diagrams:

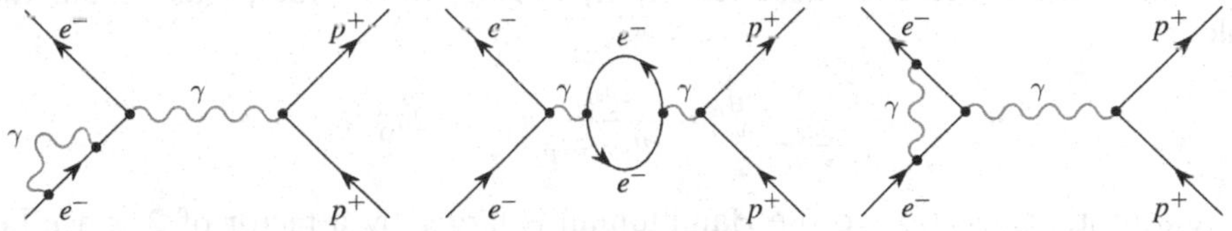

(4.39)

The first of these Feynman diagrams describes the fact that, during "free" motion through "empty vacuum," the electron interacts with a virtual photon, which changes its mass. The second diagram shows the reciprocal effect, whereby the photon mediating the interaction between the electron and the proton *en route* interacts with a virtual e^-–e^+ pair (is absorbed and then re-emitted by the pair), which effectively screens the electric charge of the nucleus and the electron in the orbit. The third diagram describes a correction to the nature of interaction of the orbiting electron and the mediating photon; this effectively changes the magnetic dipole moment of the electron and contributes to the gyromagnetic ratio (4.24b) by an amount proportional to α_e.

Suffice it here just to cite the resulting correction [407, 243, 150]:

$$E_n^{(QED)} = \begin{cases} \alpha^5 m_e c^2 \frac{1}{4n^3}\, k(n,0) & \ell = 0; \\ \alpha^5 m_e c^2 \frac{1}{4n^3}\Big[k(n,\ell) \pm \frac{1}{\pi(j+\frac{1}{2})(\ell+\frac{1}{2})}\Big], \quad j = \ell \pm \frac{1}{2}, & \ell \neq 0, \end{cases} \tag{4.40}$$

where $k(n,0)$ varies mildly from about 12.7 for $n = 1$ to about 13.2 for $n \to \infty$, while $k(n,\ell) \lesssim 0.05$ and also varies very mildly with n, ℓ. Note that, unlike the corrections considered in Sections 4.1.2–4.1.3 that are all proportional to an even power of the fine structure constant α_e, this quantum-electrodynamical contribution is proportional to α_e^5.

This contribution to the energies is called the Lamb shift. Comparing

$$E_n^{(1,r_2)} : E_n^{(2,r_1)} : E_n^{(1,SpO)} : E_n^{(1,S_eS_p)} : E_n^{(QED)} \approx n\alpha_e^2 : \alpha_e^2 : g_p\Big(\frac{m_e}{m_p}\Big) : g_p\Big(\frac{m_e}{m_p}\Big) : \alpha_e, \tag{4.41}$$

$$\approx (5.33\times 10^{-5}\cdot n) : (5.33\times 10^{-5}) : (1.52\times 10^{-3}) : (1.52\times 10^{-3}) : (7.30\times 10^{-3}). \tag{4.42}$$

This simple dimensional analysis suggests the Lamb shift to be almost five times *larger* than the hyperfine splitting; the precise numerical results are however comparable. For example, the dipole–dipole interaction (4.35) produces the "21 cm hydrogen line" at about 1.42 GHz, while the Lamb shift permits the $2^2p_{1/2} \to 2^2s_{1/2}$ transition at about 1.06 GHz.

4.1.5 Positronium

The analysis of the hydrogen atom in Sections 4.1.1–4.1.4 is easy to adapt to many two-particle bound states, where the proton or the electron (or both) are replaced by other particles. Such systems are collectively called exotic atoms. Such systems include: muonic hydrogen $(p^+\mu^-)$, pionic hydrogen $(p^+\pi^-)$, muonium (μ^+e^-), etc. Amongst these, consider positronium, (e^+e^-). Together with the hydrogen atom, this gives a good foundation for understanding "quarkonium," i.e., mesons: positronium is an adequate template for mesons composed of a quark and an antiquark of roughly the same mass, while the hydrogen atom is an adequate template for mesons where the masses of the quark and the antiquark significantly differ.

Since $m_{e^+} = m_{e^-}$, the reduced mass is $\frac{m_{e^+}m_{e^-}}{m_{e^+}+m_{e^-}} = \frac{1}{2}m_e$. By the simple $m_e \mapsto \frac{1}{2}m_e$ substitution, we obtain the Bohr-like formula:

$$E_n(e^+e^-) = \tfrac{1}{2}E_n(H) = -\alpha_e^2 m_e c^2 \frac{1}{4n^2}. \tag{4.43}$$

The wave-functions look identical to those for the hydrogen atom (4.8c), except that the Bohr radius is doubled:

$$a_0^{(pos)} = \frac{\frac{m_e m_p}{m_e+m_p}}{\frac{m_e m_e}{m_e+m_e}} a_0^{(H)} = \frac{2\,m_p}{m_e+m_p} a_0^{(H)} \approx 2a_0^{(H)}. \tag{4.44}$$

The first relativistic correction to the Hamiltonian is larger by a factor of 2, since both the electron and the positron contribute equally. However, $\langle(\vec{p}^2)^2\rangle \propto (m_e c)^4$, which is then diminished by a factor of $(\frac{1}{2})^4$ because of the smaller reduced mass. In total, the relativistic correction for positronium is an eighth of the corresponding correction for the hydrogen atom.

Significant differences from the contributions that provide the hyperfine structure to the spectrum of the hydrogen atom are: the ratio $\frac{m_{e^-}}{m_{e^+}} = 1$, the values $g_{e^+} = g_{e^-}$, and that the Thomas precession is now symmetric. The contributions analogous to (4.37) are now of the same order of magnitude as the fine structure contributions (4.31). The Lamb shift remains suppressed by a factor of α_e as compared with the contributions analogous to (4.31) and (4.37).

There exist, however, also two entirely novel effects, with no analogues in the analysis of the hydrogen atom:

Field latency In positronium, the center of the Coulomb field that acts on the electron moves with the positron, and *vice versa*. Since the changes in the Coulomb field propagate with the finite speed of light, this "tarrying" effect of field *latency* must be taken into account. This field latency may be computed in classical electrodynamics, and its contribution to the Hamiltonian is [59]

$$H_{\text{lat}} = -\frac{e^2}{4\pi\epsilon_0}\frac{1}{2m_e^2c^2}\frac{1}{r}\big(p^2 + (p\cdot\hat{r})^2\big), \tag{4.45}$$

which gives the first-order perturbative contribution:

$$E_n^{(\text{lat})} = \langle H_{\text{lat}}\rangle = \alpha_e^4 m_e c^2 \frac{1}{2n^3}\left[\frac{11}{32n} - \frac{2+\epsilon}{\ell+\frac{1}{2}}\right], \tag{4.46a}$$

where ϵ is a function of the electron and the positron spins:

$$\epsilon = \begin{cases} 0 & \text{for } j = \ell, \qquad s = 0, \\ -\frac{3\ell+4}{(\ell+1)(2\ell+3)} & \text{for } j = \ell+1, \\ \frac{1}{\ell(\ell+1)} & \text{for } j = \ell, \\ \frac{3\ell-1}{\ell(2\ell-1)} & \text{for } j = \ell-1, \end{cases} \Bigg\} \; s = 1. \tag{4.46b}$$

All spins contribute equally in positronium, so it seems reasonable to define $\vec{S} := \vec{S}_{e^-} + \vec{S}_{e^+}$, where $\vec{S}^2$ has eigenvalues $s(s+1)$ with $s = 0, 1$. Then we define $\vec{J} := \vec{L} + \vec{S}$, where $\vec{L}^2$ and $\vec{J}^2$ have eigenvalues $\ell(\ell+1)$ and $j(j+1)$.

Virtual annihilation In positronium, the electron and the positron may temporarily annihilate into a virtual photon which then, before the time alotted by Heisenberg indeterminacy relations, decays into an electron and a positron. Since the electron and the positron must be at the same location for this process, the contribution of the virtual annihilation must be proportional to $|\Psi(0)|^2$, and so can happen only when $\ell = 0$. *[✎ Why?]* Then, since the photon has spin 1, positronium also must have spin 1, i.e., it must be in the triplet state with $s = 1$ and parallel spins. The contribution to the energy of positronium is [243]

$$E_n^{(ann)} = \alpha_e^4 m_e c^2 \frac{1}{4n^3}, \qquad \ell = 0,\ s = 1. \tag{4.47}$$

Note that both new contributions (4.46) and (4.47) are of the same order of magnitude as the analogues of the fine and hyperfine structure contributions. The Lamb shift, as well as the analogues of the corrections (4.32)–(4.33) are then consistently negligible in comparison with the analogues of (4.31), (4.37), (4.40) and (4.46). The Lamb shift was shown to be $O(\alpha_e^5)$, and so contributes less than 1 % of the listed contributions, which are all $O(\alpha_e^4)$.

Real annihilation Positronium is an unstable bound state, as the comprising parts may also really annihilate and produce two or more photons. Just as in the previous discussion, since the electron and the positron must be at the same place to annihilate each other, the decay rate must be proportional to $|\Psi(\vec{0}, t)|^2$. For a two-photon decay we have the Feynman diagram

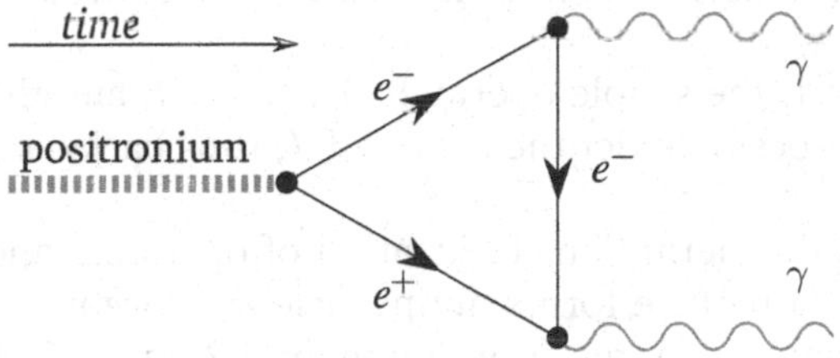

Strictly speaking, only one of the right-hand vertices is an annihilation; here we pick the lower one. The computation shows the result to be independent of this choice.

(4.48)

and it follows that $\mathfrak{M}$ must be proportional to $\frac{e^2}{4\pi\epsilon_0}$ and then also $\hbar^{-1}c^{-1}$, since $\mathfrak{M}$ is dimensionless. We will compute (5.179) in Section 5.3.2: $\mathfrak{M} = -\frac{4e^2}{\epsilon_0 \hbar c}$. Two-photon annihilation of positronium may also be interpreted as an $e^- + e^+ \to 2\gamma$ scattering, for which the effective cross-section, following the general result (3.127),

$$\sigma = \int d^2\Omega \left(\frac{\hbar c}{8\pi}\right)^2 \frac{|\mathfrak{M}|^2}{(E_{e^-} + E_{e^+})^2} \left|\frac{\vec{p}_f}{\vec{p}_i}\right| = 4\pi \frac{(\alpha_e \hbar c)^2}{(m_e c^2)^2} \frac{c}{v} = 4\pi \frac{\alpha_e^2 \hbar^2}{m_e^2 c v}, \tag{4.49}$$

where v is the relative speed of the electron and the positron. By Conclusion 3.2 on p. 113, we have

$$\Gamma = \sigma v \, |\Psi(\vec{0}, t)|^2 = \left(4\pi \frac{\alpha_e^2 \hbar^2}{m_e^2 c v}\right)\left(\frac{1}{\pi a_0^{(pos)} n^3}\right) = \left(4 \frac{\alpha_e^2 \hbar^2}{m_e^2 c v}\right)\left(\frac{\alpha_e (\frac{1}{2} m_e) c}{\hbar n}\right)^3 = \frac{\alpha_e^5 m_e c^2}{2 \hbar n^3}, \tag{4.50}$$

where the familiar expression for the $\ell = 0$ wave-function of the hydrogen atom is adapted for positronium by replacing the reduced mass, $m_e \to \frac{1}{2}m_e$. The positronium lifetime is then

$$\tau = \frac{1}{\Gamma} = \frac{2\hbar\, n^3}{\alpha_e^5 m_e c^2} = (1.24494\times 10^{-10}\,\mathrm{s}) \times n^3, \tag{4.51}$$

which is in excellent agreement with experiments [309].

4.1.6 Exercises for Section 4.1

✎ ***4.1.1*** *Compute the second-order perturbative energy correction E_n due to H'_{rel}.*

✎ ***4.1.2*** *Compute the second-order perturbative energy correction E_n due to H'_{S_eO}.*

✎ ***4.1.3*** *Compute the first-order perturbative energy correction E_n due to H''_{rel}.*

✎ ***4.1.4*** *Why can the electron and the positron making up positronium not annihilate into a single photon, i.e., why does the annihilation result in at least two photons?*

4.2 Finite symmetries

Symmetries with the structure of finite groups are widely used in solid state physics and crystallography. There are many such groups, and their structures may be very involved, and the applications very detailed and technically demanding.

In relativistic field theory, however, we are in general interested only in three rather simple finite symmetries:[5]

Parity P, which may be thought of simply as the mirror reflection of *one* of the Cartesian coordinates. In 3-dimensional space, this operation may always be followed by a 180° rotation in the mirror plane, which collectively flips the sign of all three coordinates. We then typically use this "more democratic" version of the parity operation: $P : \vec{r} \to -\vec{r}$ as well as $P : \vec{p} \to -\vec{p}$.

Time reversal T, which may be conceived classically as the simple operation $T : t \to -t$, and the physical meaning of which is simply that the process, under the action of T, runs backward in time.

Charge conjugation C, which may be thought of as the Hermitian conjugation of operators and (wave-)functions, and which physically swaps a particle for its antiparticle and *vice versa*. "Charge" here, foremost, means the electromagnetic charge (see Comment 5.2 on p. 169), but also the color in quantum chromodynamics, the *weak isospin* in weak nuclear interactions, and any and all charges related to symmetries other than spacetime coordinate transformations; see Chapters 5–7.

All three symmetries are of order 2, i.e., their successive applications (as defined here) result in the identity

$$P^2 = \mathbb{1}, \qquad T^2 = \mathbb{1}, \qquad C^2 = \mathbb{1}. \tag{4.52}$$

However, the application in quantum theory requires a little more care, as indicated in the subsequent three sections.

---

[5] As in Conclusion 2.6 on p. 78, the characteristics of the abstract mathematical model are also assigned to the concrete physical system that the model faithfully represents. This makes the symmetries of a state in the Hilbert space of the system also symmetries of the represented concrete physical system when in that state.

4.2.1 Parity

In 1956, Tsung-Dao Lee and Chen-Ning Yang studied the so-called "τ–θ" problem: two strange mesons that by this time had become known as τ and θ,[6] had all the same characteristics, except the difference in their decays:

$$``\theta^{+}" \to \pi^{+} + \pi^{0}, \qquad ``\tau^{+}" \begin{cases} \nearrow \pi^{+} + \pi^{0} + \pi^{0} \\ \searrow \pi^{+} + \pi^{+} + \pi^{-} \end{cases} \tag{4.53}$$

The parity of a system of particles is the product of *intrinsic* parities of the individual particles times a factor $(-1)^{\ell}$, where ℓ is the total angular momentum of the system. It follows that the parity of the "θ"-particle is $+1$, and the parity of the "τ"-particle is -1, since the pion's parity is -1, and 2- and 3-pion states in the processes (4.53) have total angular momentum equal to the spin of the "θ^{+}"- and the "τ^{+}"-particles, i.e., $\ell = 0$. The existence of two particles that were identical in all except their parity characteristics was very unusual. Lee and Yang proposed that they are in fact one and the same particle, but that the *P*-symmetry is violated in these weak interactions. On a second glance, they realized that parity conservation has not been experimentally confirmed in weak processes, so they recommended several experimental tests.

The same year, Chien-Shiung Wu (known as "Madam Wu") successfully completed the first of such experiments, working with E. Ambler, R. W. Hayward, D. D. Hoppes and R. P. Hudson. This experiment proved that there really do exist processes in Nature that are not invariant under the action of the parity operation. In the β-decay,

$${}^{60}_{27}\mathrm{Co} \to {}^{60}_{28}\mathrm{Ni} + e^{-} + \bar{\nu}_e, \tag{4.54}$$

Madam Wu's group showed that most electrons are emitted in high correlation with the spin of the cobalt-60 nucleus. If $\vec{p}_e$ and $\vec{S}$, respectively, are the operators of the electron's linear momentum and the spin of the cobalt-60 nucleus, and $|\Psi\rangle$ the state of this nucleus before the decay, it follows that $\langle\Psi|\vec{p}_e{\cdot}\vec{S}|\Psi\rangle \neq 0$. Now, since parity P flips the sign of the linear momentum but not of spin, it follows that

$$\begin{aligned} \langle\Psi|\vec{p}_e{\cdot}\vec{S}|\Psi\rangle &= \langle\Psi|\mathbb{1}\,\vec{p}_e{\cdot}\mathbb{1}\,\vec{S}\,\mathbb{1}|\Psi\rangle = \langle\Psi|P^{-1}P\vec{p}_e{\cdot}P^{-1}P\vec{S}\,P^{-1}P|\Psi\rangle \\ &= \big(\langle\Psi|P^{-1}\big)\big(P\vec{p}_eP^{-1}\big){\cdot}\big(P\vec{S}P^{-1}\big)\big(P|\Psi\rangle\big) = \langle\Psi'|\vec{p}_e^{\,\prime}{\cdot}\vec{S}'|\Psi'\rangle \\ &= \langle\Psi'|(-\vec{p}_e){\cdot}(+\vec{S})|\Psi'\rangle = -\langle\Psi'|\vec{p}_e{\cdot}\vec{S}|\Psi'\rangle, \end{aligned} \tag{4.55}$$

where $|\Psi'\rangle = P|\Psi\rangle$ – whatever that action on the state $|\Psi\rangle$ may be[7] – and, of course

$$\vec{p}_e^{\,\prime} := P\vec{p}_eP^{-1} = -\vec{p}_e, \qquad \text{and} \qquad \vec{S}_e^{\,\prime} := P\vec{S}P^{-1} = +\vec{S}. \tag{4.56}$$

If we assume that $[H, P] = 0$ (that parity is a symmetry of the system), it follows that:

1. either $|\Psi'\rangle = c|\Psi\rangle$ and so $\langle\Psi'| = c^{*}\langle\Psi|$, whereby the relation (4.55) would have to imply that $\langle\Psi|\vec{p}_e{\cdot}\vec{S}|\Psi\rangle = 0$, which was proven wrong by Madam Wu's experiment;
2. or $|\Psi\rangle \not\propto |\Psi'\rangle$ are degenerate states – which does not follow from otherwise successful nuclear models applicable to the cobalt-60 nucleus.

In fact, were we even to allow the possibility that the nuclear models err and $|\Psi'\rangle \neq c|\Psi\rangle$, so that $|\Psi'\rangle$ and $|\Psi\rangle$ are two distinct but degenerate states of the cobalt-60 nucleus, it may be shown [29]

[6] The early notation for the "τ"-particle must not be confused with the τ-lepton, which was experimentally detected only in 1975, almost two decades later.

[7] In any concrete representation, P acts on the wave-function by changing the argument of that wave-function, and also by changing both the integration measure and limits in expressions such as $\langle\Psi|\cdots|\Psi\rangle$. However, the result (4.55) is independent of its representation, and holds abstractly, as written here.

that the oscillation $|\Psi\rangle \leftrightarrow |\Psi'\rangle$ would be sufficiently fast to make the expectation value $\langle\Psi|\vec{p}_e\cdot\vec{S}|\Psi\rangle$ much smaller than was experimentally measured. Thus, $[H, P] \neq 0$ remains as the only possibility, i.e., that P is not a symmetry of Nature.

Once uncovered, P-violation was experimentally confirmed in more and more processes – and exclusively in processes mediated by the weak nuclear interaction. Moreover, it was discovered that *all* weak processes exhibit P-violation!

The most stunning consequence of P-violation is the fact that the "right-handed neutrinos" – if they even exist – behave significantly differently than the "left-handed" ones. That is, for every particle one may define its "helicity" [☞ Section 5.2.1 on p. 172] as the projection of the spin of the particle along the direction of its motion. For particles with non-vanishing mass, helicity cannot be Lorentz-invariant. A Lorentz boost can always transform into the particle's own coordinate system, wherein it does not move at all so the projection is undefined. It is, of course, also possible to "pass" the particle into a coordinate system in which the particle now moves in the opposite direction. All the while the spin remains unchanged, which then flips the sign of the helicity. However, a massless particle cannot be "passed," nor does there exist a coordinate system in which it is at rest, so that the helicity of a massless particle is Lorentz-invariant.

Particles with positive helicity are called "right-handed" (their spin – fictively – rotates in the direction of the fingers of the right hand when the thumb indicates the direction of motion), and the "left" particles have negative helicity. Experimental evidence to date shows that no more than about $1/10^{10}$ of all detected neutrinos are right-handed, and the mass of the observed (so almost entirely "left-handed") neutrinos is close to zero [☞ Ref. [293] and Section 7.3.2]. This extremely convincing asymmetry in Nature is of crucial importance to the structure of weak interactions [☞ Section 7.2].

For example, in the decay

$$\pi^- \to \mu^- + \bar{\nu}_\mu \tag{4.57}$$

analyzed in the pion's rest-frame, the muon and the antineutrino move in opposite directions. *[✎ Why?]* The relative angular momentum of the muon and the antineutrino must be orthogonal to the motion of the muon and the antineutrino, and so does not affect the definition of their helicities. The pion spin is zero, so the spin of the muon and the antineutrino must be antiparallel, which means that the antineutrino helicity is the same as that of the muon. Experiments confirm that all muons – and so also the antineutrinos – emerge with a right-handed helicity; the nearly complete absence (less than 1-in-10^{10}) of left-handed antineutrinos then provides for *maximal* parity violation.

By the way, in 1929, soon after the publication of Dirac's equation and the theory of the electron, Hermann Weyl gave a simpler equation suitable for spin-$\frac{1}{2}$ particles with no mass, and which uses the property that the helicity of such particles is Lorentz-invariant. Weyl's theory was neglected since the photon was the only known particle with no mass, and photons have spin 1. When Pauli, a year later, proposed the neutrino to preserve the energy conservation law, he ironically did not use Weyl's equation: Although he knew that the mass of the neutrino is small or even zero, this equation permits parity violation, which Pauli believed also to be a symmetry of Nature! Twenty-six years later, experiments proved him right about energy conservation but wrong about parity. The reason for such a convincing difference between left-handed and right-handed neutrinos remains one of the significant unexplained characteristics in elementary particle physics☝.

4.2.2 Charge conjugation and time reversal

Although the actions of the operations T and C clearly satisfy the relation (4.52), unlike P, these two operations are *anti-linear*, for example,

$$C\big(c\,\Psi(\vec{r},t)\big) = c^*\,C\big(\Psi(\vec{r},t)\big), \tag{4.58}$$

so that the proof of Conclusion A.1 on p. 461, does not apply. This makes the precise analysis of the C- and T-action nontrivial in quantum mechanics [29]; fortunately, we need not be concerned with these details. Note instead that in many cases the invariance of the dynamics with respect to the T- and/or C-operation simply implies a degeneracy, so pairs of states $|\Psi\rangle$ and $T|\Psi\rangle$, as well as $|\Psi\rangle$ and $C|\Psi\rangle$ have the same energy and the same lifetime. For example, if $|\Psi\rangle$ is used for the description of the electron, $C|\Psi\rangle$ must be assigned to the positron. The degeneracy of $|\Psi\rangle$ and $C|\Psi\rangle$ then means that the electron mass equals the positron mass, which is indeed true.

Of course, only chargeless particles may be eigenstates of the C-operation: it follows from the defining property (4.52) that, if $|\Psi\rangle$ is an eigenstate of the C-operation, then $|\overline{\Psi}\rangle = C|\Psi\rangle = \pm|\Psi\rangle$, so $|\Psi\rangle$ and $|\overline{\Psi}\rangle$ differ, at most, in the sign; one says that $|\Psi\rangle$ is its own anti-state, i.e., that the particle is its own antiparticle.

It may be shown [422] that the bound state of the spin-$\frac{1}{2}$ particle and its antiparticle is an eigenstate of the operator C with the eigenvalue $(-1)^{\ell+s}$, where ℓ is the angular momentum of the particle–antiparticle system, and s is their composite spin. For positronium, which at least virtually may annihilate into a single photon, it follows that $\ell + s = 1$ since the photon spin is 1. Since C is conserved in strong and electromagnetic processes and the electron–positron pair annihilation is evidently an electromagnetic process, it follows that the photon is a C-eigenstate, with the eigenvalue -1. Similarly, in the electromagnetic decay of the pion,

$$\pi^0 \to \gamma + \gamma, \tag{4.59}$$

there can be only an even number of photons, and the C-eigenvalue of the π^0-particle is $+1$. According to the quark model,

$$|\pi^0\rangle = \tfrac{1}{\sqrt{2}}\big(|u,\overline{u}\rangle + |d,\overline{d}\,\rangle\big) \tag{4.60}$$

is a linear combination of two particle–antiparticle bound states, so the formula $(-1)^{\ell+s}$ for the C-eigenvalue holds. Also, an n-photon system has the C-eigenvalue equal to $(-1)^n$.

On the other hand, the C conservation law is violated in weak interactions: The muons always emerge from the process (4.57) with right-handed helicity. Then

$$C(\pi^- \to \mu^-_R + \bar{\nu}_\mu) \quad = \quad \pi^+ \to \mu^+_R + \nu_\mu, \tag{4.61}$$

and the anti-muons would have to emerge also with 100% right-handed helicity from the π^+-meson decay, since the C-operation has no effect on the coordinate system, direction of particle motion and their spins. However, the same analysis as for the process (4.57) shows that the neutrino would have to have right-handed helicity – but such neutrinos may exist no more than 1-in-10^{10}! The neutrinos as well as the muons in the process (4.61) emerge with left-handed helicity. Thus, weak processes such as the $\pi^\pm$-meson decays (4.57) and (4.61) indicate that weak interactions maximally violate both P- and C-symmetry.

— ❦ —

Direct experimental verification of T-conservation or T-violation is much harder: No physical state can be a T-eigenstate, [✎ *why?*] so we cannot simply check the products of the eigenvalues on one side and the other of a process. The most direct verification would involve detailed measurement of parameters for a process $A + B + \cdots \to C + D + \cdots$, as well as the reverse process, $C + D + \cdots \to A + B + \cdots$, and then – taking into account the kinematic differences – comparing the effective cross-sections. That has indeed been done in a large number of electromagnetic and strong nuclear processes, and no trace of T-violation was found. The resulting equality, up to kinematic factors, between the "reversed" pairs of processes such as $A + B + \cdots \to C + D + \cdots$ and $C + D + \cdots \to A + B + \cdots$ is called the "principle of detailed balance."

On the other hand, the verification of the principle of detailed balance in weak nuclear processes is very hard to carry through: For example, the reversal of the weak decay $\Lambda^0 \to p^+ + \pi^-$

would require the fusion $p^+ + \pi^- \to \Lambda^0$. However, the collision $p^+ + \pi^-$ and its outcomes are (by far) so dominated by the strong nuclear interaction that the $p^+ + \pi^-$ fusion into Λ^0 simply cannot be experimentally detected, among the sea of all the different results obtained via the strong nuclear interaction. The only type of weak processes where neither the original nor the reversed process is swamped by strong and electromagnetic side-processes are the processes involving neutrinos. However, experiments with neutrinos are already very hard. Unlike other particles, neutrinos are very difficult to control in the lab, in part owing to their electromagnetic neutrality, and in part owing to the extremely small effective cross-section of their interaction with other matter.

4.2.3 The CPT-theorem and CP-violation

Note first that the operation P in 3-dimensional space is a mirror reflection of one of the Cartesian coordinates, e.g., $z \to -z$, so that the (x, y)-plane serves as the mirror. The Lorentz boost in the z-direction then mixes the z-coordinates and time, t. There exists an analytical continuation of this transformation that flips the sign of the z-coordinate and time t, and so turns the operation P into a T-operation. Finally, following the Feynman–Stückelberg interpretation of antiparticles as particles that travel backwards in time, T is equivalent to the C-operation (charge conjugation). The detailed treatment then shows that this connects the C-, P- and T-operations in any *local* field theory that (**1**) is Lorentz-invariant, (**2**) has a Lorentz-invariant ground state (vacuum), and (**3**) has a lower bound on the energy. Conversely, it follows that every non-invariance with respect to the combined CPT-transformation implies a violation of Lorentz-symmetry and/or of locality [350, 413, 300, 230, 102].

An alternative argument starts by noticing that

$$CPT(e^{\pm i(\vec{k}\cdot\vec{r}-\omega t)}) = CP(e^{\pm i(\vec{k}\cdot\vec{r}-\omega(-t))}) = C(e^{\pm i(\vec{k}\cdot(-\vec{r})+\omega t)}) = (e^{\mp i(-\vec{k}\cdot\vec{r}+\omega t)})$$
$$= e^{\pm i(\vec{k}\cdot\vec{r}-\omega t)}, \tag{4.62}$$

and then using the fact that plane waves form a complete set and are Lorentz-invariant, to argue that all spacetime-dependent Lorentz-invariant expressions (observables) are also CPT-invariant. The difficulty in this latter, seemingly much simpler argument lies in proving that there is no loss of generality and that the whole required linear combination of plane waves is also both CPT- and Lorentz-invariant.

Standard (Lagrangian) quantum field theory texts [425, 554] prove that the handful of typically used Lorentz-invariant Lagrangians are also CPT-invariant. Recently, however, a fully rigorous proof has been presented within the familiar framework of Lagrangian quantum field theories [220], applicable to all Lagrangians that depend polynomially on the fields and their derivatives.

— ❦ —

On the other hand, once it was proven in 1956 that P is not a symmetry of Nature, in 1957, Lev D. Landau suggested that the combined CP-transformation should be a symmetry of Nature. As we have seen, weak processes such as the $\pi^\pm$-meson decays (4.57) and (4.61) maximally violate both the P- and the C-symmetry, so it was reasonable to suppose that perhaps the *combined* CP-operation is a true symmetry of Nature. Thus, every naturally occurring process involving a collection of particles would have to have a “mirror image” involving antiparticles instead of particles.

However, James Cronin and Val Fitch surprised the physics community in 1964 by publishing their results showing unambiguously that, in neutral kaon decays, not even the combined CP-transformation is a symmetry of Nature.

Ironically, the possibility of CP-violation follows from a work by Murray Gell-Mann and Abraham Pais back in 1954, when they noticed that the K^0-meson cannot be its own antiparticle as its strangeness charge must equal $+1$; there then must exist a $\overline{K}^0$-meson with a strangeness charge of -1, which is evident from the following scattering events (the parenthetical indices denote strangeness):

$$\pi^-_{(0)} + p^+_{(0)} \to \Lambda^0_{(-1)} + K^0_{(+1)} \qquad \text{and} \qquad \pi^+_{(0)} + p^+_{(0)} \to p^+_{(0)} + \overline{K}^0_{(-1)} + K^+_{(+1)}. \tag{4.63}$$

These processes occur mediated by the strong interaction (so identified owing to the speed of the process), which one knows do preserve both the C- and the P-symmetry. Gell-Mann and Pais then noticed the possibility of the K^0–$\overline{K}^0$ transmutation with an explanation that is, today, easier to replace with the display of relevant Feynman diagrams shown in Figure 4.1.

We also know that the neutral kaons are pseudo-scalars, so:

$$CP|K^0\rangle = -C|K^0\rangle = -|\overline{K}^0\rangle \qquad CP|\overline{K}^0\rangle = -C|\overline{K}^0\rangle = -|K^0\rangle, \tag{4.64}$$

whereby the eigenstates of the CP-symmetry are

$$|K^0_+\rangle := \tfrac{1}{\sqrt{2}}\big(|K^0\rangle - |\overline{K}^0\rangle\big) \qquad \text{and} \qquad |K^0_-\rangle := \tfrac{1}{\sqrt{2}}\big(|K^0\rangle + |\overline{K}^0\rangle\big), \tag{4.65}$$

where

$$CP|K^0_\pm\rangle = (\pm 1)|K^0_\pm\rangle. \tag{4.66}$$

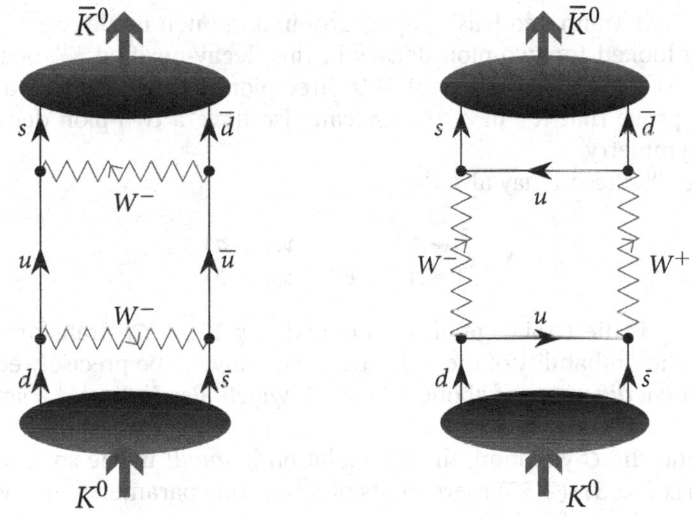

Figure 4.1 The $K^0 \to \overline{K}^0$ transmutation.

Now, neutral kaons decay (among other ways) into two or three pions, and we will neglect all other decay modes. Pions are pseudo-scalars so their intrinsic parity is -1; the parity of a two-pion system is then $+1$, and of a three-pion system, -1. Because of charge conservation, the total charge of both the two- and the three-pion systems in the neutral kaon decays must be zero, so their C-eigenvalue must be $+1$. It follows that $CP(2\pi) = +1$ but $CP(3\pi) = -1$, and it must be that

$$K^0_+ \to 2\pi, \qquad \text{and} \qquad K^0_- \to 3\pi, \tag{4.67}$$

if these (weak interaction) decays preserve the CP-symmetry. Since the two-pion decay has more energy,[8] one expects the K^0_+-state lifetime to be shorter than the K^0_--state lifetime. Indeed, the result (3.124) states that $\Gamma_{K^0_+} \propto \sqrt{1-(2m_{\pi^0}/m_{K^0})^2}$ in a decay into two particles of equal masses.

[8] From the data (2.31) it follows that about $(497.6 - 2{\times}135.0) = 227.6$ MeV remains in the two-pion decay, and only about $(497.6 - 3{\times}135.0) = 92.6$ MeV in the three-pion decay for the kinetic energy of the pions.

Analogously, $\Gamma_{K^0_-} \propto \sqrt{1-(3m_{\pi^0}/m_{K^0})^2}$ in a decay into three particles of equal masses. Thus, one expects $\Gamma_{K^0_+} > \Gamma_{K^0_-}$, and then also $\tau_{K^0_+} < \tau_{K^0_-}$. Although the ratio of these two lifetimes is not as simple, the experiments nevertheless indicate a significant difference and in the same direction as in this simplified estimate:

$$\tau^+ := \tau_{K^0_+} = 0.895\,8\times 10^{-10}\,\text{s} \qquad \text{and} \qquad \tau^- := \tau_{K^0_-} = 5.114\times 10^{-8}\,\text{s}. \tag{4.68}$$

Because of this difference, K^0_+ is called the "short" kaon ($K^0_S := K^0_+$), and K^0_- the "long" kaon ($K^0_L := K^0_-$).

Since K^0_- "lives" about 570 times longer than K^0_+, within a beam of neutral kaons (created by strong interactions, and so with a 50–50% K^0_+–K^0_- distribution) the "short" kaons quickly decay, leaving the beam as a "pure" K^0_--beam. Recall that the number of undecayed kaons diminishes exponentially, so that

$$\frac{N(K^0_+)}{N(K^0_-)} = \frac{e^{-t/\tau^+}}{e^{-t/\tau^-}} = \exp\Big\{-\frac{t}{\tau^+}+\frac{t}{\tau^-}\Big\} \approx \exp\Big\{-569.9\,\frac{t}{\tau^-}\Big\}, \tag{4.69}$$

which drops to 1.447×10^{-5} after just 1 ns.

Ten years after Gell-Mann and Pais's paper, Cronin and Fitch made use of this extraordinary property, and simply looked for two-pion decays in this decay-purified K^0_--beam. Although they only found about 1 two-pion decay to about 500 three-pion decays, this was a sufficient ($\approx$138-fold) discrepancy to prove that K^0_- nevertheless can also have a two-pion decay and so indicate the violation of *CP*-symmetry.

In addition, the K^0_--meson may also decay as

$$K^0_- \begin{cases} \to \pi^+ + e^- + \bar{\nu}_e, & (a) \\ \to \pi^- + e^+ + \nu_e, & (b) \end{cases} \tag{4.70}$$

where the a-type decay is the *CP*-image of the b-type decay. If the *CP*-transformation were a true symmetry of Nature, the probability of these decays would have to be precisely equal. Experiments, however, show a relative *difference* of about 3.3×10^{-3}, which also indicates a small but significant *CP*-violation.

Unlike the *P*- and the *C*-violation, the *CP*-violation is *small*: in the so-called Cabibbo–Kobayashi–Maskawa matrix (2.53)–(2.55) there exists precisely one parameter, δ_{13}, which parametrizes the *CP*-violation. The values of the parameter $\delta_{13} = (1.20 \pm 0.08)^\circ$ (as compared to $\max(\delta_{13}) = 180^\circ$) and its indirect appearance in computations result in the smallness of *CP*-violation, such as the $\sim \frac{1}{500}$ two-pion decays of what should be a $\leqslant \frac{1}{69\,000}$-pure $K^0_- = K^0_L$ beam. This smallness is hard to explain theoretically and remains one of the unsolved problems of elementary particle theory☝.

— ❦ —

There also exists the so-called "strong *CP*-problem": to wit, it is theoretically possible also for the strong nuclear interaction to violate *CP*-symmetry, but this is not the case. The theory of quantum chromodynamics has a parameter, ϑ,[9] which parametrizes possible strong *CP*-violation, whereupon it is a puzzle that $\vartheta \approx 0$, in all known experiments and to a high degree of precision [☞ Section 6.3.1].

Finally, the violation of *CP*-symmetry in the first seconds of the Big Bang is one of the three necessary requirements (as shown by Andrei Saharov) for an explanation of the fact that the

[9] This parameter indeed is an angle, but has no relation with the spherical coordinate of the same name.

universe that we observe today consists of matter, and not also of antimatter. This gives an unambiguous definition of the positive electric charge as that of the lepton emerging in the (somewhat but notably) more frequent "semi-leptonic" decay of the long-living neutral kaon (4.70). Thus, the existence of *CP*-violation is in fact a boon for us: If the *CP*-transformation were an exact symmetry of Nature, there could be no difference in the universe between matter and antimatter, the two would have annihilated in the first few seconds of the universe's existence, and we would not be here to notice this.

— ❦ —

It is worth noting that the *C*-, *P*-, *T*-, *CP*-, *PT*- and *CT*-symmetries are violated only in experiments that involve the weak interaction, and that these indeed are exact symmetries in all electromagnetic and strong nuclear processes.

4.2.4 Exercises for Section 4.2

✎ **4.2.1** *Suppose that the parity operation acts as* $P : |a\rangle \to |b\rangle$ *and* $P : |b\rangle \to |a\rangle$ *upon some two orthonormalized states,* $|a\rangle, |b\rangle$*. From these, try to construct the eigenfunctions of the P-operator and normalize them. Discuss the physical meaning of these eigenfunctions if they exist, or explain why an eigenstate of the P-operator cannot make sense physically.*

✎ **4.2.2** *Suppose that the time reversal operation acts as* $T : |\alpha\rangle \to |\beta\rangle$ *and* $T : |\beta\rangle \to |\alpha\rangle$ *upon some two orthonormalized states,* $|\alpha\rangle, |\beta\rangle$*. From these, try to construct the eigenfunctions of the T-operator and normalize them. Discuss the physical meaning of these eigenfunctions if they exist, or explain why an eigenstate of the T-operator cannot make sense physically.*

✎ **4.2.3** *Assuming that the CPT-transformation is an exact order-2 symmetry, prove that the eigenfunctions of the CP-operation are also T-eigenfunctions.*

4.3 Isospin

In 1932, Werner Heisenberg noticed that, for the purposes of describing atomic nuclei, it is possible to neglect the minute difference between the neutron mass and the proton mass:

$$\frac{m_n - m_p}{m_p} = \frac{939.566 - 938.272}{938.272} = 0.001\,379\,13. \tag{4.71}$$

It is even possible to ignore the fact that the proton is charged and the neutron is not: the strong nuclear interaction, which keeps the nucleus as a bound state, must be many times stronger than the electromagnetic repulsion of the protons in the nucleus. Thus, the proton and the neutron are regarded as two states of one particle, a *nucleon* (denoted *N*), just as the spin-($\pm\frac{1}{2}$) electrons are both regarded as two polarizations of the *same* particle. In analogy with spin, Heisenberg then introduced a conserved quantity that Eugene Wigner named *isospin* in 1937 and for which he employed the corresponding mathematical formalism:

$$\vec{I}: \quad [I_j, I_k] = i\varepsilon_{jk}{}^m I_m, \tag{4.72}$$

just like the $\vec{J}$ in Appendix A.3. Following the digressions A.2 on p. 465 and A.3 on p. 467, we know that there exist eigenstates

$$|I, I_3\rangle: \quad I^2|I, I_3\rangle = I(I+1)|I, I_3\rangle, \qquad I_3|I, I_3\rangle = I_3|I, I_3\rangle, \qquad |I_3| \leqslant I, \tag{4.73a}$$

$$I_\pm|I, I_3\rangle = \sqrt{I(I+1) - I_3(I_3\pm1)}\,|I, I_3\pm1\rangle, \qquad 2I,\ \triangle I \in \mathbb{Z}. \tag{4.73b}$$

4.3.1 Isospin, nucleons and pions

Heisenberg and Wigner introduced the isospin formalism for the purposes of nuclear physics, and here we identify

$$|p^+\rangle = |\tfrac{1}{2},+\tfrac{1}{2}\rangle, \qquad |n^0\rangle = |\tfrac{1}{2},-\tfrac{1}{2}\rangle. \tag{4.74}$$

Moreover, if the isospin "rotations" are a symmetry of strong interactions, then it follows that isospin is a conserved quantity in all strong nuclear processes, following Conclusion A.1 on p. 461. In 1932, the proposition of introducing such an *ad-hoc* and abstract symmetry as a further exact symmetry of strong interactions was an unusually bold move. However, such reliance on symmetries and the quantum version of Noether's theorem [☞ Conclusion A.1 on p. 461] has become one of the basic principles of fundamental physics in the twentieth century, and even grew into the gauge principle, which is the basis of contemporary understanding of interactions in general [☞ Chapters 5 and 6].

A few other (then known) hadrons are identified as

$$|\pi^+\rangle = |1,+1\rangle, \qquad |\pi^0\rangle = |1,0\rangle, \qquad |\pi^-\rangle = |1,-1\rangle, \tag{4.75}$$

$$|\Delta^{++}\rangle = |\tfrac{3}{2},+\tfrac{3}{2}\rangle, \quad |\Delta^{+}\rangle = |\tfrac{3}{2},+\tfrac{1}{2}\rangle, \quad |\Delta^{0}\rangle = |\tfrac{3}{2},-\tfrac{1}{2}\rangle, \quad |\Delta^{-}\rangle = |\tfrac{3}{2},-\tfrac{3}{2}\rangle, \quad \text{etc.} \tag{4.76}$$

The relationship between the electric charge Q, the isospin "charge" I_3, the baryon number B and strangeness S for all hadrons was found before 1974 to be in agreement with the GNN formula (2.30). Thus, isospin symmetry, soon extended into the $SU(3)_f$ approximate symmetry, offers an excellent classification tool.

— ❦ —

However, isospin is also useful in dynamics: We know from quantum mechanics that addition of spins $\frac{1}{2}$ and $\frac{1}{2}$ produces the following possibilities, here applied to isospin:

$$\begin{cases} |1,+1\rangle_S = |\tfrac{1}{2},+\tfrac{1}{2}\rangle|\tfrac{1}{2},+\tfrac{1}{2}\rangle & = |p^+,p^+\rangle, \quad (4.77a)\\ |1,0\rangle_S = \tfrac{1}{\sqrt{2}}\Big(|\tfrac{1}{2},+\tfrac{1}{2}\rangle|\tfrac{1}{2},-\tfrac{1}{2}\rangle + |\tfrac{1}{2},-\tfrac{1}{2}\rangle|\tfrac{1}{2},+\tfrac{1}{2}\rangle\Big) & = \tfrac{1}{\sqrt{2}}\big(|p^+,n^0\rangle + |n^0,p^+\rangle\big), \quad (4.77b)\\ |1,-1\rangle_S = |\tfrac{1}{2},-\tfrac{1}{2}\rangle|\tfrac{1}{2},-\tfrac{1}{2}\rangle & = |n^0,n^0\rangle; \quad (4.77c) \end{cases}$$

$$|0,0\rangle_A = \tfrac{1}{\sqrt{2}}\Big(|\tfrac{1}{2},+\tfrac{1}{2}\rangle|\tfrac{1}{2},-\tfrac{1}{2}\rangle - |\tfrac{1}{2},-\tfrac{1}{2}\rangle|\tfrac{1}{2},+\tfrac{1}{2}\rangle\Big) = \tfrac{1}{\sqrt{2}}\big(|p^+,n^0\rangle - |n^0,p^+\rangle\big), \tag{4.77d}$$

where the subscript "S" denotes that the state is symmetric with respect to swapping the two nucleons, and "A" that it is antisymmetric. However, there exists only one two-nucleon bound state: the deuteron, the deuterium nucleus, which consists of a proton and a neutron. This implies that the isospin factor in the wave-function of the deuteron must be antisymmetric with respect to swapping the two nucleons. Were this factor symmetric, isospin "rotations" would guarantee the existence of all three symmetric states (4.77a)–(4.77c) – and it is well known that the bound state of neither two protons nor two neutrons exists in Nature.

This identifies the deuteron as the isospin $|0,0\rangle$ state. Also, since $|0,0\rangle$ is antisymmetric with respect to swapping the two nucleons and since the whole wave-function must be antisymmetric with respect to the swapping of any two (otherwise identical) fermions, it follows that the product of the remaining "spatial" and "spin" factors in the wave-function of the deuteron bound state must be symmetric. So, if the proton and neutron spins are parallel (evidently symmetric) or antiparallel and symmetrized, then the spatial factor in the wave-function also must be symmetric with respect to the exchange of two nucleons. If the spins are antiparallel and antisymmetrized, the spatial factor in the wave-function must also be antisymmetric.

Without the isospin formalism, which permits treating the proton and the neutron as two polarizations of the same particle, this indirect correlation of spins and spatial factors in the wavefunction could not have been derived.

— ❦ —

Finally, isospin also easily produces *relative* effective cross-sections of various processes mostly by way of the Wigner–Eckart theorem A.3 on p. 475. Consider, e.g., the three two-nucleon collisions:

$$\begin{array}{llll} & & \underline{\quad N_1 \qquad N_2 \qquad\qquad d \quad \pi \quad} & \\ a: & p^+ + p^+ \to d + \pi^+ & \leftrightarrow \quad |\tfrac{1}{2},+\tfrac{1}{2}\rangle|\tfrac{1}{2},+\tfrac{1}{2}\rangle \to |0,0\rangle|1,+1\rangle = |1,+1\rangle, & (4.78a) \\ b: & p^+ + n^0 \to d + \pi^0 & \leftrightarrow \quad |\tfrac{1}{2},+\tfrac{1}{2}\rangle|\tfrac{1}{2},-\tfrac{1}{2}\rangle \to |0,0\rangle|1,0\rangle \;\;= |1,0\rangle, & (4.78b) \\ c: & n^0 + n^0 \to d + \pi^0 & \leftrightarrow \quad |\tfrac{1}{2},-\tfrac{1}{2}\rangle|\tfrac{1}{2},-\tfrac{1}{2}\rangle \to |0,0\rangle|1,-1\rangle = |1,-1\rangle, & (4.78c) \end{array}$$

where d denotes the deuteron, the deuterium nucleus. On the other hand, combining (4.77a)–(4.77d), we have

$$|p^+,p^+\rangle = |\tfrac{1}{2},+\tfrac{1}{2}\rangle|\tfrac{1}{2},+\tfrac{1}{2}\rangle = |1,+1\rangle, \tag{4.79a}$$

$$|p^+,n^0\rangle = |\tfrac{1}{2},+\tfrac{1}{2}\rangle|\tfrac{1}{2},-\tfrac{1}{2}\rangle = \tfrac{1}{\sqrt{2}}\big(|1,0\rangle + |0,0\rangle\big), \tag{4.79b}$$

$$|n^0,n^0\rangle = |\tfrac{1}{2},-\tfrac{1}{2}\rangle|\tfrac{1}{2},-\tfrac{1}{2}\rangle = |1,-1\rangle. \tag{4.79c}$$

Then it follows that, up to factors independent of isospin and which are equal [✎ *why?*],

$$\mathfrak{M}_a \propto \langle d,\pi^+|p^+,p^+\rangle = \langle 1,+1||\tfrac{1}{2},+\tfrac{1}{2}\rangle|\tfrac{1}{2},+\tfrac{1}{2}\rangle = 1, \tag{4.80a}$$

$$\mathfrak{M}_b \propto \langle d,\pi^0|p^+,n^0\rangle = \langle 1,0||\tfrac{1}{2},+\tfrac{1}{2}\rangle|\tfrac{1}{2},-\tfrac{1}{2}\rangle = \langle 1,0|\Big(\tfrac{1}{\sqrt{2}}\big(|1,0\rangle + |0,0\rangle\big)\Big) = \tfrac{1}{\sqrt{2}}, \tag{4.80b}$$

$$\mathfrak{M}_c \propto \langle d,\pi^-|n^0,n^0\rangle = \langle 1,-1||\tfrac{1}{2},-\tfrac{1}{2}\rangle|\tfrac{1}{2},-\tfrac{1}{2}\rangle = 1. \tag{4.80c}$$

Since $\sigma \propto |\mathfrak{M}|^2$, it follows that

$$\sigma_a : \sigma_b : \sigma_c = 2 : 1 : 2. \tag{4.81}$$

That is, it is twice as probable for a deuteron (and a pion) to emerge from the collision of two protons than from the collision of a proton and a neutron! (The collision of two neutrons is hard to arrange experimentally.)

Even more dramatic is the situation with pion–nucleon scattering. Listing all the possibilities consistent with charge conservation, we find six elastic pion–nucleon collisions:

$$(a)\ \pi^+ + p^+ \to \pi^+ + p^+,\quad (b)\ \pi^0 + p^+ \to \pi^0 + p^+,\quad (c)\ \pi^- + p^+ \to \pi^- + p^+, \tag{4.82a}$$

$$(d)\ \pi^+ + n^0 \to \pi^+ + n^0,\quad (e)\ \pi^0 + n^0 \to \pi^0 + n^0,\quad (f)\ \pi^- + n^0 \to \pi^- + n^0, \tag{4.82b}$$

and four inelastic collisions resulting in a pion and a nucleon:

$$(g)\ \pi^+ + n^0 \to \pi^0 + p^+,\qquad (h)\ \pi^0 + p^+ \to \pi^+ + n^0, \tag{4.82c}$$

$$(i)\ \pi^0 + n^0 \to \pi^- + p^+,\qquad (j)\ \pi^- + p^+ \to \pi^0 + n^0. \tag{4.82d}$$

Since $I(\pi) = 1$ and $I(N) = \frac{1}{2}$, the isospin of the incoming (initial) and of the outgoing (final) system may be either $\frac{3}{2}$ or $\frac{1}{2}$, and let $\mathfrak{M}_{3/2}$ and $\mathfrak{M}_{1/2}$ denote the corresponding so-called "reduced" amplitudes.[10] Using tables of Clebsch–Gordan coefficients we compute:

$$\pi^+ + p^+: \qquad |1,1\rangle|\tfrac{1}{2},+\tfrac{1}{2}\rangle = |\tfrac{3}{2},+\tfrac{3}{2}\rangle, \tag{4.83a}$$

[10] By the Wigner–Eckart theorem, every amplitude may be factorized as a product of a reduced amplitude and a Clebsch–Gordan coefficient [☞ Section A.3.4 and Theorem A.3 on p. 475, as well as the textbooks [362, 363, 328, 471, 480, 134, 391, 407, 472, 360, 29, 242, 3, 110, for example] and the handbook [294]].

$$\pi^0 + p^+ : \quad |1,0\rangle|\tfrac{1}{2},+\tfrac{1}{2}\rangle = \sqrt{\tfrac{2}{3}}|\tfrac{3}{2},+\tfrac{1}{2}\rangle - \tfrac{1}{\sqrt{3}}|\tfrac{1}{2},+\tfrac{1}{2}\rangle, \tag{4.83b}$$

$$\pi^- + p^+ : \quad |1,-1\rangle|\tfrac{1}{2},+\tfrac{1}{2}\rangle = \tfrac{1}{\sqrt{3}}|\tfrac{3}{2},-\tfrac{1}{2}\rangle - \sqrt{\tfrac{2}{3}}|\tfrac{1}{2},-\tfrac{1}{2}\rangle, \tag{4.83c}$$

$$\pi^+ + n^0 : \quad |1,1\rangle|\tfrac{1}{2},-\tfrac{1}{2}\rangle = \tfrac{1}{\sqrt{3}}|\tfrac{3}{2},+\tfrac{1}{2}\rangle + \sqrt{\tfrac{2}{3}}|\tfrac{1}{2},+\tfrac{1}{2}\rangle, \tag{4.83d}$$

$$\pi^0 + n^0 : \quad |1,0\rangle|\tfrac{1}{2},-\tfrac{1}{2}\rangle = \sqrt{\tfrac{2}{3}}|\tfrac{3}{2},-\tfrac{1}{2}\rangle + \tfrac{1}{\sqrt{3}}|\tfrac{1}{2},-\tfrac{1}{2}\rangle, \tag{4.83e}$$

$$\pi^- + n^0 : \quad |1,-1\rangle|\tfrac{1}{2},-\tfrac{1}{2}\rangle = |\tfrac{3}{2},-\tfrac{3}{2}\rangle. \tag{4.83f}$$

For example, the processes (a) and (f) both have $I = \frac{3}{2}$ and the Clebsch–Gordan coefficients are 1:

$$(a)\ \pi^+ + p^+ \to \pi^+ + p^+ \quad \leftrightarrow \quad \mathfrak{M}_a = \langle\tfrac{3}{2},+\tfrac{3}{2}||\tfrac{3}{2},+\tfrac{3}{2}\rangle \times \mathfrak{M}_{3/2} = \mathfrak{M}_{3/2}, \tag{4.84}$$

$$(f)\ \pi^- + n^0 \to \pi^- + n^0 \quad \leftrightarrow \quad \mathfrak{M}_f = \langle\tfrac{3}{2},-\tfrac{3}{2}||\tfrac{3}{2},-\tfrac{3}{2}\rangle \times \mathfrak{M}_{3/2} = \mathfrak{M}_{3/2}, \tag{4.85}$$

and so we have that $\mathfrak{M}_a = \mathfrak{M}_f = \mathfrak{M}_{3/2}$. The remaining processes are a mixture of $\mathfrak{M}_{3/2}$ and $\mathfrak{M}_{1/2}$, such as

$$(c)\ \pi^- + p^+ \to \pi^- + p^+ \tag{4.86}$$

$$\begin{aligned} \mapsto \quad \mathfrak{M}_c &= \Big(\tfrac{1}{\sqrt{3}}\langle\tfrac{3}{2},-\tfrac{1}{2};\mathfrak{a}_3| - \sqrt{\tfrac{2}{3}}\langle\tfrac{1}{2},-\tfrac{1}{2};\mathfrak{a}_1|\Big)\Big(\tfrac{1}{\sqrt{3}}|\tfrac{3}{2},-\tfrac{1}{2};\mathfrak{a}_3\rangle - \sqrt{\tfrac{2}{3}}|\tfrac{1}{2},-\tfrac{1}{2};\mathfrak{a}_1\rangle\Big) \\ &= \tfrac{1}{3}\langle\tfrac{3}{2},-\tfrac{1}{2};\mathfrak{a}_3||\tfrac{3}{2},-\tfrac{1}{2};\mathfrak{a}_3\rangle + \tfrac{2}{3}\langle\tfrac{1}{2},-\tfrac{1}{2};\mathfrak{a}_1||\tfrac{1}{2},-\tfrac{1}{2};\mathfrak{a}_1\rangle = \tfrac{1}{3}\mathfrak{M}_{3/2} + \tfrac{2}{3}\mathfrak{M}_{1/2} \end{aligned} \tag{4.87}$$

$$(j)\ \pi^- + p^+ \to \pi^0 + n^0 \tag{4.88}$$

$$\begin{aligned} \mapsto \quad \mathfrak{M}_j &= \Big(\tfrac{1}{\sqrt{3}}\langle\tfrac{3}{2},-\tfrac{1}{2};\mathfrak{a}_3| - \sqrt{\tfrac{2}{3}}\langle\tfrac{1}{2},-\tfrac{1}{2};\mathfrak{a}_1|\Big)\Big(\sqrt{\tfrac{2}{3}}|\tfrac{3}{2},-\tfrac{1}{2};\mathfrak{a}_3\rangle + \tfrac{1}{\sqrt{3}}|\tfrac{1}{2},-\tfrac{1}{2};\mathfrak{a}_1\rangle\Big) \\ &= \tfrac{\sqrt{2}}{3}\langle\tfrac{3}{2},-\tfrac{1}{2};\mathfrak{a}_3||\tfrac{3}{2},-\tfrac{1}{2};\mathfrak{a}_3\rangle - \tfrac{\sqrt{2}}{3}\langle\tfrac{1}{2},-\tfrac{1}{2};\mathfrak{a}_1||\tfrac{1}{2},-\tfrac{1}{2};\mathfrak{a}_1\rangle \\ &= \tfrac{\sqrt{2}}{3}\mathfrak{M}_{3/2} - \tfrac{\sqrt{2}}{3}\mathfrak{M}_{1/2}. \end{aligned} \tag{4.89}$$

The labels "$\mathfrak{a}_3$" and "$\mathfrak{a}_1$" are arrays of all other quantifiers of the isospin-$\frac{3}{2}$ and isospin-$\frac{1}{2}$ states, respectively. The effective cross-sections of these processes are then related as

$$\sigma_a : \sigma_c : \sigma_f : \sigma_j \;=\; 9|\mathfrak{M}_{3/2}|^2 : |\mathfrak{M}_{3/2} + 2\mathfrak{M}_{1/2}|^2 : 9|\mathfrak{M}_{3/2}|^2 : 2|\mathfrak{M}_{3/2} - \mathfrak{M}_{1/2}|^2. \tag{4.90}$$

In a collision regime where either $\mathfrak{M}_{3/2} \gg \mathfrak{M}_{1/2}$ or $\mathfrak{M}_{3/2} \ll \mathfrak{M}_{1/2}$, this relationship simplifies:

$$\mathfrak{M}_{3/2} \gg \mathfrak{M}_{1/2} \Rightarrow \qquad \sigma_a : \sigma_c : \sigma_f : \sigma_j \approx 9 : 1 : 9 : 2, \tag{4.91}$$

$$\mathfrak{M}_{3/2} \ll \mathfrak{M}_{1/2} \Rightarrow \qquad \sigma_a : \sigma_c : \sigma_f : \sigma_j \approx 0 : 4 : 0 : 2. \tag{4.92}$$

4.3.2 Isospin in the quark model

In processes where it suffices to track only the u and d quarks, the application of the isospin formalism is very similar within the quark model. Writing only isospin factors,

$$|u\rangle = |\tfrac{1}{2},+\tfrac{1}{2}\rangle, \qquad |d\rangle = |\tfrac{1}{2},-\tfrac{1}{2}\rangle \tag{4.93}$$

and so we have the three-quark bound states:

$$|\Delta^{++}\rangle = |u\rangle\otimes|u\rangle\otimes|u\rangle = |\tfrac{1}{2},+\tfrac{1}{2}\rangle|\tfrac{1}{2},+\tfrac{1}{2}\rangle|\tfrac{1}{2},+\tfrac{1}{2}\rangle = |\tfrac{3}{2},+\tfrac{3}{2}\rangle, \tag{4.94a}$$

$$|\Delta^{+}\rangle = |u\rangle\otimes|u\rangle\otimes|d\rangle = |\tfrac{1}{2},+\tfrac{1}{2}\rangle|\tfrac{1}{2},+\tfrac{1}{2}\rangle|\tfrac{1}{2},-\tfrac{1}{2}\rangle = |\tfrac{3}{2},+\tfrac{1}{2}\rangle, \quad |p^+\rangle = |\tfrac{1}{2},+\tfrac{1}{2}\rangle, \tag{4.94b}$$

$$|\Delta^0\rangle = |u\rangle\otimes|d\rangle\otimes|d\rangle = |\tfrac{1}{2},+\tfrac{1}{2}\rangle|\tfrac{1}{2},-\tfrac{1}{2}\rangle|\tfrac{1}{2},-\tfrac{1}{2}\rangle = |\tfrac{3}{2},-\tfrac{1}{2}\rangle, \quad |n^0\rangle = |\tfrac{1}{2},-\tfrac{1}{2}\rangle, \tag{4.94c}$$

$$|\Delta^-\rangle = |d\rangle\otimes|d\rangle\otimes|d\rangle = |\tfrac{1}{2},-\tfrac{1}{2}\rangle|\tfrac{1}{2},-\tfrac{1}{2}\rangle|\tfrac{1}{2},-\tfrac{1}{2}\rangle = |\tfrac{3}{2},-\tfrac{3}{2}\rangle. \tag{4.94d}$$

The Δ^+ particle is not identical with the proton: The isospin factors differ in the value of I, but the full wave-functions also differ in the *spin* factors: Δ^+ has spin $\frac{3}{2}\hbar$ and the proton spin is $\frac{1}{2}\hbar$. Similarly, Δ^0 and n^0 have similar isospin factors – which is identified as the "bookkeeping" notation of the u-d content. So, for example, there exists the decay[11]

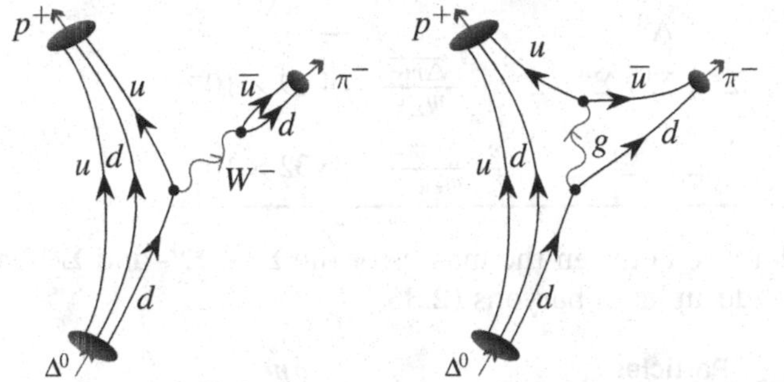

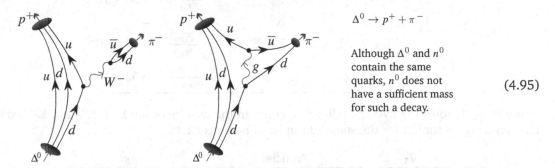

$\Delta^0 \to p^+ + \pi^-$

Although Δ^0 and n^0 contain the same quarks, n^0 does not have a sufficient mass for such a decay. (4.95)

From these Feynman diagrams we estimate that the right-hand contribution to the process amplitude is proportional to the square of the strong charge (because of the two gluon vertices) and the left-hand contribution is proportional to the square of the weak charge (because of the two W^--vertices). Owing to the immense difference in the strength of these interactions, the weak contribution is negligible and is not calculated. The right-hand diagram may be cut by any curve into the "initial" and "final" states, and the total isospin factor for the so-defined "initial" and "final" state vector computed. Across each such cut of the diagram, isospin is conserved.

The ease of application of this combination of Feynman diagrams, isospin factors in state vectors and of quick estimates of relative strengths of the contributions to the amplitude of the process is the basic reason for the Feynman diagrams' popularity. Owing to the relative simplicity of the $SU(2)$ group, isospin factors here do not give much more information than the u-d content of the diagrams, but it is clear that they are nevertheless useful in estimates using the Wigner–Eckart theorem, just as in the results (4.78)–(4.90).

4.3.3 Exercises for Section 4.3

✎ **4.3.1** *Using equation (4.83) and following the derivation of equation (4.90), find the ratios between the probabilities for all ten pion–nucleon scattering processes (4.82).*

✎ **4.3.2** *Evaluate the result (4.90) in the limit when $\mathfrak{M}_{3/2} = \mathfrak{M}_{1/2}$.*

✎ **4.3.3** *Evaluate your solution to the problem 4.3.1 in the limit when $\mathfrak{M}_{3/2} \gg \mathfrak{M}_{1/2}$.*

✎ **4.3.4** *Evaluate your solution to the problem 4.3.1 in the limit when $\mathfrak{M}_{3/2} = \mathfrak{M}_{1/2}$.*

✎ **4.3.5** *Evaluate your solution to the problem 4.3.1 in the limit when $\mathfrak{M}_{3/2} \ll \mathfrak{M}_{1/2}$.*

---

[11] Section 4.4.2 will discuss the isospin details of π^- as a bound state of a d- and an anti-u-quark.

4.4 The eightfold way, the $SU(3)_f$ group and the u, d, s quarks

Some two decades after Heisenberg and Wigner introduced the isospin formalism and the $SU(2)$ group of symmetries, several elementary particle physics researchers realized that similar benefits might be derived from grouping the eight baryons in the plot (2.32). They all have spin $\frac{1}{2}$, and their masses are "layered" in isospin multiplets [293]:

S	I_3	Particles	δm
0	$-\frac{1}{2}, +\frac{1}{2}$	$n^0,\ p^+$	$\frac{m_n - m_p}{m_p} = 1.38 \times 10^{-3}$
-1	0	Λ^0	—
	$-1, 0, +1$	$\Sigma^-,\ \Sigma^0,\ \Sigma^+$	$\frac{\overline{\Delta m_\Sigma}}{m_{\Sigma^+}} = 4.49 \times 10^{-3}$
-2	$-\frac{1}{2}, +\frac{1}{2}$	$\Xi^-,\ \Xi^0$	$\frac{m_{\Xi^-} - m_{\Xi^0}}{m_{\Xi^0}} = 5.32 \times 10^{-3}$

(4.96)

where $\overline{\Delta m_\Sigma}$ denotes the average difference between the masses of the Σ^+-, Σ^0- and Σ^--baryons. The "layering" is similar for the spin-$\frac{3}{2}$ decuplet of baryons (2.35):

S	I_3	Particles	δm
0	$-\frac{3}{2}, -\frac{1}{2}, +\frac{1}{2}, +\frac{3}{2}$	$\Delta^-,\ \Delta^0,\ \Delta^+,\ \Delta^{++}$	$\frac{\max(\Delta m_\Delta)}{m_\Delta} = 8.117 \times 10^{-4}$
-1	$-1, 0, +1$	$\Sigma^{*-},\ \Sigma^{*0},\ \Sigma^{*+}$	$\frac{\max(\Delta m_{\Sigma^*})}{m_{\Sigma^*}} = 3.117 \times 10^{-3}$
-2	$-\frac{1}{2}, +\frac{1}{2}$	$\Xi^{*-},\ \Xi^{*0}$	$\frac{m_{\Xi^{*-}} - m_{\Xi^{*0}}}{m_{\Xi^*}} = 2.087 \times 10^{-3}$
-3	0	Ω^-	—

(4.97)

The relative difference between the average masses in any one layer is some 2 orders of magnitude bigger than the in-layer relative mass differences. For the octet (4.96):

$$\frac{\overline{m_{\Lambda,\Sigma}} - \overline{m_N}}{\overline{m_N}} = 0.2223, \qquad \frac{\overline{m_\Xi} - \overline{m_{\Lambda,\Sigma}}}{\overline{m_{\Lambda,\Sigma}}} = 0.1162; \tag{4.98}$$

and for the decuplet (4.97):

$$\frac{\overline{m_{\Sigma^*}} - \overline{m_\Delta}}{\overline{m_\Delta}} = 0.1238, \qquad \frac{\overline{m_{\Xi^*}} - \overline{m_{\Sigma^*}}}{\overline{m_{\Sigma^*}}} = 0.1075, \qquad \frac{m_\Omega - \overline{m_{\Xi^*}}}{\overline{m_{\Xi^*}}} = 0.0907. \tag{4.99}$$

Thus, the approximate isospin $SU(2)$ symmetry (which in the tabulations (4.96) and (4.97) mixes the baryons horizontally) is about a hundred times better than the $SU(3)_f$ symmetry that also includes strangeness (varying vertically in these tables). This agrees with the fact that the mass of the s-quark (as measured by deep inelastic scattering) is 2 orders of magnitude bigger than that of the u- and d-quarks; see Table 4.1 on p. 152, below.

However, not only was group theory practically unknown amongst physicists in the 1950s and 1960s, but there also existed an open animosity towards group theory as representative of "abstract mathematics." Wolfgang Pauli supposedly [577] even called group theory *Gruppenpest* (group pestilence, in German). Many results of angular momentum and isospin symmetry were obtained not using the abstract methods of group theory, but by direct computations.[12] Following

[12] Even today, the Clebsch–Gordan coefficients and the Wigner–Eckart theorem with concrete applications – the main tool in using $SU(2)$ symmetry in the past three-quarters of a century – are very rarely even mentioned in mathematical group theory textbooks. The computational methods developed mainly by physicists [565, 258, 581, 105] still have not penetrated the "mathematicians' circles." On the other hand, the "abstract mathematics" does slowly seep into fundamental physics, especially in superstring theory, and here often finds unexpected uses.

the same practice, Gell-Mann derived most results in the same, "pedestrian" way, and only later discovered the elegant arguments and derivations in the then known group theory. The "eight baryon problem," i.e., the problem of finding the right generalization of the isospin $SU(2)$ symmetry that would encompass these eight baryons, thus had a thorny path. Murray Gell-Mann's "eightfold way" is in fact a collection of results that were obtained by such pedestrian methods, mostly using isospin $SU(2)$-results, as well as several phenomenological relations between strangeness, the baryon number, charge, and other properties of particles that were observed in experiments.[13]

Today, of course, we know that the relevant group is $SU(3)_f$, which indeed has the isospin $SU(2)$ as a subgroup. In hindsight, the identification of the group was obstructed by the fact that there are no three baryons that would span the fundamental 3-dimensional representation of the $SU(3)_f$ group. The early proposal by Shoichi Sakata, whereby the Λ^0-baryon extends the isospin doublet p^+, n^0 into the $SU(3)$ triplet, could not replicate the success of the isospin $SU(2)$ classification, and was soon abandoned.

Digression 4.2 In some version of Sakata's proposal, the three baryons (p^+, n^0, Λ^0) – so-called "sakatons" – were supposed to "form" all other baryons and mesons: The mesons would be obtained resulting from sakaton–antisakaton combinations, and baryons resulting from a combination of three sakatons, in all possible combinations. Formally, that indeed does produce a reasonable list of hadronic states identified by their charges, isospin, strangeness, etc. However, it was not at all clear in what sense such "products" of baryons and antibaryons could represent much lighter mesons, or – even more puzzling – how the baryon states with the quantum numbers of the sakaton triplet, (p^+, n^0, Λ^0), could also be found within the list obtained by combining three sakatons. That would imply that any one of these three baryons could be represented as a system of three copies of these very same baryons – which clearly leads to an infinite regression and obstructs the identification of this scheme as a model in which hadrons are "really" bound states of "real" particles. Many of the supporters of the so-called "S-matrix approach" to strong interactions even openly accepted this infinitely regressive interpretation of Sakata's classification scheme.

Recall that in the 1960s – especially in the southwestern parts of the USA – variants of eastern philosophies became very popular and mixed with science [☞ e.g., Refs. [91, 487, 591]]. This additionally contributed to the prejudices against group theory and to the mystique of hadron classification. As one of the most prominent advocates of $SU(3)_f$ classification, Gell-Mann contributed to this confusion both by nomenclature ("eightfold way") and by avoiding to categorically decide for or against the infinitely regressive interpretation. In sharp contrast, Richard Feynman openly advocated the "real" particle-physicist approach, whereby hadrons are really bound states of more elementary particles, which he called *partons*, avoiding Gell-Mann's "quarks." During the 1960s, Gell-Mann gradually accepted Feynman's intuitive image – they were both at CalTech (California Institute of Technology, Pasadena) – which ultimately led to the final formulation and application of the quark model.

[13] Gell-Mann spent the 1959/60 academic year in Collège de France, looking for the right generalization of $SU(2)$, and never thought of asking the resident mathematicians – amongst whom was the world-famous Jean-Pierre Serre – for help. Only late in 1960, back at CalTech, did Gell-Mann get the help of a mathematician (Richard Block) in realizing that this generalization, the $SU(3)$ group, was already very well known amongst mathematicians [119, 577].

Gell-Mann's successful prediction of the Ω^--baryon's existence [☞ Section 4.4.3], complete with its quantum numbers and its mass approximately given by the relation (2.37), was essential for accepting his "eightfold way," i.e., the classifying application of the $SU(3)_f$ group of "flavors." The original idea for the eightfold way stemmed from "finding a home" for the isospin doublet p^+–n^0 not in the direct generalization – such as Sakata's triplet – but in the *octet* of spin-$\frac{1}{2}$ baryons (2.32). In comparison, it was clear that the nine spin-$\frac{3}{2}$ baryons (4Δ, $3\Sigma^*$ and $2\Xi^*$) had to form a bigger multiplet (2.35), so that the classification scheme also had to contain in a natural way multiplets bigger than the octet, but to *not contain* multiplets such as 4-plets, 5-plets, etc.

Besides, the classification of hadrons turned out much simpler upon accepting the quark model, where the u-, d- and s-quarks span the fundamental 3-dimensional representation [☞ Section A.1.4], and mesons and baryons are bound states of quarks. Using the $SU(3)_f$ group, it is fairly easy to show that the meson and baryon multiplets must have 8, 10, 27, 28, 35, ... particles, and not some other numbers – although the $SU(3)$ group also has representations of dimensions 3, 6, 15, 21, 24, ...

To wit, representations of the $SU(3)$ group also have the so-called "triality," which is additive modulo 3 [☞ Section A.4]. The elements of the fundamental, 3-dimensional representation – i.e., the u-, d- and s-quarks – have triality 1, antiquarks triality $-1 \cong 2$, and states with n quarks and $\overline{n}$ antiquarks then have triality $(n-\overline{n})$ (mod 3). So, if both mesons and baryons must have triality of 0, this immediately rules out the $SU(3)$ representations of dimensions 3, 6, 15, 21, 24, ..., which were indeed never observed. The "triality-0" condition selects the representations of dimensions 1, 8, 10, 27, 28, 35, etc., of which, however, only the first three groupings have ever been observed.

In turn, using that mesons are quark–antiquark (**3**-**3***) bound states and since $\mathbf{3}\otimes\mathbf{3}^* = \mathbf{1}\oplus\mathbf{8}$, mesons may only form singlets and octets of the $SU(3)_f$ classification group. Similarly, using that baryons are three-quark (**3**-**3**-**3**) bound states and since $\mathbf{3}\otimes\mathbf{3}\otimes\mathbf{3} = \mathbf{1}\oplus\mathbf{8}\oplus\mathbf{8}\oplus\mathbf{10}$, baryons may only form singlets, octets and decuplets; see example A.6. Consequently, the triality-0 representations **27**, **28**, **35**, etc., may only appear as metastable multi-baryon and multi-meson states [☞ Appendix A.4.2].

Of course, just like the isospin $SU(2)$ symmetry, the $SU(3)_f$-transformations are only approximate symmetries, and with a bigger tolerance.[14] With the discovery of the J/ψ-particle and the c-quark, the $SU(3)_f$-symmetry was extended into the $SU(4)_f$-symmetry, which implies even bigger tolerance, etc. This progressively growing tolerance – i.e., measure of imprecision – of the classifying $SU(n)_f$-symmetry is reflected in the *effective masses* of quarks:[15] see Table 4.1 The basic

Table 4.1 Quark masses in MeV/c^2 [☞ Figure 2.1 on p. 76]

			Effective masses in	
	Quark	**Mass**	**Mesons**	**Baryons**
Light	u	4.2	310	363
	d	7.5		
	s	150	483	538
Heavy	c	1,100	1,500	
	b	4,200	4,700	
	t	174,200	$\gtrsim$ 174,200	

[14] In this context, the tolerance of an approximate symmetry is the margin of permitted difference between the masses of the particles linked by the purported symmetry.

[15] By its definition, mass is the measure of the object's inertia. Since quarks cannot be isolated, neither can their inertia be measured as for free particles. Their effective mass is the measure of their inertia within the bound state (meson or baryon), and so is always affected by the interactions with the "rest" of that bound state, i.e., with the other quarks and gluons and depends on the specific bound state. For the "effective mass," one then always cites average values.

idea in applications of such a *phenomenologically* defined $SU(n)_f$-symmetry is simple: Let G be the group of approximate symmetries with a given tolerance, and $H \subset G$ a subgroup of approximate symmetries with a finer tolerance. The contributions to the Hamiltonian that are H-invariant but not G-invariant are treated as "corrections" to the initial Hamiltonian that is G-invariant. A larger tolerance level implies a larger group of approximate symmetries, and a smaller (finer) tolerance level reduces the group of operations that are accepted as approximate symmetries.

The best known example for this idea is the so-called Zeeman effect: non-relativistic treatment of the hydrogen atom with neglected spins (4.1)–(4.8e) is subjected to an *external* magnetic field $\vec{B}$, which adds to the Hamiltonian the "correction"

$$H_Z = -\vec{\mu} \cdot \vec{B} = \mu_B \vec{B} \cdot (g_\ell L + g_s S), \qquad g_\ell = 1, \;\; g_s = 2\left(1 + \tfrac{\alpha}{2\pi} + \cdots\right). \tag{4.100}$$

The basic Hamiltonian, without this correction, has $Spin(4)$ symmetry [☞ Section 4.1.1], whereas the Hamiltonian with the Zeeman addition only has $Spin(2) \subset Spin(4)$ symmetry. In particle-physicist parlance, one says that the Zeeman interaction with the external magnetic field – and so that external magnetic field itself – explicitly breaks the $Spin(4)$ symmetry of the hydrogen atom. By analogy, and because of the quark mass-hierarchy given in Table 4.1 on p. 152, the Hamiltonian terms, i.e., the mass contributions for mesons and baryons may be organized as:

1. the $SU(6)_f$-symmetric, original Hamiltonian;
2. the $SU(5)_f$-symmetric "corrections," where the t-quark is separated by contributions of order $m_t \approx 174.2\,\text{GeV}/c^2$;
3. the $SU(4)_f$-symmetric "corrections," where also the b-quark is separated by contributions of order $m_b \approx 4.7\,\text{GeV}/c^2$;
4. the $SU(3)_f$-symmetric "corrections," where also the c-quark is separated by contributions of order $m_c \approx 1.5\,\text{GeV}/c^2$;
5. the $SU(2)_I$-symmetric "corrections," where also the s-quark is separated by contributions of order $m_s \approx 0.5\,\text{GeV}/c^2$;
6. the final "corrections" also break the isospin $SU(2)_I$-symmetry, and finally separate the u- and d-quarks.

In practice, this approach is used in combination with other, more directly physics-inspired ideas. The next few sections will peek into some of those estimates.

4.4.1 Quarkonium

Mesons are bound states of a quark and an antiquark, so their analysis should follow the analysis of two-body bound states, akin to the hydrogen atom and positronium [☞ Section 4.1]. There is, however, a huge difference! In the hydrogen atom, the ratio of the binding energy and the rest energy of (either of) the bound particles is $13.6\,\text{eV}/510.999\,\text{keV} \approx 2.66 \times 10^{-5}$. In contradistinction, the binding energy of the quarks in mesons and baryons is in fact infinitely large, since the quarks cannot be extracted from these bound states. This involves the fact that in attempting to extract a quark one must invest amounts of energy that are at least comparable with the rest energy of the quarks themselves, whereby it is energetically more favorable to convert the invested energy into new–quark antiquark pairs

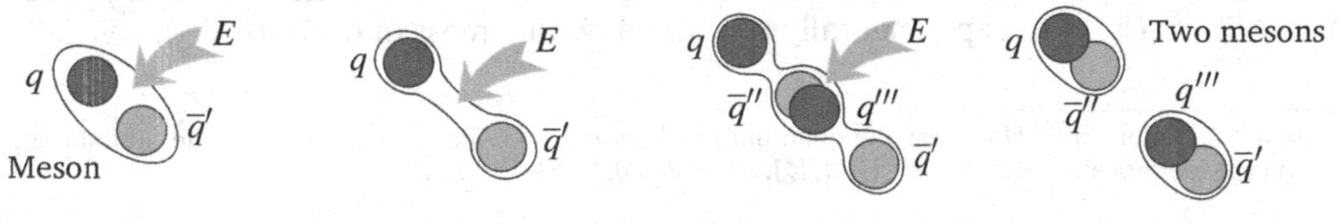

(4.101)

rather than further deforming the original bound state. Because of this possibility of creating quark–antiquark pairs, the process is essentially relativistic and definitely within the domain of field theory, where the number of particles is not conserved, as it is in standard quantum mechanics.

Besides, in the case of the hydrogen atom and exotic "atoms" such as muonium and positronium, the basic – Coulomb – potential is well known. In the case of strong interactions, however, there is no well-defined potential in the same sense: Recall that the Coulomb potential is a field that extends around the given electrically charged particle. In all points of space, it gives the information as to how the electrostatic force at that point would act upon a probing electric charge – *if and when* such a probing electric charge is placed at that point. In the case of electrodynamics, this mental construction has an excellent physical meaning, since it is physically possible to test the Coulomb field of a given particle with probing electric charges, which we really can move and place at will. Upon quarks, which are forever confined within mesons and baryons, we exert far less control.

In the case of mesons built of "heavy" quarks: c, b and t, it is possible to apply the analysis following the positronium template. Just before the discovery of the J/ψ-particle, Hugh David Politzer and Thomas Appelquist concluded that the c-quark – were it to exist following the logic of the so-called Glashow–Iliopoulos–Maiani (GIM) mechanism – would have to have non-relativistic $(c\,\bar{c})$ bound states akin to positronium, and which they called "charmonium." When the J/ψ-particle was experimentally discovered, it was immediately identified as the $1\,^3S_2$-state of charmonium,[16] and soon the other $n = 1, 2$ states (except $2\,^1P_1$, the detection of which poses exceptional experimental difficulties) were found.

The charmonium states are very well approximated following the positronium template, if the potential is modeled as

$$V_c = -\frac{4}{3}\frac{\alpha_s \hbar c}{r} + F_0\, r, \tag{4.102}$$

where F_0 is a coefficient of about 16 tons, and α_s is the strong interaction analogue of the fine structure constant; the coefficient $\frac{4}{3}$ will be computed in equation (6.68). This potential, of course, grows infinitely and so gives infinitely many bound states, the energy E_n of which asymptotically grows as $n^{3/2}$. [✎ *Why?*] However, the masses of the bound states with $n \geqslant 3$ are bigger than the "$D\overline{D}$-threshold" (the masses of the lightest D–$\overline{D}$ meson pair), so that such $(c\,\bar{c})$-states very quickly decay and are regarded as quasi-bound states. Table 4.2 lists a few lightest mesons that contain the c-quark.

— ❦ —

The story was repeated a few years later: in 1976, E. Eichten and K. Gottfried predicted that "bottomonium" would have to have even more true bound states than charmonium. When the first Υ-particle was detected in 1977, it was identified as the $1\,^3S_1$ state of the $(b\,\bar{b})$ system and during the next few years the existence of the $(b\,\bar{b})$-bound states with $n \leqslant 4$ was experimentally confirmed.

Finally, the $(t\,\bar{t})$ system was only recently detected owing to the much larger t-quark mass, and the "toponium" states are still relatively unexplored. Also, the "mixed" $(c\bar{b})$-, $(c\bar{t})$-, $(b\bar{t})$-states and their conjugates may be analyzed akin to the muonium $(\mu^+\,e^-)$. The first of these particles, $B_c^+ = (c\,\bar{b})$ and $B_c^- = (b\,\bar{c})$ are experimentally confirmed, with a mass of 6.276 GeV/c^2.

[16] The spectroscopic notation "$n\,^{2S+1}L_J$" gives the quantum numbers n, ℓ, s, j, where the letter gives the orbital angular momentum via the identification of $S, P, D, F, G, H, I, J, \ldots$ as $\ell = 0, 1, 2, 3, 4, 5, 6, 7, \ldots$

Table 4.2 Lightest mesons containing the c-quark; masses in MeV/c^2

Name	$n\,^{2S+1}L_J$	J^{PC}	Mass
η_c	$1\,^1S_0$	0^{-+}	2,980.3
J/ψ	$1\,^3S_1$	1^{--}	3,096.9
χ_{c0}	$1\,^3P_0$	0^{++}	3,414.8
χ_{c1}	$1\,^3P_1$	0^{++}	3,510.7
χ_{c2}	$1\,^3P_2$	0^{++}	3,556.2
η_c	$2\,^1S_0$	0^{-+}	3,637
J/ψ	$2\,^3S_1$	1^{--}	3,686.1

Charmonium: $(c\,\bar{c})$ **states**

Name	$(q\,\bar{q})$	J^P	Mass
D^+	$(c\,\bar{d})$	0^-	1,869.6
D^-	$(d\,\bar{c})$		
D^0	$(c\,\bar{u})$	0^-	1,864.8
$\overline{D}{}^0$	$(u\,\bar{c})$		
D^{*0}	$(c\,\bar{u})$	1^-	2,007.0
$\overline{D}{}^{*0}$	$(u\,\bar{c})$		
D^{*+}	$(c\,\bar{d})$	1^-	2,010.3
D^{*-}	$(d\,\bar{c})$		

4.4.2 Light mesons

Mesons that contain a light quark or antiquark automatically must be analyzed as relativistic bound states – for which there is no complete theoretical description✍.[17] Therefore, we remain content herein with classification.

The first fact worth noting is that although there are three quarks at our disposal, and so *nine* possible $(q\,\bar{q})$ bound states (fully neglecting spins, orbital angular momentum and dynamical details), mesons appear in groups of *eight* [☞ plot (2.31) as well as the result (A.76c)].

The reason for this is similar to the fact that with only two quarks, u and d, there exist not four but only three pions, $\pi^\pm, \pi^0$. The $SU(2)_I$ symmetry solves this puzzle by the method of isospin "addition." But, before that, the isospin of $\bar{u}$ and $\bar{d}$ must be established. In tensor notation, we have

$$\{t^1, t^2\} = \{u,\ d\} \quad \Rightarrow \quad \{t_1, t_2\} = \{\bar{u},\ \bar{d}\}. \tag{4.103}$$

However, since $\varepsilon_{\alpha\beta}$ is $SU(2)$-invariant [✎ *why?*], we may identify $(t^\alpha)^\dagger = t_\alpha = \varepsilon_{\alpha\beta}t^\beta$, so that (regarding isospin properties only!)

$$t_1 = \varepsilon_{12}t^2 = t^2 \ \Rightarrow\ |\bar{u}\rangle = |d\rangle \quad = |\tfrac{1}{2}, -\tfrac{1}{2}\rangle, \tag{4.104a}$$

$$\text{and} \qquad t_2 = \varepsilon_{21}t^1 = -t^1 \ \Rightarrow\ |\bar{d}\rangle = -|u\rangle = -|\tfrac{1}{2}, +\tfrac{1}{2}\rangle. \tag{4.104b}$$

Thus,

$$\{|\bar{u}\rangle,\ |\bar{d}\rangle\} \otimes \{|u\rangle,\ |d\rangle\} = \{|\tfrac{1}{2}, -\tfrac{1}{2}\rangle, -|\tfrac{1}{2}, +\tfrac{1}{2}\rangle\} \otimes \{|\tfrac{1}{2}, +\tfrac{1}{2}\rangle, |\tfrac{1}{2}, -\tfrac{1}{2}\rangle\} \tag{4.105}$$

$$= \{|1, \pm1\rangle,\ |1, 0\rangle\} \oplus \{|0, 0\rangle\} = \{|\pi^\pm\rangle,\ |\pi^0\rangle\} \oplus \{|\eta\rangle\}, \tag{4.106}$$

where

$$\begin{aligned}
|1, +1\rangle &= |\tfrac{1}{2}, +\tfrac{1}{2}\rangle|\tfrac{1}{2}, +\tfrac{1}{2}\rangle &&= -|\bar{d}\rangle|u\rangle,\\
|1, 0\rangle &= \tfrac{1}{\sqrt{2}}\big(|\tfrac{1}{2}, +\tfrac{1}{2}\rangle|\tfrac{1}{2}, -\tfrac{1}{2}\rangle + |\tfrac{1}{2}, -\tfrac{1}{2}\rangle|\tfrac{1}{2}, +\tfrac{1}{2}\rangle\big) &&= \tfrac{1}{\sqrt{2}}\big(-|\bar{d}\rangle|d\rangle + |\bar{u}\rangle|u\rangle\big),\\
|1, -1\rangle &= |\tfrac{1}{2}, -\tfrac{1}{2}\rangle|\tfrac{1}{2}, -\tfrac{1}{2}\rangle &&= |\bar{u}\rangle|d\rangle,\\
|0, 0\rangle &= \tfrac{1}{\sqrt{2}}\big(|\tfrac{1}{2}, +\tfrac{1}{2}\rangle|\tfrac{1}{2}, -\tfrac{1}{2}\rangle - |\tfrac{1}{2}, -\tfrac{1}{2}\rangle|\tfrac{1}{2}, +\tfrac{1}{2}\rangle\big) &&= \tfrac{1}{\sqrt{2}}\big(-|\bar{d}\rangle|d\rangle - |\bar{u}\rangle|u\rangle\big).
\end{aligned} \tag{4.107}$$

[17] N.B. One of the approaches is the so-called (MIT) "bag model": One approximates that quarks are free particles while within the meson, where they are confined by outside "pressure," which produces an impenetrable "bag." This "bag" is a 3-dimensional infinitely deep potential, the walls of which have a time-variable shape. Although the model is phenomenologically successful, it is clear that this is an *ad hoc* fiction where one needs to explain the dynamical origin of that "pressure." Another approach uses the so-called Gribov version of gauge theory of strong interactions and quark confinement [388].

Then[18]

$$\begin{cases} |\pi^+\rangle = -|\bar{d}\,u\rangle, \\ |\pi^0\rangle = \frac{1}{\sqrt{2}}(|\bar{u}\,u\rangle - |\bar{d}\,d\rangle), \\ |\pi^-\rangle = |\bar{u}\,d\rangle, \end{cases} \qquad \text{and} \qquad |\eta\rangle = -\frac{1}{\sqrt{2}}(|\bar{u}\,u\rangle + |\bar{d}\,d\rangle). \tag{4.108}$$

Note the signs that stem from the $SU(2)_I$-identification $|\bar{d}\rangle = -|\frac{1}{2},+\frac{1}{2}\rangle$, whereby $|\pi^0\rangle$ *looks* like an antisymmetric combination, but is not: a quark and an antiquark cannot be thought of as particles that are "identical up to some 'polarization' (or other selectable property)," so as to define the exchange (anti)symmetry. Instead, if we use u, d as the basis and $\bar{u}, \bar{d}$ as its conjugate basis, the three pion states (4.108) form a Hermitian matrix with no trace, whereas the η-state represents the trace of a Hermitian matrix.

So, define q^α so that $q^1 = u$ and $q^2 = d$, and

$$\pi^+ = (\bar{q}_\beta\,(\sigma^+)_\alpha{}^\beta\,q^\alpha) \quad = (\bar{d}u), \tag{4.109a}$$

$$\pi^0 = \tfrac{1}{\sqrt{2}}(\bar{q}_\beta\,(\sigma^3)_\alpha{}^\beta\,q^\alpha) = \tfrac{1}{\sqrt{2}}(\bar{u}u - \bar{d}d), \tag{4.109b}$$

$$\pi^- = (\bar{q}_\beta\,(\sigma^-)_\alpha{}^\beta\,q^\alpha) \quad = (\bar{u}d), \tag{4.109c}$$

where $\sigma^\pm := \frac{1}{2}[\sigma^1 \pm i\sigma^2]$ and $\sigma^1, \sigma^2, \sigma^3$ are Pauli matrices:

$$\sigma^1 = \begin{bmatrix} 0 & 1 \\ 1 & 0 \end{bmatrix}, \qquad \sigma^2 = \begin{bmatrix} 0 & -i \\ i & 0 \end{bmatrix}, \qquad \sigma^3 = \begin{bmatrix} 1 & 0 \\ 0 & -1 \end{bmatrix}, \tag{4.109d}$$

the halves of which satisfy the defining relations (A.38a) of the $SU(2)$ algebra.

However, comparison with experiments does not single out an unambiguous candidate for $|\eta\rangle$: in fact, there exist two spin-0 (pseudo-scalar, $J^{PC} = 0^{-+}$) particles with isospin $|0,0\rangle$:

$$\eta: \ 547.853\,\text{MeV}/c^2, \qquad \eta': \ 957.66\,\text{MeV}/c^2, \tag{4.110}$$

as well as two spin-1 (vectorial, $J^{PC} = 1^{--}$) excitations:

$$\omega: \ 782.65\,\text{MeV}/c^2, \qquad \phi: \ 1019.455\,\text{MeV}/c^2. \tag{4.111}$$

Since the masses of the η- and ω-mesons are larger than the kaon masses (2.31), it is clear that the classification must also include the mesons that contain the s-quark, and also that η and η' (and similarly ω and ϕ) are linear combinations that also contain (s-$\bar{s}$)-contributions.

Generalizing the identification (4.109), by means of including the third quark, $q^3 = s$, and using Gell-Mann's matrices (A.71), we have

$$\pi^+ = (\bar{d}u), \quad \pi^- = (\bar{u}d), \quad \pi^0 = \tfrac{1}{\sqrt{2}}(\bar{u}u - \bar{d}d), \quad \eta = \tfrac{1}{\sqrt{6}}(\bar{u}u + \bar{d}d - 2\bar{s}s), \tag{4.112a}$$

$$K^+ = (\bar{s}u), \quad K^0 = (\bar{s}d), \quad \overline{K}^0 - = (\bar{d}s), \quad K^- = (\bar{u}s), \tag{4.112b}$$

and

$$\eta' = \tfrac{1}{\sqrt{3}}(\bar{u}u + \bar{d}d + \bar{s}s). \tag{4.112c}$$

These $8+1$ states with antiparallel ($S = 0$) quark spins and orbital angular momentum $\ell = 0$ then have their total angular momentum $j = 0$.

---

[18] In fact, the *real* η^0 particle is a linear combination of not only $|u\rangle\otimes|\bar{u}\rangle$ and $|d\rangle\otimes|\bar{d}\rangle$, but also of $|s\rangle\otimes|\bar{s}\rangle$; see equations (4.112).

Combining the antiquark triplet $\{\bar{q}_\alpha,\ \alpha = 1,2,3\}$ and quark triplet $\{q^\alpha,\ \alpha = 1,2,3\}$ into 8+1 mesons (4.112) then precisely follows the $SU(3)$ decomposition (A.76c):

$$\mathbf{3}^* \otimes \mathbf{3} = \mathbf{8} \oplus \mathbf{1}. \tag{4.113}$$

Of course, this $SU(3)_f$-symmetry is approximate: in reality, kaons are heavier than the pions, as they contain the heavier s-quark instead of the lighter d-quark. Besides, the η-meson is an integral part of the $SU(3)_f$ octet and its mass is just barely larger than the kaon mass, reflecting the approximate nature of the $SU(3)_f$-symmetry. On the other hand, the η'-meson does not belong in the $SU(3)_f$ octet, but is a "singlet" – i.e., an $SU(3)_f$-invariant.

Excitation of these states where the sum of quark spins and orbital angular momentum equals 1 then produces the "vector" (spin-1) $\rho^\pm$-, ρ^0-, $K^{*\pm}$-, K^{*0}-, $\overline{K}^{*0}$- and ϕ-mesons. The total angular momentum of the bound state – with all contributions, orbital and spin – is the spin of the bound state as a particle. While the masses of the charged vector-mesons follow those of the charged pseudo-scalar mesons, ϕ- and ω-mesons mix "maximally." Experiments indicate that

$$\omega \neq \tfrac{1}{\sqrt{6}}(\bar{u}u + \bar{d}d - 2\bar{s}s), \qquad \text{but} \qquad \omega \approx \tfrac{1}{\sqrt{2}}(\bar{u}u + \bar{d}d); \tag{4.114a}$$

$$\phi \neq \tfrac{1}{\sqrt{3}}(\bar{u}u + \bar{d}d + \bar{s}s), \qquad \text{but} \qquad \phi \approx (\bar{s}s). \tag{4.114b}$$

The vector and pseudo-scalar mesons turn out to differ predominantly in the relative orientation of quark spins and both are well described as S-states, with no relative angular momentum. The difference in their masses should then stem from the spin–spin interaction, akin to the S_e–S_p contribution (4.27c) to hyperfine structure in the spectrum of the hydrogen atom. Thus, the meson mass is parametrized as

$$M(\text{meson}) \approx m_q + m_{\bar{q}} + \frac{A}{m_q m_{\bar{q}}} \langle S_q \cdot S_{\bar{q}} \rangle, \tag{4.115}$$

where the coefficient A is some multiple of $|\Psi(\vec{0},t)|^2$ that cannot be computed reliably for a relativistic system, and so is determined by comparing with experimental data. Using the well-known "trick":

$$\vec{S} := \vec{S}_q + \vec{S}_{\bar{q}} \quad \Rightarrow \quad \vec{S}_q \cdot \vec{S}_{\bar{q}} = \tfrac{1}{2}(S^2 - S_q^2 - S_{\bar{q}}^2), \tag{4.116}$$

the difference between the observed average masses of pseudo-scalar and vector mesons is rather well explained:

$$\vec{S}_q \cdot \vec{S}_{\bar{q}} = \begin{cases} \tfrac{1}{4}\hbar^2, & \text{for } S = 1 \text{ (vector mesons)}, \\ -\tfrac{3}{4}\hbar^2, & \text{for } S = 0 \text{ (pseudo-scalar mesons)}. \end{cases} \tag{4.117}$$

Using the effective (so-called "constituent") masses of quarks inside mesons from Table 4.1 on p. 152, and the best value of the parameter $A \approx \frac{4m_u^2}{\hbar^2}\,160\,\text{MeV}/c^2$, the masses of pseudo-scalar and vector mesons are obtained to within 1% from the experimental value [☞ Table 4.3] – except for the η'-meson, the mass of which poses an exceptional problem for the quark model [☞ commentary in Ref. [445]].

In this way the quark model with the $SU(3)_f$-symmetry predicts an infinitely growing ladder of meson $(\mathbf{8} + \mathbf{1})$ nonets, in good agreement with experiments up to the indicated discrepancies (4.114); Table 4.4 lists the first few nonets.

4.4.3 Baryons

The number of experimentally detected baryons composed of the u-, d- and s-quarks grows faster with mass than is the case with mesons. Foremost, this happens because of the fact that baryons are three-particle bound states, so that there exist combinatorially more different interactive contributions to the mass – such as (4.16) and (4.27a)–(4.27c), as well as (4.46a)

Table 4.3 Average masses of pseudo-scalar and vector mesons, in MeV/c^2. The η'-meson mass poses an exceptional problem for the quark model; see commentary in Ref. [445].

Meson	Computed	Measured	Meson	Computed	Measured
π	140	138	ρ	780	776
K	485	496	ω	780	783
η	559	549	K^*	896	892
η'	303	958	ϕ	1,032	1,020

Table 4.4 Lightest meson nonets in the $SU(3)_f$ quark model

			Nonet content			Mass*
ℓ	S	J^{PC}	$I=1$	$I=\frac{1}{2}$	$I=0$	(MeV/c^2)
0	0	0^{-+}	π	K	η,η'	500
	1	1^{--}	ρ	K^*	ω,ϕ	800
1	0	1^{+-}	B	Q_2	$H,?$	1,250
	1	0^{++}	δ	κ	ϵ, S^*	1,150
		1^{++}	A_1	Q_1	D,E	1,300
		2^{++}	A_2	K^*	f,f'	1,400

* Rough averages; see plot (2.31) and Ref. [293]

and (4.47) [☞ plots (2.32) and (2.35)]. These significantly complicate the computations and even just the estimates, hindering the experimental identification as to which baryon belongs to which multiplet.

Classification

As three-particle systems, baryons have *two* orbital angular momenta: the orbital angular momentum of any two of the three quarks about their center of mass, and then the orbital angular momentum of that two-quark system and the third quark about their joint center of mass. We will consider only states with $n = 1$ and $\ell = 0 = \ell'$, the masses of which are easily shown to be the lowest.[19] In this case, the baryon spin stems exclusively from the sum of the quark spins, for which the addition of three spins of magnitude $\frac{1}{2}$ we have

$$\left\{|\tfrac{3}{2},\pm\tfrac{3}{2}\rangle,\ |\tfrac{3}{2},\pm\tfrac{1}{2}\rangle\right\}_S, \qquad \left\{|\tfrac{1}{2},\pm\tfrac{1}{2}\rangle_{[12]}\right\}, \qquad \left\{|\tfrac{1}{2},\pm\tfrac{1}{2}\rangle_{[23]}\right\}. \tag{4.118}$$

The subscript S denotes total symmetry and, following Ref. [243], we use the basis

$$|\tfrac{1}{2},+\tfrac{1}{2}\rangle_{[12]} = \tfrac{1}{\sqrt{2}}\big(|{\uparrow\downarrow\uparrow}\rangle - |{\downarrow\uparrow\uparrow}\rangle\big), \qquad |\tfrac{1}{2},-\tfrac{1}{2}\rangle_{[12]} = \tfrac{1}{\sqrt{2}}\big(|{\uparrow\downarrow\downarrow}\rangle - |{\downarrow\uparrow\downarrow}\rangle\big); \tag{4.119}$$

$$|\tfrac{1}{2},+\tfrac{1}{2}\rangle_{[23]} = \tfrac{1}{\sqrt{2}}\big(|{\uparrow\uparrow\downarrow}\rangle - |{\uparrow\downarrow\uparrow}\rangle\big), \qquad |\tfrac{1}{2},-\tfrac{1}{2}\rangle_{[23]} = \tfrac{1}{\sqrt{2}}\big(|{\downarrow\uparrow\downarrow}\rangle - |{\downarrow\downarrow\uparrow}\rangle\big), \tag{4.120}$$

which are antisymmetric with respect to the exchange of the particles indicated in the subscript. It is not hard to show that

$$|\tfrac{1}{2},+\tfrac{1}{2}\rangle_{[13]} = \tfrac{1}{\sqrt{2}}\big(|{\uparrow\uparrow\downarrow}\rangle - |{\downarrow\uparrow\uparrow}\rangle\big) = |\tfrac{1}{2},+\tfrac{1}{2}\rangle_{[12]} + |\tfrac{1}{2},+\tfrac{1}{2}\rangle_{[23]}, \tag{4.121}$$

[19] The mass of a baryon as a bound state of three quarks equals the sum of the masses of the constituent quarks, minus the mass equivalent of the binding energy. Then, the strongest-bound baryons are also the lightest amongst the possible bound states of the given quarks.

and similarly for $|\frac{1}{2}, -\frac{1}{2}\rangle_{[13]}$. We introduced the abbreviations:

$$|\uparrow\rangle := |\tfrac{1}{2}, +\tfrac{1}{2}\rangle, \quad |\downarrow\rangle := |\tfrac{1}{2}, -\tfrac{1}{2}\rangle, \quad |\uparrow\downarrow\uparrow\rangle := |\tfrac{1}{2}, +\tfrac{1}{2}\rangle|\tfrac{1}{2}, -\tfrac{1}{2}\rangle|\tfrac{1}{2}, +\tfrac{1}{2}\rangle, \quad \text{etc.} \tag{4.122}$$

Using the approximate $SU(3)_f$-symmetry, u-, d- and s-quarks are treated as if they were different polarizations of the same fermion, so that Pauli's exclusion principle must be applied. That is, the *entire* wave-function of the bound state of three quarks must be antisymmetric with respect to the exchange of any two of the three quarks. The wave-function for the baryon then factorizes:

$$\Psi(\text{baryon}) = \Psi(\vec{r}, t)\, \chi(\text{spin})\, \chi(\text{flavor})\, \chi(\text{color}). \tag{4.123}$$

For states with $\ell = 0 = \ell'$, $\Psi(\vec{r}, t)$ must be a totally symmetric function since it cannot depend on angles, and so neither on the quarks' relative positions. On the other hand, the color factor depends on the additional degree of freedom: each quark has a linear combination of the three primary colors [☞ Section 2.3.13]. That is, every quark is in fact a triple of quarks that span the 3-dimensional representation of the $SU(3)_c$-symmetry,[20] and a bound state of three quarks *must* be $SU(3)_c$-invariant. Group theory applies to the $SU(3)_c$-symmetry as well as for $SU(3)_f$, and the decomposition (A.76f) provides for the fact that the $SU(3)_c$-invariant factor $\chi(\text{color})$ is totally antisymmetric.

Since the entire product (4.123) must be totally antisymmetric by Pauli's exclusion principle, and $\Psi(\vec{r}, t)$ is totally symmetric while $\chi(\text{color})$ is totally antisymmetric, it follows that the product $\chi(\text{spin})\chi(\text{flavor})$ must be totally symmetric.

Since the $\chi(\text{flavor})$ factor for the decuplet of the $SU(3)_f$-symmetry is totally symmetric [☞ decomposition (A.76f)], it follows that the $\chi(\text{spin})$ factor must also be totally symmetric. Writing out the first two of the kets (4.118):

$$|\tfrac{3}{2}, +\tfrac{3}{2}\rangle = |\uparrow\uparrow\uparrow\rangle, \qquad |\tfrac{3}{2}, +\tfrac{1}{2}\rangle = \tfrac{1}{\sqrt{3}}\big(|\uparrow\uparrow\downarrow\rangle + |\uparrow\downarrow\uparrow\rangle + |\downarrow\uparrow\uparrow\rangle\big), \tag{4.124a}$$

$$|\tfrac{3}{2}, -\tfrac{3}{2}\rangle = |\downarrow\downarrow\downarrow\rangle, \qquad |\tfrac{3}{2}, -\tfrac{1}{2}\rangle = \tfrac{1}{\sqrt{3}}\big(|\uparrow\downarrow\downarrow\rangle + |\downarrow\uparrow\downarrow\rangle + |\downarrow\downarrow\uparrow\rangle\big), \tag{4.124b}$$

we know that these four spin states $|\frac{3}{2}, m_s\rangle$ are totally symmetric, so that these ten baryons *must* have spin-$\frac{3}{2}$. That is Gell-Mann's decuplet $(4\Delta, 3\Sigma^*, 2\Xi^*, \Omega^-)$, where the fast experimental confirmation of the predicted Ω^- baryon induced most researchers to finally accept the quark model.

The construction of the octet is a little more complicated, as we must find a totally symmetric (linear combination) of products of $\chi(\text{spin})$ and $\chi(\text{flavor})$ that, separately, have a mixed symmetry. Notice first that the product $\chi_{[12]}(\text{spin})\chi_{[12]}(\text{flavor})$ is *symmetric* with respect to the exchange of the first two particles, since each of the two factors is antisymmetric. Then, it follows that the linear combination

$$\chi_{[12]}(\text{spin})\, \chi_{[12]}(\text{flavor}) + \chi_{[13]}(\text{spin})\, \chi_{[13]}(\text{flavor}) + \chi_{[23]}(\text{spin})\, \chi_{[23]}(\text{flavor}) \tag{4.125}$$

is totally symmetric and provides the spin-flavor factor in the wave-function (4.123) for the octet of spin-$\frac{1}{2}$ baryons. In spite of the relationship (4.121) – whereby $|\frac{1}{2}, m_s\rangle_{[13]}$ is linearly dependent on $|\frac{1}{2}, m_s\rangle_{[12]}$ and $|\frac{1}{2}, m_s\rangle_{[23]}$, the bilinear terms in the expression (4.125) are linearly independent, and the full expression does not simplify.

—❦—

Without the additional color degree of freedom for quarks, i.e., without the totally antisymmetric $\chi(\text{spin})$ factor in the product (4.123), the product $\Psi(\vec{r}, t)\, \chi(\text{spin})\, \chi(\text{flavor})$ would have to be

[20] Unlike the approximate $SU(3)_f$-symmetry, the $SU(3)_c$-symmetry is exact.

totally antisymmetric. For the state with smallest mass where $n = 1$ and $\ell = 0 = \ell'$, the product $\chi(\text{spin})\,\chi(\text{flavor})$ would have to be totally antisymmetric. For spin-$\frac{1}{2}$ octets, one could construct such a state, but the spin-$\frac{3}{2}$ baryons would have to have a totally antisymmetric $\chi(\text{flavor})$ factor, which would have to be the $SU(3)_f$-invariant – and so a single spin-$\frac{3}{2}$ baryon, instead of the ten experimentally detected ones (2.35).

This is the core of the problem noticed by Oscar W. Greenberg in 1964. As a resolution, he proposed [229] that the quark annihilation and creation operators should satisfy para-fermionic rules (2.41c)–(2.41d). This turns out to effectively introduce an additional degree of freedom – the same as the one called "color" in the 1965 independent proposal by Han and Nambu, where quarks had integral electric charges, with values that depend on the color [☞ Digression 5.14 on p. 214]. Their model also predicted particles (that would soon be called gluons) that mediate transformations of the color charge in quarks, where these transformations have the structure of the exact $SU(3)_c$ symmetry group and are the source of the strong interaction [☞ Section 6.1]. The current version of this model with fractionally (and color-independently) charged quarks was proposed by Harald Fritzsch and Murray Gell-Mann in 1971, and was finalized in collaboration with William A. Bardeen in 1973 [32].

Masses

By the reasoning used so far that led to the approximation (4.115), for baryons we have

$$M(\text{baryon}) \approx m_1 + m_2 + m_3 + A' \sum_{i\neq j} \frac{1}{m_i m_j} \langle S_i \cdot S_j \rangle, \tag{4.126}$$

where the spin–spin contributions (leading to the hyperfine structure in the hydrogen atom spectrum) must be computed separately for baryons in isospin groups. Indeed, in the general case, the three masses are different and every spin–spin pair – which stems from the dipole–dipole magnetic interaction – must be considered separately.

In the decuplet case, the situation is simpler, as the $\chi(\text{flavor})$ factor and therefore also the $\chi(\text{spin})$ factor are both totally symmetric. Thus, the spins of any two quarks are parallel, and the well-known "trick"

$$\vec{S}_1 \cdot \vec{S}_2 = \tfrac{1}{2}\Big((\vec{S}_1 + \vec{S}_2)^2 - S_1^2 - S_2^2\Big) \tag{4.127}$$

provides for each pair of quarks in the baryon decuplet:

$$\Big\langle \vec{S}_i \cdot \vec{S}_j \Big\rangle = \tfrac{1}{2}\big(2 - \tfrac{1}{2}(\tfrac{1}{2}+1) - \tfrac{1}{2}(\tfrac{1}{2}+1)\big)\hbar^2 = \tfrac{1}{4}\hbar^2, \qquad \text{for the decuplet.} \tag{4.128}$$

The cases

$$M(\Delta) \approx 3m_u + \frac{3A'\hbar^2}{4m_u^2} \qquad \text{and} \qquad M(\Omega^-) \approx 3m_s + \frac{3A'\hbar^2}{4m_s^2} \tag{4.129}$$

are particularly simple, where the first estimate applies to $\Delta^{++}, \Delta^{+}, \Delta^{0}$ and Δ^{-} since $m_u \approx m_d$ [☞ Table 4.1 on p. 152]. The results

$$M(\Sigma^*) \approx 2m_u + m_s + \frac{A'\hbar^2}{4}\Big(\frac{1}{m_u^2} + \frac{2}{m_u m_s}\Big), \tag{4.130a}$$

$$M(\Xi^*) \approx m_u + 2m_s + \frac{A'\hbar^2}{4}\Big(\frac{2}{m_u m_s} + \frac{1}{m_s^2}\Big) \tag{4.130b}$$

are just a little more involved.

For the baryon octet, we first must look at the isospin sub-multiplets. For example, we know the Λ^0-baryon, a (u, d, s) bound state, has the isospin $|0,0\rangle$-factor antisymmetric with respect to the $u \leftrightarrow d$ exchange. For the χ(spin) χ(flavor) product to be symmetric, it must be that the spin factor is also antisymmetric, and it then follows that the u- and d-quark spins in the Λ^0-baryon are antiparallel. Similarly, we know that the $\Sigma^\pm$- and Σ^0-baryons, also (u, d, s) bound states, form an isospin triplet, $\{|1, \pm 1\rangle, |1, 0\rangle\}$, so that the u- and d-quark spins in the Σ-baryons must be parallel. Thus,

$$\langle \vec{S}_u \cdot \vec{S}_d \rangle = \begin{cases} \frac{1}{4}\hbar^2 & \text{in the } \Lambda^0\text{-baryon,} \\ -\frac{3}{4}\hbar^2 & \text{in the } \Sigma\text{-baryons.} \end{cases} \tag{4.131}$$

Also the generalization of the relation (4.127) gives

$$\vec{S}_1 \cdot \vec{S}_2 + \vec{S}_1 \cdot \vec{S}_3 + \vec{S}_2 \cdot \vec{S}_3 = \tfrac{1}{2}\Big((\vec{S}_1 + \vec{S}_2 + \vec{S}_3)^2 - S_1^2 - S_2^2 - S_3^2\Big), \tag{4.132}$$

and

$$\langle \vec{S}_1 \cdot \vec{S}_2 + \vec{S}_1 \cdot \vec{S}_3 + \vec{S}_2 \cdot \vec{S}_3 \rangle = \begin{cases} \frac{3}{4}\hbar^2 & \text{for the spin-}\frac{3}{2}\text{ decuplet,} \\ -\frac{3}{4}\hbar^2 & \text{for the spin-}\frac{1}{2}\text{ octet.} \end{cases} \tag{4.133}$$

Adding the corresponding terms and using that $m_d \approx m_u$, we have

$$M(p^+, n^0) \approx 3m_u - \frac{3A'\hbar^2}{4m_u^2}, \tag{4.134}$$

$$M(\Lambda) \approx 2m_u + m_s - \frac{3A'\hbar^2}{4m_u^2}, \tag{4.135}$$

$$M(\Sigma) \approx 2m_u + m_s + \frac{A'\hbar^2}{4}\Big(\frac{1}{m_u^2} - \frac{4}{m_u m_s}\Big), \tag{4.136}$$

$$M(\Xi) \approx 2m_u + m_s + \frac{A'\hbar^2}{4}\Big(\frac{1}{m_s^2} - \frac{4}{m_u m_s}\Big). \tag{4.137}$$

With the effective quark masses taken from Table 4.1 on p. 152, and choosing the constant $A' = (2m_u/\hbar)^2 \cdot 50\,\text{MeV}/c^2$, one obtains excellent approximations for the measured masses [☞ Table 4.5].

Table 4.5 The lightest baryon masses in MeV/c^2

Baryon	Computed	Measured	Baryon	Computed	Measured
p^+, n^0	939	939	Δ	1,239	1,232
Λ	1,116	1,114	Σ^*	1,381	1,384
Σ	1,179	1,193	Ξ^*	1,529	1,533
Ξ	1,327	1,318	Ω	1,682	1,672

Magnetic moments

In the absence of orbital angular momenta, $\ell = 0 = \ell'$, the baryon magnetic moment is – up to corrections of the type (4.24b) – simply the sum of the constituent quarks' magnetic moments:

$$\vec{\mu}(\text{baryon}) = \vec{\mu}^{(1)} + \vec{\mu}^{(2)} + \vec{\mu}^{(3)}. \tag{4.138}$$

For a spin-$\frac{1}{2}$ particle, with charge q and mass m, we have

$$\langle \mu_3 \rangle = \Big\langle \frac{q}{m_e c} S_3 \Big\rangle = \pm \frac{q\hbar}{2m_e c}, \tag{4.139}$$

and so

$$\mu_u := \langle \mu_3^{(u)} \rangle = \pm \frac{e\hbar}{3m_u c}, \qquad \mu_d := \langle \mu_3^{(d)} \rangle = \mp \frac{e\hbar}{6m_d c}, \qquad \mu_s := \langle \mu_3^{(s)} \rangle = \mp \frac{e\hbar}{6m_s c}, \tag{4.140}$$

and

$$\langle \mu_3(\text{baryon}) \rangle = \frac{2}{\hbar} \sum_{i=1}^{3} \Big\langle \text{baryon} | \mu_i \, S_3^{(i)} | \text{baryon} \Big\rangle . \tag{4.141}$$

So, to compute the baryon magnetic moment, one must compute the right-hand side contribution to the relation (4.141) for each baryon separately, by

1. writing out the baryon state explicitly as a linear combination (4.125), using the results (4.124) and (4.118),
2. substituting this in the right-hand side sum (4.141), term by term and for both the ket and the bra,
3. evaluating each term in the so-expanded sum,
4. and finally adding the partial results.

There are clearly many contributions, but the so-obtained values are in very good agreement with the experimental measurements, as shown in Table 4.6 [243].

Table 4.6 The baryon magnetic dipole magnitudes in the first octet, expressed in units of nuclear magneton, $\frac{e\hbar}{2m_p c} = 3.152 \times 10^{-13}$ MeV/c^2/T

Baryon	$\langle \mu_3 \rangle$	Computed	Measured
p^+	$\frac{1}{3}(4\mu_u - \mu_d)$	2.79	2.793
n^0	$\frac{1}{3}(4\mu_d - \mu_u)$	−1.86	−1.913
Ξ^0	$\frac{1}{3}(4\mu_s - \mu_u)$	−1.40	−1.253
Ξ^-	$\frac{1}{3}(4\mu_u - \mu_s)$	−0.47	−0.69
Λ^0	μ_s	−0.58	−0.61
Σ^+	$\frac{1}{3}(4\mu_u - \mu_s)$	2.68	2.33
Σ^0	$\frac{1}{3}(2\mu_u + \mu_d - \mu_s)$	0.82	—
Σ^-	$\frac{1}{3}(4\mu_d - \mu_s)$	−1.05	−1.41

As a final note on hadron spectroscopy, the Reader should recall that most hadrons decay within $\sim 10^{-23}$ s of their creation within clusters of hundreds and thousands of simultaneous collision processes. The fact that such measurements on individual particles are in fact being performed is an impressive feat of resourcefulness, ingenuity and hard work.

4.4.4 Exercises for Section 4.4

✎ **4.4.1** *Estimate the relative magnitudes of the contributions analogous to (4.8b), (4.22), (4.28), (4.32), (4.33), (4.34), (4.35), (4.40), (4.46a) and (4.47), as functions of α_s, the strong interaction constant, for $(c\bar{c})$, $(b\bar{b})$ and $(t\bar{t})$ systems.*

✎ **4.4.2** *Estimate the relative magnitudes of the contributions analogous to (4.8b), (4.22), (4.28), (4.32), (4.33), (4.34), (4.35), (4.40), (4.46a) and (4.47), as functions of α_s, the strong interaction constant, for $(c\bar{b})$, $(c\bar{t})$ and $(b\bar{t})$ systems.*[21]

✎ **4.4.3** *From the fact that the lifetime of charmonium states above the D–$\overline{D}$ threshold is $\sim 10^{-23}$ s and by comparing with the positronium lifetime (4.51), estimate α_s and show that $\alpha_s \sim O(\frac{1}{10}) - O(1)$.*

✎ **4.4.4** *Write out all collisions of the $\pi + \pi \to \pi + \pi$ type.*

✎ **4.4.5** *Find the relation between the probabilities of the four collisions:*

$$\pi^+ + \pi^+ \to \pi^+ + \pi^+, \qquad \pi^+ + \pi^0 \to \pi^+ + \pi^0, \tag{4.142a}$$

$$\pi^+ + \pi^- \to \pi^+ + \pi^- \quad \text{and} \quad \pi^+ + \pi^- \to \pi^0 + \pi^0. \tag{4.142b}$$

✎ **4.4.6** *Use the Wenzel–Brillouin–Kramers (WKB) approximation to prove that, for the potential (4.102), the bound-state energies $E_n \approx E_1 n^{3/2}$ for large enough n.*

✎ **4.4.7** *Determine the degeneracy of the states predicted by the non-relativistic treatment of the potential (4.102). (Hint: try verifying the maximal symmetry of this non-relativistic Hamiltonian using the relations (4.13) and explicit computation.)*

✎ **4.4.8** *Upon fully expanding the expression (4.125) and by explicitly tracing the action of swapping quarks, show that the complete expression is symmetric with respect to the exchange of any two of the three quarks.*

✎ **4.4.9** *Derive the relations (4.130).*

[21] N.B. The details of these systems are subject to contemporary research.

5 Gauge symmetries and interactions

It is well known that the overall phase of a wave-function in quantum mechanics is not measurable. On the other hand, the so-called Aharonov–Bohm effect [☞ e.g., the texts [407, 471, 480, 472, 29, 324]] is based on the interference of two wave-functions and measures the *relative* phase, which proves that it is not possible to circumnavigate the complex nature of wave-functions. This then shows: (1) phases of wave-functions *are* physically relevant variables, and (2) any change in the *overall* (common) phase in a wave-function of the whole system must be a symmetry. This and the next chapters focus on this symmetry, and the corresponding conserved charge guaranteed to exist by Noether's theorem.

Moreover, this phase should be variable *locally*: in one way in one spacetime point, in another way in another spacetime point. It turns out that this seemingly simple (gauge) principle is actually the foundation of the contemporary understanding of all fundamental interactions [☞ [31] for the most complete review to date]. These five chapters (5–9) are dedicated to the application of this gauge idea, from technically simple examples towards more complex and realistic applications, and not following the history of its development but using the benefit of hindsight and the lessons of that history. For a flippant introduction of this idea, see also Refs. [33, 275, 269].

5.1 The non-relativistic $U(1)$ example

Start with the well-known non-relativistic quantum-mechanical description of a particle under the influence of a potential $V(\vec{r}\,)$, the wave-function of which is determined by the Schrödinger equation:

$$i\hbar\frac{\partial}{\partial t}\Psi(\vec{r},t) = \Big[-\frac{\hbar^2}{2m}\vec{\nabla}^2 + V(\vec{r},t)\Big]\Psi(\vec{r},t), \tag{5.1}$$

and by the boundary conditions. In part, the boundary conditions follow from the shape of the potential and the chosen energy E of the system, and are in part specified by choice. For example, in directions/regions where $V(\vec{r},t) > E$ as $r \to \infty$, we require $\lim_{r\to\infty}\Psi(\vec{r},t) = 0$; we also require that both $\int_V \mathrm{d}^3\vec{r}\;|\Psi(\vec{r},t)|^2$ and $\int_V \mathrm{d}^3\vec{r}\;\Psi^*(\vec{r},t)H\Psi(\vec{r},t)$ integrals are finite for every choice of the volume V. In the direction $\hat{\mathbf{e}}_k$ where $r \to \infty$ is not obstructed by a boundary condition, we may require that

$$\Psi(\vec{r},t) \sim \exp\Big\{+i\int \mathrm{d}(\hat{\mathbf{e}}_k\cdot\vec{r})\sqrt{2m[E-V(\vec{r},t)]}/\hbar\Big\}, \qquad r \to \infty. \tag{5.2}$$

Such a particle may freely "escape to infinity" in the direction $\hat{e}_k$, along which the kinetic energy remains positive, $\lim_{r\to\infty}\big(E - V(\vec{r})\big) > 0$.

It is very well known that in this formalism the complex wave-function $\Psi(\vec{r},t)$ in its entirety does not correspond to any measurable quantity, but that $|\Psi(\vec{r},t)|^2$ *is* a physically measurable probability density of finding the particle in an infinitesimal volume $d^3\vec{r}$ at the point $\vec{r}$ in space and t in time. It follows that the *phase* of the complex function $\Psi(\vec{r},t)$ is not measurable,[1] so that no transformation

$$\Psi(\vec{r},t) \quad \to \quad e^{i\varphi}\,\Psi(\vec{r},t) \tag{5.3}$$

can have any physical (measurable) consequence. The transformation (5.3) is a symmetry of the Schrödinger equation (5.1) if and only if the phase φ is a constant. In other words, the transformation (5.3) is a symmetry of the Schrödinger equation of the physical system described by the equation if and only if the identical transformation is applied to all points in space and each moment of time. Such a symmetry transformation is called *global*. Its existence is the necessary and sufficient condition for the application of Noether's theorem, and – therefore – for the existence of a corresponding conserved charge.

However, there should exist no physical obstacle for a transformation such as (5.3) to be performed with the phase φ in one point of space and at one moment in time, and a completely different phase in another point of space and at another moment in time. Indeed, the choice of the wave-function phase should be a completely arbitrary choice of an unmeasurable degree of freedom, with no measurable consequence. In other words, the transformation (5.3) would have to be an exact symmetry of the physical system even if the phase φ is an arbitrary function of $\mathrm{x} = (ct,\vec{r})$. Such transformations and symmetries are called *local*.

Digression 5.1 A rather formal justification for the transformation (5.3) to be a symmetry of the system is provided by noting that the formulation (5.3) in fact unnecessarily relies on the coordinate representation of the abstract state $|\Psi(t)\rangle$. Furthermore, it is known that only *pure* quantum states may be represented by a state vector $|\Psi(t)\rangle$, while a general state must be represented by a real, convex, normalized linear combination

$$\boldsymbol{\rho} = \sum_n r_n\,|n\rangle\langle n|, \quad \text{such that} \quad r_n \in \mathbb{R},\ \ 0 \leqslant r_n \leqslant 1,\ \ \sum_n r_n = 1. \tag{5.4a}$$

This is called the *state operator* [29], a.k.a. the *density matrix/operator* [471, 472, 360, for example]. Equivalently, $\boldsymbol{\rho}^\dagger = \boldsymbol{\rho}$, $\mathrm{Tr}[\boldsymbol{\rho}] = 1$ and $\langle u|\boldsymbol{\rho}|u\rangle \geqslant 0$ for every $|u\rangle$. A state operator (5.4a) represents a *pure* state if there exists a $|\Psi\rangle = \sum_n c_n|n\rangle$ such that $\boldsymbol{\rho} = |\Psi\rangle\langle\Psi|$; otherwise, $\boldsymbol{\rho}$ represents a *mixed* state.

The phase transformation (5.3) of the state vectors $|n\rangle$, written as $|n\rangle \to e^{i\boldsymbol{\varphi}}|n\rangle$, leaves the state operator $\boldsymbol{\rho}$ invariant:

$$\boldsymbol{\rho} \to \sum_n r_n\Big(e^{i\boldsymbol{\varphi}}|n\rangle\Big)\Big(\langle n|e^{-i\boldsymbol{\varphi}^\dagger}\Big) = \sum_n r_n|n\rangle\langle n| = \boldsymbol{\rho}, \qquad \text{if and only if} \ \ \big[e^{i\boldsymbol{\varphi}},\,|n\rangle\langle n|\big] = 0 \ \ \text{and} \ \ \boldsymbol{\varphi}^\dagger = \boldsymbol{\varphi}. \tag{5.4b}$$

---

[1] Here, we have in mind only the overall phase. In the transformation of the linear combination $\Psi = \Psi_1 + \Psi_2 \to e^{i\varphi_1}\Psi_1 + e^{i\varphi_2}\Psi_2$, the phase $(\varphi_1+\varphi_2)$ is the unmeasurable overall phase, while the *relative* phase $(\varphi_1-\varphi_2)$ is measurable by means of *interference*. This overwhelmingly reminds us of the fact that the absolute values of coordinates (and the phase is indeed a kind of coordinate) are not measurable quantities, while coordinate *differences* – i.e., distances – are.

In turn, the information about the change of this (or any other) choice cannot be transported instantly, and there will have to exist some physical mechanism for transporting this information from point to point in space and time.

It is not hard to verify that the transformation (5.3) with $\varphi = \varphi(\vec{r}, t)$ is *not* a symmetry of the Schrödinger equation (5.1):

$$i\hbar \frac{\partial}{\partial t}\Psi = \Big[-\frac{\hbar^2}{2m}\vec{\nabla}^2 + V(\vec{r},t)\Big]\Psi, \tag{5.5}$$

transformation (5.3), with $\varphi = \varphi(\vec{r}, t)$

$$i\hbar \frac{\partial}{\partial t}\Big(e^{i\varphi}\Psi\Big) = \Big[-\frac{\hbar^2}{2m}\vec{\nabla}^2 + V(\vec{r},t)\Big]\Big(e^{i\varphi}\Psi\Big),$$

$$i\hbar\, e^{i\varphi}\Big(i\frac{\partial\varphi}{\partial t}\Big)\Psi + i\hbar\, e^{i\varphi}\frac{\partial\Psi}{\partial t} = -\frac{\hbar^2}{2m}\vec{\nabla}\cdot\Big(e^{i\varphi}(i\vec{\nabla}\varphi)\Psi + e^{i\varphi}\vec{\nabla}\Psi\Big) + V(\vec{r},t)e^{i\varphi}\Psi,$$

$$i\hbar\, e^{i\varphi}\Big(i\frac{\partial\varphi}{\partial t}\Big)\Psi + i\hbar\, e^{i\varphi}\frac{\partial\Psi}{\partial t}$$
$$= -\frac{\hbar^2}{2m}\Big(e^{i\varphi}(i\vec{\nabla}\varphi)^2\Psi + e^{i\varphi}(i\vec{\nabla}^2\varphi)\Psi + 2e^{i\varphi}(i\vec{\nabla}\varphi)\cdot(\vec{\nabla}\Psi) + e^{i\varphi}\vec{\nabla}^2\Psi\Big) + V(\vec{r},t)e^{i\varphi}\Psi,$$

so, using the original equation (5.1) and upon dividing by $\Psi(\vec{r}, t)$, we obtain

$$\frac{\partial\varphi}{\partial t} = \frac{\hbar}{2m}\Big(i(\vec{\nabla}^2\varphi) + 2i(\vec{\nabla}\varphi)\cdot(\vec{\nabla}\ln(\Psi)) - (\vec{\nabla}\varphi)^2\Big). \tag{5.6}$$

This result is *absolutely unacceptable*! Not only did the (*unmeasurable!*) phase $\varphi(\vec{r}, t)$ turn out not to be an arbitrarily selectable function of space and time, but it would have to satisfy a differential equation (5.6) that depends on the particular state of the system represented by the wave-function $\Psi(\vec{r}, t)$!

The resolution of this seeming paradox can only lie in changing the Schrödinger equation, but in a way that does not ruin any of the many confirmed results obtained from this equation. Evidently, this is a very demanding request.

Following the computation (5.5)–(5.6) closely, one notices that the ultimate – and *absolutely unacceptable* – result stems from the fact that *derivatives* of the "new" wave-function $e^{i\varphi}\Psi(\vec{r}, t)$ differ from the $e^{i\varphi}$-multiples of the derivative of the "old" wave-function $\Psi(\vec{r}, t)$. With this hint, introduce a new kind of derivative:

$$\frac{\partial}{\partial t} \to D_t := \frac{\partial}{\partial t} + X, \qquad \vec{\nabla} \to \vec{D} := \vec{\nabla} + \vec{Y}, \tag{5.7a}$$

where the quantities X and $\vec{Y}$ will be determined so that these newfangled D-derivatives satisfy the relations

$$D'_t\Psi' = D'_t(e^{i\varphi}\Psi) = e^{i\varphi}(D_t\Psi), \qquad \vec{D}'\Psi' = \vec{D}'(e^{i\varphi}\Psi) = e^{i\varphi}(\vec{D}\Psi). \tag{5.7b}$$

By writing $\Psi = e^{-i\varphi}\Psi'$, these requirements show that

$$(D'_t \cdots) = e^{i\varphi}(D_t e^{-i\varphi}\cdots), \qquad (\vec{D}'\cdots) = e^{i\varphi}(\vec{D}e^{-i\varphi}\cdots), \tag{5.7c}$$

where $D'_t, \vec{D}'$ denotes these new derivatives *after* the $\Psi \to e^{i\varphi}\Psi$ transformation. In turn, with these newfangled derivatives, the Schrödinger equation becomes

$$i\hbar D_t\Psi = \Big[-\frac{\hbar^2}{2m}\vec{D}^2 + V(\vec{r})\Big]\Psi, \quad \text{or} \quad \Big[i\hbar D_t + \frac{\hbar^2}{2m}\vec{D}^2 - V(\vec{r})\Big]\Psi = 0, \tag{5.8}$$

and changes under the transformation (5.3) as

$$0=\Big[i\hbar D'_t+\frac{\hbar^2}{2m}\vec{D}'\cdot\vec{D}'-V(\vec{r})\Big](e^{i\varphi}\Psi)=\Big[i\hbar e^{i\varphi}D_t+\frac{\hbar^2}{2m}\vec{D}'\cdot e^{i\varphi}\vec{D}-e^{i\varphi}V(\vec{r})\Big]\Psi$$

$$=e^{i\varphi}\Big[i\hbar D_t+\frac{\hbar^2}{2m}\vec{D}\cdot\vec{D}-V(\vec{r})\Big]\Psi, \tag{5.9}$$

which is satisfied precisely when equation (5.8) is. Thus, with these newfangled derivatives D_t and $\vec{D}$, which themselves change via the transformation (5.3), the new Schrödinger equation (5.8) remains invariant.

Comment 5.1 *It is not at all unreasonable that the procedure for computing a rate of change (the derivative operator) needed adjustment. Recall that the total derivative* $\frac{d}{dt}f(t,g(t))=[\frac{\partial}{\partial t}+\frac{\partial g}{\partial t}\frac{\partial}{\partial g}]f$ *may be viewed as the partial derivative* $\frac{\partial}{\partial t}$ *corrected for the fact that the function* f *also depends on* t *implicitly, via its dependence on* $g(t)$*. By the same token, complex wave-functions also depend on the spacetime coordinates implicitly, via their dependence on the choice of a spacetime variable phase.*

It remains to examine the nature of these newfangled derivatives (5.7), as well as the differences between the new Schrödinger equation (5.8) and the old one (5.1). The newfangled derivatives satisfy (5.7c)

$$\Big[\Big(\frac{\partial}{\partial t}+X'\Big)\cdots\Big]=e^{i\varphi}\Big[\Big(\frac{\partial}{\partial t}+X\Big)e^{-i\varphi}\cdots\Big],$$
$$[(\vec{\nabla}+\vec{Y}')\cdots]=e^{i\varphi}[(\vec{\nabla}+\vec{Y})e^{-i\varphi}\cdots]; \tag{5.10}$$

which yields

$$X'=X-i\frac{\partial\varphi}{\partial t}\qquad\text{and}\qquad\vec{Y}'=\vec{Y}-i(\vec{\nabla}\varphi). \tag{5.11}$$

The relations (5.11) ought to be familiar to all Students who have successfully completed a standard electrodynamics course! With the definitions

$$\Phi:=\frac{\hbar}{iq}X,\qquad\vec{A}:=\frac{i\hbar}{q}\vec{Y},\qquad\lambda:=\frac{\hbar}{q}\varphi, \tag{5.12}$$

the definitions (5.7a) become

$$D_t:=\frac{\partial}{\partial t}+i\frac{q}{\hbar}\Phi,\qquad\vec{D}:=\vec{\nabla}-i\frac{q}{\hbar}\vec{A}, \tag{5.13}$$

and are called the *covariant derivatives*. Combining, we have

$$\Phi\to\Phi'=\Phi-\frac{\partial\lambda}{\partial t},\qquad\vec{A}\to\vec{A}'=\vec{A}+(\vec{\nabla}\lambda), \tag{5.14a}$$

$$\Psi(\vec{r},t)\to\Psi'(\vec{r},t)=e^{iq\lambda(\vec{r},t)/\hbar}\,\Psi(\vec{r},t). \tag{5.14b}$$

The first two relations are the standard gauge transformations of the vector and the scalar electromagnetic potentials. The third relation is the corresponding gauge transformation of the wave-function $\Psi(\vec{r},t)$ of a particle with the electric charge q, which is evidently a translation of the phase of this function.

***Comment 5.2** The action of the gauge transformation (5.14b) implies that the complex conjugate wave-function $\Psi^*(\vec{r},t)$ represents a particle with the charge that is opposite to the particle represented by $\Psi(\vec{r},t)$: $q(\Psi^*) = -q(\Psi)$.*

The first two transformation equations (5.14a) clearly imply that the effect of $\lambda(\vec{r},t)$ is indistinguishable from that of $\lambda(\vec{r},t)+\lambda_0$, where $\lambda_0 = const.$, and the single-valuedness of $\Psi(\vec{r},t)$ then implies that λ_0 must be an integral multiple of $2\pi\cdot\frac{\hbar}{q_0}$, where q_0 then must be a minimal, unit electric charge. That is, the transformation function $\lambda(\vec{r},t)$ takes $2\pi\cdot\frac{\hbar}{q_0}$-periodic distinct values, i.e., along a circle of radius $\frac{\hbar}{q_0}$. In turn, the exponential $U_\lambda := e^{iq\lambda(\vec{r},t)/\hbar}$ is unitary: $(U_\lambda)^\dagger = (U_\lambda)^{-1}$, and such λ-parametrized exponentials form the *gauge group*, called $U(1)$ [☞ Appendix A, and especially A.2].

More precisely, note that the transformation function, $\lambda = \lambda(\vec{r},t)$, remains an unrestricted, arbitrary function of space and time[2] – true to the original insight and definition as discussed above. The combined transformation (5.14) is then the true and complete *local* symmetry: a *continuous family of* $U(1)$ *gauge groups of symmetries*, one independent $U(1)$ symmetry in every point of space and time!

Owing to the identity $\vec{\nabla}\times(\vec{\nabla} f) \equiv 0$ valid for any scalar function f, it follows that $(\vec{\nabla}\times\vec{A})$ is invariant with respect to the transformations (5.14). Similarly, since the transformation of $\vec{\nabla}\Phi$ is precisely opposite of the transformation of $\frac{\partial}{\partial t}\vec{A}$, the sum $(\vec{\nabla}\Phi + \frac{\partial}{\partial t}\vec{A})$ is also invariant. These expressions are, of course, familiar:

$$\vec{B} := \vec{\nabla}\times\vec{A} \qquad \text{and} \qquad \vec{E} := -\Big(\vec{\nabla}\Phi + \frac{\partial \vec{A}}{\partial t}\Big) \tag{5.15}$$

are the magnetic and the electric fields, expressed in terms of the electromagnetic potentials. The ability to define gauge-*invariant* fields $\vec{B}$ and $\vec{E}$ will be shown to be an exceptional consequence of the abelian (commutative) nature of the $U(1)$ gauge transformation (5.14).

Digression 5.2 The term "gauge transformation" for the relations (5.14) is a historical atavism: It is a derivative of the literally translated German original coinage by Hermann Weyl, *Eichinvarianz*, by which he denoted the invariance with respect to transformations (5.14) [564]. Weyl noticed that Einstein's general theory of relativity is invariant with respect to complex rescalings. His original idea that the imaginary part of the rescaling function $\varphi(\vec{r},t)$ in the transformation (5.14b) may unite gravity with electromagnetism turned out unphysical. Such a rescaling symmetry would permit fixing a length unit in Nature, for which Weyl used the German verb *eichen*, meaning to gauge, to calibrate. The word *gauge* and its derivatives that are used in the English literature, *jauge* in French, βαθμίδας in Greek, *mérték* in Hungarian, (simply imported) *gauge* in Italian, *калибро́вочная* in Russian, *de gauge* in Spanish, etc. are all literal translations of the German verb *eichen*.

Soon, Vladimir A. Fok (first, according to Professor Okun [394], in 1926), Hermann Weyl, Erwin Schrödinger and Fritz London noticed that quantum mechanics, as governed by the Schrödinger equation, has a symmetry with respect to the combined transformations (5.14) using a real function $\varphi(\vec{r},t)$.[3] This was derived here as a

[2] Well, yes: $\lambda(\vec{r},t)$ clearly must be differentiable, at least once with respect to both t and $\vec{r}$ for the equations (5.14a) to be well defined; see however also Section 5.2.3.

[3] Woit recounts [577, pp. 61–62] that Schrödinger hinted at this in a 1922 paper, but was chidingly reminded of this neglected "tidbit" in December of 1926 by the young London; see also the account in Ref. [119].

transformation stemming from the innate property of wave-functions that their phases are not measurable.

Fundamental physics is indubitably quantum. Equations (5.1)–(5.14) and their logic then indicate the fundamental nature of this principle to be that of a ***local symmetry***, emphasizing that a spacetime variable parameter $\lambda(\vec{r},t)$ in the transformations (5.14) parametrizes a spacetime continuum of $U(1)$ symmetries. ***Local symmetry*** is then used as a conceptually correct alternative for the historically well-entrenched term *gauge symmetry* or the descriptive but rarely used modifier *phase symmetry*.

Comment 5.3 *Note that the transformation (5.14) may be understood as a spacetime-dependent translation of sorts in the (abstract, target) space of values of the functions defined over the spacetime; a translation of the electromagnetic potentials and a* **phase-shift** *of the wave-function:*

$$\textit{Eq. (5.14)} \Rightarrow \operatorname{Arg}\big[\Psi(\vec{r},t)\big] \to \operatorname{Arg}\big[\Psi(\vec{r},t)\big] + \varphi(\vec{r},t), \quad \operatorname{Arg}[z] := \tfrac{1}{2i}\ln\big(\tfrac{z}{z^*}\big). \tag{5.16}$$

The electromagnetic potentials and the phase of the wave-function are all physically **unmeasurable** *variables, the existence of which is however necessary for the consistency of the model. Lorentz symmetry requires the gauge potentials to be 4-vectors, although only two polarizations (components) have a physical meaning; the complex-analytic structure of the Schrödinger and Dirac equations requires the use of complex wave-functions, although the (overall) phase is not physically measurable.*

With the definitions (5.12), the Schrödinger equation (5.8) becomes

$$i\hbar\Big[\frac{\partial}{\partial t} + \frac{iQ}{\hbar}\Phi\Big]\Psi = \Big[-\frac{\hbar^2}{2m}\Big(\vec{\nabla} + \frac{Q}{i\hbar}\vec{A}\Big)^2 + V(\vec{r},t)\Big]\Psi. \tag{5.17}$$

That is,

$$i\hbar\frac{\partial}{\partial t}\Psi(\vec{r},t) = H_{EM}\,\Psi(\vec{r},t), \tag{5.18}$$

where

$$H_{EM} = \frac{1}{2m}\Big(\frac{\hbar}{i}\vec{\nabla} - Q\vec{A}(\vec{r},t)\Big)^2 + \Big[V(\vec{r},t) + Q\Phi(\vec{r},t)\Big] \tag{5.19}$$

is the Hamiltonian for a particle of mass m and electric charge $q := Q(\Psi)$. The dynamics of this particle is affected by the interaction with the potential $V(\vec{r},t)$, as well as the electromagnetic potentials $\vec{A}(\vec{r},t)$ and $\Phi(\vec{r},t)$.

Conclusion 5.1 *The transformation (5.14) with (5.16) is the fundamental assertion that we are at liberty to* **arbitrarily** *change the quantities that were introduced in the (mathematical) model of the physical system for its consistency, but which* **on principle** *represent no physically measurable quantity.*

It is worth noticing that the quantum description of the interaction of a charged particle with the electromagnetic field is inherently described in terms of the electromagnetic potentials $\vec{A},\Phi$ and not in terms of the electric and magnetic field, $\vec{E},\vec{B}$. Moreover, the Hamiltonian (5.19) cannot, in the general case, be expressed locally (without integration) as an interaction of a charged particle with the $\vec{E}$- and $\vec{B}$-fields.

The following facts will be shown to be consequences of the abelian (commutative) nature of the $U(1)$ symmetry group: (1) the Maxwell equations[4] (5.72) as well as the corresponding Lagrangian and Hamiltonian for the electromagnetic field *can* be expressed exclusively in terms of the electric and the magnetic field, and (2) electromagnetic potentials *can* be fully eliminated from the equations of motion, the Lagrangian and the Hamiltonian except if there is matter that interacts with the electromagnetic field.

Indeed, the transformations (5.14) are parametrized by one function, $\lambda(\vec{r},t)$, which defines the local unitary *operator*

$$\varphi(\vec{r},t) \mapsto U_\varphi := \exp\{i\varphi(\vec{r},t)\,Q\} \tag{5.20}$$

as in equation (A.37), where the operator Q may be regarded:

1. from the mathematical vantage point, as the generator of the $U(1)$ symmetry,
2. from the physical vantage point, the electric charge operator. The electric charge of a particle is then the eigenvalue and the wave-function of the particle the eigenfunction of the operator Q.

At every point of spacetime $\mathbf{x} = (ct, \vec{r})$ separately, the (continuously many) operators U_φ defined by equation (5.20) form an abelian (commutative) group, denoted $U(1)$. Since the function in the exponent manifestly satisfies $\varphi \simeq \varphi + 2\pi$, this group is sometimes identified with the circle, S^1. To repeat: Since $\varphi = \varphi(\vec{r},t)$ gives an independent "angle"-transformation at every point in space and time, we have a 4-dimensional *continuum* of $U(1)$ symmetry groups.

Comment 5.4 *The full space of "coordinates" in electrodynamics is therefore of the form (spacetime $\times\, S^1$) – a 5-dimensional topological space, equipped with a particular geometry; this was clear as early as in 1914 to Gunnar Nordstrøm [☞ Digression 11.5 on p. 414].*

5.1.1 Exercises for Section 5.1

✎ **5.1.1** *Fill in the details of the computation (5.7)–(5.14).*

✎ **5.1.2** *From the definitions (5.15), derive Gauss's law for the magnetic field and Faraday's law of induction. (This proves that the equations (5.72b) are* **consequences** *of Maxwell's definitions (5.15).)*

✎ **5.1.3** *Show that the gauge-invariant scalar functions of ϵ_0, μ_0, $\vec{E}$ and $\vec{B}$ with the dimensions of (volume) energy density and which are analytic functions of the components of the vectors $\vec{E}$ and $\vec{B}$ must be of the form*

$$c_1\big(\epsilon_0\,\vec{E}^2\big) + c_2\big(\tfrac{1}{\mu_0}\,\vec{B}^2\big) + c_3\big(\sqrt{\tfrac{\epsilon_0}{\mu_0}}\,\vec{E}\cdot\vec{B}\big). \tag{5.21}$$

The results in Table C.4 on p. 527, should be useful.

✎ **5.1.4** *Determine the constants c_1, c_2, c_3, c_4, c_5 so that*

$$\int \mathrm{d}t\,\mathrm{d}^3\vec{r}\,\Big\{c_1\big(\epsilon_0\,\vec{E}^2\big) + c_2\big(\tfrac{1}{\mu_0}\,\vec{B}^2\big) + c_3\big(\sqrt{\tfrac{\epsilon_0}{\mu_0}}\,\vec{E}\cdot\vec{B}\big) + c_4\,\rho\,\Phi + c_5\,\vec{\jmath}\cdot\vec{A}\Big\} \tag{5.22}$$

---

[4] James Clerk Maxwell described electrodynamics, originally in 1873, as a system of equations which would today be written as $\vec{E} := -\vec{\nabla}\Phi - \frac{\partial\vec{A}}{\partial t}$ and $\vec{B} := \vec{\nabla}\times\vec{A}$, and then $\vec{\nabla}\cdot(\epsilon_0\vec{E}) = \rho$ and $\vec{\nabla}\times(\vec{B}/\mu_0) - \frac{\partial(\epsilon_0\vec{E})}{\partial t} = \vec{\jmath}$. By the Maxwell equations (5.72) today, one understands the *consequences* of the first two of these equations together with the latter two, expressed exclusively in terms of the electric and the magnetic field, where the electromagnetic potentials, $\vec{A}$ and Φ, are eliminated, and where there are neither (monopole) magnetic charges nor magnetic currents: $\rho_m = 0 = \vec{\jmath}_m$.

is the Hamilton action the variation of which by Φ and $\vec{A}$, using the relations (5.15), produces Gauss and Ampère's law (5.72a).

5.2 Electrodynamics with leptons

By quantum electrodynamics one understands the relativistic theory that describes the interaction of photons and electrically charged particles. Unlike leptons, quarks and hadrons also interact via the much stronger strong nuclear interaction, so the analysis of their interactions is considerably more complicated. This section is limited to the electromagnetic interactions of leptons, and the next one turns to the electromagnetic interactions of the hadrons.

It follows from the relations (A.43d)–(A.43f) that the components of the radius-vector, and then also any other vector quantity, span a spin-1 representation of the rotation group. One thus says that the photon (represented by the vector potential,[5] $\vec{A}$) has spin 1. On the other hand, it is well known that electrically charged particles such as the electron and the quarks, which make up all tangible matter, have spin $\frac{1}{2}$.

Thus, we must first establish the relativistic generalization of the Schrödinger equation for particles of spin $\frac{1}{2}$ and 1, as well as the argument from the previous section, which specifies the interaction between them.

5.2.1 Relativistic spinors and the Dirac equation

The Schrödinger equation

$$i\hbar\frac{\partial}{\partial t}\Psi(\vec{r},t) = H\Psi(\vec{r},t) \qquad \Leftrightarrow \qquad \Psi(\vec{r},t) = e^{-i\hbar^{-1}\int_{t_0}^{t}\mathrm{d}t'\,H(t')}\,\Psi(\vec{r},t'), \quad t > t_0 \tag{5.23}$$

is simply the statement that the Hamiltonian generates the time evolution of the wave-function $\Psi(\vec{r},t)$. In non-relativistic physics (here, without electromagnetic potentials),

$$i\hbar\frac{\partial}{\partial t} = H = \frac{1}{2m}\Big(\frac{\hbar}{i}\vec{\nabla}\Big)^2 + V(\vec{r},t) \qquad \Leftrightarrow \qquad E = \frac{\vec{p}^{\,2}}{2m} + V(\vec{r},t), \tag{5.24}$$

the combination of which with equation (5.23) is the diffusion equation: of second order in spatial derivatives, but first order in the time derivative. This also implies the "quantization correspondence" (in the coordinate representation)

$$\vec{p} \leftrightarrow \vec{p} = \frac{\hbar}{i}\vec{\nabla}, \qquad \text{and} \qquad E \leftrightarrow H = i\hbar\frac{\partial}{\partial t}. \tag{5.25}$$

Instead of the non-relativistic relation (5.24), the relativistic version of the Schrödinger equation would have to correspond to the relativistic relation (3.37), and using the correspondences (5.25) we obtain

$$\vec{p}^{\,2}c^2 + m^2c^4 = E^2 \qquad \leftrightarrow \qquad \Big[c^2\Big(\frac{\hbar}{i}\vec{\nabla}\Big)^2 + m^2c^4\Big]\Psi(\vec{r},t) = \Big(i\hbar\frac{\partial}{\partial t}\Big)^2\Psi(\vec{r},t),$$
$$\Rightarrow \qquad \Big[\Box + \Big(\frac{mc}{\hbar}\Big)^2\Big]\Psi(\vec{r},t) = 0. \tag{5.26}$$

This is the so-called Klein–Gordon equation, where

$$\Box := \Big[\frac{1}{c^2}\frac{\partial^2}{\partial t^2} - \vec{\nabla}^2\Big] \tag{5.27}$$

is called the d'Alembertian or the wave operator.

[5] It will soon be shown that, as a consequence of the $U(1)$ gauge symmetry, the four functions $\Phi, \vec{A}$ represent only two physical degrees of freedom, which may be identified with two components of the vector $\vec{A}$ that are perpendicular to the direction of the photon motion.

Digression 5.3 Ironically, Schrödinger seems to have known [243] about the equation (5.26) before publishing the equation that soon acquired his name, but discarded it in the belief that the double-valuedness of the solution (3.37), $E = \pm c\sqrt{\vec{p}^2 + m^2c^2}$, precludes a probabilistic interpretation $|\Psi(\vec{r},t)|^2$. Wolfgang Pauli and Victor Weisskopf proved in 1934 that the essential obstacle to this interpretation of the quantity $|\Psi(\vec{r},t)|^2$ in relativistic physics is the fact that relativistic physics *must* contain the possibility of creating and annihilating particles, as permitted by conservation of energy, linear and angular momentum, charge, etc. This implies that the number of particles in relativistic physics is not a conserved quantity, and contradicts the elementary consequence of the Schrödinger equation:

$$(5.1) \;\Rightarrow\; \frac{\partial \varrho}{\partial t} = -\vec{\nabla}\cdot\vec{\mathcal{J}} + \frac{2}{\hbar}\,\Im m\,(V(\vec{r}))\,\varrho, \tag{5.28a}$$

$$\varrho(\vec{r},t) := |\Psi(\vec{r},t)|^2, \qquad \vec{\mathcal{J}}(\vec{r},t) := \frac{\hbar}{m}\,\Im m\,\big[\Psi^*(\vec{r},t)\vec{\nabla}\Psi(\vec{r},t)\big]. \tag{5.28b}$$

This shows that

$$\frac{\mathrm{d}}{\mathrm{d}t}\int_{\mathcal{V}} \mathrm{d}^3\vec{r}\,\varrho(\vec{r},t) = -\oint_{\partial\mathcal{V}} \mathrm{d}^2\vec{\sigma}\cdot\vec{\mathcal{J}} + \frac{2}{\hbar}\int_{\mathcal{V}} \mathrm{d}^3\vec{r}\;\Im m\,(V(\vec{r}))\,\varrho(\vec{r},t). \tag{5.28c}$$

The probability of finding the particle (represented by Ψ) within the volume $\mathcal{V}$ changes only by the probability flowing through $\partial\mathcal{V}$ (the boundary of the volume $\mathcal{V}$) – if and only if the potential $V(\vec{r},t)$ is a real function where $\varrho(\vec{r},t)$ is nonzero. The number of particles is then also conserved, and this is indeed the case in standard quantum mechanics.

Motivated by the fact that the non-relativistic Schrödinger equation is of first order in time derivatives, while the Klein–Gordon equation is of the second order, Paul Dirac found a way to factorize the Klein–Gordon equation and so obtain a differential equation that is of first order both in spatial and in time derivatives. Indeed, in the rest-frame of the particle, $\vec{p} = 0$, so that the relativistic relation (3.37) reduces to

$$E^2 - m^2c^4 = 0 \qquad\Rightarrow\qquad (E + mc^2)(E - mc^2) = 0, \tag{5.29}$$

which is the desired factorization. With $\vec{p} \neq 0$, the desired factorization of the equivalent equation (3.36) is of the form

$$\begin{aligned} \mathrm{p}^2 - m^2c^2 = 0 \qquad\Rightarrow\qquad 0 &= (\beta^\mu p_\mu + mc)(\gamma^\nu p_\nu - mc), \\ &= \beta^\mu\gamma^\nu\, p_\mu p_\nu + mc(\gamma^\mu - \beta^\mu)p_\mu - m^2c^2. \end{aligned} \tag{5.30}$$

As the original equation $\mathrm{p}^2 - m^2c^2 = 0$ has no linear terms in the 4-momentum p, it must be that $\beta^\mu = \gamma^\mu$. Equating the quadratic terms one then obtains that

$$\gamma^\mu\gamma^\nu\, p_\mu p_\nu = \mathrm{p}^2 \equiv \eta^{\mu\nu}\, p_\mu p_\nu. \tag{5.31}$$

Since $p_\mu p_\nu = p_\nu p_\mu$, we in fact have the conditions

$$\{\gamma^\mu, \gamma^\nu\} = 2\eta^{\mu\nu}, \tag{5.32}$$

where $[\eta^{\mu\nu}] = \mathrm{diag}(1,-1,-1,-1)$ is the matrix-inverse of the metric tensor (3.19) of empty spacetime. This yields

$$\mathrm{p}^2 - m^2c^2 \;=\; 0 \;=\; (\gamma^\mu p_\mu - mc)(\gamma^\mu p_\mu + mc). \tag{5.33}$$

Using the relativistic combination of the correspondences (5.25), this produces the Dirac equation:

$$p_\mu \to \frac{\hbar}{i}\partial_\mu \qquad \Rightarrow \qquad [i\hbar\gamma^\mu\partial_\mu - mc]\Psi(\mathrm{x}) = 0, \tag{5.34}$$

where the standard abbreviation [☞ Digression 3.6 on p. 93]

$$\partial_\mu := \frac{\partial}{\partial x^\mu}, \qquad \to (-\tfrac{1}{c}\partial_t, \vec{\nabla}), \tag{5.35}$$

was introduced. The choice of the second of the two factors in equation (5.33) for defining the Dirac equation is an arbitrary, but standard choice.

The question remains, what sort of objects the γ^μ are so as to satisfy the relations (5.32).

The Dirac spinor

Relations of the type (5.32) define so-called Clifford algebras. Their abstract structure, properties and representation theory had been established by mathematicians William Kingdon Clifford and Hermann Grassmann back in the second half of the nineteenth century. However, in the first half of the twentieth century, this was unknown among physicists, and Dirac independently found the smallest matrix realization of the γ^μ objects, which today we call Dirac matrices; relation (5.32) then implicitly contains the unit 4×4 matrix in the right-hand side. There exist several "standard" choices of Dirac matrices; here we follow the traditional sources [64, 63] and use the so-called Dirac basis:

$$\gamma^0 = \begin{bmatrix} \mathbb{1} & \mathbb{0} \\ \mathbb{0} & -\mathbb{1} \end{bmatrix}, \qquad \gamma^i = \begin{bmatrix} \mathbb{0} & \sigma^i \\ -\sigma^i & \mathbb{0} \end{bmatrix}, \qquad i = 1,2,3. \tag{5.36}$$

To satisfy the relations (5.32), γ^μ cannot be "ordinary" numbers but *can* be matrices. This implies that the operator that acts upon $\Psi(\mathrm{x})$ in the Dirac equation (5.34) also has to be a 4×4 matrix,[6] so $\Psi(\mathrm{x})$ must be a column-matrix with four components!

Recall that the solutions of the Schrödinger equation, e.g., for the hydrogen atom (4.8), yield $\Psi(\vec{r},t)$ as an expansion over spherical harmonics, $Y_\ell^m(\theta,\phi)$, which correspond to components of the "spin-ℓ" representation[7] of the $SO(3) \overset{1-2}{\approx} SU(2)$ rotation group [☞ Table A.2 on p. 469]. For example, the hydrogen atom states with $\ell = 1$ and $m = \pm 1, 0$ span the 3-vector representation of the rotation group, where it is also easy to define the Cartesian basis:

$$\begin{aligned} (\Psi_n)_x &:= \tfrac{1}{2}(\Psi_{n,1,+1} + \Psi_{n,1,-1}), \\ (\Psi_n)_y &:= \tfrac{1}{2i}(\Psi_{n,1,+1} - \Psi_{n,1,-1}), \end{aligned} \qquad (\Psi_n)_z := \Psi_{n,1,0}. \tag{5.37}$$

The elements of the $(2\ell+1)$-dimensional vector space $\{\Psi_{n,\ell,m}, \text{ for } |m| \leqslant \ell,\ \triangle m \in \mathbb{Z}\}$ may just as easily be represented as $(2\ell+1)$-component column-matrices.

However, the 4-component nature of the solutions to the Dirac equation represents an *additional* degree of freedom, a relativistic generalization of the "spin" factor that we used in Section 4.4.2, such as in the factorization (4.123), for example. Even for $\ell = 0$, the Dirac equation has four linearly independent solutions. In the simple case when $\vec{p} = 0$, the Dirac equation reduces to

$$\Big[\frac{i\hbar}{c}\gamma^0\frac{\partial}{\partial t} - mc\mathbb{1}\Big]\Psi = 0. \tag{5.38}$$

[6] Since the γ^μ's are 4×4 matrices, the Dirac equation should, pedantically, be written as $[i\hbar\gamma^\mu p_\mu - mc\mathbb{1}]\Psi = 0$.

[7] When it denotes a rotation group representation, the term "spin-j" is simply short for "the total angular momentum where the eigenvalue of the quadratic operator J^2 equals $j(j+1)$," regardless of the physical original and composition of this total angular momentum.

The solutions in Dirac's basis of $\boldsymbol{\gamma}$-matrices are

$$\Psi_A(t) = e^{-i(mc^2/\hbar)t}\begin{bmatrix}\Psi_1(0)\\ \Psi_2(0)\end{bmatrix}, \qquad \text{and} \qquad \Psi_B(t) = e^{+i(mc^2/\hbar)t}\begin{bmatrix}\Psi_3(0)\\ \Psi_4(0)\end{bmatrix}, \tag{5.39}$$

where $\Psi_B(t)$ represents the solutions with negative energy; i.e., *anti*-solutions with positive energy that move backwards in time, according to the Stückelberg–Feynman interpretation that is by now the standard understanding: $\Psi_B(t) \to \overline{\Psi}_B(-t)$ [☞ definition (5.49)].

Using the redefinition of solutions (wave-functions for particles) with negative energy as anti-solutions (wave-functions for antiparticles) with positive energy, the standard solutions (following the conventions of Ref. [243]) are

$$u^\uparrow \propto \begin{bmatrix}1\\ 0\\ \frac{p_z c}{E+mc^2}\\ \frac{(p_x+ip_y)c}{E+mc^2}\end{bmatrix}, \qquad u^\downarrow \propto \begin{bmatrix}0\\ 1\\ \frac{(p_x-ip_y)c}{E+mc^2}\\ \frac{p_z c}{E+mc^2}\end{bmatrix}, \tag{5.40}$$

$$v^\downarrow \propto \begin{bmatrix}\mathbb{O} & \mathbb{1}\\ \mathbb{1} & \mathbb{O}\end{bmatrix} u^\uparrow \propto \begin{bmatrix}\frac{p_z c}{E+mc^2}\\ \frac{(p_x+ip_y)c}{E+mc^2}\\ 1\\ 0\end{bmatrix}, \qquad v^\uparrow \propto \begin{bmatrix}\mathbb{O} & \mathbb{1}\\ \mathbb{1} & \mathbb{O}\end{bmatrix} u^\downarrow \propto \begin{bmatrix}\frac{(p_x-ip_y)c}{E+mc^2}\\ \frac{p_z c}{E+mc^2}\\ 0\\ 1\end{bmatrix}, \tag{5.41}$$

$$\tag{5.42}$$

where $E = +\sqrt{\vec{p}^2c^2 + m^2c^4}$ always, and the solutions with negative energy are

$$u^\uparrow_-(E,\vec{p}) = -v^\downarrow(-E,-\vec{p}) \qquad \text{and} \qquad u^\downarrow_-(E,\vec{p}) = v^\uparrow(-E,-\vec{p}). \tag{5.43}$$

Note that $u^\uparrow, u^\downarrow, u^\uparrow_-, u^\downarrow_-$ are four linearly independent solutions to the Dirac equation (5.34), whereas $v^\uparrow, v^\downarrow$ satisfy the Dirac equation with $p_\mu \to -p_\mu$ – which precisely holds for the complementary factor in equation (5.33). The solutions to the Dirac equation may then be written as

$$\Psi(\mathrm{x}) = \sum_{s=\uparrow,\downarrow}\Big[N_u\, e^{-(i/\hbar)\mathrm{x}\cdot\mathrm{p}}\, u^s(\mathrm{p}) + N_v\, e^{(i/\hbar)\mathrm{x}\cdot\mathrm{p}}\, v^s(\mathrm{p})\Big], \tag{5.44}$$

which represents the "plane wave" of a spin-$\frac{1}{2}$ particle, free of the influence of any potential. This $\Psi(\mathrm{x})$, however, is not a 4-vector in the 4-dimensional spacetime, but the so-called Dirac spinor, which we will see transforms with respect to Lorentz transformations, in an intrinsic fashion, distinct from 4-vectors.

Lorentz transformations of the Dirac spinor

From relation (A.121c), we see that the antisymmetrized products of two Dirac gamma matrices, $\boldsymbol{\gamma}^{\mu\nu} := \frac{i}{4}[\boldsymbol{\gamma}^\mu, \boldsymbol{\gamma}^\nu]$, close a Lie algebra:

$$[\boldsymbol{\gamma}^{\mu\nu}, \boldsymbol{\gamma}^{\rho\sigma}] = \eta^{\mu\rho}\boldsymbol{\gamma}^{\nu\sigma} - \eta^{\mu\sigma}\boldsymbol{\gamma}^{\nu\rho} + \eta^{\nu\sigma}\boldsymbol{\gamma}^{\mu\rho} - \eta^{\nu\rho}\boldsymbol{\gamma}^{\mu\sigma}. \tag{5.45}$$

It is not hard to verify that the definitions $J_j := \frac{1}{2i}\varepsilon_{jkl}\boldsymbol{\gamma}^{kl}$ and $K_j := i\boldsymbol{\gamma}^{0j}$ result in the commutation relations (5.45) written as

$$[J_j, J_k] = i\varepsilon_{jk}{}^m J_m, \qquad [J_j, K_k] = i\varepsilon_{jk}{}^m K_m, \qquad [K_j, K_k] = -i\varepsilon_{jk}{}^m J_m. \tag{5.46}$$

While the J_j elements generate $SO(3)$ rotations, the K_j elements generate Lorentz boosts. The elements of the Lorentz group – in the representation that acts upon 4-component Dirac spinors – are obtained as exponential functions of the linear combinations of these six generators:

$$g(\vec{\varphi},\vec{\beta}) := \exp\{-i(\varphi^j J_j + \beta^j K_j)\} = \exp\{\beta_j \boldsymbol{\gamma}^{0j} - \varepsilon_{jkm}\varphi^j \boldsymbol{\gamma}^{km}\} = \exp\{\lambda_{\mu\nu}\boldsymbol{\gamma}^{\mu\nu}\}. \tag{5.47}$$

These may be shown to actually form a double covering of the $SO(1,3)$ group, denoted $Spin(1,3)$: to each non-identity element of the $SO(1,3)$ group there correspond precisely two elements of the $Spin(1,3)$ group. For example, the 360°-rotations of $SO(1,3)$ correspond to the $\pm\mathbb{1}$ elements of $Spin(1,3)$, and only the $SO(1,3)$ 720°-rotation corresponds to the unique element $\mathbb{1} \in Spin(1,3)$.

Let us just cite here [64] that the Lorentz boost in the x^1-direction causes the transformation

$$\Psi(x) \to \left[\sqrt{\tfrac{1}{2}(\gamma+1)}\mathbb{1} - \sqrt{\tfrac{1}{2}(\gamma-1)}\boldsymbol{\gamma}^{01}\right]\Psi(x), \tag{5.48}$$

where γ with no index denotes the familiar relativistic factor $\gamma := \frac{1}{\sqrt{1-v^2/c^2}}$. It is then easy to verify that $\Psi^\dagger\Psi$ is not Lorentz-invariant, but that $\Psi^\dagger\boldsymbol{\gamma}^0\Psi$ is. One thus defines

$$\overline{\Psi} := \Psi^\dagger\boldsymbol{\gamma}^0 \tag{5.49}$$

as the Dirac-conjugate of the Dirac spinor Ψ, and note that $\overline{\Psi}\Psi$ is Lorentz-invariant.

Using the results from Appendix A.6.1, the following bilinear[8] functions may be constructed from a Dirac spinor and its Dirac-conjugate spinor:

Expression	Lorentz representation	Number of independent components
$\overline{\Psi}\Psi =$	scalar,	1
$\overline{\Psi}\boldsymbol{\gamma}^\mu\Psi =$	4-vector,	4
$\overline{\Psi}\boldsymbol{\gamma}^{\mu\nu}\Psi =$	antisymmetric rank-2 tensor,	6
$\overline{\Psi}\boldsymbol{\gamma}^\mu\widehat{\boldsymbol{\gamma}}\Psi =$	axial (i.e., pseudo-) 4-vector,	4
$\overline{\Psi}\widehat{\boldsymbol{\gamma}}\Psi =$	pseudo-scalar,	1

(5.50)

Since every complex 4×4 matrix may be written as a complex linear combination of 16 matrices (A.124) [580], the 16 functions (5.50) also form a complete system of bilinear functions of the Dirac spinor, Ψ. It is important to note that in the functions (5.50), the $\boldsymbol{\gamma}$-matrices do not transform with respect to the Lorentz transformations, but $\overline{\Psi}$ and Ψ do, and in fact just so that each bilinear product as a whole transforms in the indicated fashion. For example, $\overline{\Psi}\boldsymbol{\gamma}^\mu\Psi$ really transforms, as a whole, as the components of any other contravariant 4-vector.

Comment 5.5 *The careful Reader may have questioned the identification of the matrices J_j and K_j as the rotation and Lorentz boosts. The list (5.50) gives unambiguous confirmation, in the form of the correct Lorentz transformations of the listed bilinear expressions.*

The notation (5.50) is standard, and supposes that one consistently uses that the $\boldsymbol{\gamma}^\mu, \boldsymbol{\gamma}^{\mu\nu}$ and $\widehat{\boldsymbol{\gamma}}$ are all 4×4 matrices, $\overline{\Psi}$ is a 4-component row-matrix, and Ψ a 4-component column-matrix. Instead, one may also use the index notation, so the Ath element of the column-matrix Ψ is written Ψ^A, the Ath element of the row-matrix $\overline{\Psi}$ is $\overline{\Psi}_A$, and similarly for the $\boldsymbol{\gamma}$-matrices, so that the expressions (5.50) become

$$\overline{\Psi}_A\Psi^A, \quad \overline{\Psi}_A\,(\gamma^\mu)^A{}_B\,\Psi^B, \quad \overline{\Psi}_A\,(\gamma^{\mu\nu})^A{}_B\,\Psi^B, \quad \overline{\Psi}_A\,(\gamma^\mu)^A{}_B\,(\widehat{\gamma})^B{}_C\,\Psi^C, \quad \overline{\Psi}_A\,(\widehat{\gamma})^A{}_B\,\Psi^B. \tag{5.51}$$

[8] It is understood that "bilinear" here means "*anti*-linear + linear in $\overline{\Psi}$ and Ψ, respectively."

Similarly, instead of column-matrices (5.40)–(5.41), we may write[9] $u^{\uparrow A}$, etc., where, for example,

$$u^{\uparrow 1} = N, \quad u^{\uparrow 2} = 0, \quad u^{\uparrow 3} = N\frac{p_z c}{E+mc^2}, \quad u^{\uparrow 4} = N\frac{(p_x + ip_y)c}{E+mc^2}, \quad \text{etc.} \tag{5.52}$$

The normalizing factors in equation (5.44) are chosen so that

$$\overline{u^\uparrow}\, u^\uparrow = 2mc = \overline{u^\downarrow}\, u^\downarrow \quad \text{and} \quad \overline{v^\uparrow}\, v^\uparrow = -2mc = \overline{v^\downarrow}\, v^\downarrow. \tag{5.53}$$

The solutions of equations (5.40)–(5.41) are also complete, in the sense that[10]

$$\sum_{s=\uparrow,\downarrow} u^s\, \overline{u^s} = \not{p} + mc\mathbb{1} \quad \text{and} \quad \sum_{s=\uparrow,\downarrow} v^s\, \overline{v^s} = \not{p} - mc\mathbb{1}, \tag{5.54}$$

that is,

$$\sum_{s=\uparrow,\downarrow} u^{s,A}\, \overline{u^s}_B = (\gamma^\mu)^A{}_B p_\mu + mc\,\delta^A_B \quad \text{and} \quad \sum_{s=\uparrow,\downarrow} v^{s,A}\, \overline{v^s}_B = (\gamma^\mu)^A{}_B p_\mu - mc\,\delta^A_B. \tag{5.55}$$

The matrix (5.54) and the (explicit) index notation (5.55) may be used interchangeably, as needed and for the sake of compactness and clarity. Also, by the general Dirac spinor Ψ one understands a general linear combination

$$\Psi := \hat{\mathbf{e}}_A\, \Psi^A, \tag{5.56}$$

just as we write $\mathbf{x} = \hat{\mathbf{e}}_\mu\, x^\mu$ for a 4-vector. However, one must keep in mind that the $\hat{\mathbf{e}}_\mu$ are (Cartesian) unit vectors in the 4-dimensional spacetime in which we too move, whereas the $\hat{\mathbf{e}}_A$ are unit vectors in an abstract vector space of solutions to the Dirac equation.

Helicity, chirality and the Weyl equation

It is useful to note a very important difference between two seemingly similar properties of spin-$\frac{1}{2}$ particles: helicity and chirality. Much of the analysis here may be found in standard texts on particle physics and field theory as cited in the preface, but there is also a book dedicated to all matters of spin in particle physics [334]. The generalization of this analysis of course also exists for particles with arbitrary spin, subject however to the Weinberg–Witten theorem 6.1 on p. 249, as well as to higher-dimensional spacetime as needed in string theory.

Using the projectors (A.121b)

$$\gamma_\pm := \tfrac{1}{2}[\mathbb{1} \pm \widehat{\gamma}], \tag{5.57}$$

one defines in a fully Lorentz-invariant way:

$$\Psi_\pm := \gamma_\pm \Psi, \quad \text{so} \quad \Psi_+ + \Psi_- = \Psi, \quad \gamma_\pm \Psi_\pm = \Psi_\pm, \quad \gamma_\pm \Psi_\mp = 0. \tag{5.58}$$

For Ψ_+ (also written as Ψ_R) one says that it has right-handed chirality, and Ψ_- (also Ψ_L) has left-handed chirality. To this end, Weyl's basis (A.132) of Dirac matrices is particularly convenient. The complex 2-component projections $\Psi_\pm$ are Weyl spinors.

Independently of chirality, for particles with linear momentum $\vec{p}$ and spin $\vec{S}$, one defines the *helicity* operator, $h := \hat{p}\cdot\vec{S}/\hbar$, the eigenvalue of which is the helicity of the particle. With the mental (mnemonic and entirely fictitious!) image of the intrinsic angular momentum (spin) of the particle represented as the rotation of the particle itself, helicity may be represented as the "projection of the spin in the direction of motion." For example, a spin-$\frac{1}{2}$ particle may have helicity $+\frac{1}{2}$ or $-\frac{1}{2}$,

[9] Caution: the Dirac 4-spinors $u^\uparrow, u^\downarrow, v^\uparrow$ and $v^\downarrow$ are linearly independent and each has four components. Only a total of four of these components are linearly independent.

[10] Caution: the normalizations (5.53) and (5.54) differ from the standard quantum mechanical ones.

depending on whether, respectively, it "spins" about the direction of motion in the right-hand sense or the left-hand sense.

Helicity is not defined in a Lorentz-invariant manner. Indeed, a particle with a nonzero mass always has a rest-frame wherein it does not move, and where $\vec{p} = 0$, so the eigenvalues of h vanish. Also, it is always possible to pass such a particle, i.e., Lorentz-boost, into a coordinate system wherein the particle moves in the direction opposite to the original $\vec{p}$. Since this changes the sign of $\vec{p}$ but not of $\vec{S}$, the eigenvalues of h also change their sign. It follows that helicity cannot be Lorentz-invariant for particles with nonzero mass.

For particles with no mass, helicity *is* Lorentz-invariant, and coincides with chirality.

— ❦ —

The solutions (5.40)–(5.43) of the Dirac equation (5.34) indicate that the upper and lower components of the Dirac spinor are not independent and it is not possible to separate them in a Lorentz-invariant way. The relations (A.121b) define the projectors $\boldsymbol{\gamma}_\pm$ that are Lorentz-invariant since the $\boldsymbol{\gamma}$-matrices do not change with respect to Lorentz transformations, which gives rise to the hope that the Dirac 4-component spinor may be separated into two 2-component spinors in a Lorentz-invariant way.

Digression 5.4 One often finds a "quick" argument in the literature that $\boldsymbol{\gamma}$-matrices are Lorentz-invariant: supposedly, in the product $\boldsymbol{\gamma}^\mu p_\mu$, the Lorentz transformations act upon the physical quantity, the 4-momentum, and not on the $\boldsymbol{\gamma}$-matrices. This recalls the view that rotations of a vector $\vec{v} = \hat{\mathrm{e}}_i v^i$ act upon the basis elements $\hat{\mathrm{e}}_i$, not on the components, which are "only numerical values" in a given coordinate system. However, it is equally reasonable to adopt the vantage point where the *inverse* rotations act upon the components v^i, and not upon the basis elements $\hat{\mathrm{e}}_i$. Both applications of the transformations produce a net change in the physical quantity $\hat{\mathrm{e}}_i v^i$, which is regarded as the "active" transformation. By contrast, the "passive" transformation *simultaneously* rotates both the basis vectors $\hat{\mathrm{e}}_i$ as well as the components v^i (in the inverse sense), so that the physical quantity $\vec{v}$ remains invariant.

However, this is not a case of active/passive action of the Lorentz transformations: The Dirac $\boldsymbol{\gamma}$-matrices indeed are components of a 4-vector, but those components are *matrices*, the rows of which are in the basis of the Dirac 4-component spinor Ψ, and the columns of which are in the basis of the Dirac-conjugated spinor $\overline{\Psi}$. The Lorentz transformations act upon all three bases, and those actions mutually cancel so that the $\boldsymbol{\gamma}$-matrices remain invariant. In other words, the product $\overline{\Psi}\boldsymbol{\gamma}^\mu\Psi\, p_\mu$ is evidently Lorentz-invariant: $\overline{\Psi}\boldsymbol{\gamma}^\mu\Psi$ is a contravariant 4-vector and p_μ a covariant one, so $\overline{\Psi}\boldsymbol{\gamma}^\mu\Psi\, p_\mu$ is the scalar product of a contravariant 4-vector and a covariant 4-vector. By adapting the index notation so as to also count the components of the Dirac spinor (5.51), we have

$$(\overline{\Psi}\boldsymbol{\gamma}^\mu\Psi)\, p_\mu = \big(\overline{\Psi}_A\, (\boldsymbol{\gamma}^\mu)^A{}_B\, \Psi^B\big)\, p_\mu, \tag{5.59a}$$

so that the numerical values $(\boldsymbol{\gamma}^\mu)^A{}_B$ for each fixed μ, A, B are simply the Clebsch–Gordan coefficients in the expansion of the product $\overline{\Psi} \times \Psi$ in a spacetime 4-vector basis. In turn, the coefficients $(\boldsymbol{\gamma}^\mu)^A{}_B$ also appear in the tri-linear Lorentz-*invariant* contraction of the basis vectors $\hat{\mathrm{e}}_A (\boldsymbol{\gamma}^\mu)^A{}_B\, \hat{\mathrm{e}}_\mu\, \hat{\mathrm{e}}^B$ [☞ Section A.6]. Lastly, rewriting the above equation as

$$(\overline{\Psi}\boldsymbol{\gamma}^\mu\Psi)\, p_\mu = (\boldsymbol{\gamma}^\mu)^A{}_B\, \big(\overline{\Psi}_A\, \Psi^B\, p_\mu\big) \tag{5.59b}$$

re-interprets the matrices $\boldsymbol{\gamma}^\mu$ as the (in general) $\mathbb{C}$-valued projection of the direct product of Dirac-conjugate Dirac-spinors, Dirac-spinors and 4-momenta to Lorentz-invariant (in general) complex numbers: $\boldsymbol{\gamma} : \{\overline{\Psi}\} \times \{\Psi\} \times \{p\} \to \mathbb{C}$.

However, it is not hard to show that helicity projections do not commute with the Dirac matrices:

$$[\boldsymbol{\gamma}_\pm, \boldsymbol{\gamma}^\mu] \neq 0 : \qquad [\mathbb{1}, \boldsymbol{\gamma}^\mu] = 0 = \{\widehat{\boldsymbol{\gamma}}, \boldsymbol{\gamma}^\mu\} \quad \Rightarrow \quad \boldsymbol{\gamma}_\pm \boldsymbol{\gamma}^\mu = \boldsymbol{\gamma}^\mu \boldsymbol{\gamma}_\mp. \tag{5.60}$$

Owing to this, an attempt to use the projections (5.58) on the Dirac equation yields

$$\begin{aligned} \boldsymbol{\gamma}_\pm [i\hbar\boldsymbol{\gamma}^\mu \partial_\mu - mc\mathbb{1}]\Psi = [i\hbar\boldsymbol{\gamma}_\pm\boldsymbol{\gamma}^\mu \partial_\mu - mc\boldsymbol{\gamma}_\pm\mathbb{1}]\Psi &= [i\hbar\boldsymbol{\gamma}^\mu\boldsymbol{\gamma}_\mp \partial_\mu - mc\mathbb{1}\boldsymbol{\gamma}_\pm]\Psi \\ &= i\hbar\boldsymbol{\gamma}^\mu(\partial_\mu\Psi_\mp) - mc\Psi_\pm, \end{aligned} \tag{5.61}$$

which is a system of differential equations that couples Ψ_+ and Ψ_- precisely when $m \neq 0$. Conversely,

$$\boldsymbol{\gamma}^\mu \partial_\mu \Psi_\pm = 0 \qquad \Leftrightarrow \qquad m\Psi_\pm = 0. \tag{5.62}$$

Conclusion 5.2 (Weyl) *The Dirac spinor Ψ separates in a Lorentz-invariant way into the* ***right-handed*** *$\Psi_+ \equiv \Psi_R := \boldsymbol{\gamma}_+\Psi$ and* ***left-handed*** *$\Psi_- \equiv \Psi_L := \boldsymbol{\gamma}_-\Psi$ 2-component Weyl spinor (the eigen-spinors of the $\widehat{\boldsymbol{\gamma}}$ matrix) precisely when the mass of the particle vanishes.*

These (Weyl) spinors satisfy the simpler differential equations, $\boldsymbol{\gamma}^\mu\partial_\mu\Psi_\pm = 0$. Indeed, the Dirac differential equation (5.34) is a system of four coupled differential equations for four components of the Dirac spinor Ψ. By contrast, $\boldsymbol{\gamma}^\mu\partial_\mu\Psi_\pm = 0$ is a system of two coupled differential equations for two components of the Weyl spinor Ψ_+ and separately for Ψ_-.

Hermann Weyl noticed and published the characteristics of this special case of the Dirac equation in 1929. Yet, when Pauli invented the neutrino so as to preserve the energy conservation law, he did not want to use Weyl's equations on the grounds that they permit violating the symmetry of parity.[11] To wit, the Lorentz-invariant separation of Ψ_+ and $\Psi_- \propto P(\Psi_+)$ permits an independent – and different – treatment of these two halves of the Dirac spinor of opposite chirality. This is quite ironic, since Pauli did correctly predict the mass of the neutrino to be either very teeny or vanishing, and even during his own life it became clear that Nature really treats the left-handed neutrino very differently from the right-handed one. Until the discovery of the see-saw mechanism [☞ Section 7.3.2], the Weyl equations provided a much better model for neutrinos, and describe the maximal parity violation as observed in Nature.

— ❦ —

The frequent confusion of helicity and chirality has been fostered by the fact that massless particles are a specially simple case both for chirality and for helicity, where these two different physical quantities coincide. On the other hand, the Lorentz invariance of chirality is of fundamental importance in the contemporary formulation of weak and electroweak interactions, while helicity is easier to measure. The Reader should strive to conceptually differentiate and carefully distinguish between these two inherently different quantities.

[11] Up to the experimental confirmations of parity violation in weak interactions [☞ Sections 2.4.2 and 4.2.1], Pauli had, just as many other renowned physicists of the time, ardently advocated against ideas that include parity violation; see, e.g., A. Salam's Nobel lecture [473].

The Dirac Lagrangian density

The construction of the Dirac Lagrangian density is straightforward, if we only require the variation of the Hamilton action with that Lagrangian density to produce the Dirac equation. First, note that Ψ and $\overline{\Psi}$ may *formally* be treated as independent quantities. The Dirac equation (5.34) is then simply multiplied *from the left* by $\overline{\Psi}$ (and by c for units) and we identify

$$\mathscr{L}_D = \beta\overline{\Psi}(x)\left[c\not{p} + mc^2\mathbb{1}\right]\Psi(x) = -\beta\overline{\Psi}(x)\left[i\hbar c\boldsymbol{\gamma}^\mu\partial_\mu - mc^2\mathbb{1}\right]\Psi(x), \tag{5.63}$$

where β is an arbitrary overall sign, since the variation by $\overline{\Psi}$ yields a β-multiple of equation (5.34). Variation by Ψ yields the Hermitian conjugate of equation (5.34), i.e., nothing new (and nothing unneeded).

Digression 5.5 The Dirac spinor Ψ is a 4-tuple of formally anticommutative variables. In the general case, if ψ and χ are anticommutative and f and g are commutative variables, we have that

$$[f,g]=0, \qquad [f,\psi]=[f,\chi]=0=[g,\psi]=[g,\chi], \qquad \text{but} \qquad \{\psi,\chi\}=0; \tag{5.64a}$$

$$\Big[\frac{\partial}{\partial f},\frac{\partial}{\partial g}\Big]=0, \quad \Big[\frac{\partial}{\partial f},\frac{\partial}{\partial \psi}\Big]=\Big[\frac{\partial}{\partial f},\frac{\partial}{\partial \chi}\Big]=0=\Big[\frac{\partial}{\partial g},\frac{\partial}{\partial \psi}\Big]=\Big[\frac{\partial}{\partial g},\frac{\partial}{\partial \chi}\Big], \quad \text{but} \quad \Big\{\frac{\partial}{\partial \psi},\frac{\partial}{\partial \chi}\Big\}=0. \tag{5.64b}$$

Also,

$$\frac{\partial}{\partial\psi}\chi = -\chi\frac{\partial}{\partial\psi} \quad \text{and} \quad \frac{\partial}{\partial\chi}\psi = -\psi\frac{\partial}{\partial\chi}, \tag{5.64c}$$

which the Student must keep in mind when deriving the equations of motion from Lagrangian densities that also contain fermionic (anticommutative) variables. It is convenient to define the *right-derivative*:

$$\psi\frac{\overleftarrow{\partial}}{\partial\psi}=1, \quad (\psi\chi)\frac{\overleftarrow{\partial}}{\partial\psi}=-\Big(\psi\frac{\overleftarrow{\partial}}{\partial\psi}\Big)\chi=-\chi, \quad (\psi\chi)\frac{\overleftarrow{\partial}}{\partial\chi}=\psi\Big(\chi\frac{\overleftarrow{\partial}}{\partial\chi}\Big)=\psi, \quad \text{etc.,} \tag{5.64d}$$

and diligently apply derivatives *either* from right *or* from left.

The definition of the Lagrangian allows us to identify the components of Ψ as the *canonical coordinates*, so we may also define the canonically conjugate momentum densities:

$$\pi_\Psi := \mathscr{L}_D\frac{\overleftarrow{\partial}}{\partial\dot{\Psi}} = \big(-\beta\overline{\Psi}[i\hbar c\boldsymbol{\gamma}^\mu\partial_\mu - mc^2\mathbb{1}]\Psi\big)\frac{\overleftarrow{\partial}}{\partial(c\partial_0\Psi)} = -i\beta\hbar\overline{\Psi}\boldsymbol{\gamma}^0 = -i\beta\hbar\Psi^\dagger, \tag{5.65}$$

where we applied the right-derivative [☞ Digression 5.5 on p. 180]. The Hamiltonian then becomes

$$\begin{aligned}\mathscr{H}_D &= \pi_\Psi\dot{\Psi} - \mathscr{L}_D = (-i\beta\hbar\Psi^\dagger)(\dot{\Psi}) + \beta\overline{\Psi}[i\hbar c\boldsymbol{\gamma}^\mu\partial_\mu - mc^2\mathbb{1}]\Psi\\ &= -\beta\Psi^\dagger H\Psi + \beta\overline{\Psi}[i\hbar c\boldsymbol{\gamma}^\mu\partial_\mu - mc^2\mathbb{1}]\Psi, \qquad H \equiv i\hbar\tfrac{\partial}{\partial t}.\end{aligned} \tag{5.66}$$

The sign β in the computation (5.65) may now be determined as follows: For an on-shell Dirac fermion, i.e., one that satisfies the equations of motion (5.34), the second term in the expression (5.66) vanishes, and we obtain

$$\mathscr{H}_D\big|_{(5.34)} = -\beta\Psi^\dagger H\Psi, \qquad \text{where} \quad [\Psi] = \tfrac{1}{L^{3/2}}. \tag{5.67}$$

We thus choose $\beta = -1$ for the total energy (the Hamiltonian) of the Hamilton–Jacobi canonical formalism on-shell and the expectation value of the quantum-mechanical operator H to have the same sign. To sum up:

$$\mathscr{L}_D = -\overline{\Psi}(x)\,[c\not{p} + mc^2\mathbb{1}]\,\Psi(x) = \overline{\Psi}(x)\,[i\hbar c\gamma^\mu\partial_\mu - mc^2\mathbb{1}]\,\Psi(x), \tag{5.68a}$$

$$\pi_\Psi = i\hbar\Psi^\dagger, \tag{5.68b}$$

$$\mathscr{H}_D = \pi_\Psi\dot{\Psi} - \mathscr{L}_D = \Psi^\dagger H\Psi - \overline{\Psi}[i\hbar c\gamma^\mu\partial_\mu - mc^2\mathbb{1}]\Psi. \tag{5.68c}$$

Also, since the Dirac equation (5.34) may be written as

$$i\hbar\dot{\Psi} = H_D\Psi := [(i\hbar c\vec{\gamma}\cdot\vec{\nabla} + mc^2)\gamma^0]\Psi, \tag{5.69}$$

and H_D is the on-shell Dirac Hamiltonian operator: $\mathscr{H}_D \overset{(5.34)}{=} \overline{\Psi}H_D\Psi$, as arranged in equation (5.67).

Digression 5.6 The Dirac Lagrangian densities are often "antisymmetrized" using the identity

$$\int d^4x\,\overline{\Psi}\gamma^\mu\partial_\mu\Psi = \tfrac{1}{2}\int d^4x\,\overline{\Psi}\gamma^\mu\partial_\mu\Psi + \tfrac{1}{2}\int d^4x\,\big[\partial_\mu(\overline{\Psi}\gamma^\mu\Psi) - (\partial_\mu\overline{\Psi})\gamma^\mu\Psi\big] \tag{5.70a}$$

$$= \tfrac{1}{2}\int d^4x\,[\overline{\Psi}\gamma^\mu\partial_\mu\Psi - (\partial_\mu\overline{\Psi})\gamma^\mu\Psi] + \underbrace{\tfrac{1}{2}\oint_{\mathscr{V}} d^3(x)_\mu\,(\overline{\Psi}\gamma^\mu\Psi)}_{=0}, \tag{5.70b}$$

where the third, 3-dimensional integral is computed over the 3-dimensional boundary of spacetime, which is "at infinity." Physical fields are required to vanish there. We thus write

$$\overline{\Psi}\gamma^\mu\partial_\mu\Psi \simeq \tfrac{1}{2}[\overline{\Psi}\gamma^\mu\partial_\mu\Psi - (\partial_\mu\overline{\Psi})\gamma^\mu\Psi] =: \tfrac{1}{2}(\overline{\Psi}\gamma^\mu\overleftrightarrow{\partial}_\mu\Psi), \tag{5.70c}$$

where the middle expression defines the symbol $\overleftrightarrow{\partial}_\mu$. So antisymmetrized, we have that

$$\mathscr{L}_D \simeq -\overline{\Psi}(x)\,[\tfrac{1}{2}c\overleftrightarrow{\not{p}} + mc^2\mathbb{1}]\,\Psi(x) = \overline{\Psi}(x)\,[\tfrac{i}{2}\hbar c\overleftrightarrow{\not{\partial}} - mc^2\mathbb{1}]\,\Psi(x). \tag{5.70d}$$

Finally, the components of the canonically conjugate momentum density (5.68b) are constantly proportional to the Hermitian conjugates of the components of the Dirac spinor itself. Roughly speaking, one half of the Dirac (4-component) spinor are canonical coordinates of the system, the other half are conjugate momenta. The choice of which particular components are regarded as coordinates and which are momenta is, of course, arbitrary – up to the condition that the relations

$$\{\Psi, (i\hbar\Psi^\dagger)\} = i\hbar\mathbb{1} \quad \Rightarrow \quad \{\Psi, \Psi^\dagger\} = \mathbb{1} \tag{5.71}$$

produce the canonical *anticommutation* relations between the canonical momenta and the canonical coordinates. This arbitrariness is identical to that in classical physics.

5.2.2 The $U(1)$ gauge symmetry and photons

Classical electrodynamics builds on the Maxwell equations,

$$
\text{(Gauss)}\begin{cases}
\vec{\nabla}\cdot\vec{E} = \frac{1}{4\pi\epsilon_0}4\pi\,\rho_e, & \vec{\nabla}\times(c\vec{B}) - \frac{1}{c}\frac{\partial\vec{E}}{\partial t} = \frac{1}{4\pi\epsilon_0}\frac{4\pi}{c}\vec{\jmath}_e, \quad \text{(Ampère)} \qquad (5.72a)\\[2mm]
\vec{\nabla}\cdot(c\vec{B}) = \frac{\mu_0}{4\pi}4\pi\,\rho_m, & -\vec{\nabla}\times\vec{E} - \frac{1}{c}\frac{\partial(c\vec{B})}{\partial t} = \frac{\mu_0}{4\pi}\frac{4\pi}{c}\vec{\jmath}_m, \quad \text{(Faraday)} \qquad (5.72b)
\end{cases}
$$

that encompass the indicated laws, and where $c = \frac{1}{\sqrt{\epsilon_0\,\mu_0}}$ is the speed of propagation of light in vacuum. The densities of the magnetic (monopole!) charges, ρ_m, and currents, $\vec{\jmath}_m$, are included for later discussion of electro-magnetic duality [☞ Section 11.4]. No experiment indicates their existence, so that the equations (5.72b) are cited in the literature almost exclusively with $\rho_m \to 0$ and $\vec{\jmath}_m \to 0$. However, note that the units satisfy $[\rho_e/\epsilon_0] = [\mu_0\,\rho_m]$, as well as $[\vec{\jmath}_e/\epsilon_0] = [\mu_0\,\vec{\jmath}_m]$.

The relativistic description

For the purposes of a relativistic description of electrodynamics [☞ also Comment 8.1 on p. 294], we introduce[12]

$$A_\mu := (\Phi, -c\vec{A}), \quad \text{(gauge potential)} \qquad A^\mu := \eta^{\mu\nu}A_\nu = (\Phi, c\vec{A}); \tag{5.73a}$$

$$F_{\mu\nu} := \partial_\mu A_\nu - \partial_\nu A_\mu, \quad \begin{pmatrix}\text{antisymmetric}\\ \text{rank-2 tensor}\end{pmatrix} \qquad F^{\mu\nu} := \eta^{\mu\rho}F_{\rho\sigma}\eta^{\sigma\nu}; \tag{5.73b}$$

and identify

$$F_{00} = 0, \qquad F^{00} = 0, \tag{5.73c}$$

$$F_{0i} = \partial_0 A_i - \partial_i A_0 = \frac{1}{c}\frac{\partial(-cA_i)}{\partial t} - \frac{\partial\Phi}{\partial x^i} = E_i, \qquad F^{0i} = \eta^{00}F_{0j}\eta^{ji} = -E_i, \tag{5.73d}$$

$$
\begin{aligned}
F_{ij} &= \partial_i A_j - \partial_j A_i = \frac{\partial(-cA_j)}{\partial x^i} - \frac{\partial(-cA_i)}{\partial x^j}\\
&= c\Big(\frac{\partial A_i}{\partial x^j} - \frac{\partial A_j}{\partial x^i}\Big) = c\varepsilon_{ji}{}^k\,B_k \quad = -c\varepsilon_{ij}{}^k B_k, \qquad F^{ij} = \eta^{ik}F_{kl}\eta^{jl} = -c\varepsilon^{ijk}B_k
\end{aligned} \tag{5.73e}
$$

and, of course, $F_{\mu\nu} = -F_{\nu\mu}$. In matrix form, we have

$$
[F_{\mu\nu}] = \begin{bmatrix} 0 & E_1 & E_2 & E_3\\ -E_1 & 0 & -cB_3 & cB_2\\ -E_2 & cB_3 & 0 & -cB_1\\ -E_3 & -cB_2 & cB_1 & 0\end{bmatrix}, \quad
[F^{\mu\nu}] = \begin{bmatrix} 0 & -E_1 & -E_2 & -E_3\\ E_1 & 0 & -cB_3 & cB_2\\ E_2 & cB_3 & 0 & -cB_1\\ E_3 & -cB_2 & cB_1 & 0\end{bmatrix}. \tag{5.74}
$$

Since $F_{\mu\nu}$ are components of a rank-2 tensor, it follows that the Lorentz transformations act by [☞ Digression 3.5 on p. 91]

$$y^\mu = \mathrm{L}^\mu{}_\nu x^\nu \quad \Rightarrow \quad F_{\mu\nu}(\mathrm{y}) = \mathrm{L}^\rho{}_\mu\,F_{\rho\sigma}(\mathrm{x})\,\mathrm{L}^\sigma{}_\nu, \quad \text{i.e.} \quad \mathbf{F}(\mathrm{y}) = \mathbf{L}^T\,\mathbf{F}(\mathrm{x})\,\mathbf{L}. \tag{5.75}$$

The familiar Lagrangian [☞ also Exercises 5.1.3 and 5.1.4] for the electromagnetic field may thus be written as

$$\mathscr{L}_{EM} = -\tfrac{4\pi\epsilon_0}{4}F_{\mu\nu}\,F^{\mu\nu}. \tag{5.76}$$

---

[12] The negative relative sign in the definition of A_μ cancels the difference in signs in the definition (5.13), an additional factor of c equates the units of Φ and $\vec{A}$, which stem from the difference between D_t and $\vec{D}$.

Example 5.1 Let $\vec{E} = \hat{e}^2 E_2$ and $\vec{B} = 0$ be given in an inertial Cartesian coordinate system S, and let the inertial system $\widetilde{S}$ move with respect to S with the constant speed $\hat{e}^1 v_1$. The relations (5.75) then yield

$$\widetilde{E}_2 = \gamma E_2, \qquad \text{and also} \qquad \widetilde{B}_3 = \gamma \frac{v_1}{c^2} E_2. \tag{5.77}$$

A field that in one inertial coordinate system looks like a "purely" electric field, can in another inertial system easily be represented by a *combination* of electric and magnetic fields. Notice, however, that the equation $\vec{E}\cdot\vec{B} = 0$ remains valid. Indeed, this is a Lorentz-invariant characteristic of the specified field [☞ relations (5.80a)].

The Maxwell equations (5.72a) may then also be written as

$$\partial_\mu F^{\mu\nu} = \frac{1}{4\pi\epsilon_0} \frac{4\pi}{c} j_e^\nu, \tag{5.78}$$

where $\mathrm{j}_e = (c\rho_e, \vec{j}_e)$ is the 4-vector of electric charge and current densities. Analogously, the Maxwell equations (5.72b) may be written also as

$$\tfrac{1}{2}\varepsilon^{\mu\nu\rho\sigma}\partial_\mu F_{\nu\rho} = \frac{\mu_0}{4\pi} \frac{4\pi}{c} j_m^\sigma, \tag{5.79}$$

where $\mathrm{j}_m = (c\rho_m, \vec{j}_m)$ is the 4-vector of (monopole) magnetic charge and current densities.

Digression 5.7 Direct substitution yields

$$\tfrac{1}{2}\Gamma_{\mu\nu}F^{\mu\nu} = \vec{E}^2 - c^2\vec{B}^2 \qquad \text{and} \qquad \tfrac{1}{4}\varepsilon^{\mu\nu\rho\sigma}F_{\mu\nu}F_{\rho\sigma} = -c\vec{E}\cdot\vec{B}, \tag{5.80a}$$

which, using the transformations (5.75), shows that these two bilinear expressions in $\vec{E}$ and $\vec{B}$ are Lorentz-invariant. Evidently, these are the only linearly independent Lorentz-invariant bilinear expressions in $F_{\mu\nu}$ and $F^{\mu\nu}$, and so then also in $\vec{E}$ and $\vec{B}$. Since the Lagrangian density for electrodynamics must be a scalar (invariant) density and quadratic in electric and magnetic fields, we find that the Lagrangian density must be of the form

$$\mathscr{L}_{EM} = C_1\, F_{\mu\nu}F^{\mu\nu} + C_2\, \varepsilon^{\mu\nu\rho\sigma}\, F_{\mu\nu}F_{\rho\sigma}. \tag{5.80b}$$

The coefficients C_1, C_2 are chosen so that the variation of the Hamilton action, $\delta \int \mathrm{d}^4\mathrm{x}\, \mathscr{L}_{EM} = 0$, reproduces the Maxwell equations. The fact that this renders $C_2 = 0$ then poses the (unanswered☝) question: Why is, in the possible "addition"

$$\mathscr{L}_{\vartheta,EM} = \vartheta\, \tfrac{4\pi\epsilon_0}{4}\varepsilon^{\mu\nu\rho\sigma}\, F_{\mu\nu}F_{\rho\sigma}, \tag{5.80c}$$

to the standard Lagrangian density (5.76) of the parameter $\vartheta = 0$, either identically or up to experimental error (i.e., $\vartheta \lll 1$)?

Direct substitution of $F_{\mu\nu} = \partial_\mu A_\nu - \partial_\nu A_\mu$ on the left-hand side of equation (5.79) yields

$$\tfrac{1}{2}\varepsilon^{\mu\nu\rho\sigma}\partial_\mu\left(\partial_\nu A_\rho - \partial_\rho A_\nu\right) = \tfrac{1}{2}\varepsilon^{\mu\nu\rho\sigma}\partial_\mu\partial_\nu A_\rho - \tfrac{1}{2}\varepsilon^{\mu\nu\rho\sigma}\partial_\mu\partial_\rho A_\nu, \tag{5.81}$$

where both terms vanish separately, since

$$\varepsilon^{\mu\nu\rho\sigma}\partial_\mu\partial_\nu = \underbrace{\varepsilon^{\mu\nu\rho\sigma}\partial_\nu\partial_\mu}_{\mu\leftrightarrow\nu} = \varepsilon^{\nu\mu\rho\sigma}\partial_\mu\partial_\nu = -\varepsilon^{\mu\nu\rho\sigma}\partial_\mu\partial_\nu. \tag{5.82}$$

That is,

$$F_{\mu\nu} = \partial_\mu A_\nu - \partial_\nu A_\mu \quad \overset{(5.79)}{\Longleftrightarrow} \quad 0 = \frac{\mu_0}{4\pi}\frac{4\pi}{c} j^\sigma_m. \tag{5.83}$$

The existence of magnetic charges and currents would then be an *obstruction* for equating $F_{\mu\nu}$ with $\partial_\mu A_\nu - \partial_\nu A_\mu$, i.e., the electric and the magnetic fields could not be expressed in terms of an *unambiguously specified* 4-vector potential (5.15) [☞ Section 5.2.3], and conversely: if $F_{\mu\nu} = \partial_\mu A_\nu - \partial_\nu A_\mu$ for an unambiguously specified 4-vector potential $A_\mu(\mathbf{x})$, then no monopole magnetic charge or current can exist. We thus have:

Conclusion 5.3 *Electric and magnetic charges and currents exist* ***simultaneously*** *if and only if there can be no* ***unambiguously*** *specified 4-vector potential $A_\mu(\mathbf{x})$ for which the electromagnetic field would be $F_{\mu\nu} = \partial_\mu A_\nu - \partial_\nu A_\mu$ [☞ Section 5.2.3].*

Digression 5.8 Define a "differential 2-form" $\mathbf{F} := F_{\mu\nu}\,\mathrm{d}x^\mu\wedge\mathrm{d}x^\nu$, where "$\wedge$" denotes the antisymmetric product of the differentials, as well as the operator $\mathbf{d} := \mathrm{d}x^\mu\partial_\mu$. Then

$$\mathbf{d}\wedge\mathbf{d} \equiv \tfrac{1}{2}(\partial_\mu\partial_\nu - \partial_\nu\partial_\mu)\,\mathrm{d}x^\mu\wedge\mathrm{d}x^\nu \equiv 0. \tag{5.84a}$$

The Maxwell equations (5.72b), i.e., (5.79), are then equivalent to

$$\mathbf{d}\wedge\mathbf{F} = \mathbf{j}_m, \qquad \mathbf{j}_m := \frac{\mu_0}{4\pi}\frac{4\pi}{c} j^\sigma_m\,\varepsilon_{\mu\nu\rho\sigma}\,\mathrm{d}x^\mu\wedge\mathrm{d}x^\nu\wedge\mathrm{d}x^\rho, \tag{5.84b}$$

and the differential 3-form $\mathbf{j}_m$ is the *obstruction* for equating the differential 2-form $\mathbf{F}$ with $\mathbf{d}\wedge\mathbf{A}$, for any differential 1-form $\mathbf{A} = A_\mu\mathrm{d}x^\mu$. $\mathbf{F}$ is said to be a *nontrivial* (non-exact) 2-form.[13]

On the other hand, equations (5.72a), i.e., (5.78), may also be written in the form (5.84b). To this end, however, we need one more item of notation: in tensorial notation, any antisymmetric rank-r tensor may be turned into an antisymmetric rank-$(4-r)$ tensor by contracting with $\varepsilon_{\mu\nu\rho\sigma}$ or $\varepsilon^{\mu\nu\rho\sigma}$. Thus, a 4-vector j^μ_m is "translated" into a rank-3 tensor $j^\mu_m \to (j^\mu_m\varepsilon_{\mu\nu\rho\sigma})$ and a 3-form $\mathbf{j}_m$. A double use of this operation yields $\frac{1}{2}\varepsilon^{\mu\nu\rho\sigma}\partial_\nu(\frac{1}{2}\varepsilon_{\rho\sigma\alpha\beta}F^{\alpha\beta}) = \partial_\nu F^{\mu\nu}$. The corresponding operation with differential forms is the so-called "Hodge star," which turns an r-form into a $(4-r)$-form: $*\mathbf{A}$ is a 3-form, $*\mathbf{j}_m$ a 1-form, etc. The Maxwell equations (5.72a) and (5.78) are thus equivalent to

$$\mathbf{d}\wedge *\mathbf{F} = \mathbf{j}_e, \qquad \mathbf{j}_e := \frac{1}{4\pi\epsilon_0}\frac{4\pi}{c} j^\mu_e\,\varepsilon_{\mu\nu\rho\sigma}\mathrm{d}x^\nu\wedge\mathrm{d}x^\rho\wedge\mathrm{d}x^\sigma. \tag{5.84c}$$

Equations (5.84b) and (5.84c) respectively provide a compact form of the Maxwell equations:

$$\mathbf{d}\wedge\mathbf{F} = \mathbf{j}_m \quad \text{and} \quad \mathbf{d}\wedge *\mathbf{F} = \mathbf{j}_e. \tag{5.84d}$$

Since $\mathbf{d}\wedge\mathbf{d} \equiv 0$, $\mathbf{d}\wedge\mathbf{d}\wedge(*\mathbf{F}) = \mathbf{d}\wedge\mathbf{j}_e$ produces $\mathbf{d}\wedge\mathbf{j}_e = 0$, which is the well-known continuity equation (2.66), the integral of which yields the electric charge conservation law [☞ also Section 6.1.2]. Similarly, $\mathbf{d}\wedge\mathbf{d}\wedge\mathbf{F} = \mathbf{d}\wedge\mathbf{j}_m$ implies $\mathbf{d}\wedge\mathbf{j}_m = 0$, the continuity equation, and thus the (monopole) magnetic charge conservation law.

---

[13] By the same token is "$đQ$," in thermodynamics in general, a nontrivial 1-form and not an exact differential.

Comment 5.6 *The fact that the existence of (monopole) magnetic charges and currents obstructs the expression of the electromagnetic field $F_{\mu\nu}$ as an antisymmetric derivative of an* ***unambiguously*** *specified 4-vector potential A_μ points to a significant difference between electric and magnetic charges and currents – in spite of the fact that the Maxwell equations (5.72)* ***look*** *"symmetric." This "symmetry" – a* ***duality****, more precisely – is the mapping*

$$\varpi_{EM} : F^{\mu\nu} \longleftrightarrow (*\mathbf{F})^{\mu\nu} = [\tfrac{1}{2}\varepsilon^{\mu\nu\rho\sigma}F_{\rho\sigma}] = \begin{bmatrix} 0 & -cB_1 & -cB_2 & -cB_3 \\ cB_1 & 0 & E_3 & -E_2 \\ cB_2 & -E_3 & 0 & E_1 \\ cB_3 & E_2 & -E_1 & 0 \end{bmatrix}, \tag{5.85}$$

which swaps the roles of $\vec{E}$ and $c\vec{B}$. This implies that the vanishing of ρ_e and $\vec{j}_e$ is a necessary and sufficient condition for the existence of some unambiguously specified 1-form $\widetilde{\mathbf{A}}$ such that $*\mathbf{F} = \mathbf{d}\wedge\widetilde{\mathbf{A}}$; here, $\widetilde{\mathbf{A}} = dx^\mu \widetilde{A}_\mu$ is the 1-form of the ***dual*** *4-vector of gauge potentials.*

Conclusion 5.4 *The difference between $\mathbf{F}$ and $*\mathbf{F}$, i.e., $F_{\mu\nu}$ and $\frac{1}{2}\varepsilon_{\mu\nu\rho\sigma}F^{\rho\sigma}$, i.e., $\vec{E}$ and $c\vec{B}$, i.e., $(c\rho_e, \vec{j}_e)$ and $(\rho_m/c, \vec{j}_m/c^2)$ – and so also the whole formalism – is however fully* ***conventional****.*

The discrete transformation (5.85) is equivalent to $\varpi_{EM} : (\vec{E}, c\vec{B}) \to (c\vec{B}, -\vec{E})$. Since $\varpi_{EM}^2 = -\mathbb{1}$ and $\varpi_{EM}^4 = \mathbb{1}$, ϖ_{EM} is equivalent to a 90°-rotation. In fact, one may define even a continuous duality rotation

$$\varpi_{EM}(\vartheta) : \begin{bmatrix} \vec{E} \\ c\vec{B} \end{bmatrix} \to \begin{bmatrix} \vec{E}' \\ c\vec{B}' \end{bmatrix} = \begin{bmatrix} \cos\vartheta & \sin\vartheta \\ -\sin\vartheta & \cos\vartheta \end{bmatrix} \begin{bmatrix} \vec{E} \\ c\vec{B} \end{bmatrix} \tag{5.86}$$

and correspondingly for electric and magnetic charges and currents. The statement that there are no magnetic monopoles is then equivalent to stating that, using this "rotation," the variables $\vec{E}$ and $c\vec{B}$ (i.e., $F_{\mu\nu}$) may always be chosen so that $\rho_m = 0 = \vec{j}_m$, so that $\mathbf{F} = \mathbf{d}\wedge\mathbf{A}$, i.e., $F_{\mu\nu} = \partial_\mu A_\nu - \partial_\nu A_\mu$ – simultaneously in the whole universe and for all particles in Nature.

The standard electrodynamics

In agreement with experiments, we set $\rho_m = 0 = \vec{j}_m$, so that the relations (5.15) and (5.73b) hold, as does the so-called Bianchi identity, as a consequence of the now applicable definition (5.73b),

$$\varepsilon^{\mu\nu\rho\sigma}\partial_\nu F_{\rho\sigma} = 0, \tag{5.87}$$

and instead of equation (5.79); equations (5.74)–(5.78) remain unchanged.

In classical electrodynamics, one primarily uses the electromagnetic field $F_{\mu\nu}$, i.e., $\vec{E}$ and $\vec{B}$, and the potentials are secondary. However, in the non-relativistic formulation of the interaction (5.19) of the electromagnetic field with substance in quantum theory, the potentials had already been proved to be the fundamental quantities. Besides, the assumption that the electromagnetic field is *defined* in relations (5.73b) makes the relation (5.87) – and then also the laws (5.72b) – a trivial consequence. Thus, in electrodynamics expressed in terms of the 4-vector potential A_μ, the dynamics reduces to the equation (5.78):

$$\partial_\mu(\partial^\mu A^\nu - \partial^\nu A^\mu) = \partial_\mu\partial^\mu A^\nu - \partial^\nu(\partial_\mu A^\mu) = \frac{1}{4\pi\epsilon_0}\frac{4\pi}{c}j_e^\nu. \tag{5.88}$$

The number of independent degrees of freedom in the electromagnetic field is thereby reduced from six in the rank-2 tensor $F_{\mu\nu}$ (the components of electric and magnetic field) to four in the 4-vector A_μ.

However, the 4-vector potential, A_μ, is well known not to be unambiguously determined, as we are free to change

$$A_\mu \quad\to\quad A'_\mu = A_\mu - c\partial_\mu\lambda. \tag{5.89}$$

This is precisely the gauge transformation of the scalar and vector potential (5.14a), as it was derived in Section 5.1. The physical meaning of the transformation (5.89) may be seen from the Fourier transform:

$$A_\mu \quad\to\quad A'_\mu = A_\mu - c\partial_\mu\lambda \quad\xrightarrow{\mathscr{F}}\quad \widetilde{A}'_\mu = \widetilde{A}_\mu + ick_\mu\widetilde{\lambda}, \tag{5.90}$$

where $k_\mu := p_\mu/\hbar$ is the wave 4-vector of electromagnetic radiation. The component of the 4-vector potential in the direction of motion (in the 4-dimensional spacetime!) of the electromagnetic beam is arbitrary, and may be cancelled by a judicious choice of the gauge function λ. In that sense one frequently imposes the Lorenz gauge[14]:

$$\partial_\mu A^\mu = 0 \quad\leftrightarrow\quad k_\mu\widetilde{A}^\mu = 0. \tag{5.91}$$

Notice that this gauge is Lorentz-invariant. Using it, the dynamical part of the Maxwell equations (5.88), simplifies to

$$\Box A^\mu = \frac{1}{4\pi\epsilon_0}\frac{4\pi}{c}j_e^\mu, \tag{5.92}$$

which is the wave-equation for the gauge potentials $A^\mu(\mathbf{x})$, with the *sources* j_e^μ.

The gauge (5.91) reduces the number of degrees of freedom in the electromagnetic field (which is determined by the relation (5.73b) in terms of the 4-vector potential) from four to three. But, that's not all: the FitzGerald–Lorentz length contraction applies to all physical quantities, and so also to the components of the 4-vector potential. Since, in vacuum, the changes in the electromagnetic field propagate at the speed of light, it follows that the longitudinal component of the 4-vector potential $A_\mu(\mathbf{x})$ equals zero, that is, its Fourier transform satisfies $\vec{k}\cdot\vec{\widetilde{A}} = 0$. The inverse transformation then gives $\vec{\nabla}\cdot\vec{A} = 0$, the so-called Coulomb gauge. The combination of the Lorenz and the Coulomb gauge produces $\dot{A}_0 = 0$, so that the temporal component of the 4-vector gauge potential is an arbitrary constant.

This reduction of the number of degrees of freedom from three to two cannot be described in a Lorentz-invariant way, so there are essentially two different approaches:

1. in addition to the Lorentz-invariant gauge, impose another gauge – such as the Coulomb gauge $\vec{\nabla}\cdot\vec{A} = 0$, which explicitly violates Lorentz symmetry, or
2. leave A_μ "ungauged" and having more than two degrees of freedom. Subsequently, systematically track and subtract the contributions of the nonphysical degrees of freedom in the 4-vector A_μ.

In the absence of free carriers of electric charge, $j_e^\mu = 0$. The equation (5.92) then becomes

$$\Box A^\mu = 0, \tag{5.93}$$

which is the d'Alembert equation, i.e., the Klein–Gordon equation with $m_\gamma = 0$. The solutions are found in the form

$$A^\mu(\mathbf{x}) = a\,e^{-(i/\hbar)\mathbf{p}\cdot\mathbf{x}}\,\epsilon^\mu(\mathbf{p}), \qquad \begin{cases} p_\mu p^\mu = 0 & \Rightarrow \quad E = |\vec{p}|c, \\ p_\mu\epsilon^\mu = 0 & \to \quad \epsilon^0 = 0 = \vec{p}\cdot\vec{\epsilon}, \end{cases} \tag{5.94}$$

[14] This *gauge* (as in "condition," of "specification") bears the name of Ludvig Valentin Lorenz (1829–91), not of Hendrik Antoon Lorentz (1853–1928) after whom the Lorentz transformations, (FitzGerald–)Lorentz length contraction, and Lorentz group were named.

where in the second row we see the joint effect of the (Lorentz-invariant) Lorenz and (Lorentz-violating) Coulomb gauge, and where a is the photon amplitude.

In quantum theory, $A^\mu(\mathbf{x})$ could serve as the wave-function of the photon: the 4-vector potential that has two physical degrees of freedom, which are transversal to the direction of the photon's propagation. In a Cartesian coordinate system where the photon moves along the $(x^3 = z)$-axis, the two transversal polarizations are

$$\vec{e}_1 = (1,0,0) \qquad \text{and} \qquad \vec{e}_2 = (0,1,0), \tag{5.95}$$

and

$$\vec{e}_+ = \tfrac{1}{\sqrt{2}}(\vec{e}_1 + i\vec{e}_2) \quad \text{and} \quad \vec{e}_- = \tfrac{1}{\sqrt{2}}(\vec{e}_1 - i\vec{e}_2) \tag{5.96}$$

are the so-called right- and left-circular polarizations, the eigenvectors of the rotation generator, J_3, with the eigenvalues ± 1, respectively.

5.2.3 The magnetic monopole sneaks in

The immediate interpretation of Conclusion 5.3 on p. 184, notwithstanding, Paul Dirac found in 1931:

1. There does exist a way to include magnetic monopole charges and currents into the standard electrodynamics, i.e., the physical system described by the equations

$$\partial_\mu F^{\mu\nu} = \frac{1}{4\pi\epsilon_0}\frac{4\pi}{c} j_e^\nu, \qquad \tfrac{1}{2}\varepsilon^{\mu\nu\rho\sigma}\partial_\mu F_{\nu\rho} = 0, \qquad F_{\mu\nu} = \partial_\mu A_\nu - \partial_\nu A_\mu. \tag{5.97}$$

2. The quantum nature of Nature forces the magnetic and the electric charges to satisfy a mutual, so-called Dirac (dual charge), quantization law:

$$q_e\, q_m = 2\pi\hbar\, n, \qquad n \in \mathbb{Z}. \tag{5.98}$$

From here,

$$\alpha_e := \frac{1}{4\pi\epsilon_0}\frac{e^2}{\hbar c} \approx \frac{1}{137} \qquad \Rightarrow \qquad \alpha_m := \frac{1}{4\pi\mu_0}\frac{g^2}{\hbar c} = \frac{n^2}{4}\frac{4\pi\epsilon_0\hbar c}{e^2} \approx \frac{137}{4}n^2, \tag{5.99}$$

so that the interaction intensity with magnetic monopole charges and their currents must be very large ($\frac{\alpha_m}{\alpha_e} \approx 4{,}690\, n^2$), reciprocally to the relatively weak interaction with (electric) monopole charges and their currents, $\alpha_e \approx 1/137$.

The magnetic monopole gauge potential

Dirac's quasi-realistic model of a magnetic monopole stems from the very well known fact about magnets, that the magnetic field is strongest near the ends of a magnetic (physical) dipole and weakest near its middle. Take one such magnet – a cylindrical solenoid, for example – and affix the coordinate origin to the "north" pole of the magnet, squeeze the cross-section of the solenoid and stretch it so that the "south" pole is pulled out towards $z \to -\infty$. In the limit when the cross-section of the solenoid is negligible and the "south" pole is infinitely far, the magnetic field of such a magnet is spherically symmetric and has a source (the "north" pole) at the coordinate origin, with the "south" pole nowhere in sight.

This thought-construction evidently shows that part of the space (the negative z-semi-axis) is physically inaccessible: Every test-magnet detects a spherically symmetric (Coulomb-esque) magnetic field $\vec{B} \propto q_m\vec{r}/r^3$ in all of space around the coordinate origin – except along the negative z-semi-axis, where the test magnet cannot be placed as that is where the infinitely long and infinitely thin solenoid is. This "forbidden zone" is called the Dirac string.

Dirac showed that the vector potential [296]

$$\vec{A}(\vec{r}): \quad \text{so that} \quad \vec{\nabla} \times \vec{A} = \vec{B} = \frac{q_m}{4\pi} \frac{\vec{r}}{r^3} \tag{5.100}$$

must be singular, as a function of the position $\vec{r}$, along *some* line (the Dirac string) that begins at the coordinate origin and extends out to infinity – which is the location of the infinitely thin Dirac solenoid. However, in 1975, T. T. Wu and C. N. Yang showed that there is no need to exclude this line from the physically accessible space – paying the price in accepting that the vector potential $\vec{A}$ then cannot be an unambiguously specified (vector) function. However, since the vector potential is not directly measurable, this ambiguity (non-single valuedness) has no physically measurable repercussion.

Indeed, define [536, 210]

$$\vec{A}_N = \frac{q_m}{4\pi} \frac{x\,\hat{\mathbf{e}}_y - y\,\hat{\mathbf{e}}_x}{r(z+r)}, \qquad \vec{A}_S = \frac{q_m}{4\pi} \frac{x\,\hat{\mathbf{e}}_y - y\,\hat{\mathbf{e}}_x}{r(z-r)}, \tag{5.101a}$$

$$= -\frac{q_m}{4\pi} \frac{\cos(\theta)-1}{r\sin(\theta)}\,\hat{\mathbf{e}}_\phi, \qquad = -\frac{q_m}{4\pi} \frac{\cos(\theta)+1}{r\sin(\theta)}\,\hat{\mathbf{e}}_\phi, \tag{5.101b}$$

and notice that the function $\vec{A}_N$ is well defined everywhere except along the ("southern") z-semi-axis, while the function $\vec{A}_S$ is well defined everywhere except along the ("northern") z-semi-axis. Also, define

$$\vec{B}_N := \vec{\nabla} \times \vec{A}_N = \frac{q_m}{4\pi} \frac{\vec{r}}{r^3}, \quad \text{and} \quad \vec{B}_S := \vec{\nabla} \times \vec{A}_S = \frac{q_m}{4\pi} \frac{\vec{r}}{r^3}. \tag{5.102}$$
(except where $x = 0 = y$ and $z \leqslant 0$) (except where $x = 0 = y$ and $z \geqslant 0$)

Since $\vec{B}_N$ and $\vec{B}_S$ perfectly coincide as functions everywhere where both are defined, the "true" magnetic field $\vec{B}$ is defined to be equal to $\vec{B}_N$ or $\vec{B}_S$, using that "auxiliary" magnetic field function that is well-defined in the region of interest.[15]

Since

$$\vec{A}_S - \vec{A}_N = 2\frac{q_m}{4\pi}\left(\frac{y\hat{\mathbf{e}}_x - x\hat{\mathbf{e}}_y}{x^2+y^2}\right) = -2\frac{q_m}{4\pi}\vec{\nabla}\big[\mathrm{ATan}(x,y)\big], \tag{5.103}$$

where

$$\mathrm{ATan}(x,y) := \begin{cases} \arctan(y/x) & \text{for} \quad x \geqslant 0, \\ \pi + \arctan(y/x) & \text{for} \quad x \leqslant 0, \end{cases} \tag{5.104a}$$

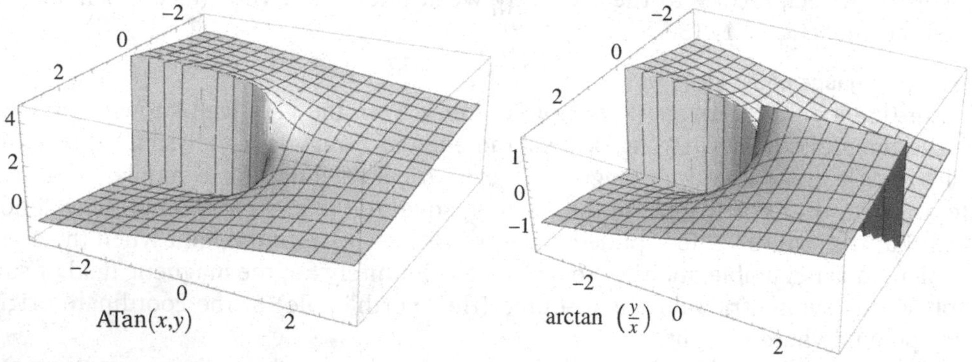

(5.104b)

[15] This is the same "trick" that cartographers use when they carve up the map of the Earth's globe (which cannot be depicted accurately on a single flat sheet of paper) into a sufficiently large number of sufficiently small maps, each of which depicts adequately a sufficiently small region of the Earth surface. These maps are then bound into an atlas where "adjacent" maps overlap sufficiently to provide the traveller with connecting information along any – of course continuous – voyage.

it follows that $\vec{A}_N$ and $\vec{A}_S$ differ by a gauge transformation (5.89) with the gauge parameter $\lambda(x) = 2q_m\,\mathrm{ATan}(x,y)$. Since the potentials $\vec{A}_N$ and $\vec{A}_S$ are not themselves measurable, but provide the same (measurable) magnetic field, the gauge transformation

$$\vec{A}_N \to \vec{A}_S = \vec{A}_N + \vec{\nabla}\lambda_{NS}, \quad \lambda_{NS}(\mathbf{x}) = -2\frac{q_m}{4\pi}\,\mathrm{ATan}(x,y) \tag{5.105}$$

is then really a symmetry of the physical system.

Dirac's dual quantization of charges

As relations (5.14a)–(5.14b) show, the gauge transformation (5.105) induces the change in the phase of the electron wave-function:

$$\Psi(\mathbf{x}) \to \Psi'(\mathbf{x}) = e^{iq_e\lambda_{NS}(\mathbf{x})/\hbar}\,\Psi(\mathbf{x}). \tag{5.106}$$

As the value of the function $\mathrm{ATan}(x,y)$ is the azimuthal angle $\phi \simeq \phi{+}2\pi$, the relation (5.105) yields

$$\exp\left\{iq_e\lambda_{NS}(\mathbf{x})/\hbar\right\} = \exp\left\{-i\frac{q_m q_e}{2\pi\hbar}\,\phi\right\}. \tag{5.107}$$

No gauge transformation – and so not this one – can *change* the single-valuedness of the wave-function, which of course is chosen single-valued to begin with. Thus the phase (5.107) also must be a single-valued function of ϕ, and $\frac{q_m q_e}{2\pi\hbar}$ must be an integer:

$$\frac{q_e q_m}{2\pi\hbar} \stackrel{!}{=} n \in \mathbb{Z}, \qquad \text{i.e.,} \qquad q_m \stackrel{!}{=} n\left(\frac{2\pi\hbar}{q_e}\right), \tag{5.108}$$

which is called the Dirac (dual charge) quantization of the magnetic charge, and where $(2\pi\hbar/q_e)$ is the elementary (unit) amount of magnetic charge.

It will prove useful to rewrite this argument by direct integration of the relation (5.105):

$$\int_1^2 \mathrm{d}\vec{r}{\cdot}\vec{A}_N = \int_1^2 \mathrm{d}\vec{r}{\cdot}\vec{A}_S + \left[\int_1^2 \mathrm{d}\vec{r}{\cdot}\vec{\nabla}\lambda_{NS} = \lambda_{NS}(\vec{r}_2) - \lambda_{NS}(\vec{r}_1)\right], \tag{5.109}$$

that is,

$$\int_1^2 \mathrm{d}\vec{r}{\cdot}\vec{A}_N - \int_1^2 \mathrm{d}\vec{r}{\cdot}\vec{A}_S = \lambda_{NS}(\vec{r}_2) - \lambda_{NS}(\vec{r}_1). \tag{5.110}$$

Dirac's quantization of the magnetic charge thus stems from the requirement that $iq_e\int_C \mathrm{d}\vec{r}{\cdot}\vec{A}$ may depend on the choice of the concrete line integration contour only up to an integral multiple of 2π:

$$e^{iq_e\int_{C_1}\mathrm{d}\vec{r}\cdot\vec{A} - q_e\int_{C_2}\mathrm{d}\vec{r}\cdot\vec{A}} = e^{iq_e\oint_{C_1-C_2}\mathrm{d}\vec{r}\cdot\vec{A}} \stackrel{!}{=} e^{2\pi i n} = 1, \tag{5.111}$$

where $(C_1{-}C_2)$ is a closed contour since C_1 and C_2 have the same end-points: $\partial C_1 = \partial C_2$.[16] Using Stokes' theorem, $\oint_C \mathrm{d}\vec{r}{\cdot}\vec{A} = \int_S \mathrm{d}^2\vec{\sigma}{\cdot}(\vec{\nabla}{\times}\vec{A})$ where S is some surface bounded by the contour C, i.e., $C = \partial S$, and the definition of the magnetic field, $\vec{B} := (\vec{\nabla}{\times}\vec{A})$, we have that

$$2\pi n \stackrel{!}{=} q_e\int_{C_1}\mathrm{d}\vec{r}{\cdot}\vec{A} - q_e\int_{C_2}\mathrm{d}\vec{r}{\cdot}\vec{A} = q_e\oint_{(C_1-C_2)=\partial S}\mathrm{d}\vec{r}{\cdot}\vec{A} = q_e\int_S \mathrm{d}^2\vec{\sigma}{\cdot}\vec{B}. \tag{5.112}$$

Applied to the magnetic field of a (hypothetical) magnetic monopole charge, this condition produces the quantization (5.108). However, the same condition also represents a reason for the existence of the so-called Aharonov–Bohm effect [☞ textbooks [407, 471, 480, 472, 29, 324],

[16] For any space $\mathscr{X}$, the symbol $\partial\mathscr{X}$ denotes the "boundary of $\mathscr{X}$."

for example], which is experimentally verified, and which should therefore be called the "Dirac–Aharonov–Bohm effect."

Today, several additional, alternative arguments are known to infer the same mutual quantization. One of them was published by Alfred S. Goldhaber in 1965. The magnetic field (5.100) of a magnetic monopole exerts a force upon a particle of electric charge q_e that passes through the magnetic field at the velocity $\vec{v}$. This so-called Lorentz force,

$$\vec{F}_L = q_e \vec{v} \times \vec{B}, \tag{5.113}$$

is perpendicular to the plane containing $\vec{v}$ and $\vec{B}$. Select a coordinate system so that $\vec{v} = v\hat{\mathbf{e}}_z$, where $\theta := \measuredangle(\vec{r},\vec{v}) = \measuredangle(\vec{B},\vec{v})$ since for the magnetic monopole $\vec{B} \propto \vec{r}$ [☞ relation (5.100)]. The distance $b := |r\sin(\theta)|$ is called the "impact parameter," just as in the set-up for the collision of two marbles, in Example 3.2 on p. 111. Select the x-axis to be in the direction of this parameter and $\vec{B}$ is in the (x,z)-plane. For sufficiently large values of b, the deflection (in the direction of the y-axis) from the trajectory (along the z-axis) will be small enough to be accurately estimated by the integral

$$(\triangle\vec{p})_y \approx \int_{-\infty}^{+\infty} \mathrm{d}t\ (\vec{F}_L)_y = \frac{q_e v q_m b}{4\pi} \int_{-\infty}^{+\infty} \frac{\mathrm{d}t}{(b^2+v^2t^2)^{3/2}} = \frac{q_e\, q_m}{2\pi b}, \tag{5.114}$$

so that

$$(\triangle\vec{L})_z = b(\triangle\vec{p})_y = \frac{q_e q_m}{2\pi}. \tag{5.115}$$

It remains to conclude – because of the quantum nature of Nature – that the change in the angular momentum must be an integral multiple of $\hbar$. This immediately reproduces equation (5.108).

Finally, let us also mention the fact that the electromagnetic field has a linear momentum density $\epsilon_0 \vec{E} \times \vec{B}$. For the field near point-like electric and magnetic charges that are separated by the vector $\vec{R}$, it may be shown that the total (integrated) linear momentum of the total field vanishes, whereby the total (integrated) angular momentum is independent of the choice of the coordinate origin and has the value [☞ [296]; this result was published by J. J. Thomson, in 1904]

$$\vec{L}_{EM} = \frac{q_e q_m}{4\pi} \frac{\vec{R}}{R}. \tag{5.116}$$

The quantization of this angular momentum in (integral) units of $\hbar$ also indicates a quantization of the magnetic charge in units that are inversely proportional to the elementary electric charge, but gives a value that is twice as large as the result (5.108). That is, the previous two arguments produce a stricter result. One could have obtained this as early as 1904 from equation (5.116), but only by adopting the quantization of angular momentum in *half-integral* units of $\hbar$ – thus foreshadowing spin-$\frac{1}{2}$ particles and systems. At the time, no one thought of it.

5.2.4 Exercises for Section 5.2

✎ **5.2.1** *Using the stated definitions of J_i, K_j and the ensuing relations (5.45), prove equation (5.46).*

✎ **5.2.2** *Using the relation (5.45) with the choice $\varphi^i = 0$, $\beta^2 = 0 = \beta^3$ and $\beta^1 = \beta$, prove relation (5.48) by expanding the exponential function, then re-summing the result after using the relation (5.32).*

✎ **5.2.3** *Prove the equivalence of results (5.34) and (5.69), as well as that $\mathscr{H}_D = \overline{\Psi} H_D \Psi$.*

✎ **5.2.4** *Using the relation (5.74)–(5.75) and (3.13b), derive equations (5.77).*

5.3 Quantum electrodynamics with leptons

The description of electrodynamics in the previous section is classical. Quantum computations are consistently derived from quantum field theory – of photons and leptons – and this derivation is outside the scope of this book. Instead, following Ref. [243] and the introductory material in Section 3.3, we will consider several examples of computations with Feynman diagrams that depict interactions of charged leptons and photons.

5.3.1 Quantum electrodynamics calculation

We have already seen Feynman diagrams that depict electromagnetic processes: $O(e^4)$ contributions to the e^-p^+ scattering are depicted by the diagrams (4.39), and the two-photon e^-e^+ annihilation is depicted by diagram (4.48). Modeled on Section 3.3.4, we first assign a mathematical expression to every graphical element, and by adapting Procedure 3.1 on p. 116, we will compute the amplitude $\mathfrak{M}$, which we will then insert into the formulae (3.112) and (3.114) for decays and scattering, respectively.

Although we will not derive the Feynman rules for electrodynamics from the Lagrangian, we present this Lagrangian density. By combining the results (5.76) and (5.68a), changing

$$\partial_\mu \to D_\mu := \partial_\mu + \tfrac{i}{\hbar c} A_\mu Q \qquad \text{so that} \qquad D'_\mu\big(e^{i\varphi(x)}\Psi(x)\big) = e^{i\varphi(x)}\big(D_\mu \Psi(x)\big), \tag{5.117}$$

in accord with the definitions (5.13) and (5.73a), and where $Q\Psi = q_\Psi \Psi$ produces the electric charge of the particle represented by Ψ, we have

$$\begin{aligned}
\mathscr{L}_{QED} &= \overline{\Psi}(x)\,[i\hbar c\not{D} - mc^2]\,\Psi(x) - \tfrac{4\pi\epsilon_0}{4} F_{\mu\nu}F^{\mu\nu}\\
&= \overline{\Psi}(x)\left[\gamma^\mu(\hbar c\, i\partial_\mu - q_\Psi A_\mu) - mc^2\right]\Psi(x)\\
&\quad - \tfrac{4\pi\epsilon_0}{4}(\partial_\mu A_\nu - \partial_\nu A_\mu)\eta^{\mu\rho}\eta^{\nu\sigma}(\partial_\rho A_\sigma - \partial_\sigma A_\rho).
\end{aligned} \tag{5.118}$$

By construction, this Lagrangian is invariant under the gauge transformation

$$A'_\mu(x) = A_\mu(x) - c\partial_\mu\varphi(x) \qquad \text{and} \qquad \Psi'(x) = e^{i\varphi Q/\hbar}\,\Psi(x). \tag{5.119}$$

Digression 5.9 The equation of motion for $A_\mu(x)$ is obtained by varying either the Lagrangian density (5.118) or the Hamilton action $\int d^4x\, \mathscr{L}_{QED}$ with respect to $A_\mu(x)$. Using so-called functional derivative generalization of partial derivatives:

$$\frac{\delta}{\delta A_\rho(y)} \mathscr{F}(A_\mu(x), (\partial_\mu A_\nu(x)), \dots) := \delta^4(x-y)\, \frac{\partial}{\partial A_\rho(x)} \mathscr{F}(A_\mu(x), (\partial_\mu A_\nu(x)), \dots), \tag{5.120a}$$

$$\frac{\delta}{\delta(\partial_\rho A_\sigma(y))} \mathscr{F}(A_\mu(x), (\partial_\mu A_\nu(x)), \dots) = \delta^4(x-y)\, \frac{\partial}{\partial(\partial_\rho A_\sigma(y))} \mathscr{F}(A_\mu(x), (\partial_\mu A_\nu(x)), \dots), \tag{5.120b}$$

we obtain the general result

$$\begin{aligned}
\frac{\delta}{\delta A_\rho(x)} \int d^4y\, \mathscr{F}(A_\mu(y), (\partial_\mu A_\nu(y))) &= \int d^4y\, \frac{\delta A_\sigma(y)}{\delta A_\rho(x)} \frac{\partial}{\partial A_\sigma(y)} \mathscr{F}(A_\mu(y), (\partial_\mu A_\nu(y)))\\
&= \int d^4y\, \delta^4(x-y)\delta^\sigma_\rho \frac{\partial}{\partial A_\sigma(y)} \mathscr{F}(A_\mu(y), (\partial_\mu A_\nu(y))) = \frac{\partial}{\partial A_\rho(x)} \mathscr{F}(A_\mu(x), (\partial_\mu A_\nu(x))).
\end{aligned} \tag{5.120c}$$

Using

$$\frac{\partial}{\partial A_\rho(\mathrm{x})} A_\mu(\mathrm{x}) = \delta^\rho_\mu, \qquad \frac{\partial}{\partial A_\rho(\mathrm{x})} (\partial_\mu A_\nu(\mathrm{x})) = 0, \tag{5.120d}$$

$$\frac{\partial}{\partial(\partial_\rho A_\sigma(\mathrm{x}))} A_\mu(\mathrm{x}) = 0, \qquad \frac{\partial}{\partial(\partial_\rho A_\sigma(\mathrm{x}))} (\partial_\mu A_\nu(\mathrm{x})) = \delta^\rho_\mu \delta^\sigma_\nu, \tag{5.120e}$$

where we need not write the arguments "(x)," we obtain

$$\partial_\mu \frac{\partial \mathscr{L}_{QED}}{\partial(\partial_\mu A_\nu)} = \frac{\partial \mathscr{L}_{QED}}{\partial A_\nu} \qquad \Rightarrow \qquad \partial_\mu F^{\mu\nu} = \frac{q_\Psi}{4\pi\epsilon_0} \overline{\Psi} \boldsymbol{\gamma}^\nu \Psi. \tag{5.120f}$$

Comparing result (5.120f) with equation (5.78) identifies

$$j^\mu_e := \frac{q_\Psi c}{4\pi} \overline{\Psi} \boldsymbol{\gamma}^\mu \Psi \tag{5.120g}$$

as the 4-vector of the electric current density. The combined Lagrangian density (5.118) shows that, while the dynamics of photons alone may be described in terms of the $F_{\mu\nu}$ field, i.e., $\vec{E}$ and $\vec{B}$, the Lagrangian description of the interaction with charged particles requires the use of the gauge 4-vector potential A_μ – although the derived equations of motion (5.120f) and the obvious (Bianchi) consequence (5.87) *may* be expressed fully in terms of the $\vec{E}$ and $\vec{B}$ fields.

Digression 5.10 Varying the Lagrangian density $\mathscr{L}_{\mathrm{QED}}$, as in equation (5.118), with respect to A_μ and $\overline{\Psi}$ (from the left), we obtain the complementary and coupled system of Euler–Lagrange equations of motion:

$$\partial_\mu F^{\mu\nu} = \frac{q_\Psi}{4\pi\epsilon_0} \overline{\Psi} \boldsymbol{\gamma}^\nu \Psi, \qquad \Big[i\hbar c\, \boldsymbol{\gamma}^\mu \partial_\mu - mc^2 \mathbb{1} \Big] \Psi = q_\Psi A_\mu \boldsymbol{\gamma}^\mu \Psi. \tag{5.121}$$

The procedure given in Digression 5.9 is equally applicable to interactions of arbitrary charged particles with photons: for a particle of a spin other than $\frac{1}{2}$, the Dirac Lagrangian density must be replaced by a corresponding Lagrangian density but where the "gauge covariant derivatives" $\partial_\mu \to D_\mu$ (5.117) are used. As an introduction and because of immediate application, the formulae will be written for a lepton/antilepton, i.e., electron/positron. The computations, however, are easy to adapt for other charged spin-$\frac{1}{2}$ particles – one should only substitute the appropriate charges and masses. Also, it should not be too hard to also adapt the computations to include charged particles without spin. This is usually called "scalar electrodynamics" in the literature, but we leave this aside.

Because of the difference in units and numerical simplification, the notation

$$g_e := \sqrt{4\pi\,\alpha_e} = \frac{|e|}{\sqrt{\epsilon_0 \hbar c}} \qquad (= |e|\sqrt{4\pi/\hbar c}, \text{ in Gauss's units}) \tag{5.122}$$

is useful. On one hand, g_e gives a dimensionless measure of the interaction strength; on the other, many electrodynamics computations may then be relatively easily adapted for weak nuclear and chromodynamics computations by changing $g_e \to g_w$ and $g_e \to g_c$, respectively, and inserting a few additional factors [☞ Chapter 6].

The practical use of most concrete models in quantum field theory reduces to the prescription (see also Procedure 11.1 on p. 416 and discussion in Section 11.2.4):

Procedure 5.1 *Start with a concrete model defined within classical field theory.*

1. *For any considered process, list the possible sub-processes, as discussed in Section 3.3.1 and in the form of a sequence of Feynman diagrams partially ordered:*
   (a) *by the number of closed loops [☞ Comment 3.5 on p. 122],*
   (b) *by the powers of a characteristic interaction parameter,*
   (c) *by the powers of $\hbar$.*
2. *Compute the amplitude $\mathfrak{M}_i$ for each (sub)process, as described by the specific Feynman calculus rules of the model; see for example Procedures 5.2 on p. 193 and 6.1 on p. 232, below.*
3. *Add the amplitudes, with a negative relative sign between sub-processes that differ only by the exchange of two identical fermions.*
4. *Compute the corresponding scattering cross-section or decay constant as discussed in Section 3.3.3, and illustrated there for a simple toy-model.*

The specific Feynman calculus rules mentioned in step 2 above are derived from the same classical action and rely on the correspondences discussed in Section 3.3.1 and in particular the listing on page 106. As stated there, that task is deferred to proper field-theory texts [64, 63, 48, 257, 307, 221, 159, 422, 423, 538, 250, 389, 243, 45, 580, 238, 241, 239, 240].

For the particular case at hand, the model describing the interaction of electrically charged spin-$\frac{1}{2}$ fermions (such as electrons) and the electromagnetic field, the classical model is described by the Lagrangian (5.118), and the specific Feynman calculus rules are as follows:

Procedure 5.2 *The contribution to the amplitude $\mathfrak{M}$ corresponding to a given Feynman diagram for an electrodynamics process with electrons and positrons is computed following the algorithm [☞ textbooks [445, 425, 586] for a derivation]:*

1. **Notation**
   (a) *Energy–momentum: Denote incoming and outgoing 4-momenta by $\mathrm{p}_1, \mathrm{p}_2, \ldots$, and the spins by $s_1, s_2, \ldots$ Denote the "internal" 4-momenta (assigned to lines that connect two vertices inside the diagram) by $\mathrm{q}_1, \mathrm{q}_2, \ldots$*
   (b) *Orientation: For a spin-$\frac{1}{2}$ particle, orient the line in the 4-momentum direction, oppositely for antiparticles. Orient external photon lines in the direction of time (herein, upward). Orient the internal photon lines arbitrarily, but use the so-chosen orientation consistently.*
   (c) *Polarization: Assign every external line the polarization factor:*

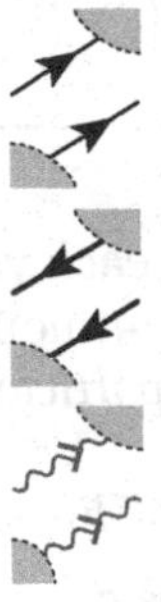

Spin-$\frac{1}{2}$ particle	*incoming*	u^s	*s = spin projection $= \uparrow, \downarrow$*
	outgoing	$\overline{u}_s$	
Spin-$\frac{1}{2}$ antiparticle	*incoming*	$\overline{v}_s$	*($\simeq$ spin-$\frac{1}{2}$ particle, travels backwards in time)*
	outgoing	v^s	
Photon	*incoming*	ϵ^μ	$\epsilon^\mu p_\mu = 0$ *and* $\epsilon^0 = 0$
	outgoing	$\epsilon^{\mu *}$	

(5.123)

2. **Vertices** *To each vertex assign the factor*

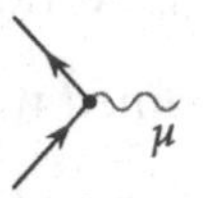

$$\longrightarrow \quad -ig_e\gamma^\mu. \tag{5.124}$$

Even without derivation, this factor clearly corresponds to the term $-q_\Psi\overline{\Psi}\not{A}\Psi$ in equation (5.118), and so represents the elementary interaction of the photon with the current of the charged particle that Ψ represents.

3. **Propagators** *To each internal line with the jth 4-momentum assign the factor:*

spin-$\frac{1}{2}$ particle: 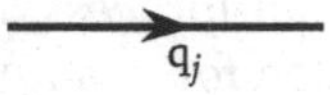

$$\longrightarrow \quad \frac{i}{\not{q}_j - m_j c} = i\frac{\not{q}_j + m_j c\mathbb{1}}{q_j^2 - m_j^2 c^2}, \tag{5.125}$$

photon: 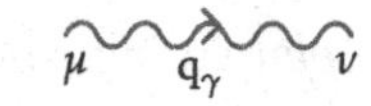

$$\longrightarrow \quad -i\frac{\eta_{\mu\nu}}{q_\gamma^2}. \tag{5.126}$$

As internal lines depict virtual particles, $\not{q}_j \neq m_j c$ and $q_\gamma^2 \neq 0$, respectively [☞ Tables C.7 on p. 529 and C.8 on p. 529]. Up to multiplicative coefficients, these factors also stem from (5.118); these are Fourier transforms of the Green functions for the differential operators $\not{D}$ and $D^{\mu\nu}$, in $\overline{\Psi}\,\not{D}\Psi := -\overline{\Psi}[i\hbar c\not{\partial} - mc^2]\Psi$ and $A_\mu D^{\mu\nu} A_\nu :\simeq -\frac{4\pi\epsilon_0}{4} F_{\mu\nu}F^{\mu\nu}$, respectively, where "$\simeq$" denotes effective equality (equivalence) under the integral, after integration by parts and "$:\simeq$" defines the left-hand side by means of such an effective equality.

Digression 5.11 *Integration by parts is used rather often, so that, e.g.,*

$$\int d^4x\,(\partial_\mu A_\nu)(\partial^\mu A^\nu) = \int d^4x\,\partial_\mu(A_\nu\partial^\mu A^\nu) - \int d^4x\,A_\nu(\partial_\mu\partial^\mu A^\nu)$$
$$= \oint_{\mathscr{V}_{(\mu)}} (d^3x)_\mu\,(A_\nu\partial^\mu A^\nu) - \int d^4x\,A_\nu(\partial_\mu\partial^\mu A^\nu), \tag{5.127a}$$

where $\mathscr{V}_{(\mu)}$ is a closed 3-dimensional hypersurface that bounds the 4-dimensional spacetime and $(d^3x)_\mu$ is the volume element of $\mathscr{V}_{(\mu)}$. As the domain of 4-dimensional integrals is typically **all of** *spacetime, $\mathscr{V}_{(\mu)}$ is a hypersurface "at infinity" where all fields are required to vanish, so the integrated term also vanishes. With this in mind, the relation (5.127a) is written as*

$$\int d^4x\,(\partial_\mu A_\nu)(\partial^\mu A^\nu) \simeq -\int d^4x\,A_\nu(\partial_\mu\partial^\mu A^\nu), \tag{5.127b}$$

which defines the relation "$\simeq$," in this context, as "equality under spacetime integral up to integrated terms that are assumed to vanish," or "equivalence up to integrals of total derivatives."

---

4. **Energy–momentum conservation** *To each vertex assign a factor $(2\pi)^4\delta^4(\sum_j k_j)$, where k_j are 4-momenta that enter the vertex. 4-momenta that* **leave** *the vertex have a negative sign – except for external spin-$\frac{1}{2}$ antiparticles, since they are equivalent to particles that move backwards in time.*
5. **Integration over 4-momenta** *Internal lines correspond to virtual particles and their 4-momenta are unknown; these variables must be integrated: $\int \frac{d^4q_j}{(2\pi)^4}$.*

6. ***Reading off the amplitude*** *The foregoing procedure yields the result*

$$-i\,\mathfrak{M}\,(2\pi)^4\delta^4(\sum_j \mathrm{p}_j), \tag{5.128}$$

where the factor $(2\pi)^4\delta^4(\sum_j \mathrm{p}_j)$ represents the 4-momentum conservation for the entire process, and where the amplitude (matrix element) $\mathfrak{M}$ is read off.

7. ***Fermion loops*** *To each fermion loop (closed line) assign a factor -1. A mathematically rigorous derivation of this rule follows from Feynman's approach using path integrals, which is far beyond the scope of this book. See however Digression 2.4 on p. 52 and especially statement 4a therein; see also the booklet [166] for an intuitive albeit not entirely rigorous explanation, Ref. [434, Vol. 1, Appendix A] for a serious introduction, and Ref. [165] for the original reference.*
8. ***Antisymmetrization*** *Since the amplitude of the process must be antisymmetric in pairs of identical (external) fermions, the partial amplitudes that differ only in the exchange of two identical external fermions must have the relative sign -1.*

As in Section 3.3.4, one draws all Feynman diagrams that contribute at the desired order in g_e, and then computes the (partial) amplitudes for each of the diagrams. The algebraic sum of these contributions yields the total amplitude, which is then inserted in formulae (3.112) and (3.114) for decays and scatterings, respectively.

In the remaining part of Section 5.3, the contributions of the following 12 Feynman diagrams will be examined, where we follow the treatment in Refs. [243] [☞ also Refs. [64, 580, 241]]: Each of these diagrams depicts a separate contribution to some $O(g_e^2)$ process and, exceptionally, $O(g_e^4)$ for the last diagram. Processes are identified by the "external" particles, whereby diagram (a) in Figure 5.1, all by itself represents one process, while diagrams (b) and (c) in Figure 5.1 represent two contributions to the same process.

Denote the external lines so that incoming are bottom-left=1 and bottom-right=2, and outgoing are top-left=3 and top-right=4. So, e.g.,

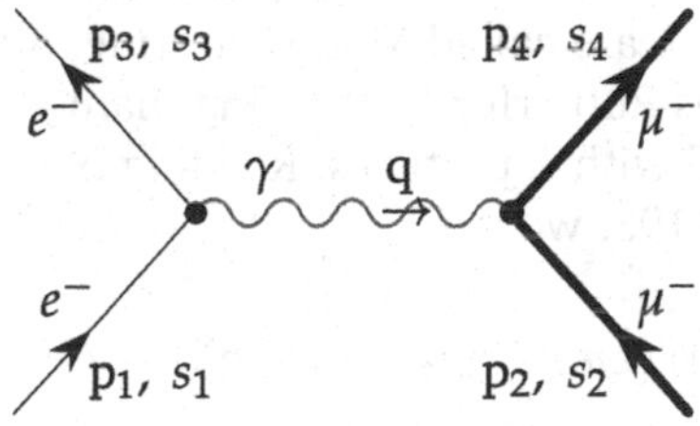

(5.129)

depicts the elastic scattering of an electron and a muon via the exchange of a photon. In fact, the incoming (and so also the outgoing) pair of fermions in the diagrams in Figure 5.1 (a)–(d), p. 196, could be identified as any other pair of different spin-$\frac{1}{2}$ particles, including the electron–proton pair in the hydrogen atom. It is, however, important to keep in mind that the relativistic description in terms of the perturbative expansion in the degree of the interaction constant g_e is appropriate for scatterings and for decays but not for bound states, the description of which is inherently non-perturbative in this sense.

To see this, note that the bound states of the hydrogen atom are determined by the Coulomb field, which results from summing over all possible exchange processes including one to infinitely many photons. The static electromagnetic field, known as the Coulomb field, may be identified with

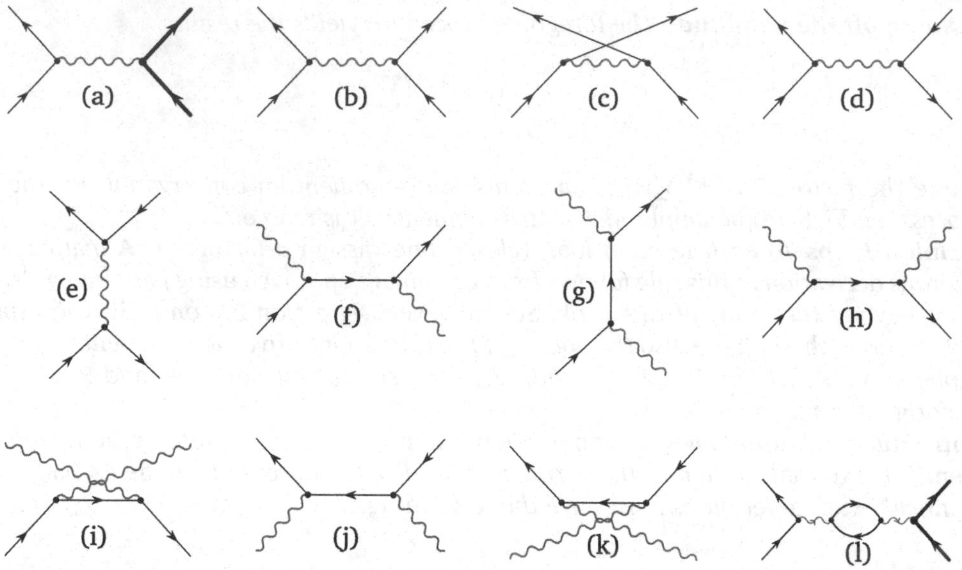

Figure 5.1 The first 12 Feynman diagrams that depict the quantum-electrodynamical processes between spin-$\frac{1}{2}$ particles and the photon. The last diagram depicts one of the corrections for the process (a).

the (Bose) *condensation* of infinitely many photons,[17] and is a phenomenon that is inherently non-perturbative in the number of exchanged particles, and so inherently unreachable in the analysis using elementary processes depicted by Feynman diagrams.

In turn, in scatterings and decays, the exchange of a single photon produces the dominant contribution, while multi-particle exchanges produce ever smaller corrections: scatterings and decays are inherently perturbative in the number of exchanged particles.

Electron–muon scattering

Scatterings of the type $e^- + \mu^- \to e^- + \mu^-$, where the muon is a "target" that is significantly heavier than the "probe" (here, e^-), are called Mott scattering, after Sir Nevill Francis Mott. In the non-relativistic regime one obtains Rutherford scattering, named after Ernest Rutherford's experiment of bombarding a foil of gold with α-particles. Reading off of the diagram in Figure 5.1, and following the Procedure 5.2 on p. 193, we get

$$
\begin{aligned}
&\int \frac{\mathrm{d}^4\mathrm{q}}{(2\pi)^4}\,(2\pi)^4\delta^4(\mathrm{p}_1-\mathrm{p}_3-\mathrm{q})\,(2\pi)^4\delta^4(\mathrm{p}_2-\mathrm{p}_4+\mathrm{q})\\
&\quad\times\big[\overline{u^{s_3}}{}_A(\mathrm{p}_3)(ig_e\,\gamma^{\mu A}{}_B)u^{s_1,B}(\mathrm{p}_1)\big]\Big(\frac{-i\eta_{\mu\nu}}{\mathrm{q}^2}\Big)\big[\overline{U^{s_4}}{}_C(\mathrm{p}_4)(ig_e\,\gamma^{\nu C}{}_D)U^{s_2,D}(\mathrm{p}_2)\big]\\
&=\frac{ig_e^2(2\pi)^4}{(\mathrm{p}_1-\mathrm{p}_3)^2}\,\delta^4(\mathrm{p}_2-\mathrm{p}_4+\mathrm{p}_1-\mathrm{p}_3)\big[\overline{u^{s_3}}{}_A(\mathrm{p}_3)\,\gamma^{\mu A}{}_B\,u^{s_1,B}(\mathrm{p}_1)\big]\big[\overline{U^{s_4}}{}_C(\mathrm{p}_4)\,\gamma_\mu{}^C{}_D\,U^{s_2,D}(\mathrm{p}_2)\big],
\end{aligned}
\tag{5.130}
$$

and comparison with the diagram (5.129) shows that $u^{s_1,b}(\mathrm{p}_1)$ represents the incoming electron, and $\overline{U^{s_4}}{}_C(\mathrm{p}_4)$ the outgoing muon, etc.

[17] And the other way round, photons are the quanta of the electromagnetic field in the sense that they are the smallest "packet" of a *change* in the electromagnetic field. These quanta – oscillations in the electromagnetic field – move at the speed of light; once established, the electrostatic or magnetostatic field does not move at all and extends through the whole available space.

From there, using the expression (5.128), we get

$$\mathfrak{M}_{(a)} = -\frac{g_e^2}{(\mathrm{p}_1-\mathrm{p}_3)^2}\left[\overline{u^{s_3}}_A(\mathrm{p}_3)\,\gamma^{\mu A}{}_B\,u^{s_1,B}(\mathrm{p}_1)\right]\left[\overline{U^{s_4}}_C(\mathrm{p}_4)\,\gamma_\mu{}^C{}_D\,U^{s_2,D}(\mathrm{p}_2)\right]. \tag{5.131}$$

If the spins of the incoming and outgoing particles are known, the polarization spinors $u^{s_1,B}$, $U^{s_2,D}$, $\overline{u^{s_3}}_A$ and $\overline{U^{s_4}}_C$ are selected as given in equations (5.40), one computes the components of the 4-vectors $[\overline{u^{s_3}}\,\boldsymbol{\gamma}^\mu\,u^{s_1}]$ and $[\overline{U^{s_4}}\,\boldsymbol{\gamma}_\mu\,U^{s_2}]$, and then the sum of the products.

When in turn the spins of the interacting particles are not measured, and we are interested in the *inclusive effective cross-section* of the scattering, i.e., the *inclusive decay constant*, summing over all spins produces an important simplification. Indeed, the formulae (3.112) and (3.114) need $|\mathfrak{M}|^2 = \overline{\mathfrak{M}}\,\mathfrak{M}$. On the other hand, $\overline{\mathfrak{M}}_{(a)}$ contains a factor

$$\begin{aligned}
\left[\overline{u}_A(\mathrm{p}_3)\,\gamma^{\mu A}{}_B\,u^B(\mathrm{p}_1)\right]^\dagger &= \left[u^\dagger(\mathrm{p}_3)\,\boldsymbol{\gamma}^0\,\boldsymbol{\gamma}^\mu\,u(\mathrm{p}_1)\right]^\dagger = \left[u^\dagger(\mathrm{p}_1)\,(\boldsymbol{\gamma}^\mu)^\dagger\,(\boldsymbol{\gamma}^0)^\dagger\,u(\mathrm{p}_3)\right]\\
&= \left[u^\dagger(\mathrm{p}_1)\mathbb{1}\,(\boldsymbol{\gamma}^\mu)^\dagger\,\boldsymbol{\gamma}^0\,u(\mathrm{p}_3)\right] = \left[u^\dagger(\mathrm{p}_1)\boldsymbol{\gamma}^0\boldsymbol{\gamma}^0\,(\boldsymbol{\gamma}^\mu)^\dagger\,\boldsymbol{\gamma}^0\,u(\mathrm{p}_3)\right]\\
&= \left[\overline{u}(\mathrm{p}_1)\,\overline{\boldsymbol{\gamma}}^\mu\,u(\mathrm{p}_3)\right], \qquad \overline{\boldsymbol{\gamma}}^\mu := \boldsymbol{\gamma}^0(\boldsymbol{\gamma}^\mu)^\dagger\boldsymbol{\gamma}^0,
\end{aligned} \tag{5.132}$$

so that $|\mathfrak{M}_{(a)}|^2$ contains the factor

$$\left[\overline{u}_A(\mathrm{p}_3)\,\gamma^{\mu A}{}_B\,u^b(\mathrm{p}_1)\right]\left[\overline{u}_C(\mathrm{p}_1)\,\gamma^{\nu C}{}_D\,u^D(\mathrm{p}_3)\right]. \tag{5.133}$$

Digression 5.12 The *physical* requirement (A.127) implies that

$$\overline{\boldsymbol{\gamma}}^\mu := \boldsymbol{\gamma}^0(\boldsymbol{\gamma}^\mu)^\dagger\boldsymbol{\gamma}^0 \overset{(A.127)}{=} \boldsymbol{\gamma}^\mu. \tag{5.134}$$

Finally, summing over spins permits using the relations (5.54):

$$\begin{aligned}
&\sum_{s_1,s_3}\left[\overline{u^{s_3}}_A(\mathrm{p}_3)\,\gamma^{\mu A}{}_B\,u^{s_1 B}(\mathrm{p}_1)\right]\left[\overline{u^{s_1}}_C(\mathrm{p}_1)\,\overline{\gamma}^{\nu C}{}_D\,u^{s_3 D}(\mathrm{p}_3)\right]\\
&\quad= \sum_{s_3}\gamma^{\mu A}{}_B\left[\sum_{s_1}u^{s_1 B}(\mathrm{p}_1)\,\overline{u^{s_1}}_C(\mathrm{p}_1)\right]\overline{\gamma}^{\nu C}{}_D\left[u^{s_3 D}(\mathrm{p}_3)\,\overline{u^{s_3}}_A(\mathrm{p}_3)\right]\\
&\quad\overset{(5.54)}{=} \gamma^{\mu A}{}_B\,(\not{p}_1 + m_e c\mathbb{1})^B{}_C\,\overline{\gamma}^{\nu C}{}_D\,(\not{p}_3 + m_e c\mathbb{1})^D{}_A\\
&\quad= \mathrm{Tr}\left[\boldsymbol{\gamma}^\mu\,(\not{p}_1 + m_e c\mathbb{1})\,\overline{\boldsymbol{\gamma}}^\nu\,(\not{p}_3 + m_e c\mathbb{1})\right],
\end{aligned} \tag{5.135}$$

which is independent of the spins s_1, s_3 that are not being measured.

It then follows that

$$\begin{aligned}
\langle|\mathfrak{M}_{(a)}|^2\rangle &= \frac{g_e^4}{(\mathrm{p}_1-\mathrm{p}_3)^4}\sum_{s_1,s_3}\mathrm{Tr}\left[\overline{u^{s_3}}(\mathrm{p}_3)\,\boldsymbol{\gamma}^\mu\,u^{s_1}(\mathrm{p}_1)\right]\mathrm{Tr}\left[\overline{u^{s_1}}(\mathrm{p}_1)\,\overline{\boldsymbol{\gamma}}^\nu\,u^{s_3}(\mathrm{p}_3)\right]\\
&\quad\times\sum_{s_2,s_4}\mathrm{Tr}\left[\overline{U^{s_4}}(\mathrm{p}_4)\,\boldsymbol{\gamma}_\mu\,U^{s_2}(\mathrm{p}_2)\right]\mathrm{Tr}\left[\overline{U^{s_2}}(\mathrm{p}_2)\,\overline{\boldsymbol{\gamma}}_\nu\,U^{s_4}(\mathrm{p}_4)\right]\\
&= \frac{g_e^4}{(\mathrm{p}_1-\mathrm{p}_3)^4}\,\mathrm{Tr}\left[\boldsymbol{\gamma}^\mu\,(\not{p}_1 + m_e c\mathbb{1})\,\overline{\boldsymbol{\gamma}}^\nu\,(\not{p}_3 + m_e c\mathbb{1})\right]\\
&\quad\times\mathrm{Tr}\left[\boldsymbol{\gamma}_\mu\,(\not{p}_2 + m_\mu c\mathbb{1})\,\overline{\boldsymbol{\gamma}}_\nu\,(\not{p}_4 + m_\mu c\mathbb{1})\right]
\end{aligned} \tag{5.136}$$

$$= \frac{g_e^4}{(\mathrm{p}_1-\mathrm{p}_3)^4}\,X^{\mu\nu}(1,3;e^-)\,X_{\mu\nu}(2,4;\mu^-). \tag{5.137}$$

Digression 5.13 This result – in fact, the entire procedure (5.131)–(5.137) – may also be depicted graphically:

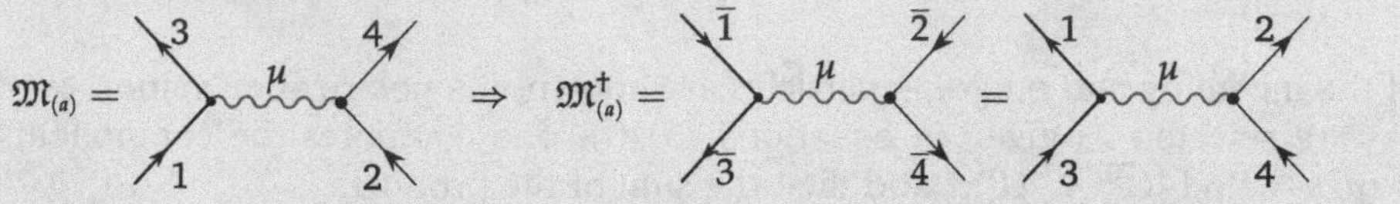

$$\mathfrak{M}_{(a)} = \quad \Rightarrow \quad \mathfrak{M}_{(a)}^{\dagger} = \quad = \quad \tag{5.138a}$$

where the diagram labels were simplified, so "1" stands for $u^{s_1,a}(\mathrm{p}_1)$ and "$\bar{1}$" for $\bar{u}_a^{s_1}(\mathrm{p}_1)$, etc. The product $\mathfrak{M}_{(a)}^{\dagger}\mathfrak{M}_{(a)}$ is then simply depicted by putting two diagrams next to each other. However, the summation of the product $\mathfrak{M}_{(a)}^{\dagger}\mathfrak{M}_{(a)}$ over spin (and, in general, all other unmeasured degrees of freedom) of, say, particle 1 is graphically depicted by connecting (concatenating) the two lines labeled "1" into a single line. Thus,

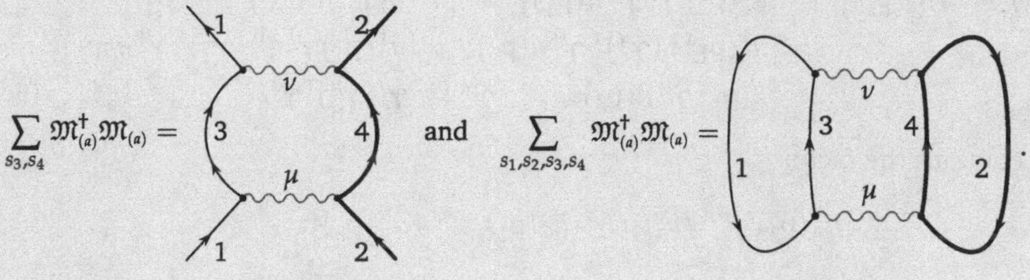

$$\sum_{s_3,s_4}\mathfrak{M}_{(a)}^{\dagger}\mathfrak{M}_{(a)} = \quad \text{and} \quad \sum_{s_1,s_2,s_3,s_4}\mathfrak{M}_{(a)}^{\dagger}\mathfrak{M}_{(a)} = \quad . \tag{5.138b}$$

By cutting the photon lines, we arrive at the graphical depiction (5.137):

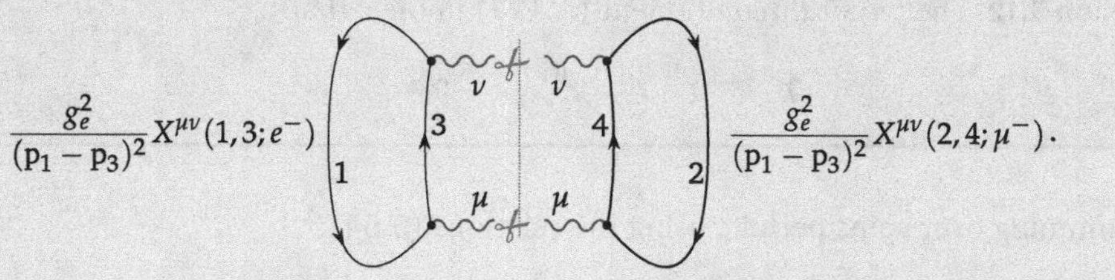

$$\frac{g_e^2}{(\mathrm{p}_1-\mathrm{p}_3)^2}X^{\mu\nu}(1,3;e^-) \qquad \frac{g_e^2}{(\mathrm{p}_1-\mathrm{p}_3)^2}X^{\mu\nu}(2,4;\mu^-). \tag{5.138c}$$

This graphical rendition of the computation of $\langle|\mathfrak{M}|^2\rangle$ is further detailed in Ref. [64].

The computation of the tensors

$$X^{\mu\nu}(1,3;e^-) := \mathrm{Tr}\left[\boldsymbol{\gamma}^{\mu}\left(\not{p}_1 + m_e c\mathbb{1}\right)\overline{\boldsymbol{\gamma}}^{\nu}\left(\not{p}_3 + m_e c\mathbb{1}\right)\right], \tag{5.139a}$$

$$X_{\mu\nu}(2,4;\mu^-) := \mathrm{Tr}\left[\boldsymbol{\gamma}_{\mu}\left(\not{p}_2 + m_{\mu} c\mathbb{1}\right)\overline{\boldsymbol{\gamma}}_{\nu}\left(\not{p}_4 + m_{\mu} c\mathbb{1}\right)\right]. \tag{5.139b}$$

reduces to writing out the $\boldsymbol{\gamma}$-polynomials in the square brackets (5.136), and simplifying using the identities (A.125). The final result is [☞ Ref. [241] for the "factorization" (5.137) and derivation]

$$\langle|\mathfrak{M}_{(a)}|^2\rangle = \frac{8g_e^4}{(\mathrm{p}_1-\mathrm{p}_3)^4}\big[(\mathrm{p}_1\!\cdot\!\mathrm{p}_2)(\mathrm{p}_3\!\cdot\!\mathrm{p}_4) + (\mathrm{p}_1\!\cdot\!\mathrm{p}_4)(\mathrm{p}_3\!\cdot\!\mathrm{p}_2) + 2(m_e m_{\mu} c^2)^2$$
$$- (m_{\mu}c)^2(\mathrm{p}_1\!\cdot\!\mathrm{p}_3) - (m_e c)^2(\mathrm{p}_2\!\cdot\!\mathrm{p}_4)\big]. \tag{5.140}$$

Electron–electron scattering

The computation (5.140) used that $e^- \neq \mu^-$. However, the result may be adapted also to the elastic $e^- + e^- \to e^- + e^-$ scattering, named after Christian Møller. This, however, does not reduce to a simple replacement $m_{\mu} \to m_e$ in the final expression (5.140), since when the two outgoing particles are identical, we must take into account another, equally possible process, depicted by the

Feynman diagram (c) in Figure 5.1 on p. 196. The total amplitude is then the *difference* between the amplitudes for (b) and for (c) in Figure 5.1 on p. 196. Indeed, since the electrons are fermions, the total amplitude must be antisymmetric with respect to the exchange of any two, and so also the two outgoing electrons. Thus,

$$\mathfrak{M}_{2e^-\to 2e^-} = \mathfrak{M}_{(b)} - \mathfrak{M}_{(c)}$$

$$= -\frac{g_e^2}{(\mathrm{p}_1-\mathrm{p}_3)^2}\big[\overline{u_3}\,\gamma^\mu\,u_1\big]\big[\overline{u_4}\,\gamma_\mu\,u_2\big] + \frac{g_e^2}{(\mathrm{p}_1-\mathrm{p}_4)^2}\big[\overline{u_4}\,\gamma^\mu\,u_1\big]\big[\overline{u_3}\,\gamma_\mu\,u_2\big] \tag{5.141a}$$

$=$ 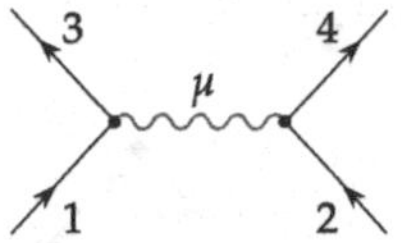

$-$ 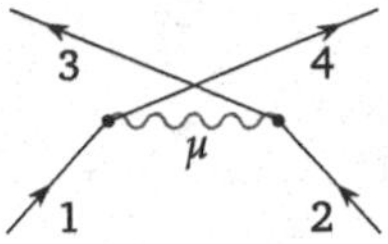

(5.141b)

where the expression is simplified by not writing the indices indicating the spin or those for the Dirac spinor components, and all these arguments are denoted by a single subscript: $u^{s_i,A}(\mathrm{p}_i) \to u_i$. Computing $\langle|\mathfrak{M}|^2\rangle$ in this case complicates as compared to (5.131)–(5.140): squaring the expression (5.141) by absolute value, we obtain

$$|\mathfrak{M}_{2e^-\to 2e^-}|^2 = |\mathfrak{M}_{(b)}|^2 + |\mathfrak{M}_{(c)}|^2 - 2\,\Re e\left(\mathfrak{M}^\dagger_{(b)}\,\mathfrak{M}_{(c)}\right). \tag{5.142}$$

The first two summands may be copied from equation (5.140), upon changing $m_\mu \to m_e$ and swapping $3 \leftrightarrow 4$ for $|\mathfrak{M}_{(c)}|^2$. The remaining, "interference" summand[18] is "a little" more complicated:

$$\mathfrak{M}^\dagger_{(b)}\mathfrak{M}_{(c)} \propto [\overline{u}_2\overline{\gamma}^\mu u_4][\overline{u}_1\overline{\gamma}_\mu u_3][\overline{u}_4\gamma^\nu u_1][\overline{u}_3\gamma_\nu u_2] = [\overline{u}_2\overline{\gamma}^\mu u_4][\overline{u}_4\gamma^\nu u_1][\overline{u}_1\overline{\gamma}_\mu u_3][\overline{u}_3\gamma_\nu u_2]$$
$$= \big[\overline{u}_2\overline{\gamma}^\mu u_4\overline{u}_4\gamma^\nu u_1\overline{u}_1\overline{\gamma}_\mu u_3\overline{u}_3\gamma_\nu u_2\big], \tag{5.143}$$

summing over spins produces

$$\langle\mathfrak{M}^\dagger_{(b)}\mathfrak{M}_{(c)}\rangle \propto \Big\langle\big[\overline{u}_2\overline{\gamma}^\mu u_4\overline{u}_4\gamma^\nu u_1\overline{u}_1\overline{\gamma}_\mu u_3\overline{u}_3\gamma_\nu u_2\big]\Big\rangle = \Big\langle\mathrm{Tr}\big[\overline{\gamma}^\mu u_4\overline{u}_4\gamma^\nu u_1\overline{u}_1\overline{\gamma}_\mu u_3\overline{u}_3\gamma_\nu u_2\overline{u}_2\big]\Big\rangle$$
$$= \mathrm{Tr}\big[\overline{\gamma}^\mu(\not{p}_4 + m_e c\mathbb{1})\gamma^\nu(\not{p}_1 + m_e c\mathbb{1})\overline{\gamma}_\mu(\not{p}_3 + m_e c\mathbb{1})\gamma_\nu(\not{p}_2 + m_e c\mathbb{1})\big], \tag{5.144}$$

for the computation of which one needs identities like (A.125), but up to and including the eighth degree in the γ-matrices. However, using the *matrix* identities (A.121)–(A.122) these may always be reduced to the listed identities (A.125) [☞ Theorem A.5 on p. 487].

Electron–positron scattering

The elastic scattering $e^- + e^+ \to e^- + e^+$ is known as Bhabha scattering, after Homi Jehangir Bhabha. Again there are contributions from two sub-processes:

$\mathfrak{M}_{(d)} =$ 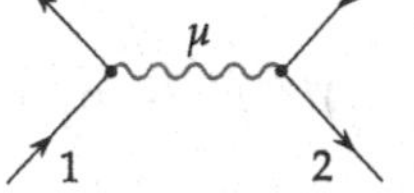

and $\mathfrak{M}_{(e)} =$ 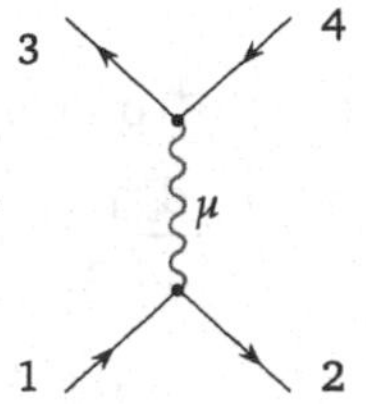

(5.145)

[18] The existence of such interference summands is the hallmark of quantum mechanics: the basic principle is that in classical physics one adds probabilities of the partial contributions to a process, whereas in quantum physics one adds the amplitudes of those probabilities and then squares this sum to obtain the probability.

It is not hard to show that $\mathfrak{M}_{(d)}(1,2,3,4) = \mathfrak{M}_{(c)}(1,3,2,4)$; i.e., by exchanging the incoming positron with the outgoing electron: the incoming positron, labeled "2," is equivalent to the outgoing electron, labeled "3," together with the $\mathrm{p}_2 \leftrightarrow -\mathrm{p}_3$ swap:

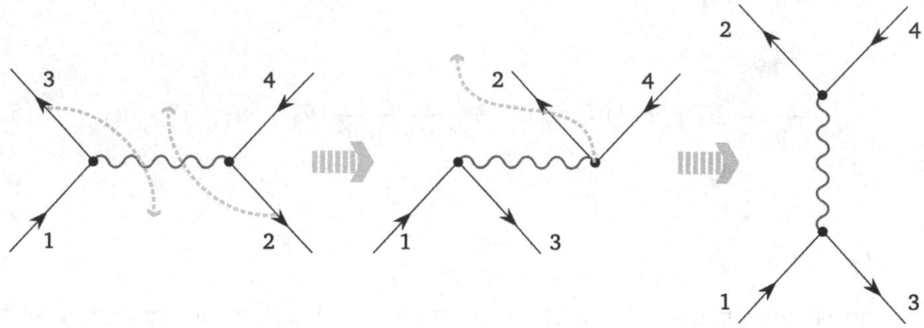

(5.146)

Antisymmetrizing with respect to this exchange of two fermions, we then have

$$\mathfrak{M}_{e^-e^+\to e^-e^+} = \mathfrak{M}_{(d)} - \mathfrak{M}_{(e)} \tag{5.147a}$$

$$= -\frac{g_e^2}{(\mathrm{p}_1-\mathrm{p}_3)^2}[\overline{u}_3\gamma^\mu u_1][\overline{v}_2\gamma_\mu v_4] + \frac{g_e^2}{(\mathrm{p}_1+\mathrm{p}_2)^2}[\overline{v}_2\gamma^\mu u_1][\overline{u}_3\gamma_\mu v_4]. \tag{5.147b}$$

The expression for $\mathfrak{M}_{(d)}$ was obtained from equation (5.131), swapping $U_2 \to v_4$: incoming muon into the incoming (backwards in time!) positron, as well as $\overline{U}_4 \to \overline{v}_2$: outgoing muon into the outgoing (backwards in time!) positron.

Compton scattering

For electron–photon scattering, there are again two diagrams:

$\mathfrak{M}_{(f)} =$ 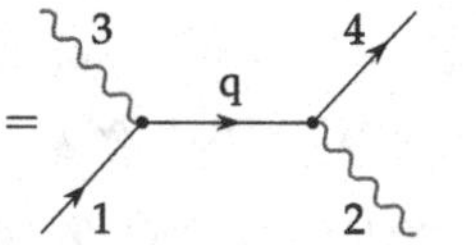

and $\mathfrak{M}_{(g)} =$ 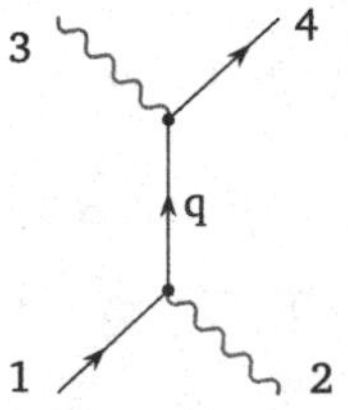

(5.148)

This time, the diagrams do not differ in an exchange of two fermions – we evidently do differentiate between the incoming and the outgoing electron, so there is no antisymmetrization; these amplitudes are therefore being added,

$$\mathfrak{M}_{e^-\gamma\to e^-\gamma} = \mathfrak{M}_{(f)} + \mathfrak{M}_{(g)}, \tag{5.149a}$$

where, following Procedure 5.2 on p. 193, we obtain

$$\begin{aligned} -i\mathfrak{M}_{(f)}(2\pi)^4\delta^4(\mathrm{p}_1+\mathrm{p}_2-\mathrm{p}_3-\mathrm{p}_4) &= \int \frac{\mathrm{d}^4\mathrm{q}}{(2\pi)^4}(2\pi)^4\delta^4(\mathrm{p}_1-\mathrm{p}_3-\mathrm{q})(2\pi)^4\delta^4(\mathrm{p}_2-\mathrm{p}_4+\mathrm{q}) \\ &\quad \times \overline{u}_4\,\epsilon_2^\mu(-ig_e\gamma_\mu)\frac{i(\not{q}+m_e c\mathbb{1})}{\mathrm{q}^2-m_e^2c^2}(-ig_e\gamma_\nu)\epsilon_3^{\nu*}\,u_1, \end{aligned} \tag{5.149b}$$

$$\mathfrak{M}_{(f)} = \frac{g_e^2}{(\mathrm{p}_1-\mathrm{p}_3)^2-m_e^2c^2}\big(\epsilon_2^\mu[\overline{u}_4\,\gamma_\mu(\not{p}_1-\not{p}_3+m_e c\mathbb{1})\gamma_\nu\,u_1]\epsilon_3^{\nu*}\big), \tag{5.149c}$$

$$\mathfrak{M}_{(g)} = \frac{g_e^2}{(\mathrm{p}_1 + \mathrm{p}_2)^2 - m_e^2 c^2} \big(\epsilon_3^{\mu *} [\overline{u}_4 \boldsymbol{\gamma}_\mu (\not{p}_1 - \not{p}_3 + m_e c \mathbb{1}) \boldsymbol{\gamma}_\nu u_1] \epsilon_2^\nu \big). \tag{5.149d}$$

We obtained the amplitude $\mathfrak{M}_{(g)}$ from $\mathfrak{M}_{(f)}$ by swapping

$$\mathrm{p}_2 \leftrightarrow -\mathrm{p}_3 \quad \text{and} \quad (\epsilon_2^\mu, \epsilon_3^{\nu *}) \leftrightarrow (\epsilon_3^{\mu *}, \epsilon_2^\nu). \tag{5.150}$$

This is easy to depict diagrammatically:

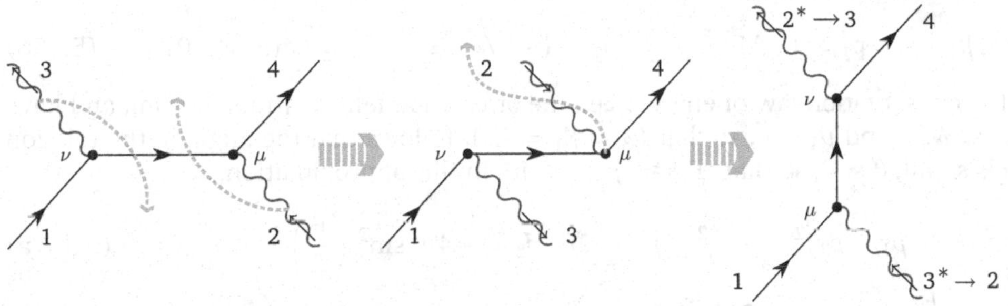

(5.151)

Electron–positron pair annihilation and creation

For inelastic scattering $e^- + e^+ \to 2\gamma$, there are again two diagrams:

$$\mathfrak{M}_{e^- + e^+ + \to 2\gamma} = \mathfrak{M}_{(h)} + \mathfrak{M}_{(i)} =$$

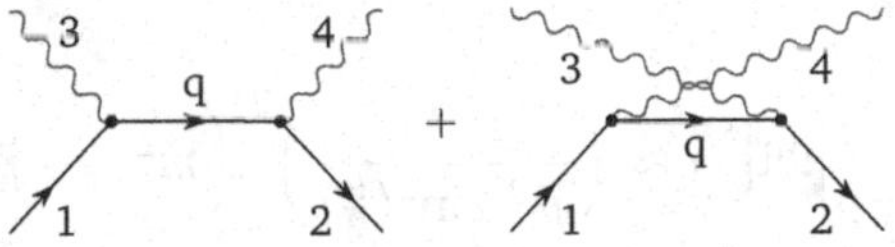

(5.152)

which are being added: They differ in the exchange of two photons $3 \leftrightarrow 4$, which are bosons, so the total amplitude is being symmetrized. As in the previous examples, we obtain

$$\begin{aligned} &= \frac{g_e^2}{(\mathrm{p}_1 - \mathrm{p}_3)^2 - m_e^2 c^2} \big(\epsilon_4^{\nu *} [\overline{v}_2 \boldsymbol{\gamma}_\nu (\not{p}_1 - \not{p}_3 + m_e c \mathbb{1}) \boldsymbol{\gamma}_\mu u_1] \epsilon_3^{\mu *}\big) \\ &\quad + \frac{g_e^2}{(\mathrm{p}_1 - \mathrm{p}_4)^2 - m_e^2 c^2} \big(\epsilon_3^{\nu *} [\overline{v}_2 \boldsymbol{\gamma}_\nu (\not{p}_1 - \not{p}_4 + m_e c \mathbb{1}) \boldsymbol{\gamma}_\mu u_1] \epsilon_4^{\mu *}\big). \end{aligned} \tag{5.153}$$

For the process of *pair-creation* $2\gamma \to e^- + e^+$, there are again two diagrams:

$$\mathfrak{M}_{2\gamma \to e^- + e^+} =$$

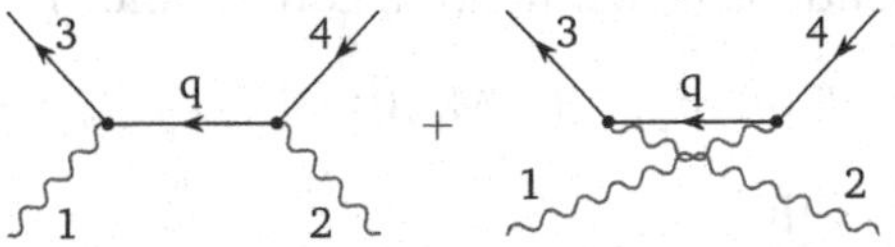

(5.154)

These contributions to the amplitude are being added as they again differ in the exchange of the two incoming photons. Owing to the evident (time-reversal) symmetry between the results (5.154) and (5.152), we have that $\mathfrak{M}_{2\gamma \to e^- + e^+} = \mathfrak{M}^\dagger_{e^- + e^+ \to 2\gamma}$.

5.3.2 *Effective cross-sections and lifetimes*

The results for $\mathfrak{M}$ and $\langle |\mathfrak{M}|^2 \rangle$ from the previous section may now be used in the above formulae (3.112) and (3.114), (3.122) and (3.127), as well as (3.159), (3.161) and (3.162).

Mott and Rutherford scattering

For the scattering of a light "probe" on a heavy "target" where the spins are not measured, we may use the results (5.136). In addition, in the approximation where the target mass ($m_B = M$) is sufficiently larger than the probe mass ($m_A = m$) so that the target recoil is negligible – which is easily realistic if the target is affixed in the lab – we use the result (3.161):

$$\frac{d\sigma}{d\Omega} \approx \left(\frac{\hbar}{8\pi Mc}\right)^2 \langle |\mathfrak{M}|^2 \rangle. \tag{5.155}$$

Since the target is immovable, we have

$$\mathrm{p}_1 = (-E/c, \vec{p}_1), \qquad \mathrm{p}_2 = (-Mc, \vec{0}), \qquad \mathrm{p}_3 \approx (-E/c, \vec{p}_3), \qquad \mathrm{p}_4 \approx (-Mc, \vec{0}), \tag{5.156}$$

where we used the conservation law of energy, i.e., the 0th component of 4-momentum, and have approximated $E_4 \approx Mc^2$ and $\vec{p}_4 \approx \vec{0}$, so that $E_3 \approx E_1 = E$. It follows that the angle in the relation $\vec{p}_1 \cdot \vec{p}_3 = \vec{p}^2 \cos\theta$ is small, $\theta \approx 0$, so that $|\vec{p}_1| \approx |\vec{p}_3| =: |\vec{p}|$. In this approximation,

$$(\mathrm{p}_1 - \mathrm{p}_3)^2 \approx -(\vec{p}_1 - \vec{p}_3)^2 = -\vec{p}_1^{\,2} - \vec{p}_3^{\,2} + 2\vec{p}_1 \cdot \vec{p}_3 = -4\vec{p}^2 \sin^2\left(\frac{\theta}{2}\right), \tag{5.157a}$$

$$(\mathrm{p}_1 \cdot \mathrm{p}_3) \approx \frac{E^2}{c^2} - \vec{p}_1 \cdot \vec{p}_3 = \vec{p}^2 + m^2c^2 - \vec{p}^2 \cos\theta = m^2c^2 + 2\vec{p}^2 \sin^2\left(\frac{\theta}{2}\right), \tag{5.157b}$$

$$(\mathrm{p}_1 \cdot \mathrm{p}_2) = ME \approx (\mathrm{p}_2 \cdot \mathrm{p}_3) \approx (\mathrm{p}_1 \cdot \mathrm{p}_4) \approx (\mathrm{p}_3 \cdot \mathrm{p}_4), \qquad (\mathrm{p}_2 \cdot \mathrm{p}_4) \approx M^2c^2. \tag{5.157c}$$

Thus,

$$\langle |\mathfrak{M}|^2 \rangle \approx \left(\frac{g_e^2\, Mc}{\vec{p}^{\,2} \sin^2(\frac{\theta}{2})}\right)^2 \left(m^2c^2 + \vec{p}^{\,2} \cos^2\left(\frac{\theta}{2}\right)\right), \tag{5.158}$$

$$\frac{d\sigma}{d\Omega} \approx \left(\frac{\alpha\hbar}{2\,\vec{p}^{\,2} \sin^2(\frac{\theta}{2})}\right)^2 \left(m^2c^2 + \vec{p}^{\,2} \cos^2\left(\frac{\theta}{2}\right)\right). \tag{5.159}$$

This is Mott's formula, which is a very good approximation of the differential cross-section for e^-–p^+ scattering, and even better for electron scattering on heavy ions. In the approximation where $\vec{p}^2 \ll m^2c^2$, we obtain

$$\frac{d\sigma}{d\Omega} \approx \left(\frac{\alpha\hbar}{2\,\vec{p}^{\,2} \sin^2(\frac{\theta}{2})}\right)^2 m^2c^2 = \left(\frac{\alpha\hbar c}{2\,m\vec{v}^{\,2} \sin^2(\frac{\theta}{2})}\right)^2, \tag{5.160}$$

which is the classical Rutherford formula (2.3).

The system of equations (4-momentum conservation)

$$(-E_1/c, \vec{p}_1) + (-Mc, \vec{0}) = (-E_3/c, \vec{p}_3) + (-E_4/c, \vec{p}_4) \tag{5.161a}$$

produces, denoting $p_i := |\vec{p}_i|$,

$$\sqrt{m^2c^2 + p_1^2} + Mc = \sqrt{m^2c^2 + p_3^2} + \sqrt{M^2c^2 + p_4^2}, \tag{5.161b}$$

$$p_1 = p_3 \cos\theta + p_4 \cos\phi, \tag{5.161c}$$

$$0 = p_3 \sin\theta - p_4 \sin\phi. \tag{5.161d}$$

Eliminating the angle ϕ from the last two equations produces

$$p_4 = |\vec{p}_1 - \vec{p}_3| = \sqrt{p_1^{\,2} - 2p_1p_3\cos(\theta) + p_3^{\,2}} \tag{5.161e}$$

which, together with relation (5.161b) gives $\left(\text{with } E_1 = c\sqrt{m^2c^2 + \vec{p}_1^{\,2}}\right)$

$$p_{3\pm} = \frac{p_1(E_1 M + m^2c^2)\cos(\theta) \pm p_1(E_1 + Mc^2)\sqrt{M^2 - m^2\sin^2(\theta)}}{(M^2+m^2)c^2 + 2E_1 M + p_1^{\,2}\sin^2(\theta)} \tag{5.162}$$

This is why simplifying approximations such as Mott's are convenient.

Electron–positron pair annihilation

In a model that has only electrons (and positrons) and photons, the decay – strictly speaking – is not possible: neither can a fermion (an electron or a positron) decay into any number of photons, nor can a photon decay into a real electron–positron pair [☞ Exercise 5.3.6]. However, the well-studied decay $\pi^0 \to 2\gamma$ is actually the process $\pi^0 = (q+\overline{q}) \to \gamma + \gamma$, which is in fact an inelastic scattering of a quark–antiquark pair that were, originally, bound into the state π^0. This process has contributions not only from the electromagnetic interaction, but also from weak and strong nuclear interactions, which complicates the estimate.

Instead, consider the decay of positronium, which is most conveniently computed in the positronium rest-frame, i.e., in the electron–positron CM system. It is known that in this system the electron and the positron move rather slowly, so we compute in the approximation where the electron and the positron are static immediately before their annihilation. The two photons created in the annihilation carry the same energy and so have linear momenta of the same magnitude and opposite direction. Thus we choose

$$\mathrm{p}_{e^-} = \mathrm{p}_1 = m_e c(-1,0,0,0), \qquad \mathrm{p}_{e^+} = \mathrm{p}_2 = m_e c(-1,0,0,0), \tag{5.163a}$$

$$\mathrm{p}_{\gamma_1} = \mathrm{p}_3 = m_e c(-1,0,0,1), \qquad \mathrm{p}_{\gamma_2} = \mathrm{p}_4 = m_e c(-1,0,0,-1), \tag{5.163b}$$

whereby it follows that

$$(\mathrm{p}_1 - \mathrm{p}_3)^2 - m_e^2c^2 = -2m_e^2c^2 = (\mathrm{p}_2 - \mathrm{p}_4)^2 - m_e^2c^2. \tag{5.164}$$

Besides, for the photons we use both the Lorenz gauge (5.91), whereby

$$\epsilon_3 \cdot \mathrm{p}_3 = 0 = \epsilon_4 \cdot \mathrm{p}_4, \tag{5.165}$$

as well as the Coulomb gauge, whereby the polarization 4-vectors ϵ_3 and ϵ_4 have no temporal component. Since p_1 and p_2 only have temporal components, it follows that

$$\epsilon_3 \cdot \mathrm{p}_1 = 0 = \epsilon_4 \cdot \mathrm{p}_1 \quad \text{and} \quad \epsilon_3 \cdot \mathrm{p}_2 = 0 = \epsilon_4 \cdot \mathrm{p}_2. \tag{5.166}$$

The expressions (5.153) may, after a little simplifying, be written as

$$\mathfrak{M}_{(h)} = \frac{g_e^2}{(\mathrm{p}_1 - \mathrm{p}_3)^2 - m_e^2c^2}\big([\overline{v}_2\, \not{\epsilon}_4^*(\not{\mathrm{p}}_1 - \not{\mathrm{p}}_3 + m_e c\mathbb{1})\not{\epsilon}_3^* u_1]\big), \tag{5.167}$$

$$\mathfrak{M}_{(i)} = \frac{g_e^2}{(\mathrm{p}_1 - \mathrm{p}_4)^2 - m_e^2c^2}\big([\overline{v}_2\, \not{\epsilon}_3^*(\not{\mathrm{p}}_1 - \not{\mathrm{p}}_4 + m_e c\mathbb{1})\not{\epsilon}_4^* u_1]\big), \tag{5.168}$$

where $\not{\epsilon}_i^* := \epsilon_i^{*\mu}\gamma_\mu$ – the gamma-matrix is not conjugated. Consider first $\mathfrak{M}_{(h)}$, where

$$\not{\mathrm{p}}_1\not{\epsilon}_3^* \overset{(A.126a)}{=} -\not{\epsilon}_3^*\not{\mathrm{p}}_1 + 2\epsilon_3^*\cdot\mathrm{p}_1 \overset{(5.166)}{=} -\not{\epsilon}_3^*\not{\mathrm{p}}_1, \tag{5.169a}$$

$$\not{\mathrm{p}}_3\not{\epsilon}_3^* \overset{(A.126a)}{=} -\not{\epsilon}_3^*\not{\mathrm{p}}_3 + 2\epsilon_3^*\cdot\mathrm{p}_3 \overset{(5.165)}{=} -\not{\epsilon}_3^*\not{\mathrm{p}}_3, \tag{5.169b}$$

$$(\not{p}_1 - \not{p}_3 + m_e c\mathbb{1})\not{\epsilon}_3^* u_1 = \not{\epsilon}_3^*(-\not{p}_1 + \not{p}_3 + m_e c\mathbb{1})u_1 = \not{\epsilon}_3^* \not{p}_3 u_1, \tag{5.169c}$$

where the last equality holds as the incoming electron, u_1, is on-shell, i.e., it satisfies the Dirac equation, $(\not{p}_1 - m_e c\mathbb{1})u_1 = 0$. $\mathfrak{M}_{(i)}$ is similarly simplified so that, using the choices (5.163b), we obtain

$$\mathfrak{M}_{e^-+e^+\to 2\gamma} = -\frac{g_e^2}{2m_e^2c^2}\overline{v}_2[\not{\epsilon}_4^*\not{\epsilon}_3^*\not{p}_3 + \not{\epsilon}_3^*\not{\epsilon}_4^*\not{p}_4]u_1 \tag{5.170}$$

$$= -\frac{g_e^2}{2m_e c}\overline{v}_2[\not{\epsilon}_4^*\not{\epsilon}_3^*(\gamma^0+\gamma^3) + \not{\epsilon}_3^*\not{\epsilon}_4^*(\gamma^0-\gamma^3)]u_1$$

$$= -\frac{g_e^2}{2m_e c}\overline{v}_2[(\not{\epsilon}_4^*\not{\epsilon}_3^* + \not{\epsilon}_3^*\not{\epsilon}_4^*)\gamma^0 + (\not{\epsilon}_4^*\not{\epsilon}_3^* - \not{\epsilon}_3^*\not{\epsilon}_4^*)\gamma^3]u_1$$

$$= -\frac{g_e^2}{2m_e c}\overline{v}_2[2\epsilon_{4\mu}^*\eta^{\mu\nu}\epsilon_{3\nu}^*\gamma^0 + 4i\epsilon_{4\mu}^*\gamma^{\mu\nu}\epsilon_{3\nu}^*\gamma^3]u_1$$

$$= -\frac{g_e^2}{2m_e c}\overline{v}_2[-2\vec{\epsilon}_4^*\cdot\vec{\epsilon}_3^*\gamma^0 + i(\vec{\epsilon}_4^*\times\vec{\epsilon}_3^*)\cdot\vec{\Sigma}\gamma^3]u_1, \tag{5.171}$$

where we used again that $\epsilon_i^0 = 0$ and where we defined

$$\Sigma_i := 2\varepsilon_{ijk}\gamma^{jk} = \tfrac{i}{2}\varepsilon_{ijk}[\gamma^j,\gamma^k]. \tag{5.172}$$

Finally, we use that the spins of the electron and the positron are antiparallel, and use

$$u_1^\uparrow = \sqrt{2}mc\begin{bmatrix}1\\0\\0\\0\end{bmatrix}, \quad u_1^\downarrow = \sqrt{2}mc\begin{bmatrix}0\\1\\0\\0\end{bmatrix}, \quad \overline{v}_2^\downarrow = \sqrt{2}mc[0\,0\,1\,0], \quad \overline{v}_2^\uparrow = \sqrt{2}mc[0\,0\,0\,1]. \tag{5.173}$$

Thus, using the concrete matrices in Appendix A.6.1,

$$\mathfrak{M}_{\uparrow\downarrow} = -2ig_e^2(\vec{\epsilon}_3^*\times\vec{\epsilon}_4^*)_z = -\mathfrak{M}_{\downarrow\uparrow}, \tag{5.174}$$

from which it follows that the symmetric state of the electron–positron system, $(|\uparrow\downarrow\rangle + |\downarrow\uparrow\rangle)/\sqrt{2}$, cannot decay into two photons. However, since the process $e^+ + e^- \to \gamma$ is kinematically forbidden, it follows that the symmetric state of positronium may only decay into three or more photons.

On the other hand, the *antisymmetric* state, $(|\uparrow\downarrow\rangle - |\downarrow\uparrow\rangle)/\sqrt{2}$, *can* decay into two photons. Thus we have

$$\mathfrak{M}_{|0,0\rangle} = \tfrac{1}{\sqrt{2}}(\mathfrak{M}_{\uparrow\downarrow} - \mathfrak{M}_{\downarrow\uparrow}), \tag{5.175}$$

where $|0,0\rangle = (|\uparrow\downarrow\rangle - |\downarrow\uparrow\rangle)/\sqrt{2}$ is the so-called *singlet* state of positronium before decay.

Next, re-insert the polarization vectors[19]

$$\vec{\epsilon}_{|1,+1\rangle} = -\tfrac{1}{\sqrt{2}}(1,i,0) \quad \text{and} \quad \vec{\epsilon}_{|1,-1\rangle} = \tfrac{1}{\sqrt{2}}(1,-i,0), \tag{5.176}$$

so that

$$(\vec{\epsilon}_3^*\times\vec{\epsilon}_4^*)_{\uparrow\downarrow} = (\vec{\epsilon}_{3,|1,+1\rangle}^*\times\vec{\epsilon}_{4,|1,-1\rangle}^*) = -\frac{1}{2}\begin{vmatrix}\hat{e}_1 & \hat{e}_2 & \hat{e}_3\\ 1 & -i & 0\\ 1 & i & 0\end{vmatrix} = -i\,\hat{e}_3 = -(\vec{\epsilon}_3^*\times\vec{\epsilon}_4^*)_{\downarrow\uparrow}, \tag{5.177}$$

whereby the photon polarization too must be in the antisymmetric superposition

$$\tfrac{1}{\sqrt{2}}\big(|1,+1\rangle_3|1,-1\rangle_4 - |1,-1\rangle_3|1,+1\rangle_4\big). \tag{5.178}$$

[19] The signs are chosen so that $\{\vec{\epsilon}_{|1,+1\rangle}, \vec{\epsilon}_{|1,-1\rangle}, \hat{e}_3\}$ would form a right-handed coordinate system.

Finally, adding the contributions to the amplitude as in the superposition (5.178),

$$\mathfrak{M}_{e^-+e^+\to 2\gamma} = -4g_e^2. \tag{5.179}$$

Although the final numeric value of this result seems disproportionately simple in comparison with the length and details of the derivation, note that we have also derived that the antiparallel spins in the electron–positron system imply that:

1. the positronium spin before the two-photon decay equals zero, i.e., the positronium is in the so-called singlet state $|0,0\rangle = (|\uparrow\downarrow\rangle - |\downarrow\uparrow\rangle)/\sqrt{2}$;
2. the spin of the two-photon state produced in the positronium decay equals zero, and the state itself is the antisymmetric superposition (5.178);
3. the triplet state of positronium, $|1,0\rangle = (|\uparrow\downarrow\rangle + |\downarrow\uparrow\rangle)/\sqrt{2}$, may only decay into three or more photons.

Given the amplitude (5.179), we may compute: First of all, using the result (3.127), we have the effective cross-section of the electron–positron annihilation in the CM system:

$$\frac{d\sigma}{d\Omega} = \Big(\frac{\hbar c}{8\pi(E_1+E_2)}\Big)^2 \frac{|\vec{p}_f|}{|\vec{p}_i|} |\mathfrak{M}|^2 = \Big(\frac{\hbar c}{16\pi(m_e c)}\Big)^2 \frac{|E_\gamma/c|}{|m_e v|} \big| -4g_e^2 \big|^2, \tag{5.180}$$

where we used that, because of (5.163a)–(5.163b), $E_1 = mc^2 = E_2$ and $E_\gamma = m_e c^2$. Simplifying, we obtain

$$\frac{d\sigma}{d\Omega} = \frac{1}{cv}\Big(\frac{\hbar\alpha}{m_e}\Big)^2, \qquad \text{and} \qquad \sigma = \frac{4\pi}{cv}\Big(\frac{\hbar\alpha}{m_e}\Big)^2, \tag{5.181}$$

since $d\sigma/d\Omega$ does not depend on angles.

For the decay constant and the lifetime of positronium, use relation (3.109), where the total number of scatterings equals $N = L\sigma$, and luminosity is $L = v\rho$, with ρ the probability density of finding the electron and the positron at the decay location. For an individual positronium "atom," $\rho = |\Psi(\vec{0},t)|^2$ and N represents the decay probability in unit time, i.e., the decay constant. Thus,

$$\Gamma = v\sigma\,|\Psi(\vec{0},t)|^2 = \frac{4\pi}{c}\Big(\frac{\hbar\alpha}{m_e}\Big)^2 |\Psi(\vec{0},t)|^2, \tag{5.182}$$

in agreement with Conclusion 3.2 on p. 113, and the relation (3.110). Recall: $[|\Psi(\vec{0},t)|^2] = L^{-3}$; at the end of Section 4.1.5, the result from analyzing the hydrogen atom was adapted, $|\Psi(\vec{0},t)|^2 = \big(\frac{\alpha m_e c}{\hbar n}\big)^3$. With this result, we finally get

$$\Gamma = \frac{4\pi}{c}\Big(\frac{\hbar\alpha}{m_e}\Big)^2 \Big[\frac{1}{\pi}\Big(\frac{\alpha(\frac{1}{2}m_e)c}{\hbar n}\Big)^3\Big] = \frac{\alpha^5\, m_e c^2}{2\hbar\, n^3}, \tag{5.183}$$

and the positronium lifetime becomes

$$\tau = \frac{1}{\Gamma} = \frac{2\hbar\, n^3}{\alpha^5\, m_e c^2} \approx (1.24494\times 10^{-10}\,\text{s}) \times n^3. \tag{4.51}$$

5.3.3 Renormalization

When discussing electron–muon scattering (5.129)–(5.140), we took into account only the Feynman diagram of lowest order in the g_{e^-}, i.e., α-power expansion. The results (5.131)–(5.140) produce $\mathfrak{M}_{(a)} = \sqrt{\langle|\mathfrak{M}_{(a)}|^2\rangle} = O(g_e^2) = O(\alpha)$. Next order corrections stem from the following diagrams:

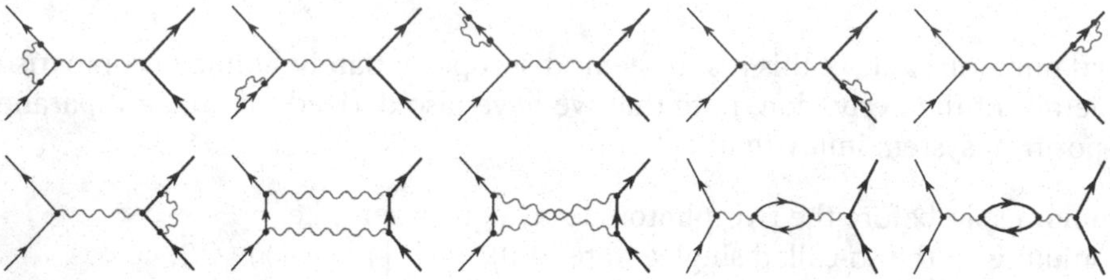

(5.184)

of which we will consider the last two. Denote

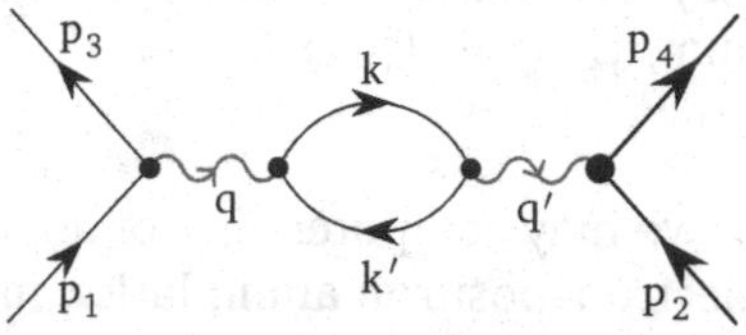

(5.185)

Calculation

The amplitude for this process is obtained following Procedure 5.2 on p. 193:

$$\begin{aligned}
&\int \frac{d^4q}{(2\pi)^4}\frac{d^4q'}{(2\pi)^4}\frac{d^4k}{(2\pi)^4}\frac{d^4k'}{(2\pi)^4}\,(2\pi)^4\delta^4(p_1-p_3-q)\,(2\pi)^4\delta^4(q-k+k')\,(2\pi)^4\delta^4(k-k'-q') \\
&\quad\times (2\pi)^4\delta^4(p_2-p_4+q')\,[\overline{u}_3(ig_e\boldsymbol{\gamma}^\mu)u_1]\Big(\frac{-i\eta_{\mu\nu}}{q^2}\Big) \\
&\quad\times(-1)\,\mathrm{Tr}\Big[(ig_e\boldsymbol{\gamma}^\nu)\frac{i}{\not{k}-m_ec}(ig_e\boldsymbol{\gamma}^\rho)\frac{i}{\not{k}-m_ec}\Big]\Big(\frac{-i\eta_{\rho\sigma}}{(q')^2}\Big)\,[\overline{U}_4(ig_e\boldsymbol{\gamma}^\sigma)U_2] \qquad (5.186a)\\
&= -g_e^4\int\frac{d^4q}{(2\pi)^4}\frac{d^4k}{(2\pi)^4}\,(2\pi)^4\delta^4(p_1-p_3-q)\,(2\pi)^4\delta^4(p_2-p_4+q) \\
&\quad\times[\overline{u}_3\,\boldsymbol{\gamma}^\mu\,u_1]\Big(\frac{\eta_{\mu\nu}}{q^2}\Big)\,\mathrm{Tr}\Big[\boldsymbol{\gamma}^\nu\frac{1}{\not{k}-m_ec}\boldsymbol{\gamma}^\rho\frac{1}{\not{k}-\not{q}-m_ec}\Big]\Big(\frac{\eta_{\rho\sigma}}{q^2}\Big)\,[\overline{U}_4\,\boldsymbol{\gamma}^\sigma\,U_2] \\
&= -i(2\pi)^4\,\delta^4(p_1+p_2-p_3-p_4) \\
&\quad\times\Big[\frac{-ig_e^4}{q^4}\int\frac{d^4k}{(2\pi)^4}\,[\overline{u}_3\,\boldsymbol{\gamma}^\mu\,u_1]\,\frac{\mathrm{Tr}[\boldsymbol{\gamma}_\mu(\not{k}+m_ec)\boldsymbol{\gamma}_\rho(\not{k}-\not{q}+m_ec)]}{(k^2-m_e^2c^2)[(k-q)^2-m_e^2c^2]}\,[\overline{U}_4\,\boldsymbol{\gamma}^\rho\,U_2]\Big]_{q=p_1-p_3}, \qquad (5.186b)
\end{aligned}$$

where the factor (-1) in the expression (5.186a) reflects rule 7 in Procedure 5.2 on p. 193. With the abbreviation $q := p_1 - p_3$, we have

$$\mathfrak{M}_{(a')} = \frac{-ig_e^4}{q^4}\,[\overline{u}_3\,\boldsymbol{\gamma}^\mu\,u_1]\Big\{\int\frac{d^4k}{(2\pi)^4}\,\frac{\mathrm{Tr}[\boldsymbol{\gamma}_\mu(\not{k}+m_ec)\boldsymbol{\gamma}_\rho(\not{k}-\not{q}+m_ec)]}{(k^2-m_e^2c^2)[(k-q)^2-m_e^2c^2]}\Big\}\,[\overline{U}_4\,\boldsymbol{\gamma}^\rho\,U_2]. \qquad (5.187)$$

Comparing with equation (5.131), we see that the inclusion of this $O(g_e^4)$ contribution[20]

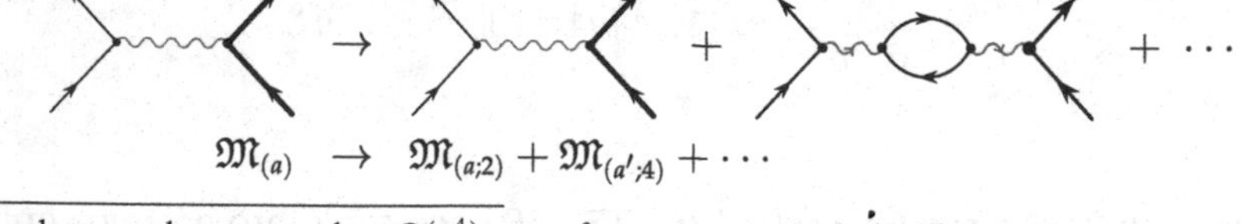

(5.188)

$$\mathfrak{M}_{(a)} \;\to\; \mathfrak{M}_{(a;2)} + \mathfrak{M}_{(a';4)} + \cdots$$

[20] For the complete result computed to $O(g_e^4)$, we of course must include all contributions (5.184) [☞ Refs. [243, 45, 580, 241]]; for brevity and pedagogical focus, only the last two are considered here.

is equivalent to replacing the photon propagator in Procedure 5.2:

$$\frac{-i\eta_{\mu\rho}}{\mathrm{q}^2} \quad\to\quad \frac{-i\eta_{\mu\rho}}{\mathrm{q}^2}+\frac{-i\,H_{\mu\rho}}{\mathrm{q}^4}+\cdots = \frac{-i}{\mathrm{q}^2}\Big[\eta_{\mu\rho}+\frac{H_{\mu\rho}}{\mathrm{q}^2}+\cdots\Big], \tag{5.189}$$

where

$$H_{\mu\rho} \;:=\; i g_e^2 \int \frac{\mathrm{d}^4\mathrm{k}}{(2\pi)^4}\,\frac{\mathrm{Tr}[\gamma_\mu(\not{\mathrm{k}}+m_e c)\gamma_\rho(\not{\mathrm{k}}-\not{\mathrm{q}}+m_e c)]}{(\mathrm{k}^2-m_e^2c^2)[(\mathrm{k}-\mathrm{q})^2-m_e^2c^2]}. \tag{5.190}$$

Since $H_{\mu\rho}$ may only depend on the (rank-2) metric tensor $\eta_{\mu\nu}$ and the 4-momentum q_μ, it must be that (as a rank-2 tensor and with $[H_{\mu\nu}]=2[\mathrm{q}]$)

$$H_{\mu\rho} = -\eta_{\mu\rho}\,\mathrm{q}^2\,I(\mathrm{q}^2) + q_\mu\, q_\rho\, J(\mathrm{q}^2). \tag{5.191}$$

Here, $I(\mathrm{q}^2)$ and $J(\mathrm{q}^2)$ are two Lorentz-invariant functions of the 4-momentum q, so they must be functions of the Lorentz-invariant square q^2. Since q is the 4-momentum of the *virtual* photon, q^2 need not be restricted to the mass shell ($\mathrm{q}^2=0$, for the massless photon) and may attain arbitrary values.

The function $J(\mathrm{q}^2)$ contributes nothing to the final result, as it occurs, within the amplitude (5.187), only contracted with the 4-momentum:

$$[\overline{u}_3\gamma^\mu\, u_1]q_\mu = [\overline{u}_3\,\not{q}\,u_1] = [\overline{u}_3\,(\not{p}_1-\not{p}_3)\,u_1] = 0. \tag{5.192}$$

This last equality holds since both the incoming and the outgoing electrons are on the mass shell:

$$\not{p}_1\, u_1 = m_e c\, u_1 \qquad\text{and}\qquad \overline{u}_3\not{p}_3 = \overline{u}_3\, m_e c. \tag{5.193}$$

(Recall: $\not{p}_i = \gamma^\mu p_{i\,\mu}$ are 4×4 matrices, u_i 4-component column-matrices and $\overline{u}_i$ 4-component row-matrices.) It remains to compute the function $I(\mathrm{q}^2)$, which may be brought into the shape [243]:

$$I(\mathrm{q}^2) = \frac{g_e^2}{12\pi^2}\Big\{\int_{m_e^2}^{\infty}\frac{\mathrm{d}\xi}{\xi} - 6\int_0^1 \mathrm{d}\zeta\;\zeta(1-\zeta)\ln\Big(1-\tfrac{\mathrm{q}^2}{m_e^2c^2}\zeta(1-\zeta)\Big)\Big\}. \tag{5.194}$$

While the first integral diverges logarithmically,

$$\int_{m_e^2}^{\infty}\frac{\mathrm{d}\xi}{\xi} = \lim_{\mu\to\infty}\int_{m_e^2}^{\mu^2}\frac{\mathrm{d}\xi}{\xi} = 2\lim_{\mu\to\infty}\ln\Big(\frac{\mu}{m_e}\Big) = \infty, \tag{5.195a}$$

the second term, in curly brackets in relation (5.194), equals

$$\begin{aligned} f(x) &:= \frac{(12-5x)\sqrt{x(x+4)} - 6(x-2)(x+4)\tan^{-1}\Big(\sqrt{\frac{x}{x+4}}\Big)}{3\sqrt{x^3(x+4)}}\\ &= \frac{4}{x}-\frac{5}{3}-\frac{2(x-2)}{x}\sqrt{\frac{x+4}{x}}\,\tan^{-1}\Big(\sqrt{\frac{x}{x+4}}\Big), \end{aligned} \tag{5.195b}$$

$$x := -\frac{\mathrm{q}^2}{m_e^2c^2} = -\frac{(\mathrm{p}_1-\mathrm{p}_3)^2}{m_e^2c^2} \approx 4\frac{\vec{p}_{e,\text{in}}^{\,2}}{m_e^2c^2}\sin^2\big(\tfrac{\theta}{2}\big) = 4\frac{\vec{v}_{e,\text{in}}^{\,2}}{c^2}\gamma_e^2\sin^2\big(\tfrac{\theta}{2}\big), \tag{5.195c}$$

where $\vec{p}_{e,\text{in}}$ and $\vec{v}_{e,\text{in}}$ are, respectively, the 3-vectors of linear momentum and the velocity of the incoming electron, γ_e the corresponding relativistic factor, and θ their deflection angle. Note that $f(x)$ varies relatively slowly:

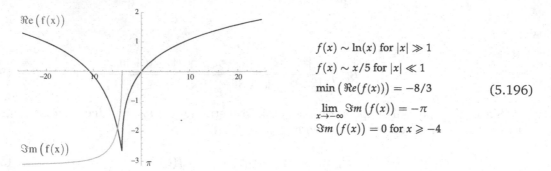

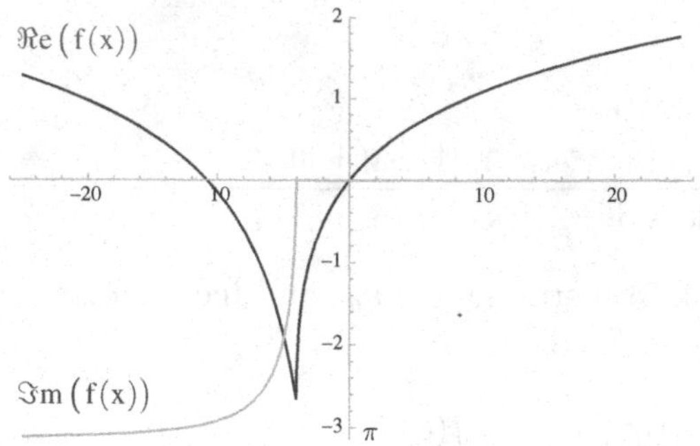

$$\begin{aligned} &f(x) \sim \ln(x) \text{ for } |x| \gg 1\\ &f(x) \sim x/5 \text{ for } |x| \ll 1\\ &\min\big(\Re e(f(x))\big) = -8/3\\ &\lim_{x\to-\infty} \Im m\,(f(x)) = -\pi\\ &\Im m\,(f(x)) = 0 \text{ for } x \geqslant -4 \end{aligned} \tag{5.196}$$

The total amplitude is

$$\mathfrak{M}_{(a)} = \lim_{\mu\to\infty} \mathfrak{M}_{(a)}(\mathrm{q}^2,\mu) + \cdots, \tag{5.197a}$$

where

$$\mathfrak{M}_{(a)}(\mathrm{q}^2,\mu) = -g_R^2(\mu)\,[\overline{u}_3\,\boldsymbol{\gamma}^\mu\,u_1]\Big(\frac{\eta_{\mu\nu}}{\mathrm{q}^2}\Big)\Big\{1+\frac{g_R^2(\mu)}{12\pi^2}\,f\Big(\frac{-\mathrm{q}^2}{m_e^2c^2}\Big)\Big\}\,[\overline{U}_4\,\boldsymbol{\gamma}^\nu\,U_2] + \cdots, \tag{5.197b}$$

$$g_{e,R}(\mu) := g_e\sqrt{1-\frac{g_e^2}{6\pi^2}\ln\Big(\frac{\mu}{m_e}\Big)}, \tag{5.197c}$$

and where "$\cdots$" denotes omitted contributions from the other diagrams (5.184), as well as $O(g_e^6)$ contributions, and the equality (5.197b) with definition (5.197c) holds up to $O(g_e^6)$ corrections.

Physical meaning

The definition (5.197a) actually uncovers a conceptual error in the original set-up of the computation: The identification of the $e^- + \mu^- \to e^- + \mu^-$ elastic scattering amplitude of course depends on the strength of the interaction of the electron and the muon with the photons that mediate the electromagnetic interaction. The measure of the strength of that interaction was initially identified [☞ definition (5.122)] with the dimensionless parameter $g_e = \frac{e}{\sqrt{\epsilon_0\hbar c}}$ used in the assignment (5.124), which in turn is derived (within a field theory course) from the *classical* Lagrangian (5.118), with $q_\Psi \to -e$ (for the electron). However, the electric charge is of course a *measured* parameter, and elastic scatterings such as $e^- + \mu^- \to e^- + \mu^-$ in fact *define* the quantity that we call the (physical) electric charge. In other words, the original parameter $q_\Psi \to -e = -g_e\sqrt{\epsilon_0\hbar c}$ used in (5.118) is neither independently nor directly measurable, and should never have been identified identically with the physical electric charge of the electron.

Conclusion 5.5 *The quantity that* **is** *measurable and which* **is** *being compared with experimental data may in turn be identified with the symbol $g_{e,R}$, as defined by the relation (5.197c) as a function of the auxiliary (intermediate and, essentially, arbitrary) parameters g_e, μ – and up to $O(g_e^4)$ contributions, which were omitted in the expansion (5.197a).*

From the form of equation (5.197c), taking the $\mu \to \infty$ limit and the physical fact that the measurable charge $g_{e,R}$ is of course a finite quantity, it follows that the original and unmeasurable variable g_e must be a function of the variable μ, so that

$$g_{e,R} := \lim_{\mu\to\infty} g_e(\mu)\sqrt{1-\frac{g_e^2(\mu)}{6\pi^2}\ln\Big(\frac{\mu}{m_e}\Big)+\cdots} < \infty \tag{5.198}$$

This perhaps fussy "detailing" is in fact logical, given that both g_e and μ are auxiliary (intermediate) variables that serve only to connect the mathematical model (developed in a perturbative way from the classical physics model) to the physical quantities that this model describes.

In field theory (a course that should follow this introduction), the rules in Procedure 5.2 on p. 193 are *derived* from the Lagrangian for electrodynamics of charged spin-$\frac{1}{2}$ particles. The parameter g_e should show up in this Lagrangian. However, just like that Lagrangian, the parameter by itself is not measurable, but defines the measurable charge by means of the iterative relation the beginning of which is given by equation (5.198). Thus, relation (5.197b) may be written as

$$\mathfrak{M}_{e^-+\mu^-\to e^-+\mu^-}(\mathrm{q}^2) = -g_{e,R}^2(\mathrm{q}^2)\,[\overline{u}_3\,\boldsymbol{\gamma}^\mu\,u_1]\Big(\frac{\eta_{\mu\nu}}{\mathrm{q}^2}\Big)[\overline{U}_4\,\boldsymbol{\gamma}^\nu\,U_2]+\cdots, \tag{5.199a}$$

$$g_{e,R}(\mathrm{q}^2) = g_{e,R}(0)\sqrt{1+\frac{g_{e,R}^2(0)}{12\pi^2}\,f\Big(\frac{-\mathrm{q}^2}{m_e^2c^2}\Big)}, \tag{5.199b}$$

that is,

$$\begin{aligned}\alpha_{e,R}(\mathrm{q}^2) &= \alpha_{e,R}(0)\Big\{1+\frac{\alpha_{e,R}(0)}{3\pi}\,f\Big(\frac{-\mathrm{q}^2}{m_e^2c^2}\Big)\Big\},\\ &\approx \alpha_{e,R}(0)\Big\{1+\frac{\alpha_{e,R}(0)}{3\pi}\,\ln\Big(\frac{\mathrm{q}^2}{m_e^2c^2}\Big)\Big\}, \qquad \mathrm{q}^2 \gg m_e^2c^2,\end{aligned} \tag{5.199c}$$

where the electric charge is defined as the renormalized parameter of the electromagnetic interaction, $g_{e,R}(\mathrm{q}^2)$, as is then defined the parameter of the electromagnetic fine structure, $\alpha_{e,R}(\mathrm{q}^2)$, as a function of the Lorentz-invariant intensity of the 4-momentum transfer, from the "probe" (here e^-) to the "target" (here μ^-).

The quantities $g_{e,R}(0)$ and $\alpha_{e,R}(0)$ are the limiting values of the functions $g_{e,R}(\mathrm{q}^2)$ and $\alpha_{e,R}(\mathrm{q}^2)$, when the 4-momentum transfer between the "probe" and "target" is negligible, and in that limit we have $\alpha_{e,R}(0) \approx \frac{1}{137}$. The numerical values of the corrections (5.199c) are relatively small, e.g., $O(6\times10^{-6})$ for a direct collision at $c/10$ speed, so that the value $\frac{1}{137}$ is used as a first approximation for $\alpha_{e,R}(\mathrm{q}^2)$ as if it were a constant. However, precise measurements of electromagnetic processes, such as in the Lamb shift (1.38a), indeed verify the corrections (5.199c).

The fact (5.199c) that the numerical value of the electric charge depends on the 4-momentum of the interaction with which that electric charge is being measured indicates the conceptual error in classical physics, where the parameters in the model of the physical system or process have *a priori* identified physical meaning and concrete value. The quantum nature of Nature teaches us that only those particular combinations and functions of the model parameters for which the values really can be measured must in fact have concrete (real and finite) values.

The contribution of the last diagram (5.184) equals the second term in equation (5.199c), only with a virtual muon in the central closed loop. That induces the replacement $m_e \to m_\mu$ in the result (5.199c), which reduces the contribution since

$$\ln\Big(\frac{\mathrm{q}^2}{m_\mu^2c^2}\Big) = \ln\Big(\frac{\mathrm{q}^2}{m_e^2c^2}\Big) - \big[\,2\ln(206) \approx 10.6558\,\big]. \tag{5.200}$$

In fact, since the fermionic closed loop in the center of the diagram (5.185) depicts a *virtual* spin-$\frac{1}{2}$ fermion – which by definition is not observed – the contribution of the same diagram should be summed over all electrically charged spin-$\frac{1}{2}$ fermions. The electron's contribution is however dominant, since the electron is the lightest of all electrically charged spin-$\frac{1}{2}$ fermions, and the corrections (5.199c) [☞ relations (5.195b)–(5.196)] are scaled by the logarithm of the inverse mass of the particle in this central loop.

Finally, the relation (5.199c) is a result of the $O(g_e^4)$ contributions, which is depicted by the diagrams (5.188). It is not hard to show that the infinite series of diagrams of growing order:

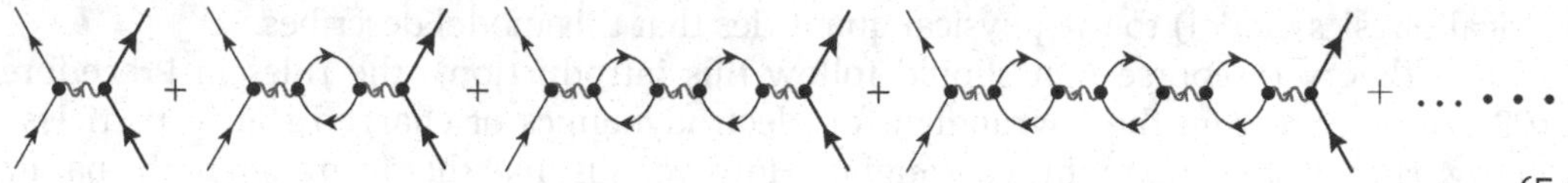

(5.201)

summing the geometric series result in

$$\alpha_{e,R}(|q^2|) \approx \frac{\alpha_{e,R}(0)}{1 - \frac{\alpha_{e,R}(0)}{3\pi} \ln\left(\frac{|q^2|}{m_e^2 c^2}\right)}, \qquad |q^2| \gg m_e^2 c^2. \tag{5.202}$$

In the domain $m_e^2 c^2 \ll q^2 \ll m_e^2 c^2 \exp\left\{\frac{3\pi}{2\alpha(0)}\right\}$, $\alpha_{e,R}(q^2)$ is a very slowly growing function, and the approximation $\alpha_{e,R}(q^2) \approx \alpha_{e,R}(0) \approx \frac{1}{137}$ is very good.

The various diagrams that are not shown in the series (5.201) [☞ collection (5.184)] either provide significantly smaller contributions than those shown (comparing diagrams of the same order in g_e^2) or their contribution may be absorbed by *renormalizing* parameters such as the mass of the electron, m_e. The contributions (5.201) are usually called the "leading logarithm" contributions.

The renormalization group

Note that the result (5.197c) was obtained by including the quantum correction of only the lowest order, and the result (5.202) includes the dominant corrections. Evidently, these corrections – computed iteratively and sequentially – may be organized in a quantitative sequence, so that from one iteration to the next one there is a "flow":

$$\left(\alpha_{e,R}^{(0)}(|q^2|) := \alpha_{e,R}(0)\right) \mapsto \cdots \mapsto \alpha_{e,R}^{(k)}(|q^2|) \mapsto \alpha_{e,R}^{(k+1)}(|q^2|) \mapsto \cdots \mapsto \alpha_{e,R}^{(\infty)}(|q^2|), \tag{5.203}$$

where only the limiting result, $\alpha_{e,R}^{(\infty)}(|q^2|)$, may be identified with the real physical quantity. The precise specification of the ordering of this renormalization "flow" depends on the concrete application – and this is one of those conceptual ideas that are applied in almost all branches of physics! The formal transformations that lead from one step in this renormalization flow into the next form a structure called the "renormalization group" – although it in fact does not satisfy the group axioms: The transformation $\mathcal{R}_{(k)}^{(k+1)}$ that takes the kth into the $(k+1)$th "step" has no binary operation defined with most other such transformations; only the consecutive "products" of the form $\mathcal{R}_{(k)}^{(k+1)} \circ \mathcal{R}_{(k-1)}^{(k)}$ are defined [☞ Comment 9.2 on p. 323].

In field theory, the application of this procedure and its structure was discovered by Ernst Stückelberg and Andre Petermann back in 1953 [502, 146]. The contemporary practice in field theory varies, but by now mostly relies on Kenneth Wilson's approach (1982 Nobel Prize), further developed by Joseph Polchinski [431]; see also Ref. [425] by Michael Peskin, who was Wilson's student, and who in turn was Gell-Mann's student. In this approach, the renormalization flow is organized by means of a varying upper limit in otherwise divergent integrals, i.e., by the energy/mass values up to which particles and excitations are included. The earlier approach, after

Murray Gell-Mann and Francis Low, varies the value of the renormalization 4-momentum μ in the computations that lead to the results such as (5.198) and is still being used.

Essentially two types of behavior can result from this renormalization flow, which motivates:

Definition 5.1 *A quantum system is* ***renormalizable*** *if merely the parameters used in the classical Lagrangian of a system change owing to quantum corrections, but the* ***functional form*** *of this Lagrangian remains the same. Otherwise, a system is* ***non-renormalizable.***

See also Definition 11.1 on p. 419 for a more precise statement. Suffice it here to say that all possible Yang–Mills type gauge theory models interacting with any spin-0 and spin-$\frac{1}{2}$ matter fields – including the Standard Model – are renormalizable.

5.3.4 Exercises for Section 5.3

✎ **5.3.1** *For* $\Gamma \in \{\mathbb{1}, \gamma^\mu, \gamma^{\mu\nu}, (\gamma^\mu \widehat{\gamma}), \widehat{\gamma}\}$*, compute for which* Γ *is* $\overline{\Gamma} := (\gamma^0 \Gamma^\dagger \gamma^0) = \Gamma$*, and for which* Γ *is* $\overline{\Gamma} = -\Gamma$.

✎ **5.3.2** *Derive the equations of motion (5.120f).*

✎ **5.3.3** *Derive equation (5.140).*

✎ **5.3.4** *Derive equation (5.144), and then equation (5.142).*

✎ **5.3.5** *Find the diagram that, as (5.138c), depicts the result (5.144) and represents it by a graphical depiction as in Digression 5.13 on p. 198.*

✎ **5.3.6** *From the 4-vector equation* $\mathrm{p}_\gamma = \mathrm{p}_{e^-} + \mathrm{p}_{e^+}$ *and the symmetry* $\vec{p}_\gamma \cdot \vec{p}_{e^-} = \vec{p}_\gamma \cdot \vec{p}_{e^+}$*, as well as* $|\vec{p}_{e^-}| = |\vec{p}_{e^+}|$*, show that the decay of a real photon into a real electron–positron pair is kinematically forbidden.*

✎ **5.3.7** *Prove relation (5.195c), i.e., that* $\mathrm{q}^2 = -4\vec{p}_i^{\,2} \sin^2(\frac{\theta}{2})$.

✎ **5.3.8** *Compute the collision energy for which the expression (5.202) diverges.*

5.4 Quantum electrodynamics of hadrons

The interaction between photons and quarks is described by the same theory as the interaction between photons and leptons, discussed in Sections 5.2.1 and 5.3.1. However, individual quarks are not available for experimenting. They are always within bound states, so-called hadrons: mesons, which are $(q\bar{q})$-systems, and baryons, which are (qqq)-systems. The interaction between a lepton (as a "probe") and a hadron (as the "target"), as well as between two hadrons, reduces to the interaction with individual (anti)quarks *within* the hadron, and so necessarily depends on the distribution of these individual (anti)quarks within the hadron. This distribution is described by so-called form-factors, which effectively[21] describe the *strong nuclear interactions* that bind the (anti)quarks into hadron bound states.

The second difficulty stems from the fact that the number of new hadrons is limited only by the available energy: as the collision energy grows, more and more new hadrons may be produced

[21] Here, "effective" means "successfully and with no a-priori detailed fundamental basis/derivation"; the adjective "phenomenological" is used in the literature, in the same sense.

in inelastic collisions, and the analysis very quickly becomes a combinatorially growing nightmare. Catalogues of hadrons such as Ref. [293] provide data about hundreds and hundreds of hadrons.

In principle, in the interaction between leptons and hadrons as well as between two hadrons, there are also contributions from weak nuclear interactions. However, that (third) source of complications is in most cases negligible, as the weak nuclear interaction is much weaker [☞ discussion on p. 67].

The two types of processes that are significant in hadronic experiments are production from electron–positron annihilation,

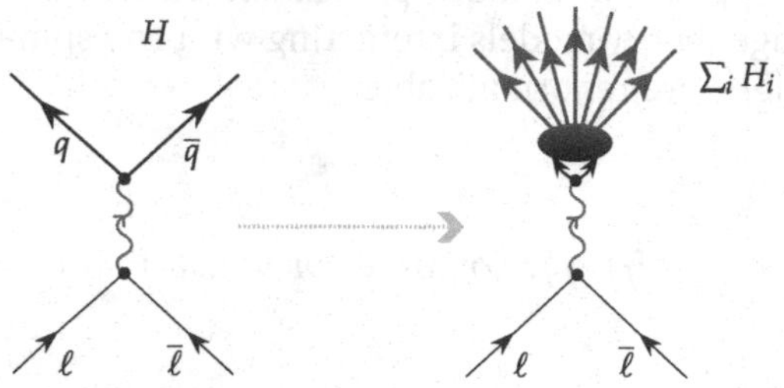

(5.204)

and so-called deep inelastic lepton–hadron collisions,

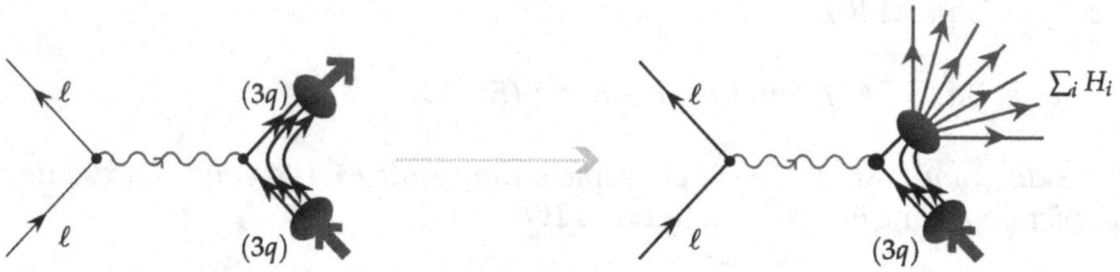

(5.205)

In both cases, increasing collision energy (indicated by the dotted arrow) gives rise to the production of a large number of outgoing hadrons. The strong nuclear interaction dominates this "hadronization," which in these diagrams is represented by the dark oval where the quark–antiquark pair (i.e., the three quarks) bind into a palette of bound states (hadrons). However, the electromagnetic part of the interaction may be separated as the interaction between the lepton and the individual (anti)quarks. Thus, computations of the amplitudes from the previous sections may be adapted also to these collisions, but the kinematic part of the analysis is significantly more complicated. Herein, we consider only the part of this analysis that is determined by the symmetries and general requirements.

5.4.1 Hadron production in electron–positron annihilation

Even with enough energy for the final collision results to include many hadrons, the electromagnetic part of the process (5.204) primarily reduces to transforming a $\ell\bar{\ell}$-pair, by way of a virtual photon, into a $q\bar{q}$-pair. That quark and antiquark then decay into lighter quarks, emit gluons and so produce a palette of various hadrons. This second stage of the process contains all the complications from strong interactions.

To describe the first stage, let m be the lepton and antilepton mass, M the mass of the produced quark and antiquark, and Q the electric charge of the quark (so $-Q$ is the electric charge of the antiquark) in units of elementary electric charge, e, so that $Q(u) = +\frac{2}{3}$, $Q(d) = -\frac{1}{3}$, etc. Adapting the second term in the result (5.147b) we have

$$\mathfrak{M}_{\ell\bar{\ell}\to q\bar{q}} = \frac{Q\,g_e^2}{(\mathrm{p}_1+\mathrm{p}_2)^2}[\overline{v}_2\gamma^\mu u_1][\overline{U}_3\gamma_\mu V_4], \tag{5.206}$$

where u_1, $\overline{v}_2$, $\overline{U}_3$ and V_4 are Dirac spinors for the incoming lepton and antilepton, and outgoing quark and antiquark, respectively. Averaging as in (5.132)–(5.137) we obtain

$$\begin{aligned}\left\langle|\mathfrak{M}_{\ell\bar{\ell}\to q\bar{q}}|^2\right\rangle &= \frac{1}{4}\Big(\frac{Q g_e^2}{(\mathrm{p}_1+\mathrm{p}_2)^2}\Big)^2 \operatorname{Tr}\left[\gamma^\mu(\not{p}_1+mc)\gamma^\nu(\not{p}_2-mc)\right]\\ &\qquad\times \operatorname{Tr}\left[\gamma^\mu(\not{p}_4-Mc)\gamma^\nu(\not{p}_3+Mc)\right]\\ &= 8\Big(\frac{Q g_e^2}{(\mathrm{p}_1+\mathrm{p}_2)^2}\Big)^2\left[(\mathrm{p}_1\cdot\mathrm{p}_3)(\mathrm{p}_2\cdot\mathrm{p}_4)+(\mathrm{p}_1\cdot\mathrm{p}_4)(\mathrm{p}_2\cdot\mathrm{p}_3)+2(mc)^2(Mc)^2\right.\\ &\qquad\left.+(mc)^2(\mathrm{p}_3\cdot\mathrm{p}_4)+(Mc)^2(\mathrm{p}_1\cdot\mathrm{p}_2)\right]\\ &= Q^2 g_e^4\Big\{1+\Big(\frac{mc^2}{E}\Big)^2+\Big(\frac{Mc^2}{E}\Big)^2+\Big[1-\Big(\frac{mc^2}{E}\Big)^2\Big]\Big[1-\Big(\frac{Mc^2}{E}\Big)^2\Big]\cos^2\theta\Big\},\end{aligned} \tag{5.207}$$

where E is the energy of the incoming lepton in the CM system and θ the angle between the incoming lepton and the outgoing quark. Treating the outgoing quark–antiquark as if they were free particles, the differential effective cross-section is given by the relation (3.127), whereupon angular integration yields

$$\sigma = \frac{\pi}{3}\Big(\frac{Q\hbar c\alpha}{E}\Big)^2\sqrt{\frac{1-(Mc^2/E)^2}{1-(mc^2/E)^2}}\Big[1+\frac{1}{2}\Big(\frac{mc^2}{E}\Big)^2\Big]\Big[1+\frac{1}{2}\Big(\frac{Mc^2}{E}\Big)^2\Big]. \tag{5.208}$$

For energies below Mc^2, the effective cross-section becomes imaginary, i.e., the process is kinematically forbidden: $E < Mc^2$ is not enough energy to produce a quark–antiquark pair of mass M each. In turn, if $E > Mc^2 \gg mc^2$, expanding the square-roots and multiplying the factors yields

$$\sigma = \frac{\pi}{3}\Big(\frac{Q\hbar c\alpha}{E}\Big)^2 F(m,M,E), \tag{5.209a}$$

$$\begin{aligned}F(m,M,E) &= \Re e\sqrt{\frac{1-(Mc^2/E)^2}{1-(mc^2/E)^2}}\Big[1+\frac{1}{2}\Big(\frac{mc^2}{E}\Big)^2\Big]\Big[1+\frac{1}{2}\Big(\frac{Mc^2}{E}\Big)^2\Big]\\ &\approx \Re e\Big[1+\Big(\frac{mc^2}{E}\Big)^2+\frac{5}{8}\Big(\frac{mc^2}{E}\Big)^4+\cdots\Big]\Big[1-\frac{3}{8}\Big(\frac{Mc^2}{E}\Big)^4+\cdots\Big].\end{aligned} \tag{5.209b}$$

In typical experiments $mc^2 \lll E$, so that the first factor in $F(m,M,E)$ is negligibly different from 1. The second factor, however, gives a significant contribution when the energy suffices to produce a quark of mass M but is not much larger than Mc^2. The behavior of the step-like function $F(m,M,E)$ near a threshold $E \sim Mc^2$, where the approximating condition $E > Mc^2$ of the expansion (5.209b) is not satisfied, is shown in Figure 5.2.

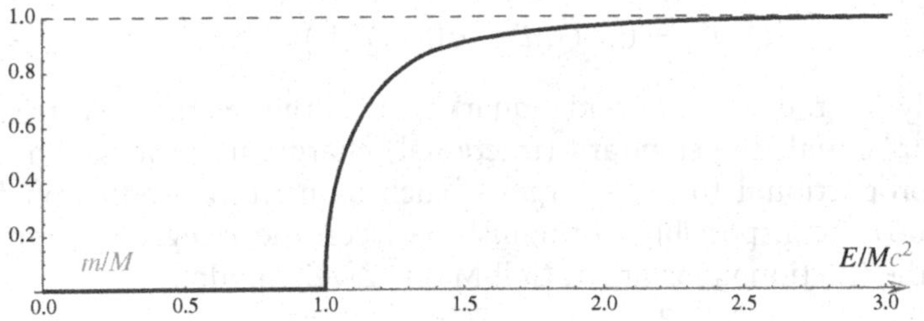

Figure 5.2 A sketch of the function $\Re e\left[F(m,M,E)\right]$ near the value $E \sim Mc^2$.

As the collision energy is increased, heavier and heavier quarks may be produced in the process. These quarks are in fact virtual particles, in the sense that they cannot be observed directly in the detectors, since they decay, emit gluons and finally bind into hadrons – dominated by strong interactions, and so very fast, $\sim 10^{-23}$ s. To avoid the need for estimating the details of this "hadronization," consider the ratio

$$R(E) := \frac{\sigma(e^- + e^+ \to \text{hadrons})}{\sigma(e^- + e^+ \to \mu^- + \mu^+)} \approx \Big[3\sum_i Q_i^2\Big]_{M_i < E/c^2}. \tag{5.210}$$

Here the universal factors such as $\frac{\pi}{3}(\frac{\hbar c\alpha}{E})^2$ cancel, and the contributions to the function $F(m, M_i, E)$ from individually created quarks give only small corrections to the result given. As a function of the collision energy, $R(E)$ is step-like,[22] increasing suddenly when the energy E reaches a threshold to produce a new quark, and is approximately constant between these thresholds. For example,

$$R(E) \approx 3[(\tfrac{2}{3})^2 + (-\tfrac{1}{3})^2] = \tfrac{5}{3}, \qquad E \leqslant M_{u,d}c^2, \tag{5.211a}$$
$$R(E) \approx 3[(\tfrac{2}{3})^2 + (-\tfrac{1}{3})^2 + (-\tfrac{1}{3})^2] = 2, \qquad E \leqslant M_s c^2, \tag{5.211b}$$
$$R(E) \approx 3[(\tfrac{2}{3})^2 + (-\tfrac{1}{3})^2 + (-\tfrac{1}{3})^2 + (\tfrac{2}{3})^2] = \tfrac{10}{3}, \qquad E \leqslant M_c c^2, \tag{5.211c}$$

and so forth. At energies between $M_c c^2 \approx 1{,}270$ MeV and about 2,000 MeV, a significant discrepancy from this simple form of $R(E)$ showed. However, it was soon discovered that this was due to resonance effects related to the production of the third (and mostly unexpected) electrically charged lepton ($m_\tau = 1{,}784\,\text{MeV}/c^2$), which decays mostly into hadrons. Considering the simplicity of the approximation and when the τ-lepton contributions are correctly accounted for, the agreement of the simple relation (5.210) with experiments is very good.

Besides, the overall factor of 3 in the relation (5.210) stems from the fact that every quark has three colors, i.e., that for every mass and electric charge there actually exist three quarks – one of each color.

Conclusion 5.6 *The very good agreement of the simple approximation (5.210) with experiments is then the direct experimental proof of the existence of color – or at least the fact that every quark exists in (otherwise unexplained) triplicate.*

Digression 5.14 At least one curious Student asked why do quarks have to have fractional electric charges, as they have been standardly assigned since 1974–5.

The original model by Han and Nambu indeed proposed quarks that were to have integral charges, dependent on color. For example, we may choose, following result (5.206),

$$Q(u^r) = +1, \;\; Q(u^y) = +1, \;\; Q(u^b) = 0,$$
$$Q(d^r) = 0, \;\; Q(d^y) = 0, \;\; Q(d^b) = -1, \tag{5.212a}$$

and similarly for the s-, c-, b- and t-quarks. Since the average electric charge for each quark "flavor" equals the standard (fractional) charge, no process where the amplitude is linearly proportional to the charges – such as hadron production from lepton collisions (5.206) – can possibly distinguish between the integrally charged Han–Nambu model and the fractionally charged Gell-Mann–Zweig model.

[22] The shape of these "steps" is described by the function $F(m, M, E)$, but it also exhibits resonant effects in the form of very narrow peaks of large intensity $\gg 1$, immediately above the $E = Mc^2$ threshold, which is not shown in the sketch in Figure 5.2 on p. 213 and the analysis of which is omitted herein.

However, a process that depends on the *square* of the electric charge – such as (5.185), which contributes to the renormalization and so ultimately to the dependence of the interaction intensity on the interaction energy (5.197c) – can distinguish the integrally charged quarks from the fractionally charged quarks. Indeed, the "corrections" under the square-root symbol are in the result (5.197c) given for a single electrically charged particle – the electron – since we assumed that the virtual particle in the closed loop in the diagram (5.185) is in fact the electron. However, with energies μc^2, one must add the contributions from all the quarks and charged leptons with masses not larger than μ, and which are proportional to the cumulative factor

$$\Big[\sum_i Q_i^2 \ln\big(\tfrac{\mu}{m_i}\big)\Big]_{m_i\leqslant\mu}, \tag{5.212b}$$

which grows differently for integrally charged quarks than for fractionally charged ones, and which give (one possible) experimentally measurable difference. Similarly, the results (5.211) would differ quantitatively for integrally charged quarks:

$$\widetilde{R}(E)\big|_{E\leqslant M_{u,d}c^2} \approx 3, \quad \widetilde{R}(E)\big|_{E\leqslant M_s c^2} \approx 4, \quad \widetilde{R}(E)\big|_{E\leqslant M_c c^2} \approx 6, \quad \text{etc.} \tag{5.212c}$$

In the early 1970s, such comparisons with experiments confirmed the fractional electric charges of the quarks in the model of Gell-Mann and Zweig; see also Footnote 25 on p. 220.

5.4.2 The electrodynamics contribution in lepton–hadron scattering

A lepton–hadron collision occurs, to a first approximation, between the lepton and one of the (anti)quarks in the hadron, and by way of exchanging a single photon. Of course, when the hadron is a baryon, one must sum over all three quarks in the baryon, and if the hadron is a meson, one must sum the contributions from the interaction with the quark and with the antiquark. In this process, the strong nuclear force field, which keeps the (anti)quark state bound, receives part of the 4-momentum transfer, but this usually produces minor corrections to this initial approximation.

Elastic lepton–hadron scattering

If the proton were a point-like spin-$\frac{1}{2}$ Dirac spinor with no additional structure, relations (5.131) and (5.137) would be valid for *elastic* collisions

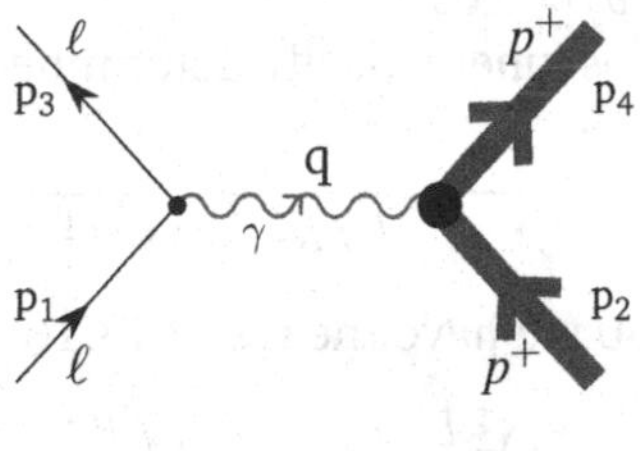

(5.213)

with only the small change, $m_\mu \to M := m_p$ in the function $X_{\mu\nu}(2,4;\mu^-) \to K_{\mu\nu}(2,4;p^+)$:

$$\Big\langle |\mathfrak{M}_{\ell p\to\ell p}|^2 \Big\rangle = \frac{g_e^4}{(\mathrm{p}_1-\mathrm{p}_3)^4} X^{\mu\nu}(1,3;\ell)\, K_{\mu\nu}(2,4;p^+), \tag{5.214}$$

$$\begin{aligned} X^{\mu\nu}(1,3;\ell) &= \mathrm{Tr}\left[\gamma^\mu\,(\not{p}_1 + m_\ell c\mathbb{1})\,\overline{\gamma}^\nu\,(\not{p}_3 + m_\ell c\mathbb{1})\right] \\ &= 2\big[p_1^\mu p_3^\nu + p_1^\nu p_3^\mu + \eta^{\mu\nu}[m_\ell^2 c^2 - (\mathrm{p}_1\cdot\mathrm{p}_3)]\big]. \end{aligned} \tag{5.215}$$

The tensors $X^{\mu\nu}$ and $X_{\mu\nu}$ were computed for point-like (elementary) spin-$\frac{1}{2}$ Dirac spinors, but protons are not point-like (elementary) and their structure causes deviations from the results (5.136)–(5.140), which produce the function $K_{\mu\nu}(2,4;p^+)$. These deviations reflect the effects of strong interaction that bind the quarks into the proton. A description of that structure in quantum chromodynamics [☞ Section 6.1] is too complex for a serious analysis here. However, we do know that $K_{\mu\nu}(2,4;p^+)$ is a rank-2 tensor, and may depend only on the 4-vectors p_2, p_4 and $\mathrm{q} := (\mathrm{p}_1 - \mathrm{p}_3) = (\mathrm{p}_4 - \mathrm{p}_2)$ and, of course, the metric $\eta_{\mu\nu}$. Following tradition, we'll use the following 4-vectors: incoming probe 4-momentum, $\mathrm{p} \equiv \mathrm{p}_2$, and the transfer 4-momentum, q, and write $\mathrm{p}_4 = (\mathrm{p}_2 + \mathrm{q})$. It is therefore possible to parametrize this corrected tensor as

$$K_{\mu\nu}(2,4;p^+) = -K_1\,\eta_{\mu\nu} + \frac{K_2}{M^2c^2}\,p_\mu p_\nu + \frac{K_4}{M^2c^2}\,q_\mu q_\nu + \frac{K_5}{M^2c^2}\,(p_\mu q_\nu + q_\mu p_\nu), \tag{5.216}$$

where K_i are functions of the only scalar variable,[23] $\mathrm{q}^2 = (\mathrm{p}_4 - \mathrm{p}_2)^2$. Since $X^{\mu\nu}(1,3;\ell)$ is a symmetric tensor (5.139a), the antisymmetric part in $K_{\mu\nu}(2,4;p^+)$ – if it even exists – does not contribute to the expression (5.214). This restricts the expansion (5.216) to be symmetric with respect to the $\mu \leftrightarrow \nu$ exchange.

Next, it may be shown [243, p. 277] and [257, Sections 8.2–8.3] that $q^\mu K_{\mu\nu} = 0$, so that [☞ Exercise 5.4.1]

$$K_4 = \frac{M^2c^2}{\mathrm{q}^2} K_1 + \tfrac{1}{4} K_2 \qquad \text{and} \qquad K_5 = \tfrac{1}{2} K_2. \tag{5.217}$$

Thus, $K^{\mu\nu}$ may be parametrized by only two form-factors:

$$K^{\mu\nu}(2,4;p^+) = -K_1(\mathrm{q}^2)\Big(\eta^{\mu\nu} - \frac{q^\mu q^\nu}{\mathrm{q}^2}\Big) + \frac{K_2(\mathrm{q}^2)}{M^2c^2}\Big(p^\mu + \tfrac{1}{2}q^\mu\Big)\Big(p^\nu + \tfrac{1}{2}q^\nu\Big). \tag{5.218}$$

Combining the results (5.214), (5.215) and (5.218), we arrive at

$$\Big\langle |\mathfrak{M}_{\ell p \to \ell p}|^2 \Big\rangle = \Big(\frac{2g_e{}^2}{\mathrm{q}^2}\Big)^2 \Big[K_1[(\mathrm{p}_1\cdot\mathrm{p}_3) - 2m_\ell^2c^2] + K_2\Big(\frac{(\mathrm{p}_1\cdot\mathrm{p})(\mathrm{p}_3\cdot\mathrm{p})}{M^2c^2} + \frac{\mathrm{q}^2}{4}\Big)\Big] \tag{5.219}$$

$$\approx \frac{g_e{}^4c^2}{4EE'\sin^4(\theta/2)}\Big(2K_1\sin^2\Big(\frac{\theta}{2}\Big) + K_2\cos^2\Big(\frac{\theta}{2}\Big)\Big) \tag{5.220}$$

where we switched to the lab frame, where the proton is initially at rest, $\mathrm{p} = (Mc,0,0,0)$, the lepton has initial energy E and is deflected, with energy E', at an angle θ from its initial direction of motion. We have also assumed that $E, E' \gg m_\ell c^2$, and have approximated $m_\ell \approx 0$. Then $\mathrm{p}_1 = E(1,\hat{p}_i)/c$, $\mathrm{p}_3 = E(1,\hat{p}_f)/c$, and $\hat{p}_i\cdot\hat{p}_f = \cos\theta$.

The outgoing lepton energy E' is kinematically determined:

$$E' = \frac{E}{1 + (2E/Mc^2)\sin^2(\frac{\theta}{2})}. \tag{5.221}$$

Besides, in the approximation $m_\ell \approx 0$ we have the result (3.160), and so

$$\frac{d\sigma}{d\Omega} = \Big(\frac{\alpha\hbar}{4ME\sin^2(\frac{\theta}{2})}\Big)^2 \frac{E'}{E}\Big(2K_1\sin^2\Big(\frac{\theta}{2}\Big) + K_2\cos^2\Big(\frac{\theta}{2}\Big)\Big)$$

$$= \Big(\frac{\alpha\hbar}{4ME\sin^2\big(\frac{\theta}{2}\big)}\Big)^2 \frac{2K_1\sin^2(\frac{\theta}{2}) + K_2\cos^2(\frac{\theta}{2})}{1 + (2E/Mc^2)\sin^2(\frac{\theta}{2})}, \tag{5.222}$$

[23] Indeed, $\mathrm{p}_2^2 = \mathrm{p}_4^2 = M^2c^2$ is a constant, and $\mathrm{q}\cdot\mathrm{p}_2 = -\frac{1}{2}\mathrm{q}^2$. Also, K_3 is the standard notation for the term that appears in the analysis of neutrino–proton collisions, but not for electrically charged leptons.

which is the so-called (Marshall Nicholas) Rosenbluth formula from 1950. By measuring the angular dependence of the electrons scattered *elastically* on initially stationary protons (or heavy positive ions), one determines experimentally the form-factors $K_1(\mathrm{q}^2)$ and $K_2(\mathrm{q}^2)$, also known as "structure functions."

Deep inelastic (light) lepton–hadron scattering

In the case of the *inelastic* collisions

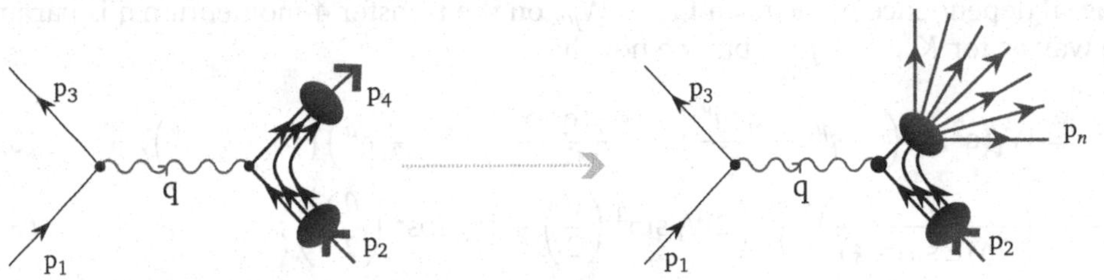

(5.223)

especially where more than one hadron emerges from the collision, the analysis must be adapted more thoroughly. Fermi's golden rule [☞ p. 113] yields

$$d\sigma = \frac{\hbar^2\left\langle|\mathfrak{M}|^2_{\ell p\to\ell X}\right\rangle}{4\sqrt{(\mathrm{p}_1\cdot\mathrm{p}_2)-(m_1m_2c^2)^2}}\prod_{i=3}^{n}\Big(\frac{c\,d^3\vec{p}_i}{(2\pi)^3 2E_i}\Big)(2\pi)^4\delta^4\Big(\mathrm{p}_1+\mathrm{p}_2-\sum_{j=3}^{n}\mathrm{p}_j\Big), \tag{5.224a}$$

$$\left\langle|\mathfrak{M}|^2_{\ell p\to\ell X}\right\rangle = \frac{g_e^4}{\mathrm{q}^4}X^{\mu\nu}(1,3;\ell\text{ (lepton)})K_{\mu\nu}(2,4;X\text{ (hadrons)}). \tag{5.224b}$$

If collisions of this type are taken inclusively and we only measure the deflection angle of the scattered lepton and its energy (and so effectively know p_3), the result (5.224) must be summed over all possible hadron results and their momenta, so we have

$$d\sigma = \frac{4\pi M\hbar^2 g_e^4\,X^{\mu\nu}(1,3;\ell)}{4\mathrm{q}^4\sqrt{(\mathrm{p}_1\cdot\mathrm{p}_2)^2-(m_1m_2c^2)^2}}\Big(\frac{c\,d^3\vec{p}_3}{(2\pi)^3 2E_3}\Big)W_{\mu\nu}, \tag{5.225}$$

$$W_{\mu\nu} := \frac{1}{4\pi M}\sum_X\int\cdots\int\prod_{i=4}^{n}\Big(\frac{c\,d^3\vec{p}_i}{(2\pi)^3 2E_i}\Big)(2\pi)^4\delta^4\Big(\mathrm{p}_1+\mathrm{p}_2-\sum_{j=3}^{n}\mathrm{p}_j\Big)K_{\mu\nu}(2,4;X). \tag{5.226}$$

For an initially stationary proton, $\mathrm{p}_2 \equiv \mathrm{p} = (-Mc,\vec{0})$ and the incoming lepton energy E, we have $\mathrm{p}_1 = (-E/c,\vec{p}_i)$. Therefore,

$$(\mathrm{p}_1\cdot\mathrm{p}_2) = ME, \quad\Rightarrow\quad \sqrt{(\mathrm{p}_1\cdot\mathrm{p}_2)^2-(m_1m_2c^2)^2} = \sqrt{M^2(E^2-m_\ell^2c^4)} \approx ME \tag{5.227}$$

since, in typical experiments of this type and with $\ell = e^\pm$, we have $m_\ell \ll E/c^2$. We therefore approximate $m_\ell \approx 0$, so that $\mathrm{p}_1 = E(-1,\hat{p}_i)/c$ and $\mathrm{p}_3 = E'(-1,\hat{p}_f)/c$. Then,

$$d^3\vec{p}_f = |\vec{p}_f|^2 d|\vec{p}_f|d\Omega \approx c^{-3}(E')^2\,dE'd\Omega, \tag{5.228}$$

and

$$\frac{d\sigma}{dE'\,d\Omega} = \Big(\frac{\alpha\hbar}{c\mathrm{q}^2}\Big)^2\frac{E'}{E}X^{\mu\nu}(1,3;\ell)\,W_{\mu\nu}. \tag{5.229}$$

Unlike in elastic collisions, $\mathrm{p}_{\text{tot}} = \sum_{i=4}^{n}\mathrm{p}_i$ with $n > 4$ in inelastic collisions where multiple hadrons emerge, so $\mathrm{p}_{\text{tot}}^2 \neq M^2c^2$ [☞ Tables C.7 on p. 529 and C.8 on p. 529]. There then exists no relation like (5.221) between E' and E,θ for inelastic collisions; E' is independent of E and θ. The result (5.229) then provides the differential effective cross-section in the span of outgoing lepton energies $[E', E'+dE']$, as is reasonable for a free and continuous variable E'.

The second consequence of $p_{tot}^2 \neq M^2c^2$ is that also $q{\cdot}p \neq -q^2/2$, and one defines the variable

$$x := -\frac{q^2}{2q{\cdot}p}, \quad x \in [0,1]. \tag{5.230}$$

The general dependence of the form-factor $W_{\mu\nu}$ on the transfer 4-momentum q is parametrized the same way as for $K_{\mu\nu}(2,4;p^+)$, but we now have

$$W^{\mu\nu} = W_1(q^2,x)\Big(-\eta^{\mu\nu} + \frac{q^\mu q^\nu}{q^2}\Big) + \frac{W_2(q^2,x)}{M^2c^2}\big(p^\mu + \tfrac{1}{2x}q^\mu\big)\big(p^\nu + \tfrac{1}{2x}q^\nu\big), \tag{5.231}$$

$$\frac{d\sigma}{dE'd\Omega} = \Big(\frac{\alpha\hbar}{2ME\sin^2(\frac{\theta}{2})}\Big)^2 \frac{E'}{E}\Big(2W_1\sin^2\Big(\frac{\theta}{2}\Big) + W_2\cos^2\Big(\frac{\theta}{2}\Big)\Big). \tag{5.232}$$

The Rosenbluth formula (5.222) is the special case obtained by substituting

$$W_i(q^2,x) = -\frac{K_i(q^2)}{2Mq^2}\,\delta(x-1), \qquad i = 1,2. \tag{5.233}$$

Note that the $\delta(x-1)$ factor not only formally fixes $x = -q^2/2q{\cdot}p \overset{!}{=} 1$, but also implies the relation (5.221).

Finally, the Rosenbluth formula (5.222) may further be specialized to an ideally point-like (elementary) proton by substituting

$$K_1 \to -q^2 \quad \text{and} \quad K_2 \to 4M^2c^2. \tag{5.234}$$

This idealization is not a bad approximation when the electron energy is sufficiently small and the electron does not come too close to the proton ($\sim 10^{-15}$ m), so that the proton internal structure has negligible influence on the scattering.

Experimental verification of the parton model

For the elastic collision $A + B \to A' + B'$, we have in the lab frame

$$p_1 = (-E/c, \vec{p}_i), \quad p_2 = (-Mc, \vec{0}), \quad p_3 = (-E'/c, \vec{p}_f), \quad p_4 = (-E''/c, \vec{P}_f). \tag{5.235}$$

Then, with $m_A = m_{A'} = m$ and $m_B = m_{B'} = M$,

$$q = (p_1 - p_3) = \big((E'-E)/c, (\vec{p}_i - \vec{p}_f)\big) = (p_4 - p_2) = \big(Mc - E''/c, \vec{P}_f\big), \tag{5.236}$$

$$q{\cdot}p_2 = M(E-E'), \tag{5.237}$$

$$q^2 \approx -4\frac{EE'}{c^2}\sin^2\Big(\frac{\theta}{2}\Big) \qquad \text{when} \qquad mc^2 \ll E, E'. \tag{5.238}$$

Note that q^2/c^2 is proportional to the Mandelstam variable t [☞ definitions (3.62)].

In the late 1960s, James Bjorken computed within the quark model that the expressions

$$F_1(x) := M\,W_1(q^2,x) \qquad \text{and} \qquad F_2(x) := \frac{-q^2}{2Mc^2x}W_2(q^2,x) \tag{5.239}$$

become asymptotically independent of q^2 at very high energies (5.238). Here the values of both $|q^2|$ and $|q{\cdot}p|$ are large (as compared to mc^2), but their ratio (5.230) remains small; $x \in [0,1]$.

This asymptotic independence from the magnitude of the 4-momentum transfer, $\sqrt{\mathrm{q}^2}$, is called "Bjorken scaling," and was soon confirmed in deep inelastic collisions, mostly of electrons and protons. May it suffice here to mention that this phenomenon confirms that in the deep inelastic collisions the 4-momentum transfer is mostly to one of the three quarks, that these quarks are much smaller than the proton and that they may be treated as point-like particles.

In 1969, Curtis Callan and David Gross proved an additional relationship between Bjorken's functions:

$$2x\,F_1(x) = F_2(x). \tag{5.240}$$

This relation was also quickly confirmed experimentally. Suffice it to say, this relation between $F_1(x)$ and $F_2(x)$ depends on the quark spins, and the Callan–Gross relation (5.240) indicates that quarks have spin $\frac{1}{2}$.

Digression 5.15 Both relations (5.239) and (5.240) may be derived [☞ Ref. [243], str. 271–277] by treating the quarks as point-like particles, and writing $f_i(x)$ for the probability that the ith quark receives the xth fraction of the 4-momentum transfer. Using the assumptions (5.233) and (5.234), write

$$W_1^i := \frac{Q_i^2}{2m_i}\delta(x_i{-}1), \quad W_2^i := -\frac{2m_i c^2 Q_i^2}{\mathrm{q}^2}\delta(x_i{-}1), \quad x_i := -\frac{\mathrm{q}^2}{2\mathrm{q}{\cdot}\mathrm{p}_i}, \tag{5.241a}$$

where m_i is the mass of the ith quark, and p_i its 4-momentum. Since the quarks *mostly* move together as the proton, suppose that

$$\mathrm{p}_i =: z_i\,\mathrm{p} \tag{5.241b}$$

is the 4-momentum of the ith quark, and equals the z_ith fraction of the 4-momentum of the whole proton. It follows that

$$\mathrm{p}_i^{\,2} = m_i^{\,2}c^2 \quad\text{and}\quad \mathrm{p}^2 = M^2c^2 \quad\Rightarrow\quad m_i = z_i M. \tag{5.241c}$$

So, if z_i varies depending on the dynamics within the proton as a bound state of three quarks, then so do the *effective* quark masses.[24] Relation (5.241b) implies that $x_i = x/z_i$, so that

$$W_1^i = \frac{Q_i^2}{2M}\,\delta(x{-}z_i) \quad\text{and}\quad W_2^i = -\frac{2Mc^2x^2Q_i^2}{\mathrm{q}^2}\,\delta\big(x{-}z_i\big). \tag{5.241d}$$

From this, we have

$$W_1 = \sum_i \int_0^1 \mathrm{d}z_i\; W_1^i = \frac{1}{2M}\sum_i Q_i^{\,2} f_i(x) \quad\Rightarrow\quad F_1(x) = \frac{1}{2}\sum_i Q_i^{\,2} f_i(x), \tag{5.241e}$$

$$W_2 = \sum_i \int_0^1 \mathrm{d}z_i\; W_2^i = -\frac{2Mc^2}{\mathrm{q}^2}x^2\sum_i Q_i^{\,2} f_i(x) \quad\Rightarrow\quad F_2(x) = x\sum_i Q_i^{\,2} f_i(x), \tag{5.241f}$$

which agrees with Bjorken's assertion that $F_1(x)$ and $F_2(x)$ are independent of q^2. Besides, the results (5.241e)–(5.241f) clearly also imply the Callan–Gross relation (5.240).

---

[24] This definition of the masses intuitively takes into account that quarks are bound within the proton, whereby their inertia, i.e., their response to interaction, differs from what it would be were they free particles.

Besides the facts that:

1. Bjorken (asymptotic) independence of the functions (5.239) from the magnitude of the 4-momentum transfer, $\sqrt{q^2}$, provides experimental confirmation that the proton consists of three electrically charged point-like (elementary) "ingredients,"
2. the Callan–Gross relation (5.240) – also experimentally confirmed – indicates that these "ingredients" have spin $\frac{1}{2}$,

the analysis in Digression 5.15 also leads to the experimental confirmation of the existence of so-called gluons, the particles that mediate the strong nuclear interaction and bind the quarks into bound states, the hadrons.

Indeed, the relation (5.241c) is too naive: quarks are bound within the hadron and their parameters are not directly measurable, whereby the quarks within the hadrons (and free ones do not exist!) are virtual particles. As such, they need not satisfy the equations of motion, i.e., they are not on the mass shell. However, the *average* fraction of 4-momentum carried by the u-quarks would have to be about twice larger than the average fraction of 4-momentum carried by the d-quarks simply because there are twice as many u-quarks as d-quarks in the proton, and their masses are approximately the same. Therefore,

$$\int_0^1 \mathrm{d}x\; x\, f_u(x) = 2\int_0^1 \mathrm{d}x\; x\, f_d(x). \tag{5.242}$$

Using the result (5.241f) as a first approximation, we obtain

$$F_2(x) \approx x\big[(\tfrac{2}{3})^2 f_u(x) + (-\tfrac{1}{3})^2 f_d(x)\big] \quad \overset{(5.242)}{\Longrightarrow} \quad \int_0^1 \mathrm{d}x\; F_2(x) \approx \int_0^1 \mathrm{d}x\; x f_d(x). \tag{5.243}$$

However, the *measured* average values of the form-factor $F_2(x)$ give

$$\int_0^1 \mathrm{d}x\; x f_d(x) \approx 0.18 \quad \text{and} \quad \int_0^1 \mathrm{d}x\; x f_u(x) \approx 0.36. \tag{5.244}$$

The sum of these average values – the fraction of the 4-momentum carried, on average, by *either one of the two u-quarks or the d-quark* – adds up to 0.54. In other words, this analysis indicates that the proton must also contain some electrically neutral "ingredients," which carry 46% of the transfer 4-momentum.

Quantum chromodynamics gives a much better estimate for the form-factors W_1 and W_2, and then also of Bjorken's functions $F_1(x)$ and $F_2(x)$, and thereby also the probabilities $f_u(x)$ and $f_d(x)$. However, the essence of the conclusion remains unchanged: A non-negligible fraction of the transfer 4-momentum is not carried by the quarks, but by electrically neutral "ingredients" of the proton. These "ingredients" of the proton must interact with the quarks by means of the strong nuclear interaction simply because that is the strongest type of interaction and, being electrically neutral, they cannot interact electromagnetically. The transfer 4-momentum is thus equally fast and uniformly shared among the three quarks, as well as these electrically neutral "ingredients."

On the other hand, the strong nuclear interaction may be described analogously to electromagnetic interactions – by means of a mediating particle. Analogously to the exchange of photons in electromagnetic interactions, strong nuclear interactions are mediated by *gluons*. As the strong nuclear interaction is independent of the electromagnetic interaction and the electric charge of the particles that interact with strong nuclear interactions, it follows that gluons must be electromagnetically neutral. This then permits the electrically neutral "ingredients" of the proton that carry about 46% of the transfer 4-momentum to be identified with gluons[25] [☞ Section 6.1].

[25] This additionally [☞ Digression 5.14 on p. 214] rules out the Han–Nambu model with integrally charged quarks: As those charge assignments are color-dependent, the gluons of the Han–Nambu model must be electrically charged and would also have to interact electromagnetically.

Detailed estimates based on quantum chromodynamics indeed provide a very good agreement with experimental data, and this then is the third significant result of deep inelastic scattering: Besides the experimental confirmation that the proton consists of three point-like spin-$\frac{1}{2}$ electrically charged quarks with charges $Q_u = \frac{2}{3}$ and $Q_d = -\frac{1}{3}$, deep inelastic scattering also confirms experimentally the existence of electromagnetically neutral gluons, which interact with quarks by means of the strong nuclear interaction.

5.4.3 Exercises for Section 5.4

✎ **5.4.1** *Derive the relation (5.217), using that* $\mathrm{q} := (\mathrm{p}_1 - \mathrm{p}_3) = (\mathrm{p}_4 - \mathrm{p}_2)$ *and* $\mathrm{p}_2^2 = \mathrm{p}_4^2 = -M^2c^2$*, as well as that* $\mathrm{p} := \mathrm{p}_2$ *and* q *are two linearly independent 4-vectors.* (*Hint: it should prove useful to first prove that* $\mathrm{q}^2 = -2\,\mathrm{q}\cdot\mathrm{p}$.)

✎ **5.4.2** *Compare the result (5.229) with the Rutherford (5.160), Mott (5.159) and Rosenbluth (5.222) formulae, as well as the limiting (5.234) case of the latter, and its reduction under the additional condition* $|\mathrm{q}^2| \ll M^2c^2$*. Exhibit the hierarchy of approximations (and their physical meaning) that relate these results.*

✎ **5.4.3** *Derive equation (5.221).*

✎ **5.4.4** *Derive equation (5.241d).*

6

Non-abelian gauge symmetries and interactions

The previous chapter showed how the fact that the phase of the electron wave-function is not an observable quantity leads to the concept of gauge symmetry, which in turn introduces the gauge potentials, and which then provides the basic framework for describing gauge interactions. The chapter before that showed that the classification of mesons and hadrons in the quark model uncovers that quarks have an additional degree of freedom – dubbed *color*, and the corresponding symmetry with the structure of the $SU(3)_c$ group. Since the physical states that may be detected must be "colorless," i.e., $SU(3)_c$-invariant, it follows that the color of any individual quark cannot be detected either, and so can be changed arbitrarily. This arbitrariness – as a function of space and time! – of the color change in quarks while maintaining the hadron composed of those quarks "colorless" is the essence of the so-called gauge principle. When applied to the local changes in the (matrix-valued) phases of wave-functions, the resulting theories are called "Yang–Mills theories"; Chapter 9 will show that the application of the same idea to local changes of parametrization of the spacetime itself leads to Einstein's theory of gravity.

6.1 The gauge symmetry of color

The first non-abelian (non-commutative) gauge theory was proposed by Oskar Klein in 1938, but that proposal was too early and remained undeveloped, unapplied, and forgotten. Non-abelian gauge theories were taken seriously only after 1954–5, after the publication of the works of C.-N. Yang and R. L. Mills, and independently, R. Shaw's dissertation mentored by A. Salam. However, both of the (currently) well-known applications, the $SU(2)_w$ theory of weak nuclear interactions and the $SU(3)_c$ theory of strong nuclear interactions, required several decades of development and novel ideas for a general acceptance and contemporary formulation of these theories, as well as their embedding in the "Standard Model."

The next few sections consider the $SU(3)_c$ gauge theory as a theoretical system for describing strong nuclear interactions. Chapter 7 will focus on the $SU(2)_w$ gauge theory of weak nuclear interactions, the $SU(2)_w \times U(1)_Q$ theory of electroweak interactions and finally the "Standard Model," based on the $SU(3)_c \times SU(2)_w \times U(1)_Q$ gauge theory. Generally, all gauge theories based on a group of symmetries that act by local changes of some generalized phases [☞ e.g., relation (6.2)] are called "Yang–Mills" gauge theories.

6.1.1 The $SU(3)_c$ gauge symmetry and gluons

Section 2.3.13 [☞ discussion on p. 61] showed that quarks have an additional 3-dimensional degree of freedom called "color." That is, the wave-function of any quark is a superposition

$$\boldsymbol{\Upsilon}_n(\mathrm{x}) = \hat{\mathrm{e}}_\alpha \Psi_n^\alpha(\mathrm{x}) = \hat{\mathrm{e}}_\mathrm{r} \Psi_n^\mathrm{r}(\mathrm{x}) + \hat{\mathrm{e}}_\mathrm{y} \Psi_n^\mathrm{y}(\mathrm{x}) + \hat{\mathrm{e}}_\mathrm{b} \Psi_n^\mathrm{b}(\mathrm{x}) = \begin{bmatrix} \Psi_n^\mathrm{r}(\mathrm{x}) \\ \Psi_n^\mathrm{y}(\mathrm{x}) \\ \Psi_n^\mathrm{b}(\mathrm{x}) \end{bmatrix}, \quad n = u, d, s, c, b, t. \tag{6.1}$$

This matrix representation of the quark wave-functions makes it evident that a local change of the phase (5.14b) of quark wave-functions, in general, becomes

$$\boldsymbol{\Upsilon}_n(\mathrm{x}) \to e^{ig_c \boldsymbol{\varphi}(\mathrm{x})/\hbar}\, \boldsymbol{\Upsilon}_n(\mathrm{x}), \qquad \boldsymbol{\varphi}(\mathrm{x}) := \varphi^a(\mathrm{x})\, Q_a, \tag{6.2}$$

where Q_a, $j = 1, \ldots, 8$ are eight 3×3 matrices that generate the $SU(3)_c$ *gauge group* [☞ Appendix A.4], and equation (6.2) is the *gauge transformation*. This $SU(3)_c$ symmetry is exact, and must not be confused with the approximate $SU(3)_f$ symmetry discussed in Section 4.4; the group structure of $SU(3)_c$ is identical with that of $SU(3)_f$ but the application is quite different. One usually uses the Gell-Mann matrices (A.71) although, of course, any other basis of Hermitian 3×3 traceless matrices serves just as well.

Digression 6.1 The non-abelian analogue of the simple formal argument in Digression 5.1 on p. 166 shows that the relation (6.2) changes the state operator for a quark by a similarity transformation, rather than leaving it invariant as in the abelian case of Digression 5.1 on p. 166. It is thus not as obvious that all generalized, non-abelian phase transformations (6.2) should be symmetries. Nevertheless, the physical motivation for requiring the transformation (6.2) to be a symmetry remains – and not just because quarks and their particular color states are not directly observable; see Conclusion 11.8 on p. 444.

By the way, the ninth linearly independent matrix generator is proportional to the unit 3×3 matrix and simply produces an overall, diagonal, phase-change

$$\boldsymbol{\Upsilon}_n(\mathrm{x}) \to e^{ig_c \varphi^0(\mathrm{x}) \mathbb{1} Q/\hbar}\, \boldsymbol{\Upsilon}_i(\mathrm{x}) = e^{iq_n (g_c/g_e) \varphi^0(\mathrm{x})/\hbar}\, \boldsymbol{\Upsilon}_n(\mathrm{x}), \tag{6.3}$$

which *looks* like the transformation (5.14b). Here, $\boldsymbol{\Upsilon}_n(\mathrm{x})$ is the eigenfunction of the operator Q, of which the eigenvalue equals the electric charge of the quark q_n represented by $\boldsymbol{\Upsilon}_n(\mathrm{x})$.[1] This provides essentially the same representation of the gauge transformation of the electrodynamics and chromodynamics interaction. It is clear that $\mathbb{1}$ commutes with all Q_a, whereby the matrices $\{\mathbb{1}; Q_1, \ldots, Q_8\}$ generate the $U(1) \times SU(3)_c$ group – except that the first factor cannot be identified with the gauge symmetry of electrodynamics straightforwardly because of the difference in the magnitudes of the respective charges and the corresponding factor (g_c/g_e) in the exponent (6.3). In addition, one *implicitly* considers the phase-change (6.2) to be limited to quarks, whereby the

[1] Unfortunately, the letter q is standardly used for charge, for the transfer 4-momentum, and for the general symbol-synonym for "quark." Herein, the 4-vector is denoted by q, and the electric charge of the quark q_n is denoted by q_n – the eigenvalue of the operator Q of the eigenfunctions that are identified with the quark q_n.

transformation (6.3) would correspond to an interaction that is limited to the hadrons and excludes leptons. Although that is a perfectly consistent possibility, such an interaction does not exist in Nature.

Akin to the analysis (5.5)–(5.7), replace $\partial_\mu \to D_\mu$ in the Dirac equation (5.34), and require that with respect to the gauge transformation (6.2) this new Dirac equation should remain unchanged:

$$[i\hbar \not{D} - mc]\mathbf{\Upsilon}_n(x) = 0 \;\to\; [i\hbar \not{D}' - mc]\mathbf{\Upsilon}'_n(x) = 0, \tag{6.4a}$$
$$D_\mu \;\to\; D'_\mu := U_{\boldsymbol{\varphi}}\, D_\mu\, U_{\boldsymbol{\varphi}}^{-1}, \tag{6.4b}$$
$$U_{\boldsymbol{\varphi}} := e^{ig_c \boldsymbol{\varphi}/\hbar}, \tag{6.4c}$$

where the matrix representation (6.1) is understood, so that $\not{D} = \gamma^\mu D_\mu$ acts as a *double* matrix derivative operator: both as a 4×4-matrix upon the spinor components (because of the $\boldsymbol{\gamma}$-matrices) and as a 3×3-matrix upon colors:

$$\not{D}\mathbf{\Upsilon}_n \equiv \boldsymbol{\gamma}^\mu D_\mu \mathbf{\Upsilon}_n, \quad \text{i.e.,} \quad (\not{D}\mathbf{\Upsilon}_n)^\alpha \equiv \boldsymbol{\gamma}^\mu D_{\mu}{}^{\alpha}_{\beta}\, \mathbf{\Upsilon}^\beta_n, \quad \text{i.e.,} \quad (\not{D}\mathbf{\Upsilon}_n)^{\alpha A} \equiv (\gamma^\mu)^A{}_B D_{\mu}{}^{\alpha}_{\beta}\, \Upsilon^{\beta B}_n, \tag{6.5}$$

where repeated α, β indices are summed over colors, red–yellow–blue. The indices A, B, which indicate the Dirac spinor components (5.51), have been written out explicitly only in the third version (6.5).

As in the procedure (5.5)–(5.14a), one finds that

$$D_\mu := \mathbb{1}\partial_\mu + \tfrac{ig_c}{\hbar c} A^a_\mu\, Q_a \qquad \text{so that} \qquad \not{D}'(e^{ig_c\boldsymbol{\varphi}(x)/\hbar}\,\mathbf{\Upsilon}_n) = e^{ig_c\boldsymbol{\varphi}(x)/\hbar}\,(\not{D}_\mu \mathbf{\Upsilon}_n), \tag{6.6a}$$
$$A'^a_\mu Q_a = A^a_\mu\, U_{\boldsymbol{\varphi}}\, Q_a U_{\boldsymbol{\varphi}}^{-1} + \tfrac{\hbar c}{ig_c} U_{\boldsymbol{\varphi}}(\partial_\mu U_{\boldsymbol{\varphi}}^{-1}) = A^a_\mu\, U_{\boldsymbol{\varphi}}\, Q_a U_{\boldsymbol{\varphi}}^{-1} - c(\partial_\mu \varphi^a) Q_a, \tag{6.6b}$$
$$\text{i.e.,} \qquad \mathbb{A}'_\mu = U_{\boldsymbol{\varphi}}\, \mathbb{A}_\mu U_{\boldsymbol{\varphi}}^{-1} - c(\partial_\mu \boldsymbol{\varphi}), \qquad \mathbb{A}_\mu := A^a_\mu Q_a \text{ and } \boldsymbol{\varphi} := \varphi^a Q_a, \tag{6.6c}$$

where Q_a are 3×3 Hermitian matrices that close the $SU(3)_c$ algebra:

$$[Q_a, Q_b] = i\, f_{ab}{}^c\, Q_c. \tag{6.6d}$$

Since the single electric charge operator Q in electrodynamics is here replaced by eight operators Q_a, it follows that the photon 4-vector gauge potential $A_\mu(x)$ must be replaced by eight *gluon* 4-vector gauge potentials, $A^a_\mu(x)$, $a = 1, \ldots, 8$. Of course, the parameter of electromagnetic interaction, g_e, is also replaced by the parameter of chromodynamic interactions, g_c.

Linearization of equation (6.6b) – the first-order expansion in $\varphi^a(x)$ – produces the gauge transformation of the gluon 4-vector gauge potentials:

$$\delta A^a_\mu = -(D_\mu\, \boldsymbol{\varphi})^a := -c(\partial_\mu \varphi^a) + \tfrac{ig_c}{\hbar c} A^b_\mu\, (\widetilde{Q}_b)_c{}^a\, \varphi^c = -(\partial_\mu \varphi^a) - \tfrac{g_c}{\hbar c} A^b_\mu\, f_{bc}{}^a\, \varphi^c, \tag{6.6e}$$
$$(\widetilde{Q}_b)_c{}^a = i\, f_{bc}{}^a, \qquad \text{where} \qquad [\widetilde{Q}_b, \widetilde{Q}_c] = i\, f_{bc}{}^a\, \widetilde{Q}_a. \tag{6.6f}$$

Here, $\widetilde{Q}_a$ are Hermitian 8×8 traceless matrices that close the same $SU(3)_c$ algebra as the 3×3 matrices Q_a, i.e., the $\frac{1}{2}\boldsymbol{\lambda}_j$'s in relations (A.71). Thus, the operators Q_a – and, in particular, the matrices $\frac{1}{2}\boldsymbol{\lambda}_a$ – provide a 3-dimensional representation of the $SU(3)_c$ group, i.e., a matrix action of the $SU(3)_c$ group upon the 3-dimensional vector space (6.1). In turn, $\widetilde{Q}_a$ provide an 8-dimensional representation of the $SU(3)_c$ group – a matrix action of the $SU(3)_c$ group upon the 8-dimensional vector space $\{\boldsymbol{\varphi}(x) := \varphi^a(x)\, Q_a\}$.[2]

[2] To be precise, this matrix-valued function $\boldsymbol{\varphi}(x)$ represents a vector space in every spacetime point x; their union forms a so-called *vector bundle* over spacetime. Gauge theories are therefore properly described by the geometry of vector bundles and their connections, here represented by the matrix 4-vector gauge potentials $\mathbb{A}_\mu$.

Comment 6.1 *In the general case, when the G-covariant derivative D_μ acts upon functions $f_{\mathcal{A}}(\mathrm{x})$ that span the d-dimensional representation of the group G, so $\mathcal{A} = 1, \ldots, d$, we have that*

$$(D_\mu f(\mathrm{x}))_{\mathcal{A}} := (\partial_\mu f_{\mathcal{A}}(\mathrm{x})) + \tfrac{ig_G}{\hbar c} A^a_\mu(\mathrm{x})\, [Q_a]_{\mathcal{A}}{}^{\mathcal{B}} f_{\mathcal{B}}(\mathrm{x}), \tag{6.7}$$

where the $d \times d$ matrices Q_a generate the group G, the 4-vectors $A^a_\mu(\mathrm{x})$ are the gauge potentials, and g_G is the magnitude of the charge of the corresponding gauge interaction.

Comparing with electrodynamics, recall that the transformation operators U_φ commute and imply the result (5.89)

$$D'_\mu = U_\varphi\, D_\mu\, U_\varphi^{-1} \quad \Rightarrow \qquad A'_\mu = A_\mu - c(\partial_\mu \lambda). \tag{6.8}$$

For the non-abelian group of chromodynamics, $SU(3)_c$, we have (expanding equation (6.6c) only to first order in φ^a)

$$D'_\mu = U_{\boldsymbol{\varphi}}\, D_\mu\, U_{\boldsymbol{\varphi}}^{-1} \quad \Rightarrow \quad (A')^a_\mu = A^a_\mu - c(D_\mu \varphi^a) = A^a_\mu - c(\partial_\mu \varphi^a) + \tfrac{g_c}{\hbar} A^b_\mu\, f_{bc}{}^a\, \varphi^c. \tag{6.9}$$

Also, in electrodynamics we have

$$F_{\mu\nu}(A') = F_{\mu\nu}(A), \tag{6.10}$$

because the fields $\vec{E}, \vec{B}$ are *invariant* with respect to the action of the electromagnetic $U(1)$ gauge transformation (5.14a) [☞ discussion of the definitions (5.15)]. In the non-abelian case, however, direct computation shows that

$$(\partial_\mu (A')^a_\nu - \partial_\nu (A')^a_\mu) \neq (\partial_\mu A^a_\nu - \partial_\nu A^a_\mu), \tag{6.11}$$

and even

$$(\partial_\mu (A')^a_\nu - \partial_\nu (A')^a_\mu) \neq U_{\boldsymbol{\varphi}}(\partial_\mu A^a_\nu - \partial_\nu A^a_\mu) U_{\boldsymbol{\varphi}}^{-1}. \tag{6.12}$$

Note, however, that both in electrodynamics and in chromodynamics the derivatives D_μ are by definition covariant:

$$U(1):\ D'_\mu = U_\varphi\, D_\mu\, U_\varphi^{-1}, \qquad SU(3):\ D'_\mu = U_{\boldsymbol{\varphi}}\, D_\mu\, U_{\boldsymbol{\varphi}}^{-1}, \tag{6.13}$$

that is, the change by means of a similarity transformation. It then follows that arbitrary (operatorial) polynomials in the D_μ's are also covariant. Finally, in electrodynamics we have that

$$[\, D_\mu\, ,\, D_\nu\,] = [\, \partial_\mu + \tfrac{iq}{\hbar c} A_\mu\, ,\, \partial_\nu + \tfrac{iq}{\hbar c} A_\nu\,] = + \tfrac{iq}{\hbar c}(\partial_\mu A_\nu - \partial_\nu A_\mu) = \tfrac{iq}{\hbar c} F_{\mu\nu}. \tag{6.14}$$

This result provides an interpretation of the fields $\vec{E}, \vec{B}$ (components of the $F_{\mu\nu}$ tensor) as curvatures in the geometry followed by electrically charged particles.[3] Indeed, in the presence of an electromagnetic field, electrically charged particles move in trajectories of which the curvature is determined by the fields $\vec{E}, \vec{B}$, i.e., the components of the $F_{\mu\nu}$ tensor. Nudged by this result, and the fact that all formal operatorial functions of $SU(3)_c$-covariant derivatives D_μ will also be $SU(3)_c$-covariant, we define

$$\begin{aligned} \mathbb{F}_{\mu\nu} &:= \tfrac{\hbar c}{ig_c}[\, D_\mu\, ,\, D_\nu\,] = \tfrac{\hbar c}{ig_c}[\, \partial_\mu + \tfrac{ig_c}{\hbar c} A^b_\mu\, Q_b\, ,\, \partial_\nu + \tfrac{ig_c}{\hbar c} A^c_\nu\, Q_c\,] \\ &= (\partial_\mu A^a_\nu - \partial_\nu A^a_\mu) Q_a + \tfrac{\hbar c}{ig_c}(\tfrac{ig_c}{\hbar c})^2\, A^b_\mu\, A^c_\nu\, [Q_b, Q_c] \;=\; F^a_{\mu\nu} Q_a, \end{aligned} \tag{6.15a}$$

$$F^a_{\mu\nu} := (\partial_\mu A^a_\nu - \partial_\nu A^a_\mu) - \tfrac{g_c}{\hbar c}\, f^a{}_{bc}\, A^b_\mu\, A^c_\nu, \tag{6.15b}$$

where we used the defining relation (A.70) of the $SU(3)$ group generators.

[3] Many a Student may find this interpretation unusual. However, Chapter 9 about gravity will, hopefully, clarify: The commutator of G-covariant derivatives provides the curvature stemming from G-gauge symmetry.

Comment 6.2 *The interaction parameter, g_c, is, in the literature, often absorbed by redefining the gluon 4-vector gauge potential, $g_c A^a_\mu \mapsto A^a_\mu$, for visibility and ease of computing. In final expressions, however, factors of g_c must be returned for comparison with experiments.*

It follows that this *matrix* $\mathbb{F}_{\mu\nu}$ transforms covariantly, as expected:

$$\begin{aligned}\mathbb{F}_{\mu\nu} \;\to\; \mathbb{F}'_{\mu\nu} &:= \tfrac{i\hbar c}{g_c}\big[\,D'_\mu\,,\,D'_\nu\,\big] = \tfrac{i\hbar c}{g_c}\big[\,U_{\boldsymbol{\varphi}}D_\mu U^{-1}_{\boldsymbol{\varphi}}\,,\,U_{\boldsymbol{\varphi}}D_\nu U^{-1}_{\boldsymbol{\varphi}}\,\big] = \tfrac{i\hbar c}{g_c}U_{\boldsymbol{\varphi}}\big[\,D_\mu\,,\,D_\nu\,\big]U^{-1}_{\boldsymbol{\varphi}}\\ &= U_{\boldsymbol{\varphi}}\,\mathbb{F}_{\mu\nu}U^{-1}_{\boldsymbol{\varphi}}.\end{aligned} \tag{6.16}$$

Thus, the $\mathbb{F}_{\mu\nu}$ tensor is in general covariant – but not invariant – with respect to the action of non-abelian (non-commutative) symmetries such as $SU(3)_c$. Only the field tensor of an abelian symmetry is invariant with respect to the action of this symmetry, as is the case with the electromagnetic field tensor $F_{\mu\nu}$, which is invariant with respect to the $U(1)_Q$ symmetry.

The relation (6.15) also implies that these matrix-represented covariant operators D_μ act upon other matrix-represented quantities by means of commutation. This implies that the covariant derivative of the gauge field, $\mathbb{F}_{\mu\nu}$, itself equals

$$D_\mu(\mathbb{F}_{\nu\rho}) = [D_\mu, \mathbb{F}_{\mu\nu}] = \tfrac{\hbar c}{i g_c}\big[\,D_\mu\,,\big[\,D_\nu\,,\,D_\rho\,\big]\big]. \tag{6.17}$$

Using the so-called Jacobi identity,

$$\big[\,A\,,\big[\,B\,,\,C\,\big]\big] + \big[\,B\,,\big[\,C\,,\,A\,\big]\big] + \big[\,C\,,\big[\,A\,,\,B\,\big]\big] \equiv 0, \tag{6.18}$$

the relation (6.17) implies that

$$\varepsilon^{\mu\nu\rho\sigma}D_\mu(\mathbb{F}_{\nu\rho}) = \tfrac{\hbar c}{i g_c}\varepsilon^{\mu\nu\rho\sigma}\big[\,D_\mu\,,\big[\,D_\nu\,,\,D_\rho\,\big]\big] = 0, \tag{6.19}$$

which generalizes the Bianchi identity (5.87) for electrodynamics.

6.1.2 The Lagrangian density for chromodynamics

Since the $SU(3)$ generators are Hermitian traceless matrices, $\mathrm{Tr}[Q_a] = 0$, it is also true that

$$\mathrm{Tr}[\mathbb{F}_{\mu\nu}] = F^a_{\mu\nu}\,\mathrm{Tr}[Q_a] = 0. \tag{6.20}$$

However, as the trace of a product of two (Hermitian or not) traceless matrices need not be zero, there is no group-theoretical reason for $\mathrm{Tr}[\mathbb{F}_{\mu\nu}\mathbb{F}^{\mu\nu}]$ to vanish. In turn, the "trace" function is invariant with respect to similarity transformations of its argument:

$$\mathrm{Tr}[\mathbb{X}] \;\to\; \mathrm{Tr}[\mathbb{S}\,\mathbb{X}\,\mathbb{S}^{-1}] = \mathrm{Tr}[\mathbb{X}\,\mathbb{S}^{-1}\,\mathbb{S}] = \mathrm{Tr}[\mathbb{X}]. \tag{6.21}$$

It follows that

$$\begin{aligned}\mathrm{Tr}[\mathbb{F}_{\mu\nu}\mathbb{F}^{\mu\nu}] \;\to\; \mathrm{Tr}[\mathbb{F}'_{\mu\nu}\mathbb{F}'^{\mu\nu}] &= \mathrm{Tr}[U_{\boldsymbol{\varphi}}\mathbb{F}_{\mu\nu}U^{-1}_{\boldsymbol{\varphi}}U_{\boldsymbol{\varphi}}\mathbb{F}^{\mu\nu}U^{-1}_{\boldsymbol{\varphi}}] = \mathrm{Tr}[\mathbb{F}_{\mu\nu}\mathbb{F}^{\mu\nu}U^{-1}_{\boldsymbol{\varphi}}\,U_{\boldsymbol{\varphi}}]\\ &= \mathrm{Tr}[\mathbb{F}_{\mu\nu}\mathbb{F}^{\mu\nu}]\end{aligned} \tag{6.22}$$

is invariant with respect to $SU(3)_c$ transformations. Up to a suitably chosen sign and coefficient [☞ Exercises 5.1.3 and 5.1.4 on p. 171], this then provides a Lagrangian density for $SU(3)_c$ gluons, analogous to the Lagrangian density for photons (5.76). Lorentz-invariance is evident since $\mathbb{F}_{\mu\nu}$ is a rank-2 tensor, and $\mathbb{F}_{\mu\nu}\mathbb{F}^{\mu\nu}$ is a scalar contraction, just as in electrodynamics [☞ Digression 5.7 on p. 183].

A Lagrangian density that, via Hamilton's principle of minimal action, produces the Dirac equation for quarks that interact with gluons is obtained by direct generalization of the Lagrangian density (5.118). That is, the $\Psi(x)$ representing an electron is replaced by $\Psi^{\alpha}_{n}(x)$, which represents the nth species (flavor) of quark and of the color α. Using the notation (6.5), we then have

$$\begin{aligned}\mathscr{L}_{QCD} &= \sum_n \mathrm{Tr}\left[\overline{\mathbf{\Psi}}_n(x)\left[i\hbar c\,\not{D} - m_n c^2\right]\mathbf{\Psi}_n(x)\right] - \tfrac{1}{4}\,\mathrm{Tr}\left[\mathbb{F}_{\mu\nu}\mathbb{F}^{\mu\nu}\right]\\ &= \sum_n \overline{\Psi}_{\alpha\, n}(x)\left[i\gamma^\mu\left(\hbar c\,\delta^\alpha_\beta\partial_\mu + ig_c\,A^a_\mu(\tfrac{1}{2}\lambda_a)^\alpha{}_\beta\right) - m_n c^2\delta^\alpha_\beta\right]\Psi^\beta_n(x) - \tfrac{1}{4}F^a_{\mu\nu}F^{\mu\nu}_a.\end{aligned} \tag{6.23}$$

As in QED, variation by A^a_μ yields the equations of motion akin to Gauss's law:

$$D_\mu F^{a\,\mu\nu} = g_c\sum_n \overline{\Psi}_{n\alpha A}(\gamma^\nu)^A{}_B(\tfrac{1}{2}\lambda^a)^\alpha{}_\beta\Psi^{\beta B}_n, \tag{6.24}$$

where the right-hand side expression may be identified as the quark contribution to the color current:

$$j^{a\,\mu}_{(q)} := g_c\sum_n \overline{\Psi}_{n\alpha A}(\gamma^\mu)^A{}_B(\tfrac{1}{2}\lambda^a)^\alpha{}_\beta\Psi^{\beta B}_n. \tag{6.25}$$

However, the left-hand side of equation (6.24) contains terms nonlinear in A^a_μ, by which this differs fundamentally from equation (5.78). For example, in electrodynamics it is true that

$$\partial_\nu j^\nu_e = \frac{4\pi\epsilon_0 c}{4\pi}\partial_\nu\partial_\mu F^{\mu\nu} \equiv 0, \qquad \text{since} \qquad F_{\mu\nu} = -F_{\nu\mu}, \tag{6.26}$$

which then immediately produces the continuity equation, i.e., charge conservation. For the equations of motion (6.24) this argument is not true:

$$\partial_\nu j^{a\,\nu}_{(q)} = \partial_\nu D_\mu F^{a\,\mu\nu} \neq 0. \tag{6.27}$$

Digression 6.2 The quark current $j^{a\,\nu}_{(q)}$ does satisfy a gauge-covariant version of the continuity equation:

$$D_\nu\, j^{a\,\nu}_{(q)} \overset{(6.24)}{=} D_\nu D_\mu F^{a\,\mu\nu} = -\tfrac{1}{2}[D_\mu, D_\nu]F^{a\,\mu\nu} \overset{(6.15)}{=} -\tfrac{1}{2}f^a{}_{bc}F^b_{\mu\nu}F^{c\,\mu\nu} = 0, \tag{6.28a}$$

where we used the definition of the quark current (6.25), that $D_\nu D_\mu F^{a\,\mu\nu} = -D_\mu D_\nu F^{a\,\mu\nu}$ because of the antisymmetry $F^{a\,\mu\nu} = -F^{a\,\nu\mu}$, as well as that $f^a{}_{bc} = -f^a{}_{cb}$ is antisymmetric with respect to the exchange $b \leftrightarrow c$, whereas $F^a_{\mu\nu}F^{b\,\mu\nu} = F^{a\,\mu\nu}F^b_{\mu\nu} = F^b_{\mu\nu}F^{a\,\mu\nu}$ is symmetric. However, the equation (6.28a) does not imply (purely quark) color conservation; following the computation (2.67) now produces

$$\begin{aligned} 0 = D_\mu j^{a\,\mu}_{(q)} &= \partial_\mu j^{a\,\mu}_{(q)} - \tfrac{g_c}{\hbar c}f^a{}_{bc}\,A^b_\mu\, j^{c\,\mu}_{(q)}\\ \Rightarrow\quad \frac{\mathrm{d}}{\mathrm{d}t}\Big(\int_V \mathrm{d}^3\vec{r}\, j^{a\,0}_{(q)}\Big) &= -\oint_{\partial V}\mathrm{d}^2\vec{r}\cdot\vec{j}^a_{(q)} + \tfrac{g_c}{\hbar c}f^a{}_{bc}\Big(\int_V \mathrm{d}^3\vec{r}\, A^b_\mu\, j^{c\,\mu}_{(q)}\Big),\end{aligned} \tag{6.28b}$$

where the additional right-hand side term does not simplify and certainly does not vanish in general.

However, by the example of (6.6e)–(6.6f),

$$D_\mu F^{a\,\mu\nu} = \partial_\mu F^{a\,\mu\nu} - \tfrac{g_c}{\hbar c}f_{bc}{}^a\,A^b_\mu F^{c\,\mu\nu}, \tag{6.29}$$

and moving the second term, $-\frac{g_c}{\hbar c} f_{bc}{}^a A^b_\mu F^{c\,\mu\nu}$, from the so-written left-hand side of the relation (6.24) to its right-hand side, we obtain

$$D_\mu F^{a\,\mu\nu} = j^{a\,\nu}_{(q)} \quad \Rightarrow \quad \partial_\mu F^{a\,\mu\nu} = J^{a\,\nu}_{(c)} \quad \Rightarrow \quad \partial_\nu J^{a\,\nu}_{(c)} = 0, \tag{6.30}$$

since $F^{a\,\mu\nu} = -F^{a\,\nu\mu}$ but $\partial_\nu\partial_\mu = +\partial_\mu\partial_\nu$. Here,

$$J^{a\,\nu}_{(c)} := j^{a\,\mu}_{(q)} + \tfrac{g_c}{\hbar c} f_{bc}{}^a A^b_\mu F^{c\,\mu\nu}, \tag{6.31}$$

$$Q^a_{(c)} := \int \mathrm{d}^3\vec{r}\, J^{a\,0}_{(c)} = g_c \int \mathrm{d}^3\vec{r}\,\Big(\sum_n [\overline{\Psi}_n \gamma^\mu \tfrac{1}{2}\lambda^a \Psi_n] + \tfrac{1}{\hbar c} f_{bc}{}^a A^b_\mu F^{c\,\mu\nu}\Big) \tag{6.32}$$

are, respectively, the chromodynamical (gauge) current density for which the continuity equation, i.e., color charge conservation, holds, and the corresponding chromodynamics (gauge) charge $Q^a_{(c)}$ that is conserved in time according to Noether's theorem.

> **Conclusion 6.1** *The continuity equation for the (chromodynamics) current (6.31), i.e., the conservation law for the (chromodynamical "color") charge (6.32) is guaranteed by the antisymmetry of the tensor (of chromodynamics) fields (6.15). This conclusion holds for all gauge theories.*

In contrast to this qualitative and conceptual similarity in all gauge theories, the specific results (6.15)–(6.32) also indicate two fundamental differences in comparison with electromagnetism:

> **Conclusion 6.2** *The chromodynamics (non-abelian gauge) field tensor $F^a_{\mu\nu}$ is nonlinear in the gluon 4-vector potentials A^a_μ. By contrast, the electromagnetic field tensor $F_{\mu\nu}$ is a linear function of the photon 4-vector potential.*

> **Conclusion 6.3** *The chromodynamics (non-abelian gauge) current (6.31) and corresponding charge (6.32) have contributions both from quarks and from gluons! By contrast, photons have no electromagnetic charge and do not contribute to the electric current.*

Example 6.1 For illustration, consider the $SU(3)_c$-covariant equations of motion (6.24) where we fix $\nu = 0$, and where we use equation (6.29):

$$\partial_\mu F^{a\,\mu 0} - \tfrac{g_c}{\hbar c} f^a{}_{bc} A^b_\mu F^{c\,\mu 0} = j^{a\,0}_{(q)}, \qquad a,b,c = 1,\ldots,8. \tag{6.33}$$

Just as in electrodynamics [☞ Section 5.2.2], we define

$$\vec{E}^a := \hat{\mathbf{e}}_i F^{a\,i0}, \qquad \rho^a_{(q)} := j^{a\,0}_{(q)}, \qquad \vec{A}^a := -\hat{\mathbf{e}}^i A^a_i, \tag{6.34}$$

where we are free to absorb all numerical factors in these definitions, and where we may fix $A^a_0 = 0$, $a = 1,\ldots,8$. The equations (6.33) then reduce to

$$\vec{\nabla}\cdot\vec{E}^a = \rho^a_{(q)} - \tfrac{g_c}{\hbar c} f^a{}_{bc} \vec{A}^b\cdot\vec{E}^c, \tag{6.35}$$

where $\rho^a_{(q)}$ is evidently a source of the chromo-electric field $\vec{E}^a$, but where the chromodynamics vector potentials and fields – *of other colors* – themselves contribute to the source! Equation (6.35) is the generalization of Gauss's law for non-commutative (non-abelian) charges – here, of the chromodynamics "color." For example, using the concrete values (A.71) of $f^a{}_{bc}$, we see that

$$\vec{\nabla}\cdot\vec{E}^1 = \rho^1_{(q)} - \tfrac{g_c}{\hbar c}\big(2\vec{A}^{[2}\cdot\vec{E}^{3]} + \vec{A}^{[4}\cdot\vec{E}^{7]} + \vec{A}^{[5}\cdot\vec{E}^{6]}\big), \tag{6.36}$$

where $X^{[a}Y^{b]} := \frac{1}{2}(X^aY^b - X^bY^a)$. That is, the indicated (nonlinear!) coupling of the chromodynamics potentials and fields serves as an *additional* source (or sink, depending on the overall sign) for chromodynamics fields, besides the quark source $\rho^a_{(q)}$.

It follows that the equations of motion for non-abelian gauge theory – obtained by varying the Lagrangian density $\mathscr{L}_{QCD}$ (6.23) with respect to A^a_μ and Ψ^α_n:

1. cannot be expressed without explicit use of the 4-vector gauge potentials, A^k_μ,
2. form a nonlinearly coupled system of differential equations.

Because of Conclusion 6.2 above, the chromodynamics generalizations of the electric and the magnetic fields are rarely used in quantum chromodynamics. Chromodynamics is expressed using the 4-vector potentials A^a_μ, the quanta of which are interpreted as gluons. Conclusion 6.3 indicates a fundamentally larger complexity and technical demand in chromodynamics – and then also the relative simplicity of electrodynamics in comparison with chromodynamics.[4]

With this in mind, the fact that the exploration of chromodynamics is still a very active research field☞ should not come as a surprise. In about four decades, many different approaches in this exploration have been developed from the need to "extract" from this conceptually successful theoretical model concrete quantitative predictions for precise comparison with experiments, but also for better theoretical understanding of the model itself. Among these approaches, may it suffice here to mention three:

Lattice QCD In this approach, the otherwise continuous spacetime is replaced by a lattice of a small spacing. The equations of motion are then solved numerically, and one estimates the forms of those solutions in the limit when the spacetime lattice spacing tends to zero.

Large-*N* QCD Since no experiment can identify any one of the colors in any one real physical process, the contributions to the physical processes must be summed and averaged over all colors. If N is to denote the number of "colors," summing over colors tends to produce factors of N while averaging tends to incur factors of $\frac{1}{N}$. The contributions to the various processes may thus be classified according to the exponent in the overall factor N^ν. Such re-organizing of the computations sometimes permits summing contributions that are all $O(N^\nu)$, albeit from different orders of perturbation as counted by powers of g_s or $\hbar$, and this produces results not derivable otherwise.

QCD strings The original motivation for introducing strings into the physics of elementary particles was the fact that hadrons (mesons and baryons) in collisions at sufficiently high energies show a structure that appears filamentary in a first approximation. The results from the quark model and quantum chromodynamics soon surpassed the precision of this filamentary approximation. However, recent results in the mathematical analysis of superstrings – and

[4] This insight will hopefully not discourage the Students who are already acquainted with this "relative simplicity" of electrodynamics, and so also the "relative simplicity" of the exam problems in that course.

especially the so-called AdS/CFT, i.e., gravitation/gauge duality – led to new methods, the application of which to the original problem of hadronic physics produces new results and new avenues for exploration.

To sum up: chromodynamics exhibits the generalization of the Maxwell equations:

$$D_\mu \mathbb{F}^{\mu\nu} = \mathbb{J}^\nu_{(q)} \qquad \text{and} \qquad \varepsilon^{\mu\nu\rho\sigma} D_\mu(\mathbb{F}_{\nu\rho}) = 0, \tag{6.37}$$

where

$$\mathbb{J}^\nu_{(q)} := g_c \Big(\sum_n \overline{\Psi}_{n\alpha A} (\gamma^\mu)^A{}_B (\tfrac{1}{2}\lambda^a)^\alpha{}_\beta \Psi^{\alpha A}_n \Big) Q_a \tag{6.38}$$

is the quark contribution to the chromodynamics current. This current density, however, does not satisfy the continuity equation and $\int \mathrm{d}^3\vec{r}\, \mathbb{J}^0_{(q)}$ is not a conserved color charge. Instead, there exists a redefinition of the (Gaussian) first half of equations (6.37):

$$\partial_\mu \mathbb{F}^{\mu\nu} = \mathbb{J}^\nu_{(c)}, \qquad \text{where} \qquad \mathbb{J}^\nu_{(c)} := \mathbb{J}^\nu_{(q)} + \tfrac{ig_c}{\hbar c}[\mathbb{A}_\mu, \mathbb{F}^{\mu\nu}], \tag{6.39}$$

so that

$$\partial_\nu \mathbb{J}^\nu_{(c)} = 0, \qquad \text{and so also} \qquad \frac{\mathrm{d}}{\mathrm{d}t}\int_V \mathrm{d}^3\vec{r}\, \mathbb{J}^0_{(c)} = -c \oint_{\partial V} \mathrm{d}^2\vec{\sigma} \cdot \vec{\mathbb{J}}_{(c)}. \tag{6.40}$$

It should be clear that in the application to an abelian (*commutative*) group where $f_{ab}{}^c = 0$, the relations (6.37)–(6.30) reduce to the Maxwell equations, the definition of the electric current density, the continuity equation and the electric charge conservation in electrodynamics, respectively.

6.1.3 Exercises for Section 6.1

✎ ***6.1.1*** *Expanding the $\sigma = 0$ component of the system of equations (6.19), obtain the chromodynamic equivalent of Gauss's law for the chromomagnetic field, and show that the nonlinear coupling of gluons also provides an effective chromomagnetic source term in this equation.*

✎ ***6.1.2*** *Determine the gauge covariance of equation (6.24); prove that the left-hand side and the right-hand side of the equality both transform the same.*

✎ ***6.1.3*** *As in the previous exercise, determine separately the gauge covariance of the left-hand side and the right-hand side of the equality $\partial_\mu F^{a\,\mu\nu} = J^{a\,\nu}_{(c)}$ in the result (6.30), where $J^{a\,\nu}_{(c)}$ is defined by the equation (6.31).*

✎ ***6.1.4*** *As in the previous exercise and using the definition (6.31), determine the gauge covariance of the chromodynamics continuity equation $\partial_\nu J^{a\,\nu}_{(c)} = 0$ in the result (6.30).*

6.2 Concrete calculations

Conceptually, quantum-chromodynamics processes are analyzed in the same way as the quantum-electrodynamics ones, via computations that begin with the rules for Feynman diagrams. Adapting Procedure 5.2 on p. 193, we then have the analogous algorithm. However, QCD computations for diagrams with closed loops require exceptional care, additional rules and even additional, so-called ghost fields with so-called BRST symmetry[5] – which is beyond our present scope. The presently given algorithm therefore suffices only for chromodynamics diagrams with no gluon loops.

Procedure 6.1 *The contribution to the amplitude $\mathfrak{M}$ corresponding to a given Feynman diagram (with no gluon loops) for the chromodynamics processes with quarks, antiquarks and gluons is computed following the algorithm [☞ textbooks [445, 425, 586] for a derivation]:*

1. **Notation**
   (a) *Energy–momentum: Denote incoming and outgoing 4-momenta by* $\mathrm{p}_1, \mathrm{p}_2, \ldots$, *and the spins* $s_1, s_2, \ldots$ *Denote the "internal" 4-momenta (assigned to lines that connect two vertices inside the diagram) by* $\mathrm{q}_1, \mathrm{q}_2, \ldots$
   (b) *Orientation: For a spin-$\frac{1}{2}$ particle, orient the line in the 4-momentum direction, oppositely for antiparticles. Orient external gluon lines in the direction of time (herein, upward). Orient the internal gluon lines arbitrarily, but use the so-chosen orientation consistently.*
   (c) *Polarization: Assign every external line the polarization factor:*

Spin-$\frac{1}{2}$ quark	incoming		$u^s_f \chi^\alpha$	$s =$ spin projection $= \uparrow, \downarrow$	
	outgoing		$\overline{u}_{f,s}\, \chi^\dagger_\alpha$	$\alpha =$ quark color $= r, y, b$; $f =$ quark flavor: $u, d, s, \ldots$	
Spin-$\frac{1}{2}$ antiquark	incoming		$\overline{v}_{f,s}\, \chi^\dagger_\alpha$	($\simeq$ spin-$\frac{1}{2}$ quark, travels backwards in time)	(6.41)
	outgoing		$v^s_f \chi^\alpha$		
Gluon	incoming		$\epsilon^\mu \chi^a$	$\epsilon^\mu p_\mu = 0$ and $\epsilon^0 = 0$	
	outgoing	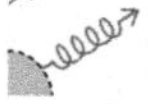	$\epsilon^{\mu *} \chi^{a*}$		

2. **Vertices** *To each vertex assign the factor according to type:*
   (a) *Quark–gluon vertex:*[6]

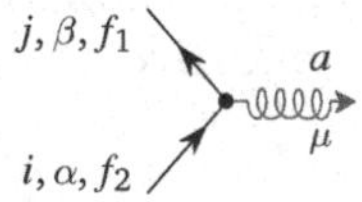

$$\rightarrow \quad -ig_c \gamma^\mu\, \delta^{f_1}_{f_2}\, (\tfrac{1}{2}\lambda_a)^\beta{}_\alpha. \tag{6.42}$$

This factor clearly corresponds to the term $-g_c \overline{\Psi}_\alpha \gamma^\mu A^a_\mu (\lambda_a)^\alpha{}_\beta \Psi^\beta$ *in the Lagrangian (6.23), and represents the elementary gluon–quark interaction.*

   (b) *3-gluon vertex:*

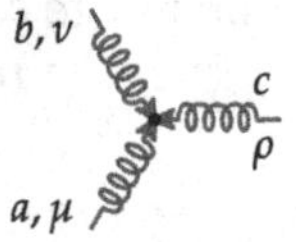

$$\rightarrow \quad -g_c f^{abc}\big[\eta_{\mu\nu}(k_1-k_2)_\rho + \eta_{\nu\rho}(k_2-k_3)_\mu + \eta_{\rho\mu}(k_3-k_1)_\nu\big]. \tag{6.43}$$

---

[5] The name of this symmetry is an acronym from the names of the discoverers: C. M. Becchi, A. Rouet and R. Stora [44], and I. V. Tyutin [526]. The method of using ghost fields itself is usually called after L. Faddeev and V. Popov, the physicists who were among the first to use the method; B. DeWitt, who published very similar ideas at the same time but in a technically much more demanding fashion, is unfortunately almost never cited in the invention of this method. Unlike electrodynamics, the non-abelian nature of the gluon interactions in quantum chromodynamics unavoidably couples all degrees of freedom in the gluon 4-vectors A^a_μ, so that the unphysical degrees of freedom (the longitudinal polarization and the temporal component) cannot be consistently eliminated. However, the *contributions* of these unphysical degrees of freedom may be consistently eliminated by introducing ghost fields and reducing correspondingly the gauge symmetry to the BRST nilpotent symmetry. This level of technical details is beyond the scope of this book; see, e.g., Refs. [425, 123, 586, 316, 277] and especially the texts [268, 555, 484, 496, 589, 590].

[6] In traditional normalization, just as *halves* of Pauli matrices generate the $SU(2)$ group, so do *halves* of Gell-Mann matrices (A.71) generate the $SU(3)$ group.

This factor corresponds to the terms in the Lagrangian (6.23) that are 3-linear in A^a_μ, and represents the elementary interaction of three gluons.

(c) *4-gluon vertex:*

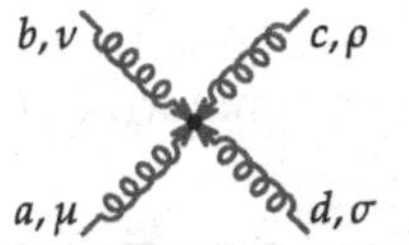

$$\rightarrow \quad -ig_c^2[f^{abe}f^{cd}{}_e(\eta_{\mu\sigma}\eta_{\nu\rho}-\eta_{\mu\rho}\eta_{\nu\sigma}) + f^{ace}f^{db}{}_e(\eta_{\mu\sigma}\eta_{\nu\rho}-\eta_{\mu\nu}\eta_{\rho\sigma}) + f^{ade}f^{bc}{}_e(\eta_{\mu\nu}\eta_{\rho\sigma}-\eta_{\mu\rho}\eta_{\nu\sigma})]. \tag{6.44}$$

This factor corresponds to the terms in the Lagrangian (6.23) that are 4-linear in A^a_μ, and represents the elementary interaction of four gluons.

3. **Propagators** *To each internal line with the jth 4-momentum assign the factor*

quark: 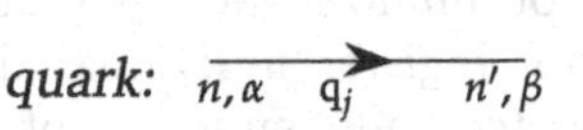

$$\rightarrow \quad \frac{i\delta^{n,n'}\delta^\beta_\alpha}{\not{q}_j - m_j c} = i\delta^{n,n'}\delta^\beta_\alpha\frac{\not{q}_j + m_j c\mathbb{1}}{q_j^2 - m_j^2c^2}, \tag{6.45}$$

gluon: 

$$\rightarrow \quad -i\frac{\eta_{\mu\nu}}{q_g^2}\delta^{ab}. \tag{6.46}$$

As internal lines depict virtual particles, $\not{q}_j \neq m_j c$ and $q_g^2 \neq 0$, respectively [☞ Tables C.7 on p. 529 and C.8 on p. 529]. Up to multiplicative coefficients, these factors also stem from the Lagrangian (6.23); these are Fourier transforms of the Green functions for the differential operators $\not{D}$ and $D^{\mu\nu}_{ab}$ in $\overline{\Psi}\not{D}\Psi := -\sum_n \overline{\Psi}_{n,\alpha}[i\hbar c\not{\partial} - mc^2]\Psi^\alpha_n$ and $A^a_\mu D^{\mu\nu}_{ab}A^b_\nu :\simeq -\frac{1}{4}\mathring{F}^a_{\mu\nu}\mathring{F}_a^{\mu\nu}$, respectively, where $\mathring{F}^a_{\mu\nu} := (\partial_\mu A^a_\nu - \partial_n A^a_\nu)$ is the so-called linearization of the field $F^a_{\mu\nu}$.

4. **Energy–momentum conservation** *To each vertex assign a factor $(2\pi)^4\delta^4(\sum_j k_j)$, where k_j are 4-momenta that enter the vertex. 4-momenta that* **leave** *the vertex have a negative sign except for external spin-$\frac{1}{2}$ antiparticles, since they are equivalent to particles that move backwards in time.*
5. **Integration over 4-momenta** *Internal lines correspond to virtual particles and their 4-momenta are unknown; these variables must be integrated: $\int\frac{d^4q_j}{(2\pi)^4}$.*
6. **Reading off the amplitude** *The foregoing procedure yields the result*

$$-i\mathfrak{M}\,(2\pi)^4\delta^4(\sum_j p_j), \tag{6.47}$$

where the factor $(2\pi)^4\delta^4(\sum_j p_j)$ represents the 4-momentum conservation for the entire process, and where the amplitude (matrix element) $\mathfrak{M}$ is read off.

7. **Fermion loops** *To each fermion loop (closed line) assign a factor −1. A mathematically rigorous derivation of this rule follows from Feynman's approach using path integrals, which is far beyond the scope of this book [☞ booklet [166] for an intuitive albeit not entirely rigorous explanation, [434, Vol. 1, Appendix A] for a serious introduction, and [165] for the original reference].*
8. **Antisymmetrization** *Since the amplitude of the process must be antisymmetric in pairs of identical (external) fermions, the partial amplitudes that differ only in the exchange of two identical external fermions must have the relative sign −1.*

As in Section 3.3.4, one draws all Feynman diagrams that contribute at the desired order in g_c, and then computes the (partial) amplitudes for each of the diagrams. The algebraic sum of these contributions yields the total amplitude, which is then inserted in formulae (3.112) and (3.114) for decays and scatterings, respectively.

Fermion loops (closed lines) will be discussed at the end of this section.

Digression 6.3 Fermion wave-functions, as a whole, must mutually anticommute. When "factorizing" (4.123), $\boldsymbol{\Psi} = \sum_i \Psi_i(\vec{r},t)\,\chi_i(\text{spin})\,\chi_i(\text{flavor})\,\chi_i(\text{color})$, an odd number of factors is anticommuting, and the choice is in principle arbitrary. However, because of the spin-statistics theorem, herein we consistently choose $\chi_i(\text{spin})$ to be anticommuting for half-integral spin, and the other factors to be commutative functions [☞ also Digression 10.2 on p. 360].

Unlike the computations in Sections 3.3.4 and 5.3.1–5.3.3 where the ultimate goal was to compute the lifetime for decays or the differential and total effective cross-section for collisions, for chromodynamics interactions we cannot finalize the computation. Since quarks cannot be extracted from hadrons, detectors cannot register individual quarks, so that, e.g., an elastic collision $u + d \to u + d$ cannot be detected independently of the hadronic bound states of which these quarks are the building blocks. Thus, the (differential) effective cross-section for this collision does not have a physical meaning, as it cannot be compared with experiments.

However, the amplitude for chromodynamics processes does have a physical meaning and may easily be used to compare with concrete experiments, somewhat akin to familiar application of the Wigner–Eckart theorem A.3 on p. 475. That is, in a hadronic process such as the elastic collision

$$[p^+ = (u,u,d)] + [n^0 = (u,d,d)] \to [p^+ = (u,u,d)] + [n^0 = (u,d,d)], \tag{6.48}$$

the dominant, $O(g_s^2)$ contribution is chromodynamical and stems from the quark–quark interaction: the dominant contributions to the amplitude of hadronic processes stem from the interaction of one quark from each of the two hadrons; these contributions then add algebraically, depending on the symmetries of the bound states (6.48). Since the u- and d-quark have approximately the same mass, and may have any of the same spin states ($|\frac{1}{2},+\frac{1}{2}\rangle$ and $|\frac{1}{2},-\frac{1}{2}\rangle$) and any of the same colors (red, yellow, blue), the chromodynamics interaction does not distinguish between u–u, u–d and d–d interactions. For the purely QCD contributions, we have

$$\mathfrak{M}^{(\text{QCD})}_{u+u\to u+u} \approx \mathfrak{M}^{(\text{QCD})}_{u+d\to u+d} \approx \mathfrak{M}^{(\text{QCD})}_{d+d\to d+d} \tag{6.49a}$$

up to $O(\frac{m_u - m_d}{m_u + m_d})$ and $O(g_c^4)$ corrections and up to non-QCD contributions such as were discussed in the computations (4.82)–(4.92). Also,

$$\mathfrak{M}^{(\text{QCD})}_{p^+ + p^+ \to p^+ + p^+} \approx \mathfrak{M}^{(\text{QCD})}_{p^+ + n^0 \to p^+ + n^0} \approx \mathfrak{M}^{(\text{QCD})}_{n^0 + n^0 \to n^0 + n^0}, \tag{6.49b}$$

in the same approximation. The chromodynamic interaction (up to corrections of the next order in magnitude) thus does not differentiate between protons and neutrons, and the result (6.49) is in excellent agreement with concrete experimental data in nuclear physics. Thus, the differences in the binding energy of protons and neutrons within a nucleus may be reduced to differences in spin values,[7] in the spatial factors that also include the orbital angular moments, as well as in the isospin factors [☞ Section 4.3.1].

[7] The (anti)symmetrization is fairly complex in baryons: recall the argument for relations (4.123) and (4.125).

6.2.1 Quark–quark interaction

To describe the interaction between two quarks, assume that they are different, so that only one $O(g_s^2)$ diagram exists:

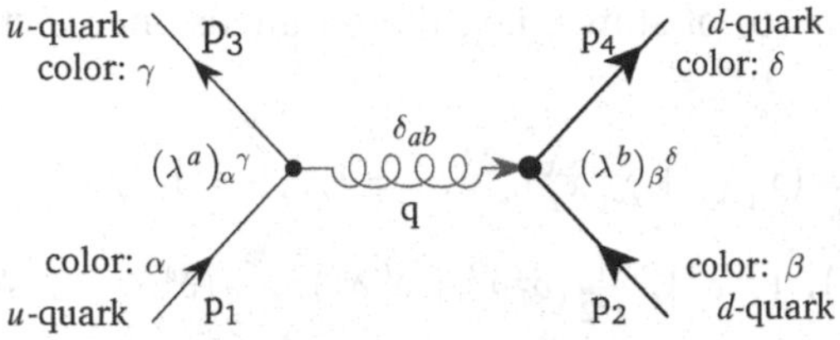

(6.50)

for which the amplitude is obtained following Procedure 6.1 on p. 232,

$$\mathfrak{M}_{u+d\to u+d} = -\frac{g_s^2}{2}\frac{1}{q^2}\left[\overline{u}_3\,\boldsymbol{\gamma}^\mu\,u_1\right]\left[\overline{u}_4\,\boldsymbol{\gamma}_\mu\,u_1\right](\chi_3^\dagger\,\boldsymbol{\lambda}^a\,\chi_1)\,(\chi_4^\dagger\,\boldsymbol{\lambda}_a\,\chi_2), \tag{6.51}$$

which is analogous to amplitude (5.131), except that:

1. g_e is replaced with g_c,
2. the color factor, $f_c(3,4|1,2) = \frac{1}{4}(\chi_3^\dagger\boldsymbol{\lambda}^a\chi_1)(\chi_4^\dagger\boldsymbol{\lambda}_a\chi_2)$, is inserted.

The fact that Feynman calculus from quantum electrodynamics is fairly easy to adapt to quantum chromodynamics as well as other kinds of non-abelian gauge interactions has contributed to the popularity of the technique.[8]

Since the electromagnetic interaction of two charged particles of the type (5.129) is known to lead to the Coulomb potential $\frac{\alpha_e\hbar c}{r} = \frac{1}{4\pi\epsilon_0}\frac{e^2}{r}$ and since the result (6.51) differs from (5.131) only in $g_e \to g_c$ and the inserted factor f_c, we conclude that the quantum-chromodynamics interaction of the type (6.50) also leads to a Coulomb-like potential:

$$V_{qq}(r) = f_c\frac{\alpha_s\hbar c}{r}, \tag{6.52}$$

and it only remains to determine the color factor:

$$f_c(3,4|1,2) = \tfrac{1}{4}(\chi_3^\dagger\boldsymbol{\lambda}^a\chi_1)(\chi_4^\dagger\boldsymbol{\lambda}_a\chi_2) = \tfrac{1}{4}\,\chi_{3\gamma}^\dagger\,\chi_{4\delta}^\dagger\,(\lambda^a)_\alpha{}^\gamma\,(\lambda_a)_\beta{}^\delta\,\chi_1^\alpha\,\chi_2^\beta. \tag{6.53}$$

Use the correspondence between the index- and matrix-notation:

$$\chi^r \leftrightarrow \delta_1^\alpha \leftrightarrow \begin{bmatrix}1\\0\\0\end{bmatrix}, \qquad \chi^y \leftrightarrow \delta_2^\alpha \leftrightarrow \begin{bmatrix}0\\1\\0\end{bmatrix}, \qquad \chi^b \leftrightarrow \delta_3^\alpha \leftrightarrow \begin{bmatrix}0\\0\\1\end{bmatrix}. \tag{6.54}$$

From the $SU(3)$ group representation theory result (A.76a) [☞ Appendix A.4.2], we know that the color factors for two (incoming) quarks, $\chi_1^\alpha\chi_2^\beta$, must belong to one of two vector spaces:

1. The antisymmetric triplet ($\mathbf{3}^*$) of states, i.e., the 3-dimensional vector space spanned by two-quark color factors:

$$\begin{aligned}&\{\chi_{12}{}^{[\alpha\beta]} := \tfrac{1}{\sqrt{2}}(\chi_1^\alpha\chi_2^\beta - \chi_1^\beta\chi_2^\alpha),\ \ \alpha,\beta = red, yellow, blue = 1,2,3\}\\ &= \Big\{\tfrac{1}{\sqrt{2}}(\delta_1^\alpha\delta_2^\beta - \delta_1^\beta\delta_2^\alpha),\ \tfrac{1}{\sqrt{2}}(\delta_1^\alpha\delta_3^\beta - \delta_1^\beta\delta_3^\alpha),\ \tfrac{1}{\sqrt{2}}(\delta_2^\alpha\delta_3^\beta - \delta_2^\beta\delta_3^\alpha)\Big\},\end{aligned} \tag{6.55}$$

[8] The insolubility of quantum chromodynamics stems from the fact that α_s varies with energy much faster than the electrodynamics fine structure parameter, α_e, and oppositely, α_s *diminishes* with energy [☞ Section 6.2.4]. Moreover, perturbative computations indicate that below about 200 MeV, α_s becomes larger than 1, so the perturbative approach to quantum chromodynamics where α_s is the perturbative parameter makes neither practical nor conceptual sense when the interaction energy is less than about 200 MeV [☞ Section 6.3]. Here we then focus on sufficiently high energies.

where the black subscripts 1 and 2 in the first row indicate the first and second quark, respectively. In the second row, we dispensed with these subscripts,[9] so as not to confuse them with the color labels $1 = r$, $2 = y$ and $3 = b$, which were needed in the second row.

2. The symmetric 6-tuplet (**6**) of states, i.e., the 6-dimensional vector space spanned by two-quark color factors:

$$\begin{aligned}\{\chi_{12}{}^{(\alpha\beta)} &:= \tfrac{1}{\sqrt{(1+\delta_{\alpha\beta})}}(\chi_1^\alpha \chi_2^\beta + \chi_1^\beta \chi_2^\alpha),\ \ \alpha,\beta = r,y,b = 1,2,3\} \\ &= \Big\{(\delta_1^\alpha \delta_1^\beta),\ (\delta_2^\alpha \delta_2^\beta),\ (\delta_3^\alpha \delta_3^\beta),\ \tfrac{1}{\sqrt{2}}(\delta_1^\alpha \delta_2^\beta + \delta_1^\beta \delta_2^\alpha),\ \tfrac{1}{\sqrt{2}}(\delta_1^\alpha \delta_3^\beta + \delta_1^\beta \delta_3^\alpha),\ \tfrac{1}{\sqrt{2}}(\delta_2^\alpha \delta_3^\beta + \delta_2^\beta \delta_3^\alpha)\Big\}.\end{aligned} \tag{6.56}$$

The quantum-mechanical normalization[10] of the color factors was used (so that $\|\chi_{12}{}^{[\alpha\beta]}\|^2 = 1$ as well as $\|\chi_{12}{}^{(\alpha\beta)}\|^2 = 1$ for every choice of α,β) and the numerical identification $\alpha,\beta = r,y,b = 1,2,3$ for the basis in which the Gell-Mann matrices (A.71) are given. For outgoing quarks the Hermitian conjugate factors (6.55)–(6.56) must be used, but note that Hermitian conjugation preserves the (anti)symmetry of the two-particle color factors.

That is, in the process (6.50), the color factor for the incoming quarks (with colors α and β) may be in any linear combination of either the antisymmetrized elements (6.55), or the symmetrized elements (6.56). The color factor for the outgoing quarks (with colors γ and δ) may be – independently of the incoming quarks – in any one of the Hermitian conjugates of those states. In principle then, one must compute the color factors (6.53) for each of the combinations

$$f_c(3,4|1,2) \;=\; f_c(\mathbf{3}_A^*|\mathbf{3}_A^*),\ f_c(\mathbf{3}_A^{*\prime}|\mathbf{3}_A^*),\ f_c(\mathbf{6}_S|\mathbf{3}_A^*),\ f_c(\mathbf{3}_A^*|\mathbf{6}_S),\ f_c(\mathbf{6}_S|\mathbf{6}_S),\ f_c(\mathbf{6}_S'|\mathbf{6}_S), \tag{6.57}$$

where $\mathbf{3}_A^*$ denotes some concrete antisymmetrized state, $\mathbf{6}_S$ denotes some concrete symmetrized state, and prime simply indicates some *other* such concrete state.

Example 6.2 A concrete computation of the first type (6.57), i.e., $f_c(\mathbf{3}_A^*|\mathbf{3}_A^*)$ is done taking, e.g., the red–blue $\frac{1}{\sqrt{2}}(\delta_\gamma^1 \delta_\delta^3 - \delta_\delta^1 \delta_\gamma^3) \in \mathbf{3}^*$ element:

$$\begin{aligned}\{\tfrac{1}{4}\,(\chi_{3\gamma}^\dagger \chi_{4\delta}^\dagger)_{\mathbf{3}}\,(\lambda^a)_\alpha{}^\gamma\,(\lambda_a)_\beta{}^\delta\,(\chi_1^\alpha \chi_2^\beta)_{\mathbf{3}^*}\} &\supset \tfrac{1}{4}\,\tfrac{1}{\sqrt{2}}(\delta_\gamma^1 \delta_\delta^3 - \delta_\delta^1 \delta_\gamma^3)\,(\lambda^a)_\alpha{}^\gamma\,(\lambda_a)_\beta{}^\delta\,\tfrac{1}{\sqrt{2}}(\delta_1^\alpha \delta_3^\beta - \delta_1^\beta \delta_3^\alpha) \\ &= \tfrac{1}{8}\big[\lambda^a{}_1{}^1\,\lambda_{a3}{}^3 - \lambda^a{}_3{}^1\,\lambda_{a1}{}^3 - \lambda^a{}_1{}^3\,\lambda_{a3}{}^1 + \lambda^a{}_3{}^3\,\lambda_{a1}{}^1\big] = \tfrac{1}{4}\big[\lambda^a{}_1{}^1\,\lambda_{a3}{}^3 - \lambda^a{}_3{}^1\,\lambda_{a1}{}^3\big].\end{aligned} \tag{6.58a}$$

The sums over Gell-Mann matrices, $a = 1,\dots,8$ simplify, as there is only one matrix for which $(\lambda_{a1}{}^1 \neq 0 \neq \lambda_{a3}{}^3)$, and only two matrices for which $(\lambda_{a3}{}^1 \neq 0 \neq \lambda_{a1}{}^3)$:

$$= \tfrac{1}{4}\big[\lambda^8{}_1{}^1\,\lambda_{83}{}^3 - \lambda^4{}_3{}^1\,\lambda_{41}{}^3 - \lambda^5{}_3{}^1\,\lambda_{51}{}^3\big] = \tfrac{1}{4}\big[\tfrac{1}{\sqrt{3}}\cdot\tfrac{-2}{\sqrt{3}} - 1\cdot 1 - i\cdot(-i)\big] = -\tfrac{2}{3}. \tag{6.58b}$$

[9] We imply that the first factor in every monomial – whether formally χ^α or the Kronecker symbol – refers to the first quark, and the second factor to the second quark.

[10] In mathematical sources, if such explicit constructions are given at all, one mostly finds combinatorial functions, the normalization of which refers to their use in probability theory. However, wave-functions are not probabilities but probability amplitudes, so the desired normalization mostly requires factors of the type $\frac{1}{\sqrt{2}}$ (for a probability amplitude) instead of $\frac{1}{2}$ (for a probability), etc.

Similarly, the type $f_c(\mathbf{3}^{*\prime}_A|\mathbf{3}^*_A)$ computation yields

$$\{\tfrac{1}{4}(\chi^\dagger_{3\gamma}\chi^\dagger_{4\delta})_{3'}(\lambda^a)_\alpha{}^\gamma(\lambda_a)_\beta{}^\delta(\chi^\alpha_1\chi^\beta_2)_{3^*}\}\supset\tfrac{1}{4}\tfrac{1}{\sqrt{2}}(\delta^1_\gamma\delta^2_\delta-\delta^1_\delta\delta^2_\gamma)(\lambda^a)_\alpha{}^\gamma(\lambda_a)_\beta{}^\delta\tfrac{1}{\sqrt{2}}(\delta^\alpha_1\delta^\beta_3-\delta^\beta_1\delta^\alpha_3)$$
$$=\tfrac{1}{8}[\lambda^a{}_1{}^1\lambda_{a3}{}^2-\lambda^a{}_3{}^1\lambda_{a1}{}^2-\lambda^a{}_1{}^2\lambda_{a3}{}^1+\lambda^a{}_3{}^2\lambda_{a1}{}^1]$$
$$=\tfrac{1}{4}[\lambda^a{}_1{}^1\lambda_{a3}{}^2-\lambda^a{}_3{}^1\lambda_{a1}{}^2]=0, \tag{6.58c}$$

since there is no Gell-Mann matrix for which $(\lambda_{a1}{}^1\neq0\neq\lambda_{a3}{}^2)$ or $(\lambda_{a3}{}^1\neq0\neq\lambda_{a1}{}^2)$. Also, for the $f_c(\mathbf{6}_S|\mathbf{3}^*_A)$ type, one checks

$$\{\tfrac{1}{4}(\chi^\dagger_{3\gamma}\chi^\dagger_{4\delta})_6(\lambda^a)_\alpha{}^\gamma(\lambda_a)_\beta{}^\delta(\chi^\alpha_1\chi^\beta_2)_{3^*}\}\supset\text{ four characteristic cases:}$$
$$\Big\{\tfrac{1}{4}(\delta^1_\gamma\delta^1_\delta)(\lambda^a)_\alpha{}^\gamma(\lambda_a)_\beta{}^\delta\tfrac{1}{\sqrt{2}}(\delta^\alpha_1\delta^\beta_2-\delta^\beta_1\delta^\alpha_2)=\tfrac{1}{4\sqrt{2}}[\lambda^a{}_1{}^1\lambda_{a2}{}^1-\lambda^a{}_2{}^1\lambda_{a1}{}^1]=0, \tag{6.58d}$$
$$\tfrac{1}{4}(\delta^3_\gamma\delta^3_\delta)(\lambda^a)_\alpha{}^\gamma(\lambda_a)_\beta{}^\delta\tfrac{1}{\sqrt{2}}(\delta^\alpha_1\delta^\beta_2-\delta^\beta_1\delta^\alpha_2)=\tfrac{1}{4\sqrt{2}}[\lambda^a{}_1{}^3\lambda_{a2}{}^3-\lambda^a{}_2{}^3\lambda_{a1}{}^3]=0, \tag{6.58e}$$
$$\tfrac{1}{4}\tfrac{1}{\sqrt{2}}(\delta^1_\gamma\delta^3_\delta+\delta^1_\delta\delta^3_\gamma)(\lambda^a)_\alpha{}^\gamma(\lambda_a)_\beta{}^\delta\tfrac{1}{\sqrt{2}}(\delta^\alpha_1\delta^\beta_3-\delta^\beta_1\delta^\alpha_3)$$
$$=\tfrac{1}{8}[\lambda^a{}_1{}^1\lambda_{a3}{}^3-\lambda^a{}_3{}^1\lambda_{a1}{}^3+\lambda^a{}_1{}^3\lambda_{a3}{}^1-\lambda^a{}_3{}^3\lambda_{a1}{}^1]=0, \tag{6.58f}$$
$$\tfrac{1}{4}\tfrac{1}{\sqrt{2}}(\delta^1_\gamma\delta^2_\delta+\delta^1_\delta\delta^2_\gamma)(\lambda^a)_\alpha{}^\gamma(\lambda_a)_\beta{}^\delta\tfrac{1}{\sqrt{2}}(\delta^\alpha_1\delta^\beta_3-\delta^\beta_1\delta^\alpha_3)$$
$$=\tfrac{1}{8}[\lambda^a{}_1{}^1\lambda_{a3}{}^2-\lambda^a{}_3{}^1\lambda_{a1}{}^2+\lambda^a{}_1{}^2\lambda_{a3}{}^1-\lambda^a{}_3{}^2\lambda_{a1}{}^1]=0\Big\}. \tag{6.58g}$$

The complete collection of values of the function $f_c(\mathbf{6}_S|\mathbf{3}^*_A)$ of course consists of $6\times3=18$ cases, but these may all be obtained from the above four concrete cases by permuting the values $\alpha,\beta,\gamma,\delta=1,2,3$. It follows that the outgoing pair of quarks is always in the same concrete antisymmetric state as was the incoming pair. (That also follows from the $SU(3)_c$ color conservation, but it is reassuring to confirm this by direct computation.)

Direct computation [☞ Example 6.2 on p. 236, and Exercise 6.2.1] confirms that

$$f_c(\mathbf{3}^*_A|\mathbf{3}^*_A)=-\tfrac{2}{3}\quad\text{and}\quad f_c(\mathbf{6}_S|\mathbf{6}_S)=+\tfrac{1}{3}, \tag{6.59}$$

while $f_c(\mathbf{3}^{*\prime}_A|\mathbf{3}^*_A)$, $f_c(\mathbf{6}_S|\mathbf{3}^*_A)$, $f_c(\mathbf{3}^*_A|\mathbf{6}_S)$ and $f_c(\mathbf{6}'_S|\mathbf{6}_S)$ vanish for all cases.

Conclusion 6.4 *These results indicate that a gluon exchange between two quarks does not change the color combination for two-quark states.*[11] *Besides, the sign in the result (6.59) indicates the one-gluon exchange chromodynamics force, computed in the standard fashion as $\vec{F}_{qq}=-\vec{\nabla}(V_{qq})$ from the relation (6.52), to be:*

1. *attractive if the quark colors are antisymmetrized,*
2. *repulsive if the quark colors are symmetrized.*

Comment 6.3 *The emphasis that this amounts to only a single-gluon exchange contribution to the chromodynamics force is very important: It does not follow that the exchange of*

[11] The computation is of course shown only for the exchange of a single gluon, but its direct iteration is applicable to the exchange of an arbitrary finite number of gluons. Extending this to a formally infinite number of exchanged gluons, including gluon condensation, remains an open issue☝.

more gluons follows the same regularity, and so it does not follow that the total chromodynamics force follows the same regularity. Several further contributions, however, have been computed and they preserve the qualitative character of the result (6.59).

A baryon, of course, has three quarks, and the options for the color factor are (A.78):

1. totally symmetric, so-called "**10**" (10-dimensional) representation,
2. mixed symmetric, so-called "**8**" (8-dimensional) representation (in two distinct ways),
3. totally antisymmetric, so-called "**1**" (1-dimensional) representation

of the $SU(3)_c$ group, where only the last one is $SU(3)_c$-invariant. Also, only in the last case is the system antisymmetric (i.e., the wave-function of the baryon as a three-particle bound state is antisymmetric) with respect to the exchange of any two quarks. Conclusion 6.4 then indicates that this is the only case in which the chromodynamics force between all quarks in the baryon is attractive.

Also, since $O(g_c^2)$ computations indicate that the chromodynamic interaction is binding (attractive) only when the factor $\chi(\text{color})$ in the factorization (4.123) is totally antisymmetric, it follows that the bound state (i.e., its wave-function) for every baryon must be totally symmetric in the remaining three factors:

$$\Psi(\text{baryon}) = \left[\Psi(\vec{r},t)\,\chi(\text{spin})\,\chi(\text{flavor})\right]_S \chi_A(\text{color}).$$

Since the factor $\chi(\text{flavor})$ is determined by the choice of the hadron [☞ Section 4.4] as totally symmetric for the **10**-plet of flavors and mixed symmetric [☞ relation (4.125)] for the **8**-plet, for each of these baryons the symmetries determine the correlation between spin and orbital angular momentum. In ground states, the angular momenta in the three-quark system all vanish, so the spin factor is unambiguously determined to be:

1. spin-$\frac{3}{2}$ and totally symmetric for the decuplet $\{\Delta, \Sigma^*, \Xi^*, \Omega\}$,
2. spin-$\frac{1}{2}$, with a rather more complicated symmetry (4.125) with (4.119)–(4.120) for the octet $\{p^+, n^0, \Lambda, \Sigma, \Xi\}$ of baryons.

Conclusion 6.5 *Furthermore, the chromodynamics interaction between two $SU(3)_c$-invariant bound states cannot happen via the exchange of a single gluon [☞ Example 6.3 on p. 240], but must involve a simultaneous exchange of at least two gluons, and so is of the order of at least $O(g_s^4)$, or a gluon and a quark pair; see process (6.77). Indeed, if the baryon that emits any particle is to remain $SU(3)_c$-invariant both before and after emitting, it follows that the emitted intermediary itself must be $SU(3)_c$-invariant. As none of the eight gluons are $SU(3)_c$-invariant, the intermediary must be an $SU(3)_c$-invariant state composed of at least two gluons or a quark–antiquark pair.*

It then follows that the simplest chromodynamics interaction between two nucleons within an atomic nucleus is about $O(g_c^2)$ times weaker than the strong interaction between two quarks.[12] (This reminds us a little of the fact that the dipole–dipole interaction between two neutral hydrogen atoms is weaker than the Coulomb interaction between the electron and the proton within one atom.)

---

[12] At this introductory level, we have no means of assessing the contribution to the effective strength of interaction provided by the exchange of quarks between two hadrons. However, the $SU(3)_c$-invariance requirement on the particle mediating the strong interaction between two hadrons clearly forces it to be of higher order than the direct, $SU(3)_c$-variant one-gluon-mediated interaction between quarks.

Digression 6.4 $SU(3)_c$-invariant states composed entirely of gluons are called "*glueballs*" and in principle may be observed, but no such state has so far been reliably detected. However, all quantum numbers of such purely gluon $SU(3)_c$-invariant bound states are identical to quantum numbers of electrically neutral mesons such as π^0, ρ^0, etc., with which they mix. This mixing makes experimental differentiation of "glueballs" from ordinary mesons extremely difficult, and no "glueball" state has yet been conclusively detected.

6.2.2 Quark–antiquark interaction

Mesons are much easier to study than baryons, as they are bound states of a quark and an antiquark. However, with this simplification also comes a complication – at least when the meson is neutral with respect to all interactions, so the bound state is of the type

$$\bar{u}\,u + \bar{d}\,d + \bar{s}\,s + \cdots. \tag{6.60}$$

Indeed, now the quark and the antiquark may mutually annihilate. We first consider differently flavored quark–antiquark mesons, where not even virtual annihilation can happen; the next section will consider the possible annihilation in a type (6.60) system.

The amplitude of a single-gluon exchange has a contribution only from one Feynman diagram:

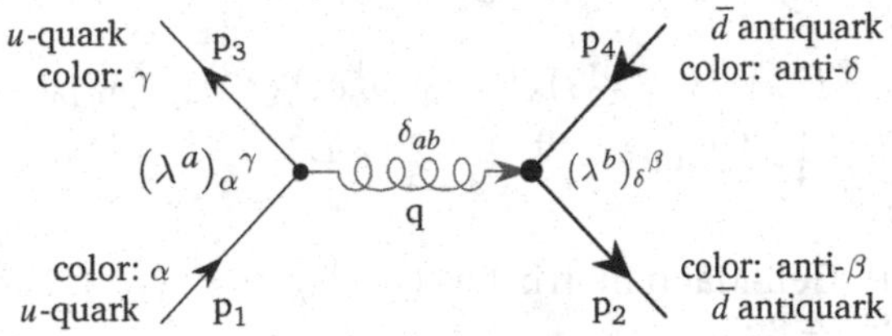

(6.61)

Following Procedure 6.1 on p. 232, and analogously to the result for the first part of (5.147b), we have

$$\mathfrak{M}_{u+\bar{d}\to u+\bar{d}} = -\frac{g_c^2}{4\mathrm{q}^2}[\bar{u}_3\boldsymbol{\gamma}^\mu u_1][\bar{v}_2\boldsymbol{\gamma}_\mu v_4]\,(\chi_3^\dagger\boldsymbol{\lambda}^a\chi_1)\,(\chi_2^\dagger\boldsymbol{\lambda}_a\chi_4), \tag{6.62}$$

where $\mathrm{q} = (\mathrm{p}_1 - \mathrm{p}_3)$ is the 4-momentum exchange, and the result differs from the electrodynamics one only in that:

1. g_e is replaced by g_c,
2. the color factor, $f_c(3,\bar{4}|1,\bar{2}) = \frac{1}{4}(\chi_3^\dagger\boldsymbol{\lambda}^a\chi_1)(\chi_2^\dagger\boldsymbol{\lambda}_a\chi_4)$, is inserted.

The color factor for the incoming quark–antiquark pair again must belong to one of the two vector spaces:

1. The Hermitian octet (**8**) of states, i.e., the 8-dimensional vector space spanned by the color factors,

$$\begin{aligned}\{\chi_{12}{}^\alpha{}_\beta &= \sqrt{1+\tfrac{1}{2}\delta^\alpha_\beta}\,(\chi_1^\alpha\chi_{2\beta}^\dagger - \tfrac{1}{\sqrt{3}}\delta^\alpha_\beta\,\mathring{\boldsymbol{\chi}}),\ \ \alpha,\beta = \mathit{red}, \mathit{yellow}, \mathit{blue} = 1,2,3\}\\ &= \Big\{\sqrt{\tfrac{3}{2}}(\delta^\alpha_1\delta^1_\beta - \mathring{\boldsymbol{\chi}}),\ \sqrt{\tfrac{3}{2}}(\delta^\alpha_2\delta^2_\beta - \mathring{\boldsymbol{\chi}}),\ \sqrt{\tfrac{3}{2}}(\delta^\alpha_3\delta^3_\beta - \mathring{\boldsymbol{\chi}}),\\ &\qquad (\delta^\alpha_1\delta^2_\beta),\ (\delta^\alpha_1\delta^3_\beta),\ (\delta^\alpha_2\delta^1_\beta),\ (\delta^\alpha_2\delta^3_\beta),\ (\delta^\alpha_3\delta^1_\beta),\ (\delta^\alpha_3\delta^2_\beta)\Big\},\end{aligned} \tag{6.63}$$

which form a traceless Hermitian matrix, where $\mathring{\boldsymbol{\chi}} := \frac{1}{\sqrt{3}}\,\mathrm{Tr}(\chi_1\chi_2^\dagger) = \frac{1}{\sqrt{3}}(\chi_1^\alpha\chi_{2\alpha}^\dagger)$.

2. The $SU(3)_c$-invariant (**1**), where ${\chi_{12}}^\alpha{}_\beta = \delta^\alpha_\beta \mathring{\boldsymbol{\chi}}$ is a multiple of the unit matrix.

Normalization is again quantum mechanical, so $\|{\chi_{12}}^\alpha{}_\beta\|^2 = 1$ for every choice α, β.

Similarly to the result (6.57), for $u + \bar{d} \to u + \bar{d}$ we have

$$f_c(3, \bar{4}|1, \bar{2}) = f_c(\mathbf{8}|\mathbf{8}),\ f_c(\mathbf{8}'|\mathbf{8}),\ f_c(\mathbf{8}|\mathbf{1}),\ f_c(\mathbf{1}|\mathbf{8}),\ f_c(\mathbf{1}|\mathbf{1}). \tag{6.64}$$

Also, just as in electrodynamics, the gluon exchange gives rise to a potential of the form

$$V_{q\bar{q}}(r) = -f_c \frac{\alpha_c \hbar c}{r}, \tag{6.65}$$

where the sign is now negative, since the color charges of a quark and an antiquark are "opposite": one is the (chromodynamics) "color" the other the "anticolor."[13]

Example 6.3 To compute the functions $f_c(\mathbf{8}|\mathbf{8})$, $f_c(\mathbf{8}'|\mathbf{8})$ and $f_c(\mathbf{1}|\mathbf{1})$, we pick the simplest particular cases for each; the diligent Student will convince themselves by direct computation that all cases give the same quantitative results.

For $f_c(\mathbf{8}|\mathbf{8})$, the incoming and the outgoing quark–antiquark pair have the same combination of color–anticolor; take, e.g., the red–antiblue $(\delta^1_\gamma \delta^\delta_3) \in \mathbf{8}$ element:

$$\begin{aligned}\{\tfrac{1}{4}\,(\chi^\dagger_{3\gamma}\,\chi^\delta_4)_{\mathbf{8}}\,(\lambda^a)_\alpha{}^\gamma\,(\lambda_a)_\delta{}^\beta\,(\chi^\alpha_1\,\chi^\dagger_{2\beta})_{\mathbf{8}}\} &\supset \tfrac{1}{4}\,(\delta^1_\gamma\delta^\delta_3)\,(\lambda^a)_\alpha{}^\gamma\,(\lambda_a)_\delta{}^\beta\,(\delta^\alpha_1\delta^3_\beta)\\ &= \tfrac{1}{4}\,{\lambda^a}_1{}^1\,{\lambda_{a3}}^3 = \tfrac{1}{4}\,{\lambda^8}_1{}^1\,{\lambda_{83}}^3 = \tfrac{1}{4}\,\tfrac{1}{\sqrt{3}}\,\tfrac{-2}{\sqrt{3}} = -\tfrac{1}{6},\end{aligned} \tag{6.66}$$

since only the eighth Gell-Mann matrix has $({\lambda^a}_1{}^1 \neq 0 \neq {\lambda_{a3}}^3)$. For $f_c(\mathbf{8}'|\mathbf{8})$ we take, e.g., $(\delta^1_\gamma\delta^\delta_3) \in \mathbf{8}$ and $(\delta^3_\gamma\delta^\delta_1) \in \mathbf{8}'$:

$$\begin{aligned}\{\tfrac{1}{4}\,(\chi^\dagger_{3\gamma}\,\chi^\delta_4)_{\mathbf{8}'}\,(\lambda^a)_\alpha{}^\gamma\,(\lambda_a)_\delta{}^\beta\,(\chi^\alpha_1\,\chi^\dagger_{2\beta})_{\mathbf{8}}\} &\supset \tfrac{1}{4}\,(\delta^3_\gamma\delta^\delta_1)\,(\lambda^a)_\alpha{}^\gamma\,(\lambda_a)_\delta{}^\beta\,(\delta^\alpha_1\delta^3_\beta)\\ &= \tfrac{1}{4}\,{\lambda^a}_1{}^3\,{\lambda_{a1}}^3 = \tfrac{1}{4}({\lambda^4}_1{}^3\,{\lambda_{41}}^3 + {\lambda^5}_1{}^3\,{\lambda_{51}}^3) = \tfrac{1}{4}\,(1{\cdot}1 + (-i){\cdot}(-i)) = 0.\end{aligned} \tag{6.67}$$

Since the representation **1** has only one dimension, for $f_c(\mathbf{1}|\mathbf{1})$ there is a single contribution:

$$\begin{aligned}&\tfrac{1}{4}\,(\chi^\dagger_{3\gamma}\,\chi^\delta_4)_{\mathbf{1}}\,(\lambda^a)_\alpha{}^\gamma\,(\lambda_a)_\delta{}^\beta\,(\chi^\alpha_1\,\chi^\dagger_{2\beta})_{\mathbf{1}}\\ &\quad= \tfrac{1}{4}\,\tfrac{1}{\sqrt{3}}(\delta^1_\gamma\delta^\delta_1 + \delta^2_\gamma\delta^\delta_2 + \delta^3_\gamma\delta^\delta_3)\,(\lambda^a)_\alpha{}^\gamma\,(\lambda_a)_\delta{}^\beta\,\tfrac{1}{\sqrt{3}}(\delta^\alpha_1\delta^1_\beta + \delta^\alpha_2\delta^2_\beta + \delta^\alpha_3\delta^3_\beta)\\ &\quad= \tfrac{1}{12}\,{\lambda^a}_\alpha{}^\gamma\,{\lambda_{a\gamma}}^\alpha = \tfrac{1}{12}\,\delta_{ab}\,\mathrm{Tr}(\boldsymbol{\lambda}^a\,\boldsymbol{\lambda}^b) = \tfrac{1}{12}\,\delta_{ab}\,2\delta^{ab} = \tfrac{1}{6}\,8 = \tfrac{4}{3},\end{aligned} \tag{6.68}$$

where we used the relation (A.72). This coefficient, $f_c(\mathbf{1}|\mathbf{1}) = \frac{4}{3}$, has shown up in the relation (4.102).

---

[13] In electrodynamics, of course, there is only one kind of charge – electric – and the opposite charge is simply the negative charge. For chromodynamics colors, "anticolor" is not simply negative "color," but the opposite "color"; i.e., the color that together with the original one produces a colorless, i.e., an $SU(3)_c$-invariant whole. This we may write, e.g., $(\chi_{\alpha\,(\text{red})})^\dagger = (\chi^\dagger)^{\alpha\,(\text{green})}$. We will not use this notational possibility, as it additionally complicates the tensor algebra rules and necessitates printing in color; with the current convention, computations may be followed even in monochromatic printout.

Direct computation shows also that $f_c(\mathbf{8}|\mathbf{1}), f_c(\mathbf{1}|\mathbf{8}) = 0$, and we have:

Conclusion 6.6 *These results show that the single-gluon exchange*[14] *between a quark and an antiquark preserves the color state: incoming and outgoing quark–antiquark pairs have the same color combination. Besides, the chromodynamics force (6.65) between a quark and an antiquark is*

1. *attractive when both the incoming and the outgoing pair are in the $SU(3)_c$-invariant state, and*
2. *repulsive otherwise.*

6.2.3 Quark–antiquark annihilation

The single-gluon exchange amplitude now has two contributions, corresponding to the two Feynman diagrams:

$$\mathfrak{M}_{u+\bar{u}\to u+\bar{u}} = \quad (6.69)$$

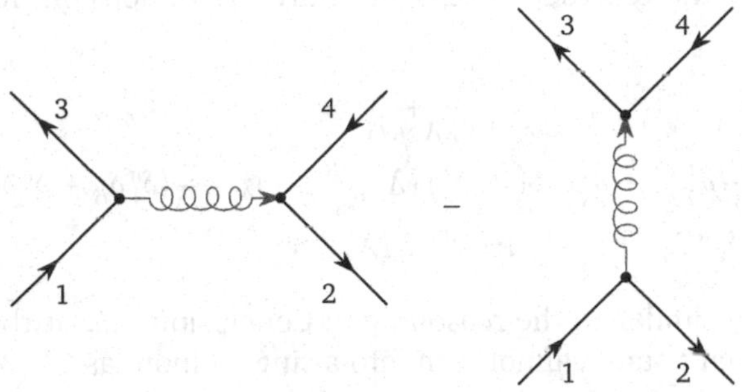

where the relative minus sign follows from the fact that the amplitude for the second sub-process (the virtual annihilation and re-creation of the $u\,\bar{u}$ pair) equals the first, upon exchanging the incoming antiquark, 2, and the outgoing quark 3 [☞ discussion of the Bhabha scattering and procedure (5.145)]. Adapting the result (5.147b), we have that the amplitude of this process equals

$$\begin{aligned}\mathfrak{M}_{u+\bar{u}\to u+\bar{u}} &= -\frac{g_c^2}{4(\mathrm{p}_1-\mathrm{p}_3)^2}[\overline{u}_3\gamma^\mu u_1][\overline{v}_2\gamma_\mu v_4](\chi_3^\dagger\boldsymbol{\lambda}^a\chi_1)(\chi_2^\dagger\boldsymbol{\lambda}_a\chi_4)\\ &\quad+\frac{g_c^2}{4(\mathrm{p}_1+\mathrm{p}_2)^2}[\overline{v}_2\gamma^\mu u_1][\overline{u}_3\gamma_\mu v_4](\chi_2^\dagger\boldsymbol{\lambda}^a\chi_1)(\chi_3^\dagger\boldsymbol{\lambda}_a\chi_4),\end{aligned} \quad (6.70)$$

where we used that the color factor, f_c, for the first diagram is identical to the factor in the result (6.62), and the factor for the second diagram, $\widetilde{f}_c$, is obtained by swapping $2 \leftrightarrow 3$.

Example 6.4 We will compute one sample value of each of $\widetilde{f}_c(\mathbf{8}|\mathbf{8})$, $\widetilde{f}_c(\mathbf{8}'|\mathbf{8})$ and $\widetilde{f}_c(\mathbf{1}|\mathbf{1})$, and we choose the simplest cases to this end; the diligent Student should verify by direct computation that all cases produce quantitatively the same results. Alternatively, this may also be proven by $SU(3)_c$ group action from the results presented here [☞ Exercise 6.2.2].

For $\widetilde{f}_c(\mathbf{8}|\mathbf{8})$, the incoming and outgoing quark–antiquark pair have the same color–anticolor combination; fix this to be the red–antiblue combination:

$$\begin{aligned}\{\tfrac{1}{4}\,(\chi^\dagger_{3\gamma}\,\chi_4^\delta)_\mathbf{8}\,(\lambda^a)_\alpha{}^\beta\,(\lambda_a)_\delta{}^\gamma\,(\chi_1^\alpha\,\chi^\dagger_{2\beta})_\mathbf{8}\} &\supset \tfrac{1}{4}\,(\delta^1_\gamma\delta^\delta_3)\,(\lambda^a)_\alpha{}^\beta\,(\lambda_a)_\delta{}^\gamma\,(\delta^\alpha_1\delta^3_\beta)\\ = \tfrac{1}{4}\,\lambda^a{}_1{}^3\,\lambda_{a3}{}^1 = \tfrac{1}{4}\,(\lambda^4{}_1{}^3\,\lambda_{43}{}^1+\lambda^5{}_1{}^3\,\lambda_{53}{}^1) &= \tfrac{1}{4}\,(1{\cdot}1+(-i){\cdot}(i)) = \tfrac{1}{2},\end{aligned} \quad (6.71)$$

---

[14] See Footnote *11* on p. 237.

since only $\boldsymbol{\lambda}^4$ and $\boldsymbol{\lambda}^5$ have $({\lambda^a}_1{}^3 \neq 0 \neq {\lambda_{a3}}^1)$. For $\widetilde{f}_c(\mathbf{8}'|\mathbf{8})$ we have, e.g.,

$$\begin{aligned}\{\tfrac{1}{4}\,(\chi^{\dagger}_{3\gamma}\,\chi^{\delta}_{4})_{\mathbf{8}'}\,(\lambda^a)_\alpha{}^\beta\,(\lambda_a)_\delta{}^\gamma\,(\chi^{\alpha}_{1}\,\chi^{\dagger}_{2\beta})_{\mathbf{8}}\} &\supset \tfrac{1}{4}\,(\delta^3_\gamma\delta^\delta_1)\,(\lambda^a)_\alpha{}^\beta\,(\lambda_a)_\delta{}^\gamma\,(\delta^\alpha_1\delta^3_\beta)\\ &= \tfrac{1}{4}\,{\lambda^a}_1{}^3\,{\lambda_{a1}}^3 = \tfrac{1}{4}({\lambda^4}_1{}^3\,{\lambda_{41}}^3+{\lambda^5}_1{}^3\,{\lambda_{51}}^3) = \tfrac{1}{4}\,(1{\cdot}1+(-i){\cdot}(-i)) = 0,\end{aligned} \tag{6.72}$$

or, e.g.,

$$\begin{aligned}\{\tfrac{1}{4}\,(\chi^{\dagger}_{3\gamma}\,\chi^{\delta}_{4})_{\mathbf{8}'}\,(\lambda^a)_\alpha{}^\beta\,(\lambda_a)_\delta{}^\gamma\,(\chi^{\alpha}_{1}\,\chi^{\dagger}_{2\beta})_{\mathbf{8}}\} &\supset \tfrac{1}{4}\,(\delta^2_\gamma\delta^\delta_1)\,(\lambda^a)_\alpha{}^\beta\,(\lambda_a)_\delta{}^\gamma\,(\delta^\alpha_1\delta^3_\beta)\\ &= \tfrac{1}{4}\,{\lambda^a}_1{}^3\,{\lambda_{a1}}^2 = 0,\end{aligned} \tag{6.73}$$

as no Gell-Mann matrix has a nonzero 1st entry in both the 2nd and 3rd row (or column). In turn, since the representation **1** has only one dimension, for $\widetilde{f}_c(\mathbf{1}|\mathbf{1})$ there is a single case,

$$\begin{aligned}&\tfrac{1}{4}\,(\chi^{\dagger}_{3\gamma}\,\chi^{\delta}_{4})_{\mathbf{1}}\,(\lambda^a)_\alpha{}^\beta\,(\lambda_a)_\delta{}^\gamma\,(\chi^{\alpha}_{1}\,\chi^{\dagger}_{2\beta})_{\mathbf{1}}\\ &\quad= \tfrac{1}{4}\,\tfrac{1}{\sqrt{3}}(\delta^1_\gamma\delta^\delta_1+\delta^2_\gamma\delta^\delta_2+\delta^3_\gamma\delta^\delta_3)\,(\lambda^a)_\alpha{}^\beta\,(\lambda_a)_\delta{}^\gamma\,\tfrac{1}{\sqrt{3}}(\delta^\alpha_1\delta^1_\beta+\delta^\alpha_2\delta^2_\beta+\delta^\alpha_3\delta^3_\beta)\\ &\quad= \tfrac{1}{12}\,{\lambda^a}_\alpha{}^\alpha\,{\lambda_{a\gamma}}^\gamma = \tfrac{1}{12}\,\mathrm{Tr}(\boldsymbol{\lambda}^a)\,\mathrm{Tr}(\boldsymbol{\lambda}_a) = 0,\end{aligned} \tag{6.74}$$

which is very similar to the reasoning in Conclusion 6.5, at the end of Section 6.2.1: an $SU(3)_c$-invariant state cannot turn into a single gluon, as $SU(3)_c$-invariant gluons do not exist.

Using the direct computations from Example 6.3, we have that

$$\begin{aligned}\mathfrak{M}_{u+\overline{u}\to u+\overline{u}} = &-\frac{g_c^2}{(\mathrm{p}_1-\mathrm{p}_3)^2}\left\{\begin{matrix}-\frac{1}{6}\\ +\frac{4}{3}\end{matrix}\right\}[\overline{u}_3\boldsymbol{\gamma}^\mu u_1][\overline{v}_2\boldsymbol{\gamma}_\mu v_4]\\ &+\frac{g_c^2}{(\mathrm{p}_1+\mathrm{p}_2)^2}\left\{\begin{matrix}\frac{1}{2}\\ 0\end{matrix}\right\}[\overline{v}_2\boldsymbol{\gamma}^\mu u_1][\overline{u}_3\boldsymbol{\gamma}_\mu v_4], \qquad \text{if}\ \begin{cases}\boldsymbol{\chi}_{12}\subset\mathbf{8},\\ \boldsymbol{\chi}_{12}=\mathbf{1}.\end{cases}\end{aligned} \tag{6.75}$$

If a concrete incoming quark–antiquark pair in fact form a meson, the color factors $\boldsymbol{\chi}_{12}=\boldsymbol{\chi}_{34}$ must be $SU(3)_c$-invariant, so that the second diagram (6.69) in fact contributes nothing because of color conservation. However, in hadronic elastic collisions of the type $n^0+\pi^-\to n^0+\pi^-$, both diagrams contribute:

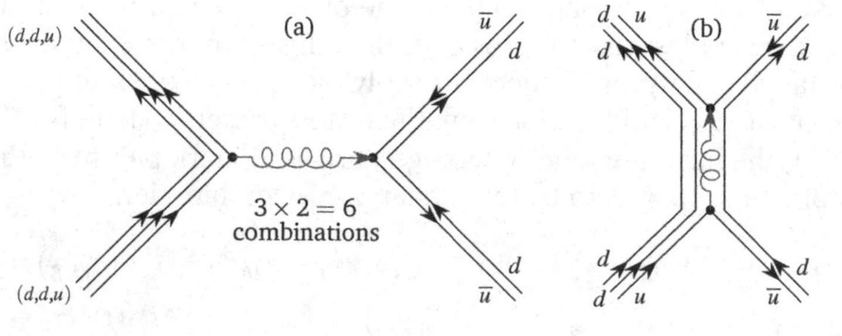

(6.76)

The diagram (a) contributes in six ways (either of the three quarks in the neutron may exchange a gluon with either the $\overline{u}$ antiquark, or the d quark within the pion); the diagram (b) contributes in only one way. Except, the processes depicted in diagram (a) are prohibited by Conclusion 6.5, i.e., either at least one more gluon and/or a d quark (being common to both incoming hadrons) must be exchanged, as for example in

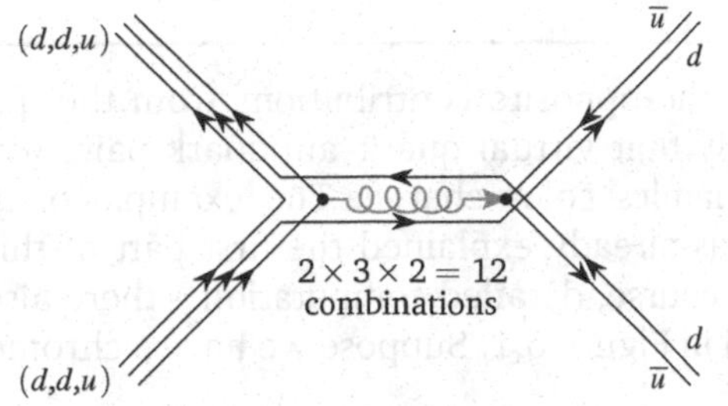

(6.77)

This then is still $O(g_c^2)$, up to the undetermined d-quark exchange factor; additional gluon exchange would increase the order. Note that the exchanged state (propagating in the horizontal direction in this Feynman diagram) may well be interpreted as the exchange of a virtual pion, π^0 – vindicating in part Yukawa's original proposal for strong interactions.

6.2.4 Renormalization and asymptotic freedom

In Section 5.3.3, we obtained the relation (5.202),

$$\alpha_{e,R}(|\mathrm{q}^2|) \approx \frac{\alpha_{e,R}(0)}{1 - \frac{\alpha_{e,R}(0)}{3\pi}\ln\left(\frac{|\mathrm{q}^2|}{m_e^2c^2}\right)}, \qquad |\mathrm{q}^2| \gg m_e^2c^2, \tag{5.202}$$

which indicates the electromagnetic fine structure constant to in fact be a variable, and to depend on the transfer 4-momentum q at which the measurement takes place.

In the analogous analysis of $O(g_s^4)$ corrections to the amplitude of the collision (6.50) new diagrams appear, precisely because of the non-abelian (non-commutative) nature of the chromodynamics interaction. Ignoring diagrams that only correspond to renormalizing the parameters of the incoming and outgoing particles, for $O(g_s^4)$ contributions we have

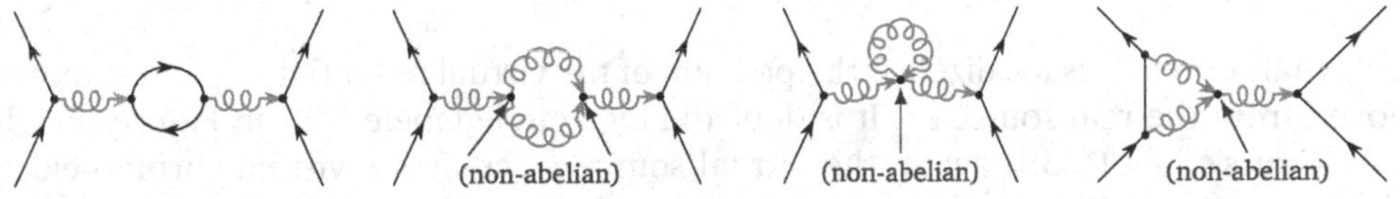

(6.78)

The computation of the contributions depicted by the last three diagrams requires additional rules that involve the introduction of ghost fields and the so-called BRST nilpotent symmetry [☞ Footnote 5 on p. 232]. That level of technical details is beyond the scope of this book, and we simply cite [445] the final result:

$$\alpha_{s,R}(|\mathrm{q}^2|) \approx \frac{\alpha_{s,R}(\mu^2c^2)}{1 + \frac{\alpha_{s,R}(\mu^2c^2)}{3\pi}\frac{11n-2n_f}{4}\ln\left(\frac{|\mathrm{q}^2|}{\mu^2c^2}\right)}, \qquad |\mathrm{q}^2| \gg \mu^2c^2. \tag{6.79}$$

This holds for all $SU(n)$-gauge interactions, where n_f is the total number of Dirac spin-$\frac{1}{2}$ fermions that possess such an n-dimensional $SU(n)$ charge. The fermion loop (6.78) contributions are the opposite of the gauge boson loop contributions.[15] The precise computation produces the coefficient

[15] Recall that fermion loops require an additional -1 factor in the amplitude, as well as that both quarks and gluons contribute to the chromodynamics color charge (6.32) [☞ Conclusion 6.3 on p. 229].

$\frac{11n-2n_f}{4}$, which in our case is $+5\frac{1}{4}$: we have $n=3$ colors and $n_f=6$ quark flavors. Since the relative sign in the denominator (6.79) is opposite from the relative sign in the denominator (5.202), it follows that $\alpha_s(|q^2|)$ *diminishes* as $|q^2|$ grows.

Example 6.5 Effectively, the opposite contributions from the quarks and the gluons in the relation (6.79) imply that virtual quark–antiquark pairs *screen*, and virtual gluons *enhance* the chromodynamics color charge. The example of quantum electrodynamics [☞ Section 5.3.3] has already explained the first part of this phenomenon. For the second part – except, of course, detailed computation – there also exists a qualitative argument [425], depicted in Figure 6.1. Suppose we have a chromodynamic charge source

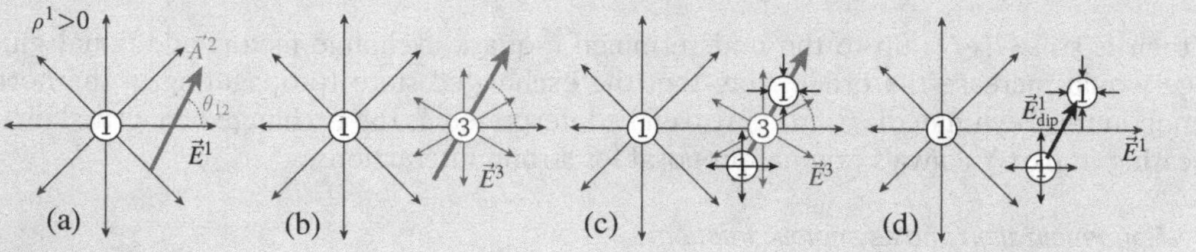

Figure 6.1 A qualitative depiction of the mechanism by which virtual gluons enhance the chromodynamic charge

of the color ρ^1, depicted by the "central" circle labeled "1" in Figures 6.1(a)–(d). By Gauss's law (6.35), this creates a chromo-electric field $\vec{E}^1$; see Figure 6.1(a). Let somewhere nearby a virtual quantum of the chromodynamics field appear; in Figures 6.1(a)–(c), this is depicted by the vector $\vec{A}^2$, at an angle of $\theta_{12} = 60°$ from the positive direction of $\vec{E}^1$. This virtual quantum $\vec{A}^2$ couples to the pre-existing field $\vec{E}^1$ and produces via the non-abelian (non-commutative) part of equation (6.35) a virtual source for the field $\vec{E}^3$:

$$\vec{\nabla}\cdot\vec{E}^3 = -\tfrac{g_c}{\hbar c} f^3{}_{21}\vec{A}^2\cdot\vec{E}^1 = -\tfrac{g_c}{\hbar c}(-1)|\vec{A}^2||\vec{E}^1|(\cos\theta_{12} = +\tfrac{1}{2}) = +\tfrac{g_c}{2\hbar c}|\vec{A}^2||\vec{E}^1|. \tag{6.80}$$

This virtual source ρ^3 is localized at the position of the virtual potential $\vec{A}^2$, i.e., somewhat removed from the real source ρ^1. It is depicted by a circle labeled "3" in Figures 6.1(b)–(c). By Gauss's law (6.35) again, the virtual source ρ^3 creates a virtual chromo-electric field $\vec{E}^3$, depicted in Figures 6.1(b)–(c). Iterating, the coupling of this virtual field $\vec{E}^3$ and the virtual potential $\vec{A}^2$ serves as an *additional* source (or sink) for the field $\vec{E}^1$. Indeed, just outside the location of the "bare" source ρ^1 and near the virtual source ρ^3, we have

$$\vec{\nabla}\cdot\vec{E}^1 = -\tfrac{g_c}{\hbar c} f^1{}_{23}\vec{A}^2\cdot\vec{E}^3 = -\tfrac{g_c}{\hbar c}(+1)|\vec{A}^2||\vec{E}^3|\cos\theta_{32}, \tag{6.81}$$

where θ_{32} is the angle between the virtual potential $\vec{A}^2$ and the virtual field $\vec{E}^3$. In Figure 6.1(c), p. 244, we see that:

1. $\cos\theta_{32} > 0$ north-east from the virtual source ρ^3, and
2. $\cos\theta_{32} < 0$ south-west from the virtual source ρ^3.

Thus, the coupling of the virtual field $\vec{E}^3$ and the virtual potential $\vec{A}^2$ serves as an *additional sink* for $\vec{E}^1$ near the virtual source ρ^3 and a little further away from the "bare" source ρ^1, and as an *additional source* for $\vec{E}^1$ near the virtual source ρ^3 and a little closer to the "bare" source ρ^1. This additional source-and-sink form a small dipole of the $\vec{E}^1_{\rm dip}$ field, at the location of the virtual potential $\vec{A}^2$. Such additional dipoles result in a vacuum polarization owing to the nonlinear coupling of chromodynamics fields and potentials.

For clarity, Figure 6.1(d), p. 244, depicts the contributions of only the chromo-electric field $\vec{E}^1$, where we see that the coupling of the virtual potentials $\vec{A}^2$ with the induced virtual field $\vec{E}^3$ has produced the additional field $\vec{E}^1_{\rm dip}$, and just so that the "bare" source "1" is effectively *enhanced* (anti-screened) rather than screened, i.e., diminished: In the induced virtual dipole, the source is closer to the "bare" source, and the sink is further away.

Repeating the analysis with any other combination of distribution and value of the initial "bare" source and the virtual potential, as well as the further iterations of this nonlinear coupling, confirms this qualitative conclusion. The virtual quanta of the chromodynamics field of course appear with a random distribution around the "bare" source, but the so-induced vacuum polarization uniformly enhances the "bare" source.

The transfer 4-momentum $\mathrm{q} := (\mathrm{p}_1 - \mathrm{p}_2)$ between the left- and the right-hand particles in each diagram (6.78) is

$$\mathrm{q}^2 = \mathrm{p}_1^2 + \mathrm{p}_2^2 - 2\mathrm{p}_1\cdot\mathrm{p}_2 = (m_1^2 + m_2^2)c^2 + 2\vec{p}_1\cdot\vec{p}_2 - 2\frac{E_1 E_2}{c^2} = \frac{(E_1 - E_2)^2}{c^2} - (\vec{p}_1 - \vec{p}_2)^2. \tag{6.82}$$

The distance covered by the virtual particles, which occurs predominantly in the horizontal, mediating portion of the diagram (6.78), is inversely proportional to this transfer momentum. Thus, $\alpha_{s,R}(|\,q^2|)$ *grows* with the distance at which the interaction occurs, which confirms earlier given qualitative arguments and is in full accord with experimental observations; see Section 2.3.14.

Digression 6.5 The careful Reader will have noticed that the relation (6.79) gives the chromodynamics fine structure parameter at the energy $c\sqrt{|\mathrm{q}^2|}$ as a function of two quantities: the mass μ and the value of the chromodynamics fine structure parameter at the transfer momentum μc. These two quantities may be "collected," by defining

$$\Lambda_{QCD}: \quad \ln(\Lambda^2_{QCD}) := \ln(\mu^2 c^2) - \frac{12\pi}{(11n - 2n_f)\alpha_{s,R}(\mu^2 c^2)}, \tag{6.83}$$

the substitution of which into the result (6.79) yields

$$\alpha_{s,R}(|\mathrm{q}^2|) \approx \frac{12\pi}{(11n - 2n_f)\ln\big(\frac{|\mathrm{q}^2|}{\Lambda^2_{QCD}}\big)}, \tag{6.84}$$

where Λ_{QCD} is the magnitude of the transfer 4-momentum at which $\alpha_{s,R}(|\mathrm{q}^2|)$ diverges; this divergence is called the Landau pole, after L. D. Landau. The importance of this divergence is only formal, since perturbative computations fail to make sense before that, when $\alpha_{s,R}(|\mathrm{q}^2|) \lesssim 1$. Experimental estimates give only an approximate region $100\,\mathrm{MeV}/c < \Lambda_{QCD} < 500\,\mathrm{MeV}/c$, and one typically uses $\Lambda_{QCD} \approx 220\,\mathrm{MeV}/c$ as the approximate value of the geometric mean of the experimental bounds.

Finally, it is worth noting that for quantum electrodynamics, in relation (5.202), the reference value of $\alpha_{e,R}(0) \approx \frac{1}{137}$ is an excellent choice. That is the value of the fine structure parameter – and so also the intensity of the electromagnetic interaction (5.122), $g_e = \sqrt{4\pi\alpha_e} = |e|/\sqrt{\epsilon_0 \hbar c}$ – that is measurable in experiments where the interacting electric charges are at a distance much larger than the typical (sub-)atomic distances. Those are, of course, all "classical" experiments with electric charges.

By contrast, in quantum chromodynamics, $\alpha_{s,R}(0)$ makes no sense. Perturbative computations wherein the parameter $\alpha_{s,R}(|\mathrm{q}^2|)$ and the relation (6.79) are defined fails to be valid at 4-momenta *below* $\sim$200 MeV/c, i.e., at distances bigger than $\sim 10^{-15}$ m. Perturbative computations in quantum chromodynamics make sense only at distances smaller than $\sim 10^{-15}$ m, i.e., at energies larger than $\sim$200 MeV. That makes the introduction of an arbitrary reference value, such as $\alpha_{s,R}(\mu^2 c^2)$, necessary. For sufficiently large μ, $\alpha_{s,R}(\mu^2 c^2)$ can even be measured, whereupon the relation (6.79) is of better practical use than the simpler relation (6.84).

A moment's thought reveals that this striking difference between $\alpha_{s,R}(|\mathrm{q}^2|)$ and $\alpha_{e,R}(|\mathrm{q}^2|)$ in their dependence on the transfer 4-momentum fully supports two of the experimentally noted properties of quarks:

Asymptotic freedom the limit $\lim_{|\mathrm{q}^2|\to\infty} \alpha_{s,R}(|\mathrm{q}^2|) = 0$ agrees with the experimentally observed fact that the strong interaction between quarks is vanishingly small at vanishingly small distances.

Confinement the limit $\lim_{|\mathrm{q}^2|\to\Lambda_{QCD}} \alpha_{s,R}(|\mathrm{q}^2|) = \infty$ agrees with the experimentally observed fact that the strong interaction between quarks grows as the distance between two quarks is being increased, e.g., so as to separate them.

Note that this is *not* a proof of confinement, since perturbation theory, used to compute $\alpha_{s,R}(|\mathrm{q}^2|)$, breaks down as $|\mathrm{q}^2| \to \Lambda_{QCD}$; nevertheless, this perturbative result is encouraging and gives good hope that other methods will eventually provide a rigorous proof☝.

6.2.5 Exercises for Section 6.2

✎ **6.2.1** *Following Example 6.2 on p. 236, compute all possible cases of the color factors* $f_C(\mathbf{3}^*_A|\mathbf{6}_S)$, $f_C(\mathbf{6}_S|\mathbf{6}_S)$ *and* $f_C(\mathbf{6}'_S|\mathbf{6}_S)$.

✎ **6.2.2** *Using all elements (6.63) and by explicit computation – or using the* $SU(3)_c$ *action – show that the results of Example 6.3 on p. 240 are independent of the choice of the concrete case(s).*

✎ **6.2.3** *Following Example 6.3 on p. 240, compute all possible cases of* $f_C(\mathbf{8}|\mathbf{1})$.

✎ **6.2.4** *For all six elastic collisions of a nucleon* (p^+, n^0) *and a pion* $(\pi^\pm, n^0)$*, determine the relative contribution of the diagrams of type (a) and type (b) in the display (6.76).*

✎ **6.2.5** *Redoing the analysis of Example 6.5 on p. 244, verify that a virtual gauge vector* $\vec{A}^2$ *oriented, however, at an angle* $120°$ *will produce the same effect of anti-screening of the initial source* ρ^1.

6.3 Non-perturbative comments

Field theory is – in practice – a perturbative discipline, and most of the detailed work in quantum chromodynamics indeed relies on perturbative computations. Because the fine structure parameter and the interaction intensity depend on the distance at which the interaction takes place (mediated by exchange of gluons) [☞ result (6.79)–(6.84)], perturbative computations do not suffice. A complete solution of quantum chromodynamics must include essentially non-perturbative effects. Here, we mention a few themes that appear in attempts at non-perturbative analysis.

6.3.1 Strong CP-violation, "topological" solutions and the ϑ-vacuum

The chromodynamics analogue of the question at the end of Digression 5.7 on p. 183, about the expression (5.80c), is as follows: In the most general (both gauge- and Lorentz-invariant) Lagrangian density for quantum chromodynamics,

$$\mathscr{L}_{\text{QCD}+} = -\sum_n \text{Tr}\left[\overline{\mathbf{\Psi}}_n(x)\,[i\hbar c\not{D} - m_n e^{i\vartheta'\gamma}c^2]\,\mathbf{\Psi}_n(x)\right] \\ -\tfrac{1}{4}\,\text{Tr}\left[\mathbb{F}_{\mu\nu}\mathbb{F}^{\mu\nu}\right] - \tfrac{n_f g_s^2\vartheta}{32\pi^2}\varepsilon^{\mu\nu\rho\sigma}\,\text{Tr}\left[\mathbb{F}_{\mu\nu}\mathbb{F}_{\rho\sigma}\right], \tag{6.85}$$

why are the parameters $\vartheta' \sim \vartheta < 3 \times 10^{-10}$? The most reliable bound follows from the fact that the presence of the ϑ, ϑ'-dependent terms would provide the neutron with an electric dipole momentum. Experimentally, the electric dipole momentum of the neutron vanishes, and the bounds then follow from the limits on the experimental error in that measurement. Unlike the *CP*-violation as discussed in Section 4.2.3, *CP*-violation that follows from this so-called "ϑ_{QCD}-problem" is also called the "strong *CP*-violation."

The additional term, $\varepsilon^{\mu\nu\rho\sigma}\,\text{Tr}\left[\mathbb{F}_{\mu\nu}\mathbb{F}_{\rho\sigma}\right]$ is the 4-divergence of the so-called Loos–Chern–Simons "current" [555],

$$\mathcal{K}^\mu = \tfrac{n_f g_s^2}{32\pi^2}\varepsilon^{\mu\nu\rho\sigma}\big(\delta_{ab}\,A^a_\nu\,F^b_{\rho\sigma} - \tfrac{1}{3}g_s f_{abc}\,A^a_\nu\,A^b_\rho\,A^c_\sigma\big). \tag{6.86}$$

Then, a formal ϑ-transformation $\exp\{i\vartheta\mathcal{Q}\}$ exists with $\mathcal{Q} := \int d^3\vec{r}\,\mathcal{K}^0$ that transforms the vacuum $|0\rangle$ with $\vartheta = 0$ into the vacuum $|\vartheta\rangle = e^{i\vartheta\mathcal{Q}}|0\rangle$ with the $\vartheta \neq 0$ value. Since all operators transform as $H(\vartheta) = e^{i\vartheta\mathcal{Q}}H(0)e^{-i\vartheta\mathcal{Q}}$, it follows that all physics with $\vartheta \neq 0$ is identical to the physics with $\vartheta = 0$. The vacua with distinct values of ϑ define "sectors" in the Hilbert space of quantum chromodynamics, and "our World" could easily be one such sector, which is physically indistinguishable from the sector with $\vartheta = 0$.

On the other hand, the equations of motion for quantum chromodynamics, derived from the Lagrangian density (6.23), are nonlinear equations, and have solutions that cannot be obtained by perturbative methods. To a large degree, such solutions are similar to magnetic monopoles that were discussed in Section 5.2.3; because of the nonlinear nature of the coupled system of equations of motion (6.37)–(6.40), one expects the set of solutions to be more complex and varied than in the case of electrodynamics. Suffice it to mention here the fact that such solutions are often determined by "global geometry," i.e., by boundary conditions at infinity, which often includes (but is not limited to) topology.

In the physics jargon, such solutions are often called "topological." This typically implies that the solutions are parametrized (also) by some characteristic integers. As such integer characterization cannot continuously vary, this provides a degree of stability to such solutions. With the benefit of hindsight, we see that the stationary states of the hydrogen atom – counted by the "quantum numbers" $n, \ell, m \in \mathbb{Z}$ and $m_s = \pm\frac{1}{2}$ – are also stable precisely because of the (half-)integrality of these numbers. In spite of this qualitative similarity, it is important to note that such "topological" solutions – which also include the Dirac monopole from Section 5.2.3 – also exist for the gauge field alone, i.e., for the electromagnetic field without charged particles, the chromodynamics field without quarks, and so on.

Using geometrical and topological methods that are beyond our present scope, it may be shown that non-perturbative solutions of the system (6.37)–(6.40) may be counted by an integer index. These solutions are similar to the vacua of the various ϑ-sectors as discussed in the previous paragraph. However, Alexander Belavin, Alexander Polyakov, Albert S. Schwartz and Yuri Tyupkin [484, 555] showed in 1975 that there is "tunneling" (via so-called BPST *instantons*[16]) from one sector into another, and that the true vacuum is a linear combination $|\vartheta\rangle := \sum_N e^{i\vartheta N}|N\rangle$, for $N \in \mathbb{Z}$. This effectively cancels the conclusion of the discussion about the result (6.86), as it proves that different ϑ-sectors are not independent.

In gauge theories with the Higgs field [☞ Chapter 7] the same role is played by the so-called 't Hooft–Polyakov monopole, and Polyakov also showed that instanton effects in quantum electrodynamics where photons interact with a scalar field (e.g., with the Higgs field) provide the photon with a mass – which is simply unacceptable.

The question why $\vartheta, \vartheta' = 0$, therefore remains unanswered.

— ❦ —

On the other hand, the discussion of the Dirac monopole and its Wu–Yang construction (5.101)–(5.105) as well as 't Hooft and Polyakov's constructions for non-abelian Yang–Mills theories with a Higgs field extends the gauge principle, which originated from the observation that the phases of complex wave-functions are fundamentally unmeasurable quantities, just as are the generalized, matrix-valued phases of wave-function n-tuples such as the chromodynamics triples of quarks (6.1).

> **Conclusion 6.7** *Since the gauge 4-vectors $\mathbb{A}_\mu$ in all Yang–Mills gauge theories are themselves fundamentally unobservable quantities, they may well be multi-valued or otherwise ambiguously defined as functions over spacetime. It is necessary and sufficient only that the gauge fields, the tensor components $\mathbb{F}_{\mu\nu} := [D_\mu, D_\nu]$ (up to a conventional multiplicative constant), are well-defined functions over spacetime.*

As an immediate corollary of this conclusion and the Wu–Yang and then the t'Hoof–Polyakov construction where gauge transformations connect differently specified gauge 4-vector potentials into a class, it follows that Yang–Mills theories, even without appropriately charged matter, may have a class of nontrivial "topological" solutions to their equations of motion. Here "topological" refers to the fact that the existence and the counting of such solutions may be determined by topological methods, depending on the gauge symmetry groups and boundary conditions [☞ also the nontrivial geometries of "empty spacetime" in Section 9.3].

In quantum theories, all (and so also the topologically nontrivial) solutions of the equations of motion may be used as "vacua." Particles – the quanta of all fields, including the gauge fields the nontrivial classical solution of which defines the vacuum – then move through this vacuum, to a first approximation without disturbing it. We thus have:

> **Conclusion 6.8 (background fields)** *Each (and so also the topologically nontrivial) solution of a system of classical equations of motion for all fields defines a "vacuum" in which the quanta of those fields move, to a first approximation, without changing these classical,* ***background fields****.*

6.3.2 The Weinberg–Witten theorem

On the heels of the quark model success, theories of preons and of technicolor became popular in the 1980s. At least some particles among the quarks, leptons, gauge and Higgs bosons were

[16] Instantons in general denote special particle-like objects in field theory, which are well localized not only in (position) space but also in time. That is, instantons are particles that exist but for an instant in time. They were first discovered in non-abelian Yang–Mills theory, but can appear generally in all nonlinearly coupled field theories.

represented as composite states in these models. In an attempt to disqualify such models with a general argument, Stephen Weinberg and Edward Witten [564] proved a theorem now bearing their names:

Theorem 6.1 (Weinberg–Witten) *No quantum field theory in* $(3+1)$*-dimensional spacetime with a Poincaré-covariant and gauge-invariant 4-vector current* J^μ *that satisfies a continuity equation may have a massless particle with a helicity bigger than* $\frac{1}{2}$ *and a non-vanishing charge of* $\int d^3\vec{r}\, J^0$.

No quantum field theory in $(3+1)$*-dimensional spacetime with a Poincaré-covariant and gauge-invariant rank-2 tensor that satisfies a continuity equation may have a massless particle with a helicity bigger than 1.*

Comment 6.4 *The expression "Poincaré-covariant" means that it transforms properly with respect to the Lorentz transformations and translations in spacetime, regardless of whether co- or contra-variant and how many times; the continuity equation for a 4-vector is the usual* $\partial_\mu J^\mu = 0$*, and for a rank-2 tensor,* $T^{\mu\nu}$*, it is* $\partial_\mu T^{\mu\nu} = 0$.

The proof of the theorem is non-perturbative and very general, but the assumptions of the theorem are very stringent. Indeed, it turns out that the theorem in fact does not apply precisely in the models that were meant to be disqualified. For example, in at least several preonic models and in the technicolor theory, there exists an additional non-abelian gauge interaction, the purpose of which is to bind the states that in such models replace some of the particles that are regarded as elementary in the Standard Model. As shown in relations (6.24)–(6.32), the non-abelian (non-commutative) current that is conserved, i.e., satisfies a continuity equation (6.30), is not gauge-invariant, whereby the (prerequisite) assumption of the Weinberg–Witten theorem is not satisfied and the theorem does not apply.

This is related to another unresolved question. Indeed, in a regime where the quark masses are negligible, the chromodynamics Lagrangian density has a doubly larger symmetry: the Dirac spinors representing quarks may be projected into the left- and the right-handed $\Psi_\pm$ (5.58). This Lagrangian density is invariant with respect to an *independent* and global (constant in spacetime) $SU(n_f)$ flavor transformation of the left- and right-handed quarks, so that the full symmetry of this Lagrangian density is $SU(n_f)_L \times SU(n_f)_R \times U(1)_L \times U(1)_R$. Quantum effects in quantum chromodynamics break this symmetry into the "diagonal"[17] $SU(n_f) \times U(1)_B$, where the $U(1)_B$ charge is the baryon number, and which has two significant consequences for the complete understanding of which one also needs material from Section 7.1:

1. The quantum (not spontaneous) breaking of the classical $U(1)_A$ symmetry (the complement of $U(1)_B$ in the product $U(1)_L \times U(1)_R$) is an *anomaly*; instanton solutions from the previous section contribute to this effect as well as to the "strong" *CP*-violation and connect these two unexplained characteristics of quantum chromodynamics☝. Generally, anomalies are an indicator of an inconsistency in the model, but as $U(1)_A$ is an *approximate* symmetry, anomalies indicate an inconsistency in the model only in the unphysical limiting case when the quark masses vanish.
2. The eight spin-0 mesons $(\pi^\pm, \pi^0, K^\pm, K^0, \bar{K}^0, \eta)$ could be identified as the Goldstone bosons [☞ Section 7.1.2] of the symmetry breaking $SU(3)_L \times SU(3)_R \to SU(3)_f$. Of course, quark

[17] For groups of the form $G_L \times G_R$, where the two factors have the same structure but act upon different objects or different aspects of a given object, the "diagonal" subgroup $G \subset G_L \times G_R$ again has the same structure but acts simultaneously both as G_L and as G_R. Only when G_L and G_R are abelian (commutative) does there also exist an "anti-diagonal" complement. Thus $U(1)_L \times U(1)_R = U(1)_D \times U(1)_A$, where the first factor is the diagonal subgroup, so the $U(1)_D$ charge is the sum of $U(1)_L$ and $U(1)_R$ charges; the $U(1)_A$ charge is their difference.

masses are not zero, the broken symmetry was never exact, and neither are the masses of these spin-0 mesons zero, but they are significantly lower than the $SU(3)_f$ singlet meson η'.

3. The eight spin-1 mesons $(\rho^{\pm}, \rho^0, K^{*\pm}, K^{*0}, \bar{K}^{*0}, \phi)$ could be identified as the gauge bosons of the remaining symmetry $SU(3)_f$. Of course, quark masses are not zero, this symmetry is not exact, and neither are the masses of these spin-1 mesons zero. It is not clear, however, if the masses of these mesons (and especially their lightness) may be explained completely as an *explicit* $SU(n_f)$ symmetry-breaking effect, or if there exists a generalization of the Weinberg–Witten theorem that would apply.

7 The Standard Model

The data and facts about elementary particles introduced so far almost completely define the so-called Standard Model of elementary particles; the few missing pieces are:

1. a detailed description of the weak interactions as a gauge theory with the $SU(2)_L$ symmetry group, the gauge bosons of which interact only with fermions of left chirality,
2. a mechanism of providing mass to gauge bosons as well as other particles, and
3. a unification of the weak and the electromagnetic interaction.

Straightforwardly adding an $m^2\|\mathbb{A}_\mu\|^2 := m^2\,\mathrm{Tr}[\mathbb{A}_\mu\eta^{\mu\nu}\mathbb{A}_\nu]$ term to the Lagrangian density would certainly provide the 4-vector potential $\mathbb{A}_\mu$ with the mass m. However, that term is not invariant under the gauge transformation, and explicitly breaks precisely that symmetry because of which $\mathbb{A}_\mu$ was introduced. On the other hand, particles that mediate the weak interaction must have a mass [☞ discussion in the passage on the weak processes (2.56)]. Thus, finding a hopefully more skillful, and certainly gauge-invariant mechanism for providing gauge bosons with a mass is absolutely indispensable for consistency, and we first attend to that matter.

7.1 Boundary conditions and solutions of symmetric equations

Simply inserting an explicit mass term, $m^2\|\mathbb{A}_\mu\|^2$, into the Lagrangian density would destroy precisely that symmetry which is gauged by the 4-vector gauge potential $\mathbb{A}_\mu$ and would thus be self-contradicting. The equations of motion, and so also the Lagrangian density and the Hamilton action, therefore, must remain gauge invariant. Recall, however, that *concrete solutions* of a given system of equations need not have all the symmetries of the system that they solve [☞ Appendix A.1.3 and Comment A.2 on p. 458]. However, if a symmetry X of a system of equations is not a symmetry of a concrete solution f so $X(f) \neq f$, then $X(f)$ is nevertheless a (different) solution of the system, and X is the transformation that maps one solution into the other. Finally, recall that the solutions of a model are not determined only by the system of equations, but also by the boundary (initial, analyticity, etc.) conditions, so it must be that at least some of those conditions distinguish f from $X(f)$.

It is then – in principle – possible to find a solution of gauge-invariant equations of motion that represent massive gauge bosons, i.e., concrete solutions that break precisely the gauge symmetry of the system. This desired solution wherein gauge bosons have a mass breaks the gauge symmetry, and this "boundary" condition must be in the abstract space of gauge phases [☞ Comment 5.3 on p. 170], and this cannot be imposed "by hand" without destroying precisely the symmetry that we are trying to describe.

The mechanism in which a choice of such a condition in the space of gauge phases can be imposed does exist, and it is based on a concatenation of ideas:

1. In 1950, L. D. Landau and V. L. Ginzburg analyzed *phenomenologically* ferromagnet magnetization, following Landau's early work in 1937 [330, 207].[1]
2. P. Anderson's comment and Y. Nambu's research (1960), where the BCS (J. Bardeen, L. N. Cooper and J. R. Shiffer, 1957) model of superconductance is adapted to the description of vacuum in quantum field theory.
3. J. Goldstone's theorem (1961–2) about the Nambu–Goldstone modes (1961), the final proof of which within special relativistic theoretical systems was provided by J. Goldstone, A. Salam and S. Weinberg [214].
4. P. Anderson's work (1963 [15]) about *non-relativistic* plasmons, gauge symmetry and the emergence of effective mass.
5. Independent proposals (1964) by:
   (a) R. Brout and F. Englert,
   (b) P. Higgs,
   (c) G. Guralnik, C. R. Hagen and T. Kibble.
6. In 1971, G. 't Hooft (PhD work advised by M. J. G. Veltman) showed the renormalizability of models where a non-abelian gauge symmetry is broken *via* the Higgs mechanism.
7. 1973: S. Coleman and E. Weinberg analyzed the effect of quantum corrections.

Owing to this complex genesis of this group of ideas, I will not delve into the historical details, but will focus on the description of the effect and its technical details, leaving out the discussion of the individual contributions. Also, I will use the simple expressions such as the "Goldstone mode," the "Higgs mechanism" and the "Higgs particle," with no intention to downplay the relevance of others' contributions. Ever since the LHC at CERN started the experiment of which one of the aims is the detection of the Higgs particle, historical reminders have been (re-)emerging; see, for example, Ref. [252]. For more information, beyond the scope of this book, see Refs. [311, 499, 359, 368].

7.1.1 The Landau–Ginzburg phenomenological description of magnetization

To describe the magnetization of a magnet, introduce the vector function $\vec{M}(\vec{r},t)$, the direction and magnitude of which describe the state of magnetization in the object in an infinitesimally small volume (and which we regard as a tiny domain) at the point $\vec{r}$ at the time t. As the direction and magnitude of magnetization in nearby domains affects the magnetization in a given domain, one expects that the change in the magnetization spreads, at least in the first approximation, as a wave. One therefore expects that the magnetization satisfies an equation of the form

$$\Big[\vec{\nabla}^2 - \frac{1}{v^2}\frac{\partial^2}{\partial t^2}\Big]\vec{M} = \cdots \tag{7.1}$$

where v is the propagation speed of the magnetization wave and where one must supply the missing terms on the right-hand side. If we temporarily define $\mathrm{x} := (vt, \vec{r})$, akin to the relativistic practice, this equation would follow from a Lagrangian density with the "kinetic" term

[1] With the benefit of hindsight, this analysis may today be viewed disparagingly as "fitting" the potential to describe the observed effect. However, this analysis is valuable as it indicates the essential result – precisely the effective potential – that every fundamental, so-called microscopic model *must* reproduce. This then presents an extraordinarily effective criterion to filter the possible, more fundamental models.

$\frac{A}{2}\eta^{\mu\nu}(\partial_\mu\vec{M})\cdot(\partial_\nu\vec{M})$, where A is a constant with appropriate physical units. Adding a potential $V(\vec{M})$, one obtains the classical equations of motion

$$\delta_{ij}\Big[\vec{\nabla}^2 - \frac{1}{v^2}\frac{\partial^2}{\partial t^2}\Big]M^j = -\frac{1}{A}\frac{\partial V}{\partial M^i}, \tag{7.2}$$

where M^i are the components of the magnetization vector in some arbitrary Cartesian coordinate system. In quantum theory, one must of course switch to operators and define an adequate Hamiltonian by integrating (in space) the Hamiltonian density obtained from the Lagrangian density. However, the essence of this procedure is that the ground state of the quantum model is defined by the global minimum of the potential $V(\vec{M})$. This phenomenological approach (based on Landau's theory of phase transitions [☞ for example, Ref. [340]]) then reduces to choosing an appropriate potential function $V(\vec{M})$.

In the familiar example of the harmonic oscillator, the potential $V(x) = \frac{1}{2}m\omega^2x^2$ has a unique minimum, $x = 0$. Correspondingly, the model has a unique ground state for all physically acceptable values of the parameters $\omega, m > 0$. Landau's essential insight, which provides the basis for the Landau–Ginzburg description of magnetization, is that a more complicated potential may well have several distinct minima, depending on the choice of its parameters. Thus, e.g., the anharmonic generalization of the linear harmonic potential, $V(x) = \frac{1}{2}\mu x^2 + \frac{1}{4}\lambda x^4$, has two phases:

1. **when $\mu > 0$:** the minimum of the potential $V(x)$ is at $\check{x}_0 = 0$;
2. **when $\mu < 0$:** the minima of the potential $V(x)$ are at $\check{x}_\pm = \pm\sqrt{-\mu/\lambda}$,

where $\check{x} := \min(V(x))$. The quantum-mechanical expectation value of the observable x (the position of the oscillator) is the average value, $\langle x\rangle = 0$, but in the second case may be "localized" at $\check{x}_\pm$. For the Hilbert space to consist of normalizable bound states and so that the above local minima would in fact be global minima, one requires $\lambda > 0$. [✎ *Why?*] (The $\lambda < 0$ choice implies that $\lim_{x\to\infty} V(x) \to -\infty$, which is unphysical as it prevents the existence of a stable ground state.)

The application of this idea in the Landau–Ginzburg phenomenological model also uses the fact that the potential is a scalar function of the vector $\vec{M}(\vec{r},t)$, and so can depend only on the magnitude $|\vec{M}| = \sqrt{\delta_{ij}M^i(\vec{r},t)M^j(\vec{r},t)}$. If one also requires that the potential is an analytic function, it must be that

$$V(\vec{M}) = \tfrac{1}{2}\mu|\vec{M}|^2 + \tfrac{1}{4}\lambda(|\vec{M}|^2)^2 + \cdots \tag{7.3}$$

It then follows that the ground state of the quantum-mechanical description of magnetization is determined by minimizing the potential:

1. **if $\mu > 0$:** the minimum of the potential $V(\vec{M})$ is at $\langle\vec{M}\rangle_0 = 0$;
2. **if $\mu < 0$:** the minima of the potential $V(\vec{M})$ are at $\langle\vec{M}\rangle_> = \sqrt{-\mu/\lambda}\,\hat{M}$,

where $\hat{M}$ is one of a continuum of *arbitrary* unit vectors in the 3-dimensional space in which the magnetization $\vec{M}(\vec{r},t)$ is a 3-vector – and which *coincides* with the "actual," real space.

Before we return to the question: "*Which* arbitrary direction $\hat{M}$?," let us finish the parametrization of the Landau–Ginzburg model by noting that one of course knows that the magnet loses its magnetization when heated to a sufficiently high temperature. It then follows that μ must be a function of temperature, and so that $\mu < 0$ for $T < T_c$, whereas $\mu > 0$ for $T > T_c$. The concrete choice of the $\mu = \mu(T)$ dependence, as well as the presence of an additional $(|\vec{M}|^2)^3$ term in the expansion of the potential (7.3) in the original Landau–Ginzburg potential stems from additional requirements to also describe successfully physical characteristics such as the susceptibility – which may be ignored for the present purposes. We then simply adopt

$$V(\vec{M}) = \tfrac{1}{2}\mu_0(T^2 - T_c^2)|\vec{M}|^2 + \tfrac{1}{4}\lambda(|\vec{M}|^2)^2 + \cdots, \qquad \mu_0, \lambda > 0. \tag{7.4}$$

It then follows that the ground state of the quantum-mechanical description of the magnetization is determined by the minimum of the potential:

$$\min\big[V(\vec{M})\big] = \begin{cases} \langle\vec{M}\rangle_> = \sqrt{\mu_0(T_c^2-T^2)/\lambda}\,\hat{M} & \text{when } T < T_c; \\ \langle\vec{M}\rangle_0 = 0 & \text{when } T \geqslant T_c. \end{cases} \tag{7.5}$$

Notice that $\min[V(\vec{M})]$ is a continuous (but not smooth) function of the temperature.

Thus, at a sufficiently high temperature, the object has no magnetization, $\langle\vec{M}\rangle_0 = 0$, whereas lowering of the temperature below T_c causes the object to *spontaneously magnetize*. That is, we have that $\langle\vec{M}\rangle_> = \sqrt{\mu_0(T_c^2-T^2)/\lambda}\,\hat{M}$, in the direction $\hat{M}$ – which remains undetermined by the dynamics of the model.

In an actual, real situation, there *always* exists some small *external* magnetic field, which "chooses" a preferred direction: The interactions of the tiny domains with this external magnetic field then direct them opposite to this external magnetic field, which removes the arbitrariness of the choice of $\hat{M}$.

Comment 7.1 *In the Landau–Ginzburg description of magnetization, the 3-dimensional space in which the magnetization vector $\vec{M}(\vec{r},t)$ has magnitude and direction is in fact the "actual," real space in which we ourselves live and move. In the other applications of this idea, this need not be so.*

Classical analysis straightforwardly shows the following properties:

1. As temperature decreases through the critical value T_c, the character of the potential $V(\vec{M})$ changes suddenly. However, the gradient of the potential (the generalized "force" that moves the magnetization of the object) in fact always vanishes at the point $\vec{M} = 0$; that is a consequence of the fact that an analytic potential function must depend on $|\vec{M}|^2$ and not on $|\vec{M}|$. This necessitates an influence to "move" the system from $\vec{M} = 0$, and this external influence then also fixes the ultimate direction of the magnetization $\hat{M}$. This may literally be an external influence (a small external magnetic field), or also a simply random (quantum) fluctuation within the object/system itself.
2. Immediately after the transition, when $T \lesssim T_c$, the potential has a very mild "slope" near $\vec{M} = 0$, the "inclination" of which grows with the distance from the $\vec{M} = 0$ point. The global minimum of the potential function moves from $\vec{M} = 0$ to a circle of "radius" $|\vec{M}(T)| = \sqrt{\mu_0(T_c^2-T^2)/\lambda}$, which grows as the temperature decreases: $T < T_c$ and $T \to 0$.
3. Even if moved by an external influence, the actual value $\langle\vec{M}\rangle$ will lag behind the growing value of the "radius" $M(T)$. The change in the magnetization from $\langle\vec{M}\rangle_0$ towards $\langle\vec{M}\rangle_>$ will be accelerated, just as with rolling down a steepening slope.
4. When the system reaches $\langle\vec{M}\rangle_>$, the motion regime turns oscillatory around $\langle\vec{M}\rangle_>$, where the loss of energy through interaction with the environment leads to a stabilization of the value $\langle\vec{M}\rangle \to \langle\vec{M}\rangle_>$.
5. In this entire process, the system has (through interaction with the environment) lost the energy

$$\Delta V := V\Big(|\vec{M}|{=}0\Big) - V\Big(|\vec{M}|{=}\sqrt{\mu_0(T_c^2-T^2)/\lambda}\Big), \tag{7.6}$$

which somewhat akin to the latent heat of a first-order (discontinuous) phase transition such as freezing of water. To be precise, magnetization is however a second-order (continuous) transition, where $\langle\vec{M}\rangle$ continuously changes between its values.

7.1.2 The Goldstone theorem

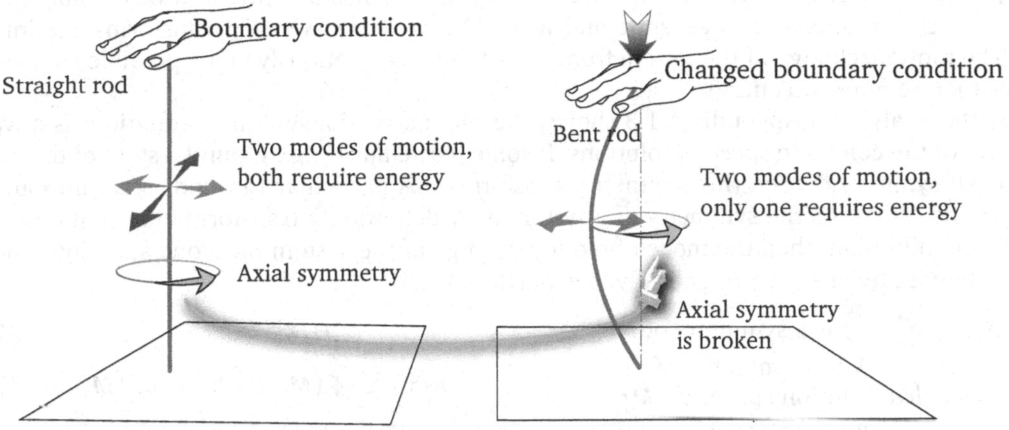

Figure 7.1 The example of a straight and a bent rod. Notice that the mode of motion of the bent rod that (ignoring friction) requires no energy is identical with the symmetry that is broken by bending. The difference is induced by changing the boundary conditions.

Before we apply the ideas from the previous section to a scalar field in a relativistic theory, consider a simple model, shown in Figure 7.1. This model illustrates several of the characteristics of symmetry breaking, with a faithful analogy in the case of spontaneous magnetization as the temperature drops.

The straight rod has a manifest axial symmetry. Analogously, at temperatures above T_c, a magnetic material has $\langle \vec{M} \rangle = 0$, i.e., the magnetic domain orientation distribution is spherically symmetric. The bent rod does not have the axial symmetry, but its horizontal rotation requires no energy if we neglect friction. Analogously, at temperatures below T_c, the magnetic material has $\langle \vec{M} \rangle \neq 0$, i.e., the magnetic domain orientation distribution is no longer spherically symmetric. However, fluctuations in the magnetization orientation form a wave (dubbed a *magnon*), the propagation of which in the magnet requires very little energy, which fails to vanish only because of imperfections and finiteness of a real, physical magnet.

Similarly, the molecular velocity distribution in any fluid is spherically symmetric. When the temperature of the fluid drops below the freezing point, the material can form a crystal, in which molecules move only in modes permitted by the crystalline geometry; this breaks the continuous spherical symmetry into the discrete crystalline symmetry. Correspondingly, there appear *phonons* in the crystal, the propagation of which in the crystal requires very little energy, which fails to vanish because of the imperfection and finiteness of the real, physical crystal.

These examples exhibit the essence of the Goldstone theorem, a technically simplified form of which is:

> ***Theorem 7.1*** *For every continuous (and local) symmetry of a system (and for which there then exists a current that satisfies the continuity equation and a conserved charge) that is not a symmetry of the vacuum (ground state), there exists an excitation (a motion/fluctuation mode) of the system that requires no energy.*

The idea of the proof is very simple: the ground state that breaks the continuous symmetry is only one of continuously many possible such states. In the example of a bent rod, the direction of bending is one of continuously many arbitrary directions; similarly, the ultimate direction of magnetization $\langle\vec{M}\rangle$ and of the crystalline lattice represent arbitrary choices from among continuously many possibilities. Thus, a system with a broken symmetry has a continuum of possible ground states – which are of course degenerate and which the broken symmetry transforms one into another. The motion/change of the system from one of these continuously many possible choices into another then requires no energy.

As the analysis in Appendix A.1.3 shows, the symmetry of a system of equations is always a symmetry of the complete space of solutions. If some particular – e.g., ground – state of the system is not itself symmetric, then this symmetry transforms this particular (ground) state into another (also ground) state. As the symmetry of a system is by definition a transformation that commutes with the Hamiltonian, then the mode of motion/change of the system from one state into another cannot require any energy. Symbolically (see Section A.1.3):

$X\,(=(X^\dagger)^{-1})$ is a symmetry of $\boldsymbol{M}$.	$\leftrightarrow$	$[H,X]=0.$	(7.7a)
X is a symmetry of the *complete* solution space, $\mathscr{X}(\boldsymbol{M})$.	$\leftrightarrow$	$X\lvert\Psi\rangle\in\mathscr{X}(\boldsymbol{M})\Leftrightarrow\lvert\Psi\rangle\in\mathscr{X}(\boldsymbol{M}).$	(7.7b)
$\lvert\Psi\rangle\in\mathscr{X}(\boldsymbol{M})$ breaks X.	$\leftrightarrow$	$(X\lvert\Psi\rangle=\lvert\Psi'\rangle)\neq\lvert\Psi\rangle.$	(7.7c)
$\Rightarrow$ The "motion" $\lvert\Psi\rangle\to\lvert\Psi'\rangle$ in $\mathscr{X}(\boldsymbol{M})$ requires no energy.	$\leftrightarrow$	$\langle\Psi\lvert H\rvert\Psi\rangle-\langle\Psi'\lvert H\rvert\Psi'\rangle=0.$	(7.7d)

The final result follows since

$$\langle\Psi'\lvert H\rvert\Psi'\rangle=\langle\Psi\lvert X^\dagger HX\rvert\Psi\rangle\overset{(7.7a)}{=}\langle\Psi\lvert X^{-1}XH\rvert\Psi\rangle=\langle\Psi\lvert H\rvert\Psi\rangle. \tag{7.8}$$

Since $\langle\Psi'\lvert H\rvert\Psi'\rangle=\langle\Psi\lvert H\rvert\Psi\rangle$, the transformation/motion $\lvert\Psi\rangle\to\lvert\Psi'\rangle$ cannot possibly require any energy. ☑

The careful Reader must have noticed the minor differences in the implied definitions and concepts in the above several paragraphs, and a technically complete treatment of the Goldstone theorem requires a consistent and technically precise connection between these ideas. Besides, one must keep in mind the finiteness of the resolution of any concrete measurement, which then implies limitations in the definition of physical quantities. For example, the "bare" electric charge is not distinguishable from a system consisting of that same electric charge but together with the electromagnetic field created by that charge, the intensity of which is below the threshold of observability. That is, the "bare" electric charge is indistinguishable from the electric charge surrounded by a sea of photons that are either sufficiently "soft" (of small frequency) or are reabsorbed too fast to permit detection.

7.1.3 *The Higgs effect for gauge symmetry*

As we begin analyzing the gradual development of a model based on the ideas from the previous Sections 7.1.1–7.1.2, note that in field theory the quadratic term provides a field with a mass, as was mentioned in the beginning of this section.

Field shift

A simple relativistic Lagrangian density (constructed in the spirit of the discussion in Section 7.1.1) for one, real, scalar field, $\phi(x)$ is

$$\mathscr{L}_0=\tfrac{1}{2}\eta^{\mu\nu}(\partial_\mu\phi)(\partial_\nu\phi)-\tfrac{\varkappa}{2}\big(\tfrac{mc}{\hbar}\big)^2\phi^2-\tfrac{1}{4}\lambda\phi^4, \tag{7.9}$$

so that the classical, Euler–Lagrange equation of motion is

$$0=\partial_\mu\frac{\partial\mathscr{L}_0}{\partial(\partial_\mu\phi)}-\frac{\partial\mathscr{L}_0}{\partial\phi}=\partial_\mu(\eta^{\mu\nu}\partial_\nu\phi)+\varkappa\big(\tfrac{mc}{\hbar}\big)^2\phi+\lambda\phi^3, \tag{7.10}$$

that is,[2]

$$\big[\tfrac{1}{c^2}\partial_t^2 - \vec{\nabla}^2 + \varkappa\big(\tfrac{mc}{\hbar}\big)^2\big]\phi = -\lambda\phi^3 \quad\Leftrightarrow\quad \big[-E^2 + \vec{p}^2c^2 + \varkappa m^2c^4\big]\phi = (\hbar^2c^2\lambda)\phi^3, \tag{7.11}$$

identifying $\sqrt{\varkappa}\,m$ as the mass of the ϕ field, so we fix $\varkappa \to 1$ for now.

Comment 7.2 *Since $[\int d^4x\,\mathscr{L}_0] = [\hbar] = ML^2/T$, then $[\mathscr{L}_0] = M/L^2T$. As the metric tensor $\eta_{\mu\nu}$ and its inverse $\eta^{\mu\nu}$ are dimensionless and $[\partial_\mu] = L^{-1}$, it follows that the units of the so-defined scalar field $[\phi] = \sqrt{M/T}$ and the units of λ are $[\lambda] = T/ML^2$. In turn, comparing the ϕ^2-terms in the Lagrangian density (7.9), $[\partial_\mu] = [\frac{mc}{\hbar}]$ and m is really a mass, $[m] = M$.*

The potential energy density in the Lagrangian density (7.9) is $\mathscr{V}_0 = \frac{\varkappa}{2}\big(\frac{mc}{\hbar}\big)^2\phi^2 + \frac{1}{4}\lambda\phi^4$, and the Hamiltonian density is

$$\mathscr{H}_0 := (\dot\phi)\frac{\partial\mathscr{L}_0}{\partial(\dot\phi)} - \mathscr{L}_0 = \tfrac{1}{2}\big[\tfrac{1}{c^2}\dot\phi^2 + (\vec\nabla\phi)\cdot(\vec\nabla\phi)\big] + \tfrac{\varkappa}{2}\big(\tfrac{mc}{\hbar}\big)^2\phi^2 + \tfrac{1}{4}\lambda\phi^4. \tag{7.12}$$

The expressions (7.11) and (7.12) indicate that changing $\varkappa = 1 \to 1$, aiming to describe a symmetry breaking as in Section 7.1.1, implies that the mass of the ϕ field has become imaginary ($\sqrt{\varkappa}m = m \to im$) – which is nonsensical in classical physics.

However, recall that the parameters in the classical Lagrangian are only auxiliary, helping parameters, and that the true, physically measurable values are obtained only after an adequate redefinition of those parameters, i.e., after renormalization [☞ Sections 5.3.3 and 6.2.4]. With that idea, in 1973 Sydney Coleman and Erick Weinberg analyzed the effect of the electromagnetic field on the mass of an electrically charged scalar particle [112] and found that there exists a regime (phase) of the parameter m, λ choices where the effective mass of the field (owing to renormalization effects) really does become imaginary and so induces the breaking of a symmetry, i.e., indicates an instability of the state with the unbroken symmetry. With this in mind, we now simply change $m^2 \to -m^2$ without delving into the detailed reasons and dynamics of this change.

With the potential energy density $\widetilde{\mathscr{V}}_0 = -\frac{1}{2}\big(\frac{mc}{\hbar}\big)^2\phi^2 + \frac{1}{4}\lambda\phi^4$, the system is unstable at $\phi_0 = 0$, and the global minima appear at the values $\phi \to \pm\frac{mc}{\hbar\sqrt{\lambda}}$. One thus expects that, after enough time, the system settles at either $\langle\phi\rangle = +\frac{mc}{\hbar\sqrt{\lambda}}$ or $\langle\phi\rangle = -\frac{mc}{\hbar\sqrt{\lambda}}$. Feynman's perturbative computation would then have to be adapted so that all fields vanish at the chosen classical minimum, i.e., that the fields represent fluctuations around that minimum. It is thus convenient to introduce one of the two substitutions:

$$\text{either}\quad \varphi_+ := \phi - \frac{mc}{\hbar\sqrt{\lambda}}, \qquad \text{when } \langle\phi\rangle = +\frac{mc}{\hbar\sqrt{\lambda}}, \qquad \text{so } \langle\varphi_+\rangle = 0, \tag{7.13a}$$

$$\text{or}\quad \varphi_- := \phi + \frac{mc}{\hbar\sqrt{\lambda}}, \qquad \text{when } \langle\phi\rangle = -\frac{mc}{\hbar\sqrt{\lambda}}, \qquad \text{so } \langle\varphi_-\rangle = 0, \tag{7.13b}$$

whereby the Lagrangian density (7.9), with the sign in the mass-term changed by hand,

$$\widetilde{\mathscr{L}}_0 = \tfrac{1}{2}\eta^{\mu\nu}(\partial_\mu\phi)(\partial_\nu\phi) + \tfrac{1}{2}\big(\tfrac{mc}{\hbar}\big)^2\phi^2 - \tfrac{1}{4}\lambda\phi^4, \tag{7.14}$$

becomes – corresponding to the choice (7.13) – one of the two Lagrangian densities:

$$\text{either}\quad \mathscr{L}_+ = \tfrac{1}{2}\eta^{\mu\nu}(\partial_\mu\varphi_+)(\partial_\nu\varphi_+) - \big(\tfrac{mc}{\hbar}\big)^2\varphi_+^2 - \tfrac{mc\sqrt{\lambda}}{\hbar}\varphi_+^3 - \tfrac{1}{4}\lambda\varphi_+^4 + \tfrac{m^4c^4}{4\lambda\hbar^4}, \tag{7.15a}$$

[2] Identification of the operator ∂_μ with the components of the 4-momentum is obtained fastest by using the quantum-mechanical relations in the coordinate representation, $H = i\hbar\partial_t = i\frac{\hbar}{c}\partial_0$ and $\vec{p} = -i\hbar\vec\nabla$, so that substituting the eigenvalues, $\hbar^2\partial_t^2 \mapsto -E^2$ and $\hbar^2\vec\nabla^2 \mapsto -\vec{p}^2$ [☞ Digression 3.6 on p. 93, and the relation (3.37) that holds when $\lambda \to 0$].

$$\text{or}\quad \mathscr{L}_- = \tfrac{1}{2}\eta^{\mu\nu}(\partial_\mu\varphi_-)(\partial_\nu\varphi_-) - \left(\tfrac{mc}{\hbar}\right)^2\varphi_-^2 + \tfrac{mc\sqrt{\lambda}}{\hbar}\varphi_-^3 - \tfrac{1}{4}\lambda\varphi_-^4 + \tfrac{m^4c^4}{4\lambda\hbar^4}. \tag{7.15b}$$

For these "shifted" fields (7.13), the mass is real, $m_\pm = \sqrt{2}m$, as the sign of the quadratic term turned negative, and other than the anharmonic term $\varphi_\pm^4$, now there is also a cubic term, $\varphi_\pm^3$. Finally, the total value of the Lagrangian density shifted by the constant $+\frac{m^4c^4}{4\lambda\hbar^4}$, which means that the value of the total energy density (Hamiltonian density) of the system *decreased* by $-\frac{m^4c^4}{4\lambda\hbar^4}$; recall, $\mathscr{H} = p_i\dot{q}^i - \mathscr{L}$. This contribution to the energy of the system is the excess energy density of the phase transition from the phase where the effective mass is real and $\langle\phi\rangle = 0$ into the phase where the effective mass is imaginary and $\langle\phi\rangle = \pm\frac{mc}{\hbar\sqrt{\lambda}}$.[3]

As the minimum of the total energy in the phase with $\langle\phi\rangle = \pm\frac{mc}{\hbar\sqrt{\lambda}}$ is lower than that in the phase with $\langle\phi\rangle = 0$, it follows that the ground state of the system after the sign change of the quadratic term must have one of the two possible values: $\langle\phi\rangle = \pm\frac{mc}{\hbar\sqrt{\lambda}}$, and the choice between these two values is arbitrary.

Conclusion 7.1 *The Lagrangian density (7.9) describes two phases of the system:*

1. *the symmetric phase, where $\varkappa > 0$ and $\langle\phi\rangle = 0$,*
2. *the broken symmetry phase, where $\varkappa < 0$ and $\langle\phi\rangle = \pm\frac{mc}{\hbar\sqrt{\lambda}} \neq 0$.*

Typically, the parameter $\varkappa$ is a function of temperature and turns negative when the temperature drops below some critical value. The change $\varkappa > 0 \;\leftrightarrow\; \varkappa < 0$ is, of course, a ***phase transition****, for which the excess energy density equals $\frac{m^4c^4}{4\lambda\hbar^4}$, as seen in the expressions (7.15).*

The Lagrangian density (7.9), and then also the equations of motion (7.11), have the symmetry $\varpi : \phi \to -\phi$. However, when the parameter m^2 turns into $-m^2$ and the mass becomes unphysically imaginary, as in the Lagrangian density (7.14), the state where $\langle\phi\rangle = 0$ becomes unstable. Instead, one chooses one of the two states where $\langle\phi\rangle = \pm\frac{mc}{\hbar\sqrt{\lambda}}$ and, corresponding to the change in the notation (7.13), one uses one of the two Lagrangian densities (7.15). The transformation ϖ is still a symmetry of the system:

$$\varpi : \phi \to -\phi \quad\Rightarrow\quad \varphi_\pm \to -\varphi_\pm \mp 2\tfrac{mc}{\hbar\sqrt{\lambda}} \quad\Rightarrow\quad \mathscr{L}_\pm \to \mathscr{L}_\pm. \tag{7.16}$$

As this is a discrete transformation, the Goldstone theorem does not apply. However, there evidently exists a mapping $\varphi_\pm \to -\varphi_\mp$ that turns $\mathscr{L}_\pm(\varphi_\pm) \to \mathscr{L}_\mp(\varphi_\mp)$; i.e., that connects the two existing and degenerate vacua.[4] By breaking discrete symmetries, the existence of such a discrete mapping is a property that is closest to the existence of a Goldstone mode. Although this "goldstonesque" transformation $\varphi_\pm \to -\varphi_\mp$ is not identically equal to the initial symmetry (7.16), the two transformations are isomorphic: both are reflections, albeit across different points in the field space.

Basic building blocks of Feynman diagrams correspond to the terms in the Lagrangian density (7.15). Terms that are quadratic in φ define the "propagator," i.e., the Green function. Its physical meaning is that the change in the φ field in one spacetime point correlates with a change in a neighboring point. For a scalar field, this function is represented in Feynman diagrams by a

[3] The contribution to the total energy is, of course, $-\int d^3\vec{r}\,\frac{m^4c^4}{4\lambda\hbar^4}$, which diverges because of the integral over the infinitely large space. However, this is but one example of the need to renormalize the reference energy level of the "empty spacetime."

[4] When the number of degenerate states is finite, as here, it makes sense to construct (anti)symmetrized linear combinations $\mathscr{L}_+$ and $\mathscr{L}_-$. However, we will be interested in the breaking of *continuous* symmetries, where this is not possible – or at least does not have the same physical meaning.

dashed line. Cubic and quartic terms, respectively, represent correlated changes in three and four spacetime points, and are represented by vertices where, respectively, three and four dashed lines meet:

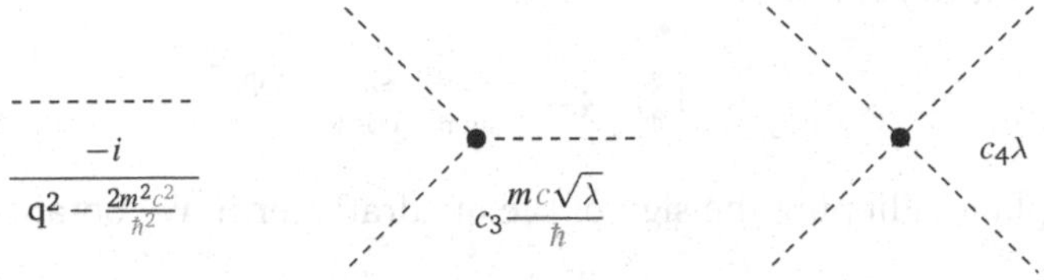

(7.17)

where the concrete choice of (combinatorial and normalizing) constants c_3, c_4 is not relevant here. Note, however, that the triple vertex may be obtained from the quadruple one by formally "ending" one of its four edges – as if the field ϕ here *sinks* into the vacuum:

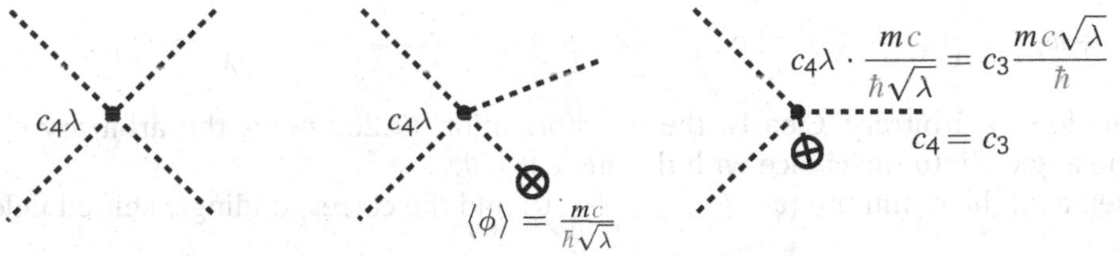

(7.18)

or *wells* up from it [☞ discussion about the diagram (3.82)]. The nonzero value $\langle\phi\rangle$ indicates that the number of ϕ-quanta is not conserved in the vacuum with the broken symmetry. In contrast, the number of $\varphi_\pm$-quanta is conserved as $\langle\varphi_\pm\rangle = 0$, and this is the normal mode for describing the system in vacuum with the broken symmetry. After the substitution $\phi \to \varphi_\pm$, the system has only elementary diagrams of the type (7.17), from which one can, of course, construct much more complex Feynman diagrams, and so also much more complex processes. However, in the φ_+- or φ_--description (depending on the choice of the vacuum), there are no diagrams with "sinks" or "sources" such as in the ϕ-description (7.18).

Finally, the Feynman diagrams represent corresponding perturbative contributions, understanding that the fields fluctuate about their classical solutions. Thereby, the choice of the Lagrangian density $\mathscr{L}_\pm$ implies that the ϕ field fluctuates about the expectation value $\langle\phi\rangle = \pm\frac{mc}{\hbar\sqrt{\lambda}}$, so $\langle\varphi_\mp\rangle = 0$. Similarly, just as the ground state of the linear harmonic oscillator is centered at $x = 0$, so is the ground state of the model with the Lagrangian density $\mathscr{L}_+$ centered at $\varphi_+ = 0$, and for $\mathscr{L}_-$ at $\varphi_- = 0$. These then are two distinct models of the system, which the "goldstonesque" transformation $\varphi_\pm \to -\varphi_\mp$ maps one into the other, and proves them to be physically equivalent descriptions of the same system.

Conclusion 7.2 *In the* ***symmetric*** *phase, one uses the Lagrangian density (7.9) with $\varkappa > 0$ and the ϕ field, so that the Feynman diagrams (7.17) – without the triple vertex – correspond to the so-described processes. When $\varkappa < 0$, for the description of this* ***non-symmetric*** *phase one picks either the Lagrangian density (7.15a) or (7.15b) and, correspondingly, either φ_+ or φ_-; correspondingly, the Feynman diagrams (7.17) change their meaning although the mechanics of the computations remains the same.*

Breaking of continuous symmetry

One of the simplest generalizations of the above results to the case where a continuous symmetry is broken by the choice of the ground state uses two real scalar fields in place of one: $\phi(\mathbf{x}) \to (\phi_1(\mathbf{x}), \phi_2(\mathbf{x}))$. The Lagrangian density is chosen akin to (7.9)

$$\mathscr{L}_{2d} = \tfrac{1}{2}\eta^{\mu\nu}\delta^{ij}(\partial_\mu\phi_i)(\partial_\nu\phi_j) - \tfrac{1}{2}\big(\tfrac{mc}{\hbar}\big)^2(\delta^{ij}\phi_i\phi_j) - \tfrac{1}{4}\lambda(\delta^{ij}\phi_i\phi_j)^2. \tag{7.19}$$

Owing to the specific choice of the potential function, the Lagrangian density (7.19) is invariant under the action of an arbitrary rotation

$$\varpi_\vartheta : \begin{bmatrix}\phi_1\\ \phi_2\end{bmatrix} \to \begin{bmatrix}\phi'_1\\ \phi'_2\end{bmatrix} := \begin{bmatrix}\cos\vartheta & -\sin\vartheta\\ \sin\vartheta & \cos\vartheta\end{bmatrix}\begin{bmatrix}\phi_1\\ \phi_2\end{bmatrix} \tag{7.20}$$

in the abstract (ϕ_1, ϕ_2)-plane. Flipping the sign of the quadratic term, we obtain

$$\widetilde{\mathscr{L}}_{2d} = \tfrac{1}{2}\eta^{\mu\nu}\delta^{ij}(\partial_\mu\phi_i)(\partial_\nu\phi_j) + \tfrac{1}{2}\big(\tfrac{mc}{\hbar}\big)^2(\delta^{ij}\phi_i\phi_j) - \tfrac{1}{4}\lambda(\delta^{ij}\phi_i\phi_j)^2, \tag{7.21}$$

where the potential energy density is easily found to have continuously many minima, forming the *circle*

$$(\phi_1^2 + \phi_2^2)\big|_{\min} = \tfrac{m^2c^2}{\hbar^2\lambda}, \qquad \text{i.e.,} \qquad (\phi_1, \phi_2)_{\min} = \Big(\tfrac{mc}{\hbar\sqrt{\lambda}}\cos\theta, \tfrac{mc}{\hbar\sqrt{\lambda}}\sin\theta\Big), \tag{7.22}$$

where the angle θ is arbitrary. Clearly, the transformation (7.20) maps the arbitrary choice of minima at the angle θ into the choice with the angle $\theta + \vartheta$.

Consider, e.g., the minimum $(\phi_1, \phi_2) \to (\tfrac{mc}{\hbar\sqrt{\lambda}}; 0)$ and the correspondingly shifted fields:

$$\varphi_1 = \phi_1 - \tfrac{mc}{\hbar\sqrt{\lambda}}, \qquad \varphi_2 = \phi_2. \tag{7.23}$$

With these, the Lagrangian density (7.19) becomes

$$\begin{aligned}\widetilde{\mathscr{L}}_{2d} &= \tfrac{1}{2}\eta^{\mu\nu}\delta^{ij}(\partial_\mu\varphi_i)(\partial_\nu\varphi_j) - \big(\tfrac{mc}{\hbar}\big)^2\varphi_1^2\\ &\quad - \tfrac{mc\sqrt{\lambda}}{\hbar}\varphi_1(\varphi_1^2 + \varphi_2^2) - \tfrac{1}{4}\lambda(\varphi_1^2 + \varphi_2^2)^2 + \tfrac{m^4c^4}{4\lambda\hbar^4} \end{aligned}\tag{7.24a}$$

$$\begin{aligned} &= \tfrac{1}{2}\eta^{\mu\nu}(\partial_\mu\varphi_1)(\partial_\nu\varphi_1) - \big(\tfrac{mc}{\hbar}\big)^2\varphi_1^2 - \tfrac{mc\sqrt{\lambda}}{\hbar}\varphi_1^3 - \tfrac{1}{4}\lambda\varphi_1^4\\ &\quad + \tfrac{1}{2}\eta^{\mu\nu}(\partial_\mu\varphi_2)(\partial_\nu\varphi_2) - \tfrac{1}{4}\lambda\varphi_2^4\\ &\quad - \tfrac{mc\sqrt{\lambda}}{\hbar}\varphi_1\varphi_2^2 - \tfrac{1}{2}\lambda\varphi_1^2\varphi_2^2 + \tfrac{m^4c^4}{4\lambda\hbar^4}, \end{aligned}\tag{7.24b}$$

where we separated the terms that produce the dynamics of the φ_1 and the φ_2 fields into two separate rows, and left the coupling terms and the excess energy density for the last row.

Just as in the one-dimensional example (7.9)–(7.16), the φ_1 field has acquired a real mass $m_1 = \sqrt{2}|m|$, as well as an additional cubic term, besides the φ_1^4 term. However, the $\varphi_2 = \phi_2$ field has *lost* its mass, and only has a φ_2^4 term in the potential! Finally, the φ_1 and the φ_2 fields are coupled via the $\varphi_1\varphi_2^2$ and the $\varphi_1^2\varphi_2^2$ terms, in the sense that the Euler–Lagrange equations of motion form a coupled system. The transformation

$$\varpi_\vartheta : \begin{bmatrix}\varphi_1\\ \varphi_2\end{bmatrix} \to \begin{bmatrix}\varphi'_1\\ \phi'_2\end{bmatrix} := \begin{bmatrix}\cos\vartheta & -\sin\vartheta\\ \sin\vartheta & \cos\vartheta\end{bmatrix}\begin{bmatrix}\varphi_1\\ \varphi_2\end{bmatrix} + \begin{bmatrix}\tfrac{mc}{\hbar\sqrt{\lambda}}(\cos\vartheta - 1)\\ \tfrac{mc}{\hbar\sqrt{\lambda}}\sin\vartheta\end{bmatrix} \tag{7.25}$$

is still a symmetry of the system – and is merely rewritten into the new coordinates, making it evident that this is not a rotation about the coordinate origin in the (φ_1, φ_2)-plane. Since the rotations in the (φ_1, φ_2)-plane about the point $(0,0)$ are not symmetries, the fact that the φ_2 field has lost its mass indicates that (in this choice of the parametrization of the system) φ_2 represents the Goldstone boson.

Conclusion 7.3 *Following Conclusion 7.2 on p. 259, one may use the Lagrangian densities (7.19) and (7.24), respectively, to describe the symmetric ($\varkappa > 0$) and the non-symmetric ($\varkappa < 0$) phases of the system. Unlike in the situation in Conclusion 7.2, the non-symmetric phase now contains a* **continuous degeneracy**: *any one of continuously many scalar fields that satisfy the relations (7.22) represents a minimum of the potential in the non-symmetric phase. Any one concrete choice, such as (7.23), then represents one concrete spontaneous breaking of the original symmetry, from among continuously many such choices.*

Digression 7.1 Note that after ad hoc changing the sign from the Lagrangian density (7.19) into the Lagrangian density (7.21), varying the ϕ field produces the change in the equation of motion:

$$[\Box + (\tfrac{mc}{\hbar})^2]\phi_j = -\lambda\phi_j\|\phi\|^2 \qquad \to \qquad [\Box - (\tfrac{mc}{\hbar})^2]\phi_j = -\lambda\phi_j\|\phi\|^2, \tag{7.26a}$$

where $\Box = \frac{1}{c^2}\frac{\partial^2}{\partial t^2} - \vec{\nabla}^2$ is the wave operator, a.k.a., the d'Alembertian. In the absence of interactions ($\lambda \to 0$), the Klein–Gordon operator $[\Box + (\frac{mc}{\hbar})^2]$ thus changes into $[\Box - (\frac{mc}{\hbar})^2]$. Since the standard Klein–Gordon operator corresponds to the relation (3.36), we have

$$[\Box + (\tfrac{mc}{\hbar})^2]\phi_j = 0 \;\Leftrightarrow\; E^2 = \vec{p}^2c^2 + m^2c^4 \;\Leftrightarrow\; \frac{\vec{p}^2}{E^2/c^4} = v^2 = c^2\Big(1 - \frac{m^2c^4}{E^2}\Big) < c^2. \tag{7.26b}$$

However, flipping the sign of the $m^2\phi^2$ term, by hand, would produce

$$[\Box - (\tfrac{mc}{\hbar})^2]\phi_j = 0 \;\Leftrightarrow\; E^2 = \vec{p}^2c^2 - m^2c^4 \;\Leftrightarrow\; \frac{\vec{p}^2}{E^2/c^4} = v^2 = c^2\Big(1 + \frac{m^2c^4}{E^2}\Big) > c^2. \tag{7.26c}$$

Thus, simply flipping the sign of the $m^2\phi^2$ term in the Lagrangian density would have two correlated consequences:

1. The vacuum where $\langle\phi\rangle = 0$ would become unstable, as a local *maximum* of the potential energy density, which indicates the tendency of the system to transition into a state where $\langle\phi\rangle = \frac{mc}{\hbar\sqrt{\lambda}} \neq 0$.
2. The ϕ field would become tachyonic (superluminal): it would propagate faster than the speed of light in the "false" vacuum where $\langle\phi\rangle = 0$; by transitioning into the "true" vacuum where $\langle\phi\rangle = \frac{mc}{\hbar\sqrt{\lambda}} \neq 0$, ϕ (i.e., now φ) becomes again a physical, tardionic (subluminal) field.

However, the sign of the (quadratic) mass term is in reality a continuously variable parameter, and the evolution of the system is considerably more involved than could be shown here; see for example [81, 20]. Nevertheless, the appearance of a tachyonic particle/state in a simple analysis as shown here does signal vacuum instability.

The correspondence between the broken symmetry and the Goldstone boson is not perfectly evident in this parametrization, since φ_2 does not represent rotations. This correspondence becomes clearer by using, instead of (7.23), the nonlinear transformation

$$\phi_1 = \rho\cos\theta, \qquad \phi_2 = \rho\sin\theta, \tag{7.27}$$

with which the Lagrangian density with the flipped sign of the quadratic term becomes

$$\widetilde{\mathscr{L}}_{2d} = \tfrac{1}{2}\eta^{\mu\nu}\left[(\partial_\mu\rho)(\partial_\nu\rho) + \rho^2(\partial_\mu\theta)(\partial_\nu\theta)\right] + \tfrac{1}{2}\left(\tfrac{mc}{\hbar}\right)^2\rho^2 - \tfrac{1}{4}\lambda\rho^4. \tag{7.28a}$$

Finding the minimum on the circle of radius $\rho = \frac{mc}{\hbar\sqrt{\lambda}}$ and after the substitution

$$\varrho := \rho - \tfrac{mc}{\hbar\sqrt{\lambda}}, \tag{7.28b}$$

we obtain

$$\begin{aligned}\widetilde{\mathscr{L}}_{2d} &= \tfrac{1}{2}\eta^{\mu\nu}(\partial_\mu\varrho)(\partial_\nu\varrho) - \left(\tfrac{mc}{\hbar}\right)^2\varrho^2 - \tfrac{mc\sqrt{\lambda}}{\hbar}\varrho^3 - \tfrac{1}{4}\lambda\varrho^4 + \tfrac{m^4c^4}{4\lambda\hbar^4}\\ &\quad + \tfrac{1}{2}\left(\varrho + \tfrac{mc}{\hbar\sqrt{\lambda}}\right)^2\eta^{\mu\nu}(\partial_\mu\theta)(\partial_\nu\theta), \qquad \varpi_\vartheta : \theta \xrightarrow{(7.20)} \theta + \vartheta.\end{aligned} \tag{7.28c}$$

This makes it evident that the rotations (7.20) map the system from a parametrization where the Feynman calculus is defined about the ground state with $(\varrho,\theta) = (0,0)$ into a parametrization centered at $(\varrho,\theta) = (0,\theta_*)$, and where the θ field has no mass – nor in fact any potential – and so represents the Goldstone mode.

In turn, by shifting the fields in a ϑ-dependent fashion:

$$\varphi_1 = \phi_1 - \tfrac{mc}{\hbar\sqrt{\lambda}}\cos(\vartheta), \qquad \varphi_2 = \phi_2 - \tfrac{mc}{\hbar\sqrt{\lambda}}\sin(\vartheta), \tag{7.29}$$

we obtain

$$\begin{aligned}\widetilde{\widetilde{\mathscr{L}}}_{2d} &= \tfrac{1}{2}\eta^{\mu\nu}\delta^{ij}(\partial_\mu\phi_i)(\partial_\nu\phi_j) - \left(\tfrac{mc}{\hbar}\right)^2\Big(\cos(\vartheta)\,\phi_1 + \sin(\vartheta)\,\phi_2\Big)^2 + \tfrac{m^4c^4}{4\lambda\hbar^4}\\ &\quad + \sqrt{\lambda}\left(\tfrac{mc}{\hbar}\right)\Big(\cos(\vartheta)\,\phi_1 + \sin(\vartheta)\,\phi_2\Big)(\phi_1^2 + \phi_2^2) - \tfrac{1}{4}\lambda(\phi_1^2 + \phi_2^2)^2,\end{aligned} \tag{7.30}$$

which evidently interpolates between (7.23)–(7.24) and a continuum of equivalently shifted Lagrangian densities.

Notice the extraordinary similarity between the descriptions (7.27)–(7.30) and the illustration in Figure 7.1 on p. 255, whereby it is possible to identify the pair of fields (ϕ_1,ϕ_2) with the motion denoted by the dark/light arrows on the left-hand side, and ϱ with the radial motion (dark arrows) on the left-hand side, and where the ϖ_ϑ rotation evidently perfectly corresponds to the rotational motion denoted by the light and outlined arrow. Unfortunately, the nonlinear coupling in the kinetic term, between $(\partial_\mu\theta)$ and ϱ, is the "price" of making this relationship between the Goldstone mode and the broken symmetry evident. This "polar" parametrization of the model is therefore rather cumbersome for defining Feynman diagrams and the perturbative computations, and is not used except to identify symmetries.

The Higgs effect for gauge *U*(1) symmetry

The 2-dimensional model from the previous section may be combined with gauge symmetry. One only needs to reinterpret the pair of real scalar fields, ϕ_1,ϕ_2, as one complex scalar field, $\boldsymbol{\phi} = \phi_1 + i\phi_2$. This complex field then has a phase, and the description from Sections 5.1 and 5.3 may be adapted. Start therefore with the Lagrangian density

$$\mathscr{L}_{CW} = \tfrac{1}{2}\eta^{\mu\nu}(D_\mu\boldsymbol{\phi})^*(D_\nu\boldsymbol{\phi}) - \tfrac{1}{2}\left(\tfrac{mc}{\hbar}\right)^2|\boldsymbol{\phi}|^2 - \tfrac{1}{4}\lambda\left(|\boldsymbol{\phi}|^2\right)^2 - \tfrac{4\pi\epsilon_0}{4}F_{\mu\nu}F^{\mu\nu}, \tag{7.31}$$

where

$$D_\mu\boldsymbol{\phi} = \partial_\mu\boldsymbol{\phi} + \tfrac{iq_\phi}{\hbar c}A_\mu\,\boldsymbol{\phi}, \tag{7.32}$$

is the (electromagnetically) $U(1)$-covariant derivative, and q_ϕ is the electric charge of the complex field $\boldsymbol{\phi}$. Varying by $\boldsymbol{\phi}$ and $\boldsymbol{\phi}^*$, we obtain the Euler–Lagrange equations of motion, varying by $\dot{\boldsymbol{\phi}}$ and $\dot{\boldsymbol{\phi}}^*$ produces the canonical momenta, etc. However, we are here concerned with the breaking of the gauge symmetry

$$\boldsymbol{\phi}(\mathbf{x}) \to \exp\left\{\tfrac{i}{\hbar} q_\phi\, \chi(\mathbf{x})\right\} \boldsymbol{\phi}(\mathbf{x}), \qquad A_\mu(\mathbf{x}) \to A_\mu(\mathbf{x}) - c\,\partial_\mu \chi(\mathbf{x}), \tag{7.33}$$

and will, as before, consider the Lagrangian density (7.31), the one with the "wrong" sign of the quadratic term:

$$\widetilde{\mathscr{L}}_{\text{CW}} = \tfrac{1}{2}\eta^{\mu\nu}(D_\mu\boldsymbol{\phi})^*(D_\nu\boldsymbol{\phi}) + \tfrac{1}{2}\big(\tfrac{mc}{\hbar}\big)^2|\boldsymbol{\phi}|^2 - \tfrac{1}{4}\lambda\big(|\boldsymbol{\phi}|^2\big)^2 - \tfrac{4\pi\epsilon_0}{4} F_{\mu\nu}F^{\mu\nu}. \tag{7.34}$$

It is not hard to show, e.g., by parametrizing $\boldsymbol{\phi} = R\,e^{i\Theta}$, [✎ *do it]* that the potential energy density

$$\widetilde{\mathscr{V}}_{\text{CW}} = -\tfrac{1}{2}\big(\tfrac{mc}{\hbar}\big)^2|\boldsymbol{\phi}|^2 + \tfrac{1}{4}\lambda\big(|\boldsymbol{\phi}|^2\big)^2 = -\tfrac{1}{2}\big(\tfrac{mc}{\hbar}\big)^2 R^2 + \tfrac{1}{4}\lambda R^4 \tag{7.35}$$

has a minimum when $R := |\boldsymbol{\phi}| = \frac{mc}{\hbar\sqrt{\lambda}}$ and the "angle" $\Theta \in [0, 2\pi]$ is arbitrary, which parametrizes a circle of radius $\frac{mc}{\hbar\sqrt{\lambda}}$ – in the complex field plane of $\boldsymbol{\phi} = \phi_1 + i\phi_2$. The classical solutions, i.e., the quantum-expectation values $|\langle\boldsymbol{\phi}\rangle| = \frac{mc}{\hbar\sqrt{\lambda}}$, are equally probable for every choice of the "angle" Θ, and the ultimate value $\langle\Theta\rangle$ is determined by the initial conditions and external influences. (As per Conclusion 1.1, perfect initial conditions do not exist.)

Choosing, e.g., $\Theta = 0$ for the ground state and in the Feynman diagrammatic calculus,[5] we must redefine the fields so that they describe fluctuations about the chosen classical solution. We thus define $\boldsymbol{\varphi} = \boldsymbol{\phi} - \frac{mc}{\hbar\sqrt{\lambda}}$, but are free to return to the Cartesian basis, with $\varphi_1 := \Re e(\boldsymbol{\phi}) - \frac{mc}{\hbar\sqrt{\lambda}}$ and $\varphi_2 := \Im m(\boldsymbol{\phi})$. This substitution yields

$$\begin{aligned}
\widetilde{\mathscr{L}}_{\text{CW}} &= \tfrac{1}{2}\eta^{\mu\nu}\big[D_\mu\big((\varphi_1 + \tfrac{mc}{\hbar\sqrt{\lambda}}) + i\varphi_2\big)\big]^*\big[D_\nu\big((\varphi_1 + \tfrac{mc}{\hbar\sqrt{\lambda}}) + i\varphi_2\big)\big] - \tfrac{4\pi\epsilon_0}{4}F_{\mu\nu}F^{\mu\nu}\\
&\quad + \tfrac{1}{2}\big(\tfrac{mc}{\hbar}\big)^2\big|(\varphi_1 + \tfrac{mc}{\hbar\sqrt{\lambda}}) + i\varphi_2\big|^2 - \tfrac{1}{4}\lambda\big(\big|(\varphi_1 + \tfrac{mc}{\hbar\sqrt{\lambda}}) + i\varphi_2\big|^2\big)^2\\
&= \Big[\tfrac{1}{2}(\partial_\mu\varphi_1)(\partial^\mu\varphi_1) - \tfrac{m^2c^2}{\hbar^2}{\varphi_1}^2\Big] + \Big[\tfrac{1}{2}(\partial_\mu\varphi_2)(\partial^\mu\varphi_2)\Big] - \Big[\tfrac{4\pi\epsilon_0}{4}F_{\mu\nu}F^{\mu\nu} - \tfrac{1}{2}\tfrac{q_\varphi^2 m^2}{\hbar^4\lambda}A_\mu A^\mu\Big]\\
&\quad + \underline{\tfrac{q_\varphi\, m}{\hbar^2\sqrt{\lambda}}A_\mu(\partial^\mu\varphi_2)} + \tfrac{q_\varphi}{c\hbar}A_\mu\big[\varphi_1(\partial^\mu\varphi_2) - (\partial^\mu\varphi_1)\varphi_2\big]\\
&\quad + \tfrac{q_\varphi^2 m}{c\hbar^3\sqrt{\lambda}}\varphi_1 A_\mu A^\mu - \tfrac{mc\sqrt{\lambda}}{\hbar}\varphi_1({\varphi_1}^2 + {\varphi_2}^2)\\
&\quad + \tfrac{1}{2}\tfrac{q_\varphi^2}{c^2\hbar^2}A_\mu A^\mu({\varphi_1}^2 + {\varphi_2}^2) - \tfrac{1}{4}\lambda({\varphi_1}^2 + {\varphi_2}^2)^2 + \tfrac{m^4c^4}{4\lambda\hbar^4}.
\end{aligned} \tag{7.36}$$

The appearance of the underlined "mixed" quadratic term $\frac{q_\varphi\, m}{\hbar^2\sqrt{\lambda}}A_\mu(\partial^\mu\varphi_2)$ indicates that the functions $\varphi_1, \varphi_2, A_0, A_1, A_2$ and A_3 are not the normal modes of this system. [✎ *Why?]* However, instead of pursuing the diagonalization procedure, we may use the gauge transformation

$$\begin{aligned}
\boldsymbol{\phi} \to e^{i\vartheta}\boldsymbol{\phi} &= (\cos\vartheta + i\sin\vartheta)(\phi_1 + i\phi_2)\\
&= (\phi_1\cos\vartheta - \phi_2\sin\vartheta) + i(\phi_1\sin\vartheta + \phi_2\cos\vartheta)
\end{aligned} \tag{7.37}$$

where we select [☞ definition (5.104a)]

$$\vartheta = -\,\text{ATan}(\phi_1, \phi_2) = -\,\text{ATan}\Big(\varphi_1 + \frac{mc}{\hbar\sqrt{\lambda}}, \varphi_2\Big), \tag{7.38}$$

[5] In classical physics, where $\phi_1 = \Re e(\boldsymbol{\phi})$ and $\phi_2 = \Im m(\boldsymbol{\phi})$ would be functions of (only) time, a similar choice would be convenient for describing small oscillations. Feynman's diagrammatic calculus is indeed a generalization of small oscillations in field theory, in a quite general sense.

so that $\varphi'_2 := \Im m(e^{i\vartheta}\boldsymbol{\phi}) = 0$. Also, $\varphi'_1 := \Re e\,(e^{i\vartheta}(\boldsymbol{\phi} - \frac{mc}{\hbar\sqrt{\lambda}}))$ and, of course, $A'_\mu := A_\mu + (\hbar c\,\partial_\mu\vartheta)$. The Lagrangian density (7.36) being invariant with respect to gauge transformations, it follows that *the same* Lagrangian density may also be expressed in terms of these new, gauge-transformed fields:

$$\begin{aligned}\mathscr{L}_{\mathrm{CW}} = &\Big[\tfrac{1}{2}(\partial_\mu\varphi'_1)(\partial^\mu\varphi'_1) - \tfrac{m^2c^2}{\hbar^2}\varphi_1'^2\Big] - \Big[\tfrac{4\pi\epsilon_0}{4}F'_{\mu\nu}F'^{\mu\nu} - \tfrac{1}{2}\tfrac{q_\varphi^2 m^2}{\hbar^4\lambda}A'_\mu A'^\mu\Big]\\ &+ \tfrac{q_\varphi^2 m}{c\hbar^3\sqrt{\lambda}}\varphi'_1 A'_\mu A'^\mu - \tfrac{mc\sqrt{\lambda}}{\hbar}\varphi_1'^3 + \tfrac{1}{2}\tfrac{q_\varphi^2}{c^2\hbar^2}A'_\mu A'^\mu\varphi_1'^2 - \tfrac{1}{4}\lambda\varphi_1'^4 + \tfrac{m^4c^4}{4\lambda\hbar^4},\end{aligned} \tag{7.39}$$

where we note that φ'_2 no longer appears. The same result could, of course, have been obtained by the standard diagonalization procedure.

It must be kept in mind that the three Lagrangian densities (7.34), (7.36) and (7.39) all describe the same system, only in slightly different parametrization, and where the ultimate version (7.39) achieves the most concise description. Varying the Lagrangian density (7.39) by A_μ produces the Euler–Lagrange equations of motion:

$$\Box A'^\nu - \partial^\nu(\partial_\mu A'^\mu) + \frac{q_\varphi^2 m^2}{4\pi\epsilon_0\hbar^4\lambda}A'^\nu = -\frac{q_\varphi^2}{4\pi\epsilon_0 c^2\hbar^2}\big(\varphi'_1 + \tfrac{mc}{2\hbar\sqrt{\lambda}}\big)\varphi'_1 A'^\nu. \tag{7.40}$$

This proves that the gauge field A'_μ acquired the mass

$$m_A = \frac{q_\varphi\, m}{\sqrt{4\pi\epsilon_0\hbar c}\sqrt{\lambda}} = \frac{q_\varphi}{\sqrt{4\pi\epsilon_0}}\frac{1}{c^2}\langle\phi_1\rangle, \qquad \langle\phi_1\rangle = \frac{mc}{\hbar\sqrt{\lambda}}, \tag{7.41}$$

since by using the Lorenz gauge, $\partial_\mu A'^\mu = 0$, the equation of motion (7.40) becomes

$$\Big[\Box + \big(\tfrac{q_\varphi\, m}{\sqrt{4\pi\epsilon_0}\hbar^2\sqrt{\lambda}}\big)^2\Big]A'^\nu = -\frac{q_\varphi^2}{4\pi\epsilon_0 c^2\hbar^2}\big(\varphi'_1 + \tfrac{mc}{2\hbar\sqrt{\lambda}}\big)\varphi'_1 A'^\nu, \tag{7.42}$$

where the operator in the square brackets is the same as in the Klein–Gordon equation (5.26).

The algebraic substitutions and operations that turn the Lagrangian densities (7.34)–(7.36) into (7.39) may also be represented graphically, since the various homogeneous terms[6] unambiguously correspond to the Feynman diagrams. So, e.g., the gauge boson mass stems from the interaction of these bosons with the Higgs field, where both "scalar" legs of this 2+2-leg vertex sink into the vacuum, or well from it:

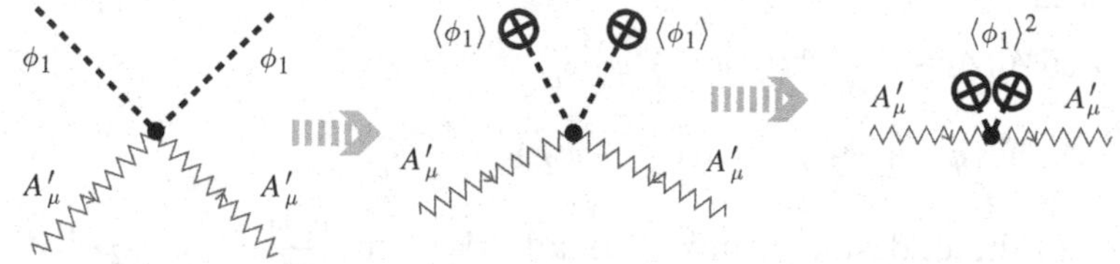

(7.43)

The incessant sinking into the vacuum and welling from it of the ϕ_1-field acts as an effective "viscosity" for all the fields interacting with ϕ_1. This is what impedes the propagation of gauge fields A'_μ, so the quanta of this field acquire an (increased) inertia, i.e., mass. It is not hard to show that, in the (ϕ_1, ϕ_2)-picture, the Feynman diagrams of all "additional" terms in the Lagrangian density (7.39) have dashed lines that sink into the vacuum or well from it, as shown in diagrams (7.18) and (7.43). After the substitution $\phi_1 \to \varphi_1 + \frac{mc}{\hbar\sqrt{\lambda}}$, all diagrams that contain sinks/sources $\langle\phi_1\rangle$ are simply drawn as new, independent diagrams.

[6] By "homogeneous terms" one understands all the terms that have the same power of the various fields of the model. For example, $(\partial_\mu\phi_1)(\partial^\mu\phi_1)$ and $\frac{m^2c^2}{\hbar^2}\phi_1^2$ are homogeneous and together contribute to the propagator, i.e., the first Feynman diagram (7.17). Similarly, the Lagrangian density (7.39) has only one cubic term, $-\frac{mc\sqrt{\lambda}}{\hbar}\varphi_1'^3$, and this is the only term that contributes to the triple vertex Feynman diagram, shown as the middle diagram (7.17).

Conclusion 7.4 *In a diagram such as (7.18) or (7.43), the crucial role is played by the property of Higgs bosons that they have a non-vanishing vacuum expectation value. The direct interpretation of these diagrams is that the Higgs bosons mediate the interaction of other particles with the* ***true*** *vacuum, so that the Higgs bosons in fact also* ***mediate*** *a type of interaction.*

Supporting the claim that these are but different descriptions of the same system, let us count the degrees of freedom in the Lagrangian density:

Equation (7.34) The complex scalar field $\boldsymbol{\phi}$ has two real functions, $\phi_1(\mathrm{x})$ and $\phi_2(\mathrm{x})$. The $U(1)$-gauge potential $A_\mu(\mathrm{x})$ has four real components, but only two are physical, as the gauge symmetry permits the imposition of the Lorenz and the Coulomb gauge, which leave only the two components (those orthogonal to the photon's direction of motion) having a physical meaning. Jointly, these count as four real functions.

Equation (7.39) The real scalar (Higgs) field $\varphi_1'(\mathrm{x})$ is of course just one real function. The vector potential $A_\mu(\mathrm{x})$ here has a mass, and so also has, besides the two components that are orthogonal to the direction of motion, the longitudinal component.[7] Jointly, these again count as four real functions.

By rewriting the Lagrangian density from its form (7.34) into the form (7.39), the imaginary part of the scalar field $\boldsymbol{\phi}$ became the *physical, longitudinal component* of the 4-vector gauge potential A_μ, whereby that gauge boson acquired the mass (7.41), proportional to the charge and the vacuum expectation value of the Higgs field $\boldsymbol{\phi}$. One says that the gauge boson "ate" the imaginary part of the Higgs field, φ_2, which had no mass in the Lagrangian density (7.36) and so represented the Goldstone boson. Suffice it here then to state, without a detailed proof [257, 307, 159, 422, 423, 538, 250, 389, 243]:

Conclusion 7.5 *In the general case of non-abelian (non-commutative) gauge symmetry breaking via the Higgs effect, there exists a symmetric ($\varkappa > 0$) phase, where the complete gauge symmetry is exact, and all Higgs fields are "accounted for" and have the same, real mass.*

There also exists a non-symmetric ($\varkappa < 0$), i.e., Higgs phase, where the gauge symmetry is broken so that from the original group of symmetries G only a subgroup H of symmetries is exact. For each generator of the so-called coset G/H [☞ the lexicon entry, in Appendix B.1] and corresponding to each broken symmetry:

1. *one Higgs scalar field turns into*
2. *the longitudinal component of one gauge 4-vector potential,*
3. *and the particle represented by that 4-vector potential becomes massive.*

The choice between the symmetric or non-symmetric phase is made by the sign $\varkappa$, which is a function of the ***order parameter*** *(typically, the temperature T), so that*

$$\varkappa(T) = \begin{cases} \varkappa > 0 & \text{for} \quad T > T_c \quad \text{symmetric phase,} \\ \varkappa < 0 & \text{for} \quad T < T_c \quad \text{non-symmetric phase.} \end{cases} \tag{7.44}$$

Comment 7.3 *All the Lagrangian densities involving a Higgs field such as (7.39) exhibit an excess energy density, $\frac{m^4c^4}{4\lambda\hbar^4}$. This quantity must contribute to the vacuum energy density of our universe (there is no external reservoir to siphon it away), the $8\pi G_N/c^4$-multiple of*

[7] See the discussion on p. 186, as well as the explanation in Footnote 29 on p. 67.

which is the cosmological constant, and which is known to be some 55 orders of magnitude smaller than $\frac{m^4c^4}{4\lambda\hbar^4}$; whence the term "excess energy density." This discrepancy only becomes worse with the grand-unifying attempts that we will explore in the next chapter. Ultimately, a theory also including gravity would – based on dimensional arguments alone – predict a vacuum energy density that is some 122 orders of magnitude larger than what is observed. This is often cited as the "vacuum catastrophe" and the "worst theoretical prediction in the history of physics" [272]. However, this is not the first time dimensional analysis alone presented a manifestly wrong answer; see Section 1.2.5.

Comment 7.4 *In processes where the energies of the involved particles are bigger (smaller) than $k_B T_c$, one expects the system to be in the symmetric (non-symmetric) phase. In practice therefore, the energy available to the particles in observed processes is identified with the order parameter, i.e., temperature. Finally, the critical energy then must be proportional to the value $\langle\phi\rangle$, and dimensional analysis dictates that*

$$E_c = \hbar c\sqrt{\lambda}\langle\phi\rangle \;= k_B T_c. \tag{7.45}$$

7.1.4 Exercises for Section 7.1

✎ ***7.1.1*** *Confirm the results (7.15) by explicit computation.*

✎ ***7.1.2*** *Confirm the results (7.22) by explicit computation.*

✎ ***7.1.3*** *Expanding the Lagrangian density (7.21) about $(\varphi_1, \varphi_2) = (\phi_1, \phi_2 - \frac{mc}{\hbar\sqrt{\lambda}})$, verify that now φ_1 plays the role of the Goldstone boson.*

✎ ***7.1.4*** *Confirm the results (7.24) by explicit computation.*

✎ ***7.1.5*** *Confirm the results (7.36) by explicit computation.*

7.2 The weak nuclear interaction and its consequences

Interactions of gauge 4-vector potentials and spin-$\frac{1}{2}$ fermions studied in Chapter 5 faithfully describe the interactions of electromagnetic and strong interactions, the gauge bosons of which are massless. The Higgs effect, described in the previous section, provides a correct description of massive $W^{\pm}$- and Z^0-bosons. However, for the description of the interaction of these bosons with spin-$\frac{1}{2}$ fermions, we need one additional detail, to which we now turn.

7.2.1 The asymmetry in weak interactions

Chapter 5 describes interactions of gauge bosons with 4-component Dirac fermions, which were shown in Section 5.2.1 on p. 172 to decompose in a Lorentz-invariant way into the eigenstates of $\gamma_{\pm}$ [☞ Conclusion 5.2 on p. 179], the so-called Weyl spinors:

$$\Psi = \Psi_+ + \Psi_-, \qquad \Psi_{\pm} := (\gamma_{\pm}\Psi), \qquad \gamma_{\pm} = \tfrac{1}{2}[\mathbb{1} \pm \hat{\gamma}]. \tag{7.46}$$

Using the relations (A.121a)–(A.121b) and (A.130), we obtain that

$$\begin{aligned}&\overline{\Psi}\big[i\hbar c\gamma^{\mu}D_{\mu} - \tfrac{mc}{\hbar}\mathbb{1}\big]\Psi \\ &\quad = \overline{\Psi_+}\big[i\hbar c\gamma^{\mu}D_{\mu}\big]\Psi_+ + \overline{\Psi_-}\big[i\hbar c\gamma^{\mu}D_{\mu}\big]\Psi_- - \tfrac{mc}{\hbar}\big[\overline{\Psi_-}\Psi_+ + \overline{\Psi_+}\Psi_-\big].\end{aligned} \tag{7.47}$$

That is, the interaction of a spin-$\frac{1}{2}$ fermion with the gauge field as described in Chapter 5 includes both "left-handed" ($\Psi_- \equiv \Psi_L$) and "right-handed" ($\Psi_+ \equiv \Psi_R$) fermions.[8]

Note that the Lagrangian term that defines the mass, $-\frac{mc}{\hbar}\overline{\Psi}_-\Psi_+$, couples Ψ_+ and Ψ_-. This is the so-called Dirac mass. By contrast, the previous two terms in the expression (7.47) "link" fermions of *the same* chirality. This property permits massless spin-$\frac{1}{2}$ particles to satisfy the simpler, Weyl equation (5.62) instead of the more complicated Dirac equation (5.34).

As was discussed in Section 4.2.1, the weak interactions *maximally* break the parity symmetry as the interaction of the $W^{\pm}$ boson with a charged lepton and a neutrino exclusively couples the "left-handed" fermions. Thus, e.g., the interaction $e^- \to W^- + \nu_e$ in the Lagrangian density must correspond to the term

$$\begin{aligned}
&\overline{\Psi^{(\nu,e)}_-}\big[i\hbar c\boldsymbol{\gamma}^\mu(\mathbb{1}\partial_\mu + \tfrac{ig_w}{\hbar c}\mathbf{W}_\mu)\big]\Psi^{(\nu,e)}_-, \qquad \mathbf{W}_\mu := \tfrac{1}{2}\sigma_a W^a_\mu,\\
&= \overline{\Psi^{(\nu,e)}\boldsymbol{\gamma}_-}\big[i\hbar c\boldsymbol{\gamma}^\mu(\mathbb{1}\partial_\mu + \tfrac{ig_w}{\hbar c}\mathbf{W}_\mu)\big](\boldsymbol{\gamma}_-\Psi^{(\nu,e)})\\
&\overset{(A.130)}{=} \overline{\Psi^{(\nu,e)}}\boldsymbol{\gamma}_+\big[i\hbar c\boldsymbol{\gamma}^\mu(\mathbb{1}\partial_\mu + \tfrac{ig_w}{\hbar c}\mathbf{W}_\mu)\big]\boldsymbol{\gamma}_-\Psi^{(\nu,e)}\\
&= \overline{\Psi^{(\nu,e)}}\big[i\hbar c\boldsymbol{\gamma}_+\boldsymbol{\gamma}^\mu\boldsymbol{\gamma}_-(\mathbb{1}\partial_\mu + \tfrac{ig_w}{\hbar c}\mathbf{W}_\mu)\big]\Psi^{(\nu,e)}\\
&= \overline{\Psi^{(\nu,e)}}\big[i\hbar c\boldsymbol{\gamma}^\mu\boldsymbol{\gamma}_-^2(\mathbb{1}\partial_\mu + \tfrac{ig_w}{\hbar c}\mathbf{W}_\mu)\big]\Psi^{(\nu,e)}\\
&\overset{(A.121b)}{=} \overline{\Psi^{(\nu,e)}}\big[i\hbar c\boldsymbol{\gamma}^\mu\boldsymbol{\gamma}_-(\mathbb{1}\partial_\mu + \tfrac{ig_w}{\hbar c}\mathbf{W}_\mu)\big]\Psi^{(\nu,e)}.
\end{aligned} \tag{7.48}$$

That is, the first term in the left–right symmetric expression (7.47) must not appear in the Lagrangian density for weak interactions. As the key terms in the Lagrangian density must include factors of the type

$$\begin{aligned}
\overline{\Psi^{(\nu,e)}_-}\boldsymbol{\gamma}^\mu\mathbf{W}_\mu\Psi^{(\nu,e)}_- &= \overline{\Psi^{(\nu,e)}}\boldsymbol{\gamma}^\mu\boldsymbol{\gamma}_-\mathbf{W}_\mu\Psi^{(\nu,e)} = \tfrac{1}{2}\overline{\Psi^{(\nu,e)}}\boldsymbol{\gamma}^\mu[\mathbb{1} - \widehat{\boldsymbol{\gamma}}]\mathbf{W}_\mu\Psi^{(\nu,e)}\\
&= \tfrac{1}{2}\Big[\underbrace{\overline{\Psi^{(\nu,e)}}\boldsymbol{\gamma}^\mu\tfrac{1}{2}\sigma_a\Psi^{(\nu,e)}}_{\text{vector}} - \underbrace{\overline{\Psi^{(\nu,e)}}\boldsymbol{\gamma}^\mu\widehat{\boldsymbol{\gamma}}\tfrac{1}{2}\sigma_a\Psi^{(\nu,e)}}_{\text{axial vector}}\Big]W^a_\mu,
\end{aligned} \tag{7.49}$$

one says that weak interactions are of the "$V-A$" type – contrary to the electrodynamics and chromodynamics interactions that are of purely "V" (vector) type.

Thus, the Lagrangian density describing the interactions of gauge bosons $W^{\pm}$ may be written with the projectors $\boldsymbol{\gamma}_-$ consistently inserted for all fermions; interactions with the Z^0-boson are still more complicated [☞ Sections 7.2.4 and 7.2.5].

7.2.2 The GIM mechanism

Section 2.3.14 showed that the quark states that interact by weak interaction are not the eigenstates of the "free" Hamiltonian that defines the mass: The quark states that are detected as d-, s- and b-quarks primarily differ in mass [☞ Figure 2.1 on p. 76, and Table 4.1 on p. 152]. However, the eigenstates of the Hamiltonian term describing the interaction with the $W^{\pm}$- and Z^0-bosons are nontrivial linear combinations (2.53) of these mass-identified states.

The first-order effect

When Nicola Cabibbo suggested the first variation of this phenomenon in 1963, only the u-, d- and s-quarks were known. Proposing that the states that interact with the $W^{\pm}$- and Z^0-bosons are in fact

$$|u\rangle, \quad |d_w\rangle := \cos\theta_c|d\rangle + \sin\theta_c|s\rangle, \qquad |s_w\rangle := \cos\theta_c|s\rangle - \sin\theta_c|d\rangle, \tag{7.50}$$

$$\text{so}\quad |d\rangle = \cos\theta_c|d_w\rangle - \sin\theta_c|s_w\rangle, \qquad |s\rangle = \cos\theta_c|s_w\rangle + \sin\theta_c|d_w\rangle, \tag{7.51}$$

[8] It is standard to use the adjectives "left/right-handed" regarding both the chirality eigenstates and the helicity eigenstates of spin-$\frac{1}{2}$ fermions – although these coincide only for massless particles. The context usually makes it clear which of these two characteristics is meant; herein, we have in mind only chirality.

Cabibbo explained the existence of processes of the type

$$d \to W^- + u \qquad \text{and} \qquad s \to W^- + u. \tag{7.52}$$

Since the s-quark carries strangeness, and u- and d-quarks do not, the first process is assigned $\triangle S = 0$ and the second one $\triangle S = 1$. In these processes the W^--boson is said to interact with the quark "current" $d \to u$ (which preserves strangeness), and respectively $s \to u$ (where strangeness is broken). Using the principle of detailed balance [☞ Section 2.14], we also have the processes $W^- \leftrightarrow d_w + \overline{u}$, and akin to the expression (7.49), we define the quark 4-current density that interacts with the weak gauge bosons:

$$W^\mu_+ : \quad \mathfrak{J}^\mu_+ = \overline{d}_{w\,L}\gamma^\mu u_L \qquad \to \cos\theta_c\,\overline{d}u + \sin\theta_c\,\overline{s}u, \tag{7.53a}$$

$$W^\mu_- : \quad \mathfrak{J}^\mu_- = \overline{u}_L\gamma^\mu d_{w\,L} \qquad \to \cos\theta_c\,\overline{u}d + \sin\theta_c\,\overline{u}s, \tag{7.53b}$$

whereby it follows that

$$Z^0 : \quad \mathfrak{J}^\mu_0 = \overline{u}_L\gamma^\mu u_L - \overline{d}_{w\,L}\gamma^\mu d_{w\,L} \to \overline{u}u - \cos^2\theta_c\,\overline{d}d - \tfrac{1}{2}\sin 2\theta_c(\overline{d}s + \overline{s}d) - \sin^2\theta_c\,\overline{s}s. \tag{7.53c}$$

This implies the existence and relative strength of the following processes:

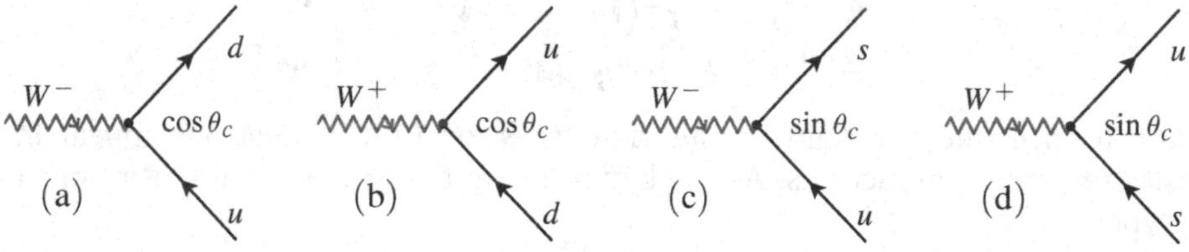

(7.54)

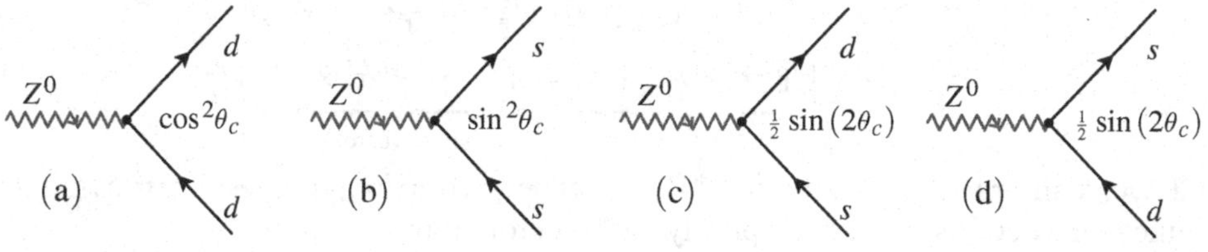

(7.55)

as well as their variations obtained through the principle of detailed balance, and where the relative θ_c-dependent factors for the amplitudes of these processes are written next to the vertices. The processes (7.54a,b) and (7.55a,b) have $\triangle S = 0$, and the processes (7.54c,d) and (7.55c,d) have $\triangle S = \pm 1$.

Combining the processes (7.54d) and (7.55d) with similar processes where the $W^\pm$- and Z^0-bosons create a lepton–antilepton pair, we obtain the Feynman diagrams

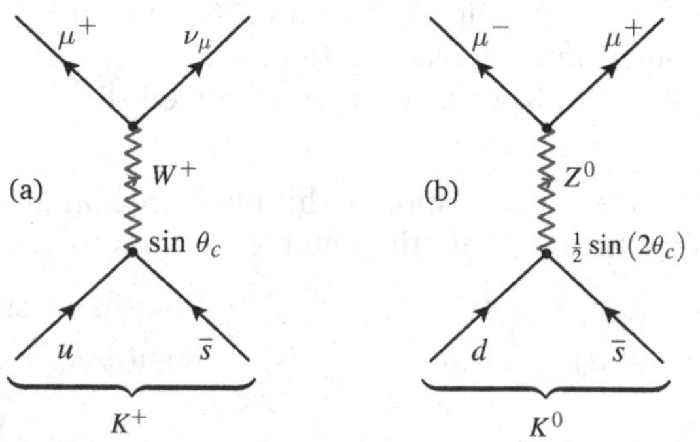

(7.56)

Except for the θ_c-dependent factor and the dependence on the particle masses, the amplitude of these processes would have to be approximately the same since

$$\left|\frac{\frac{1}{2}\sin(2\theta_c)}{\sin(\theta_c)}\right|^2 = \cos^2(\theta_c) \sim O(\tfrac{1}{2})\text{–}O(1). \tag{7.57}$$

However, experiments confirm that the first of these two processes really happens and with the expected probability, but the second of these two processes practically does not occur [293]:[9]

$$\frac{\Gamma(K^+ \to \mu^+ + \nu_\mu)}{\Gamma(K^+ \to \text{all})} \approx 64\%, \qquad \frac{\Gamma(K^0 \to \mu^- + \mu^+)}{\Gamma(K^0 \to \text{all})} < 9\times10^{-9}. \tag{7.58}$$

In the general case, it is experimentally verified that the processes with $\triangle S = \pm 1$ mediated by the Z^0-boson occur many orders of magnitude less frequently than other weak processes that can be described using the diagrams (7.54)–(7.55) and their equivalents with leptons instead of quarks. Cabibbo's original parametrization (7.50) thus implies the result (7.53c), which – besides the experimentally confirmed processes of the type (7.56a) – also predicts the flavor-changing neutral current processes, such as (7.56b), which do not occur. According to the discussion that led to Conclusion 1.1 on p. 6, Cabibbo's then model *must* be corrected.

To explain the tremendous difference (7.58), Glashow, Iliopoulos and Maiani (GIM) proposed in 1970 that there exists a fourth quark, c, so that the quark current densities that interact with the $W^{\pm}$- and Z^0 bosons are

$$W^+_\mu: \quad \mathfrak{J}^\mu_+ = \overline{d}_{wL}\gamma^\mu u_L + \overline{s}_{wL}\gamma^\mu c_L \to \cos\theta_c\,\overline{d}u + \sin\theta_c\,\overline{s}u - \sin\theta_c\,\overline{d}c + \cos\theta_c\,\overline{s}c, \tag{7.59a}$$

$$W^-_\mu: \quad \mathfrak{J}^\mu_- = \overline{u}_L\gamma^\mu d_{wL} + \overline{c}_L\gamma^\mu s_{wL} \to \cos\theta_c\,\overline{u}d + \sin\theta_c\,\overline{u}s - \sin\theta_c\,\overline{c}d + \cos\theta_c\,\overline{c}s, \tag{7.59b}$$

$$Z^0: \quad \mathfrak{J}^\mu_0 = \overline{u}_L\gamma^\mu u_L - \overline{d}_{wL}\gamma^\mu d_{wL} + \overline{c}_L\gamma^\mu c_L - \overline{s}_{wL}\gamma^\mu d_{sL} \to \overline{u}u + \overline{c}c - \overline{d}d - \overline{s}s. \tag{7.59c}$$

This proposal corrects Cabibbo's model in that it does not alter the results for the processes of the type (7.56a), but – in agreement with the experimental non-observation – prohibits processes of the type (7.56b). That is, in contrast to the quark current density (7.53c) that contains mixing terms $\overline{d}s$ and $\overline{s}d$, the quark current density (7.59c) contains no mixing term. The "price" for so diagonalizing the Z^0-boson interaction in the flavor space was the postulate of the existence of the fourth quark, and that proposal and its consequences are usually called the GIM mechanism.

> ***Comment 7.5*** *The Reader should notice the conceptual parallel between Glashow, Iliopoulos and Maiani's postulation of a new quark so as to preserve the logical consistency of the model and Pauli's postulation of the neutrino so as to preserve the energy conservation law [☞ Section 2.3.9].*

The second-order effect

Now, even if the decay $K^0 \to \mu^+ + \mu^-$ by way of a simple $O(g_w^2)$ process (7.56b) is forbidden, it does not follow that this physical process cannot happen by way of a more complex interaction, i.e., by way of a more complex Feynman diagram. Indeed, one straightforwardly constructs the $O(g_w^4)$ diagrams:

[9] Processes mediated by Z^0-bosons are usually labeled by the FCNC acronym, standing for *flavor-changing neutral current*.

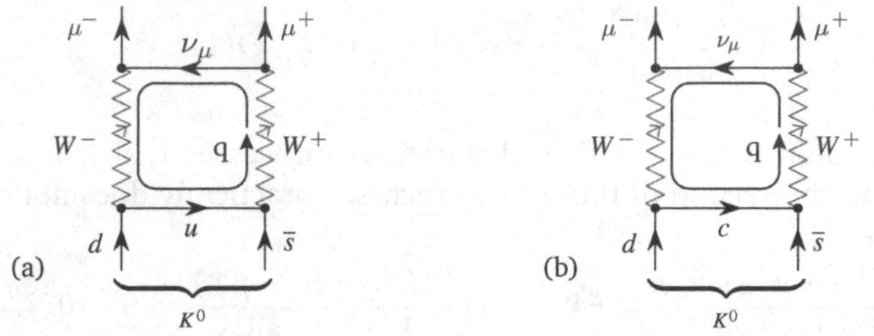

(7.60)

The sub-processes described by these two diagrams are identical, except that the u-quark in the left-hand diagram is replaced by a c-quark in the right-hand one. As these quarks are virtual in these diagrams, according to Conclusion 2.3 on p. 56, both sub-processes contribute to the decay $K^0 \to \mu^+\mu^-$. However, the diagram (7.54) implies that the amplitude of the diagram (7.60a) is proportional to $(\cos\theta_c)(-\sin\theta_c)$, while the amplitude for the diagram (7.60b) is proportional to $(\sin\theta_c)(\cos\theta_c)$. Since the amplitudes of these sub-processes are being added, these two contributions would exactly cancel if the u- and c-quark masses were equal.

That is, the application of the 4-momentum conservation in all vertices straightforwardly implies that one of the (internal) 4-momenta remains undetermined, and its integration remains unrestricted. We may always choose this to be the 4-momentum shown as circulating in the central loop/box and which was denoted "q." The $\int d^4q$-integral is dominated by contributions that stem from the $|q| \gtrsim (m_W c) = 80.403\,\text{GeV}/c$ regime, which is far in excess of m_u, m_c. The u- and c-quark mass dependence of the amplitudes must therefore be fairly soft, causes a very small ultimate difference between the two amplitudes, and guarantees their approximate cancellation. One expects the amplitude $\mathfrak{M}$ to be a function of $m_c - m_u$, and $\mathfrak{M} \propto (m_W)^{-2}$, owing to the two W-propagators. Thus, this estimate $|\mathfrak{M}|^2 \propto |\frac{(m_c-m_u)^2}{m_w^2}|^2 \sim 10^{-8}$ is already amazingly close to the experimental result (7.58) [293].

It may further be shown that the GIM mechanism actually guarantees the approximate cancellation of all possible contributions to the Z^0-mediated weak processes where $\triangle S = \pm 1$, and so guarantees good agreement between the Cabibbo–GIM model with four quarks and the experimental data. Nevertheless, the postulation of a new particle so as to preserve the logical consistency of the model was still regarded an extravagant "solution" of a problem of the otherwise (in the early 1970s) experimentally insufficiently justified quark model [243].

7.2.3 $U(1)_A$ anomaly

The existence of the fourth, c-quark was experimentally confirmed in 1974, but even before that, an extraordinarily strong but "purely theoretical" argument for its existence was known – separate from the GIM mechanism, but just as often ignored as "idle theory."

In the classical (non-quantum) version of the quark model, the functions used to represent the various particles satisfy their equations of motion:

$$\begin{aligned} i\hbar\partial_\mu[\overline{\Psi}_1\boldsymbol{\gamma}^\mu\Psi_2] &= (i\hbar\partial_\mu\overline{\Psi}_1\boldsymbol{\gamma}^\mu)\Psi_2 + \overline{\Psi}_1\boldsymbol{\gamma}^\mu(i\hbar\partial_\mu\Psi_2) = -\overline{(i\hbar\not{\partial}\Psi_1)}\Psi_2 + \overline{\Psi}_1(i\hbar\not{\partial}\Psi_2) \\ &= -\overline{(m_1c\Psi_1)}\Psi_2 + \overline{\Psi}_1(m_2c\Psi_2) = (m_2-m_1)c\,\overline{\Psi}_1\Psi_2, \end{aligned} \tag{7.61}$$

since the quark functions Ψ satisfy the Dirac equation (5.34). Analogously,

$$\begin{aligned} i\hbar\partial_\mu[\overline{\Psi}_1\widehat{\boldsymbol{\gamma}}\boldsymbol{\gamma}^\mu\Psi_2] &= (i\hbar\partial_\mu\overline{\Psi}_1(-\boldsymbol{\gamma}^\mu\widehat{\boldsymbol{\gamma}}))\Psi_2 + \overline{\Psi}_1\widehat{\boldsymbol{\gamma}}\boldsymbol{\gamma}^\mu(i\hbar\partial_\mu\Psi_2) = \overline{(i\hbar\not{\partial}\Psi_1)}\widehat{\boldsymbol{\gamma}}\Psi_2 + \overline{\Psi}_1\widehat{\boldsymbol{\gamma}}(i\hbar\not{\partial}\Psi_2) \\ &= \overline{(m_1c\Psi_1)}\widehat{\boldsymbol{\gamma}}\Psi_2 + \overline{\Psi}_1\widehat{\boldsymbol{\gamma}}(m_2c\Psi_2) = (m_1+m_2)c\,\overline{\Psi}_1\widehat{\boldsymbol{\gamma}}\Psi_2. \end{aligned} \tag{7.62}$$

We then have:

Theorem 7.2 *For spinors Ψ_i that satisfy the Dirac equation $[i\hbar\partial\!\!\!/ - m_i c]\Psi_i = 0$:*

1. *the 4-vector current $\mathfrak{J}^\mu_{ij} := [\overline{\Psi}_i\gamma^\mu\Psi_j]$ satisfies the continuity equation*

$$\partial_\mu \mathfrak{J}^\mu_{ij} = 0 \qquad \textit{precisely when} \quad m_i = m_j. \tag{7.63}$$

2. *The pseudo (axial) 4-vector current $\widehat{\mathfrak{J}}^\mu_{ij} := [\overline{\Psi}_i\widehat{\gamma}\gamma^\mu\Psi_j]$ satisfies the continuity equation*

$$\partial_\mu \widehat{\mathfrak{J}}^\mu_{ij} = 0 \qquad \textit{precisely when} \quad m_i = m_j = 0. \tag{7.64}$$

The continuity equations (7.63) and (7.64) guarantee that the "charges"

$$Q_{ij} := \int \mathrm{d}^3\vec{r}\, \mathfrak{J}^0_{ij} \qquad \text{and} \qquad \widehat{Q}_{ij} := \int \mathrm{d}^3\vec{r}\, \widehat{\mathfrak{J}}^0_{ij} \tag{7.65}$$

are conserved in all classical processes. For example, if we select i, j to count all quarks, then let $j = i$ and sum, $\sum_i Q_{ii}$ represents the quark number, and the expression $(3\sum_i Q_{ii})$ equals the baryon number [☞ Section 2.4.2, especially p. 76]. *Conversely* to Noether's theorem A.1 on p. 461, each current density that satisfies the equation of continuity defines a symmetry, and the "charges" Q_{ij} and $\widehat{Q}_{ij}$ are the formal generators of these corresponding symmetries. These are the *classical* symmetries of the system.

However, quantum effects in principle need not preserve classical symmetries, which then causes the appearance of quantum contributions that "spoil" the continuity equations

$$\partial_\mu \mathfrak{J}^\mu_{ij} = \mathfrak{A}_{ij} \qquad \text{and} \qquad \partial_\mu \widehat{\mathfrak{J}}^\mu_{ij} = \widehat{\mathfrak{A}}_{ij}, \tag{7.66}$$

where $\mathfrak{A}_{ij}$ and $\widehat{\mathfrak{A}}_{ij}$ are (quantum) *anomalies* of the current 4-vector densities $\mathfrak{J}^\mu_{ij}$ and $\widehat{\mathfrak{J}}^\mu_{ij}$, respectively, i.e., of the symmetries corresponding to these currents, whereby the *anomalies* $\mathfrak{A}_{ij}$ and $\widehat{\mathfrak{A}}_{ij}$ measure the quantum breaking of these symmetries.

It is paramount to realize the general nature of this phenomenon! We distinguish the following cases:

Approximate symmetries, as is the case with the "axial" currents (7.64), which are approximately conserved only in the specific regime of energies, 3-momenta and precision where we may neglect the differences between the masses of the particles amongst which the considered approximate symmetries operate. Even classically, such a current satisfies the continuity equation only approximately; its breaking produces a so-called pseudo-Goldstone mode, the mass of which is of the order of the resolution of the assumed approximation.

Global symmetries, such as the baryon number, for which the formal charge (7.65) is given by $(3\sum_i Q_{ii})$ and where the sum extends over all quark flavors. For that case, quantum chromodynamics yields $\mathfrak{A} \propto \vartheta\varepsilon^{\mu\nu\rho\sigma}\,\mathrm{Tr}[\mathbb{F}_{\mu\nu}\mathbb{F}_{\rho\sigma}]$, where ϑ is a free parameter for which experiments indicate $\vartheta < 3{\times}10^{-10}$, the tininess of which has no complete theoretical explanation [☞ Section 6.3.1] ☡.

Gauge symmetries, for which the appearance of anomaly indicates an essential contradiction. That is, models with anomalous gauge symmetry simply make no sense – unless they can be extended so as to cancel all gauge anomalies.

The analysis of precisely this last type of anomaly (S. Adler, and independently J. S. Bell and R. Jackiw) in 1969 pointed to the appearance of an anomalous quantum contribution in the continuity equation to the familiar electromagnetic current, owing to the coupling with the axial current [425, 586]. All amplitude contributions for any concrete process that leads to the appearance of an anomaly are products of a single, characteristic and incurably divergent type of integral and an indicative numerical factor. The algebraic sum of these contributions, the amplitude is thus a product of this characteristic and divergent integral and the sum of these indicative numeric factors. Such a result makes sense only if the sum of the indicative numeric factors identically cancels, as is the case, e.g., with the sum of electric charges within the family $\{u, d; \nu_e, e^-\}$ of fermions[10]

$$\sum_i Q_i = 3\big[\big(+\tfrac{2}{3}\big) + \big(-\tfrac{1}{3}\big)\big] + (0) + (-1) = 0, \tag{7.67}$$

where the explicit pre-factor "3" stems from summing over the three colors of the u- and d-quarks. The identical cancellation of this sum – and the corresponding absence of the quantum anomaly in electric charge conservation – has the following implications:

1. Every lepton pair $\{\nu_\ell, \ell^-\}$ requires a corresponding quark pair with (color-averaged) charges $+\frac{2}{3}$ and $-\frac{1}{3}$.
2. Quarks with electric charges $+\frac{2}{3}$ and $-\frac{1}{3}$ *must* occur in triples. Alternatively, the integrally charged quarks of the Han–Nambu model (5.212a) also occur in triples.

The latter of these two consequences confirms the necessity of the existence of quark colors.

However, more importantly, the first of these two consequences implies that the existence of the muon necessarily predicts the existence not only of the s-quark (with $-\frac{1}{3}$ charge) but also of the c-quark (with $+\frac{2}{3}$ charge). Since the neutrino has no electric charge, the unavoidable need for a consistent and complete cancellation of the quantum anomaly of the electric current had by 1969 predicted the existence of the fourth quark. However, it was not clear at the time that this conclusion was absolutely inevitable, and even the theoretical motivations for predicting the fourth quark, such as the GIM mechanism, originally did not include the anomaly analysis.

Digression 7.2 The lesson from Pauli's prediction of the neutrino [☞ Section 2.3.9] so as to save the 4-momentum conservation law seems not to have been learned well enough. Between 1969 and 1974, several separate theoretical considerations indicated that inconsistency and contradiction within the theoretical models of particle physics could only be avoided by introducing a new particle, the c-quark. Nevertheless, few particle physicists took these arguments seriously, since the discovery of the J/ψ particle, the lowest-energy $c\bar{c}$-bound state, came as a surprise to most.

It behooves us to finally learn that logical consistency and absence of self-contradiction is a terrific tool of theoretical physics.

The benefit of hindsight today of course permits complete certainty in limiting to quark models that include only complete quark–lepton fours (so-called "families"):

$$\underbrace{\underbrace{\underbrace{\begin{bmatrix} u \\ d \end{bmatrix}, \begin{bmatrix} \nu_e \\ e^- \end{bmatrix};} \quad \begin{bmatrix} c \\ s \end{bmatrix}, \begin{bmatrix} \nu_\mu \\ \mu^- \end{bmatrix};} \quad \begin{bmatrix} t \\ b \end{bmatrix}, \begin{bmatrix} \nu_\tau \\ \tau^- \end{bmatrix}}. \tag{7.68}$$

Including the s-quark without the c-quark or the b-quark without the t-quark is simply inconsistent, as it causes the quantum effects to ruin the $U(1)$ symmetry of quantum electrodynamics and the corresponding electric charge conservation – contradicting experiments, as well as contradicting the gauge symmetry of electromagnetism and the corresponding interactions with gauge bosons.

[10] The fundamental Standard Model fermions are typically divided into three copies of the first four: $\{u, d; \nu_e, e^-\}$, $\{c, s; \nu_\mu, \mu^-\}$ and $\{t, b; \nu_\tau, \tau^-\}$. These copies are called – figuratively – either *generations* or *families*. Without any implication or judgement about the former of these – or indeed any filial/paternal, sororal or fraternal relations, I will herein use the latter name.

Example 7.1 The anomaly analysis from Section 7.2.3 may be applied to the pair of Feynman diagrams where $f_L \in \{u_L, d_L; \nu_{eL}, e^-_L; \overline{u}_L = \overline{u_R}, \overline{d}_L = \overline{d_R}; e^+_L = \overline{e^-_R}; \cdots\}$:

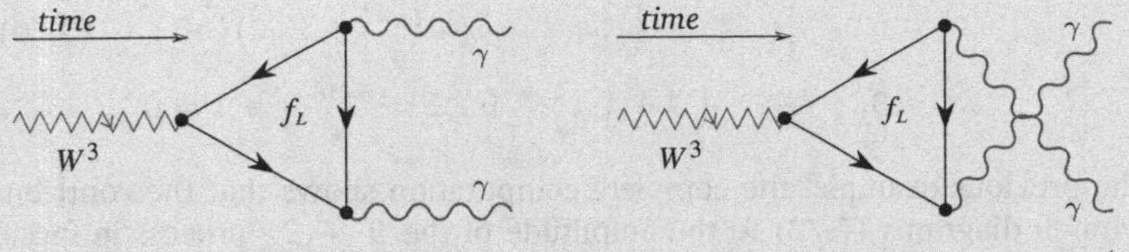

(7.69)

the amplitudes of which contain terms proportional to the sum $\sum_{f_L} I_w(f_L)\,(Q(f_L))^2$. Summing over the fermions of only the first family [☞ Table 7.1 on p. 275, as well as Refs. [425, 586, Chapter 19] for details],

$$\sum_{f_L} I_w(f_L)(Q(f_L))^2 = 3\Big[(+\tfrac{1}{2})(+\tfrac{2}{3})^2 + (-\tfrac{1}{2})(-\tfrac{1}{3})^2\Big] + (+\tfrac{1}{2})(0)^2 + (-\tfrac{1}{2})(-1)^2$$
$$= 3\Big[+\tfrac{2}{9} - \tfrac{1}{18}\Big] - \tfrac{1}{2} = 3\Big(+\tfrac{3}{18}\Big) - \tfrac{1}{2} = 0. \tag{7.70}$$

The complete computation shows that the contributions of the Feynman diagrams (7.69) in fact diverge. Thus, the contributions of the Feynman diagrams (7.69) to the amplitudes that contain the $W^3 \to 2\gamma$ factor are finite (and in fact vanish) if and only if the virtual fermions forming the triangle loops include complete families $\{u, d; \nu_e, e^-\}$, $\{c, s; \nu_\mu, \mu^-\}$, etc. Without the cancellation (7.70), models that include these Feynman diagrams simply make no sense. Notice that the same computation for the Han–Nambu model (5.212a) of integrally charged quarks,

$$\sum_{f_L} I_w(f_L)(Q(f_L))^2$$
$$= \Big[(+\tfrac{1}{2})\big((+1)^2{+}(+1)^2{+}(0)^2\big) + (-\tfrac{1}{2})\big((-1)^2{+}(0)^2{+}(0)^2\big)\Big]$$
$$+ (+\tfrac{1}{2})(0)^2 + (-\tfrac{1}{2})(-1)^2$$
$$= \Big[(+\tfrac{1}{2})2 + (-\tfrac{1}{2})1\Big] + (-\tfrac{1}{2})(-1)^2 = [+1 - \tfrac{1}{2}] - \tfrac{1}{2} = 0, \tag{7.71}$$

implies that it too is free of this gauge anomaly.

Example 7.2 Akin to Example 7.1, we may analyze the pair of Feynman diagrams where the unobserved fermion in the loop is again $f_L \in \{u_L, d_L; \nu_{eL}, e^-_L; \overline{u_R} = \overline{u}_L, \overline{d}_L; e^+_L; \ldots\}$:

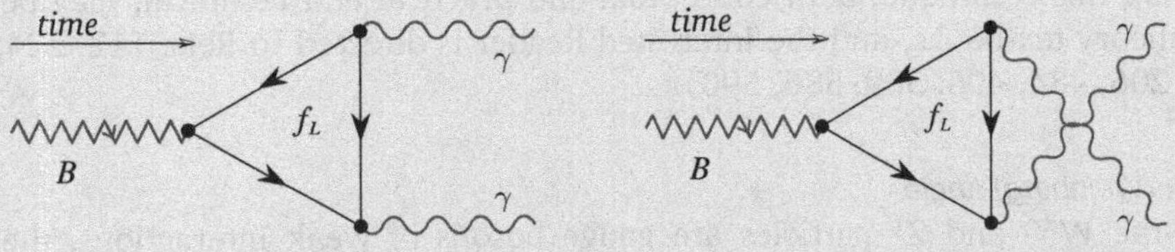

(7.72)

the amplitudes of which contain terms proportional to the sum $\sum_{f_L} Y_w(f_L)\,(Q(f_L))^2$. Summing over the fermions of only the first family [☞ Table 7.1 on p. 275, as well as Refs. [425, 586, Chapter 19] for details],

$$\sum_{f_L} Y_w(f_L)(Q(f_L))^2 = 3\Big[(+\tfrac{1}{3})\big((+\tfrac{2}{3})^2+(-\tfrac{1}{3})^2\big)\Big] + (-1)(0)^2 + (-1)(-1)^2$$
$$+3\Big[(-\tfrac{4}{3})(-\tfrac{2}{3})^2 + (+\tfrac{2}{3})(+\tfrac{1}{3})^2\Big] + (+2)(+1)^2 + (0)(0)^2$$
$$= 3\big(\tfrac{1}{3}\cdot\tfrac{4+1}{9} - \tfrac{4}{3}\cdot\tfrac{4}{9} + \tfrac{2}{3}\cdot\tfrac{1}{9}\big) - 1 + 2 = \tfrac{5-16+2}{9} + 1 = 0. \tag{7.73}$$

As in the previous example, the complete computation shows that the contributions of the Feynman diagrams (7.72) to the amplitude of the $B \to 2\gamma$ process in fact diverge. Again, this result makes sense only if the virtual fermions depicted by the triangular loops include complete families $\{u, d; \nu_e, e^-\}$, $\{c, s; \nu_\mu, \mu^-\}$, etc. Without a cancellation such as in (7.73), models that include these Feynman diagrams simply make no sense. Notice that the same computation for the Han–Nambu model (5.212a) of integrally charged quarks,

$$\sum_{f_L} Y_w(f_L)(Q(f_L))^2$$
$$= \Big[(+\tfrac{1}{3})\big((+1)^2+(+1)^2+(0)^2 + (-1)^2+(0)^2+(0)^2\big)\Big] + (+1)(0)^2 + (-1)(-1)^2$$
$$+ \Big[(-\tfrac{4}{3})\big((+1)^2+(+1)^2+(0)^2\big) + (+\tfrac{2}{3})\big((-1)^2+(0)^2+(0)^2\big)\Big]$$
$$+ (+2)(+1)^2 + (0)(0)^2$$
$$= \big[\tfrac{1}{3}\cdot 3 - \tfrac{4}{3}\cdot 2 + \tfrac{2}{3}\cdot 1\big] - 1 + 2 = \tfrac{3-8+2}{3} + 1 = 0, \tag{7.74}$$

implies that it is also free of this gauge anomaly.

Conclusion 7.6 *Since the joint contributions of the Feynman diagram pairs (7.69) vanish, as they also do for the diagram pair (7.72), the joint contributions then also vanish for the linear combination* $Z^0 = \cos(\theta_w)W^3 - \sin(\theta_w)B$.[11] *The same holds if in these diagrams the* W^3*- and* B*-particle, respectively (which are the normal modes in the* $SU(2)_w \times U(1)_y$ *symmetric phase) are replaced with the* Z^0*-particle, one of the two normal modes after the* $SU(2)_w \times U(1)_y \to U(1)_Q$ *symmetry breaking.*

In the general case, the anomaly of any symmetry must remain conserved through any phase transition, and so also through the $SU(2)_w \times U(1)_y \to U(1)_Q$ *electroweak symmetry breaking. Anomalies of gauge symmetries of course must vanish (cancel), but the conservation of anomalies of other (including approximate, and exact but global) symmetries is a useful "sum rule" in the study of all phase transitions.*

Further details on this technique, both conceptual and practical and technical, may be found in standard field theory textbooks, and the interested Reader is directed to Refs. [12, 224, 75, 261, 425, 554, 555, 206, 484, 496, 589, 586, 590].

7.2.4 The weak (Weinberg) angle

Although both the $W^\pm$- and Z^0-particles are gauge bosons of weak interactions, their masses are not equal [☞ Table C.2 on p. 526]. This is a consequence of the fact that the Z^0-boson

[11] The angle θ_w is usually called "weak" or the Weinberg angle (although it was Glashow who introduced it); experimentally, $\theta_w \approx 28.75°$.

and the photon are linear combinations of the $SU(2)_w$-partner of the $W^{\pm}$-boson and the $U(1)_y$-gauge boson. This effect is well described in the Glashow–Weinberg–Salam model of electroweak interactions.

The conclusions of Sections 7.2.1–7.2.3 indicate a finer structure among the particles in Table 2.3 on p. 67, of which all matter consists. That is, weak interactions may be described by a non-abelian (non-commutative) gauge model in which, owing to the relation (7.49), the left- and the right-handed fermions are treated differently. Akin to the GNN formula (2.44b), the weak isospin I_w and the weak hypercharge Y_w are defined so as to satisfy the relation

$$Q = I_w + \tfrac{1}{2} Y_w. \tag{7.75}$$

Table 7.1 The weak isospin, the weak hypercharge and the electric charge of the elementary fermions are related by equation (7.75). The values are, however, different for fermions of left-handed and right-handed chirality.

	Fermion family			**Charges**		
	1	2	3	Q	I_w	Y_w
$\Psi_- = \gamma_- \Psi$ (left-handed)	$\begin{bmatrix} u \\ d \end{bmatrix}_L$	$\begin{bmatrix} c \\ s \end{bmatrix}_L$	$\begin{bmatrix} t \\ b \end{bmatrix}_L$	$+\frac{2}{3}$ $-\frac{1}{3}$	$+\frac{1}{2}$ $-\frac{1}{2}$	$+\frac{1}{3}$ $+\frac{1}{3}$
	$\begin{bmatrix} \nu_e \\ e^- \end{bmatrix}_L$	$\begin{bmatrix} \nu_\mu \\ \mu^- \end{bmatrix}_L$	$\begin{bmatrix} \nu_\tau \\ \tau^- \end{bmatrix}_L$	0 -1	$+\frac{1}{2}$ $-\frac{1}{2}$	-1 -1
$\Psi_+ = \gamma_+ \Psi$ (right-handed)	u_R	c_R	t_R	$+\frac{2}{3}$	0	$+\frac{4}{3}$
	d_R	s_R	b_R	$-\frac{1}{3}$	0	$-\frac{2}{3}$
	e_R^-	μ_R^-	τ_R^-	-1	0	-2
	ν_{eR}	$\nu_{\mu R}$	$\nu_{\tau R}$	0	0	0

It must be emphasized that the weak isospin and the weak hypercharge are defined akin to the previously defined and similarly named quantities, and so that they satisfy the familiar formula (2.44b). However, Table 7.1 shows that these quantities coincide with the "old" values (2.44a) only for the left-handed eigenfunctions of chirality and not for the right-handed ones – which have no weak isospin and so are invariant with respect to $SU(2)_w$. In this way, the weak isospin and $SU(2)_w$ play the role, respectively, of the charge and the symmetry for the gauge model of weak interactions.

In the gauge $SU(2)_w \times U(1)_y$ model (Glashow, Weinberg and Salam) one introduces the gauge bosons $W_\mu^{\pm}$ and W_μ^3 for the $SU(2)_w$ factor, and B_μ for the $U(1)_y$ factor. The weak isospin and the weak hypercharge [☞ Table 7.1] determine the interaction intensity between these gauge bosons and the fermions $\{u, d; \nu_e, e^-; c, s; \nu_\mu, \mu^-; \dots\}$, so we know that the interaction terms in the Lagrangian density are, in order:

$$\mathscr{L}_{\text{GWS}} \ni g_w \big(W_\mu^+ J_+^\mu + W_\mu^- J_-^\mu + W_\mu^3 J_3^\mu \big) + g_y B_\mu J_y^\mu, \tag{7.76a}$$

$$J_+^\mu := \big\{ [\overline{u_L}\, \boldsymbol{\gamma}^\mu\, d_{wL}] + [\overline{c_L}\, \boldsymbol{\gamma}^\mu\, s_{wL}] + [\overline{t_L}\, \boldsymbol{\gamma}^\mu\, b_{wL}] \big\}, \tag{7.76b}$$

$$J_-^\mu := \big\{ [\overline{d_{wL}}\, \boldsymbol{\gamma}^\mu\, u_L] + [\overline{s_{wL}}\, \boldsymbol{\gamma}^\mu\, c_L] + [\overline{b_{wL}}\, \boldsymbol{\gamma}^\mu\, t_L] \big\}, \tag{7.76c}$$

$$\begin{aligned} J_3^\mu := \Big\{ &\tfrac{1}{2} \Big([\overline{u_L}\, \boldsymbol{\gamma}^\mu\, u_L] + [\overline{c_L}\, \boldsymbol{\gamma}^\mu\, c_L] + [\overline{t_L}\, \boldsymbol{\gamma}^\mu\, t_L] + [\overline{\nu_{eL}}\, \boldsymbol{\gamma}^\mu\, \nu_{eL}] + [\overline{\nu_{\mu L}}\, \boldsymbol{\gamma}^\mu\, \nu_{\mu L}] + [\overline{\nu_{\tau L}}\, \boldsymbol{\gamma}^\mu\, \nu_{\tau L}] \Big) \\ &- \tfrac{1}{2} \Big([\overline{d_L}\, \boldsymbol{\gamma}^\mu\, d_L] + [\overline{s_L}\, \boldsymbol{\gamma}^\mu\, s_L] + [\overline{b_L}\, \boldsymbol{\gamma}^\mu\, b_L] + [\overline{e_L^-}\, \boldsymbol{\gamma}^\mu\, e_L^-] + [\overline{\mu_L^-}\, \boldsymbol{\gamma}^\mu\, \mu_L^-] + [\overline{\tau_L^-}\, \boldsymbol{\gamma}^\mu\, \tau_L^-] \Big) \Big\}, \end{aligned} \tag{7.76d}$$

$$\begin{aligned}
J_y^\mu := \Big\{ &\tfrac{1}{6}\Big([\overline{u_L}\,\gamma^\mu\,u_L] + [\overline{c_L}\,\gamma^\mu\,c_L] + [\overline{t_L}\,\gamma^\mu\,t_L] + [\overline{d_L}\,\gamma^\mu\,d_L] + [\overline{s_L}\,\gamma^\mu\,s_L] + [\overline{b_L}\,\gamma^\mu\,b_L] \Big) \\
&- \tfrac{1}{2}\Big([\overline{\nu_{eL}}\,\gamma^\mu\,\nu_{eL}] + [\overline{\nu_{\mu L}}\,\gamma^\mu\,\nu_{\mu L}] + [\overline{\nu_{\tau L}}\,\gamma^\mu\,\nu_{\tau L}] + [\overline{e^-_L}\,\gamma^\mu\,e^-_L] + [\overline{\mu^-_L}\,\gamma^\mu\,\mu^-_L] + [\overline{\tau^-_L}\,\gamma^\mu\,\tau^-_L] \Big) \\
&+ \tfrac{2}{3}\Big([\overline{u_R}\,\gamma^\mu\,u_R] + [\overline{c_R}\,\gamma^\mu\,c_R] + [\overline{t_R}\,\gamma^\mu\,t_R] \Big) - \tfrac{1}{3}\Big([\overline{d_R}\,\gamma^\mu\,d_R] + [\overline{s_R}\,\gamma^\mu\,s_R] + [\overline{b_R}\,\gamma^\mu\,b_R] \Big) \\
&- \Big([\overline{e^-_R}\,\gamma^\mu\,e^-_R] + [\overline{\mu^-_R}\,\gamma^\mu\,\mu^-_R] + [\overline{\tau^-_R}\,\gamma^\mu\,\tau^-_R] \Big) \Big\},
\end{aligned} \tag{7.76e}$$

where d_w, s_w and b_w are the quark states defined by the Cabibbo–Kobayashi–Maskawa (CKM) mixing (2.53)–(2.55), the subscript "L" denotes the projection to the left-handed chirality, and where the expression for J_y^μ includes the factor $\frac{1}{2}$ from the formula $Q = I_w + \frac{1}{2}Y_w$, modeled on the original GNN formula (2.44b).

For the purposes of $SU(2)_w \times U(1)_y \to U(1)_Q$ symmetry breaking, Weinberg and Salam[12] introduced a doublet of complex Higgs fields:

$$\mathbb{H} = \begin{bmatrix} H_1 \\ H_2 \end{bmatrix}, \quad \text{with} \quad \begin{cases} I_w(H_1) = +\frac{1}{2} & Y_w(H_1) = +1 \quad Q(H_1) = +1, \\ I_w(H_2) = -\frac{1}{2} & Y_w(H_2) = +1 \quad Q(H_2) = 0. \end{cases} \tag{7.77}$$

We thus identify $H_1 = H^+$, $(H_1)^\dagger = H^-$, $H_2 = H^0$ and $(H_2)^\dagger = \overline{H}^0$.

Besides, $W^\pm_\mu, W^3_\mu$ and B_μ also interact with the complex Higgs field doublet, $\mathbb{H}$,

$$\mathscr{L}_{\mathbb{H}} = \big\|(\partial_\mu - ig_w W^\alpha_\mu \tfrac{1}{2}\boldsymbol{\sigma}_\alpha - ig_y B_\mu \tfrac{1}{2}\mathbb{1})\mathbb{H}\big\|^2_\eta + \tfrac{1}{2}\big(\tfrac{\mu c}{\hbar}\big)^2(\mathbb{H}^\dagger\mathbb{H}) - \tfrac{1}{4}\lambda(\mathbb{H}^\dagger\mathbb{H})^2, \tag{7.78}$$

where the index α is summed over the values $1, 2, 3$, and where

$$\boldsymbol{\sigma}_1 = \begin{bmatrix} 0 & 1 \\ 1 & 0 \end{bmatrix}, \quad \boldsymbol{\sigma}_2 = \begin{bmatrix} 0 & -i \\ i & 0 \end{bmatrix}, \quad \boldsymbol{\sigma}_3 = \tfrac{1}{2}\begin{bmatrix} 1 & 0 \\ 0 & -1 \end{bmatrix}. \tag{7.79}$$

With the sign of the quadratic term as in equation (7.78), the minimum of the potential lies in the values of the field $\mathbb{H}$ that satisfy

$$|H_1|^2 + |H_2|^2 = H_{1r}^2 + H_{1i}^2 + H_{2r}^2 + H_{2i}^2 = \big(\tfrac{\mu c}{\lambda\hbar}\big)^2 \tag{7.80}$$

and which form a 3-sphere $S^3 \subset \mathbb{R}^4 \approx \mathbb{C}^2$. One such value is $\mathbb{H} = \big(\frac{\mu c}{\lambda\hbar}\big)\begin{bmatrix} 0 \\ 1 \end{bmatrix}$.

Digression 7.3 That is, with the standard choice of the Higgs field (7.77), $\langle H_1\rangle = \langle H^+\rangle \neq 0$ would imply that the vacuum has the electric charge $+1$ and that the $U(1)$ gauge symmetry of the electromagnetic interaction is broken – which is not the case! Of

[12] S. L. Glashow had already in 1958, in his PhD dissertation mentored by J. Schwinger, proposed an electro-weak unification based only on the $SU(2)_w$ group, where the photon corresponds to the diagonal generator J_3, and the $W^\pm$-bosons correspond to the generators $J_\pm$ [☞ relations (A.38)]. The model was worked out in collaboration with H. Georgi and it turned out that this cannot be made to agree with experiments [209]. It became clear in the early 1960s that the gauge group $SU(2)_w \times U(1)_y$ is a better choice, so that the photon (7.85) would interact with fermions with an intensity equal to the electric charge obtained from the GNN formula (7.75). The mass of the $W^\pm$- and the Z^0-bosons had, however, remained a mystery: Simply added "by hand" (as Glashow advocated), the mass of the gauge bosons explicitly breaks the gauge invariance but also the renormalizability (and then also the self-consistency) of the model. In 1967–8, Weinberg and, independently, Salam showed that the Higgs mechanism may be applied and produces the desired mass. G. 't Hooft (1971), B. W. Lee and J. Zinn-Justin, and finally G. 't Hooft and M. Veltman (1972) proved the renormalizability of the Glashow–Weinberg–Salam model of electroweak interactions, and D. J. Gross and R. Jackiw, and then C. Bouchiat, J. Iliopoulos and P. Meyer showed the same year (1972) that all anomalies cancel in this model [209, 552, 473].

course, the choice $\langle H_1\rangle \neq 0$ and $\langle H_2\rangle = 0$ would only imply that the remaining massless field is not A_μ but Z_μ amongst the linear combinations (7.85)–(7.86), and that the corresponding $U(1) \subset SU(2)_w \times U(1)_y$ remains the exact gauge symmetry. This group $U(1)$ and this field would then have to be identified, respectively, with the gauge symmetry of electromagnetism and the photon.

After redefining the Higgs field,

$$\widetilde{\mathbb{H}} := \mathbb{H} - \langle \mathbb{H}\rangle, \qquad \langle \mathbb{H}\rangle = \left(\tfrac{\mu c}{\lambda\hbar}\right)\left[\begin{smallmatrix}0\\1\end{smallmatrix}\right], \tag{7.81}$$

it follows that $\widetilde{H}_{1r}, \widetilde{H}_{1i}, \widetilde{H}_{2r}, \widetilde{H}_{2i}, W^1_\mu, W^2_\mu, W^3$ and B_μ are not the normal modes – just as in the Lagrangian density (7.36)–(7.39) – and one must again diagonalize the fields. The identification of normal modes is fairly simple. From equation (7.78), we have

$$\begin{aligned}
&\left[(\partial_\mu - ig_w W^\alpha_\mu\,\tfrac{1}{2}\sigma_\alpha - ig_y B_\mu\,\tfrac{1}{2}\mathbb{1})\mathbb{H}\right]^\dagger \eta^{\mu\nu}\left[(\partial_\nu - ig_w W^\beta_\nu\,\tfrac{1}{2}\sigma_\beta - ig_y B_\nu\,\tfrac{1}{2}\mathbb{1})\mathbb{H}\right]\\
&\quad = \cdots + \left(\tfrac{\mu c}{\lambda\hbar}\right)^2\left[(-ig_w W^\alpha_\mu\,\tfrac{1}{2}\sigma_\alpha - ig_y B_\mu\,\tfrac{1}{2}\mathbb{1})\left[\begin{smallmatrix}0\\1\end{smallmatrix}\right]\right]^\dagger \eta^{\mu\nu}\left[(-ig_w W^\beta_\nu\,\tfrac{1}{2}\sigma_\beta - ig_y B_\nu\,\tfrac{1}{2}\mathbb{1})\left[\begin{smallmatrix}0\\1\end{smallmatrix}\right]\right] + \cdots\\
&\quad = \cdots + \tfrac{1}{4}\left(\tfrac{\mu c}{\lambda\hbar}\right)^2 (g_w W^3_\mu - g_y B_\mu)^\dagger \eta^{\mu\nu}(g_w W^3_\nu - g_y B_\nu) + \cdots.
\end{aligned} \tag{7.82}$$

Using the "weak angle"

$$\theta_w = \arccos\left(\frac{g_w}{\sqrt{g_w^2+g_y^2}}\right), \qquad \text{so} \quad \cos\theta_w = \frac{g_w}{\sqrt{g_w^2+g_y^2}} \quad \text{and} \quad \sin\theta_w = \frac{g_y}{\sqrt{g_w^2+g_y^2}}, \tag{7.83}$$

the expression (7.82) becomes

$$\cdots + \tfrac{1}{2}\left(\tfrac{\mu c}{\sqrt{2}\lambda\hbar}\right)^2 (g_w^2+g_y^2)\left\|\left(\cos(\theta_w)W^3_\mu - \sin(\theta_w)B_\mu\right)\right\|^2_\eta + \cdots. \tag{7.84}$$

The normal modes then are the linear combinations

$$A_\mu := \cos(\theta_w)B_\mu + \sin(\theta_w)W^3_\mu, \qquad \text{with the mass } = 0, \tag{7.85}$$

$$Z_\mu := -\sin(\theta_w)B_\mu + \cos(\theta_w)W^3_\mu, \qquad \text{with the mass } = \tfrac{\mu c}{\sqrt{2}\lambda\hbar}\sqrt{g_w^2+g_y^2}. \tag{7.86}$$

The gauge boson represented by the 4-vector A_μ is identified as the photon, and the gauge boson represented by the 4-vector Z^0_μ acquired a mass and is identified with the massive Z^0-particle. Similarly,

$$\begin{aligned}
&\left[(\partial_\mu - ig_w W^\alpha_\mu\,\tfrac{1}{2}\sigma_\alpha - ig_y B_\mu\,\tfrac{1}{2}\mathbb{1})\mathbb{H}\right]^\dagger \eta^{\mu\nu}\left[(\partial_\nu - ig_w W^\beta_\nu\,\tfrac{1}{2}\sigma_\beta - ig_y B_\nu\,\tfrac{1}{2}\mathbb{1})\mathbb{H}\right]\\
&\quad = \cdots + \left(\tfrac{\mu c}{\lambda\hbar}\right)^2\left[-ig_w(W^1_\mu\,\tfrac{1}{2}\sigma_1 + W^2_\mu\,\tfrac{1}{2}\sigma_2)\left[\begin{smallmatrix}0\\1\end{smallmatrix}\right]\right]^\dagger \eta^{\mu\nu}\left[-ig_w(W^1_\nu\,\tfrac{1}{2}\sigma_1 + W^2_\nu\,\tfrac{1}{2}\sigma_2)\left[\begin{smallmatrix}0\\1\end{smallmatrix}\right]\right] + \cdots\\
&\quad = \cdots + \tfrac{1}{2}g_w^2\left(\tfrac{\mu c}{\lambda\hbar}\right)^2\left[(W^+_\mu\sigma_- + W^-_\mu\sigma_+)\left[\begin{smallmatrix}0\\1\end{smallmatrix}\right]\right]^\dagger \eta^{\mu\nu}\left[(W^+_\nu\sigma_- + W^-_\nu\sigma_+)\left[\begin{smallmatrix}0\\1\end{smallmatrix}\right]\right] + \cdots\\
&\quad = \cdots + g_w^2\left(\tfrac{\mu c}{\sqrt{2}\lambda\hbar}\right)^2 W^+_\mu \eta^{\mu\nu} W^-_\nu + \cdots,
\end{aligned} \tag{7.87}$$

where

$$W^\pm_\mu := \tfrac{1}{\sqrt{2}}(W^1_\mu \pm iW^2_\mu) \qquad \text{and} \qquad \sigma_+ = \left[\begin{smallmatrix}0&1\\0&0\end{smallmatrix}\right], \quad \sigma_- = \left[\begin{smallmatrix}0&0\\1&0\end{smallmatrix}\right]. \tag{7.88}$$

This shows that the mass of the $W^{\pm}$-bosons equals $g_w(\frac{\mu c}{\sqrt{2}\lambda\hbar})$, and using the definition (7.83) and the results (7.84) and (7.87) we have

$$M_W = \cos(\theta_w)\, M_Z, \tag{7.89a}$$

$$\text{since} \quad \Big[g_w(\tfrac{\mu c}{\sqrt{2}\lambda\hbar})\Big] = \frac{g_w}{\sqrt{g_w^2+g_y^2}}\Big[\sqrt{g_w^2+g_y^2}(\tfrac{\mu c}{\sqrt{2}\lambda\hbar})\Big]. \tag{7.89b}$$

Conclusion 7.7 *Note that the gauge fields B_μ and W^3_μ couple, respectively, to the corresponding "charges" Y_w and I_w, and that the gauge field A_μ – the photon – couples to the electric charge Q. The linear relation (7.85) then corresponds to the "weak" version of the GNN formula, $Q = I_w + \frac{1}{2}Y_w$, which holds for the values of these charges as they are given in Table 7.1 on p. 275.*

The fermion currents that interact with the gauge fields $W^{\pm}$ remain the same as in (7.76b)–(7.76c), and the A_μ and the Z^0_μ fields respectively interact with the fermion currents:

$$J^\mu_{\text{em}} := [J^\mu_3 + J^\mu_y] \qquad = [J^\mu_{\text{em}\,L} + J^\mu_{\text{em}\,R}], \tag{7.90}$$

$$J^\mu_Z := \frac{1}{\cos(\theta_w)}\big[J^\mu_3 - \sin^2(\theta_w) J^\mu_{\text{em}\,L}\big] = \frac{1}{\cos(\theta_w)}\big[\cos^2(\theta_w) J^\mu_3 - \sin^2(\theta_w) J^\mu_y\big], \tag{7.91}$$

where

$$J^\mu_{\text{em}\,i} := \sum_{i=L,R}\Big\{+\tfrac{2}{3}\Big([\overline{u_i}\,\boldsymbol{\gamma}^\mu\,u_i] + [\overline{c_i}\,\boldsymbol{\gamma}^\mu\,c_i] + [\overline{t_i}\,\boldsymbol{\gamma}^\mu\,t_i]\Big) - \tfrac{1}{3}\Big([\overline{d_i}\,\boldsymbol{\gamma}^\mu\,d_i] + [\overline{s_i}\,\boldsymbol{\gamma}^\mu\,s_i] + [\overline{b_i}\,\boldsymbol{\gamma}^\mu\,b_i]\Big)$$
$$-1\Big([\overline{e_i^-}\,\boldsymbol{\gamma}^\mu\,e_i^-] + [\overline{\mu_i^-}\,\boldsymbol{\gamma}^\mu\,\mu_i^-] + [\overline{\tau_i^-}\,\boldsymbol{\gamma}^\mu\,\tau_i^-]\Big)\Big\}. \tag{7.92}$$

Digression 7.4 That is, we have that

$$g_w\,W^3_\mu\,J^\mu_3 + g_y\,B_\mu\,J^\mu_y = g_w\Big[\sin(\theta_w)A_\mu + \cos(\theta_w)Z_\mu\Big]J^\mu_3 + g_y\Big[\cos(\theta_w)A_\mu - \sin(\theta_w)Z_\mu\Big]J^\mu_y$$
$$= \Big[g_w\sin(\theta_w)\,J^\mu_3 + g_y\cos(\theta_w)\,J^\mu_y\Big]A_\mu + \Big[g_w\cos(\theta_w)\,J^\mu_3 - g_y\sin(\theta_w)\,J^\mu_y\Big]Z_\mu, \tag{7.93a}$$

where, of course, we know that

$$\Big[g_w\sin(\theta_w)\,J^\mu_3 + g_y\cos(\theta_w)\,J^\mu_y\Big] = \Big[\frac{g_w\,g_y}{\sqrt{g_w^2+g_y^2}}J^\mu_3 + \frac{g_y\,g_w}{\sqrt{g_w^2+g_y^2}}J^\mu_y\Big] = g_e\,J^\mu_{\text{em}}. \tag{7.93b}$$

This recovers the original GNN formula (2.30), i.e., (2.44b):

$$J^\mu_{\text{em}} = J^\mu_3 + J^\mu_y, \tag{7.93c}$$

since the $\frac{1}{2}$ factor in the GNN formula (2.30) is built into the definition of J^μ_y (7.76e). Also,

$$\frac{g_w\,g_y}{\sqrt{g_w^2+g_y^2}} \overset{(7.83)}{=} g_w\sin(\theta_w) \overset{(7.83)}{=} g_y\cos(\theta_w) \overset{(7.90)}{=} g_e. \tag{7.93d}$$

In turn,

$$g_w\cos(\theta_w)\,J^\mu_3 - g_y\sin(\theta_w)\,J^\mu_y = g_z\Big[\cos^2(\theta_w)\,J^\mu_3 - \sin^2(\theta_w)\,J^\mu_{\text{em}}\Big] \tag{7.93e}$$

recovers equation (7.91), where

$$g_z = g_w/\cos(\theta_w) = \sqrt{g_w^2+g_y^2}. \tag{7.93f}$$

Note that $g_z = g_w/\cos(\theta_w) > g_w\sin(\theta_w) = g_e$, and $\frac{g_e}{g_z} = \frac{1}{2}\sin(2\theta_w)$.

Already from the expansions (7.76) and (7.90)–(7.91), we see that the complete Lagrangian density contains very many terms. There exist several different "economical" ways of writing that "pack" of the myriads of summands in different ways. For example, we may write

$$J^{\mu}_{\text{em}} = \sum_{n} \Big\{ \tfrac{2}{3}[\overline{U}_n \gamma^{\mu} U_n] - \tfrac{1}{3}[\overline{D}_n \gamma^{\mu} D_n] - [\overline{\ell}_n \gamma^{\mu} \ell_n] \Big\}, \qquad n = 1,2,3, \tag{7.94}$$

where $U_1 = u$, $U_2 = c$, $U_3 = t$, $D_1 = d$, $D_2 = s$, $D_3 = b$, $\ell_1 = e^-$, $\ell_2 = \mu^-$ and $\ell_3 = \tau^-$, and omitting the projections to left-handed chirality of a particle indicates the inclusion of both left- and right-handed particles in the sum.

For concrete computations, it is however more convenient to simply list the amplitude contributions of each possible vertex and line, as done in the next section.

7.2.5 Feynman's rules for weak interactions

Interactions of the $W^{\pm}$-bosons with elementary Standard Model fermions are simple as compared to the interactions of the Z^0-boson. It is important, however, to keep in mind that the d_w-, s_w- and b_w-quark states, which interact by weak interactions, are defined as the CKM combinations (2.53)–(2.55):

$$\begin{bmatrix} |d_w\rangle \\ |s_w\rangle \\ |b_w\rangle \end{bmatrix} := \begin{bmatrix} V_{ud} & V_{us} & V_{ub} \\ V_{cd} & V_{cs} & V_{cb} \\ V_{td} & V_{ts} & V_{tb} \end{bmatrix} \begin{bmatrix} |d\rangle \\ |s\rangle \\ |b\rangle \end{bmatrix}, \tag{7.95a}$$

$$= \begin{bmatrix} c_{12}c_{13} & s_{12}c_{13} & s_{13}\, e^{-i\delta_{13}} \\ -s_{12}c_{23} - c_{12}s_{23}s_{13}\, e^{i\delta_{13}} & c_{12}c_{23} - s_{12}s_{23}s_{13}\, e^{i\delta_{13}} & s_{23}c_{13} \\ s_{12}s_{23} - c_{12}c_{23}s_{13}\, e^{i\delta_{13}} & -c_{12}s_{23} - s_{12}c_{23}s_{13}\, e^{i\delta_{13}} & c_{23}c_{13} \end{bmatrix} \begin{bmatrix} |d\rangle \\ |s\rangle \\ |b\rangle \end{bmatrix}, \tag{7.95b}$$

$$\text{where} \qquad c_{ij} := \cos(\theta_{ij}), \qquad s_{ij} := \sin(\theta_{ij}), \quad i,j = 1,2,3 = d,s,b,$$

and where $|d\rangle$, $|s\rangle$ and $|b\rangle$ are the eigenstates of the "free" Hamiltonian, i.e., the states with the well-defined mass.[13] This permits writing

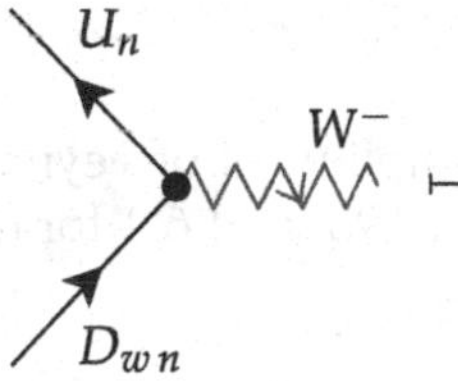

$$\longmapsto \; -\frac{i g_w}{2\sqrt{2}} \gamma^{\mu} [\mathbb{1} - \hat{\gamma}]\, \delta_n^{n'} \qquad \begin{bmatrix} U_n \\ D_{w\,n} \end{bmatrix} = \underset{n=1}{\begin{bmatrix} u \\ d_w \end{bmatrix}}, \underset{n=2}{\begin{bmatrix} c \\ s_w \end{bmatrix}}, \underset{n=3}{\begin{bmatrix} t \\ b_w \end{bmatrix}}, \tag{7.96}$$

and

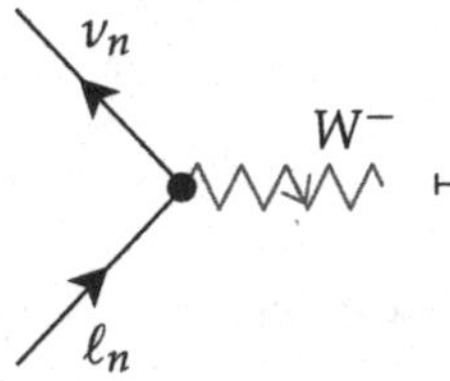

$$\longmapsto \; -\frac{i g_w}{2\sqrt{2}} \gamma^{\mu} [\mathbb{1} - \hat{\gamma}]\, \delta_n^{n'} \qquad \begin{bmatrix} \nu_n \\ \ell_n \end{bmatrix} = \underset{n=1}{\begin{bmatrix} \nu_e \\ e^- \end{bmatrix}}, \underset{n=2}{\begin{bmatrix} \nu_\mu \\ \mu^- \end{bmatrix}}, \underset{n=3}{\begin{bmatrix} \nu_\tau \\ \tau^- \end{bmatrix}}, \tag{7.97}$$

which of course implies all processes that may be obtained from $D_{w\,n} \to W^- + U_n$ and $\ell_n \to W^- + \nu_n$ using the crossing symmetry and the principle of detailed balance [☞ Section 2.3.8].

[13] These are the *stationary states*, well known to the Student who successfully covered quantum mechanics, the eigenstates of the "free" Hamiltonian, i.e., the one where the mixing and interaction terms are omitted.

That is, using the CKM definitions (7.95), the interactions with the $W^{\pm}$-bosons do not mix the CKM-redefined "families" of quarks.

Although A_μ is a linear combination of the W^3_μ-field with the "$V{-}A$" type of interaction with elementary Standard Model fermions and of the B_μ-field that interacts with the fermions of both left- and right-handed chirality, the values of I_w and Y_w in Table 7.1 on p. 275 ensure that the resulting interaction with the A_μ-field is purely of the "V" type. That is, the A_μ-field interacts equally with fermions of both left- and right-handed chirality, and of course, precisely as the photon in electrodynamics [☞ Procedure 5.2 on p. 193].

The neutral Z_μ-field is the complementary linear combination of the neutral W^3_μ- and B_μ-fields, and the interactions of this Z_μ-field with the elementary Standard Model fermions are not as simple as those of the A_μ-field. Following the textbook [243], we may write

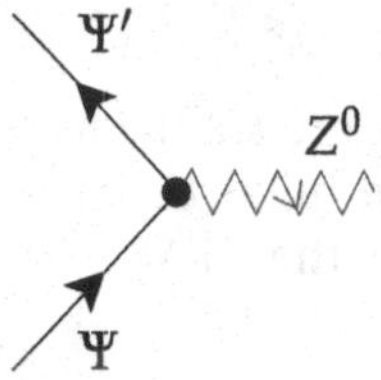

$$\longmapsto\quad -\frac{i g_w}{2\sqrt{2}}\boldsymbol{\gamma}^\mu[c_V\mathbb{1}-c_A\widehat{\boldsymbol{\gamma}}]\delta^{\Psi'}_{\Psi} \tag{7.98}$$

Ψ	c_V	c_A
ν_n	$\frac{1}{2}$	$\frac{1}{2}$
ℓ_n	$-\frac{1}{2}+2\sin^2(\theta_w)$	$-\frac{1}{2}$
U_n	$\frac{1}{2}-\frac{4}{3}\sin^2(\theta_w)$	$\frac{1}{2}$
D_n	$-\frac{1}{2}+\frac{2}{3}\sin^2(\theta_w)$	$-\frac{1}{2}$

As regards the internal lines that correspond to $W^{\pm}$- and Z^0-boson exchanges, analogously to step 3 in the procedures 5.2 on p. 193, and 6.1 on p. 232, we assign

$$\longmapsto\quad -\frac{i(\eta_{\mu\nu}-q_\mu q_\nu/M^2c^2)}{\mathrm{q}^2-M^2c^2}, \tag{7.99}$$

where $M = M_W$ or $M = M_Z$, depending on whether the propagator corresponds to the $W^{\pm}$- or the Z^0-boson exchange. When the exchange energies are sufficiently smaller than Mc^2, we have

$$\lim_{(|\mathrm{q}^2|/M_W^2c^2)\to 0} -\frac{i(\eta_{\mu\nu}-q_\mu q_\nu/M^2c^2)}{\mathrm{q}^2-M^2c^2} \approx \frac{i\eta_{\mu\nu}}{M^2c^2}, \tag{7.100}$$

which is usually a good first approximation.

In addition to these definitions, the procedure for computing amplitudes of Feynman diagrams is identical to Procedures 5.2 for quantum electrodynamics on p. 193, and 6.1 for quantum chromodynamics on p. 232.

Example 7.3 The elastic collision $\nu_\mu + e^- \to \nu_\mu + e^-$ may occur, to $O(g_w^2)$ order, only mediated by a Z^0-boson exchange:

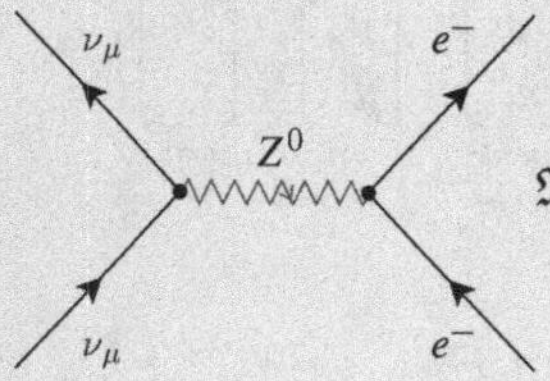

$$\mathfrak{M} = \frac{g_z{}^2}{8M_Z^2c^2}\big[\overline{v}_3\boldsymbol{\gamma}^\mu(\mathbb{1}-\widehat{\boldsymbol{\gamma}})v_1\big]\big[\overline{e}_4\boldsymbol{\gamma}^\mu(c_V\mathbb{1}-c_A\widehat{\boldsymbol{\gamma}})e_1\big], \tag{7.101}$$

where $v_i := \Psi_{\nu_\mu}(\mathrm{p}_i)$ and $e_i := \Psi_{e^-}(\mathrm{p}_i)$. Computing as in the case (5.131)–(5.140) we obtain

$$\langle|\mathfrak{M}|^2\rangle = \tfrac{1}{2}\Big(\frac{g_Z}{4M_Z c}\Big)^4 \Big\{(c_V+c_A)^2(\mathrm{p}_1\cdot\mathrm{p}_2)(\mathrm{p}_3\cdot\mathrm{p}_4) + (c_V-c_A)^2(\mathrm{p}_1\cdot\mathrm{p}_4)(\mathrm{p}_3\cdot\mathrm{p}_2) - m_e^2c^2(c_V^2-c_A^2)(\mathrm{p}_1\cdot\mathrm{p}_3)\Big\}. \tag{7.102}$$

In the CM-system and neglecting the electron mass, $m_e \to 0$, we obtain the simpler relation

$$\langle|\mathfrak{M}|^2\rangle = \tfrac{1}{2}\Big(\frac{g_Z\,E}{M_Z c^2}\Big)^4 \Big\{(c_V+c_A)^2 + (c_V-c_A)^2\cos^4(\tfrac{1}{2}\theta)\Big\}, \tag{7.103}$$

where E is the energy of the electron (as well as the neutrino) in the CM-system, and θ is the electron deflection angle. Then

$$\frac{d\sigma}{d\Omega} = 2\Big(\frac{\hbar c}{\pi}\Big)^2\Big(\frac{g_Z}{4M_Z c^2}\Big)^4 E^2\,\Big\{(c_V+c_A)^2 + (c_V-c_A)^2\cos^4(\tfrac{1}{2}\theta)\Big\}, \tag{7.104}$$

$$\begin{aligned}\sigma &= \frac{2}{3\pi}(\hbar c)^2\Big(\frac{g_Z}{2M_Z c^2}\Big)^4 E^2\,(c_V^2+c_A^2+c_V\,c_A)\\ &= \frac{2}{\pi}(\hbar c)^2\Big(\frac{g_Z}{2M_Z c^2}\Big)^4 E^2\,\Big(\tfrac{1}{4}-\sin^2(\theta_w)+\tfrac{4}{3}\sin^4(\theta_w)\Big).\end{aligned} \tag{7.105}$$

Comparing with the similar process $\nu_\mu + e^- \to \nu_e + \mu^-$ that involves the exchange of a W-boson:

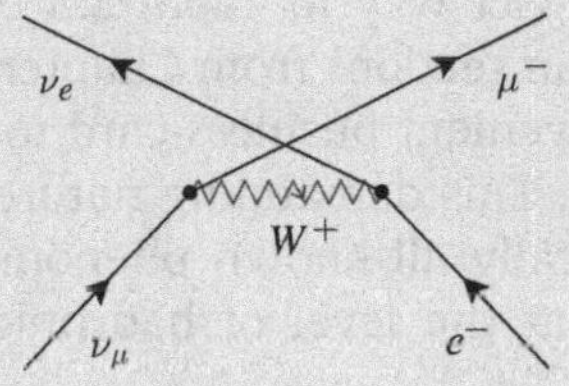

$$\sigma = \frac{1}{8\pi}\Big[\Big(g_w M_W c^2\Big)^2\,\hbar c E\Big]^2\Big[1-\Big(\frac{m_\mu c^2}{2E}\Big)^2\Big]^2 \tag{7.106}$$

and at energies $E \gg m_\mu c^2$, we have (using $\theta_w = 28.75°$, from the ratio of the measured masses M_W/M_Z)

$$\frac{\sigma(\nu_\mu + e^- \to \nu_\mu + e^-)}{\sigma(\nu_\mu + e^- \to \nu_e + \mu^-)} \approx \tfrac{1}{4} - \sin^2(\theta_w) + \tfrac{4}{3}\sin^4(\theta_w) = 0.0900. \tag{7.107}$$

This agrees with the experimental value $0.11 \pm 10\%$ fairly well.

7.2.6 Exercises for Section 7.2

✎ ***7.2.1*** *Following Example 7.1, show that the sum of all amplitudes for both diagrams of the type (7.69) but with two (three) W^3-particle and one (no) photon also vanishes.*

✎ ***7.2.2*** *For the potential process $\gamma \to 2\gamma$ described by an appropriate algebraic sum of diagrams of the type (7.69) but with a photon in place of W^3, show that the symmetrization of the outgoing photons (as bosons) guarantees that the sum of the contributions of these Feynman diagrams vanishes.*

✎ ***7.2.3*** *Following Example 7.2, show that the sum of all amplitudes for both diagrams of the type (7.72) but with n B-particles and $(3-n)$ photons also vanishes, for every $n = 0, 1, 2, 3$.*

✎ ***7.2.4*** *Complete the computation in Example 7.3.*

7.3 The Standard Model

The elaborate structure called the *Standard Model* of elementary particle physics has the following components:

1. the elementary spin-$\frac{1}{2}$ fermions in Table 7.1 on p. 275, and the data (2.44);
2. electromagnetic interactions with the $U(1)_Q$ gauge symmetry [☞ Section 5.3];
3. chromodynamic interactions with the $SU(3)_c$ gauge symmetry [☞ Section 6.1];
4. the asymmetric treatment of particles with left- and right-handed chirality [☞ discussion around the expressions (5.57)–(5.62), then Sections 7.2.1 and 7.2.4];
5. the GIM mechanism, anomaly cancellation and generalization of the GIM mechanism with the Cabibbo–Kobayashi–Maskawa quark mixing [☞ Section 7.2.2];
6. the $SU(2)_w \times U(1)_y$ gauge symmetry of the electroweak interactions, in the symmetric phase;
7. the spontaneous $SU(2)_w \times U(1)_y \to U(1)_Q$ gauge symmetry breaking of electroweak interactions in the Higgs phase [☞ Sections 7.1 and 7.2.4];
8. the very intricate and detailed structure of fermion masses [☞ Tables 4.1 on p. 152, and C.2 on p. 526].

This structure is presented in an extremely short and ultra-compact way in Table 2.3 on p. 67. However, the incremental development of the material presented in sections from Chapter 2 up to now clearly indicates that this short compactness is merely a convenient business-card to an otherwise technically very demanding and intricate Standard Model. This demanding nature should not be surprising, since this model successfully describes practically all known phenomena not only at the fundamental level of quarks and leptons, but also at the level of hadronic bound states [☞ Section 2.4.1 and Conclusion 2.4 on p. 71].

Undoubtedly, the most complex parts of the Standard Model pertain to the aspects of weak interactions, which are roughly presented in the foregoing part of this chapter. It remains to discuss (1) the general mechanism in the Standard Model by which fermions in Table 7.1 on p. 275 acquire a mass, and (2) neutrino mixing.

7.3.1 Fermion masses

The argument at the very beginning of Chapter 7 shows that the gauge bosons are massless by construction – except, as we have seen here, those corresponding to symmetries spontaneously broken via the Higgs mechanism [☞ Section 7.1.3]. The mass of these gauge bosons stems from the interaction with the Higgs field [☞ expressions (7.84) and (7.87)] and owing to the shift $\mathbb{H} \to \widetilde{\mathbb{H}} + \langle\mathbb{H}\rangle$, which is dictated by the fact that the "flipped" sign of the quadratic term in the Lagrangian density (7.78) puts the minimum of the potential energy at one of the points with $\mathbb{H}^\dagger\mathbb{H} = (\frac{\mu c}{\lambda\hbar})^2 > 0$, so that the vacuum expectation value of the two-component Higgs field is not zero, $\langle\mathbb{H}\rangle \neq 0$.

Similarly, one expects that the fermion masses also stem from the Higgs field shift $\mathbb{H} \to \widetilde{\mathbb{H}} + \langle\mathbb{H}\rangle$. The expression (7.47) shows that a typical term in the Lagrangian density that provides the fermion fields with a mass must be of the form (with the customary notation $\Psi_+ = \Psi_L$ and $\Psi_- = \Psi_R$)

$$\overline{\Psi_L}\,\mathbb{H}\,\Psi_R \quad \text{and} \quad \overline{\Psi_R}\,\mathbb{H}\,\Psi_L. \tag{7.108}$$

Such terms are possible precisely because $\mathbb{H}$ is an $SU(2)_w$-doublet, just as are the wave-functions for all fermions of left-handed chirality, whereas all right-handed fermions are invariant under $SU(2)_w$ transformations. Therefore, terms such as[14]

$$\begin{aligned} h_e\, \overline{e_R^-}\, \mathbb{H}^\dagger \left[\begin{smallmatrix} \nu_e \\ e^- \end{smallmatrix}\right]_L + h.c. &= h_e\, \overline{e_R^-}\, \left[\begin{smallmatrix} H_1^* & H_2^* \end{smallmatrix}\right] \left[\begin{smallmatrix} \nu_e \\ e^- \end{smallmatrix}\right]_L + h.c. = h_e\, \overline{e_R^-}\, (H_1^* \nu_{eL} + H_2^* e_L^-) + h.c. \\ &= \Re e(h_e \langle H_2 \rangle^*)\, (\overline{e_R^-}\, e_L^- + \overline{e_L^-}\, e_R^-) + \cdots \end{aligned} \tag{7.109}$$

are $SU(2)_w \times U(1)_y$-invariant and produce the electron mass, $m_e = \Re e\,(h_e \langle H_2 \rangle)/c^2$. Similarly, for d-quarks one has

$$\begin{aligned} h_d\, \overline{d_R}\, \mathbb{H}^\dagger \left[\begin{smallmatrix} u \\ d \end{smallmatrix}\right]_L + h.c. &= h_d\, \overline{d_R}\, \left[\begin{smallmatrix} H_1^* & H_2^* \end{smallmatrix}\right] \left[\begin{smallmatrix} u \\ d \end{smallmatrix}\right]_L + h.c. = h_d\, \overline{d_R} (H_1^* u_L + H_2^* d_L) + h.c. \\ &= \Re e(h_d \langle H_2 \rangle^*)\, (\overline{d_R}\, d_L + \overline{d_L}\, d_R) + \cdots, \end{aligned} \tag{7.110}$$

which are also $SU(2)_w \times U(1)_y$-invariant and produce $m_d = \Re e\,(h_d \langle H_2 \rangle)/c^2$, the d-quark mass. For u-quarks, an additional definition [☞ discussion of the relation (A.49)] is needed:

$$C : \mathbb{H} = \begin{bmatrix} H_1 \\ H_2 \end{bmatrix} \longmapsto \mathbb{H}^c := -\boldsymbol{\varepsilon}\mathbb{H}^* = \left[\begin{smallmatrix} 0 & -1 \\ 1 & 0 \end{smallmatrix}\right] \begin{bmatrix} H_1^* \\ H_2^* \end{bmatrix} = \begin{bmatrix} -H_2^* \\ H_1^* \end{bmatrix}, \tag{7.111}$$

which transforms, under $SU(2)_w$, the same as $\mathbb{H}$. We can therefore add to the Lagrangian density also the terms

$$\begin{aligned} -h_u\, \overline{u_R}\, (\mathbb{H}^c)^\dagger \left[\begin{smallmatrix} u \\ d \end{smallmatrix}\right]_L + h.c. &= -h_u\, \overline{u_R}\, \left[\begin{smallmatrix} -H_2 & H_1 \end{smallmatrix}\right] \left[\begin{smallmatrix} u \\ d \end{smallmatrix}\right]_L + h.c. = -h_u\, \overline{u_R} (-H_2 u_L + H_1 d_L) + h.c. \\ &= \Re e(h_u \langle H_2 \rangle)\, (\overline{u_R}\, u_L + \overline{u_L}\, u_R) + \cdots, \end{aligned} \tag{7.112}$$

which are also $SU(2)_w \times U(1)_y$-invariant and produce $m_u = \Re e\,(h_u \langle H_2 \rangle)/c^2$, the u-quark mass.

The structure of the Standard Model neither requires nor prohibits adding the neutrino of right-handed chirality, which is noted in Table 7.1 on p. 275: ν_{iR} (with $i = e, \mu, \tau$) are included in the table but are separated from the other fermions. If one includes these right-handed neutrinos, one can include in the Lagrange density also the terms

$$\begin{aligned} -h_\nu\, \overline{\nu_{eR}}\, (\mathbb{H}^c)^\dagger \left[\begin{smallmatrix} \nu_e \\ e^- \end{smallmatrix}\right]_L + h.c. &= -h_\nu\, \overline{\nu_{eR}}\, \left[\begin{smallmatrix} -H_2 & H_1 \end{smallmatrix}\right] \left[\begin{smallmatrix} \nu_e \\ e^- \end{smallmatrix}\right]_L + = -h_\nu\, \overline{\nu_{eR}} (-H_2 \nu_{eL} + H_1 e_L^-) + h.c. \\ &= \Re e(h_\nu \langle H_2 \rangle)\, (\overline{\nu_{eR}}\, \nu_{eL} + \overline{\nu_{eL}}\, \nu_{eR}) + \cdots, \end{aligned} \tag{7.113}$$

which are also $SU(2)_w \times U(1)_y$-invariant and produce $m_\nu = \hbar\, \Re e\,(h_\nu \langle H_2 \rangle)/c$, the neutrino mass.

The quantities defined by the relations (7.109), (7.110), (7.112) and (7.113) are the so-called Dirac masses, since the variation of the Lagrangian density by fermion fields produces the Dirac equation (5.34), with the indicated masses. In addition, terms that were omitted in the expressions (7.109), (7.110), (7.112) and (7.113) are of the general form

$$h_i\, \Re e\,(H_2)\, (\overline{\Psi_{iR}}\, \Psi_{iL} + \overline{\Psi_{iL}}\, \Psi_{iR}), \tag{7.114}$$

which define interactions of the Higgs particle, $\Re e(H_2)$, with the Standard Model fermions. The remaining components of the complex Higgs doublet, $H_1 = H^+$, $H_1^* = H^-$ and $\Im m(H_2)$ have become the longitudinal components of the $W^\pm$- and the Z^0-bosons; see Section 7.1.3, Conclusion 7.5 on p. 265, and equation (7.49).

The so-obtained fermion masses (7.109), (7.110), (7.112) and (7.113) as well as the masses of the Z^0- and the $W^\pm$-bosons (7.82)–(7.87) are all proportional to the mass $\Re e\,(\langle H_2 \rangle)/c^2$. The

[14] The abbreviation "$+h.c.$" is standard for adding the Hermitian conjugate terms.

Yukawa parameters h_e, h_d, h_u, h_ν (and similarly for the remaining two families) are, however, completely arbitrary parameters of the Standard Model and, besides in the fermion masses, appear only in the terms of the type (7.114) that describe the Yukawa interactions of the fermions with the Higgs particle. This then links the intensity of this interaction with the fermion masses. Of course, until the details of the interactions of the Higgs particle with the Standard Model fermions are measured sufficiently precisely, the choice of the parameters h_e, h_d, h_u, h_ν, etc., is determined only in terms of the measured fermion masses – except for the neutrinos; see the next section.

Since the Standard Model fermion masses [☞ Tables 4.1 on p. 152, and C.2 on p. 526] differ significantly from the masses of the $W^\pm$- and the Z^0-bosons, it follows that the parameters h_e, h_d, h_u, h_ν, etc., are quite far from numbers of order 1, and the structure represented by this list of parameters ought to be explained somehow. However, that is a task beyond the Standard Model.

Digression 7.5 Let us mention a non-standard version of the Standard Model [169], where one introduces a Higgs field that is $SU(2)_w \times U(1)_y$-invariant, but has Yukawa interactions $(\overline{\Psi}\,\widetilde{H}\,\Psi)$ with the Standard Model fermions. Shifting $\widetilde{H} \to \widetilde{H}' + \langle\widetilde{H}\rangle$, the fermions acquire a mass just as by the previously described standard method (7.114). As $SU(2)_w \times U(1)_y$ gauge bosons do not interact directly with this Higgs boson, their masses stem from perturbative corrections of the type

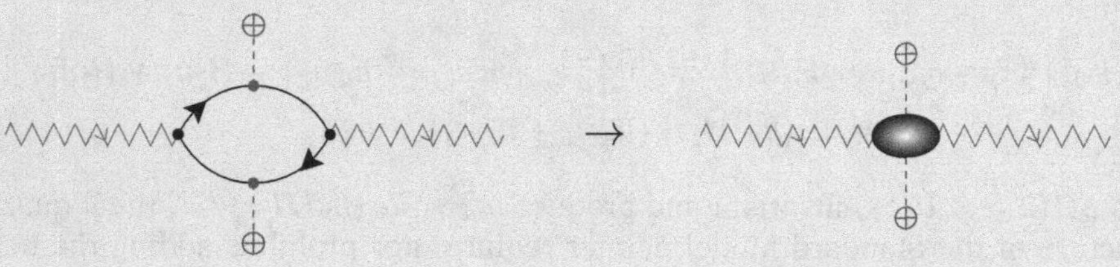

(7.115)

where the shaded oval in the right-hand diagram represents the resulting effective (indirect) interaction between $SU(2)_w \times U(1)_y$ gauge bosons and the Higgs field $\widetilde{H}$ that sinks into the vacuum, i.e., $\langle\widetilde{H}\rangle \neq 0$; compare with the illustration (7.43). Effectively, the so-obtained mass for the gauge bosons produces a model that differs from the Standard Model results only at energies significantly larger than $m_W, m_Z \sim 100\,\text{GeV}/c^2$. Since these masses are radiatively induced, the mass of the Higgs particle itself is expected to be larger than $100\,\text{GeV}/c^2$ – in agreement with the recent LHC results at CERN [25, 109, 293]. Only detailed measurements of the interactions of the Higgs particle with the other Standard Model particles can distinguish this possibility from the original version, or other generalizations and extensions.

7.3.2 Neutrino mixing

It was noted in the early 1990s that amongst the neutrinos that arrive at the Earth's surface there are fewer muon neutrinos, ν_μ, than expected. That is, neutrinos are produced in the atmosphere mainly through the decay of pions and muons:

$$\pi^+ \to \mu^+ + \nu_\mu, \quad \to (e^+ + \nu_e + \overline{\nu}_\mu) + \nu_\mu, \tag{7.116}$$

$$\pi^- \to \mu^- + \overline{\nu}_\mu, \quad \to (e^- + \overline{\nu}_e + \nu_\mu) + \overline{\nu}_\mu. \tag{7.117}$$

Evidently, one expects about twice as many muon (anti)neutrinos than electron (anti)neutrinos to reach the Earth's surface. However, experimental results of the KamiokaNDE installation showed that the atmospheric muon-to-electron (anti)neutrino number ratio depends on the direction of their arrival: Among the (anti)neutrinos arriving at the Earth's surface well-nigh vertically, the ratio was really close to 2:1. However, amongst the neutrinos arriving at a large angle from the vertical, this ratio is closer to 1:1. This indicates that the muon (anti)neutrinos somehow vanish whilst passing through the atmosphere, much faster than the electron (anti)neutrinos, and certainly much faster than would be expected from the known fact that the effective cross-section of the interaction between neutrinos and other matter is extremely small. These experimental results were later confirmed in the Super-KamiokaNDE installation.

In turn, the mechanisms that produce the enormous energy of a star such as our Sun had been subject to research from the beginning of the nineteenth century, when Lord Rayleigh showed that – with the *then* generally accepted assumption that the Sun's energy stems from gravitational contraction – the Sun could not be as old as the geological finds (of Earth) indicate and as needed for the process of evolution. However, Becquerel discovered radioactivity in 1896, and by about 1920 the atomic weights were measured sufficiently precisely to make it possible for Arthur Eddington to notice that four hydrogen atoms are a little heavier than the helium atom. According to Einstein's relation $E_0 = mc^2$, the difference $(4m_{\rm H}-m_{\rm He})$ indicates that fusing four hydrogen atoms into an atom of helium should release energy.

In the early 1930s Chadwick discovered the neutron, Pauli postulated the existence of the neutrino and Fermi described the basic process of weak nuclear interaction, $n^0 \to p^+ + e^- + \overline{\nu}_e$. This opened the possibility for a realistic description of the nuclear processes that produce most of the radiation energy of the Sun. By 1938, Hans Bethe had worked out the details of the so-called carbon cycle, where the process of fusion is catalyzed by carbon, nitrogen and oxygen, and which is the dominant process in very large stars. In the Sun, which is a relatively smaller and lighter star, the basic mechanism is the so-called *pp*-process:

1. $$p^+ + p^+ \to d^+ + e^+ + \nu_e, \qquad \text{(continuous spectrum)} \tag{7.118a}$$
$$p^+ + p^+ + e^- \to d^+ + \nu_e, \qquad \text{(discrete spectrum)} \tag{7.118b}$$

2. $$d^+ + p^+ \to {}^3\text{He}^{++} + \gamma, \tag{7.118c}$$

3. $${}^3\text{He}^{++} + p^+ \to \alpha^{++} + e^+ + \nu_e, \qquad \text{(continuous spectrum)} \tag{7.118d}$$
$${}^3\text{He}^{++} + {}^3\text{He}^{++} \to \alpha^{++} + p^+ + p^+, \tag{7.118e}$$
$${}^3\text{He}^{++} + \alpha^{++} \to {}^7\text{Be}^{4+} + \gamma, \tag{7.118f}$$

4. $${}^7\text{Be}^{4+} + e^- \to {}^7\text{Li}^{3+} + \nu_e, \qquad \text{(discrete spectrum)} \tag{7.118g}$$
$${}^7\text{Li}^{3+} + p^+ \to \alpha^{++} + \alpha^{++}, \tag{7.118h}$$
$${}^7\text{Be}^{4+} + p^+ \to {}^8\text{B}^{5+} + \gamma, \tag{7.118i}$$
$${}^8\text{B}^{5+} \to ({}^8\text{Be}^{4+})^* + e^+ + \nu_e, \qquad \text{(continuous spectrum)} \tag{7.118j}$$
$${}^8\text{B}^{5+} + e^- \to ({}^8\text{Be}^{4+})^* + \nu_e, \qquad \text{(discrete spectrum)} \tag{7.118k}$$
$$({}^8\text{Be}^{4+})^* \to \alpha^{++} + \alpha^{++}. \tag{7.118l}$$

The processes (7.118a), (7.118d) and (7.118j) produce neutrinos with a continuous distribution of energies, while the neutrinos produced in the processes (7.118b), (7.118g) and (7.118k) have a fixed energy [☞ Section 3.2: when a collision or a decay produces only two particles, their energies are completely determined]. Most of the neutrinos are created in the process (7.118a) as the concentration of input "ingredients" (protons) is much larger than the concentration of input "ingredients" in the other processes. However, the energy of the so-produced neutrinos is no larger than about 400 keV, which makes their detection harder. In turn, neutrinos produced in the processes (7.118d) and (7.118j) have energies reaching over 1 MeV, where the detectors are far more sensitive.

John Bahcall's additional and detailed computations of the resulting distribution and total neutrino flux were finally verified in 1968 [355, 369]: Ray Davis's group monitored a giant tank (4,850 feet underground, in the *Homestake* gold mine in South Dakota) containing a dry-cleaning fluid with a large content of chlorine, seeking the results of the reaction

$$\nu_e + n^0 \to p^+ + e^-, \qquad \text{by way of} \qquad \nu_e + {}^{37}\text{Cl} \to {}^{37}\text{Ar} + e^-. \tag{7.119}$$

The detection of argon-37 indicated that only about one-third of electron neutrinos that the Sun emits actually arrive at the surface of the Earth. This discrepancy in the number of solar electron neutrinos was dubbed the "neutrino problem."

A little earlier, in 1967, Bruno Pontecorvo proposed (following up on a decade-earlier proposal) a simple solution of the neutrino problem, by postulating that the electron neutrinos produced in the Sun at least partially transform during their flight to the Earth into another type (muon and tau) of neutrinos or even antineutrinos. As the Davis experiment could detect only electron neutrinos, the transformed neutrinos would show up as "missed." This mechanism is, in general, called "neutrino oscillation," as it is based on an essentially simple quantum-mechanical effect.

To wit, with two eigenstates of the Hamiltonian

$$H|1\rangle = E_1|1\rangle \qquad \text{and} \qquad H|2\rangle = E_2|2\rangle, \tag{7.120}$$

the evolution of a linear combination of these two stationary states is described as

$$|``1{+}2"; t\rangle = C_1 e^{-iE_1 t/\hbar}|1\rangle + C_2 e^{-iE_2 t/\hbar}|2\rangle, \tag{7.121}$$

where the constants C_1, C_2 are determined from the initial condition. The probability that this linear combination is after the amount of time t in the state $\cos(\alpha)|1\rangle + \sin(\alpha)|2\rangle$ equals

$$\begin{aligned} P_\alpha &:= \Big|\big[\cos(\alpha)\langle 1| + \sin(\alpha)\langle 2|\big]|``1{+}2"; t\rangle\Big|^2 \\ &= |C_1|^2\cos^2(\alpha) + |C_2|^2\sin^2(\alpha) + \sin(2\alpha)\,\Re e\big[C_1 C_2^* e^{-i(E_1-E_2)t/\hbar}\big]. \end{aligned} \tag{7.122}$$

If the system was originally in the "opposite" linear combination, $\cos(\alpha)|2\rangle - \sin(\alpha)|1\rangle$ so $C_1 = -\sin(\alpha)$ and $C_2 = \cos(\alpha)$, we have that

$$P_{|\alpha+\frac{\pi}{2}\rangle\to|\alpha\rangle} = \sin^2(2\alpha)\sin^2(\tfrac{1}{2}\omega_{12}t), \qquad \omega_{12} := \tfrac{E_1-E_2}{\hbar}. \tag{7.123}$$

Therefore, the system oscillates:

$$\big(|\alpha{+}\tfrac{\pi}{2}\rangle = -\sin(\alpha)|1\rangle + \cos(\alpha)|2\rangle\big) \quad\longleftrightarrow\quad \big(|\alpha\rangle = \cos(\alpha)|1\rangle + \sin(\alpha)|2\rangle\big) \tag{7.124}$$

under the conditions that

1. the two stationary states are not degenerate: $E_1 \neq E_2$, so that $\omega_{12} \neq 0$, and
2. the system is initially in a nontrivial ($\alpha \neq 0$) linear combination of the two stationary states.

It is evident that the conceptually same phenomenon occurs in a system with three non-degenerate stationary states, but the oscillations are more complicated.

For relativistic particles, we have that (using that $\vec{p}_1 = \vec{p}_2 = \vec{p}$)

$$E_1 - E_2 = \sqrt{|\vec{p}|^2c^2 + m_1{}^2c^4} - \sqrt{|\vec{p}|^2c^2 + m_2{}^2c^4} \approx |\vec{p}|c\left[\frac{1}{2}\frac{(m_1{}^2 - m_2{}^2)c^2}{|\vec{p}|^2} + \cdots\right]$$
$$\approx \frac{(m_1{}^2 - m_2{}^2)c^3}{2|\vec{p}|} + \cdots \approx \frac{(m_1{}^2 - m_2{}^2)c^4}{2\overline{E}}, \tag{7.125}$$

where $\overline{E}$ is the average value of energies E_1 and E_2.

Just as $|d\rangle$, $|s\rangle$ and $|b\rangle$ – eigenstates of the free, kinetic Hamiltonian and thus characterized by their well-defined masses – are not the eigenstates of weak interactions, suppose that the electron-, muon- and tau-neutrinos (identified as the eigenstates of weak interactions) are not the eigenstates of the free Hamiltonian, $|\nu_i\rangle$. Then,

$$|\nu_e\rangle = -\sin(\theta_\nu)|\nu_1\rangle + \cos(\theta_\nu)|\nu_2\rangle, \qquad |\nu_\mu\rangle = \cos(\theta_\nu)|\nu_1\rangle + \sin(\theta_\nu)|\nu_2\rangle, \tag{7.126}$$

neglecting the third family. From this,

$$P_{\nu_e\to\nu_\mu} \approx \sin^2(2\theta_\nu)\sin^2\Big(\tfrac{(m_1{}^2-m_2{}^2)c^4}{4\overline{E}\hbar}t\Big) = \sin^2(2\theta_\nu)\sin^2\Big(\tfrac{(m_1{}^2-m_2{}^2)c^3}{4\overline{E}\hbar}z\Big), \tag{7.127}$$

where $z = ct$ is approximately equal to the distance that neutrinos traverse (the masses m_1, m_2 are very small, so the neutrinos propagate with speeds that are close to c). This shows that after a traversed distance of

$$(2n+1)\,z_*, \qquad \text{where} \quad z_* = \frac{2\pi\overline{E}\hbar}{(m_1^2 - m_2^2)c^3}, \qquad n = 0, 1, 2, \ldots \tag{7.128}$$

all electron neutrinos have converted into muon neutrinos, and at distances $2n\,z_*$ all electron neutrinos have turned back into their initial state. In other words, $2z_*$ is the wavelength of the simple oscillation between two types of neutrinos.

Of course, there do exist *three* types of neutrinos, and the oscillations are more complicated. Besides, traversing matter additionally changes the parameters of neutrino mixing. This was first described by Lincoln Wolfenstein, Stanislav Mikheyev and Alexei Smirnov, and this additional effect is named after then, the MSW effect. In 2001, the first results were published from Super-KamiokaNDE, which uses water in the detector, and which can detect all three types of neutrinos, albeit with different levels of efficiency. Independently, in the same year, the first results were published also from SNO (Sudbury Neutrino Observatory), which uses heavy water in the detector. Because of the presence of the neutron in the deuterium nuclei, SNO detects two additional processes with neutrinos that are not detected in Super-KamiokaNDE.

By April 2002, the combination of these experimental results unambiguously showed that the neutrino oscillations exist and solved the so-called "neutrino problem," showing clearly that the neutrino stationary states, ν_1, ν_2, ν_3 have nonzero and different masses, and that the weak interaction eigenstates, the particles ν_e, ν_μ, ν_τ, are linear combinations of the stationary states ν_1, ν_2, ν_3. Experiments also give the *difference* of the squares of masses:

$$\triangle_{12}(m_\nu^2) \approx 8\times10^{-5}\,(\text{eV}/c^2)^2, \qquad \triangle_{23}(m_\nu^2) \approx 3\times10^{-3}\,(\text{eV}/c^2)^2, \tag{7.129}$$

but cannot show if the pattern of masses is two similar masses significantly smaller than the third one, or two similar masses significantly larger than the third one [☞ book [369], and [370] for a more recent and thorough review].

Finally, Section 2.3.10 discussed the research of R. Davis and D. S. Harmer, who concluded that ν_e and $\overline{\nu}_e$ are distinct elementary particles. However, a detailed analysis of the non-occurring process (2.24), i.e., $\overline{\nu}_e + n^0 \not\to p^+ + e^-$, shows that it may well be possible for ν_e and $\overline{\nu}_e$ to be the same particle – that the neutrino is its own antiparticle – but that this process (2.24) is forbidden by helicity/chirality: whereas $\nu_e + n^0 \to p^+ + e^-$ could happen with a *left-handed* neutrino, the absence of a left-handed *anti*neutrino would then prevent the process (2.24).

A more direct consequence of the logically possibility that $\nu_e = \overline{\nu}_e$ would be the neutrino-less double β-decay:

$$2d \to 2u + 2e^- + (2\overline{\nu}_e \to \overline{\nu}_e + \nu_e \to 0) \to 2u + 2e^-, \tag{7.130}$$

which has never been observed. Nevertheless, the logical possibility that $\nu_e = \overline{\nu}_e$ still attracts considerable interest as it is necessary for the so-called *see-saw mechanism*. This mechanism uses the fact that the (left-handed) neutrinos are the $I_w = +\frac{1}{2}$ components of the lepton doublets that interact by means of weak interactions, and so also with the doublets of Higgs fields. In turn, one may always add to the Standard Model the right-handed neutrino, which has no weak charge (isospin):

$$I_w(\nu_{e\,L}) = +\tfrac{1}{2}, \qquad I_w(H_2) = -\tfrac{1}{2}, \qquad I_w(\nu_{e\,R}) = 0. \tag{7.131}$$

The Standard Model Lagrangian density may then contain the terms[15]

$$m_\nu\left(\overline{\nu_{e\,R}}\,\nu_{e\,L} + \overline{\nu_{e\,L}}\,\nu_{e\,R}\right) + \tfrac{1}{2}M_\nu\,\overline{\nu_{e\,R}}\,\nu^c_{e\,R}, \tag{7.132a}$$

where m is the mass that stems from the (so-called Yukawa) interaction term (7.113), where $H_2 \to \widetilde{H}_2 + \langle H_2\rangle$ produces $m_\nu = h_\nu\langle H_2\rangle$. In the basis $(\nu_{e\,L}, \nu_{e\,R})$, the Lagrangian terms (7.132a) produce the mass matrix

$$\begin{bmatrix} 0 & m_\nu \\ m_\nu & M_\nu \end{bmatrix} \quad \overset{\text{diag.}}{\longmapsto} \quad m_\pm = \frac{1}{2}\left|M_\nu \pm \sqrt{4m_\nu^2 + M_\nu^2}\right| \approx \begin{cases} M_\nu, \\ m_\nu^2/M_\nu. \end{cases} \tag{7.132b}$$

One expects that $m_\nu \sim 10^2\,\text{GeV}/c^2$, while experiments indicate that the neutrino masses are $m_{\nu,\text{exp}} < 2\,\text{eV}$ [293]. Therefore, $M_\nu \sim (m_\nu^2/m_\pm) \gtrsim 10^{13}\,\text{GeV}/c^2$.

A mass parameter such as $M_\nu \gtrsim 10^{13}\,\text{GeV}/c^2$ must stem from effects that are beyond the Glashow–Weinberg–Salam theory of the electroweak interactions,[16] and also beyond the Standard Model, but are probably related to the so-called Grand Unification or some other phenomena expected to occur at such high characteristic energies.

It is worth mentioning that in 1962 Ziro Maki, Masami Nakagawa and Shoichi Sakata proposed a general neutrino mixing, akin to the CKM mixing of the "lower" quarks and extending a similar proposal by Bruno Pontecorvo [353]. The analogous general neutrino mixing matrix is thus called the PMNS-matrix [369, 370].

7.3.3 The Standard Model, summarized

We are finally ready to summarize the Lagrangian density for the Standard Model, using the list on p. 282:

---

[15] For any fermion, $\overline{\Psi}\Psi^c$ has twice every charge of $\overline{\Psi}$, i.e., Ψ^c, a Majorana mass term $M\,\overline{\Psi}\Psi^c$ requires a mass parameter M that has twice every charge of Ψ. All gauge symmetries corresponding to these charges must therefore be broken; either explicitly by introducing such a term by hand, or spontaneously if the mass parameter is the vacuum expectation of a scalar field. The Majorana mass term $\frac{1}{2}M_\nu\,\overline{\nu_{e\,R}}\,\nu^c_{e\,R}$ is possible exclusively because all the charges of a right-handed neutrino vanish, so that $\nu^c_{e\,R} := C(\nu_{e\,R})$ transforms identically to $\nu_{e\,L}$ with respect to all *unbroken* Standard Model symmetries.

[16] Since the mass scale of the GWS-model is of the order of magnitude of $W^\pm$- and Z^0-bosons, $\sim 10^2\,\text{GeV}/c^2$, a mass parameter of the order of magnitude $\sim 10^{13}\,\text{GeV}/c^2$ would require a numerical coefficient of the order $\sim 10^{11}$, the kind of which never occurs in typical computations. That is, although the Standard Model contains dimensionless coefficients such as h_e, h_d, h_u, h_ν in the expressions (7.109), (7.110), (7.112) and (7.113), all these dimensionless coefficients are smaller than 1 and there is no systematic computation where a combination of them would emerge to be of the order $\sim 10^{11}$. This situation here is very similar to the discussion of the hydrogen atom in Sections 1.2.5 and 4.1, where *negative powers* of the fine structure constant (and so also of dimensionless coefficients larger than 1) do not occur.

$$\mathscr{L}_{\rm SM} = \mathscr{L}_{\rm F} + \mathscr{L}_{\rm G} + \mathscr{L}_{\rm H} + \mathscr{L}_{\rm Y} + \mathscr{L}_{M_\nu}, \tag{7.133a}$$

$$\mathscr{L}_{\rm F} = i\hbar c \sum_n \Big[\overline{\Psi_{nL}}\not{D}\Psi_{nL} + \overline{\Psi_{nR}}\not{D}\Psi_{nR}\Big], \tag{7.133b}$$

$$D_\mu := \partial_\mu + \tfrac{ig_c}{\hbar c} G^a_\mu Q_{c\,a} + \tfrac{ig_w}{\hbar c} W^\alpha_\mu \mathbb{V}^\dagger_w I_{w\,\alpha} \mathbb{V}_w + \tfrac{ig_y}{\hbar c} B_\mu Y_w, \tag{7.133c}$$

$$\mathscr{L}_{\rm G} = -\tfrac{1}{4}\sum_{a=1}^{8} G^a_{\mu\nu} G^{a\,\mu\nu} - \tfrac{1}{4}\sum_{a=1}^{3} W^a_{\mu\nu} W^{a\,\mu\nu} - \tfrac{1}{4} B_{\mu\nu} B^{\mu\nu}, \tag{7.133d}$$

$$\mathscr{L}_{\rm H} = \big\|[\partial_\mu - \tfrac{ig_w}{\hbar c} W^a_\mu \sigma_a - \tfrac{ig_y}{\hbar c}\mathbb{1}]\mathbb{H}\big\|^2_\eta - \tfrac{\varkappa}{2}\big(\tfrac{\mu c}{\hbar}\big)^2(\mathbb{H}^\dagger\mathbb{H}) - \tfrac{1}{4}\lambda(\mathbb{H}^\dagger\mathbb{H})^2, \tag{7.133e}$$

$$\mathscr{L}_{\rm Y} = \sum_n \big(h_n \overline{\Psi_{nR}}(\mathbb{H}^\dagger\Psi_{nL}) + h^*_n(\overline{\Psi_{nL}}\mathbb{H})\Psi_{nR}\big), \tag{7.133f}$$

$$\mathscr{L}_{M_\nu} = \tfrac{1}{2} M_\nu c^2\, \overline{\nu_{e\,R}}\, \nu^c_{e\,R}. \tag{7.133g}$$

Here, the summands in the Lagrangian density (7.133d) were written akin to (5.118) and (6.23), but the gauge field tensors were denoted

$$G^a_{\mu\nu} = \partial_\mu G^a_\nu - \partial_\nu G^a_\mu - \tfrac{g_c}{\hbar c} f^a{}_{bc} G^b_\mu G^c_\nu, \qquad a,b,c = 1,2,\dots,8, \tag{7.134}$$

for the $SU(3)_c$ gluon field,

$$W^\alpha_{\mu\nu} = \partial_\mu W^\alpha_\nu - \partial_\nu W^\alpha_\mu - \tfrac{g_w}{\hbar c}\epsilon^\alpha{}_{\beta\gamma} W^\beta_\mu W^\gamma_\nu, \qquad W^\pm = W^1 \pm iW^2, \quad \alpha,\beta,\gamma = 1,2,3, \tag{7.135}$$

for the $SU(2)_w$ gauge field, and

$$B_{\mu\nu} = \partial_\mu B_\nu - \partial_\nu B_\mu, \tag{7.136}$$

for the $U(1)_y$ gauge field. As customary, convention-dependent coefficients such as $4\pi\epsilon_0$ for electromagnetism have been absorbed in the definition of the gauge field tensors and are not explicitly shown. In the expressions (7.133b), the derivative D_μ (7.133c) is covariant with respect to the *complete* $SU(3)_c \times SU(2)_w \times U(1)_y$ Standard Model gauge group action:

1. The operator Q_a is the ath generator of the chromodynamics $SU(3)_c$ gauge symmetry (6.6d), which annihilates $SU(3)_c$-invariant fields and wave-functions.
2. The operator $I_{w\,\alpha}$ is the αth (isospin) generator of the weak $SU(2)_w$ gauge symmetry. The fermions in Table 7.1 on p. 275 are the eigenstates of the generator $I_{w\,3}$, with the eigenvalues I_w; the operators $I_{w\,\pm}$ raise and lower the values of I_w by 1 [☞ relations (A.38) for the general $SU(2)$ algebra].
3. The operator Y_w produces the weak hypercharge of the field or wave-function on which it acts.

The $\mathbb{V}_w$ matrix encodes the CKM mixing of the lower, d-, s- and b-quarks [☞ relations (2.53)] and leaves the other fermions unchanged. The sum in the expression (7.133b) contains all the elementary fermions from Table 7.1 on p. 275.

As in Section 7.1.3, the parameter $\varkappa$ in the expression (7.133e) separates the symmetric ($\varkappa = +1$) and the "non-symmetric" ($\varkappa = -1$) phases. For $\varkappa = +1$, $\langle\mathbb{H}\rangle = 0$ and the $SU(3)_c \times SU(2)_w \times U(1)_y$ gauge symmetry is unbroken; for $\varkappa = -1$, $\langle\mathbb{H}\rangle \neq 0$ and the gauge symmetry is broken to $SU(3)_c \times U(1)_Q$, the normal modes of the gauge 4-vector potentials are (7.85)–(7.86) and the $W^{\pm}$- and Z^0-bosons acquired a mass [☞ expressions (7.84) and (7.87)]. In the non-symmetric phase, the linear combination of gauge bosons (7.85) is identified with the photon. On one hand, this linear combination remains massless; on the other, this linear combination interacts with the Standard Model elementary fermions proportionally to the g_e-multiple of the combination of charges $Q = I_w + \frac{1}{2}Y_w$, which is by construction equal to the electric charge.

Similarly, in the symmetric phase ($\varkappa = +1$), the terms (7.133f) describe only the interaction between elementary fermions and the Higgs doublet of scalar fields. In the non-symmetric phase, owing to the shift $\mathbb{H} \to \widetilde{\mathbb{H}} + \langle\mathbb{H}\rangle$ where $\langle\mathbb{H}\rangle \neq 0$, the terms (7.133f) also provide the Standard Model elementary fermions with mass. Finally, the last term (7.133g) is needed for the "see-saw mechanism" [☞ Section 7.3.2]. This models the left-handed neutrino masses – many orders of magnitude below other Standard Model elementary fermion masses – by means of new physics expected at energies that are many orders of magnitude above the Standard Model masses; for example, the masses of the right-handed neutrinos, which thereby remain not observable directly for now.

As has been widely reported, the search for the Higgs particle has been on for the past decade or so, with most of the meticulous analyses centering on the LEP (Large Electron–Positron collider) and more recently the LHC (Large Hadron Collider) experiments at CERN. These culminated recently with the "5-σ" (99.999,9%) confirmation by the ATLAS and CMS collaborations from the LHC at CERN of a new, $\approx 125.9\,\text{GeV}/c^2$ particle [293], consistent with the Standard Model Higgs particle [25, 109]. However, it is important to realize that the Higgs particle is hard to identify unambiguously in experiments, since its mass, decay modes and their branching ratios all strongly depend on the details of the Standard Model – and its variations. The data compiled from the pertinent experiments are found to be compatible with the Standard Model as described above, but do not exclude several generalizations. For a review of recent experimental results, including also supersymmetric variants of the Standard Model and models wherein the Higgs field is a composite bound state, see Refs. [25, 109], the references therein, and in particular also Refs. [160, 493, 494, 475] ☡.

7.3.4 Exercises for Section 7.3

✎ **7.3.1** *For the expressions (7.109), (7.110), (7.112) and (7.113) to be Lagrangian density terms, compute the physical unit-dimensions of the Yukawa coupling coefficients h_U, h_D, h_ν and h_ℓ in the $M^x L^y T^z$ format.*

✎ **7.3.2** *Confirm the result (7.122) by explicit computation, using equation (7.121).*

✎ **7.3.3** *Confirm the result (7.125) by explicit computation.*

✎ **7.3.4** *Confirm the result (7.132b) by explicit diagonalization.*

✎ **7.3.5** *Compute the simplified neutrino oscillation wavelength $2z_*$ (7.128), using one and then the other value in equations (7.129).*

✎ **7.3.6** *Compute the order of magnitude of M_ν so that equation (7.132b) would produce $m_- \sim 1\,\text{eV}/c^2$.*

Part III
Beyond the Standard Model

8

Unification: the fabric of understanding Nature

Just as done several times in the previous chapters, we reconsider the historical development of the key aspects of modern physics, using the benefit of hindsight to perceive the character of this development. Throughout the foregoing material, the Democritean atomism provided the *warp*, complemented by the gauge principle as its *weft*. Along the way, however, this fabric reveals the ubiquity of the third conceptual strand (*woof*, as it were [☞ lexicon entry on p. 508, in Appendix B.1]) – unification; we now turn to explore this more closely.

8.1 Indications

The Newtonian theory of gravity unites the mechanics of the so-called terrestrial and the celestial objects and so eradicates this difference postulated within the Aristotelian philosophy of Nature, which the Roman Catholic clergy (sanctioned by the AD 313 Edict of Milan) imposed as exclusive of all other world views from the pre-hellenic and the hellenic cultures. According to Newton's law of gravity – as a descriptive model of this natural phenomenon – gravity is obeyed equally by both the Sun and the Moon, and the planets and the stars, by the communication satellites and the rockets as well as the Rockettes, by both the basketball and the baseball balls as well as the players in those games, and of course also by the apple that supposedly fell from the tree under which Newton sat...

Of course, Newton's unified description of Nature is not a unification of pre-existing theoretical models – in the contemporary sense of the word "model." It is, however, one of the first rigorous applications of the principle that Nature is one and that it can be understood in a unified fashion, and not as a (jury rigged) patchwork of different and diverse ideas – each with but a very narrow aim and applicability.

It behooves us then to examine also the nature of our ideas about unification.

8.1.1 Unification of relativistic and quantum physics

Modern fundamental physics is based on the requirement that a description of Nature include both its quantumness and its relativity, in the senses of the general theory of relativity.

Special relativistic unification

The first example of unification of *existing* scientific models is provided by the Maxwell equations: Indeed, Ampère's and Faraday's laws and Gauss's laws (5.72) were already known, as well as experimentally verified in situations for which those laws of Nature had been identified.[1] Their combination into one unified, *electromagnetic system* – and the extension of Ampère's law for the sake of agreement with the continuity equation and general consistency in the non-static/stationary case – has far-reaching consequences:

1. The electric field and the magnetic field (viewed as two distinct physical phenomena) are the limiting cases of a unified electromagnetic field, in the formal limit $c \to \infty$; accordingly, the Maxwell equations (5.72), without magnetic (monopole) charges and currents, become (using the relation $1/\epsilon_0 c^2 = \mu_0$)

$$\vec{\nabla}\cdot\vec{E} = \frac{4\pi\rho_e}{4\pi\epsilon_0}, \qquad \vec{\nabla}\times\vec{B} = \frac{4\pi}{4\pi\epsilon_0 c^2}\vec{\jmath}_e + \frac{1}{c^2}\frac{\partial\vec{E}}{\partial t} \;\to\; \mu_0\,\vec{\jmath}_e, \tag{8.1a}$$

$$\vec{\nabla}\cdot\vec{B} = 0, \qquad -\vec{\nabla}\times\vec{E} = \frac{\partial\vec{B}}{\partial t}. \tag{8.1b}$$

   Thus, in the formal limit $c \to \infty$, only the last relation (Faraday's law) still relates the electric and the magnetic fields, and only when the magnetic field varies in time and the electric field varies in space, so as to have a nonzero curl: $\vec{\nabla}\times\vec{E} \neq 0$. In turn, the full electrodynamics (5.72) is then the extension of this electro-and-magneto-*static* system, both self-consistent and consistent with Nature.
2. Changes in the electromagnetic field propagate with the speed of light, in the form of waves. Using the Lorenz gauge, the Maxwell equations (5.72) without magnetic (monopole) charges and currents produce the wave equation for the 4-vector gauge potential (5.92), so that their changes propagate at the speed of light in vacuum, c. The electromagnetic field, as gauge-invariant derivatives of the gauge potentials (5.15), also satisfies the wave equation

$$\Box\vec{B} = \vec{\nabla}\times(\mu_0\vec{\jmath}_e), \qquad \Box\vec{E} = -\vec{\nabla}\Big(\frac{\rho_e}{\epsilon_0}\Big) - \frac{\partial(\mu_0\vec{\jmath}_e)}{\partial t}. \tag{8.2}$$

3. The system of Maxwell equations (5.72) has symmetries:
   (a) Lorentz transformations of spacetime (3.1), i.e., (3.13) and corresponding transformations of the electromagnetic field (5.75),
   (b) duality (5.86) between the electric and the magnetic field.
4. The existence of magnetic (monopole) charges and currents would obstruct the (unambiguous) expression of the electromagnetic field in terms of a gauge 4-vector potential [☞ Comment 5.6 on p. 185].
5. The regime where the unified electrodynamics may be regarded as a collection of separate subjects of electro-*statics*, magneto-*statics* and wave optics is the "$c \to \infty$" formal limit.

Comment 8.1 *Since c is a natural constant, the formal limit "$c \to \infty$" makes sense only in the form of dimensionless ratios $v_{ij}/c \to 0$, where v_{ij} ranges over all relative speeds observable in the considered system. This has three significant consequences:*

1. *Non-relativistic physics is a special,* ***limiting case*** *of relativistic physics, which is in turn an* ***extension*** *of non-relativistic physics. For any given system, in the*

[1] Maxwell noticed that without the displacement current, $-\mu_0\partial(\epsilon_0\vec{E})/\partial t$, the divergence of Ampère's original law, $\vec{\nabla}\times\vec{B} = \mu_0\vec{\jmath}_e$, produces $\vec{\nabla}\cdot\vec{\jmath}_e = 0$, which holds only in the restricted cases when the free charge density in the entire observed space is unchanging in time, i.e., only in the static/stationary situations for which Ampère originally identified the law.

space of all possible relative speeds $\{v_{12}, v_{13}, \dots\}$, the non-relativistic regime involves only the lowest-order nonzero results in the $v_{ij}/c \ll 1$ approximation, i.e., ***near the point****: $v_{ij} = 0$ for all i, j; everything else is relativistic physics.*

2. *By non-relativistic systems one may understand only the cases where the relativistic corrections are* ***negligible*** *– for which the limits of precision are necessarily subject to* ***convention****.*
3. *Since the changes in the electromagnetic field propagate at the speed of light, all systems with variable electromagnetic fields are unavoidably relativistic.*

The property that changes in the electromagnetic field propagate as waves, at the speed of light, unifies the (electro- and magneto-)*static* phenomena with the wave phenomena (ultraviolet radiation, light, heat radiation and radio-waves, which were known by the end of the nineteenth century to be but different types of electromagnetic radiation), and then also the high-frequency limit of wave optics known as geometric optics.

Digression 8.1 From the contemporary, symmetry vantage point, the symmetries of the Maxwell equations are the Lorentz transformations [☞ Section 3.1]. The symmetries of Newtonian mechanics are the Galilean transformations, which differ from Lorentz transformations in that the boost transformations do not change time:

$$\textbf{Galileo}\quad \vec{r}\,' = \vec{r} - \vec{v}t, \qquad t' = t, \tag{8.3a}$$

$$\textbf{Lorentz}\quad \vec{r}\,' = \vec{r} - \gamma\vec{v}t + (\gamma - 1)(\hat{v}\cdot\vec{r})\,\hat{v}, \qquad t' = \gamma\Big(t - \frac{\vec{v}\cdot\vec{r}}{c^2}\Big). \tag{8.3b}$$

In Newtonian physics, time is absolute. Since charged particles interact with the electromagnetic fields and when they move, it is necessary that the theoretical model of those interactions is a single, coherent and consistent theoretical system – which can happen only if one can either:

1. adapt the Maxwell equations so as to exhibit Galilean symmetries of Newtonian physics,
2. or adapt Newtonian laws so as to exhibit Lorentz symmetries of relativistic physics.

As is well known, Nature picks the second, and not the first of these logical possibilities.

General relativistic unification

Chapter 9 will provide a telegraphic review of the general theory of relativity, but let us note here that the "general theory of relativity" (and then also its special case, the special theory of relativity) is in fact a ***theoretical system*** [☞ Section 8.3.1]. The pivotal idea in the theory of relativity is also the gauge principle, but applied to the "real," i.e., concrete spacetime, rather than to an abstract space of phases as was the case with electroweak and strong interactions [☞ Chapter 5]:

1. To describe physical systems, one uses coordinate systems the points of which are the points of spacetime in which the parts of that system move. To this end, one uses the 4-vector of spacetime coordinates, x.
2. The coordinates in such coordinate systems are not themselves physically observable, i.e., they cannot be measured. Indeed, absolute positions of various objects cannot be measured, but distances between them can.

3. To measure distances,

$$s(\mathrm{x}_i,\mathrm{x}_f) := \int_{\mathrm{x}_i}^{\mathrm{x}_f} \mathrm{d}s, \qquad \text{where} \quad \mathrm{d}s^2 := g_{\mu\nu}(\mathrm{x})\,\mathrm{d}x^\mu \mathrm{d}x^\nu, \tag{8.4}$$

one must know the metric tensor $g(\mathrm{x})$ in the chosen coordinate system, represented by $\mathrm{x} = (x^0, x^1, x^2, x^3)$. The components $g_{\mu\nu}(\mathrm{x})$ are – in principle, and definitely in the general case corresponding to the *general* theory of relativity – arbitrary functions of the coordinates x. For special relativity, $g_{\mu\nu}(\mathrm{x}) = -\eta_{\mu\nu}$; see (3.17)–(3.19).

4. Since the coordinates x cannot be observed directly, it ought be possible to change the coordinate system – through the substitution $\mathrm{x} \to \mathrm{y}$, but so that

$$\mathrm{d}s^2_{(x)} = g_{\mu\nu}(\mathrm{x})\,\mathrm{d}x^\mu \mathrm{d}x^\nu \stackrel{!}{=} g_{\mu\nu}(\mathrm{y})\,\mathrm{d}y^\mu \mathrm{d}y^\nu = \mathrm{d}s^2_{(y)}, \tag{8.5}$$

from which it follows that [☞ Digression 3.2 on p. 88, and Chapter 9]

$$g_{\mu\nu}(\mathrm{y}) = \frac{\partial x^\rho}{\partial y^\mu}\frac{\partial x^\sigma}{\partial y^\nu}\,g_{\rho\sigma}(\mathrm{x}), \tag{8.6}$$

that is, that the metric tensor is indeed a tensor, of rank 2 and of type $(0,2)$.

5. Chapter 9 shows how invariance with respect to general coordinate transformations implies the existence of gauge potentials and the gravitational interaction – exactly the way invariance with respect to local phase transformations implies the Yang–Mills type gauge interactions [☞ Chapters 5–6].

Comment 8.2 *For the special theory of relativity, we have $g_{\mu\nu}(\mathrm{x}) \to -\eta_{\mu\nu}$,*[2] *which is the constant metric tensor (3.19) of the "flat" spacetime. In this sense, the special theory of relativity is a "**limit-point**" in the space of all possible general coordinate systems and corresponding metric tensors described in the general theory of relativity. In turn, the general theory of relativity is then an **extension** of the special theory of relativity.*

The practical demarcation between special and general theories of relativity may thus naively be estimated by considering the departure of the actual metric tensor $g_{\mu\nu}(\mathrm{x})$ from the metric tensor of flat spacetime, $-\eta_{\mu\nu}$. This, however, is not well defined. Indeed, owing to the relation (8.6), neither is specifying any particular component of the metric tensor nor is its comparison with the same component from another metric tensor independent of the choice of coordinates. However, there do exist so-called curvature invariants, the values of which are independent of coordinate choices, and these then may serve for demarcation purposes. In $(3+1)$-dimensional spacetime, there are 20 such invariants, of which the simplest one is the so-called scalar curvature, $R := g^{\mu\rho}R_{\mu\nu\rho}{}^{\nu}$, where $R_{\mu\nu\rho}{}^{\sigma}$ is the so-called Riemann tensor [☞ Chapter 9]. Suffice it to say, if any one of these 20 curvature invariants cannot be neglected (in the considered processes and in comparison with some earlier specified precision limits), the system is generally relativistic.

The general theory of relativity contains (Einstein's) model of gravity, while the special theory of relativity pertains to flat spacetime, with no gravitational effects. Thereby, the special theory of relativity may be regarded as the formal $G_N \to 0$ limit of the general theory.

Comment 8.3 *As in the case of the formal limit "$c \to \infty$" and since G_N is a natural constant, the formal limit "$G_N \to 0$" may be understood only as a statement that all characteristic quantities of the system commensurate with G_N (of the same physical units) are much larger*

[2] The expression (8.4) defines the metric tensor $g_{\mu\nu}$ by way of defining the **distance**, while the expression (3.17) defines the proper **time** in spacetime. The signs in $\eta_{\mu\nu}$ are therefore opposite from the signs in $g_{\mu\nu}$.

than G_N. Intuitively, these characteristic quantities ought to be some invariant measures of the spacetime curvature, but all such invariants are computable from the Riemann tensor, the dimensions of which are $[R_{\mu\nu\rho}{}^{\sigma}] = L^{-2}$. On the other hand, $[G_N] = L^3T^{-2}M^{-1}$ and curvature invariants (obtained as various contractions of various tensor products of the Riemann tensor) cannot be compared with G_N, but can be compared with the constant $\ell_P := \sqrt{\hbar G_N/c^3}$, the Planck length [☞ Table 1.1 on p. 24].

Thus, for the purposes of estimating the "non-gravitational" limiting case, it is more convenient to use the natural constant ℓ_P **instead of** Newton's gravitational constant. *This limiting case then may be written formally as "$\ell_P \to 0$," understanding here relations of the type $|\mathbf{R}_i|\ell_P \ll 1$, where:*

1. *$|\mathbf{R}_i|$ is the norm of the ith curvature invariant, defined so as to have dimensions $[\mathbf{R}_i] = L^{-1}$;*
2. *the relation "$\ll$" here means "smaller than a previously set limit of precision."*

In this sense, the notation "$G_N \to 0$" is being used as a synonym for the formal limit "$\ell_P \to 0$", while keeping $\hbar$ and c constant [☞ Comments 8.1 on p. 294 and 8.4 on p. 298].

Quantum unification

As the Maxwell equations – the theoretical model of the electromagnetic field – indicate that the electro-static and the magneto-static fields are only limiting cases of the electromagnetic field whereby the descriptions of these natural phenomena are unified, so does quantum mechanics unite the notion of a particle and that of a wave.

The very notion of a particle presupposes that the position of the observed object in "ordinary" space may be localized arbitrarily well, i.e., that the object is ideally located in a perfectly well-specified (mathematically dimensionless) point of "ordinary" space: the position of this object is perfectly precisely specified. In a complementary fashion, the very notion of a (plane) wave presupposes that the position of the observed object in *momentum space* may be localized arbitrarily well, i.e., that the object is ideally located in a perfectly well-specified (mathematically dimensionless) point of momentum space: the wave vector of the object is perfectly precisely specified.

However, the Heisenberg indeterminacy relations, $\triangle x \triangle p_x \geqslant \frac{1}{2}\hbar$, imply that a *quantum* object cannot be localized more precisely than within a region in the phase space,[3] the "surface area" of which is never smaller than $\frac{1}{2}\hbar$. This gives the phase space in quantum physics a "granular" structure. In turn, it is also known that functions (or, more generally, distributions) over the phase space that may be used to represent *classical* observables cannot reproduce consistently and completely all properties of the *quantum* state operator; see quantum mechanics textbooks such as Ref. [29]. Thus, quantum physics cannot be described simply as classical physics with the additional requirement of a "granular" phase space. Quantum mechanics teaches us that real "things" are neither ideal particles nor ideal waves, but "something else"; something that in appropriate circumstances may be *approximated* by the limiting case of a point-particle, while in other circumstances an *approximation* by the limiting case of a wave is more precise.

The conceptual analogy with electro-*static* and magneto-*static* fields on one hand, and electromagnetic waves (always moving) on the other should be manifest. It should then come as no surprise that field theory in this conceptual sense interpolates between particles and waves. However, field theory is not a theory of a collection of wave packets – that literally interpolate between particles represented by the Dirac δ-function as one limiting case, and plane waves as the other

[3] The geometric shape, and even connectedness of this region remains a-priori undetermined, regardless of the choice of a system, and its evolution during the passage of time.

limiting case. Field theory contains both wave packets as limiting cases – less special than particles and plane waves, but limiting cases nevertheless.[4]

Finally, the transition from quantum to classical physics is often cited as the formal limiting process $\hbar \to 0$, which identifies classical physics as a limiting case of quantum physics.

> ***Comment 8.4*** *Since $\hbar$ is a natural constant, this limiting process makes sense only as the limit $(\hbar/S_i) \to 0$ for $i = 1, 2 \ldots$, where S_i are various physical observable quantities characteristic for the given system and with units $\frac{ML^2}{T}$, such as the angular momentum and Hamilton's action, $S = \int dt\, L$, where L is the Lagrangian of the system. In this precise sense, classical (non-quantum) physics is a* ***limiting case*** *of quantum physics, which is in turn an* ***extension*** *of classical physics.*

The theoretical system of relativistic quantum physics

The combination of limiting processes described in Comments 8.1, 8.2 and 8.4 in this section then provides the complete depiction (see Figure 8.1) of the *theoretical system* within which the Standard Model of elementary particle physics is formulated [☞ Section 8.3].

For all Standard Model purposes, suppose that the spacetime curvature and corresponding gravitational effects are negligible, i.e., that a full sequence of conditions of the form $\boldsymbol{R}_i \ell_P \to 0$ is satisfied, as discussed in Comment 8.3 [☞ also Chapter 9, as well as Refs. [508, 62, 367, 548, 66]], and that reduces Einstein's general theory of relativity to the special theory of relativity with no gravitation; the Newtonian theory of gravity may be derived as a lowest-order nonzero effect *near* this limit [95, 96, 271, 58]; see also Section 9.2.4. In individual interaction processes between elementary particles, the gravitational interaction is many orders of magnitude weaker than the strong or even the electroweak interactions, whereby special relativity suffices for all Standard Model purposes.

With this assumption, the schematic diagram in Figure 8.1 reduces to the first quadrant in the coordinate $(\frac{1}{c}, \hbar)$-plane, which represents (specially) relativistic quantum physics, i.e., field theory.

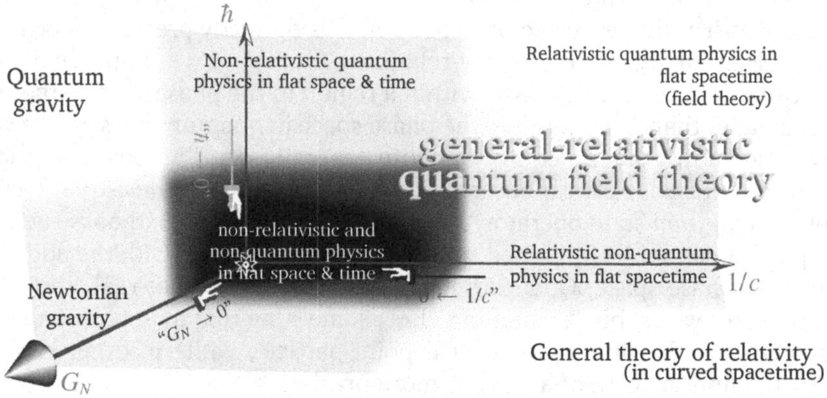

Figure 8.1 A sketch of the limiting cases of the general and special theory of relativity as well as quantum physics. The boundaries of the formal transitions into the approximations "$c \to \infty$," "$\hbar \to 0$" and "$G_N \to 0$" (i.e., "$\ell_P \to 0$"; see text) are conventionally defined, as depicted by gradual shading. General-relativistic quantum field theory is up front, well inside the first octant.

[4] It may help to imagine the palette of possibilities covered by field theory as a multi-dimensional geometric object with an "edge." The points of this "edge" correspond to various wave-packets, and its two end-points correspond to the particle and the plane wave, respectively.

The demarcations that determine the negligibility of characteristic quantities $(\hbar/S_i)$ and (v_{jk}/c) of the system are *conventional*, and this is represented in Figure 8.1 as a gradual change in the shading. Here, the non-relativistic physics is "sufficiently near" the vertical axis, and non-quantum physics is "sufficiently near" the horizontal axis. The practical criteria for this "nearness" – i.e., the boundary where the non-relativistic or non-quantum approximation is no longer sufficiently good – depends on the adopted *conventions* regarding the required precision of computational results.

Let us then emphasize the conceptual differences:

1. in phase transitions, the boundary between the symmetric and the non-symmetric phase is precisely determined by the system: see Conclusion 7.5 and relation (7.45);
2. the transition from quantum (or relativistic) physics into the non-quantum (non-relativistic) *approximation* is conditioned by the convention of computation precision. Strictly speaking (with absolute precision), non-quantum and non-relativistic physics are merely idealized limiting cases.

The transition from the regime where the electroweak interaction is united into the regime where electrodynamics essentially differs from weak interactions (photons are massless, $W^\pm$-, Z^0-bosons are massive) is manifestly a phase transition and not a conventional approximation. In turn, the transition from the regime of electrodynamics into the regime where we – practically and pragmatically – separate electro-statics from magneto-statics is conditioned by the convention of computational precision, i.e., whether or not relativistic corrections may be neglected.

However, there do exist significant similarities. The conceptual similarity is reflected in the facts that both electrodynamics and electroweak interactions have both a "unified" and a "separated" regime, as well as that the symmetries of the system in the unified regime are larger than the symmetries in the separated regime; see Table 8.1.

Table 8.1 Conceptual similarities and differences between the unification of the electric and the magnetic fields into the electromagnetic (EM) one, and the electromagnetic and weak fields into the electroweak (EW) field. $Po(1,3)$ is the Poincaré group of linear transformations of spacetime: Lorentz transformations and translations.

	United regime	**Separated regime**
Electromagnetism	The relative speed between at least two subsystems is not negligibly small, $v_{ij}/c \not\ll 1$.	The relative speed between at least two subsystems is negligibly small, $v_{ij}/c \ll 1$.
	The transition demarcation is ***specified by a convention*** in resolution.	
	Separation and differentiation between the $\vec{E}$- and the $\vec{B}$-fields depends on the choice of the coordinate system; see Example 5.1 on p. 183, and relations (5.75) and (5.77).	In a system where the free charges are static and the idealized currents stationary, the electric and the magnetic fields are static and perfectly separated.
	The symmetries of the Maxwell equations form the Lorentz group, together with spacetime translations, i.e., the Poincaré group, $Po(1,3)$.	The symmetries of electro- and magneto-static systems are limited to rotations in space, Galilean boosts and translations in space and time, $Ga(1,3) \subsetneq Po(1,3)$.
Electroweak int.	Particles in a process have energies $E_i > \hbar c\sqrt{\lambda}\langle\mathbb{H}\rangle|_{\varkappa<0} \sim M_{W^\pm}c^2$.	Particles in a process have energies $E_i < \hbar c\sqrt{\lambda}\langle\mathbb{H}\rangle|_{\varkappa<0} \sim M_{W^\pm}c^2$.
	The transition demarcation (the order parameter critical value) is ***determined by the system***.	
	$W^\pm, W^3_\mu$ and B_μ are the normal modes, and are all massless.	B_μ and W^3_μ are not normal modes; A_μ (massless) and Z_μ (massive) are; see relations (7.85)–(7.86).
	Local (gauge) symmetries of electroweak interactions form the $SU(2)_w \times U(1)_y$ group.	Local (gauge) symmetries of electroweak interactions reduce $U(1)_Q \subset SU(2)_w \times U(1)_y$.

Conclusion 8.1 (unification) *Since Newton's* Principia *(1687) and through the unification of electroweak interactions (Glashow, Weinberg and Salam, 1979 Nobel Prize), three distinct notions of unification have grown into fundamental physics:*

(a) conceptual *in the sense that Nature is one and that its scientific descriptions (models) should be conceptually uniform, and not a patchwork (hodgepodge) of diverse and disparate ideas;*
(b) limiting *in the sense that one marked "regime" of behavior of a system is, strictly speaking, merely a special limiting case (i.e., approximation) of another, more general and/or more exact description;*
(c) phase/regime *where the description of a system contains a definition of an order parameter and its critical value that divides two phases, i.e., regimes of a system.*

Note the double duty pulled by the word "regime," used in two different senses in the second and third notions of unification as listed here. Similarly, the word "phase" is used here in the sense exemplified by solids vs. liquids – very different from its use in Chapters 5–7.

8.1.2 Indications for exploring beyond the Standard Model

The Standard Model explains a lot, but also indicates the unknown source of some of the basic characteristics of this model and the state of understanding Nature that this model represents:

Spacetime For Standard Model purposes, one *assumes* the spacetime to be a continuous topological real 4-dimensional space with a flat metric tensor $-\eta_{\mu\nu}$ of signature $(1,3)$, i.e., that one of the four dimensions is of a time-like and three are of a space-like character. We do not know why this is so.

The interaction hierarchy The fundamental interactions in the Standard Model emerge from the gauge principle and the local (gauge) symmetry group $SU(3)_c \times SU(2)_w \times U(1)_y$. The dependence of the interaction strength on the 4-momentum transfer involved where this strength is measured as well as the electroweak symmetry-breaking $SU(2)_w \times U(1)_y \to U(1)_Q$ are described within the Standard Model. However, the relative intensities of the concrete values of the parameters α_s, α_w and α_y (i.e., α_e) – obtained by measuring at any one concrete energy – are not determined within the Standard Model and may only be regarded as given (and unexplained) "initial data."

The scale and the mass hierarchy structure All Standard Model fermions acquire their mass via interaction with the Higgs field, through the field shift $\mathbb{H} \to \mathbb{H} + \langle \mathbb{H} \rangle$ [☞ relations (7.109)–(7.113)]. However, nothing in the Standard Model determines the concrete values of the specific constants h_Ψ that describe the intensity of the direct (Yukawa) interaction of the Standard Model fermions with the Higgs boson – and thus also the masses of these fermions [☞ Tables 4.1 on p. 152 and C.2 on p. 526]. Since $\langle \mathbb{H} \rangle$ is determined from the experimental data for $m_Z = 91.1876\,\text{GeV}/c^2$ [☞ relation (7.81) and (7.86)], it follows that

$$\langle \mathbb{H} \rangle \sim 10^2\,\text{GeV}/c^2, \quad \text{and} \quad h_u, h_d \sim 10^{-5}, \quad h_s \sim 10^{-3}, \quad h_c, h_b \sim 10^{-2}, \quad h_t \sim 1. \tag{8.7}$$

Neither the general smallness h_Ψ (except h_t) nor the hierarchy of these parameters is explained in the Standard Model. Until the Higgs particle is fully confirmed and its characteristics (including all the coupling parameters h_Ψ) are measured, the fermion masses remain without explanation in the Standard Model.

CKM quark mixing The very fact that the eigenstates of the free Hamiltonian are also the eigenstates of the strong, electromagnetic and gravitational interactions, but not also of weak interaction is not unusual: there is no a-priori theoretical reason for a coincidence of eigenstates of all various interaction terms in the Hamiltonian. However, the origin and the concrete values of the Cabibbo–Kobayashi–Maskawa parameters (angles) that control the quark mixing in weak interactions (2.53) are not determined at all by the Standard Model and remain unknown☝.

Neutrino mixing and oscillations Similarly to quarks, there is no a-priori theoretical reason for a coincidence of the eigenstates of the free and the (only) weak interactive term in the Hamiltonian for neutrinos. However, the origin and the concrete value of the parameter M_ν in equation (7.132) and, more generally, the origin and the concrete values of the parameters (in the PMNS-matrix [☞ Section 7.3.2]) that control the neutrino mixing in free propagation (as compared the neutrinos defined by the weak interactions) are also not determined at all by the Standard Model, and this remains an open problem☝.

The number of fermion families The Standard Model simply includes the fact that there exist three families of fundamental fermions [☞ Table 7.1 on p. 275], but this fact is neither mandated nor explained and remains one of the puzzles of the Standard Model☝.

***CP*-violation** The combined discrete *CPT*-operation must be a symmetry in all Lorentz-invariant models [☞ Section 4.2.3]. However, the combined *CP*-operation need not be (and is not) a symmetry of Nature, and neither need then the time reversal operation be. On the other hand, *T*-violation is necessary for the irreversible creation of a sufficient surplus of matter (as compared to antimatter) in the first seconds of the Big Bang, and *CP*-violation via weak interactions is, roughly and little as it is, of just the sufficient amount. However, nothing in the Standard Model explains the concrete value of the angle δ_{13} in the CKM matrix (2.53), nor the complete absence of the otherwise perfectly possible – and many orders of magnitude larger – *CP*-violation through strong nuclear interactions [☞ Section 6.3.1], which remains a complete mystery☝.

Cosmological constant Phase transitions always have excess energy density [☞ Conclusion 7.1 on p. 258 and Comment 7.3 on p. 265]. For water to freeze, an external heat reservoir must remove this excess energy. However, when the entire Universe undergoes a phase transition, there is no "external heat reservoir," and this energy remains as a homogeneous and isotropic background energy. The recent discovery that the expansion of the Universe is in fact accelerating implies the existence of some kind of background "*dark energy*" – however, the observed value of even the so-called cosmological constant is many tens of orders of magnitude smaller than the excess energy density of the electroweak phase transition; the origin and the concrete value of this astoundingly extravagant discrepancy remains a puzzle; see Comment 7.3 on p. 265☝.

Dark matter Observations of the distribution of rotation speeds of stars about their galactic centers imply the existence of an invisible source of gravity (mass), the quantity and volume of which surpasses the mass and volume of the visible matter in galaxies. The Standard Model contains no adequate candidate for such matter, the origin and nature of which then remain a puzzle☝. The variants of the cosmological "inflationary model" require that the total amount of matter in the Universe should even be ten times more than the best estimates for the amount of visible matter. For these models – which successfully describe most cosmically large-scale properties of our Universe – the existence of dark matter is crucial.

— ❦ —

The questions that the Standard Model uncovers may in many cases be formulated only based on the description of Nature and insights into its properties given precisely by that same Standard Model. It is then not inappropriate to regard the Standard Model as a tool for systematizing our questions about Nature that are conceptually beyond reach of the Standard Model. We thus speak of research "beyond the Standard Model."

Theoretical, experimental – and even aesthetic – successes of the electro-weak unification inspired many a researcher in the last quarter of the twentieth century to formulate a model that would unify the strong with the electroweak interaction, as well as explain at least some of the Standard Model puzzles. This idea receives significant support from the fact that the coupling parameters α_s, α_w and α_y change with the magnitude of the 4-momentum transfer at which these parameters are measured – and in a, roughly, convergent fashion. That is, if we suppose that above the energies $\sim 10^2$ GeV there exist no new fundamental fermions as well as no new interactions – which is referred to as the "grand desert hypothesis" – the functions $\alpha_s(\mathrm{q})$, $\alpha_w(\mathrm{q})$ and $\alpha_y(\mathrm{q})$ converge and meet approximately at the energy $|\mathrm{q}|c \sim 10^{15-17}$ GeV. The details of this convergence and of this "merging" depend on the concrete model and additional assumptions and so are necessarily of a speculative nature; finally, one talks about an extrapolation over 15–16 orders of magnitude, with no precedent in the history of physics!

Consider the relation (5.202), as well as (6.79), which holds for the general case of $SU(n)$-gauge interactions of n_f fermion flavors, and note that the reciprocals of the fine structure parameters are approximately linear functions of the logarithm of the magnitude of the 4-momentum transfer $|\mathrm{q}|$ at which the parameters are measured:

$$\left.\begin{array}{ll} U(1): & \dfrac{1}{\alpha_{1,R}(|\mathrm{q}^2|)} \approx \dfrac{1}{\alpha_{1,R}(\mu^2c^2)} - \dfrac{4}{12\pi}\ln\Big(\dfrac{|\mathrm{q}^2|}{\mu^2c^2}\Big) \\ SU(n): & \dfrac{1}{\alpha_{n,R}(|\mathrm{q}^2|)} \approx \dfrac{1}{\alpha_{n,R}(\mu^2c^2)} + \dfrac{11n-2n_f}{12\pi}\ln\Big(\dfrac{|\mathrm{q}^2|}{\mu^2c^2}\Big) \end{array}\right\} \quad |\mathrm{q}^2| \gg \mu^2c^2, \tag{8.8}$$

where μ is the largest fermion mass that can occur in the loops such as (5.201), the total number of which equals n_f. At energies over $\mu c^2 = m_\tau c^2 = 174.2$ GeV, we have

$$SU(3)_c: \quad n_f = 3\times 2_{(w)}, \qquad 11n - 2n_f = +21, \tag{8.9}$$

$$SU(2)_w: \quad n_f = 3\times(3_{(c)}+1), \qquad 11n - 2n_f = -2, \tag{8.10}$$

for $SU(2)_w$ and the same μ, and where the number of $SU(3)_c$-interacting quarks equals 6 (one doublet of quark $SU(3)_c$-triplets in each of the three families) and the number of $SU(2)_w$-interacting fermions equals 12: one (color) triplet of quark $SU(2)_w$-doublets and one lepton $SU(2)_w$-doublet in each of the three families. One thus obtains

$$U(1)_y: \quad \frac{1}{\alpha_{y,R}(|\mathrm{q}^2|)} \approx \frac{1}{\alpha_{y,R}(\mu^2c^2)} - \frac{4}{12\pi}\ln\left(\frac{|\mathrm{q}^2|}{\mu^2c^2}\right), \tag{8.11a}$$

$$SU(2)_w: \quad \frac{1}{\alpha_{w,R}(|\mathrm{q}^2|)} \approx \frac{1}{\alpha_{w,R}(\mu^2c^2)} - \frac{2}{12\pi}\ln\left(\frac{|\mathrm{q}^2|}{\mu^2c^2}\right), \tag{8.11b}$$

$$SU(3)_c: \quad \frac{1}{\alpha_{s,R}(|\mathrm{q}^2|)} \approx \frac{1}{\alpha_{s,R}(\mu^2c^2)} + \frac{21}{12\pi}\ln\left(\frac{|\mathrm{q}^2|}{\mu^2c^2}\right). \tag{8.11c}$$

where the values of $\alpha_{y,R}(\mu^2c^2)$, $\alpha_{w,R}(\mu^2c^2)$ and $\alpha_{s,R}(\mu^2c^2)$ are experimentally determined. The depiction of the system (8.11) in Figure 8.2 is very suggestive: the magnitudes of the $SU(3)_c$-, $SU(2)_w$- and $U(1)_y$-interactions converge and become approximately equal somewhere around $|\mathrm{q}| \sim 10^{15}$ GeV$/c$. The details that ensure that the three functions (8.11) really merge in one point include an increasing precision of the measurements of the "initial" values, as well as the assumption of possible new particles with masses between $m_t \sim 174.2$ GeV$/c^2$ and the energy where the functions (8.11) acquire the same value.

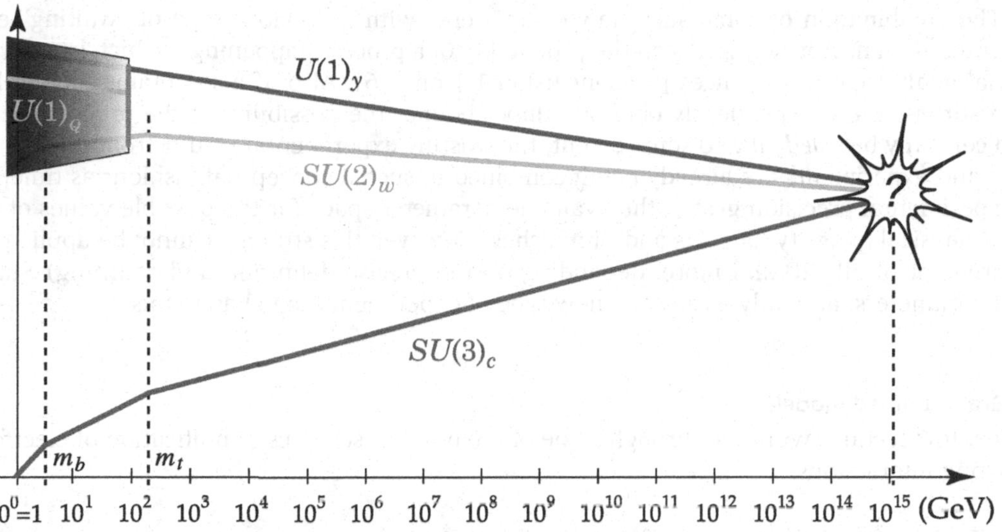

Figure 8.2 The convergence of the $SU(3)_c \times SU(2)_w \times U(1)_y$ gauge interaction strengths in the Standard Model. The slope changes indicate energy thresholds where new real quarks may be produced. The shaded area indicates the $SU(2)_w \times U(1)_y \to U(1)_Q$ phase transition.

The simplest assumption – that in this enormous span of energies nothing new exists – in fact *does not* lead to a precise merging of all three functions. In turn, in some of the possible and explored extensions of the Standard Model, this agreement is much better. One such extension is the so-called Minimally (extended) Supersymmetric Standard Model (MSSM), where this "grand desert" is populated by new particles: one superpartner for each Standard Model particle.

Of course, only concrete experiments may decide and provide the ultimate conclusion about the best model of unification of gauge interactions – as well as whether such a unification even takes place at all. As is known from even the popular literature and daily newspapers, the installations that such experiments require have in the twentieth century grown ever larger and more complex, and so are subject to both financial and political difficulties – already of international proportions. A glance into the past and the much more modest requirements of epoch-making experiments at the turn of the nineteenth into the twentieth century implies the practical impossibility of continuing one of the two pillars of experimental physics (and Rutherford's legacy): colliders (where beams of particles are accelerated and then collided, and where real collision processes are observed to happen) are becoming prohibitively expensive and complex.

The other conceptual type of experiments is based on the quantum essence of natural processes: Even if the energy in a system is insufficient for the interaction mediator in the process to be produced as a real particle, the process may nevertheless occur by exchanging virtual mediating particles. Although this significantly diminishes the probability for the observed process to happen, one then observes an enormous amount of matter (an enormous ensemble of particles) where such a process may happen, and then... waits for an unambiguous signal that the process really did happen. Until a concrete event is registered, the experiment produces *only an upper bound* for the probability for this process to happen, and cannot show if the process is in fact forbidden.

A new epoch-making advance in experimental physics will most probably require the invention of a radically new conceptual set-up of the experiment that would, in lieu of an opportunity to produce the concrete process or interaction as a real process, give a *lower bound* for the probability of this process occurring – complementary to the "waiting" experiments.

The combination of some such new experiments with a previous type of "waiting" experiment could then narrow the limits on the probability of a process happening – which is anyway the essential goal of natural sciences [☞ Conclusion 1.1 on p. 6]. Also, if a so-obtained lower bound should surpass the independently obtained upper bound, the possibility of the process occurring would certainly be *ruled out*. To some extent, the existing experimental results from diverse installations and experiments are already being combined in such a conceptual fashion; as time passes and experimental precision grows, the available parameter space for the possible values of a considered physical quantity narrows and diminishes. However, this strategy cannot be applied to the measurement of all (20 and more, depending on the precise definition and counting) Standard Model parameters, and only a radically new type of experiments can change this.

8.2 Grand unified models

The next few sections will skim through some of the possible schemes of unification of electroweak and strong interactions.

8.2.1 The Pati–Salam $SU(4)_c \times SU(2)_L \times SU(2)_R$ model

In a series of papers [411, 410, 412] in 1973–4, Jogesh C. Pati and Abdus Salam proposed a unification scheme based on two simple ideas:

1. that "lepton-ness" is the fourth color (extending the three quark colors), and
2. that there exists a phase in which parity is an exact, i.e., restored symmetry.

These two ideas may be presented rather effectively in the form of a table:

Electroweak interaction		Chirality	$SU(4)_c$: $SU(3)_c$: r	y	b	ℓ
$SU(2)_L$	$+\frac{1}{2}$	L	u^r	u^y	u^b	ν_e^ℓ
	$-\frac{1}{2}$		d^r	d^y	d^b	$e^{-\ell}$
$SU(2)_R$	$+\frac{1}{2}$	R	u^r	u^y	u^b	ν_e^ℓ
	$-\frac{1}{2}$		d^r	d^y	d^b	$e^{-\ell}$

plus two more "families" of fundamental fermions, each with an identical structure. (8.12)

In the fully symmetric phase, the "Pati–Salam" group $SU(4)_c \times SU(2)_L \times SU(2)_R$ specifies the gauge symmetries of the model, and this certainly contains the Standard Model gauge symmetry group $SU(3)_c \times SU(2)_L \times U(1)_y$ as a subgroup. Reference [412] describes several variants of this unification, but over the subsequent years this concrete model was singled out as the most successful. The 16 fermion states in the table (8.12) are denoted typically as the

$$(\mathbf{4},\mathbf{2},\mathbf{1})_L \oplus (\mathbf{4},\mathbf{1},\mathbf{2})_R \tag{8.13}$$

representation of the $SU(4)_c \times SU(2)_L \times SU(2)_R \rtimes \mathbb{Z}_2$ group, where $\mathbb{Z}_2 = \{\mathbb{1}, P\}$ and P is the operation of parity; the symbol "$\rtimes$" denotes the semidirect product [☞ the lexicon entry, in Appendix B.1]. With respect to this complete symmetry of the Pati–Salam model, the representation (8.13) is *irreducible*, i.e., there is no proper subset of the fermions in the table (8.12), which all elements of the complete symmetry $\big(SU(4)_c \times SU(2)_L \times SU(2)_R\big) \rtimes \mathbb{Z}_2$ transform into that same subset only. In turn, since parity is a symmetry of the model, the $SU(2)_L$ and $SU(2)_R$ coupling parameters must be equal, but the $SU(4)_c$ coupling parameter is independent.

With respect to the $SU(3)_c \times SU(2)_w \times U(1)_y \subset SU(4)_c \times SU(2)_L \times SU(2)_R \rtimes \mathbb{Z}_2$ subgroup, the representation (8.13) decomposes into

$$\big[(\mathbf{4},\mathbf{2},\mathbf{1}) \to (\mathbf{3},\mathbf{2})_{\frac{1}{3}} \oplus (\mathbf{1},\mathbf{2})_{-1}\big]_L \oplus \big[(\mathbf{4},\mathbf{1},\mathbf{2}) \to (\mathbf{3},\mathbf{1})_{\frac{4}{3}} \oplus (\mathbf{3},\mathbf{1})_{-\frac{2}{3}} \oplus (\mathbf{1},\mathbf{1})_{-2} \oplus (\mathbf{1},\mathbf{1})_0\big]_R. \tag{8.14}$$

This is the "physics-standard" notation, where the group representations are denoted by their dimensions [☞ Appendix A].[5] In particular, $(\mathbf{m},\mathbf{n},\mathbf{p})$ denotes the $SU(4)_c \times SU(2)_L \times SU(2)_R$-representation, which is the tensor product of the $\mathbf{m}$-dimensional representation of the $SU(4)_c$ group, the $\mathbf{n}$-dimensional representation of the $SU(2)_L$ group and the $\mathbf{p}$-dimensional representation of the $SU(2)_R$ group. Thus, $(\mathbf{4},\mathbf{2},\mathbf{1})$ is an $SU(4)_c$-quartet of $SU(2)_L$-pairs of quark-leptons, which decompose (8.14) into

$$(\mathbf{4},\mathbf{2},\mathbf{1}) = \begin{Bmatrix} u^r, u^y, u^b, \nu_e^\ell \\ d^r, d^y, d^b, e^{\ell-} \end{Bmatrix}_L \to \left[(\mathbf{3},\mathbf{2})_{\frac{1}{3}} = \begin{Bmatrix} u^r, u^y, u^b \\ d^r, d^y, d^b \end{Bmatrix}\right]_L \oplus \left[(\mathbf{1},\mathbf{2})_{-1} = \begin{Bmatrix} \nu_e^\ell \\ e^{\ell-} \end{Bmatrix}\right]_L, \tag{8.15a}$$

$$(\mathbf{4},\mathbf{1},\mathbf{2}) = \begin{Bmatrix} u^r, u^y, u^b, \nu_e^\ell \\ d^r, d^y, d^b, e^{\ell-} \end{Bmatrix}_R \to \begin{cases} \big[(\mathbf{3},\mathbf{1})_{\frac{4}{3}} = \{u^r, u^y, u^b\}\big]_R \oplus \big[(\mathbf{1},\mathbf{1})_{-2} = \{e^{\ell-}\}\big]_R \\ \big[(\mathbf{3},\mathbf{1})_{-\frac{2}{3}} = \{d^r, d^y, d^b\}_R\big]_R \oplus \big[(\mathbf{1},\mathbf{1})_0 = \{\nu_e^\ell\}\big]_R \end{cases}. \tag{8.15b}$$

In distinction from the left–right asymmetric interactions in the Standard Model [☞ Table 7.1 on p. 275], the extended electroweak interaction with the $SU(2)_L \times SU(2)_R$ gauge symmetry is *universal*. That is, the table (8.12) makes it clear that this model unavoidably predicts the existence of the right-handed neutrino. The right-handed neutrinos were indeed listed in Table 7.1 on p. 275, but the Standard Model does not mandate their existence. The right-handed neutrinos are invariant under the action of the Standard Model gauge symmetries and those symmetries do not link them with any other particles. In fact, all right-handed fermions in Table 7.1 on p. 275 do not partake in weak interactions and are invariant under its gauge symmetry $SU(2)_L$. In stark contrast, the left–right symmetric gauge group $SU(4)_c \times SU(2)_L \times SU(2)_R$ in the Pati–Salam model includes right-handed neutrinos in the $SU(2)_R$-doublets, extends weak interactions to left-handed particles, and thus provides the system a phase with a weak interaction that is universal (and not restricted to left-handed particles only) and where the symmetry of parity is restored.

In turn, this model then also makes it possible to describe the spontaneous breaking of the parity symmetry.

In the early 1970s, one could only suppose that there should exist a method of endowing the left- and the right-handed neutrino with masses non-symmetrically. The so-called see-saw model [☞ discussion of the relation (7.132a)–(7.132b)] was discovered only much later, and this model – the only one known – requires a mass parameter $M_\nu \gtrsim 10^{15}\,\text{GeV}/c^2$. This then cannot stem from the Standard Model but may easily be the consequence of some symmetry breaking in the diagram (8.16); the critical energy of such symmetry breaking must be many orders of magnitude larger than $m_{W^\pm}c^2, m_Z c^2 \sim 10^2$ GeV. The technical method for parity breaking, so as to reproduce the experimentally observed phenomena, still remains insufficiently understood in left–right symmetric constructions such as the Pati–Salam model☝. In principle, one expects to be able to come up with some variant of spontaneous symmetry breaking *à la* Sections 7.1.1–7.1.2, but none of the explored models seems to be able to reproduce all experimental details.

[5] In the general case, this is not sufficiently precise, as all Lie groups except $SU(2) \cong Spin(3)$ have distinct representations of equal dimensions, but this ambiguity turns up very rarely within the examples of interest, and in those exceptional cases those distinct representations of equal dimensions are distinguished by additional decorations such as $\mathbf{15}$ and $\mathbf{15'}$ in $SU(3)$.

The complete phase diagram – which in the 1970s was not discussed in detail – contains (at least) the five regimes ("phases")

$$
\begin{array}{c|c}
SU(4)_c \times SU(2)_L \times SU(2)_R \rtimes \mathbb{Z}_2 & SU(3)_c \times U(1)_{y'} \times SU(2)_L \times SU(2)_R \rtimes \mathbb{Z}_2 \\
\hline
SU(4)_c \times SU(2)_L \times U(1)_R & SU(3)_c \times U(1)_{y'} \times SU(2)_L \times U(1)_R \\
\cline{2-2}
& \multicolumn{1}{|c}{\to SU(3)_c \times SU(2)_L \times U(1)_y} \\
\cline{2-2}
\end{array}
\tag{8.16}
$$

and the existence or absence of these (and possibly many other) regimes (and phase transitions between them) depends on the choice of the Higgs field(s) that control the symmetry-breaking process. In the display (8.16), the "vertical" phase transition (between the first two rows) is characterized by the breaking of the right-handed copy of the weak isospin gauge group, $SU(2)_R \to U(1)_R$, as well as the parity $\mathbb{Z}_2$. The "horizontal" transition is characterized by the breaking of the extended color symmetry, $SU(4)_c \to SU(3)_c \times U(1)_{y'}$. Finally, the phase transition from the regime in the right-hand side region in the second row into the regime in the last row is characterized by the breaking of the abelian (commutative) symmetries $U(1)_{y'} \times U(1)_R \to U(1)_y$. The original work on this model [412] indicated that there exists a choice of (by now reduced to eight) Higgs fields that can describe the required symmetry breaking, $SU(4)_c \times SU(2)_L \times SU(2)_R \to SU(3)_c \times SU(2)_L \times U(1)_y$.

Besides, this model also predicts the possibility of proton decay! Namely, the symmetry $SU(4)_c$ also contains transformations of any one quark into a corresponding lepton, such as $u^{r,y,b} \to \nu_e$ and $d^{r,y,b} \to e^-$. The gauge bosons that mediate such interactions, collectively named X, must violate the baryon and the lepton number, but preserve the fermion (i.e., "quark+lepton") number. Nevertheless, the decay of the proton into lighter particles – electrons, positrons, (anti)neutrinos and pions – is not possible in this model by way of exchanging only the gauge bosons, but requires also the exchange of some Higgs field(s). Owing to the conservation of the fermion number, the simplest such proton decay could be of the form $p^+ \to 3\nu_e + \pi^+$ or $p^+ \to 4\nu_e + e^+$.

The possibility that the proton is not stable was first seriously considered within this Pati–Salam model, but proton decay has not been experimentally confirmed to date.

8.2.2 *The Georgi–Glashow SU(5) model*

Almost at the same time, Howard Georgi and Sheldon Lee Glashow suggested a competing unification model, based on the gauge group $SU(5)$ [202]. This model explicitly contains the left–right asymmetry of the Standard Model. Also, the Standard Model fermions appear within two distinct representations of the gauge group, $SU(5)$. However, since the gauge group has a single factor, unlike the Pati–Salam group, there is only one coupling parameter, and this model explains the relative ratio of the coupling parameters α_s, α_w and α_y (i.e., α_e).

The Standard Model fermions of each family are herein grouped:

$$
(f_{10})^{[AB]} = \left[\begin{array}{cc|ccc}
0 & e^+ & u^r & u^y & u^b \\
 & 0 & d^r & d^y & d^b \\
\hline
 & & 0 & \overline{u}_b & \overline{u}_y \\
\multicolumn{2}{l}{\text{anti-}} & & 0 & \overline{u}_r \\
\multicolumn{2}{l}{\text{symmetric}} & & & \\
\multicolumn{2}{l}{\text{rank-2 tensor}} & & & 0
\end{array}\right]_L, \qquad
(\overline{f}_{5^*})_A = \left[\begin{array}{c}
e^- \\ \nu_e \\ \hline \overline{d}_r \\ \overline{d}_y \\ \overline{d}_b
\end{array}\right]_L, \qquad
f_1 = (\overline{\nu}_e)_L,
\tag{8.17a}
$$

$$
\mathbf{10} \to (\mathbf{3},\mathbf{2})_{\frac{1}{3}} \oplus (\mathbf{3}^*,\mathbf{1})_{-\frac{4}{3}} \oplus (\mathbf{1},\mathbf{1})_2, \quad \mathbf{5}^* \to (\mathbf{3}^*,\mathbf{1})_{\frac{2}{3}} \oplus (\mathbf{1},\mathbf{2})_{-1}, \quad \mathbf{1} \to (\mathbf{1},\mathbf{1})_0,
\tag{8.17b}
$$

where only the left-handed fermions are listed; clearly, $(\overline{\nu}_e)_L = \overline{\nu_{e,R}}$, so the anti-fermions of left-handed chirality represent fermions of right chirality. The indices $A, B = 1, \cdots, 5$ here count the components of the fundamental, 5-dimensional representation of the $SU(5)$ group, and $(f_{10})^{[AB]}$

denotes the components of the antisymmetric matrix that represents the 10-dimensional representation. $\overline{f}_{5^*}$ represents the conjugate fundamental, 5-dimensional representation but is here shown as a column-matrix rather than a row-matrix to save space.

The $SU(3)_c \times SU(2)_w \times U(1)_y$ gauge subgroup representations (8.17b) are identified akin to the decomposition (8.15), and were already indicated in the decomposition (8.17):

$$(\mathbf{1},\mathbf{1})_2 \leftrightarrow \{e^+\}_L, \qquad (\mathbf{3},\mathbf{2})_{\frac{1}{3}} \leftrightarrow \begin{Bmatrix} u^r, u^y, u^b \\ d^r, d^y, d^b \end{Bmatrix}_L, \qquad (\mathbf{3}^*,\mathbf{1})_{-\frac{4}{3}} \leftrightarrow \{\overline{u^r}, \overline{u^y}, \overline{u^b}\}_L, \tag{8.18a}$$

$$(\mathbf{1},\mathbf{2})_{-1} \leftrightarrow \begin{Bmatrix} \nu_e \\ e^- \end{Bmatrix}_L, \qquad (\mathbf{3}^*,\mathbf{1})_{\frac{2}{3}} \leftrightarrow \{\overline{d^r}, \overline{d^y}, \overline{d^b}\}_L, \qquad (\mathbf{1},\mathbf{1})_0 \leftrightarrow \{\overline{\nu_e}\}_L. \tag{8.18b}$$

The $SU(5)$ gauge bosons in this model contain the $SU(3)_c \times SU(2)_L \times U(1)_y$ Standard Model gauge bosons, and also six additional gauge bosons, which form an $SU(2)_L$-symmetry doublet, and an $SU(3)_c$-symmetry triplet:

$$\begin{Bmatrix} X^r \\ Y^r \end{Bmatrix}, \begin{Bmatrix} X^y \\ Y^y \end{Bmatrix}, \begin{Bmatrix} X^b \\ Y^b \end{Bmatrix}: \qquad \begin{matrix} I_3(X) = +\frac{1}{2}, & Q(X) = \frac{4}{3}, \\ I_2(Y) = -\frac{1}{2}, & Q(Y) = \frac{1}{3}. \end{matrix} \tag{8.19}$$

It is easy to find X- and Y-mediated processes in this model whereby the proton decays; for example,

$$\begin{aligned} p^+ = (u+u+d) &\to (u+u+(\overline{X}+e^+)) \to (u+(u+\overline{X})+e^+) \to (u+\overline{u}+e^+) \\ &\to \pi^0 + e^+ \quad \to \quad 2\gamma + e^+. \end{aligned} \tag{8.20}$$

Estimates of the proton lifetime then give the basic bounds for the X and Y gauge boson masses, and thus also the critical energy of the $SU(5) \to SU(3)_c \times SU(2)_L \times U(1)_y$ phase transition. Conversely, using the results $M_X, M_Y \sim 10^{15}\,\text{GeV}/c^2$ from estimates such as Figure 8.2 on p. 303, it follows that the proton lifetime is $\tau_p \sim 10^{28}$–10^{29} years, which is too short: Experiments have by now raised the lower bounds to about 6.6×10^{33} years [293].

In turn, although the right-handed neutrino may be added to the fermions $f_{10} \oplus \overline{f}_{5^*}$, as in the decomposition (8.17), it is an $SU(5)$-invariant, i.e., neutral (chargeless) with respect to all $SU(5)$-gauge interactions. Thus, the right-handed neutrino may only have interactions of the Yukawa type (a product of two fermions and a scalar in the Lagrangian density), the coefficients of which are completely free parameters.

8.2.3 More complex models

Since the Pati–Salam $SU(4)_c \times SU(2)_L \times SU(2)_R$ model and the Georgi–Glashow $SU(5)$ unification model leave some of the Standard Model questions unanswered, it is reasonable to seek models with a gauge group that contain both the Pati–Salam and the Georgi–Glashow gauge group. It is interesting that the model built using the $SO(10)$ gauge group[6] contains both:

$$SO(10) \begin{matrix} \nearrow SU(4)_c \times SU(2)_L \times SU(2)_R, \\ \searrow SU(5) \times U(1)', \end{matrix} \qquad \mathbf{16}_L \begin{matrix} \nearrow (\mathbf{4},\mathbf{2},\mathbf{1})_L \oplus (\mathbf{4}^*,\mathbf{1},\mathbf{2})_L; \\ \searrow (\mathbf{10}_{-1})_L \oplus (\mathbf{5}^*{}_3)_L \oplus (\mathbf{1}_{-5})_L. \end{matrix} \tag{8.21}$$

In this model, all Standard Model fermions of one family – together with the right-handed neutrino – form the 16-dimensional irreducible spinor representation of the gauge group. The

[6] To be precise, this is in fact the $Spin(10)$ group, the double covering of the $SO(10)$ group, so that the spinor representations are faithful, i.e., single-valued. However, in the physics literature one usually writes $SO(10)$, implicitly understanding the single-valuedness requirement.

model is explicitly left–right symmetric and the coupling parameters α_s, α_w and α_y (i.e., α_e) all stem from a single coupling parameter of the $SO(10)$-gauge interaction. The $SO(10)$ unification thus contains the unification characteristics of both the Pati–Salam and the Georgi–Glashow models.

The number of both principal and practical puzzles of the Standard Model is thus reduced, but some of the questions still remain unanswered. Amongst them is the question: Why are there three fundamental fermion families? It would then seem reasonable to extend the gauge symmetry so as to also include a symmetry that mixes these fundamental fermion "families," the breaking of which should also explain the differences in the average masses of the fermions in the first, second and third "families." The simplest suggestion is the addition of another $SU(3)$ factor,[7] but this is evidently ad hoc, and it would be more desirable if this "familial" symmetry were a subgroup of some grand-unifying group.

The extension of the $SO(10)$ symmetry that would suffice in unifying all three fundamental fermion families into one irreducible gauge group representation, and where there exists a symmetry-breaking possibility such that precisely the three known families remain relatively light while all others (if any) acquire masses of the order $\gtrsim 10^{15}\,\text{GeV}/c^2$ must be $SO(18)$. In models with such a large gauge group the number of additional particles (additional fundamental fermions, additional gauge bosons and Higgs fields) reaches many thousands, and such models are not easy to take seriously [☞ Refs. [104, 285], and the references cited therein].

Researchers of so-called GUT[8] models have explored most of the Lie groups that are sufficiently large to contain the Standard Model, but are in one way or the other minimal. In other words, since this research is mostly speculative owing to the extrapolations over enormous energies, the researchers mostly adhere to the Ockham principle, whereby the symmetry structure and the content (the fundamental particles list) of the Standard Model is extended only if this extension offers an explanation for one of the Standard Model puzzles.

Superstring theory revived interest in some of the earlier explored exotic unifying models, and foremost in a model based on the E_6 gauge symmetry group. In this model, the fermions of one family fit into the smallest (27-dimensional!) irreducible representation of the E_6 group, so each family of E_6-fermions also contains 11 completely new fermions, the absence of which from experiments must be explained separately. With the E_6-model, one often mentions a model based on the $SU(3)_c \times SU(3)_L \times SU(3)_R \subset E_6$ subgroup, dubbed "trinification."[9]

8.3 On the formalism and characteristics of scientific systems

The unification of our knowledge about Nature into a single, coherent, comprehensive and logically consistent system with as few as possible basic concepts and ideas is the leitmotiv of the foregoing exposition. The same guiding idea also permeates the remainder of this book, where the understanding of Nature so far acquired will be expanded with considerations about gravity and the geometrization of physics, aspects of a possible unification of bosons and fermions, as well as a final unification of matter, all its interactions and even spacetime.

It behooves us then to summarize the hierarchical structure that is usually referred to as a "scientific system," somewhat as a reprise of the introductory thoughts of Chapter 1, but now with the background of Chapters 2–7.

---

[7] This "familial" factor in the symmetry group must have a 3-dimensional representation to represent the three "families," and this 3-dimensional representation must be complex, as are the wave-functions of the fundamental fermions.

[8] GUT stands for "Grand Unified Theory."

[9] This term is indeed the amalgamation of "trinity" and "unification". Herein, trinity indicates the three $SU(3)$ factors in the gauge group; the double entendre allusion to the Holy Trinity may well be on purpose.

8.3.1 The hierarchical structure of scientific systems

First of all, following the discussion in Section 1.1.2, by "scientific systems" one understands systems of understanding Nature that are based on iterating the cycle of observing–predicting–checking, which asymptotically improves this understanding. During this iterative process, the mathematical models and the apparatus we use to describe natural phenomena are extended and become technically more complex, and also describe Nature ever better and indicate an ever-increasing wealth of detail. No Student could fail to notice that the mathematical language that sufficed in the introductory hours of the first physics course quickly became inadequately scant, and that mastering new material in physics made it necessary to develop this mathematical language.

In retrospect, both the material mastered in other courses and that presented in Chapters 2–7 indicate the following categories of descriptive structures:

A model provides a mathematical description (surrogate) for a concrete physical system, whether this concerns a description of a concrete and simple physical system such as the pendulum or the lever, or a similarly concrete but complex system such as the Standard Model of elementary particle physics. In this description, every parameter quantifies a characteristic of the given and concrete physical system. For a more precise definition, see Procedure 11.1 on p. 416.

A theory is an axiomatic system in which a small number of physically motivated and logically consistent axioms (postulates) determines an infinite sequence of consequences that ensue with logical and mathematical rigor. Of course, we are interested in *physics* theories, of which one also expects that neither its axioms nor any of their consequences contradict Nature; the logically and mathematically incontrovertible consequences that (as yet) have not been tested are thus the predictions of the theory.

A theoretical system is a coherent and logically consistent axiomatic system that contains several distinct and otherwise independently defined and separately applicable theories.

Comment 8.5 *Here, we are primarily interested in the* **theoretical** *approach, hence we speak of* **theoretical systems**. *The analogous category of* **scientific systems** *of course includes both theoretical and experimental aspects of the system.*

During the second half of the twentieth century a subfield emerged within elementary particle physics that is usually referred to as phenomenology, and which effectively connects the ever more separated theoretical and experimental research. The scientific system then of course includes this bridging subfield.

Strictly in form, a theoretical system is indistinguishable from a theory; the difference stems from the physics application that dictates the source/motivation and justification of the axioms, as well as whether a sub-system can be applied separately. The following concrete example of two well-known theories as well as two theoretical systems containing those may serve to illustrate this.

Example 8.1 The special theory of relativity is based on two well-known postulates [☞ introductory part of Section 3.1, and in particular Definition 3.1 on p. 84, and comments], and of course the requisite mathematical apparatus that is well known from various earlier courses and was used in Chapter 3.

Similarly, quantum physics may also be introduced axiomatically. Various Authors cite different numbers of axioms: six [110], four [480], three [391] or two [29], mostly because the longer lists also contain some purely mathematical results, whereas the shorter lists presuppose the mathematical apparatus as independent (prerequisite) material [29].

In turn, there exist two relatively well distinguished theoretical systems, both of which include both the special theory of relativity and quantum theory:

1. relativistic quantum mechanics, and
2. relativistic quantum field theory.[10]

Both theoretical systems satisfy the above-cited requirement to contain (at least) two distinct and otherwise independently defined and separately applicable theories. The latter theoretical system is however more general: relativistic quantum field theory contains relativistic mechanics. The precise distinction between these two systems is beyond the scope of this book, but suffice it to state that an axiomatic approach to relativistic quantum field theory also requires the system of so-called Wightman axioms [572] or the Haag–Kastler alternative approach to *local quantum physics*, dubbed *algebraic quantum field theory* [254], or some other effective substitute for these.

The difference is simplest to see by comparing two standard texts, by the same Authors: Ref. [64] for relativistic quantum mechanics and Ref. [63] for relativistic quantum field theory.

The objective of distinguishing these categories of descriptive structures is to identify the important characteristics that distinguish models from theories and from theoretical systems. Consider again a concrete example: the success of Bohr's model of the hydrogen atom is oft cited as the turning point in adopting quantum physics.

Simplified, one says that classical physics cannot describe the hydrogen atom. Notice that classical physics is certainly a theoretical system, even if by classical physics one understands only classical mechanics with the additional, and simplest description of the Coulomb interaction.

We are now in a position to note the finesse (and trap) of Popper's falsifiability criterion [☞ Digression 1.1 on p. 9] – and so also of Conclusion 1.2 on p. 9: The necessity of quantizing the angular momentum indicates the falsifiability of *one concrete model* of classical physics – the classical planetary model of the atom by Rutherford, implicitly including the assumption of continuously variable angular momentum. This does not speak of the theoretical system (called classical physics) as a whole. As the classical planetary model predicts that the electrons in the orbit must lose energy via Bremsstrahlung – which of course is not the property of true atoms in Nature – one faces the format of a proof by contradiction. The logic of that type of proof indicates that at least one of the concrete assumptions (premisses) of the model must be at fault. As this includes all implicit/tacit assumptions, it is not a-priori clear that the non-classical quantization of the angular momentum is the only resolution of the disagreement between the model and Nature.

The fact that no one came up with a construction of a classical but stable planetary model of the atom[11] and many other results (Planck's black body spectrum formula, Einstein's explanation of the photoelectric effect, Compton's explanation of the effect that is now named after him) *jointly* indicate that the quantumness is convincingly indispensable in the description of Nature. That is,

---

[10] In practice, one always understands "quantum field theory" to be relativistic. As our present aim is to explicitly emphasize both relativity and quantumness of this theoretical system, both adjectives are explicitly stated.

[11] Note that Bohr's postulate of orbital angular momentum quantization all by itself does not necessarily preclude a purely classical explanation. For example, the complex system of Saturn's rings exhibits resonance phenomena that do provide excellent explanations for the stability of at least some of them. Similarly, the Titius–Bode rule, $\frac{1}{10}(4+3{\cdot}2^n)$ for $n=-\infty, 0, 1, 2, 3\dots$ and in units of Earth's semi-major axis, specifies the semi-major axis of each solar planet to within a small percentage except for Neptune. Neither this rule nor its generalization, Stanley Dermott's law (which then also applies to major satellites of solar planets), have a known theoretical explanation, although simulations support the belief that the regularity stems from many-body resonant phenomena [262]. It therefore simply does not follow that some as yet unknown but purely classical resonance phenomena could not in principle provide for the stability of certain select – quantized as it were – atomic orbits.

the quantum description of Nature is the only *known* one, wherein models are *as best as known* consistent with all these and a vast many other phenomena observed in Nature.

However, the quantum description of Nature cannot possibly *refute* (or *falsify*) classical theory, since classical theory is a limiting case (i.e., an approximation [☞ Figure 8.1 on p. 298]) and so also an integral part of the quantum theory. It makes no sense to state that the whole refutes one of its integral parts or limiting cases. Quantum theory *extends* classical theory and is applicable to concrete systems where classical theory is no longer sufficiently accurate: recall the gradual transitions in the sketch in Figure 8.1 on p. 298 and the dependence of this feature on the adopted conventions of accuracy.

The analogous situation holds for the theory of relativity which of course does not refute its "non-relativistic" limiting case. The situation is analogous also with the (desired, but not yet existing) generally relativistic quantum field theory – i.e., the theory that coherently and consistently unifies both quantumness and general relativity in Nature. All so far known models that faithfully describe the various natural phenomena and aspects of Nature must be integral parts (as limiting or special cases, or as concrete applications) of this all-encompassing theoretical system.

Some of these objections to the ideas regarding falsifiability also emerge upon examining more closely the helicoidal cycle that Popper uses to describe advances in scientific knowledge [442]:

$$\boxed{\begin{array}{l}\text{Problem}\\ \text{situation}\end{array}}_1 \to \boxed{\begin{array}{l}\text{Tentative}\\ \text{theory}\end{array}}_1 \to \boxed{\begin{array}{l}\text{Error}\\ \text{elimination}\end{array}}_1 \to \boxed{\begin{array}{l}\text{Problem}\\ \text{situation}\end{array}}_2 \to \cdots \tag{8.22}$$

Here the appearance of a "problem situation" (such as an unexplained observation) triggers the creation of several competing "tentative theories" that do explain the problem situation. These are then subject to increasingly more rigorous testing (attempts at falsification), which eliminates those that turn out to be erroneous in this third step. The remaining (unfalsified) theory is then upheld until the next "problem situation" emerges and the cycle repeats.

This reminds us of the three-step iterative process "observe–model–predict" described in Section 1.1.2, which may be recast into the above format for comparison:

$$\boxed{\text{Observation}}_1 \to \boxed{\text{Model}}_1 \to \boxed{\text{Predict}}_1 \to \boxed{\text{Observe}}_2 \to \cdots \tag{8.23}$$

The following observations are immediate on comparison:

1. The outcome of an observation need not pose a "problem situation," i.e., a conflict with the previously established/trusted theory. It could be anywhere between complete confirmation and outright conflict, including indications for minor corrections. But most importantly, new observations may well imply wholly new phenomena, the qualitative *separateness* of which may not be fully understood until much later. For example, both electric and magnetic phenomena had been noticed some 24 centuries before they were systematically represented in mathematical models by Coulomb, Gauss, Ampère, Faraday, etc.
2. Models are neither theories, nor tentative theories nor conjectures, but concrete mathematical surrogates of a predefined accuracy and tolerance, and constructed *within* the framework provided by one or more pertinent theories or theoretical systems.
3. Comparisons of model predictions against Nature rarely have a binary outcome of either-true-or-false, and so can rarely lead to outright falsification of the model at hand. This is even more true of theories and theoretical systems, as discussed above and in Digression 1.1 on p. 9, but cannot be overemphasized:
   (a) relativistic physics does not falsify non-relativistic physics, but *extends* it;
   (b) quantum physics does not falsify classical physics, but *extends* it.

Conclusion 8.2 *It is a historical fact that the contemporary description of Nature is a growing and integrated theoretical system, based on a number of postulates that is relatively small as compared to the scope and span of this description, and where the whole system as well as the candidates for additions are continually* ***filtered*** *by comparison with Nature and also by the need to form a coherent and logically consistent integration.*

In this sense is the contemporary description of Nature a growing organism.

We will return to this discussion in Section 11.5.

8.3.2 Inside indications of limitations

One of the systemically interesting characteristics of classical field theory, which includes the special theory of relativity, is that the field theory (theoretical system) indicates the limits of its own applicability.

The electrodynamics of charged particles

Start with the fact that both the formulation and the understanding of electrodynamics – as the basic example of a concrete classical field theory and with concrete application in mind – has essentially changed since the original, James Clerk Maxwell description in 1873.

In this original description, the electromagnetic field represented the deformation of aether, just as sound is a wave-like deformation of the medium through which it passes. For the aether itself, one supposed that it is at rest in Newton's absolute space and time. In this description, charged matter appears as a discontinuity in aether and in this sense is of secondary meaning.

In 1892, Hendrik A. Lorentz reformulated electrodynamics as a theory of the interactions between atomistic material particles and the all-permeating electromagnetic field, which permeates even the interior of the material particles. Lorentz initially had in mind the ions as these basic charged particles, but upon Thomson's discovery of electrons, in 1897, Lorentz's reformulation of electrodynamics focused on interactions of the electrons and the electromagnetic field. Following Lorentz's description, the charged particles are represented as little pellets of a finite size [☞ Digressions 4.1 on p. 132, 3.13 on p. 123, and 8.2 on p. 313], the electric charge of which is distributed over the surface and possibly also in the interior. Einstein's special theory of relativity – introduced as the basis for a description of electrodynamics of charged objects in motion – demands that the energy and momentum of a particle under the action of the Lorentz transformations change as components of a 4-vector, that the mass of the particle is Lorentz-invariant [☞ Chapter 3], and that they are related by equation (3.36), i.e., that the mass is the Lorentz-invariant magnitude of the 4-momentum, i.e., energy–momentum.

A way to satisfy this requirement in the Abraham–Lorentz model of an electron was never found, and all indications are that the 4-momentum, and then also the mass obtained from the relation (3.36), may be defined independently from the interactions of the electron with the electromagnetic field. From a contemporary, symmetry vantage point, the electric charge is a conserved Noether charge that corresponds to the continuous gauge symmetry (5.14), while the 4-momentum is the conserved Noether charge corresponding to spacetime translations [☞ Section 2.4.2, as well as Conclusion 9.6 on p. 329]. Since the gauge transformations (5.14) and spacetime translations are both logically and functionally independent, it follows that the mass of a particle must be independent of its electric charge.

Thus, it follows that classical electrodynamics is not complete in the sense that it does not seem capable of producing a consistent and complete model for charged material particles. As the well-known Michelson–Morley, Fizeau and other experiments imply that the concept of aether does not describe the experimental facts, it follows that one cannot go back to the original Maxwell view either, wherein material particles are "merely" discontinuities in aether.

As noted in Digression 3.13 on p. 123, the total energy (mass) of the electric field of a point-like electron diverges. Paul A. M. Dirac, in 1938, suggested a covariant procedure for separating the finite portion of this energy.[12] This procedure, however, results in a reactive force that is proportional to the derivative of acceleration, changes the familiar expression for the Lorentz force in electrodynamics, and causes the pre-acceleration effect, where a particle starts accelerating before a force is applied [35]! It would seem to be possible to avoid the effects of pre-acceleration only if the electron were large enough so that the (changes in the) field would need enough time to permeate the particle. The current experimental bounds are orders of magnitude smaller than the so-obtained estimates; see, however, Refs. [420, 421, 336, 17, 464, 78, 79] for a rather more complex non-point-like model, which is argued to be consistent with contemporary experiments.

Digression 8.2 Digression 3.13 on p. 123 showed that the energy of the electric field of an electron – which is in the Abraham–Lorentz model to be thought of as a rotating sphere of radius r_e – equals $\alpha_e b \frac{\hbar c}{r_e}$. The value $b = \frac{1}{2}$ holds if the electric charge of the electron is uniformly distributed over the surface of the sphere, and $b = \frac{3}{5}$ if it is uniformly distributed over the interior of the sphere. At any rate, b is a constant of the order ~ 1, which is true even for more complex electron charge distributions.

If this energy – by definition necessary to bring the electric charge of the electron from infinite distances into any concrete configuration – is identified with the electron rest energy, $m_e c^2$, one obtains that the electron classical radius is

$$r_e = \alpha_e b \frac{\hbar}{m_e c} = \alpha_e\, b\, \bar{\lambda}_e = 2.817\,940\,289\,4 \times 10^{-15}\, b\ \text{m}, \tag{8.24}$$

where $\bar{\lambda}_e = \frac{\lambda_e}{2\pi}$ is the so-called (reduced) Compton wavelength [☞ Table C.3 on p. 527] and which – up to the factor b – agrees with J. J. Thomson's estimates from collision processes. Namely, Thomson found the effective cross-section for electromagnetic radiation scattering off of non-relativistically moving electrons to be proportional to the area r_e^2, which agrees with the elementary analysis such as shown in Example 3.2 on p. 111. It follows that $b \lesssim 1$, and that b cannot be much smaller than 1. Interestingly, it is again Compton scattering – albeit at novel high-energy regimes – that may provide new information in this continuing quest [78, 79]; see also Refs. [420, 421, 336, 17, 464, 57].

In modern experiments, electrons are collided with energies of the order of 10^2 GeV, indicating that they come to a distance of about 10^{-18} m from each other – and do not show any sign of spatial structure. Down to such distances, electrons behave as point-like material particles, in full agreement with the relativistic quantum field theory, and the Gaussian distribution of the probability of finding the electron about this point, completely typical in quantum theory. The Abraham–Lorentz (and any other, classical) model of charged material particles thus does not agree with the experimental fact that $(r_e)_{\text{exp.}} < 10^{-18}$ m, nor with the general theoretical result about the minimal size of the charge distribution [☞ Section 11.4]. Even the proton, which is not an elementary particle, is 2–3 orders of magnitude smaller than the classical radius of a particle with the elementary unit of electric charge.

From this one concludes that, for particles of mass m and electric charge qe, the classical radius $r_{cl} \sim \alpha_e \frac{q\hbar}{m\,c}$ and the corresponding time $t_{cl} \sim \alpha_e \frac{q\hbar}{m\,c^2}$ are the (lower) bounds of applicability of this scientific system called the classical electrodynamics of charged bodies. Notice that $\hbar$ appears

[12] This type of procedure is today referred to as "regularization" and is an integral part of renormalization computations.

explicitly in these bounds only owing to the definition of α_e; writing $r_{cl} \sim \frac{qe^2}{4\pi\epsilon_0 m c^2}$ and $t_{cl} \sim \frac{qe^2}{4\pi\epsilon_0 m c^3}$ instead gives these definitions a decidedly more classical appearance.

Pointillist quantum gravity

Section 1.3, and especially 1.3.3, has already discussed the joint characteristic of the combination of quantum theory and the qualitative characteristic of the theory of relativity: the existence of a minimal, Planck length. The analysis of Section 1.3.3 indicates that it suffices to endow the quantum theory with Newtonian gravity and the relativistic requirement that no material object can travel faster than the speed of light in vacuum. However, it also indicates that the Planck length, as a (lower) bound of resolution and knowledge, is not a characteristic of the quantum theory by itself, but of the theoretical system obtained by joining the quantum theory with (at least Newtonian and also with Einsteinian) gravity, and the relativistic limit on speeds $v < c$.

One thus expects this amalgamated theoretical system not to be fundamental, but to be an approximation to a more complete theoretical system. In fact, with the view that physics theories and theoretical systems only asymptotically approach their aim, the *Final Theory* is of course just a *dream*, and even an *impossible dream* – to paraphrase Refs. [553] and [549, 338], respectively. Nevertheless, the contemporary physics en route to that dream is no less real, pragmatic and successful in describing Nature as comprehensively, coherently and consistently as possible.

The (super)string theory (in fact a theoretical system) is currently the most complete candidate, and it necessarily contains quantum general-relativistic field theory, but we do not at present know enough about this complex theoretical system for a final estimate as to the measure in which this theoretical system can contain a faithful description of (our) Nature. For the most part, this uncertainty derives from the fact that many of the questions raised within and about (super)string theory have simply never before been posed. Other attempts, such as loop gravity and spacetime foam [489], as well as some more recent attempts, are insufficiently developed even just as (merely) theories of quantum gravity, and they certainly do not include matter and other interactions as (super)string theory *does*; we will return to these issues in Chapter 11.

9 Gravity and the geometrization of physics

There exist excellent textbooks on Einstein's theory of gravity, ranging from non-technical introductions [329, 469, 187, 10, 231] to technically detailed ones [264, 367, 390, 55, 205, 414, 103, 548, 210, 131, 164, 66, 135, 96, 398, 506, 272, 315, 380, 342], as well as on tensor differential calculus in curved spaces [☞ [508, 62, 563, 210] to begin with], which is typically regarded as a prerequisite for a technical mastery of the material. The purpose of this chapter then cannot compete with these rich and detailed sources nor with textbooks on black holes and wormholes [103, 543], gravity in general and cosmology [418, 481, 419, 28, 558], and the interested Reader is wholeheartedly directed to this literature.

Complementing these resources, the general theory of relativity as a theory of (classical, i.e., non-quantum) gravity is here presented in comparison with Yang–Mills gauge theories from Chapters 5–8, thus continuing the unifying guiding idea that led us to this point; approaches to *quantum* gravity will be addressed in Chapter 11.

9.1 Einstein's equivalence principle and gauge symmetry

Most books that discuss general relativity and gravity – regardless of the technical level – start off with A. Einstein's principle of equivalence. Complementing this historically standard approach, gravity and general relativity may also be described and even "discovered" by (1) carefully examining the possible spacetime geometries as frameworks for real physical observations as done by R. Geroch [205]; (2) exploring the appearance and use of multi-valued fields [☞ magnetic monopole, Section 5.2.3] in a variety of physical models as done by H. Kleinert [315]; or (3) modeling the familiar gravitational and inertial phenomena from the point of view of a particle theory virtuoso as done by R. P. Feynman [164].

Borrowing from these approaches, we do start with Einstein's equivalence principle, but show that it is conceptually identical to the idea of gauge symmetry employed in Chapters 5 and 6. Thus it fits perfectly in the unifying "business card" of Nature, Table P.1 on xiii. Using the same concepts developed in Chapters 5 and 6, this lets us identify the analogue of the gauge vector potentials, construct a Lagrangian for them and derive Einstein's equation (9.44), below.

9.1.1 Inertial vs. gravitational mass

It was pointed out in Digression 8.1 that the full Lorentz transformations (including rotations and boosts; see Section 3.1.1) are a symmetry of the well-established Maxwell equations (5.72), while the Galilean group is a symmetry of Newtonian mechanics. While the Galilean group is the $c \to \infty$ limit of the Lorentz group, it is not a subgroup, and the two frameworks cannot be coherently combined, so as to describe the electrodynamics of moving electric charges. As is well known, the $c \to \infty$ limit of the Maxwell equations (so they would exhibit the Galilean group of symmetries) is unphysical: light propagates at finite speed. We are thus left with Nature's choice, relativistic physics.

The framework of relativistic physics, however, leaves a curious dichotomy regarding the concept of mass: On one hand, we have a simple mathematical result (3.36), which equates the Lorentz-invariant magnitude of its 4-momentum with the mass of an object, which is in turn identified (3.28)–(3.30) with the "inertial mass" familiar from non-relativistic mechanics. This mass is the ratio $m = \frac{|\vec{F}|}{|\vec{a}|}$, where $\vec{F}$ is a force applied to an object, $\vec{a}$ its resulting acceleration, where all observations are made in a coordinate system where the object was initially at rest, and we may even consider the limit where $\vec{F}$ and so also $\vec{a}$ are arbitrarily small.

On the other hand, in Newton's universal law of gravity, the mass of an object determines how strongly the gravitational attraction acts upon it – and there is no a-priori reason for this "gravitational mass" to be the same as the "inertial mass." That is to say, there remains the logical possibility that inertial effects upon an object may not be proportional to the same "mass" as are the gravitational effects, which is something Nature can – and does – decide for us: They are indeed one and the same.

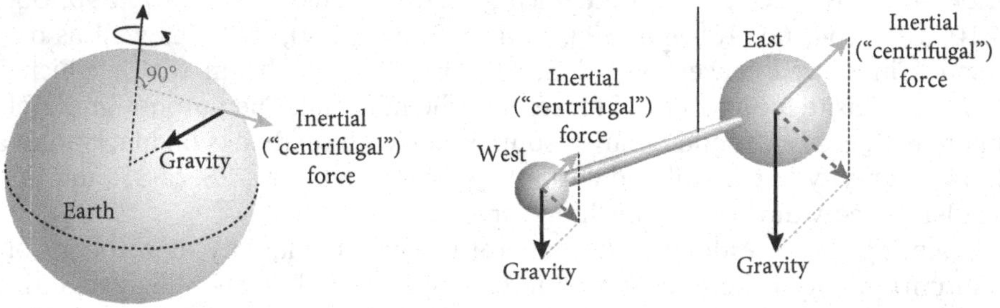

Figure 9.1 The classic Eötvös experiment: balancing the dumbbell horizontally compares the gravitational forces, while balancing it in the obtuse upward direction (grey arrows) compares inertial forces.

Experiments to this end have been carried out since around 1885, at first by Eötvös Loránd, where two substantial masses connected by a rod are balancing, suspended by a thin thread. The gravitational force acts towards the center of the Earth, while the inertial ("centrifugal") force due to Earth's rotation acts away from the axis of rotation, at an obtuse angle from the gravitational force. By aligning the horizontally balanced dumbbell initially in the east–west direction, all forces acting on each massive object are perpendicular to the connecting rod, and any difference in the sum of forces acting on one object vs. the other will produce a torque and twist the dumbbell from the initial east–west alignment. No matter what variety of the "eastern" and the "western" object in this torsion dumbbell were tried, the gravitational and the inertial forces were always found to be in the same proportion, thus proving the equality of the "inertial" and the "gravitational" mass, by now to the precision (relative error) of 10^{-11} [462].

Another logical possibility, that antiparticles [☞ Section 2.3.7] and particles *repel* each other by gravity, is easily dispelled in similarly high-precision experiments with elementary particles such as the neutral kaons [☞ Section 4.2.3]: Since the decay eigenstates (4.65), $|K^0_S\rangle := \frac{1}{\sqrt{2}}(|K^0\rangle - |\overline{K}^0\rangle)$, and $|K^0_L\rangle := \frac{1}{\sqrt{2}}(|K^0\rangle + |\overline{K}^0\rangle)$, are linear combinations of the particle and its antiparticle and beams of K^0_S and K^0_L propagate in Earth's gravitational field between creation and detection, a difference in the sign of the masses of K_0 and $\overline{K}_0$ would have to show. The experiments indeed do have the requisite precision, and indicate that K_0 and $\overline{K}_0$ have a positive (attracting) "gravitational" mass [164].

Between his seminal papers on special and general relativity, 1905–16, Einstein of course did not know about kaons, but must have been aware of the Eötvös-type experiment and its variations. He must have also been aware of the physically unnatural restriction to inertial coordinate systems within the special theory of relativity, as well as the fact that changes in the gravitational field could not propagate faster than the speed of light. To all of these issues, he came up with a single and elegant solution:

> ***Conclusion 9.1 (Principle of Equivalence)*** *Not only are the "inertial" and the "gravitating" masses equal, but inertial and gravitational physical effects are in fact* ***identical****.*

Tracing Einstein's line of thought in the popular as well as most standard textbook presentations repeatedly brings up the example of a person in an enclosure such as an elevator with no windows. While at rest at the ground floor, the person in the elevator feels Earth-normal gravity. While the elevator accelerates upward, the inertial effect is added to the gravitational effect, and the person experiences an increase in their weight – which a scale will readily verify is quite real. During the constant motion between the floors, the weight experienced returns to Earth-normal. Finally, while the elevator decelerates when reaching the destination floor above, the person experiences a decrease in their weight. In fact, this much can be easily reasoned simply from Newton's third law: the force measured by the scale on which the person in the elevator stands doesn't care whether the reaction (with which it holds the person from falling through) balances the gravitational or the kinematic acceleration.

Extrapolating from these very familiar experiences, one can easily imagine a person within an enclosure, who would not be able to tell whether the experienced weight (or lack thereof) is a consequence of the gravitational force of some nearby planet, or the fact that the enclosure (perhaps a rocket ship) is moving in an accelerated fashion. Indeed, this is clearly true as long as the considered accelerated motion and related inertial forces and the gravitational forces are confined to one direction.

Even certain simple arrangements with additional forces and accelerations in additional directions easily permit such a dual interpretation. Consider for example a person at the North Pole, observing the motion of a so-called "spherical" pendulum, such as a bundle of keys attached to a keychain that the person holds firmly. With the Earth's rotational axis passing through the person's hand holding the keychain, the keys would be moving under the influence of three types of forces:

1. the gravitational force ($\vec{F}_g$), vertically downward to a very good approximation;
2. the horizontal "centrifugal" force ($\vec{F}_{cf}$), directed away from the axis of Earth's rotation;
3. the horizontal Coriolis force ($\vec{F}_C$), at every instant perpendicular to both the axis of Earth's rotation and to the direction of motion of the keys.

Exactly the same effects would be observed by a person in an accelerating rocket ship that additionally rotates about the direction of its linear motion – such "co-rotating" non-inertial coordinate systems were considered on p. 84, so as to exclude them from the Definition 3.1 on p. 84 of inertial coordinate systems; see the left-hand pair of illustrations in Figure 9.2.

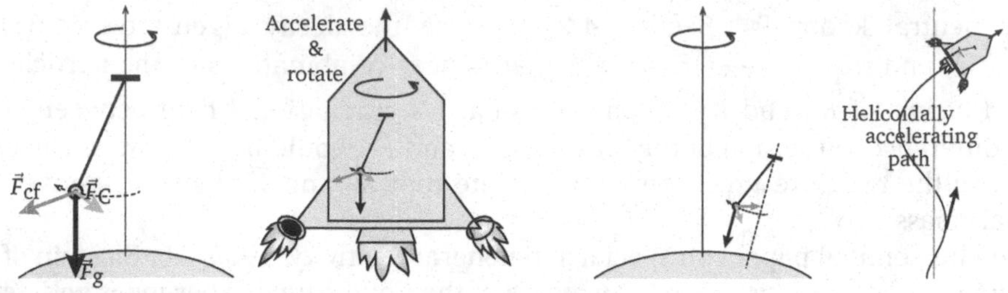

Figure 9.2 Two rotating pendula and the corresponding co-rotating accelerating coordinate systems.

Finding an appropriately accelerating coordinate system to be equivalent to an arrangement with more and more complicated systems of forces and accelerations of course becomes more and more complicated. For example, if the person with the swinging keychain were to move away from the North Pole, the direction of the gravitational acceleration would no longer coincide with the axis of Earth's rotation – as is the case with Foucault's pendulum in Paris, France. Effectively, the direction of gravitational acceleration for that person co-rotates about the axis of Earth's rotation, with which it also forms a nonzero angle. The corresponding accelerating coordinate system would then have to accelerate in a direction that forms the complementary angle with the Earth's axis of rotation, and precesses about it, thus accelerating along an expanding helicoidal path; see the right-hand pair of illustrations in Figure 9.2.

Any mechanical system under the influence of a *homogeneous* gravitational field is already perfectly equivalent in Newtonian mechanics to making the same mechanical system *uniformly* accelerate. Einstein's equivalence principle (Conclusion 9.1) is, however, fully general and applies to all physical phenomena, not just mechanics. W. Pauli then showed in his inimitable swift (and parsimonious) fashion, that this principle implies [414, Section 53]:

1. The influence of Newtonian (weak-field) gravity on a slowly moving object is determined by a scalar potential.
2. The gravitational field of stars causes a red shift in their spectral lines.
3. Even in a static gravitational field, light rays do not follow a geodesic in the 3-dimensional sense, but in the 4-dimensional spacetime sense: light rays are bent by gravity.

We will discuss the first of these results below, after introducing the requisite technical details.

9.1.2 Spacetime geometry and general coordinate transformations

As Geroch shows in detail [205, pp. 67–165], for every arrangement and scenario of particles moving in gravitational fields, there is a co-moving *spacetime geometry*. These are coordinate systems, each with four coordinates x^μ, $\mu = 0,1,2,3$, and a specified metric, $g_{\mu\nu}(\mathrm{x})$ of signature $(1,3)$; see Definition 3.2 on p. 89, we will explore some of the more interesting ones in some detail in Section 9.3. However, unlike in Chapter 3, these coordinates are inherently curvilinear in most applications, as should be clear from the example in the right-hand illustrations of Figure 9.2.

Away from certain exceptional locations (singularities) to be discussed in Section 9.3 and in sufficiently small regions (so-called patches) of spacetime, these inherently curvilinear coordinates can always be related to the Cartesian coordinates, much as every smooth curve can be approximated by its tangent. In Cartesian coordinates, the generalization of Pythagoras' theorem to spacetime [☞ relations (3.15)–(3.17)] defines the (spatial) so-called line element:

$$ds^2 := -c^2 d\tau^2 = dx^\mu(-\eta_{\mu\nu})dx^\nu. \tag{9.1}$$

The relation (3.11c) then provides the expression in arbitrary coordinates $x^\mu \mapsto y^\mu = y^\mu(x)$:

$$ds^2 := dx^\mu(-\eta_{\mu\nu})dx^\nu = dy^\rho \underbrace{\Big(\frac{\partial x^\mu}{\partial y^\rho}\Big)(-\eta_{\mu\nu})\Big(\frac{\partial x^\nu}{\partial y^\sigma}\Big)} dy^\sigma = dy^\rho\, g_{\rho\sigma}(y)\, dy^\sigma, \tag{9.2}$$

$$g_{\rho\sigma}(y) := \Big(\frac{\partial x^\mu}{\partial y^\rho}\Big)(-\eta_{\mu\nu})\Big(\frac{\partial x^\nu}{\partial y^\sigma}\Big), \quad \text{the metric tensor.} \tag{9.3}$$

Comment 9.1 *Note that the overall sign of the metric tensor (9.2) is opposite from the overall sign of the metric tensor (3.19). This unfortunate difference in conventions stems from the fact that the metric tensor (9.2) in general relativity defines a* ***distance****, while the expression (3.17) defines the proper* ***time*** *of a particle that moves in spacetime.*

The analogous computation for an *arbitrary* invertible coordinate substitution $y^\mu \to z^\mu(y)$ produces

$$g_{\mu\nu}(y) = \frac{\partial z^\rho}{\partial y^\mu}\frac{\partial z^\sigma}{\partial y^\nu}\, g_{\rho\sigma}(z), \tag{9.4}$$

proving that the metric tensor $g_{\mu\nu}$ is a rank-2, type-(0, 2) tensor.[1] More precisely, we define:

Definition 9.1 *Coordinate system transformations $x^\mu \to y^\mu(x)$ that are (1) unambiguously invertible, and (2) preserve the space/time character (signature) of spacetime [☞ Definitions 3.2 on p. 89 and 3.3 on p. 90] are* ***general coordinate transformations****.*

Unless otherwise stated, we only consider coordinate transformations that belong to this class.

Using the matrix notation, relation (9.4) may be written as

$$[g_{..}(x)] = \Big[\frac{\partial z}{\partial x}\Big][g_{..}(z)]\Big[\frac{\partial z}{\partial x}\Big]^T, \tag{9.5}$$

where the superscript T denotes matrix transposition.[2] Computing the determinants produces

$$g(x) = \Big(\det\Big[\frac{\partial z}{\partial x}\Big]\Big)^2 g(z), \quad \text{where} \quad g(x) := \det\big[g_{..}(x)\big]. \tag{9.6}$$

Since the metric tensor in spacetime has an odd number of negative eigenvalues,[3] it follows that the determinant of the metric tensor is negative, and

$$\sqrt{-g(x)} = \det\Big[\frac{\partial z}{\partial x}\Big]\sqrt{-g(z)} \tag{9.7}$$

---

[1] According to definition (9.2) of the quantity ds as a *distance* – which for purely spatial 4-vectors must agree with the familiar notion of the Euclidean distance – and owing to the "particle" convention (3.19) features the relative difference in the overall sign between $\eta_{\mu\nu}$ and $g_{\mu\nu}$: in flat spacetime, $g_{\mu\nu} \to -\eta_{\mu\nu}$. Both quantities are, however, called metric tensors, and the Reader is expected to read from the context which of the two conventions are used.

[2] The careful Reader will note that in the matrix representation of the components $g_{\rho\sigma}(z)$ one of the two indices must be counting rows while the other then must be counting columns. In the contraction with the matrices of partial derivatives in relation (9.4), the upper index (on the z-coordinate) in one of these two matrices must count columns (being contracted with the rows of $[g_{\rho\sigma}]$), but in the other it must count rows, whence the matrix representation of one of these matrices of partial derivatives is necessarily transposed in comparison with the other one.

[3] The general coordinate transformations, by Definition 9.1, preserve the signature, i.e., the numbers of positive and negative eigenvalues of the metric matrix.

is a real ***scalar density*** of ***weight*** -1. The weight of $\sqrt{-g}$ being -1 signifies that it transforms oppositely from the 4-fold differential (which then is a scalar density of weight $+1$):

$$\mathrm{d}^4\mathrm{x} = \det\left[\frac{\partial \mathrm{x}}{\partial \mathrm{y}}\right] \mathrm{d}^4\mathrm{y}, \tag{9.8}$$

which is computed straightforwardly (B.37) in Appendix B.2.1.

Conclusion 9.2 *The result (9.7) and the computation (B.37) in Appendix B.2.1 then imply that*

$$\sqrt{-g(\mathrm{x})}\,\mathrm{d}^4\mathrm{x} = \sqrt{-g(\mathrm{z})}\,\mathrm{d}^4\mathrm{z} \tag{9.9}$$

*is an invariant with respect to the general coordinate transformations [☞ Definition 9.1 on p. 319], and provides the **invariant (differential) 4-volume element**.*

Given the metric tensor $g_{\mu\nu}(\mathrm{y})$, the *inverse metric tensor* is defined by matrix inversion:

$$g^{\mu\nu}(\mathrm{y}): \quad g^{\mu\nu}(\mathrm{y})\, g_{\nu\rho}(\mathrm{y}) \overset{!}{=} \delta^\mu_\rho \overset{!}{=} g_{\rho\nu}(\mathrm{y})\, g^{\rho\mu}(\mathrm{y}), \tag{9.10}$$

point-by-point $\mathrm{y} = (y^0, \dots, y^3)$ in spacetime. Since

$$0 = \partial_\sigma(\delta^\mu_\rho) = \partial_\sigma(g_{\rho\nu}\, g^{\rho\mu}) = (\partial_\sigma g_{\rho\nu})g^{\rho\mu} + g_{\rho\nu}(\partial_\sigma g^{\rho\mu}), \tag{9.11}$$

it follows that

$$(\partial_\sigma g^{\lambda\mu}) = -g^{\rho\mu} g^{\lambda\nu}(\partial_\sigma g_{\rho\nu}). \tag{9.12}$$

In turn, derivatives of the determinant $g = \det[g_{..}]$ are computed using the Jacobi relation:

$$\partial_\rho g = g\, g^{\mu\nu}\, \partial_\rho g_{\mu\nu}, \tag{9.13}$$

from which it follows that

$$\partial_\rho \sqrt{-g} = -\tfrac{1}{2}\frac{\partial_\rho g}{\sqrt{-g}} = -\tfrac{1}{2}\sqrt{-g}\,(g_{\mu\nu}\,\partial_\rho g^{\mu\nu}) = \tfrac{1}{2}\sqrt{-g}\,(g^{\mu\nu}\,\partial_\rho g_{\mu\nu}). \tag{9.14}$$

For more detail, see Appendix B.2.3.

9.1.3 Einstein's equivalence principle as a gauge principle

Reconsider an object such as $\Psi(\mathrm{x})$, the wave-function used to describe an electron in Chapter 5. As discussed in detail in Section 5.1 and employed throughout Chapters 5–7, the (complex) function $\Psi(\mathrm{x})$ perforce contains unphysical information and is physically equivalent to $e^{i\varphi(\mathrm{x})}\Psi(\mathrm{x})$, where the phase $\varphi(\mathrm{x})$ is an undetermined function over spacetime. Consequently, the rate of change of $\Psi(\mathrm{x})$ in spacetime is computed not using partial derivatives, but using gauge-covariant derivatives (5.13), i.e., (5.117): $D_\mu := \partial_\mu + \frac{i}{\hbar c}A_\mu Q$. Since $\Psi(\mathrm{x})$ depends on the spacetime point both explicitly and also through the undetermined phase $\varphi(\mathrm{x})$, the partial derivative in D_μ computes the rate of change in spacetime owing to the explicit dependence on spacetime, while the $\frac{i}{\hbar c}A_\mu Q$ terms provides the "correction" owing to the indirect dependence via the undetermined phase, $\varphi(\mathrm{x})$.

The discussion in Section 9.1.1 showed that Einstein's principle of equivalence is itself equivalent to the statement that the difference between gravitational and inertial effects is purely a difference in the mathematical description, i.e., a difference in the *choice* of the coordinate system. Section 9.1.2 then formalizes the notion of spacetime geometry, as a spacetime coordinate system together with the corresponding metric, and *changing* this choice is accomplished by means of a general coordinate transformation.

As discussed by Pauli [414, p. 150], besides the technical aspects of the general coordinate transformations as formalized by tensor calculus, the key physical import of Einstein's principle of equivalence as provided in Conclusion 9.1 on p. 317 is its universal nature. That is, the equality of the various gravitational and inertial effects holds not only for certain (say, mechanical) phenomena, but for all physical phenomena. Therefore, there can be no physical distinction between them, and gravitational and inertial effects are not merely equal, but identical.

However, this insistence on universality is implied by the completely general (applicable to all of fundamental physics!) first, "conceptual" notion of unification as specified in part (a) of Conclusion 8.1 on p. 300. Under the umbrella of this overarching unifying principle, Einstein's equivalence principle is equivalent to

> **Conclusion 9.3 (Gauge principle of coordinate equivalence)** *General coordinate transformations [☞ Definition 9.1 on p. 319] can have no physically measurable consequences – and so must be symmetries [☞ Appendix A.1.3].*

In turn, this is conceptually identical to the gauge principle as employed in Chapters 5–7, except that the principle is here applied to the parametrization of spacetime, rather than to the abstract phases of wave-functions as in Chapters 5–7. Also, general coordinate transformations are typically nonlinear; this renders any gauge theory relating to general coordinate transformations intrinsically more complicated than the gauge theories considered in Chapters 5–7. We will explore the parallels and the differences between Yang–Mills gauge theory as discussed in Chapters 5–7 and general relativity, and will develop a selection of topics within general relativity specifically to that end. The Reader should, however, be aware of other possible approaches to gravity (some of them not entirely unrelated to the approach adopted herein), such as "gauge gravity" [451, 276] or "emergent gravity" [486, 315], to name a few.

Nevertheless, the *conceptual* similarity between the gauge principle as employed in Chapters 5–7 and the gauge equivalence principle (Conclusion 9.3 on p. 321) is striking:

1. Positions (in space of phases vs. in spacetime):
   (a) The choice of the overall phase of a wave-function is not observable; relative phases of different summands in a linear combination of wave-functions *are* observable.
   (b) The position of an object in spacetime is not observable; relative positions of different objects – distances between them – *are* observable.
2. Local (gauge) symmetry (changing the "position"):
   (a) Changing the choice of the overall phase of a wave-function locally in spacetime, i.e., by amounts that differ from point to point in spacetime.
   (b) Changing the choice of the coordinate system locally in spacetime, i.e., by (nonlinear) general coordinate transformations.
3. Gauge-covariant derivative operators (see below):
   (a) Correct the computation of the rate of change in spacetime to compensate for the spacetime variations in the choice of the undetermined phase.
   (b) Correct the computation of the rate of change in spacetime to compensate for the spacetime variations (nonlinearity) in the spacetime coordinate system itself.
4. Gauge interactions and curving trajectories (see below):
   (a) Gauge potentials and fields interact with test particles and curve their trajectories.
   (b) Spacetime is curved by the presence of matter, and curves the trajectories of test particles (including light).

9.1.4 Exercises for Section 9.1

✎ **9.1.1** *Show that, when y^μ are also Cartesian spacetime coordinates, the relation (9.3) implies that $g_{\rho\sigma}(y) = -\eta_{\rho\sigma}$.*

✎ **9.1.2** *Show that, when both x^μ and y^μ are Cartesian spacetime coordinates, $\frac{\partial x^\mu}{\partial y^\rho}$ must be a Lorentz transformation as discussed in Section 3.1.1.*

✎ **9.1.3** *Prove (9.9).*

✎ **9.1.4** *Prove the result (9.14).*

9.2 Gravity vs. Yang–Mills interactions

Having identified in Section 9.1.2 the key elements by which tensor algebra as used in Chapter 3 generalizes to the general spacetime geometries (Appendix B.2 has more details), we turn to employing the gauge symmetry concept from Chapters 5–7 to general coordinate transformations. In particular, given a 4-tuple of contravariant components $A^\mu(x)$ of a vector field as well as a 4-tuple of covariant components $B_\mu(x)$ of another vector field, we quote the definition of the covariant derivatives:

$$\text{result (B.55):}\quad D_\mu A^\rho := \left[\partial_\mu A^\rho + \Gamma^\rho_{\mu\nu} A^\nu\right] \quad\text{and}\quad D_\mu B_\nu := \left[\partial_\mu B_\nu - \Gamma^\rho_{\mu\nu} B_\rho\right]. \tag{9.15}$$

As shown in Appendix B.2, the second term in these derivatives compensates for the fact that the frame of reference, i.e., system basis vectors in a curvilinear coordinate system, varies point-to-point in spacetime. They also ensure that these derivatives are covariant with respect to general coordinate transformations:

$$\big(\widetilde{D_\mu A^\rho(y)}\big) = \frac{\partial x^\nu}{\partial y^\mu}\frac{\partial y^\rho}{\partial x^\sigma}\big(D_\nu A^\sigma(x)\big) \quad\text{and}\quad \big(\widetilde{D_\mu B_\rho(y)}\big) = \frac{\partial x^\nu}{\partial y^\mu}\frac{\partial x^\sigma}{\partial y^\rho}\big(D_\nu B_\sigma(x)\big), \tag{9.16}$$

and covariant derivatives of vectors transform as rank-2 proper tensors. That is, these covariant derivatives behave with respect to general coordinate transformations *identically* as do the gauge-covariant derivatives (5.7), (5.117) and (6.6) with respect to the local (gauge) symmetry of Yang–Mills type models described in Chapters 5–7.

It should then present no surprise that the necessary introduction of the $\Gamma^\rho_{\mu\nu}$-dependent "correcting" terms in the covariant derivatives (9.15) – to accommodate for the spacetime variable coordinatization of the spacetime geometry – will result in a gauge interaction. Furthermore, the results (9.48)–(9.49) below will identify this interaction as gravity.

9.2.1 The metric connection and the Christoffel symbol

The formal characterization (B.66) of the covariant derivative is formally identical to the general form (5.10), i.e., (6.6); its action on a type-(p,q) tensor is given by the general relation owing to the definition (B.40):

$$(D_\mu\,\mathbb{T})_{\rho_1\cdots\rho_q}^{\nu_1\cdots\nu_p} = \big(\partial_\mu T_{\rho_1\cdots\rho_q}^{\nu_1\cdots\nu_p}\big) + \sum_{i=1}^{p}\Gamma^{\nu_i}_{\mu\sigma_i}\,T_{\rho_1\cdots\cdots\cdots\rho_q}^{\nu_1\cdots\sigma_i\cdots\nu_p} - \sum_{i=1}^{q}\Gamma^{\sigma_i}_{\mu\rho_i}\,T_{\rho_1\cdots\sigma_i\cdots\rho_q}^{\nu_1\cdots\cdots\cdots\nu_p}; \tag{9.17}$$

see also Appendix B.2.3. The well-known special cases are (9.15) and the rank-2 case:

$$(D_\mu\,\mathbb{T})_{\nu\rho} = \partial_\mu T_{\nu\rho} - \Gamma^\sigma_{\mu\nu} T_{\sigma\rho} - \Gamma^\sigma_{\mu\rho} T_{\rho\sigma}. \tag{9.18}$$

Notice: the precise index notation of the covariant derivative action on tensor densities depends on the rank and type of those tensor densities, as then also does the action of the Levi-Civita connection 4-vector $\mathbb{\Gamma}_\mu$, i.e., the Christoffel symbol $\Gamma^\rho_{\mu\nu}$.

It follows that the symbol $\Gamma^\rho_{\mu\nu}$ transforms *inhomogeneously* – and so is not a tensor:

$$\Gamma^\rho_{\mu\nu}(\mathrm{y}) = \frac{\partial x^\sigma}{\partial y^\mu}\frac{\partial x^\tau}{\partial y^\nu}\frac{\partial y^\rho}{\partial x^\kappa}\Gamma^\kappa_{\sigma\tau}(\mathrm{x}) + \frac{\partial y^\rho}{\partial x^\sigma}\frac{\partial^2 x^\sigma}{\partial y^\mu \partial y^\nu}, \tag{9.19}$$

exactly as in the case of gauge 4-vector potentials in the (abelian) electrodynamics (5.89) and non-abelian chromodynamics (6.6b). At a first glance, the inhomogeneous term in the expressions (5.89) and (6.6b) is proportional to $(\partial_\mu \boldsymbol{\varphi}) = (\partial_\mu U)U^{-1}$, which may seem different from the second term in the result (9.19). However, using the matrix notation

$$[U]^\rho{}_\sigma = \frac{\partial y^\rho}{\partial x^\sigma}, \qquad \text{we have} \qquad \frac{\partial y^\rho}{\partial x^\sigma}\frac{\partial^2 x^\sigma}{\partial y^\mu \partial y^\nu} = [U]^\rho{}_\sigma \frac{\partial}{\partial y^\mu}[U^{-1}]^\sigma{}_\nu, \tag{9.20}$$

which then fully agrees with $(\partial_\mu \boldsymbol{\varphi}) = (\partial_\mu U)U^{-1} = -U(\partial_\mu U^{-1})$, up to a conventional sign of the phase "angle" $\boldsymbol{\varphi}$.

Comment 9.2 *The transformations $U = [\frac{\partial \mathrm{y}}{\partial \mathrm{x}}]$ employed here are general coordinate transformations [☞ Definition 9.1 on p. 319], which form a (gauge) group only in a restricted sense.*[4] *The physical manifestations of the theory in which $\mathbb{\Gamma}_\mu$ is the gauge potential and U the gauge transformation will be identified below as gravity; see equations (9.48)–(9.49).*

One may also construct the so-called the connection (differential) 1-forms[5]

$$\mathbb{A} := dx^\mu \mathbb{A}_\mu, \qquad \text{i.e.,} \qquad \mathbb{\Gamma} := dx^\mu \mathbb{\Gamma}_\mu. \tag{9.21}$$

Since $\mathbb{A} = dx^\mu A^a_\mu Q_a$ and Q_a are elements of the *algebra* of the gauge group, one says that $\mathbb{A}$ is valued in the gauge algebra. Similarly, $\mathbb{\Gamma}$ is a differential 1-form with values in the algebra of the group of transformations (B.41); the covariant differential $dx^\mu D_\mu$ is also-called the Levi-Civita connection.

Conclusion 9.4 *As the algebra of a group is essentially specified by linearizing (A.9), it follows that $\mathbb{\Gamma}$ may be regarded as a differential 1-form that takes values in the algebra of transformations of the tangent 4-plane (at any given spacetime point) into itself, which is the algebra of the Lorentz group, Spin(1, 3). Although no spinor appears in this discussion, the Lorentz group of course must act unambiguously on spinors also, whereupon we write Spin(1, 3) instead of SO(1, 3) [☞ discussion about relations (5.45)–(5.48)].*

However, note the difference: For the Yang–Mills gauge symmetries in Chapters 5–7, the unitary operator of the symmetry transformation, $U := \exp\{ig_c\varphi^a(\mathrm{x})Q_a/\hbar\}$, depends on (the coordinates of) the spacetime point $\mathrm{x} = (x^0, \ldots, x^3)$ but describes a change in parametrizing another, abstract space of generalized phases of wave-functions. Within our present context, $[U]^\rho{}_\sigma = \frac{\partial y^\rho}{\partial x^\sigma}$ depends on the spacetime point x, but simultaneously describes the change in the coordinate parametrization (basis elements) of that *very same* spacetime. Besides, the coordinate transformations $x^\mu \to y^\nu = y^\nu(\mathrm{x})$ are nonlinear in general. This conceptual as well as literal nonlinearity

[4] The binary combination of two transformations exists only when they "concatenate": $\frac{\partial x^\mu}{\partial y^\nu}\frac{\partial z^\rho}{\partial x^\mu} = \frac{\partial z^\rho}{\partial y^\nu}$ and $\frac{\partial x^\mu}{\partial y^\nu}\frac{\partial y^\nu}{\partial z^\rho} = \frac{\partial x^\mu}{\partial z^\rho}$, but a product such as $\frac{\partial x^\mu}{\partial y^\nu}\frac{\partial z^\nu}{\partial w^\rho}$ does not simplify as a closed binary operation. This structure curiously reminds us of the so-called "renormalization group," see Section 5.3.3 on p. 210.

[5] Instead of dx^μ, one may of course use any arbitrary basis elements, e^μ, resulting also in 1-forms, albeit not differential. The use of the dx^μ-basis is however standard, as it provides a connection with differential and integral calculus.

provides the root of all differences between (Yang–Mills) gauge theories and the general theory of relativity, viewed as a gauge theory.

This difference also reflects in the following: The gauge vector potential (6.6c) has a matrix representation:

$$\mathbb{A}_\mu := A^a_\mu\, Q_a \qquad \to \qquad [\mathbb{A}_\mu]_\alpha{}^\beta. \tag{9.22}$$

The gauge vector potential for general coordinate transformations (9.20) is the Levi-Civita connection 4-vector,

$$\mathbb{\Gamma}_\mu \qquad \to \qquad [\mathbb{\Gamma}_\mu]_\nu{}^\rho, \tag{9.23}$$

that acts upon a vector according to the relations (9.15), in perfect analogy with the action of the chromodynamics gauge vector potential (9.22) upon a quark wave-function:

$$[\mathbb{A}_\mu \cdot \Psi]^\alpha = [\mathbb{A}_\mu]_\beta{}^\alpha\, \Psi^\beta \qquad \leftrightarrow \qquad [\mathbb{\Gamma}_\mu \cdot V]^\rho = \Gamma^\rho_{\mu\nu}\, V^\nu. \tag{9.24}$$

Note, however, that the chromodynamics gauge potentials are matrices in the abstract space of (color) phases and covariant vectors in real spacetime. By contrast, the Christoffel symbol is a matrix in the very same spacetime wherein it is also a connection 4-vector. What is more, it is not hard to show that (see, e.g., the derivation of (B.59) in Appendix B.2.3)

$$\Gamma^\rho_{\mu\nu} = \tfrac{1}{2} g^{\rho\sigma} \Big[\frac{\partial g_{\sigma\nu}}{\partial x^\mu} + \frac{\partial g_{\mu\sigma}}{\partial x^\nu} - \frac{\partial g_{\mu\nu}}{\partial x^\sigma} \Big]. \tag{9.25}$$

That is, the gauge potential for general coordinate transformations, the Levi-Civita connection 4-vector $\mathbb{\Gamma}_\mu$, can be derived from the metric tensor (9.3),[6] which thereby serves as a gauge "*pre*-potential." In Yang–Mills gauge theories, no such thing exists.

In turn, relation (9.25) is equivalent to the result

$$D_\mu\, g_{\nu\rho} = 0 \qquad \Leftrightarrow \qquad D_\mu\, g^{\nu\rho} = 0. \tag{9.26}$$

That is, the metric tensor and its inverse are "covariantly constant," so (9.25) may just as well be derived from either of the two relations (9.26). Again, Yang–Mills gauge theories contain no such nontrivial "covariantly constant" object.

Thus, while the electric and magnetic fields may be obtained as derivatives of an electromagnetic potential (5.15)–(5.73) A_μ, this potential cannot be obtained as a derivative of some more fundamental prepotential. Similarly, chromodynamics fields $\mathbb{F}_{\mu\nu} = F^a_{\mu\nu} Q_a$ can also be expressed in terms of a chromodynamics potential (6.15) $\mathbb{A}_\mu = A^a_\mu Q_a$, but these potentials cannot be expressed in terms of something more fundamental yet. In sharp contrast, the Christoffel symbol $\Gamma^\rho_{\mu\nu}$ *may be and is* expressed in terms of a derivative of the metric tensor (9.25) and the inverse metric tensor. From relations (9.25) it also follows that the Christoffel symbol is symmetric with respect to the exchange of the indices

$$\Gamma^\rho_{\mu\nu} = +\Gamma^\rho_{\nu\mu}. \tag{9.27}$$

In the Yang–Mills gauge vector potentials $[A_\mu]_\alpha{}^\mu$, an analogous symmetrization (here, for $\mu \leftrightarrow \alpha$) simply makes no sense at all: μ and α indicate basis elements in completely different spaces.

[6] Strictly speaking, this is true only in the absence of fermions. With fermions present, one uses the so-called Palatini formalism, wherein the metric tensor and the Levi-Civita connection 4-vector $\mathbb{\Gamma}_\mu$ are independent.

Digression 9.1 Some useful consequences of the relations (9.25)–(9.26) are

$$\frac{\partial g_{\mu\nu}}{\partial x^\sigma} = \Gamma^\rho_{\mu\sigma} g_{\rho\nu} + \Gamma^\rho_{\nu\sigma} g_{\rho\mu}, \qquad \frac{\partial g^{\mu\nu}}{\partial x^\sigma} = -g^{\mu\rho}\Gamma^\nu_{\sigma\rho} - g^{\nu\rho}\Gamma^\mu_{\sigma\rho}, \tag{9.28a}$$

$$\Gamma^\mu_{\mu\nu} = \frac{\partial}{\partial x^\nu}\ln\big(\sqrt{-g}\big), \qquad g := \det[g_{..}]; \quad g < 0 \text{ because of signature } (1,3), \tag{9.28b}$$

where we used the relation

$$\frac{\partial g}{\partial g_{\mu\nu}} = g\, g^{\mu\nu}, \quad \text{so that} \quad \frac{\partial g}{\partial x^\rho} = g\, g^{\mu\nu}\frac{\partial g_{\mu\nu}}{\partial x^\rho}. \tag{9.28c}$$

The signature is the number of positive and negative eigenvalues of the metric tensor [☞ discussion about the expression (3.19) and Definition 3.3 on p. 90].

Digression 9.2 Also, definition (9.17) produces the following oft-used results:

$$\operatorname{grad}(f)_\mu := D_\mu f = (\partial_\mu f); \tag{9.29a}$$

$$\operatorname{curl}(V_.)^{\rho\sigma} := \varepsilon^{\mu\nu\rho\sigma} D_\nu V_\mu = \varepsilon^{\mu\nu\rho\sigma}(\partial_\nu V_\mu); \tag{9.29b}$$

$$\operatorname{curl}(V^\cdot)^{\rho\sigma} := \varepsilon^{\mu\nu\rho\sigma} D_\mu(g_{\nu\lambda}V^\lambda) = \varepsilon^{\mu\nu\rho\sigma}\partial_\mu(g_{\nu\lambda}V^\lambda); \tag{9.29c}$$

$$\operatorname{div}(V^\cdot) := D_\mu V^\mu = \frac{1}{\sqrt{-g}}\big(\partial_\mu(\sqrt{-g}\,V^\mu)\big); \tag{9.29d}$$

$$\operatorname{div}(V_.) := D_\mu(g^{\mu\nu}V_\nu) = \frac{1}{\sqrt{-g}}\big(\partial_\mu(\sqrt{-g}\,g^{\mu\nu}V_\nu)\big); \tag{9.29e}$$

$$\Box f := D_\mu(g^{\mu\nu}D_\nu f) = \frac{1}{\sqrt{-g}}\Big[\partial_\mu\big(\sqrt{-g}\,g^{\mu\nu}(\partial_\nu f)\big)\Big]. \tag{9.29f}$$

Note that, in 1+3-dimensional spacetime, the curl of a 4-vector is a rank-2 tensor. On the other hand, the spacetime analogue of $\vec\nabla^2\vec A \equiv \vec\nabla(\vec\nabla\cdot\vec A) - \vec\nabla\times(\vec\nabla\times\vec A)$ may be used to compute

$$\Box A^\mu = \left[g^{\mu\nu}\partial_\nu\left(\frac{\partial_\rho(\sqrt{-g}\,A^\rho)}{\sqrt{-g}}\right) + \frac{1}{\sqrt{-g}}\varepsilon^{\mu\nu\rho\sigma}\varepsilon^{\alpha\beta\kappa\lambda}\partial_\nu\left(\frac{(\partial_\alpha A_\beta)}{\sqrt{-g}}\,g_{\kappa\rho}g_{\lambda\sigma}\right)\right]. \tag{9.29g}$$

9.2.2 The curvature of spacetime

Finally, just as the gauge field $\mathbb{F}_{\mu\nu}$ is defined in relation (6.15) as the commutator of covariant derivatives, so too may the Riemann curvature tensor be defined:

$$\begin{aligned} R_{\mu\nu\rho}{}^\sigma &:= [\,D_\mu\,,\,D_\nu\,]_\rho{}^\sigma = [(\delta^\sigma_\lambda\partial_\nu + \Gamma^\sigma_{\nu\lambda})\Gamma^\lambda_{\mu\rho}] - [(\delta^\sigma_\lambda\partial_\mu + \Gamma^\sigma_{\mu\lambda})\Gamma^\lambda_{\nu\rho}] \\ &= \partial_\nu\Gamma^\sigma_{\mu\rho} - \partial_\mu\Gamma^\sigma_{\nu\rho} + \Gamma^\sigma_{\nu\lambda}\Gamma^\lambda_{\mu\rho} - \Gamma^\sigma_{\mu\lambda}\Gamma^\lambda_{\nu\rho}. \end{aligned} \tag{9.30}$$

Note the formal similarity of the defining expression (9.30) and the definition of the gauge field for non-abelian gauge symmetry (6.15). However, unlike $\mathbb{F}_{\mu\nu}$ which is an antisymmetric rank-2 tensor and the components of which are matrices in the abstract space of phases, the Riemann tensor is a rank-4, type-$(1,3)$ tensor. Besides, it may be shown that [508, 62, 367, 548, 66, 96]

$$R_{\mu\nu\rho}{}^\rho = 0, \tag{9.31}$$

and that the closely related tensor

$$R_{\mu\nu\,\rho\sigma} := R_{\mu\nu\rho}{}^{\lambda}\, g_{\lambda\sigma} \tag{9.32a}$$

satisfies the relations:

$$R_{\mu\nu\,\rho\sigma} = -R_{\nu\mu\,\rho\sigma}, \tag{9.32b}$$
$$R_{\mu\nu\,\rho\sigma} = -R_{\mu\nu\,\sigma\rho}, \tag{9.32c}$$
$$R_{\mu\nu\,\rho\sigma} = +R_{\rho\sigma\,\mu\nu}, \tag{9.32d}$$
$$\varepsilon^{\lambda\nu\rho\sigma} R_{\mu\nu\,\rho\sigma} = 0, \qquad \text{1st Bianchi identity,} \tag{9.32e}$$
$$\varepsilon^{\kappa\lambda\mu\nu} D_{\lambda} R_{\mu\nu\,\rho\sigma} = 0, \qquad \text{2nd Bianchi identity.} \tag{9.32f}$$

This 2nd Bianchi identity (9.32f) is both formally and conceptually analogous to the Bianchi identity (5.87) in electrodynamics and (6.19) for non-abelian gauge fields.

Relation (9.31) is analogous to the requirement that in the expansion $\mathbb{F}_{\mu\nu} = F^a_{\mu\nu}\, Q_a$, the generators Q_a of non-abelian factors in the gauge group are traceless: $\mathrm{Tr}[Q_a] = [Q_a]_\alpha{}^\alpha = 0$. This is certainly true of the gauge field of the $SU(3) \times SU(2)_w$ group, and is not true precisely for the *abelian* electromagnetic $U(1)$ field $F_{\mu\nu}$. The Riemann tensor $R_{\mu\nu\rho}{}^\sigma$ may be regarded as a special rank-2 and type-$(0,2)$ tensor, the components of which are matrices and traceless rank-2 and type-$(1,1)$ tensors, $R_{\mu\nu\rho}{}^\sigma = [R_{\mu\nu}]_\rho{}^\sigma$, subject to the additional constraints (9.32b)–(9.32f). The fact that both $\mathbb{F}_{\mu\nu}$ and $R_{\mu\nu\rho}{}^\sigma$ are defined as commutators of appropriate covariant derivatives then guarantees the first of the relations, (9.32b). This similarity permits the interpretation of the Riemann tensor as a general coordinate transformation analogue of the tensor $\mathbb{F}_{\mu\nu}$. The components $R_{\mu\nu\rho}{}^\sigma$ are then interaction fields associated with general coordinate transformations, and in fact represent the general-relativistic generalization of the gravitational field; see below.

The very existence of the definition (9.32a) points to the difference between $R_{\mu\nu\rho}{}^\sigma$ and $[F_{\mu\nu}]\alpha^\beta$. For orthogonal and symplectic gauge groups,[7] their invariant quadratic forms would play the role of $g_{\lambda\sigma}$ and produce $[F_{\mu\nu}]_{\alpha\beta}$. Unitary groups (such as $SU(3)_c$) have no such tensor, and for them there can exist nothing analogous to definition (9.32a). Also, for unitary gauge groups there exist no analogues of the relations (9.32c)–(9.32e).

Furthermore, for Yang–Mills gauge fields, $[\mathbb{F}_{\mu\nu}]_\alpha{}^\beta$, there is no way to perform the contraction between one of the "matrix" indices α or β and one of the "tensor" indices μ or ν. In turn, the contractions that can be performed,

$$g^{\mu\nu}\mathbb{F}_{\mu\nu} \equiv 0, \qquad \begin{cases} \mathrm{Tr}[\mathbb{F}_{\mu\nu}] &= [\mathbb{F}_{\mu\nu}]_\alpha{}^\alpha = 0, \quad \text{for semisimple Lie groups,} \\ \mathrm{Tr}[F_{\mu\nu}] &= F_{\mu\nu}, \quad \text{for } U(1) \text{ factors,} \end{cases} \tag{9.33}$$

are trivial: The first equality holds owing to the fact that $g_{\mu\nu} = +g_{\nu\mu}$ but $\mathbb{F}_{\mu\nu} = -\mathbb{F}_{\nu\mu}$. The second one follows from the fact that $\mathrm{Tr}[Q_a] \neq 0$ only for $U(1)$ factors.

The situation is, however, different for the Riemann tensor: neither is

$$\text{the Ricci tensor:} \qquad R_{\mu\rho} := R_{\mu\nu\rho}{}^{\nu}, \tag{9.34}$$

trivial, nor is its trace,

$$\text{the scalar curvature:} \qquad R := g^{\mu\rho}\, R_{\mu\rho} = g^{\mu\rho}\, R_{\mu\nu\rho}{}^{\nu}. \tag{9.35}$$

[7] Orthogonal and symplectic groups may be defined as the groups of linear transformations of some specified real vector space that preserve a (pseudo-)Euclidean, i.e., symplectic quadratic form, respectively [☞ Appendix A]. However, this invariant quadratic form does not determine the gauge potential of Yang–Mills theories with orthogonal and symplectic group of symmetries, unlike the fact that the relation (9.25) does determine the Christoffel symbol in terms of the metric.

It is also useful to know that, following Conclusion 9.4 on p. 323, we have that the differential 2-form[8]

$$\mathbf{R} := \big[\, \mathrm{d}x^\mu D_\mu \,,\, \mathrm{d}x^\nu D_\nu \,\big], \qquad \text{i.e.,} \qquad [\mathbf{R}]_\rho{}^\sigma := \mathrm{d}x^\mu\, \mathrm{d}x^\nu\, R_{\mu\nu\rho}{}^\sigma \tag{9.36}$$

also has values in the algebra of the Lorentz group *Spin*(1, 3).

Definition (9.30) shows that the components of the Riemann tensor $R_{\mu\nu\rho}{}^\sigma$ are derivatives of the second order (or are quadratic in derivatives of the first order) of the metric tensor components,[9] but it contains also the inverse metric tensor. $R_{\mu\nu\rho}{}^\sigma$ is therefore a nonlinear function of the metric tensor components, $g_{\mu\nu}$, but precisely of second order in spacetime derivatives of those components.[10] The same is then true also of the Ricci tensor (9.34), as well as the scalar curvature (9.35).

Yang–Mills gauge theories have nothing analogous to the expressions (9.34)–(9.35). There, the Lagrangian density (6.23) is found in the form $-\frac{1}{4}\,\mathrm{Tr}[\mathbb{F}_{\mu\nu}\,\mathbb{F}^{\mu\nu}]$, which is quadratic in the derivatives of $\mathbb{A}_\mu$. This Lagrangian density then yields equations of motion (6.24) that are analogous to Gauss's law for the electric field and Ampère's law for the electromagnetic field (6.37).

Analogously to the expression $-\frac{1}{4}\,\mathrm{Tr}[\mathbb{F}_{\mu\nu}\mathbb{F}^{\mu\nu}]$ in the Lagrangian density (6.23), the Hamilton action with the Riemann tensor would be proportional to the integral

$$\int \sqrt{-g}\,\mathrm{d}^4\mathrm{x}\; R_{\mu\nu\rho}{}^\sigma\, g^{\mu\kappa} g^{\nu\lambda}\, R_{\kappa\lambda\sigma}{}^\rho. \tag{9.37}$$

Since both $\sqrt{-g}\,\mathrm{d}^4\mathrm{x}$ and $R_{\mu\nu\rho}{}^\sigma\, g^{\mu\kappa} g^{\nu\lambda}\, R_{\kappa\lambda\sigma}{}^\rho$ are scalar quantities, this integral is invariant under general coordinate transformations. Varying this action by the components of the Christoffel symbol would, in the standard fashion, produce Euler–Lagrange equations of the second order in derivatives of the Christoffel symbol, $\mathbb{\Gamma}$. However, the Christoffel symbol is itself a derivative of the metric tensor, and varying this action by components of the metric tensor (which is more fundamental than the Christoffel symbol) would produce Euler–Lagrange equations of motion for the metric tensor components that are of the *fourth order* in spacetime derivatives, which agrees with neither classical (non-quantum) theory of gravity nor with experimental facts about gravity.

Fortunately – and completely unlike in Yang–Mills gauge theory – with the Riemann tensor it is possible to define another, so-called Einstein–Hilbert action:

$$\frac{c^3}{16\pi\, G_N} \int \sqrt{-g}\,\mathrm{d}^4\mathrm{x}\; R, \qquad \text{where } R \overset{(9.35)}{=} g^{\mu\rho}\, R_{\mu\nu\rho}{}^\nu. \tag{9.38}$$

The powers of the natural constants $c, \hbar$ and G_N in the prefactor are determined:

1. by requiring the Hamilton action to have the dimensions $\frac{ML^2}{T}$ [☞ Sections 1.2.3 and 1.2.2],
2. by definition (3.10) whereby $[\mathrm{d}^4\mathrm{x}] = L^4$ (note: $\mathrm{d}^4\mathrm{x} = c\,\mathrm{d}t\,\mathrm{d}^3\vec{r}$),[11]
3. by definitions (9.2), (9.25) and (9.30), from which it follows that $[g_{\mu\nu}] = 1$, $[\Gamma^\rho_{\mu\nu}] = L^{-1}$ and $[R_{\mu\nu\rho}{}^\sigma] = L^{-2}$, respectively.

The conventional numerical prefactor $\frac{1}{16\pi}$ simplifies many derivations and many final results. Varying this action by the metric tensor components produces [508, 62, 367, 548, 66, 96]

[8] When defining differential p-forms, one automatically uses the antisymmetric product of basis elements and without any notational distinction: $(\cdots \mathrm{d}x^\mu \mathrm{d}x^\nu \cdots) = -(\cdots \mathrm{d}x^\nu \mathrm{d}x^\mu \cdots)$.

[9] All told, every summand in the defining expression (9.30) contains precisely two spacetime derivatives.

[10] Unlike the quadratic, cubic or another expression of a relatively low degree, the components of the inverse metric tensor are by definition ratios of the determinants of various cofactors and the determinant of the entire metric tensor. A Taylor expansion in the components of the original metric tensor is then an infinite series, containing arbitrarily high powers of the components of the original metric tensor. This makes the inverse metric tensor, and then also the Riemann and other curvature tensors, *very* nonlinear.

[11] Some Authors imply $\mathrm{d}^4\mathrm{x} := \mathrm{d}t\,\mathrm{d}^3\vec{r}$, so that the prefactor in the action (9.38) has c^4 instead of c^3 as given here.

$$R_{\mu\nu} - \tfrac{1}{2}g_{\mu\nu}R = 0. \tag{9.39}$$

This system of differential equations, the Einstein equations, determines the metric tensor components as functions of the spacetime coordinates, and in the absence of all matter, i.e., in empty space. The combination $G_{\mu\nu} := R_{\mu\nu} - \frac{1}{2}g_{\mu\nu}R$ is called the Einstein tensor.

Already, writing the Einstein equations (9.39), with definitions (9.30) and (9.25), indicates the essential differences from Yang–Mills gauge theories: The differential equations (6.37) are at most cubic in the 4-vector potentials $\mathbb{A}_\mu$, while the Einstein equations (9.39) are *very* nonlinear in the metric tensor components. The definition of the Christoffel symbol and the scalar curvature involve the inverse metric tensor, the components of which are ratios of cubic polynomials in the components $g_{\mu\nu}$ and the determinant $\det[g_{\mu\nu}]$. This much more radical nonlinearity of the differential equations (9.39) – and also the action (9.38) from which the Einstein equations follow – is the root of the technical differences between the general theory of relativity and Yang–Mills gauge theories.

9.2.3 Coupling of gravity and matter

Finally, the operations so far defined may be combined and produce a relevant result for our present purposes:

Conclusion 9.5 *In the general case, Hamilton's action is*

$$S[\phi_i(x)] := \int \sqrt{-g}\,\mathrm{d}^4x\,\mathscr{L}(\phi_i(x), (D_\mu\phi_i(x)), \dots; x), \tag{9.40}$$

$$g := \det[g(x)], \quad \mathrm{d}^4x := \tfrac{1}{4!}\varepsilon_{\mu\nu\rho\sigma}\mathrm{d}x^\mu\mathrm{d}x^\nu\mathrm{d}x^\rho\mathrm{d}x^\sigma, \tag{9.41}$$

where $\mathscr{L}$ is the "Lagrangian density" (in the sense of "Lagrangian per unit 4-volume"). In turn, both $\sqrt{-g}\,\mathrm{d}^4x$ and $\mathscr{L}$ are scalars, i.e., invariants with respect to general coordinate transformations [☞ Definition 9.1 on p. 319].

Comment 9.3 Lagrangian densities $\mathscr{L}(\phi_i(x), (\partial_\mu\phi_i(x)), \dots; x)$ constructed within the *special-relativistic field theory may continue to be used, but "covariantizing" the derivatives, $\partial_\mu \mapsto D_\mu := \partial_\mu + \mathbb{\Gamma}_\mu$, where $\mathbb{\Gamma}_\mu$ is the formal Levi-Civita* **connection** *4-vector, which when acting on tensors may be represented by the Christoffel symbol (9.17).*

In the general case, the covariant derivative is $D_\mu = \partial_\mu + \mathbb{\Gamma}_\mu + \sum_k \frac{ig_k}{\hbar c}A^{(k)}_\mu\cdot Q^{(k)}$, where $Q^{(k)}_{a_k}$ are generators of the kth factor in the Yang–Mills group of gauge symmetries with the coupling parameter g_k and gauge 4-vector potentials $A^{(k)}_\mu{}^{a_k}$.

In the general case, let $\mathscr{L}_M$ be the Lorentz-invariant Lagrangian density for any type of matter – here, "matter" denotes everything except the metric tensor $g_{\mu\nu}$, the Levi-Civita connection 4-vector potential $\mathbb{\Gamma}_\mu$, and the Riemann tensor $R_{\mu\nu\rho}{}^\sigma$ and quantities constructed from these. The corresponding model that is invariant with respect to general coordinate transformations has the Hamilton action

$$\int \sqrt{-g}\,\mathrm{d}^4x \left[\frac{c^3}{16\pi G_N}R - \mathscr{L}_M\right], \tag{9.42}$$

where all the derivatives in the Lagrangian density $\mathscr{L}_M$ are "covariantized" as discussed in Comment 9.3 on p. 328. Varying this action by the components of the inverse metric tensor yields

$$\frac{\delta R}{\delta g^{\mu\nu}} + \frac{R}{\sqrt{-g}}\frac{\delta(\sqrt{-g})}{\delta g^{\mu\nu}} = -\frac{16\pi G_N}{c^3}\frac{1}{\sqrt{-g}}\frac{\delta(\sqrt{-g}\,\mathscr{L}_M)}{\delta g^{\mu\nu}}, \tag{9.43}$$

that is [508, 62, 367, 548, 66, 96],

$$R_{\mu\nu} - \tfrac{1}{2}g_{\mu\nu}R = \frac{8\pi G_N}{c^4}T_{\mu\nu}, \tag{9.44}$$

where the rank-2 and type-(0, 2) tensor

$$T_{\mu\nu} := -\frac{2c}{\sqrt{-g}} \frac{\delta(\sqrt{-g}\,\mathscr{L}_M)}{\delta g^{\mu\nu}} \tag{9.45}$$

has the physical meaning of the energy–momentum tensor density for the physical system described by the Lagrangian density $\mathscr{L}_M$.

Digression 9.3 Note that the inverse metric tensor and the metric tensor of course are not independent quantities, since the inverse metric tensor is defined so as to satisfy

$$g_{\mu\nu}\, g^{\rho\nu} = \delta^{\rho}_{\mu}, \qquad g_{\mu\nu} = +g_{\nu\mu} \;\Rightarrow\; g^{\mu\nu} = +g^{\nu\mu}. \tag{9.46a}$$

It then follows that varying the inverse metric tensor is not independent of varying the metric tensor itself:

$$0 = \delta(\delta^{\rho}_{\mu}) = \delta(g_{\mu\nu}\, g^{\rho\nu}), \tag{9.46b}$$

$$\Rightarrow \qquad \delta g^{\mu\nu} = -g^{\mu\rho} g^{\nu\sigma}\,(\delta g_{\rho\sigma}), \quad \text{and} \quad \frac{\delta}{\delta g^{\mu\nu}} = -g_{\mu\rho} g_{\nu\sigma} \frac{\delta}{\delta g_{\rho\sigma}}. \tag{9.46c}$$

Varying the action (9.42) by various fields that represent various "matter" degrees of freedoms produces the Euler–Lagrange equations of motion for these fields. As all the derivatives in the Lagrangian density $\mathscr{L}_M$ are covariantized, the resulting Euler–Lagrange equations of motion will, in the general case, depend on the Levi-Civita connection 4-vector $\mathbb{\Gamma}_\mu$ as well as on the metric $g_{\mu\nu}$. The Euler–Lagrange equations of motion and the Einstein equations (9.44) then form a coupled system of differential equations, which are certainly nonlinear in the metric tensor components.

Although such coupled systems of differential equations most often are not soluble in closed form, the geometric meaning of the Einstein equations (9.44) is very clear:

1. On the left-hand side, $R_{\mu\nu} - \frac{1}{2} g_{\mu\nu} R$ is a nonlinear expression in the metric tensor components, which is of precisely second order in spacetime derivatives; the left-hand side depends only on the metric tensor components and their spacetime derivatives.
2. On the right-hand side, $T_{\mu\nu}$ is the energy–momentum tensor density, which describes the spacetime (and general-relativistic) generalization of mass of the matter.

The differential equation (9.44) thus determines the metric tensor, for which the energy–momentum tensor density plays the role of the "source" – just as the differential equation representing Gauss's law determines the electric field for which electric charge density plays the role of the source, and Ampère's law determines the electromagnetic field for which the electric current density plays the role of the source.

What's more, comparing the Einstein equations with the differential equations representing the Gauss–Ampère laws is more than suggestive: it may be shown that the energy–momentum tensor density, $T_{\mu\nu}$, is indeed the Noether "current" density that corresponds to the continuous symmetry of spacetime translations.

Since the metric tensor is the quantity that determines the spacetime geometry, we have:

Conclusion 9.6 *Conceptually, the Einstein equations are perfectly analogous to Gauss's law for the electric field and Ampère's law for the electromagnetic field, and they determine the spacetime geometry, for which the energy–momentum tensor density of the present matter is the "source," i.e., the "driving force."*

That is: by virtue of its presence, matter curves spacetime.

Digression 9.4 Relation (9.24) gives a *formal* correspondence between Yang–Mills gauge theories and the general theory of relativity:

$$[\mathbb{A}_\mu]_\alpha{}^\beta \quad\longleftrightarrow\quad \Gamma^\rho_{\mu\nu}, \qquad \text{and so also} \qquad [\mathbb{F}_{\mu\nu}]_\alpha{}^\beta \quad\longleftrightarrow\quad R_{\mu\nu\rho}{}^\sigma. \tag{9.47a}$$

This formal correspondence is also qualitatively correct, and foremost in its geometric sense, where the tensors $\mathbb{F}_{\mu\nu}$ and $R_{\mu\nu\rho}{}^\sigma$ represent the curvature of the effective spacetime for the purposes of field propagation and particle motion.

However, in a strictly practical sense – the so-called "engineering" spirit of Section 9.3.4 that also permeates the discussion leading to Conclusion 9.6 – the formal correspondence (9.47a) is not appropriate.[12] The Einstein equations (9.44) identify the differential expression that is of second order in spacetime derivatives of the metric tensor with the energy–momentum tensor density $T_{\mu\nu}$ for that distribution of matter:

$$\Big\{R_{\mu\nu}-\tfrac{1}{2}g_{\mu\nu}\,R = \tfrac{1}{2}g^{\rho\sigma}\big(\partial_\mu\partial_\rho g_{\nu\sigma}+\partial_\nu\partial_\rho g_{\mu\sigma}\big)+\cdots\Big\} = \frac{8\pi\,G_N}{c^4}T_{\mu\nu}. \tag{9.47b}$$

That system of differential equations is formally analogous to the Gauss–Ampère laws (5.88), expressed in terms of the gauge potential:

$$\Big\{(\Box A^\mu) - \eta^{\mu\nu}(\partial_\nu\partial_\rho A^\rho)\Big\} = \frac{1}{4\pi\epsilon_0}\frac{4\pi}{c}j_e^\nu. \tag{9.47c}$$

Comparing equations (9.47b) and (9.47c) implies the correspondence

$$A_\mu \;\longleftrightarrow\; g_{\mu\nu}, \qquad F_{\mu\nu} \;\longleftrightarrow\; \Gamma^\rho_{\mu\nu}, \qquad j_e^\mu \;\longleftrightarrow\; T_{\mu\nu}, \tag{9.47d}$$

which better fits this "engineering" sense. The differences between the correspondences (9.47a) and (9.47d) stem from the already mentioned differences, and foremost from the following facts:

1. Both in Yang–Mills gauge theories and in the general theory of relativity, the covariant derivative is defined so that $D_\mu - \partial_\mu \propto \mathbb{A}_\mu$, i.e., $D_\mu - \partial_\mu \propto \mathbb{\Gamma}_\mu$. However, $\mathbb{A}_\mu$ cannot be expressed as the derivative of anything "more fundamental," whereas $\mathbb{\Gamma}_\mu$ can: see equation (9.25).
2. Both in Yang–Mills gauge theories and in general theory of relativity, the curvature is defined as the commutator $[D_\mu, D_\nu]$. However, the Hamilton action for Yang–Mills gauge theory is quadratic in the curvature, while the Einstein–Hilbert action is linear in the (scalar) curvature (9.35).

Finally, the identity

$$R = -g^{\mu\nu}\big(\Gamma^\sigma_{\mu\rho}\Gamma^\rho_{\nu\sigma} - \Gamma^\rho_{\mu\nu}\Gamma^\sigma_{\rho\sigma}\big) + \partial_\mu K^\mu \tag{9.47e}$$

shows the Einstein–Hilbert Lagrangian to be quadratic in $\mathbb{\Gamma}_\mu$, making it similar – though definitely not identical – to the Yang–Mills type Lagrangians (5.76) and (6.23), in further support of the "engineering" correspondence (9.47d).

[12] This practical sense is regarded "engineering" in the sense that the Gauss–Ampère laws may be used to find the desired electromagnetic field, by constructing the appropriate distribution of charges and currents. Analogously, the Einstein equations (9.44) may be used so that by constructing a particular distribution of matter one produces the desired gravitational field, and so also the spacetime of the desired curvature.

9.2.4 Geometry and Newtonian limit

In turn, if we take $\mathscr{L}_M = m\sqrt{g_{\mu\nu}\frac{\partial x^\mu}{\partial t}\frac{\partial x^\nu}{\partial t}}$,[13] which is the Lagrangian density [☞ definition L_0 in Digression 3.7 on p. 93, and defining equation (9.2)] for a particle that moves in spacetime with the metric tensor $g_{\mu\nu}$, then varying the action (9.42) by x^μ yields

$$\frac{d^2x^\rho}{dt^2} + \Gamma^\rho_{\mu\nu}\frac{dx^\mu}{dt}\frac{dx^\nu}{dt} = 0. \tag{9.48}$$

These are the differential equations that determine the so-called geodesic (extremal) lines. In flat spacetime, $g_{\mu\nu} = -\eta_{\mu\nu}$ and the Christoffel symbol vanishes, so equation (9.48) gives $\ddot{x}^\mu = 0$, i.e., $x^\mu = x^\mu_0 + v^\mu_0 t$ gives straight lines in spacetime. Rearranging the second term we obtain the analogue of Newton's second law:

$$m\frac{d^2x^\rho}{dt^2} = F^\rho_{\text{grav}} := -m\,\Gamma^\rho_{\mu\nu}\frac{dx^\mu}{dt}\frac{dx^\nu}{dt}, \tag{9.49}$$

where the right-hand side provides the gravitational force that curves the trajectory of the particle, the acceleration of which appears on the left-hand side.

Conclusion 9.7 *The possibility of reinterpreting essentially geometric information as essentially physical information*

$$\left.\begin{array}{r}\textit{spacetime curvature}\\ \textit{appearing in equation (9.48)}\end{array}\right\} \Leftrightarrow \left\{\begin{array}{l}\textit{definition of the force and}\\ \textit{interaction in equation (9.49)}\end{array}\right. \tag{9.50}$$

points to the fundamental equivalence of these two ways of thinking and explaining natural phenomena.

Of course, this is merely one of the simplest examples, but it should be clear that now even in the most general context – including also the Yang–Mills type of gauge interactions[14] [☞ Chapters 5 and 6] – the coupled system of the Einstein equations and the general-relativistically covariant Euler–Lagrange equations of motion may be reinterpreted:

1. either in a purely geometric sense, where objects move along geodesic (extremal) trajectories defined by the (charge/color/isospin-sensing) curvature of spacetime,[14]
2. or in a purely "physicsy" sense, where objects move under the influence of forces with which these objects affect one another.

It behooves us to keep in mind that this latter way of interpreting natural phenomena *implicitly* presupposes the existence of an "empty" spacetime in which these objects move. Therefore, the first, geometric way of interpretation is more economical, and represents the basis of "geometrizing" physics: the notion of force may be replaced by the notion of curvature in the (appropriately generalized) spacetime; see also Comment 3.2.

Starting from (9.48), following Pauli [414], we focus on a spatial component of x, $x^\rho \to x^k$, use that $x^0 = ct$, and assume that $g_{\mu\nu}$ deviates only slightly from its flat-space value, $-\eta_{\mu\nu}$, and obtain

$$\frac{d^2x^k}{dt^2} \approx -c^2\,\Gamma^k_{00}, \tag{9.51}$$

[13] Here, t denotes an arbitrary parameter of the dimension of time, which grows monotonously along the worldline of the given particle.

[14] From this "geometrized" point of view, the various phases that are subject to gauge transformation are to be included in the "total spacetime." Since these phases vary over the usual spacetime, the resulting structure is a called a fiber bundle, where the spacetime-variable phases span the *fibers* over the *base* spacetime. The fiber-wise curvature is measured by the $\mathbb{F}_{\mu\nu}$-type tensors, and is detected only by particles that have the appropriate type of charge: electromagnetic, weak isospin or chromodynamic color.

where terms quadratic in the small deviations $\gamma_{\mu\nu} := (g_{\mu\nu}+\eta_{\mu\nu})$ have been dropped. Assuming furthermore that the components of the metric $g_{\mu\nu}$ are slowly varying in time so that time derivatives may be neglected,

$$\Gamma^k_{00} = \tfrac{1}{2}g^{k\sigma}(\partial_0 g_{\sigma 0} + \partial_0 g_{0\sigma} - \partial_\sigma g_{00}) \approx -\tfrac{1}{2}g^{k\ell}(\partial_\ell g_{00}). \tag{9.52}$$

In fact, since we must keep $\partial_\ell g_{00} = \partial_\ell(\gamma_{00}-1) = (\partial_\ell \gamma_{00})$, where γ_{00} is the small deviation, dropping terms that are second order in $\gamma_{\mu\nu}$ allows us to drop the (also small) contributions from:

1. off-diagonal terms from the ℓ-summation, and
2. the deviations in g_{kk} from $(-\eta_{kk} = 1)$, whereby $g^{kk} \to 1$.

This produces

$$\frac{d^2x^k}{dt^2} \approx \tfrac{1}{2}c^2(\partial_k \gamma_{00}), \qquad \text{i.e.,} \qquad \frac{d^2\vec{r}}{dt^2} \approx \tfrac{1}{2}c^2(\vec{\nabla}\gamma_{00}) \stackrel{!}{=} -\vec{\nabla}\Phi_N, \tag{9.53}$$

and allows us to identify $-\frac{1}{2}c^2\gamma_{00} := -\frac{1}{2}c^2(g_{00}+1)$ with Newton's gravitational potential, such as $\Phi_N = -G_N\frac{M}{r}$ for a point-like source of gravity of mass M, so the potential energy of the considered particle with mass m at a distance r from the gravitational source is $m\Phi_N = -G_N\frac{mM}{r}$.

Much more detailed derivations of the Newtonian weak-field limit of gravity may be found in the literature; see for example Refs. [96, 95, 271, 58].

9.2.5 Exercises for Section 9.2

✎ **9.2.1** *Prove the relations in Digression 9.1 on p. 325.*

✎ **9.2.2** *Prove the relations in Digression 9.2 on p. 325.*

✎ **9.2.3** *Prove that the Riemann tensor has 20 independent degrees of freedom. (Hint: the rank-4 tensor itself of course has $4^4 = 256$ components. Show that the relations (9.32b) reduce this to 36, the relation (9.32d) further to 21, and relation (9.32e) to 20.)*

✎ **9.2.4** *Prove the relation (9.32f) using the definition (9.30) of $R_{\mu\nu\rho\sigma}$.*

✎ **9.2.5** *Prove that the Ricci tensor is symmetric: $R_{\mu\nu} = R_{\nu\mu}$.*

✎ **9.2.6** *Prove that the equations (9.48) are covariant, i.e., that a coordinate substitution changes these equations only up to a non-vanishing overall multiplicative factor.*

✎ **9.2.7** *Derive the Euler–Lagrange equations of motion for the n-plet of scalar fields $\phi_i(\mathbf{x})$ with the Lagrangian density*

$$\mathscr{L}[\phi_i] = \frac{1}{2}g^{\mu\nu}\delta_{ij}\,(D_\mu\phi^i)(D_\nu\phi^j) - \frac{m^2c^2}{2\hbar^2}\delta_{ij}\phi^i\phi^j. \tag{9.54}$$

(Hint: since ϕ^i are Lorentz-scalars, determine first the action of $D_\mu\phi^i$ from relation (9.17).)

✎ **9.2.8** *From the Lagrangian density (9.54), derive the energy–momentum tensor density, $T_{\mu\nu}$, and the system of Euler–Lagrange equations from the previous exercise coupled with the Einstein equations.*

9.3 Special solutions

Solutions of the Einstein equations (9.44) represent various spacetime geometries – various universes[15] – of which each one may serve as the background/arena in which all "other" physics happens, including the elementary particle physics as analyzed so far. Besides, the Einstein equations – as a system of differential equations for the metric tensor components – are nonlinear, making the existence of a growing class of exact solutions all the more interesting.

9.3.1 The Schwarzschild solution

Only a month after the publication of Einstein's general theory of relativity and gravitation, in 1915, Karl Schwarzschild published the first and best known exact solution to the Einstein equations. Six years later, the mathematician George David Birkhoff proved a theorem[16] whereby any spherically symmetric solution of the Einstein equations without matter (9.39) must be stationary and asymptotically flat, i.e., the geometry of the outer region of spacetime must be described by the Schwarzschild metric tensor (see Refs. [367, 264, 103, 548, 131] and also [128, 587, 127]), given here in spherical coordinates:

$$\textbf{Schwarzschild}\quad \begin{cases} [g_{\mu\nu}] = \text{diag}\big(-f_S(r), \frac{1}{f_S(r)}, r^2, r^2\sin^2(\theta)\big), \\ ds^2 = -f_S(r)c^2 dt^2 + \frac{1}{f_S(r)}dr^2 + r^2\big(d\theta^2 + \sin^2(\theta)\, d\varphi^2\big), \end{cases} \tag{9.55a}$$

where

$$f_S(r) := \Big(1 - \frac{r_S}{r}\Big), \qquad r_S = \frac{2G_N M}{c^2}. \tag{9.55b}$$

As the metric tensor (9.55) satisfies the Einstein equations with $T_{\mu\nu} = 0$, it follows that the Schwarzschild solution describes *empty spacetime*, in the sense that this is a possible geometry of spacetime in the *absence* of any matter. The mass $M := \frac{c^2 r_S}{2G_N}$ that may be ascribed to the point-like object at the origin of the coordinate system then does not represent a particle of matter that is placed there, but is a characteristic of spacetime itself [☞ Digression 9.5 on p. 340], which for observers outside r_S is curved as if there existed an object of mass M.

The meaning of the Schwarzschild radius, r_S, is as follows: The well-known expression for the (first) escape velocity, i.e., the velocity of separation from a planet of mass M at a distance r from the center of the planet is

$$v_1 = \sqrt{\frac{2G_N M}{r}}. \tag{9.56}$$

It follows that the separation velocity at the Schwarzschild radius becomes $v_1(r_S) = c$. This literally means that Schwarzschild's solution (9.55) holds for $r \geqslant r_S$. For observers that are outside the Schwarzschild radius, objects that pass through the surface of the sphere of radius r_S can no longer return. This sphere is thus called the "event horizon" and effectively separates the exterior from the interior. As the same conclusion holds also for light, classical physics predicts that the interior of this horizon is completely black for observers in the exterior – whence the popular name "black hole." Formally, the metric tensor (9.55) is applicable also in the interior of the event horizon, but

---

[15] The distinction between a "spacetime geometry" and a "universe" – as the latter word is used in this chapter – is far from strict: the latter term is used merely to emphasize its *global* meaning. A "universe," after all, has an all-encompassing ring to it and so allows "spacetime geometry" to have either just a local reference, if desired, or a fully global one. In recent times however, the terms "multiverse" and "metaverse" came into vogue, denoting a collection – sometimes infinitely large – of universes [513, 514, 515, 557, for starters]. Especially when these universes within a multiverse are connected, the connotation of globalness of a single universe is restricted in some way or another, at the least. Herein, in turn, a "universe" will be used to denote a closed, isolated and geodesically complete spacetime, unless explicitly stated otherwise.

[16] It was recently discovered that this theorem, many years known under Birkhoff's name, was proven two years earlier (in 1919) by the Norwegian physicist Jørg Tofte Jebsen [297].

here the coordinate t becomes space-like and r becomes time-like; the physical meaning of this change remains uncertain, foremost because – at least within classical physics[17] – it is not possible to design an experiment (even a thought-experiment) with which one could compare the evolution of physical phenomena outside the event horizon with those unfolding within the horizon.

Singularities

The functional dependence of the Schwarzschild metric on the radius indicates that there exist two special places within the space with the geometry (9.55):

1. the Schwarzschild radius, where $f_S(r) = 0$, so the metric tensor has a singularity: the coefficient of the $\mathrm{d}t^2$ term vanishes, and the coefficient of the $\mathrm{d}r^2$ term diverges;
2. the coordinate origin, where $f_S(r)$ diverges, so the coefficient of the $\mathrm{d}t^2$ term diverges, and the coefficient of the $\mathrm{d}r^2$ term vanishes.

However, the metric tensor transforms under general coordinate transformations as a rank-2 and type-$(0,2)$ tensor, and it is not clear a priori if these special places are indeed singularities. As the metric tensor is of type $(0,2)$, this transformation has the form [☞ Definition B.2 on p. 511]

$$g_{\mu\nu}(\xi) = \frac{\partial\zeta^\rho}{\partial\xi^\mu}\, g_{\rho\sigma}(\zeta)\, \frac{\partial\zeta^\sigma}{\partial\xi^\sigma} \qquad \Longleftrightarrow \qquad g' = U^T g\, U \quad \text{(in matrix form)}, \tag{9.57}$$

which is *not* a similarity transformation. Thus, neither the characteristic polynomial, $\det[\boldsymbol{g} - \lambda\mathbb{1}]$, nor the eigenvalues of the matrix $\boldsymbol{g}$ are invariants. The only invariant that can be constructed from the metric tensor is $\delta^\rho_\mu = g_{\mu\nu} g^{\rho\nu}$, which produces no information about possible singularities.

However, depending on the first and second derivatives of the metric tensor components, the Riemann curvature tensor does contain information about their (non)analyticity, and one only needs to find a way to extract that information in an invariant fashion. The scalar curvature (9.35) is one such invariant. As the Riemann tensor has 20 independent degrees of freedom [☞ Exercise 9.2.3], this leaves precisely 19 independent invariants, but an explicit listing of such invariants remains an open problem✌. Now, there do exist two simple quadratic invariants

$$\|R_{\mu\nu}\|^2 := R_{\mu\nu}\, g^{\mu\rho} g^{\nu\sigma}\, R_{\rho\sigma} \qquad \text{and} \qquad \|R_{\mu\nu\rho}{}^\sigma\|^2 := R_{\mu\nu\rho}{}^\sigma\, g^{\mu\alpha} g^{\nu\beta} g^{\rho\gamma} g_{\sigma\delta}\, R_{\alpha\beta\gamma}{}^\delta, \tag{9.58}$$

of which the second, the so-called Kretschmann invariant for the Schwarzschild metric, equals

$$\|R_{\mu\nu\rho}{}^\sigma\|^2 = \frac{48 G_N{}^2 M^2}{c^4\, r^6}, \tag{9.59}$$

and is indeed divergent at the coordinate origin, $r = 0$. This proves that the coordinate origin is really a singularity of the geometry. The fact that neither the scalar curvature (9.35) nor the quadratic curvature invariants (9.58) diverge on the event horizon does not prove that the location $r = r_S$ is not a singularity. It remains, in principle, to check 17 other independent invariants; the divergence of any one of those invariants on the sphere $r = r_S$ would prove that the event horizon is a singularity. Unfortunately, as no list of 20 independent invariants is known, such a direct verification is not available in practice.[18]

[17] The quantum theory of gravity is not a complete theory, and this analysis is not without debate. However, in the early 1970s, Stephen Hawking was among the first to apply the "semi-classical" analysis and so discover that black holes radiate, emitting the so-called Hawking radiation. The same methods led to the derivation of the Bekenstein–Hawking formula according to which the entropy of a black hole is proportional to the surface area of the event horizon. A recent application of stringy methods and the gravity–gauge duality [☞ p. 443] discovered newer, and not just semi-classical results.

[18] Nor may this suffice even in principle: As discussed in Ref. [264, Section 8.1], because of the non-definiteness of the metric $g_{\mu\nu}$, there could exist singular solutions to the Einstein equations for which all invariant curvature polynomials (constructed from $g_{\mu\nu}$, $g^{\mu\nu}$, $\varepsilon_{\mu\nu\rho\sigma}$ and $R_{\mu\nu\rho}{}^\sigma$) are finite. Also, there do exist special solutions such as the Taub-NUT (Newman, Unti and Tamburino) solution, where the invariant curvature polynomials remain bounded but the spacetime contains incomplete geodesics within a compact neighborhood of the horizon.

Fortunately, Georges Lemaître discovered in 1933 that the coordinate substitution (introduced by Arthur Eddington in 1924, without noting the significance)

$$\mathrm{d}\tau := \mathrm{d}t + \sqrt{\frac{r_S}{r}}\frac{\mathrm{d}r/c}{(1-\frac{r_S}{r})}, \qquad \mathrm{d}\varrho := \mathrm{d}t + \sqrt{\frac{r}{r_S}}\frac{\mathrm{d}r/c}{(1-\frac{r_S}{r})} \tag{9.60a}$$

changes the appearance of the Schwarzschild metric tensor into

$$\mathrm{d}s^2 = -c^2\mathrm{d}\tau^2 + \Big(\frac{2r_S}{3(\varrho - c\tau)}\Big)^{\frac{2}{3}}\mathrm{d}\varrho^2 + r^2\big(\mathrm{d}\theta^2 + \sin^2(\theta)\mathrm{d}\varphi^2\big) \tag{9.60b}$$

and so clearly shows that the sphere $r = r_S$, i.e., $\varrho = \varrho_S := \big(\frac{2}{3}r_S + c\tau\big)$ is free of singularities.

Thus, the event horizon is a completely non-singular location in spacetime and the unlucky observer who drifts through it would notice nothing unusual in his immediate vicinity – except that he would not be able to return outside the event horizon. This phenomenon is often compared with the fact that fish that arrive too close to a waterfall can no longer return upstream.

In turn, the $r = 0$ location is indeed a real singularity [☞ equations (9.58)–(9.59)], and its existence explains the fact that the Schwarzschild solution describes empty space with no matter located within the event horizon, although the coordinate origin may be ascribed the mass $M = \frac{r_S c^2}{2G_N}$. More precisely, any Gaussian sphere that fully encloses the event horizon will detect a gravitational field as if within it there existed a mass M. However, such a Gaussian sphere can be shrunk down only as far as the event horizon; beyond that, no information could be extracted from the gravitational field detectors (scales) bedecking the Gaussian sphere. Mathematically, this unusual property stems from the nonlinearity of the Einstein equations and the singularity of the Schwarzschild solution of those equations. Physically, this indicates the self-interaction of the gravitational field – which is conceptually very similar to the self-interaction of non-abelian Yang–Mills gauge fields [☞ so-called "glueballs," discussed on p. 239], and this self-interaction mimics a material particle located at the origin. In fact, the formation of black holes may be described as a phase transition [148, 147] and even have a Landau–Ginzburg effective description [149], much like the Higgs effect [☞ Section 7.1]. However, unlike the fact that black holes have mass, no self-interacting non-abelian Yang–Mills gauge field configuration could exist that would exhibit a non-vanishing charge (color, isospin,...) at the origin.

There is, however, another important conceptual difference in describing and modeling Yang–Mills interactions and gravity:

1. The standard models of Yang–Mills interactions [☞ Chapters 5–7] are formulated in flat and infinitely large spacetime, which has the geometry of $\mathbb{R}^{1,3}$, i.e., real 4-dimensional spacetime with the flat metric $g_{\mu\nu} = -\eta_{\mu\nu}$.
2. Models of gravity generally involve a *choice* of a nontrivial metric $g_{\mu\nu} \neq -\eta_{\mu\nu}$, defined on a spacetime that need not at all have the simple structure of $\mathbb{R}^{1,3}$.

When modeling gravity, we *are* free to chose a spacetime where portions – such as singular points – are excised. If all singular points are excised, the remaining spacetime will be singularity-free, but this typically comes at a price: there will exist geodesic paths, solutions to equations (9.48), which tend towards the points that have been excised or are otherwise absent from the given spacetime. It then may or may not be possible to "fill in" (complete) this spacetime in a way that renders all geodesics complete and also (re-)introduces no singular points. Already this observation should make it clear that the (non-)singularity of spacetime is a rather delicate issue that cannot be resolved simply by identifying whether or not all curvature invariants (were one even able to enlist them all) are (non-)singular.

In addition, geodesic incompleteness is not the only way of detecting an incompleteness in the spacetime, and it is standard [367, 264] to distinguish at least three a-priori different notions of completeness and incompleteness as its logical negative:

> **Definition 9.2** *A spacetime is* ***geodesically complete*** *if every geodesic path can be extended infinitely within the given spacetime. One may further specify geodesic (in)completeness by restricting to* ***time-like, null*** *or* ***space-like*** *geodesics.*

This permits the logical possibility that a given spacetime with a given choice of metric is both time-like and null-geodesically complete, but contains incomplete space-like geodesics.

Besides considering geodesic paths as a continuous sequence of points, one may consider any other (discrete) Cauchy sequence of points; this leads to:

> **Definition 9.3** *A spacetime is* ***metrically complete*** *if every Cauchy sequence converges to a point within the given spacetime.*

For a positive-definite metric, it turns out that the geodesic and metric notions of (in)completeness are equivalent [317, 318]. However, the physically interesting case involves the Lorentzian metric of signature $(1,3)$, which is not positive-definite, and where this equivalence does not hold.

There is also another definition, due to C. Ehresmann (1957) and B. G. Schmidt (1971), which generalizes geodesic completeness: One considers all possible smooth (once differentiable) curves in a given spacetime and shows that the length of any such curve is finite in a given parametrization if and only if it is also finite in any other parametrization obtained by parallel transport. Variables parametrizing such curves in a 1–1 fashion are called *(generalized) affine parameters*. Curves with this class of parametrization define a *bundle*, which is then used in the definition [264]:

> **Definition 9.4** *A spacetime is* ***b-complete*** *if every once-differentiable curve of finite length as measured by a generalized affine parameter is within the given spacetime.*

If a finite once-differentiable curve with its end-point(s) contained in the spacetime is a geodesic, this geodesic is complete in the sense of Definition 9.2. If the metric is positive-definite, b-completeness coincides with metric completeness.

The metric is of course not positive-definite in the physically interesting Lorentzian spacetime, in which case it turns out that b-completeness of spacetime implies its geodesic completeness, but the converse is not true [264]. This prompts Hawking and Ellis to *define* a spacetime to be singularity free if it is b-complete, and concede that:

> ...one might possibly wish to weaken this condition slightly, to say that space-time is singularity-free if it is only *non-spacelike b-complete*, i.e., if there is an end-point for all non-spacelike C^1 [once-differentiable] curves with finite length as measured by a generalized affine parameter.

Needless to say, a detailed analysis of singularities in spacetime geometry and the theory of gravity is much more involved than the purely algebraic considerations around equation (9.59) and certainly beyond our present scope. In addition, the study of gravitation, spacetime geometry, astrophysics and cosmology brings up the questions whether a singularity could dynamically develop within an initially non-singular spacetime, whether an initially singular spacetime could dynamically de-singularize, and how various singularities might interact with each other. The interested Reader is therefore directed to standard references [367, 264, 548, 66, 96], to begin with.

9.3.2 Charged and rotating solutions

In 1916–18, Hans Reissner and Gunnar Nordstrøm generalized the Schwarzschild solution to electrically charged black holes:

Reissner–Nordstrøm
$$\begin{cases} [g_{\mu\nu}] = \mathrm{diag}\big(-f_{RN}(r), \frac{1}{f_{RN}(r)}, r^2, r^2\sin^2(\theta)\big), \\ \mathrm{d}s^2 = -f_{RN}(r)c^2\mathrm{d}t^2 + \frac{1}{f_{RN}(r)}\mathrm{d}r^2 + r^2\big(\mathrm{d}\theta^2 + \sin^2(\theta)\,\mathrm{d}\varphi^2\big), \end{cases} \tag{9.61a}$$

where

$$f_{RN}(r) := \Big(1 - \frac{r_S}{r} + \frac{r_q^2}{r^2}\Big), \qquad r_q := \sqrt{\frac{q^2\,G_N}{4\pi\epsilon_0\,c^4}}. \tag{9.61b}$$

This solution has a horizon at the location where $g_{rr} \to \infty$, i.e., where $f_{RN}(r) = 0$:

$$r_\pm = \tfrac{1}{2}\Big(r_S \pm \sqrt{r_S^2 - 4r_q^2}\Big). \tag{9.62}$$

For $2r_q < r_S$, the concentric spheres of radii r_+ and r_- are two concentric horizons. When $2r_q = r_S$, the two horizons coincide, and this case is called the *extremal Reissner–Nordstrøm* solution. Using equations (9.55b) and (9.61b), the extremal case is characterized by the relation $q = \sqrt{4\pi\epsilon_0 G_N}\,M$. For two extremal Reissner–Nordstrøm solutions of the same-sign electric charge, the gravitational attraction precisely cancels the electrostatic repulsion and there is effectively no interaction. In the case when $2r_q > r_S$, i.e., when $q > \sqrt{4\pi\epsilon_0 G_N}M$ and the black hole is "overcharged," there are no horizons and the singularity at the coordinate origin would be visible to the observer at any distance.

Comment 9.4 *A singularity that is not enclosed by an event horizon is called "naked." The existence of naked singularities would violate Roger Penrose's cosmic censorship hypothesis (to wit, that every singularity is enclosed within an event horizon and is accessible to no "outside" observer). In accord with this hypothesis, it is* ***believed*** *that the gravitational collapse of matter cannot create naked singularities.*

The exact solution for a chargeless, static, spinning black hole was discovered by Roy Kerr only in 1963, and is now most often specified in the coordinates given by Robert H. Boyer and Richard W. Lindquist in 1967:

Kerr
$$\begin{cases} \mathrm{d}s^2 = -\Big(1 - \frac{r_S\,r}{\rho^2}\Big)c^2\mathrm{d}t^2 + \rho^2\Big(\frac{1}{\Delta}\mathrm{d}r^2 + \mathrm{d}\theta^2\Big) \\ \qquad + \Big(r^2 + \ell^2 + \frac{r_S\,r\,\ell^2}{\rho^2}\sin^2(\theta)\Big)\sin^2(\theta)\,\mathrm{d}\varphi^2 - \frac{2r_S\,r\,\ell\,\sin^2(\theta)}{\rho^2}c\,\mathrm{d}t\,\mathrm{d}\varphi, \end{cases} \tag{9.63a}$$

where

$$\ell := \frac{L}{Mc}, \qquad \rho := \sqrt{r^2 + \ell^2\cos^2(\theta)}, \qquad \Delta := r^2 - r_S\,r + \ell^2, \tag{9.63b}$$

and L is the angular momentum. Note that – unlike in the Schwarzschild (9.55) and Reissner–Nordstrøm (9.61) solutions – the Kerr metric tensor is not diagonal: the (ct, r, θ, φ) coordinates are not orthogonal in the Kerr geometry. This solution possesses two event horizons at the location where $g_{rr} \to \infty$, which gives two concentric spheres of radii

$$r_H^\pm = \tfrac{1}{2}\Big(r_S \pm \sqrt{r_S^2 - 4\ell^2}\Big), \tag{9.64}$$

of which r_H^+ is clearly *the* relevant event horizon for outside observers. In turn $g_{tt} \to 0$ occurs on the ellipsoids (adopting Visser's nomenclature [540]):

$$\textbf{ergosurface} \quad r_E^\pm = \tfrac{1}{2}\Big[r_S \pm \sqrt{r_S^2 - 4\ell^2 \cos^2(\theta)}\Big]. \tag{9.65}$$

The space between the outer one of these ellipsoids and the outer one of the spherical event horizons is called the *ergoregion*. Objects that enter through the outer ergosurface (9.65) must co-rotate with an angular speed of at least

$$\Omega = -\frac{g_{t\varphi}}{g_{\varphi\varphi}} = \frac{2r_S\, r\, \ell\, c}{\rho^2(r^2+\ell^2) + r_S\, r\, \ell^2 \sin^2(\theta)}, \tag{9.66}$$

even if this implies that they move faster than c, in reference to outside observers. Such superluminal motion, however, does not contradict the theory of relativity, as in a real sense the spacetime itself inside the ergoregion co-rotates akin to a radially accelerating conveyor belt, and objects are – in reference to this co-rotating spacetime – not moving faster than c.

However, since the ergosurface (9.65) is not a "one-way" event horizon, objects can dip into the ergoregion and come back out of it. As the motion during the passage through the ergoregion is faster than a "parallel" motion outside the ergoregion, such an object will draw energy from the spinning black hole. Indeed, consider a conveyer belt that passes through the ergoregion but loops back outside the ergoregion. The co-rotation within the ergoregion will thus drive the conveyor belt outside the ergoregion and so do useful work. This process of drawing energy from a spinning black hole is called the Penrose process, after Roger Penrose, who discovered this possibility. Also, there exist trajectories that pass through the ergoregion, which make it possible to travel backwards in time.

Two years later, in 1965, Ezra Newman found a generalization of the Kerr metric tensor, for an electrically charged spinning black hole:

$$\textbf{Kerr–Newman} \quad \left\{\begin{aligned} \mathrm{d}s^2 &= -\frac{\Delta}{\rho^2}\Big(c\,\mathrm{d}t - \ell \sin^2(\theta)\,\mathrm{d}\varphi\Big)^2 + \rho^2\Big(\frac{1}{\Delta}\mathrm{d}r^2 + \mathrm{d}\theta^2\Big) \\ &\quad + \frac{\sin^2(\theta)}{\rho^2}\Big((r^2+\ell^2)\,\mathrm{d}\varphi - \ell c\,\mathrm{d}t\Big)^2, \end{aligned}\right. \tag{9.67a}$$

where

$$\ell := \frac{L}{Mc}, \quad \rho := \sqrt{r^2 + \ell^2\cos^2(\theta)}, \quad \Delta := r^2 - r_S\, r + \ell^2 + r_q^2, \quad r_q := \sqrt{\frac{q^2\, G_N}{4\pi\epsilon_0\, c^4}}, \tag{9.67b}$$

and L, M, and q are the angular momentum, the mass and the electric charge of the black hole. Just as the Kerr metric tensor (9.63), the Kerr–Newman metric tensor (9.67) is also not diagonal, and the (ct, r, θ, φ) coordinates are not orthogonal in the Kerr–Newman geometry.

Furthermore, direct computation proves that the location $\rho = 0$ is a true coordinate singularity for both the Kerr geometry (9.63) and the Kerr–Newmann solution (9.67), since the Kretschmann curvature invariant $\|R_{\mu\nu\rho}{}^{\sigma}\|^2$ defined in equations (9.58) diverges there. Given that the location $r = 0$ within the standard interpretation of the coordinates (r, θ, φ) is a single point, the result

$$\rho = 0 \quad \Leftrightarrow \quad r = 0, \text{ and } (\theta = \tfrac{\pi}{2} \text{ if } \ell \neq 0), \tag{9.68}$$

may appear puzzling, in that the coordinate location "$r = 0$ and ($\theta \neq \frac{\pi}{2}$ if $\ell \neq 0$)" is singular in neither the Kerr geometry nor the Kerr–Newmann geometry. This indicates that the coordinate locations "within the coordinate origin,"

$$O_* := \{r=0,\ \theta=\tfrac{\pi}{2},\ \varphi\in[0,2\pi]\} \quad\text{and}\quad O_\circ := \{r=0,\ \theta\neq\tfrac{\pi}{2},\ \varphi\in[0,2\pi]\}, \tag{9.69}$$

must be distinguished. This makes it obvious that the coordinates (r,θ,φ) must not be interpreted literally as the standard spherical coordinates for the Kerr and the Kerr–Newmann geometries, (9.63) and (9.67), respectively; R. Wald provides the standard argument for O_* to be interpreted as a ring-shaped singularity in these geometries [548, pp. 314–315]; see also [540]. Consequently, the whole coordinate region

$$O := \{r=0,\ \theta\in[0,\pi],\ \varphi\in[0,2\pi],\ \ \varphi\simeq\varphi+2\pi\} \tag{9.70}$$

must be regarded as a *null* 2-sphere standing in the place of the standard coordinate origin, and the singularity of the Kerr and the Kerr–Newmann geometry is then located on the equator of this null 2-sphere. This recalls the process of "blowing up a singularity," where the null 2-sphere is the "exceptional divisor" [279, for starters].

Not even a decade later, in 1972–3, Akira Tomimatsu and Humitaka Sato discovered a class of exact solutions [523, 524, 270] [☞ also [200] for a recent review and applications] that generalize the Kerr solution (with polar coordinates $\rho := \sqrt{x^2+y^2}$ and φ):

Kerr–Tomimatsu–Sato
$$ds^2 = -Fc^2\big[dt-\omega\, d\varphi\big]^2 + F^{-1}\big[E\,(d\rho^2+dz^2)+\rho^2 d\varphi^2\big], \tag{9.71a}$$

where the functions E, F and G are most easily expressed in terms of prolonged spheroidal coordinates (ξ,η,φ):

$$x=\rho_0\sqrt{(\xi^2-1)(1-\eta^2)}\cos\varphi,\quad y=\rho_0\sqrt{(\xi^2-1)(1-\eta^2)}\sin\varphi,\quad z=\rho_0\,\xi\eta, \tag{9.71b}$$

so $\rho=\rho_0\sqrt{(\xi^2-1)(1-\eta^2)}$:

$$E(\xi,\eta) := \frac{A(\xi,\eta)}{p^{2\delta}(\xi^2-\eta^2)^{\delta^2}},\quad F(\xi,\eta) := \frac{A(\xi,\eta)}{B(\xi,\eta)},\quad G(\xi,\eta) := \frac{2L/mc}{A(\xi,\eta)}(1-\eta^2)C(\xi,\eta), \tag{9.71c}$$

where $A(\xi,\eta)$, $B(\xi,\eta)$ and $C(\xi,\eta)$ are polynomials of degree $2\delta^2$, $2\delta^2$ and $(2\delta^2-1)$, respectively, and where the constants ρ_0 and p are algebraic functions of the mass m, angular momentum L, the integral parameter δ and the natural constants [523, 524]

$$\rho_0 := \frac{G_N}{c^2}\frac{p}{\delta}m \qquad\text{and}\qquad p=\sqrt{1-\frac{c^2}{G_N{}^2}\frac{L^2}{m^4}}. \tag{9.72}$$

The Tomimatsu–Sato solutions depend on the parameter δ, so that $\delta=1$ gives the Kerr solution, but for $\delta\neq1$ the Tomimatsu–Sato solutions contain naked singularities.

— ❦ —

It is important to understand that the very nontrivial solutions (9.55), (9.61), (9.63), (9.67) and (9.71) are but a few special – and physically very interesting – representatives of a general class of solutions of the Einstein equations without matter. In other words, solutions to the Einstein equations (9.39) include very nontrivial geometries that even contain locations (in the presented case, the so-called black holes) that have the appearance of a particle: they have a mass, and may have electric charge and intrinsic angular momentum.

Digression 9.5 It is reasonable then to inquire whether, e.g., the electron could be simply a charged black hole. However, with the mass and the charge of the electron, one easily obtains

$$r_S(e^-) = 1.353 \times 10^{-57}\,\text{m} \ll \ell_P \qquad \text{and} \qquad r_q(e^-) = 9.152 \times 10^{-37}\,\text{m} < \ell_P. \tag{9.73a}$$

Since $r_S(e^-) < r_q(e^-)$, this black hole has no event horizon, and represents a naked singularity. However, as both characteristic radii are smaller than the Planck length, Conclusion 1.5 on p. 30 indicates that this model is unverifiable. That is, because of Conclusion 1.5, it simply is not possible to determine any concretely verifiable consequence of representing the electron by a charged miniature (classical!) black hole.

Strictly speaking, the complete theory of quantum gravity does not exist,[19] so that only estimates exist that indicate that – contrary to the name – *quantum* black holes radiate. This radiation is named after Stephen Hawking, who in 1974 explained the quantum process that enables this radiation, and without violating the "one-way" nature of the event horizon. These estimates indicate that black holes lose mass via the Hawking radiation, and so have an "evaporation time" [403]:

$$t_{\text{evap.}} \approx 5{,}120\pi \frac{G_N{}^2}{\hbar c^4} M^3 \approx 8.407 \times 10^{-17} (M/\text{kg})^3\,\text{s}. \tag{9.73b}$$

For charged leptons and quarks, electric charge conservation would have to obstruct their evaporation, but for neutrinos with a mass $m_\nu < 2\,\text{eV} \sim 3 \times 10^{-36}\,\text{kg}$ the evaporation time is of the order $< 4 \times 10^{-127}\,\text{s}$, which is some 83 orders of magnitude shorter than the Planck time. Conservation of angular momentum ($\frac{1}{2}\hbar$) of all fundamental fermions would also have to obstruct their evaporation – including neutrinos – when represented by a miniature black hole: Indeed, the Hawking radiation may consist only of particles that are lighter than the black hole that is evaporating by means of this radiation; only photons are lighter than neutrinos, but photons have integral spin.

In principle, therefore, miniature black hole models for quarks and leptons would have to be stable, but such models would seem to be essentially unverifiable owing to the result (9.73a); see however also Refs. [420, 421, 336, 17, 464, 57, 78, 79]. In particular, it has been known since 1968 [98] that a Kerr–Newmann black hole has no electric dipole moment, but does have a magnetic dipole moment with a gyromagnetic ratio equal to 2, just like the Dirac electron without the field theory $O(\alpha)$ corrections [☞ Digression 4.1 on p. 132].

Finally, the general solutions to the Einstein equations without matter (9.39), including the Schwarzschild, the Reissner–Nordstrøm, the Kerr and the Kerr–Newman geometry, define a class of macroscopic geometries of various possible vacua; i.e., empty spacetimes. In such models, the central objects such as black holes are not to be treated as matter, but as a geometric property (defect) of spacetime itself. Even qualitatively, this recalls the "topological" solutions discussed in Section 6.3.1, including also the Dirac magnetic monopole from Section 5.2.3, other similar solutions [☞ Conclusion 6.7 on p. 248], and the "glueball" solutions in non-abelian Yang–Mills gauge theories, discussed on p. 239.

[19] String theory is known to be a quantum theory that contains gravity; the technical development of this theory suffices to confirm these estimates but not yet to compute any corrections.

9.3.3 Other interesting solutions

This section will explore some known solutions to the Einstein equations. As in the previous section, the solutions are specified by providing the line element $\mathrm{d}s^2 = g_{\mu\nu}\mathrm{d}x^\mu\mathrm{d}x^\nu$. This determines the "background" spacetime geometry [☞ Conclusion 6.8 on p. 248] in which one may analyze the motion of particles, the presence of which one supposes is a small perturbation to the energy–momentum tensor density and so also the Einstein equations, so that the spacetime geometry is not significantly changed. Such solutions are often called "universes" or "worlds," understanding that this is an extremely simplified picture where this "world" consists only of the background spacetime geometry, the matter/energy required to stabilize this geometry, and the test particles the effect of which upon the geometry may be neglected. The interested Reader is directed to the catalogues [497, 372] for starters.

Standard geometries in cosmology

The definition of the geometry that is most often used in cosmology and is understood to be the standard was provided by Alexander Friedman, Georges Henri Joseph Édouard Lemaître, Howard Percy Robertson and Arthur Geoffrey Walker, and this we will refer to as the FLRW geometry. (Depending on the historical precision and socio-political accent, Authors in the research literature not infrequently omit one or more of these names and initials.) The metric tensor of the FLRW geometry is given in terms of the "reduced-circumference polar coordinates":

$$\textbf{FLRW}\quad \mathrm{d}s^2 = -c^2\mathrm{d}t^2 + a^2(t)\mathrm{d}\Sigma^2, \qquad \begin{cases} \mathrm{d}\Sigma^2 := \Big[\dfrac{\mathrm{d}r^2}{1-Kr^2} + r^2\mathrm{d}\Omega^2\Big], \\ \mathrm{d}\Omega^2 := \mathrm{d}\theta^2 + \sin^2(\theta)\mathrm{d}\varphi^2, \end{cases} \tag{9.74}$$

where $a(t)$ is a dimensionless "scale function" of time, and K is the Gauss curvature of the space at the time when $a(t) = 1$. Alternatively, one writes $k := \frac{K}{|K|} = \pm 1$ a $k = 0$ when $K = 0$, whereby r is a dimensionless variable in the direction of the distance from the coordinate origin and $a(t)$ has the physical dimensions of length. In the case of positive curvature, space is a 3-sphere and the coordinates (r, θ, φ) cover only half of this space in a single-valued fashion, whereupon they are called the "reduced-circumference polar coordinates": in analogy with the cylindrical distance from the z-axis on the surface of a 2-sphere, the radial variable r grows from the north pole up to the equator but then decreases towards the south pole and this results in the two-valuedness of the coordinate system. Instead of (r, θ, φ), we may use the hyper-spherical coordinates:

$$\mathrm{d}\Sigma^2 = \mathrm{d}r^2 + S_K^2(r)\mathrm{d}\Omega^2, \qquad S_K(r) := \begin{cases} \frac{1}{\sqrt{K}}\sin(r\sqrt{K}) & K > 0, \\ r & K = 0, \\ \frac{1}{\sqrt{|K|}}\sinh(r\sqrt{|K|}) & K < 0, \end{cases} \tag{9.75}$$

which do not have this drawback.

This metric tensor solves the Einstein equations in the case when the matter has a homogeneous and isotropic energy–momentum tensor density, so that the Einstein equations reduce to the pair:

$$\Big(\frac{\dot{a}}{a}\Big)^2 + \frac{Kc^2}{a^2} - \frac{\Lambda c^2}{3} = \frac{8\pi G_N}{3}\varrho, \tag{9.76a}$$

$$2\frac{\ddot{a}}{a} + \Big(\frac{\dot{a}}{a}\Big)^2 + \frac{Kc^2}{a^2} - \Lambda c^2 = -\frac{8\pi G_N}{c^2}p, \tag{9.76b}$$

where ϱ and p denote the density and pressure of matter, and Λ is the cosmological constant. Since the redefinitions

$$\varrho \to \varrho - \frac{\Lambda c^2}{8\pi G_N} \quad\text{and}\quad p \to p + \frac{\Lambda c^4}{8\pi G_N} \tag{9.77}$$

effectively eliminate the cosmological constant, it follows that the presence of the cosmological constant may be simulated by *something* that (1) permeates the universe, (2) is homogeneous and isotropic, and (3) the pressure and the density of which satisfy the relation

$$p = -\varrho c^2. \tag{9.78}$$

Generally, *anything* that has a negative pressure ($p/\varrho < 0$) is called *dark energy* and its presence in the FLRW cosmology induces the universe to expand. For an accelerated expansion of the universe it would suffice were the dark energy to satisfy the relation

$$p < -\tfrac{1}{3}\varrho c^2. \tag{9.79}$$

A scalar field with this property is dubbed *quintessence*, and the ratio p/ϱ is then not necessarily a constant. Finally, one obtains an extremely accelerated expansion of the universe if

$$p < -\varrho c^2, \tag{9.80}$$

which is then referred to as phantom energy. Note that these are phenomenological definitions:

> **Definition 9.5** *Anything homogeneous and isotropic throughout the whole spacetime is called:*
>
> **dark energy** *if the pressure is negative:* $p/\varrho < 0$;
> **quintessence** *if the density and the pressure satisfy (9.79):* $p/\varrho < -c^2/3$;
> **cosmological constant** *if the density and the pressure satisfy (9.78):* $p/\varrho = -c^2$;
> **phantom energy** *if the density and the pressure satisfy (9.80):* $p/\varrho < -c^2$.

Dark energy is thus an umbrella term including its three more specific types. The demarcations are determined by the qualitative differences in the induced evolution of the universe: The cosmological constant causes the spacetime geometry to accelerate its expansion, while phantom energy causes this expansion to diverge in finite time. In turn, models of quintessence typically involve at least one dynamical field, which then varies over spacetime; moduli fields in superstring theory are natural and oft-tried candidates [☞ Footnote *34* on p. 443].

Of particular interest are the special cases of the FLRW geometry [367]:

$$\mathrm{d}s^2 = \begin{cases} -c^2\mathrm{d}t^2 + a_0^2\, e^{+2c\sqrt{\Lambda/3}\,t}\, \mathrm{d}\vec{r}^{\,2}, & \textbf{de Sitter}, \\ -c^2\mathrm{d}t^2 + \mathrm{d}\vec{r}^{\,2}, & \textbf{Minkowski}, \\ -c^2\mathrm{d}t^2 + a_0^2\, e^{-2c\sqrt{\Lambda/3}\,t}\, \mathrm{d}\vec{r}^{\,2}, & \textbf{anti de Sitter}, \end{cases} \tag{9.81}$$

where $H := 2\sqrt{\Lambda/3} > 0$ is the so-called Hubble constant,[20] and $\mathrm{d}\vec{r}^{\,2} = \mathrm{d}\vec{r}\cdot\mathrm{d}\vec{r}$ is the familiar Euclidean norm of the spatial differential $\mathrm{d}\vec{r}$. Because of using the familiar (flat) Euclidean norm for the spatial part of the differential, the coordinates in equation (9.81) are also called the "flat coordinates." There also exists the "static" parametrization

$$\mathrm{d}s^2 = -c^2\big(1 \mp \tfrac{1}{3}\Lambda\rho^2\big)\mathrm{d}\tau^2 + \big(1 \mp \tfrac{1}{3}\Lambda\rho^2\big)^{-1}\mathrm{d}\rho^2 + \rho^2\big(\mathrm{d}\theta^2 + \sin^2(\theta)\mathrm{d}\phi^2\big), \tag{9.82}$$

where ρ is a suitable "radial" coordinate; for a precise relation between equations (9.81) and (9.82), see Ref. [367]; the upper (negative) sign produces the metric for the de Sitter spacetime, and the lower (positive) sign for the anti de Sitter spacetime. Finally, there also exists the quotient parametrization

$$\mathrm{d}s^2_{\text{AdS}} = \tfrac{L^2}{z^2}\big(-c^2\mathrm{d}t^2 + \mathrm{d}x^2 + \mathrm{d}y^2 + \mathrm{d}z^2\big). \tag{9.83}$$

[20] The proposal that the universe is expanding and with a rate now called the Hubble constant was made by Georges Lemaître in 1927, two years before Edwin Hubble confirmed and more precisely determined the expansion rate; see Refs. [550, 532, 68] and the references therein.

Expression (9.82) should make it clear that the de Sitter spacetime has a spherical horizon with the radius $\rho_H = \sqrt{3/\Lambda}$. In turn, the $z^2 \to 0$ limiting case of the expression (9.83) defines the flat metric $-\mathrm{d}z^2 = -c^2\mathrm{d}t^2 + \mathrm{d}x^2 + \mathrm{d}y^2$ on a (2+1)-dimensional space with the Minkowski metric, and that forms the "conformal limit" of the anti de Sitter spacetime.

Finally, the $(n+1)$-dimensional de Sitter spacetime may be defined also as the orthogonal group coset $O(1,n+1)/O(1,n)$, and the anti de Sitter spacetime equals $O(2,n)/O(1,n)$.

— ❦ —

Note that the $g^{\mu\nu}$-trace of the Einstein equations (9.44) produces $R = -\frac{8\pi G_N}{c^4}T$. Substituting this into equation (9.44) yields

$$R_{\mu\nu} = \frac{8\pi G_N}{c^4}\Big[T_{\mu\nu} - \tfrac{1}{2}g_{\mu\nu}T\Big]. \tag{9.84}$$

This makes it clear that every solution where the energy–momentum tensor density of matter vanishes, $T_{\mu\nu} = 0$, the Ricci tensor also must vanish. And the other way around, the vanishing of the Ricci tensor, via the Einstein equations (9.44), implies that $T_{\mu\nu} = 0$ also. The geometries (choices of the metric tensor) for which the Ricci tensor vanishes (and so $T_{\mu\nu} = 0$) are called Ricci-flat geometries. This of course includes the *flat* geometry, where the metric tensor $g_{\mu\nu} = -\eta_{\mu\nu}$ is a constant, and all components of both the Christoffel symbol and the Riemann tensor vanish.

In turn, neither the vanishing of the Ricci tensor – nor even of the entire Riemann tensor – implies that the metric is flat. For example, the Kasner geometry has the metric tensor defined as [367, generalized]

$$\textbf{Kasner}\quad \mathrm{d}s^2 = -c^2\mathrm{d}t^2 + \sum_{i=1}^{3}\big(\tfrac{t}{T_i}\big)^{2p_i}(\mathrm{d}x^i)^2, \tag{9.85}$$

where T_i are arbitrary constants with units of time. If the parameters p_i are chosen to satisfy the Kasner conditions

$$\sum_{i=1}^{3} p_i = 1 = \sum_{i=1}^{3}(p_i)^2, \tag{9.86}$$

the Einstein tensor ($G_{\mu\nu} := R_{\mu\nu} - \frac{1}{2}g_{\mu\nu}R$) and even the Ricci tensor vanish. If we further set any two of three parameters p_i to be zero and the third to be 1, then the entire Riemann tensor vanishes, although the metric tensor is not equal to $-\eta_{\mu\nu}$, the constant metric tensor of flat spacetime.

One of the unusual properties of the Kasner geometry inexorably follows from Kasner's conditions (9.86) themselves: one of the three parameters must be non-positive. That is, we have

$$\text{equation (9.86)} \;\Rightarrow\; \begin{cases} p_2^{\pm} &= \frac{1}{2}\Big(1 - p_1 \pm \sqrt{1 + 2p_1 - 3p_1{}^2}\Big),\\ p_3^{\pm} &= 1 - p_1 - \frac{1}{2}\Big(1 - p_1 \pm \sqrt{1 + 2p_1 - 3p_1{}^2}\Big), \end{cases} \tag{9.87}$$

where $-\frac{1}{3} \leqslant p_i \leqslant 1$ for $i = 1,2,3$. It is easy to verify that the only non-negative solutions are the permutations of the triple $\vec{p} = (0,0,1)$. In turn, if one of the parameters is maximally negative, we have permutations of the triple $\vec{p} = (-\frac{1}{3},\frac{2}{3},\frac{2}{3})$. A few examples with rational values are the permutations of $\vec{p} = (-\frac{2}{7},\frac{3}{7},\frac{6}{7}), (-\frac{3}{13},\frac{4}{13},\frac{12}{13}), (-\frac{6}{19},\frac{10}{19},\frac{15}{19}), (-\frac{4}{21},\frac{5}{21},\frac{20}{21}), (-\frac{5}{31},\frac{6}{31},\frac{30}{31})$, etc.

Since $\sqrt{-g} = ct/(T_1{}^{p_1}T_2{}^{p_2}T_3{}^{p_3})$, the volume of Kasner geometry expands linearly in time. However, except for the class where the values (p_1,p_2,p_3) are permutations of the triple $(0,0,1)$ and where the Kasner geometry stagnates in two directions and expands in the third, the Kasner geometry expands in two spatial directions but shrinks in the third in all other cases.

Gödel's universe

One of the most unusual solutions to the Einstein equations was discovered in 1949 by Kurt Gödel; the metric tensor for the geometry of the so-called Gödel universe is specified as [372, 219]

$$\textbf{Gödel}\quad ds^2 = -c^2 dt^2 + \frac{dr^2}{1+\left(\frac{r}{r_g}\right)^2} + r^2\left[1-\left(\frac{r}{r_g}\right)^2\right] d\phi^2 + dz^2 - c\frac{2\sqrt{2}\,r^2}{r_g} dt\, d\phi, \tag{9.88}$$

where r_g is the Gödel radius. These cylindrical coordinates (t, r, ϕ, z) co-rotate with the entire universe, which results in the additional non-diagonal $dt\, d\phi$-term.

In this universe and with reference to the coordinate system (t, r, ϕ, z), a light ray that starts from the coordinate origin in the horizontal (r, ϕ)-plane follows an elliptical path that bends in the counter-clockwise direction. At the point where it reaches the distance r_g from the coordinate origin, the light ray is moving in the $+\hat{e}_\phi$ direction and begins to return to the coordinate origin, where it closes the elliptic path. Thus, observers that are at rest in the coordinate origin cannot see outside the cylinder of the horizontal radius of r_g, which then defines an optical horizon for these observers. This curious property is a consequence of the fact that the light cones (generated by light-like vectors) at every point of the (x, y)-plane tilt in the $+\hat{e}_\phi$ direction at an angle (away from the coordinate t-axis) that grows with the distance from the origin. At the distance $r_g \ln(1+\sqrt{2}) \approx 0.88\, r_g$, the light cone tilts over so much that one of the light-like vectors becomes parallel with $+\hat{e}_\phi$ and generates a circular light-like path in the (x, y)-plane: a beam of light can be emitted so as to travel on a closed circle of radius $\approx 0.88\, r_g$ – without advancing in coordinate time, t [264].

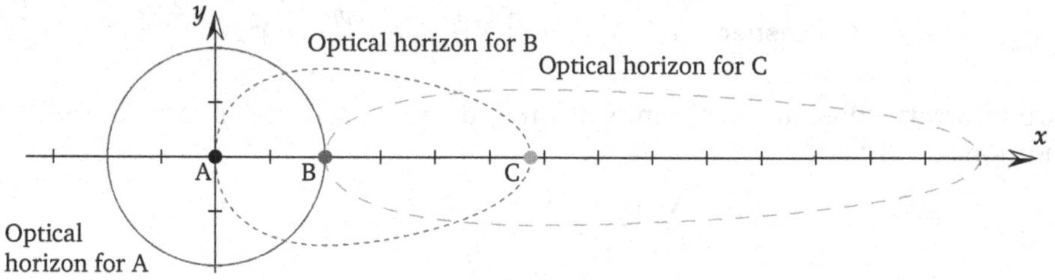

Figure 9.3 Optical horizons for observers A, B and C in the Gödel universe.

For an observer located outside the coordinate origin there exists a similar optical horizon, of an ovoid shape where the ovoid is narrower and longer in the region further away from the coordinate origin, as shown in Figure 9.3. Note that, by the definition of the optical horizon in the Gödel universe, light returns owing to the Doppler effect and the co-rotation of the entire universe and not owing to gravity. This optical horizon is thus of an essentially different nature from the event horizon in the Schwarzschild geometry.

In turn, a particle with a non-vanishing mass that is at some point at rest with respect to this coordinate system remains in that resting state, i.e., moves only in time. Thus, the coordinates (t, r, ϕ, z) are called co-rotating, and the radius r_g presents the effective optical horizon. For the Gödel universe it is convenient to define the angular speed

$$\Omega_g := \frac{\sqrt{2}\,c}{r_g} \tag{9.89}$$

with which the matter "at rest" and the entire Gödel universe rotate.

In spite of this rotation, the geometry (9.88) is homogeneous. From the form of the metric tensor, it should be clear that translations in the $\hat{e}_t$ and $\hat{e}_z$ directions as well as rotations in the $\hat{e}_\phi$

directions are isometries (symmetries of the metric tensor) and that they are respectively generated by the differential operators

$$X_0 := \tfrac{1}{\Omega_g}\,\partial_t, \qquad X_3 := r_g\,\partial_z \qquad \text{and} \qquad X_2 := \partial_\phi. \tag{9.90}$$

As for the radial direction, ∂_r is clearly not a symmetry as this would translate $r \to r + r_0$, leaving a cylindrical "hole" of radius r_0, whereas a $r \to r - r_0$ translation would map points near the z-axis into a nonexistent domain with the absurd value $r < 0$. However, it turns out that there do exist *two* differential operators,

$$X_1 := \frac{1}{\sqrt{1+(\frac{r}{r_g})^2}}\Big[\tfrac{r}{\sqrt{2}c}\cos\phi\,\partial_t + \tfrac{r_g}{2}\big[1+(\tfrac{r}{r_g})^2\big]\sin\phi\,\partial_r + \tfrac{r_g}{2r}\big[1+2(\tfrac{r}{r_g})^2\big]\cos\phi\,\partial_z\Big], \tag{9.91}$$

$$X_4 := \frac{1}{\sqrt{1+(\frac{r}{r_g})^2}}\Big[\tfrac{r}{\sqrt{2}c}\cos\phi\,\partial_t - \tfrac{r_g}{2}\big[1+(\tfrac{r}{r_g})^2\big]\cos\phi\,\partial_r + \tfrac{r_g}{2r}\big[1+2(\tfrac{r}{r_g})^2\big]\sin\phi\,\partial_z\Big], \tag{9.92}$$

that do generate isometries. Gödel, in his original work in 1949, already used four of these five isometries to show that this geometry is homogeneous, and it was shown in 1992 [167] that the complete set of *five* isometries closes the $\mathfrak{so}(3) \oplus \mathfrak{tr}(\mathbb{R}^{1,1})$ algebra:

$$L_1 := X_4, \quad L_2 := X_1, \quad L_3 := -i(X_0{+}X_2), \qquad \begin{cases} [\,L_j\,,\,L_k\,] = i\varepsilon_{jk}{}^{\ell}L_\ell, \\ [\,L_j\,,\,X_0\,] = 0 = [\,L_j\,,\,X_3\,], \end{cases} \tag{9.93}$$

where $\mathfrak{tr}(\mathbb{R}^{1,1})$ is the abelian algebra of translations in the (t,z)-plane. These symmetries can then be used to map points, paths, vectors and other tensors from one point of the Gödel universe into another, so that it suffices to work out the geometric properties with reference to the given coordinate system (9.88) and with the origin of the spatial coordinates as the reference point.

The coordinate time t and the proper time τ are identical for the observer "at rest" at the coordinate origin. Very near the z-axis, so for $r \sim 0$, the Gödel geometry is approximately flat (in cylindrical coordinates). Using the homogeneity and the action of the algebra (9.93), this then holds locally for any observer.

In the co-rotating basis the Einstein tensor ($G_{\mu\nu} := R_{\mu\nu} - \frac{1}{2}g_{\mu\nu}R$) is given as

$$[G_{\mu\nu}] = \Omega_g^2\,\mathrm{diag}(-1,1,1,1) + 2\Omega_g^2\,\mathrm{diag}(1,0,0,0). \tag{9.94}$$

The Einstein equations then dictate that this geometry is maintained by a type of matter for which the energy–momentum tensor density has the same value. The first contribution describes the so-called lambda-vacuum, i.e., the solution with the cosmological constant [☞ relations (9.77)–(9.79)]. The second contribution describes a co-rotating perfect (and all-permeating) fluid, i.e., a co-rotating dust.[21] Note that the coefficients of the two contributions must be in the precise proportion as given in equation (9.94).

> **Conclusion 9.8** *The Gödel geometry of spacetime may be understood as the result of an even permeation of the whole spacetime with dark energy (cosmological constant) and a perfect fluid, and in the precise proportion provided in the expression (9.94).*

The Gödel geometry is a relatively rare example of a geodesically complete and non-singular geometry [☞ the lexicon entry, in Appendix B.1]: The coordinate system (9.88) covers the entire Gödel spacetime, and contains no singularity; it also has an unusually symmetric structure (9.93).

[21] The cylindrical solution with co-rotating dust was discovered in 1924 by Cornelius Lanczos [325], but the solutions is better known after Willem Jacob van Stockum, who analyzed it in 1937 [534].

Traveling through time

Of course, every particle incessantly travels – through time, in the direction of time flow. However, the Lanczos–Stockum solution of co-rotating dust and Gödel's co-rotating universe were amongst the first solutions to contain so-called closed time-like curves; the Kerr solution (9.63) also has such curves. Those are curves for which the tangent vector is always time-like [☞ Definitions 3.3 on p. 90], but which are closed in spacetime. An ordinary particle with nonzero mass may travel along such a curve, and so can return into its own past!

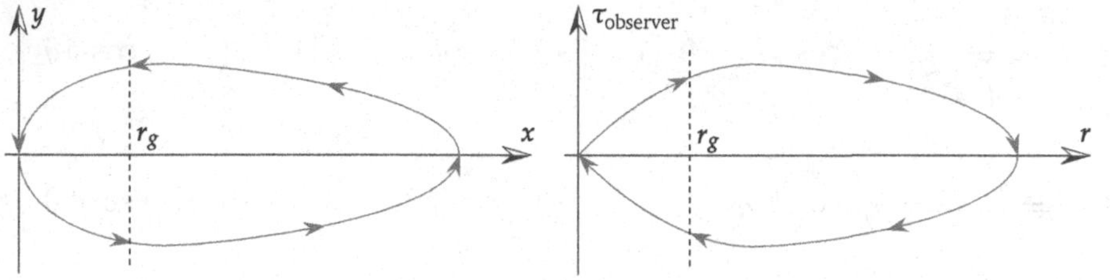

Figure 9.4 A time-like closed path in the Gödel universe.

The simplest such closed time-like curve in the Gödel universe is an ovoid path in the (x, y)-plane, e.g., with the x-axis as its symmetry axis, as shown (following the analysis of Ref. [219]) in Figure 9.4. Similar, but much more complicated closed time-like curves may then be found both in the Gödel universe, and in the Lanczos–Stockum solution with co-rotating dust, and also in many other exact solutions [☞ catalogues [497, 372] as well as the texts [367, 548]]. Note that, following the path in Figure 9.4, the particle moves backwards in time only outside the optical horizon of the observer at the coordinate origin. Also, Ref. [219] gives the necessary conditions: For a particle to move along such a closed time-like curve, it must be launched with the initial speed $v \gtrsim 0.98c$ (measured in the co-rotating coordinate system) and from a location $r \lesssim 1.7r_g$, as well as any other initial conditions that are obtained from these by isometry algebra transformations (9.93).

These concrete, exactly solved examples are particularly important to indicate the fact that many intuitively clear and acceptable characteristics of flat spacetime – including also the perhaps beguiling but precisely resolved situations in the special theory of relativity[22] – simply need not hold in the general theory of relativity. For details about closed time-like curves, the ambitious Reader should consult the books [519, 265].

Digression 9.6 Most typical scenarios of reversing the direction of traveling in time contradict energy conservation: Suppose an object X were to travel forward in time from $t < t_0$ to $t_1 > t_0$, then "turn around" to travel in time from t_1 back to $t_0 < t_1$, and then continue traveling in time forward as usual, through t_1 and beyond. Figure 9.5 depicts this process in two versions, to the left where the object X travels continuously backward in time, and to the right where it "jumps." So, in version (a), the

[22] The so-called paradoxes most often mentioned are the twin-paradox, and those of the ladder and the barn, the ruler and hole in the table, but there exist many others [512]. Not one of these puzzling situations is a real paradox and merely indicates that many of our notions acquired in everyday life are approximations that are really fit only for flat spacetime or at most locally, and so must be reconsidered and adapted beyond such local applications. For example, simultaneity becomes a relative notion, and the rigid body makes sense only at non-relativistically small speeds, since the action of the force cannot propagate through the body faster than the speed of light in vacuum, so that each body bends under the influence of non-simultaneous forces.

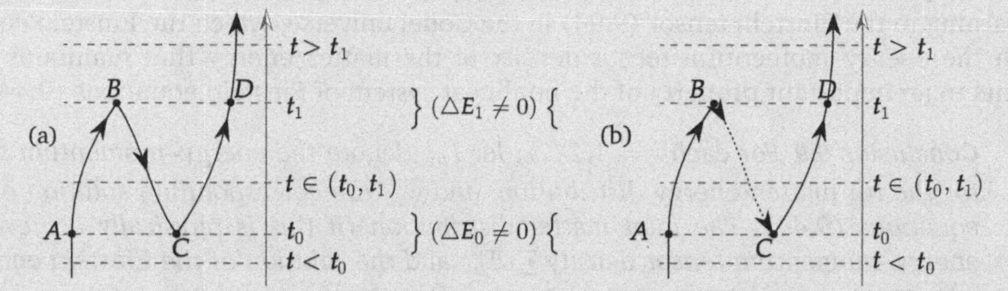

Figure 9.5 Two typical scenarios of time-travel: (a) with continuous backward travel, and (b) with "instantaneous" backward travel. Energy is measured by adding up all contributions at spacetime points simultaneous to a given observer and connected by the dashed lines. In both scenarios, energy fails to be conserved.

change $\triangle E_0 = (m_X + m_{\overline{X}})c^2$ occurs as time passes from before t_0 to after t_0, and then $\triangle E_1 = -(m_X + m_{\overline{X}})c^2$ as time passes through t_1. In version (b), $\triangle E_{0,1} = \pm m_X c^2$ at these two points in time. In diagrams with elementary particles similar to (a), another kind of particle is emitted from the point B and absorbed at the point C to balance 4-momentum conservation.

However, in the general, nontrivial geometries (and topologies) necessary to describe gravity in all generality, energy and 3-momentum are not globally well defined. These quantities are spatial 3-dimensional integrals of the $T_{\mu\nu}$ components of the energy–momentum density tensor, where the domain of integration is a 3-dimensional space-like hypersurface of simultaneous points in the 4-dimensional spacetime, as chosen by a specific class of observers. Most admissible 4-dimensional spacetime geometries admit a wide variety of such 3-dimensional space-like hypersurfaces, over which the required integrals produce widely differing results; the analysis is improved by restricting to coordinate systems satisfying the de Donder gauge condition, $\partial_\mu(\sqrt{-g}g^{\mu\nu}) = 0$ [2]. This exhibits the close relationship between energy conservation and time-travel, so the simple energy-conservation argument in Figure 9.5 need not hold. In fact, no general argument preventing time-travel can exist.

Counter-intuitively, and using the isometry algebra (9.93), it was shown [219] that the closed time-like curves in the Gödel universe nevertheless do not violate causality. In other cases, such as the closed time-like curves through the ergoregion of the Kerr geometry, paths that go through "wormholes" [☞ below] and many others [519, 265], where causality may be violated in principle, semi-classical arguments indicate that the quantum physics *probably* precludes violations of causality. However, based on such semi-classical arguments, Stephen Hawking hypothesized in 1992 that there exists a general chronology protection principle, except within the indeterminacy specified by Heisenberg's relations. Much milder is the hypothesis proposed by Igor Novikov back in 1975, whereby only self-consistent paths are permitted; this also includes traveling backwards in time if this does not cause a change in the existing history. A survey of these hypotheses and other practical, technical and conceptual questions related to closed time-like curves may be found in Refs. [542, 544]. Of course, as no complete theory of quantum gravity exists as yet, physical realizations of traveling along closed time-like curves and the physical realization of even chronology violation remain an open question☝.

9.3.4 Engineering spacetime, wormholes and topological bridges

Returning to the Einstein tensor (9.94) in the Gödel universe, which the Einstein equations equate with the energy–momentum tensor density of the matter/energy that maintains this geometry, points to an important property of the nonlinear system of Einstein equations (9.44):

> **Conclusion 9.9** *For each $i = 1, 2\ldots$, let $T^{(i)}_{\mu\nu}$ denote the energy–momentum tensor density for the ith matter/energy distribution, and $g^{(i)}_{\mu\nu}$ the corresponding solution of the Einstein equations (9.44). The joint matter distribution (if this is physically achievable) has the energy–momentum tensor density $\sum_i T^{(i)}_{\mu\nu}$ and the solution of the Einstein equations (9.44) $g^{(\Sigma)}_{\mu\nu}$. However, $g^{(\Sigma)}_{\mu\nu}$ is most often significantly different from either of the "partial" solutions $g^{(i)}_{\mu\nu}$, as well as from their sum.*

This property is intuitively acceptable: It should be the case that we can always freely combine different types of matter/energy (except that two macroscopic material objects, of course, cannot exist in the same place at the same time) and to add them to any initially given spacetime. The presence of additional matter/energy then must change the geometry of spacetime again in a way determined by the Einstein equations. However, the resulting metric tensor, in general, is not simply an analogous linear combination of metric tensors that follow from the presence of one or the other component energy–momentum tensor density. Succinctly,

> **Conclusion 9.10** *Energy–momentum density tensors of matter/energy distributions and their Einstein tensors are additive; the corresponding metric tensors are not.*

These conclusions rely on the usual interpretation of the Einstein equations as a differential system that determines the metric tensor as a function of a provided energy–momentum tensor density and initial and boundary data.

The converse approach partially follows from the logical sequence in Conclusion 9.8 on p. 345, and is sometimes referred to as the "engineering approach," wherein:

1. specify a desired geometry by specifying the corresponding metric tensor;
2. compute the Einstein tensor $G_{\mu\nu} = R_{\mu\nu} - \frac{1}{2} g_{\mu\nu} R$ for this metric tensor;
3. this specifies the required energy–momentum tensor density $T_{\mu\nu}$ of the matter/energy distribution that produces/maintains the desired geometry by its presence.
4. Finally, explore:
   (a) What (physical/engineering) characteristics should this matter/energy distribution have, so as to have the required $T_{\mu\nu}$?
   (b) Is it (at least in principle) possible to construct a structure with the matter/energy distribution and the required $T_{\mu\nu}$?

For the purpose of classifying the types of matter/energy, the characterizing "energy conditions" were introduced. To define these conditions, we need:[23]

1. a time-like 4-vector field with components $\xi^\mu(\mathrm{x})$, i.e., $g_{\mu\nu}\xi^\mu\xi^\nu < 0$, $\forall\mathrm{x}$;
2. a light-like (or null) 4-vector field with components $k^\mu(\mathrm{x})$, i.e., $g_{\mu\nu}k^\mu k^\nu = 0$, $\forall\mathrm{x}$;
3. a causal 4-vector field with components $\zeta^\mu(\mathrm{x})$, i.e., $g_{\mu\nu}\zeta^\mu\zeta^\nu \leqslant 0$, $\forall\mathrm{x}$.

Since ξ^μ may be interpreted as a 4-vector that is tangential to the worldline of a massive particle, it follows that $\varrho := T_{\mu\nu}\xi^\mu\xi^\nu$ is the total mass–energy density (of the material particle as well as all

[23] Recall that the signature of the metric tensor $g_{\mu\nu}$ in the relativistic tradition followed in this chapter is the reverse of the signature of the metric tensor of flat spacetime, $\eta_{\mu\nu}$, used in the particle physics tradition; $g_{tt} < 0$ while $\eta_{tt} > 0$.

non-gravitational fields that act upon this particle in this spacetime point). Similarly, the quantity $\varrho_0 := T_{\mu\nu}k^\mu k^\nu$ is the limiting value of the mass–energy density ϱ for a massless particle/field.

The following "energy conditions" are used to typify matter/energy:

	Condition	For all
Dominant	$g^{\mu\nu}T_{\mu\rho}T_{\nu\sigma}\zeta^\rho\zeta^\sigma \leqslant 0$ and $g^{0\mu}T_{\mu\nu}\zeta^\nu < 0$	$g_{\mu\nu}\zeta^\mu\zeta^\nu \leqslant 0,\ (\zeta^0 > 0)$
Weak	$T_{\mu\nu}\xi^\mu\xi^\nu \leqslant 0$	$g_{\mu\nu}\xi^\mu\xi^\nu < 0$
Null*	$T_{\mu\nu}k^\mu k^\nu \leqslant 0$	$g_{\mu\nu}k^\mu k^\nu = 0$
Strong	$\big[T_{\mu\nu} - \frac{1}{2}g_{\mu\nu}T\big]\xi^\mu\xi^\nu \leqslant 0$	$g_{\mu\nu}\xi^\mu\xi^\nu < 0$

(9.95)

* The null condition is also often referred to as "light-like."

The relationship between these conditions is

$$\textbf{Dominant} \Rightarrow \textbf{Weak} \Rightarrow \textbf{Null} \Leftarrow \textbf{Strong}, \tag{9.96}$$

where it is important to note that, nomenclature to the contrary, the strong condition does not imply the weak, nor vice versa. These conditions also have their "averaged" version, where the integral of the condition over some spacetime region is satisfied although the condition is violated somewhere within the given region.

The Einstein–Rosen "bridge"

The Schwarzschild metric tensor (9.55) exhibits two pathological properties at the distance $r = r_S$:

1. the time component, $g_{00} = g_{tt} = -\big(1-\frac{r_S}{r}\big)c^2$ vanishes,
2. the radial component, $g_{rr} = -\big(1-\frac{r_S}{r}\big)^{-1}$ diverges.

In turn, as discussed in Section 9.3.1 on p. 334, the divergence or vanishing of an individual component of the metric tensor does not necessarily imply a real singularity in the geometry. Moreover, Lemaître's coordinates (9.60) prove that the location $r = r_S$ is not singular. This supports the nagging doubt that the familiar spherical coordinates (t, r, θ, φ) – and so maybe even Lemaître's – do not in fact describe the complete spacetime geometry in the vicinity of the black hole.

Also, a detailed analysis of the various trajectories of massive particles and light rays that pass through the event horizon [367] points to a very bizarre property, sketched in the diagram on the left-hand side of Figure 9.6: Particles directed towards the black hole follow spacetime paths that are seemingly disconnected when passing through the event horizon and require the coordinate time to diverge to $t \to +\infty$ (whereas the proper time remains finite), and the path segment within the event horizon to move backwards in coordinate time while computation proves that the proper time continues to pass forward for massless particles and to stagnate for light.

In Figure 9.6(a), follow a light ray directed at the black hole from an initial point A, as it passes through the point C in the coordinate time $t = 0$, passes through the horizon $(r = r_S)$ in coordinate time $t = +\infty$ at the "point" D, then *returns* in coordinate time, within the horizon, and falls into the $r = 0$ singularity in the spacetime point F. Namely,

$$\text{for } r < r_S, \qquad f_S(r) < 0 \quad \text{so } g_{tt} = -f_S(r) > 0 \text{ and } g_{rr} = \big(f_S(r)\big)^{-1} < 0. \tag{9.97}$$

Thus, within the horizon, the coordinate t has a space-like character (particles may move in both directions of t) and the coordinate r has a time-like character, and particles may move only in the direction $r \to 0$, i.e., towards the singularity. Similarly to a light ray, a massive particle directed towards the black hole from the initial spacetime point B, passes through the point C in coordinate

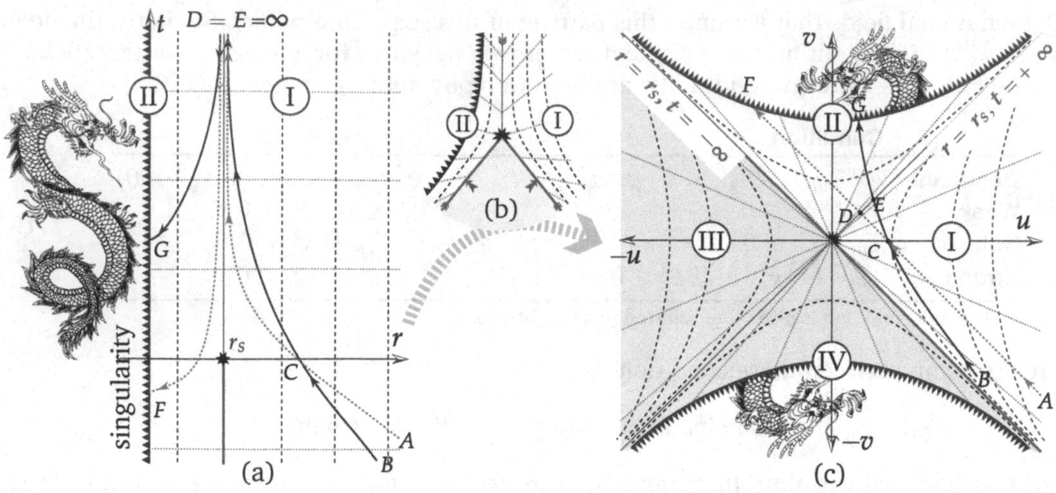

Figure 9.6 The Schwarzschild geometry (a) in the original (t, r) coordinates, (c) in Kruskal–Szekeres coordinates and (b) the transitionary stadium of the mapping from (a) to (c). A light ray directed toward the black hole follows the *A-C-D-F* path, while a particle with a non-vanishing mass directed toward the black hole follows the *B-C-E-G* path. The depiction (9.55) is spherically symmetric and angular coordinates θ, φ are not shown; every point in the figure lies on a sphere of the given radius. The diagram (b) shows how the diagram (a) "opens" in the mapping to the diagram (c).

time $t=0$, passes through the horizon ($r=r_S$) in coordinate time $t=+\infty$ (at the "point" E), and then returns *retrograde* in coordinate time within the horizon, and falls into the $r=0$ singularity (G). Throughout, the proper time of a massive particle passes forward, and remains finite.

Besides the appearance of a fictitious singularity at $r = r_S$, the discontinuity of the path – along which we know that the proper time is not discontinuous – also indicates that the Schwarzschild coordinates (t, r, θ, φ) are not appropriate. The Eddington–Lemaître coordinates (9.60) do remove the first but not also the second of these two problems. In 1950, John L. Synge discovered the incompleteness of the Schwarzschild coordinates, as well as a system of coordinates that is complete. Independently and unaware of Synge's results, Christian Fronsdal again proved the incompleteness of Schwarzschild coordinates in 1959 (at CERN), and found a complete analytical description of the Schwarzschild geometry in the form of a higher-dimensional coordinate system with an algebraic constraint.[24] His solution turned out to be very similar to the solution that Martin Kruskal (at Princeton University) found a little earlier but did not publish, and of which D. Finkelstein and J. A. Wheeler (then professors at Princeton University) knew and to whom Fronsdal, in his original work [181], gave thanks for the communication. Independently from this group of explorers, the same solution was discovered also by Szekeres György, in Australia; the independent works by Kruskal and Szekeres were published in 1960 and this finite – and explicit – version of the description is today called the Kruskal–Szekeres diagram, and u and v in Figure 9.6(c), p. 350, are the Kruskal–Szekeres coordinates [367]. In turn, Fronsdal's implicit description is today rarely mentioned.

[24] By definition, spaces of solutions of systems of algebraic equations are called *algebraic varieties* and form a major subject in the mathematical discipline of *algebraic geometry*. This connection between mathematics and physics will recur later, and much more vigorously, with the exploration of (super)strings.

The Schwarzschild and Kruskal–Szekeres coordinates are related as follows:

K–Sz		**Schwarzschild**	**K–Sz**		**Schwarzschild**	
$u_I,\ -u_{III}$	$=$	$\sqrt{\frac{r}{r_S}-1}\, e^{r/r_S} \cosh\left(\frac{ct}{2r_S}\right)$	$u_{II},\ -u_{IV}$	$=$	$\sqrt{1-\frac{r}{r_S}}\, e^{r/r_S} \sinh\left(\frac{ct}{2r_S}\right)$	(9.98a)
$v_I,\ -v_{III}$	$=$	$\sqrt{\frac{r}{r_S}-1}\, e^{r/r_S} \sinh\left(\frac{ct}{2r_S}\right)$	$v_{II},\ -v_{IV}$	$=$	$\sqrt{1-\frac{r}{r_S}}\, e^{r/r_S} \cosh\left(\frac{ct}{2r_S}\right)$	

$$\left(\tfrac{r}{r_S}-1\right) e^{r/r_S} = u^2 - v^2, \quad t = \begin{cases} \frac{2r_S}{c}\,\mathrm{arth}\left(\frac{v}{u}\right) & \text{in regions I and III;} \\ \frac{2r_S}{c}\,\mathrm{arth}\left(\frac{u}{v}\right) & \text{in regions II and IV;} \end{cases} \tag{9.98b}$$

where the subscript to Kruskal–Szekeres coordinates denotes the region in which the stated relation holds. By definition, $r \geqslant 0$, so the half-plane $(t,r)_{r<0}$ has no physical meaning. However, the half-plane $(t,r)_{r\geqslant 0}$ with the boundary $(r = 0)$ is not geodesically complete – as was shown: paths that start outside the horizon, pass through the horizon and then fall into the singularity "pass" through the point at infinity and come back from it. In turn, the domain of Kruskal–Szekeres coordinates (shown in Figure 9.6(c), p. 350, as the part of the (u,v)-plane bounded by the singularity hyperbolas) is geodesically complete: All geodesic lines are either completely contained within this region or have a limiting point at infinity and outside the singularity hyperbolas. Also, every finite part of every geodesic path is entirely contained within the domain of Kruskal–Szekeres coordinates.

Figure 9.6(c), p. 350, is the Schwarzschild geometry presented in Kruskal–Szekeres coordinates (u,v): the half-plane $(t,r)_{r\geqslant 0}$ from Figure 9.6(a) is mapped into the region bounded by the "$r = r_S$, $t = -\infty$" diagonal and the upper singularity hyperbola. Figure 9.6(b) shows the "intermediate phase" between the Schwarzschild picture and the Kruskal–Szekeres picture, where one sees that:

1. the diagonal "$r = r_S$, $t = -\infty$" appears by "splitting" the lower Schwarzschild semi-axis $r = r_S$, $t \in (-\infty, 0]$ into two semi-axes that then open into the "$r = r_S$, $t = -\infty$" diagonal;
2. the "splitting" of the lower Schwarzschild semi-axis $r = r_S$, $t \in (-\infty, 0]$ provides the space of regions III and IV;
3. the upper Schwarzschild semi-axis $r = r_S$, $t \in [0, +\infty)$ becomes the semi-axis that divides the regions I and II, and its copy divides the regions III and IV.

The comparative examination of these two coordinates of the Schwarzschild geometry clearly demonstrates that the mapping $(t,r)_{r\geqslant 0} \xrightarrow{1-2} (u,v)$ is two-valued, i.e., that the Kruskal–Szekeres picture is a double covering of the Schwarzschild picture.

This double covering implies that every spacetime region with the Schwarzschild geometry there automatically must have an exact copy, and these two regions touch along the "$r = r_S$, $t = -\infty$" diagonal in the Kruskal–Szekeres picture. By means of Figure 9.6(b), p. 350, we see that in the Schwarzschild picture this means that the two copies of spacetime touch along the event horizon, but only up to the coordinate time $t = 0$. As the coordinates may be changed by arbitrary general coordinate transformations [☞ Definition 9.1 on p. 319], the time $t = 0$ of course has no invariant meaning and the moment when the two spacetime regions separate depends on the choice of the observer; the text [367] shows the detailed history of this process from the vantage point of two different observers.

Since $\mathrm{arth}(x) = \tanh^{-1}(x) = \sum_{k=0}^{\infty} \left(\frac{x}{k}\right)^{2k+1}$, in regions I and III and for sufficiently large but fixed u_*, we have that $t(u_*) \approx \frac{2r_S}{cu_*} v$, and the Kruskal–Szekeres coordinate v approximates the Schwarzschild time t. Thus, the Schwarzschild-simultaneous points in Kruskal–Szekeres Figure 9.6(c), p. 350, all lie on predominantly horizontal and approximately straight lines when "sufficiently deep" within the regions I and III;[25] in passing through the regions II and IV, these Schwarzschild-simultaneous points are depicted by the nonlinear curves in the Kruskal–Szekeres coordinates.

Figure 9.6 (c), p. 350, then clearly indicates that this depicts a dynamical process where, from the vantage point of a fixed observer outside the event horizon, a "bridge" (or tunnel) appears that connects the spacetime regions I and III. This process was discovered by Albert Einstein and Nathan Rosen in 1935, hence its name. However, only in 1962 did John A. Wheeler and Robert W. Fuller discover that this bridge is in fact an unstable configuration and that neither material objects nor light can pass through it. Because of this impassability and topological form $S^2 \times \mathbb{R}^1$ that is a 3-dimensional generalization of the cylinder ($S^1 \times \mathbb{R}^1$), these configurations became known as *wormholes*.

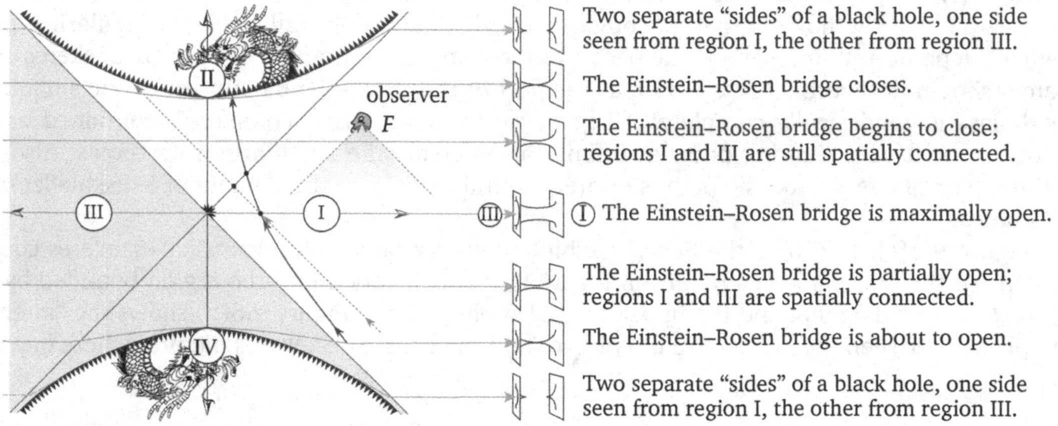

Figure 9.7 The Einstein–Rosen "bridge" as a dynamical process. The ($\approx$ time) v coordinate distance between the lower (earlier) and upper (later) singularity has no physical meaning: Particles directed towards the "bridge" end up in the upper (future) singularity: massive particles follow the path depicted by the solid line, light follows the dashed one. The physically accessible regions I and III meet only at the Kruskal–Szekeres coordinate origin, usually thought of as the circumference of the "throat" of the bridge.

Figure 9.7 shows the Schwarzschild geometry in the Kruskal–Szekeres coordinates, where the Schwarzschild-simultaneous hypersurfaces are depicted as predominantly horizontal lines, which indicate to the right the status of the Einstein–Rosen bridge by a sketch of its cross-section. The lines $\mathscr{C}$ that connect the regions I and III through the Einstein–Rosen bridge always have a spatial character, i.e., tangent 4-vectors $V \in T_x(\mathscr{C})$ along these lines ($x \in \mathscr{C}$) are space-like, $g_{\mu\nu}(x)V^\mu(x)V^\nu(x) > 0$ for every $x \in \mathscr{C}$. The diagram in Figure 9.7 shows that not even light rays – in the Kruskal–Szekeres coordinate system, light travels along straight 45° lines – can reach either from the inside of region I into the inside of region III, or the other way. The same is true of real, massive particles.

Only light rays that are entirely within the event horizon (diagonal lines that intersect in the center of the diagram in Figure 9.7) pass from the boundary of region I into the boundary of region

[25] Recall that the angular coordinates θ and φ are not depicted in the diagrams in Figure 9.6 on p. 350, so every point represents an entire sphere of indicated radius, and every line is then a 3-dimensional space of the $\mathbb{R}^1 \times S^2$ topology, where the radius of the sphere S^2 varies along the line $\mathbb{R}^1$, collapsing to a point only where this line $\mathbb{R}^1$ touches the singularity.

III and the other way around. However, these paths (of light-like character) are forever trapped in the event horizon.

Conclusion 9.11 *In spite of the existence of spatial connections (by paths to which all tangent vectors are of spatial character) between regions I and III, the Einstein–Rosen "bridge" is forever closed for real particles (which travel along paths of time-like character), including here light and all other gauge fields.*

Comment 9.5 *The Einstein–Rosen bridge, however, is* **not** *closed to virtual particles. This in principle permits an interference of wave-functions that permeate through the Einstein–Rosen bridge, and provides a form of Aharonov–Bohm effect: The spacetime for Feynman-esque integration over paths (histories) [☞ Procedure 11.1 on p. 416] is multiply connected and connects otherwise unreachable portions of the universe.*

Notice that the topology of spacetime is necessarily a dynamical concept since one of the dimensions is time-like. Abstractly, the 4-dimensional mathematical space of the physical spacetime is multiply connected, and the bridge is "always" present. However, the simultaneous points for any real physical observer, F, form a 3-dimensional subspace, $\mathscr{P}_{F,t}$, of space-like character, so that all tangent vectors $\mathrm{V} \in T_{\mathrm{x}}(\mathscr{P}_{F,t})$ to this subspace (for each $\mathrm{x} \in \mathscr{P}$) are space-like 4-vectors: $g_{\mu\nu}(\mathrm{x})V^{\mu}(\mathrm{x})V^{\nu}(\mathrm{x}) > 0$ for every $\mathrm{x} \in \mathscr{P}_{F,t}$. As the time t of the physical observer F passes, the topology of this subspace $\mathscr{P}_{F,t}$ varies, as sketched in the right-hand half of Figure 9.7 on p. 352. In the example of the Einstein–Rosen "bridge" in the Schwarzschild geometry, the two separated regions of space:

1. have a black hole each;
2. these two black holes connect in a moment;
3. the connection of these black holes opens into a space-like "bridge" (wormhole) of the $S^2 \times \mathbb{R}^1$ topology;
4. this "bridge" closes before even light can pass through it;
5. there remain two separated regions, with a black hole each.

It can, however, not be overstated that every real physical observer, F, can see only the events that can signal F from the interior of the "past" light cone $\mathscr{C}^{\wedge}_{F,t}$, the vertex ("here, now") of which is in the spacetime point $\mathrm{x}_{F,t}$. Figure 9.7 on p. 352 then makes it clear that no real physical observer can even see through an Einstein–Rosen bridge. Owing to the somewhat "instantaneous" nature of the Einstein–Rosen bridge, it vaguely recalls the instantons mentioned in Chapter 6 and the tunneling through them; see Footnote *16* on p. 248.

Stabilization of traversable wormholes

Recall that the Schwarzschild geometry solves the Einstein equations without an energy–momentum tensor density on the right-hand side. The above description of the Einstein–Rosen "bridge" shows that even the topology and geometry of otherwise empty spacetime may be highly nontrivial.

The geodesically complete picture of the Schwarzschild geometry [☞ Figures 9.6 on p. 350 and 9.7 on p. 352] indicates that the Einstein equations have solutions where the spacetime is topologically nontrivial. Namely, the regions I and III may be either regions in otherwise separate universes, or regions in the same universe, which are however arbitrarily far from one another as measured along any path that does not pass through the Einstein–Rosen "bridge." Concretely, suppose in a given moment one such "bridge" opens temporarily between a black hole near Earth and some black hole in this same spacetime, but in the Andromeda Galaxy. In this case our spacetime would become multiply connected, and the space would become momentarily multiply connected, as there exist closed paths that do pass through that "bridge" from Earth to Andromeda, and then return to Earth along a (much) longer way. In the moment when such a path (space-like, for the Einstein–Rosen "bridge" is impassable) exists, such a path cannot be continuously shrunk to a point. Alternatively, such a path is not the boundary of any surface that is entirely contained in the given spacetime.

This topological property is identical to the property of the surface of a torus, which contains closed paths that traverse the "big" or the "small" circle at least once, so they cannot be continuously deformed into a point. In contrast to such non-contractible paths, there also exist of course closed paths that are the boundaries of surfaces that are completely contained in the given space, and which then may be continuously contracted to a point. Thus "topologically" seen, the surface of a torus is equivalent to the surface of a 2-sphere to which was added a cylindrical "handle" (wormhole), as shown in Figure 9.8.

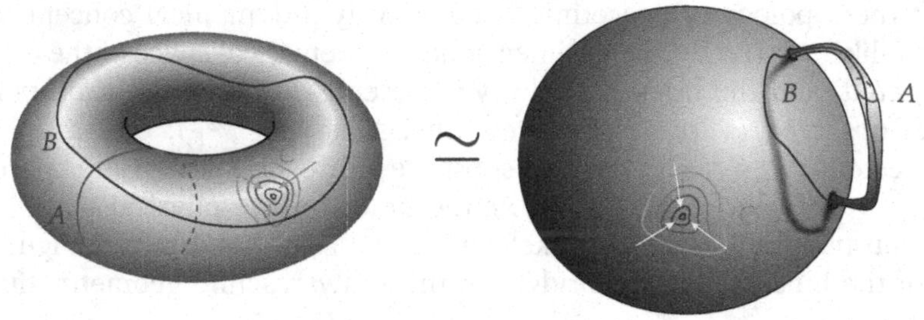

Figure 9.8 The torus surface with three topologically distinct closed paths: Neither A nor B can be continuously deformed into a point as can be done with the path C. Besides, the path A cannot be continuously deformed into the path B. The same holds for the "sphere with a handle" to the right, which is topologically equivalent to the torus.

In turn, that multiple connectedness – for real particles, fields and objects – has no practical meaning as the Einstein–Rosen "bridge" is impassable for them.

It is then reasonable to ask if there may exist some deformation of the Schwarzschild (or similar) geometry in which some such bridge between otherwise distant spacetime regions could exist and which would be traversable by real particles, fields and objects.

The metric tensor that exactly describes such a geometry evidently must have elements that are at least quadratic functions of at least some spatial coordinates, so that the spacetime solution would have two "branches," i.e., "sheets," which would then be connected by a tunnel, and so that in an adequate geodesically complete spacetime diagram (such as the Kruskal–Szekeres diagram for the Schwarzschild geometry) the otherwise separated regions of spacetime are connected through that tunnel by time-like paths. For solutions of this type the popular name "wormhole" was kept, but unlike the Einstein–Rosen space-like "bridge," these time-like wormholes are named Lorentzian wormholes [541, 543].

The simplest example is provided by the metric tensor

$$ds^2 = -c^2 dt^2 + d\ell^2 + (k^2+\ell^2)\big(d\theta^2 + \sin^2(\theta) d\varphi^2\big), \tag{9.99}$$

where $r = \pm\sqrt{k^2+\ell^2}$ is the "true" radial coordinate, and $k > 0$ is a constant. For this metric tensor one computes the Einstein tensor, in spherical coordinates:

$$[G_{\mu\nu} = R_{\mu\nu} - \tfrac{1}{2} g_{\mu\nu} R] = \frac{k^2}{(k^2+\ell^2)^2} \mathrm{diag}\big[-c^2, -1, (k^2+\ell^2), (k^2+\ell^2)\sin^2(\theta)\big]. \tag{9.100}$$

The Einstein equations then equate this tensor with the energy–momentum tensor density of the matter/energy that is necessary at the connection of the two "branches" of the solution to maintain this geometry.

This use of the Einstein equations is identical to the use of the Gauss–Ampère equations in electrodynamics. There, the spherically symmetric electric field, for example, with a magnitude that decays as $1/r^2$ implies that there must exist an electric charge at the coordinate origin that maintains this field.

As the physical meaning of the T_{tt} component of the energy–momentum tensor density is the usual matter/energy density (including all non-gravitational fields), and T_{rr} is the radial pressure of this matter density, we see that the energy–momentum tensor density that is being equated with the result (9.100) must represent a very unusual matter/energy: both its density and its radial pressure are negative. However, in the original paper in 1989, Matt Visser [541] pointed out that there do exist physical systems that have been realized in laboratories, such as for the Casimir effect, and which exhibit at least some of these exotic properties. Later research in this respect discovered several other physical systems, the combinations of which could – in principle – be used to open and stabilize such Lorentz wormholes.

The fact that the matter/energy that maintains a traversable wormhole *must* have exotic properties follows from the simple insight [519]: When light enters a traversable wormhole, the rays are being focused towards a fictitious center, following the spacetime curvature caused by the gravitational effect of the energy/matter that maintains the wormhole traversable. The incoming rays therefore behave precisely as if they are gravitationally focused by the gravitational field of a massive object. In turn, when the light comes out on the "other side" of a traversable wormhole, the rays must be emanating as if they were welling from a center, following the spacetime curvature caused by the gravitational effects of the energy/matter that maintains the wormhole traversable. Effectively, these rays are then *refracted* by the gravitational field, indicating that the matter/energy density that maintains the wormhole traversable must be less than the density of empty, flat spacetime, i.e., must be negative.

The interested Reader should consult Refs. [546, 544] for additional examples and literature.

9.3.5 Exercises for Section 9.3

✎ ***9.3.1*** *Verify that the substitutions (9.77) eliminate the cosmological constant from the equations (9.76).*

✎ ***9.3.2*** *Adapting the relation (9.94), specify the proportion of cosmological constant and co-rotating perfect fluid that can emulate* (**a**) *dark energy,* (**b**) *quintessence, and* (**c**) *phantom energy.*

✎ ***9.3.3*** *Estimate the energy conditions (9.95) for* (**a**) *dark energy,* (**b**) *quintessence,* (**c**) *cosmological constant, and* (**d**) *phantom energy.*

✎ ***9.3.4*** *Determine which of the energy conditions (9.95) are violated by the matter/energy distribution required to support the Lorentzian wormhole (9.100).*

10

Supersymmetry: boson–fermion unification

The previous chapters, and foremost Chapter 8, show that the development of fundamental physics is inherently based on the idea of unification, and in three related but distinct ways [☞ Conclusion 8.1 on p. 300]. However, one aspect remains in which the objects in fundamental physics, as discussed so far, remain separated:

1. The basic building blocks of matter – quarks and leptons – have spin $\frac{1}{2}\hbar$ and so are *fermions*: they are subject to Pauli's exclusion principle (no two fermions can coexist in the same state) and an ensemble of fermions obeys the Fermi–Dirac statistics.
2. Interaction mediators – gauge and Higgs[1] fields – have integral spin and so are *bosons*: not subject to Pauli's exclusion principle, their ensemble obeys the Bose–Einstein statistics; infinitely many bosons in the same state form a Bose *condensate*.

Digression 10.1 The following parallel practically suggests itself:

1. Subject to Pauli's exclusion principle, two fermions cannot be simultaneously in the same quantum state, i.e., "in the same place" in the Hilbert space – just as in classical physics two material objects cannot be simultaneously in the same place in the real space.
2. Not subject to Pauli's exclusion principle, two bosons *can* be simultaneously in the same quantum state, i.e., "in the same place" in the Hilbert space – just as in classical physics two interaction fields can be simultaneously in the same place in the real space.

Also, matter (substance) elementary particles are fermions, and mediating elementary particles of interaction fields are bosons [☞ Table 2.3 on p. 67]. As if, by extending classical physics into quantum, we transported the "events" of physics from spacetime into the Hilbert space.

[1] Recall Conclusion 7.4 on p. 265: Higgs bosons mediate the interaction of other particles with the *true* vacuum.

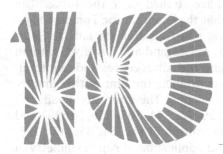

This chapter offers a brief review of the *only possible* way to bridge this last divide: the symmetry transformations that change bosons into fermions and back. The so-extended symmetries of spacetime are called supersymmetries.

The mathematical structure of supersymmetry is a kind of *superalgebras*, i.e., of *supergroups*, which are abstract algebraic structures that mathematicians have studied since the 1960s. The special property of supersymmetries among superalgebras is that they contain the Poincaré algebra (i.e., group) in flat spacetime, as well as the corresponding generalization for anti de Sitter spacetime[2] or with so-called conformal symmetry. In 1971, Yuri A. Gol'fand and Evgeny Likhtman discovered that supersymmetry [☞ Section 10.3] helps in dealing with divergences and renormalization computations in field theory. Besides the conceptual importance, the aim of this chapter is then also to show this practical aspect of supersymmetry application. The interested Reader is, besides texts and monographs in physics [189, 387, 562, 560, 129, 76, 344, 308, 556, 516, 8] and mathematics [178, 125, 535, 461], also directed to the on-line sources [144, 351, 356, 60, 19]; finally, Refs. [115, 186] give a detailed review of the effects and application of supersymmetry in quantum mechanics.

— ❦ —

Supersymmetry that will be considered here is a *global*, i.e., *rigid* symmetry: the symmetry transformation parameters [☞ definition (10.62)] are constants over all spacetime. Of course, there also exists a gauge generalization of supersymmetry, where the supersymmetry transformation parameters are free functions over spacetime, in perfect analogy with the procedure in Section 5.1. Such a gauge, i.e., *local* supersymmetry, turns out *necessarily* to include gravitation, as well as interactions that are mediated by spin-$\frac{3}{2}$ *gravitinos*, the superpartners of spin-2 gravitons. The structure of these models is a fascinating unification of gravitation and the gravitons with particles of lower spin – including gauge 4-vectors, Dirac fermions and scalars, but is also technically much more demanding than the material covered so far, so the interested Reader is directed to the abundant literature, and especially to the textbooks [189, 562, 560, 76]. Besides, it turns out that these "supergravity" models are – by themselves – neither renormalizable nor can they include all the delicate details of the Standard Model without extension within superstring theory, which will be reviewed in Chapter 11.

10.1 The linear harmonic oscillator and its extensions

Before delving into a review of concrete applications of supersymmetry in field theory and elementary particle physics, consider the appearance of supersymmetry in one of the simplest and most familiar quantum-mechanical systems, in the supersymmetric extension of the linear harmonic oscillator.

10.1.1 The harmonic oscillator

The linear harmonic oscillator is very well known and studied in full within every quantum mechanics course, so we recall only the basic relations, to set up the notation. With the standard notation

$$[A,B] := AB-BA \quad \text{and} \quad \{A,B\} := AB+BA, \tag{10.1}$$

[2] The generalization of empty spacetime when the cosmological constant is positive (as is the case with the real spacetime in which we live) is called de Sitter geometry, whereas the empty spacetime with a negative cosmological constant is called anti de Sitter geometry [☞ relations (9.81)]. Supersymmetry turns out not to be definable in spacetimes with de Sitter geometry ($\Lambda > 0$). Thus, the value of the cosmological constant is an indirect measure of supersymmetry breaking, if the fundamental description of Nature indeed is supersymmetric.

in the "excitation representation," we have

$$H_{\rm LHO} := \tfrac{1}{2}\hbar\omega\{a^\dagger, a\} = \hbar\omega(a^\dagger a + \tfrac{1}{2}), \qquad [a, a^\dagger] = 1; \tag{10.2a}$$

$$\mathscr{H}_{\rm LHO} = \Big\{ |n\rangle : \langle n|n'\rangle = \delta_{n,n'},\ \sum_n |n\rangle\langle n| = \mathbb{1},\ n, n' \in 0, 1, 2, \dots \Big\}, \tag{10.2b}$$

$$a|n\rangle = \sqrt{n}|n-1\rangle, \qquad a^\dagger|n\rangle = \sqrt{n+1}|n+1\rangle, \tag{10.2c}$$

as well as

$$H_{\rm LHO}|n\rangle = E_n|n\rangle, \qquad E_n = \hbar\omega(n + \tfrac{1}{2}). \tag{10.2d}$$

The ground state, $|0\rangle$ is characterized by the fact that

$$|0\rangle : \quad a|0\rangle = 0 \quad \text{and} \quad E_0 = \tfrac{1}{2}\hbar\omega \neq 0. \tag{10.3}$$

The Hilbert space (10.2b) is sketched in Figure 10.1(a), on p. 362. Since every observable physical quantity $\widetilde{\mathcal{F}}$ for the linear harmonic oscillator may be expressed as a function of operators $a, a^\dagger$, *[✎ why?]* the relations (10.2a) and (10.2c) suffice to compute every matrix element $\langle n'|\widetilde{\mathcal{F}}|n\rangle$:

$$\widetilde{\mathcal{F}} = \sum_{p,q=0}^{\infty} c_{p,q}(a^\dagger)^p(a)^q, \qquad \langle n'|(a^\dagger)^p(a)^q|n\rangle = \begin{cases} N_{p,q}\,\delta_{n'-p,n-q}, & q \leqslant n \text{ and } p \leqslant n', \\ 0 & \text{otherwise,} \end{cases} \tag{10.4a}$$

$$N_{p,q} = \sqrt{\underbrace{n(n-1)\cdots(n-q+1)}_{q}\,\underbrace{(n-q+1)(n-q+2)\cdots(n-q+p)}_{p}}. \tag{10.4b}$$

The linear harmonic oscillator is said to be completely solved.

10.1.2 The fermionic extension

Now extend the oscillator (10.2) with a degree of freedom represented by the operators $b, b^\dagger$, which obey

$$\{b, b^\dagger\} = 1 \quad \text{and} \quad \{b, b\} = 0 = \{b^\dagger, b^\dagger\} \quad \Rightarrow \quad b^2 = 0 = b^{\dagger 2}, \tag{10.5}$$

$$[a, b] = 0, \quad [a, b^\dagger] = 0, \quad [a^\dagger, b] = 0, \quad [a^\dagger, b^\dagger] = 0, \tag{10.6}$$

and where the Hamiltonian for the extended system is

$$H_{\rm LHO^+} = \tfrac{1}{2}\hbar\omega\{a^\dagger, a\} + \tfrac{1}{2}\hbar\tilde{\omega}[b^\dagger, b] \;=\; \hbar(\omega\, a^\dagger a + \tilde{\omega}\, b^\dagger b) + \tfrac{1}{2}\hbar(\omega - \tilde{\omega}). \tag{10.7}$$

Just as in the well-known algebraic analysis of the linear harmonic oscillator, suppose that the operator $b^\dagger b$ (as it occurs in the Hamiltonian) has eigenstates

$$b^\dagger b|\nu\rangle_f = \nu|\nu\rangle_f. \tag{10.8}$$

Then,

$$b^\dagger b(b^\dagger|\nu\rangle_f) = b^\dagger(1 - b^\dagger b)|\nu\rangle_f = \begin{cases} b^\dagger(1-\nu)|\nu\rangle_f & = (1-\nu)(b^\dagger|\nu\rangle_f), \\ b^\dagger|\nu\rangle_f - \underline{\underline{b^{\dagger 2}}}\, b|\nu\rangle_f & = (b^\dagger|\nu\rangle_f), \quad b^{\dagger 2} \equiv 0, \end{cases} \tag{10.9}$$

computed in two different ways, produces the relation $(1-\nu)b^\dagger|\nu\rangle_f = b^\dagger|\nu\rangle_f$. That is, $\nu\, b^\dagger|\nu\rangle_f = 0$, so that

$$\text{either} \quad b^\dagger|\nu\rangle_f \equiv 0, \qquad \text{or} \quad \nu = 0 \ \text{and} \ b^\dagger|0\rangle_f \propto |1\rangle_f. \tag{10.10}$$

Similarly,

$$b^\dagger b(b|\nu\rangle_f) = \begin{cases} b^\dagger \underline{\underline{b^2}}\, b|n\rangle_f & \equiv 0, \quad b^2 \equiv 0, \\ (1 - bb^\dagger)b|\nu\rangle_f & = b(1 - b^\dagger b)|\nu\rangle_f = b(1-\nu)|\nu\rangle_f = (1-\nu)(b|\nu\rangle_f), \end{cases} \tag{10.11}$$

computed in two different ways, produces the relation $(1-\nu)b|\nu\rangle_f = 0$. Thus,

$$\text{either} \quad b|\nu\rangle_f \equiv 0, \qquad \text{or} \quad \nu = 1 \text{ and } b|1\rangle_f \propto |0\rangle_f. \tag{10.12}$$

Consistently with these results, we have that

$$b|0\rangle_f \equiv 0, \qquad b^\dagger|0\rangle_f = |1\rangle_f, \qquad b|1\rangle_f = |0\rangle_f, \qquad b^\dagger|1\rangle_f \equiv 0. \tag{10.13}$$

We define for the extended system:

$$|n,\nu\rangle := |n\rangle \otimes |\nu\rangle_f, \qquad n = 0,1,2,3,\dots, \qquad \nu = 0,1, \tag{10.14a}$$

which defines the $b, b^\dagger$-extended Hilbert space:

$$\mathscr{H}_{\text{LHO}^+} := \Big\{ |n,\nu\rangle \ : \ \langle n,\nu|m,\mu\rangle = \delta_{n,m}\delta_{\nu,\mu}, \ \sum_{n,\nu} |n,\nu\rangle\langle n,\nu| = \mathbb{1} \Big\}, \tag{10.14b}$$

where $n, n' = 0,1,2,3\dots$ and $\nu, \nu' = 0,1$, and where the energy levels are given as

$$H_{\text{LHO}^+}|n,\nu\rangle = E_{n,\nu}|n,\nu\rangle, \qquad E_{n,\nu} = \hbar\big[\omega(n+\tfrac{1}{2}) + \widetilde{\omega}(\nu - \tfrac{1}{2})\big]. \tag{10.14c}$$

The energy of the ground state, $|0,0\rangle$, is

$$E_{0,0} = \tfrac{1}{2}\hbar(\omega - \widetilde{\omega}). \tag{10.15}$$

Since $n = 0,1,2,3\dots$, it follows that the $a^\dagger$-excitations of the familiar linear harmonic oscillator are not limited by Pauli's exclusion principle, and so are identified as bosonic excitations/particles. Since $\nu = 0,1$, it follows that the (single possible) $b^\dagger$-excitation does obey Pauli's exclusion principle, and so is identified as a fermionic excitation/particle with which the linear harmonic oscillator is extended.

The Hilbert space of this fermion-extended linear harmonic oscillator is sketched in Figure 10.1(b), where the white nodes represent bosonic states and the black ones are fermionic states. In that figure, $\widetilde{\omega}$ is chosen to be equal to $\frac{4}{5}\omega$, so that the difference in the energies of the ground state and the first fermionic excitation, $|0,1\rangle$, is $\frac{4}{5}$ of the energy gap between the ground state and the first bosonic excitation, $|1,0\rangle$.

Digression 10.2 By the way, there exist two distinct conventions for **Hermitian conjugation:**

1. the physicists' rule [189, 76], where $(XY)^\dagger = Y^\dagger X^\dagger$ regardless whether "X" and "Y" are commuting or anticommuting objects;
2. the mathematicians' rule [178, 124], where $(XY)^\dagger = (-1)^{\pi(X)\pi(Y)} Y^\dagger X^\dagger$ and where $\pi(X) = 0$ for commuting X and $\pi(X) = 1$ for anticommuting X.

These rules coincide except for anticommuting (fermionic) objects, $\chi\psi = -\psi\chi$:

$$\text{physicists' rule:} \quad (\psi\chi)^\dagger = +\chi^\dagger\psi^\dagger, = -\psi^\dagger\chi^\dagger, \tag{10.16a}$$

$$\text{mathematicians' rule:} \quad (\psi\chi)^\dagger = -\chi^\dagger\psi^\dagger, = +\psi^\dagger\chi^\dagger. \tag{10.16b}$$

The product of two real fermions is imaginary by the physicists' rule, but real by the mathematicians' rule. Herein, we adopt the physicists' practice and rule.

10.1.3 The supersymmetric oscillator

With the operators $a, a^\dagger, b$ and $b^\dagger$, we define the bilinear operators $(b^\dagger a)$ and $(a^\dagger b)$, for which we compute

$$[H_{\text{LHO}^+}, b^\dagger a] = \hbar(\widetilde{\omega} - \omega) b^\dagger a, \qquad [H_{\text{LHO}^+}, a^\dagger b] = \hbar(\omega - \widetilde{\omega}) a^\dagger b, \tag{10.17}$$

$$\{a^\dagger b, b^\dagger a\} = a^\dagger a + b^\dagger b. \tag{10.18}$$

This shows that the choice $\widetilde{\omega} \to \omega$ gives a special case, where the operators

$$H := \hbar\omega(a^\dagger a + b^\dagger b), \qquad Q := \sqrt{2\hbar\omega}\, a^\dagger b, \qquad Q^\dagger := \sqrt{2\hbar\omega}\, b^\dagger a, \tag{10.19}$$

define the so-called supersymmetry algebra, for which

$$\{Q^\dagger, Q\} = 2H, \qquad [H, Q] = 0 = [H, Q^\dagger] \tag{10.20}$$

are the defining relations. The last two relations show that the operators Q and $Q^\dagger$ generate symmetries of this specially tuned ($\widetilde{\omega} \to \omega$) fermion-extended oscillator. The first relation identifies the operators Q and $Q^\dagger$ as square-roots of this specially tuned fermion-extended Hamiltonian H.

Finally, we compute

$$Q^\dagger|n{+}1, 0\rangle = \sqrt{2\hbar\omega(n{+}1)}\,|n, 1\rangle, \qquad \text{and} \qquad Q|n, 1\rangle = \sqrt{2\hbar\omega(n{+}1)}\,|n{+}1, 0\rangle, \tag{10.21}$$

$$\tfrac{1}{2}\{Q^\dagger, Q\}|n, \nu\rangle = H|n, \nu\rangle = \hbar\omega(n{+}\nu)|n, \nu\rangle, \tag{10.22}$$

so that

$$E_{n,\nu} = \hbar\omega(n{+}\nu). \tag{10.23}$$

Thus, for every $n = 0, 1, 2, 3 \ldots$, the states $|n{+}1, 0\rangle$ and $|n, 1\rangle$ form a degenerate pair of states that the operators Q and $Q^\dagger$ map one into another, as is shown in Figure 10.1(c).

It is now clear that the ground state, $|0, 0\rangle$, is the only non-degenerate state and that it has a vanishing energy; the spectrum in Figure 10.1(c) fully exhausts the Hilbert space (10.14b) for this specially tuned ($\widetilde{\omega} = \omega$) extended harmonic oscillator. The action of the operators $Q, Q^\dagger$ on the Hilbert space (10.14b) is manifestly a symmetry. With respect to this symmetry, only the ground state $|0, 0\rangle$ is invariant, while for every $n = 1, 2, 3 \ldots$, $(|n{+}1, 0\rangle; |n, 1\rangle)$ is a boson–fermion pair of superpartner states, a so-called *supermultiplet*.

Definition 10.1 *A symmetry is called* ***supersymmetry*** *if (1) it maps bosonic states into fermionic ones and vice versa, and (2) it is generated by operators Q and $Q^\dagger$ the anticommutator of which contains the Hamiltonian H of the system.*

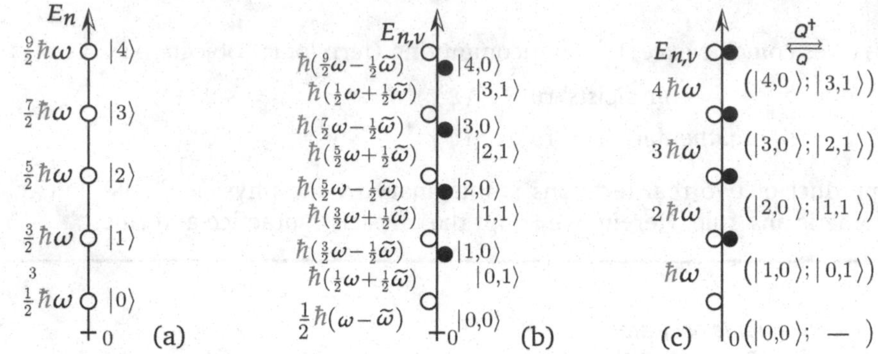

Figure 10.1 A sketch of Hilbert spaces: (a) the linear harmonic oscillator, (b) its fermionic extension with $\widetilde{\omega} \approx \frac{4}{5}\omega$, (c) its supersymmetric fermionic extension.

Digression 10.3 The dimensions (units) of the quantum-mechanical supersymmetry generator follow directly from relations (10.20), and are given as $[Q] = \frac{\sqrt{ML}}{T}$.

The system described by the creation and annihilation operators, $a^\dagger, b^\dagger$ and a, b respectively, for which the (anti)commutation relations (10.2a) and (10.5)–(10.6) hold and the Hamiltonian is specified by the first of equations (10.19), is the supersymmetric harmonic oscillator. In the general case the states are represented by wave-functions, which are functions of time and of the general form:

$$\begin{aligned} \phi(t) &:= \textstyle\sum_n \phi_n(t)|n,0\rangle, & \text{and} \qquad \psi(t) &:= \textstyle\sum_n \psi_n(t)|n-1,1\rangle, \\ &= \textstyle\sum_n \phi_n(t)\frac{(a^\dagger)^n}{\sqrt{n!}}|0,0\rangle, & &= \textstyle\sum_n \psi_n(t)\frac{(a^\dagger)^{n-1}b^\dagger}{\sqrt{(n-1)!}}|0,0\rangle, \end{aligned} \tag{10.24}$$

where $\phi(t)$ is a bosonic state and $\psi(t)$ a fermionic one. Let $\mathscr{B}$ and $\mathscr{F}$ be the vector spaces spanned by bosonic and fermionic wave-functions, respectively. Then the operators Q and $Q^\dagger$ map

$$Q \oplus Q^\dagger : \quad \mathscr{B} := \Big\{\sum_n \phi_n(t)|n,0\rangle\Big\} \leftrightarrows \mathscr{F} := \Big\{\sum_n \psi_n(t)|n-1,1\rangle\Big\}, \tag{10.25}$$

except for the ground state, $|0,0\rangle$, which both Q and $Q^\dagger$ annihilate. The ground state thus forms the *kernel* of the supersymmetry mapping (10.25) [☞ the lexicon entry for "kernel," in Appendix B.1]. Since the mapping $Q \oplus Q^\dagger$ acts both ways, the kernel could – in general – have both a bosonic and a fermionic component, so the precise statement is that

$$\{\phi_0(t)|0,0\rangle\} = \ker(Q \oplus Q^\dagger) \cap \mathscr{B}. \tag{10.26}$$

The function $\phi_0(t)|0,0\rangle$, as a special mode in the expansion (10.24), is often referred to as the "zero mode."

In the general supersymmetric case[3] it is possible that the mapping (10.25) has both bosonic and fermionic components in the kernel, i.e., it is possible that there exist n_B bosonic and n_F fermionic states that are annihilated by both Q and $Q^\dagger$. With such a generalization in mind, we have:

[3] A quantum-mechanical system with the general Hamiltonian for which there exist adequately general operators Q and $Q^\dagger$ so that the relations (10.20) hold is supersymmetric [☞ Refs. [115, 186] for a classification and examples].

Definition 10.2 (the Witten index) *For a quantum-mechanical system with a Hamiltonian H and a Hermitian-conjugate pair of operators $(Q, Q^\dagger)$ that satisfy the relations (10.20), define*

$$\iota_W := n_B - n_F \qquad \text{(the Witten index)}, \tag{10.27}$$

$$n_B = \dim\Big(\ker(Q \oplus Q^\dagger) \cap \mathscr{B}\Big), \qquad n_F = \dim\Big(\ker(Q \oplus Q^\dagger) \cap \mathscr{F}\Big), \tag{10.28}$$

where $\mathscr{B}$ and $\mathscr{F}$ are the vector spaces of bosonic and fermionic states, so that the Hilbert space of the system is $\mathscr{H} = \mathscr{B} \oplus \mathscr{F}$, and $(Q \oplus Q^\dagger) : \mathscr{B} \leftrightarrows \mathscr{F}$.

In 1981, Edward Witten showed that this *index* – by definition integral – can change only with radical changes in the Hamiltonian, such as the radical change in the potential from the harmonic $\frac{1}{2}m\phi^2$ to the anharmonic $\frac{1}{4}\lambda\phi^4$. For example, if the potential is given as

$$V(\phi) = \tfrac{1}{2}\big(m\phi + \lambda\phi^2\big)^2, \quad \text{with } |m|, |\lambda| < \infty, \tag{10.29}$$

the Witten index continues to have the constant value ($\iota_W = 2$) for arbitrary finite values of the parameter m while $\lambda \neq 0$. The value of the index changes discontinuously (into $\iota_W = 1$) in the parameter subspace where $\lambda = 0$. The Witten index is similarly constant with almost all continuous changes in parameters such as the parameters in the Lagrangian density (7.9). Using this stability, Witten proved the theorem within field theory [573]:

Theorem 10.1 (Witten) *Supersymmetry may be broken spontaneously only if $\iota_W = 0$. Conversely, supersymmetry must remain an exact symmetry while $\iota_W \neq 0$.*

This theorem then automatically also holds within quantum mechanics (adequate for this section), and within statistical physics.

That is, the Witten index ι_W is an *obstruction* for supersymmetry breaking. By definition integral, ι_W cannot change continuously with continuous changes in parameters and so can change only abruptly. This property makes the Witten index one of the first examples of *quasi-topological* invariants in physics, after Dirac's quantization of the magnetic monopole (5.98) charges. However, the relationship between the Witten index and (super)symmetry breaking is definitely the first example where such an invariant plays the role of an obstruction for a physical process such as the breaking of a symmetry and the accompanying phase transition.

Digression 10.4 It proves useful to list the parameters of a model, then designate the subspaces of this parameter space according to the values of the Witten index; this produces the first, rough sketch of the phase diagram for the system.

If the parameter space has at least two subspaces (two phases), each labeled by "its" value of the Witten index, then a change of the parameters that moves from one into the other subspace describes a phase transition. In a phase transition, the Hilbert space of the model changes radically: if we treat the potential (10.29) quantum mechanically, so $\phi = \phi(t)$, the radical change is seen from the fact that:

1. For $\lambda \neq 0$, the Hilbert space $\mathscr{H}_{\lambda\neq 0}$ consists of wave-functions that must decay asymptotically as $\exp\{-\alpha|\phi|^3\}$, for $\phi \to \pm\infty$ and a suitable $\alpha > 0$.
2. For $\lambda = 0$, the Hilbert space $\mathscr{H}_{\lambda=0}$ consists of wave-functions that must decay asymptotically as $\exp\{-\beta|\phi|^2\}$, for $\phi \to \pm\infty$ and a suitable $\beta > 0$.

Since $\exp\{-\alpha|\phi|^3\}$ decays faster than $\exp\{-\beta|\phi|^2\}$, then $\mathscr{H}_{\lambda\neq 0} \subsetneq \mathscr{H}_{\lambda=0}$, and the Hilbert space over the generic part of the parameter space (where $\lambda \neq 0$) is thus *more limited* than the Hilbert space over the special subspace where $\lambda = 0$, and where the Hilbert space is strictly larger.

10.1.4 Exercises for Section 10.1

✎ ***10.1.1*** *Compute the results (10.4).*

✎ ***10.1.2*** *Find an alternative to equation (10.13), or prove that this is the only possibility.*

✎ ***10.1.3*** *Using the definitions (10.19), compute equations (10.20).*

✎ ***10.1.4*** *Compute equation (10.23).*

✎ ***10.1.5*** *Verify (or disprove) the claims made in Digression 10.4.*

10.2 Supersymmetry in descriptions of Nature

The previous section introduced and defined supersymmetry as a symmetry of a very simple model, which may perhaps appear to be an artificial toy, an abstract example that is not applicable in the "real world." However, the early history of the discovery and application of supersymmetry is a meandering and branching story that indicates both a wide applicability, as well as the fact that many ideas in physics are conceived of in one area, but are then applied more successfully and notably in another area. Something like that was already seen in the telegraphic review of the discovery of spontaneous symmetry breaking, on p. 252.

10.2.1 Applications of supersymmetry

While supersymmetry in fundamental physics is still awaiting experimental confirmation [182], this fermion–boson symmetry has found rather successful applications elsewhere. In fact, novel applications of supersymmetry are still being discovered, so that this review is, at best, a starting point for the interested Reader.

Supersymmetry and hadrons Already in 1966–8, Hironari Miyazawa had discovered the (approximate) boson–fermion symmetry as a formal mapping between mesons (bosons) and baryons (fermions). Miyazawa's approach required the use of the $\mathfrak{su}(6|21)$ *superalgebra*, which was a very unfamiliar structure at the time, and this phenomenological approach did not gain much acceptance. Recall Pauli's denigrating stance towards group theory and its methods [☞ p. 150], which remained well-entrenched until Gell-Mann and Ne'emann used $SU(3)_f$ in hadron classification – seven or eight years after Miyazawa! Much later, it turned out [☞ e.g., Ref. [100]] that Miyzawa's approach together with the quark model (which was accepted only several years after Miyazawa's work) yields quite good results, and is useful in hadron phenomenology.

Supersymmetry and strings In 1971, Jean-Loup Gervais and Bunji Sakita [549] discovered the boson–fermion symmetry in fermionic string theories, which is actually a superconformal symmetry – a combination of supersymmetry and conformal symmetry. At the time, string theory competed with the quark model in attempting to describe hadrons and strong interactions. As the quark model soon (1973–4) proved to be superior in describing hadrons and strong interactions, this application of supersymmetry also fell by the wayside until 1984, when (super)string theory was revived as a theory of fundamental physics, and not of hadronic bound states [☞ Chapter 11].

Supersymmetry and field theory In the same year, 1971, Yuri A. Gol'fand and Evgeny Likhtman discovered that the use of supersymmetry in field theory removes a large number of divergent results and markedly simplifies (and sometimes even trivializes) the problem of renormalization [☞ Sections 5.3.3 and 6.2.4]. Similar conclusions were soon – and independently – published by Dmytro V. Volkov and V. P. Akulov, in 1972, as well as Julius Wess and Bruno Zumino in 1974. Also in 1974, Abdus Salam and John A. Strathdee introduced the notion of *superspace* as a supersymmetric extension of spacetime, and *superfield* as fields defined over superspace, and which contain both bosonic and fermionic fields as components. These ideas soon generated significant interest, and in less than ten years, Marcus Grisaru, S. James Gates, Jr., Martin Roček and Warren Siegel had already published the first textbook on supersymmetry, superspace, superfields and supergravity [189]; for more details and topically organized original references, see Ref. [76].

Supersymmetry and nuclear structure On the other, phenomenological side, supersymmetry is used also in the analysis of nuclear structure; see Ref. [364] for experimental confirmation, a recent article [185], the review [399] and references therein. Indeed, atomic nuclei of adjacent isotopes and elements, which differ only in one neutron or proton, may be treated as superpartners: Suppose a particular atomic nucleus ${}^{A}_{Z}X$ has an even atomic number (the number of protons and neutrons together) and so is a boson. Then the nuclei that have one neutron more or less, ${}^{A\pm1}_{\ Z}X'$, or one proton more or less, ${}^{A\pm1}_{Z\pm1}X''$, are fermions. The formal boson–fermion (supersymmetric) transformations

$$\begin{array}{ccccc} & & {}^{A+1}_{\ Z}X' & & \\ & & \downarrow\uparrow & & \\ {}^{A-1}_{Z-1}X'' & \leftrightarrows & {}^{A}_{Z}X & \leftrightarrows & {}^{A+1}_{Z+1}X'' \\ & & \downarrow\uparrow & & \\ & & {}^{A-1}_{\ Z}X' & & \end{array} \tag{10.30}$$

may all be used to predict the structure and the energy levels of the ${}^{A\pm1}_{\ Z}X'$ and ${}^{A\pm1}_{Z\pm1}X''$ nuclei, starting with the known properties of the ${}^{A}_{Z}X$ nucleus. This approximate supersymmetry may even be used for estimating information about nuclei that in comparison to a well-known ${}^{A}_{Z}X$ nucleus have both an additional proton and an additional neutron, ${}^{A\pm2}_{Z\pm1}X'''$ [401], which fit in the corners of the diagram (10.30), as well as the so-called *hypernuclei*, which are short-lived nuclei that captured a Λ^0 baryon [400] and which extend the diagram (10.30) in a third dimension. This application of supersymmetry is similar to Gell-Mann's application of $SU(3)$ algebra in classifying hadrons.

Supersymmetry as an approximate, phenomenological symmetry Supersymmetry may be applied in a similar, approximate and phenomenological fashion wherever bosonic states clearly differ from fermionic but have (approximately) the same energy [☞ Theorem 10.3 on p. 369 and Eq. (10.20)], and where the process by which a bosonic state may be transformed into a fermionic one and back is easy to identify. The simplest example in atomic physics would be the simple ionization of any neutral atom. Indeed,

1. a neutral atom has $A+Z$ spin-$\frac{1}{2}$ particles: Z protons, $(A-Z)$ neutrons and Z electrons;
2. simple ionization removes a single electron, leaving the atom with one fewer electrons.

If $A+Z$ is even, the original neutral atom was a boson, and the once-ionized atom is a fermion, and vice versa. In any case, the ionization process turns a bosonic state into a fermionic one or the other way around. The same holds for molecules, and the question is only whether the application of supersymmetry may help to discover anything new about these relatively well studied systems. Leaving this to the interested Reader✍, we return to the supersymmetry in field theory and as a possible fundamental symmetry.

Supersymmetry in lower-dimensional systems By far the majority of the real physical systems extend through all three dimensions of real space. However, there do exist physical systems that may be regarded, to a good approximation, as 2-dimensional (such as the monolayer systems in *solid state physics*: crystals and materials that consists of mostly a single layer of atoms, molecules or ions) or even just 1-dimensional (such as the enormously long molecules of DNA in *biophysics*).

Supersymmetry may, of course, also be discovered in such systems, as is the case with the monolayer system of graphene, where supersymmetry and the Witten index successfully describe the appearance of the unconventional quantum Hall effect; see, e.g., the articles [402, 153, 5, 408, 347] as well as the references cited therein.

Three levels of fundamental physics Even in fundamental physics, supersymmetry [☞ Definition 10.1 on p. 361] may occur in either of the three very different (albeit closely related) levels; see also Section 11.2 for a slightly different layering of the (super)string theoretical system, and so also the layered appearance and application of supersymmetry. These are:

1. The description of the physical system itself – whether in the classical Hamiltonian formalism, or in the formalism of quantum mechanics or field theory – in the real $(3+1)$-dimensional spacetime. If supersymmetric, the list of supersymmetry generators contain the Hamiltonian density for the given physical system, and also the linear momentum densities. The algebra of operators that are assigned to these physical quantities is then given by relations that contain the algebra (10.20), but are typically rather more complicated (10.63).
2. In analyzing any physical system, the dynamics and the evolution in time are important, and the so-called dimensional reduction to the worldline offers a frequently used approach to analysis. In this approach, for every physical quantity:[4]
   (a) First neglect the dependence on spatial coordinates, and treat the result as a (relativistic or non-relativistic, as needed) quantum-mechanical system.
   (b) All symmetries of the higher-dimensional theory remain to be symmetries of the dimensionally reduced quantum-mechanical "shadow," but the dynamics of the 1-dimensional system – and of the supersymmetry algebra (10.20) or (10.31) too – is simpler to analyze.
   (c) A dynamical solution to the 1-dimensional system and its symmetries (which contain the "shadows" of the Lorentz symmetries of the original higher-dimensional system) are used to reconstruct a corresponding dynamical solution to the original higher-dimensional system.
3. In the Schrödinger picture, every quantum description of any model has a Hilbert space of state functions (or state operators), upon which the Hamiltonian of the system has an induced action. If the system has a supersymmetry, it then manifests as an (induced) supersymmetry in the Hilbert space. Owing to the separate role of time in the Schrödinger picture, this supersymmetry always has the 1-dimensional algebra (10.31).

Conclusion 10.1 *Every supersymmetric model always contains an inherently 1-dimensional (induced) supersymmetry (10.31) in the Hilbert space, which is physically distinct from the dimensional reduction in the second item of the above list, even if they turn out to be mathematically isomorphic. See also Digression 10.1 on p. 357.*

[4] Although many researchers intuitively use this conceptual approach, to the best of my knowledge, the first formal description of this conceptual approach to the research program appeared in Ref. [197].

10.2.2 Additional (super)symmetry

The introductory form of supersymmetry, given in relations (10.20), and Definition 10.1 on p. 361, suggests some simple generalizations.

On one hand, it is evidently possible to find systems with several pairs of supersymmetry generators, e.g., proton and neutron ones in supersymmetric models of nuclear structure; see the diagram (10.30). Denote such replicas by $Q_i, Q^{\dagger j}$, so that the defining relations (10.20) become

$$\{ Q^{\dagger i}, Q_j \} = 2\delta^i{}_j H, \qquad [H, Q_i] = 0 = [H, Q^{\dagger j}], \qquad i,j = 1,2,\dots,N. \tag{10.31}$$

Equivalently, it is possible to introduce a real basis

$$\mathcal{Q}_j := Q_j + Q^{\dagger j} \qquad \text{and} \qquad \mathcal{Q}_{N+j} := i(Q^{\dagger j} - Q_j), \qquad j = 1,2,\dots,N, \tag{10.32a}$$

$$\{ \mathcal{Q}_I, \mathcal{Q}_J \} = 2\delta_{IJ} H, \qquad [H, \mathcal{Q}_I] = 0, \qquad I,J = 1,2,\dots,2N, \tag{10.32b}$$

and then generalize to a supersymmetric algebra (10.32b) with an *odd* number of real generators $\mathcal{Q}_I$. In this real (Hermitian) basis, $\mathcal{Q}_I^2 = H$ holds, and $\mathcal{Q}_I$ may literally be treated as square-roots of the Hamiltonian. On the other hand, the supersymmetry algebra may be defined *starting* with the relations (10.32b), including the case of an odd number of real operators $\mathcal{Q}_I$. The supersymmetry (10.31)–(10.32) is referred to as "2N-extended supersymmetry."

Superalgebras (10.31) and (10.32) may be further extended by adding bosonic operators (with various possible actions upon the considered physical system), as well as by adding commutation relations among these additional bosonic operators and the operators given by (10.31), i.e., (10.32). For example, to the relations (10.32b) we may add a matrix of operators Z_{IJ}, so that the relations (10.32b) are replaced with

$$\{ \mathcal{Q}_I, \mathcal{Q}_J \} = 2\delta_{IJ} H + Z_{IJ}, \qquad [H, \mathcal{Q}_I] = 0, \qquad I,J = 1,2,\dots,2N, \tag{10.33}$$

where

$$\delta^{IJ} Z_{IJ} = 0, \qquad [\mathcal{Q}_I, Z_{JK}] = 0 = [H, Z_{IJ}], \qquad [Z_{IJ}, Z_{KL}] = 0, \tag{10.34}$$

which represents a *central* extension of the superalgebra (10.32b). On the other hand, the last group of commutation relations, $[Z_{IJ}, Z_{KL}] = 0$, may also be replaced by

$$[Z_{IJ}, Z_{KL}] = f_{IJKL}{}^{MN} Z_{MN}, \tag{10.35}$$

so that the operators Z_{IJ} generate some nontrivial Lie algebra [☞ Appendix A]. The physical meaning of some of the operators Z_{IJ} may well be spacetime (such as translations, rotations and Lorentz boosts), in which case at least some of the commutators $[H, Z_{IJ}]$ become non-vanishing. The remaining Z_{IJ}'s may generate "internal" symmetries such as the gauge symmetries corresponding to changes in the phases of complex wave-functions, weak isospin and color in the Standard Model. In the equations (10.33)–(10.35), it was assumed that $[Z_{IJ}] = [H] = \frac{ML^2}{T^2}$, so that the coefficients $f_{IJKL}{}^{MN}$ must have these same dimensions (units) – or be scaled by an appropriate constant of such dimensions. In a concrete application, this may well need to be modified by introducing appropriate constants ($\hbar, c$, etc.) in these equations.

Extending this analysis to include *fermionic* (super)symmetry operators, and correspondingly to superalgebras where the binary operation is the *supercommutator*:

$$[X, Y\} := XY - (-1)^{|X||Y|} YX, \qquad |X| = \begin{cases} 0 & \text{if } X \text{ is a boson,} \\ 1 & \text{if } X \text{ is a fermion.} \end{cases} \tag{10.36}$$

The general theory of algebraic structure imposes only the requirement that the various (anti)commutation relations (10.20)–(10.35) be self-consistent, for which the verification of the generalization of the Jacobi identities is necessary and sufficient:

$$0 \equiv \big[B_1,[B_2,B_3]\big] + \big[B_2,[B_3,B_1]\big] + \big[B_3,[B_1,B_2]\big], \tag{10.37a}$$
$$0 \equiv \big[B_1,[B_2,F_3]\big] + \big[B_2,[F_3,B_1]\big] + \big[F_3,[B_1,B_2]\big], \tag{10.37b}$$
$$0 \equiv \big\{F_1,[F_2,B_3]\big\} + \big\{F_2,[F_1,B_3]\big\} + \big[B_3,\{F_1,F_2\}\big], \tag{10.37c}$$
$$0 \equiv \big[F_1,\{F_2,F_3\}\big] + \big[F_2,\{F_3,F_1\}\big] + \big[F_3,\{F_1,F_2\}\big], \tag{10.37d}$$

where B_1, B_2, B_3 are any three bosonic operators and F_1, F_2, F_3 are any three fermionic operators from the considered superalgebra.

Digression 10.5 A superalgebra $\mathfrak{S}$ is the generalization of the algebraic structure of *algebra*, the elements of which are either *even* (bosonic) $B_1, B_2, \ldots \in \mathfrak{S}^0$, or *odd* (fermionic) $F_1, F_2, \ldots \in \mathfrak{S}^1$. The binary "multiplication" operation is called the "supercommutator," denoted $[\ ,\ \}$, such that:

$$[B_1,B_2\} := [B_1,B_2] \in \mathfrak{S}^0, \quad [B_1,F_1\} := [B_1,F_2] \in \mathfrak{S}^1, \quad [F_1,F_2\} := \{F_1,F_2\} \in \mathfrak{S}^0. \tag{10.38a}$$

The supersymmetry algebra is then specified by the defining relations

$$[X_a, X_b\} = i f_{ab}{}^c X_c, \tag{10.39}$$

which define the Killing–Cartan metric tensor:

$$g_{ab} := f_{ac}{}^d\, f_{bd}{}^c. \tag{10.40}$$

For example, in the supersymmetry algebra (10.32b), define $X_0 = H$ and $X_I = \mathcal{Q}_I$ where $I = 1, 2, \ldots, 2N$. Then

$$f_{00}{}^0 = 0 = f_{00}{}^I, \qquad f_{IJ}{}^0 = -i\delta_{IJ}, \qquad f_{0I}{}^J = 0 = f_{IJ}{}^K, \tag{10.41}$$

so the complete Killing–Cartan metric tensor vanishes identically:

$$g_{00} = f_{00}{}^0 f_{00}{}^0 + f_{0K}{}^0 f_{00}{}^K + f_{00}{}^L f_{0L}{}^0 + f_{0K}{}^L f_{0L}{}^K = 0, \tag{10.42a}$$
$$g_{0J} = f_{00}{}^0 f_{J0}{}^0 + f_{0K}{}^0 f_{J0}{}^K + f_{00}{}^L \underline{f_{JL}{}^0} + f_{0K}{}^L f_{JL}{}^K = 0, \tag{10.42b}$$
$$g_{IJ} = f_{I0}{}^0 f_{J0}{}^0 + \underline{f_{IK}{}^0} f_{J0}{}^K + f_{I0}{}^L \underline{f_{JL}{}^0} + f_{IK}{}^L f_{JL}{}^K = 0, \tag{10.42c}$$

where the only nonzero factors are underlined. This high level of degeneracy prevents an effective application of standard (Lie-algebraic) methods of classification and study.

Also, representations of supersymmetry algebras are vector spaces of the form $\mathscr{B} \oplus \mathscr{F}$, where $\mathscr{B}$ denotes the vector space of bosonic wave-functions and $\mathscr{F}$ is the vector space of fermionic wave-functions, which the supersymmetry transformations map into each other, generalizing the relation (10.25). Note that $\mathscr{H} = \mathscr{B} \oplus \mathscr{F}$ is actually a complete Hilbert space for the considered model, and in supersymmetric theories one automatically and by definition considers the (super)symmetries of this complete Hilbert space. Automatically, we obtain results of the form

$$\langle b|F|b\rangle \equiv 0 \equiv \langle f|F|f\rangle, \quad \langle f|B|b\rangle \equiv 0 \equiv \langle f|B|f\rangle, \qquad \forall|b\rangle \in \mathscr{B}, \quad \forall|f\rangle \in \mathscr{F}, \tag{10.43}$$

where B is any bosonic operator and F any fermionic operator; they are called super-selection rules and hold in all models with supersymmetry. This result is consistent with the definition of the "fermionic number," which is 1 for all fermions (states, functions, operators, ...) and 0 for all bosons. In products, this number is added, and it is defined modulo 2. Thus, e.g.,

$$\mathrm{F}(\langle b|F|b\rangle) = \mathrm{F}(\langle b|) + \mathrm{F}(F) + \mathrm{F}(|b\rangle) = 0+1+0 = 1 \neq 0, \tag{10.44a}$$

$$\mathrm{F}(\langle f|F|f\rangle) = \mathrm{F}(\langle f|) + \mathrm{F}(F) + \mathrm{F}(|f\rangle) = 1+1+1 = 3 \simeq 1 \ (\mathrm{mod}\ 2), \ \neq 0, \tag{10.44b}$$

and so on. It turns out that this "fermionic number" may be defined consistently in spacetimes of all dimensions, and that it differentiates spinorial from tensorial representations of the Lorentz group. Also, the Witten index (10.27) may be formally defined as

$$\iota_W = \operatorname*{Tr}_{\mathscr{H}} \big[(-1)^{\mathrm{F}} \big]. \tag{10.45}$$

10.2.3 Exercises for Section 10.2

✎ **10.2.1** *By explicit computation show that the operators $Q_i, Q^{\dagger j}$ and H that satisfy the algebra (10.33) also satisfy the Jacobi identities (10.37).*

✎ **10.2.2** *By explicit computation show that the operators $\mathcal{Q}_i, Z_{IJ}$ and H that satisfy the algebra (10.31) also satisfy the Jacobi identities (10.37).*

10.3 Supersymmetric field theory

In the 1960s (before the experimental confirmation and consequent wide acceptance of the quark model!), many elementary particle physics researchers explored how much and what may all be proven and established about the behavior of leptons and hadrons – without a detailed knowledge of their dynamics, i.e., without knowing the "microscopic" theory of these interactions. Also, attempts were made to combine the symmetries of spacetime, such as the rotational (i.e., angular momentum or spin) $SU(2)$ group of symmetries, with the so-called *internal* symmetries of elementary particles, such as isospin and its $SU(3)_f$ generalization by Gell-Mann and Ne'emann. The successful non-relativistic combination $SU(2) \times SU(3)_f \subset SU(6)$ surprisingly turned up the frustration:[5] a fully relativistic generalization could not be found, rousing suspicions of a profound obstruction.

Indeed, in 1965, Lochlainn O'Raifeartaigh published a proof [396, 397] of the theorem that today bears his name, and which may be paraphrased simply as [344]:

Theorem 10.2 (O'Raifeartaigh) *The Hilbert space of the states of a particle with finite and non-vanishing mass is invariant with respect to the action of the Lie group of transformations that contains the Poincaré group (Lorentz transformations and spacetime translations).*

A sharper version of one key aspect of this theorem was provided by P. Roman and C. J. Koh the same year [463]:

Theorem 10.3 (Roman–Koh) *Distinct particles and states transformed into each other by a Lie group have the same Lorentz-invariant mass $m := \sqrt{\mathrm{p}\cdot\mathrm{p}}$.*

Only two years later, Sidney Coleman and Jeffrey Mandula (in 1967) proved the theorem [111] for all relativistic field theories:

[5] This $SU(6)$ is indeed part of Miyazawa's $\mathfrak{su}(6|21)$ superalgebra framework mentioned in Section 10.2.1.

Theorem 10.4 (Coleman–Mandula) *In any model of particles with finite and non-vanishing masses and which (directly or indirectly) interact with each other, the only permissible symmetries form the Poincaré group and some Lie group, the elements of which commute with all of the Poincaré group of symmetries.*

It then follows that no (bosonic) symmetry transformation can change the fermionic number of any state or particle upon which the operator acts, i.e., the fermionic number of the wave-function that represents this state or particle.

In 1975, Rudolf Haag, Jan Łopuszanski and Martin Sohnius noticed the "hole" in these results: It was tacitly assumed that the symmetry operators were bosonic, so that the symmetries form a Lie group, and the group generators satisfy a Lie algebra where the operation of multiplication is a commutator. The more general algebraic structures defined by bosonic *as well as fermionic* operators together are *superalgebras*, where the binary operation is a supercommutator (10.36). Within this extension of the Lie algebras, Haag, Łopuszanski and Sohnius proved the theorem [255]:

Theorem 10.5 (Haag–Łopuszanski–Sohnius) *In every model with a (1) finite number of distinct types of particles, (2) each of which has a finite and non-vanishing mass, and (3) with an asymptotically complete S-matrix,*[6] *the only permissible symmetries form a so-called supersymmetric extension of the product of the Poincaré group and some Lie group, the elements of which commute with all of the Poincaré group of symmetries [☞ Definition 10.3].*

Definition 10.3 *The Poincaré algebra, $\mathfrak{po}(1,3) = \mathfrak{spin}(1,3) \mathbin{\ltimes} \mathfrak{tr}(\mathbb{R}^{1,3})$, [☞ Section A.5.3] is generated by Lorentz transformations (A.110) and spacetime translations (A.109), i.e., the operators $L_{\mu\nu}$ and P_μ, respectively, which satisfy the relations schematically given as [☞ also the definition (10.64)]*

$$[L, L'] = L'', \qquad [L, P] = P', \qquad [P, P'] = 0. \tag{10.46}$$

The supersymmetric extension of the Poincaré algebra then has the additional spin-$\frac{1}{2}$ generators Q, which satisfy the relations schematically given as

$$\{Q, Q'\} = P \oplus Z, \qquad [L, Q] = \tfrac{1}{2} Q', \qquad [P, Q] = 0, \tag{10.47a}$$

$$[Z, Z'] = Z'', \qquad [L, Z] = 0, \qquad [Z, P] = 0, \qquad [Z, Q] = 0. \tag{10.47b}$$

The generators Q are called **supercharges**, *and Z are* **central charges**.

Comment 10.1 *Theorem 10.5 also guarantees that in all relativistic field theories only the supersymmetric generators, and exclusively with spin $\frac{1}{2}$, may change the spin (and also the fermionic number) of the particles upon which they act, and to extend the symmetries into supersymmetry. Also, it is known that the inclusion of massless particles does not change the conclusion of the theorem if those particles are Yang–Mills gauge bosons and their superpartners (**gauginos**).*

Digression 10.6 Without inserting any dimensionful constants such as $\hbar$ or c in the equations (10.46)–(10.47), the implied dimensions of these field theory supersymmetry generators Q and central charges Z are $[Q] = \sqrt{\frac{ML}{T}}$ and $[Z] = \frac{ML}{T}$, which differ from their quantum-mechanical counterparts because of $P_0 = -H/c$; see Digressions 10.3 on p. 362 and 10.7 on p. 378.

[6] The S-matrix by definition maps all possible incoming states of the system into all possible outgoing states.

10.3.1 Supersymmetry stabilizes the vacuum

Upon review – with the benefit of a century's worth of hindsight – the need to include ***quantumness*** in the description of Nature may be understood as the only ***universal property*** that stabilizes the atoms and so also all the tangible matter [☞ Digression 2.3 on p. 45]. Besides, the quantumness of physics unifies the concepts (our idealized mnemonic imagery) of particles and waves [☞ Section 8.1.1 on p. 297].

On the other hand, the need to include (general) ***relativity*** in the description of Nature was seen in Chapters 5, 6 and 9 to be part of a ***universal gauge principle*** that connects the existence of unmeasurable degrees of freedom in the description of Nature with *local* symmetries, then interactions and curvature of spacetime in which the physical particles move and fields extend. Besides, the special theory of relativity unifies space and time into spacetime, energy and momentum into 4-momentum, rotations and boosts into the Lorentz group, etc. The general theory of relativity unifies the notion of gravitation and acceleration, and provides the inherent relation between the curvature of spacetime and the presence of matter.

On the third hand, already the classical and certainly the quantum field theory indicate that the precise definition of observable quantities is not infrequently a very delicate task – the naive expressions even for the energy of empty spacetime not infrequently diverge [☞ Digression 3.13 on p. 123, and Sections 5.3.3 and 6.2.4]. Besides, in interactive field theories that include gravity even the ground state of a system is not guaranteed to have a non-negative energy, nor in fact is it guaranteed to have a globally defined energy bounded from below.

Regarding this last issue, supersymmetry helps (which is stated here with no detailed and mathematically strict justification and proof):

> ***Conclusion 10.2*** ***Supersymmetry*** *offers (as best as known) the only* ***universal mechanism*** *for stabilizing the vacuum: in* ***every*** *system without gravity [☞ Ref. [73] for energy positivity conditions without supersymmetry],*
>
> 1. *the minimum of energy is zero if and only if the system is supersymmetric;*
> 2. *the minimum of energy is positive if the system has a spontaneously broken supersymmetry.*

> ***Comment 10.2*** *If the description of the system includes the general theory of relativity (to describe gravity), the energy is not a globally well-defined quantity, and statements of non-negativity of energy do not have an invariant meaning.*

Besides, supersymmetry is the only property that may unify bosons and fermions [☞ Theorems 10.2 on p. 369, and 10.4 on p. 370]. For a compact and comprehensive summary of these properties, see Table P.1 on p. xiii, i.e., Table 11.1 on p. 409.

Technical advantages of supersymmetry

Even when spontaneously broken, supersymmetry also has two technically very advantageous consequences:

1. it significantly lessens (or even eliminates) the need for renormalization of parameters in field theory;
2. it prevents the "mixing" of characteristic energies.

That is, in any model (in field theory without supersymmetry) where in the classical version there exist two distinct characteristic energies (such as the energy of electro-weak unification, $m_W c^2 \sim 10^2$ GeV and the energy of grand unification $m_X \sim 10^{15}$ GeV), quantum effects "spoil" results such as masses of the order $10^2\text{GeV}/c^2$ via renormalization "corrections" of order $10^{15}\text{GeV}/c^2$.

Since masses are by definition invariant with respect to the action of Lorentz symmetries as well as all gauge symmetries, no fundamental symmetry principle – except supersymmetry – can "protect" them from such catastrophic quantum corrections.[7] Thus, in models without supersymmetry we may expect only one (the largest) effective characteristic energy, which in theories with gravity must be the Planck mass, $M_P \sim 10^{19}\,\text{GeV}/c^2$. All other masses then would have to be a multiple of this big mass and there is no reason for the existence of minuscule dimensionless coefficients such as [☞ result (7.132b), and Tables 4.1 on p. 152, and C.2 on p. 526]

$$\frac{m_{\nu_e}}{M_P} \lesssim 10^{-28}, \qquad \frac{m_e}{M_P} \sim 10^{-23}, \qquad \frac{m_u}{M_P} \sim 10^{-22}, \tag{10.48}$$

for them to remain stable with respect to quantum corrections, and much smaller than $O(1)$ numbers. It follows that in models without supersymmetry there is neither a fundamental reason for the masses of the elementary particles to be so many orders of magnitude smaller than the Planck mass, nor a mechanism that would "protect" such minuscule masses (were we to choose them so "by hand") from quantum corrections.

The presence of supersymmetry in any theoretical model (and so too in the Standard Model), has an important effect on the appearance (and stability with respect to quantum corrections) of experimentally established minuscule parameters such as (10.48) [189, 562, 560, 76]:

Theorem 10.6 *In any supersymmetric model, quantum effects do not change the part of the Lagrangian density that stems from the so-called* ***superpotential*** *[☞ Section 10.3.2].*

Corollary 10.1 *Although – all by itself – supersymmetry cannot* **cause** *minuscule parameters such as (10.48), supersymmetry does "protect" them if they enter via Lagrangian terms that stem from the superpotential, and in particular owing to the shift in the Higgs field in the process of spontaneous symmetry breaking. In practice, that includes all masses.*

This property of supersymmetry is exceptionally advantageous in the *technical* sense, because of the fact that most field theory models are analyzed and used in practice within perturbative computational frameworks described in Procedure 5.1. Renormalization is inherently a feature of iterative additions of ever higher contributions within a perturbative computational framework. Therefore, the appearance of divergences, the need for renormalization as well as the property of softening and limiting this need via supersymmetry is – by definition – a technical and not a conceptual property. This characterization holds even if some of the "non-perturbative" results and properties of a particular model are known [☞ Section 6.3], and they are:

1. statements about the existence of alternative vacua which cannot be computed by perturbative methods defined about the usual vacuum, but where the results are again obtained by some kind of perturbative computation about some such alternative vacuum,
2. general statements about the whole Hilbert space.

— ❦ —

In all Yang–Mills type gauge field theories [☞ Chapters 5 and 6], the divergences can be removed from precisely defined expressions for measurable physical quantities [☞ Section 3.3.4, especially the closing part and the discussion about Digression 3.11 on p. 122, to begin with]. In as much as the renormalization procedure has not satisfied the intuition and conceptual insight of some of the most influential twentieth-century physicists, the number of *live* physicists who do not accept

[7] The notable exceptions to this reasoning are the "pseudo-Goldstone modes" mentioned in Section 7.2.3.

renormalization pragmatically as a "procedure that works" is ever smaller [☞ paraphrasing Max Planck, on p. 124 and Digression 3.11 on p. 122]. However, the renormalization *procedure* is, indubitably, a technical detail of the current understanding of Nature, and not a fundamental principle of this (not even the current) understanding.

It should then be clear that the original motivation for supersymmetry stemmed from the very practical fact, of which Gol'fand and Likhtman discovered the first glimpses in 1971, that this peculiar type of symmetry automatically removes many of the divergences that occur in field theory. A detailed analysis of this general procedure is far outside the scope of this book, although some of the simplest aspects will nevertheless be made visible.

Sections 5.3.3 and 6.2.4 showed concrete (albeit the simplest) Feynman computations with diagrams where the need for renormalization appears. In the remainder of this section we will consider one (the simplest) conceptual problem in field theory, and then also the mechanism by which supersymmetry completely removes this problem.

Vacuum energy

Consider, for example, a scalar field with the Lagrangian density (7.9), where we set for simplicity $\lambda \to 0$:

$$\mathscr{L}_{\text{KG}} = \tfrac{1}{2}\eta^{\mu\nu}(\partial_\mu\phi)(\partial_\nu\phi) - \tfrac{1}{2}\big(\tfrac{mc}{\hbar}\big)^2\phi^2 \;=\; \tfrac{1}{2c^2}\dot{\phi}^2 - \tfrac{1}{2}\big[\vec{\nabla}^2 + \big(\tfrac{mc}{\hbar}\big)^2\big]\phi^2. \tag{10.49}$$

The Euler–Lagrange equation of motion derived from this Lagrangian density is

$$\big[\tfrac{1}{c^2}\partial_t^2 - \vec{\nabla}^2 + \big(\tfrac{mc}{\hbar}\big)^2\big]\phi(\mathbf{x}) = 0, \tag{10.50}$$

the so-called Klein–Gordon equation. If we expand $\phi(\mathbf{x})$ in plane waves,

$$\phi(\mathbf{x}) = \frac{1}{(2\pi)^{3/2}}\int \mathrm{d}^3\vec{k}\;\phi_{\vec{k}}(\mathbf{x}), \qquad \phi_{\vec{k}}(\mathbf{x}) := f_{\vec{k}}(t)\,e^{i\vec{k}\cdot\vec{r}}, \tag{10.51}$$

the $\vec{\nabla}^2$-term produces the eigenvalue $-\vec{k}^2$, and the equation of motion becomes

$$\big[\partial_t^2 + \big(\vec{k}^2c^2 + \tfrac{m^2c^4}{\hbar^2}\big)\big]f_{\vec{k}}(t) = 0. \tag{10.52}$$

The wave-modes $\phi_{\vec{k}}(\mathbf{x})$ are linearly independent, so every plane wave $\phi_{\vec{k}}(\mathbf{x})$ behaves as an independent degree of freedom, "counted" by the vectors $\vec{k}$, and with the dynamics of the harmonic oscillator with the frequency $c\sqrt{\vec{k}^2 + \frac{m^2c^2}{\hbar^2}}$. The presence of interactions (as would be produced by the $\frac{1}{4}\lambda\phi^4$ term in the Lagrangian density (7.9) and which we have omitted for simplicity) couples these independent oscillators but does not reduce their number nor does it destroy their linear independence. Every such *quantum* oscillator has its stationary states with energies [☞ relation (3.37)]

$$E_{n,\vec{k}} = E_{\vec{k}}(n + \tfrac{1}{2}), \qquad E_{\vec{k}} := \hbar c\sqrt{\vec{k}^2 + \tfrac{m^2c^2}{\hbar^2}} = \sqrt{(\hbar\vec{k})^2c^2 + m^2c^4}, \tag{10.53}$$

and the energy of the entire field (summed over all oscillators, of course) in the ground state is

$$E_{\text{vacuum}} = \tfrac{1}{2}\int \mathrm{d}^3\vec{k}\;E_{\vec{k}} = 2\pi\int_0^\infty k^2\mathrm{d}k\,\sqrt{\hbar^2k^2c^2 + m^2c^4}. \tag{10.54}$$

This evidently diverges $\sim k^4$ as $k \to \infty$: there are (continuously) infinitely many vectors $\vec{k}$ and all except $\vec{k} = \vec{0}$ have a positive magnitude $\vec{k}^2 > 0$.

For the free electromagnetic field, the result is virtually identical, only with the ultra-relativistic expression $E_{\vec{k}} = |\hbar\vec{k}|c$, since $m_\gamma \equiv 0$, so the result for E_{vacuum} diverges again.

However, modeling after the supersymmetric harmonic oscillator in Section 10.1.3, we may construct a supersymmetric model beginning with:

1. a pair of fields (10.51) combined into a complex scalar field $\phi(x)$;
2. the Lagrangian density of the type (7.19), but with $\lambda \to 0$;
3. adding a complex Weyl fermion $\Psi_+(x)$ of left chirality [☞ Section 5.2.1 on p. 172] and an auxiliary complex field $F(x)$;
4. adding Lagrangian (counter)terms that are specially tuned so that:
   (a) the Hamilton action for the whole system $(\phi; \Psi_+; F)$ is invariant with respect to the linear action of supersymmetry;
   (b) the Euler–Lagrange equations of motion form a system of:
      i. one differential equation of the second order for the complex field $\phi(x)$,
      ii. one pair of differential equations of the first order for the two components of the complex Weyl fermion $\Psi_+(x)$ – which also means that one linear combination of these components is the canonical coordinate while another is the canonically conjugate momentum,
      iii. one *non-differential* equation or the auxiliary complex field $F(x)$.

The non-differential equation obtained in step 4(b)iii holds point-by-point in all of spacetime *separately*, and so can be used – at least in principle – to substitute its solution back into the Lagrangian density, whereupon the differential equations in steps 4(b)i and 4(b)ii need to be re-derived from the so-substituted Lagrangian density. These *differential* equations of motion, however, express the values of the fields $\phi(x)$ and $\Psi_+(x)$ at any one point in spacetime in terms of the values of those fields at infinitesimally nearby points, and so describe dynamical (continually propagating) fields. A detailed analysis of the physical degrees of freedom then shows that all states in the Hilbert space (except for the ground states, with a vanishing energy) occur in boson–fermion pairs, generalizing the situation shown in Figure 10.1(c), on p. 362.

In so-constructed models the result (10.23) guarantees that the equivalent computation for the vacuum energy gives $E_{\text{vacuum}} = 0$. This, in fact, is a direct (and so universal) consequence of the algebra (10.31), and up to a factor c^{-1} also of the algebra (10.47) [☞ Digression 10.6 on p. 370], where

$$\sum_i \{Q^{\dagger i}, Q_i\} = 2NH, \qquad \text{since} \quad \mathrm{Tr}[\delta^i_j] = N, \tag{10.55}$$

and where it is easy to show that the left-hand side is non-negative. The algebraic details of all consistent generalizations of supersymmetry – as long as the trace of the coefficient in front of the Hamiltonian (δ^i_j) on the right-hand side of equation (10.31) is positive – guarantee the non-negativity of the Hamiltonian spectrum, so that $\langle H \rangle \geqslant 0$ is a universal result in all (rigidly) supersymmetric theories.

Supersymmetric states, supersymmetry breaking and details

The states $|\Omega\rangle$ with vanishing energy, for which $\langle\Omega|H|\Omega\rangle = 0$, must in turn satisfy

$$\begin{aligned} 0 = \langle\Omega|H|\Omega\rangle &= \Big\langle\Omega\Big|\tfrac{1}{N}\sum_i \{Q^{\dagger i}, Q_i\}\Big|\Omega\Big\rangle = \tfrac{1}{N}\sum_i \Big\{\langle\Omega|Q^{\dagger i}Q_i|\Omega\rangle + \langle\Omega|Q_i Q^{\dagger i}|\Omega\rangle\Big\} \\ &= \tfrac{1}{N}\sum_i \Big\{\big|Q_i|\Omega\rangle\big|^2 + \big|Q^{\dagger i}|\Omega\rangle\big|^2\Big\}, \end{aligned} \tag{10.56}$$

which is a sum of non-negative contributions, so each must vanish separately, whereupon

$$\text{both} \quad Q_i|\Omega\rangle = 0 \quad \text{and} \quad Q^{\dagger i}|\Omega\rangle = 0 \quad \text{for all } i. \tag{10.57}$$

From there, it follows that

$$U_{\epsilon,\bar{\epsilon}}|\Omega\rangle = |\Omega\rangle, \qquad U_{\epsilon,\bar{\epsilon}} := \exp\big\{-i(\epsilon\cdot Q + \epsilon^\dagger\cdot Q^\dagger)\big\}, \tag{10.58}$$

whereby the states $|\Omega\rangle$ are supersymmetric, i.e., unchanged under the supersymmetry transformation.

There may exist several supersymmetric states, and even continuously many. In the general case, when the bosonic (fermionic) states form a space $\mathscr{V}_B$ (i.e., $\mathscr{V}_F$), the Witten index is given by the relation

$$\iota_W := \chi_E(\mathscr{V}_B) - \chi_E(\mathscr{V}_F), \tag{10.59}$$

where $\chi_E(\mathscr{X})$ is the Euler characteristic of the space $\mathscr{X}$, which reduces to the previous definition (10.27) since the Euler characteristic of a point equals $\chi_E(\cdot) = 1$ – and which also holds for any space that contracts (continuously) to a point, such as $\mathbb{R}^n$ and $\mathbb{C}^n$.

From this analysis it follows that the Hilbert space of every supersymmetric model can only consist of:

1. supersymmetric states (of zero energy, so these are the ground states of the system),
2. supersymmetric boson–fermion pairs of states with positive energy.

In supersymmetric models, every $E > 0$ energy level must be evenly degenerate. That is, for each bosonic state, $|b_a\rangle$ with $E_a \neq 0$, we construct the fermionic state $|f_{a,I}\rangle := \mathcal{Q}_I|b_a\rangle$ and vice versa:

$$H|b_a\rangle = E_a|b_a\rangle \quad \Rightarrow \quad |b_a\rangle = \frac{H}{E_a}|b_a\rangle = \frac{\mathcal{Q}_I\mathcal{Q}_I}{E_a}|b_a\rangle = \frac{1}{E_a}(\mathcal{Q}_I|f_{a,I}\rangle), \tag{10.60}$$

which is evidently possible if and only if $E_a \neq 0$. (It is possible to prove further also that the total number of bosonic and fermionic states with a given energy $E_a \neq 0$ must be the same [560].) Thus, only the degeneracy of the ground state(s) (where $E = 0$) is not determined and only the ground state(s) may be non-degenerate, and only if the Witten index is nonzero, $\iota_W \neq 0$.

In supersymmetric models, the Hilbert space is of the form of a direct sum of so-called "sectors," of which every one consists of one ground state (with $E = 0$) and an infinite ladder of boson–fermion pairs of states (with $E > 0$), formally obtained by acting with operators of creation on the given ground state, just as is the case in Figure 10.2(a), and which generalizes the situation shown in Figure 10.1(c), p. 362.

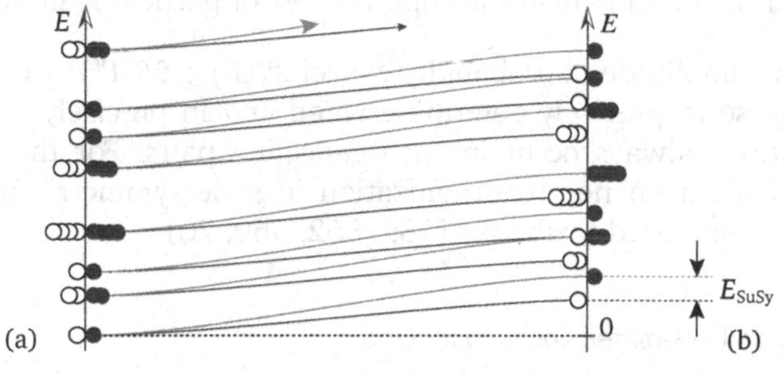

Figure 10.2 A sketch of a sector in the Hilbert space of a supersymmetric system, before (a) and after (b) spontaneous supersymmetry breaking; E_{SuSy} is the supersymmetry-breaking parameter. For supersymmetry to be broken, the Witten index must vanish, which means that the ground states must occur in boson–fermion degenerate pairs.

If there are no supersymmetric (ground) states with $E = 0$ energy, supersymmetry is broken: The states in the Hilbert space are formally obtained as a direct sum of sectors, each of which is obtained by choosing a state with lowest, albeit positive, energy and upon which one acts with creation operators. These sectors of the Hilbert space in the general case form semi-infinite ladders of bosonic and (independently) fermionic states, and those states are not guaranteed to be ordered in pairs as shown in Figure 10.1 on p. 362, and in Figure 10.2.

Spontaneous supersymmetry breaking If supersymmetry is broken because the system of Euler–Lagrange equations of motion in the list on p. 374 does not have a solution for which the potential energy minimum[8] equals zero, and Hamilton's action functional continues to be supersymmetric, supersymmetry is said to be *spontaneously broken*. In such cases, every sector of the Hilbert space looks like a semi-infinite ladder of states, as shown in Figure 10.2(b), and where the difference between the masses of adjacent bosonic and fermionic states, E_{SuSy}, is the supersymmetry-breaking parameter; in the limiting case $E_{\text{SuSy}} \to 0$, this sector returns from the shape in Figure 10.2(b) into the shape in Figure 10.2(a). In practice, this case is confirmed by analyzing the subsystem of non-differential equations in step 4(b)iii on p. 374, and the simplest model (named after the physicist who discovered this possibility, Lochlainn O'Raifeartaigh) where supersymmetry is spontaneously broken requires at least three distinct super-multiplets $(\phi^a; \Psi^a_+; F^a)$ $a = 1,2,3$ [189, 560]; see Digression 10.11 on p. 385.

Explicit supersymmetry breaking If supersymmetry of the Hamilton action is breaking because of the occurrence (or addition "by hand") of a term in the Lagrangian density, supersymmetry is said to be *explicitly broken* by that term. The effect of the explicit supersymmetry breaking on the Hilbert space of course depends on the concrete Lagrangian term that breaks supersymmetry.

— ❦ —

A detailed analysis of the mechanism whereby supersymmetry removes the need for renormalizing parameters in the Lagrangian density that stem from the superpotential [☞ Section 10.3.2] is far beyond the scope of this book. However, at least intuitively, the source of this property is seen from the fact listed in Rule 7 for Feynman calculus with the diagrams in quantum electrodynamics [☞ Procedure 5.2 on p. 193]. That is, each fermionic loop (closed path) in a given Feynman diagram requires an additional factor of (-1) as compared to an otherwise identical diagram where that same loop is bosonic. If then the Feynman diagrams Γ and Γ' differ only by:

1. the loop $\mathscr{C} \subset \Gamma$ is a closed path of particle X in the diagram Γ,
2. the loop $\mathscr{C} \subset \Gamma'$ is a closed path of the superpartner of particle X in the diagram Γ',

the contributions to the amplitude of probability cancel $\mathfrak{M}(\Gamma) + \mathfrak{M}(\Gamma') = 0$.

It remains of course to precisely determine when and in precisely which perturbative calculations do contributions always occur in such canceling pairs. For the details and a precise formulation of this theorem on non-renormalization in supersymmetric models, the interested Reader is directed to the standard textbooks [189, 562, 560, 76].

10.3.2 Supersymmetry in 1+3-dimensional spacetime

In spacetime, the Hamiltonian is the time component of the 4-momentum operator, so the relation (10.33) may be adapted by replacing the operator Z_{IJ} with the operators of linear momentum. Also, since the supersymmetry generators Q_I, according to the Haag–Łopuszanski–Sohnius theorem, transform as spin-$\frac{1}{2}$ representations of the Lorentz group, the index I must count the components of the corresponding spinor, or several copies of it.

For the purposes of this introductory text, we restrict to *simple* (unextended) supersymmetry, the generators of which form a single Dirac spinor. Generalizations are described in the literature;

[8] By potential energy we mean the value of the total energy, i.e., Hamiltonian where all derivatives of all fields are set to zero.

see e.g., Refs. [189, 562, 560, 129, 76, 308, 178, 535, 461, 144, 351, 356, 60, 19, 115, 186] for starters.

Superalgebra and supersymmetry

We are interested in the Lorentz group in 1+3-dimensional spacetime, with the very convenient fact (A.5.2) that the algebra of the Lorentz group $Spin(1,3) \cong SL(2;\mathbb{C})$ is isomorphic with the direct sum of two copies of the algebra $\mathfrak{spin}(3) = \mathfrak{su}(2)$, which is in turn very familiar both from classical as well as from quantum mechanics as the group of rotations, i.e., spin. It is then convenient to use the notation that expresses this mathematical structure [☞ Section A.5 to begin with, and the textbooks [189, 560, 76] for more precise and abundant details].

The 4-component Dirac spinor may be decomposed, in a Lorentz-invariant way, into two 2-component Weyl spinors [☞ Section 5.2.1 on p. 172, Appendix A.6 and especially A.6.2], which in Weyl's (chiral) basis of the Dirac matrices (A.132) is

$$\Psi \equiv \Psi_+ + \Psi_- = (\gamma_+\Psi) + (\gamma_-\Psi), \qquad \Psi_+ \mapsto \begin{bmatrix}\psi_\alpha\\ 0\end{bmatrix} \quad \text{and} \quad \Psi_- \mapsto \begin{bmatrix}0\\ \overline{\chi}_{\dot\alpha}\end{bmatrix}, \quad \alpha,\dot\alpha = 1,2. \tag{10.61}$$

In simple supersymmetry, the total number of generators, Q_α and $\overline{Q}_{\dot\alpha}$, is minimal and itself forms a Dirac spinor. The supersymmetry transformation operator, following the definitions (5.20) and (6.2), then is

$$U_{\epsilon,\bar\epsilon} := e^{\delta_Q(\epsilon)} = \mathbb{1} + \delta_Q(\epsilon) + \cdots, \qquad \delta_Q(\epsilon) := -i\big(\epsilon\cdot Q + \bar\epsilon\cdot\overline{Q}\big). \tag{10.62}$$

The defining relations of supersymmetry without any central extension are[9]

$$\{\, Q_\alpha\,,\, \overline{Q}_{\dot\alpha}\,\} = -2\,\sigma^\mu_{\alpha\dot\alpha}\, P_\mu, \qquad \big[\, L_{\mu\nu}\,,\, Q_\alpha\,\big] = i\hbar\,(\sigma_{\mu\nu})_\alpha{}^\beta Q_\beta, \tag{10.63a}$$

$$\big[\, L_{\mu\nu}\,,\, P_\rho\,\big] = i\hbar\,(\eta_{\mu\rho}P_\nu - \eta_{\nu\rho}P_\mu), \qquad \big[\, L_{\mu\nu}\,,\, \overline{Q}_{\dot\alpha}\,\big] = i\hbar\,(\overline{\sigma}_{\mu\nu})_{\dot\alpha}{}^{\dot\beta}\overline{Q}_{\dot\beta}, \tag{10.63b}$$

$$\big[\, L_{\mu\nu}\,,\, L_{\rho\sigma}\,\big] = i\hbar\,(\eta_{\mu\rho}L_{\nu\sigma} - \eta_{\mu\sigma}L_{\nu\rho} + \eta_{\nu\sigma}L_{\mu\rho} - \eta_{\nu\rho}L_{\mu\sigma}), \tag{10.63c}$$

with all other (anti)commutators vanishing, and where the matrices $\boldsymbol{\sigma}_{\mu\nu}$ are defined in relations (A.158) [☞ Appendix A.6.2 in more detail]. The generators P_μ and $L_{\mu\nu}$ have the well-known differential operator representation over spacetime [☞ also relations (A.111)],

$$P_\mu = \tfrac{\hbar}{i}\partial_\mu \quad \text{and} \quad L_{\mu\nu} := \tfrac{\hbar}{i}(\eta_{\mu\rho}x^\rho\partial_\nu - \eta_{\nu\rho}x^\rho\partial_\mu), \tag{10.64}$$

while Q_α and $\overline{Q}_{\dot\alpha}$ are at this point abstract operators. For them to acquire a differential operator representation, spacetime itself must be extended, and we now turn to this.

Superspace

In 1974, Abdus Salam and John Strathdee postulated superspace, as an extension of spacetime and in which spacetime is contained as a subspace. Since then, supersymmetry researchers mostly form two schools: those who fully rely on superspace and the methods of super-functional analysis, and those who regard superspace as an irrelevant crutch. However, it has been proven recently [282] that the canonical relation[10] $[H,t] = i\hbar$ and self-consistency of the supersymmetry algebra (10.31) via Jacobi identities (10.37) implies the existence of superspace. Although the very existence of

---

[9] The negative sign in the first of relations (10.63) follows from that in relations (3.35) and (3.38).

[10] In spacetime, this is $[p_\mu, x^\nu] = -i\hbar\delta^\nu_\mu$, where the negative sign in the right-hand side stems from the definition $(p_\mu) = (-E/c, \vec{p})$ [☞ the derivation of equation (3.35)], as well as the identification of $E \to i\hbar\frac{\partial}{\partial t}$ and $\vec{p} \to \frac{\hbar}{i}\vec\nabla$ in the coordinate representation. Keep in mind that time t in quantum mechanics and spacetime coordinates x^μ in quantum field theory are not eigenvalues of any Hermitian operators but parameters [☞ Ref. [29] for a detailed discussion in quantum mechanics].

superspace does not *force* us to use it, this will be convenient for the purposes of this book, since it explicitly represents and effectively uses the unification of bosons and fermions.

The standard superspace is the extension of spacetime that in addition to the four bosonic coordinates has another four fermionic, *anticommuting* coordinates:

$$\mathbf{x} \mapsto (x^\mu;\theta^\alpha,\bar{\theta}^{\dot{\alpha}}) = (ct,x^1,x^2,x^3;\theta^1,\theta^2,\bar{\theta}^{\dot{1}},\bar{\theta}^{\dot{2}}), \qquad \{\theta^\alpha,\theta^\beta\} = 0 = \{\theta^\alpha,\bar{\theta}^{\dot{\alpha}}\}. \tag{10.65}$$

The corresponding derivatives also anticommute:

$$\partial_\alpha := \frac{\partial}{\partial\theta^\alpha}, \quad \bar{\partial}_{\dot{\alpha}} := \frac{\partial}{\partial\bar{\theta}^{\dot{\alpha}}}, \qquad \{\partial_\alpha,\partial_\beta\} = \{\partial_\alpha,\bar{\partial}_{\dot{\beta}}\} = \{\bar{\partial}_{\dot{\alpha}},\bar{\partial}_{\dot{\beta}}\} = 0. \tag{10.66}$$

It is not hard to verify that the combined operators

$$Q_\alpha := i\partial_\alpha + \hbar\sigma^\mu_{\alpha\dot{\alpha}}\bar{\theta}^{\dot{\alpha}}\partial_\mu \qquad \text{and} \qquad \bar{Q}_{\dot{\alpha}} := i\bar{\partial}_{\dot{\alpha}} + \hbar\sigma^\mu_{\alpha\dot{\alpha}}\theta^\alpha\partial_\mu \tag{10.67}$$

satisfy the relations (10.63) and so, together with the definitions (10.64), give a differential representation of the abstract operators in the algebra (10.63). Newcomers in this field usually find it surprising that there exists a second pair of operators

$$D_\alpha := \partial_\alpha + i\hbar\sigma^\mu_{\alpha\dot{\alpha}}\bar{\theta}^{\dot{\alpha}}\partial_\mu \qquad \text{and} \qquad \bar{D}_{\dot{\alpha}} := \bar{\partial}_{\dot{\alpha}} + i\hbar\sigma^\mu_{\alpha\dot{\alpha}}\theta^\alpha\partial_\mu \tag{10.68}$$

that satisfy

$$\{\, D_\alpha\,,\, \bar{D}_{\dot{\alpha}}\,\} = -2\sigma^\mu_{\alpha\dot{\alpha}}P_\mu = 2i\hbar\sigma^\mu_{\alpha\dot{\alpha}}\partial_\mu, \tag{10.69}$$

as well as the other relations (10.63) upon substituting $Q \to D$ and $\bar{Q} \to \bar{D}$, and finally that

$$\left.\begin{array}{l} \{\, D_\alpha\,,\, Q_\beta\,\} = 0 = \{\, D_\alpha\,,\, \bar{Q}_{\dot{\beta}}\,\} \\ \{\, \bar{D}_{\dot{\alpha}}\,,\, Q_\beta\,\} = 0 = \{\, \bar{D}_{\dot{\alpha}}\,,\, \bar{Q}_{\dot{\beta}}\,\} \end{array}\right\} \quad \Leftrightarrow \quad \left\{\begin{array}{l} U^{-1}_{\epsilon,\bar{\epsilon}}D_\alpha U_{\epsilon,\bar{\epsilon}} = D_\alpha, \\ U^{-1}_{\epsilon,\bar{\epsilon}}\bar{D}_{\dot{\alpha}} U_{\epsilon,\bar{\epsilon}} = \bar{D}_{\dot{\alpha}}. \end{array}\right. \tag{10.70}$$

Recalling the property (10.70), D_α and $\bar{D}_{\dot{\alpha}}$ are usually called *super-covariant derivatives*, although they are in fact *invariant* with respect to supersymmetry transformations; for brevity and to avoid this imprecision, we use "super-derivative" instead. Note that

$$-iQ_\alpha = D_\alpha - 2i\hbar\sigma^\mu_{\alpha\dot{\alpha}}\bar{\theta}^{\dot{\alpha}}\partial_\mu \qquad \text{and} \qquad -i\bar{Q}_{\dot{\alpha}} = \bar{D}_\alpha - 2i\hbar\sigma^\mu_{\alpha\dot{\alpha}}\theta^\alpha\partial_\mu. \tag{10.71}$$

Digression 10.7 The definitions (10.64) and relations (10.63) and (10.69) imply that the physical dimensions (units) of the operators in the supersymmetry algebra are

$$[P_\mu] = \frac{M\,L}{T}, \quad [L_{\mu\nu}] = \frac{M\,L^2}{T}, \quad [Q_\alpha] = [\bar{Q}_{\dot{\alpha}}] = \sqrt{\frac{M\,L}{T}} = [D_\alpha] = [\bar{D}_{\dot{\alpha}}]. \tag{10.72a}$$

Also,

$$[\theta^\alpha] = [\bar{\theta}^{\dot{\alpha}}] = \sqrt{\frac{T}{M\,L}}, \quad \text{so} \quad [\hbar\theta^\alpha\sigma^\mu_{\alpha\dot{\alpha}}\bar{\theta}^{\dot{\alpha}}] = [x^\mu]. \tag{10.72b}$$

In turn, using the high energy particle physics convention where powers of $\hbar$ and c are implied and unwritten, the dimensions of these field theory operators are expressed by specifying the appropriate *power* of energy:

$$\begin{gathered} [P_\mu] = 1 = [H], \quad [L_{\mu\nu}] = 0, \quad [Q_\alpha] = [\bar{Q}_{\dot{\alpha}}] = \tfrac{1}{2} = [D_\alpha] = [\bar{D}_{\dot{\alpha}}], \\ [\theta^\alpha] = [\bar{\theta}^{\dot{\alpha}}] = -\tfrac{1}{2}, \quad [x^\mu] = -1, \end{gathered} \tag{10.72c}$$

implying, e.g., that $\sqrt{\frac{\text{MeV}}{c}}$ are units for Q_α and $\frac{\hbar c}{\text{MeV}}$ for x^μ; see Digressions 10.3 on p. 362 and 10.6 on p. 370, as well as Table C.5 on p. 528.

Superfields

Since all the operators in the algebra (10.63) are now realized as differential operators (10.64), (10.67) and (10.68) with respect to the coordinates of superspace (10.65), it is natural to introduce functions over superspace, so-called "superfields," $\mathbb{F}(x;\theta,\bar{\theta})$. The very definition of coordinates (10.65) implies that they are nilpotent:

$$\{\theta^\alpha,\theta^\beta\}=0 \quad \overset{\alpha=\beta}{\Longrightarrow} \quad 0=\{\theta^\alpha,\theta^\alpha\}=2(\theta^\alpha)^2 \;\Rightarrow\; (\theta^\alpha)^2=0,\;\alpha=1,2; \tag{10.73a}$$

$$\{\bar{\theta}^{\dot{\alpha}},\bar{\theta}^{\dot{\beta}}\}=0 \quad \overset{\dot{\alpha}=\dot{\beta}}{\Longrightarrow} \quad 0=\{\bar{\theta}^{\dot{\alpha}},\bar{\theta}^{\dot{\alpha}}\}=2(\bar{\theta}^{\dot{\alpha}})^2 \;\Rightarrow\; (\bar{\theta}^{\dot{\alpha}})^2=0,\;\dot{\alpha}=1,2. \tag{10.73b}$$

Therefore, every function of the variables $\theta^\alpha,\bar{\theta}^{\dot{\alpha}}$ has a formal Taylor expansion that terminates and gives a finite polynomial:

$$\mathbb{F}(x;\theta,\bar{\theta})=\phi(x)+\theta^\alpha\psi_\alpha(x)+\bar{\theta}^{\dot{\alpha}}\bar{\chi}_{\dot{\alpha}}(x)+\cdots+\theta^2\bar{\theta}^2\mathcal{F}(x), \tag{10.74}$$

where the coefficients in the expansion are ordinary functions over ordinary spacetime and where $\theta^2:=\frac{1}{2}\varepsilon_{\alpha\beta}\theta^\alpha\theta^\beta$ and $\bar{\theta}^2:=\frac{1}{2}\varepsilon_{\dot{\alpha}\dot{\beta}}\bar{\theta}^{\dot{\alpha}}\bar{\theta}^{\dot{\beta}}$ [☞ Appendix A.6.2 and especially Comment A.3 on p. 490, for notation]. If $\mathbb{F}(x;\theta,\bar{\theta})$ is given as a commuting, scalar superfield and since the $\theta,\bar{\theta}$ coordinates anticommute, the coefficient functions – called *component fields* – alternate between being commuting and anticommuting:

0. $\phi(x)$ is a commuting function and a scalar,
1. $\psi_\alpha(x)$ and $\bar{\chi}_{\dot{\alpha}}(x)$ are anticommuting functions and spin-$\frac{1}{2}$ spinors,

   ⋮

4. $\mathcal{F}(x)$ is a commuting function and a scalar.

Alternatively, the component fields may be defined as the coefficients in the Taylor expansion over $(\theta^\alpha,\bar{\theta}^{\dot{\alpha}})$, using the super-derivatives projected to the spacetime subspace of superspace:

$$\phi(x):=\mathbb{F}(x;\theta,\bar{\theta})\big|; \tag{10.75a}$$
$$\psi_\alpha(x):=\big[D_\alpha\mathbb{F}(x;\theta,\bar{\theta})\big]\big|; \tag{10.75b}$$
$$\bar{\chi}_{\dot{\alpha}}(x):=\big[\bar{D}_{\dot{\alpha}}\mathbb{F}(x;\theta,\bar{\theta})\big]\big|; \tag{10.75c}$$
$$F(x):=-\tfrac{1}{4}\big[D^2\mathbb{F}(x;\theta,\bar{\theta})\big]\big|; \tag{10.75d}$$
$$V_{\alpha\dot{\alpha}}(x):=-\tfrac{1}{2}\big[[D_\alpha,\bar{D}_{\dot{\alpha}}]\mathbb{F}(x;\theta,\bar{\theta})\big]\big|, \qquad V_\mu:=\tfrac{1}{2}\bar{\sigma}_\mu^{\dot{\alpha}\alpha}V_{\alpha\dot{\alpha}}; \tag{10.75e}$$
$$G(x):=-\tfrac{1}{4}\big[\bar{D}^2\mathbb{F}(x;\theta,\bar{\theta})\big]\big|; \tag{10.75f}$$
$$\lambda_\alpha(x):=-\tfrac{1}{4}\big[\bar{D}^2D_\alpha\mathbb{F}(x;\theta,\bar{\theta})\big]\big|; \tag{10.75g}$$
$$\bar{\kappa}_{\dot{\alpha}}(x):=-\tfrac{1}{4}\big[D^2\bar{D}_{\dot{\alpha}}\mathbb{F}(x;\theta,\bar{\theta})\big]\big|; \tag{10.75h}$$
$$\mathcal{F}(x):=\tfrac{1}{32}\big[(D^2\bar{D}^2+\bar{D}^2D^2)\mathbb{F}(x;\theta,\bar{\theta})\big]\big| \tag{10.75i}$$

where the vertical right-delimiter denotes the projection to the "ordinary" spacetime:

$$(X)\big| \;:=\; \lim_{\theta,\bar{\theta}\to 0}(X). \tag{10.76}$$

These definitions use super-derivatives instead of ordinary partial derivatives:

1. Since all definitions (10.75) contain a projection $\theta,\bar{\theta}\to 0$ that annihilates contributions that include $i\hbar\sigma^\mu\!\cdot\!\bar{\theta}\partial_\mu$ and $i\hbar\theta\!\cdot\!\sigma^\mu\partial_\mu$, the end result is the same as if $D_\alpha\to\partial_\alpha$ and

$\overline{D}_{\dot{\alpha}} \to \overline{\partial}_{\dot{\alpha}}$ were used – up to spacetime derivatives of component fields of "lower" physical dimensions (units), obtained upon multiple application of super-derivatives, where one of these derivatives acts on the $\theta, \overline{\theta}$ coordinates in the other.

2. The advantage of using super-derivatives in the definitions (10.75) follows from the relations (10.70): Q and $\overline{Q}$ may be freely anticommuted with the super-derivatives $D, \overline{D}$, which is not true of ordinary partial derivatives $\partial_\alpha, \overline{\partial}_{\dot{\alpha}}$.
3. The super-covariant derivatives may be used for imposing superdifferential constraints, which are then evidently covariant with respect to supersymmetry transformations, implemented by the operator (10.62) [☞ Section 10.3.3].

Note: when acting upon superfields and superdifferential expressions of superfields, whereas the super-derivatives $D_\alpha, \overline{D}_{\dot{\alpha}}$ act as usual, from the left, the supersymmetry generators $Q_\alpha, \overline{Q}_{\dot{\alpha}}$ act *from the right* [189, 76]. Thus,

$$Q_\alpha(\mathbb{F}) = \mathbb{F}\overleftarrow{Q_\alpha} = +(Q_\alpha\mathbb{F}), \qquad \text{but} \qquad Q_\alpha(D_\beta\mathbb{F}) = (D_\beta\mathbb{F})\overleftarrow{Q_\alpha} \overset{(10.70)}{=} -(Q_\alpha \circ D_\beta\mathbb{F}), \tag{10.77}$$

where in the final expressions, both $+Q_\alpha\mathbb{F}$ and $-Q_\alpha D_\beta\mathbb{F}$ act as usual, from the left. It is useful to note that the operators used in the definitions (10.75) form a hierarchy of super-derivatives:

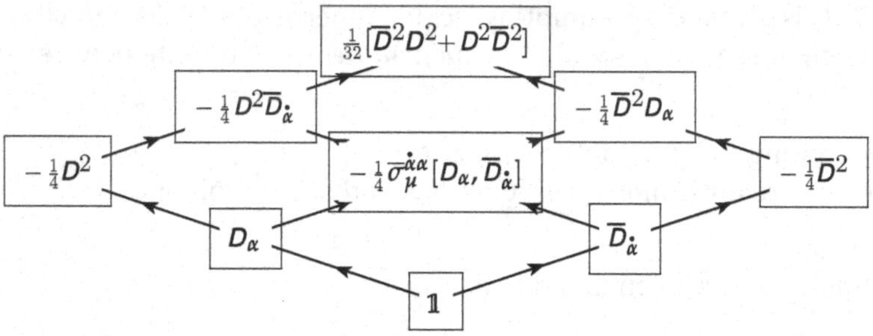

(10.78)

This structure is partially ordered by the physical dimension [☞ Digression 10.7 on p. 378], which grows upward in the diagram (10.78), and by successive application of D_α and $\overline{D}_{\dot{\alpha}}$ denoted by arrows in the diagram (10.78).

Example 10.1 The infinitesimal supersymmetry transformations of any component field may be obtained by computing the projection

$$\begin{aligned}\mathscr{D}[\delta_Q(\epsilon)\mathbb{F}]\big| &\overset{(10.70)}{=} \delta_Q(\epsilon)(\mathscr{D}\mathbb{F})\big| = \big[-i(\epsilon\cdot Q+\overline{\epsilon}\cdot\overline{Q})(\mathscr{D}\mathbb{F})\big]\big| \\ &= \big(\epsilon\cdot(D-2i\hbar\sigma^\mu\cdot\overline{\theta}\,\partial_\mu)+\overline{\epsilon}\cdot(\overline{D}-2i\hbar\theta\cdot\sigma^\mu\partial_\mu)\big)(\mathscr{D}\mathbb{F})\big| = (\epsilon\cdot D+\overline{\epsilon}\cdot\overline{D})(\mathscr{D}\mathbb{F})\big|,\end{aligned} \tag{10.79}$$

where $\mathscr{D}$ is the specific D-operator from the basis (10.78) that projects on the desired component field within the superfield $\mathbb{F}$. For example,

$$\delta_Q(\epsilon)\phi = (\epsilon^\alpha D_\alpha + \overline{\epsilon}^{\dot{\alpha}}\overline{D}_{\dot{\alpha}})\mathbb{F}\big| = \epsilon^\alpha\psi_\alpha + \overline{\epsilon}^{\dot{\alpha}}\overline{\chi}_{\dot{\alpha}}; \tag{10.80}$$

$$\begin{aligned}\delta_Q(\epsilon)\psi_\alpha &= (\epsilon\cdot D+\overline{\epsilon}\cdot\overline{D})(D_\alpha\mathbb{F})\big| = \tfrac{1}{2}\epsilon^\beta\varepsilon_{\beta\alpha}D^2\mathbb{F}\big| + \overline{\epsilon}^{\dot{\alpha}}\big(\tfrac{1}{2}\{D_\alpha,\overline{D}_{\dot{\alpha}}\} - \tfrac{1}{2}[D_\alpha,\overline{D}_{\dot{\alpha}}]\big)\mathbb{F}\big| \\ &= \tfrac{1}{2}\epsilon^\beta\varepsilon_{\beta\alpha}(-4F) + i\hbar\sigma^\mu_{\alpha\dot{\alpha}}\overline{\epsilon}^{\dot{\alpha}}(\partial_\mu\phi) - \tfrac{1}{4}\sigma^\mu_{\alpha\dot{\alpha}}\overline{\epsilon}^{\dot{\alpha}}\big(\overline{\sigma}_\mu^{\dot{\beta}\beta}[D_\beta,\overline{D}_{\dot{\beta}}]\mathbb{F}\big)\big| \\ &= 2\varepsilon_{\alpha\beta}\epsilon^\beta F + i\hbar\sigma^\mu_{\alpha\dot{\alpha}}\overline{\epsilon}^{\dot{\alpha}}(\partial_\mu\phi) + \sigma^\mu_{\alpha\dot{\alpha}}\overline{\epsilon}^{\dot{\alpha}}F_\mu; \qquad \text{etc.}\end{aligned} \tag{10.81}$$

Another key property is that every function of the form

$$[D^2\overline{D}^2 f(\mathbb{F}_1,\mathbb{F}_2,\dots)]| \tag{10.82}$$

is automatically an invariant under the supersymmetry transformation. More precisely, we have the standard result:

Theorem 10.7 *For every analytic functional expression $f(\mathbb{F}_1,\mathbb{F}_2,\dots)$ constructed from superfields $\mathbb{F}_1$, $\mathbb{F}_2$, etc., the Hamilton action of the form $\int \mathrm{d}x\,[D^2\overline{D}^2 f]|$ is supersymmetric:*

$$\delta_Q(\epsilon)\int \mathrm{d}^4x\,[D^2\overline{D}^2 f(\mathbb{F}_1,\mathbb{F}_2,\dots)]| = \int \mathrm{d}^4x\,\partial_\mu \mathcal{K}^\mu = 0, \tag{10.83}$$

where the functional expression and the component fields of the superfields $\mathbb{F}_i$ satisfy the restrictions that are usual in field theory, and which guarantee that the spacetime integrals (10.83) are well defined and convergent.

Comment 10.3 *The concrete choice of the functional expression $f(\mathbb{F}_1,\mathbb{F}_2,\dots)$ of course depends on which concrete terms one desires in the Lagrangian density:*

$$\mathscr{L} := [D^2\overline{D}^2 f(\mathbb{F}_1,\mathbb{F}_2,\dots)]|. \tag{10.84}$$

By definition, the Lagrangian density is said to be supersymmetric if it defines a supersymmetric Hamilton action, which means that $\delta_Q(\epsilon)\mathscr{L} = \partial_\mu\mathcal{K}^\mu$ suffices.

Proof The result (10.83) follows from direct computation with the two terms:

$$\begin{aligned}
\delta_Q(\epsilon)\int \mathrm{d}^4x\,[D^2\overline{D}^2 f(\mathbb{F}_1,\mathbb{F}_2,\dots)]| &\overset{(10.71)}{=} \int \mathrm{d}x\Big\{(\epsilon\cdot D_\alpha + \bar\epsilon\cdot\overline{D} + \dots)D^2\overline{D}^2 f(\mathbb{F})\Big\}\Big| \\
&\overset{\theta,\bar\theta\to 0}{=} \int \mathrm{d}^4x\Big\{\epsilon^\alpha \underbrace{D_\alpha D^2\,\overline{D}^2}_{\equiv 0} f(\mathbb{F}) + \bar\epsilon^{\dot\alpha}\overline{D}_{\dot\alpha}D^2\overline{D}^2 f(\mathbb{F})\Big\}\Big| \\
&\overset{(A.164)}{=} \int \mathrm{d}^4x\Big\{\bar\epsilon^{\dot\alpha}\big[-4i\hbar\sigma^\mu_{\alpha\dot\alpha}\partial_\mu\varepsilon^{\alpha\beta}D_\beta + D^2\overline{D}_{\dot\alpha}\big]\overline{D}^2 f(\mathbb{F})\Big\}\Big| \\
&= \int \mathrm{d}^4x\Big\{\partial_\mu\underbrace{\big[-4i\hbar\bar\epsilon^{\dot\alpha}\sigma^\mu_{\alpha\dot\alpha}\varepsilon^{\alpha\beta}D_\beta\overline{D}^2 f(\mathbb{F})\big]|}_{:=\mathcal{K}^\mu} + \bar\epsilon^{\dot\alpha}D^2\underbrace{\overline{D}_{\dot\alpha}\overline{D}^2}_{\equiv 0} f(\mathbb{F})|\Big\} \\
&= \int_{\text{spacetime}} \mathrm{d}^4x\;\partial_\mu\mathcal{K}^\mu = \oint_{\partial(\text{spacetime})} (\mathrm{d}^3x)_\mu\;\mathcal{K}^\mu,
\end{aligned} \tag{10.85}$$

where the last integral vanishes, since the "boundary" of spacetime is an infinitely distant 3-sphere, where the fields, and also the integral of $\mathcal{K}^\mu$, are routinely required to vanish. ☑

Comment 10.4 *As it is only necessary for the entire integral $\oint_{\partial(\text{spacetime})} (\mathrm{d}^3x)_\mu\,\mathcal{K}^\mu$ to vanish, it would suffice for the expression $\mathcal{K}^\mu$ to vary "at infinity" so that the sum over the infinitely distant (spacetime) 3-sphere should evaluate to zero.*

Digression 10.8 Given an anticommuting (Grassmann) variable θ, integration over θ that is invariant with respect to constant translations $\theta\to\theta+\epsilon$ must in fact be functionally identical to the partial θ-derivative: $\int \mathrm{d}\theta\, f(\theta) \equiv \frac{\partial}{\partial\theta} f(\theta)$; this is called Berezin integration, after Felix Alexandrovich Berezin. Since θ must be nilpotent, the result of such integration must be θ-independent, and no loss is incurred by appending the $\theta \to 0$ projection. However, delimited by this trailing projection, the action of a super-derivative such as $D := \frac{\partial}{\partial\theta} + i\hbar\theta\frac{\partial}{\partial x}$ (where x is a commuting variable) is identical to the action of a partial θ-derivative, which in turn is identical to Berezin integration:

$$[D\, f(\theta)]\big| = \Big[\frac{\partial}{\partial\theta} f(\theta)\Big]\Big| = \int \mathrm{d}\theta\, f(\theta). \tag{10.86}$$

The $\theta, \overline{\theta} \to 0$ projection (10.76) of the 4-fold super-derivative (10.84) of any superfield function is thus equal to its $\mathrm{d}^2\theta\mathrm{d}^2\overline{\theta}$-integral. In turn, this re-interprets the integral–super-derivative combination such as in (10.83) as a $\mathrm{d}^4x\mathrm{d}^2\theta\mathrm{d}^2\overline{\theta}$-integration over the whole superspace and so provides a completely uniform and geometrical treatment. In practice, however, one evaluates these integrals by means of projections of super-derivatives, which is why they are so indicated throughout this chapter.

10.3.3 The chiral and the gauge superfield

The superfield $\mathbb{F}(x;\theta,\overline{\theta})$ may also be regarded a partially ordered set of component fields, which may be partially ordered by growing physical dimensions (units) akin to the diagram (10.78). Preserving this structure, it is possible to impose constraints on some of the component fields, which is most effectively achieved using super-derivatives.

Super-constraints and the chiral superfield

Using the superfields and super-derivatives D_α and $\overline{D}_{\dot\alpha}$, it is possible to specify superdifferential equations that – because of the relations (10.70) – transform covariantly with respect to the action of supersymmetry transformations $U_{\epsilon,\overline{\epsilon}}$, given by equation (10.62).

One of the simplest such superdifferential equations defines the so-called chiral (and the conjugate, anti-chiral) superfield:[11]

$$\textbf{chiral}\quad \overline{D}_{\dot\alpha}\,\Phi = 0 \qquad \text{and} \qquad \textbf{anti-chiral}\quad D_\alpha\,\overline{\Phi} = 0. \tag{10.87}$$

It is then not hard to show that

$$\phi := [\Phi]\big|, \qquad \psi_\alpha := [D_\alpha\Phi]\big|, \qquad F := -\tfrac{1}{4}[D^2\Phi]\big| \tag{10.88}$$

are the only non-trivial component fields: two complex scalar fields ϕ and F, and one 2-component complex spin-$\frac{1}{2}$ field ψ_α; the physical dimensions (units) of these two scalar fields, however, are not equal: $[F] = [\phi]\cdot\frac{ML}{T}$. The remaining components either vanish or do not include new fields; for example,

$$[\overline{D}_{\dot\alpha}\Phi]\big| \overset{(10.87)}{=} 0, \qquad [\overline{D}^2\Phi]\big| \overset{(10.87)}{=} 0, \tag{10.89}$$

$$[\overline{D}_{\dot\alpha}D_\alpha\Phi]\big| \overset{(10.69)}{=} [(2\sigma^\mu_{\alpha\dot\alpha}P_\mu - D_\alpha\overline{D}_{\dot\alpha})\Phi]\big| \overset{(10.87)}{=} [(-2i\sigma^\mu_{\alpha\dot\alpha}\hbar\partial_\mu)\Phi]\big| = -2i\hbar\sigma^\mu_{\alpha\dot\alpha}(\partial_\mu\phi). \tag{10.90}$$

Supersymmetric transformations are easily derived following Example 10.1 on p. 380:

$$\delta_Q(\epsilon)\phi = (\epsilon\cdot D + \overline{\epsilon}\cdot\overline{D})\Phi\big| = \epsilon\cdot\psi; \tag{10.91a}$$

[11] The analogy with complex-analytic functions is fully justified and valuable.

$$\delta_Q(\epsilon)\psi_\alpha = (\epsilon\cdot D + \bar{\epsilon}\cdot\bar{D})D_\alpha\Phi\big| = \big(\tfrac{1}{2}\epsilon^\beta\varepsilon_{\beta\alpha}D^2 + 2i\hbar\bar{\epsilon}^{\dot{\alpha}}\sigma^\mu_{\alpha\dot{\alpha}}\partial_\mu\big)\Phi\big|,$$

$$= 2\varepsilon_{\alpha\beta}\epsilon^\beta\, F + 2i\hbar\sigma^\mu_{\alpha\dot{\alpha}}\bar{\epsilon}^{\dot{\alpha}}(\partial_\mu\phi); \tag{10.91b}$$

$$\delta_Q(\epsilon)F = (\epsilon\cdot D + \bar{\epsilon}\cdot\bar{D})(-\tfrac{1}{4}D^2\Phi)\big| = -\tfrac{1}{4}\bar{\epsilon}^{\dot{\alpha}}(4i\hbar\sigma^\mu_{\alpha\dot{\alpha}}\varepsilon^{\alpha\beta}\partial_\mu D_\beta)\Phi\big|,$$

$$= -i\hbar\sigma^\mu_{\alpha\dot{\alpha}}\bar{\epsilon}^{\dot{\alpha}}\varepsilon^{\alpha\beta}(\partial_\mu\psi_\beta). \tag{10.91c}$$

Digression 10.9 Iterating the result (10.91), one can show that

$$[\delta_Q(\epsilon_{(1)}),\delta_Q(\epsilon_{(2)})](\phi;\psi_\alpha;F) = 2i\hbar\,(\epsilon_{(2)}\cdot\boldsymbol{\sigma}^\mu\cdot\epsilon_{(1)} - \epsilon_{(1)}\cdot\boldsymbol{\sigma}^\mu\cdot\epsilon_{(2)})\partial_\mu(\phi;\psi_\alpha;F). \tag{10.92a}$$

That is, the commutator of two supersymmetry transformations formally equals a translation in spacetime. However, notice that this translation parameter,

$$\epsilon^\mu_{(1,2)} := (\epsilon_{(2)}\cdot\boldsymbol{\sigma}^\mu\cdot\bar{\epsilon}_{(1)} - \epsilon_{(1)}\cdot\boldsymbol{\sigma}^\mu\cdot\bar{\epsilon}_{(2)}), \tag{10.92b}$$

is not an ordinary spacetime vector! The supersymmetry transformation parameters, $\epsilon^\alpha_{(1)}, \epsilon^\alpha_{(2)}$ anticommute, and so are nilpotent [☞ relations (10.73)]; the vector (10.92b) is therefore itself (degree-4) *nilpotent*: $(\epsilon^\mu_{(1,2)})^4 \equiv 0$ for any $\mu = 0,1,2,3$. Similarly, $\epsilon\cdot\psi$ is only formally a "shift" in the scalar field ϕ, according to the transformation relation (10.91a), since the expression $\epsilon\cdot\psi(x)$ is in every spacetime point (degree-4) nilpotent and the function $\phi(x)$ in every spacetime point has values that are *ordinary*, i.e., non-nilpotent commuting complex numbers.

Conclusion 10.3 *Although the (symmetrized) iterative application of the supersymmetry generators Q_α and $\bar{Q}_{\dot{\alpha}}$ is equivalent to the application of the spacetime translation generator P_μ, supersymmetry transformations (10.62) do not produce transformations in "real" spacetime.*

The fact that supersymmetry transformations map the fields $\phi(x) \leftrightarrow \psi_\alpha(x) \leftrightarrow F(x)$ (and their derivatives) in every spacetime point, however, remains.

Notice that chiral superfields (at the same spacetime point) form the "ring" algebraic structure [☞ the lexicon entry, in Appendix B.1]:

Conclusion 10.4 *The product of two chiral superfields is again a chiral superfield:*

$$\bar{D}_{\dot{\alpha}}\Phi_1 = 0 = \bar{D}_{\dot{\alpha}}\Phi_2, \quad \Rightarrow \quad \bar{D}_{\dot{\alpha}}(\Phi_1\Phi_2) = 0, \tag{10.93}$$

*with the usual rules of distribution between multiplication and addition. It follows that chiral fields (at the same spacetime point) form the "**chiral ring**." Moreover, it follows that an arbitrary analytic function of chiral superfields (defined by its Taylor expansion) is also a chiral superfield.*

Theorem 10.8 *The most general supersymmetric Lagrangian density for a chiral superfield Φ must be of the form*

$$\mathscr{L}[\Phi] = [D^2\bar{D}^2\,K(\Phi^\dagger,\Phi)]\big| \;+\; [D^2\,W(\Phi)]\big| + [\bar{D}^2\,\overline{W}(\Phi^\dagger)]\big|. \tag{10.94}$$

Proof Since the Lagrangian density must be real, for the first term of the general form (10.84) one selects a real function $K(\Phi^\dagger,\Phi)$, and adds the second term and its Hermitian conjugate where

$W(\Phi)$ is an arbitrary analytic function. Indeed, this second term is also (and independently!) supersymmetric:

$$
\begin{aligned}
&(\epsilon{\cdot}Q+\bar{\epsilon}{\cdot}\overline{Q})\, D^2\, W(\Phi)\big| \\
&\quad\overset{(10.71)}{=} \epsilon^{\alpha}(iD_{\alpha}-2\hbar\sigma^{\mu}_{\alpha\dot{\alpha}}\bar{\theta}^{\dot{\alpha}}\partial_{\mu})\, D^2\, W(\Phi)\big| + \bar{\epsilon}^{\dot{\alpha}}(i\overline{D}_{\dot{\alpha}}-2\hbar\sigma^{\mu}_{\alpha\dot{\alpha}}\theta^{\alpha}\partial_{\mu})\, D^2\, W(\Phi)\big| \\
&\quad\overset{\theta,\bar{\theta}\to 0}{=} i\epsilon^{\alpha}\underbrace{D_{\alpha}D^2}_{\equiv 0}\, W(\Phi)\big| + i\bar{\epsilon}^{\dot{\alpha}}(-2i\hbar\sigma^{\mu}_{\alpha\dot{\alpha}}\partial_{\mu}\varepsilon^{\alpha\beta}D_{\beta}+D^2\overline{D}_{\dot{\alpha}})W(\Phi)\big| \\
&\quad= \partial_{\mu}\underbrace{\big[2\hbar\bar{\epsilon}^{\dot{\alpha}}\sigma^{\mu}_{\alpha\dot{\alpha}}\varepsilon^{\alpha\beta}D_{\beta}W(\Phi)\big|\big]}_{\mathcal{K}^{\mu}} + i\bar{\epsilon}^{\dot{\alpha}}D^2\,\underbrace{\overline{D}_{\dot{\alpha}}W(\Phi)}_{=0}\big| = \partial_{\mu}\mathcal{K}^{\mu},
\end{aligned}
\tag{10.95}
$$

so the $\int d^4x$-integral vanishes again, owing to the usual restrictions on the fields. Listing all possible Lorentz-invariant terms, one shows that the expression (10.94) is the most general form of a supersymmetric Lagrangian density. ☑

The standard choice $K(\Phi^{\dagger},\Phi) = \Phi^{\dagger}\Phi$ gives (after some "D-gymnastics" [☞ relations (A.162)–(A.165)]) the standard Lagrangian density for a scalar ϕ and a fermion ψ_{α}, and the total resulting Lagrangian density is – up to integration by parts for symmetrization of the expression,

$$
\begin{aligned}
\mathscr{L}[\Phi] = &-(\partial_{\mu}\phi^*)\eta^{\mu\nu}(\partial_{\nu}\phi) - \tfrac{i}{2}\bar{\sigma}^{\mu\dot{\alpha}\alpha}\big[\overline{\psi}_{\dot{\alpha}}(\partial_{\mu}\psi_{\alpha}) - (\partial_{\mu}\overline{\psi}_{\dot{\alpha}})\psi_{\alpha}\big] + F^*F \\
&+ F\,W'(\phi) + \tfrac{1}{2}\varepsilon^{\alpha\beta}\psi_{\alpha}\psi_{\beta}\,W''(\phi) \;+\; F^*\,\overline{W}'(\phi^*) + \tfrac{1}{2}\varepsilon^{\dot{\alpha}\dot{\beta}}\overline{\psi}_{\dot{\alpha}}\overline{\psi}_{\dot{\beta}}\,\overline{W}''(\phi^*).
\end{aligned}
\tag{10.96}
$$

Since the Euler–Lagrange equations of motion for the component fields F and F^*,

$$
F^* = -W'(\phi) \qquad\text{and}\qquad F = -\overline{W}'(\phi^*),
\tag{10.97}
$$

are non-differential equations in F and F^*, they may be used to substitute F and F^*:

$$
\begin{aligned}
\mathscr{L}[\Phi] = &-\tfrac{i}{2}\bar{\sigma}^{\mu\dot{\alpha}\alpha}\big[\overline{\psi}_{\dot{\alpha}}(\partial_{\mu}\psi_{\alpha}) - (\partial_{\mu}\overline{\psi}_{\dot{\alpha}})\psi_{\alpha}\big] - (\partial_{\mu}\phi^*)\eta^{\mu\nu}(\partial_{\nu}\phi) \\
&- |W'(\phi)|^2 + \tfrac{1}{2}\varepsilon^{\alpha\beta}\psi_{\alpha}\psi_{\beta}\,W''(\phi) + \tfrac{1}{2}\varepsilon^{\dot{\alpha}\dot{\beta}}\overline{\psi}_{\dot{\alpha}}\overline{\psi}_{\dot{\beta}}\,\overline{W}''(\phi^*).
\end{aligned}
\tag{10.98}
$$

The constant $\hbar$ has been eliminated in the expressions such as (10.94)–(10.98) by redefining the component fields to emphasize the similarity with the Lagrangian density (7.34) [☞ Exercise 10.3.6 on p. 388]. A similar redefinition of the fermion fields $\psi_{\alpha},\overline{\psi}_{\dot{\alpha}}$ and the use of the basis (A.132) for the Dirac γ-matrices shows the first term (10.98) to give the standard Lagrangian density (5.68a) for Dirac fermions.

The computation (10.94)–(10.98) clearly shows that the $D^2\overline{D}^2\Phi^{\dagger}\Phi|$ term produced the standard "kinetic" part of the Lagrangian density, while the terms $D^2W(\Phi)| + \overline{D}^2\overline{W}(\Phi^{\dagger})|$ produce, after eliminating F and F^* via their equations of motion (10.97), the potential

$$
V(\phi) = |W'(\phi)|^2 \;\geqslant\; 0.
\tag{10.99}
$$

Finally, the terms $-\psi^2W''(\phi^*) - \overline{\psi}^2\overline{W}''(\phi^*)$ provide the supersymmetric completion of the potential $|W'(\phi)|^2$. Owing to the relation (10.99) with the potential, the function $W(\Phi)$ is called the *superpotential*.

Digression 10.10 On one hand, owing to Theorem 10.7 on p. 381, and the similar result (10.95), the Lagrangian density (10.96) is known to be supersymmetric, i.e., its infinitesimal supersymmetry transformation $\delta_Q(\epsilon) = -i(\epsilon \cdot Q + \bar{\epsilon} \cdot \overline{Q})$ changes the Lagrangian density (10.96) into a total derivative. This may also be directly verified by substituting the supersymmetry transformations of the component fields

$$\delta_Q(\epsilon)\phi = \epsilon^\alpha \psi_\alpha, \quad \delta_Q(\epsilon)\psi_\alpha = 2\varepsilon_{\alpha\beta}\epsilon^\beta F + 2i\hbar\sigma^\mu_{\alpha\dot{\alpha}}\bar{\epsilon}^{\dot{\alpha}}(\partial_\mu \phi), \qquad \delta_Q(\epsilon)F = i\hbar\sigma^\mu_{\alpha\dot{\alpha}}\varepsilon^{\alpha\beta}\bar{\epsilon}^{\dot{\alpha}}(\partial_\mu \psi_\beta) \tag{10.100a}$$

into the Lagrangian density (10.96).

However, the Lagrangian density (10.98) is *not* invariant with respect to the supersymmetry transformations (10.100a)! These transformations represent the original (and linear) supersymmetry action upon the superfield Φ, i.e., upon the component fields – including F. The elimination of F by (10.97) changes this action, so that the transformation rules (10.100a) also change, and the Lagrangian density (10.98) is invariant with respect to the so-changed transformations. As $W'(\phi)$ is nonlinear in the general case, these changed supersymmetry transformation rules are also nonlinear.

Digression 10.11 The simplest model in which supersymmetry is spontaneously broken was found by O'Raifeartaigh, and has the superpotential

$$D^2[\lambda\Phi_0 + m\Phi_1\Phi_2 + g\Phi_0{\Phi_1}^2]\big| + h.c., \tag{10.101a}$$

where Φ_0, Φ_1 and Φ_2 are three chiral superfields. The non-differential equations of motion for the auxiliary components F_0, F_1 and F_2 are (10.98)

$$F_0 = -\lambda - g{\phi_1}^2, \quad F_1 = -m\phi_1 - 2g\phi_0\phi_1, \quad F_2 = -m\phi_2, \tag{10.101b}$$

which make the potential in this model into (10.99)

$$V = \sum_{k=0}^{2} |F_k|^2 = \big|\lambda + g{\phi_1}^2\big|^2 + \big|m\phi_2 + 2g\phi_0\phi_1\big|^2 + \big|m\phi_1\big|^2. \tag{10.101c}$$

This cannot possibly vanish: the last term can vanish only where $\phi_1 = 0$ and where the potential becomes $V = |\lambda|^2 + |m\phi_2|^2 > 0$. Supersymmetry is therefore broken spontaneously as there is no solution to the equations of motion where $V = 0$.

One of the original ideas for the application of supersymmetry was to find superfields where the bosons and fermions of the Standard Model [☞ Table 2.3 on p. 67] would all be component fields of the *same* superfields. However, the component fields of the same superfield differ only by spin, and not by charges (electric, weak isospin or color), so this is not possible: the Standard Model fermions have completely different charges from the bosons.

It follows that identifying the 2-component Weyl fermion $\psi_\alpha(x) \in \Phi(x;\theta,\bar{\theta})$ with a left-handed chiral half of a Dirac 4-component wave-function of the electron, the complex scalar field $\phi(x) \in \Phi(x;\theta,\bar{\theta})$ is the electron superpartner, the so-called *selectron*, which must be added to the list in Table 2.3 on p. 67. Bosons with the charges given in Table 7.1 on p. 275 have not been detected experimentally, while supersymmetry implies their existence; one jokes that supersymmetry is already 50% experimentally verified. However, this means that supersymmetry – in Nature – must be broken, and in such a way that the masses of the bosonic superpartners of the elementary fermions from Table 2.3 on p. 67 are sufficiently big, bigger than the masses of the elementary fermions by the amount E_{SuSy}/c^2, so that this explains why they have not been detected experimentally so far [☞ examples in Figure 10.2 on p. 375].

The gauge superfield

Consider now the real superfield $\mathbb{A}^\dagger = \mathbb{A}$. The component fields may be found by means of Taylor-esque projections (10.75); specifically, the projection (10.75e) finds the real 4-vector field $A_\mu(\mathrm{x}) \in \mathbb{A}(\mathrm{x};\theta,\bar{\theta})$. On the other hand, the same component of the combined superfield $i(\Phi - \Phi^\dagger)$ gives

$$\tfrac{1}{4}\bar{\sigma}_\mu^{\dot{\alpha}\alpha}\big[[D_\alpha, \bar{D}_{\dot{\alpha}}]i(\Phi - \Phi^\dagger)\big]\big| = 2\hbar\,(\partial_\mu\, \Re e(\phi)), \tag{10.102}$$

so that the superfield transformation

$$\mathbb{A} \to \mathbb{A}' = \mathbb{A} + i(\Phi - \Phi^\dagger) \quad \ni \quad A_\mu \to A'_\mu = A_\mu + \partial_\mu (2\hbar\, \Re e(\phi)) \tag{10.103}$$

contains the gauge transformation (5.89) of the vector component field, where $-\frac{\hbar}{c}\,\Re e(\phi)$ plays the role of the gauge local parameter. Of course, if the chiral superfields Φ_i are intended one for each Standard Model elementary fermion and we introduce the real superfield $\mathbb{A} \ni A_\mu$ for the electromagnetic field, to parametrize the gauge transformation of the electromagnetic field we must introduce a separate chiral superfield $\boldsymbol{\Lambda}$, the scalar component of which plays the role of the gauge local parameter.

A detailed analysis [189, 562, 560, 76] of the component fields in the combination $\mathbb{A} + i(\boldsymbol{\Lambda} - \bar{\boldsymbol{\Lambda}})$ shows that a suitable choice of the superfield $\boldsymbol{\Lambda}$ eliminates the component fields in the "lower half" of the superfield $\mathbb{A}$, as per diagram (10.78). However, it is more practical to define the *chiral–anti-chiral* pair of fermionic superfields:

$$\mathbb{A}_\alpha := (\bar{D}^2 D_\alpha \mathbb{A}) \qquad \text{and} \qquad \bar{\mathbb{A}}_{\dot{\alpha}} := (D^2 \bar{D}_{\dot{\alpha}} \mathbb{A}), \tag{10.104}$$

which satisfy

$$\mathbb{A} = \mathbb{A}^\dagger \quad \Rightarrow \quad \varepsilon^{\alpha\beta} D_\alpha \mathbb{A}_\beta = \varepsilon^{\dot{\alpha}\dot{\beta}} \bar{D}_{\dot{\alpha}} \bar{\mathbb{A}}_{\dot{\beta}}, \tag{10.105}$$

and the components of which include

$$\mathbb{A}_\alpha\big| =: \lambda_\alpha, \qquad \bar{\mathbb{A}}_{\dot{\alpha}}\big| =: \bar{\lambda}_{\dot{\alpha}}, \tag{10.106a}$$

$$D_\alpha \mathbb{A}_\beta\big| =: \varepsilon_{\alpha\beta}\mathrm{D} + i(\boldsymbol{\sigma}^{\mu\nu})_\alpha{}^\gamma\, \varepsilon_{\beta\gamma} F_{\mu\nu}, \qquad \bar{D}_{\dot{\alpha}} \bar{\mathbb{A}}_{\dot{\beta}}\big| =: \varepsilon_{\dot{\alpha}\dot{\beta}}\mathrm{D} + i(\bar{\boldsymbol{\sigma}}^{\mu\nu})^{\dot{\gamma}}{}_{\dot{\alpha}}\, \varepsilon_{\dot{\beta}\dot{\gamma}} F_{\mu\nu}, \tag{10.106b}$$

$$D^2 \mathbb{A}_\alpha\big| = -i\hbar\sigma^\mu_{\alpha\dot{\alpha}}\varepsilon^{\dot{\alpha}\dot{\beta}}(\partial_\mu \bar{\lambda}_{\dot{\beta}}), \qquad \bar{D}^2 \bar{\mathbb{A}}_{\dot{\alpha}}\big| = -i\hbar\sigma^\mu_{\alpha\dot{\alpha}}\varepsilon^{\alpha\beta}(\partial_\mu \lambda_\beta). \tag{10.106c}$$

Here,

$$F_{\mu\nu} := (\partial_\mu A_\nu - \partial_\nu A_\mu), \tag{10.107}$$

and the component fields from the "lower half" of the original superfield $\mathbb{A}$ show up neither in the expressions (10.106) nor in any other projection of the superfields $\mathbb{A}_\alpha$ and $\bar{\mathbb{A}}_{\dot{\alpha}}$. A supersymmetric Lagrangian density that includes the standard $-\frac{1}{4}F_{\mu\nu}F^{\mu\nu}$ Lagrangian density is then obtained from the expression

$$\begin{aligned}
\mathscr{L}[\mathbb{A}] &= -\tfrac{1}{4}[D^2\, \varepsilon^{\alpha\beta}\mathbb{A}_\alpha \mathbb{A}_\beta]\big| - \tfrac{1}{4}[\bar{D}^2\, \varepsilon^{\dot{\alpha}\dot{\beta}}\bar{\mathbb{A}}_{\dot{\alpha}}\bar{\mathbb{A}}_{\dot{\beta}}]\big| \\
&= -\tfrac{1}{4}F_{\mu\nu}F^{\mu\nu} - \tfrac{i\hbar}{2}\bar{\sigma}^{\mu\,\dot{\alpha}\alpha}\big[\bar{\lambda}_{\dot{\alpha}}(\partial_\mu \lambda_\alpha) - (\partial_\mu \bar{\lambda}_{\dot{\alpha}})\lambda_\alpha\big] + 2\mathrm{D}^2.
\end{aligned} \tag{10.108}$$

The first term is – up to a (re)scaling of the field A_μ – the Lagrangian density that is identical to the density in the expression (5.76) for electromagnetic fields. The equations of motion for the spinor fields $(\lambda_\alpha, \bar{\lambda}_{\dot{\alpha}})$ are the Dirac equations, which is typical for spin-$\frac{1}{2}$ fermions, but the mass of these spinors vanishes. These then are the superpartners of the gauge fields and are in general called *gauginos*.[12] Notice that the relation (10.105) equates the component functions that occur in D- and $\overline{D}$-projections (10.106b), but leaves $\bar{\lambda}_{\dot{\alpha}}$ formally independent of λ_α. The condition (10.105), however, guarantees that the Dirac spinor $(\lambda_\alpha, \bar{\lambda}_{\dot{\alpha}})$ has four *real* independent components.

Supersymmetric electrodynamics

The minimal supersymmetric Lagrangian density where the chiral field $\mathbb{F}$ interacts with the gauge superfield $\mathbb{A}$, for the supersymmetric version of electrodynamics for example, is obtained in the form

$$\mathscr{L} = -\tfrac{1}{4}[D^2\,\varepsilon^{\alpha\beta}\mathbb{A}_\alpha\mathbb{A}_\beta]\big| - \tfrac{1}{4}[\overline{D}^2\,\varepsilon^{\dot{\alpha}\dot{\beta}}\overline{\mathbb{A}}_{\dot{\alpha}}\overline{\mathbb{A}}_{\dot{\beta}}]\big| + \big[D^2\overline{D}^2\,\overline{\Phi}\,e^{q_\Phi\mathbb{A}}\,\Phi\big]\big|. \tag{10.109}$$

This Lagrangian density is invariant with respect to the gauge transformations

$$\mathbb{A} \to \mathbb{A} + i(\Lambda - \overline{\Lambda}), \qquad \Phi \to e^{iq_\Phi\Lambda}\Phi, \qquad \overline{\Phi} \to e^{-iq_\Phi\overline{\Lambda}}\overline{\Phi}, \tag{10.110}$$

which coincide with the transformations (5.14a) for the component fields A_μ, and that of the $\psi_\alpha \in \Phi$ with the left-handed chiral projection of the transformation (5.14b). Expanding the expression (10.109) produces the Lagrangian density for supersymmetric electrodynamics, where the additional terms involve the superpartners of both the photon (itself represented by the 4-vector potential A_μ), and the left-handed chiral electron (represented by the fermion field ψ_α).

To extend this minimal model so as to include also the right-handed chiral electron, we must introduce another chiral field, $\Phi^c := C(\Phi)$, which is defined so that ψ^c_α is the left-handed chiral spin-$\frac{1}{2}$ fermionic field with the electric charge opposite to that of the electron. Then, $\overline{\psi}^c_{\dot{\alpha}} \in \overline{\Phi^c}$ is the right-handed chiral spin-$\frac{1}{2}$ fermions field with the electric charge equal to the electron charge. Therefore, the Lagrangian density for electrodynamics with a massive electron must be of the form

$$\begin{aligned}\mathscr{L} = &-\tfrac{1}{4}[D^2\,\varepsilon^{\alpha\beta}\mathbb{A}_\alpha\mathbb{A}_\beta]\big| - \tfrac{1}{4}[\overline{D}^2\,\varepsilon^{\dot{\alpha}\dot{\beta}}\overline{\mathbb{A}}_{\dot{\alpha}}\overline{\mathbb{A}}_{\dot{\beta}}]\big| + \big[D^2\overline{D}^2\,\overline{\Phi}\,e^{q_\Phi\mathbb{A}}\,\Phi\big]\big| + \big[D^2\overline{D}^2\,\overline{\Phi^c}\,e^{-q_\Phi\mathbb{A}}\,\Phi^c\big]\big|\\ &+ m\Big\{[D^2\Phi\Phi^c]\big| + [\overline{D}^2\overline{\Phi}\,\overline{\Phi^c}]\big|\Big\},\end{aligned} \tag{10.111}$$

where we added terms in the second row, which produce the Lagrangian terms $m(\psi\cdot\psi^c + \overline{\psi}\cdot\overline{\psi^c})$ for the electron, as well as (after eliminating the auxiliary scalar fields F and F^c using their non-differential equations of motion) $m^2\phi\phi^c \equiv m^2|\phi|^2$ for the selectron.

The minimal supersymmetric Standard Model

The construction of the complete supersymmetric Standard Model is now seen as a generalization of the procedure that led us to the Lagrangian density (10.111). For the details of this construction, the interested Reader is directed to the abundant literature [☞ textbooks [189, 562, 560, 76] to begin with]. However, note:

1. On one hand, supersymmetry conceptually unites bosons and fermions – and requires that every boson has a fermion superpartner, and *vice versa*.
2. On the other hand, the concrete bosons and fermions of the Standard Model cannot be each others' superpartners, since the (gauge and Higgs) bosons in the Standard Model transform differently from the fundamental fermions in the Standard Model with respect to the action of the gauge group $SU(3)_c \times SU(2)_w \times U(1)_Q$.

[12] Superpartners of bosonic particles are named using the boson's name with an attached *-ino* suffix, such as *photino*, *gluino* and *higgsino*. The superpartners of fermionic particles are named by attaching an *s-* prefix to the fermion's name, such as *selectron*, *sneutrino* and *squark*.

Also, it turns out that the details of the fermion mass hierarchy require introducing not one but two chiral superfields for each Higgs field in the Standard Model, and it follows:

> ***Conclusion 10.5*** *The so-called Minimally Supersymmetric Standard Model (MSSM) requires a little over twice as many particles as the Standard Model.*

Considering this simple counting of degrees of freedom used to describe Nature, the reason for supersymmetrizing the Standard Model certainly is not economy. However, recall the conceptual and practical (technical) consequences of supersymmetry [☞ Section 10.3.1]:

1. vacuum stabilization,
2. mass hierarchy stabilization, and
3. simplification of the renormalization procedure.

Note also the fact that before the invention of supersymmetry, which successfully solves these problems of the Standard Model, these problems were hardly mentioned. Of course, that owes partly to the approach of describing Nature pragmatically and axiomatically:

> ***Comment 10.5*** *Theoretical models are constructed with the aim to describe, in a logically coherent and consistent theoretical system, the known phenomena without predicting nonexistent phenomena, and while keeping the necessary assumptions as few as possible.*
>
> *These assumptions (axioms) are re-examined only when the resulting theoretical system "paints" the development of the model "into a corner" and when within this theoretical system it is not possible to construct a model that does not err in a concrete aspect of the description of Nature, or when an opportunity emerges to explain it in a conceptually more fundamental or practically simpler system of assumptions.*

Of course, the question remains: In what measure is supersymmetry of help in models of quantum physics that contain the general theory of relativity? Considering that the complete theory of quantum gravity does not exist yet, a final answer to this question then does not exist either. However, the next chapter will permit us to say a little more about this.

10.3.4 Exercises for Section 10.3

✎ ***10.3.1*** *Prove that the left-hand side of the relation (10.55) is non-negative.*

✎ ***10.3.2*** *Show that the operators (10.67) and (10.68) satisfy the operatorial relations (10.69) and (10.70).*

✎ ***10.3.3*** *Confirm the relations (10.91).*

✎ ***10.3.4*** *Confirm the result (10.95).*

✎ ***10.3.5*** *By iterative and consistent use of relations (10.69) and definitions (10.88) (a.k.a. "D-gymnastics"), derive equation (10.96) from (10.94).*

✎ ***10.3.6*** *Return the proper factors of $\hbar$ and c in the expressions (10.96)–(10.98).*

10.4 Classification of off-shell supermultiplets

Recall Procedure 5.1 on p. 193, which is generally accepted as the only systematic procedure applicable in all known models of quantum field theory [☞ also Procedure 11.1 on p. 416]. For the purposes of this procedure we must define an integral of the general type

$$\int \mathbf{D}[\phi]\, e^{iS[\phi]/\hbar} \tag{10.112}$$

to be computed over all possible fields, here represented by the symbol ϕ. $S[\phi]$ is the classical Hamilton action, which according to Hamilton's principle has a minimum for the choice of fields ϕ that satisfy the classical (Euler–Lagrange) equations of motion, i.e., for fields ϕ that are *on shell*. For such *classical* fields, $S[\phi]$ is minimal, and the integrand $\exp\{iS[\phi]/\hbar\}$ oscillates minimally. By contrast, a choice of the fields that are "far" from such classical fields then causes the integrand to oscillate very fast, so that the contributions mostly cancel. The naive reasoning then is that the contributions of the classical fields dominate the formal integral (10.112). This is in no way proven rigorously, as the space of the choices of the fields ϕ is infinite-dimensional: although the contribution to the integral (10.112) from any one non-classical field is infinitesimally small, there are continuously many such fields and the sum over them may well even diverge.

However, we certainly know that the fields over which the integration (10.112) is to be performed must a priori be *off-shell*, i.e., not subject to any differential equation, and foremost not the classical (Euler–Lagrange) equations of motion: that would be outright contradictory. *Quantum* supersymmetric models then must be constructed using *off-shell* supermultiplets (collections of particles and their superpartners); in models of supersymmetric quantum field theory, both the known particles and all their superpartners must be represented by *off-shell* fields.

With that in mind it is then surprising that four decades after the introduction of supersymmetry in field theory there is still no complete theory of *off-shell* representation of supersymmetry algebras☡. Recent research in this direction [139, 140, 141, 142] indicates a fantastic and combinatorially complex multitude of possibilities, very different from the well-known theory of the finite-dimensional unitary representations of Lie algebras, and even the supersymmetric *on-shell* representations, which are well known.

The remainder of this section gives a telegraphic description of this research, mostly so as to indicate some open possibilities for research. However, this introduction is restricted to *intact* supermultiplets[13] – those that have not been constrained or gauged in any way. Constrained and gauged (gauge-equivalences of) supermultiplets are indeed very widely used, and the interested Reader is directed to the textbooks [189, 562, 560, 76].

10.4.1 One-dimensional supersymmetry as the common denominator

Recall the three levels of theoretical analysis of physical systems [☞ description on p. 366] where supersymmetry may show up, and especially the second and third levels of analysis, where supersymmetry reduces to supersymmetric quantum mechanics, with the algebra

$$\{\, Q_I\,,\, Q_J \,\} = 2\delta_{IJ}H, \qquad [\, H\,,\, Q_I \,] = 0, \qquad I, J = 1, \ldots, N, \tag{10.113}$$

where $H = i\hbar\partial_\tau$ is the Hamiltonian and τ the proper time, and where in the general case one does not require N to be even as in the relations (10.32). Note that we revert to the quantum-mechanical normalization, $[Q_I] = \frac{\sqrt{ML}}{T}$.

---

[13] The adjective "intact" is simply shorter than the detailed "unconstrained and ungauged."

Following the lesson from the conclusion in Digression 10.9 on p. 383, or even just simply the pragmatic application of supersymmetry as a transformation that maps bosons into fermions and back, the goal of classifying representations of supersymmetry is to classify all possible supermultiplets, i.e., collections of bosons and fermions that supersymmetry maps one into the other. To this end, it is convenient to introduce a graphical notation as described in Table 10.1.[14]

Table 10.1 The correspondence between Adinkras and supersymmetry transformations gives: node ↔ component field; white/black node ↔ boson/fermion; Ith color/index edge ↔ $\mathcal{Q}_I$; dashed edge ↔ − sign; edge direction ↔ 1 (∂_τ when following an edge in the opposite direction). In addition, the Adinkras are drawn putting the nodes at levels proportional to their relative units, so the implicit edge directions are upward.

Adinkra	Supersymmetry transf.	Adinkra	Supersymmetry transf.
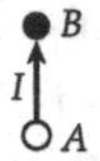	$\mathcal{Q}_I \begin{bmatrix}\psi_B\\ \phi_A\end{bmatrix} = \begin{bmatrix}i\dot{\phi}_A\\ \psi_B\end{bmatrix}$		$\mathcal{Q}_I \begin{bmatrix}\psi_B\\ \phi_A\end{bmatrix} = \begin{bmatrix}-i\dot{\phi}_A\\ -\psi_B\end{bmatrix}$
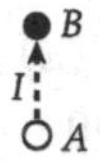	$\mathcal{Q}_I \begin{bmatrix}\phi_A\\ \psi_B\end{bmatrix} = \begin{bmatrix}\dot{\psi}_B\\ i\phi_A\end{bmatrix}$		$\mathcal{Q}_I \begin{bmatrix}\phi_A\\ \psi_B\end{bmatrix} = \begin{bmatrix}-\dot{\psi}_B\\ -i\phi_A\end{bmatrix}$

Edges are here labeled by the index I; for a fixed I, they are drawn in the Ith color.

The next two examples of supermultiplets of $N = 2$ supersymmetric quantum mechanics should clarify the application of the rules in Table 10.1.

Example 10.2 The simplest supermultiplet is of the general form that reflects the basis of the type (10.78):

$$\begin{aligned} \mathcal{Q}_1\,\phi &= \psi_1, & \mathcal{Q}_2\,\phi &= \psi_2, && (10.114a)\\ \mathcal{Q}_1\,\psi_1 &= i\dot{\phi}, & \mathcal{Q}_2\,\psi_1 &= -iF, && (10.114b)\\ \mathcal{Q}_1\,\psi_2 &= iF, & \mathcal{Q}_2\,\psi_2 &= i\dot{\phi}, && (10.114c)\\ \mathcal{Q}_1\,F &= \dot{\psi}_2, & \mathcal{Q}_2\,F &= -\dot{\psi}_1. && (10.114d) \end{aligned}$$

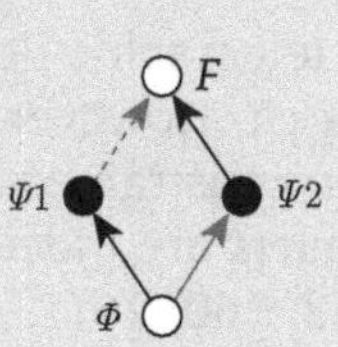

The black edges depict the action of the $\mathcal{Q}_1$ supercharge and the gray edges the $\mathcal{Q}_2$-action. The fact that in a two-colored quadrangle (10.114) an odd number of edges must be dashed (i.e., the corresponding supercharge action has an additional −1 sign) follows from the fact that

$$\left.\begin{aligned} \phi \xrightarrow{\mathcal{Q}_1} \psi_1 \xrightarrow{-\mathcal{Q}_2} F: \quad F &= -\mathcal{Q}_2(\mathcal{Q}_1(\phi)),\\ \phi \xrightarrow{\mathcal{Q}_2} \psi_1 \xrightarrow{\mathcal{Q}_1} F: \quad F &= \mathcal{Q}_1(\mathcal{Q}_2(\phi)), \end{aligned}\right\} \quad \Rightarrow \quad \mathcal{Q}_1\mathcal{Q}_2 = -\mathcal{Q}_2\mathcal{Q}_1, \tag{10.115}$$

in agreement with equation (10.31).

[14] A graphical representation of a system of equations offers the evident advantage of heuristic insight and is not at all a new idea [177]; the formalization of such graphs – called *Adinkras* – for the purposes of supersymmetry, however, is of recent origin [139]. They are particularly useful in depicting intact supermultiplets.

Example 10.3 Another example of an $N = 2$ supermultiplet is

$$\mathcal{Q}_1\,\varphi_1 = \chi_1, \qquad \mathcal{Q}_2\,\varphi_1 = \chi_2, \tag{10.116a}$$
$$\mathcal{Q}_1\,\varphi_2 = \chi_2, \qquad \mathcal{Q}_2\,\varphi_2 = -\chi_1, \tag{10.116b}$$
$$\mathcal{Q}_1\,\chi_1 = i\dot{\varphi}_1, \qquad \mathcal{Q}_2\,\chi_1 = -i\dot{\varphi}_2, \tag{10.116c}$$
$$\mathcal{Q}_1\,\chi_2 = i\dot{\varphi}_2, \qquad \mathcal{Q}_2\,\chi_2 = i\dot{\varphi}_1. \tag{10.116d}$$

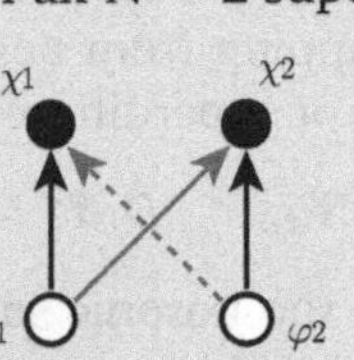

Formally, equating $(\phi;\psi_1\psi_2;F) = (\varphi_1;\chi_1,\chi_2;\dot{\varphi}_2)$ identifies the two supermultiplets, but this implies the relation $F = \dot{\varphi}_2$ and so $\varphi_2 = \int\mathrm{d}\tau\,F$, which is evidently non-local. The two supermultiplets, (10.114) and (10.116), thus cannot be considered equivalent off-shell supermultiplets.

Both examples, 10.2 and 10.3, depict supermultiplets that consist of two bosons and two fermions. The difference is indicated by the fact that in Example 10.2 $[F] = [\phi]\cdot\frac{ML^2}{T^2}$, whereas in Example 10.3 $[\varphi_1] = [\varphi_2]$; see Table C.5 on p. 528. It is then evident that the supersymmetric Lagrangian of the form

$$\mathscr{L}_2 := \tfrac{1}{2}\mu\big[(\dot{\varphi}_1)^2 + (\dot{\varphi}_2)^2 + \tfrac{2i}{\hbar}(\chi_1\dot{\chi}_2 - \dot{\chi}_1\chi_2)\big], \tag{10.117}$$

with an appropriate characteristic constant μ, produces the familiar equations of motion: second order in time derivatives for the bosons φ_1, φ_2 and first order for fermions χ_1, χ_2. By contrast, the analogous supersymmetric Lagrangian

$$\mathscr{L}_1 := \tfrac{1}{2}\mu\big[(\dot{\phi})^2 + \tfrac{1}{\hbar^2}F^2 + \tfrac{2i}{\hbar}(\psi_1\dot{\psi}_2 - \dot{\psi}_1\psi_2)\big] \tag{10.118}$$

produces the usual equations of motion for the boson ϕ and the fermions ψ_1, ψ_2, but an algebraic equation for the boson F [☞ step 4(b)iii on p. 374, as well as the equations of motion (10.97)].

This *dynamical* information is thus encoded by the "*height arrangement*" of the nodes in the Adinkra, which defines the relative physical units of the component fields in the depicted supermultiplet.

Digression 10.12 The formal difference between the supermultiplets (10.114) and (10.116) is seen by analyzing the identifications

$$\begin{aligned} (\phi;\psi_1,\psi_2;F) &\xrightarrow{=} (\varphi_1;\chi_1,\chi_2;(\dot{\varphi}_2)),\\ (\phi,(\textstyle\int\mathrm{d}\tau\,F);\psi_1,\psi_2) &\xleftarrow{=} (\varphi_1,\varphi_2;\chi_1,\chi_2). \end{aligned} \tag{10.119}$$

This gives a formal bijection between the two supermultiplets. However, since $\partial_\tau : \varphi_2 \mapsto (\dot{\varphi}_2)$ annihilates the constant term in a power expansion of the function $\varphi_2(\tau)$ and $\partial_\tau^{-1} : F \mapsto (\int\mathrm{d}\tau\,F)$ adds an arbitrary (integration) constant, this formal bijection is not a perfect 1–1 mapping in both ways, and the supermultiplets (10.116) and (10.114) must be considered different.

Digression 10.13 For the *Lagrangians* (not Lagrangian densities!) $\mathscr{L}_1$ and $\mathscr{L}_2$ to have the units of energy and μ to be identifiable as a mass, $[\phi] = [\varphi_i] = L$, $[\psi_i] = [\chi_i] = \frac{\sqrt{ML^2}}{T}$ and $[F] = \frac{ML^3}{T^2}$.

Supermultiplets that can be depicted with Adinkras (graphs that are constructed based on the rules in Table 10.1 on p. 390 [☞ article [139] for the appropriate theorems and details]) have the property that the supersymmetric mapping from bosonic to fermionic component fields and back may also be represented by a system of superdifferential relations:

$$D_I \boldsymbol{\Phi}_i = (L_I)_i{}^{\alpha} \boldsymbol{\Upsilon}_\alpha, \qquad D_I \boldsymbol{\Upsilon}_\alpha = i\hbar (R_I)_\alpha{}^{i} \dot{\boldsymbol{\Phi}}_i, \tag{10.120}$$

where the index I counts supercharges, i the bosonic superfields $\boldsymbol{\Phi}_i$, α the fermionic superfields $\boldsymbol{\Upsilon}_\alpha$, chosen so that:

1. component fields $\phi_i = \boldsymbol{\Phi}_i|$ and $\psi_\alpha = \boldsymbol{\Upsilon}_\alpha|$ (up to a ∂_τ- or ∂_τ^{-1}-prefactor as needed) are the complete system of component fields for the desired supermultiplet, and
2. in every row and every column, the numerical matrices $(L_I)_i{}^{\alpha}$ and $(R_I)_\alpha{}^{i}$ have precisely one nonzero entry, which equals ± 1.

Because of the relations (10.71), the system of superdifferential relations specifies the supersymmetric transformations within the supermultiplet.

Although there exist supermultiplets that do not satisfy these requirements, all worldline off-shell supermultiplets may be constructed starting with such "*adinkraic*" supermultiplets [284, 143]. *Adinkras* for a few such supermultiplets for small N (in the variant where neither ∂_τ- nor ∂_τ^{-1}-prefactors were used) are

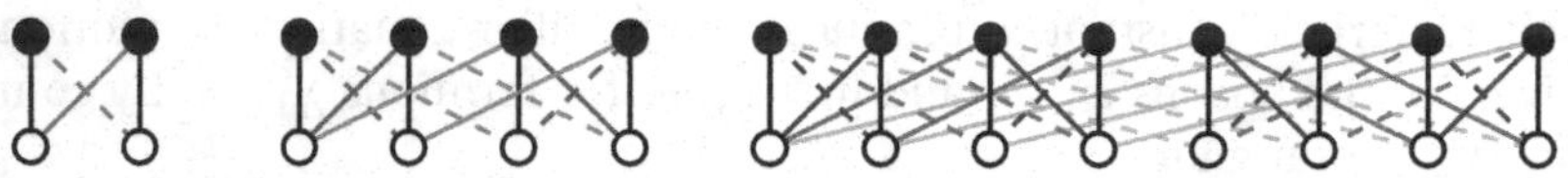

etc. (10.121)

It should now be clear that there exist a combinatorially (hyper-exponentially) growing number of different node-height arrangements in Adinkras with growing N. Every new node-height arrangement corresponds to a new application of ∂_τ- and ∂_τ^{-1}-prefactors, which then specifies a new supermultiplet, which in turn results in a number of different supermultiplets that grows combinatorially with a growing N.

In turn, the matrices $\mathbb{L}_I$ and $\mathbb{R}_I$ in the equations (10.120) satisfy the relations

$$(L_I)_i{}^{\alpha}\,(R_J)_\alpha{}^{k} + (L_J)_i{}^{\alpha}\,(R_I)_\alpha{}^{k} = 2\delta_{IJ}\,\delta_i^k, \tag{10.122}$$

$$(R_I)_\alpha{}^{j}\,(L_J)_j{}^{\beta} + (R_J)_\alpha{}^{j}\,(L_I)_j{}^{\beta} = 2\delta_{IJ}\,\delta_\alpha^\beta, \tag{10.123}$$

which define a double cover of the Clifford algebre $\mathfrak{Cl}(0, N)$.[15] In the original articles [195, 197, 196, 198, 199, 194, 193] the algebra (10.123) was denoted $\mathcal{GR}(d, N)$, where it is assumed that, as needed, the superfields $\boldsymbol{\Phi}_i, \boldsymbol{\Upsilon}_\alpha$ may be replaced by their ∂_τ-derivatives. Indeed, this formal ∂_τ-mapping connects all supermultiplets with the same "chromo-topology" [139]. For the relatively simple case of quantum-mechanical $N = 2$ supersymmetry, iterations of such ∂_τ-mapping yield the cyclic sequence

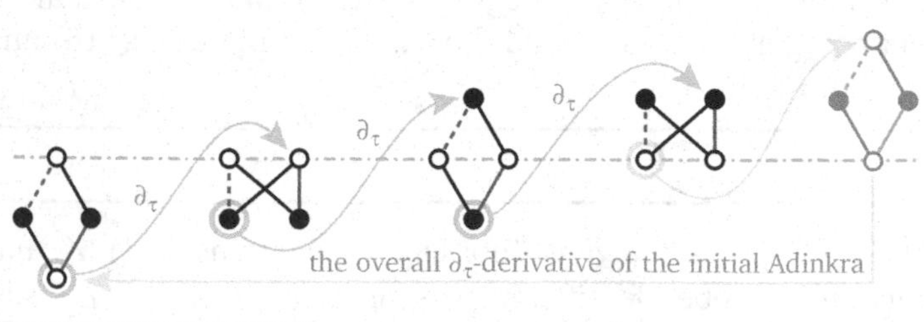

(10.124)

[15] The double-covered Clifford algebra is obtained by identifying $\mathbb{L}_I, \mathbb{R}_I \overset{2\text{–}1}{\longmapsto} e_I$.

For quantum-mechanical $N = 3$ supersymmetry, the analogous cyclic sequence is

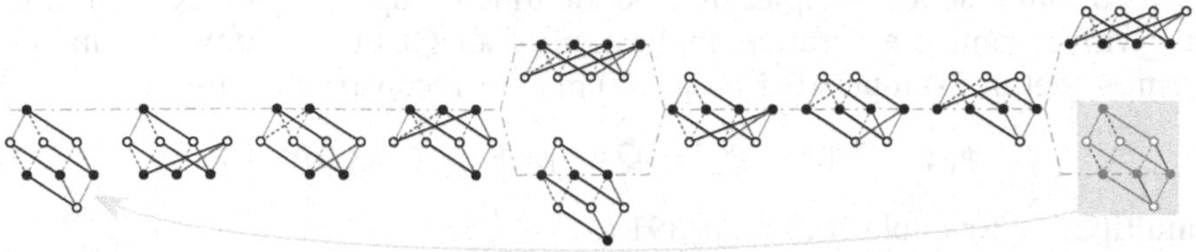

(10.125)

where the gray-highlighted Adinkra (bottom, right) is identical (up to the overall level, indicated by the gray dot-n-dash line) to the initial one, at far left, thus repeating the cycle. These illustrations show that for the (even just adinkraic) finite-dimensional representations of quantum-mechanical N-extended supersymmetries the number of possible node-height arrangements – and so the number of different supermultiplets – grows combinatorially with the growing number of supersymmetries, N.

In addition, starting with $N = 4$, there emerges a new possibility – "projections" – of which more in the next section. May it suffice here to show but one example:

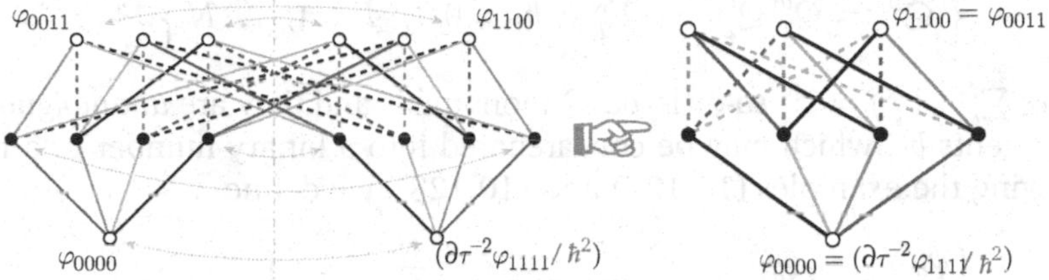

(10.126)

The dashed double-ended arrows indicate some of the pairs of component fields in the left-hand supermultiplet that are identified so as to obtain the component fields of the right-hand supermultiplet. The naming convention of the labeled component fields is explained in the next section.

It has been proven that the number of such "*adinkraic*" off-shell supermultiplets grows fantastically fast with the number of supersymmetries, and one expects about 10^{47} distinct supermultiplets for $N \leqslant 32$, which are expected to form about 10^{12} equivalence classes [141, 142]. Finally, it has been shown that an infinite number of ever larger (and non-adinkraic) supermultiplets can be constructed as networks of adinkraic supermultiplets, connected by one-way supersymmetry transformations [284]; this is also the structure of some rather well-known supermultiplets of simple supersymmetry in 4-dimensional spacetime [190].

— ❦ —

For such a (worldline) supermultiplet to be the 1-dimensional "shadow" of a supermultiplet from a 4-dimensional supersymmetric field theory, it is necesary that both the component fields and the supercharge action are compatible with Poincaré symmetry in 4-dimensional spacetime. One expects this to be a rather nontrivial requirement [197, 157, 158, 191, 283], which should drastically reduce the number of possible supermultiplets in higher-dimensional spacetime, but this verification (dimensional reconstruction) is far from solved in general☝; see Refs. [157, 158, 191, 283, 409].

10.4.2 Supermultiplets and binary encryption

It is fascinating that the classification of off-shell quantum-mechanical supermultiplets [140, 142] is closely related to the classification of doubly even binary linear block codes, which may be used in error-detecting and error-correcting encryption [286].

That is, in the quantum-mechanical N-extended supersymmetry (10.113) we have N real supercharges $\mathcal{Q}_I$, so that a supermultiplet may be identified – up to an application of ∂_τ- and/or ∂_τ^{-1}-prefactors – with a complete iterative application of all $\mathcal{Q}_I$'s upon some starting component field. The supermultiplet in Example 10.2 on p. 390 may be reconstructed also as

$$\{\,\phi,\ \psi_1 := \mathcal{Q}_1(\phi),\ \psi_2 := \mathcal{Q}_2(\phi),\ F := \mathcal{Q}_1\big(\mathcal{Q}_2(\phi)\big)\,\}, \tag{10.127}$$

and the supermultiplet in Example 10.3 on p. 391 as

$$\{\,\varphi_1,\ \chi_1 := \mathcal{Q}_1(\varphi_1),\ \chi_2 := \mathcal{Q}_2(\varphi_1),\ \varphi_2 := \partial_\tau^{-1}\big(\mathcal{Q}_1(\mathcal{Q}_2(\varphi_1))\big)\,\}. \tag{10.128}$$

As the defining relations of the supersymmetry algebra (10.113) imply that

$$\mathcal{Q}_I\mathcal{Q}_J = -\mathcal{Q}_J\mathcal{Q}_I, \qquad I \neq J, \tag{10.129a}$$

$$\big(\mathcal{Q}_I\big)^2 = H, \qquad I = 1,\dots,N, \tag{10.129b}$$

it follows that every formal $\mathcal{Q}$-monomial can be expressed as a linear combination of H-multiples of lexicographically ordered monomials from the basis

$$\Big\{\,\boldsymbol{\mathcal{Q}}^{\boldsymbol{b}} := \mathcal{Q}_1^{b_1}\mathcal{Q}_2^{b_2}\cdots\mathcal{Q}_N^{b_N},\quad b_I = 0,1,\ I = 1,\dots,N\Big\}. \tag{10.130}$$

Evidently, there are $\sum_{k=0}^{N}\binom{N}{k} = 2^N$ so-ordered $\mathcal{Q}$-monomials and they are *unambiguously* encoded by the binary exponents b_I, which may be concatenated into a binary number of a formal binary exponent $\boldsymbol{b}$. Following the examples (10.127) and (10.128), we define

$$\begin{Bmatrix}\phi_{\boldsymbol{b}}\\ \psi_{\boldsymbol{b}}\end{Bmatrix} := \boldsymbol{\mathcal{Q}}^{\boldsymbol{b}}(\phi_{00\cdots}),\quad \text{when } |\boldsymbol{b}| := \sum_{I=1}^{N} b_I \text{ is } \begin{cases}\text{even,}\\ \text{odd.}\end{cases} \tag{10.131}$$

The field identification in the relation between the two Adinkras (10.126) requires the imposition of the operatorial conditions[16]

$$\mathcal{Q}_1\mathcal{Q}_2 \simeq +\mathcal{Q}_3\mathcal{Q}_4, \quad \mathcal{Q}_1\mathcal{Q}_3 \simeq -\mathcal{Q}_2\mathcal{Q}_4, \quad \mathcal{Q}_1\mathcal{Q}_4 \simeq +\mathcal{Q}_2\mathcal{Q}_3, \tag{10.132a}$$

in addition to the relations (10.113), i.e., (10.129). Indeed, acting (always only from the right!) by the operators $\mathcal{Q}_1$, $\mathcal{Q}_2$, $\mathcal{Q}_3$ and $\mathcal{Q}_4$ on the relations (10.132a) produces

$$H\mathcal{Q}_1 \simeq +\mathcal{Q}_2\mathcal{Q}_3\mathcal{Q}_4, \quad H\mathcal{Q}_2 \simeq -\mathcal{Q}_1\mathcal{Q}_3\mathcal{Q}_4, \quad H\mathcal{Q}_3 \simeq +\mathcal{Q}_1\mathcal{Q}_2\mathcal{Q}_4, \quad H\mathcal{Q}_4 \simeq -\mathcal{Q}_1\mathcal{Q}_2\mathcal{Q}_3, \tag{10.132b}$$

and then, finally, also

$$\big(H^2 = -\hbar^2\partial_\tau^2\big) \simeq -\mathcal{Q}_1\mathcal{Q}_2\mathcal{Q}_3\mathcal{Q}_4. \tag{10.132c}$$

This last relation corresponds to the identification of the component fields:

$$\Big(H^2\phi_{0000} = -\hbar^2\big(\partial_\tau^2\phi_{0000}\big)\Big) = \Big(-\mathcal{Q}_1\mathcal{Q}_2\mathcal{Q}_3\mathcal{Q}_4(\phi_{0000}) =: -\phi_{1111}\Big). \tag{10.133}$$

Similarly, other relations (10.132) encode all other identifications (10.126), and so also the projection of the bigger, left-hand side supermultiplet to the smaller, right-hand side supermultiplet. It is essential to note that the relations (10.132) do not impose any ∂_τ-differential equation upon any of the component fields, and each field – and so the entire supermultiplet – remains off-shell.

[16] By operatorial conditions/relations one implies conditions/relations between two operatorial expressions, and which conditions/relations must hold when the left-hand and the right-hand sides of the equality are applied on any object upon which the operation of the given operators is defined.

Digression 10.14 Note that – *up to additional H-factors* – each of the eight relations (10.132) may be obtained from any other one. For example,

$$HQ_2 \simeq -Q_1Q_3Q_4 \xrightarrow{\cdot Q_3} HQ_2Q_3 \simeq -Q_1Q_3Q_4Q_3 = +HQ_1Q_4. \tag{10.134a}$$

In that sense are the three relations (10.132a) "basic" since all other relations (10.132b)–(10.132c) follow with no additional H-factors, whereas the converse does not follow. Jointly, the relations (10.132a) may be written as

$$Q_IQ_J - \tfrac{1}{2}\varepsilon_{IJ}{}^{KL}Q_KQ_L \simeq 0, \tag{10.134b}$$

which indicates the need for the Levi-Civita symbol ε_{IJKL}, where all four indices have precisely one of the four possible values – corresponding to the binary number b = 1111 [☞ Ref. [141] for analogous relations that correspond to other codes].

All the relations (10.132), and so also all the identifications (10.126) are almost unambiguously encoded by the binary number $b = 1111$,[17] which generates a so-called "binary doubly even linear block code" d_4 [286], which is also the simplest such code. These codes are used in binary encryption that helps in communications by enabling the detection of transmission errors and even some corrections, and without re-transmitting the original message. Once projected, as in the example (10.126), the smaller supermultiplet may be connected with various "node-height rearrangements" by applying the formal ∂_τ- and ∂_τ^{-1}-prefactors, which then generates all possible supermultiplets with that chromo-topology [142].

Thus, the classification of off-shell worldline supermultiplets is closely related to the classification of "binary doubly even linear block codes," and gives a close relationship between supersymmetry and encryption – which is a fully unexpected and fascinating result in this research. Numerically, it is even more fascinating that there are at least $\sim 10^{47}$ such codes for $N \leqslant 32$ (which is a limit suggested by the M-theoretic extension of superstrings), and that they form at least $\sim 10^{12}$ equivalence classes; moreover, the number of supermultiplets of which the "chromo-topology" [142] is defined by any one such code itself grows combinatorially with N, which further increases the "menagerie."

The construction and classification of off-shell supermultiplets in higher-dimensional spacetimes starting from the so far discussed worldline off-shell supermultiplets is in progress [157, 158, 191, 283, 409] ☝. In addition, other approaches and methods can complement these efforts, even if in more specific setting (such as for a fixed number of supersymmetries, N): see, for example, Refs. [281, 292, 47], to begin with.

10.4.3 Exercises for Section 10.4

✎ **10.4.1** *Prove that the Lagrangian terms (10.117) and (10.118) are invariant with respect to the supersymmetry transformations (10.114)–(10.116).*

✎ **10.4.2** *Derive and solve the equations of motion defined by the Lagrangian density (10.117).*

✎ **10.4.3** *Complete the Lagrangian term* $\mathscr{L}_3 = \omega(\varphi_1\dot{\varphi}_2 - \varphi_2\dot{\varphi}_1) + \cdots$ *so it is invariant with respect to the supersymmetry transformations (10.116).*

✎ **10.4.4** *Derive and solve the equations of motion defined by the Lagrangian density* $\mathscr{L}_1 + \mathscr{L}_3$*, as defined in the expression (10.117) and the solutions of Exercise 10.4.3.*

[17] Except for the choice of the relative sign in equation (10.134b), for cases with a total of $N = 4k$ supersymmetries.

11

Strings: unification of all foundations of reality

The theory (i.e., theoretical *system*) of strings was already mentioned in Chapter 1 and Section 10.2, and in a very fleeting way also in the historical review in Chapter 2 – which must be supplemented before we turn to even a non-technical review of the theoretical system as well as some of the lessons from this research. The historical review will therefore introduce a few new terms, which will thereafter be clarified in the remainder of this chapter.

By now, there exist complete and pedagogically well organized texts on string theory [225, 224, 417, 483, 434, 124, 594, 398, 298, 46, 312, 251, 510], lecture notes [424, 381, 432, 170, 430], in relation to Yang–Mills gauge theories [439], recent reviews [373] and even popular books at the "guide for complete idiots" level [375, 358, 299]. This, final chapter – and the entire book – can then possibly serve only as an aperitif and prerequisite to most of this growing literature.

11.1 Strings: recycling, recycling...

Only about four decades old, the string theoretical system is based on the fundamental idea that the elementary particles – the basic building blocks of Nature and the Democritean ideal indivisibles – are not point-like. As we have concluded (with the benefit of hindsight!) already in Section 1.3.3, there exist natural limits to the tininess of elementary objects, and the "material point" is merely an idealization.[1] For students who have successfully completed a course in electrodynamics, it will be natural to regard these basic building blocks as 1-dimensional elementary objects in the next order of approximation (multipole expansion). Except, unlike a *rigid* dipole (rigid bodies being a non-relativistic idealization), strings are relativistic one-dimensional objects that possess "internal" dynamics, which essentially stems from their 1-dimensionality. However, these fundamental strings do not consist of anything "more elementary," and it is precisely the dynamics of this irreducible

[1] As was mentioned in the Preface, what exactly is identified as an elementary object is historically qualified. Chemists of the nineteenth century had rightly considered atoms as elementary; in the transition into the twentieth century, electrons and atomic nuclei appeared elementary, but it soon became evident that the nuclei consist of more elementary nucleons, and up until the last quarter of the twentieth century the list of elementary particles consisted of several leptons and a combinatorially growing list of hadrons. In the last quarter of the twentieth century, the list of elementary particles shortened to the compact Table 2.3 on p. 67, to a dozen or so elementary fermions and another dozen or so mediators of their interactions.

non-point-likeness that produces the unexpected complexity of the string theoretical system as well as many other properties with no precedent in the physics of elementary particles.

11.1.1 The original idea and application

The basic ideas (roots) of the string theoretical system date back to 1943, when Werner Heisenberg introduced the idea of the *S*-matrix. Namely, Heisenberg noticed that the familiar notions used in the classical description of physics (space, time, particle, etc.) need not be well defined in quantum physics, and tried to design a formalism that deals only with observable quantities [☞ quote on p. xi].[2] By definition, the *S*-matrix of a physical process maps precisely every incoming state into every possible outgoing state, and depends only on the positions, momenta, energies, etc., defined and measured sufficiently far from the location of all interactions. Thus designed, the incoming and the outgoing states are called *asymptotic*, and the *S*-matrix theory is maximally non-local in the sense that it specifies relations only between events that are sufficiently separated in spacetime.

In the late 1950s and the 1960s, this approach grew into a program, the so-called "*S*-matrix theory," the most notable advocates of which were Stanley Mandelstam and Geoffrey Chew. Hendrik Kramers and Ralph Kronig discovered that assuming the *S*-matrix to be *analytic* allowed one to derive dispersion relations, which in turn imply causality between the asymptotic states even when causality is not microscopically well defined.

String theory – or, more precisely, the "dual resonant model" – was originally invented to describe hadrons, in the late 1960s. Namely, in collisions at sufficiently high energies, mesons become spatially elongated, somewhat akin to Figure 1.5 on p. 28, i.e., in display (4.101). In 1968, Gabrielle Veneziano discovered the formula (soon to be dubbed the Veneziano amplitude) that describes very well the amplitude of the effective cross-section of mesonic $A+B \to C+D$ collisions [☞ [434, Vol. 1] as well as [594]]:

$$\mathfrak{M}(p_A, p_B; p_C, p_D) \propto \frac{g^2}{\alpha'} I(s,t), \tag{11.1a}$$

$$I(s,t) := \int_0^1 d\lambda\, \lambda^{-\alpha' s-2}(1-\lambda)^{-\alpha' t-2} = \frac{\Gamma(-\alpha' s-1)\Gamma(-\alpha' t-1)}{\Gamma(-\alpha'(s+t)-2)}, \tag{11.1b}$$

where $\Gamma(z)$ is the Euler gamma function, and the ratio $B(a,b) := \frac{\Gamma(a)\Gamma(b)}{\Gamma(a+b)}$ that appears in the Veneziano formula (11.1) is the Euler beta function. The variables s, t, u in the Veneziano amplitude (11.1) are the Mandelstam variables (3.62), which appear in the analysis of the lowest order Feynman calculus for the process $A+B \to C+D$:

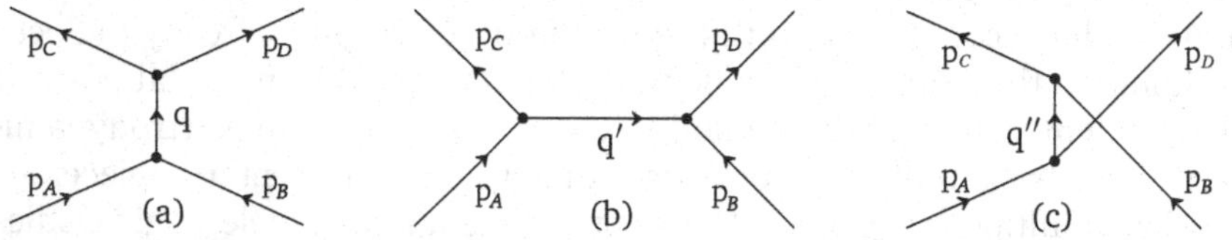

$$s := -q^2 = -(p_A^2 + p_B^2) \qquad t := -(q')^2 = -(p_A^2 - p_C^2) \qquad u := -(q'')^2 = -(p_A^2 - p_D^2) \tag{11.2}$$

The amplitudes for these sub-processes evidently satisfy the relations

$$\mathfrak{M}_{(a)}(p_A, p_B; p_C, p_D) = \mathfrak{M}_{(b)}(p_A, -p_C; -p_B, p_D) = \mathfrak{M}_{(c)}(p_A, -p_D; p_C, -p_D). \tag{11.3}$$

[2] One knew that protons and neutrons have finite sizes, about 10^{-15} m, and that the strength of the interaction between them at such distance was without precedent. As one after another of the attempts to model this force failed to correctly account for its peculiarities, Heisenberg believed the properties of space and time to radically change at nuclear and smaller distances.

However, for $A+B \to C+D$ meson collisions, experiments show that the first two amplitudes are in fact *equal* – which agrees with the Veneziano formula (11.1). In other words, the first two diagrams are only two different depictions of the very same physical process, so that these two depictions are equivalent, i.e., dual to one another.

Generalizations for amplitudes of collisions where more than four incoming and outgoing mesons appear were soon discovered by Yoichiro Nambu (1968), Holger Bech Nielsen (1969) and Leonard Susskind (1969) [549]. All these results indicated that the mediating state that carries the transfer 4-momenta q, q' and q'' in the diagrams (11.2), may be represented as a linear superposition of infinitely many linear harmonic oscillators (*resonances*) with masses and frequencies that were determined by the poles of the Euler gamma function, and which are all integral multiples of a fundamental mass, i.e., frequency. This property evidently points to the possibility of interpretation of the mediating state as something filamentary, of which these linear harmonic oscillators are the Fourier modes.

This is where the original identification of mesons with open *strings* (akin to the letter "I") emerges from the so-called "dual resonant model," where *duality* refers to the equivalence in the description of the first two collision processes (11.2) using either one of the two string-diagrams:

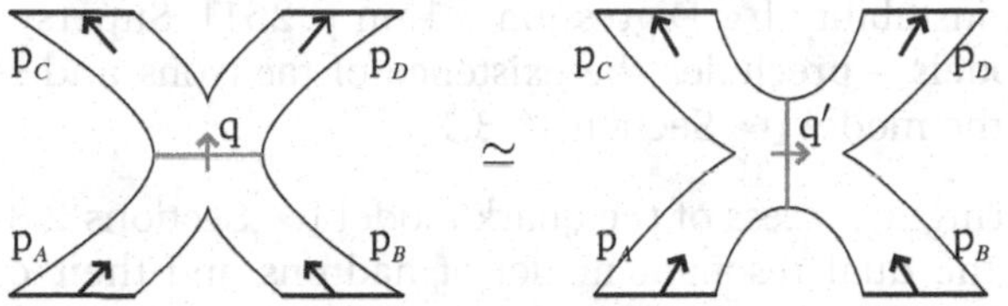

(11.4)

It is evident that the two surfaces in the Feynman diagrams (11.4) are equal, merely specified in somewhat different parametrization and with a differing interpretation of the "mediating" state, here denoted by the gray line. In the model name, "dual resonant model," the adjective *resonant* refers to the infinite sequence of harmonic *resonances* – the Fourier modes of the mediating state – whether these are identified with a virtual string that propagates upward (time-like) in the left-hand diagram, or to the right (space-like) in the right-hand diagram. In this model, each concrete meson is identified with one of the Fourier modes of the string, whereby the incoming and the outgoing mesons are also represented as strings, fixed into the configuration of the particular Fourier mode that corresponds to the given meson.

These diagrams make it evident that strings interact by joining the end-points and splitting in two, so the two incoming open strings in the left-hand diagram (11.4) join into one, mediating, which subsequently splits into the two outgoing strings. The dual resonant model (of strings) was very popular and intensively explored in the period 1968–74, by which time certain essential properties of this model were discovered:

1. Open strings (akin to the letter "I") may join into closed strings (akin to the letter "O"), the Fourier modes of which have no charge and for which the effective cross-section grows with the energy. In the late 1960s, Vladimir Gribov dubbed this subset of mesons the "pomeron sector," after Isaak Pomerančuk, who proved their necessary existence in all string models.[3]
2. In the pomeron sector, the Veneziano amplitude is fully (s,t,u)-symmetric, so that

$$\mathfrak{M}(\mathrm{p}_A, \mathrm{p}_B; \mathrm{p}_C, \mathrm{p}_D) \propto \frac{g^2}{\alpha'}\big[I(s,t) + I(t,u) + I(u,s)\big], \tag{11.5}$$

[3] Geoffrey Chew and Steven Frautschi brought many of the results obtained by (then) Soviet Union physicists to the West within their "democratic" theory, in which "hadrons consist of hadrons" and have no other, "more elementary" constituent factors and which is sometimes also referred to as the "bootstrap model," alluding to a story involving Karl Friedrich Hieronymus, Freiherr von Münchhausen [65]. This reminds us of fractals – an idea that Benoit Mandelbrot would introduce in mathematics several years later, in 1975.

which is depicted by the fact that the three string-diagrams for these three sub-processes:

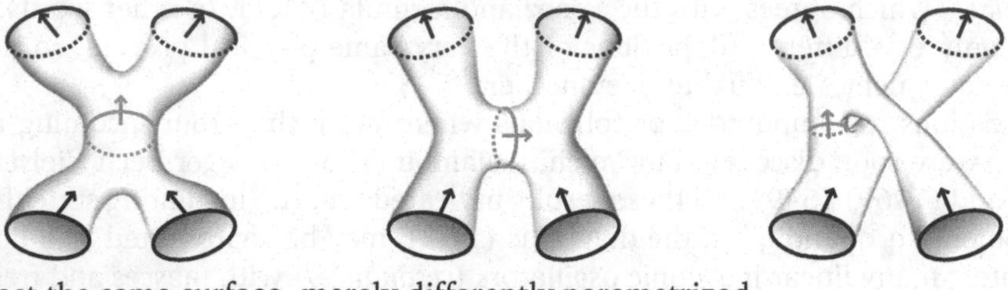

(11.6)

are in fact the same surface, merely differently parametrized.

3. The pomeron sector always contains also a *massless* spin-2 Fourier mode (this particle was dubbed the *pomeron*), whereas such a hadron was never experimentally detected.
4. The consistency of the dynamics of (bosonic) strings, which must move relativistically, requires them to propagate through a flat spacetime of $25+1$ dimensions. Supersymmetric strings, which Jean-Loup Gervais and Bunji Sakita discovered in 1971 [551], consistently propagate through $(9+1)$-dimensional flat spacetime.
5. All string models without supersymmetry contain tachyons, the presence of which in the spectrum indicates an instability [☞ Digression 7.1 on p. 261]. Supersymmetry – in any model, not just string models – precludes the existence of tachyons and so stabilizes the vacuum (ground state) of the model [☞ Section 10.3.1].

The well-nigh overnight success of the quark model [☞ Sections 2.3.12 and 2.3.13] in 1974 completely suppressed the dual resonant model of hadrons and their depiction as strings, and string theory became ignored.

In the same year, 1974, Tamiaki Yoneya and independently Joël Scherk and John Schwarz noticed that string theory can be applied not as a model for hadrons, but as a model of gravity [☞ Footnote *13* on p. 13]. The pomeron (the nonexistent massless spin-2 hadron) was thus identified with the graviton, and the string *tension* (T_0) and parameter α' in the Veneziano formulae (11.1) and (11.5) were now related through the extended equality

$$\alpha' = \frac{1}{2\pi T_0 \hbar c} = \frac{G_N}{\hbar c^5}, \tag{11.7}$$

with the Newton constant, where the units are

$$[\alpha'] = \frac{T^4}{M^2 L^4} = (\text{energy})^{-2}, \qquad [T_0] = \frac{ML}{T^2} = \left[\frac{\text{energy}}{\text{length}}\right]. \tag{11.8}$$

This changes the characteristic length of strings from the characteristic size of hadrons to the Planck length:

$$\ell_S := \sqrt{\alpha'}\hbar c = \sqrt{\frac{\hbar c}{2\pi T_0}} \sim 10^{-15}\,\text{m} \quad \to \quad \ell_S = \ell_P \sim 10^{-35}\,\text{m}. \tag{11.9}$$

This suggestion remained mostly ignored, partly because of the sudden focus on the quark model, and partly owing to the fantastic requirement that spacetime in string theory would have to have 25+1 dimensions (9+1 with supersymmetry). Scherk and Schwarz supposed that the so-called (Nordstrøm)–Kałuża–Klein compactified geometry could reduce the effective spacetime dimension to 3+1.[4]

[4] The idea that space may be more than 3-dimensional, and that the additional dimensions are periodic and with too small a radius to be detected stems from Gunnar Nordstrøm, in 1914. However, this model was based on his model of gravity, which differs from the general theory of relativity, and (as determined by 1919) also from Nature. Thus, this compactification idea was mostly forgotten together with his theory of gravity, until Theodor Kałuża in 1919 and Oscar Klein in 1926 independently revived it.

By the early 1980s, relatively few physicists were working on string theory, but even in this period of relative isolation, several significant results were derived:

1. In 1979, Daniel Friedan proved a fascinating fact about the (1+1)-dimensional field theory defined on the worldsheet swept out in time by strings. Requiring that quantum corrections do not renormalize the background metric of the spacetime through which the string propagates reproduces and generalizes the Einstein equations for gravity. That is, the quantum dynamics of strings implies gravity in the spacetime through which the strings move.
2. In 1981, Alexander Polyakov proved that the Hamilton action

$$S[X] = \frac{1}{4\pi\alpha'\hbar c^2}\int_\Sigma \mathrm{d}^2\xi\; g^{\alpha\beta}(\xi)\,(\partial_\alpha X^\mu)\,G_{\mu\nu}(X)\,(\partial_\beta X^\nu) \tag{11.10}$$

is classically equivalent to the Hamilton action

$$S[X] = -\frac{T_0}{c}\int_\Sigma \mathrm{d}^2\xi\,\sqrt{-\det\left[(\partial_\alpha X^\mu)\,G_{\mu\nu}(X)\,(\partial_\beta X^\nu)\right]}, \tag{11.11}$$

which Yoichiro Nambu and independently Tetsuo Goto originally formulated as the surface area of the worldsheet Σ that the string sweeps in spacetime $\mathscr{X}$. Here $X^\mu(\tau,\sigma)$ are the coordinates that indicate where in spacetime $\mathscr{X}$ is the point $\xi^\alpha = (c\tau,\sigma)$ of the string worldsheet Σ. The advantage of the Polyakov theory for the quantum theory of strings is evident: The quantization of the Hamilton action (11.11) is far more complicated (and without a generally accepted treatment) than that of the action (11.10). In turn, the geometric interpretation of equation (11.11) as the surface area of the string's worldsheet, which the Hamilton's minimal action principle minimizes, is retained by the virtue of the fact that the matrix

$$(X^*G)_{\alpha\beta} := (\partial_\alpha X^\mu)\,G_{\mu\nu}(X)\,(\partial_\beta X^\nu) \tag{11.12}$$

is the metric tensor on the worldsheet, induced ("pulled back") by the mapping $X : \Sigma \to \mathscr{X}$ from the metric tensor $G_{\mu\nu}(X)$ in spacetime.
3. Michael B. Green and John Schwarz discovered that two string models (II A and II B) are the T-dual of one another. This "T-duality" and its generalizations later proved to be one of the most important properties in the string theoretical system, by which all stringy models essentially differ from all *pointillist* models.[5]
4. In 1984, Louis Alvarez-Gaumé and Edward Witten published Ref. [12] (which had in its preprint form circulated since August 1983) with a detailed analysis of anomalies [☞ Section 7.2.3] in interactions with gravity, and with the result that the only models in (9+1)-dimensional spacetime where these anomalies do not destroy self-consistency are the string models II A (which is not chiral and so has no anomalies) and II B (where the anomalies cancel).

11.1.2 The string revolution

In summer of 1984, a preprint by Michael B. Green and John Schwarz started the "first string revolution," by publishing the proof [223] that Alvarez-Gaumé and Witten had omitted an important anomaly-cancellation possibility, and then demonstrated a concrete – and unexpected – mechanism (now called the Green–Schwarz mechanism) whereby anomalies cancel in the particular cases of $SO(32)$ and $E_8 \times E_8$ gauge symmetries! (Open strings with $SO(32)$ gauge symmetry were known; no string model with the $E_8 \times E_8$ gauge symmetry was then known.)

[5] To emphasize the fact that picturing elementary particles as idealized point-particles is but an approximation, in this chapter I will use the suggestive adjective *pointillist*.

Early in 1985, the papers (in circulation since late in 1984) by David J. Gross, Jeffrey A. Harvey, Emil Martinec and Ryan Rohm [247, 246, 248] were published, wherein the so-called heterotic string models were constructed: one with $SO(32)$ and another with $E_8 \times E_8$ gauge symmetry. Only a few months later, Philip Candelas, Gary Horowitz, Andy Strominger and Edward Witten showed [88], following the just-published work by Candelas with Derek Reine [90], that the $(9+1)$-dimensional spacetime of the $E_8 \times E_8$ heterotic string model may be compactified (*à la* Nordstrøm–Kałuża–Klein) on a complex Calabi–Yau 3-fold $\mathscr{Y}$,[6] so as to produce an effective model with $E_6 \times E_8$ gauge symmetry in (3+1)-dimensional spacetime and with $\frac{1}{2}\chi_E(\mathscr{Y})$ families of left-handed (chiral) fermions in the **27**-dimensional representation of the E_6 gauge group! ($\chi_E(\mathscr{Y})$ is the so-called Euler characteristic of the space $\mathscr{Y}$.) Since E_6 contains the $SU(5)$, the $SO(10)$ and the $SU(3)_c \times SU(3)_L \times SU(3)_R$ grand-unifying groups, it was clear that *there exist* string models that can contain the Standard Model of elementary particles.

Candelas soon showed that the details of the geometry of the complex Calabi–Yau 3-fold used in the compactification correlate with dynamical parameters in the Standard Model [114], which indicates that many if not all details of the Standard Model very likely may be derived from details of the geometry of so-compactified string models. This correlates many (if not all) of the physics properties of superstring models with the geometry of (generalized) spacetime!

Between spring 1984 and spring 1985, the attitude of most researchers in elementary particle physics completely changed, from totally ignoring string theory to fully focusing on constructing and exploring its models, including their compactifications.[7]

Soon, string models were constructed that at first blush had no geometric interpretation [377, 378]. Namely, in pointillist models, the configuration space is directly generated from the spacetime $\mathscr{X}$ through which these points move, and the dynamics of these material points is determined by the familiar geometry of the spacetime $\mathscr{X}$. In string models, the configuration space has a more complex structure. The Fourier spectrum of strings contains modes that propagate both in one and in the other direction around the closed string, these two classes of modes are independent and may satisfy different boundary conditions. This not only effectively doubles the configuration space but also permits constructions that are simply impossible in pointillist theories. This also requires a generalization of the usual structures in geometry – where the research is still developing. In 1993 Edward Witten constructed, and Jacques Distler and Shamit Kachru asymmetrically generalized the first class of models [574, 136] that interpolate between "geometric" Calabi–Yau models and "non-geometric" Landau–Ginzburg orbifold models first proposed by Cumrun Vafa. The name of these latter models indeed points to the fact that these are generalizations of the Landau–Ginzburg model as we have seen in Section 7.1.1. In these interpolating models, the "geometric" and "non-geometric" constructions appear as different phases of the same physical system defined on the worldsheet.

11.1.3 The second string revolution

It was already known in 1976 [482] that string models also include p-dimensional hypersurfaces where the open strings end if they are to satisfy Dirichlet boundary conditions [☞ Digression 11.6 on p. 415]. The integral parameter $p = 0$ here denotes a point, $p = 1$ refers to a string (filament),

[6] The term "compactification" implies that the spacetime geometry changes $\mathscr{X} = \mathbb{R}^{1,9} \to \mathbb{R}^{1,3} \times \mathscr{Y}$, where $\mathscr{Y}$ is a compact real 6-dimensional space, and $\mathbb{R}^{1,3}$ is the (3+1)-dimensional flat spacetime. Here, $\mathscr{Y}$ is a complex 3-dimensional subspace of some better known space, typically specified by a system of algebraic equations.

[7] In April 1983, I was present at a lecture by John Schwarz about strings at ICTP, in Trieste, when no one from the audience of some 300 or so active researchers in elementary particle physics asked a question after the lecture, except for Herman Nicolai who hosted the event.

$p = 2$ to a membrane, etc. During 1989–95, Joe Polchinski showed that these so-called Dp-branes,[8] for consistency, must be treated as independent dynamical objects. They are then as fundamental as the strings from which one may have started [☞ articles [438, 433] or books [434, vol. 2] and [298]]. In this nomenclature, in the original model of open bosonic strings with $SO(32)$ gauge group, the full (25+1)-dimensional spacetime is a $D25$-brane, and the end-points of the open strings are restricted to this 25-dimensional space but move freely in it. Many of these various elementary p-branes are, in the string theoretical system, in fact elementary (extremely charged) black objects – a realization (recycling [☞ Footnote *13* on p. 13]) of the idea in Digression 9.5 on p. 340, just not for electrons but for these new, p-dimensional objects. This makes it clear that the name "string theory," and even "string theoretical system," is a misnomer: this theoretical system *must* include objects of various dimensions[9] [☞ Section 11.4]!

By 1995, Leonard Susskind had included Gerardus 't Hooft's holography principle in string theory, whereby the high-energy excitations of strings coincide with the thermal states of black holes and by which the fluctuations of the event horizon describe not only the degrees of freedom of the black hole itself but also of the nearby objects. That same year, Edward Witten showed that the five basic string models (open, the $SO(32)$ heterotic, the $E_8 \times E_8$ heterotic, the Type II A and the Type II B) as well as their various compactifications may be regarded as limiting cases of a more fundamental, so-called "M-theory," which he showed to also have a sixth limiting case, which contains the (otherwise unique) point-particle supergravity in (10+1)-dimensional spacetime [575].

Digression 11.1 M-theory extended string theory thus *incorporates* (rather than *falsifies* in a Popperian sense) point-particle field theory, and requires it to have the specific symmetries and structure of the 11-dimensional supergravity, enriched by including also specific 5-branes.

The unification of these various models into a theoretical system of strings (by now extended by various p-branes), i.e., M-theory, around 1995, is regarded as the "second string revolution," whereby the revolution of 1984 is in retrospect counted as the first. Some Authors regard the change of application of string models in 1974 (from hadrons to gravity) as the first revolution, so their counting is shifted by one from the one used herein.

Following Witten's implicit "definition" of M-theory, Tom Banks, Willy Fischler, Stephen Shenker and Leonard Susskind generalized the holography principle to the whole M-theory. Juan Maldacena then noticed [354] the following sequence of relationships:

1. Low-energy excitations near a black hole are represented by physical objects that are localized near the event horizon of the black hole.
2. In the case of extremely charged Reissner–Nordstrøm black holes, the event horizon is of the form $AdS^d \times S^{9-d}$, where AdS^d denotes the d-dimensional anti de Sitter space (9.81) and the sphere S^{9-d} carries the flux of some gauge field.
3. This latter configuration may be described by an $N = 4$ supersymmetric (and conformally symmetric) version of Yang–Mills field theory [☞ Chapter 6 and 10].

[8] The coinage "p-brane" appears in the literature, as a back-formation from *mem*brane, and where the number p denotes the number of spatial dimensions. Continuing this back-formation, the term "brane" is used even all by itself as a dimensionally non-specific collective name.

[9] This provides an opportunity for a neatly rhyming recap: *the theoretical system of strings and things.*

This sequence of relationships was soon generalized, worked out and theoretically confirmed in detail (by Edward Witten, Steven Gubser, Igor Klebanov and Alexander Polyakov, among others [510]), and is today called the *AdS/CFT* (or, more generally, the *gravity/gauge*) *correspondence* and represents a concrete theoretical realization of the holography principle. In a roundabout way, the string theoretical system has thus returned to its original application, to the description of interactions via gauge fields.

On the other hand, soon after Witten's proposal of *M*-theory, Cumrun Vafa generalized this proposal into the so-called *F*-theory, which indicates the existence of a phase of this united theory in which the spacetime has 12 dimensions [530] – and for which the effective 12-dimensional field theory is known only in various (partially) compact variants, and for which it is not determined if it has 10+2 or 11+1 space+time dimensions (there exist arguments for both cases)☝.

11.1.4 The third string revolution

The numerologically inclined Reader must have noticed the approximate cycle of about 11 years (just like Sun-spots?) between:

1. since 1974: strings are used to describe gravity and not hadrons;
2. since 1985: the five basic string models and their compactifications;
3. since 1995–7: indirect hints of *M*- and *F*-theory as the completion of the string theoretical system, as well as the establishment of the *AdS/CFT* correspondence.

However, 2007 arrived after an unhurried percolation of several ideas that fused into the picture of *the landscape* of string theories and *the swamp* of other models [☞ the book [505], a partial critique [152], the works [531, 395], as well as the rest of this chapter for starters]. For some participants and observers, the shift in understanding the task, the purpose, and even the power of physics – with the backdrop of this landscape and swamp – represents an anti-catharsis. Namely, the very idea of the existence of an enormous and *connected* "web" of all possible string models is not new,[10] and draws its roots from the phase diagrams in grand-unified models [☞ Chapter 8], which in turn conceptually remind us of the phase diagrams in the physics of bulk materials.

The open question is, however:

1. is there a principle (such as minimization of free energy in statistical mechanics) that singles out one of all those models – hopefully such that it describes Nature just the way we observe it?
2. or is this Universe of ours selected by the fact that we – such as we are – could not even exist in some different Universe; which is the so-called "anthropic principle?"

Some regard the adoption of this anthropic principle as a revolution in understanding Nature, while others on the opposite end of a continuous palette of opinions regard this as a sign of intellectual capitulation; yet others regard it as a signal of the hopelessness of "string theory" within the science of physics [☞ paraphrasing Planck, on p. 124]. And then, there is also the infrequently adopted, but to my mind important view, best stated by Douglas Adams:

> This is rather as if you imagine a puddle waking up one morning and thinking, "This is an interesting world I find myself in – an interesting hole I find myself in – fits me rather neatly, doesn't it? In fact it fits me staggeringly well, must have been made to have me in it!" This is such a powerful idea that as the sun rises in the sky and the air heats up

[10] Early results in this direction [227, 226, 86, 87] came before their time: seven years later, their physics aspects started clearing in cases with more supersymmetry [22, 23, 24]. Two years after that, the original application of this physical process was shown to require some "isolated" models [304]; see however also recent works such as Refs. [492, 582, 290, 365] and references therein for some recent developments☝.

and as, gradually, the puddle gets smaller and smaller, it's still frantically hanging on to the notion that everything's going to be alright, because this world was meant to have him in it, was built to have him in it; so the moment he disappears catches him rather by surprise. I think this may be something we need to be on the watch out for. [6]

Somewhat in response to the realization of this seemingly unprecedented and complex opulence of the landscape and the swamp,[11] and the development in the vantage point (the ambivalence between the anthropic and the minimizing principle) within the string theoretical system, and partly also as a reply from the shadow of the fantastic activity around string models, there emerged critical reviews [490, 577], which are not infrequently kibitzing and criticize even just the interest in "string theory" since it is "too general to be experimentally falsified" [paraphrased, T.H.].[12] The basis of this critique stems from the fact that the string *theoretical system* has an immense (perhaps even infinite-dimensional [453]) continuous parameter space, which makes it seem impossible to refute, i.e., demonstrate that none of those choices can produce a realistic theory – or, more precisely, a concrete model that describes the observed Nature.

However, this is a very naive view of both the refutation process, as well as the logical justification of refuting. For example, the $N = 8$ supergravity in (3+1)-dimensional spacetime also contains a continuum of parameter choices, but since the 1980s it has been known with certainty that all the models within this theory possess the (non-realistic) unbroken parity symmetry; as of recently, we also know that all these models are most probably non-renormalizable [☞ Definition 5.1 on p. 211]. In fact, before the Green–Schwarz discovery (in 1984) of the mechanism of anomaly cancellation that was missed by the otherwise complete analysis of Alvarez-Gaumé and Witten, it was trusted that the analysis of anomalies offers a systematic characteristic of all string models (except for Type IIA and Type IIB, which in turn do not contain the Standard Model) that disqualified them all [☞ Section 11.1.2]. This definitely proves that even infinite-dimensional collections of models *can* in principle be ruled out. By virtue of the Green–Schwarz mechanism of anomaly cancellation in specific cases, the number of possible string models is significantly restricted.

In turn, even the Standard Model of elementary particles contains a continuum of parametric possibilities [☞ Chapter 7], and is reliably known to contain neither explanation nor a selection mechanism from among the possible choices of these parameters. By contrast, the last decade of research – and the imagery of the landscape and the swamp – indicates that the parameters in the string theoretical system in fact form a *discretuum* – a sufficiently dense but countable (and perhaps even finite) subset of parametric choices that is fixed among others also by a generalization of the Dirac mutual quantization of the electric and magnetic charges [☞ originally [74] and also the more recent general review [31]].

More generally and paraphrasing Refs. [505, 531, 395], there exists an immense landscape of perfectly quantum-consistent string models that contain the characteristics of Nature as we know it and which we inhabit. It is then believable that amongst those models there exists *one* that simultaneously features *all* of the characteristics of precisely *our* Nature, and with no "surplus."

---

[11] In reality, the plethora of theoretical models could always have been seen, but was not since many of the properties such as the dimension of spacetime or the choice of the gauge groups were taken for granted.

[12] For example, Lee Smolin [489, 490] promotes loop (quantum) gravity [☞ below] as competing with strings. Peter Woit [577] applies Pauli's denigrating "not even wrong" to "string theory" (which he openly calls a "failure") and lobbies theorists to do something else – without himself contributing to any concrete research. Bert Schroer, amid numerous historical-philosophical essays, lobbies for a very non-standard formalism that also contains a "delocalization" of point-particles [478, 479, and references quoted therein], which is "mistakenly interpreted as a string" (whatever that might mean). The critical review of Smolin's and Woit's books [490, 577] that Joe Polchinski published in the bimonthly *American Scientist* (Jan./Feb. 2007) caused a highly ramified tree of internet blog-debates that are freely accessible [☞ [436, 491, 429, 435] as well as [576], for starters].

There also exists a much larger swamp of classical, semi-classical and otherwise partially quantum-consistent models, which are however not completely quantum-consistent (and do not belong to the theoretical system of strings with its M- and F-theoretic extensions). The landscape models emerge from this swamp like islands.

> **Conclusion 11.1** *The (M- and F-theory extended) (super)string theoretical system is the first one in the history of fundamental physics that can predict a countability – and maybe even finiteness – of the number of perfectly quantum-consistent models with an adequate content of matter, broken P- and CP-symmetries, gravity and gauge interactions, and the first one that still maintains the hope that one of its models faithfully and completely describes* **our** *Nature.*

Given the fact that theoretical systems are by construction axiomatic systems that seem complex enough to be subject to Gödel's incompleteness theorem [☞ the lexicon entry, in Appendix B.1, and Appendix B.3], this feature of the string theoretical system is even surprising!

Besides, "string theory" is actually a theoretical system, which is pointless to falsify much as the *theoretical system* of classical mechanics is not falsified, nor can it be falsified in the naive Popperian sense [☞ Digression 1.1 on p. 9, Section 8.3.1 and Digression 11.2 on p. 408]. The string theoretical system contains models that are very successful in describing various natural phenomena, amongst which are also some characteristics that previously had been simply taken for granted [☞ Section 11.2]! Finally, as understood nowadays, the string theoretical system [434, Vol. 2, Figure 14.4] contains various (super)string models that are in fact limiting cases of a more fundamental theory (provisionally identified with the hints of M- or even F-theory) of which we so far know only what can be discerned from the vantage points of known special limiting cases. It is simply too early to tell.

11.1.5 Quantum gravity

The string theoretical system is by no means the only attempt at a rigorous definition and construction of both a qualitative and a quantitative description of quantum gravity [367, 322, 291, 495, 310, 67, 465, 342].

Namely, unlike Yang–Mills gauge theories, the (gauge) general theory of relativity is not renormalizable [217], nor is any field theory that includes gravity.[13] The technical part of the problem indubitably stems from the essentially nonlinear nature of the general theory of relativity [☞ Chapter 9]. Thus, a "complete theory of quantum gravity" simply (so far) does not exist (and even less existent is its quantum-consistent unification with the Standard Model). There exist the following more or less well developed candidates:

1. the (M- and F-theory extended) (super)string theoretical system,
2. loop (quantum) gravity (LQG),
3. various modifications of gravity.

The original idea in the loop (quantum) gravity approach is that the quantization procedure should be applied starting from a classical Hamilton action for gravity, which is classically equivalent to the Einstein–Hilbert action (9.38), but offers some (technical) advantage in the quantization procedure. By contrast, the Einstein equations (9.44) and the Einstein–Hilbert action (9.38) are only approximate results in the string theoretical system [☞ Section 11.1.1].

[13] So-called *perturbative gravity* is an effective description of quantum gravity where the perturbative contributions are systematically suppressed by powers of ratios of the form $E/(M_P c^2)$, where E is the characteristic energy of the considered process. In this formulation, there evidently is no chance of obtaining convergent results when the energy E approaches $M_P c^2 \sim 10^{19}\text{GeV}/c^2$, but the description becomes ever better at ever lower energies [343, and the references therein].

There exist more or less critical and relatively contemporary reviews of these alternative approaches (2 and 3), and the interested Reader is directed to these resources, e.g., starting with [11, 386, 385, 467, 466, 468] and the plentiful references in those works.

The alternative approach to quantizing gravity, the so-called loop (quantum) gravity, draws on the earlier *geometrodynamics* approach [310] and the use of Abhay Ashtekar's variables [21]. These were defined (in 1986) as certain contour integrals and so correspond to the (homotopy) classes of these contours. The *practice* within this approach regarding the treatment of symmetries and the subsequent constraints upon quantum states, however, radically differs from canonical (and so also Dirac, BRST, BV and BVF) quantum treatment of constraints [386, 385, 11], which is the essential building block element in the whole fundamental physics that led to the Standard Model, grand-unified models, and also the string theoretical system [☞ texts such as [554, 555, 484, 496]]. In this respect, it is not clear whether the approach to quantization in loop (quantum) gravity is in agreement with Nature as canonical quantization is known to be. Namely, the formalism of loop (quantum) gravity does not seem to detect any of the anomalies,[14] so it is not clear how this procedure could possibly be consistent with the Standard Model wherein anomalies play a crucial role.

However, the very definition (and evident possibility of identifying these contours with strings) indicates a possible connection between loop (quantum) gravity and string models. At any rate, however, loop (quantum) gravity for now does not even try seriously to unify gravity with other interactions and matter, and this approach is in this sense in a very different (and much more modest) category than the string theoretical system.

In other conceptual approaches, one postulates that on small enough distances either the Lorentz symmetry is no longer valid [☞ [273] and the recent review [545]], and perhaps even the space ceases to be a continuous topological space as one otherwise usually regards it and becomes something akin to foam [☞ Footnote 2 on p. 398], or the law of gravity changes on cosmic scales [155, and references therein], or some other characteristic of gravity and/or spacetime varies. In the spacetime foam approach, the macroscopic and well-known characteristics of spacetime (continuity, smoothness) are simply a result of averaging over an enormous ensemble of structures that are, each by itself, of a wholly different nature. For astrophysical considerations that prompt such modifications of the "ΛCDM" (Einstein gravity with a cosmological constant, Λ, and cold dark matter) model and a review of modified Newtonian dynamics (MOND) proposals, see recent reviews such as Refs. [180, 498, 156] and references therein. In loop (quantum) gravity, spacetime also emerges as a dynamical and produced structure: The space itself is produced from so-called spin networks, which in time sweep out the so-called spin foam, whereby these two a-priori independent approaches turn out to have a common point.

Somewhat more recent are approaches wherein gravity emerges from a simpler theory that included neither general coordinate invariance nor a version of Einstein's equivalence principle [486, 315]. It is also possible to apply the gauge principle to different subsets of general coordinate transformations or treat them in somewhat different ways, and so obtain differently gauged theories of gravity [451, 276].

Finally, the common characteristic of all these approaches to quantum gravity is that, so far, no feasible experiment is known that would rule out or confirm any one of them. For example, the stringy corrections to the Einstein equations are too small to be measurable except in strongly curved spacetime, as should be the case near a singularity.

---

[14] See Digression 7.2. Also, LQG fails to explain the ultimate fate of the well-known Goroff–Sagnotti two-loop divergence [217] that signals the failure of conventional renormalization within quantum gravity; see also Refs. [386, 385, 11] for a more detailed discussion of shortcomings of the LQG approach.

11.2 The theoretical system of (super)strings

The subject matter of elementary particle physics had by the end of the twentieth century grown beyond the usual confines of a physics discipline such as, e.g., atomic physics, molecular physics or astrophysics. On one hand, the subject matter is not a single, relatively bounded domain of natural phenomena and structures, but forms a hierarchy of at least two levels of such structures[15] [☞ Table 2.5 on p. 71]:

1. hadron physics,
2. quark–lepton physics,

and a third, essentially different level, temporarily identified as

3. (*M*- and *F*-theory extended) (super)string physics, including alternative approaches such as loop (quantum) gravity, spacetime foam and other modifications of gravity.

On the other hand, (super)string theory is no longer a concrete theory of one concrete (our) reality, but a theoretical system that we hope will be able to describe our reality *also*. In the same sense, nor is classical (non-relativistic and non-quantum) physics a single, particular theory of a single, particular mechanical system, but a theoretical system applicable to a broad (continuous, in fact!) spectrum of phenomena, both natural and also completely artificial.[16]

Digression 11.2 Classical mechanics – as a very well known theoretical *system* – is perfectly applicable to force laws: $F = \alpha x^{\sqrt{17}}$, $F = \beta x^{x/x_0}$, $F = \gamma^{\frac{\arctan(x/x_0)}{\ln(1+kx)}}$, etc. From among *uncountably many* functional forms, *our* Nature chose $F = -kx + O(x^3)$ for springs, but $F = \frac{k'}{r}$ for gravity. In 3-dimensional space (and with one dimension of time), Bertrand's theorem [327, 213] guarantees that only these two force laws provide for stable orbits, but nothing – within the theoretical system of classical mechanics – prevents the existence of $F \propto \frac{1}{x+x_0}$ springs, or $F \propto -r$ gravity. Yet, no one deems classical mechanics any less "scientific" because of its inability to "predict" the correct force law from first principles.

Amusingly, extending classical mechanics by including the gauge principle permits one to derive the $\propto 1/r$ force law for gravity (in $(3+1)$-dimensional spacetime). Extending it by including some ideas about elasticity (either by postulate or as derived from the microscopic structure and interactions within materials) permits one to derive the $\propto -x$ restoring force law of springs. This supports the expectation that many of the particular but unexplained characteristics of the Standard Model will be explained by (and derived from) developments *beyond* the Standard Model itself.

The next step in the evolution of that theoretical system is provided by the theory of relativity and quantum theory; the coherent unification (but without gravity, and so without acceleration) is known as quantum field theory. Conceptually, this is a well-defined theoretical system,

[15] The delineation of these "levels" is of course *practical* but artificial; Nature is one. Just as there exist chemical processes that belong both in organic and in inorganic chemistry, so does the structure of small nuclei (those of deuterium, tritium, helium,...) belong both in nuclear and in hadron physics, and the structure and dynamics of quark bound states both in hadron and in "quark–lepton physics." Finally, the electromagnetic (and also gravitational) interaction of course appears through all this physics, from microscopic to macroscopic scales.

[16] It is not difficult to see that even a small change in the concrete value of some of the natural constants would have significant repercussions with the end result that such a World would be significantly different from ours – and, thus, very *unnatural*. Numerous humorous examples of such ilk form the scientific basis for the popular books for laymen [183].

in that it is known that all field theory necessary in the Standard Model of elementary particle physics [☞ Chapters 3–7] is renormalizable and has no anomalies.

The remaining step in this evolution then must include gravity, and for this the (M- and F-theory extended) (super)string theoretical system is the most successful candidate.

One hopes that the *true* fundamental *Theory of Everything* is simply finite, and that no renormalization is needed; there do exist indications that this may well be the case with (super)string models. Finally, string models as a class of theoretical constructions are still the most likely milieu for approaching the Theory of Everything. It is literally too early to judge, owing to the simple reason that the class of (M- and F-theory extended) (super)string models is by far not known sufficiently well [505] – although one knows things about string models that one would not even have thought of asking of the physics before strings.

11.2.1 General requirements

The basic characteristics of Nature, as uncovered by the physics of the twentieth century, are summarized in Table 11.1. Any theoretical system with the ambition to describe Nature must contain these characteristics as its integral properties. Alternatively, were we to wish to substitute any of the listed characteristics with something else, we would have to prove not only that the alternative equally well generates models of natural phenomena, but also that it fits equally well with all other characteristics of Nature.

Table 11.1 Characteristics of describing Nature, key properties/purpose and the resulting unifications

	Characteristic	Universal property	Unifies/describes
	Quantumness	Stabilizes atoms	Waves and particles
Gauge principle	Special relativity	Links symmetries, conservation laws, forces/interactions and geometry	Spacetime, energy–momentum
	General relativity		Acceleration-gravitation, mass-inertia
	Relativity of phases (of wave-functions)		(Electro-magneto)+weak, and strong interactions
	Supersymmetry*	Stabilizes vacuum	Bosons and fermions

* Supersymmetry is the only characteristic listed here that is not yet experimentally verified, but is the only (known!) universal characteristic of which the consequences include vacuum stabilization.

Conclusion 11.2 *Nature is one; the fragments of our description of Nature sooner or later must fit into a single, coherent and consistent whole.*

11.2.2 The spacetime perspective

Start with the simplest example, where an open string moves through flat space. Let σ be a coordinate along the string, so that $\sigma = 0$ and $\sigma = \ell_S$ are the end-points (and $\sigma \in [0, 2\ell_S]$ for a closed string, where we identify the ends to form a circle of circumference $2\ell_S$), and let τ be the proper time of the string.[17] If X^μ are the coordinates in the *target* spacetime $\mathscr{X}$ through which that string moves, then $(X^0(\tau,\sigma), X^1(\tau,\sigma), \dots, X^{n-1}(\tau,\sigma))$ is the n-plet of *coordinate functions* that specify where the point σ on the string, at the proper time τ, is located in the target spacetime $\mathscr{X}$, where $\dim(\mathscr{X}) = n$.

[17] Since a string is not a point and different parts of the string may move with different velocities, every point on the string has its own proper time, and we only require that the parametrization of proper time vary continuously from point to point along the string so that the pair of variables (τ, σ) provides a coordinate system covering the worldsheet swept out by the string.

As the proper time τ passes, the string sweeps out the worldsheet on which we may introduce the general coordinates (ξ^0, ξ^1). Suppose that the string is at rest so $c\tau = \xi^0 = X^0$ and that it extends along the coordinate $X^1 = \xi^1 = \sigma$. Then, the surface of the worldsheet is obtained by the integral $\int dX^0 dX^1$. In the general case, when the string moves arbitrarily through the target spacetime, a coordinate change must include all coordinates X^μ as well as the metric tensor of the target spacetime, and the result is the Nambu–Goto action (11.11).

Geometrically, the n-tuplet of functions $X^\mu(\xi)$ provides the mapping $X : \Sigma \to \mathscr{X}$ of the worldsheet Σ (that the string sweeps out in the process of moving) in the target spacetime $\mathscr{X}$, and the Hamilton action (11.11) characterizes this mapping.

Comment 11.1 *To be precise, $X(\Sigma) \subset \mathscr{X}$ denotes the **image** of the worldsheet in the target spacetime, to be distinguished from the abstract worldsheet Σ. For example, a **constant** mapping X produces the image $X(\Sigma)$ that is a single **point** in $\mathscr{X}$.*

Namely, for every possible image of the worldsheet that connects the string in any given initial position and the string in any given final position, one can compute the action $S[X]$. The classical worldsheet is the one that minimizes the Hamilton functional $S[X]$, and provides a textbook example of the application of Hamilton's variation principle [☞ Figure 11.1]. From this vantage

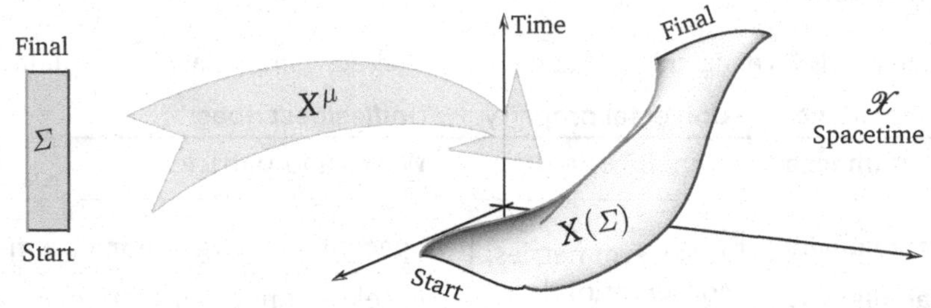

Figure 11.1 An image of the worldsheet of an open string in spacetime.

point, of primary importance is the motion of the string from its initial into its final position and the image of the worldsheet that this string sweeps out in the spacetime $X(\Sigma) \in \mathscr{X}$. The canonical quantization of the dynamics of this motion – using the Hamiltonian defined originally from the Nambu–Goto action (11.11) and subsequently from the Polyakov action (11.10) – gave the original results such as the critical dimension $\dim(\mathscr{X}) = 26$ for the ordinary string, i.e., $\dim(\mathscr{X}) = 10$ for the supersymmetric string, where the string oscillators are accompanied also by supersymmetric partners. The string end-points (i.e., the "side" edges of the surface $X(\Sigma)$) may be assigned charges. Strings interact by splitting in two and by joining ends [☞ Figure 11.2] – recall that

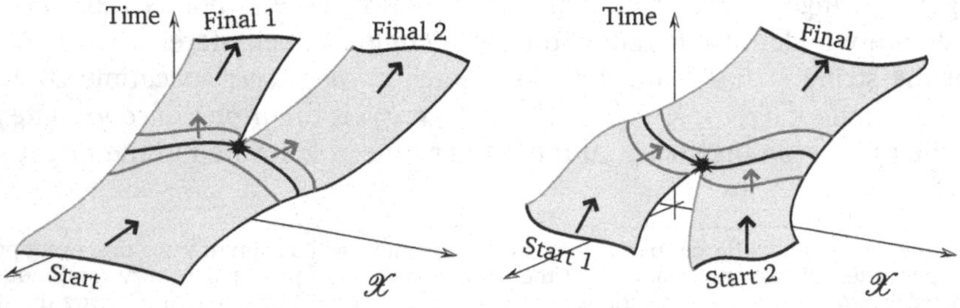

Figure 11.2 String interactions: splitting of one string into two, and joining two strings into one.

the original inspiration for strings were mesons.[18] In mesons, the string end-points are identified as the locations of the quark and the antiquark that carry charges (isospin, electric charge, ...), while the string itself represents the continuous and two-way flux of the chromodynamic field that binds the quark and the antiquark. It follows that one end-point of a string may join with another end-point of that or another string only if all charges on one of the two end-points are of the *opposite type* from the charges on the other of the two end-points. Also, when a string splits, the two newly created end-points must have opposite charges. Thus, open strings with charged end-points represent the combination $(q, \overline{q})$, where q is the n-dimensional "charge."[19] For either of the two end-points of one string to be able to join with either of the two end-points of another, their interactions must be governed by the group $SO(n)$; quantum consistence then requires $n = 32$ [225].

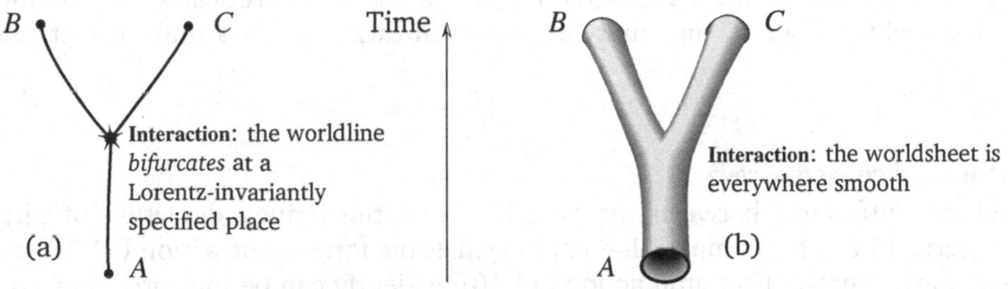

Figure 11.3 The $A \to B + C$ decay in the pointillist description (a), and in the stringy description (b).

The conceptual difference between pointillist and stringy descriptions of a simple process, e.g., the decay of a particle (3.130) $A \to B{+}C$, may be seen in Figure 11.3, for an example with a closed string. In the case of the interaction of open strings [☞ illustrations (11.4)], in certain subsets of coordinate systems there still exist special points at the edge of the string's worldsheet [☞ the cusp-like edge-points of the surfaces in the illustrations (11.4)]. However, it may be shown that this worldsheet parametrization is, via a conformal mapping, equivalent to another, where these particular points are not singled out.

Since the parametrization of the worldsheet that connects the starting and the final positions of the strings in any process cannot even in principle be measured, the measurable quantities in the stringy description must be averaged over all possible parametrizations. In doing so, not only do we have to average over all parametrizations of a fixed abstractly specified worldsheet that connects the starting with the final positions of the strings, but we must average also over all possible worldsheets that satisfy those boundary conditions. The decay process $A \to B + C$ is thus described by Feynman diagrams such as in Figure 11.3 only in the lowest approximation. The next approximation requires adding the probability amplitudes depicted by the Feynman diagrams such as in Figure 11.4. In the pointillist description of elementary particles, the two Feynman diagrams contribute differently and separately – and in fact contribute to the renormalization of different physical quantities. In the stringy variant the contributions of the two Feynman diagrams are equal, since the worldsheets of those diagrams are equivalent: one of these worldsheets may be mapped into the other by means of a continuous deformation.

[18] It is certainly not clear how to represent baryons in this naive picture, which is the additional reason for neglecting strings as a model for hadrons.

[19] In the sense that the electric charge is a 1-dimensional charge, isospin is a 2-dimensional charge, color is a 3-dimensional charge, etc.

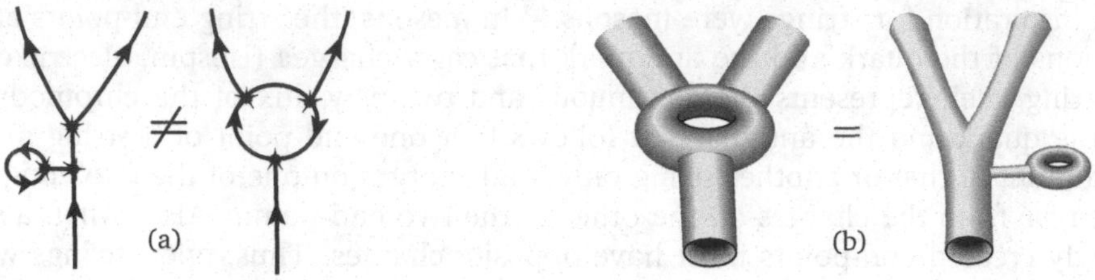

Figure 11.4 Feynman diagrams for first-order corrections to the $A \to B + C$ decay, in the pointillist description (a), and in the stringy description (b).

This engenders the intuitive impression that the string theoretical system is much better defined in the technical sense, and that the renormalization process may in fact not even be necessary.

11.2.3 The worldsheet perspective

The geometrical difference between the pointillist and the stringy depiction of physical processes [☞ Figures 11.3–11.4 on pp. 411–412] as well as the form of the action (11.10) points to an alternative perspective: the Hamilton action (11.10) evidently can be interpreted as an action for a physical system that "lives" in the (1+1)-dimensional spacetime of the *abstract* worldsheet of the string. The function $X^\mu(\xi)$ is here simply the μth field that is a scalar with respect to the (1+1)-dimensional Lorentz symmetry transformations. From this perspective, the (1+1)-dimensional worldsheet spacetime Σ is the fundamental spacetime, in which the geometry is specified by the metric tensor $g_{\alpha\beta}(\xi)$. The spacetime $\mathscr{X}$ is here simply the (abstract) *target space* in which the fields $X^\mu(\xi)$ take values and in which the metric tensor $G_{\mu\nu}(X)$ specifies how to compute the kinetic Lagrangian term for the 26-ple of fields $(X^0, \ldots, X^{25})$.

From the worldsheet perspective, the choice $\mathscr{X} = \mathbb{R}^{1,25}$ simply produces a model with 26 scalar (coordinate) fields, subject to the Polyakov action

$$\begin{aligned} S[\mathscr{X};\eta_{\mu\nu}] &= \frac{1}{4\pi\alpha'\hbar c^2}\int_\Sigma \mathrm{d}^2\xi\; g^{\alpha\beta}(\xi)\,\eta_{\mu\nu}\,(\partial_\alpha X^\mu)(\partial_\beta X^\nu) \\ &= \frac{1}{4\pi\alpha'\hbar c^2}\int_\Sigma \mathrm{d}^2\xi\; g^{\alpha\beta}(\xi)\Big[(\partial_\alpha X^0)(\partial_\beta X^0) - \sum_{i=1}^{25}(\partial_\alpha X^i)(\partial_\beta X^i)\Big], \end{aligned} \tag{11.13}$$

where we note that the contributions of 25 scalar functions $X^1, \ldots, X^{25}$ have the "wrong" sign of this generalized "kinetic" term,[20] which is specified by the choice of *parameters*:

$$\big[G_{\mu\nu}(X)\big]_{\text{from (11.10)}} \longmapsto \big[\eta_{\mu\nu}\big] = \mathrm{diag}[1,-1,\ldots,-1], \qquad \mu,\nu = 0,\ldots,25, \tag{11.14}$$

which in the target space $\mathscr{X} = \mathbb{R}^{1,25}$ represents the metric tensor.

[20] The sign of the whole Hamilton action depends on the choice of whether the metric tensor $g_{\alpha\beta}(\xi)$ on the worldsheet follows the "particle" or "relativist" convention – compare Chapters 3 and 9 – which is not essential here. The relative negative sign in the Lagrangian density (11.13), however, *is* essential and follows from the choice of the *signature* of the metric tensor $G_{\mu\nu}(X)$ in the target space $\mathscr{X}$.

Digression 11.3 In the familiar application of field theory in elementary particle physics in (3+1)-dimensional spacetime, the fields in the Lagrangian densities such as (5.118), (6.23) and (7.78) represent quarks, leptons, gauge and Higgs fields. Each of these (classical!) fields takes values in a corresponding number of copies of the real or complex space, $\mathbb{R}$ or $\mathbb{C}$. The geometry – and topology – of these "target" spaces is trivial: they are *contractible*; they can be continuously contracted (by scaling) to a point, and so also contain no subspace that could *not* be continuously contracted to a point.

Digression 11.4 The essential reason for the existence of a critical dimension is precisely the non-point-like nature of strings. Owing to their spatial extension, each string itself has an infinite sequence of harmonic resonances, each of which contributes to the Hamiltonian. Each of the infinitely many such quantum oscillators contributes a non-vanishing "zero energy" [☞ the constant term in equation (10.2a)], making a formally divergent sum – but one that may be unambiguously assigned a unique, finite and definite value [259]. These contributions to the Hamiltonian are offset by the freedom of general coordinate transformations [☞ Definition 9.1 on p. 319] on the worldsheet, $\xi^{\alpha}(\tau,\sigma) \to \zeta^{\alpha}(\xi(\tau,\sigma))$. For the ground state and the observables of a quantum theory of strings to be well defined, the net "zero energy" must in fact vanish. This cancellation limits the ways in which the string can oscillate, and thereby the structure of the (generalized) spacetime probed/spanned by those oscillations [224, 434, 594, 46, 312].

In particular, the types of oscillation possible in flat spacetime limits this spacetime to have the "critical" 25+1 dimensions. Including fermionic and other types of oscillators reduces the critical dimension of the spacetime at the expense of adding structure to it. In particular, to reduce the critical dimension of this flat "target" spacetime to 9+1, one may include oscillators that generate $N = 1$ supersymmetry and either $E_8 \times E_8$ or $SO(32)$ gauge symmetry, or one may include oscillators that generate $N = 2$ supersymmetry in the (9+1)-dimensional spacetime.

Herein, we cannot delve into the details of such computations as existing texts do [225, 224, 434, 594, 46, 312]. Suffice it here to mention that Polchinski [434] demonstrates the existence and computes the value of the critical dimension in seven different ways; one of these is detailed accessibly by Zwiebach [594], another by Kiritsis [312]. In turn, it seems that the quantization approach in loop (quantum) gravity does not identify the anomalies of which the cancellation produces the critical dimension [386, 385], whereby the results of this quantization approach do not agree with the results of standard methods.

Amusingly, the shift in perspective, from spacetime to the worldsheet, provided inspiration to explore the analogous shift in perspective in pointillist models: from spacetime to the worldline. Evidently, instead of a field theory in (1+1)-dimensional worldsheet spacetime, here we have a field theory in (0+1)-dimensional *time* – i.e., ordinary mechanics!

Indeed, this formalism is much better understood, and the analysis ought in fact to be simpler! However, even a swift glance at the Feynman diagrams in the left-hand side of Figures 11.3 on p. 411 and 11.4 on p. 412 indicate serious difficulties: The worldline on which one is to construct the (quantum and relativistic) mechanical model *bifurcates* and so is not a Hausdorff space! In the so-designed model, one would have to define functions such as scalar fields $X^{\mu}(\xi)$ in the expression (11.13), but now as functions of one argument, $X^\mu(\tau)$. Here, τ stands for the proper time τ, which is however not even unambiguously defined: its domain space – the worldline – is in fact not a simple line. Also, at the bifurcation points even the first derivatives with respect to the variable τ (necessary in Lagrangian dynamics) are multi-valued, and already the set-up of this approach indicates serious technical difficulties.

On the other hand, the worldsheet swept out by interacting strings is everywhere smooth (and so also all the worldsheet derivatives, not just $\partial_\alpha X^\mu(\xi)$ needed in Lagrangian dynamics) and each Hamilton action such as (11.10) is perfectly well defined – even with arbitrarily many "handles" (the right-hand diagrams in Figure 11.4 on p. 412, both have one "handle"), i.e., even for an arbitrarily high order contribution in the stringy version of the usual Feynman perturbation theory.

Digression 11.5 The shift in perspective – from spacetime, understanding this to be the "real" spacetime in which *we* live,[21] into the worldsheet spacetime – inexorably leads to the question: "Can the spacetime dimension, $n = \dim(\mathscr{X})$, be something other than $n = 4$?" In all of the "pre-stringy" development of fundamental physics through the Standard Model and beyond [☞ Chapter 8], the "obvious" dimension of spacetime, $3+1$, was taken for granted. The Nordstrøm–Kałuża–Klein model (in 1914, and 1919–26) was a small exception to this fact, but one that was forgotten owing to its initial lack of success in unifying gravity and electromagnetism – which solidified the opinion (prejudice?) that the 4-dimensionality and even uniqueness of spacetime were obvious. Also, all particle research implicitly assumed spacetime to be *flat*, open and infinitely large, $\mathscr{X} = \mathbb{R}^{1,3}$. Nontrivial geometries [☞ Chapter 9] had occupied the attention of the separate team of researchers, mostly "relativists," who for the most part did not follow the contemporary developments in elementary particle physics; in turn, neither did "particle physicists" follow the contemporary developments in the research of nontrivial solutions in the general theory of relativity.

The stringy shift in perspective irrevocably erased that chasm.

There is another property of the string theoretical model that is easiest to see from this perspective. Consider the Hamilton action (11.13) and simplify it by choosing coordinates $\xi^0 = \tau$ and $\xi^1 = \sigma$, so that $(g_{\alpha\beta}) = \left[\begin{smallmatrix} c^2 & 0 \\ 0 & -1 \end{smallmatrix}\right]$; to describe a closed string, take the coordinate σ to be periodic $\sigma \simeq \sigma + 2\pi R$. Varying the action (11.13) then produces the equations of motion

$$\partial_+\partial_- X^\mu = 0, \qquad \mu = 0,\dots,25, \quad \partial_\pm := \tfrac{1}{2}\left[\tfrac{\partial}{\partial\sigma} \pm \tfrac{1}{c}\tfrac{\partial}{\partial\tau}\right], \tag{11.15}$$

the general solutions of which are

$$X^\mu(\tau,\sigma) = X^\mu_L(\sigma^+) + X^\mu_R(\sigma^-), \qquad \partial_+ X^\mu_R(\sigma^-) = 0 = \partial_- X^\mu_L(\sigma^+), \qquad \sigma^\pm := (c\tau \pm \sigma). \tag{11.16}$$

That is, the general solution of the D'Alembert equation in (1+1)-dimensional spacetime is a linear combination of two *arbitrary* functions, X^μ_L and X^μ_R, each of which depends however only on one variable, σ^+ and σ^-, respectively. Since $X^\mu_L(\sigma^+)$ remains constant when after the time $\triangle\tau > 0$

[21] This manifest *subjectivity* is the first indication that the (3+1)-dimensional spacetime in which we live is neither the only one nor is it uniquely defined in physics models. Namely, this "real" or "true" spacetime is no more "real" than the (1+1)-dimensional spacetime of the string worldsheet, or even the 1-dimensional worldline that a point-particle sweeps out as time passes. The spacetime in which objects (particles, strings, ...) move is typically called the target spacetime, in the sense of the mapping depicted in Figure 11.1 on p. 410, and as will become clearer in Section 11.2.4.

the σ coordinate shifts to the *left* ($\triangle\sigma = -c\triangle\tau < 0$), the function $X^\mu_L(\sigma^+)$ "moves" to the left; analogously, the function $X^\mu_R(\sigma^-)$ "moves" to the right.

It is of paramount importance that the Lorentz group *Spin*(1,1) in (1+1)-dimensional spacetime is abelian (commutative), whereby all irreducible representations are 1-dimensional. It is not hard to show that the Lorentz group does not mix $X^\mu_L(\sigma^+)$ and $X^\mu_L(\sigma^-)$. The only linear transformation that swaps $X^\mu_L(\sigma^+) \leftrightarrow X^\mu_L(\sigma^-)$ is the discrete transformation of parity, $P : \sigma^+ \leftrightarrow \sigma^-$. Thus, every string model automatically has for every scalar field $X^\mu(\xi)$ two independent functions $X^\mu_L(\sigma^+)$ and $X^\mu_L(\sigma^-)$, which may well be treated completely independently. The generalization of this phenomenon to worldsheets with arbitrary metric is only technically more complex, but conceptually remains the same, precisely owing to the signature and commutativity of the Lorentz group on the string worldsheet.

This type of "doubling" of degrees of freedom evidently does not exist in pointillist models. In turn, in models where the material points are replaced by p-dimensional objects with $p > 1$, the Lorentz group on the world-"$(p+1)$-volume," $Spin(1,p)$, for $p > 1$ is no longer abelian (commutative), and there is no Lorentz-invariant separation of the fields X^μ into two or more independent functions. The separation of fields on the string worldsheet into left- and right-moving functions is therefore a unique phenomenon in string models, and ensures the uniqueness of some of the features in string models.

Digression 11.6 The combination of these two perspectives may even offer a useful insight into a phenomenon that is harder to understand from either one of the two perspectives.

Let $X^{\hat\mu}(\tau,\sigma)$ be the $\hat\mu$th string coordinate for some fixed $\hat\mu$; as a function of the string proper time, τ, and coordinate σ along the string, the function $X^{\hat\mu}(\tau,\sigma)$ has the classical equation of motion (11.15) and one may consider two basic types of boundary conditions in the spatial coordinate σ:

$$\textbf{Dirichlet condition} \qquad X^{\hat\mu}(\tau,0) = x_0^{\hat\mu}, \quad X^{\hat\mu}(\tau,\ell_s) = x_1^{\hat\mu}, \tag{11.17a}$$

$$\textbf{von Neumann condition} \qquad \partial_\sigma X^{\hat\mu}(\tau,0) = \acute{x}_0^{\hat\mu}, \quad \partial_\sigma X^{\hat\mu}(\tau,\ell_s) = \acute{x}_1^{\hat\mu}, \tag{11.17b}$$

where $x_0^{\hat\mu}$ and $x_1^{\hat\mu}$ are constant positions along the $\hat\mu$-axis, and $\acute{x}_0^{\hat\mu}$ and $\acute{x}_1^{\hat\mu}$ are dimensionless constants. The von Neumann condition at either end of the string imposes no restriction on either the position of that end-point or the velocity of its motion. Thus, string end-points with von Neumann conditions are simply free. However, imposing the Dirichlet condition to an end of a string fixes the target space position of that end, at the location $x_0^{\hat\mu}$, i.e., $x_1^{\hat\mu}$.

So, if a certain string is to satisfy the boundary conditions, say,

$$\text{von Neumann} \quad \partial_\sigma X^\mu(\tau,0) = 0, \quad \partial_\sigma X^\mu(\tau,\ell_s) = 0, \qquad \mu = 0,\dots,9, \tag{11.17c}$$

$$\text{Dirichlet} \quad X^\mu(\tau,0) = 0, \quad X^\mu(\tau,\ell_s) = L, \qquad \mu = 10,\dots,25, \tag{11.17d}$$

the $\sigma = 0$ and $\sigma = \ell_s$ ends of that string are trapped on the $(9+1)$-dimensional coordinate hypersurfaces specified, respectively, by the conditions $x_0^\mu = 0$ and $x_1^\mu = L$ for $\mu = 10,\dots,25$. These two $(9+1)$-dimensional spacetime hypersurfaces are two *D*9-branes (each with 9 space-like and 1 time-like dimension).

The existence of such p-dimensional $(0 \leqslant p \leqslant 25)$ objects – called "*Dp*-branes" – in string theory was already known in 1976 [482]; their dynamics was explored and emphasized only much later [434, 433, 438, 298].

11.2.4 Of models, again

It will be useful to reflect on the importance of the shift in perspective from Section 11.2.2 to Section 11.2.3. Note the common denominator in both perspectives – and with the benefit of hindsight we see that this is the case also in *all* theoretical systems of contemporary physics – the fact that scientific models have the following *canonical* geometrical content:

Procedure 11.1 *Theoretical models in general are constructed by specifying the following structural elements [☞ also the Procedure 5.1 on p. 193]:*

1. *A domain space, $\mathfrak{D}$, with local coordinates ξ.*
2. *A target space, $\mathfrak{T}$.*
3. *Maps $\varphi : \mathfrak{D} \to \mathfrak{T}$, values of which serve as local coordinates in $\mathfrak{T}$. In physics parlance, φ represents the generalized coordinates, i.e., the fields of the model.*
4. *The dynamical functional $S[\varphi;C] = \int_{\mathfrak{D}} \mathscr{L}(\varphi,\dot{\varphi},\dots;C)$ and boundary conditions, where C denote auxiliary parameters that specify the model.*
5. *Probing **currents/sources** ϑ, which are fields over the domain space $\mathfrak{D}$, chosen so that $\int_{\mathfrak{D}} \vartheta\cdot\varphi$ is invariant under the action of all symmetries and general coordinate transformations [☞ Definition 9.1 on p. 319] in the model.*

The classical version *of the so-specified model is immediate: varying the action $S[\varphi;C]$ produces the equations of motion, to which one must find solutions that satisfy the given initial/boundary conditions.*

Instead of a formal proof of the identifications made in Procedure 11.1, suffice it here to consider the following examples:

Example 11.1 Non-relativistic classical mechanics of a massive point-particle
The domain space is $\mathfrak{D} = \mathbb{R}^1$ (time), the target space is $\mathfrak{T} = \mathbb{R}^3$ (space),

$$S[\vec{r}] = \int \mathrm{d}t\, L(\vec{r},\dot{\vec{r}},\dots), \qquad L(\vec{r},\dot{\vec{r}},\dots) = \tfrac{m}{2}\dot{\vec{r}}^2 - V(\vec{r}), \tag{11.18}$$

where $V(\vec{r})$ is the potential energy and $\vec{r}(t)$ the mapping $\vec{r} : (\mathfrak{D}=\mathbb{R}^1) \to (\mathfrak{T}=\mathbb{R}^3)$. Choosing the coordinates $\vec{r} = \xi^i\hat{e}_i$, Hamilton's variational principle produces the Euler–Lagrange equations of motion

$$\sum_{k=0}^{\infty}(-1)^k \frac{\mathrm{d}^k}{\mathrm{d}t^k}\left[\frac{\partial L}{\partial\big(\frac{\mathrm{d}^k\xi^i}{\mathrm{d}t^k}\big)}\right] = 0, \tag{11.19}$$

which are to be solved subject to the boundary conditions for the classical solution.

Relativistic theory is conceptually the same, but with $L = -mc^2\sqrt{1-\frac{\vec{v}^2}{c^2}} - V(\vec{r})$.

Example 11.2 Non-relativistic classical mechanics of n massive point-particles
The domain space is $\mathfrak{D} = \mathbb{R}^1$ (time), the target spaces is $\mathfrak{T} = \mathbb{R}^{3n} = \oplus_{a=1}^{n}(\mathbb{R}^3)_i$,

$$S[\vec{r}_a] = \int \mathrm{d}t\, L(\vec{r}_a,\dot{\vec{r}}_a,\dots), \qquad L(\vec{r}_a,\dot{\vec{r}}_a,\dots) = \tfrac{1}{2}\sum_{a=1}^{n} m_a\dot{\vec{r}}_a^{\,2} - V(\vec{r}_1,\cdots,\vec{r}_n), \tag{11.20}$$

where the n-tuple of 3-vectors $\{\vec{r}_a(t), a = 1,\dots,n\}$ is the mapping $\vec{r}_a : (\mathfrak{D}=\mathbb{R}^1) \to (\mathfrak{T}=\mathbb{R}^{3n})$. Choosing the coordinates $\vec{r}_a = \xi^i_a\hat{e}_i$, Hamilton's variational principle produces the Euler–Lagrange equations of motion

$$\sum_{k=0}^{\infty}(-1)^k \frac{\mathrm{d}^k}{\mathrm{d}t^k}\left[\frac{\partial L}{\partial\left(\frac{\mathrm{d}^k \xi_a^i}{\mathrm{d}t^k}\right)}\right] = 0, \quad a = 1,\dots,n, \tag{11.21}$$

which are to be solved subject to the boundary conditions for the classical solution.

Example 11.3 Classical electromagnetic field without free charges and currents
The domain space is $\mathfrak{D} = \mathbb{R}^{1,3}$ (the (3+1)-dimensional spacetime), the target space is now the quotient space [☞ Appendix A.1.1] $\mathfrak{T} = (\mathbb{R}^{1,3}/\mathbb{R}^{1,1}) \cong \mathbb{R}^{0,2}$ (the physical polarizations of all gauge fields are 4-vector potentials modulo gauge transformations),

$$S[A_\mu] = \int \mathrm{d}^4\mathrm{x}\, \mathscr{L}((\partial_\mu A_\nu),\dots), \qquad \mathscr{L}((\partial_\mu A_\nu),\dots) = -\tfrac{4\pi\epsilon_0}{4} F_{\mu\nu}F^{\mu\nu}, \tag{11.22}$$

where $F_{\mu\nu} := (\partial_\mu A_\nu - \partial_\nu A_\mu)$, and the 4-vector A_μ is the function that maps $A : \mathbb{R}^{1,3} \to \mathbb{R}^{1,3}$, but owing to gauge invariance all four components $A_0, \dots, A_3$ are not independent. Imposing the Lorenz and the Coulomb gauge [☞ discussion about equation (5.91)] the temporal and longitudinal component are eliminated, which leaves the 2-dimensional space $\mathbb{R}^{0,2}$. Hamilton's variational principle produces the Euler–Lagrange equations of motion

$$\frac{\partial L}{\partial A_\mu} - \partial_\nu \frac{\partial L}{\partial(\partial_\nu A_\mu)} + \partial_\nu\partial_\rho \frac{\partial L}{\partial(\partial_\nu\partial_\rho A_\mu)} - \cdots = 0, \tag{11.23}$$

which are to be solved subject to the boundary conditions for the classical solution.

Example 11.4 Classical electromagnetic field with free charges and currents
The previous example is easily modified by adding the "probing currents/sources." These are – reading off from the Lagrangian density (5.22) – simply the *probing/test* electric charge density and current density, $\rho, \vec{\jmath}$. These should be distinguished from any charge and current density that are provided so as to produce a particular desired field. Notice that the gauge transformations (5.14a) change

$$(\rho\Phi - \vec{\jmath}\cdot\vec{A}) \;\to\; \rho\Phi - \rho\dot{\lambda} - \vec{\jmath}\cdot\vec{A} - \vec{\jmath}\cdot\vec{\nabla}\lambda = (\rho\Phi - \vec{\jmath}\cdot\vec{A}) - (\vec{\jmath}\cdot\vec{\nabla}\lambda + \rho\dot{\lambda}). \tag{11.24}$$

Under the spacetime integral, partial integrations yields

$$\delta_{\text{gauge}} \int \mathrm{d}^4\mathrm{x}\,(\rho\,\Phi - \vec{\jmath}\cdot\vec{A}) = \int \mathrm{d}^4\mathrm{x}\,(\vec{\nabla}\cdot\vec{\jmath} + \dot{\rho})\,\lambda - \int \mathrm{d}t \oint_{S^2_\infty} \mathrm{d}^2\vec{r}\,(\vec{\nabla}\cdot\vec{\jmath}\lambda) - \int \mathrm{d}^3\vec{r}\,\big[\rho\,\lambda\big]_{t=-\infty}^{t=+\infty}. \tag{11.25}$$

The requirement that $(\rho\Phi - \vec{\jmath}\cdot\vec{A})$ remain gauge-invariant for all *arbitrary* gauge parameter functions $\lambda(\mathrm{x})$ imposes the conditions on the probing charge and current densities: **(1)** they must vanish sufficiently fast at the boundary of the spacetime domain ($\vec{\jmath}$ at spatial infinity, ρ at time-like infinity) for the first integral to vanish after the application of Gauss's theorem, and **(2)** they must satisfy the continuity equation, $\dot{\rho} = -\vec{\nabla}\cdot\vec{\jmath}$.

The quantum version of the model described in Procedure 11.1 may be obtained – among a historical sequence of various approaches that conceptually follow the so-called canonical quantization[22] – using the Feynman–Hibbs path-integral formalism. A proper and detailed introduction to the path-integral formalism and its applications is best deferred to the texts on the subject [459, 537, 509, 165, 123, 316, 277]. Suffice it here to provide a heuristic motivation, based on Feynman's original intuitive imagery of summing over all possible histories of a system and extending the analysis familiar from quantum mechanics.

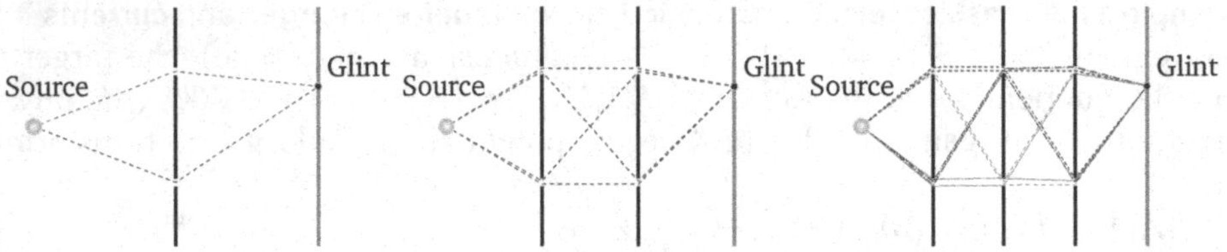

Figure 11.5 Three progressively more complicated double-slit experiments: Left, with two options, passing either through one or the other slit; middle with $2^2 = 4$ options; right, with $2^3 = 8$ options.

Consider the standard two-slit experiment, where a quantum particle is emitted from one side of the two-slitted screen and is then detected on the other side, at a specific location on a scintillating screen, as shown on the left-hand side of Figure 11.5. This arrangement is discussed in most texts of quantum mechanics, and the intensity of the glint on the scintillating screen is determined by computing the interference between the particle/wave traveling along the two distinct types of paths, one passing through the top slit, the other through the bottom one. The middle arrangement, with two consecutive two-slit experiments offers $2^2 = 4$ options: top–top, top–bottom, bottom–top and bottom–bottom. The right-most arrangement offers $2^3 = 8$ options, and so on. With n consecutive screens, each with p slits, there are p^n options for the particle passing through them, and the intensity of the glint on the scintillating screen is determined by the interference of the particle/wave traveling along all possible paths.

By letting both the number of screens and the number of slits in them grow infinitely, the collection of possible paths grows to include all possible paths that start at the source and end at the glint.[23] This is precisely as if there were no screens at all, but we expressly avoid presupposing which way a particle/wave moves from one point to the next. That in turn is precisely the situation with quantum physics, where we expressly avoid relying on classical equations of motion to determine the path of a particle between the only two points where it has been observed!

The intensity of the glint is thus determined by the total interference of the particle/wave traveling along each one from the ensemble of all possible paths from the source to the glint. Along each path, one should calculate the net phase of the particle/wave, as the difference between them determines the result of interference between particles/waves arriving along any two of the paths. This is effectively accomplished by summing phase factors $e^{i\,\text{phase}}$ over the ensemble of all possible paths. Feynman and Hibbs proved [165] that (1) this reasoning can be extended to all models of quantum physics, and (2) the correct phase-factor is in fact $\exp\{-iS[\varphi;C]/\hbar\}$, with $S[\varphi;C]$ the classical action for the system considered, as in Procedure 11.1.

[22] The contemporary BRST- and ZJBV-quantization in the Lagrangian formalism and the BFV-quantization in the Hamiltonian formalism are direct and universal generalizations of canonical quantization (as it is traditionally called), and are thus just as canonical. See Ref. [554] for pre-BRST methods, and [268, 555, 484, 496, 589, 590] for contemporary treatments.

[23] In fact, there is no a-priori reason to exclude back-tracking paths either.

With the data provided in Procedure 11.1, one then defines the partition functional

$$Z[\vartheta;C] := \int \mathbf{D}[\varphi]\; e^{-\frac{i}{\hbar}(S[\varphi;C]+\int d^4x\; \vartheta\cdot\varphi)}, \tag{11.26}$$

where $\int \mathbf{D}[\varphi]$ is the functional integral summing over all (independent and unconstrained, i.e., *free* and certainly off-shell) fields φ. This quantity turns out to be a generating function for so-called n-point correlation functions [459, 165, 123, 316, 277]

$$G(\xi_1,\xi_2,\dots) := \langle \varphi(\xi_1)\,\varphi(\xi_2)\cdots\rangle \tag{11.27}$$

$$= \frac{1}{Z[\vartheta;C]}\Big[\frac{\delta}{\delta\vartheta(\xi_1)}\frac{\delta}{\delta\vartheta(\xi_2)}\cdots Z[\vartheta;C]\Big]\Big|_{\vartheta\to 0'}, \tag{11.28}$$

which produce the probability amplitude for the correlation of perturbations in the field φ at the n points in the domain space, $\xi_i \in \mathfrak{D}$. They generalize the (2-point) Green function. From (11.26), one can also easily define the so-called effective action

$$S_{\text{eff.}}[\varphi_*;C'] := i\hbar\,\log\Big(\int \mathbf{D}[\varphi]\; e^{-\frac{i}{\hbar}S[(\varphi_*+\varphi);C]}\Big). \tag{11.29}$$

The fields φ_* are called "background," and provide the interpretation of φ as the (quantum) fluctuations around this background. The effective action $S_{\text{eff.}}[\varphi_*;C']$ may well be used as if a classical action, producing equations of motion for the background fields, φ_*. However, owing to the integration over all "fluctuations" in the right-hand side of the definition (11.29), $S_{\text{eff.}}$ effectively includes quantum corrections.

The concrete and technically precise use of the general framework based on (11.26)–(11.29) requires the expanse afforded by quantum field theory texts such as Refs. [459, 165, 123, 316, 277], but the above intuitive motivation, formal definitions and concepts still permit making two key observations. First, note that the definition in no way guarantees that $S_{\text{eff.}}[\varphi_*;C']$ would even resemble the original action $S[\varphi;C']$. We thus revisit Definition 5.1 on p. 211 and make it more precise:

> **Definition 11.1** *A quantum system is* ***renormalizable*** *if the effective action, $S_{\text{eff.}}[\varphi_*;C']$, has the same functional form and dependence of its parameters as the original action, $S[\varphi;C]$, and where the formal transformation $S[\varphi;C] \to S_{\text{eff.}}[\varphi_*;C']$ is fully described by the* ***renormalization*** *of the system parameters, $C \to C'$.*

Second, we note that the quantities such as (11.26), (11.28) and (11.29) are in practice most often calculated perturbatively, and individual contributions turn out to be calculable using Feynman diagrams – the same tools we have seen in Sections 3.3, 5.3–5.4 and 6.2. It turns out that the partition functional $Z[\vartheta;C]$ receives contributions from all possible Feynman diagrams, whereas the effective action $S_{\text{eff.}}[\varphi_*;C']$ receives contributions only from the connected diagrams. That is, contributions to the effective action are those of the partition functional, however with the contributions from disconnected Feynman diagrams subtracted [425, 586]. Both formally and in their physical meaning, these subtractions depicted by disconnected Feynman diagrams generalize the "excisions" that appear in the quantum-mechanical perturbative calculations (3.93)–(3.102).

— ❦ —

The space of all mappings $\mathfrak{T}^{\mathfrak{D}} := \{\mathfrak{D} \to \mathfrak{T}\}$ over which the integral (11.26) is to be computed – in the general case – is not the same as the configuration space. Namely, a concrete physical system is very often limited by constraints, $\chi(\varphi) = 0$, which "restrict" the mappings. The integral (11.26) is defined over *free* mappings, and constraints are to be included into the dynamics by means of Lagrange multipliers $S[\varphi; C] \to S[\varphi; C] + \int_{\mathfrak{D}} \Lambda \cdot \chi(\varphi)$, so that varying the Lagrange multipliers Λ imposes the constraints $\chi(\varphi) = 0$. Among other cases, every symmetry of a system is a constraint since the symmetry transformation, S, does not change the physical state $|\Psi\rangle$, whereby it is always true that

$$S|\Psi\rangle = |\Psi\rangle \quad \text{i.e.,} \quad \chi_a|\Psi\rangle = 0, \quad \text{where} \quad \chi_a := -i\frac{\partial S}{\partial \epsilon^a}\Big|_{\epsilon=0} \quad \text{and} \quad S = e^{i\epsilon^a \chi_a}, \qquad (11.30)$$

and where ϵ^a denotes the ath symmetry transformation parameter.

Digression 11.7 In 1950–1, P. A. M. Dirac showed [132, 133] that it is essential to ensure that the dynamics of the system (in the Hamiltonian formalism, generated by the Hamiltonian) *preserves* the constraints χ_i. In classical physics, this means that all Poisson brackets $\{\chi_i, \chi_j\}$ and $\{H, \chi_i\}$ must automatically equal a linear combination of the constraints, and the model is consistent only if this can be achieved by a combination of:

1. extending the collection of all constraints, $\{\chi_1, \chi_2, \dots\}$,
2. redefining the Poisson brackets (into Dirac brackets),

whereby the (redefined or not) brackets must satisfy the Jacobi identity (6.18).

   In quantum physics, the analogous situation must hold for commutators, i.e., anticommutators between spinor operators, including the generalization of the Jacobi identities (10.37). So far, the most general known procedures that ensure this are the Zinn–Justin–Batalin–Vilkovisky (BV) quantization [40, 41, 555, 484] in the Lagrangian formalism, and the Batalin–Fradkin–Vilkovisky (BFV) quantization [174, 39, 172, 36] in the Hamiltonian formalism.

Considerations and statements such as these, in this section and also in Sections 8.1.1 and 8.3, actually are not themselves part of physics (the subject matter of which is *Nature*), but of a discipline the subject matter of which is the *scientific discipline* "physics" and its structure. In the analogous situation, the discipline of which the subject matter is mathematics is called *metamathematics* [314]. Analogously, the discipline of which the subject matter is physics should be called *metaphysics*, but this name is already the standard moniker for a branch of philosophy concerned with the nature of existence and of the world.

   As this discipline (about the formal structure of physics) is by its nature (just as in the description of the Procedure 11.1 and in Appendix B.3) rather mathematical, perhaps *metamatephysics* would not be too inappropriate?

11.2.5 *Reconstructing the spacetime perspective*

The shift in perspective from the spacetime (in Section 11.2.2) to the worldsheet (in Section 11.2.3) of course has its inverse process that reconstructs the *effective* spacetime field theory from the original worldsheet field theory.

   Namely, the worldsheet Hamilton action and partition functional

$$S[X; G_{\mu\nu}, \dots] = \int_{\Sigma} \mathscr{L}(X^{\mu}, (\partial_{\alpha} X^{\mu}), \dots; \underbrace{G_{\mu\nu}, \dots}_{\text{parameters}}),$$

$$Z[Y; G] = \int \mathbf{D}[X]\, e^{\frac{i}{\hbar} S[X; G_{\mu\nu}, \dots] + \int_{\Sigma} Y \cdot X}, \qquad (11.31)$$

where the parameters specify the concrete model in question. In the concrete actions (11.13) and (11.31), for example, there appears the metric tensor

$$[G_{\mu\nu}(X)]_{\text{from (11.10)}} \longmapsto [\eta_{\mu\nu}] = \mathrm{diag}(1,-1,\ldots,-1), \qquad \mu,\nu = 0,\ldots,25. \tag{11.32}$$

Using the partition functional $Z[Y;G]$, we may compute how quantum fluctuations alter these specified parameters, and in 1979, Daniel Friedan showed that the condition – that these quantum fluctuations *do not alter* the chosen metric tensor – reproduces (at the lowest order in perturbative computations) the Einstein equations for the given metric tensor.

Similar results hold for all parameters in the Hamilton action (11.31): the condition for quantum stability of every parameter is an equation that in the lowest order of perturbative computation looks like a classical equation of motion for the physical quantity represented by this parameter. Thus, quantum stability of models defined on the worldsheet Σ defines the *effective* field theory in the spacetime $\mathscr{X}$, such that the classical equations of motion equal the condition for quantum stability of the original worldsheet model. Let $\{\phi^a \simeq \delta G_{\mu\nu},\ldots\}$ be the collection of all parameter fluctuations in the Hamilton action (11.31).[24]

Next, construct a Hamilton action (of the second level):

$$\int_{\mathscr{X}} \mathscr{L}(\phi,(\partial_\mu\phi),\ldots;\mathcal{G}_{ab},\ldots), \qquad \mathscr{L} = \mathcal{G}_{ab}\,G^{\mu\nu}\,(\partial_\mu\phi^a)(\partial_\nu\phi^b) + \cdots, \tag{11.33}$$

where $\mathcal{G}_{ab}$ (and similar) parameters in the Lagrangian density (11.34) are chosen so that the classical equations of motion for the $\phi^a \simeq \delta G_{\mu\nu},\ldots$ variations of the parameters in the Hamilton action of the first level precisely produce the conditions for the quantum stability of the model (11.31).

Of course, this new Hamilton action defines the quantum model

$$S[\phi;\mathcal{G}_{ab},\ldots] = \int_{\mathscr{X}} \mathscr{L}(\phi^a,(\partial_\mu\phi^a),\ldots;\underbrace{\mathcal{G}_{ab},\ldots}_{\text{parameters}}),$$
$$Z[J;\mathcal{G}] = \int \mathbf{D}[\phi]\, e^{\frac{i}{\hbar}S[\phi;\mathcal{G}_{ab},\ldots]+\int_{\mathscr{X}} J\cdot\phi}, \tag{11.34}$$

where $\mathscr{F}$ is the target space, where ϕ^a take values. This quantum model is then the *effective quantum field theory* in spacetime $\mathscr{X}$ of which a part (in a realistic model, see Section 11.3) is identified with the "real" spacetime in which we live. In this model, $\phi^a \simeq \delta G_{\mu\nu},\ldots$ are fields that are identified with the "real" fields such as the graviton (for $\delta G_{\mu\nu}$), the photon, the electron, the quark,... To keep the notation simple, only the graviton is explicitly written (in $\phi^a \simeq \delta G_{\mu\nu},\ldots$), but each of the fields in the Standard Model may be identified as the variation of some parameter in the Hamilton action (11.31).

Evidently, the concept of generating the Hamilton action (11.34) from the previous, worldsheet Hamilton action (11.31) can be repeated: The quantum model (11.34) itself depends on parameters $\mathcal{G}_{ab},\ldots$ The quantum stability of the model (11.34) produces conditions for the variations $\mathbf{\Phi}^A \sim \delta\mathcal{G}_{ab},\ldots$ One then constructs a Hamilton action (of the *third level*) for the variables $\mathbf{\Phi}^A$, which is chosen so that the equations of motion for $\mathbf{\Phi}^A$ are precisely the conditions for quantum stability of the model (11.34) from the previous (second) level. This Hamilton action of the third level then defines the effective quantum field theory that "lives" in the target space of the quantum model (11.34) of the previous, second level. In principle, this iterative construction of

---

[24] The presentation here is drastically simplified! In practice, one must first construct the Hilbert space where the states are constructed akin to the linear harmonic oscillator, and with the creation operators from the expansion (11.38). In this Hilbert space there exist, e.g., states such as $G^{(m,n)}_{\mu\nu}(X)a^{\mu}_{m,R}a^{\nu}_{n,L}|0\rangle$, amongst which the expectation values with $m = -1 = n$ define the metric tensor $G_{\mu\nu}(X)$ on the spacetime $\mathscr{X}$ in which the coordinate fields $X^\mu(\xi)$ take values. The variables $\{\phi^a \simeq \delta G_{\mu\nu},\ldots\}$ therefore parametrize the fluctuations in the Hilbert space of the worldsheet model.

ever higher levels of effective field theories never stops; the iterative scheme may be presented formally:

$$\{\,\mathfrak{D}_{(k+1)} \xrightarrow{\varphi_{(k+1)}} \mathfrak{T}_{(k+1)}\,\};\quad \delta S_{(k+1)}[\varphi_{(k+1)};C_{(k+1)}]=0;\quad \delta_{\text{qu.}}C_{(k+1)}=0; \tag{11.35a}$$

$$\Vert \qquad \nearrow\ \varphi_{(k+1)}:=\delta C_{(k)} \qquad \Big\downarrow \text{ choice of } S_{(k+1)}[\varphi_{(k+1)};C_{(k+1)}]$$

$$\{\,\mathfrak{D}_{(k)} \xrightarrow{\varphi_{(k)}} \mathfrak{T}_{(k)}\,\};\quad \underbrace{\delta S_{(k)}[\varphi_{(k)};C_{(k)}]=0;}_{\text{classical physics}}\quad \underbrace{\delta_{\text{quantum}}C_{(k)}=0;}_{\text{quantum stability}} \tag{11.35b}$$

where the scheme begins with the level where $\mathfrak{D}_{(1)} = \Sigma$ is the string worldsheet, $\mathfrak{T}_{(1)}$ the (extended) target spacetime (in the $(3+1)$-dimensional portion of which we seem to live), $\varphi_{(1)}$ are the coordinate fields immersing $\mathfrak{D}_{(1)} \to \mathfrak{T}_{(1)}$, and $C_{(1)}$ are the "coupling constants" of this first level field theory, including quantities that are in turn identified as structural characteristics of $\mathfrak{T}_{(1)}$, such as the metric tensor.

> ***Conclusion 11.3*** *String models contain an infinitely iterative hierarchy of (effective) field theories, defined by iteratively following the construction of the Hamilton action (11.34) from (11.31) [☞ scheme (11.35)].*

Of these, (at least) the first three "levels" are used routinely: the first level describes the dynamics of the (super)strings themselves (11.31), the second level describes the dynamics of the fields such as the quarks and leptons (11.34), the third level is used to explore the so-called *modular* spaces. Namely, the parameters $\mathcal{G}_{ab}$ in the second level Hamilton action, in the expressions (11.34), determine the geometry of the domain spacetime of this (second) level, and represent points in the space of possible geometries. Variations of these parameters then represent variations of these geometries and so represent local coordinates in the space of possible geometries, the so-called *modular* space. Such a modular space is then the target space in the third level and the Hamilton action in this third level then contains parameters that correspond to the structure of this modular space. In this way, the third level field theory within (super)string theories serves also as a "laboratory" for studying the structure of this modular space. It is interesting to mention that the physically motivated choice of the Zamolodchikov metric tensor on modular spaces of so-defined models coincides with the mathematically "natural" choice of the Weil–Petersson metric tensor [89], whereby the successful applications of these physical models in mathematics – amongst which some original works are collected in Refs. [85, 84] – were a fascinating surprise.

11.3 Towards realistic string models

The choice (11.13) is clearly but the simplest case, when the strings move through flat, empty and infinitely large spacetime. However, it is fairly simple to change this geometry in this model. For example, some of the scalar fields $X^\mu(\xi)$ may be required to satisfy a periodicity condition, and constantly so over the worldsheet Σ, for simplicity:

$$X^i(\xi) \simeq X^i(\xi) + 2\pi R_i, \qquad i = 4,\dots,25, \quad \forall \xi \in \Sigma. \tag{11.36}$$

As a result, the scalar fields $X^0(\xi),\dots,X^3(\xi)$ still take values in an open, flat and infinitely large space, $\mathbb{R}^{1,3}$. However, each of the scalar fields $X^4(\xi),\dots,X^{25}(\xi)$ now takes values in what is seen to be a closed and finite (*compact!*) circle of radius $R_4,\dots,R_{25}$. The shape of the target space (in which the functions $X^0(\xi),\dots,X^{25}(\xi)$ take values) has through the imposition of the conditions (11.36) turned into

$$\mathscr{X} = \mathbb{R}^{1,25} \xrightarrow{(11.36)} \mathscr{X}' = \mathbb{R}^{1,3} \times T^{22}, \quad T^{22} := S^1_{(R_4)} \times \cdots \times S^1_{(R_{25})}. \tag{11.37}$$

The space $\mathscr{X}'$ is compact in 22 directions, but remains non-compact in the $X^0, \dots, X^3$ directions. The conditions (11.37) represent a direct application of the Nordstrøm–Kałuża–Klein compactification.

Besides, two additional modifications may be introduced owing to the special properties of the (1+1)-dimensional worldsheet:

1. For the scalar "coordinate" fields, such as $X^4, \dots, X^{25}$, one may use the opportunity described in the discussion of the relations (11.15)–(11.16): Each of these 22 fields harbors two independent functions, upon which boundary and/or periodicity conditions may be imposed independently – and so *differently*.
2. The oscillators obtained through Fourier decomposition of functions (11.16) are bosonic and in the quantum variant correspond to creation and annihilation operators, just as with linear harmonic oscillators. Exclusively in physics defined on the (1+1)-dimensional worldsheet, every pair of bosonic creation/annihilation operators may be substituted with two pairs of *fermionic* creation/annihilation operators.[25]

The above two peculiarities of field theory in (1+1)-dimensional worldsheet spacetime (the independence of left- and right-moving modes in fields and the possibility of fermionization of bosons – and reciprocally of bosonizing fermions) makes the following construction possible:

Construction 11.1 (heterotic string) *Replace the 16 right-moving functions $X^{10}_R, \dots, X^{25}_R$ with 32 right-moving fermions, $\lambda^1_R, \dots, \lambda^{32}_R$. Impose periodicity to the left-moving functions $X^{10}_L, \dots, X^{25}_L$, so that this 16-tuple of functions $X^{10}_L, \dots, X^{25}_L$ takes values on a 16-dimensional torus that is identical with the so-called maximal torus of either the $E_8 \times E_8$ or the $D_{16} = \mathfrak{so}(32)$ algebra. The Hilbert space in such a model is built akin to the Hilbert space in Section 10.1.3, applying creation operators from the Fourier expansion following Refs. [225, 224, 434, 594],*

$$X^\mu(\tau,\sigma) = x^\mu + \frac{p_\mu}{p_-} c\tau + i\sqrt{\frac{\alpha'}{2}} \sum_{\substack{n=-\infty \\ n\neq 0}}^{+\infty} \left[\frac{a^\mu_{n,R}}{n} e^{-2\pi i n \varsigma^+} + \frac{a^\mu_{n,L}}{n} e^{2\pi i n \varsigma^-} \right], \tag{11.38}$$

where

$$\varsigma^\pm := (\sigma \pm c\tau)/\ell_s, \qquad \text{and} \qquad a^\mu_{-n,R} = (a^\mu_{n,R})^\dagger, \quad a^\mu_{-n,L} = (a^\mu_{n,L})^\dagger, \tag{11.39}$$

and where $p_\pm$ are the momenta corresponding to the coordinates $x^\pm := \frac{1}{\sqrt{2}}(x^0 \pm x^1)$. By choosing x^+ to be along proper time, $x^+ = \tau$, x^+ and p_+ are to be treated as simple parameters, but $x^-, x^2, \dots, x^9$ and $a^\mu_{n,R}, a^\mu_{n,L}$ and their canonically conjugate variables $p_-, p_2, \dots, p_9$ and $a^\mu_{-n,R}, a^\mu_{-n,L}$ are all the familiar quantum operators. The traditional normalization is provided by the canonical commutation relations

$$[x^-, p_-] = -i\hbar, \qquad [x^{\hat\mu}, p_{\hat\nu}] = i\hbar\delta^{\hat\mu}_{\hat\nu}, \qquad \hat\mu, \hat\nu = 2, \dots, 9; \tag{11.40a}$$

$$[a^\mu_{m,R}, a^\nu_{n,R}] = n\delta^{\mu\nu}\delta_{n,-m}, \quad [a^\mu_{m,L}, a^\nu_{n,L}] = n\delta^{\mu\nu}\delta_{n,-m}. \tag{11.40b}$$

A similar oscillator expansion also exists for $X^{10}_L, \dots, X^{25}_L$ and $\lambda^1_R, \dots, \lambda^{32}_R$. Gross, Harvey, Martinec and Rohm showed [247, 246, 248] that such a model:

---

[25] This transformation and its unique existence within field theory in (1+1)-dimensional spacetime was noticed in 1935–7 by Pascual Jordan, who attempted to explain a photon as a bound state of a neutrino and an antineutrino [301]; owing to the failure of this application, the basic idea of *fermionization* was itself neglected until its successful recycling [☞ Footnote *13* on p. 13] within string theory.

1. *effectively describes the propagation of strings through a (9+1)-dimensional, flat and infinitely large spacetime, $\mathbb{R}^{1,9}$, parametrized by the values of the scalar fields $X^0(\xi),\dots,X^9(\xi)$;*
2. *the geometric and physical meaning of $x^0,\dots,x^9$ is that they are the center of mass coordinates of the string;*
3. *with a suitable choice of the relative coefficients in the Hamilton action, the system exhibits a supersymmetry owing to the presence of fermionic modes that replaced the bosonic functions $X_R^{10},\dots,X_R^{25}$;*
4. *with a suitable choice of the radii $R_{10},\dots,R_{25}$ and angles between the coordinates $X^{10},\dots,X^{25}$, the system also has either the $E_8\times E_8$ or the $SO(32)$ gauge symmetry.*

These two specifically stringy constructions are called *heterotic* strings and proffer the possibility to construct models that contain (much more than) enough gauge symmetry and matter to describe the "real world" [247, 246, 248].

This telegraphic synopsis is a far cry from describing the details of the construction of these string models, whereby the construction certainly appears to be rather ad hoc. The technical details and consistency conditions in these constructions are, however, very rigorous; for example, these conditions single out the gauge group to be either $E_8\times E_8$ or $SO(32)$, and the number of "flat and infinitely large" spacetime dimensions to be precisely 9+1 and only if the model is supersymmetric.

11.3.1 *Partially compact topology and geometry*

Somewhat in the manner of the Nordstrøm–Kałuża–Klein compactification (11.37), Philip Candelas, Gary Horowitz, Andy Strominger and Edward Witten showed that by substituting the spacetime geometry

$$\mathbb{R}^{1,9} \longrightarrow \mathbb{R}^{1,3}\times\mathscr{Y} \tag{11.41}$$

in the $E_8\times E_8$ heterotic string model, the system remains minimally (simply) supersymmetric precisely if the real 6-dimensional space $\mathscr{Y}$ is chosen to be in fact a complex 3-dimensional, compact and so-called Calabi–Yau space,[26] a.k.a., 3-fold. The hallmark feature of such spaces is that they admit a metric tensor $g_{i\bar{\imath}}$ for which:

1. $g_{i\bar{\imath}}$ is Kähler, i.e., $g_{i\bar{\imath}}=\frac{\partial}{\partial z^i}\frac{\partial}{\partial z^{\bar{\imath}}}K(z,\bar{z})$, where $K(z,\bar{z})$ is the Kähler potential;
2. the Ricci tensor computed from this metric is a total derivative, so $\oint_S \mathrm{d}z\cdot R\cdot\mathrm{d}\bar{z}=0$ for every closed real 2-dimensional (complex 1-dimensional) surface $S\in\mathscr{Y}$.

Here (z^1,z^2,z^3) are complex local coordinates for $\mathscr{Y}$, and $z^{\bar{\imath}}=(z^i)^*$. Since the space $\mathbb{R}^{1,3}\times\mathscr{Y}$ is compact in the $\mathscr{Y}$-directions, such constructions are referred to as *Calabi–Yau compactifications*. Note that, unlike the original Nordstrøm–Kałuża–Klein compactification, here the metric tensor components in the directions of the compact space $\mathscr{Y}$ do not produce any gauge fields, as Calabi–Yau 3-folds have no isometries.

Without delving into the details of such constructions (to which end the interested Reader is directed to the book [279] and the references therein), may it suffice here to mention that the complex 3-dimensional compact Calabi–Yau spaces have two variable characteristic numbers, denoted $h^{1,1}$ and $h^{2,1}$, and that Calabi–Yau compactifications of the $E_8\times E_8$ heterotic string models produce:

---

[26] Eugenio Calabi's conjecture, that the necessary and sufficient criterion for a complex 3-dimensional, compact space to admit a Kähler metric is that its first Chern class should vanish, was proven in 1974 by Shing-Tung Yau, for which he was awarded the Fields Medal. For a detailed history of both the related mathematical discoveries as well as their applications in physics and especially in (super)string theory, see Ref. [584].

1. an effectively (3+1)-dimensional, flat and infinitely large spacetime, $\mathbb{R}^{1,3}$;
2. minimal (simple) supersymmetry, as described in Section 10.3.2;
3. the gauge symmetry group is reduced to $E_6 \times E_8$;
4. matter fields, in the following collections:
   (a) $h^{2,1}$ copies of the **27**-representation of the group E_6,
   (b) $h^{1,1}$ copies of the $\overline{\mathbf{27}}$-representation of the group E_6;
5. a connection between the otherwise arbitrary Standard Model parameters, such as the Yukawa coupling parameters h_e, h_u, h_d, h_ν [☞ Section 7.3.1] and the geometry of the selected Calabi–Yau space $\mathscr{Y}$.

Every **27**-representation of the (compactification-reduced) gauge group E_6 contains one family of Standard Model fundamental fermions with the usual $SU(3)_c \times SU(2)_w \times U(1)_Q \subset E_6$ charges, and the $\overline{\mathbf{27}}$-representation contains the same particles but with wrong (opposite) charges. Ideally, one would like to construct a Calabi–Yau space with $h^{2,1} = 3$ and $h^{1,1} = 0$, or the other way around.[27] Besides, if the given Calabi–Yau space is not simply connected, it is possible to establish a flux of "background" gauge field along a closed contour that cannot be continuously contracted to a point. Such a non-contractible closed-contour integral of such a flux effectively serves as a Higgs field: It breaks the gauge symmetry and can "pair" fields from the **27**-representation with fields from the $\overline{\mathbf{27}}$-representation and provide them with a mass of the order of $10^{17-19}\text{GeV}/c^2$. For further details about constructing Calabi–Yau spaces and analyzing the models obtained by compactifying on Calabi–Yau spaces, the interested Reader is directed to search the contemporary literature (at `www.arXiv.org`), perhaps with some help from the by now two decades old Ref. [279] for starters.

11.3.2 Mirror symmetry

The analysis of the application of Calabi–Yau spaces in compactification of string models discovered the phenomenon that for every model with $h^{2,1}$ "families" and $h^{1,1}$ anti-"families" one may construct a "mirror-dual" model in which the number of "families" and anti-"families" is flipped (Brian Greene and Ronen Plesser, 1990 [236]):

$$(\mathscr{Y},\mathscr{Y}') : \qquad h^{2,1}(\mathscr{Y}) = h^{1,1}(\mathscr{Y}'), \quad h^{1,1}(\mathscr{Y}) = h^{2,1}(\mathscr{Y}'). \tag{11.42}$$

It was soon proved [51, 83] that the phenomenon is rather typical, which provided support for a research field that may rightly be called *experimental mathematics*. Namely, between 1984 and 2002, the catalogue of constructions grew from a handful to nearly half a billion [☞ review [321] and references therein], and the statistically significant tendencies in this collection acquire significant probabilities of being systemic, and then are well worth exploring as candidates for mathematically rigorous theorems. Besides, the insight into the physical qualities of the models in which these tendencies are noted may provide an argumentation that is fully alien to the mathematical tradition, whence these tendencies may appear surprising or even "magical" from the mathematicians' vantage point.

One such example is precisely the "mirror duality," where the identifications are in fact not only on the level of numerical characteristics (11.42), but also among certain physically motivated algebraic structures that correspond to Yukawa interactions (7.133f). Using this physics insight, mirror duality may be used to compute certain mathematical characteristics of Calabi–Yau spaces [85, 84, 49, 50], to which end the "purely mathematical" methods are still not known, and which at first seemed fantastic and unbelievable.

However, intrigued by the manifest computational efficiency, mathematicians had already in the same year, 1993–4, proved the phenomenon of "mirror duality" within a well-defined class of

[27] It is always possible to *completely* flip the construction, so that *all* Yang–Mills gauge charges are reversed.

constructions [42, 71, 43]. In turn, this gave rise to a whole "industry" of research [583, 235, 426, 117, 124, 274], and then also the proof that the "mirror duality" – in the general case – is an example of the so-called T-duality [501].

> **Definition 11.2** *Two spacetimes, $\mathscr{X}$ and $\mathscr{X}'$, are T-dual if the string model that describes string propagation through the spacetime $\mathscr{X}$ is physically identical to the string model that describes string propagation through the spacetime $\mathscr{X}'$.*

In other words, the spacetimes $\mathscr{X}$ and $\mathscr{X}'$ are T-dual if strings, by propagating through them, do not distinguish between them. In the case when $\mathscr{X} = \mathbb{R}^{1,3} \times \mathscr{Y}$ a $\mathscr{X}' = \mathbb{R}^{1,3} \times \mathscr{Y}'$, T-duality is naturally referring to the $\mathscr{Y}$ and $\mathscr{Y}'$ ("purely" space-like) factors. This relation between the spaces $\mathscr{Y}$ and $\mathscr{Y}'$ is thus indirect, as it is based on an identification of structures between observable quantities, as schematically presented in Figure 11.6. Precisely because of this indirectness are

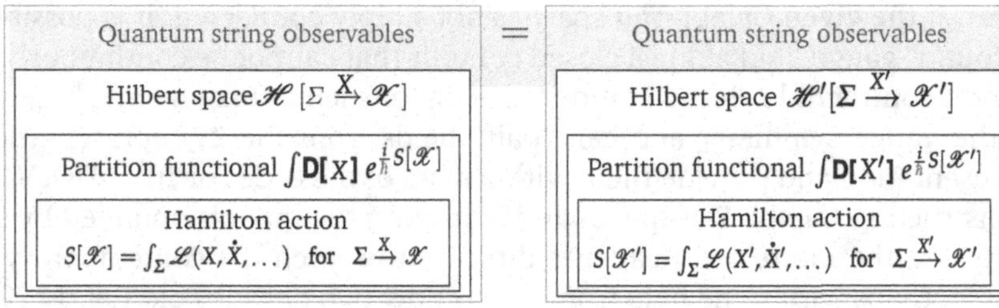

Figure 11.6 A depiction of the indirect relation of "stringy duality" between spacetimes $\mathscr{X}$ and $\mathscr{X}'$.

the so-obtained relations very unexpected, so that the construction and exploration of relations between string models may be regarded also as a machine for generating mathematically non-trivial conjectures, the final proof of which then significantly advances both mathematics and, reciprocally, also physics [455].

Recently, one such physically motivated and unusual general construction [51] of "mirror dual" Calabi–Yau spaces from 1993 was re-examined, providing it with a mathematically more natural formulation 16 years later, and then also with a rigorous proof [320, 108, 72]. It seems worthwhile to note that this construction of "mirror dual" spaces holds even when not all of the defining conditions of Calabi–Yau spaces are fulfilled, and so points to a much more general phenomenon in algebraic geometry, for which there is no other indication, and the mathematical implications of which are only now being explored.

Notice also that the so-defined "stringy duality" is being identified at the level of quantum observables, which are understood to act upon a Hilbert space, and a partition functional for concrete computations. In particular, it is logically not necessary for a model to have a Lagrangian formulation and the associated geometric interpretation. In fact, models have been constructed for which at the time only the partition function was known, and neither a Lagrangian formulation nor a geometrical interpretation was known or needed; see Ref. [203], for starters, and [26, 503, 335] for related recent references.

11.3.3 Variable geometry and cosmology

The basic idea in the Nordstrøm–Kałuża–Klein compactification was the assumption that the entire spacetime has the structure of a product such as $\mathscr{X} = \mathbb{R}^{1,3} \times \mathscr{Y}$. That is, at every point of the first factor there exists an entire copy of the second factor and *vice versa*; also, all copies of the second factor "along" the first factor are identical, and *vice versa*.

Of course, this may be generalized so that one factor varies from point to point of the other factor. That is, the compact Calabi–Yau space $\mathscr{Y}$ may depend on the location in the non-compact space $\mathbb{R}^{1,3}$. In the general case, such variations define a structure called fibration in mathematics, and the type of the variation provides a finer classification of such constructions. Two examples of fibration are shown in Figure 11.7. In the case of fibering a complex 3-dimensional compact

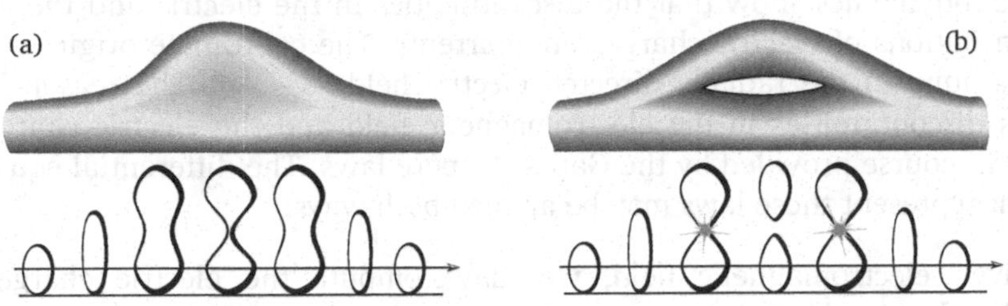

Figure 11.7 Two simple examples of fibration of a loop ($\sim S^1$) along a line: (a) the loop changes its geometry but not its topology, (b) the loop changes both its geometry and its topology ($S^1 \to 2S^1 \to S^1$). In the latter case, there necessarily exist points in the base (horizontal) space where the loop is singular (indicated by the stars).

Calabi–Yau space "along" a (3+1)-dimensional spacetime, the situation is of course much more complicated than the simple examples in Figure 11.7 [237, 228]. However, if we suppose that the variations of the compact complex 3-dimensional Calabi–Yau space $\mathscr{Y}$ along the spacetime $\mathbb{R}^{1,3}$ occur subject to certain complex-analytic limitations, it follows that [228]:

1. The compactification space $\mathscr{Y}$ *must* become singular at some spacetime points $\mathbf{x}_* \in \mathbb{R}^{1,3}$, similarly to the situation in Figure 11.7(b).
2. The spacetime locations $\mathbf{x}_* \in \mathbb{R}^{1,3}$ where the Calabi–Yau space $\mathscr{Y}$ becomes singular for a typical (3+1)-dimensional observer look like massive objects.
3. The metric tensor in spacetime in the vicinity of these objects has an additional contribution
$$\delta g_{\mu\nu} = (\partial_\mu \phi^a)\, \mathcal{R}_{ab}(\phi)\, (\partial_\nu \phi^b), \tag{11.43}$$
where $\phi^a(x)$ are the scalar fields in spacetime $\mathbb{R}^{1,3}$ (not on the worldsheet Σ), which represent the changes in the complex-analytical structure of the compactifying space $\mathscr{Y}$. The tensor $\mathcal{R}_{ab}$ is the Ricci tensor, computed from the metric tensor $\mathcal{G}_{ab}(\phi)$ given on the target (modular) space in which the scalar field $\phi^a(x)$ takes values.
4. The total number and degree of singularizations may be computed exactly for any concrete model, and is a topological characteristic of the model.
5. With the analytic limitations specified in Refs. [237, 228], these massive objects are lines of cosmic proportions – *cosmic strings* – and affect the distribution of matter (galaxies) in the universe. The gravitational field of filamentary objects in 3-dimensional space decreases as $\sim r^{-1}$, and so dominates in accreting matter from which stars, stellar systems, galaxies and clusters form.
6. Relaxing the analytic limitations [237, 228], the dynamics of these cosmic strings may be analyzed perturbatively, but the total number of interactions (joining and splitting) of these cosmic strings is an exactly computable topological invariant for every model.

Thus, in this rather unexpected way, the details of the (microscopic!) string models also have direct cosmological consequences. The connection between the physics of elementary particles and cosmology is already known even in the popular literature [551], but contemporary research in this area is outside the scope of this book. However, this connection became much more direct in constructing stringy models, whereby the purpose of the following section is to at least offer a sampling from this rich research palette.

11.3.4 Localization of gravity

Courses in electrodynamics show that the discontinuities in the electric and the magnetic fields stem from distributions of electric charges and currents. The coordinate origin, $r = 0$, is in this sense a discontinuity in the radially directed electric field, $\frac{1}{4\pi\epsilon_0}\frac{q}{r^2}\hat{r}$. The reason behind the relation between discontinuities in the electromagnetic field and the electric charge and current distributions is of course provided by the Gauss–Ampère laws. The differential equations (5.72a), i.e., (5.78), that represent these laws may be applied both ways:

1. For a given electromagnetic field, we may compute the electric charge and current distribution that produces it.
2. For a given electric charge and current distribution, solve the differential equation and find the produced field.

Regarding gravity and the general theory of relativity, the Einstein equations (9.44) are analogous to the Gauss–Ampère equations. One then expects that discontinuities and other peculiarities in the gravitational field and spacetime curvature may be addressed in an "engineering" fashion, by assembling appropriate distributions of matter [☞ Section 9.3.4].

In typical superstring models, the consistency in dynamics requires that the spacetime through which the superstrings move has 9+1 dimensions. It is then reasonable to ask if it is possible to construct a model where the (9+1)-dimensional spacetime has (3+1)-dimensional isolated subspaces ("defects") of which some may perhaps serve as *our* universe. In such a construction, one must establish:

Q.1 Are there (9+1)-dimensional spacetimes with (3+1)-dimensional "defects"?
Q.2 Are there matter modes that are effectively "trapped" in these "defects"?
Q.3 Do Yang–Mills fields have modes that are effectively "trapped" in these "defects"?
Q.4 Does gravity have precisely one mode that is effectively "trapped" in these "defects"?
Q.5 And, of course, is there at least one (3+1)-dimensional "defect" in which all of the above-cited features occur☝?

Applying Gauss's law within the (3+1)-dimensional spacetime of such defects uses a static subspace of the 3-dimensional space that encloses the source – electric charge or mass – and may be "radially" contracted to the very location of the source. For point-like sources [☞ Section 11.4] in 3-dimensional space, these then are 2-dimensional surfaces, the surface area of which grows with the square of the linear size, whereby the electrostatic and gravitational fields decrease $\sim r^{-2}$ to maintain the constant flux – as we know is the case in Nature.

In turn, if Yang–Mills gauge fields and/or gravity are not trapped in the (3+1)-dimensional defect, the fields will decrease faster: If the field permeates an n-dimensional space, the Gaussian enclosing "wrap" is an $(n-1)$-dimensional sphere, whereupon the Yang–Mills as well as gravitational fields (and forces) decrease following a $\sim r^{1-n}$ law.

— ❦ —

It turns out that the answers to the first three of the above questions are positive, and under very general conditions [280]. Namely, the condition that the compactification space (11.41) is of the Calabi–Yau type (that it admits a metric tensor of which the Ricci tensor is a total derivative), in the worldsheet field theory perspective ("first level" [☞ Section 11.2.4]) becomes a cancellation condition for certain anomalies.

From that same perspective, this same anomaly cancellation condition holds equally for the non-compactified part of the spacetime; for the $\mathbb{R}^{1,3}$ factor (11.41), this anomaly cancellation condition is trivially satisfied. Recall that anomalies are an indication of contradictions in the quantum model – here, in the quantum field theory in the (1+1)-dimensional worldsheet spacetime.

The corresponding geometric condition in (9+1)-dimensional spacetime field theory ("second level") is that the *entire* spacetime must admit a metric tensor, the Ricci tensor of which is a total derivative. Of course, in the situation (11.41), this is trivially satisfied, since the entire Riemann tensor over the $\mathbb{R}^{1,3}$ factor vanishes. However, this implies that the Wick-rotated $t \to it$ analytic continuation of spacetime may be chosen to be a *non-compact*, complex 5-dimensional Calabi–Yau space.

Existence of large universes with isolated sub-universes

The general property of all Calabi–Yau spaces is that they typically have a large number of nontrivial subspaces, which neither have boundaries themselves nor are the boundary of some other subspace – just like the closed contours A and B in Figure 9.8 on p. 354 and the left-hand illustration in Figure 11.8. The complex 3-dimensional Calabi–Yau spaces – used in compactification (11.41) – have such real 2-, 3- and 4-dimensional subspaces [279]. Complex 5-dimensional Calabi–Yau spaces have such real k-dimensional subspaces with $k = 2, \dots, 8$. However, such subspaces are not *isolated*, just as the closed paths of type A and B on the surface of a torus are not: each of these contours has continuous deformations/shifts, such as those depicted in the left-hand side of Figure 11.8.

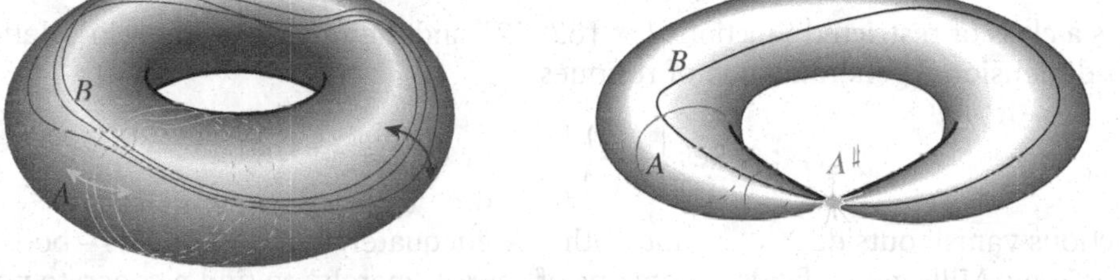

Figure 11.8 The torus surface (left) with two deformable topologically nontrivial closed paths: neither A nor B can be continuously deformed to a point but they can both be deformed into a continuum of "nearby" paths. In the "pinched" torus surface (right), however, the point $A^\sharp$ is isolated [☞ text].

In stark contrast, the surface of the "pinched" torus, on the right-hand side of Figure 11.8 contains the point $A^\sharp$, which is the limiting case of the 1-dimensional subspaces, the closed paths of type A. The point $A^\sharp$ is geometrically and even topologically singled out: any sufficiently small neighborhood of every other point on the surface of the "pinched" torus is of the form of a circle (disc); every neighborhood of the point $A^\sharp$ is of the form of two cones joined at their vertices, and this vertex is a "double point," which is the property that *isolates* this point.

The surface of a torus is in fact a compact complex 1-dimensional Calabi–Yau space, and serves as an intuitive model for higher-dimensional constructions. However, the fact that the double point is also necessarily singular happens only in spaces of one complex dimension. Within complex 3-dimensional Calabi–Yau spaces, such special isolated subspaces are of the form of non-singular real 2-dimensional spheres; they have smooth neighborhoods, but nevertheless cannot be deformed/shifted within the given complex 3-dimensional Calabi–Yau space and so are isolated [279].

In general, virtually all complex n-dimensional Calabi–Yau spaces contain special subspaces of complex dimension $\lfloor\frac{n-1}{2}\rfloor$, the integral part of the fraction $\frac{n-1}{2}$. For $n = 5$, these are complex 2-dimensional, i.e., real 4-dimensional subspaces!

Much as a real 2-dimensional surface of a torus serves as an example of a compact complex 1-dimensional Calabi–Yau space, the 2-dimensional cylinder serves as an example of a non-compact complex 1-dimensional Calabi–Yau space. Besides, note that the 2-dimensional sphere has a positive curvature, but that excising two separate points (the 0-dimensional Calabi–Yau space) leaves behind a surface that is a continuous deformation of a cylinder, which is *flat*: there exists a global and single-valued metric tensor for which the Riemann tensor vanishes.

Similarly, every *non-compact* complex n-dimensional Calabi–Yau space may be obtained by excising from a compact complex n-dimensional Fano space[28] a compact complex $(n-1)$-dimensional Calabi–Yau subspace [520, 521]. If the surgery is arranged to also excise part of a special, isolated real 4-dimensional subspace, its remainder is then also non-compact. After a "reverse" $it \to t$ analytical continuation so that one of the four real dimensions is again time-like, these now $(3+1)$-dimensional non-compact subspaces really can serve as isolated examples of $(3+1)$-dimensional spacetime [280].

Conclusion 11.4 *It follows that the analytical continuation of a typical non-compact complex 5-dimensional Calabi–Yau space is a* $(9+1)$*-dimensional spacetime that contains numerous isolated* $(3+1)$*-dimensional sub-spacetimes.*

Localization of matter and Yang–Mills gauge interactions

For every example of a complex space Z with a complex algebraic subspace $X \subset Z$, specified as the space of solutions of the system of algebraic equations

$$X \subset Z: \quad X := \{z \in Z,\ \Phi(z) = 0\}, \tag{11.44}$$

there exists a class of restricted functions [☞ [82, 52] and the references therein], defined via the complex n-dimensional generalization of residues,

$$f(x) := \operatorname*{Res}_{z\in X}\left[\frac{f(z)}{\Phi(z)}\right], \qquad x \in X = \Phi^{-1}(0). \tag{11.45}$$

These functions vanish outside $X \subset Z$, and within X adequately represent fields – both for matter, and also for Yang–Mills gauge fields. It remains of course "merely" to find a concrete non-compact complex 5-dimensional Calabi–Yau space, with a suitable isolated subspace in which (after analytic continuation so that one of the coordinates in X is time-like) the number and type of localized fields can reproduce the contents of the elementary particle physics Standard Model [☞ Table 2.3 on p. 67, as well as Conclusion 2.2 on p. 46].

In the 1990s, string models were routinely constructed containing various p-branes on which, by their very definition, end-points of open strings are trapped on the given p-brane. This then guarantees localized degrees of freedom amongst which it seems realistic to seek the particle content of the Standard Model, including the Yang–Mills gauge fields.

However, owing to significant differences between Yang–Mills gauge fields and the gravitational field [☞ Section 9.2] and since the graviton is inherently realized in string theory by closed strings [☞ discussion that leads to (11.7)], it is not clear that the analysis in Refs. [82, 52] may be adapted so as to be applied to gravity.[29] Thus, the proposal wherein the additional six spatial

---

[28] after the Italian mathematician, Gino Fano, spaces with a positive curvature are called Fano spaces.

[29] The works of Keiichi Akama and other researchers in Japan and the Soviet Union 1967–82 [☞ [9] and the references therein], where it is shown that an effective general relativity and gravity may be *induced* in $(5+1)$-dimensional models with $(3+1)$-dimensional vortices, until recently were not known outside Japan and the former Soviet Union. However, these models are not renormalizable, and cannot be part of a fundamental theoretical system.

dimensions are not compactified and unobservably small, but it is our (3+1)-dimensional spacetime that is an isolated part – "defect" – within the (9+1)-dimensional spacetime, could not be taken seriously. Compactification *à la* Nordstrøm–Kałuża–Klein remained the only known logical possibility for constructing realistic string (and so also *M*- and *F*-theory extended) models, *almost* to the end of the twentieth century.

Localization of gravity

In 1999, Lisa Randall and Raman Sundrum discovered a relatively simple situation in which gravity *is localized* at (a part of) the boundary of a space [450], which "opened vistas and paved avenues" for constructing alternatives to compactification models.

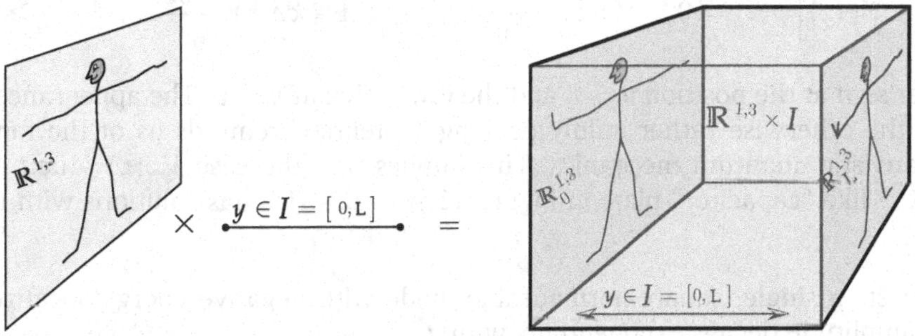

Figure 11.9 The Randall–Sundrum cosmology toy-model.

Randall and Sundrum studied the toy-model wherein spacetime is 5-dimensional geometry reminiscent of a capacitor: the 5-dimensional spacetime is of the form $\mathbb{R}^{1,3} \times I$, where $I = [0, L]$ is the closed interval, i.e., the interval together with its boundary points, as in Figure 11.9. From the definition of the coordinate y, it follows that it can only have non-negative values, and the metric tensor in the 5-dimensional spacetime $\mathbb{R}^{1,3} \times [0, L]$ must depend on $|y|$. However, since the Einstein tensor – the left-hand side of the Einstein equations (9.44) – is a differential expression of *second* order in spacetime derivatives of the metric tensor components, it follows that the Riemann, Ricci and Einstein tensors, as well as the scalar curvature, must include terms proportional to the Dirac δ-function, $\delta(y)$.

Concretely, Randall and Sundrum define

$$ds^2 = -e^{-2k|y|}\eta_{\mu\nu}dx^\mu dx^\nu + dy^2, \quad \text{i.e.,} \quad [g(x,y)] = \begin{bmatrix} -e^{-2k|y|} & 0 & 0 & 0 & 0 \\ 0 & e^{-2k|y|} & 0 & 0 & 0 \\ 0 & 0 & e^{-2k|y|} & 0 & 0 \\ 0 & 0 & 0 & e^{-2k|y|} & 0 \\ 0 & 0 & 0 & & 1 \end{bmatrix}, \tag{11.46}$$

whereby the Ricci tensor and the scalar curvature are

$$[\boldsymbol{R}] = \left[\begin{array}{c|c} -\eta_{\mu\nu}\, e^{-2k|y|} f(y) & 0 \\ \hline 0 & g(y) \end{array}\right], \qquad \begin{cases} f(y) = 2k\big[\delta(y) - 2k\ \text{sig}^2(y)\big], \\ g(y) = 4k\big[2\,\delta(y) - k\ \text{sig}^2(y)\big], \end{cases} \tag{11.47a}$$

$$R = 16\,k\,\delta(y) - 20\,k^2\ \text{sig}^2(y) = \begin{cases} 16\,k\,\delta(y) & y = 0, \\ -20\,k^2 & y \neq 0. \end{cases} \tag{11.47b}$$

Here

$$\text{sig}(y) := \begin{cases} -1, & y < 0, \\ 0, & y = 0, \\ +1, & y > 0, \end{cases} \qquad \text{so} \qquad \text{sig}^2(y) := \begin{cases} +1, & y \neq 0, \\ 0, & y = 0, \end{cases} \tag{11.48}$$

and the results hold in the vicinity of $y=0$, as if $L \to \infty$. The exact result is more complicated and of course must include terms with $\delta(y-L)$ and $\mathrm{sig}^2(y-L)$ because of the analogous effect of the $y=L$ boundary.

The Einstein equations then dictate that $T_{\mu\nu}$ must contain terms proportional to $\delta(y)$, $\delta(y-L)$, $\mathrm{sig}^2(y)$ and $\mathrm{sig}^2(y-L)$. This implies that the maintenance of such a geometry requires the existence of matter that is localized at the $y=0$ and the $y=L$ boundaries of this 5-dimensional "universe," as well as matter that permeates this universe along the fifth, y-coordinate. However, it is more important that the differential equations for the metric tensor components, after separation of variables and a suitable substitution $z=z(y)$, include the differential equation [450],

$$\Big[-\frac{1}{2}\frac{\mathrm{d}^2}{\mathrm{d}z^2}+\widetilde{V}_{\pm}(z)\Big]\hat{\psi}(z)=0,\quad \widetilde{V}_{\pm}(z)=\frac{15\,k^2}{8(k\,|z|+1)^2}\pm\tfrac{3}{2}\,k\,\delta(z)-\frac{m^2c^2}{\hbar^2},\qquad z,y\geqslant 0, \tag{11.49}$$

with the upper sign at the position $y=L$ and the lower sign at $y=0$. The appearance of the Dirac δ-function in the otherwise rather mildly peaking "potential" reminds us of the familiar system from non-relativistic quantum mechanics. This implies that the case $\widetilde{V}_{-}(z)$ – i.e., at the $y=0$ copy of the $\mathbb{R}^{1,3}$-like "capacitor" plate in Figure 11.9 on p. 431 – has solutions with the following properties:

1. There exists a single, square-normalizable mode with negative energy, localized at $y=0$, and its amplitude decays exponentially with y.
2. There is a continuum of modes:
   (a) with continuous mass/energy $m^2 \in [0,+\infty)$,
   (b) the envelopes (amplitudes) of which are very small near $z=0$ because of the $\frac{15\,k^2}{8(k|z|+1)^2}$ barrier,
   (c) which asymptotically (for $|z|\to\infty$) approach plane waves.
3. Owing to the property 2(b), the interference between the unique localized ("bound-state") mode and the continuum of modes is suppressed.

Randall and Sundrum then showed [450] that the unique localized mode in the metric tensor effectively serves as the metric tensor in the $\mathbb{R}^{1,3}_{y=0}$-boundary of their 5-dimensional model, and leads to a usual formulation of the general theory of relativity in this part of the boundary, as well as to the familiar Newton/Kepler gravitational potential $\sim r^{-1}$. The continuum of modes produces a correction of Newton's law of gravity:

$$V(r)=G_{\mathrm{N}}\frac{M_1\,M_2}{r}\Big(1+\frac{1}{(kr)^2}\Big), \tag{11.50}$$

where the parameter k is a measure of the curvature (not the size!) of the 5-dimensional spacetime along the fifth coordinate.

Conclusion 11.5 *Comparing equations (11.50) and (11.47b) shows that the correction to Newton's law of gravity is suppressed by the* ***curvature*** *of the "big" spacetime along the fifth dimension, and not the size of this fifth dimension. This result is* ***qualitatively*** *different from similar results in compactification models: There, all corrections that stem from the existence of compact dimensions are always suppressed by the* ***volume*** *of the (small) compact space.*

The fifth coordinate in the Randall–Sundrum model may well be even infinitely large(!); its existence nevertheless changes the effective (3+1)-dimensionality of physics in the $y=0$ boundary 3-brane no more than as specified in equation (11.50).

This important result started a "minor industry" of elaborations of this and similar ideas, whereby the string theoretical system again forayed into cosmology with these *brane geometry* models. In the 1990s, many details of the interactions between various p-branes and other more-or-less exotic objects that appear in the string theoretical system were worked out. Now that we know of the Randall–Sundrum mechanism for localizing gravity, it is worth exploring the possibility that some of these p-branes – as well as other $(9+1)$-dimensional spacetimes – are of cosmic proportions, and that we happen to live on one of these 3-branes, which provides the basic conceptual idea of the cosmology of so-called brane-worlds [☞ [352], for a recent review].

— ❦ —

Amongst such models an interesting possibility emerges that again links the microscopic and the macroscopic physics in unusual ways: Namely, we know that no supersymmetric partner particle of any of the known particles has ever been found, so – if the fundamental theory of Nature is supersymmetric at all, supersymmetry must be broken, and direct experimental evidence for this (e.g., a Goldstone fermion) is also lacking. On the other hand, the discovery that our universe is expanding in an accelerated fashion implies that the corresponding geometry (in large, cosmic proportions) is the de Sitter geometry. However, it is known [189, 562, 560, 76] that the de Sitter geometry does not admit supersymmetry. Nevertheless, it is possible to construct superstring models containing a $(3+1)$-dimensional sub-spacetime, i.e., 3-brane [53] [☞ also [485, 303, 302] and [16, 120] for recent works]:

1. with the de Sitter geometry,
2. with localized gravity,
3. with an exponential relation between the Planck mass and the mass of $W^{\pm}$- and Z^0-bosons,
4. where the geometry is induced by the presence of a modular field,[30]
5. with the cosmological constant related to the supersymmetry breaking [54].

It follows that it is possible to break supersymmetry by means of the spacetime geometry, which in turn is produced by the interaction of gravity with modular fields that are unique to stringy models.

However, this is but one of many possibilities; one "merely" ought to find the model in which on some of its 3-branes with localized gravity there exist enough localized matter and Yang–Mills gauge fields for the Standard Model [☞ Table 2.3 on p. 67, and Conclusion 2.2 on p. 46]. The Reader interested in this class of brane-world models is directed to the rich literature, starting for example with Ref. [452] for relations with strings, and Ref. [101] for F-theory extensions.

Exospace

Besides the two general mechanisms discussed so far,

compactified worlds the Norstrøm–Kałuża–Klein compactification, both the constant (11.41), and the variable kind as discussed in Section 11.3.3,

brane-worlds the Randall–Sundrum mechanism of localizing gravity to some of the sub-spacetimes of a big, $(9+1)$-dimensional spacetime,

there is however also a *third* possibility.

Namely, physics without strings is based on describing the motion of point-like particles ("material points") and their extension, "point-local" fields: Although a field by definition extends and

[30] This is literally a stringy "signature." The particular modular field involved here is not single-valued: rotations in a plane within the "extra" dimensions induce a so-called Möbius, or $SL(2;\mathbb{Z})$ transformation in the field. This occurs in no non-stringy theory/model, and *every* string model contains this particular modular field.

permeates the entire space, functions that are used to describe fields are fundamentally local quantities. For example, the gauge potential $\mathbb{A}_\mu(\mathrm{x})$ depends on the coordinates of a *single point* in spacetime, and the differential equations that mathematically represent the laws about such fields are local: the fluctuation of the field at any one point in spacetime causes – via the local differential equation of motion – the propagation of the fluctuation from one spacetime point to the infinitesimally neighboring points.

The physics of strings is not local in the same sense. From the worldsheet perspective, the field theory in (1+1)-dimensional worldsheet spacetime is not local in the same sense as it is in the (3+1)-dimensional spacetime. Students who have successfully passed a course in electrodynamics must know that the Green functions for the wave operator (the d'Alembertian) in $(n{+}1)$-dimensional spacetime *grow* with the distance when $n < 2$. Thus, scalar fields in (1+1)-dimensional spacetime correlate between arbitrarily distant fluctuations, and so are fundamentally global and non-local fields.

On the other hand, from the spacetime perspective in which the strings propagate, it is clear that strings are not local objects, but exist simultaneously (however this to be understood) in a continuum of space-like separated points within the spacetime – this is a property of all extended objects, including also all p-branes with $p > 0$. Besides, the following facts also hold about string interactions:

- String interaction *is* local in the spacetime in which the strings propagate: From any observer's reference system, the joining of two strings into one and the splitting of a string into two happens at one spacetime point.
- String interaction is *not* local in the configuration space of strings; if it were, two strings would be joining into one (and one splitting into two) in a single point in the configuration space – which is a particular configuration of the entire string.
- String interaction is *not* local in the string worldsheet spacetime; moreover, a string interaction represents a "cosmological" fusion of two such (1+1)-dimensional spacetimes into one, or the splitting of one into two.

Of course, these are merely picturesque indications that the motion and interactions of strings (in fact, of all p-branes for $p > 0$) differ essentially from those of point-particles (0-branes).

It turns out, however, that these differences are crucial in determining through what kinds of spacetimes strings – and more generally, p-branes with $p > 0$ – can consistently propagate. It was already known in 1985 that so-called *orbifolds* – spaces with conical singularities of the form $\mathbb{R}^n/D$ where D is the action of some finite group of rotations – pose no problem [138, 137]. A complete and final criterion to answer the question "through how singular a spacetime can strings consistently propagate" is not yet known☝, but it is known that requiring supersymmetry in stringy dynamics permits singularities of rather high degree [278]. This certainly includes both orbifold and canonical singularities [☞ the "Young Persons' Guide" [454]]. The hallmark property of these types of singularities is that they can be smoothed by means of processes called *blow-up*, *deformation* and *small resolution*, which either maintain or can be restricted to maintain all the characteristics essential for superstring dynamics, such as Ricci flatness [279, for starters].

Closely related to singular spaces are so-called *stratified pseudo-manifolds* through which strings also move consistently [24].[31] Such spaces generalize the cases shown in Figure 11.9 on p. 431, where the 5-dimensional space has 4-dimensional "boundary" parts. In general, stratified pseudo-manifolds are connected unions of several parts organized by dimension so that:

[31] The Authors of Ref. [24] have not emphasized this fact explicitly, but their Figure 19 explicitly depicts a complex 3-dimensional pseudo-manifold with a complex 1-dimensional connected additional part.

1. separately taken, every part is a space of constant dimension,
2. subspaces of the same dimension form a *stratum*,
3. there may exist more than one stratum, i.e., parts of more than one dimension.

Instead of detailed definitions, may it suffice here to consider the two examples in Figure 11.10. The left-hand example is a surface (a), defined by the equation $z = (x/y)^2$, with a self-intersection along the non-negative part of the z-axis, where the surface evidently has two-fold defined tangent vectors and has "unusual" (exotic) neighborhoods. However, this surface may be decomposed into:

0. the (excised) 0-dimensional stratum: the coordinate origin O,
1. the (excised) 1-dimensional stratum: positive z-axis denoted z_+,
2. the (remaining) 2-dimensional stratum: two surfaces, A and B.

In this example, *every* point x_* of every stratum has arbitrarily near points that belong to a higher-dimensional stratum; in general, this need not be true.

The right-hand side of Figure 11.10 shows a more unusual but also more general example (b), which may be decomposed into:

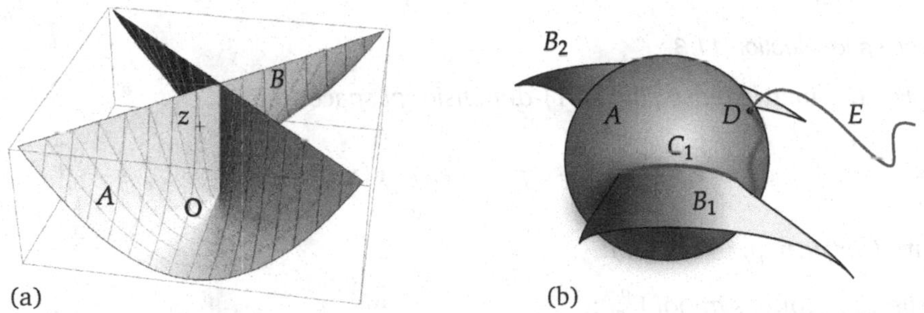

Figure 11.10 Two stratified pseudo-manifolds.

0. the (excised) 0-dimensional stratum: the point D,
1. the (excised) 1-dimensional stratum: the "seams" C_1 and C_2 (on the back side) and the "tail" E,
2. the (remaining) 2-dimensional stratum: two "wings" B_1 and B_2 and the surface of the sphere A with three holes (two line-like cuts where the "seams" C_1 and C_2 were excised, and one point-like hole where the point D was excised).

Unlike example (a), not every point of the 1-dimensional stratum has points in the higher-dimensional stratum (the surface of the sphere A and of the wings B_1 and B_2 and without the "seams" C_1 and C_2) that are arbitrarily close to it. To wit, *each* point in the "seams" C_i is arbitrarily near *some* points in both wings, B_1 and B_2, and in the sphere A. On the other hand, most points in the "tail" E are nowhere near any point in the 2-dimensional stratum. One says that the "tail" E is *outside* the 2-dimensional stratum. Extending the nomenclature of Ref. [24], let the prefix *exo-* denote the *exotic* parts that are mostly *outside* the highest(-dimensional) stratum. As the highest dimension in this example is two (2), the only exo-space (here, exo-line) is the 1-dimensional "tail" E.

Also, the complete pseudo-manifold (b) in Figure 11.10 does not have well (unambiguously) defined derivatives at the "seams" C_1 and C_2 as well as at the joining point D. This is also true of the example (a), along the non-negative semi-axis $z \geqslant 0$.

Finally, the spacetime (so-called *brane geometry*) in a string model can easily be some such stratified pseudo-manifold $\mathscr{Z}$ that contains (3+1)-dimensional exo-space $\mathscr{X} \subset \mathscr{Z}$, i.e., that $\mathscr{Z}$ admits a metric tensor[32] that gives a 4-dimensional exo-space $\mathscr{X}$ the Lorentzian signature, $(1,3)$. For example, $\mathscr{Z}$ could be caricatured by the right-hand example in Figure 11.10, where $\mathscr{X}$ could be represented by the "tail" E and which could of course easily be of cosmic proportions. To an observer who is in a part of $\mathscr{X}$ sufficiently far from the region where $\mathscr{X}$ joins with the "rest" of the complete spacetime $\mathscr{Z}$, the spacetime evidently looks 3+1 dimensional. However, an observer who is sufficiently close to the connecting region between $\mathscr{X}$ and the "rest of $\mathscr{Z}$" will of course be able to experimentally verify that by passing from $\mathscr{X}$ into the "rest of $\mathscr{Z}$" the number of spacetime dimensions changes. In such a model, all (gravitational and Yang–Mills gauge) fields must be localized, simply as they have nowhere else to propagate in the neighborhood of almost all points in $\mathscr{X}$. Then, the Gaussian surface that encloses point-like sources (charges and/or masses) may almost everywhere in $\mathscr{X}$ be chosen to be a sphere of surface $4\pi r^2$ at a distance r from the source, whereby the Coulomb and the Newton/Kepler forces indeed decrease as $\sim r^{-2}$. The exploration of such models is in its infancy, but it is clear that in many models the p-branes may well be *exo*-branes, just like in Figure 19 of Ref. [24] or in the right-hand example in Figure 11.10. For mathematical details about stratified pseudo-manifolds, the Reader is directed to the literature, e.g., starting with the book [313].

11.3.5 Exercises for Section 11.3

✎ **11.3.1** *For the metric tensor in* (2+1)*-dimensional spacetime*

$$ds^2 = -e^{-2k|y|}c^2dt^2 + e^{-2k|y|}dx^2 + dy^2, \quad \text{i.e.,} \quad [g(\mathbf{x},y)] = \begin{bmatrix} -e^{-2k|y|} & 0 & 0 \\ 0 & e^{-2k|y|} & 0 \\ 0 & 0 & 1 \end{bmatrix}, \tag{11.51}$$

compute (for $m,n,p,r = 0,1,2$):

1. *the Christoffel symbol Γ^p_{mn};*
2. *the Riemann tensor $R_{mnp}{}^r$;*
3. *the Ricci tensor R_{mn};*
4. *the scalar curvature R_{mn};*
5. *the Einstein tensor $G_{mn} := R_{mn} - \frac{1}{2}g_{mn}R$.*

✎ **11.3.2** *Show by direct computation that the change $e^{-2k|y|} \to e^{-2ky}$ in the metric tensor definition in the previous exercise "erases" all matter localized in the spacetime plane $y = 0$ and that the results for Γ^p_{mn}, $R_{mnp}{}^r$, R_{mn}, R and G_{mn} agree with the formal substitution $\text{sig}^2(y) \to 1$ in the results of the previous exercise.*

11.4 Duality and dual worldviews

Section 5.2.2 showed that there exists a symmetry between the electric field, charge and current on one hand, and the magnetic field, charge and current on the other [☞ Conclusion 5.4 on p. 185]. The fact that – as best as known – there are no magnetic monopoles in Nature, i.e., monopole magnetic charges and currents means that the duality rotation (5.85), $\varpi_{EM}(\vartheta)$, may be applied simultaneously both to electromagnetic fields and to electromagnetic sources so that this basis is – everywhere in the universe "simultaneously"! – "turned" orthogonal to $(\rho_m/c, \vec{j}_m/c^2)$ so that only $(c\rho_e, \vec{j}_e)$ remain. However, the *same* system can be equally well described in the "basis"

[32] The metric tensor may be defined separately in each part of constant dimension, and then require that these tensors coincide in places where the various parts touch.

that is "turned" orthogonal to $(c\rho_e, \vec{\jmath}_e)$ so that only $(\rho_m/c, \vec{\jmath}_m/c^2)$ remain, as was summarized in Conclusion 5.4 on p. 185. This then provides two *dual* descriptions of Nature.

Exploration of the string theoretical system discovered many other such *dual* relations between many (sometimes very) different models. In this sense, one says:

> **Definition 11.3** *Two given descriptions of Nature, O_1 and O_2, are **dual** if they are **physically indistinguishable**, i.e., if the collection of all physical observables (together with all relations between them) in the description O_1 is isomorphic to the corresponding collection in the description O_2.*
>
> *(There may be more than two dual descriptions, so one talks of **triality**, but in practice one never talks of **quadrality**, **quintality**, or any other **n-ality**.)*

Given the complexity of the relations between the physical system, the mathematical model, the solutions to that model and measurable (verifiable) results [☞ Figure A.2 on p. 457], the limitations on the mathematical model itself are evidently very indirect and roundabout. It should then be clear that the mental caricatures and images that one uses in formulating the model are merely a crutch in the construction of the model, and not "the one and true" image of "reality" – recall the Copernican legacy [☞ Section 1.1.1]. Thus, as long as two or more models (even if based on different images) equally well describe Nature, we are free to choose which of these two (or more) images to tentatively identify with Nature [☞ Section 11.3.2, and especially Figure 11.6 on p. 426]. In doing so, we must stick to the facts [☞ Example 1.1 on p. 11]:

1. The chosen image is but one of the a-priori equally "real" formulation images.
2. The choice of the formulation image is, without an experimentally verifiable difference, subjective and tentative.
3. Measurable results (and not the formulation image) of a model make up its goal, and so also its point.

The last two decades made it ever clearer that the string theoretical system integrally contains multiple differing formulation images in describing the same "thing" – which lends support to conceiving of a new (unexplored and undeveloped☝) kind of "symmetry," acting via corresponding transformations between different formulation images, and which are *not* the usual gauge symmetries [126, 77, 470].

11.4.1 T-duality

The first examples of duality were discovered more-or-less accidentally, whereby this research reminds us of experimental physics where the first task is to find as many examples as possible so as to perceive the common properties, so as to then seek the basic principles of this phenomenon. Although there exist many seemingly different examples of T-duality, it turns out that this is one of the simpler classes of duality, and that other types of duality are progressively more and more unusual – from the vantage point of the well-known general properties in previously known physical models. This of course merely points to the fact that the (M- and F-theory extended) (super)string theoretical system has radically new, unknown and unexpected properties.

$R \to 1/R$ duality

The first signal of this multiplicity is the so-called $R \to 1/R$ duality, which is described in detail in textbooks [434, 594, 46]. This is an essential consequence of the fact that strings are not point-like and that the relation between strings and the spacetime through which they move is very different from the analogous relation for point-like particles.

Let the $\hat{\imath}$th spacetime coordinate (for some fixed $\hat{\imath} > 0$) be periodic and closed into a circle of circumference $2\pi R$. To a string that propagates through such a spacetime, the coordinate field in this periodic direction must satisfy the same periodicity condition:

$$X^{\hat{\imath}}(\tau,\sigma) \simeq X^{\hat{\imath}}(\tau,\sigma) + 2\pi w\, R, \qquad w \in \mathbb{Z}, \tag{11.52}$$

where w counts how many times the string is wrapped around the circle of the periodic coordinate. This periodicity changes neither the Hamilton action nor the Lagrangian density, but imposes the linear momentum quantization in the direction of the periodic (compact) coordinate:

$$p_{\hat{\imath}\, n} = \frac{n\hbar}{R}, \qquad n \in \mathbb{Z}, \quad \text{fixed } \hat{\imath}. \tag{11.53}$$

The proper generalization of the relation (3.36) gives the square of the Lorentz-invariant mass of that string [434, 594] as

$$m^2c^2 = \frac{n^2\hbar^2}{R^2} + \frac{w^2R^2}{\alpha'^2\hbar^2c^4} + \frac{2}{\alpha' c^2}(N_L + N_R - 2), \tag{11.54}$$

where N_L and N_R are the total excitation numbers for the left-moving and the right-moving oscillators (11.38), respectively, counting only oscillators that are transversal to the worldsheet and which satisfy the condition

$$nw + N_R - N_L = 0. \tag{11.55}$$

Manifestly, the relations (11.54)–(11.55) remain unchanged under the exchange

$$n \leftrightarrow w, \qquad \text{and} \qquad R \leftrightarrow \frac{\alpha'\hbar^2c^2}{R} = \frac{{\ell_s}^2}{R}, \tag{11.56}$$

which implies that string models do not distinguish between spacetimes in which a periodic coordinate describes a circle of circumference $2\pi R$ and those with a circle of circumference $2\pi({\ell_s}^2/R)$. For a complete proof of this equivalence according to the diagram in Figure 11.6 on p. 426, see Chapter 8 in the textbook [434, Vol. 1].

Manifestly, if $R < \ell_s$, then $({\ell_s}^2/R) > \ell_s$. Thus, a spacetime with a compact dimension that is smaller than the string characteristic size, $\ell_s \sim 10^{-35}$ m, is equivalent to a spacetime where the compact dimension is reciprocally larger than this characteristic length. In this sense, compact dimensions cannot be "too small" to be "seen" by strings – which is exactly the opposite behavior from that in "pointillist" models, for which $\ell_P \sim 10^{-35}$ m is the minimal discernible distance [☞ Section 1.3].

This duality between "big" and reciprocally "small" dimensions is called the T-duality, in that the target space in which the coordinates are all periodic has the geometry of a torus, $T^n := S^1 \times \cdots \times S^1$ (with n factors).

Mirror duality, again

Mirror duality, discussed in Section 11.3.2, was first discovered as a relation between two Calabi–Yau compactification models that use two very concrete constructions [236]: the first Calabi–Yau manifold, X, is the space of solutions to the algebraic equation

$$\sum_{i=1}^{5} {z_i}^5 = 0, \qquad z_i \in \mathbb{C}, \quad (z_1,\dots,z_5) \simeq (\lambda z_1,\dots,\lambda z_5), \quad 0 \neq \lambda \in \mathbb{C}. \tag{11.57}$$

The other Calabi–Yau manifold is obtained as the quotient space $Y = X/(\mathbb{Z}_5 \times \mathbb{Z}_5 \times \mathbb{Z}_5)$, where each factor $\mathbb{Z}_5$ denotes an independent symmetry of the fifth order, such as the transformation

$$(z_1, z_2, \dots, z_5) \simeq (\omega^1 z_1, \omega^2 z_2, \dots, \omega^5 z_5), \qquad \omega := e^{2\pi i/5}, \ \omega^5 = 1. \tag{11.58}$$

It may be computed [236, 279] that $h^{1,1}(X) = 1 = h^{2,1}(Y)$ and $h^{2,1}(X) = 101 = h^{1,1}(Y)$, which is an indication of the "mirror duality" $X \leftrightarrow Y$. The complete proof [236] is much more detailed and requires showing that compactifications on the manifolds X and Y are physically equivalent, in the sense of the diagram in Figure 11.6 on p. 426. This construction of an explicit pair of mirror-dual Calabi–Yau manifolds was soon generalized in several different ways, and Andy Strominger, Shing-Tung Yau and Eric Zaslow had by 1996 proven that all such pairs are special cases of T-duality [501]. Complementary to this research, done from the perspective of the spacetime through which the string moves, models in (1+1)-dimensional worldsheet spacetime were also soon constructed that completely reproduce the mirror duality [188, 371, 192].

— ❦ —

In all these examples, many of the geometric and topological properties of the space X and its T-dual space X^T are different, but $\dim(X^T) = \dim(X)$. Even this need not be the case in other types of duality.

11.4.2 Gauss's law and its consequences

The insight that the Einstein equations are the analogue of Gauss's law for the gauge symmetry of general coordinate reparametrizations motivates a re-examination of Gauss's law for the various possible cases.

Electric and magnetic sources in 4-dimensional spacetime

Electric charge is – in principle – measured by means of measuring the total flux of the electric field through a closed (Gaussian) surface that encloses the given charge distribution. From the practical application of this (Gauss's) law in electrostatics, we know that for the electric field the smallest possible electric charge is point-like, i.e., 0-dimensional. Of course, a collection of point-like charges may well form a linear, surface or volume distribution of charges, as most often observed in Nature. However, the point here is that nothing in the structure of electrodynamics obstructs an electric charge from being as little as 0-dimensional.

Analogously, the same thought-experiment/measuring may also be set up for a magnetic field, and it follows that the smallest possible magnetic charge is also point-like – as discussed in Section 5.2.3. The fact that point-like magnetic charges (magnetic monopoles) are not observed in Nature is then a puzzling property of *our* particular Nature, even if we take into account the relativity of what is called "electric" and what "magnetic" [☞ Conclusion 5.4 on p. 185]☝.

In preparation to re-examine this fact in the more general case, consider in a bit more detail how the application of Gauss's law produces the minimal dimension (= 0) for electric and magnetic charges in (3+1)-dimensional spacetime.

Using the 4-vector notation (5.73), the components of the electric field are identified as $E_i = F_{0i}$. The flux of the electric field is maximal when measured (integrated) over a surface of which the tangent plane at every point is orthogonal to the direction of the electric field. (As the electric field is a 3-vector, in 3-dimensional space the orthogonal subspace is of course a 2-dimensional surface.) The Gaussian integral over the closed surface (2-dimensional sphere) is then written

$$\Phi_E := \oint_{S_G} \mathrm{d}^2\vec{\sigma} \cdot \vec{E} = \oint_{S_G} \mathrm{d}^2\sigma^i\, E_i = \oint_{S_G} \mathrm{d}^2\sigma^i\, F_{0i} = \oint_{S_G} \mathrm{d}x^\mu \mathrm{d}x^\nu\, \varepsilon_{\mu\nu}{}^{0i}\, F_{0i}. \tag{11.59}$$

Owing to the defining properties of the Levi-Civita symbol, $\varepsilon_{\mu\nu}{}^{\rho\sigma} := \varepsilon_{\mu\nu\kappa\lambda}\eta^{\kappa\rho}\eta^{\lambda\sigma}$, we know that $\mu, \nu \neq 0, i$, i.e., that $\mathrm{d}x^\mu$ and $\mathrm{d}x^\nu$ are differentials of coordinates that are, in every point of the Gaussian surface S_G, orthogonal to the direction of the electric field E_i, as well as to the direction of time – i.e., they are constant in time. As the electric field $\vec{E}$ is directed *from* the source (positive electric charge) or *towards* the sink (negative electric charge), the Gaussian surface S_G can then be shrunk "radially" – along the ith coordinate in equation (11.59) – to the source/sink of the $\vec{E}$-field, electric charge itself. The dimension of this source/sink then must be equal to

$$\dim(\rho_e) \geqslant \dim(\text{space}) - \dim(S_G) - \dim(\text{radius}) = 3 - 2 - 1 = 0. \tag{11.60}$$

That is, electric charges may well be as little as 0-dimensional (point-like).

The components of the magnetic field are identified as $B^i := \frac{1}{2}\varepsilon^{ik}F_{jk} = \frac{1}{2}\varepsilon^{0ik}F_{jk}$, i.e., $F_{jk} = B^i\varepsilon_{0ik}$, so unlike equation (11.59), for the magnetic flux we have

$$\begin{aligned}\Phi_M &:= \oint_{S'_G} \mathrm{d}^2\sigma_i\, B^i = \oint_{S'_G} \mathrm{d}x^i\mathrm{d}x^j\, B^k\, \varepsilon_{0ijk} = \oint_{S'_G} \mathrm{d}x^j\mathrm{d}x^k\, F_{jk}\\ &= \oint_{S'_G} \mathrm{d}x^\mu\mathrm{d}x^\nu\, \varepsilon_{\mu\nu 0i}\big(\tfrac{1}{2}\varepsilon^{0ijk}\, F_{jk}\big) \overset{(5.85)}{=} \oint_{S'_G} \mathrm{d}x^\mu\mathrm{d}x^\nu\, \varepsilon_{\mu\nu 0i}({*F}^{0i}).\end{aligned} \tag{11.61}$$

Again, the use of the Levi-Civita symbol ε^{0ijk} guarantees that $\mathrm{d}x^i$, $\mathrm{d}x^j$ and B^k are mutually orthogonal. The Gaussian surface S'_G again may be shrunk "radially" to the magnetic charge itself, the dimension of which then must be equal to

$$\dim(\rho_m) \geqslant \dim(\text{space}) - \dim(S'_G) - \dim(\text{radius}) = 3 - 2 - 1 = 0. \tag{11.62}$$

Thus, in (3+1)-dimensional spacetime the sources (or sinks) of both electric and magnetic fields may be as little as 0-dimensional (point-like). Notice the similarity between equations (11.59) and (11.61) owing to the fact that both $F_{\mu\nu}$ and its dual, ${*F}^{\mu\nu}$, are rank-2 tensors.

Sources for gauge fields in *n*-dimensional spacetime

The above analysis probably seems unnecessarily complicated for such an "obvious" result. However, in more than (3+1)-dimensional spacetime, the results are less obvious.

In n-dimensional spacetime, the electric flux is

$$\Phi_E := \oint_{S_G} \mathrm{d}x^{\mu_1}\cdots\mathrm{d}x^{\mu_{n-2}}\, \varepsilon_{\mu_1\cdots\mu_n}\eta^{\mu_{n-1}0}\eta^{\mu_n i}\, F_{0i}, \tag{11.63a}$$

$$\dim(\rho_e) \geqslant (n{-}1) - (n{-}2) - 1 = 0: \quad \Rightarrow \text{ point-like electric charges.} \tag{11.63b}$$

The magnetic flux (writing $({*F})^{\rho_1\cdots\rho_{n-2}} := \frac{1}{2}\varepsilon^{\rho_1\cdots\rho_{n-2}jk}\, F_{jk}$) is

$$\Phi_M := \oint_{S_G} \mathrm{d}x^\mu\mathrm{d}x^\nu\, \tfrac{1}{(n-2)!}\varepsilon_{\mu\nu\rho_1\cdots\rho_{n-2}}\, ({*F})^{\rho_1\cdots\rho_{n-2}}, \tag{11.64a}$$

$$\dim(\rho_m) \geqslant (n{-}1) - (2) - 1 = n{-}4: \quad \Rightarrow (n{-}4)\text{-dimensional magnetic charges.} \tag{11.64b}$$

Thus, in spacetime of more than $3+1$ dimensions, the magnetic sources/sinks may no longer be point-like. This follows from the fact that in n-dimensional spacetime the components of the magnetic field are identified with the components of the *dual* tensor:

$$E_i := F_{0i}, \qquad \text{but} \qquad B^{i_1\cdots i_{n-3}} := \varepsilon^{0i_1\cdots i_{n-3}jk}F_{jk}, \tag{11.65}$$

so the components of the electric field always form a spatial vector, but the components of the magnetic field form a spatial rank-$(n{-}3)$ tensor: A rank-r tensor "emanates" radially from the source (or towards a sink) that is therefore $(r{-}1)$-dimensional.

In string models for the first time there routinely also appear gauge fields for which the gauge potential itself is an antisymmetric rank-r spacetime (Lorentz) tensor, $A_{\mu_1\cdots\mu_r}(x)$. The gauge fields are then defined using the rank-$(r{+}1)$ tensor:

$$E_{i_1\cdots i_r} := F_{0i_1\cdots i_r}, \qquad\qquad A_{\mu_1\cdots\mu_r} := A_{[\mu_1\cdots\mu_r]}, \tag{11.66}$$

$$B^{i_1\cdots i_{n-r-2}} := \varepsilon^{0\,i_1\cdots i_{n-r-2}\,j_1\cdots j_{r+1}} F_{j_1\cdots j_{r+1}}, \qquad F_{\mu_1\mu_2\cdots\mu_{r+1}} := \partial_{[\mu_1} A_{\mu_2\cdots\mu_{r+1}]} \tag{11.67}$$

where the square brackets around the indices denote total antisymmetrization [☞ the lexicon entry, in Appendix B.1]:

$$a_{[\mu}b_{\nu]} := \tfrac{1}{2}(a_\mu b_\nu - a_\nu b_\mu) = \tfrac{1}{2!(n-2)!}\varepsilon_{\mu\nu\rho_1\cdots\rho_{n-2}}\varepsilon^{\kappa\lambda\rho_1\cdots\rho_{n-2}}a_\kappa b_\lambda, \tag{11.68a}$$

$$\begin{aligned} a_{[\mu}b_\nu c_{\rho]} &:= \tfrac{1}{3!}\big(a_{[\mu}b_{\nu]}c_\rho + a_{[\rho}b_{\mu]}c_\nu + a_{[\nu}b_{\rho]}c_\mu\big) \\ &= \tfrac{1}{3!(n-3)!}\varepsilon_{\mu\nu\rho\sigma_1\cdots\sigma_{n-3}}\varepsilon^{\iota\kappa\lambda\sigma_1\cdots\sigma_{n-2}}a_\iota b_\kappa c_\lambda, \qquad \text{etc.} \end{aligned} \tag{11.68b}$$

These imply the generalization of the relations (11.63b) and (11.64b):

Conclusion 11.6 *The minimal dimensions of the electric and magnetic charges in n-dimensional spacetime for gauge interactions with a rank-r gauge potential are*

$$\dim(\rho_e) \geqslant ((n-1)) - (n-(r+1)) - 1 = r-1: \quad \textit{electric } (r-1)\textit{-branes}; \tag{11.69a}$$

$$\dim(\rho_m) \geqslant ((n-1)) - (r+1) - 1 = n-r-3: \quad \textit{magnetic } (n-r-3)\textit{-branes}. \tag{11.69b}$$

For neither of these to become negative, it follows that

$$0 \leqslant [p := (r-1)] \leqslant (n-4). \tag{11.70}$$

The upper limit is then always taken to be $(n-4) = 7$, corresponding to $n = 11$ in the M-theory extension of string theory. It is not clear if the F-theory extension could also permit 8-branes, but 7-branes certainly do play a key role in the original definition of F-theory [530].

Comment 11.2 *Note that all branes have a tension that is determined by a relation of the type (11.7), and that the largest branes have $n-4$ spatial dimensions. Such branes can then "trap" $n-4$ dimensions of space, obstructing their expansion during the Big Bang, which provides the possibility of a* **dynamical** *explanation of the fact that only 4 dimensions of spacetime have characteristic scales of cosmic proportions, while the remaining $n-4$ spatial dimensions may have a size of the order of the Planck length, ℓ_P – thus supporting the compactification type of spacetime geometry.*

This image is supported by the example of the vibrations and motion of a closed string (as a 1-brane): The vibrations of the string are not limited by the string tension except in the "radial" direction, which would significantly change the length/circumference of the string itself. The range of motion of a closed string may thus be parametrized by (1) *a "radial" coordinate that is effectively and naturally "trapped" to be within the order of magnitude of ℓ_P, and* (2) *one or more "transversal" coordinates that are not so limited. In this perspective [☞ Section 11.2.3], spacetime is indeed* **spanned/generated** *by the modes of string motion/oscillation, and the range of these oscillations thus describes the geometry of this generated spacetime as effectively compactified in the "radial" direction but flat in the "transversal" directions.*

Thus, the electro-magnetic duality then implies that the observable dynamics of electric $(r-1)$-branes is dual to the dynamics of magnetic $(n-r-3)$-branes – although these are in most cases objects of differing dimensions. For example, for rank-2 antisymmetric tensor gauge potentials, $A_{\mu\nu} = -A_{\nu\mu}$, so that $F_{\mu\nu\rho} := (\partial_{[\mu}A_{\nu\rho]}) = \frac{1}{3}(\partial_\mu A_{\nu\rho} + \partial_\nu A_{\rho\mu} + \partial_\rho A_{\mu\nu})$, we have that $r = 2$, so

$$\dim(\rho_e) \geqslant 1, \qquad \dim(\rho_m) \geqslant (n-5). \tag{11.71}$$

Electric charges are then at least 1-dimensional (linear, filamentary distributions), and magnetic charges are at least $(n-5)$-dimensional – point-like in $(4+1)$-dimensional spacetime, but linear in $(5+1)$-dimensional spacetime, etc.

For all these generalized abelian (commutative) fluxes, one defines

$$\mathbf{d} := \mathrm{d}x^{\mu}\partial_{\mu}, \qquad \mathbf{A}_{(r)} := \mathrm{d}x^{\mu_1}\cdots\mathrm{d}x^{\mu_r}A_{\mu_1\cdots\mu_r}(\mathbf{x}), \tag{11.72a}$$

$$\mathbf{F}_{(r+1)} := \mathbf{dA}_{(r)} := \mathrm{d}x^{\mu_1}\cdots\mathrm{d}x^{\mu_{r+1}}\big[F_{\mu_1\cdots\mu_{r+1}}(\mathbf{x}) := \big(\partial_{[\mu_1}A_{\mu_2\cdots\mu_{r+1}]}(\mathbf{x})\big)\big]. \tag{11.72b}$$

Then, by direct generalization of the Dirac dual charge quantization condition (5.112), one obtains

$$q_e^{(r)}\int_S \mathbf{F}_{(r+1)} = q_e^{(r)}\oint_{\partial S}\mathbf{A}_{(r)} \stackrel{!}{=} 2\pi\, n_{(r)} \in 2\pi\mathbb{Z}, \qquad \forall S, \quad \dim(S) = (r+1). \tag{11.73a}$$

However, the duality between the magnetic and electric fields in $\mathbf{F}_{(r+1)}$ then implies also

$$q_m^{(n-r-2)}\int_{S'}(*\mathbf{F})_{(n-r-1)} = q_m^{(n-r-2)}\oint_{\partial S'}(\widetilde{\mathbf{A}})_{(n-r-2)} \stackrel{!}{=} 2\pi\, n_{(n-r-2)} \in 2\pi\mathbb{Z}, \tag{11.73b}$$

for every $(n-r-1)$-dimensional subspace $S' \subset \mathscr{X}$ of the spacetime $\mathscr{X}$ in which the magnetic charge is or moves; following Comment 5.6, we define $\widetilde{\mathbf{A}}$ to satisfy $*\mathbf{F} = \mathbf{d}\wedge\widetilde{\mathbf{A}}$.

These two quantization relations (11.73) of course produce the same generalization of the result (5.108). The important novelty follows from its application in cases when the total spacetime is a product, $\mathscr{X} = \mathscr{X}' \times \mathscr{Y}$, and the pair of quantization conditions (11.73) may be applied using (iteratively) the dualization (denoted by the symbol $*$) *independently* within either one of the three spaces, $\mathscr{X}$, $\mathscr{X}'$ and/or $\mathscr{Y}$. For example, if $\dim(\mathscr{X}) = 10$, $\dim(\mathscr{X}') = 4$ and $\dim(\mathscr{Y}) = 6$, we have

$$\mathbf{F}_{(r+1)}, \qquad (*_{\mathscr{X}}\mathbf{F})_{(10-r-1)}, \qquad (*_{\mathscr{X}'}\mathbf{F})_{(4-r-1)}, \qquad (*_{\mathscr{Y}}\mathbf{F})_{(6-r-1)} \tag{11.74}$$

at our disposal, as well as the *independent* charge-quantization for all these fields. This property was first noticed in 1996 [500], and four years later it was shown that this condition, within the string theoretical system, implies the quantization of many quantities that are continuous in pointillist theories. The resulting discrete available space of models is thus dubbed (in distinction from continuum) *discretuum* [74].

> ***Comment 11.3*** *Interestingly, string models are being applied, fully in the spirit of Section 11.1, also in areas of physics that seemingly have no relation with either (relativistic) elementary particle fundamental physics or cosmology, but where experiments are possible and even rather easily accessible [☞ for example, Refs. [529, 337], for starters].*

Swamp and landscape

In the first decade of the twenty-first century, a trend was noticed in several general properties in the (M- and F-theory extended) (super)string theoretical system, which distinguishes this theoretical system from the pointillist theoretical system.[33] Conceptual differences are mostly in favor of the stringy models, although they are technically (much more) demanding.

Wherever possible, one can compare the "volume" of the space of possible models within the string theoretical system with the corresponding result for the pointillist models. For example, in a model where a scalar field appears, one may inquire how big is the space in which the scalar field takes values. To this end, we need a preferred "volume" measure, which usually follows from

[33] The nomenclature is necessarily imprecise here: The theoretical system of strings necessarily also includes various *irreducible* p-branes (of various origins, of various properties and also for various values of $0 \leqslant p \leqslant 7$ or perhaps $\leqslant 8$), and even 0-branes that are really point-like objects; "theoretical system of strings and things" thus does seem to be a nitpickingly correct name. By contrast, in pointillist models, all spatially extended objects (and charge distributions as well as fields) may be reduced to collections of functions that depend on the coordinates of only one spacetime point.

the (preferred) choice of the metric tensor on the space of values of this scalar field, and the same choice then also dictates the dynamics in the given model. (Notice that such questions were rarely if ever raised before the advent of string theory.) The "physically preferred choice" is thus obtained by reverse reasoning:

$$\text{field dynamics} \to \begin{matrix}\text{metric on the}\\ \text{space of field values}\end{matrix} \to \begin{matrix}\text{volume element for the}\\ \text{space of field values.}\end{matrix} \tag{11.75}$$

Even in the case of a periodic dimension of radius R [☞ first part of Section 11.4.1], strings effectively "reduce" the space of possible radii, $\mathcal{M}(R)$, to the semi-infinite interval $R \in [\ell_s, \infty)$, which is equivalent to the interval $R \in (0, \ell_s]$, and the "volume" of which $\mathrm{Vol}(\mathcal{M}(R)) = |\int_0^{\ell_s} \frac{dr}{r}| = |\int_{\ell_s}^{\infty} \frac{dr}{r}|$ diverges but only logarithmically, and only at the limit $R \to 0$, which is dual to the infinitely large radius where periodicity fails to make sense. By contrast, pointillist models have no reason for excluding any part of the whole interval $R \in [0, \infty)$, except that in the limiting case $R \to \infty$ periodicity fails to make sense; however, the limiting case $R \to 0$ now effectively corresponds to a new model with spacetime of one fewer dimensions.

The interval $[0, \infty)$ is thus the parametric space of every pointillist model with one periodic coordinate – together with the limiting case $R = 0$, which in fact represents a radically different model. By contrast, the interval $[\ell_s, \infty)$ is the parametric space for stringy models with one periodic coordinate – and does not include the lower-dimensional "stowaway."

Even better motivated is the case of a so-called *modulus*[34] in Calabi–Yau compactifications, including the so-called dilaton–axion (complex) scalar field, where various dualities help to reduce the otherwise infinite space of choices of values for this field to a space of finite volume. Besides, every concrete compactification has a finite number of parameters, and so – for every compactification model – the total number of choices is described by a space of which the volume in the Weil–Petersson–Zamolodchikov metric [89, and references therein] is finite [522, 348], just as was the case for the much simpler torus compactifications, where $\mathscr{Y}$ in the decomposition (11.41) is a real 6-dimensional torus.

> **Conclusion 11.7** *From the ever more rigorously verified property that the (M- and F-theory extended) (super)string theoretical system consists of a* ***discretuum*** *(and not a continuum) of models,*[35] *it follows that the string theoretical system is a far better defined theoretical system than any pointillist theoretical system.*

This induced the picturesque vision [531] where the vast majority of models that can be constructed within classical, (in various ways) incompletely quantum and/or quantum but non general-relativistic physics form a *swamp*, from which emerge the models that are completely quantum and general-relativistically consistent, and which form the *landscape*. These latter models, or at least their non-empty subset, one believes are string models.

Of course, for *our* Nature, one believes it to be described by some model within the *landscape*, and the problem is "only" that there are so many models that one does not know where to start looking.

AdS/CFT, i.e., gravity/gauge duality

The idea of duality has, for the first time, clearly been manifested in the example of the particle–wave duality in the creation of standard quantum physics – which belongs to the pointillist

[34] *Moduli* are scalar fields, the expectation values of which parametrize the geometrical characteristics of the compactification Calabi–Yau spaces, such as the choice of the complex structure and of the complexified Kähler class. More generally, the analogous reasoning may apply to all *parameters* in stringy models [☞ the structure (11.35)].

[35] Conceptually, all limitations on the theoretical system that indicate the discreteness of the string theoretical system follow from a combination of quantumness and general relativity of Nature.

theoretical system. The electro-magnetic duality – although in fact part of the classical theory – has acquired a wide application only through its generalization within stringy models and their *M*- and *F*-theory extensions, where it indicates that neither the dimension of mathematical objects used to represent physical objects nor the dimension of spacetime in which these objects move and interact are inviolable and sacrosanct constants.

Within the theoretical system of strings, however, a duality was discovered in 1997 that relates string models with point-particle quantum field theory, the so-called AdS/CFT duality [354, 440]. By now, this duality has been generalized to many other examples, into a general gravity/gauge duality [☞ lecture notes [437], where this phenomenon was related to a loophole in the Weinberg–Witten theorem 6.1 on p. 249]. The general characteristic of the gravity/gauge dual examples is the identity between the physical observables and relations between them, for two models that were obtained as different limiting cases of the same superstring model, where

1. one limiting case represents a *superstring* model with the spacetime geometry that contains an $(n+1)$-dimensional anti de Sitter factor [☞ definition (9.81)],
2. the other limiting case is a *point-particle* supersymmetric gauge theory, the degrees of freedom of which are "trapped" at the $(n-1)+1$-dimensional conformal boundary (with the Minkowski metric) of the anti de Sitter space [☞ expression (9.83) and the related discussion].

Contemporary (at the beginning of the twenty-first century) understanding of the string theoretical system contains both the consistent and the absolutely comprehensive application of both (1) the gauge principle and (2) the principle of source/sink completeness [☞ [31] and references therein]:

> ***Conclusion 11.8 (conjecture)*** *In consistent quantum models with gravity, (1) all symmetries are gauged, and (2) all sources/sinks (electric and magnetic) for all gauge fields are included and satisfy the appropriate generalization of the Dirac dual charge quantization condition [☞ Section 5.2.3 and relation (5.108)].*

This conclusion is so far the strictest known formulation of fundamental limitations on models in the string theoretical system, and so far has the status of a very strong *conjecture*: in spite of very strong indications, there is no rigorous proof (as yet)☝ [☞ also the discussion in the Polchinski–Smolin debate [429, 435]].

The early twenty-first century understanding of the so-called string theory – and especially with its *M*- and *F*-theory extensions – is by far not "just a theory." Just as classical and statistical mechanics are (axiomatic) theoretical systems that provide a conceptual and technical framework for addressing large classes of respective phenomena, so is "string theory" also a theoretical *system* and not a (single) theory. In fact, "string theory" includes *three* major theories: quantum theory, gauge theory and the theory of relativity, and is moreover the one known framework that unifies them in a coherent, cohesive and logically consistent fashion.

While all these considerations in no way oblige Nature to be describable within the string theoretical system, they do make it our best candidate, ever.

Discrete spacetime

The whole basis of the Democritean atomistic worldview relies on the idea that – everyday sensory experiences to the contrary – matter is not continuous, but consists of an immense number of very teeny *elementary* particles. The fundamental physics of the twentieth century, and foremost quantum physics, convinces us that all existent matter, including also the interaction fields, may be described in this same atomistic fashion:

Conclusion 11.9 *All "ingredients" of the Standard Model [☞ Section 7.3.3] and including the gravitational field [☞ Chapter 9] are represented by quantum fields, i.e., elementary* ***particles*** *[☞ the lexicon entry, in Appendix B.1, entries for* ***field (physical)*** *and* ***quantum****, as well as Footnote 17 on p. 196].*

Thus, there is no logical obstruction for spacetime to also be discrete. For example, from the worldsheet perspective of the string theoretical system, target spacetime is simply the space of values spanned (dynamically generated) by the coordinate fields such as $X^\mu(\xi)$ in the Polyakov action (11.10).

The idea that spacetime is (at least in some directions) not in fact a continuous, commutative *topological space* is not new [☞ e.g., [130, 92, 361] and references therein]. One of many and various possibilities is very close to the computational method that is used in so-called "lattice QCD" [☞ description on p. 230]. The results of recent experiments in the LHC installation at CERN seem to lend support to a variant of this idea [374, 14, 13]. Here, in the simplest *model*, spacetime literally has the lattice structure of a crystal, in the sense that it consists of discrete points through which all material objects pass as if those points are ordered at constant distances and in uniform directions, just like atoms in crystalline lattices. Note that this is a very radical idea where the *auxiliary* space (the one throughout which the points of the *true* spacetime are distributed in regularly periodic and uniform fashion as a crystalline lattice) is a purely fictitious structure. If we further assume that the points in this "crystalline" spacetime are ordered akin to Cartesian directions and that the distances between the points in different directions are significantly different,

$$L_z \gg L_y \gg L_x \gg L_{ct}, \tag{11.76}$$

then, denoting $\lambda_{\rm dB}$ the de Broglie wavelength, we have that:

1. to probes with $\lambda_{\rm dB} > L_z$; spacetime appears continuous and 3+1 dimensional;
2. to probes with $L_z > \lambda_{\rm dB} > L_y$ the whole spacetime appears to be a vertical stack (uniform sequence) of horizontal and continuous $(2+1)$-dimensional (surface) spacetimes;
3. to probes with $L_y > \lambda_{\rm dB} > L_x$ the whole spacetime appears to be a two-directional stack of horizontal and continuous $(1+1)$-dimensional (linear) spacetimes;
4. to probes with $L_x > \lambda_{\rm dB} > L_{ct}$ the whole spacetime appears to be a three-directional stack of $(0+1)$-dimensional space-like arrangement of points with (still) a continuous passage of time;
5. to probes with $L_{ct} > \lambda_{\rm dB}$ the whole spacetime appears to be a four-directional stack of spacetime *points*, i.e., disconnected events.

Amusingly, such a discrete structure of spacetime is implied by the assumption that the space of conjugate momenta is compact. This concrete Cartesian "crystalline" structure is simply obtained, e.g., in the *momentum* representation of quantum mechanics where one imposes periodic conditions to the linear momenta,[36] which gives to the momentum space the geometry of a 3-dimensional torus, the radii proportional to the reciprocals of the distances L_x^{-1}, L_y^{-1} and L_z^{-1}. Evidently, more complicated (and for now ad hoc imposed) compact geometry of the 4-momentum space then implies a more complicated structure of the discrete spacetime.

11.4.3 Lessons of fundamental physics as a model of nature

The duality between different – and even different-dimensional – models reminds us that the picturesque formulation imagery at the foundation of a given model is merely a mental caricature and image – a crutch – just as the hydrogen atom is merely *represented/imagined* as a point-like

[36] Within non-relativistic quantum mechanics at least, it seems essentially contradictory to require periodicity of energy.

electron orbiting a point-like proton. The *real* atom is not two point-like charges orbiting each other, nor is the atom a charged standing wave undulating along a circle around an oppositely charged point-particle, nor is the atom a negatively charged cloud centered on a positively charged proton... Of course, such picturesque formulation imagery is very useful in describing the atom, in that it dictates the construction of a corresponding mathematical model, which is then used to "produce" predictions of the model – intending to compare those predictions with Nature.

The connection between the picturesque formulation imagery and the ultimate authority, Nature, is very indirect, and so then is the justification of the formulation imagery, however impressively picturesque it may be. It should thus come as no surprise that even very different formulation images may turn out to produce models that agree equally with Nature [☞ Definition 11.3 on p. 437, and the discussion after this definition; see also the recent work [222]].

The caution that "the map is not the territory" (Alfred Korzybsky) is perfectly in agreement with this lesson, and omits the cultural–historical and perhaps even religious connotations of the ancient Tao principle "The way you can go is not the real way. The name you can say is not the real name" [527].

11.4.4 Exercises for Section 11.4

✎ ***11.4.1*** *Verify the results (11.69).*

✎ ***11.4.2*** *Using the definitions (11.72), generalize the flux definitions (11.63a) and (11.64a) for a rank-r antisymmetric gauge potential. Show the conditions (11.73a) and (11.73b) to reproduce the same quantization condition for the product* $q_e^{(r)} q_m^{(n-r-2)}$.

11.5 Instead of an epilogue: unified theory of everything

The fluffiness of clouds and the babbling of a brook are examples of emergent phenomena, which is not within the domain of fundamental physics, but of the physics of collectives – and that is a relatively *new* and emerging discipline in physics. The subject matter here is precisely the regularity and circumstances wherein relatively simple basic rules and their theoretical systems may produce (by means of nonlinear and/or self-interactive coupling) very complex phenomena.

For example, the basic laws of chemical bonds are relatively simple and stem from elementary quantum mechanics [☞ Schrödinger's quotation on p. 13 and its discussion], but nevertheless produce a fantastically diverse palette of an uncounted number of chemical compounds, as well as different materials. These compounds and materials then, through their dynamics and interactions, produce complex structures and behaviors that can in no particular sense or case be simply and *completely* reduced to elementary quantum mechanics, although in a completely literal sense they stem – draw roots – from it.

This reminds us of the fact that the stability, functionality and beauty of a palace are not properties of its bricks, shingles and other materials of which the palace is built [☞ also the discussion in Section 1.1.4]. Similarly, neither is the evolutionary role, nor the mimicry function or the beauty of the complicated patterns on butterfly wings simply the "diffraction and coherent scattering of light," although this is the basic mechanism for the appearance of most colors in the often stunningly exquisite wings.

In a sense, akin to the term "epiphenomenon," this new discipline could be called *epiphysics*. On one hand, this new discipline would be concerned with phenomena that are *beyond* currently familiar physics, and on the other, the subject matter of this new discipline would still be the *Nature* (φύσις ≈ *physis*, Greek) of this next level of natural phenomena. However, it is important to keep in mind that the demarcation between fundamental physics and this *epiphysics* must be hazy; in the end, Nature is one [☞ Conclusion 11.2 on p. 409].

The helix of learning has thus come full circle, and hopefully one floor higher: Bohr's thought, quoted in the Preface, on p. xi, resonates through the entire development of the fundamental physics of elementary particles, appears explicitly also in the discussion around Digression 1.1 on p. 9, then again in Section 8.3, and completely permeates Chapter 11 and especially Section 11.4. At any rate, during the twentieth century, the fundamental physics of elementary particles has been developing from a discipline in which one believed to have almost everything solved to a discipline that is bound to separate into at least two or three separate disciplines within physics [☞ Section 11.2], and possibly also into a discipline the subject matter of which is the *structure* of (theoretical) physics itself. Evidently, within this development, a theoretical system has been discovered within which there is hope of finding a description of the *fundamentals* of Nature, but this has, *en route*, instructed us very pointedly about the very nature of our understanding of Nature.

Part IV

Appendices

A Groups: structure and notation

In high energy theory one has plenty of opportunity to use results from group theory, for which Ref. [488] is one of the most often used sources. We will be interested in linear *representations* of groups, i.e., the applications of abstract groups in the form of linear transformations of a vector space, V. By specifying this vector space together with a basis, the group representation is specified in the form of matrices that map vectors from V linearly into vectors that are also in V. A telegraphically brief and cursory review of some of the useful results in group theory provided here cannot possibly compete with the serious sources such as Refs. [565, 258, 287, 581, 201, 80, 333, 260, 334, 256, 447].

A.1 Groups: definitions and applications

This cluster of appendices describes the general algebraic structure of groups and in particular of Lie groups, and then discusses the general properties of the application of groups in physics. This is important for understanding the content of scientific models and their relation with Nature, for the description of which these models were invented.

A.1.1 Axioms and a rough classification

We will need several group-theoretical and algebraic structures and their concrete applications, and they are briefly described here.

Groups

A group G consists of a set of elements $\{a,b,c,\dots\}$ equipped with a binary operation $*$ that satisfies the following axioms (given here with a textual "translation" of the formal symbolism):

1. $\forall a,b \in G,\ a*b \in G;$ (A.1a)
   For each ($\forall$) two elements a,b from the group G, the result of the binary operation $a*b$ is also in ($\in$) the group G, making the operation $*$ ***closed****;*
2. $\forall a,b,c \in G,\ a*(b*c) = (a*b)*c;$ (A.1b)
   The binary operation $*$ is ***associative****, i.e., the result of a repeated application of the binary operation $*$ is independent from the order in which the two operations are computed;*
3. $\exists e \in G,\ \forall a \in G:\quad a*e = e*a = a;$ (A.1c)
   There exists ($\exists$) a ***neutral element*** *(e) of the group G, such that the results of the binary operations $a*e$ and $e*a$ equal the original element a, for each ($\forall$) a of the group G.*

4. $\forall a \in G,\ \exists a^{-1} \in G: \quad a * a^{-1} = a^{-1} * a = e.$ (A.1d)

*For each ($\forall$) element a of the group G, there exists ($\forall$) an **inverse element** a^{-1} in the group, such that the results of the binary operations $a * a^{-1}$ and $a^{-1} * a$ equal the neutral element, e.*

Pedantically, it is not necessary to require that the neutral and the inverse elements are *both-sided*: it suffices to require that there exist, say, the left-neutral element ($\mathbb{1}_L * a = a$) and the left-inverse element ($a_L^{-1} * a = \mathbb{1}$); the existence of the right-neutral element ($a * \mathbb{1}_D = a$) and the right-inverse element ($a * a_D^{-1}$), as well as the equalities ($\mathbb{1}_L = \mathbb{1}_D$ and $a_L^{-1} = a_D^{-1}$) then follow [331, 332].

A group is called abelian (commutative) if the binary operation *commutes*: $(a * b) = (b * a)$, for each two $a, b \in G$; otherwise, the group is called non-abelian (non-commutative). A group G is called *additive* if $*$ is an addition, and *multiplicative* if $*$ is a multiplication.

According to the number of their elements, groups are classified as:

1. Finite, with a finite number of elements. For example, $\mathbb{Z}_2 = \{1, -1; \cdot\}$ is the multiplicative group that consists of two elements, 1 and -1.
2. Countably infinite, with countably infinitely many elements. For example, $\{\mathbb{Z}; +\}$ is the additive group of all (countably many) integers.
3. Continuous, with a continuum of elements, which are further subdivided as:
   (a) Finite-dimensional. For example, $U(1)$ is the multiplicative group of (complex) unitary numbers,[1] i.e., numbers of the form $e^{i\varphi}$, where $\varphi \simeq \varphi + 2\pi$. The number of group elements is continuously infinite, since there is one element for each of the continuously many angles $\varphi \in [0, 2\pi]$. These angles evidently form a subset of the 1-dimensional real axis, $\mathbb{R}^1$, and $U(1)$ is a 1-dimensional group.
   (b) Infinite-dimensional.[2] For example, *Diff*(S^1) is the multiplicative group of all diffeomorphisms (continuous reparametrizations) of the circle, which is a concrete example of the group of general coordinate transformations [☞ Definition 9.1 on p. 319], useful within the theoretical system of strings.

Coset

Besides groups, we also need the concept of a ***coset***: For any group G and its subgroup H, the (right) coset G/H consists of the elements

$$\textbf{coset}: \quad G/H := \{g \simeq g * h : \quad g \in G,\ h \in H\}, \tag{A.2}$$

where $*$ is the binary operation in the group G and in the subgroup $H \subset G$. In other words, the coset elements are defined as equivalence classes "up to right 'multiplication' by elements from H." The left coset is defined similarly, and if the group G is abelian, the left and the right coset are identical, of course.

This formal definition describes some very familiar examples:

Days of the week Consider the additive group of integers $\mathbb{Z}_+$ (which is abelian, i.e., commutative), and its subgroup $7\mathbb{Z}_+$, the additive group of integers that are divisible by 7. The coset $\mathbb{Z}_7 := \mathbb{Z}_+/7\mathbb{Z}_+$ is then defined as the additive group of equivalence classes of integers $\mathbb{Z}_+$, where numbers $n \in \mathbb{Z}$ and $n + k$ (for each $k \in 7\mathbb{Z}$) are regarded as equivalent ($\simeq$). The coset $\mathbb{Z}_7$ therefore consists of elements

$$[0 \simeq 7 \simeq 14 \simeq \ldots],\ [1 \simeq 8 \simeq 15 \simeq \ldots],\ [2 \simeq 9 \simeq 16 \simeq \ldots],\ \ldots \tag{A.3}$$

which may be ***represented***:

$$\Big\{[0], [1], [2], [3], [4], [5], [6]\Big\} = \mathbb{Z}_7, \tag{A.4}$$

[1] It follows that their modulus, i.e., absolute value is 1: $z^{-1} = z^* \Rightarrow 1 = z^* z = |z|^2 \Rightarrow |z| = 1$, as $|z| \geqslant 0$.

[2] These are further subdivided into several classes, but this will not concern us here.

and where the classes $[n]$ may be identified with the days of the week, $[0]$=Sunday, $[1]$=Monday, etc. Indeed, seven days from Monday is again Monday, twenty-one days before Saturday was again Saturday, $7n$ days from Tuesday is again Tuesday, etc.

Circle Consider the additive group of real numbers $\mathbb{R}_+$ and its subgroup of additive numbers $2\pi\mathbb{Z}_+$, the elements of which are integral multiples of 2π. The coset $\mathbb{R}_+/2\pi\mathbb{Z}_+$ then may be identified with the circle S^1, as the coset $\mathbb{R}_+/2\pi\mathbb{Z}_+$ is parametrized by the equivalence classes of real number $[\phi \simeq \phi + 2n\pi]$, for each $n \in \mathbb{Z}$, known as *angles*. Thus, $\mathbb{R}_+/2\pi\mathbb{Z}_+ \cong S^1$.

It is useful to know that all n-dimensional spheres may be identified with the coset

$$S^n := \left\{ \mathbf{x} \in \mathbb{R}^{n+1} : \sum_{i=0}^{n} x_i{}^2 = r^2 \right\} \cong SO(n+1)/\,SO(n), \tag{A.5}$$

where $SO(n)$ is the group of real and orthogonal $n\times n$ matrices of determinant $+1$. For the details of the isomorphism ($\cong$), the Reader is directed to the literature on Lie groups [565, 258, 581, 256, 80, 260, 333, 447].

Quotient space

The following generalization of the coset turns out to be very useful. Let V be a vector space over the field $\Bbbk$, and $\mu : V \to V$ some mapping of that vector space into itself. One then says that

$$V/\mu := \{[\vec{v} \simeq \mu(\vec{v})] : v \in V\} \tag{A.6}$$

is a quotient space of the vector space V by the action of the mapping μ. The coset is then the special case of the quotient space, where V is regarded as an additive group,[3] and μ is a mapping that preserves this structure, e.g.:

1. Adding integral linear combinations of a specified collection of vectors $\vec{w}_i \in V$, $i = 1, 2, 3, \dots$; indeed, the subset $\{n\vec{v}_0 : n \in \mathbb{Z}\}$ evidently forms a subgroup of the additive group V.
   Example: The 2-dimensional torus $T^2 = \mathbb{R}^2/\Lambda$, where $\Lambda = \{nL_1\hat{\mathrm{e}}_1 + mL_2\hat{\mathrm{e}}_2\}$ is a Cartesian lattice with spacings L_1 and L_2, which are then the circumferences of one and the other circle in the torus.
2. (An)isotropic homothety: rescaling of the (basis) vectors

$$\mu : (\hat{\mathrm{e}}^1, \hat{\mathrm{e}}^2, \dots) \to (\lambda^{a_1}\hat{\mathrm{e}}^1, \lambda^{a_2}\hat{\mathrm{e}}^2, \dots) \in V \tag{A.7}$$

   where $0 \neq \lambda \in \Bbbk$, and since $\vec{a} = a_i\hat{\mathrm{e}}^i$ is an invariantly defined vector, the definition (A.7) is in fact independent of the choice of a basis $\{\hat{\mathrm{e}}^1, \hat{\mathrm{e}}^2, \dots\} \in V$.
   Example: The n-dimensional sphere S^n may be identified also with the quotient space $\mathbb{R}^{n+1}/\mathbb{R}^*_{>0}$, where $\mathbb{R}^*_{>0}$ is the multiplicative group of positive real numbers and the particular action on $\mathbb{R}^{n+1}$ is isotropic:

$$\mu : (x^0, x^1, \dots) \mapsto (\lambda x^0, \lambda x^1, \dots), \quad \lambda > 0. \tag{A.8}$$

   Every element $\mathbb{R}^{n+1}/\mathbb{R}^*_{>0}$ then looks like a ray in the $(n+1)$-dimensional space, starting at the coordinate origin (not including the origin itself) to infinity (not including infinity). Taking one point to represent each ray, e.g., at a *same*, fixed distance from the coordinate origin, then gives the familiar image of the n-dimensional sphere.

[3] The sum of any two vectors is again a vector; adding vectors is associative; $\vec{0}$ is the neutral element with respect to addition; $-\vec{v}$ is the "inverse" vector with respect to addition.

Besides, the physical degrees of freedom in all gauge fields and potentials (including also gravitation) always have the structure of a quotient space [☞ Examples 11.1–11.4, p. 416–417]: the number of physical polarizations of a gauge particle is always smaller than the number of components of the mathematical object (gauge 4-vector, metric tensor, etc.) that must be used to represent the particle.

A.1.2 Lie groups

Of the finite-dimensional continuous groups, of special interest are the so-called Lie groups, G, the elements of which may be written as $g(\mathrm{a}) := \exp\{i\,a^j\,T_j\}$, where summing over j is understood, $\mathrm{a} := (a^1, \ldots, a^n)$ is an n-tuple of *parameters*, n the dimension of the group, and T_j are the group generators. Conversely, the group generators, T_j, are obtained by linearizing:

$$T_j := -i\frac{\partial g(\mathrm{a})}{\partial a^j}\bigg|_{a^k=0}. \tag{A.9}$$

This means that the space of elements of every Lie group has a well-defined tangent plane in every point, whereupon this group space is a smooth manifold, which locally looks like a Euclidean n-dimensional space. The non-abelian structure of a group G reflects in the difference

$$\mathbb{1} - g(\mathrm{a})\,g(\mathrm{b})\,g(\mathrm{a})^{-1}\,g(\mathrm{b})^{-1} = a^i b^j [T_i, T_j] + \cdots \tag{A.10}$$

where "…" denotes contributions of higher order in parameters a, b.[4] Since a product of group elements must again be a group element, the product $g(\mathrm{a})\,g(\mathrm{b})\,g(\mathrm{a})^{-1}\,g(\mathrm{b})^{-1}$ must be expressible as $g(\mathrm{c}) = \mathbb{1} + ic^j\,T_j + \cdots$ for some c, from which it follows that the generators T_i must satisfy the relations

$$[\,T_j\,,\,T_k\,] = i\,f_{jk}{}^m\,T_m, \tag{A.11a}$$

where the coefficients $f_{ij}{}^k = -f_{ji}{}^k$ are the group *structure constants*, and the binary operation $[\ ,\ ]$ is called the *commutator*, or the *Lie bracket*.

Definition A.1 *Formally, the n-dimensional vector space $\mathfrak{A}$, the elements of which are of the form $a^j\,T_j$, and for which the multiplicative operation*

$$(a^j\,T_j) * (b^k\,T_k) := a^j b^k [T_j, T_k] = (i\,a^j b^k\,f_{jk}{}^m)\,T_m \in \mathfrak{A} \tag{A.11b}$$

is defined is called the **algebra** *of the group G.*

Comment A.1 *Since both the Lie groups and the Lie algebras have* **continuously** *many elements, omitting (the action of) finitely many elements does not change the formal relation between a group and its algebra, but it is important to account for such elements.*

Example A.1 For example, the Pauli matrices

$$\sigma^1 = \begin{bmatrix} 0 & 1 \\ 1 & 0 \end{bmatrix}, \quad \sigma^2 = \begin{bmatrix} 0 & -i \\ i & 0 \end{bmatrix}, \quad \sigma^3 = \begin{bmatrix} 1 & 0 \\ 0 & -1 \end{bmatrix} \tag{A.12}$$

may be used as generators of the group $SU(2)$, the elements of which are of the form $\exp\{i\,a_j\sigma^j\}$, and also as a basis for the $\mathfrak{su}(2)$ algebra, the elements of which are of the

[4] Recall that, in this book, n-vectors as a whole are denoted by upright letters, so a and b are n-vectors with components a^i and b^i, $i = 1, 2, \ldots, n$.

form $a_j\sigma^j$.[5] On the other hand, the $SU(2)$ group elements are defined (in its fundamental representation) as 2×2 unitary matrices with unit determinant. That certainly includes both the 2×2 identity matrix $\mathbb{1} = \exp\{i\,\mathbb{O}\}$ that corresponds to the coordinate origin in the a-space, $\mathrm{a} = (0,0,0)$. However, the $SU(2)$ group also includes the element $-\mathbb{1} = \exp\{i\pi\mathbb{1}\}$, which is omitted in the relation between the $SU(2)$ group and the $\mathfrak{su}(2)$ algebra, since $\pi\mathbb{1} \neq a_j\sigma^j$, and $\pi\mathbb{1} \not\subset \mathfrak{su}(2)$. Thus, although $\exp\{i\,a_j\sigma^j\}$ differs from $SU(2)$ by continuously infinitely many elements of the form $-\mathbb{1}\exp\{i\,a_j\sigma^j\}$, all the omitted elements may be recovered by multiplying (from left or from right) $\exp\{i\,a_j\sigma^j\}$ by $-\mathbb{1}$, the action of which then is the one (and so finite) difference between $SU(2)$ and $\exp\{i\,a_j\sigma^j\}$.

Together, $\mathbb{1} \subset \exp\{i\,a_j\sigma^j\}$ and this omitted element, $-\mathbb{1}$, form a multiplicative finite subgroup of $SU(2)$, denoted $\mathbb{Z}_2 = \{\mathbb{1}, -\mathbb{1}\} \subset SU(2)$. The representations of the group $SU(2)$ that are eigenspaces of the $\exp\{i\pi\mathbb{1}\}$ element of this subgroup $\mathbb{Z}_2 \subset SU(2)$ and have the eigenvalue $+1$ are called tensorial, while the ones with the eigenspace -1 are spinors. Notice that the choice $\mathrm{a} = (0,0,\phi)$ represents the rotation about the third axis; by writing the standard generator as $\frac{1}{2}\sigma^3$, we find this to represent a rotation by the angle $\frac{1}{2}\phi$ – as befits, e.g., a 2-component spin-$\frac{1}{2}$ wave-function, and which is why it changes sign upon a 2π-rotation.

Whereas every algebra $\mathfrak{A}$ gives rise to a group G by means of "exponentiating," i.e., by defining that $g := \exp\{a\} \in G$ for every $a \in \mathfrak{A}$, not infrequently the algebra $\mathfrak{A}$ also *contains* a multiplicative group $\mathfrak{A}^\times$ with the algebra "multiplication" as the binary operation in $\mathfrak{A}^\times$. We thus have the *formal* relation of these three structures $\mathfrak{A}^\times \subset \mathfrak{A} \xrightarrow{\exp} G$.

Example A.2 Note that the $\{1, i, -1, -i\}$-multiples of the 2×2 identity matrix and the Pauli matrices also form a multiplicative group of 16 elements:

$$\{\mathbb{1},\ \sigma^1,\ \sigma^2,\ \sigma^3,\ i\mathbb{1},\ i\sigma^1,\ i\sigma^2,\ i\sigma^3,$$
$$-\mathbb{1},\ -\sigma^1,\ -\sigma^2, -\sigma^3,\ -i\mathbb{1},\ -i\sigma^1,\ -i\sigma^2,\ -i\sigma^3\}. \tag{A.13}$$

Indeed, the Pauli matrices satisfy *two* relations:

$$[\,\sigma^j, \sigma^k\,] = 2i\,\varepsilon^{jk}{}_{\ell}\,\sigma^\ell, \qquad \text{as well as} \qquad \{\,\sigma^j, \sigma^k\,\} = 2\,\delta^{jk}\,\mathbb{1}, \tag{A.14}$$

where $\{A, B\} := A\,B + B\,A$ is the *anticommutator*. Thus, the formula

$$\sigma^j\,\sigma^k = \delta^{jk}\,\mathbb{1} + i\varepsilon^{jk}{}_{\ell}\,\sigma^\ell \tag{A.15}$$

has, for each $j,k = 1,2,3$, precisely one element on the right-hand side. Thus, multiplying Pauli matrices, one produces the identity matrix and i-multiples of the Pauli matrices, and these must be added to the list of elements of the group. Multiplying in that extended collection, one obtains the (-1)- and $(-i)$-multiples of the Pauli matrices as well as

[5] The standard choice of using the *halves* of the Pauli matrices makes the structure constants equal to i-fold multiples of the Levi-Civita symbol, the same algebra as the rotation generators, $L^x := i(y\frac{\partial}{\partial z} - z\frac{\partial}{\partial y})$, etc. cyclically, so that $[L^j, L^k] = i\varepsilon^{jk}{}_{\ell}L^\ell$. Using the Pauli matrices instead, we have that $[\sigma^j, \sigma^k] = 2i\,\varepsilon^{jk}{}_{\ell}\,\sigma^\ell$.

$(-\mathbb{1})$ and $(i\mathbb{1})$, which also must be added to the list of elements. Multiplication within that again-extended collection also yields the $(-i\mathbb{1})$, and this completes the procedure of closing the set: The 16 elements (A.13) form a group with respect to the familiar matrix-multiplication.

For all semisimple Lie algebras,[6] the Killing form

$$g_{jl} := -f_{jk}{}^{m}\, f_{lm}{}^{k} \tag{A.16}$$

is positive-definite, and serves as a metric tensor, and defines

$$f_{jkl} := f_{jk}{}^{m}\, g_{ml}, \tag{A.17}$$

which may be shown to be a totally antisymmetric tensor.

Digression A.1 It is worth noting that the relation (A.11a) determines only the antisymmetric product of the generators. The symmetric product, the so-called *anticommutator*, remains free to be specified separately:

$$\{\, T_j\,,\, T_k \,\} = N\,\delta_{jk}\mathbb{1} + \tfrac{1}{2}\, d_{jk}{}^{m}\, T_m, \qquad \text{if the set } \{\mathbb{1}, T_1, \dots, T_n\} \text{ is complete.} \tag{A.18a}$$

In any given representation, the vector space (representation) V of dimension $r := \dim(V)$ is given, upon which the operators T_j act as $r \times r$ matrices, and the normalization constant N depends on r. Also, these $r \times r$ matrices that play the role of the generators T_j typically satisfy certain additional conditions: they may be symmetric, Hermitian, traceless, etc. If the collection of matrix representatives $\{\mathbb{1}, T_1, \dots, T_n\}$ is complete for the specified type of matrices, the relation (A.18a) follows automatically. Otherwise, one expects that the anticommutators $\{T_j, T_k\}$ include matrices that cannot be represented as the linear combination $\mathbb{1}$ and T_j. Thus, both the existence of the relation (A.18a) and then also the constants $d_{jk}{}^{m}$ strongly depend on the representation of the generators T_j.

If the additional relation (A.18a) exists, its combination with the relation (A.11a) reduces

$$T_j\, T_k = N\,\delta_{jk}\mathbb{1} + (i f_{jk}{}^{m} + \tfrac{1}{2}\, d_{jk}{}^{m})\, T_m \tag{A.18b}$$

to a linear combination of the identity $\mathbb{1}$ and algebra generators T_j – which provides more information than the abstract defining requirement of the Lie algebra (A.11a). Thus, abstract Lie algebras include less structure than what their applications in physics not infrequently have [☞ Section A.1.4].

A.1.3 Groups in (fundamental) physics

Every model of every physical system uses some collection of variables[7] that quantify the system, and imposes relations between those variables in the form of systems of equations and conditions for those variables, appropriate for the physical system being described. That system of equations, together with all conditions on the domain of variables and the operators used to write

[6] A Lie algebra $\mathfrak{A}$ is semisimple if it has no abelian (commutative) direct summand, i.e., no abelian subalgebra that commutes with the whole algebra $\mathfrak{A}$; for the precise statement, see Refs. [581, 256].

[7] In this general description, "variables" includes every mathematical symbol that may have a value, thus, variables include both arguments of some functions, as well as those functions, and various additional parameters.

out the specified equations forms the *mathematical model*, ***M***, of the physical system. In lieu of experimental results against the model, one regards the model as adequately representing the considered physical system, and one often identifies experimental results in routine conversation. However, it is very important not to confuse in principle the components in this description of the physical system [☞ Figure A.1].

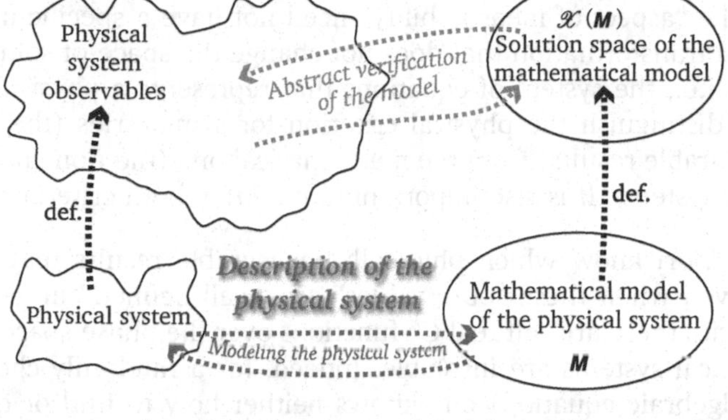

Figure A.1 Relations between the physical system and its observables, as well as the mathematical model and its space of solutions. The smoothness of the mathematical side of this image indicates the fundamental idealizations.

Symmetries of physics systems and symmetry breaking

The situation is actually more complicated than shown in Figure A.1. Namely, the observables in realistic physical systems are usually not specified "once and for all," and their improved definition is an iterative process. In turn, in realistic cases, only some of the theoretically definable observables can be measured in practice, and this subset must be marked. Besides, real physical systems often contain details that are either included in the mathematical model or neglected from it in an iterative or layered fashion. The situation in realistic cases then looks more like the diagram in Figure A.2.

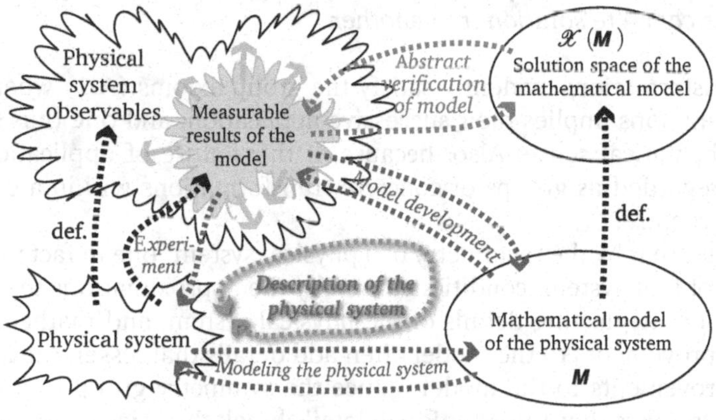

Figure A.2 Relations between the physical system and its observables, as well as the mathematical model, its space of solutions, and their comparisons via experiments.

The collection and domain of variables and operators needed for the description of the physical system usually permits certain changes of those variables and operators, without such re-definitions affecting any concrete, measurable result for the physical system, and obtained by means of this model. Alternatively, the model may be viewed as a mathematical system of equations and conditions, which defines the space of solutions of the system, i.e., the space of solutions of the model, $\mathscr{X}(\boldsymbol{M})$ – regardless whether those solutions can be computed.

The procedure of changing those variables in a way that changes no measurable aspect of the model is called a *symmetry transformation* of the model of the physical system, and the property of the system that permits such a change is called a *symmetry* of the model, i.e., of the system represented by the model. Similarly, instead of the physical system one may consider any system of equations where the "aspect of measurability" need not have a specific meaning. A symmetry is then by definition a transformation that does not change the space of solutions of the specified mathematical model, i.e., the system of equations that represents a physical system. It is important to conceptually distinguish the physical criterion for symmetries (the non-changing of the collection of all measurable results) from the mathematical one (the non-changing of the space of solutions to the given system). It is also important to note that both criteria are *hard*:

1. One cannot a priori know which physically measurable results may possibly exist in a given model, even when these "observables" are "well defined" in general; for example, in classical physics these are "all real C^k-functions over the phase space."[8]
2. Most mathematical systems are insoluble. Indeed, for a randomly chosen system of (differential and algebraic equation) one knows neither how to find or determine the exact solution ("in closed form"), nor of an algorithm of an iterative method for obtaining such a solution, and sometimes even all the known approximations do not suffice for a concrete application. Moreover, it may well be the case that many mathematical systems are not soluble even in principle.

In spite of that – in practice, and so in models that have so far been considered – it is not infrequently possible to definitively determine if a particular transformation is a symmetry of the system or not. In addition, the models used in practice of course form an "infinitesimally" teeny subset of all possible models, and they are chosen precisely so that – besides adequately representing the *interesting physical systems* – they are "sufficiently soluble" so as to be of practical use.

> **Comment A.2** *In addition, note that the concrete solutions often do not possess all the symmetries of the system that they solve. In that case, however, the symmetry of the system transforms one concrete solution into another.*

Symmetry transformations evidently satisfy the group axioms (A.1) when the binary operation of two transformations implies their successive application, and one thus speaks of *symmetry groups*, in this mathematical sense. Also, because of this nature of application of group theory, groups are always regarded as groups of concrete transformations within a concrete model, and not as an abstract structure.

That also implies that by the "symmetry of a physical system" one in fact understands the symmetry of the model of that system, conditioned also by the approximations that have been applied in the model by way of neglecting details of the physical system, and mathematical idealizations in the model. As improvements to the model often add details that lessen the number and domain of symmetries, improvements to the model reduce the symmetry group to a subgroup G_1 of the original group G_0. One says that the additional details break the original group into its subgroup, $G_1 \subset G_0$. Although only G_1 is then the "real" symmetry group, the extended structure $G_1 \subset G_0$ provides useful additional information about the model. Not infrequently, the improvements to the

[8] The choice of k and the type of functions (C^0-functions are continuous, C^1-functions are smooth, etc.) depends on the requirements in a concrete application. In classical physics, one *usually* restricts to C^2-functions, as the equations of motion are differential equations of second order, and at least the second derivatives need to be well defined. However, more detailed requirements in the analysis of deformations require higher derivatives, so the required function type must be adapted.

model may be organized iteratively, corresponding to a *chain* $G_2 \subset G_1 \subset G_0$ of subgroups, which may have an alternative $G_2 \subset G_1' \subset G_0$. The entire *web* of such chains of subgroups provides a hierarchy of model improvements, which corresponds to a hierarchy of physical phenomena and corresponding corrections to measurable results of the model, such as energy.

A simple example

As an illustration of the ideas and concepts depicted in Figure A.2 on p. 457, consider the very familiar example:

$$F = m\,a = m\frac{\mathrm{d}^2 x}{\mathrm{d}t^2}, \qquad \text{with the conditions} \qquad x|_{t=t_0} = x_0, \quad \frac{\mathrm{d}x}{\mathrm{d}t}\Big|_{t=t_0} = v_0. \tag{A.19}$$

In the familiar application of these equations, F and m are parameters in the problem; respectively, the force that acts upon a given body and the mass (measure of inertia) of that body. The function of time, $x = x(t)$, is the position of the body, and x_0, v_0 are boundary (initial) conditions.

The physical system of all bodies of mass m under the action of a force F is thus represented by the model ***M***, which is the *abstraction and simplification* of the physical system and which consists of the differential equation (A.19) together with the conditions x_0, v_0 that specify the concrete conditions of a concrete body in a concrete situation to which the model may be applied.[9]

The mathematical solution of this model (assuming that F and m are independent of time) is the function

$$x = x(t) = x_0 + v_0\,(t-t_0) + \frac{F}{2m}(t-t_0)^2, \tag{A.20}$$

so the *space of mathematical solutions* is the abstract space $\mathscr{X}(\boldsymbol{M})$, of the *four-parameter family of functions* $x = x(t; F, m, x_0, v_0)$. Since the t-dependence is determined by the equation (A.20), this space of mathematical solutions has four dimensions, with the coordinates F, m, x_0, v_0. The **phase diagram** is the partitioning of this 4-dimensional space into regions where the model behaves uniformly, and where the passage from one region into another – through some interface region – represents a phase transition in the system.

Similarly named, but something entirely different, is the **phase space**, Φ. For this system, this is the 2-dimensional space parametrized by the values of the pair of functions $(x(t), p(t))$, where $p(t) := m\frac{\mathrm{d}x}{\mathrm{d}t}$. The motion of the body sweeps a path in Φ, parametrized by time. The space of physical observables is then the infinite-dimensional space of all (continuous, and if desired perhaps also analytic and/or square-integrable, etc.) real functions $\mathscr{M}(\Phi)$ over the 2-dimensional phase space Φ.

Finally, the space of measurable model results, $\mathscr{R}(\boldsymbol{M})$, is – *in principle* – a subspace of the space $\mathscr{M}(\Phi)$; see Figure A.3. In this simple model, however, every element of the space $\mathscr{M}(\Phi)$ in fact may be represented as a model result, the question is only whether it is *experimentally possible* to directly measure that result: Namely, it may be that the result must be "factored" into factors and/or summands, which one measures directly and which are then "put together" into the *indirectly* "measured" complex result – but our goal here is not to delve into the details of *experimental* methods.[10] For the model (A.19), in fact, $\mathscr{R}(\boldsymbol{M}) = \mathscr{M}(\Phi)$, i.e., every observable of this physical system may in fact be represented by an (in principle measurable) result of the model (A.19).

Some symmetries: Assuming that the mass of the body is an absolute constant, the differential equation (A.19) has (among others, also) two independent symmetries:

$$\mathsf{P}\colon \begin{cases} x \to -x, \\ F \to -F; \end{cases} \tag{A.21a}$$

[9] Model (A.19) neglects whatever friction may exist, the resistance of the medium through which the body may be moving, etc.

[10] For example, even a relatively simple observable such as speed is *usually* not measured directly, but one measures independently the observables of "traversed distance" and "elapsed time," and speed is then *computed* as their ratio.

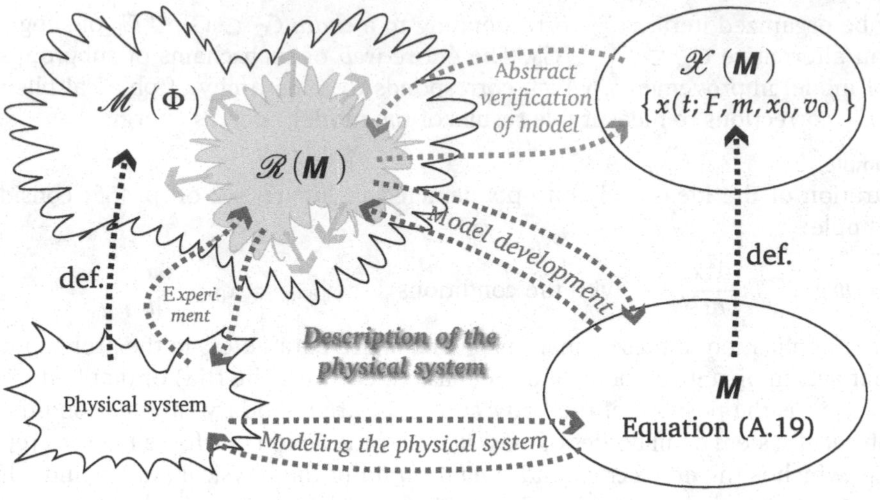

Figure A.3 Relations between the physical system (a body of mass m under the influence of the force F) and its observables, its mathematical model, the space of solutions thereof and the measurable results of this model, as well as their comparisons via experiments. The reason for the relation $\mathscr{R}(\boldsymbol{M}) \subsetneq \mathscr{M}(\Phi)$ is evident: there exist real functions over the phase space $\Phi = \{x, p_x\}$ which are therefore *observables* in the formal sense, but for which no one knows how such a function in fact might be measured, whereupon they do not belong to $\mathscr{R}(\boldsymbol{M})$.

$$T_\tau: \quad t \to t+\tau, \qquad \tau \in \mathbb{R}. \tag{A.21b}$$

The operation P is the mirror reflection of one of the spatial coordinates, which is in 3-dimensional space equivalent to the reflection of all three coordinates through the coordinate original. Its physical meaning is that one is free to pick the direction of measuring the position x either to the right or to the left, from some initially specified point identified as the coordinate origin. Of course, a change of this convention requires the sign of the force to also be changed simultaneously and correspondingly. The other symmetry, T_τ, is the time-translation. The solution (A.20) is also invariant with respect to the first of these two symmetries if the parameters (the integration constants) x_0, v_0 simultaneously satisfy

$$P: \begin{cases} x_0 & \to & -x_0, \\ v_0 & \to & -v_0, \end{cases} \tag{A.22}$$

which is in agreement with the definition x_0, v_0 as the position and speed in the $t = t_0$ moment: if the convention of measuring positions is changed from right-ward to left-ward, all quantities $x(t), x_0, v_0, F$ evidently change signs. With respect to the simultaneous action of the operation P, specified by the relations (A.21)–(A.22), both the system (A.19) and its solutions (A.20) – each a solution by itself! – are invariant with respect to P, i.e., this transformation is a symmetry in the most direct sense.

On the other hand, the physical interpretation of the time-translation is that the behavior of the system does not depend on when we begin to measure time, and this is a symmetry in a slightly indirect sense. Namely, from the fact that the function (A.20) remains a solution of the system (A.19) although it is not invariant under the action of T,[11] it follows that the solutions of a

[11] The constant t_0 does not change under the action of the operation T – indeed, t_0 is chosen so as to be an absolute constant that specifies the beginning of time measurements for the purposes of the applications of the model to a concrete physical model.

system need not possess all the symmetries of the system [☞ Comment A.2 on p. 458]. However, with t_0 as a fixed constant, we have

$$T_\tau: \qquad t_0 \to t'_0 := t_0 + \tau, \tag{A.23}$$

$$\begin{aligned} x(t;t_0) &= x_0 + v_0(t-t_0) + \frac{F}{2m}(t-t_0)^2 \\ \to x(t;t'_0) &= \underbrace{\Big(x_0 - v_0\tau + \frac{F}{2m}\tau^2\Big)}_{x'_0} + \underbrace{\Big(v_0 - \frac{F}{m}\tau\Big)}_{v'_0}(t-t_0) + \frac{F}{2m}(t-t_0)^2. \end{aligned} \tag{A.24}$$

Indeed, the symmetry transformation of the system (A.19) changes the integration constants and turns the solution where the time measurement began at t_0, into the solution where the time measurement began at t'_0. Thereby, the symmetry T_τ of the system (A.19) is not a symmetry of a concrete solution (A.20), but transforms one concrete solution into another concrete solution. Thus, the transformation (A.19) is a symmetry of the *entire space* of solutions, $\mathscr{X}(\boldsymbol{M})$.

Symmetries and conservation laws In classical physics, the implications of such symmetries are the content of the (Amalie Emmy) Noether theorem, whereby in classical physics and briefly:

> **Theorem A.1 (Amalie Emmy Noether)** *Every continuous symmetry has a corresponding current 4-vector, j^μ, which satisfies the continuity equation, $\partial_\mu j^\mu = 0$, and $\int d^3\vec{r}\, j^0$ is the corresponding "charge," conserved in time.*

Generally, *additive symmetries* (such as T_τ) have additive conserved charges, and for T_τ, this is the total energy of the system:

$$\frac{dx}{dt}\cdot(\text{A.19}) \;\Rightarrow\; m\frac{dx}{dt}\frac{d^2x}{dt^2} - \frac{dx}{dt}F = 0, \tag{A.25}$$

$$\Rightarrow\; \frac{dE}{dt} = 0, \quad \text{where} \quad E := \frac{m}{2}\Big(\frac{dx}{dt}\Big)^2 - xF. \tag{A.26}$$

The energy E therefore does not change in time, and the Noether theorem connects this property to the symmetry $T_\tau : t \to t + \tau$ of the differential equation (A.19), owing to the fact that $\frac{d}{d(t+\tau)} = \frac{d}{dt}$. The additivity of energy means that the energy of a combined system is the sum of energies of the individual sub-systems.

Similarly, the multiplicative symmetries (such as P), have multiplicative conserved "charges"; for P, this is the *parity* of the system.[12] $\mathsf{P}: f(x) = f(-x) = \mathsf{p} f(x)$, $\mathsf{p} = \pm 1$. The multiplicativity of parity means that the parity of a combined system is the product of the parities of the individual sub-systems.

—❦—

In quantum physics, the relation between symmetries and conserved quantities is even more direct:

> **Conclusion A.1** *Let P and Q be two canonically conjugate variables in the sense of the classical description of a system, and P and Q the respectively corresponding operators so $[\mathsf{P}, \mathsf{Q}] = i\hbar$. Then the operator $\frac{1}{\hbar}\mathsf{P}$* **generates** *translations of the eigenvalues of the operator Q and* **vice versa***:*
>
> $$e^{iq_0\mathsf{P}/\hbar}\,\mathsf{Q}\,e^{-iq_0\mathsf{P}/\hbar} = \mathsf{Q} + q_0. \tag{A.27}$$

[12] Of course, the eigenvalue of parity may be written $e^{i\pi\tilde{\mathsf{p}}}$, with $\tilde{\mathsf{p}} = 0$ or 1, and this $\tilde{\mathsf{p}}$ would then be a conserved mod-2 additive quantity. The "multiplicative" practice followed herein is, however, the generally accepted one.

If the translation of eigenvalues of Q is a symmetry of the system and H the Hamiltonian of the quantum description of this system, then P is a conserved quantity, and **vice versa**. *More precisely, if $Q \simeq Q + q_0$ then*

$$\frac{dP}{dt} = \frac{1}{i\hbar}[H, P] + \frac{\partial P}{\partial t} = 0, \tag{A.28}$$

and conversely: if $[H, P] = 0$ and P does not explicitly depend on time, the eigenvalues of P are conserved quantities and $Q \to Q + q_0$ is a symmetry of the system; the unitary operators $U_{q_0} := \exp\{\frac{i}{\hbar} q_0 P\}$ ***realize*** *this symmetry.*

For a proof and a detailed discussion, see standard textbooks of quantum mechanics, such as [407, 471, 328, 480, 472, 242, 360, 29, 339, 324].

The best known example of this relation is provided by the canonically conjugate pair (position, momentum). In coordinate representation, the operator $\frac{1}{\hbar} p_x = -i\frac{\mathrm{d}}{\mathrm{d}x}$ is indeed the generator of translation in the x-coordinate:

$$e^{iap_x/\hbar} f(x) = e^{a\frac{\mathrm{d}}{\mathrm{d}x}} f(x) = \sum_{k=0}^{\infty} \frac{a^k}{k!} \frac{\mathrm{d}^k}{\mathrm{d}x^k} f(x) = \sum_{k=0}^{\infty} \frac{a^k}{k!} f^{(k)}(x) = f(x+a). \tag{A.29}$$

Since p does not explicitly depend on time, the condition that p is a conserved quantity reduces to the condition that p commutes with the Hamiltonian. Since evidently $[p, \frac{1}{2m}p^2] = 0$ this condition becomes $[p, V(x)] = 0 = \frac{\hbar}{i}[\frac{\mathrm{d}}{\mathrm{d}x}, V(x)]$, i.e., that the potential is a constant. Indeed, for a constant potential, $x \to x + x_0$ is a manifest symmetry. Table A.1 lists several examples of often used symmetries and the corresponding conserved quantities. Absolutely essential is the fact that conserved quantities are eigenvalues of operators that generate corresponding transformations. So, for example, the unitary operator $U_{\vec{a}} = \exp\{i\vec{a}\cdot\vec{p}\}$ produces translation in space $\vec{r} \to \vec{r} + \vec{a}$, and the operator $U_{\vec{\xi}} = \exp\{i\vec{\xi}\cdot\vec{r}\}$ produces translation in the momentum space: $\vec{p} \to \vec{p} + \vec{\xi}$.

Table A.1 Some examples of continuous symmetries and corresponding conserved quantities. For various transformations, "charge" denotes various physical quantities; for translation of the phase of complex wave-functions, "charge" is indeed the electric charge.

Symmetry			**Conserved quantity**	
Time translation	$t \to t + t_0$	$\leftrightarrow$	Energy	E
Space translation	$\vec{r} \to \vec{r} + \vec{r}_0$	$\leftrightarrow$	Linear momentum	$\vec{p}$
Rotation (about the z-axis)	$\phi \to \phi + \phi_0$	$\leftrightarrow$	Angular momentum	L_z
Gauge transformation (general)	Phase shift	$\leftrightarrow$	Charge (general)	q
Reflection (through coordinate origin)	$\vec{r} \to -\vec{r}$	$\leftrightarrow$	Parity	P

A.1.4 Matrix groups and bilinear invariants

Application of group theory in physics always implies concrete action of the group elements upon concrete physical objects, i.e., upon the mathematical variables that represent those objects in the particular model of the physical system. The linear group action is then always in the form of a linear transformation of a vector space that those mathematical variables span, so these are always matrix groups.

The most often used matrix groups for $n, p, q \in \mathbb{N}$ are defined as follows [581, 260]:

$GL(n;\Bbbk)$ is the group of invertible $n\times n$ matrices with $\Bbbk$-elements, where $\Bbbk = \mathbb{Q}, \mathbb{R}, \mathbb{C}, \mathbb{H}$ denotes the *base field* of rational, real, complex and quaternion numbers, respectively.

$SL(n;\Bbbk)$ is the subgroup of $GL(n;\Bbbk)$, the elements, $\mathbb{A}$, of which have unit determinant, and so preserve the volume element: $\mathrm{d}^n(\mathbb{A}\,\mathrm{x}) = \mathrm{d}^n\mathrm{x}$ for $\mathrm{x} \in \Bbbk^n$.

$O(p,q;\Bbbk)$ is the subgroup $GL(p{+}q;\Bbbk)$ the elements of which are $\eta_{(p,q)}$-orthogonal,

$$\mathbf{L}^T\,\eta_{(p,q)}\,\mathbf{L} = \eta_{(p,q)}, \qquad \eta_{(p,q)} := \mathrm{diag}(\underbrace{1,\dots,1}_{p\text{ times}},\underbrace{-1,\dots,-1}_{q\text{ times}}) \tag{A.30}$$

and preserve the pseudo-Riemannian scalar product:

$$(\mathrm{x},\mathrm{y})_{(p,q)} := x^\mu\,\eta_{\mu\nu}\,y^\nu = \sum_{\mu=1}^{p} x^\mu\,y^\mu - \sum_{\mu=p+1}^{p+q} x^\mu\,y^\mu, \qquad \mathrm{x},\mathrm{y} \in \Bbbk^{p,q}. \tag{A.31}$$

$SO(p,q;\Bbbk)$ is the subgroup of $O(p,q;\Bbbk)$, the elements of which have unit determinant.

$Sp(2n;\Bbbk)$, for $\Bbbk = \mathbb{R}$ or $\mathbb{C}$, is the subgroup of $SL(2n;\Bbbk)$, the elements of which preserve the symplectic quadratic form

$$\mathrm{x}\wedge\mathrm{y} := 2\sum_{\mu-1}^{2n} x^\mu\wedge x^{\mu+n} = x^\mu\,\Omega_{\mu\nu}\,x^\nu, \qquad [\Omega_{\mu\nu}] = \begin{bmatrix} \mathbb{0} & \mathbb{1} \\ -\mathbb{1} & \mathbb{0} \end{bmatrix}, \tag{A.32}$$

and where the $2n\times 2n$ matrix $[\Omega_{\mu\nu}]$ is called the "symplectic identity."

$U(p,q)$ is the subgroup of $GL(p{+}q;\mathbb{C})$, the elements of which are unitary and preserve the Hermitian scalar product

$$\langle\mathrm{x}|\mathrm{y}\rangle_{(p,q)} := \sum_{\mu=1}^{p}(x^\mu)^*\,y^\mu - \sum_{\mu=p+1}^{p+q}(x^\mu)^*\,y^\mu, \qquad \mathrm{x},\mathrm{y} \in \mathbb{C}^{p,q}. \tag{A.33}$$

$Sp(p,q) = U(p,q;\mathbb{H})$ is the subgroup of $GL(2p{+}2q;\mathbb{H})$, the elements of which are quaternion-unitary and preserve the quaternion–Hermitian scalar product

$$\langle\mathrm{z}|\mathrm{w}\rangle_{(p,q)} := \bar{\mathrm{z}}^\mu\,\eta_{\mu\nu}\,w^\nu = \sum_{i=\mu}^{p} \bar{\mathrm{z}}^\mu\,w^\mu - \sum_{\mu=p+1}^{p+q} \bar{\mathrm{z}}^\mu\,w^\mu, \qquad \mathrm{z},\mathrm{w} \in \mathbb{H}^{p,q}, \tag{A.34}$$

where $\bar{x}^\mu$ denotes the quaternion-conjugate of x^μ. This group in fact is not *symplectic*, in the sense that it does not preserve any symplectic quadratic form. Because of this, for the previous group, $Sp(2n;\Bbbk)$, the base field is always denoted, and for $Sp(p,q)$ never, and by convention $Sp(n) \equiv Sp(n,0)$.

$SU(p,q)$ is the subgroup of $U(p,q)$, the elements of which have unit determinant.

Quaternion (also known as hyper-complex) numbers and algebra were invented by William Rowan Hamilton, in 1843. Quaternion numbers may be defined as the formal sum $\mathrm{q} = x^0 + ix^1 + jx^2 + kx^3$, where i,j,k are the formal quaternion units that satisfy $i^2 = j^2 = k^2 = ik = -1$. The quaternion-conjugate number is then equal to $\bar{\mathrm{z}} = z^0 - iz^1 - jz^2 - kz^3$. Quaternions do not commute, and $ij = k$ but $ji = -k$, etc. Quaternion "units" may be represented by the complex matrices

$$1 \to \left[\begin{smallmatrix}1&0\\0&1\end{smallmatrix}\right], \quad i \to \left[\begin{smallmatrix}i&0\\0&-i\end{smallmatrix}\right], \quad j \to \left[\begin{smallmatrix}0&1\\-1&0\end{smallmatrix}\right], \quad k \to \left[\begin{smallmatrix}0&i\\i&0\end{smallmatrix}\right], \tag{A.35}$$

so the definitions that use quaternions may be rewritten as complex-matrix definitions.

Owing to frequent use, the base field $\mathbb{R}$ is not written for orthogonal groups so that "$SO(1,3)$" means $SO(1,3;\mathbb{R})$, the base field $\mathbb{C}$ is not written for unitary groups so that "$SU(3)$" means $SU(3;\mathbb{C})$.

A.1.5 Exercises for Section A.1

✎ ***A.1.1*** *Prove relation (A.10) by explicit expansion of exponential functions.*

✎ ***A.1.2*** *Prove that the collection* $\{\mathbb{1},(i\sigma^1),(i\sigma^2),(i\sigma^3),-\mathbb{1},(-i\sigma^1),(-i\sigma^2),(-i\sigma^3)\}$ *forms a group, which is a subgroup of the group (A.13).*

✎ ***A.1.3*** *Show that scaling operations* $\{R_\rho : x \to \rho x,\ \rho \in (-\infty,+\infty)\}$ *form a group if the binary operation is consecutive application. For* R_ρ *to be a symmetry of the system (A.19), one must require that simultaneously* $R_\rho : F \to \rho F$*. Determine the action that is consistent with (a) the group structure, and (b) physical meaning of all symbols in the expressions (A.19)–(A.20).*

✎ ***A.1.4*** *Show that* $\{\mathbb{1}, P\}$ *forms a subgroup of* $\{R_\rho : x \to \rho x,\ \rho \in (-\infty,+\infty)\}$*.*

✎ ***A.1.5*** *Show that* $\{\mathbb{1}, T : t \to -t\}$ *forms a group.*

✎ ***A.1.6*** *Show that* $\{\mathbb{1}, T, P, (PT)\}$ *forms a group. Show that* $[T, P] = 0$*.*

A.2 The $U(1)$ group

The multiplicative group of unitary complex numbers, $U(1) = \{e^{i\varphi}, \cdot\}$ where $\varphi \in \mathbb{R}^1$ and $\varphi \simeq \varphi + 2\pi$, is one of the best known in theoretical physics. Representations of the group $U(1)$ are complex functions upon which the group acts by phase transformation: $f \to e^{iq_f\varphi} f$, $f \in \mathbb{C}$. In the general case, the real number q_f is called the "charge" of the particle represented by the function f, and the representation is unambiguously specified by the charge. In the case of the application in electromagnetism, the charge is the electric charge. The $U(1)$ charges are simply additive:

$$U(1) : f \to e^{iq_f\varphi} f,\ g \to e^{iq_g\varphi} g, \qquad \Rightarrow \qquad (fg) \to e^{i(q_f+q_g)\varphi}(fg). \tag{A.36}$$

It is possible – although it is rarely so denoted – to define the $U(1)$ group elements as

$$U(1) = \{\, e^{i\varphi Q},\ \varphi \simeq \varphi + 2\pi \,\}, \tag{A.37}$$

where the operator Q is the *generator* of the group, and q_f the eigenvalue of the eigenfunction: $Qf = q_f f$. Thus, $e^{i\varphi Q} f = e^{i\varphi q_f} f = e^{iq_f\varphi} f$. In complex analysis, q_f is called the winding number of the complex function f, and in the physical application of complex analysis the product $(q_f\varphi)$ is called the *phase* of the function f. This relationship between complex analysis and its application in gauge models is of special importance in string models from the worldsheet perspective [☞ Section 11.2.3], where one easily switches to the complex coordinate system $(\tau, \sigma) \to z = \sigma + i\tau$, and so also to complex analysis.

A.2.1 Exercises for Section A.2

✎ ***A.2.1*** *Given two mutually commuting* $U(1)$ *groups, generated respectively by the mutually commuting Hermitian operators* A *and* B*, show that the two-parameter family of elements* $g_{a,b} := \exp\{i(aA + bB)\}$ *form the abelian group* $U(1)_A \times U(1)_B$ *for* $a, b \in \mathbb{R}$ *with respect to the usual multiplication.*

✎ ***A.2.2*** *Show that any two linearly independent linear combinations of A and B from the previous exercise can serve as generators for the group $U(1)_A \times U(1)_B$. The particular choice $C_+ := A+B$ generates the* **diagonal** *subgroup $U(1)_+ \subset U(1)_A \times U(1)_B$, while the combination $C_- := A-B$ generates the complementary $U(1)_- \subset U(1)_A \times U(1)_B$.*

✎ ***A.2.3*** *Show that $U(1)_+$ and $U(1)_-$ as defined in the previous problem commute with each other and that, as groups, $U(1)_+ \times U(1)_- = U(1)_A \times U(1)_B$.*

A.3 The $SU(2)$ group

This group is familiar from the quantum-mechanical formalism of spin and orbital angular momentum. The group is *generated* by any three operators that satisfy the relations

$$[J_j, J_k] = i\,\varepsilon_{jkl}\,\delta^{lm}\,J_m := i\,\varepsilon_{jk}{}^m\,J_m. \tag{A.38a}$$

The $SU(2)$ group elements are then operators of the form $U_{\vec{a}} := \exp\{ia^j J_j\}$. Conversely, the generators may be formally defined by the relation

$$J_k := \frac{1}{i}\left[\frac{\partial g(\vec{a})}{\partial a^k}\right]_{\vec{a}=\vec{0}}. \tag{A.38b}$$

It follows that the quadratic J^2 operator commutes with all three J_j:

$$[J^2, J_j] = 0, \quad j = 1,2,3, \qquad \text{where} \qquad J^2 := J_1^2 + J_2^2 + J_3^2, \tag{A.38c}$$

so the operators[13] J^2 and J_3 have a simultaneous (common) basis of eigenfunctions $|j,m\rangle$:

$$J^2|j,m\rangle = j(j+1)|j,m\rangle, \qquad J_3|j,m\rangle = m|j,m\rangle, \tag{A.38d}$$

where

$$\Delta m \in \mathbb{Z}, \quad j := \max(m) \;\Rightarrow\; -j \leqslant m \leqslant +j. \tag{A.38e}$$

It follows that j and m are both either integral (tensorial) or half-integral (spinorial), and that

$$J_\pm := (J_1 \pm iJ_2), \quad J_\pm|j,m\rangle = \sqrt{j(j+1) - m(m\pm1)}\,|j,m\pm1\rangle. \tag{A.38f}$$

Note that

$$J_+|j,j\rangle \equiv 0, \qquad \text{as well as} \qquad J_-|j,-j\rangle \equiv 0 \tag{A.39}$$

by virtue of relations (A.38f), as derived in Digression A.2.

Digression A.2 (Proof of equation (A.38), following Ref. [18]) As no two operators from the collection $\{J_1, J_2, J_3\}$ commute, there is no subsystem of mutually commuting operators that would have a simultaneous eigenbasis. One thus chooses one, usually J_3, to find its eigenstates. Then, one proves by direct computation that

$$[J_i, J^2] = 0, \qquad J^2 := J_1^2 + J_2^2 + J_3^2, \quad i = 1,2,3. \tag{A.40a}$$

---

[13] The choice of J_3 is arbitrary, and is called the quantization axis choice for angular momentum in quantum mechanics.

Since J_3 and J^2 commute, they have a simultaneous (common) eigenbasis:

$$J^2|\lambda,m\rangle = \lambda|\lambda,m\rangle, \qquad J_3|\lambda,m\rangle = m|\lambda,m\rangle, \tag{A.40b}$$

which may always be ortho-normalized via the Gram–Schmidt procedure:

$$\langle\lambda',m'|\lambda,m\rangle = \delta_{\lambda',\lambda}\,\delta_{m',m}. \tag{A.40c}$$

Using the remaining two operators, J_1, J_2, we define

$$J_\pm := J_1 \pm iJ_2, \qquad (J_\pm)^\dagger = J_\mp, \tag{A.40d}$$

so that

$$J_\pm J_\mp = J_1^2 + J_2^2 \pm J_3, \qquad J^2 = J_+J_- + J_-J_+ + J_3^2, \tag{A.40e}$$

$$[J_3, J_\pm] = \pm J_\pm \qquad \text{and} \qquad [J_\pm, J_\mp] = \pm 2J_3. \tag{A.40f}$$

Next, check how $J_\pm$ act upon $|\lambda,m\rangle$:

$$J_3\big(J_\pm|\lambda,m\rangle\big) = \big(J_\pm J_3 \pm J_\pm\big)|\lambda,m\rangle = (m\pm1)\big(J_\pm|\lambda,m\rangle\big), \tag{A.40g}$$

so it must be that

$$J_\pm|\lambda,m\rangle = N_\pm(m)\,|\lambda,m{\pm}1\rangle. \tag{A.40h}$$

Thus, the operators $J_\pm$ raise/lower the second eigenvalue, m, but do not change the first, λ.

Since J_3 and J^2 are Hermitian operators, λ, m must be real numbers. Also,

$$\begin{aligned}\lambda = \langle J^2\rangle &= \langle J_+J_-\rangle + \langle J_-J_+\rangle + \langle J_3^2\rangle \\ &= \big\|J_-|\lambda,m\rangle\big\|^2 + \big\|J_+|\lambda,m\rangle\big\|^2 + m^2 \geqslant m^2.\end{aligned} \tag{A.40i}$$

Thus m^2, and so also m, has a maximum; let $j := \max(m)$. Then $J_+|\lambda,+j\rangle$ would have to be proportional to $|\lambda,j{+}1\rangle$. However, since $(j{+}1) > j = \max(m)$, $|\lambda,j{+}1\rangle$ cannot exist. It follows that

$$J_+|\lambda,j\rangle = 0. \tag{A.40j}$$

Applying $\langle\lambda,m|J_-$ to this result, we have that

$$\begin{aligned}0 = \langle\lambda,j|J_-J_+|\lambda,j\rangle &= \langle\lambda,j|(J_1^2 + J_2^2 - J_3)|\lambda,j\rangle = \langle\lambda,j|(J^{\,2} - J_3^2 - J_3)|\lambda,j\rangle \\ &= \lambda - j(j{+}1), \qquad \Rightarrow \qquad \lambda = j(j{+}1).\end{aligned} \tag{A.40k}$$

Following the analogous reasoning for $J_+ \leftrightarrow J_-$, we obtain that

$$\min(m) = -j, \qquad J_-|\lambda,-j\rangle = 0. \tag{A.40l}$$

Renaming the basis $|\lambda,m\rangle \mapsto |j,m\rangle$, we have that

$$J^2|j,m\rangle = j(j{+}1)|j,m\rangle, \qquad J_3|j,m\rangle = m|j,m\rangle. \tag{A.40m}$$

In addition, the operators $J_i, J^2, J_\pm$ can change m only in unit increments. It follows that

$$\Delta m \in \mathbb{Z}, \quad |m| \leqslant j := \max(m), \qquad \Rightarrow \qquad \begin{cases} j \;\in \mathbb{Z}_{\geqslant 0}, & \text{tensors;} \\ j \;\in \mathbb{Z}_{\geqslant 0} + \tfrac{1}{2}, & \text{spinors.}\end{cases} \tag{A.40n}$$

Similarly, we have that

$$|N_\pm(m)|^2 = \langle j,m|J_\pm J_\mp|j,m\rangle = j(j+1) - m(m\pm1), \tag{A.40o}$$

so

$$N_\pm(m) = \sqrt{j(j+1) - m(m\pm1)}. \tag{A.40p}$$

A.3.1 Representations of $SU(2)$

The relations (A.38d)–(A.38f) imply that

$$\begin{aligned} U_{\vec\varphi}|j,m\rangle &= e^{i\varphi^k J_k}|j,m\rangle = \exp\{i(\varphi^+ J_+ + \varphi^- J_- + \varphi^3 J_3)\}|j,m\rangle \\ &= \sum_{|m'|\leqslant j} c_{m,m'}|j,m'\rangle. \end{aligned} \tag{A.41}$$

That is, the action of the unitary operator $U_{\vec\varphi}$ does not change j in $|j,m\rangle$ upon which it acts, but – for a general choice of $\vec\varphi$ – transforms any one $|j,m\rangle$ into a linear combination of all $|j,m'\rangle$ with all the permitted values of m'. The abstract vector space

$$V_j := \left\{ \sum_{|m|\leqslant j} c_m|j,m\rangle,\ (c_{-j},\dots,c_j) \in \Bbbk^{2j+1} \text{ and equation (A.41)} \right\} \cong \Bbbk^{2j+1}, \tag{A.42a}$$

$$U_{\vec\varphi} : V_j \to V_j, \qquad V_j \text{ is a } \left\{ \begin{smallmatrix}\text{tensorial}\\ \text{spinorial}\end{smallmatrix} \right\} \text{ representation if } j \left\{ \begin{smallmatrix}\text{integral}\\ \text{half-integral}\end{smallmatrix} \right\} \tag{A.42b}$$

is a $(2j+1)$-dimensional (unitary) representation of the $SU(2)$ group, i.e., the $SU(2)$ group maps the vector space V_j into itself, and $SU(2)$ is a group of symmetries of the vector space V_j, for every $2j \in \mathbb{Z}_{\geqslant 0}$. Correspondingly, the same partitioning of representations into these two subclasses is also obtained by partitioning into the eigen-representations of the element $\exp\{i\pi\mathbb{1}\} \in \mathbb{Z}_2 \subset SU(2)$ [☞ Example A.1 on p. 454].

Table A.2 on p. 469 lists the first several such representations. It is important to *keep* in mind that the spaces V_j are not simply copies of $\Bbbk^{2j+1}$ (where $\Bbbk = \mathbb{Q}, \mathbb{R}, \mathbb{C}$ or $\mathbb{H}$, as required), but imply the $SU(2)$ action (A.41). It follows that no $SU(2)$ representation V_j contains a strictly smaller representation $V_{j'}$, with $j' < j$. One says that every representation V_j is *irreducible*.

Digression A.3 The results (A.40m)–(A.40n) give a *complete list* of irreducible representations of the $SU(2)$ group and its algebra (A.38a):

1. tensorial representations, of which the most familiar are:
   (a) scalars, i.e., invariants, represented by $|0,0\rangle$;
   (b) 3-vectors, represented by the basis $\{|1,-1\rangle, |1,0\rangle, |1,+1\rangle\}$;
   (c) (spin-2) quadrupoles, represented by $\{|2,-2\rangle, |2,-1\rangle, |2,0\rangle, |2,+1\rangle, |2,+2\rangle\}$;
   etc.
2. spinorial representations, of which the most familiar are:
   (a) spin-$\frac{1}{2}$ systems, represented by the basis $\{|\frac{1}{2},-\frac{1}{2}\rangle, |\frac{1}{2},+\frac{1}{2}\rangle\}$;
   (b) spin-$\frac{3}{2}$ systems, represented by the basis $\{|\frac{3}{2},-\frac{3}{2}\rangle, |\frac{3}{2},-\frac{1}{2}\rangle, |\frac{3}{2},+\frac{1}{2}\rangle, |\frac{3}{2},+\frac{3}{2}\rangle\}$;
   etc.

Note that the bases of formal vectors $\{|j,m\rangle,\ |m|\leqslant j\}$ are just a formal notation for bases of spherical harmonics $\{Y_j^m(\theta,\phi),\ |m|\leqslant j\}$, which are the coordinate representation of the formal $|j,m\rangle$. For example,

$$|1,+1\rangle \leftrightarrow Y_1^{+1}(\theta,\phi) = -\sqrt{\frac{3}{8\pi}}\sin\theta\, e^{+i\phi}, \tag{A.43a}$$

$$|1,0\rangle \leftrightarrow \quad Y_1^{0}(\theta,\phi) = +\sqrt{\frac{3}{4\pi}}\cos\theta, \tag{A.43b}$$

$$|1,-1\rangle \leftrightarrow Y_1^{-1}(\theta,\phi) = -\sqrt{\frac{3}{8\pi}}\sin\theta\, e^{-i\phi}, \tag{A.43c}$$

from which it follows that Cartesian coordinates may be expressed as

$$x = r\,\sin\theta\,\cos\phi = -r\sqrt{\frac{2\pi}{3}}\Big(Y_1^1(\theta,\phi)+Y_1^{-1}(\theta,\phi)\Big) \quad\leftrightarrow\quad |1,+1\rangle+|1,-1\rangle, \tag{A.43d}$$

$$y = r\,\sin\theta\,\sin\phi = i\,r\sqrt{\frac{2\pi}{3}}\Big(Y_1^1(\theta,\phi)-Y_1^{-1}(\theta,\phi)\Big) \quad\leftrightarrow\quad |1,+1\rangle-|1,-1\rangle, \tag{A.43e}$$

$$z = r\,\sin\phi \qquad = r\sqrt{\frac{4\pi}{3}}\,Y_1^0(\theta,\phi) \quad\leftrightarrow\quad |1,0\rangle. \tag{A.43f}$$

Similar relations exist for all bases $\{|j,m\rangle,\ |m|\leqslant j\}$ for $j\in\mathbb{Z}$. The other half of representations, the spinors $\{|j,m\rangle,\ |m|\leqslant j\}$ for $(j+\frac{1}{2})\in\mathbb{Z}$ also have an analogous representation in terms of spherical and Cartesian coordinates, but are less well known, and are double-valued and so are not determined unambiguously.

Table A.2 lists several well-known irreducible representations of the $SU(2)$ group, denoted in several alternative and oft-used forms, and Table A.3 on p. 470 lists the first few spherical harmonics $Y_j^m(\theta,\phi)$, which are the functional representation[14] (in spherical coordinates) of the abstract elements $|j,m\rangle$. Of course, the abstract operators $J_\pm, J_3$ and J^2 also have a corresponding functional representation:

$$J_\pm = \pm e^{\pm i\phi}\Big[\frac{\partial}{\partial\theta}\pm i\cot(\theta)\frac{\partial}{\partial\phi}\Big], \tag{A.44a}$$

$$J_3 = -i\frac{\partial}{\partial\phi}, \tag{A.44b}$$

$$J^2 = -\Big[\frac{1}{\sin(\theta)}\frac{\partial}{\partial\theta}\Big(\sin(\theta)\frac{\partial}{\partial\theta}\Big)+\frac{1}{\sin^2(\theta)}\frac{\partial^2}{\partial\phi^2}\Big]. \tag{A.44c}$$

Evidently and except for J_3, computations with the abstract operators and eigenstates of the $SU(2)$ group are simpler than with the functional representation of these.

— ❦ —

Other than the formal ($|j,m\rangle$) and the functional ($Y_j^m(\theta,\phi)$) notation, the matrix notation is also widely used. It is well known that halves of the Pauli matrices (A.147)

$$\mathbb{J}_1^{(1/2)} = \tfrac{1}{2}\begin{bmatrix}0&1\\1&0\end{bmatrix},\qquad \mathbb{J}_2^{(1/2)} = \tfrac{1}{2}\begin{bmatrix}0&-i\\i&0\end{bmatrix},\qquad \mathbb{J}_3^{(1/2)} = \tfrac{1}{2}\begin{bmatrix}1&0\\0&-1\end{bmatrix}, \tag{A.45a}$$

[14] Unfortunately, the word "representation" is used in two slightly different senses: here, it is in the sense of "realization," in distinction from the technical sense of an "$SU(2)$ group representation," according to the definition (A.42).

Table A.2 Several smallest representations of the $SU(2)$ group; formal ket-notation precisely corresponds to spherical harmonics $|j,m\rangle \leftrightarrow Y_j^m(\theta,\phi)$ when $j \in \mathbb{Z}$.

	Dim.	Formal ket-notation	Index[a]	Matrix
V_0	1	$\{\|0,0\rangle\}$	t	$[x]$
$V_{\frac{1}{2}}$	2	$\{\|\frac{1}{2},-\frac{1}{2}\rangle,\|\frac{1}{2},+\frac{1}{2}\rangle\}$	t^a	$\begin{bmatrix}x^1\\x^2\end{bmatrix}$
V_1	3	$\{\|1,-1\rangle,\|1,0\rangle,\|1,+1\rangle\}$	$t^{(ab)}$	$\begin{bmatrix}x=t^{(11)}\\y=t^{(12)}\\z=t^{(22)}\end{bmatrix}$
$V_{\frac{3}{2}}$	4	$\{\|\frac{3}{2},-\frac{3}{2}\rangle,\|\frac{3}{2},-\frac{1}{2}\rangle,\|\frac{3}{2},+\frac{1}{2}\rangle,\|\frac{3}{2},+\frac{3}{2}\rangle\}$	$t^{(abc)}$	$\begin{bmatrix}x^1=t^{(111)}\\ \vdots\\x^4=t^{(222)}\end{bmatrix}$
V_2	5	$\{\|2,-2\rangle,\|2,-1\rangle,\|2,0\rangle,\|2,+1\rangle,\|2,+2\rangle\}$	$t^{(abcd)}$	$\begin{bmatrix}x^1=t^{(1111)}\\ \vdots\\x^5=t^{(2222)}\end{bmatrix}$
$\vdots$	$\vdots$	$\vdots$	$\vdots$	
V_j	$2j+1$	$\{\|j,-j\rangle,\|j,1-j\rangle\cdots,\|j,j-1\rangle,\|j,+j\rangle\}$	$t^{(a_1\cdots a_{2j})}$	$\begin{bmatrix}x^1=t^{(1\cdots1)}\\ \vdots\\x^{2j+1}=t^{(2\cdots2)}\end{bmatrix}$

[a] The indices are $a,b,c\ldots \in \{1,2\}$; round parentheses denote symmetrization: $t^{(ab)} = +t^{(ba)}$.

satisfy the relations (A.38a), which identifies the eigenvectors of the $\mathbb{J}_3^{(1/2)}$-matrix with the eigenvectors of the abstract operator J_3:

$$\left[\begin{smallmatrix}1\\0\end{smallmatrix}\right] \leftrightarrow |\tfrac{1}{2},+\tfrac{1}{2}\rangle \quad\text{and}\quad \left[\begin{smallmatrix}0\\1\end{smallmatrix}\right] \leftrightarrow |\tfrac{1}{2},-\tfrac{1}{2}\rangle. \tag{A.45b}$$

In a fully identical fashion, the matrices

$$\mathbb{J}_1^{(1)} = \frac{1}{\sqrt{2}}\left[\begin{smallmatrix}0&1&0\\1&0&1\\0&1&0\end{smallmatrix}\right],\quad \mathbb{J}_2^{(1)} = \frac{1}{\sqrt{2}}\left[\begin{smallmatrix}0&-i&0\\i&0&-i\\0&i&0\end{smallmatrix}\right],\quad \mathbb{J}_3^{(1)} = \left[\begin{smallmatrix}1&0&0\\0&0&0\\0&0&-1\end{smallmatrix}\right], \tag{A.46a}$$

also satisfy the relations (A.38a), which identifies the eigenvectors of the $\mathbb{J}_3^{(1)}$-matrix with the eigenvectors of the abstract operator J_3:

$$\left[\begin{smallmatrix}1\\0\\0\end{smallmatrix}\right] \leftrightarrow |1,+1\rangle,\quad \left[\begin{smallmatrix}0\\1\\0\end{smallmatrix}\right] \leftrightarrow |1,0\rangle \quad\text{and}\quad \left[\begin{smallmatrix}0\\0\\1\end{smallmatrix}\right] \leftrightarrow |1,-1\rangle, \tag{A.46b}$$

and,

$$\mathbb{J}_1^{(3/2)} = \frac{1}{2}\begin{bmatrix}0&\sqrt{3}&0&0\\\sqrt{3}&0&2&0\\0&2&0&\sqrt{3}\\0&0&\sqrt{3}&0\end{bmatrix},\quad \mathbb{J}_2^{(3/2)} = \frac{1}{2}\begin{bmatrix}0&-\sqrt{3}i&0&0\\\sqrt{3}i&0&-2i&0\\0&2i&0&-\sqrt{3}i\\0&0&\sqrt{3}i&0\end{bmatrix},\quad \mathbb{J}_3^{(3/2)} = \frac{1}{2}\left[\begin{smallmatrix}3&0&0&0\\0&1&0&0\\0&0&-1&0\\0&0&0&-3\end{smallmatrix}\right], \tag{A.46c}$$

Table A.3 The first few spherical harmonics

$Y_0^0 = \frac{1}{\sqrt{4\pi}}$	$= \frac{1}{\sqrt{4\pi}}$	$Y_2^0 = \sqrt{\frac{15}{16\pi}}(3\cos^2\theta - 1)$	$= \sqrt{\frac{5}{16\pi}} \frac{3z^2 - r^2}{r^2}$
$Y_1^1 = -\sqrt{\frac{3}{8\pi}} \sin\theta\, e^{i\phi}$	$= -\sqrt{\frac{3}{8\pi}} \frac{x+iy}{r}$	$Y_3^3 = -\sqrt{\frac{35}{64\pi}} \sin^3\theta\, e^{3i\phi}$	$= -\sqrt{\frac{35}{64\pi}} \frac{(x+iy)^3}{r^3}$
$Y_1^0 = +\sqrt{\frac{3}{4\pi}} \cos\theta$	$= \sqrt{\frac{3}{4\pi}} \frac{z}{r}$	$Y_3^2 = \sqrt{\frac{105}{32\pi}} \sin^2\theta\, \cos\theta\, e^{2i\phi}$	$= \sqrt{\frac{105}{32\pi}} \frac{(x+iy)^2 z}{r^3}$
$Y_2^2 = \sqrt{\frac{15}{32\pi}} \sin^2\theta\, e^{2i\phi}$	$= \sqrt{\frac{3}{8\pi}} \frac{(x+iy)^2}{r}$	$Y_3^1 = \sqrt{\frac{21}{64\pi}} \sin\theta(1 - 5\cos^2\theta) e^{i\phi}$	$= -\sqrt{\frac{21}{64\pi}} \frac{(x+iy)(5z^2 - r^2)}{r^3}$
$Y_2^1 = \sqrt{\frac{15}{8\pi}} \sin\theta \cos\theta\, e^{i\phi}$	$= \sqrt{\frac{3}{4\pi}} \frac{(x+iy)z}{r}$	$Y_3^0 = \sqrt{\frac{7}{16\pi}}(5\cos^2\theta - 3)\cos\theta$	$= \sqrt{\frac{7}{16\pi}} \frac{z(5z^2 - 3r^2)}{r^3}$

similarly provide a 4-dimensional realization for spin-$\frac{3}{2}$ systems. An analogous matrix realization of the operators $\vec{J}$ and eigenvectors $|j, m\rangle$ is of course possible for all j.

Finally, in the tensor notation, we have

$$t^1 \;\leftrightarrow\; |\tfrac{1}{2}, +\tfrac{1}{2}\rangle \qquad \text{and} \qquad t^2 \;\leftrightarrow\; |\tfrac{1}{2}, -\tfrac{1}{2}\rangle, \tag{A.47a}$$

which, with the definition $(u, v) := (t^1, t^2)$, implies the definitions

$$\mathfrak{J}_1^{(1/2)} := \tfrac{1}{2}\Big(v\frac{\partial}{\partial u} + u\frac{\partial}{\partial v}\Big), \quad \mathfrak{J}_2^{(1/2)} := \tfrac{i}{2}\Big(v\frac{\partial}{\partial u} - u\frac{\partial}{\partial v}\Big), \quad \mathfrak{J}_3^{(1/2)} := \tfrac{1}{2}\Big(u\frac{\partial}{\partial u} - v\frac{\partial}{\partial v}\Big). \tag{A.47b}$$

For $j = 1$, one typically identifies the formal tensor variables $t^{(11)}, t^{(12)}, t^{(22)}$ with the Cartesian x, y, z, respectively, and we have the well-known

$$\mathfrak{J}_1^{(1)} := i\Big(x\frac{\partial}{\partial y} - y\frac{\partial}{\partial x}\Big), \quad \mathfrak{J}_2^{(1)} := i\Big(y\frac{\partial}{\partial z} - z\frac{\partial}{\partial y}\Big), \quad \mathfrak{J}_3^{(1)} := i\Big(z\frac{\partial}{\partial x} - x\frac{\partial}{\partial z}\Big). \tag{A.47c}$$

As each of these notations and representations is convenient in some but not all computations, it behooves the Reader to practice "translating" from any one of these representations into any other one.

It is useful to note that the Levi-Civita symbol,

$$\varepsilon_{ab}: \quad \varepsilon_{12} = 1 = -\varepsilon_{21}, \ \ \varepsilon_{11} = 0 = \varepsilon_{22}, \tag{A.48}$$

is invariant with respect to $SU(2)$ transformations, since, using relations (B.38) and after the computation (B.37), it follows that the change of basis $t^a \to \tau^a$ produces

$$\mathrm{d}^2 t := \tfrac{1}{2}\varepsilon_{ab}\, \mathrm{d}t^a \mathrm{d}t^b = \det\Big[\tfrac{\partial(t^1, t^2)}{\partial(\tau^1, \tau^2)}\Big]\, \tfrac{1}{2}\varepsilon_{ab}\, \mathrm{d}\tau^a \mathrm{d}\tau^b = \mathrm{d}^2\tau, \tag{A.49}$$

since the determinant of $SU(2)$ transformations equals $\det\big[\frac{\partial(t^1, t^2)}{\partial(\tau^1, \tau^2)}\big] = 1$ by definition. The analogous situation holds by definition for all $SU(n)$ groups, but for $SU(2)$, exceptionally, ε_{ab} is a rank-2 tensor, and so may also serve as an (antisymmetric!) metric tensor, which is appropriate for anticommuting variables that are used in supersymmetry [☞ Chapter 10].

A.3.2 The SU(2) and SO(3) groups

Rotations in real, 3-dimensional space may be represented as real, orthogonal 3×3 matrices with unit determinant. Their successive application may be identified with matrix multiplication, which does not commute, and this multiplicative group is denoted $SO(3)$. Its algebra, $\mathfrak{so}(3)$, is identical to the $\mathfrak{su}(2)$ algebra. However, although $\mathfrak{so}(3) = \mathfrak{su}(2)$, the groups $SO(3)$ and $SU(2)$ differ: the $SU(2)$ action upon all representations $V_j = \{|j,m\rangle,\ |m| \leqslant j\}$ is single-valued.

In distinction, the group $SO(3)$ action is single-valued upon integral (tensorial) representations $V_j = \{|j,m\rangle,\ |m| \leqslant j \in \mathbb{Z}\}$, but not upon half-integral (spinorial) representations, $V_j = \{|j,m\rangle,\ |m| \leqslant j,\ (j+\frac{1}{2}) \in \mathbb{Z}\}$. Since a φ-rotation about the x^3-axis acts by $\exp\{i\,\varphi\,J_3\}$ and eigenvalues of J_3 on elements of spinorial representations V_j are half-integral, spinors are "double-valued functions" under $SO(3)$ rotations. By an appropriate change of basis, it is easy to show that the eigenvalues of any one component $\vec{J}$, in any direction, are equal to their J_3 eigenvalues. Thus, the conclusion about double-valuedness of the elements of spinorial representations V_j holds for rotations about any axis. Thus, spinors change their sign upon any 360°-rotation; only 720°-rotations act upon them as the identity.

Since the algebras are identical, $\mathfrak{so}(3) = \mathfrak{su}(2)$, the elements of both algebras – and so also both the $SU(2)$ and the $SO(3)$ generators – are rightfully called angular momenta. Understanding this 2–1 relationship between these groups, $SU(2)$ is the two-fold covering of the $SO(3)$ group, and the elements of the $SU(2)$ group are also frequently called rotations. Pedantically, the $SU(2)$ group is the double covering of the $SO(3)$ group of rotations.

A.3.3 Addition of angular momenta

In the concrete application of the $SU(2)$ group in elementary particle physics, it is important to keep in mind that angular momentum is not a *directly* measurable quantity.

This is partly true also in classical physics of macroscopic bodies: for an ice-skater in a pirouette or a spinning top, the angular momentum cannot be measured directly. Instead, usually, one identifies a "marking" on the spinning object (the ice-skater's face or a pattern on the top), and the angular velocity is determined by following the motion of this marking. Independently, one determines the moment of inertia for the same object in some way,[15] and then *computes* the angular momentum from the so-obtained values of the moment of inertia and the angular velocity. That is, there's no such thing as an "angularmomentumometer."

With elementary particles, the situation is even more indirect: by definition, elementary particles cannot have a "marking" the motion of which one could follow even in principle, so as to measure the angular velocity, compute the moment of inertia, etc. Instead, the angular momentum is even *defined* indirectly. For example, the intrinsic angular moment of an electron – the so-called spin – is in fact a fictive rotation [☞ Digression 4.1 on p. 132] which one *computes*, by way of relation (4.24a), from the *measured* magnetic dipole momentum.

In the situation when we have several magnetic fields, it is perfectly logical to compute their vectorial sum. Conversely, since the dipole momenta of these magnetic fields define spins and orbital angular momenta,[16] to the sum of magnetic fields then corresponds a sum of angular momenta, both intrinsic ("spins") and relative ("orbital").

— ❦ —

The technique of adding angular momenta in quantum theory differs from "ordinary vectorial addition" which is expected in classical physics, and this is discussed in great detail in standard textbooks of quantum mechanics. We recall here the basic relations.

[15] In principle, this is possible by approximating the geometry of the object and its mass distribution, whereupon one *computes* the moment of inertia by integrating, or by physically applying a force, and the moment of inertia is *computed* as the ratio of the applied torque and the produced change in its angular velocity.

[16] Although Bohr's model of the atom depicts the electron as a point-particle that rotates about a point-like proton, so that the rotation of the electron's charge forms a current that produces an "orbiting magnetic field," experiments actually only measure this magnetic field, from which then – in turn – one *concludes* about the rotating of the *mental image* of the point-like electron in Bohr's atom.

Let $\{L_1, L_2, L_3\}$ and $\{S_1, S_2, S_3\}$ be two triples of operators, of which each independently satisfies the relations (A.38a) – ***regardless of their physical meaning*** – and let

$$[L_j, S_k] = 0 \quad \text{for every pair of indices } j,k = 1,2,3. \tag{A.50}$$

These two triples then generate two separate copies of the $SU(2)$ group, where elements of one commute with the elements of the other, and we have $SU(2)_L \times SU(2)_S$. One then defines

$$J_j := L_j + S_j \qquad \Rightarrow \qquad [J_j, J_k] = i\varepsilon_{jk}{}^m J_m, \tag{A.51}$$

and the triple J_i generates the *diagonal subgroup* $SU(2)_J \subset SU(2)_L \times SU(2)_S$. For each triple, one defines operators such as J^2 and $J_\pm$, yielding results akin to (A.38), repeating the computations in Digression A.2 on p. 465:

$$L^2|\ell, m_\ell\rangle = \ell(\ell{+}1)|\ell, m_\ell\rangle, \qquad L_3|\ell, m_\ell\rangle = m_\ell|\ell, m_\ell\rangle; \tag{A.52a}$$
$$S^2|s, m_s\rangle = s(s{+}1)|s, m_s\rangle, \qquad S_3|s, m_s\rangle = m_s|s, m_s\rangle; \tag{A.52b}$$
$$J^2|j, m_j\rangle = j(j{+}1)|j, m_j\rangle, \qquad J_3|j, m_j\rangle = m_j|j, m_j\rangle. \tag{A.52c}$$

The relation (A.50) implies that L^2, L_3, S^2, S_3 all mutually commute, so that the tensor product of the eigenbases (A.52a) and (A.52b),

$$|\ell, s; m_\ell, m_s\rangle := |\ell, m_\ell\rangle \otimes |s, m_s\rangle, \tag{A.53a}$$

is a simultaneous eigenbasis of all four operators:

$$L^2|\ell, s; m_\ell, m_s\rangle = \ell(\ell{+}1)|\ell, s; m_\ell, m_s\rangle, \qquad S^2|\ell, s; m_\ell, m_s\rangle = s(s{+}1)|\ell, s; m_\ell, m_s\rangle, \tag{A.53b}$$
$$L_3|\ell, s; m_\ell, m_s\rangle = m_\ell|\ell, s; m_\ell, m_s\rangle, \qquad S_3|\ell, s; m_\ell, m_s\rangle = m_s|\ell, s; m_\ell, m_s\rangle. \tag{A.53c}$$

The operator J_3 commutes with L^2, L_3, S^2, S_3, but is of course not linearly independent since equation (A.51) implies that $J_3 = L_3 + S_3$. We thus also have that

$$J_3|\ell, s; m_\ell, m_s\rangle = (m_\ell{+}m_s)|\ell, s; m_\ell, m_s\rangle. \tag{A.53d}$$

In turn,

$$[J^2, L_3] = 2i\,\varepsilon^{jk}{}_3 L_j S_k = 2i(L_1 S_2 - L_2 S_1) = -[J^2, S_3], \tag{A.54}$$

and J^2 does not commute with every operator from the collection $\{L^2, L_3, S^2, S_3\}$. Thus, the 4-plet $\{L^2, L_3, S^2, S_3\}$ is a maximal collection of linearly independent mutually commuting operators.

In turn, the operators $\{J^2, L^2, S^2, J_3\}$ also all mutually commute, and since L_3 and S_3 do not commute with J^2, this second operator quartet is also a maximal collection of linearly independent mutually commuting operators. Thus, they too have a simultaneous eigenbasis:

$$J^2|j, \ell, s; m_j\rangle = j(j{+}1)|j, \ell, s; m_j\rangle, \qquad L^2|j, \ell, s; m_j\rangle = \ell(\ell{+}1)|j, \ell, s; m_j\rangle, \tag{A.55a}$$
$$J_3|j, \ell, s; m_j\rangle = m_j|j, \ell, s; m_j\rangle, \qquad S^2|j, \ell, s; m_j\rangle = s(s{+}1)|j, \ell, s; m_j\rangle. \tag{A.55b}$$

In textbooks of quantum mechanics, L are identified with the orbital angular momentum, S with the spin and J with the "total" angular momentum, (e.g., of an electron in a hydrogen atom). Ignoring the fact that J does not include the nuclear spin, and so in reality is *not* the total angular momentum, there exist many situations where there are more than two triples of operators each of which satisfies the relations such as do L and S, and where at least some of such operators have no relation with rotations, even if fictitious. For example, there is no obstruction to add – akin to equations (A.51) – the angular momentum of a nucleon in one nucleus, say, with the isospin of that or any other nucleon.

Thus, L and S as well as their eigenbasis (A.53a) will be referred to as "constituent," and J and the eigenbasis (A.55) will be referred to as "composite."

Of course, since both bases are complete, it follows that every element of one may be expressed in terms of the elements of the other:

$$|\ell,s;m_\ell,m_s\rangle = \sum_{j=|\ell-s|}^{\ell+s} C^{j,m_j}_{\ell,s;m_\ell,m_s}\,|j,\ell,s;m_j\rangle, \tag{A.56a}$$

$$|j,\ell,s;m_j\rangle = \sum_{\substack{m_\ell=-\ell\\ |m_s|=|m_j-m_\ell|\leqslant s}}^{\ell} \big(C^{j,m_j}_{\ell,s;m_\ell,m_s}\big)^*\,|\ell,s;m_\ell,m_s\rangle, \tag{A.56b}$$

where

$$C^{j,m_j}_{\ell,s;m_\ell,m_s} := \langle j,\ell,s;m_j|\ell,s;m_\ell,m_s\rangle \equiv \langle j,m_j|\ell,s;m_\ell,m_s\rangle \tag{A.56c}$$

are the Clebsch–Gordan coefficients, which by standard convention all have real values. In addition, we have:

Theorem A.2 *For the sum of two triples of operators, $L_i + S_i = J_i$, each of which satisfies relations (A.38) and (A.50), the relations (A.52) follow, as well as:*

$$|\ell-s| \leqslant j \leqslant (\ell+s), \qquad |j-\ell| \leqslant s \leqslant (j+\ell), \qquad |j-s| \leqslant \ell \leqslant (j+s), \tag{A.57}$$

$$m_j = m_\ell + m_s, \qquad |m_j| \leqslant j, \qquad |m_\ell| \leqslant \ell, \qquad |m_s| \leqslant s, \tag{A.58}$$

where j,ℓ and s assume precisely once all the integrally separated values within the indicated limits.

Thus, using the notation from the left-most two columns of Table A.2 on p. 469, we have that

$$V_\ell \otimes V_s = \oplus_{j=|\ell-s|}^{(\ell+s)} V_j \qquad \Leftrightarrow \qquad (\mathbf{2\ell{+}1}) \otimes (\mathbf{2s{+}1}) = \oplus_{j=|\ell-s|}^{(\ell+s)} (\mathbf{2j{+}1}). \tag{A.59}$$

For example:

$$\begin{array}{lcccccc}
\underline{V_\ell \otimes V_s = V_j} & \Leftrightarrow & (\mathbf{2\ell{+}1}) & \otimes & (\mathbf{2s{+}1}) & = & (\mathbf{2j{+}1}) \\
\hline
V_{1/2} \otimes V_{1/2} = V_1 \oplus V_0 & \Leftrightarrow & \mathbf{2} & \otimes & \mathbf{2} & = & \mathbf{3} \oplus \mathbf{1} \\
V_1 \otimes V_{1/2} = V_{3/2} \oplus V_{1/2} & \Leftrightarrow & \mathbf{3} & \otimes & \mathbf{2} & = & \mathbf{4} \oplus \mathbf{2} \\
V_1 \otimes V_1 = V_2 \oplus V_1 \oplus V_0 & \Leftrightarrow & \mathbf{3} & \otimes & \mathbf{3} & = & \mathbf{5} \oplus \mathbf{3} \oplus \mathbf{1} \\
V_2 \otimes V_1 = V_3 \oplus V_2 \oplus V_2 & \Leftrightarrow & \mathbf{5} & \otimes & \mathbf{3} & = & \mathbf{7} \oplus \mathbf{5} \oplus \mathbf{3} \\
\hline
\end{array} \tag{A.60}$$

and so on. The first row here corresponds to the detailed relations

$$V_{1/2} = \big\{c_+|\tfrac12,+\tfrac12\rangle + c_-|\tfrac12,-\tfrac12\rangle\big\}, \tag{A.61}$$

$$\begin{aligned}
\big\{c_+|\tfrac12,+\tfrac12\rangle + c_-|\tfrac12,-\tfrac12\rangle\big\} &\otimes \big\{c'_+|\tfrac12,+\tfrac12\rangle' + c'_-|\tfrac12,-\tfrac12\rangle'\big\}\\
&= \big\{c_1|1,+1\rangle + c_0|1,0\rangle + c_{-1}|1,-1\rangle\big\} \oplus \big\{c'_0|0,0\rangle\big\}
\end{aligned} \tag{A.62}$$

where $\{c_+,c_-\}$, $\{c'_+c'_-\}$ and $\{c_1,c_0,c_{-1};c'_0\}$ are coefficients in the linear combinations appropriate for the vector spaces $V_{1/2}$, $V'_{1/2}$, V_1 and V_0, and where

$$V_1 : \begin{cases}
|1,+1\rangle = |\tfrac12,+\tfrac12\rangle|\tfrac12,+\tfrac12\rangle', & \text{(A.63a)}\\
|1,0\rangle = \tfrac{1}{\sqrt2}\Big(|\tfrac12,+\tfrac12\rangle|\tfrac12,-\tfrac12\rangle' + |\tfrac12,-\tfrac12\rangle|\tfrac12,+\tfrac12\rangle'\Big), & \text{(A.63b)}\\
|1,-1\rangle = |\tfrac12,-\tfrac12\rangle|\tfrac12,-\tfrac12\rangle', & \text{(A.63c)}
\end{cases}$$

$$V_0: \quad |0,0\rangle = \tfrac{1}{\sqrt{2}}\Big(|\tfrac{1}{2},+\tfrac{1}{2}\rangle|\tfrac{1}{2},-\tfrac{1}{2}\rangle' - |\tfrac{1}{2},-\tfrac{1}{2}\rangle|\tfrac{1}{2},+\tfrac{1}{2}\rangle'\Big). \tag{A.63d}$$

For bigger groups this detailed representation is also possible, but the notation becomes more complicated, so statements expressed in the "dimensional" notation, in the right-hand side of tabulation (A.60), are more often found in the physics literature.

Corollary A.1 *Every representation V_j may be assigned a* **parity**, *$\pi(V_j) := 2j \pmod 2$, so $\pi(V_j) = 0$ for tensors, and $\pi(V_j) = 1$ for spinors [☞ definition (A.42)]. Then it follows that parity is mod-2 additive: $\pi(V_\ell \otimes V_s) \equiv 2(\ell+s) \mod 2$.*

Finally, the tensor/index-notation is also used, especially for larger groups, and in that notation the relations (A.63) become

$$\mathbf{2}\otimes\mathbf{2} = \mathbf{3}\oplus\mathbf{1} \quad \leftrightarrow \quad t^\alpha \otimes u^\beta = \underbrace{v^{(\alpha\beta)}}_{\text{3 comps.}} \oplus \underbrace{\Big(v^{[\alpha\beta]} = v\,\varepsilon^{\alpha\beta}\Big)}_{\text{1 component}}, \tag{A.64}$$

where

$$V_0 = \{c_0|0,0\rangle\} = \{b_0\,t\}, \tag{A.65a}$$
$$V_{1/2} = \{c_+|\tfrac{1}{2},+\tfrac{1}{2}\rangle + c_-|\tfrac{1}{2},-\tfrac{1}{2}\rangle\} = \{b_1\,t^1 + b_2\,t^2\}, \tag{A.65b}$$
$$V_1 = \{c_1|1,+1\rangle + c_0|1,0\rangle + c_{-1}|1,-1\rangle\} = \{b_{11}\,t^{(11)} + b_{12}\,t^{(12)} + b_{22}\,t^{(22)}\}, \tag{A.65c}$$

and so on. The formal variables t for V_0, $\{t^1,t^2\}$ for $V_{1/2}$, $\{t^{(11)},t^{(12)},t^{(22)}\}$ for V_1, etc., play the role of basis vectors in the tensor notation. Also, the Levi-Civita symbol $\varepsilon_{\alpha\beta}$ is an $SU(2)$-invariant antisymmetric 2-form, so the antisymmetric rank-2 tensor may be identified with the invariant: $v^{[\alpha\beta]} \mapsto v = (\tfrac{1}{2}\varepsilon_{\alpha\beta}v^{[\alpha\beta]})$. Similarly, we have the projections

$$V_1\otimes V_{1/2}\supset V_{1/2} \quad \Leftrightarrow \quad t^{(\alpha\beta)}u^\gamma \mapsto v^\alpha := (\varepsilon_{\beta\gamma}t^{(\alpha\beta)}u^\gamma), \tag{A.66a}$$
$$V_{3/2}\otimes V_{1/2}\supset V_1 \quad \Leftrightarrow \quad t^{(\alpha\beta\gamma)}u^\delta \mapsto v^{(\alpha\beta)} := (\varepsilon_{\gamma\delta}t^{(\alpha\beta\gamma)}u^\delta), \tag{A.66b}$$
$$V_1\otimes V_1\supset V_1 \quad \Leftrightarrow \quad t^{(\alpha\beta)}u^{(\gamma\delta)} \mapsto v^{(\alpha\gamma)} := (\varepsilon_{\beta\delta}t^{(\alpha\beta)}u^{(\gamma\delta)}), \tag{A.66c}$$
$$V_1\otimes V_1\supset V_0 \quad \Leftrightarrow \quad t^{(\alpha\beta)}u^{(\gamma\delta)} \mapsto v := (\varepsilon_{\alpha\gamma}\varepsilon_{\beta\delta}t^{(\alpha\beta)}u^{(\gamma\delta)}), \tag{A.66d}$$

and so on.

A.3.4 $SU(2)$-covariant operators and the Wigner–Eckart theorem

Relations (A.38d) and (A.38f) have a very simple generalization from eigen-*vectors* to covariant/eigen-*operators*: If a $(2r{+}1)$-tuple of operators $\{T^{(r)}_\rho,\ |\rho| \leqslant r\}$ satisfies the relations

$$[\,J^2, T^{(r)}_\rho\,] = r(r{+}1)\,T^{(r)}_\rho, \qquad [\,J_3, T^{(r)}_\rho\,] = \rho\,T^{(r)}_\rho, \tag{A.67a}$$

then also

$$[\,J_\pm, T^{(r)}_\rho\,] = \sqrt{r(r{+}1) - \rho(\rho{+}1)}\,T^{(r)}_{\rho\pm1}, \tag{A.67b}$$

and the formal vector space $\{\sum_{\rho=-r}^{r} c_\rho T^{(r)}_\rho, c_\rho \in \mathbb{R}\} \cong \mathbb{R}^{2r+1}$ is also an $SU(2)$ group representation. Then we have [362, 363, 471, 328, 480, 134, 391, 407, 472, 360, 29, 339, 242, 3, 110, 324, for example]:

Theorem A.3 (Wigner–Eckart) *For the $(2r+1)$-tuple of operators $\{T^{(r)}_\rho,\ |\rho|\leqslant r\}$ that satisfy relations (A.67), for vectors $|j,m_j;\alpha\rangle$ that satisfy relations (A.38d) and if α represents additional eigenvalues of operators independent of J, we have*

$$\langle j'm'_j;\alpha'|\,T^{(r)}_\rho\,|j,m_j;\alpha\rangle = \langle j',m'_j|r,j;\rho,m_j\rangle\,\langle j';\alpha'\|\,T^{(r)}\,\|j;\alpha\rangle, \tag{A.68}$$

where $\langle j';\alpha'\|\,T^{(r)}\,\|j;\alpha\rangle$ is the so-called reduced matrix element (amplitude) that does not depend on m_j,ρ,m'_j, and $\langle j',m'_j|r,j;\rho,m_j\rangle$ is a Clebsch–Gordan coefficient.

This theorem is most often used when ratios of matrix elements are needed where the reduced matrix elements are equal and cancel in the ratio.

— ❦ —

For a practical use of relations (A.56) and the Wigner–Eckart theorem A.3, one needs the numerical values of the Clebsch–Gordan coefficients. To this end one most often uses tables [242, 105] [☞ also [294]], although there is a "closed formula" [328]:

$$C^{j,m_j}_{\ell,s;m_\ell,m_s} = \delta_{m_j,m_\ell+m_s}\,A^j_{\ell,s}\,B^{j,m_j}_{\ell,s;m_\ell,m_s}\,D^{j,m_j}_{\ell,s;m_\ell,m_s}\,, \tag{A.69a}$$

$$\delta_{m_j,m_\ell+m_s} = \begin{cases} 1 \text{ if } m_j = m_\ell+m_s, \\ 0 \text{ if } m_j\neq m_\ell+m_s; \end{cases} \tag{A.69b}$$

$$A^j_{\ell,s} := \sqrt{\frac{(\ell{+}s{-}j)!\,(j{+}\ell{-}s)!\,(s{+}j{-}\ell)!\,(2j{+}1)}{(\ell{+}s{+}j{+}1)!}}\,, \tag{A.69c}$$

$$B^{j,m_j}_{\ell,s;m_\ell,m_s} := \sqrt{(j{+}m_j)!\,(j{-}m_j)!\,(\ell{+}m_\ell)!\,(\ell{-}m_\ell)!\,(s{+}m_s)!\,(s{-}m_s)!}\,, \tag{A.69d}$$

$$D^{j,m_j}_{\ell,s;m_\ell,m_s} := \sum_r \frac{(-1)^r}{(\ell{-}m_\ell{-}r)!\,(s{+}m_s{-}r)!\,(j{-}s{+}m_\ell{+}r)!\,(j{-}\ell{-}m_s{+}r)!\,(\ell{+}s{-}j{-}r)!\,r!}\,, \tag{A.69e}$$

where the sum over r is limited by the facts that division by factors in the denominator produces a zero when

$$r>(\ell{-}m_\ell),\ (s{+}m_s),\ (\ell{+}s{-}j),\qquad r<0,\ (s{-}j{-}m_\ell),\ (\ell{+}m_s{-}j), \tag{A.69f}$$

which makes the sum finite. Evidently, formula (A.69) is not best suited for quick computations "by heart," but is appropriate for machine computation.

— ❦ —

In view of these well-known results for the $SU(2)$ group and its algebra, we have:

Conclusion A.2 *For applications of any group in physics, it is desired to have, in order of importance (and technical demand):*

1. *the complete list of finite-dimensional unitary representations, such as (A.42),*
2. *the complete list of decompositions of products, such as (A.59),*
3. *the complete list of Clebsch–Gordan coefficients, such as (A.69), or at least a method/algorithm for their computation.*

It is fascinating that for off-shell representations of supersymmetry not even the first task is solved☝, not even in N-extended supersymmetric quantum mechanics [☞ Section 10.4].

A.3.5 Exercises for Section A.3

✎ ***A.3.1*** *Using the differential representation (A.44) of J^2, J_3 and $J_\pm$ as well as the functional representations (A.43a)–(A.43c), verify the general results (A.40m), and (A.40h) with (A.40p) for the cases $j = 1$, $m = \pm 1, 0$.*

✎ ***A.3.2*** *Verify by explicit computation that the matrices (A.46a) satisfy the $\mathfrak{su}(2)$ algebra relations (A.38a). Construct the 3×3 matrix representative of $(\vec{\mathbb{J}}^{(1)})^2$.*

✎ ***A.3.3*** *Given two separate triples of Hermitian operators, $\vec{L}$ and $\vec{S}$, satisfying the $\mathfrak{su}(2)$ algebra (A.38a) and commuting mutually (A.50), prove that equation (A.51) defines the one and only nontrivial linear combination that also satisfies the $\mathfrak{su}(2)$ algebra (A.38a).*

A.4 The *SU*(3) group

The *SU*(3) group is defined as the group of 3×3 unitary matrices with unit determinant.

Digression A.4 Corollary A.1 on p. 474 defines parity for representations of the *SU*(2) group, which is additive for products of representations. Similarly, the *SU*(3) group has a *triality*: representations are either real with triality 0, or a conjugate pair of complex representations with triality 1 and $-1 \cong 2$. The triality of a product of two representations with trialities t_1 and t_2, respectively, is $(t_1+t_2) \pmod 3$. Similarly, one defines a mod-n additive "n-ality" of representations of the $SU(n)$ group for every n.

A.4.1 The $\mathfrak{su}(3)$ algebra

As a generalization of the relations (A.38a) for the *SU*(2) group generators and a special case of the general relation for all Lie algebras (A.11a), the *SU*(3) group is generated by *eight* operators Q_a that satisfy the relations

$$[\,Q_a\,,\,Q_b\,] = i\,f_{ab}{}^c\,Q_c. \tag{A.70}$$

It is useful to note the *SU*(3) analogue of the generator matrices (A.45a), i.e., the standard choice among the matrix realizations (of the *doubles*[17]) of the *SU*(3) generators in the smallest, 3-dimensional and *fundamental* representation are the so-called Gell-Mann matrices:

$$\boldsymbol{\lambda}_1 = \begin{bmatrix}0&1&0\\1&0&0\\0&0&0\end{bmatrix},\quad \boldsymbol{\lambda}_2 = \begin{bmatrix}0&-i&0\\i&0&0\\0&0&0\end{bmatrix},\quad \boldsymbol{\lambda}_3 = \begin{bmatrix}1&0&0\\0&-1&0\\0&0&0\end{bmatrix},\quad \boldsymbol{\lambda}_4 = \begin{bmatrix}0&0&1\\0&0&0\\1&0&0\end{bmatrix}, \tag{A.71a}$$

$$\boldsymbol{\lambda}_5 = \begin{bmatrix}0&0&-i\\0&0&0\\i&0&0\end{bmatrix},\quad \boldsymbol{\lambda}_6 = \begin{bmatrix}0&0&0\\0&0&1\\0&1&0\end{bmatrix},\quad \boldsymbol{\lambda}_7 = \begin{bmatrix}0&0&0\\0&0&-i\\0&i&0\end{bmatrix},\quad \boldsymbol{\lambda}_8 = \tfrac{1}{\sqrt{3}}\begin{bmatrix}1&0&0\\0&1&0\\0&0&-2\end{bmatrix}. \tag{A.71b}$$

The first three matrices evidently generate one of continuously many $SU(2) \subset SU(3)$ subgroups. This choice of matrix representations of generators shows that the structure constants $f_{abc} = f_{ab}{}^d\,g_{dc}$ are totally antisymmetric, $f_{abc} = -f_{bac} = -f_{acb} = -f_{cba}$, and we have

$$f_{123} = 1,\quad f_{458} = f_{678} = \tfrac{\sqrt{3}}{2},\quad f_{147} = -f_{156} = f_{246} = f_{257} = f_{345} = -f_{367} = \tfrac{1}{2}. \tag{A.71c}$$

It is useful to know that

$$\mathrm{Tr}(\boldsymbol{\lambda}^a\boldsymbol{\lambda}^b) = 2\delta^{ab}. \tag{A.72}$$

[17] Just as *halves* of Pauli matrices close the $\mathfrak{su}(2)$ algebra, *halves* of the Gell-Mann matrices close the $\mathfrak{su}(3)$ algebra.

A.4.2 Representations of SU(3)

One may define the ket-notation as well as the matrix notation for every Lie group, but only the dimensional and the tensor/index notation are shown here:

$$\mathbf{1} \simeq t, \qquad \mathbf{3} \simeq t^{\alpha}, \qquad \mathbf{3}^* \simeq t_{\alpha} = \tfrac{1}{2}\varepsilon_{\alpha\beta\gamma} t^{[\beta\gamma]}, \qquad \alpha,\beta,\gamma,\ldots = 1,2,3, \tag{A.73}$$

$$\mathbf{6} \simeq t^{(\alpha\beta)}, \qquad \mathbf{6}^* \simeq t_{(\alpha\beta)}, \qquad \mathbf{8} \simeq t^{\alpha}{}_{\beta},\ t^{\alpha}{}_{\alpha} \equiv 0, \qquad \mathbf{10} \simeq t^{(\alpha\beta\gamma)}, \qquad \text{etc.} \tag{A.74}$$

Here, e.g., $t^{(\alpha\beta)}$ is the symmetric 3×3 matrix, $t^{[\alpha\beta]}$ is the antisymmetric 3×3 matrix, $t^{\alpha}{}_{\beta}$ is the Hermitian 3×3 matrix the trace of which vanishes, etc. It is important to recall the identities:

$$\varepsilon_{\alpha\beta\gamma}\varepsilon^{\delta\epsilon\phi} = \delta^{\delta}_{\alpha}\delta^{\epsilon}_{\beta}\delta^{\phi}_{\gamma} - \delta^{\delta}_{\alpha}\delta^{\phi}_{\beta}\delta^{\epsilon}_{\gamma} + \delta^{\phi}_{\alpha}\delta^{\delta}_{\beta}\delta^{\epsilon}_{\gamma} - \delta^{\phi}_{\alpha}\delta^{\epsilon}_{\beta}\delta^{\delta}_{\gamma} + \delta^{\epsilon}_{\alpha}\delta^{\phi}_{\beta}\delta^{\delta}_{\gamma} - \delta^{\epsilon}_{\alpha}\delta^{\delta}_{\beta}\delta^{\phi}_{\gamma}, \tag{A.75a}$$

$$\Rightarrow \quad \varepsilon_{\alpha\beta\gamma}\varepsilon^{\delta\epsilon\gamma} = \delta^{\delta}_{\alpha}\delta^{\epsilon}_{\beta} - \delta^{\epsilon}_{\alpha}\delta^{\delta}_{\beta}, \qquad \varepsilon_{\alpha\beta\gamma}\varepsilon^{\delta\beta\gamma} = 2\delta^{\delta}_{\alpha}, \qquad \varepsilon_{\alpha\beta\gamma}\varepsilon^{\alpha\beta\gamma} = 6. \tag{A.75b}$$

Then,

$$\mathbf{3}\otimes\mathbf{3} = \mathbf{6}_S \oplus \mathbf{3}^*_A \;\Leftrightarrow\; t^{\alpha}\, s^{\beta} = t^{(\alpha}s^{\beta)} + t^{[\alpha}s^{\beta]}, \qquad \begin{cases} t^{(\alpha}s^{\beta)} &:= \tfrac{1}{2}(t^{\alpha}s^{\beta} + t^{\beta}s^{\alpha}), \\ t^{[\alpha}s^{\beta]} &:= \tfrac{1}{2}(t^{\alpha}s^{\beta} - t^{\beta}s^{\alpha}); \end{cases} \tag{A.76a}$$

where subscripts S and A, respectively, denote the symmetric and antisymmetric parts of a product. Next,

$$\begin{aligned} \mathbf{6}\otimes\mathbf{3} = \mathbf{10}\oplus\mathbf{8} \;\Leftrightarrow\; t^{(\alpha\beta)}\, s^{\gamma} &= t^{(\alpha\beta}s^{\gamma)} + \tfrac{4}{3}\, t^{(\alpha[b)}s^{\gamma]}, \\ t^{(\alpha\beta}s^{\gamma)} &:= \tfrac{1}{3}\big(t^{(\alpha\beta)}s^{\gamma} + t^{(\beta\gamma)}s^{\alpha} + t^{(\gamma\alpha)}s^{\beta}\big), \\ t^{(\alpha[\beta)}s^{\gamma]} &:= \tfrac{1}{4}\big((t^{(\alpha\beta)}s^{\gamma} - t^{(\alpha\gamma)}s^{\beta}) + (t^{(\beta\alpha)}s^{\gamma} - t^{(\beta\gamma)}s^{\alpha})\big) \\ &= \tfrac{1}{4}\big(2t^{(\alpha\beta)}s^{\gamma} - t^{(\alpha\gamma)}s^{\beta} - t^{(\beta\gamma)}s^{\alpha}\big) \end{aligned} \tag{A.76b}$$

where it follows that $t^{(\alpha[\beta)}s^{\gamma]}\,\varepsilon_{\alpha\beta\gamma} \equiv 0$;

$$\mathbf{3}^*\otimes\mathbf{3} = \mathbf{8}\oplus\mathbf{1} \;\Leftrightarrow\; t_{\alpha}\, s^{\beta} = \Big(t_{\alpha}s^{\beta} - \tfrac{1}{3}\delta^{\beta}_{\alpha}(t_{\gamma}s^{\gamma})\Big) + \tfrac{1}{3}\delta^{\beta}_{\alpha}(t_{\gamma}s^{\gamma}). \tag{A.76c}$$

Besides, we also have that

$$\tfrac{4}{3}\, t^{(\alpha[\beta)}s^{\gamma]}\,\varepsilon_{\beta\gamma\delta} = t^{(\alpha\beta)}s^{\gamma}\,\varepsilon_{\beta\gamma\delta} =: (t^{(\cdot\cdot)}s^{\cdot})^{\alpha}{}_{\delta} \quad : \quad \delta_{\alpha}{}^{\delta}\,(t^{(\cdot\cdot)}s^{\cdot})^{\alpha}{}_{\delta} \equiv 0, \tag{A.76d}$$

so that $(t^{(\cdot\cdot)}s^{\cdot})^{\alpha}{}_{\delta}$ is a Hermitian matrix with vanishing trace. Since $\varepsilon_{\beta\gamma\delta}\varepsilon^{\epsilon\phi\delta} = \delta^{\beta}_{\epsilon}\delta^{\gamma}_{\phi} - \delta^{\gamma}_{\epsilon}\delta^{\beta}_{\phi}$, we have also the "converse" relations:

$$(t^{(\cdot\cdot)}s^{\cdot})^{\alpha}{}_{\delta}\,\varepsilon^{\beta\gamma\delta} = t^{(\alpha\beta)}s^{\gamma} - t^{(\alpha\gamma)}s^{\beta}, \qquad \tfrac{2}{3}(t^{(\cdot\cdot)}s^{\cdot})^{(\alpha}{}_{\delta}\,\varepsilon^{\beta)\gamma\delta} = \tfrac{4}{3}\, t^{(\alpha[\beta)}s^{\gamma]}. \tag{A.76e}$$

Finally, we also need the combination (A.76a), (A.76b) and (A.76c):

$$\mathbf{3}\otimes\mathbf{3}\otimes\mathbf{3} = (\mathbf{6}_S\oplus\mathbf{3}^*_A)\otimes\mathbf{3} = (\mathbf{10}_S\oplus\mathbf{8})\oplus(\mathbf{8}\oplus\mathbf{1}_A), \tag{A.76f}$$

where the subscripts S and A, respectively, denote the totally symmetric and totally antisymmetric product, and two 8-plets have a mixed symmetry:

$$(\mathbf{3}\otimes\mathbf{3}\otimes\mathbf{3})_S \;\leftrightarrow\; t^{(\alpha}u^{\beta}v^{\gamma)}, \qquad (\mathbf{3}\otimes\mathbf{3}\otimes\mathbf{3})_A \;\leftrightarrow\; t^{[\alpha}u^{\beta}v^{\gamma]}. \tag{A.76g}$$

For the cubic expressions with mixed symmetry, there exist many possible choices, one of which follows from the iterative procedure (A.76f):

$$(\mathbf{3}\otimes\mathbf{3}\otimes\mathbf{3})_{\mathbf{8}_{(1)}} = \Big(\underbrace{(\mathbf{3}\otimes\mathbf{3})_S}_{\mathbf{6}}\otimes\mathbf{3}\Big)_{\mathbf{8}_{(1)}} \quad\leftrightarrow\quad (t^{(\alpha}u^{\beta)})v^{\gamma}\varepsilon_{\beta\gamma\delta}, \tag{A.76h}$$

$$(\mathbf{3}\otimes\mathbf{3}\otimes\mathbf{3})_{\mathbf{8}_{(2)}} = \Big(\underbrace{(\mathbf{3}\otimes\mathbf{3})_A}_{\mathbf{3}^*}\otimes\mathbf{3}\Big)_{\mathbf{8}_{(2)}} \quad\leftrightarrow\quad (t^{\alpha}u^{\beta}\varepsilon_{\alpha\beta\delta})v^{\gamma}\big(\delta^{\delta}_{\epsilon}\delta^{\phi}_{\gamma} - \tfrac{1}{3}\delta^{\delta}_{\gamma}\delta^{\phi}_{\epsilon}\big). \tag{A.76i}$$

These two expressions provide two linearly independent 3×3 Hermitian matrices with a vanishing trace.

These results indicate that the answer to Exercises A.3.1 and A.3.2 on 476, is given by Weyl's general construction:

> ***Construction A.1 (Weyl)*** *All finitely dimensional unitary representations of every Lie group may be constructed projecting n-fold tensor products of the* ***fundamental*** *(**spinorial** for Spin groups) representation, $V^{\otimes n}$, by means of the so-called Young symmetrizer.*

The computations (A.76a)–(A.76f) provide concrete examples of this construction:

0. For the $SU(3)$ group, the fundamental (defining) representation is the complex 3-dimensional, denoted **3**, also denoted in the tensor representation as $\mathbf{3} = \{t^{\alpha},\ \alpha = 1,2,3\}$.
1. The product $\mathbf{3}\otimes\mathbf{3}$ may be projected to:
   (a) the symmetric part of the product, **6**: $t^{(\alpha}u^{\beta)} = +t^{(\beta}u^{\alpha)}$, and
   (b) the antisymmetric part of the product, $\mathbf{3}^*$: $t^{[\alpha}u^{\beta]} = -t^{[\beta}u^{\alpha]}$, which is isomorphic to the conjugate 3-dimensional representation: $t^{[\alpha}u^{\beta]}\varepsilon_{\alpha\beta\gamma} = (t{\cdot}u{\cdot})_{\gamma}$.
2. The product $\mathbf{6}\otimes\mathbf{3}$ may be projected to:
   (a) the totally symmetric part of the product, **10**: $t^{(\alpha\beta}u^{\gamma)} = +t^{(\beta\alpha}u^{\gamma)} = +t^{(\gamma\beta}u^{\alpha)} = +t^{(\alpha\gamma}u^{\beta)}$, and
   (b) the part of the product with mixed symmetry, **8**: $t^{(\alpha[\beta)}u^{\gamma]} = +t^{(\beta[\alpha)}u^{\gamma]}$, but the $\beta\leftrightarrow\gamma$ antisymmetrization in $t^{(\alpha[\beta)}u^{\gamma]}$ is broken by imposing the $\alpha\leftrightarrow\beta$ symmetrization.

Projecting may be understood also as a linear mapping of vector spaces:

$$\mathrm{Sym}: \mathbf{3}\otimes\mathbf{3}\to\mathbf{6} \qquad\text{and}\qquad \mathbf{3}^* = \ker\big(\mathrm{Sym}(\mathbf{3}\otimes\mathbf{3})\big), \tag{A.77}$$

that is, $\mathbf{3}^*$ is the part of the product $\mathbf{3}\otimes\mathbf{3}$ that is annihilated by symmetrization. A consistent and iterative application of this procedure is called "Young symmetrization."

To decompose the triple tensor product, we may use the table of coefficients:

		$\alpha\beta\gamma$	$\alpha\gamma\beta$	$\gamma\alpha\beta$	$\gamma\beta\alpha$	$\beta\gamma\alpha$	$\beta\alpha\gamma$
10	$t^{(\alpha\beta\gamma)}$	+1	+1	+1	+1	+1	+1
8	$t^{(\alpha[\beta)\gamma]}$	+2	−1	−1	−1	−1	+2
8	$t^{[\alpha(\beta]\gamma)}$	+2	+1	−1	+1	−1	−2
1	$t^{[\alpha\beta\gamma]}$	+1	−1	+1	−1	+1	−1

(A.78)

so that, e.g.:

$$\begin{aligned} t^{(\alpha[\beta)\gamma]} &\propto (+2t^{\alpha\beta\gamma} - t^{\alpha\gamma\beta} - t^{\gamma\alpha\beta} - t^{\gamma\beta\alpha} - t^{\beta\gamma\alpha} + 2t^{\beta\alpha\gamma}) \propto (2t^{(\alpha\beta)\gamma} - t^{(\alpha\gamma)\beta} - t^{(\gamma\beta)\alpha}) \\ &\propto \big[\big(t^{(\alpha\beta)\gamma} - t^{(\alpha\gamma)\beta}\big) + \big(t^{(\alpha\beta)\gamma} - t^{(\beta\gamma)\alpha}\big)\big] \propto \big(t^{\alpha[\beta\gamma]} + t^{\beta[\alpha\gamma]}\big), \end{aligned} \tag{A.79}$$

which agrees with the result (A.76b). However, table (A.78) also provides the identity

$$t^{\alpha\beta\gamma} = t^{(\alpha\beta\gamma)} + t^{(\alpha[\beta)\gamma]} + t^{[\alpha(\beta]\gamma)} + t^{[\alpha\beta\gamma]}, \tag{A.80}$$

which reproduces the decomposition (A.76f).

To simplify decompositions such as (A.76), we use (Alfred) Young tableaux, which provide yet another alternative notation for representations of Lie groups [581, 168] and for the Young symmetrization mentioned in Construction A.1.

Construction A.2 (Young) *The fundamental, complex n-dimensional $SU(n)$ group representation is depicted by a box, $\Box$. A symmetric product Sym$(\mathbf{n}\otimes\mathbf{n})$ is depicted by placing two boxes next to each other: $\Box\Box$. An antisymmetric product* $\ker(\mathrm{Sym}\,(\mathbf{n}\otimes\mathbf{n}))$ *is depicted by placing two boxes one under the other: $\begin{smallmatrix}\Box\\\Box\end{smallmatrix}$. Therefore,*

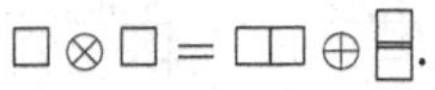

(A.81)

*A **Young tableau** is no more than n vertically stacked horizontal series of boxes, where:*

1. *all horizontal series being from the same position on the left,*
2. *no horizontal series has more boxes than the one above it;*
3. *a column of n boxes depicts the $SU(n)$-invariant tensor $\varepsilon^{\alpha_1\cdots\alpha_n}$, and may be deleted from the tableau.*

Example A.3 Decomposition (A.76f), i.e., (A.80) is then depicted as

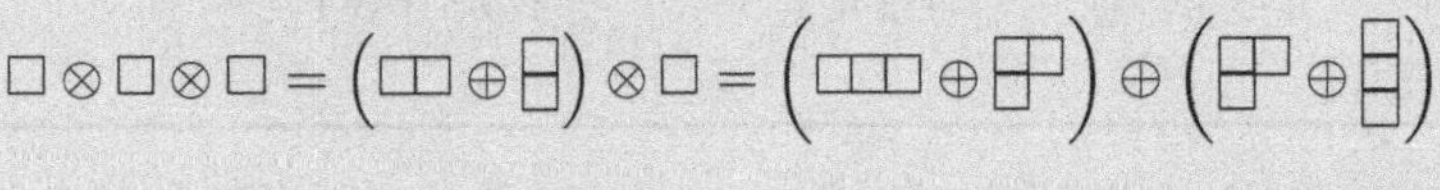

(A.82)

where multiplication and decomposition are performed iteratively, by attaching the right-hand box to the left-hand tableau in all possible and permitted ways.

For complete rules for multiplying arbitrary tableaux – and for all Lie groups – the interested Reader is directed to the literature [581, 168].

Example A.4 May it suffice here to list the following four examples:

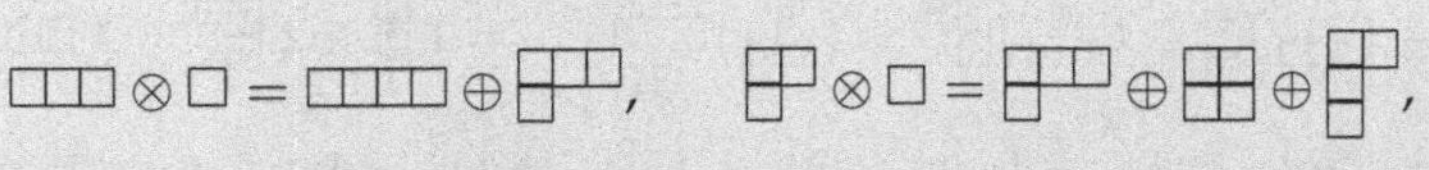

(A.83a)

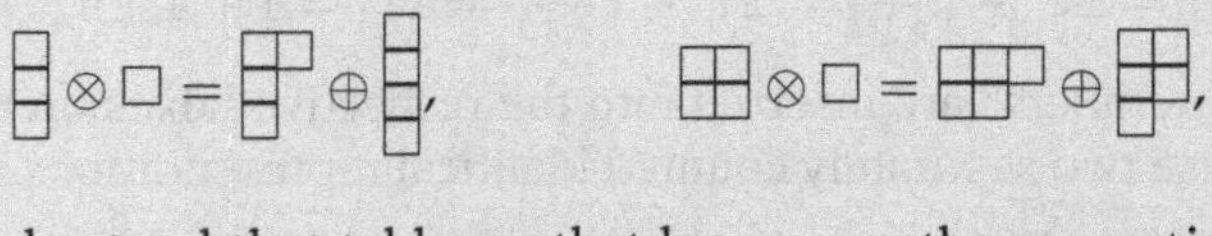

(A.83b)

where it is understood that tableaux that have more than n vertically stacked boxes are discarded for the $SU(n)$ group. Thus, in the third example, only the first tableau remains for $SU(3)$, but both summands remain for $SU(n)$, $n > 3$.

When the $SU(3)$ group structure is applied to the "flavor" of hadrons, the 3-dimensional representation, **3**, which is spanned by u-, d- and s-quarks, in the Young tableau notation, the quarks are depicted by a box and antiquarks with a column of two boxes. Then, it is clear that:

1. Mesons (bound states of a quark and an antiquark) are depicted by Young tableaux from the product $\begin{array}{|c|}\hline \\ \hline \\ \hline \end{array} \otimes \square$, i.e., $\mathbf{3}^* \otimes \mathbf{3} = \mathbf{1} \oplus \mathbf{8}$.
2. Baryons (bound states of tri quarks) are depicted by Young tableaux (A.82).
3. Other $SU(3)_f$ group representations can appear only in "exotic" bound states such as di-mesons ($\bar{q}\bar{q}qq$), di-baryons ($qqqqqq$), etc.

There exist two useful combinatorial formulae, for which we first need a function that associates to every box one more than the total number of boxes to the right and below a given box. Because of the geometric shape of the union of the counted boxes, this function is called the "hook number." In Young tableaux (A.84) the values of the "hook numbers" are inscribed into the boxes:

$$\begin{array}{|c|c|}\hline 3 & 1 \\ \hline 1 \\ \cline{1-1} \end{array}, \quad \begin{array}{|c|c|c|}\hline 4 & 2 & 1 \\ \hline 1 \\ \cline{1-1} \end{array}, \quad \begin{array}{|c|c|c|}\hline 5 & 3 & 1 \\ \hline 3 & 1 \\ \cline{1-2} 1 \\ \cline{1-1} \end{array}, \qquad \text{etc.} \tag{A.84}$$

Then, the representation depicted by the tableau YT appears

$$n_{YT} := \frac{N!}{\text{product of "hook numbers"}} \tag{A.85}$$

times in the tensor product $V^{\otimes N}$.

Example A.5 For the examples in the series (A.84) this formula yields:

$$\frac{3!}{1\cdot 3\cdot 1} = 2, \qquad \frac{4!}{1\cdot 2\cdot 4\cdot 1} = 3, \qquad \frac{6!}{1\cdot 3\cdot 5\cdot 1\cdot 3\cdot 1} = 16. \tag{A.86}$$

This number is also the dimension of the representation of the permutation group S_N represented by this Young tableau, so that $N! = \sum (n_{YT})^2$, with the sum extending over all the tableaux with N boxes.

Example A.6 For baryons represented as 3-quark bound states, we cited the fact that $\mathbf{3} \otimes \mathbf{3} \otimes \mathbf{3} = \mathbf{1} \oplus \mathbf{8} \oplus \mathbf{8} \oplus \mathbf{10}$; see the discussion around equation (2.40) and also in Section 4.4. The formula (A.85) then proves that there are two separately counted, 2-dimensional representations of permutation symmetries S_3:

$$\square\otimes\square\otimes\square = \left(\tfrac{3!}{3\cdot 2\cdot 1}\right)\begin{array}{|c|}\hline 3 \\ \hline 2 \\ \hline 1 \\ \hline \end{array} \oplus \left(\tfrac{3!}{3\cdot 1\cdot 1}\right)\begin{array}{|c|c|}\hline 3 & 1 \\ \hline 1 \\ \cline{1-1} \end{array} \oplus \left(\tfrac{3!}{3\cdot 2\cdot 1}\right)\begin{array}{|c|c|c|}\hline 3 & 2 & 1 \\ \hline \end{array} \quad = 1\cdot\begin{array}{|c|}\hline 3 \\ \hline 2 \\ \hline 1 \\ \hline \end{array} \oplus 2\cdot\begin{array}{|c|c|}\hline 3 & 1 \\ \hline 1 \\ \cline{1-1} \end{array} \oplus 1\cdot\begin{array}{|c|c|c|}\hline 3 & 2 & 1 \\ \hline \end{array}, \tag{A.87a}$$

$$3! = \left(\tfrac{3!}{3\cdot 2\cdot 1} = 1\right)^2 + \left(\tfrac{3!}{3\cdot 1\cdot 1} = 2\right)^2 + \left(\tfrac{3!}{3\cdot 2\cdot 1} = 1\right)^2 = 1 + 4 + 1, \tag{A.87b}$$

where the "hook numbers" are inserted into the respective boxes on the right, to aid the computation. These two separately counted identical representations in the middle of the expansion, depicted as $\begin{array}{|c|c|}\hline & \\ \hline \\ \cline{1-1} \end{array}$, have mixed symmetry and correspond to the baryon octets, **8**. In turn, the totally antisymmetric and the totally symmetric representations occur at the beginning and the end of the expansion, respectively; these are both 1-dimensional (unique) and occur once in the expansion.

For the second formula, for the $SU(n)$ group, inscribe into every row of boxes the ascending series of integers, starting with n in the top row, with $(n-1)$ in the second row and so on; these are called the "box numbers." The dimension of the $SU(n)$ representation depicted by the tableau YT is then given by the formula

$$d_{YT} = \frac{\text{product of "box numbers"}}{\text{product of "hook numbers"}}. \tag{A.88}$$

Example A.7 For the examples (A.84) and the $SU(4)$ group, we have the dimensions

$$\frac{(4{\cdot}5)(3)}{1{\cdot}3{\cdot}1} = 20, \qquad \frac{(4{\cdot}5{\cdot}6)(3)}{1{\cdot}2{\cdot}4{\cdot}1} = 45, \qquad \frac{(4{\cdot}5{\cdot}6)(3{\cdot}4)(2)}{1{\cdot}3{\cdot}5{\cdot}1{\cdot}3{\cdot}1} = 64, \tag{A.89}$$

while for the $SU(3)$ group, the same tableaux have the dimensions

$$\frac{(3{\cdot}4)(2)}{1{\cdot}3{\cdot}1} = 8, \qquad \frac{(3{\cdot}4{\cdot}5)(2)}{1{\cdot}2{\cdot}4{\cdot}1} = 15, \qquad \frac{(3{\cdot}4{\cdot}5)(2{\cdot}3)(1)}{1{\cdot}3{\cdot}5{\cdot}1{\cdot}3{\cdot}1} = 8. \tag{A.90}$$

Note that the formula for dimensions of $SU(n)$ tableaux (A.88) automatically returns zero if the tableau contains a column of more than n boxes: for $SU(2)$, the third tableau in the sequence (A.84) yields $\frac{(2\cdot3\cdot4)(1\cdot2)(0)}{1\cdot3\cdot5\cdot1\cdot3\cdot1} = 0$.

— ❦ —

The notational systems presented have their advantages but also their shortcomings:

1. The dimensional notation is unambiguous only for the $SU(2)$ group, and one must use additional "decorations" to distinguish the distinct representations that happen to have the same dimension.
2. The ket-notation is unambiguous, but requires specifying some complete collection of mutually commuting (Casimir) operators – such as J^2 and J_3 for $SU(2)$ – and their eigenvalues [☞ [488] for a list of Casimir operators].
3. The tensor/index notation is unambiguous, but the specification of the various symmetrization patterns using round parentheses and square brackets quickly becomes unwieldy and confusing.
4. The matrix notation requires ever bigger matrices.
5. Young tableaux are unambiguous and very compact, but products of arbitrary representations for some of the Lie groups may well require very complex rules [168].

Thus, in practice, one typically uses a combination of at least two notational systems, and so it is very important to know all the notational systems and how to successfully "translate" from any one into any other one of them.

A.4.3 Exercises for Section A.4

✎ ***A.4.1*** *Using the formula (A.88), compute the dimensions of the representations depicted by all the Young tableaux in the decomposition (A.81) and verify agreement for* $n = 3,4,5$.

✎ ***A.4.2*** *Using the formula (A.88), compute the dimensions of the representations depicted by all the Young tableaux in the decomposition (A.82) and verify agreement for $n = 3, 4, 5$.*

✎ ***A.4.3*** *Using the formula (A.88), compute the dimensions of the representations depicted by all the Young tableaux in the decomposition (A.83) and verify agreement for $n = 3, 4, 5$.*

A.5 Orthogonal and *Spin* groups

We have already encountered the rotation group $SO(3)$, and the Lorentz group $SO(1,3)$. In the general case, the group $SO(p,q)$ is the group of linear transformations of real $(p{+}q)$-dimensional vectors $(x^1, \ldots, x^{p+q})$, which preserve the bilinear scalar product [581, 260, 334]:

$$(\mathrm{x} \cdot \mathrm{y})_{p,q} := x^1 y^1 + \cdots x^p y^p - x^{p+1} y^{p+1} - \cdots x^{p+q} y^{p+q}. \tag{A.91}$$

This definition is equivalent to the statement that elements of the $SO(p,q)$ group may be represented as $(p{+}q) \times (p{+}q)$ matrices $\mathbf{L}$, which satisfy the requirement of the generalized orthogonality

$$\mathbf{L}^T \boldsymbol{\eta}_{(p,q)} \mathbf{L} \stackrel{!}{=} \boldsymbol{\eta}_{(p,q)} \qquad \Leftrightarrow \qquad \boldsymbol{\eta}_{(p,q)} \mathbf{L}^T \boldsymbol{\eta}_{(p,q)} \stackrel{!}{=} \mathbf{L}^{-1}, \tag{A.92}$$

where $\boldsymbol{\eta}_{(p,q)}$ is the diagonal matrix with the first p diagonal elements equal to $+1$, and the remaining q elements equal to -1. In the usual case $(p,q) = (1,3)$ and $\boldsymbol{\eta} := \boldsymbol{\eta}_{(1,3)}$.

A.5.1 Spinors

Just as Dirac constructed the spinorial representation $\{\hat{\mathrm{e}}_a \Psi^a\}$, starting from the 4-vector $\mathrm{p} = \hat{\mathrm{e}}^\mu p_\mu$ of the Lorentz group $SO(1,3)$, this can also be done for every $SO(p,q)$, and the $SO(p,q)$ transformation of those spinors is just as two-valued. Analogously to the double covering of the $SO(3)$ group one also defines the double covering of every $SO(p,q)$ group, denoted $Spin(p,q)$. As representations of the $Spin(p,q)$ group, both tensors and spinors are single-valued functions. The algebra of the $Spin(p,q)$ group is denoted $\mathfrak{spin}(p,q)$, and it is worth knowing that $\mathfrak{spin}(p,q) = \mathfrak{spin}(p{+}q,0) = \mathfrak{spin}(p{+}q)$. In other words, for a fixed $p{+}q$, different $Spin(p,q)$ groups differ only in the "finite part" [☞ Definition A.1 on p. 454, and Comment A.1 on p. 454] and their algebras are identical. For $p{+}q \leqslant 6$, there exist additional identities among algebras:

$$\mathfrak{spin}(3) = \mathfrak{su}(2), \quad \mathfrak{spin}(4) = \mathfrak{su}(2) \oplus \mathfrak{su}(2), \quad \mathfrak{spin}(5) = \mathfrak{sp}(4), \quad \mathfrak{spin}(6) = \mathfrak{su}(4), \tag{A.93}$$

Table A.4 Some low-dimensional $(p{+}q \leqslant 6)$ spin groups; $Spin(p,q) = Spin(q,p)$ [☞ Section A.1.4]

$Spin(1) \cong O(1) \cong \mathbb{Z}_2$	$Spin(2,2) \cong SU(1,1) \times SU(1,1)$
$Spin(2) \cong U(1) \cong SO(2)$	$Spin(5) \cong Sp(2)$
$Spin(1,1) \cong GL(1;\mathbb{R})$	$Spin(1,4) \cong Sp(1,1)$
$Spin(3) \cong SU(2) \cong Sp(1) \cong SL(1;\mathbb{H})$	$Spin(2,3) \cong Sp(2;\mathbb{R})$
$Spin(1,2) \cong SU(1,1)$	$Spin(6) \cong SU(4)$
$Spin(4) \cong SU(2) \times SU(2)$	$Spin(1,5) \cong SL(2;\mathbb{H})$
$Spin(1,3) \cong SL(2;\mathbb{C}) \cong Spin(3;\mathbb{C})$	$Spin(2,4) \cong SU(2,2)$
$SO^{\uparrow}(1,3) \cong SO(3;\mathbb{C})$	$Spin(3,3) \cong SL(4;\mathbb{R})$

$Spin(p,q)$ groups for $p{+}q > 6$ are not isomorphic to other Lie groups.

which imply identities between the corresponding groups. Spin groups are defined as double coverings of orthogonal groups, i.e., the general relation

$$SO(p,q) = Spin(p,q)/\mathbb{Z}_2. \tag{A.94}$$

For physics applications, the practical meaning of this relation is that the multiplicative group $\mathbb{Z}_2 = \{\mathbb{1}, -\mathbb{1}\}$ is a subgroup of $Spin(p,q)$. Tensorial representations do not transform under this $\mathbb{Z}_2$-action, while spinorial ones change their sign under the action of $-\mathbb{1} \in \mathbb{Z}_2$. This sign equals $(-1)^{\mathrm{F}}$, where F is the so-called "fermion number" defined in the text leading to equations (10.44): $\mathrm{F} = 0$ for bosons and $\mathrm{F} = 1$ for fermions. Thus, spinorial representations of the $Spin(p,q)$ group are double-valued with respect to the $SO(p,q)$-action, and are not "true" functions; tensorial representations are single-valued under both the $Spin(p,q)$- as well as the $SO(p,q)$-action. May it suffice here to quote without proof [565, 258, 581, 256, 80, 260, 333, 447]:

> **Theorem A.4** *If for two groups, G_1 and G_2, it is true that $G_1 = G_2/H$, then $H \subset G_2$ is a subgroup of G_2, and elements of G_1 are obtained by identifying those elements from G_2 that differ only by the action of the subgroup $H \subset G_2$. Besides, the representations of G_1 are H-invariant representations of G_2.*

The relation $SO(p,q) = Spin(p,q)/\mathbb{Z}_2$ then implies that the $SO(p,q)$ representations are $\mathbb{Z}_2$-invariant $Spin(p,q)$ representations – and those are the tensors, the fermion number of which is $\mathrm{F} = 0$.

Spinors are, however, the $Spin(p,q)$ representations that are not invariant with respect to the action of the subgroup $\mathbb{Z}_2 \subset Spin(p,q)$ – the spinors' fermion number is $\mathrm{F} = 1$ and they change their sign pod under the action of the nontrivial $\mathbb{Z}_2$ element. Let $g_+, g_- \in Spin(p,q)$ be group elements that differ only by the $\mathbb{Z}_2$-action, so g_+ does not change its sign while g_- does. Since the relation $SO(p,q) = Spin(p,q)/\mathbb{Z}_2$ implies that the group elements $g \in SO(p,q)$ are obtained by identifying $g := [g_+ \simeq g_-]$, it is clear that the $SO(p,q)$-action upon spinors is *double-valued.*

A.5.2 $Spin(1,3)$

In relativistic physics, what is physically relevant is not the Euclidean length in spacetime, but the *interval*, of the form $\sqrt{(x^0)^2 - (x^1)^2 - \cdots}$. Thus, for relativistic physics purposes, we are most often interested in Lorentz groups $SO(1,n)$ and their double coverings, $Spin(1,n)$, where n is the number of spatial dimensions. The algebras of these groups are the same as for their Euclidean counterparts, so the identities (A.93) may be used, but it is important to keep in mind that the group $Spin(1,n)$ differ from $Spin(1{+}n)$; see Table A.4 on p. 482.

From Table A.4 on p. 482, we have that

$$Spin(1,3) = SL(2,\mathbb{C}), \tag{A.95}$$

where $SL(2,\mathbb{C})$ denotes the group of complex $2{\times}2$ matrices of unit determinant. This group is generated by

$$\boldsymbol{\tau}_j := \tfrac{1}{2}\sigma^j \quad \text{and} \quad \widetilde{\boldsymbol{\tau}}_j := \tfrac{i}{2}\sigma^j, \qquad j = 1,2,3, \tag{A.96}$$

the nonzero commutation relations of which are

$$[\boldsymbol{\tau}_j, \boldsymbol{\tau}_k] = i\varepsilon_{jk}{}^m\,\boldsymbol{\tau}_m, \qquad [\boldsymbol{\tau}_j, \widetilde{\boldsymbol{\tau}}_k] = i\varepsilon_{jk}{}^m\,\widetilde{\boldsymbol{\tau}}_m, \qquad [\widetilde{\boldsymbol{\tau}}_j, \widetilde{\boldsymbol{\tau}}_k] = -i\varepsilon_{jk}{}^m\,\boldsymbol{\tau}_m. \tag{A.97}$$

On the other hand, starting from relation (5.45):

$$[\,\boldsymbol{\gamma}^{\mu\nu}, \boldsymbol{\gamma}^{\rho\sigma}\,] = \eta^{\mu\rho}\boldsymbol{\gamma}^{\nu\sigma} - \eta^{\mu\sigma}\boldsymbol{\gamma}^{\nu\rho} + \eta^{\nu\sigma}\boldsymbol{\gamma}^{\mu\rho} - \eta^{\nu\rho}\boldsymbol{\gamma}^{\mu\sigma}. \tag{A.98}$$

This shows that $J_j := \frac{1}{2i}\varepsilon_{jkl}\gamma^{kl}$ with $j,k,l = 1,2,3$ satisfy the $\mathfrak{su}(2) = \mathfrak{so}(3)$ subalgebra:

$$[J_1, J_2] = [(-i\gamma^{23}), (-i\gamma^{31})] = -\eta^{23}\gamma^{31} + \eta^{21}\gamma^{33} - \eta^{31}\gamma^{23} + \eta^{33}\gamma^{21} = (-1)(-iJ_3) \\ = iJ_3, \tag{A.99}$$

and so forth, for the remaining two permutations, $[J_2, J_3]$ and $[J_3, J_1]$. Denote the remaining elements $K_j := i\gamma^{0j}$, and find

$$[K_1, K_2] = [i\gamma^{01}, i\gamma^{02}] = -\eta^{00}\gamma^{12} + \eta^{02}\gamma^{10} - \eta^{12}\gamma^{00} + \eta^{10}\gamma^{02} = -(+1)(+iJ_3) \\ = -iJ_3, \tag{A.100}$$

and so forth, for the remaining two permutations, $[K_2, K_3]$ and $[K_3, K_1]$. Finally, the mixed commutators yield

$$[J_1, K_1] = [(-i\gamma^{23}), i\gamma^{01}] = \eta^{20}\gamma^{31} - \eta^{21}\gamma^{30} + \eta^{31}\gamma^{20} - \eta^{30}\gamma^{21} = 0, \tag{A.101}$$

$$[J_1, K_2] = [(-i\gamma^{23}), i\gamma^{02}] = +\eta^{20}\gamma^{32} - \eta^{22}\gamma^{30} + \eta^{32}\gamma^{20} - \eta^{30}\gamma^{22} = -(-1)(iK_3) \\ = iK_3, \tag{A.102}$$

and so forth, for the remaining two permutations, $[K_2, K_3]$ and $[K_3, K_1]$. We thus have the general structure of commutators:

$$[J_j, J_k] = i\varepsilon_{jk}{}^{m} J_m, \qquad [J_j, K_k] = i\varepsilon_{jk}{}^{m} K_m, \qquad [K_j, K_k] = -i\varepsilon_{jk}{}^{m} J_m, \tag{A.103}$$

which are identical in form to the relations (A.97). This shows that the groups $SL(2,\mathbb{C})$ and $Spin(1,3)$, and thus also $SO(1,3) \cong Spin(1,3)/\mathbb{Z}_2$, have identical algebras.

Finally, define

$$M_j := \tfrac{1}{2}J_j + \tfrac{i}{2}K_j \qquad \text{and} \qquad \overline{M}_j := \tfrac{1}{2}J_j - \tfrac{i}{2}K_j, \tag{A.104}$$

and find

$$[M_j, M_k] = i\varepsilon_{jk}{}^{m} M_m, \qquad [\overline{M}_j, \overline{M}_k] = i\varepsilon_{jk}{}^{m} \overline{M}_m, \qquad [M_j, \overline{M}_k] = 0, \tag{A.105}$$

which demonstrates that

$$\mathfrak{alg}(SL(2;\mathbb{C})) = \mathfrak{alg}(Spin(1,3)) = \mathfrak{alg}(Spin(3;\mathbb{C})) = \mathfrak{alg}(SO(1,3)) \\ = \mathfrak{su}(2)_L \oplus \mathfrak{su}(2)_R. \tag{A.106}$$

In the physics literature one sometimes comes across the statement that the Lorentz *group* is isomorphic (or even equals) the product $SU(2)_L \times SU(2)_R$, which is false. Luckily, the precise details of the Lorentz *group* $Spin(1,3)$ and its precise relationship with the groups $SU(2)_L$ and $SU(2)_R$ generated by the operators M_j and $\overline{M}_j$ are usually not relevant, and the relation (A.5.2) for the corresponding algebras suffices.

Note that the discrete operations P and T generate the "finite part" of the $O(1,3)$ group, the group of real 4×4-matrix transformations of spacetime 4-vectors that preserve the relativistic interval. The action of the P and T transformations may then also be represented in the form of 4×4-matrices:[18]

$$P = \begin{bmatrix} 1 & 0 & 0 & 0 \\ 0 & -1 & 0 & 0 \\ 0 & 0 & -1 & 0 \\ 0 & 0 & 0 & -1 \end{bmatrix} \quad \text{and} \quad T = \begin{bmatrix} -1 & 0 & 0 & 0 \\ 0 & 1 & 0 & 0 \\ 0 & 0 & 1 & 0 \\ 0 & 0 & 0 & 1 \end{bmatrix}, \quad \text{so} \quad PT = -\mathbb{1}. \tag{A.107}$$

[18] Caution: the matrix representation of the operations P and T evidently describes linear operations. However, in quantum theory the operation T is *anti-linear* and its action cannot be represented this way.

The definition of the $O(1,3)$ group does not include the requirement of a unit determinant, but orthogonality implies that the determinant of $O(1,3)$-matrices equals ±1. Elements with the determinant -1 do not form a group, as they exclude the identity element, while the elements with determinant $+1$ do form the $SO(1,3)$ group, which then is evidently a subgroup of $O(1,3)$.

The physical meaning of Lorentz transformations requires that the direction of the flow of time remains unchanged. Such transformations form a subgroup of $SO(1,3)$, which is called the orthochronous Lorentz group,[19] denoted $L^{\uparrow} \equiv SO^{\uparrow}(1,3)$. It may be shown that this is a connected group, i.e., every element of the orthochronous Lorentz group may be continuously "shrunk" to the identity element: Every Lorentz transformation may be factorized into the product of three rotations and three Lorentz boosts, akin to the well-known factorization of every rotation into three Euler angle rotations. Each of those six parameters, three angles and three components of velocity, may be continuously shrunk to 0, whereby every orthochronous Lorentz transformation may be continuously shrunk to $\mathbb{1}$.

Denote by $TL^{\uparrow}$ the collection of all products of elements from $L^{\uparrow}$ with the element T; since $L^{\uparrow}$ is continuously connected, so is $TL^{\uparrow}$. The analogy holds for $PL^{\uparrow}$ and for $PTL^{\uparrow}$. It should be evident that the $TL^{\uparrow}$, $PL^{\uparrow}$ and $PTL^{\uparrow}$ components cannot be continuously turned into $\mathbb{1}$, nor can an element of one of these three components be continuously turned into an element of another component. It then follows that the $O(1,3)$ group is a disconnected union of four components: $L^{\uparrow}$, $TL^{\uparrow}$, $PL^{\uparrow}$ and $PTL^{\uparrow}$, and that the disconnected unions $L^{\uparrow}$ and $PTL^{\uparrow}$ form a subgroup $SO(1,3) \subset O(1,3)$.

A.5.3 The Poincaré algebra and group in 1+3-dimensional spacetime

Transformations of the tangent space of 1+3-dimensional spacetime are linear transformations of the space $\mathbb{R}^{1,3}$, of the form

$$x^\mu \to y^\mu = \mathrm{L}^\mu{}_\nu\, x^\nu + \xi^\mu, \tag{A.108}$$

where the matrix $\mathbf{L} = [\mathrm{L}^\mu{}_\nu]$ provides the Lorentz transformations of 4-vectors in (flat) spacetime, and the 4-vector ξ^μ parametrizes translations in spacetime. These transformations have an *induced* action of functions of spacetime, by means of the differential operators

$$x^\mu \to x^\mu + \xi^\mu \quad \Rightarrow \quad f(\mathbf{x}) \to f(\mathbf{x}+\xi) = \exp\{\xi^\mu \partial_\mu\} f(\mathbf{x}); \tag{A.109}$$

$$x^\mu \to \mathrm{L}^\mu{}_\nu\, x^\nu \quad \Rightarrow \quad f(\mathbf{x}) \to f(\mathbf{L}\mathbf{x}) = \exp\{\lambda_\mu{}^\nu \mathrm{L}^\mu{}_\nu\} f(\mathbf{x}). \tag{A.110}$$

The translation generators are then differential operators ∂_μ, and the Lorentz transformation generators are

$$\mathrm{L}^\mu{}_\nu = x^\mu \partial_\nu - \eta^{\mu\rho}\eta_{\nu\sigma} x^\sigma \partial_\rho, \tag{A.111a}$$

so that:

$$\textbf{boost}\quad \mathrm{L}^0{}_i = x^0\partial_i - \eta^{00}\eta_{ij}x^j\partial_0 = ct\tfrac{\partial}{\partial x^i} + \delta_{ij}x^j\tfrac{1}{c}\tfrac{\partial}{\partial t}, \tag{A.111b}$$

$$\begin{aligned} \mathrm{L}^i{}_0 &= x^i\partial_0 - \eta^{ij}\eta_{00}x^0\partial_j = x^i\tfrac{1}{c}\tfrac{\partial}{\partial t} + \delta^{ij}ct\tfrac{\partial}{\partial x^j} = \delta^{ij}\big(ct\tfrac{\partial}{\partial x^j} + \delta_{jk}x^k\tfrac{1}{c}\tfrac{\partial}{\partial t}\big) \\ &= \delta^{ij}\,\mathrm{L}^0{}_j, \end{aligned} \tag{A.111c}$$

$$\textbf{rot.}\quad \mathrm{L}^i{}_j = x^i\partial_j - \eta^{ik}\eta_{j\ell}x^\ell\partial_k = \eta^{ik}\big(\eta_{kn}x^n\tfrac{\partial}{\partial x^j} - \eta_{jn}x^n\tfrac{\partial}{\partial x^k}\big)_i = \varepsilon^i{}_{jk}\,\varepsilon^{k\ell}{}_n\,x^n\tfrac{\partial}{\partial x^\ell}. \tag{A.111d}$$

[19] The nomenclature here is not quite standard: some Authors call the full $O(1,3)$ group the Lorentz group while others reserve this name only for the orthochronous component, $SO^{\uparrow}(1,3)$, of the $SO(1,3)$ group.

For example,

$$\begin{aligned}
\exp\{\xi^\mu\partial_\mu\}f(\mathrm{x}) &= \sum_{k=0}^{\infty}\frac{1}{k!}\bigg(\Big[\prod_{i=1}^{k}\xi^{\mu_i}\frac{\partial}{\partial y^{\mu_i}}\Big]f(\mathrm{y})\bigg)_{\mathrm{y}\to\mathrm{x}}\\
&= \frac{1}{0!}f(\mathrm{x}) + \frac{1}{1!}\xi^\mu\Big(\frac{\partial f(\mathrm{y})}{\partial y^\mu}\Big)_{\mathrm{y}\to\mathrm{x}} + \frac{1}{2!}\xi^\mu\xi^\nu\Big(\frac{\partial^2 f(\mathrm{y})}{\partial y^\mu\partial y^\nu}\Big)_{\mathrm{y}\to\mathrm{x}} + \cdots\\
&= f(\mathrm{x}+\xi).
\end{aligned}\tag{A.112}$$

Translations in 1+3-dimensional space $\mathbb{R}^{1,3}$ commute and are parametrized by the 4-vector $\xi^\mu \in \mathbb{R}^{1,3}$. As the operators $\frac{\partial}{\partial x^\mu}$ also span the vector space $\mathbb{R}^{1,3}$, we may write that $\mathfrak{tr}(\mathbb{R}^{1,3}_{\mathrm{x}}) \cong \mathbb{R}^{1,3}_{\partial}$, where $\mathfrak{tr}(\mathscr{X})$ denotes "the algebra of translations in space $\mathscr{X}$." It is not hard to verify that

$$[\lambda\cdot\mathrm{L},\,\lambda'\cdot\mathrm{L}] = \lambda''\cdot\mathrm{L}, \qquad [\lambda\cdot\mathrm{L},\,\xi\cdot\partial] = \xi'\cdot\partial, \qquad [\xi\cdot\partial,\,\xi'\cdot\partial] = 0, \tag{A.113a}$$

$$\lambda\cdot\mathrm{L} := \lambda_\mu{}^\nu\mathrm{L}^\mu{}_\nu, \qquad \xi\cdot\partial := \xi^\mu\partial_\mu, \tag{A.113b}$$

so that the Poincaré algebra is $\mathfrak{po}(1,3) = \mathfrak{spin}(1,3) \niplus \mathfrak{tr}(\mathbb{R}^{1,3})$, and the Poincaré group is $Po(1,3) = Spin(1,3) \ltimes \mathbb{R}^{1,3}$, where the asymmetric binary symbol $\niplus$ ($\ltimes$) denotes the semidirect sum (product) and recalls the fact that the left-hand summand (factor) acts upon the right-hand one [☞ the lexicon entry, in Appendix B.1].

A.5.4 Exercises for Section A.5

✎ ***A.5.1*** *Verify equations (A.103) by explicit computation, using however only the definitions (5.45).*

✎ ***A.5.2*** *Verify equations (A.105), using the definitions (A.104) and the previous results.*

✎ ***A.5.3*** *Verify equations (A.113) by explicit computation, using however only the definitions (A.111).*

✎ ***A.5.4*** *Using the definitions (A.111) and your results in the above problems, reconstruct the differential operator representation of the operators* $\overline{M}_j$ *and* $\overline{M}_k$.

A.6 Spinors and Dirac γ-matrices

$SO(p,q)$ denotes the group of homogeneous and linear transformations of (p,q)-vectors $\vec{v}$ that preserve the bilinear product

$$\vec{v}\cdot\vec{u} := \sum_{i=1}^{p} v_i u_i - \sum_{i=p+1}^{p+q} v_i u_i \stackrel{!}{=} \sum_{i=1}^{p+q} (\mathbb{M}_i{}^k u_k)\cdot\eta^{ij}(\mathbb{M}_j{}^\ell u_\ell), \tag{A.114}$$

where the $(p{+}q)\times(p{+}q)$ matrices $\mathbb{M}$ have a unit determinant and where

$$[\eta^{ij}] = \boldsymbol{\eta}^{(p,q)} = \mathrm{diag}(\underbrace{+1,\ldots,+1}_{p},\underbrace{-1,\ldots,-1}_{q}). \tag{A.115}$$

The vectors $\vec{v}$ form the defining vector space V_v. One also writes $SO(p,q) = SO(V_v;\boldsymbol{\eta}^{(p,q)})$, where the latter notation quite literally stands for "the group of unimodular orthogonal transformations of the vector space V_v that preserve the bilinear product obtained using the matrix $\boldsymbol{\eta}^{(p,q)}$."

Every (unimodular orthogonal) group $SO(p,q)$ has a double covering (that is also a group), denoted $Spin(p,q)$ [333], the single-valued spinorial representations of which are double-valued representations of the $SO(p,q)$ group. Every $Spin(p,q)$ group has the (Dirac) spinor representation V_Ψ as well as its formally dual representation $(V_\Psi)^* = V_{\overline{\Psi}}$, for which the relation

$$V_\Psi \otimes V_{\overline{\Psi}} \supset V_v, \tag{A.116}$$

holds, where $V_v\ (p,q)$ is the defining vector representation of $SO(p,q)$. For any chosen bases,

$$\hat{e}_A \in V_\Psi, \qquad \hat{e}^A \in V_{\overline{\Psi}}, \qquad \text{and} \qquad \hat{e}^\mu \in V_v, \tag{A.117}$$

and the Dirac $\boldsymbol{\gamma}$-matrices are arrays of the coefficients in the projection (A.116):

$$\hat{e}_A (\boldsymbol{\gamma}^\mu)^A{}_B\, \hat{e}^B = \hat{e}^\mu. \tag{A.118}$$

A.6.1 Dirac matrices in (3+1) dimensional spacetime

The elements $\boldsymbol{\gamma}^\mu$, $\mu = 0,1,2,3$, which satisfy

$$\{\boldsymbol{\gamma}^\mu, \boldsymbol{\gamma}^\nu\} = 2\eta^{\mu\nu}, \qquad \text{with} \qquad [\eta^{\mu\nu}] = \text{diag}(+1,-1,-1,-1), \tag{A.119}$$

form the Clifford algebra $\mathfrak{Cl}(1,3)$. Following Feynman, one defines

$$\not{p} := \boldsymbol{\gamma}^\mu p_\mu \qquad \text{for each 4-vector p.} \tag{A.120}$$

This implies the following definitions and results:

$$\widehat{\boldsymbol{\gamma}} := i\boldsymbol{\gamma}^0\boldsymbol{\gamma}^1\boldsymbol{\gamma}^2\boldsymbol{\gamma}^3 := \tfrac{i}{4!}\varepsilon_{\mu\nu\rho\sigma}\boldsymbol{\gamma}^\mu\boldsymbol{\gamma}^\nu\boldsymbol{\gamma}^\rho\boldsymbol{\gamma}^\sigma, \qquad \{\widehat{\boldsymbol{\gamma}}, \boldsymbol{\gamma}^\mu\} = 0, \qquad (\widehat{\boldsymbol{\gamma}})^2 = \mathbb{1}; \tag{A.121a}$$

$$\boldsymbol{\gamma}_\pm := \tfrac{1}{2}[\mathbb{1} \pm \widehat{\boldsymbol{\gamma}}], \qquad [\boldsymbol{\gamma}_+, \boldsymbol{\gamma}_-] = 0, \qquad \boldsymbol{\gamma}_+ + \boldsymbol{\gamma}_- = \mathbb{1}, \qquad (\boldsymbol{\gamma}_\pm)^2 = \boldsymbol{\gamma}_\pm, \tag{A.121b}$$

$$\boldsymbol{\gamma}^{\mu\nu} := \frac{i}{4}[\boldsymbol{\gamma}^\mu, \boldsymbol{\gamma}^\nu], \qquad [\boldsymbol{\gamma}^{\mu\nu}, \boldsymbol{\gamma}^{\rho\sigma}] = \eta^{\mu\rho}\boldsymbol{\gamma}^{\nu\sigma} - \eta^{\mu\sigma}\boldsymbol{\gamma}^{\nu\rho} + \eta^{\nu\sigma}\boldsymbol{\gamma}^{\mu\rho} - \eta^{\nu\rho}\boldsymbol{\gamma}^{\mu\sigma}. \tag{A.121c}$$

$$\boldsymbol{\gamma}^\mu\boldsymbol{\gamma}_\mu = 4\mathbb{1}, \qquad \boldsymbol{\gamma}^\mu\boldsymbol{\gamma}^\nu\boldsymbol{\gamma}^\rho\boldsymbol{\gamma}_\mu = 4\boldsymbol{\gamma}^\nu\boldsymbol{\gamma}^\rho, \tag{A.122a}$$

$$\boldsymbol{\gamma}^\mu\boldsymbol{\gamma}^\nu\boldsymbol{\gamma}_\mu = -2\boldsymbol{\gamma}^\nu, \qquad \boldsymbol{\gamma}^\mu\boldsymbol{\gamma}^\nu\boldsymbol{\gamma}^\rho\boldsymbol{\gamma}^\sigma\boldsymbol{\gamma}_\mu = -2\boldsymbol{\gamma}^\nu\boldsymbol{\gamma}^\rho\boldsymbol{\gamma}^\sigma, \tag{A.122b}$$

$$\boldsymbol{\gamma}^\mu\boldsymbol{\gamma}^\nu\boldsymbol{\gamma}^\rho = \eta^{\mu\nu}\boldsymbol{\gamma}^\rho - \eta^{\mu\rho}\boldsymbol{\gamma}^\nu + \eta^{\nu\rho}\boldsymbol{\gamma}^\mu + i\varepsilon^{\mu\nu\rho\sigma}\boldsymbol{\gamma}_\sigma\widehat{\boldsymbol{\gamma}}\ . \tag{A.122c}$$

Theorem A.5 *Owing to the relations (A.119), (A.121a) and (A.122c), it follows that every $\boldsymbol{\gamma}$-matrix polynomial may be reduced to the quadratic polynomial*

$$C_0\mathbb{1} + C_\mu\boldsymbol{\gamma}^\mu + \tfrac{1}{2}C_{\mu\nu}\boldsymbol{\gamma}^{\mu\nu} + \widehat{C}_\mu\boldsymbol{\gamma}^\mu\widehat{\boldsymbol{\gamma}} + \widehat{C}_0\widehat{\boldsymbol{\gamma}}. \tag{A.123}$$

That is, the basis

$$\mathbb{1},\ \boldsymbol{\gamma}^\mu,\ \boldsymbol{\gamma}^{\mu\nu},\ \boldsymbol{\gamma}^\mu\widehat{\boldsymbol{\gamma}},\ \widehat{\boldsymbol{\gamma}} \tag{A.124}$$

for the Dirac algebra (A.119) is complete.

We also have

$$\text{Tr}[\boldsymbol{\gamma}^\mu] = 0, \qquad \text{Tr}[\boldsymbol{\gamma}^\mu\boldsymbol{\gamma}^\nu\boldsymbol{\gamma}^\rho] = 0, \qquad \text{Tr}[\boldsymbol{\gamma}^\mu\boldsymbol{\gamma}^\nu\boldsymbol{\gamma}^\rho\boldsymbol{\gamma}^\sigma\boldsymbol{\gamma}^\lambda] = 0, \qquad \text{etc.} \tag{A.125a}$$

$$\text{Tr}[\boldsymbol{\gamma}^\mu\boldsymbol{\gamma}^\nu] = 4\eta^{\mu\nu}, \qquad \text{Tr}[\boldsymbol{\gamma}^\mu\boldsymbol{\gamma}^\nu\boldsymbol{\gamma}^\rho\boldsymbol{\gamma}^\sigma] = 4(\eta^{\mu\nu}\eta^{\rho\sigma} - \eta^{\mu\rho}\eta^{\nu\sigma} + \eta^{\mu\sigma}\eta^{\nu\rho}), \tag{A.125b}$$

$$\text{Tr}[\widehat{\boldsymbol{\gamma}}] = 0, \qquad \text{Tr}[\boldsymbol{\gamma}^\mu\boldsymbol{\gamma}^\nu\widehat{\boldsymbol{\gamma}}] = 0, \qquad \text{Tr}[\boldsymbol{\gamma}^\mu\boldsymbol{\gamma}^\mu\boldsymbol{\gamma}^\rho\boldsymbol{\gamma}^\sigma\widehat{\boldsymbol{\gamma}}] = -4i\varepsilon^{\mu\nu\rho\sigma}. \tag{A.125c}$$

These relations imply

$$\not{p}\not{p} = p^2\,\mathbb{1}, \qquad \not{p}\not{q} = (p{\cdot}q - 2i p_\mu \gamma^{\mu\nu} q_\nu)\mathbb{1}; \tag{A.126a}$$
$$\not{p}\not{q} + \not{q}\not{p} = 2(p{\cdot}q)\,\mathbb{1}, \qquad \not{p}\not{q} - \not{q}\not{p} = -4i(p_\mu \gamma^{\mu\nu} q_\nu)\mathbb{1}; \tag{A.126b}$$
$$\mathrm{Tr}[\not{p}\not{q}] = 4p{\cdot}q\,\mathbb{1}, \qquad \mathrm{Tr}[\not{p}\not{q}\not{r}\not{s}] = 4[(p{\cdot}q)(r{\cdot}s) - (p{\cdot}r)(q{\cdot}s) + (p{\cdot}s)(q{\cdot}r)]; \tag{A.126c}$$
$$\mathrm{Tr}[\not{p}] = 0 = \mathrm{Tr}[\not{p}\not{q}\not{r}], \quad \mathrm{Tr}[\hat{\gamma}\not{p}\not{q}\not{r}\not{s}] = 4i\varepsilon^{\mu\nu\rho\sigma} p_\mu\, q_\nu\, r_\rho\, s_\sigma; \tag{A.126d}$$
$$\gamma^\mu \not{p}\not{q}\gamma_\mu = 4p{\cdot}q\,\mathbb{1}, \qquad \gamma^\mu \not{p}\gamma_\mu = -2\not{p}, \qquad \gamma^\mu \not{p}\not{q}\not{r}\gamma_\mu = -2\not{r}\not{q}\not{p}. \tag{A.126e}$$

In physics applications, besides the relations (A.119) that define the Clifford algebra, one *additionally* requires the matrices γ^μ to satisfy

$$(\gamma^0)^\dagger = \gamma^0, \quad \text{and} \quad (\gamma^i)^\dagger = -\gamma^i,\ i = 1,2,3, \qquad \Leftrightarrow \qquad (\gamma^\mu)^\dagger = \gamma^0\gamma^\mu\gamma^0. \tag{A.127}$$

This requirement is not an integral part of the definition and structure of Clifford algebras, which one must keep in mind when using mathematical results about Clifford algebras. The use of the algebra (A.119) in the physics literature always assumes the additional conditions (A.127) – as well as their consequences.

Corresponding to Dirac conjugation of spin-$\frac{1}{2}$ fermions (5.49), we have

$$\overline{\Psi} := \Psi^\dagger\gamma^0 \qquad \Leftrightarrow \qquad \overline{\gamma^\mu} := \gamma^0(\gamma^\mu)^\dagger\gamma^0 \overset{(A.127)}{=} \gamma^\mu. \tag{A.128}$$

Therefore,

$$\begin{aligned}\overline{\hat{\gamma}} &:= \gamma^0(i\gamma^0\gamma^1\gamma^2\gamma^3)^\dagger\gamma^0 = -i\gamma^0(\gamma^3)^\dagger\gamma^0\gamma^0(\gamma^2)^\dagger\gamma^0\gamma^0(\gamma^1)^\dagger\gamma^0\gamma^0(\gamma^0)^\dagger\gamma^0 = -i\overline{\gamma^3}\,\overline{\gamma^2}\,\overline{\gamma^1}\,\overline{\gamma^0}\\ &\overset{(A.127)}{=} -i\gamma^3\gamma^2\gamma^1\gamma^0 \overset{(A.119)}{=} -i\gamma^0\gamma^1\gamma^2\gamma^3 = -\hat{\gamma},\end{aligned} \tag{A.129}$$

and so

$$\overline{\gamma_\pm} = \gamma_\mp, \qquad \text{whereby} \qquad \overline{\Psi_\pm} = \overline{\Psi}_\mp. \tag{A.130}$$

Besides the Dirac basis:

$$\gamma^0 = \begin{bmatrix}\mathbb{1} & \mathbb{O}\\ \mathbb{O} & -\mathbb{1}\end{bmatrix}, \quad \gamma^i = \begin{bmatrix}\mathbb{O} & \sigma^i\\ -\sigma^i & \mathbb{O}\end{bmatrix}, \quad \hat{\gamma} = \begin{bmatrix}\mathbb{O} & \mathbb{1}\\ \mathbb{1} & \mathbb{O}\end{bmatrix}, \tag{A.131}$$

the most often used choices are the Weyl basis:

$$\gamma^0 = \begin{bmatrix}\mathbb{O} & -\mathbb{1}\\ -\mathbb{1} & \mathbb{O}\end{bmatrix}, \quad \gamma^i = \begin{bmatrix}\mathbb{O} & \sigma^i\\ -\sigma^i & \mathbb{O}\end{bmatrix}, \quad \hat{\gamma} = \begin{bmatrix}\mathbb{1} & \mathbb{O}\\ \mathbb{O} & -\mathbb{1}\end{bmatrix}, \quad \Psi_{\text{Dirac}} = \begin{bmatrix}\Psi_+\\ \Psi_-\end{bmatrix}; \tag{A.132}$$

and the Majorana basis:

$$\begin{aligned}&\gamma^0 = \begin{bmatrix}\mathbb{O} & \sigma^2\\ \sigma^2 & \mathbb{O}\end{bmatrix}, \quad \gamma^1 = \begin{bmatrix}i\sigma^3 & \mathbb{O}\\ \mathbb{O} & i\sigma^3\end{bmatrix}, \quad \gamma^2 = \begin{bmatrix}\mathbb{O} & -\sigma^2\\ \sigma^2 & \mathbb{O}\end{bmatrix}, \quad \gamma^3 = \begin{bmatrix}-i\sigma^1 & \mathbb{O}\\ \mathbb{O} & -i\sigma^1\end{bmatrix},\\ &\hat{\gamma} = \begin{bmatrix}\sigma^2 & \mathbb{O}\\ \mathbb{O} & \sigma^2\end{bmatrix},\end{aligned} \tag{A.133}$$

in which all components of the Dirac spinor Ψ are real, while the Dirac matrices themselves are all imaginary in the Majorana basis.

A.6.2 Weyl's notation for spinors

The literature about supersymmetry [☞ [189, 562, 560, 129, 76, 308], to list only textbooks] is unfortunately replete with differences in notation and conventions. For consistency, the conventions of Ref. [76] are adopted herein, and the Reader is left to compare with other sources and correctly translate the notation and conventions.

Left and right spinors

The result (5.58) indicates the fact that the Dirac 4-component spinor may, in a Lorentz-invariant way, be separated into a pair of two-component spinors, $\Psi = (\Psi_+, \Psi_-)$, where $\Psi_\pm$ are defined by the projections $\boldsymbol{\gamma}_\pm$ (5.57). This separation reflects the fact that the Lorentz group in 1+3-dimensional spacetime is $Spin(1,3) \cong Spin(3;\mathbb{C}) \cong SL(2;\mathbb{C})$, and that the Lorentz algebra is

$$\mathfrak{spin}(1,3) = \mathfrak{su}(2)_L \oplus \mathfrak{su}(2)_R. \tag{A.134}$$

That is, Ψ_+ transforms under the $\mathfrak{su}(2)_L$-action and is invariant under the $\mathfrak{su}(2)_R$-action, while Ψ_- transforms the other way around:

$$\Psi_+ \sim (\tfrac{1}{2},0), \quad \Psi_- \sim (0,\tfrac{1}{2}) \qquad \text{with respect to } \mathfrak{spin}(1,3) = \mathfrak{su}(2)_L \oplus \mathfrak{su}(2)_R. \tag{A.135}$$

In physics literature one often encounters the statement "$Spin(1,3) \cong SU(2)_L \times SU(2)_R$," which does not hold for the group. For most all of physics purposes, however, the relation (A.135) suffices, which is true of the algebra; the Reader is directed to the literature [565, 258, 581, 256, 80, 260, 333, 447] for the precise details about these groups, their representations and differences.

For the Weyl spinors (5.58), one uses the 2-component notation:

$$\Psi_+ := \boldsymbol{\gamma}_+\Psi \mapsto \psi_\alpha, \qquad \psi_\alpha \to \psi'_\alpha = M_\alpha{}^\beta\,\psi_\beta, \tag{A.136a}$$

$$\Psi_- := \boldsymbol{\gamma}_-\Psi \mapsto \overline{\chi}_{\dot\alpha}, \qquad \overline{\chi}_{\dot\alpha} \to \overline{\chi}'_{\dot\alpha} = \overline{\chi}_{\dot\beta}\,(\overline{M}^{-1})^{\dot\beta}{}_{\dot\alpha}. \tag{A.136b}$$

Here, $M_\alpha{}^\beta$ and $\overline{M}^{\dot\beta}{}_{\dot\alpha}$ are matrix elements of $SL(2;\mathbb{C})$-matrices $\mathbb{M} = \exp\{\mathfrak{m}_L\}$ with $\mathfrak{m}_L \in \mathfrak{su}(2)_L$ and $\overline{\mathbb{M}} = \exp\{\mathfrak{m}_R\}$ with $\mathfrak{m}_R \in \mathfrak{su}(2)_R$; the matrices $\mathbb{M}$ and $\overline{\mathbb{M}}$ are independent, and one refers to independent "left" and "right" action.

The spin-$\frac{1}{2}$ wave-functions ψ and χ are used to represent *fermionic* wave-functions, so that the components ψ_α and $\overline{\chi}_{\dot\alpha}$ are *anticommuting* functions.[20] Thus the Levi-Civita symbols $\varepsilon^{\alpha\beta}$ and $\varepsilon^{\dot\alpha\dot\beta}$ serve as (antisymmetric!) metric tensors for "left" and "right" Weyl spinors, ψ, χ and $\overline{\psi}, \overline{\chi}$:

$$(\psi{\cdot}\chi) := \psi_\alpha\varepsilon^{\alpha\beta}\chi_\beta = \psi_1\chi_2 - \psi_2\chi_1 = -\chi_\beta\varepsilon^{\alpha\beta}\psi_\alpha = \chi_\beta\varepsilon^{\beta\alpha}\psi_\alpha = (\chi{\cdot}\psi), \tag{A.137}$$

$$(\overline{\psi}{\cdot}\overline{\chi}) := \overline{\psi}_{\dot\alpha}\varepsilon^{\dot\alpha\dot\beta}\overline{\chi}_{\dot\beta} = \overline{\psi}_1\overline{\chi}_2 - \overline{\psi}_2\overline{\chi}_1 = -\overline{\chi}_{\dot\beta}\varepsilon^{\dot\alpha\dot\beta}\overline{\psi}_{\dot\alpha} = \overline{\chi}_{\dot\beta}\varepsilon^{\dot\beta\dot\alpha}\overline{\psi}_{\dot\alpha} = (\overline{\chi}{\cdot}\overline{\psi}), \tag{A.138}$$

where we must pay attention to detail:

$$\varepsilon^{\alpha\gamma}\varepsilon_{\beta\gamma} = \delta^\alpha_\beta, \quad \text{but} \quad \varepsilon^{\alpha\gamma}\varepsilon_{\gamma\beta} = -\delta^\alpha_\beta; \qquad \varepsilon^{\dot\alpha\dot\gamma}\varepsilon_{\dot\beta\dot\gamma} = \delta^{\dot\alpha}_{\dot\beta}, \quad \text{but} \quad \varepsilon^{\dot\alpha\dot\gamma}\varepsilon_{\dot\gamma\dot\beta} = -\delta^{\dot\alpha}_{\dot\beta}. \tag{A.139}$$

By convention, we set $\varepsilon^{12} = 1 = \varepsilon^{\dot\alpha\dot\beta}$.

[20] To be precise, every component of the field ψ_α and χ_α may be identified with a spacetime-dependent linear combination of anticommuting operators, such as b and $b^\dagger$ in Section 10.1, where creation operators act upon a vacuum state and create states with appropriate fermionic excitations.

Products of 2-component Weyl spinors satisfy the following identities:

$$\psi_\alpha\chi_\beta = \tfrac{1}{2}\varepsilon_{\alpha\beta}(\psi\cdot\chi) - \tfrac{1}{2}\sigma^{\mu\nu}_{\alpha\beta}(\psi\,\sigma_{\mu\nu}\,\chi), \tag{A.140}$$

$$\overline{\psi}_{\dot\alpha}\overline{\chi}_{\dot\beta} = \tfrac{1}{2}\varepsilon_{\dot\alpha\dot\beta}(\overline{\psi}\cdot\overline{\chi}) - \tfrac{1}{2}\overline{\sigma}^{\mu\nu}_{\dot\alpha\dot\beta}(\overline{\psi}\,\overline{\sigma}_{\mu\nu}\,\overline{\chi}), \tag{A.141}$$

$$\psi^2 := \tfrac{1}{2}\varepsilon^{\alpha\beta}\psi_\alpha\psi_\beta, \qquad \psi_a\overline{\chi}_{\dot\alpha} = -\tfrac{1}{2}\sigma^\mu_{\alpha\dot\alpha}(\psi\,\sigma_\mu\,\overline{\chi}), \tag{A.142}$$

$$(\psi_1\cdot\psi_2)(\psi_3\cdot\psi_4) = -(\psi_1\cdot\psi_3)(\psi_2\cdot\psi_4) - (\psi_1\cdot\psi_4)(\psi_2\cdot\psi_3), \tag{A.143}$$

$$(\psi_1\cdot\psi_2)(\overline{\psi}_3\cdot\overline{\psi}_4) = -\tfrac{1}{2}(\psi_1\sigma^\mu\overline{\psi}_4)(\psi_2\sigma_\mu\overline{\psi}_3). \tag{A.144}$$

Comment A.3 *Since fermionic wave-functions are anticommuting, they must also be nilpotent:*

$$\{\,\psi_\alpha(x)\,,\,\psi_\beta(x)\,\} = 0 \qquad \Rightarrow \qquad (\psi_\alpha(x))^2 \equiv 0. \tag{A.145}$$

The notation "ψ^2" is then free for the definition:

$$\psi^2(x) := \psi_1(x)\psi_2(x) = \tfrac{1}{2}\varepsilon^{\alpha\beta}\psi_\alpha(x)\psi_\beta(x). \tag{A.146}$$

4-Vectors and Pauli's matrices

4-vectors such as the spacetime 4-vector x transform as the $(\frac{1}{2},\frac{1}{2})$ representation of the $\mathfrak{spin}(1,3) = \mathfrak{su}(2)_L \oplus \mathfrak{su}(2)_R$ algebra, i.e., of the $Spin(1,3) \cong SL(2;\mathbb{C})$ group. The $SL(2;\mathbb{C})$ group action on the 4-vector x^μ is easiest represented using Pauli matrices:

$$[\sigma^\mu]_{\alpha\dot\alpha}: \qquad \sigma^0 := \left[\begin{smallmatrix}1&0\\0&1\end{smallmatrix}\right],\quad \sigma^1 := \left[\begin{smallmatrix}0&1\\1&0\end{smallmatrix}\right],\quad \sigma^2 := \left[\begin{smallmatrix}0&-i\\i&0\end{smallmatrix}\right],\quad \sigma^3 := \left[\begin{smallmatrix}1&0\\0&-1\end{smallmatrix}\right], \tag{A.147}$$

which are identified with the index notation $\sigma^\mu_{\alpha\dot\alpha}$ so that, e.g., $\sigma^2_{12} = -i$. Using $\varepsilon^{\alpha\beta}$ and $\varepsilon^{\dot\alpha\dot\beta}$ to "raise" spinor indices and $\eta_{\mu\nu}$ to "lower" the vector index, we have

$$\overline{\sigma}^{\dot\alpha\alpha}_\mu := \varepsilon^{\dot\alpha\dot\beta}\varepsilon^{\alpha\beta}\eta_{\mu\nu}\sigma^\nu_{\beta\dot\beta}: \qquad [\overline{\sigma}_\mu] = ([\mathbb{1}],[\sigma^1],[\sigma^2],[\sigma^3]) = [\sigma^\mu]. \tag{A.148}$$

That is, the matrices σ^μ and $\overline{\sigma}_\mu$ look alike. However, the matrices

$$\sigma_\mu := \eta_{\mu\nu}\sigma^\nu \qquad \text{and} \qquad \overline{\sigma}^\mu := \eta^{\mu\nu}\overline{\sigma}_\nu \tag{A.149}$$

have a differing sign: $[\overline{\sigma}^1] = -[\sigma^1]$, $[\overline{\sigma}^2] = -[\sigma^2]$, $[\overline{\sigma}^3] = -[\sigma^3]$, as well as $[\sigma_1] = -[\overline{\sigma}_1]$, $[\sigma_2] = -[\overline{\sigma}_2]$, $[\sigma_3] = -[\overline{\sigma}_3]$.

One therefore writes

$$\boldsymbol{x} := x^\mu\overline{\boldsymbol{\sigma}}_\mu, \qquad \boldsymbol{x} \to \boldsymbol{x}' = \overline{\mathbb{M}}\,\boldsymbol{x}\,\mathbb{M}^{-1}, \qquad \mathbb{M},\overline{\mathbb{M}} \in SL(2;\mathbb{C}), \tag{A.150}$$

where the matrices $\mathbb{M} = \exp\{i\omega_\mu\sigma^\mu\}$ and $\overline{\mathbb{M}} = \exp\{i\overline{\pi}_\mu\overline{\sigma}^\mu\}$ are independent, and represent the independent "left" and "right" action, so that

$$(\overline{\chi}'\cdot\boldsymbol{x}'\cdot\psi') = (\overline{\chi}\overline{\mathbb{M}}^{-1}\cdot\overline{\mathbb{M}}\,\boldsymbol{x}\,\mathbb{M}^{-1}\cdot\mathbb{M}\psi) = (\overline{\chi}\cdot\boldsymbol{x}\cdot\psi) \tag{A.151}$$

is an $SL(2;\mathbb{C})$-invariant. In the index notation,

$$\begin{aligned}(\overline{\chi}_{\dot\alpha}\,x^{\dot\alpha\alpha}\,\psi_\alpha) \to (\overline{\chi}'_{\dot\alpha}\,x'^{\dot\alpha\alpha}\,\psi'_\alpha) &= \overline{\chi}_{\dot\beta}\,(\overline{M}^{-1})^{\dot\beta}{}_{\dot\alpha}\,\overline{M}^{\dot\alpha}{}_{\dot\gamma}\,x^{\dot\gamma\gamma}\,(M^{-1})_\gamma{}^\alpha\,M_\alpha{}^\beta\,\psi_\beta\\ &= (\overline{\chi}_{\dot\alpha}\,x^{\dot\alpha\alpha}\,\psi_\alpha).\end{aligned} \tag{A.152}$$

Finally, notice that

$$\det[\,\boldsymbol{x}\,] = (x^0)^2 - (x^1)^2 - (x^2)^2 - (x^3)^2 = x^\mu\,\eta_{\mu\nu}\,x^\nu \overset{(3.17)}{=} \mathbf{x}^2, \tag{A.153}$$

which is also an $SL(2;\mathbb{C})$-invariant:

$$\det[\,\boldsymbol{x}\,] \to \det[\,\boldsymbol{x}'\,] = \det[\,\overline{\mathbb{M}}\,\boldsymbol{x}\,\mathbb{M}^{-1}\,] = \det[\overline{\mathbb{M}}]\det[\,\boldsymbol{x}\,]\det[\mathbb{M}^{-1}] = \det[\,\boldsymbol{x}\,], \tag{A.154}$$

since the $SL(2;\mathbb{C})$ elements are unimodular, $\det[\overline{\mathbb{M}}] = 1 = \det[\mathbb{M}]$.

The Pauli matrices (A.147) and (A.148) satisfy the following useful identities:

$$(\boldsymbol{\sigma}_\mu\overline{\boldsymbol{\sigma}}_\nu + \boldsymbol{\sigma}_\nu\overline{\boldsymbol{\sigma}}_\mu)_\alpha{}^\beta = 2\eta_{\mu\nu}\,\delta_\alpha^\beta, \qquad (\overline{\boldsymbol{\sigma}}_\mu\boldsymbol{\sigma}_\nu + \overline{\boldsymbol{\sigma}}_\nu\boldsymbol{\sigma}_\mu)^{\dot\alpha}{}_{\dot\beta} = 2\eta_{\mu\nu}\,\delta_{\dot\beta}^{\dot\alpha}; \tag{A.155}$$

$$\mathrm{Tr}\,\big[\boldsymbol{\sigma}_\mu\overline{\boldsymbol{\sigma}}_\nu\big] = \mathrm{Tr}\,\big[\overline{\boldsymbol{\sigma}}_\mu\boldsymbol{\sigma}_\nu\big] = 2\eta_{\mu\nu}, \qquad \sigma^\mu_{\alpha\dot\alpha}\,\overline{\sigma}_\mu^{\dot\beta\beta} = 2\delta_\alpha^\beta\,\delta_{\dot\alpha}^{\dot\beta}, \tag{A.156}$$

and are suitable for the conversion of $Spin(1,3)$-tensors into (bi)spinor expressions:

$$V_{\alpha\dot\alpha} := \sigma^\mu_{\alpha\dot\alpha}V_\mu \quad\Leftrightarrow\quad V_\mu = \tfrac{1}{2}\overline{\sigma}_\mu^{\dot\alpha\alpha}V_{\alpha\dot\alpha}. \tag{A.157}$$

It is convenient to also define the matrices

$$(\boldsymbol{\sigma}_{\mu\nu})_\alpha{}^\beta := \tfrac{1}{4}\big(\sigma_{\mu\,\alpha\dot\alpha}\overline{\sigma}_\nu{}^{\dot\alpha\beta} - \sigma_{\nu\,\alpha\dot\alpha}\overline{\sigma}_\mu{}^{\dot\alpha\beta}\big), \qquad (\overline{\boldsymbol{\sigma}}_{\mu\nu})^{\dot\alpha}{}_{\dot\beta} := \tfrac{1}{4}\big(\overline{\sigma}_\mu{}^{\dot\alpha\alpha}\sigma_{\nu\,\alpha\dot\beta} - \overline{\sigma}_\nu{}^{\dot\alpha\alpha}\sigma_{\mu\,\alpha\dot\beta}\big), \tag{A.158a}$$

which, $\boldsymbol{\sigma}_{\mu\nu}$ and $\overline{\boldsymbol{\sigma}}_{\mu\nu}$ independently, close the $\mathfrak{spin}(1,3)$ algebra (A.121c), and for which

$$(\boldsymbol{\sigma}_{\mu\nu})_{\alpha\beta} := (\boldsymbol{\sigma}_{\mu\nu})_\alpha{}^\gamma\varepsilon_{\beta\gamma} \qquad\text{and}\qquad (\boldsymbol{\sigma}_{\mu\nu})^{\alpha\beta} := \varepsilon^{\alpha\gamma}(\boldsymbol{\sigma}_{\mu\nu})_\gamma{}^\beta, \tag{A.158b}$$

$$(\overline{\boldsymbol{\sigma}}_{\mu\nu})^{\dot\alpha\dot\beta} := (\overline{\boldsymbol{\sigma}}_{\mu\nu})^{\dot\alpha}{}_{\dot\gamma}\varepsilon^{\dot\beta\dot\gamma} \qquad\text{and}\qquad (\overline{\boldsymbol{\sigma}}_{\mu\nu})_{\dot\alpha\dot\beta} := \varepsilon_{\dot\alpha\dot\gamma}(\overline{\boldsymbol{\sigma}}_{\mu\nu})^{\dot\gamma}{}_{\dot\beta}. \tag{A.158c}$$

For these matrices (with $\varepsilon_{0123} = 1$), it is true that

$$(\boldsymbol{\sigma}_{\mu\nu})_\alpha{}^\beta(\boldsymbol{\sigma}_{\rho\sigma})_\beta{}^\alpha = \tfrac{1}{2}(\eta_{\mu\rho}\eta_{\nu\sigma} - \eta_{\mu\sigma}\eta_{\nu\rho}) + \tfrac{i}{2}\varepsilon_{\mu\nu\rho\sigma}, \tag{A.159}$$

$$(\overline{\boldsymbol{\sigma}}_{\mu\nu})^{\dot\alpha}{}_{\dot\beta}(\overline{\boldsymbol{\sigma}}_{\rho\sigma})^{\dot\beta}{}_{\dot\alpha} = \tfrac{1}{2}(\eta_{\mu\rho}\eta_{\nu\sigma} - \eta_{\mu\sigma}\eta_{\nu\rho}) - \tfrac{i}{2}\varepsilon_{\mu\nu\rho\sigma}. \tag{A.160}$$

Super-derivatives

In supersymmetry research, the so-called "super-derivatives"

$$D_\alpha := \partial_\alpha - i\sigma^\mu_{\alpha\dot\alpha}\overline{\theta}^{\dot\alpha}\partial_\mu \qquad\text{and}\qquad \overline{D}_{\dot\alpha} := \overline{\partial}_{\dot\alpha} - i\sigma^\mu_{\alpha\dot\alpha}\theta^\alpha\partial_\mu \tag{10.68$'$}$$

are of special importance. They anticommute with the generators of supersymmetry, $Q_\alpha, \overline{Q}_{\dot\alpha}$, and so commute with the operator of the supersymmetry transformation:

$$D_\alpha U_{\epsilon,\bar\epsilon} = U_{\epsilon,\bar\epsilon}D_\alpha \quad\text{and}\quad \overline{D}_{\dot\alpha}U_{\epsilon,\bar\epsilon} = U_{\epsilon,\bar\epsilon}\overline{D}_{\dot\alpha}, \qquad U_{\epsilon,\bar\epsilon} := \exp\{-i(\epsilon^\alpha Q_\alpha + \bar\epsilon^{\dot\alpha}\overline{Q}_{\dot\alpha})\}. \tag{A.161}$$

The operators $D_\alpha, \overline{D}_{\dot\alpha}$ are then, in fact, literally *invariant* with respect to the supersymmetry action, but their name, "(super)covariant," stuck in the literature; herein, the shorter and more precise term "super-derivative" is used.

The basic property of the super-derivatives,

$$\{\,D_\alpha\,,\,\overline{D}_{\dot\alpha}\,\} = 2\hbar^{-1}\sigma^\mu_{\alpha\dot\alpha}P_\mu = -2i\,\sigma^\mu_{\alpha\dot\alpha}\partial_\mu, \tag{10.69$'$}$$

is sometimes called *super-commutativity* and permits simplifying higher-order super-derivatives:

$$D_\alpha D_\beta = \tfrac{1}{2}\epsilon_{\alpha\beta}D^2, \qquad \overline{D}_{\dot\alpha}\overline{D}_{\dot\beta} = \tfrac{1}{2}\epsilon_{\dot\alpha\dot\beta}\overline{D}^2; \tag{A.162}$$

$$D_\alpha D_\beta D_\gamma = 0, \qquad \overline{D}_{\dot\alpha}\overline{D}_{\dot\beta}\overline{D}_{\dot\gamma} = 0; \tag{A.163}$$

$$[D^2, \overline{D}_{\dot\alpha}] = 4i\sigma^\mu_{\alpha\dot\alpha}\varepsilon^{\alpha\beta}\partial_\mu D_\beta, \qquad [\overline{D}^2, D_\alpha] = 4i\sigma^\mu_{\alpha\dot\alpha}\varepsilon^{\dot\alpha\dot\beta}\partial_\mu\overline{D}_{\dot\beta}; \tag{A.164}$$

$$D^2\overline{D}^2 + \overline{D}^2D^2 - 2\varepsilon^{\dot\alpha\dot\beta}\overline{D}_{\dot\alpha}D^2\overline{D}_{\dot\beta} = -16\Box, \quad \Box := \eta^{\mu\nu}\partial_\mu\partial_\nu. \tag{A.165}$$

A.6.3 Exercises for Section A.6

✎ **A.6.1** *Prove the relations (A.121) using only the anticommutation relations (A.119).*

✎ **A.6.2** *Prove the relations (A.122) using only the anticommutation relations (A.119).*

✎ **A.6.3** *Prove Theorem A.5 using only the anticommutation relations (A.119).*

✎ **A.6.4** *Prove Theorem A.5 using the Cayley–Hamilton theorem.*

✎ **A.6.5** *Prove the relations (A.125) using only the anticommutation relations (A.119).*

✎ **A.6.6** *Prove the relations (A.126) using only the anticommutation relations (A.119).*

✎ **A.6.7** *Prove the relations (A.162)–(A.165) using only the relations (10.69).*

B A lexicon

When describing previously uncharted territories, discoverers and inventors are forced to adopt and adapt previously known terms, concepts and techniques for the new phenomena, or invent wholly new ones. This appendix collects a listing of perhaps less familiar but oft-used terms in our field, then turns to the vector/tensor and even functional extension to the hopefully well-familiar rules of multivariate calculus, and closes with a brief on Gödel's incompleteness theorem.

B.1 The jargon

The jargon of theoretical and mathematical physics is very much in development and in some cases not yet standardized. With the aim of using compact but precise terms to name very specific ideas, many scientists begin using an otherwise rarely used word and, at times, their choice "catches on" and becomes standardized. At other times, different terms are used by competing (or non-communicating) research groups for the same or closely related concepts, whereupon one of the two "competing" terms may turn into a standard but only after a long period during which both terms are used. As the fundamental physics of elementary particles is still very much in development, consistency and expediency required me to make certain choices in terminology, which I have, to the best of my knowledge, indicated together with possible alternatives.

The subsequent lexicon offers brief explanations for some of the perhaps less familiar technical terms and expressions, most of which are fairly standard, but in a field other than particle physics.

Abelian (commutative, symmetric) A binary operation $\star$ is abelian if $a \star b = b \star a$. By extension, structures defined using an abelian binary operation are also called abelian. Operations that are not abelian are called non-abelian (= non-commutative, = asymmetric), as are structures defined using them.

Algebra A vector space $\mathfrak{A}$ over a field $\Bbbk$, equipped with a binary operation $*$, which satisfies the distribution law over addition: $a*(b+c) = (a*b)+(a*c)$, for all elements $a, b, c \subset \mathfrak{A}$, and for which it is true that $\alpha(a*b) = (\alpha a)*b = a*(\alpha b)$, for each $\alpha \in \Bbbk$ and $a, b \in \mathfrak{A}$. The operation $*$ is typically a type of multiplication; it is often commutative, i.e., symmetric, but in Lie algebras it is antisymmetric: $a*b = -b*a$.

Amplitude In the context of field theory and so also in (high energy) elementary particle physics, this is the matrix element $\mathfrak{M}_{i\to f} := \langle f|H_{\text{int}}|i\rangle$ where $|i\rangle$ and $|f\rangle$ are the initial and final states and H_{int} is the (algebraic sum of all) interaction operator(s) that can bring about the process $|i\rangle \to |f\rangle$. The probability for this process is then proportional to $|\mathfrak{M}_{i\to f}|^2$ [☞ display (3.85) and Section 3.3.3 on p. 113].

Analytic function A function $f(x)$ is analytic in a domain $\mathscr{D}$ if it has a (convergent) Taylor expansion $f(x) = \sum_{n=0}^{\infty} a_n(x-x_0)^n$ for every $x_0, (x-x_0) \in \mathscr{D}$.

Anomaly Structural changes in relations between observables caused by passing from classical to quantum theory. If those relations represent the algebra of symmetry transformations and anomalies obstruct the closure or change the structure of that algebra, then anomalies destroy or change the symmetry that was built into the system originally – which points to an inconsistency. Models with an anomaly in a gauge symmetry are simply inconsistent [☞ Section 7.2.3], whereupon all gauge anomaly ought to cancel. [☛ geometric quantization; canonical quantization] In turn, anomalies in global and approximate symmetries need not cancel, but are characteristic quantities that cannot be altered by field redefinitions, and so must remain conserved throughout the evolution of a system, including phase transitions. This is a direct consequence of the underlying principle in Dirac quantization. [☛ Dirac quantization]

Auxiliary field A field that has a non-differential equation of motion, which determines the field point-by-point. If this equation of motion can be solved, the solution can be reinserted in the Lagrangian density, which is classically equivalent to the original Lagrangian density but involves fewer fields. The equivalence need not hold between the quantum models defined from the two Lagrangian densities.

Baryon Since the acceptance of the quark model in 1973, a bound state of three quarks. Originally, a particle that interacts by means of the strong nuclear force (at $\sim 10^{-23}$ s), can be detected as an isolated particle, and has a mass that is not smaller than that of the proton, such as a neutron.

Bijection A mapping $f : X \to Y$ that is both (1) an ***injection*** (i.e., "1–1"), so for every $x \in X$ there is precisely one $y = f(x) \in Y$, and (2) a ***surjection***, so for every $y \in Y$ there is an $x \in X$ so that $f(x) = y$. Bijection = surjective injection, i.e., injective surjection. [☛ injection, surjection]

BFV-quantization A contemporary version (by Igor Batalin, Efim S. Fradkin and Grigori Vilkovisky) of canonical quantization in the Hamiltonian formalism, which generalizes the evolution of the canonical–Dirac–BRST quantization to the general case when the constraints do not close the structure of an algebra [174, 39, 172, 36, 37, 345, 38, and references therein]; see also the texts [268, 555, 484, 496, 589, 590] and [509]. [☛ BRST quantization; Dirac quantization; canonical quantization]

Bose condensation The state of a system where infinitely many particles (bosons) are in the same quantum state. The Coulomb static potential may be understood as a Bose condensation of infinitely many photons.

Boson By definition, a particle (as well as its mathematical representatives: wave-functions, creation and annihilation operators or fields) that obeys the Bose–Einstein statistics; Pauli's exclusion principle does not apply to bosons and bosons may condense [☛ Bose condensate]. By the spin-statistics theorem (in Lorentz-covariant models), physical particles whose mathematical representatives transform as tensor representations of the Lorentz group are bosons. The possible values of bosonic wave-functions and fields are (ordinary) commuting numbers ("c-numbers").

BRST quantization A procedure (by Carlo M. Becchi, Alain Rouet and Raymond Stora, and separately by Igor V. Tyutin) of constructing a quantum theory from an originally classical field theory with a gauge symmetry, in which the gauge symmetry reduces to a BRST symmetry and counterterms are added to the Lagrangian density that are invariant with respect to the BRST symmetry, although not with respect to the original (classical) gauge symmetry. As a gauge symmetry is realized in quantum theory by imposing constraints (that the physical states are invariant under the action of the symmetry) and these constraints close an algebra, BRST quantization is a canonical generalization of the Dirac quantization with constraints of the first class in Dirac's classification [445, 425, 345]; see also the texts [555, 484, 496, 589, 590]. [☛ Dirac quantization; canonical quantization]

BRST symmetry A reduction of a gauge symmetry where the parameters in a gauge transformation are replaced by *ghost fields*: functions of spacetime that have the opposite statistics from the original parameters but transform identically as the original parameters under the action of both the gauge and the Lorentz transformations. For example, Yang–Mills gauge theories have ordinary (commutative) scalar functions as gauge parameters. In the corresponding BRST symmetry, to the system is added a pair of canonically conjugate *anticommutative* scalar fields that otherwise, in every other aspect, transform identically as the original parameters of the given gauge transformation. Interactions of these ghost fields with other fields are determined precisely so that they cancel the contributions of the unphysical components in the gauge fields [44]; see also the texts [268, 555, 484, 496, 589, 590]. [☛ ghost fields; nonphysical components]

Bundle [☞ vector bundle]

BV-quantization A contemporary version of canonical quantization (by Jean Zinn–Justin, then by Igor Batalin and Grigori Vilkovisky) in the Lagrangian formalism, which generalizes the evolution of the canonical–Dirac–BRST quantization to the general case when the constraints do not close the structure of an algebra [41, 345]; see also the texts [555, 484]. [☛ BRST quantization; Dirac quantization; canonical quantization]

Canonical quantization Also known as the second quantization; the adjective "canonical" stems from using the canonical Hamiltonian formalism of classical physics and its quantum reinterpretation, where the relations between observables in a given model are preserved as well as possible, and with a formal replacement of the Poisson brackets by commutators. Changes in these relations, e.g., if the Poisson bracket $\{\mathcal{A},\mathcal{B}\} = \mathcal{C}$ upon canonical quantization becomes $[A,B] = C + \Delta$, the additional term Δ is one of the measures of this *anomaly*. [☛ anomaly]

Cartesian product Also known as the *direct product*: for two sets X and Y, the Cartesian product is the set of all ordered pairs:

$$X \times Y := \{(x,y) \,:\, x \in X,\ y \in Y\}. \tag{B.1}$$

Cauchy sequence Given a metric space (a set of points x_i with a well-defined distance function $d(x_i,x_j)$ between any two points), this is a sequence of points $x_1, x_2, \ldots$, where

$$d(x_i,x_j) < \epsilon, \quad \forall i,j > N, \tag{B.2}$$

for some predefined integer N and positive real number (tolerance) ϵ. That is, all points sufficiently far up the sequence are closer than ϵ to each other.

Chirality The eigenvalue of the operator $\widehat{\gamma}$. A particle is said to have a well-defined chirality if its wave-function is an eigenfunction of this operator. The operators $\frac{1}{2}[\mathbb{1} \pm \widehat{\gamma}]$, with the $\widehat{\gamma}$-matrix defined in Appendix A.6.1, project to spin-$\frac{1}{2}$ particles of chirality $\pm\frac{1}{2}$. By construction, chirality is Lorentz-invariant. However, as $\widehat{\gamma}$ anticommutes with the Dirac operator $\gamma^\mu\partial_\mu$ and commutes with the mass,

$$[i\hbar\gamma^\mu\partial_\mu - mc\mathbb{1}]\,\tfrac{1}{2}[\mathbb{1} \pm \widehat{\gamma}] \not\propto \tfrac{1}{2}[\mathbb{1} \pm \widehat{\gamma}]\,[i\hbar\gamma^\mu\partial_\mu - mc\mathbb{1}], \tag{B.3}$$

and the chirality of a massive particle is not a constant.

CM system For a system of particles located at the positions $\vec{r}_i$ and having the masses m_i, the position and velocity of the center of mass are, by definition,

$$\vec{r}_{CM} := \frac{\sum_i m_i \vec{r}_i}{\sum_i m_i}, \qquad \vec{v}_{CM} := \frac{\sum_i m_i \vec{v}_i}{\sum_i m_i}. \tag{B.4}$$

A coordinate system where $\vec{v}_{CM} = 0$ is called the center of momentum frame, where $\vec{r}_{CM}$ need not vanish; a coordinate system where additionally also $\vec{r}_{CM} = 0$ is called the center of mass system, or "CM-system" for short.

Codimension For a subspace $X \subset Y$, $\mathrm{cod}(X \subset Y) := \dim(Y) - \dim(X)$. If the subspace X is defined by means of a system of algebraic equations, near every point $x \in X$, that system must have $\mathrm{cod}(X \subset Y)$ independent equations.

Codomain For a mapping $f : X \to Y$, the collection of elements Y wherein the map points, and wherein the *values* of f and its *image* lie; $f(x) = y \in Y$ for all $x \in X$.

Cokernel For a linear mapping $f : X \to Y$ of a vector space X into Y, the *cokernel* of f consists of the equivalence classes $\mathrm{cok}(f) := \{[y \simeq y + f(x)] : x \in X, y \in Y\}$.

Color In the context of elementary particles, the 3-dimensional $SU(3)_c$ charges of quarks, such that baryons consist of three quarks with one of the three linearly independent colors ("red," "yellow," "blue") each, so that the baryon is "colorless," or more precisely, $SU(3)_c$-invariant. Owing to the ubiquity of computer graphics, the so-called subtractive color system is ever more familiar, but we adopt the familiar additive color system. Here, red and yellow produces orange, and its mix with blue produces ***colorless***, i.e., black. The opposite (anti-)colors of primary colors are: anti-red = green, anti-yellow = purple, anti-blue = orange; the mixture of any color and its anti-color produces ***colorless***. Because of this regularity the name *color* is convenient as a mnemonic crutch for adding $SU(3)_c$ vectors [☞ Appendix A.4].

Compact space A *topological space* [☛ topological space] X where every open neighborhood (and so also the whole X) may be covered by a finite number of open neighborhoods is called quasi-compact. A topological space where every two distinct points have some non-intersecting neighborhoods,

$$\forall x \neq x' \in X, \quad \exists U, U' \subset X: \quad U \ni x, \ U' \ni x', \ U \cap U' = \varnothing \tag{B.5}$$

is called Hausdorff. A Hausdorff space that is also quasi-compact is compact. In practice in theoretical physics, it is crucial that compact spaces have a well-defined *size*, so that compact spaces may be chosen to be smaller (or larger) than a given size/length.

Compactification The procedure where a non-compact topological space X is added to a topological space Y of strictly lesser dimension, so that $X^c := (X \cup Y)$ is compact. The simplest example is $S^1 = \mathbb{R}^1 \cup \{\text{point}\}$, where a point "at infinity" was added to the open line ($\mathbb{R}^1$), so as to obtain the circle (S^1).

Concrete applications of this procedure within the present subject stem from the proposal originally made by Gunnar Nordstrøm, in 1914, whereby the spatial dimension of the form of an open and infinitely large line, $\mathbb{R}^1$, is replaced by a closed, compact and small circle, S^1. The proposal was rediscovered by Theodor F. E. Kałuża in 1919 (published in 1921) and also Oscar Klein in 1921. The latter two publications being generally known, this is typically called "Kałuża–Klein compactification." Symmetries of the compactified space result in Yang–Mills gauge symmetries in the non-compact spacetime. The special case when the compact space is a Calabi–Yau manifold is called "Calabi–Yau compactification." As Calabi–Yau spaces of more than one complex dimension do not have continuous symmetries, Calabi–Yau compactification does not give rise to any gauge symmetry, and in fact typically reduces what gauge symmetry there was prior to compactification; see Section 11.3.1.

Complex structure [☛ conjugation]
Commutative [☛ abelian]
Conjugation is a mapping of one (generalized) *complex structure* into its equivalent partner. Most generally, a ***complex structure*** is specified by an operation $\hat{\mathcal{I}}$, the two-fold repetition of which results in a sign change: $\hat{\mathcal{I}} \circ \hat{\mathcal{I}} = -\mathbb{1}$. Therefore, $-\hat{\mathcal{I}}$ is always also a complex structure, distinct from $\mathcal{I}$ but equivalent to it for all purposes, and all complex structures always occur in such equivalent pairs.

Complex conjugation Every rule by which a pair of real numbers (x, y) is assigned a complex number z has a conjugate rule. For example, *relative* to the definition $z := (x + iy)$, $z^* := (x - iy)$ is the ***complex conjugation*** of the complex number z. Operatively, complex conjugation changes $i \to -i$. The analogous situation holds also for matrices, functions, operators, etc.

Hermitian conjugation of matrices is the combination of complex conjugation (of every element) of the matrix with its transposition: $(a_{ij})^\dagger := a^*_{ji}$. [☛ Digression 10.2 on p. 360]

Dirac conjugation of a Dirac spinor Ψ is the Hermitian conjugation combined with right-multiplication by the γ^0 matrix: $\overline{\Psi} := \Psi^\dagger\gamma^0$. Correspondingly, the Dirac conjugate of the operator R is $\overline{R} := (\gamma^0)^{-1}R\gamma^0$. For a Cartesian basis of γ-matrices with the metric tensor (3.19), it follows that $(\gamma^0)^{-1} = \gamma^0$ so $\overline{R} = \gamma^0 R \gamma^0$, which agrees with the definition (5.132).

Contact interaction Interaction that requires that all participants in the interaction are localized in the same spacetime point – akin to the collision of two marbles. All elementary processes in the Standard Model are contact interactions. For example, the emission and the absorption of a (virtual) photon by an electron requires that the "incoming" electron in a spacetime point turn into the "outgoing" electron and that the photon in this interaction is emitted from or absorbed at that same point. The Yukawa interaction is analogous, except that a scalar particle is emitted or absorbed instead of a photon. The Fermi interaction is also analogous, except that here two fermions collide in a spacetime point from which then two other fermions emerge, or one fermion decays into three fermions, all emitted from the same spacetime point.

Contravariant vector A vector the components of which, $A^\mu(\mathrm{x})$, are transformed as

$$A^\mu(\mathrm{x}) = \left(\frac{\partial x^\mu}{\partial y^\nu}\right) A^\nu(\mathrm{y}) \tag{3.11c}$$

by the coordinate system transformation $\mathrm{x} \to \mathrm{y}$.

Coset [☛ Appendix A.1.1.]

Cotangent bundle The vector bundle $\mathscr{T}^*_{\mathscr{X}} := E(\mathscr{X}; T^*_x(\mathscr{X}); \pi)$ where $T^*_x(\mathscr{X})$ is the cotangent space of the space $\mathscr{X}$ at the point $x \in \mathscr{X}$. If x^μ are local coordinates in the space $\mathscr{X}$ at the given point, then $T^*_x(\mathscr{X})$ may be represented as the formal vector space of linear combinations $\omega_\mu \mathrm{d}x^\mu$.

Coulomb field, potential A stationary electric charge is surrounded by the constant Coulomb electrostatic field, $\vec{E}$; $q_0\vec{E}$ is the force that acts upon the probing particle of charge q_0. For the same situation, $\vec{E} = -\vec{\nabla}\Phi$, where Φ is the Coulomb potential; $q_0\Phi$ is the potential energy of the probing charge q_0 in the field $\vec{E}$. It follows from Gauss's law that the Coulomb field of a point-like charge is $\vec{E} \propto 1/r^{d-1}$, where d is the dimension of the space and r the distance between the source of the field and the place where the field is measured; also, $\Phi \propto 1/r^{d-2}$.

Covariant derivative A measure of the amount of change in the "overall value" of a generalized function $\boldsymbol{F}$ owing to a change of one of the arguments of $\boldsymbol{F}$ in the limiting case when the change in the argument is infinitesimal and tends to zero. For a real scalar (invariant) function, the "overall value" is simply the "value" or intensity, and the covariant derivative of such a function is the same as the partial derivative. However, for more general functions $\boldsymbol{F}$ that take values in a multi-dimensional space, such as spacetime itself or some abstract space, the covariant derivative also takes into account that the space of values of $\boldsymbol{F}$ may well change over the space of arguments. This then additionally changes the "overall value" (both the intensity and the "direction") of $\boldsymbol{F}$ at an infinitesimally close neighboring value of the argument. The covariant derivatives therefore have the general form $D := \partial + \boldsymbol{\Gamma}$, where $\boldsymbol{\Gamma}$ is the gauge potential and encodes the variation in the space of values of $\boldsymbol{F}$. [☛ gauge potential, gauge field]

Covariant vector A vector the components of which, $B_\mu(\mathrm{x})$, are transformed as

$$B_\mu(\mathrm{x}) = \left(\frac{\partial y^\nu}{\partial x^\mu}\right) B_\nu(\mathrm{y}) \tag{3.11d}$$

by the coordinate system transformation $\mathrm{x} \to \mathrm{y}$.

Covering For a given (topologial) space X, the n-fold (finite) cover Y is a space for which there exists an n–1 mapping $\pi : Y \to X$ such that for every point $x \in U \subset X$, where U is any open neighborhood in X, there exist exactly n points and non-intersecting open neighborhoods $y_i \in V_i \subset Y$, such that $\pi(y_i) = x$ and $\pi(V_i) = U$. That is, π is a continuous surjection. The points y_i are called the π-inverse images of x, i.e., $\pi^{-1}(x) = \{y_1, y_2, \ldots\}$.

Curvature Given a space $\mathscr{X}$ over which the functions $f(x^\mu)$ are defined locally, i.e., in sufficiently small open neighborhoods, $f(x^\mu)$ is unambiguously defined. Let D_μ be local derivatives that (in sufficiently small open neighborhoods) correctly compute the difference $\mathrm{d}x^\mu(D_\mu f) = f(x^\mu{+}\mathrm{d}x^\mu) - f(x^\mu)$. Then, in general, the relations

$$\left[D_\mu\,,\,D_\nu\right] = \boldsymbol{T}_{\mu\nu}{}^\rho D_\rho + \boldsymbol{R}_{\mu\nu} \tag{B.6}$$

define the ***torsion*** $\boldsymbol{T}_{\mu\nu}{}^\rho$ and the ***curvature*** $\boldsymbol{R}_{\mu\nu}$ of the space $\mathscr{X}$. These two (local) structures specify the (local) geometry of the space $\mathscr{X}$ and a class of functions $f(x^\mu)$ over this space.

In examples where $\mathscr{X}$ is spacetime and $f(x^\mu)$ a complex wave-function representing a lepton or a quark, the torsion vanishes, and $\boldsymbol{R}_{\mu\nu}$ is the Yang–Mills gauge field (denoted $\mathbb{F}_{\mu\nu}$ [☞ Chapters 5 and 6]). The torsion vanishes also when $f(x^\mu)$ represents a tensor over spacetime $\mathscr{X}$, in which case is $\boldsymbol{R}_{\mu\nu}$ the Riemann tensor [☞ Chapter 9]. In turn, when $D_\alpha, \overline{D}_{\dot\alpha}$ are super-derivatives (10.68), so the commutator in the relation (B.6) is replaced by an anticommutator, the curvature vanishes and the torsion does not [☞ relation (10.69), which holds for the extended basis of super-spacetime derivatives $\{D_\alpha, \overline{D}_{\dot\alpha}, \partial_\mu\}$]. Finally, in the theory of Lie groups, the Lie group itself is a differentiable space where the derivatives are closely related to the generators Q, and their commutator, akin to (A.70), defines the *structure constants* of the Lie group as the torsion and where the curvature vanishes.

Dirac quantization The development of general canonical quantization for systems in which there exist constraints, and the specification how to treat these constraints in quantum theory so they remain satisfied throughout the evolution of the system in time; see Digression 11.7 on p. 420, the texts [64, 445, 425], as well as Dirac's book [134]. Dirac's procedure proves the fundamental equivalence between Heisenberg's "matrix mechanics" and Schrödinger's "wave mechanics" and connects the ideas from both approaches. [☛ canonical quantization]

Direct product [☛ Cartesian product]

Domain For a map $f : X \to Y$, this is X, the collection of elements that are being mapped by f; $X := \{x : f(x) \text{ is well defined}\}$.

Einstein–Rosen bridge A wormhole that connects the inside of the event horizon of one of two Schwarzschild black holes with the inside of the event horizon of another black hole of the same type. [☛ wormhole]

Energy–momentum (4-momentum) transfer In collisions $A + B \to A' + \cdots$, where B is initially a target at rest and A and A' the incoming and outgoing probe,[1] $\mathrm{q} := (\mathrm{p}_A - \mathrm{p}_{A'})$ is the 4-momentum that the probe transfers to the target. In elastic collisions, $A + B \to A' + B'$, we have that $\mathrm{q} = (\mathrm{p}_{B'} - \mathrm{p}_B)$.

Equivalence A binary relation $\sim$ between elements of a set A is an equivalence if and only if it is (1) reflexive ($a \sim a$), (2) symmetric (if $a \sim b$ then $b \sim a$), (3) transitive (if $a \sim b$ and $b \sim c$ then $a \sim c$). An equivalence class is a subset of A consisting of elements that are all equivalent to each other; different equivalence classes are disjoint subsets of A, and their union equals A.

Euler characteristic Denoted $\chi_E(\mathscr{X})$, the Euler (or the Euler–Poincaré) characteristic is the topological invariant of the topological space $\mathscr{X}$. If $\mathscr{X}$ is a real 2-dimensional surface that has a triangulation (an approximation by a network of finitely many triangles), $\chi_E(\mathscr{X}) = k_0 - k_1 + k_2$, where k_0 is the number of vertices (corners), k_1 the number of edges and k_2 the number of triangles. A generalization exists also to higher-dimensional spaces (using a generalization of triangles): $\chi_E(\mathscr{X}) = \sum_{i=0}^{\dim X} (-1)^i k_i$, where k_0, k_1, k_2 are defined as for surfaces, k_3 the number of (exclusively tetrahedral) 3-dimensional elements, etc.

Extremal black hole A nontrivial solution of the Einstein equations, such as the Reissner–Nordstrøm solutions (9.61) where the two horizons coincide, $2r_q = r_s$, and which is marginal between the solutions where the singularity is screened by the event horizon and the solutions where it is not, i.e., solutions with a naked singularity.

Fermion By definition, a particle (as well as its mathematical representatives: wave-functions, creation and annihilation operators, or fields) that obeys Pauli's exclusion principle (two fermions cannot be in the same quantum state) and therefore also the Fermi–Dirac statistics. Owing to the spin-statistics theorem (in Lorentz-covariant models), physical particles whose mathematical representatives transform as spinorial representations of the Lorentz group are fermions. Fermionic wave-functions and fields have values that are anticommuting "numbers" ("a-numbers").

Fibration The space obtained by generalizing the tensor product of two spaces, where one of the factors in the product changes "along" the other factor. The type of that change (continuous, smooth, analytic, complex-analytic, …) distinguishes the various fibrations. Even the topology, i.e., homotopy of the variable factor may change, i.e., this factor may change discontinuously. [☛ homotopy class, Figure 11.7 on p. 427]

Field (mathematics) A collection of elements, $\Bbbk$, for which two operations, # and $*$, are defined so that:

1. $(\Bbbk, \#)$ is an abelian (commutative) group, with $e \in \Bbbk$ the neutral element;
2. $(\Bbbk \smallsetminus \{e\}, *)$ is an abelian (commutative) group;
3. the distribution rules $a * (b \# c) = (a * b) \# (a * c)$ and $(a \# b) * c = (a * c) \# (b * c)$ hold.

Field (physics) A function over spacetime. A scalar field is a function the values of which are scalars, a vector field is a function the values of which are vectors, etc. By a *"gauge field,"* however, one means the concrete fields such as the electric and magnetic fields, and their generalizations to other gauge models. [☛ gauge field] Variations/perturbations in a field are quantized in quantum physics. [☛ quantum]

Flavor The type of quark – distinguished by their masses and various charges, see the tabulation (2.44a). These are eigenstate of the free (propagation) Hamiltonian, and flavor ranges over *up, down, strange, charm, beauty* and *top*.

Gauge fields In the most familiar example, electromagnetism, these are the electric and the magnetic fields, which jointly form Maxwell's tensor $F_{\mu\nu}$ [☞ relations (5.73)]. More generally,

[1] A and A' are one and the same particle, with changed kinematical parameters: energy, linear momentum and angular momentum, including spin.

Yang–Mills gauge fields are the components of the matrix-valued tensor $\mathbb{F}_{\mu\nu}$ [☞ definition (6.15)], and for gravity these are the components of the Riemann tensor (9.30). In the most general case, gauge fields are defined, up to multiplicative constants, as the result of computing $[D_\mu, D_\nu]$, where D_μ are the correspondingly gauge-covariant derivatives, so $[D_\mu, D_\nu]$ is a measure of the non-commutativity of the changes of the considered generalized (complex-, vector-, tensor-, spinor-, matrix-, Lie-algebra-, ... valued) functions, i.e., the curvature of the space of such generalized functions. [☛ covariant derivative]

Gauge potential In the most familiar example, electromagnetism, these are the scalar and the vector potentials that jointly form the 4-vector A_μ [☞ relations (5.73)] and represent the difference between the covariant and the partial derivative [☞ definition (5.13)]. More generally, Yang–Mills gauge potentials form a matrix-valued 4-vector $\mathbb{A}_\mu$ [☞ definition (6.6a)], and for gravity these are the Christoffel symbols (9.17). In the most general case, the gauge potential is the difference between the gauge-covariant and the partial derivative: $\Gamma = D - \partial$. [☛ covariant derivative, potential]

Geodesic completeness The property of a given coordinate system with the given metric tensor that the limiting points of all geodesic lines (9.48) are within the range of those coordinates. A typical nontrivial example is the surface of a torus, for which we choose the coordinates (x, y), where x parametrizes the "little circle" so $x \simeq x + 2\pi R_1$, and y parametrizes the "big circle" so $y \simeq y + 2\pi R_2$, with $R_2 \geqslant R_1$. The coordinate system (x, y) is thus geodesically complete. As a counter-example, consider the "northern" stereographic projection of a sphere to the (x, y)-plane, so that the south pole corresponds to the coordinate origin and the equator to the circle of unit radius centered at the coordinate origin. Then geodesic lines on the sphere that contain the north pole correspond to geodesic lines in the plane that contain the point at infinity – which is not within the range of the coordinates. Such geodesic lines are thus incomplete or even disconnected, so that the coordinate system (x, y) with any Euclidean metric is geodesically incomplete as a description of a sphere.

Geometric quantization The process of constructing a quantum theory from the original classical theory, which uses the symplectic structure ω of the phase space Φ of the classical theory [288, 173, 579, 56]. Observables in classical theory are simply real functions $\mathcal{A}, \mathcal{B}, \mathcal{C}, \ldots$ over Φ. Geometric quantization is based on the introduction of a ω-compatible *polarization* $\pi(\Phi)$. In physics practice, π denotes the concrete choice of the half of the coordinates in the phase space Φ, which are the canonical coordinates, q^i, for which the ω-complementary half of the coordinates over Φ play the role of canonically conjugate momenta, p_i. With that standard notation, the symplectic structure is simply given by the Poisson brackets $\omega(\mathcal{A}, \mathcal{B}) := \frac{\partial \mathcal{A}}{\partial q^i}\frac{\partial \mathcal{B}}{\partial p_i} - \frac{\partial \mathcal{A}}{\partial p_i}\frac{\partial \mathcal{B}}{\partial q^i}$. That same polarization produces the quantum observables $A = \pi(\mathcal{A})$, $B = \pi(\mathcal{B})$, etc. The difference

$$\Delta := [\pi(\mathcal{A}), \pi(\mathcal{B})] - \pi\big(\omega(\mathcal{A}, \mathcal{B})\big) \tag{B.7}$$

is one of the measures of *anomaly*. [☛ anomaly]

Geometrization of physics The process by which physics is increasingly described in terms of geometry. At its simplest, this is the dual interpretation of the geodesic equation either as a bending of trajectories owing to spacetime curvature (9.48) or owing to the action of a gravitational force (9.49). At a rather more comprehensive level, in string theory models compactified on a space $\mathscr{Y}$, many of the physical properties of the effective particle physics model are derived as geometrical and topological characteristics of $\mathscr{Y}$; see discussion on p. 402 and in Section 11.3.1.

Ghost field Of the four components of the gauge 4-vector potential $\mathbb{A}_\mu$, only two correspond to degrees of freedom with a physical meaning. It turns out that it is possible to introduce two (anticommuting scalar) "ghost fields," the detailed kinematics and dynamics of which are chosen precisely so as to cancel the extraneous contributions of the two unphysical degrees of freedom in the 4-vector $\mathbb{A}_\mu$ [441, 425, 555, 484, 496, 589, 590]. The gauge symmetry is thereby reduced to the nilpotent BRST symmetry.

Gluon The particle (quantum) that mediates the strong interaction. Gluons interact with each other as well as with quarks and antiquarks, which they bind into hadrons. The interaction between hadrons is then a residual interaction, just as the molecular forces between electrically neutral atoms are modeled as dipole–dipole and higher order electromagnetic interactions [☞ Section 6.1.1].

Gödel's incompleteness theorem This theorem proves that no axiomatic system that is sufficiently complex to contain arithmetics can be both complete and self-consistent. Gödel's proof is constructive, and shows that within all such self-consistent axiomatic systems it is explicitly possible to construct a statement that can neither be proven nor disproven within the given axiomatic system. Therefore, either that statement or its logical negation may *always* be added to the axiomatic system as a new axiom, and this extensibility never stops [211, 376]. Although Gödel constructed a particular undecidable statement in his proof, and expressly for the purpose of proving the theorem, it does follow that there exist infinitely many such undecidable statements – and some of those, within physics as a formal axiomatic system, are bound to be of interest. [☛ Appendix B.3]

Gram–Schmidt procedure In a vector space V, equipped with a finite scalar product, i.e., where $\langle a|b\rangle < \infty$ for every $a, b \in V$, the Gram–Schmidt procedure produces an *orthonormal basis*:

1. Pick an element $a \in V$ and define $\alpha_1 := a/\sqrt{\langle a|a\rangle}$ and set $k = 1$.
2. If there is some $b \in V$ that is linearly independent from $\alpha_i \in \mathscr{B}_V := \{\alpha_1, \dots, \alpha_k\}$,
   (a) Define $\alpha_{k+1} := \sum_{i=1}^{k} c_i\alpha_i + c_{k+1}b$.
   (b) Determine $\{c_1, \dots, c_{k+1}\}$ so that
      i. $\langle \alpha_{k+1}|\alpha_i\rangle = 0$, for all $i = 1, \dots, k$,
      ii. and $\langle \alpha_{k+1}|\alpha_{k+1}\rangle = 1$.
   (c) Increase k by one ($k \mapsto k+1$), and return to step 2.
3. The basis for the vector space V is $\mathscr{B}_V = \{\alpha_1, \dots, \alpha_k\}$ and $\dim(V) = k$.

Group A collection of elements G equipped with a binary operation $\star$ that satisfies the four axioms [☛ Appendix A.1.1]:
closure $\forall a, b \in G$, $(a \star b) \in G$;
associativity $\forall a, b, c \in G$, $(a \star b) \star c = a \star (b \star c)$;
neutral element $\exists e \in G$ such that $\forall a \in G$, $a \star e = a = e \star a$;
inverse element $\forall a \in G$, $\exists a^{-1} \in G$ such that $a^{-1} \star a = e = a \star a^{-1}$.
That is, a group is an invertible monoid.

Groupoid [☛ magma]

Hadron A particle that interacts by means of the strong nuclear force (at $\sim 10^{-23}$ s) and can be detected as an isolated particle; e.g., a proton or a pion.

Hausdorff space A topological space in which distinct points have disjoint neighborhoods. Most variables typically considered in physics models span/form Hausdorff spaces. Examples of non-Hausdorff spaces include bifurcating (Y-shaped) 1-dimensional lines such as the Feynman diagrams (3.130)–(3.131) and the left-hand side of Figures 11.3 on p. 411 and 11.4 on p. 412. [☛ topological space]

Helicity The eigenvalue of the operator $\hat{p}\cdot\vec{S}/\hbar$, i.e., the projection of spin in the direction of motion of the particle, in units of $\hbar$. As massless particles move at the speed of light in vacuum, their helicity is Lorentz-invariant and equals their chirality.

Hermitian conjugation [☛ Digression 10.2 on p. 360]

Homotopy class Geometric objects that can be continuously transformed one into another form a homotopy class of such objects; different objects in the same homotopy class are homotopy equivalents of each other. Continuous interpolation between two homotopy equivalent objects is called the homotopy (between those two objects). Thus is the surface of a sphere a homotopy equivalent of the surface of a cube and a tetrahedron for example, but not of a torus or a pretzel.

Hypersurface The subspace $X \subset Y$ is a hypersurface if the codimension $\mathrm{cod}(X \subset Y) = 1$; near every point $x \in X$, the subspace $X \subset Y$ is specified by a single constraint.

Image For a mapping $f : X \to Y$, the f-image of the space X is the collection of points in Y obtained by mapping the points of X: $\mathrm{im}(f) = f(X) = \{f(x) = y \in Y : x \in X\}$.

Injection A "1–1" (one-to-one) mapping $f : X \hookrightarrow Y$, such that for every $a \in A$ there is precisely one $y = f(x) \in Y$.

Isometry A symmetry of a space $\mathscr{X}$ that leaves the metric on $\mathscr{X}$ unchanged.

Isomorphism Bijective homomorphism, i.e., a bijection $f : X \to Y$ for which both f and f^{-1} preserve the algebraic structure of the objects X and Y, and so are ***homomorphisms***. For example, if X and Y are groups, the f-image of every group axiom in X results in the corresponding group axiom in Y, and vice versa. We write $X \cong Y$.

KamiokaNDE The Kamioka Nucleon Decay Experiment, run at the Kamioka Observatory, Institute for Cosmic Ray Research, near the Kamioka section of the city of Hida, Japan. KamiokaNDE was initially designed to detect proton decay, but was successfully used to detect solar and atmospheric neutrinos, through upgrades known as KamiokaNDE-II, Super-KamiokaNDE, Super-KamiokaNDE-II and -III.

Kernel Elements of a vector space X that a linear mapping $f : X \to Y$ maps to $0 \in Y$ form the ***kernel*** of the linear mapping f, denoted $\ker(f) := \{x \in X,\ f(y) = 0 \in Y\}$. In other words, $\ker(f)$ consists of the elements of the vector space X annihilated by the mapping f.

Kronecker product The special case of the *tensor product* for matrices of arbitrary size, so including also column-matrices and row-matrices. The result of the Kronecker product is the block-matrix:

$$\mathbb{A} = \begin{bmatrix} a & b & c \\ d & e & f \end{bmatrix}, \quad \mathbb{B} = \begin{bmatrix} \alpha \\ \beta \end{bmatrix}, \quad \text{then} \quad \mathbb{A} \otimes \mathbb{B} = \begin{bmatrix} a\begin{bmatrix} \alpha \\ \beta \end{bmatrix} & b\begin{bmatrix} \alpha \\ \beta \end{bmatrix} & c\begin{bmatrix} \alpha \\ \beta \end{bmatrix} \\ d\begin{bmatrix} \alpha \\ \beta \end{bmatrix} & e\begin{bmatrix} \alpha \\ \beta \end{bmatrix} & f\begin{bmatrix} \alpha \\ \beta \end{bmatrix} \end{bmatrix} = \begin{bmatrix} a\alpha & b\alpha & c\alpha \\ a\beta & b\beta & c\beta \\ d\alpha & e\alpha & f\alpha \\ d\beta & e\beta & f\beta \end{bmatrix}. \tag{B.8}$$

Note that $\mathbb{B} \otimes \mathbb{A} \neq \mathbb{A} \otimes \mathbb{B}$:

$$\mathbb{B} \otimes \mathbb{A} = \begin{bmatrix} \alpha\begin{bmatrix} a & b & c \\ d & e & f \end{bmatrix} \\ \beta\begin{bmatrix} a & b & c \\ d & e & f \end{bmatrix} \end{bmatrix} = \begin{bmatrix} \alpha a & \alpha b & \alpha c \\ \alpha d & \alpha e & \alpha f \\ \beta a & \beta b & \beta c \\ \beta d & \beta e & \beta f \end{bmatrix} = \begin{bmatrix} a\alpha & b\alpha & c\alpha \\ d\alpha & e\alpha & f\alpha \\ a\beta & b\beta & c\beta \\ d\beta & e\beta & f\beta \end{bmatrix} \neq \begin{bmatrix} a\alpha & b\alpha & c\alpha \\ a\beta & b\beta & c\beta \\ d\alpha & e\alpha & f\alpha \\ d\beta & e\beta & f\beta \end{bmatrix} = \mathbb{A} \otimes \mathbb{B}. \tag{B.9}$$

Kronecker symbol The index representation of the identity matrix

$$\delta^i_j := \begin{cases} 1, & \text{if } i = j, \\ 0, & \text{if } i \neq j, \end{cases} \tag{B.10}$$

allows the generalizations after the pattern:

$$\delta^{ij}_{[k\ell]} := \tfrac{1}{2}\big(\delta^i_k\delta^j_\ell - \delta^i_\ell\delta^j_k\big), \qquad \delta^{ij}_{(k\ell)} := \tfrac{1}{2}\big(\delta^i_k\delta^j_\ell + \delta^i_\ell\delta^j_k\big), \tag{B.11a}$$

$$\delta^{ijk}_{[\ell mn]} := \tfrac{1}{3!}\big(\delta^i_\ell\delta^j_m\delta^k_n - \delta^i_\ell\delta^j_n\delta^k_m + \delta^i_n\delta^j_\ell\delta^k_m - \delta^i_n\delta^j_m\delta^k_\ell + \delta^i_m\delta^j_n\delta^k_\ell - \delta^i_m\delta^j_\ell\delta^k_n\big), \tag{B.11b}$$

$$\delta^{ijk}_{(\ell mn)} := \tfrac{1}{3!}\big(\delta^i_\ell\delta^j_m\delta^k_n + \delta^i_\ell\delta^j_n\delta^k_m + \delta^i_n\delta^j_\ell\delta^k_m + \delta^i_n\delta^j_m\delta^k_\ell + \delta^i_m\delta^j_n\delta^k_\ell + \delta^i_m\delta^j_\ell\delta^k_n\big), \quad \text{etc.,} \tag{B.11c}$$

which are also called (anti-)symmetrized Kronecker symbols.

Lepton A particle that does not interact by means of the strong nuclear force (at $\sim 10^{-23}$ s); e.g., the electron.

Levi-Civita symbol The index representation of the permutation symbol

$$\varepsilon_{i_1\cdots i_n} := \begin{cases} +1, & \text{if the order } i_1,\ldots,i_n \text{ is an even permutation of } 1,2,\ldots,n, \\ -1, & \text{if the order } i_1,\ldots,i_n \text{ is an odd permutation of } 1,2,\ldots,n, \\ 0, & \text{otherwise.} \end{cases} \tag{B.12}$$

We also define $\varepsilon^{i_1\cdots i_n} := \varepsilon_{i_1\cdots i_n}$. (Some Authors prefer using a definition such that $\varepsilon^{i_1\cdots i_n} := -\varepsilon_{i_1\cdots i_n}$, for numerical convenience in some computations.) The key relation between the Levi-Civita and the Kronecker symbols is

$$\begin{aligned} \varepsilon^{i_1\cdots i_n}\,\varepsilon_{j_1\cdots j_n} &= \delta^{\,i_1\cdots i_n}_{[j_1\cdots j_n]}, \\ &= \tfrac{1}{n!}\Big(\delta^{i_1}_{j_1}\cdots\delta^{i_{n-1}}_{j_{n-1}}\delta^{i_n}_{j_n} - \delta^{i_1}_{j_1}\cdots\delta^{i_{n-1}}_{j_n}\delta^{i_n}_{j_{n-1}} + \cdots \;\; (n!\text{ permutations, total})\Big). \end{aligned} \tag{B.13}$$

Lie group [☛ Appendix A.1.1]

Luxon a particle that travels through vacuum at the speed of light in vacuum, c, and has no mass. All mediators of gauge interactions that correspond to unbroken gauge symmetries are luxons.

Magma (groupoid) A collection of elements M equipped with a *closed* binary operation $\star$, i.e., $\forall a,b \in M,\ (a\star b) \in M$.

Manifold A space where every sufficiently small neighborhood of every point is isomorphic to the flat space $\mathbb{R}^n$, where n is the dimension of the manifold. A manifold is everywhere smooth and the tangent space at every point is a copy of $\mathbb{R}^n$.

Mass shell In the 4-dimensional space of 4-momentum, the "mass shell" for a particle of mass m is the subspace defined by the relation $E^2 - \vec{p}^2c^2 = m^2c^4$. For $m^2 > 0$ (ordinary particles and antiparticles), this is the two-component hyperboloid, where $E = \pm\sqrt{m^2c^4 + \vec{p}^2c^3}$ on both "shells." For $m = 0$ (photons, gluons and gravitons), this is the "light cone" the two portions of which touch in the point $(E/c, \vec{p}) = (0, \vec{0})$. For $m^2 < 0$ (tachyons), this is the single-component hyperboloid.

Meson Since the acceptance of the quark model in 1973, a bound state of a quark and an anti-quark. Originally, a particle that interacts by means of the strong nuclear force (at $\sim 10^{-23}$ s), can be detected as an isolated particle, and has a mass that is between the electron mass and the proton mass; e.g., $\pi^\pm, \pi^0$.

Minimal coupling The coupling between matter and interaction field that occurs by the interaction field modifying the spacetime derivative of the matter field. The gauge principle introduces only minimal coupling [☛ Chapters 5–7 and 9].

For example, let $\Psi(\mathrm{x})$ represent the matter field and $A_\mu(\mathrm{x})$ the gauge potential of the interaction field. They are minimally coupled through replacing $\partial_\mu\Psi \to (\partial_\mu + igA_\mu)\Psi$, where g is a suitable (coupling) parameter; $g = \frac{q_\Psi}{\hbar c}$ in electromagnetism, where q_Ψ is the electric charge of the matter particle represented by $\Psi(\mathrm{x})$.

Monoid A collection of elements M equipped with a binary operation $\star$ that satisfies the three axioms [☛ Appendix A.1.1]:

closure $\forall a,b \in M,\ (a\star b) \in M$;

associativity $\forall a,b,c \in M,\ (a\star b)\star c = a\star(b\star c)$;

neutral element $\exists e \in M$ such that $\forall a \in M,\ a\star e = a = e\star a$.

That is, a monoid is a semigroup with a neutral element.

Multipole expansion The expansion of a function over 3-dimensional flat space, in which we use spherical coordinates, over the complete system of spherical harmonics [☞ relations (4.2)–(4.4)]:

$$F(r,\theta,\phi) = \sum_{\ell,m} f_\ell^m(r)\, Y_\ell^m(\theta,\phi), \tag{B.14}$$

where

$$f_\ell^m(r) := \int_0^{2\pi} \mathrm{d}\phi \int_0^{\pi} \sin\theta\, \mathrm{d}\theta\, \big(Y_\ell^m(\theta,\phi)\big)^* F(r,\theta,\phi), \tag{B.15}$$

$$\vec{\nabla}^2 F(r,\theta,\phi) = \frac{1}{r}\Big[\frac{\mathrm{d}^2}{\mathrm{d}r^2} r\, F(r,\theta,\phi)\Big] - \frac{1}{r^2}\Big[\vec{L}^2 F(r,\theta,\phi)\Big], \tag{B.16}$$

$$\vec{L}^2 Y_\ell^m(\theta,\phi) = \ell(\ell{+}1)\, Y_\ell^m(\theta,\phi); \qquad \vec{L}^2 := -\vec{\nabla}^2\big|_{r=1}, \ \ell \geqslant 0. \tag{B.17}$$

Notice that the coordinates θ,ϕ parametrize a 2-sphere, $S^2 = \mathbb{R}^3\big|_{r=1}$. More generally, for every compact Riemann space $\mathscr{K}$, the Laplacian $\vec{\nabla}^2_K|_{r=1}$ has a non-positive spectrum (collection of eigenvalues), and corresponding eigenfunctions, which generalize the spherical harmonics.

Noether theorem To every continuous symmetry of a physical system in classical physics, there corresponds an additive current density that satisfies the continuity equation, and produces an additive conserved charge. In quantum theory, the conserved charges are eigenvalues of generators of the corresponding symmetries, and these in turn are the momenta canonically conjugate to the canonical variables the (eigen)values of which the symmetries change. For example, the linear momentum $\vec{p}$ is the eigenvalue of the operator of linear momentum $\vec{p}$, and also the conserved "charge" of the corresponding translations in position $\vec{r}$, which is generated by $\vec{p} = \frac{\hbar}{i}\vec{\nabla}$ and implemented by the unitary operator $\exp\{i\vec{a}\cdot\vec{p}/\hbar\} = \exp\{\vec{a}\cdot\vec{\nabla}\}$.

Conserved "charges" of finite symmetries are multiplicative: a product of two parity eigenfunctions is also a parity eigenfunction, with the eigenvalue that is a product of eigenvalues of the factors. Although Noether's original theorem does not apply to finite symmetries, the generalization is easy to derive. However, the operators that implement discrete symmetries may be both linear (and so unitary), and anti-linear (and then anti-unitary), such as the operator of charge conjugation: $\mathsf{C}(\alpha\mathsf{A}) = \alpha^*\mathsf{C}(\mathsf{A})$, for every operator A and constant $\alpha \in \mathbb{C}$.

Non-abelian [☛ abelian]

Non-commutative [☛ abelian]

Nonphysical components Within every Lorentz-covariant formalism, one uses only fields and operators that form complete representations of the Lorentz group. Thus, for example, gauge potentials in (3+1)-dimensional spacetime are always presented by 4-vectors, $A_\mu(x)$. However, only two components of this 4-vector are physically measurable, while two are not: for example, for a freely propagating field in empty space, the temporal and the longitudinal components are nonphysical. There exists no Lorentz-covariant method of isolating them from the 4-tuple (A_0, A_1, A_2, A_3). For example, the Lorenz gauge, $\eta^{\mu\nu}\partial_\mu A_\nu = 0$ specifies one *differential* relation between the 4-vector components $A_0(x), \ldots, A_3(x)$ in a Lorentz-invariant way, which formally permits expressing one of the four components in terms of an *integral* of the derivatives of the other three components. This effectively removes one degree of freedom, but this relation is not local. However, for the removal of the other nonphysical component, there does not even exist a Lorentz-invariant gauge condition – neither algebraic nor differential. [☛ BRST quantization]

Normal subgroup A subgroup $N \subset G$ is normal if

$$\forall n \in N \subset G, \ \ \forall g \in G, \ \ gng^{-1} \in N. \tag{B.18}$$

Ockham's principle Also known as *Ockham's razor*, as well as the *principle of parsimony*, of *economy* and of *succinctness*, whereby from among two competing possible explanations one must choose the simpler. Although this principle is useful in research practice, one must recognize that its application depends strongly on the cultural "background": ideas and elements that are well known within one culture (and are therefore regarded simpler) may well be alien in another culture. Thus, there is a danger that the application of this principle is simply a façade of a prejudice.

Pauli's principle Two identical fermions cannot simultaneously be in the same quantum state, i.e., they cannot simultaneously occupy the same "place" in the Hilbert space.

Photon The particle (quantum) that mediates the electromagnetic interaction. Photons interact directly with quarks, antiquarks, (electrically) charged leptons (e^-, μ^- and τ^-) and also with the charged weak gauge bosons $W^{\pm}$ [☞ Sections 2.3.4 and 5.2.2].

Physical components In practice, physical quantities are not infrequently represented by multi-component mathematical objects such as vectors, tensors and spinors. Components that in some way may be measured experimentally (such as the transversal polarizations of the electromagnetic radiation, for example) are *physical*. [☛ nonphysical components]

Point-like The property of showing sign of neither internal structure nor spatial extension.

Potential Short for "gauge potential," this term is used as a generalization of the electro-static potential, where we have that if $\Phi(\vec{r},t)$ is the potential, then:

1. $g\,\Phi(\vec{r},t)$ is the *potential energy* of a particle with charge g when placed in the potential $\Phi(\vec{r},t)$ that interacts with this charge,
2. $-\vec{\nabla}\Phi(\vec{r},t)$ is the *(gauge) field* corresponding to the potential,
3. $-g\,\vec{\nabla}\Phi(\vec{r},t)$ is the force that the potential $\Phi(\vec{r},t)$ exerts on a particle of charge g.

In the relativistic generalization, one speaks of the "4-vector potential," $(\Phi, -c\vec{A})$, for which the *fields* are the components of the $F_{\mu\nu} := (\partial_\mu A_\nu - \partial_\nu A_\mu)$ tensor [☞ definitions (5.73)]; in the non-abelian (non-commutative) generalization the *fields* are defined as the components of the $\mathbb{F}_{\mu\nu} := [D_\mu, D_\nu]$ tensor, where $D_\mu := \partial_\mu + \frac{iq}{\hbar c}\mathbb{A}_\mu$ [☞ definition (6.15)]. Finally, in the general theory of relativity, Christoffel symbols and the connection 4-vector play the role of the *potential* and the components of the Riemann tensor are the *fields* [☞ Sections 9.2.1 and 9.2.2]. [☛ gauge potential]

Quantum In quantum physics, all material entities (matter as well as interactions thereof) are subject to quantization of the Hamilton action, which cannot vary continuously, but as integral multiples of the Planck constant, $\hbar$. Note that the "background" (settled, static, infinitely spread-out, classical, i.e., non-quantum) fields, such as the Coulomb field of a static charge distribution, are but a convenient idealization, representable by averaging over an infinite number of quanta. [☛ field (physics)]

Quotient space [☛ Appendix A.1.1]

Range (of a mapping) For a mapping $f: X \to Y$, this can variously denote either the *codomain* or the *image* of f; this ambiguity and this term are avoided herein.

Rank (of a mapping) For a mapping $f: X \to Y$, $\mathrm{rank}(f) = \dim\big(\mathrm{im}(f)\big) = \dim\big(f(X)\big)$.

Rank (of a tensor density) [☛ definition on p. 511]

Ring A collection of elements, $\Bbbk$, for which two operations, # and $*$, are defined so that:

1. $(\Bbbk, \#)$ is an abelian (commutative) group, with $e \in \Bbbk$ the neutral element;
2. $(\Bbbk, *)$ is a monoid (like a group, but without invertibility);
3. the distribution rules: $a*(b\#c) = (a*b)\#(a*c)$ and $(a\#b)*c = (a*c)\#(b*c)$ hold.

Semidirect product Some groups have the structure $G = H \ltimes N$, where $H \subset G$ is a subgroup, and $N \subset G$ is a normal subgroup [☛ normal subgroup]. This implies that the only common element is $N \cap H = \mathbb{1} \in G$, and that every group element $g \in G$ can be factorized as $g = h \circ n = n' \circ h'$, where $n, n' \in N$ and $h, h' \in H$. The group G is said to be an N-extension of the group H; it is also true that H is isomorphic to the quotient group G/N [☞ definition (A.6) for the quotient space, which here inherits the group structure].

A well-known example is the Poincaré group, $Po(1,3) = Spin(1,3) \ltimes \mathbb{R}^{1,3}$, which is the extension of the Lorentz group $Spin(1,3)$ by translations $\mathbb{R}^{1,3}$ in spacetime, and where the asymmetry of the symbol $\ltimes$ reminds us that the elements of the subgroup $Spin(1,3)$ map $\mathbb{R}^{1,3} \to \mathbb{R}^{1,3}$.

Semidirect sum Some algebras have the structure $\mathfrak{A} = \mathfrak{A}_1 \mathbin{\dot{+}} \mathfrak{A}_2$, where for every $a, b \in \mathfrak{A}_1$ and $c, d \in \mathfrak{A}_2$ it is true that

$$a * b \in \mathfrak{A}_1, \quad \text{but} \quad c * d,\ a * c,\ c * a \in \mathfrak{A}_2, \tag{B.19}$$

and where $*$ is a "multiplication" in the algebra $\mathfrak{A}$. Formally,

$$\mathfrak{A}_1 * \mathfrak{A}_1 \in \mathfrak{A}_1, \quad \text{but} \quad \mathfrak{A}_1 * \mathfrak{A}_2,\ \mathfrak{A}_2 * \mathfrak{A}_1,\ \mathfrak{A}_2 * \mathfrak{A}_2 \in \mathfrak{A}_2. \tag{B.20}$$

The algebra $\mathfrak{A}$ is said to be a $\mathfrak{A}_2$-extension of the algebra $\mathfrak{A}_1$. The asymmetry symbol "$\dot{+}$" here reminds us that $\mathfrak{A}_1$ maps $\mathfrak{A}_1 : \mathfrak{A}_2 \xrightarrow{*} \mathfrak{A}_2$, but it is not a standard notation in the literature, where mostly the uninformative symmetrical symbols $+$ and $\oplus$ are used, and it is left to the Reader to figure out from the context the direction of the inherently asymmetrical relation, i.e., whether $\mathfrak{A}_1 * \mathfrak{A}_2 \in \mathfrak{A}_2$ or $\mathfrak{A}_2 * \mathfrak{A}_1 \in \mathfrak{A}_1$.

Semigroup A collection of elements S equipped with a binary operation $\star$ that satisfies the two axioms [☛ Appendix A.1.1]:
closure $\forall a, b \in S,\ (a \star b) \in S$;
associativity $\forall a, b, c \in S,\ (a \star b) \star c = a \star (b \star c)$.
That is, a semigroup is an associative magma.

Signature In every real n-dimensional vector space V (over the scalar field $\Bbbk$) in which the scalar product $g(v_1, v_2) \in \Bbbk$ is defined for every $v_1, v_2 \in V$, one may find a basis in which $g(\ ,\ )$ is a diagonal matrix. For real vector fields (where $\Bbbk = \mathbb{R}$) the number of positive, negative and vanishing diagonal elements in the diagonalized $g(\ ,\ )$ is called the *signature*. The metric tensor (3.19), $(\eta_{\mu\nu}) = \text{diag}(1, -1, -1, -1)$, in (3+1)-dimensional spacetime has the signature $(1,3)$. A group of linear transformations is also said to have signature $(1,3)$ if those transformations preserve the scalar product (3.17) defined by the metric tensor of signature $(1,3)$; such transformations form the group $O(1,3)$; $SO(1,3)$ is the subgroup of transformations the determinant of which equals $+1$.

Span A maximal collection of linearly independent elements $\hat{\mathrm{e}}_i$, $i = 1, 2, 3 \ldots$, is said to *span* the vector space $V := \{v^i\, \hat{\mathrm{e}}_i,\ v^i \in \Bbbk\}$ over a given field of scalars $\Bbbk$.

Spin Intrinsic (albeit perhaps fictitious) angular momentum of an object (particle or physical system) X, meaning that under rotations of the coordinate system the *orientation* of the object X transforms as a representation of the rotation group with the given "angular momentum." For example, a photon has spin $1\hbar$, meaning that its orientation (i.e., polarization) transforms as a spin-$1\hbar$ (vector) representation of the rotation group, the electron as a spin-$\frac{1}{2}\hbar$ (spinor) representation of the rotation group, and the graviton as a spin-$2\hbar$ (rank-2 tensor) representation. The spin of composite systems is the vector sum of all angular momenta of its constituents,[2] but the spin of an elementary particle is not the result of any rotation: elementary particles are *point-like*.

Stückelberg–Feynman interpretation The antiparticle is identified with the particle moving backwards in time. This interpretation follows from the fact that if $\Psi(x)$ is the wave-function of the particle, then its Hermitian conjugate (and, for spin-$\frac{1}{2}$ particles, also the right multiple

[2] The spin of the hydrogen atom as a bound state of an electron and a proton is the vector sum of the orbital angular momentum of the electron in its orbit around the proton, as well as the electron's and the proton's spin.

by γ^0) produces the wave-function of the antiparticle. Expanding into a Fourier series we have that $\Psi(x) = \sum_\omega e^{i\omega t}\psi_\omega(\vec{r})$, so the Hermitian conjugation is formally identical with the reversal of time.

Surjection A mapping $f : X \twoheadrightarrow Y$, such that for every $y \in Y$ there is an $x \in X$ such that $f(x) = y$.

Symmetry breaking vs. violation A particular process is said to violate a symmetry X if either (1) the X-image of the process does not occur as frequently, i.e., with the same probability, as the original process, or (2) the conserved quantity corresponding to symmetry X is not constant (conserved) during the considered process.

In turn, the symmetry X is broken in a physical system if either (1) the symmetry does not preserve some of the conditions (such as a boundary condition) required of the concrete physical system, or (2) X does not commute with the full Hamiltonian of the system.

Tachyon A particle that propagates through vacuum faster than light, and has an imaginary mass; the appearance of tachyons indicates that the vacuum is not stable [☞ Digression 7.1 on p. 261].

Tangent bundle A vector bundle $\mathscr{T}_{\mathscr{X}} := E(\mathscr{X}; T_{\mathscr{X}}; \pi)$ where $T_x(\mathscr{X}) \cong T_{\mathscr{X}}$ is the tangent space of the space $\mathscr{X}$ at the point $x \in \mathscr{X}$. If x^μ are local coordinates in the space $\mathscr{X}$ at the given point, then $T_x(\mathscr{X})$ may be represented as the vector space of linear combinations $v^\mu \frac{\partial}{\partial x^\mu}$.

Tardion a particle that propagates through vacuum slower than light, and has a real mass; all known matter (and anti-matter) is tardionic, whereupon this term is rarely used.

Tensor product The most general bilinear product of two algebraic structures of the same type, such as vector spaces, algebras, etc. Let X and Y be two vector spaces over the same field, $\mathbb{k}$. The elements of the tensor product $\mathrm{X} \otimes \mathrm{Y}$ are $\mathbb{k}$-linear combinations of elements of the direct product of the sets of elements X and Y, where additionally one requires that the pairs of elements satisfy the relations

$$R := \begin{cases} e(x{+}x', y) \sim e(x,y) + e(x',y), \quad e(x, y{+}y') \sim e(x,y) + e(x,y'), \\ \qquad\qquad c\,e(x,y) \sim e(cx,y) \sim e(x,cy). \end{cases} \tag{B.21}$$

Then formally,

$$\mathrm{X} \otimes \mathrm{Y} = \Big\{ \sum_i c_i\, e(x_i, y_i) \;:\; c_i \in \mathbb{k},\ (x_i, y_i) \in \mathrm{X} \times \mathrm{Y} \Big\} \Big/ R, \tag{B.22}$$

which is again a vector space. Similarly, the tensor product of two algebras is again an algebra. In other words, the tensor product inherits the algebraic structure of its factors. Alternatively, Definition B.6 on p. 514 also holds – given using the components with respect to any chosen basis.

Topological space A set of elements ("points") $\mathscr{X}$ with the *topology* τ, which consists of a collection of subsets of the set $\mathscr{X}$ such that they satisfy the axioms:

1. The empty set and the whole set $\mathscr{X}$ belong to τ.
2. The union of an arbitrary number of sets in τ is also in τ.
3. The intersection of an arbitrary finite number of sets in τ is also in τ.

For this system of axioms, the sets in τ are called *open subsets* of the set $\mathscr{X}$; every point $x \in X$ is contained in at least one such open subset, which is then called the *open neighborhood* of the point x. There also exists a complementary definition of topology, using *closed subsets* of the set $\mathscr{X}$; the empty set and the set $\mathscr{X}$ itself here too belong to τ. [☛ also Hausdorff space]

Torsion [☛ curvature]

Vector bundle Let $\mathscr{X}$ be the "base" space, equipped with a copy of a vector space V_x at every point $x \in \mathscr{X}$ of the base space, so that the vector spaces V_x transform homogeneously one into another when the basis point x moves through the base space. The union $\bigcup_{x\in\mathscr{X}} V_x$ is then called the vector bundle over the base space $\mathscr{X}$.

There is also a reverse definition: the total space $E(\mathscr{X}; V; \pi)$ of a vector bundle with a given vector space V over the base space $\mathscr{X}$ is such that π is the "vertical" projection with the property that $\pi(E) = \mathscr{X}$, and $\pi^{-1}(x) = V_x \cong V$ for each $x \in \mathscr{X}$.

Vector space A collection of elements (vectors) of which every linear combination with coefficients from a field $\Bbbk$ is also an element of this collection is called a vector space V over the field $\Bbbk$.

Warp, weft and woof are the mutually transversal strands of yarn in a simply woven fabric: *warp* stretches lengthwise from beginning to end, and the strand that is woven left to right and back, weaving through the strands of warp, is variably called *weft* or *woof*.

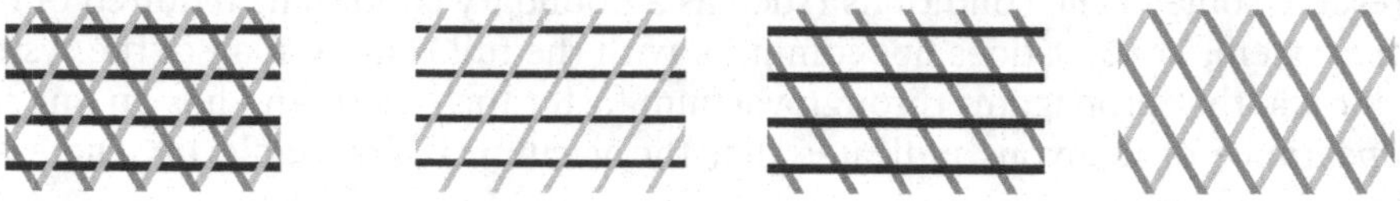

Figure B.1 The triple weave: leaving out any one of the strands dissolves the fabric.

In the theoretical fundamental physics as described herein, the three conceptual strands are provided by (1) the Democritean idea of a smallest portion of matter that shows no further, internal constituents, (2) the gauge principle of local symmetry, which provides a coherent description of all known fundamental interactions, and (3) the idea that all of Nature is to be understood within a unified, comprehensive and logically consistent framework. The (M- and F-theory extended) superstring theoretical system is a framework that conceptually unifies all matter, all of its interactions, as well as the spacetime in which they exist. [☛ Section 1.3.3; Chapters 5–7 and 9]

Wormholes The region in spacetime shaped as a "tunnel," $\mathbb{R}^r \times K^{d-r}$ for $1 \leqslant r < d$, where d denotes the total dimension of spacetime, and which either connects two otherwise distant regions of one spacetime, or two otherwise separate spacetimes; K^{d-r} is some compact space (e.g., the 2-sphere, S^2) and represents the "cross-section" of the "tunnel." In known examples, the size of the "cross-section" is typically very small, of the order of $\ell_P \sim 10^{-35}$ m and most often has a nonzero size only for a very short time, $t_P \sim 10^{-43}$ s. The matter required to keep the wormhole open for a material body or even light to pass through must have "exotic" properties (negative energy density and/or pressure). [☛ Section 9.3.4, Einstein–Rosen bridge]

Yang–Mills interaction, symmetry, theory A gauge interaction, model, symmetry and/or theory is said to be of Yang–Mills type when the gauge 4-vector potential, $\mathbb{A}_\mu \propto (D_\mu - \partial_\mu)$, is the fundamental physical degree of freedom that describes such an interaction. This is the case with electromagnetic, strong and weak nuclear interactions [☞ Chapters 5 and 6], but not with gravity: there, the Christoffel symbol, $\boldsymbol{\Gamma} \propto (D_\mu - \partial_\mu)$, may be expressed as an algebraic combination of the inverse metric tensor and the derivatives of the metric tensor [☞ Chapter 9].

Yukawa field, potential (screened Coulomb field, potential) The Yukawa potential in d-dimensional space is $\Phi_Y = e^{-r/r_0}/r^{d-2}$ and the Yukawa field is $-\vec{\nabla}\Phi_Y$; the negative sign is chosen so that the $r_0 \to \infty$ limiting case of the Yukawa field coincides with the traditional definition of the electrostatic field. Here, r_0 is the range of the Yukawa potential and the field.

Yukawa interaction (Yukawa coupling) The coupling between matter field $\Psi(\mathrm{x})$ and the Yukawa potential $\Phi(\mathrm{x})$ produced by the Lagrangian density term $h_\Psi \overline{\Psi}\Phi\Psi$, where h_Ψ is the Yukawa coupling parameter [☛ contact interaction].

ZJBV-quantization [☛ BV-quantization]

B.2 Tensor calculus basics

We start with 4-tuples of coordinates such as $\mathrm{x} = (x^0, x^1, x^2, x^3)$, two functions of such coordinates, f and g, and the well-known derivative rules in multi-variate calculus:

$$\textbf{product rule}\quad \frac{\partial}{\partial x^\mu}\big(f(\mathrm{x})\,g(\mathrm{x})\big) = \Big(\frac{\partial f(\mathrm{x})}{\partial x^\mu}\Big)g(\mathrm{x}) + f(\mathrm{x})\Big(\frac{\partial g(\mathrm{x})}{\partial x^\mu}\Big), \tag{B.23}$$

$$\textbf{chain rule}\quad \frac{\partial}{\partial x^\mu}\Big(y^\nu(\mathrm{z}(\mathrm{x}))\Big) = \Big(\frac{\partial y^\nu}{\partial z^\rho}\Big)\Big(\frac{\partial z^\rho}{\partial x^\mu}\Big). \tag{B.24}$$

Taking $\mathrm{x} = (x^0, x^1, x^2, x^3)$, $\mathrm{y} = (y^0, y^1, y^2, y^3)$ and $\mathrm{z} = (z^0, z^1, z^2, z^3)$ to provide *general* coordinate systems, these 4-tuples need not span vector spaces in general: In general coordinate systems, linear combinations $c_\mu x^\mu$ with numerical (dimensionless) constants c_μ need make no sense at all. At the very least, the constants c_μ could be equipped with appropriate physical units. For example, in the familiar spherical coordinate system (r, θ, ϕ), a linear combination such as $(\frac{\pi}{2}r - \sqrt{3}\theta)$ makes no sense since the two summands have wholly different physical units. In turn, denoting by L some suitable and constant length, the linear combination $(\frac{\pi}{2L}r - \sqrt{3}\theta)$ does make sense in general, although it does not seem to provide any physically reasonable quantity. Even so, and owing to the generally curvilinear nature of general coordinates and their diverse behavior (e.g., $\theta \simeq \theta \pm 2\pi$ while $r \geqslant 0$), linear combinations (even if adjusted for physical units) of general coordinates do not, in general, represent a point in the space parametrized by these coordinates.

However, owing to the infinitesimal nature of the differentials $\mathrm{d}x^\mu$ and the operators $\frac{\partial}{\partial x^\mu}$, the 4-tuples $(\mathrm{d}x^0, \mathrm{d}x^1, \mathrm{d}x^2, \mathrm{d}x^3)$ and $(\frac{\partial}{\partial x^0}, \frac{\partial}{\partial x^1}, \frac{\partial}{\partial x^2}, \frac{\partial}{\partial x^3})$ *do* span two vector spaces – again with the proviso that the constants in the respective linear combinations may have to be equipped with adequate physical units. The application of the chain rule to these clearly distinguishes them and permits the definition of two *distinct* types of 4-vectors:

$$\textbf{contravariant vector}\ (3.11\text{c})\quad \mathrm{d}x^\mu = \mathrm{d}y^\nu\Big(\frac{\partial x^\mu}{\partial y^\nu}\Big) \quad\leftrightarrow\quad A^\mu(\mathrm{x}) = A^\nu(\mathrm{y})\Big(\frac{\partial x^\mu}{\partial y^\nu}\Big); \tag{B.25}$$

$$\textbf{covariant vector}\ (3.11\text{d})\quad \frac{\partial}{\partial x^\mu} = \Big(\frac{\partial y^\nu}{\partial x^\mu}\Big)\frac{\partial}{\partial y^\nu} \quad\leftrightarrow\quad B_\mu(\mathrm{x}) = \Big(\frac{\partial y^\nu}{\partial x^\mu}\Big)B_\nu(\mathrm{y}), \tag{B.26}$$

simply by observing that they transform with the *opposite* partial derivatives, as was already done in Digression 3.2 on p. 88.

B.2.1 Basis elements

We then proceed as follows: Given any coordinate system $\mathrm{x} := (x^0, x^1, x^2, x^3)$ equipped with a metric tensor, $g_{\mu\nu}(\mathrm{x})$, we specify:

1. The line element $\mathrm{d}s$ provides the invariant norm of the coordinate differentials:
$$\mathrm{d}s := \sqrt{\mathrm{dx}\cdot\mathrm{dx}}, \quad \mathrm{dx}\cdot\mathrm{dx} := g_{\mu\nu}(\mathrm{x})\,\mathrm{d}x^\mu\mathrm{d}x^\nu. \tag{B.27}$$
2. The invariant Kronecker symbol
$$\delta^\mu_\nu := \frac{\partial x^\mu}{\partial x^\nu} = \begin{cases} 1 \text{ if } \mu = \nu, \\ 0 \text{ if } \mu \neq \nu, \end{cases} \tag{B.28}$$
is simply the statement that the coordinates x^μ are mutually independent.
3. The invariant Levi-Civita symbol is defined implicitly by expanding the Jacobian of a coordinate transformation $\mathrm{x} \to \mathrm{y}$:
$$\Big|\frac{\partial \mathrm{x}}{\partial \mathrm{y}}\Big| =: \varepsilon^{\mu\nu\rho\sigma}\frac{\partial x^0}{\partial y^\mu}\frac{\partial x^1}{\partial y^\nu}\frac{\partial x^2}{\partial y^\rho}\frac{\partial x^3}{\partial y^\sigma} =: \varepsilon_{\mu\nu\rho\sigma}\frac{\partial x^\mu}{\partial y^0}\frac{\partial x^\nu}{\partial y^1}\frac{\partial x^\rho}{\partial y^2}\frac{\partial x^\sigma}{\partial y^3}, \tag{B.29}$$

that is,

$$\varepsilon^{\mu\nu\rho\sigma} = \varepsilon_{\mu\nu\rho\sigma} := \begin{cases} +1 & \text{if } \mu,\nu,\rho,\sigma = \text{even permutation of } 0,1,2,3; \\ -1 & \text{if } \mu,\nu,\rho,\sigma = \text{odd permutation of } 0,1,2,3; \\ 0 & \text{otherwise.} \end{cases} \tag{B.30}$$

4. It follows that

$$\varepsilon_{\alpha\beta\gamma\delta}\varepsilon^{\mu\nu\rho\sigma} = 4!\,\delta^{\mu\nu\rho\sigma}_{[\alpha\beta\gamma\delta]}, \tag{B.31}$$

where

$$\delta^{\mu\nu}_{[\alpha\beta]} := \tfrac{1}{2}(\delta^{\mu}_{\alpha}\delta^{\nu}_{\beta} - \delta^{\mu}_{\beta}\delta^{\nu}_{\alpha}), \quad \delta^{\mu\nu\rho}_{[\alpha\beta\gamma]} := \tfrac{1}{3}(\delta^{\mu\nu}_{[\alpha\beta]}\delta^{\rho}_{\gamma} + \delta^{\mu\nu}_{[\beta\gamma]}\delta^{\rho}_{\alpha} + \delta^{\mu\nu}_{[\gamma\alpha]}\delta^{\rho}_{\beta}),$$
$$\text{and so } \delta^{\mu\nu\rho\sigma}_{[\alpha\beta\gamma\delta]} := \tfrac{1}{4}(\delta^{\mu\nu\rho}_{[\alpha\beta\gamma]}\delta^{\sigma}_{\delta} - \delta^{\mu\nu\rho}_{[\delta\alpha\beta]}\delta^{\sigma}_{\gamma} + \delta^{\mu\nu\rho}_{[\gamma\delta\alpha]}\delta^{\sigma}_{\beta} - \delta^{\mu\nu\rho}_{[\beta\gamma\delta]}\delta^{\sigma}_{\alpha}). \tag{B.32}$$

5. Owing to the reciprocal transformation rules (3.11c)–(3.11d), the *contractions*

$$\mathrm{A}(\mathrm{x})\cdot\mathrm{B}(\mathrm{x}) = A^{\mu}(\mathrm{x})\,B_{\mu}(\mathrm{x}), \quad \mathrm{A}(\mathrm{x})\cdot\partial = A^{\mu}(\mathrm{x})\partial_{\mu}, \quad \mathrm{dx}\cdot\mathrm{B}(\mathrm{x}) = \mathrm{d}x^{\mu}\,B_{\mu}(\mathrm{x}), \tag{3.12a$'$}$$

$$\text{and} \quad \mathrm{d} := \mathrm{dx}\cdot\partial := \mathrm{d}x^{\mu}\frac{\partial}{\partial x^{\mu}} \tag{B.33}$$

are all invariant under general coordinate transformations $x^{\mu} \mapsto y^{\mu}(\mathrm{x})$, as specified in Definition 9.1 on p. 319. Thus, the $\mathrm{d}x^{\mu}$ may be used as basis vectors for covariant components $B_{\mu}(\mathrm{x})$, and the ∂_{μ} may be used as basis vectors for contravariant components $A^{\mu}(\mathrm{x})$. This is the typical *choice* in the mathematics literature as it connects tensor algebra and differential geometry; see Comment B.1 on p. 512.

6. Let $\mathrm{e}(\mathrm{x})$ denote an event – a point in spacetime specified, with the coordinates x, and let the displacement to an infinitesimally near event be $\mathrm{de} = \frac{\partial \mathrm{e}}{\partial x^{\mu}}\mathrm{d}x^{\mu}$, expressed in the x^{μ} coordinates. Then, we define:

$$\textbf{covariant basis element} \quad \mathrm{e}_{\mu}(\mathrm{x}) := \frac{\partial \mathrm{e}}{\partial x^{\mu}}, \tag{B.34}$$

$$\textbf{contravariant basis element} \quad \mathrm{e}^{\mu}(\mathrm{x}) := g^{\mu\nu}\mathrm{e}_{\nu}(\mathrm{x}). \tag{B.35}$$

The scalar product of these basis elements is defined so that

$$\mathrm{e}_{\mu}(\mathrm{x})\cdot\mathrm{e}_{\nu}(\mathrm{x}) = g_{\mu\nu}(\mathrm{x}), \quad \mathrm{e}^{\mu}(\mathrm{x})\cdot\mathrm{e}^{\nu}(\mathrm{x}) = g^{\mu\nu}(\mathrm{x}) \quad \text{and} \quad \mathrm{e}_{\mu}(\mathrm{x})\cdot\mathrm{e}^{\nu}(\mathrm{x}) = \delta^{\nu}_{\mu}. \tag{B.36}$$

7. Given the contravariant components of a 4-vector, $A^{\mu}(\mathrm{x})$, the 4-vector is invariantly specified as $\mathrm{A}(\mathrm{x}) = A^{\mu}(\mathrm{x})\,\mathrm{e}_{\mu}(\mathrm{x})$. Given the covariant components of a 4-vector, $B_{\mu}(\mathrm{x})$, the 4-vector is invariantly specified as $\mathrm{B}(\mathrm{x}) = B_{\mu}(\mathrm{x})\,\mathrm{e}^{\mu}(\mathrm{x})$.

Here, "invariant," "covariant" and "contravariant" all refer to transformation properties with respect to the general coordinate transformations specified in Definition 9.1 on p. 319.

Given the definition of contravariant vectors (B.25), it is straightforward to compute the transformation rule for the differential "volume" element:

$$\begin{aligned} \mathrm{d}^4x = \mathrm{d}x^0\mathrm{d}x^1\mathrm{d}x^2\mathrm{d}x^3 &= \tfrac{1}{4!}\varepsilon_{\mu\nu\rho\sigma}\,\mathrm{d}x^{\mu}\mathrm{d}x^{\nu}\mathrm{d}x^{\rho}\mathrm{d}x^{\sigma} \\ &= \tfrac{1}{4!}\varepsilon_{\mu\nu\rho\sigma}\Big(\frac{\partial x^{\mu}}{\partial y^{\alpha}}\mathrm{d}y^{\alpha}\Big)\Big(\frac{\partial x^{\nu}}{\partial y^{\beta}}\mathrm{d}y^{\beta}\Big)\Big(\frac{\partial x^{\rho}}{\partial y^{\gamma}}\mathrm{d}y^{\gamma}\Big)\Big(\frac{\partial x^{\sigma}}{\partial y^{\delta}}\mathrm{d}y^{\delta}\Big) \\ &= \tfrac{1}{4!}\varepsilon_{\mu\nu\rho\sigma}\frac{\partial x^{\mu}}{\partial y^{\alpha}}\frac{\partial x^{\nu}}{\partial y^{\beta}}\frac{\partial x^{\rho}}{\partial y^{\gamma}}\frac{\partial x^{\sigma}}{\partial y^{\delta}}\,\mathrm{d}y^{\alpha}\mathrm{d}y^{\beta}\mathrm{d}y^{\gamma}\mathrm{d}y^{\delta} \\ &= \tfrac{1}{4!}\varepsilon_{\mu\nu\rho\sigma}\frac{\partial x^{\mu}}{\partial y^{\alpha}}\frac{\partial x^{\nu}}{\partial y^{\beta}}\frac{\partial x^{\rho}}{\partial y^{\gamma}}\frac{\partial x^{\sigma}}{\partial y^{\delta}}\,\delta^{\alpha\beta\gamma\delta}_{[\epsilon\varphi\lambda\kappa]}\,\mathrm{d}y^{\epsilon}\mathrm{d}y^{\varphi}\mathrm{d}y^{\lambda}\mathrm{d}y^{\kappa} \end{aligned} \tag{B.37a}$$

$$= \Big[\tfrac{1}{4!}\varepsilon_{\mu\nu\rho\sigma}\frac{\partial x^\mu}{\partial y^\alpha}\frac{\partial x^\nu}{\partial y^\beta}\frac{\partial x^\rho}{\partial y^\gamma}\frac{\partial x^\sigma}{\partial y^\delta}\varepsilon^{\alpha\beta\gamma\delta}\Big]\,\tfrac{1}{4!}\varepsilon_{\epsilon\varphi\lambda\kappa}\,\mathrm{d}y^\epsilon\mathrm{d}y^\varphi\mathrm{d}y^\lambda\mathrm{d}y^\kappa$$

$$= \det\Big[\frac{\partial \mathrm{x}}{\partial \mathrm{y}}\Big]\,\mathrm{d}^4 y, \tag{B.37b}$$

where the key relation (B.31) between the Levi-Civita and the Kronecker symbols was used. We have also used the general expressions for the determinants of $n \times n$ matrices representing rank-2 tensor densities:

$$\textbf{type (1,1)}\qquad \det[\mathbb{M}] := \tfrac{1}{4!}\varepsilon_{\mu_1\cdots\mu_n}\,M^{\mu_1}_{\nu_1}\cdots M^{\mu_n}_{\nu_n}\,\varepsilon^{\nu_1\cdots\nu_n}, \tag{B.38a}$$

$$\textbf{type (0,2)}\qquad \det[\mathbb{N}] := \tfrac{1}{4!}\varepsilon^{\mu_1\cdots\mu_n}\,N_{\mu_1\nu_1}\cdots N_{\mu_n\nu_n}\,\varepsilon^{\nu_1\cdots\nu_n}, \tag{B.38b}$$

$$\textbf{type (2,0)}\qquad \det[\mathbb{P}] := \tfrac{1}{4!}\varepsilon_{\mu_1\cdots\mu_n}\,P^{\mu_1\nu_1}\cdots P_{\mu_n\nu_n}\,\varepsilon_{\nu_1\cdots\nu_n}. \tag{B.38c}$$

Given the definitions of the "ingredients":

1. a contravariant vector (3.11c),
2. a covariant vector (3.11d),
3. a scalar density (9.8),

we adapt Weyl's Construction A.1 and generate representations of the group of general coordinate transformations, by taking tensor products of the "ingredients" and symmetrizing like factors in all possible ways. More precisely,

Definition B.1 *Tensor densities may be formally constructed from a scalar density U, a contravariant vector $V = V^\mu \mathrm{e}_\mu$ and a covariant vector $W = W_\mu \mathrm{e}_\mu$:*

$$U(\mathrm{y}) = \Big(\det\Big[\frac{\partial \mathrm{y}}{\partial \mathrm{x}}\Big]\Big)U(\mathrm{x}),\quad V^\mu(\mathrm{y}) = \frac{\partial y^\mu}{\partial x^\nu}V^\nu(\mathrm{x}),\quad W_\mu(\mathrm{y}) = \frac{\partial x^\nu}{\partial y^\mu}W_\nu(\mathrm{x}). \tag{B.39}$$

One constructs first the vector space of ordered products,

$$T(p,q;w) := U^w \cdot \underbrace{V\otimes\cdots\otimes V}_{p}\otimes\underbrace{W\otimes\cdots\otimes W}_{q}, \tag{B.40}$$

on which the permutation group $S_p \times S_q$ acts, where S_p permutes the V-factors and S_q permutes the W-factors. The vector space $T(p,q;w)$ may then be decomposed, in a unique fashion, into a direct sum of irreducible representations of the permutation group (index symmetrization). Finally, each summand in the so-obtained direct sum may be further decomposed by ***contracting*** *with invariant tensors δ^μ_ν, $\varepsilon_{\mu\nu\rho\sigma}$ and $\varepsilon^{\mu\nu\rho\sigma}$.*

Focusing on the structure of the transformation properties, i.e., how a quantity transforms with respect to general coordinate transformations, rather than how it may have been constructed, produces the complementary general definition:

Definition B.2 (tensor density) *A quantity that is in some coordinate system (with coordinates x^μ) specified by its* ***components*** *$\{T^{\mu_1\cdots\mu_p}_{\nu_1\cdots\nu_q}(\mathrm{x})\}$ and the components of which in some other coordinate system (with coordinates y^μ) may be computed using the relations*

$$T^{\rho_1\cdots\rho_p}_{\sigma_1\cdots\sigma_q}(\mathrm{y}) = \Big(\det\Big[\frac{\partial \mathrm{y}}{\partial \mathrm{x}}\Big]\Big)^w \frac{\partial y^{\rho_1}}{\partial x^{\mu_1}}\cdots\frac{\partial y^{\rho_p}}{\partial x^{\mu_p}}\frac{\partial x^{\nu_1}}{\partial y^{\sigma_1}}\cdots\frac{\partial x^{\nu_q}}{\partial y^{\sigma_q}}\,T^{\mu_1\cdots\mu_p}_{\nu_1\cdots\nu_q}(\mathrm{x}) \tag{B.41}$$

is called a ***tensor density*** *of weight w, type (p,q) and rank $p+q$. Weight-0 tensor densities are called* ***tensors****; rank-1 tensors are called* ***vectors****, and rank-0 tensors are* ***scalars****, i.e.,* ***invariants****. The symbol $[\frac{\partial \mathrm{y}}{\partial \mathrm{x}}]$ denotes the matrix of partial derivatives that appear in equation (B.26).*

This special meaning of the word "density" – which in this special use always follows the adjective "tensor," "vector" or "scalar" – must not be confused with the familiar notion as in "per unit of volume." Thus, for example, "Lagrangian density" literally means "Lagrangian per unit of volume." On the other hand, in the sense of Definition B.2 and in the typical practice in theoretical and mathematical physics, Lagrangian densities are – as well as the Lagrangians and Hamiltonians and Hamiltonian densities – **scalars**, i.e., weight-0 scalar densities [☞ Conclusion 9.5 on p. 328].

Comment B.1 *Specifying all components in any one concrete basis* **does** *specify the tensor density abstractly, since the relations (B.41) provide the transformation rules from one basis into any other one. In the mathematical literature one typically uses the natural basis* $\{dx^\mu, \frac{\partial}{\partial x^\mu}\}$*, whereby a tensor density is specified* ***invariantly*** *as*

$$T(\mathrm{x}) \;:\quad dx^{\nu_1}\otimes\cdots\otimes dx^{\nu_p}\, T^{\mu_1\cdots\mu_p}_{\nu_1\cdots\nu_q}(\mathrm{x})\,\frac{\partial}{\partial x^{\mu_1}}\otimes\cdots\otimes\frac{\partial}{\partial x^{\mu_q}}. \tag{B.42a}$$

Using the relations (B.25)–(B.26) and (B.41), it is then easy to show that

$$T(\mathrm{y}) = \left(\det\left[\frac{\partial \mathrm{y}}{\partial \mathrm{x}}\right]\right)^w T(\mathrm{x}). \tag{B.42b}$$

In this book, I follow the physicists' practice of specifying and manipulating components (with respect to any one particular basis) as the representatives of the whole tensor density; see Digression 3.3 on p. 88, as well as the discussion in Wald's textbook [548].

B.2.2 Tensor algebra

Scalar functions (weight-0 scalar densities) over spacetime are, in the physics nomenclature, typically called scalar fields. These scalar fields (in the physics sense) form – at every spacetime point separately – a field in the mathematical sense. That is, addition and multiplication of scalar fields – taken at any particular spacetime point – follows the usual rules of addition and multiplication of "ordinary" (real and complex) numbers. It is, however, important to note that this is not the case when adding/multiplying scalar fields where the summands/factors are taken at different spacetime points: $f(\mathrm{x})\,g(\mathrm{y})$ is not a function of either just x or just y, but of both. Thus, scalar functions (over the whole spacetime) do not form the usual algebraic structure of a field. However, restricting the binary operations to the cases when both summands/factors are taken at the same spacetime point produces an algebraic structure that minimally deviates from the standard definition of the (mathematical) field, i.e., extends this definition.[3] The corresponding generalizations of functions (and all the tensor densities as well) over general, curved spaces are called *sections* of various *bundles* [☞ [563, 210, 379, 176], to begin with].

Similarly, tensor densities $T^{\mu_1\cdots\mu_p}_{\nu_1\cdots\nu_q}(\mathrm{x})$ may be multiplied by scalar densities $f(\mathrm{x})$ by simply multiplying each component. Also, it should be clear that the tensor densities of the same type and weight may be added, which permits defining point-by-point linear combinations such as

$$f(\mathrm{x})\,T^{\mu_1\cdots\mu_p}_{\nu_1\cdots\nu_q}(\mathrm{x}) + h(\mathrm{x})\,U^{\mu_1\cdots\mu_p}_{\nu_1\cdots\nu_q}(\mathrm{x}), \tag{B.43}$$

as long as the sum of weights of f and T equals the sum of weights of h and U, and this generates a structure that minimally generalizes the structure of a vector space:

[3] The deviation pertains precisely to the general case, when the arguments of the two factors in a product are not the same. For those cases, one may simply declare that multiplication is not defined – which is already a departure from the standard definition of a field, or one may define such a product via some formal expansion into a series in powers of the difference $(\mathrm{x}-\mathrm{y})$ – when such a power series is well defined, etc.

Definition B.3 *Tensor densities of the same type form a generalization of the vector space as their linear combination is defined by specifying*

$$f(\mathbf{x})\,T^{\mu_1\cdots\mu_p}_{\nu_1\cdots\nu_q}(\mathbf{x}) + h(\mathbf{x})\,U^{\mu_1\cdots\mu_p}_{\nu_1\cdots\nu_q}(\mathbf{x}), \tag{B.44}$$

where the coefficients are scalar densities of complementary weights:

$$w\big[\,f(\mathbf{x})\,T^{\mu_1\cdots\mu_p}_{\nu_1\cdots\nu_q}(\mathbf{x})\,\big] = w\big[\,h(\mathbf{x})\,U^{\mu_1\cdots\mu_p}_{\nu_1\cdots\nu_q}(\mathbf{x})\,\big]. \tag{B.45}$$

The linearity of the definition guarantees that the result (B.44) is again a tensor density of the same rank, type and weight.

The structure of a vector space is recovered by restricting to constant coefficients and tensor densities of the same weight.

The following two operations are also important:

Definition B.4 (Contraction) *For any type-(p,q) tensor density, where $p \neq 0 \neq q$, one constructs the* ***contraction***

$$\delta^{\nu_i}_{\mu_j}\;:\;T^{\mu_1\cdots\mu_p}_{\nu_1\cdots\nu_q}(\mathbf{x}) \mapsto T^{\mu_1\cdots\widehat{\mu}_j\cdots\mu_p}_{\nu_1\cdots\widehat{\nu}_i\cdots\nu_q}(\mathbf{x}) = \Big(\delta^{\nu_i}_{\mu_j}\,T^{\mu_1\cdots\mu_j\cdots\mu_p}_{\nu_1\cdots\nu_i\cdots\nu_q}(\mathbf{x})\Big), \tag{B.46}$$

where $\widehat{\mu}_i$ denotes that the index μ_i is omitted from the sequence. The result of contracting is a type-$(p-1,q-1)$ tensor density of the same weight as the original tensor density.

Definition B.5 *For any two indices of the same type, one defines*

$$T^{(\mu\nu)\cdots}_{\cdots} := \tfrac{1}{2}\big(T^{\mu\nu\cdots}_{\cdots} + T^{\nu\mu\cdots}_{\cdots}\big), \quad \text{and} \quad T^{[\mu\nu]\cdots}_{\cdots} := \tfrac{1}{2}\big(T^{\mu\nu\cdots}_{\cdots} - T^{\nu\mu\cdots}_{\cdots}\big), \tag{B.47}$$

the so-called symmetric and antisymmetric part of the original tensor density. The linearity of the definition guarantees that both parts retain the rank, type and weight of the original tensor density.

With tensor densities of a rank higher than two, the combinatorial possibilities and wealth of various (anti)symmetrization patterns grow very quickly; some simple examples are given in relations (A.66) and (A.76). Technically more precisely, the various forms of (anti)symmetrization provide various representations of the permutation group that acts by permuting the indices of the same type (here, subscript vs. superscripts).

Comment B.2 *Every tensor density with at least two indices of the same type may always be decomposed:*

$$T^{\mu\nu\cdots}_{\cdots} \equiv 2\cdot\tfrac{1}{2}T^{\mu\nu\cdots}_{\cdots} + \tfrac{1}{2}T^{\nu\mu\cdots}_{\cdots} - \tfrac{1}{2}T^{\nu\mu\cdots}_{\cdots} = T^{(\mu\nu)\cdots}_{\cdots} + T^{[\mu\nu]\cdots}_{\cdots}, \tag{B.48}$$

where $T^{(\mu\nu)\cdots}_{\cdots}$ and $T^{[\mu\nu]\cdots}_{\cdots}$ transform the same as the original tensor density, $T^{\mu\nu\cdots}_{\cdots}$. More generally, every tensor density may be decomposed into a sum of tensor densities, each of which is an irreducible representation of the permutation group that acts by permuting indices of the same type.

The operations provided by the definitions B.3, B.4 and B.5 generate a structure that is usually called simply "linear algebra."

Finally, define also the multiplication of tensor densities:

Definition B.6 *For any two tensor densities $T^{\mu_1\cdots\mu_p}_{\nu_1\cdots\nu_q}(\mathrm{x})$ and $U^{\rho_1\cdots\rho_{p'}}_{\sigma_1\cdots\sigma_{q'}}(\mathrm{x})$, respectively of type (p,q) and $(p'q')$ and weights w and w', the* ***tensor product*** *may be specified by the relation*

$$(T\otimes U)^{\mu_1\cdots\mu_{p+p'}}_{\nu_1\cdots\nu_{q+q'}}(\mathrm{x}) := T^{\mu_1\cdots\mu_p}_{\nu_1\cdots\nu_q}(\mathrm{x})\, U^{\mu_{p+1}\cdots\mu_{p+p'}}_{\nu_{q+1}\cdots\nu_{q+q'}}(\mathrm{x}) \tag{B.49}$$

the result of which is a type-$(p{+}p', q{+}q')$ and weight-$(w+w')$ tensor density.

B.2.3 Tensor calculus

The *rate of change* of a vector such as $\mathrm{A}(\mathrm{x}) = A^\mu(\mathrm{x})\,\mathrm{e}_\mu(\mathrm{x})$ over spacetime is then

$$\frac{\partial \mathrm{A}}{\partial x^\mu} = \frac{\partial}{\partial x^\mu}\Big(A^\nu(\mathrm{x})\,\mathrm{e}_\nu(\mathrm{x})\Big) = \frac{\partial A^\nu}{\partial x^\mu}\,\mathrm{e}_\nu(\mathrm{x}) + A^\nu(\mathrm{x})\,\frac{\partial \mathrm{e}_\nu}{\partial x^\mu}. \tag{B.50}$$

Since e_ν form a complete set, the partial derivative in the second term must be expressible as a linear combination in the same basis:

$$\frac{\partial \mathrm{e}_\nu}{\partial x^\mu} =: \Gamma^\rho_{\mu\nu}(\mathrm{x})\,\mathrm{e}_\rho(\mathrm{x}), \tag{B.51}$$

where $\Gamma^\rho_{\mu\nu}(\mathrm{x})$ are, for each pair (μ,ν) and at each point x in spacetime, simply the 4-tuple of coefficient functions in the linear combination of basis vectors $\mathrm{e}_\rho(\mathrm{x})$. Combining results (B.50) and (B.51), we have

$$\frac{\partial \mathrm{A}}{\partial x^\mu} = \Big[\frac{\partial A^\rho}{\partial x^\mu} + A^\nu\,\Gamma^\rho_{\mu\nu}\Big]\mathrm{e}_\rho(\mathrm{x}). \tag{B.52}$$

It is straightforward that

$$\frac{\partial}{\partial x^\mu}\Big(\mathrm{e}_\nu{\cdot}\mathrm{e}^\rho = \delta^\rho_\nu\Big) = 0 \quad\Rightarrow\quad \frac{\partial \mathrm{e}^\rho}{\partial x^\mu} = -\Gamma^\rho_{\mu\nu}(\mathrm{x})\,\mathrm{e}^\nu(\mathrm{x}), \tag{B.53}$$

whereby

$$\frac{\partial \mathrm{B}}{\partial x^\mu} = \frac{\partial B_\nu}{\partial x^\mu}\,\mathrm{e}^\nu(\mathrm{x}) + B_\nu(\mathrm{x})\,\frac{\partial \mathrm{e}^\nu}{\partial x^\mu} = \Big[\frac{\partial B_\nu}{\partial x^\mu} - B_\rho\,\Gamma^\rho_{\mu\nu}\Big]\mathrm{e}^\nu(\mathrm{x}). \tag{B.54}$$

The quantities in the square brackets in equations (B.52) and (B.54) are then defined as the *covariant derivatives* of the components

$$D_\mu A^\rho := \big[\partial_\mu A^\rho + \Gamma^\rho_{\mu\nu} A^\nu\big] \quad\text{and}\quad D_\mu B_\nu := \big[\partial_\mu B_\nu - \Gamma^\rho_{\mu\nu} B_\rho\big]. \tag{B.55}$$

The formula (9.17) is then the straightforward iteration of these two definitions, as dictated by Weyl's Construction A.1 on p. 478, adapted here to provide Definition B.1 on p. 511.

The definition of $\Gamma^\rho_{\mu\nu}(\mathrm{x})$ in equation (B.51) and the relations (B.36) then imply several important properties of $\Gamma^\rho_{\mu\nu}(\mathrm{x})$. First,

$$\Gamma^\rho_{\mu\nu}\,\mathrm{e}_\rho = \frac{\partial\,\mathrm{e}_\nu}{\partial x^\mu} = \frac{\partial^2\,\mathrm{e}}{\partial x^\mu \partial x^\nu} = \frac{\partial^2\,\mathrm{e}}{\partial x^\nu \partial x^\mu} = \Gamma^\rho_{\nu\mu}\,\mathrm{e}_\rho \quad\Rightarrow\quad \Gamma^\rho_{\mu\nu} = \Gamma^\rho_{\nu\mu}. \tag{B.56}$$

Next, compute

$$\frac{\partial g_{\mu\nu}}{\partial x^\rho} = \frac{\partial}{\partial x^\rho}(\mathrm{e}_\mu{\cdot}\mathrm{e}_\nu) = \Gamma^\sigma_{\mu\rho}\mathrm{e}_\sigma{\cdot}\mathrm{e}_\nu + \mathrm{e}_\mu{\cdot}\Gamma^\sigma_{\nu\rho}\mathrm{e}_\sigma = \Gamma^\sigma_{\mu\rho}g_{\sigma\nu} + g_{\mu\sigma}\Gamma^\sigma_{\nu\rho}. \tag{B.57}$$

Reusing this equality with permuted indices μ,ν,ρ, we obtain

$$\frac{\partial g_{\mu\sigma}}{\partial x^\nu} + \frac{\partial g_{\nu\sigma}}{\partial x^\mu} - \frac{\partial g_{\mu\nu}}{\partial x^\sigma} = 2g_{\sigma\rho}\Gamma^\rho_{\mu\nu}, \tag{B.58}$$

which implies the standard formula [508, 62, 367, 548, 66, 96]

$$\Gamma^{\rho}_{\mu\nu} = \tfrac{1}{2} g^{\rho\sigma} \Big[\frac{\partial g_{\mu\sigma}}{\partial x^\nu} + \frac{\partial g_{\nu\sigma}}{\partial x^\mu} - \frac{\partial g_{\mu\nu}}{\partial x^\sigma} \Big]. \tag{B.59}$$

It is then straightforward to show that

$$D_\mu g_{\nu\rho} = 0 = D_\mu g^{\nu\rho}. \tag{B.60}$$

We close with a useful result and a comment. The Jacobi identity for derivatives of the determinant $g := \det[g_{..}]$ is

$$\frac{\partial g}{\partial x^\mu} = g\, g^{\nu\rho} \frac{\partial g_{\nu\rho}}{\partial x^\mu} \quad \Rightarrow \quad g^{\nu\rho} \frac{\partial g_{\nu\rho}}{\partial x^\mu} = \frac{1}{g} \frac{\partial g}{\partial x^\mu} = \frac{1}{(-g)} \frac{\partial (-g)}{\partial x^\mu} = \frac{\partial \ln(-g)}{\partial x^\mu}, \tag{B.61}$$

where the sign-change was necessary as spacetime metrics have an odd number of negative eigenvalues and so a negative determinant. Now contract the expression (B.59):

$$\Gamma^{\mu}_{\mu\nu} = \tfrac{1}{2} g^{\mu\sigma} \Big[\frac{\partial g_{\mu\sigma}}{\partial x^\nu} + \frac{\partial g_{\nu\sigma}}{\partial x^\mu} - \frac{\partial g_{\mu\nu}}{\partial x^\sigma} \Big] = \tfrac{1}{2} g^{\mu\sigma} \frac{\partial g_{\mu\sigma}}{\partial x^\nu}, \tag{B.62}$$

since $g^{\mu\sigma} \frac{\partial g_{\nu\sigma}}{\partial x^\mu} \overset{\mu:\sigma}{=} g^{\sigma\mu} \frac{\partial g_{\nu\mu}}{\partial x^\sigma} = g^{\mu\sigma} \frac{\partial g_{\mu\nu}}{\partial x^\sigma}$ and the last two terms cancel. Using then the identity (B.61) yields

$$\Gamma^{\mu}_{\mu\nu} = \tfrac{1}{2} \frac{\partial \ln(g)}{\partial x^\mu} = \frac{\partial \ln\big(\sqrt{g}\big)}{\partial x^\mu} = \frac{1}{\sqrt{g}} \frac{\partial \sqrt{g}}{\partial x^\mu}. \tag{B.63}$$

Therefore,

$$\begin{aligned} D{\cdot}\mathrm{A} = \mathrm{e}^\mu {\cdot} \frac{\partial \mathrm{A}}{\partial x^\mu} &= (D_\mu A^\nu)\, \mathrm{e}^\mu {\cdot} \mathrm{e}_\nu = (D_\mu A^\mu) = \frac{\partial A^\mu}{\partial x^\mu} + \Gamma^\mu_{\mu\nu} A^\nu, \\ &= \frac{\partial A^\nu}{\partial x^\nu} + \Big(\frac{1}{\sqrt{g}} \frac{\partial \sqrt{g}}{\partial x^\nu} \Big) A^\nu = \frac{1}{\sqrt{g}} \frac{\partial (\sqrt{g} A^\nu)}{\partial x^\nu} = \frac{1}{\sqrt{g}} \frac{\partial (\sqrt{g}\, g^{\nu\rho} A_\rho)}{\partial x^\nu} \end{aligned} \tag{B.64}$$

provides the definition of the spacetime gradient of a 4-vector, alternatively given for a vector specified in terms of contravariant and covariant components. The spacetime gradient of a type-(p,q) tensor density of weight w is then obtained by iterating this result. For example, the spacetime divergence of a type-$(2,0)$ tensor is

$$(D{\cdot}\mathbb{T})^\nu = \frac{\partial T^{\mu\nu}}{\partial x^\mu} + \Gamma^\mu_{\mu\sigma} T^{\sigma\nu} + \Gamma^\nu_{\mu\sigma} T^{\mu\sigma} = \frac{1}{\sqrt{g}} \frac{\partial (\sqrt{g} T^{\sigma\nu})}{\partial x^\sigma} + \Gamma^\nu_{\mu\sigma} T^{\mu\sigma}. \tag{B.65}$$

The general result is

$$\big(\widetilde{D_\lambda \mathbb{T}}(\mathrm{y})\big)^{\rho_1 \cdots \rho_p}_{\sigma_1 \cdots \sigma_q} = \Big(\det \Big[\frac{\partial \mathrm{y}}{\partial \mathrm{x}} \Big| \Big)^w \frac{\partial y^{\rho_1}}{\partial x^{\mu_1}} \cdots \frac{\partial y^{\rho_p}}{\partial x^{\mu_p}} \frac{\partial x^{\nu_1}}{\partial y^{\sigma_1}} \cdots \frac{\partial x^{\nu_q}}{\partial y^{\sigma_q}} \frac{\partial x^\kappa}{\partial y^\lambda} \big(D_\kappa \mathbb{T}(\mathrm{x})\big)^{\mu_1 \cdots \mu_p}_{\nu_1 \cdots \nu_q}. \tag{B.66}$$

That is, the covariant derivative of a type-(p,q) tensor density of weight w is a type-$(p,q{+}1)$ tensor density of weight w.

Finally, we note that for every μ the vector $\mathrm{e}_\mu(\mathrm{x})$ is defined infinitesimally near the point x. Using the 4-vector of partial derivatives $\frac{\partial}{\partial x^\nu}$, we may define

$$e_\mu{}^\nu(\mathrm{x}): \quad \mathrm{e}_\mu = e_\mu{}^\nu(\mathrm{x}) \frac{\partial}{\partial x^\nu}, \tag{B.67}$$

exhibiting that the basis elements $\mathrm{e}_\mu(\mathrm{x})$ span a linear vector space. This locally (infinitesimally) defined (tangent) spacetime must then be isomorphic to $\mathbb{R}^{1,3}$, and we are free to choose Cartesian coordinates in it, say ξ^m, for which $g_{mn}(\xi) = -\eta_{mn}$, so that $\mathrm{e}_\mu(\mathrm{x}) = e_\mu{}^m(\mathrm{x})\,\frac{\partial}{\partial\xi^m}$. In turn, comparing the straightforward computation

$$\mathrm{e}_\mu := \frac{\partial \mathrm{e}}{\partial x^\mu} = \frac{\partial \xi^m}{\partial x^\mu}\frac{\partial \mathrm{e}}{\partial \xi^m}, \tag{B.68}$$

with the definition of $e_\mu{}^\nu(\mathrm{x})$ given in (B.67), we see that $e_\mu{}^m(\mathrm{x}) = \frac{\partial\xi^m}{\partial x^\mu}$, when the local tangent-space derivatives $\frac{\partial}{\partial\xi^m}$ are used as covariant basis elements, instead of the curvilinear $\frac{\partial \mathrm{e}}{\partial x^\mu}$.

The so-defined 4×4 matrix of coefficients $e_\mu{}^m(\mathrm{x})$ is variously called a *tetrad*, a *Fierbein* (German: *fier* = four, *Bein* = leg), a "moving frame," or a "soldering form" [508, 62, 367, 548, 66, 96], as it relates curvilinear derivatives to the local, tangent-space, $\frac{\partial}{\partial x^\mu} = e_\mu{}^m(\mathrm{x})\frac{\partial}{\partial\xi^m}$, at every point in spacetime. Straightforwardly,

$$e_\mu{}^m(\mathrm{x})\,(-\eta_{mn})\,e_\nu{}^n(\mathrm{x}) = g_{\mu\nu}(\mathrm{x}), \tag{B.69}$$

and $e_\mu{}^m(\mathrm{x})$ may be regarded as a square-root of the metric tensor. By abuse of language, one says that $\mu,\nu,\ldots$ are "curved indices," meaning that they indicate curvilinear coordinates; in turn, $m,n,\ldots$ are dubbed "flat indices," meaning that they indicate Cartesian coordinates in the flat tangent spacetime $\cong \mathbb{R}^{1,3}$, which is defined locally (infinitesimally) at every point x of otherwise arbitrarily curved but smooth spacetime.

Clearly, at any point where the local system of partial derivatives $\frac{\partial\xi^m}{\partial x^\mu}$ is ill-defined, this construction in the specified coordinates breaks down, detecting a candidate (putative) singularity; see the discussion in Section 9.3.1, starting on p. 334.

B.2.4 *Functionals and functional derivatives*

Without delving into technical details and a rigorous definition of functionals and functional derivatives, we provide here a heuristic introduction and a few results that prove useful in computations such as done in Digression 5.9 on p. 191 or Section 11.2.4.

Consider first a 4-vector $\mathrm{x} = (x^0, x^1, x^2, x^3)$. The value of the symbol "k^μ" clearly depends on the choice of the index, which indicates one of the four components. Note that there are only a finite number of choices for μ, and thus a finite number of components of k^μ. This is conceptually similar to the notion of a function $f(x)$, the value of which depends on the choice of the argument x – except that x varies *continuously* over a range of values. For each of the permissible choices of the argument x, $f(x)$ returns a value and so the space of possible values may well also form a continuously infinite set.

We frequently consider summation over the indices – which we will write explicitly in this section, such as,

$$(\mathrm{x}\cdot\eta)_\nu := \sum_{\mu=0}^{3} x^\mu\eta_{\mu\nu} = x^0\eta_{0\nu} + x^1\eta_{1\nu} + x^2\eta_{2\nu} + x^3\eta_{3\nu}. \tag{B.70}$$

In the 4-vector quantity so defined, the index ν appears on both sides of the equation and remains *free*: it may be freely chosen and changed at will. By contrast, the index μ has been summed over, does not even appear in the right-most, expanded version of the sum, and is not free to substitute arbitrarily chosen values (from within $0,1,2,3$); it is a dummy summation variable. *Conceptually,* this is identical to the fact that in the integral

$$F[f;y] := \int_a^b \mathrm{d}x\, f(x)\,H(x,y), \tag{B.71}$$

the argument y remains free and available for substitution with any of its allowed values, while the variable x has been "used up" to compute the integral. Just as the sum (B.70) depends on the 4-vector x and its 4 components (x^0, x^1, x^2, x^3), so does the integral (B.71) depend on the choice of the function $f(x)$ and its values. Just as $(\mathrm{x}\cdot\eta)_\nu$ no longer depends on the "used-up" index μ, neither does the integral $F[f;y]$ depend on the "used-up" variable x.

Both of these expressions depend on the summation (integration) limits, they also depend on the additional rank-2 tensor (2-argument function) quantities, $\eta_{\mu\nu}$ $(H(x,y))$, but we focus here the dependence on the 4-vector x^μ vs. the function $f(x)$. In particular, we easily compute the derivative by x^α of the first of these quantities:

$$\frac{\partial}{\partial x^\alpha}(\mathrm{x}\cdot\eta)_\nu = \frac{\partial}{\partial x^\alpha}\sum_{\mu=0}^{3} x^\mu \eta_{\mu\nu} = \sum_{\mu=0}^{3}\frac{\partial}{\partial x^\alpha}x^\mu\eta_{\mu\nu} = \sum_{\mu=0}^{3}\Big(\frac{\partial x^\mu}{\partial x^\alpha}\eta_{\mu\nu} + x^\mu \underbrace{\frac{\partial \eta_{\mu\nu}}{\partial x^\alpha}}_{\text{assume }=0}\Big)$$
$$= \sum_{\mu=0}^{3}\Big(\frac{\partial x^\mu}{\partial x^\alpha} = \delta^\mu_\alpha\Big)\eta_{\mu\nu} = \eta_{\alpha\nu}. \tag{B.72}$$

In the indicated assumption, we state that the rank-2 tensor $\eta_{\mu\nu}$ is defined independently of the 4-vector x^μ. In perfect analogy with (B.72), we compute the *functional* (also called *variational*) derivative of the integral (B.71):

$$\frac{\delta}{\delta f(z)}F[f;y] = \frac{\delta}{\delta f(z)}\int_a^b \mathrm{d}x\, f(x)\,H(x,y) = \int_a^b \mathrm{d}x\,\frac{\delta}{\delta f(z)}f(x)\,H(x,y)$$
$$= \int_a^b \mathrm{d}x\,\Big(\frac{\delta f(x)}{\delta f(z)}H(x,y) + f(x)\underbrace{\frac{\delta H(x,y)}{\delta f(z)}}_{\text{assume }=0}\Big)$$
$$= \int_a^b \mathrm{d}x\,\Big(\frac{\delta f(x)}{\delta f(z)} = \delta(x-z)\Big)H(x,y) = H(z,y). \tag{B.73}$$

In the indicated assumption, we state that the 2-argument function $H(x,y)$ is defined independently of the function $f(x)$. Still more generally, consider a nonlinear functional of the function $f(x)$:

$$\mathcal{F}[f] := \int_a^b \mathrm{d}x\,\mathscr{F}(f(x)), \tag{B.74}$$

where $\mathscr{F}$ is an arbitrary functional expression involving $f(x)$, such as $\frac{(f(x))^2}{\sqrt{\log(f(x)+1)}}$. Requiring the basic chain rule to apply, we obtain

$$\frac{\delta}{\delta f(z)}\mathcal{F}[f] := \int_a^b \mathrm{d}x\,\frac{\delta}{\delta f(z)}\mathscr{F}(f(x)) = \int_a^b \mathrm{d}x\,\delta(x-z)\Big[\frac{\partial\mathscr{F}(\xi)}{\partial\xi}\Big]_{\xi\to f(x)}, \tag{B.75}$$

where the symbol f is used in the partial derivative within the square brackets as a formal argument of the function $\mathscr{F}$ and the derivative is calculated in the standard way. Once the derivative is computed, $\xi \to f(x)$ is substituted back in the resulting (derivative) functional expression. Note that the Dirac δ-function, $\delta(x-z)$, quenches the integration to an evaluation at $x \to z$.

There is, however, an aspect of functional derivatives that does not have a direct analogue in the 4-vector calculus framework of (B.72), and it has to do with cases where the definition of the functional such as (B.71) depends not only on the function, but also on its derivatives.

This is actually a fairly typical case as most Lagrangians are functional expressions involving not only fields, but also their time derivatives. The time-integral of any Lagrangian is then Hamilton's action, and it is a functional of the fields involved. For example,

$$S[\phi;C] = \int_0^T \mathrm{d}t\; L(\phi,\phi',\phi'',\dots). \tag{B.76}$$

$$\begin{aligned}\frac{\delta}{\delta\phi(\tau)}S[\phi;C] &= \int_0^T \mathrm{d}t\; \frac{\delta}{\delta\phi(\tau)}L(\phi,\phi',\phi'',\dots)\\ &= \int_0^T \mathrm{d}t\; \Big\{\frac{\delta\phi(t)}{\delta\phi(\tau)}\Big[\frac{\partial L}{\partial\phi}\Big] + \frac{\delta\phi'(t)}{\delta\phi(\tau)}\Big[\frac{\partial L}{\partial\phi'}\Big] + \frac{\delta\phi''(t)}{\delta\phi(\tau)}\Big[\frac{\partial L}{\partial\phi''}\Big] + \cdots\Big\},\end{aligned} \tag{B.77}$$

where all the expressions in the square brackets treat $\phi,\phi',\phi'',\dots$ as independent variables to perform the indicated partial derivatives, then re-submit $\phi\to\phi(t)$, $\phi'\to\phi'(t)$, etc. Next, we use without proof that $\frac{\delta\phi'(t)}{\delta\phi(\tau)} = \big(\frac{\mathrm{d}}{\mathrm{d}t}\frac{\delta\phi(t)}{\delta\phi(\tau)}\big)$:

$$\begin{aligned}&= \int_0^T \mathrm{d}t\; \Big\{\frac{\delta\phi(t)}{\delta\phi(\tau)}\Big[\frac{\partial L}{\partial\phi}\Big] + \Big(\frac{\mathrm{d}}{\mathrm{d}t}\frac{\delta\phi(t)}{\delta\phi(\tau)}\Big)\Big[\frac{\partial L}{\partial\phi'}\Big] + \Big(\frac{\mathrm{d}^2}{\mathrm{d}t^2}\frac{\delta\phi(t)}{\delta\phi(\tau)}\Big)\Big[\frac{\partial L}{\partial\phi''}\Big] + \cdots\Big\}\\ &= \int_0^T \mathrm{d}t\; \Big\{\delta(t-\tau)\Big[\frac{\partial L}{\partial\phi}\Big] + \Big(\frac{\mathrm{d}}{\mathrm{d}t}\delta(t-\tau)\Big)\Big[\frac{\partial L}{\partial\phi'}\Big] + \Big(\frac{\mathrm{d}^2}{\mathrm{d}t^2}\delta(t-\tau)\Big)\Big[\frac{\partial L}{\partial\phi''}\Big] + \cdots\Big\}.\end{aligned}$$

Next, we integrate by parts; the second term once, the third term twice and so on:

$$= \int_0^T \mathrm{d}t\; \delta(t-\tau)\Big\{\Big[\frac{\partial L}{\partial\phi}\Big] - \Big(\frac{\mathrm{d}}{\mathrm{d}t}\Big[\frac{\partial L}{\partial\phi'}\Big]\Big) + \Big(\frac{\mathrm{d}^2}{\mathrm{d}t^2}\Big[\frac{\partial L}{\partial\phi''}\Big]\Big) + \cdots\Big\} + \text{B.T.}, \tag{B.78}$$

where "B.T." denotes boundary terms stemming from the integrations by part. Finally,

$$\frac{\delta}{\delta\phi(\tau)}S[\phi;C] = \sum_{k=0}^{\infty}(-1)^k\frac{\mathrm{d}^k}{\mathrm{d}\tau^k}\frac{\partial\, L(\phi(\tau),\phi'(\tau),\phi''(\tau),\dots)}{\partial\,\phi^{(k)}(\tau)} + \text{B.T.} \tag{B.79}$$

A further generalization of this to n-tuples of fields, and to dependence on more than one variable is straightforward:

$$\frac{\delta}{\delta\phi_a(\mathrm{x})}S[\phi_\cdot;C] = \sum_{k=0}^{\infty}(-1)^k\partial^k\frac{\partial\, L(\phi_\cdot(\mathrm{x}),\partial^1\phi_\cdot(\mathrm{x}),\partial^2\phi_\cdot(\mathrm{x}),\dots)}{\partial(\partial^k\phi_a^{(k)}(\mathrm{x}))} + \text{B.T.}, \tag{B.80a}$$

$$\partial^k := \underbrace{\partial_\mu\partial_\nu\cdots\partial_\rho}_{k\text{ factors}}, \qquad \text{and} \qquad a = 1,2,\dots,n, \tag{B.80b}$$

and where a summation is implied between each spacetime partial derivative occurring within the two copies of ∂^k – one acting on the partial derivative of the Lagrangian density and the other in the specification of the derivative field with respect to which the partial derivative of the Lagrangian is computed:

$$= \frac{\partial L}{\partial\phi_a(\mathrm{x})} - \partial_\mu\frac{\partial L}{\partial(\partial_\mu\phi_a(\mathrm{x}))} - \partial_\mu\partial_\nu\frac{\partial L}{\partial(\partial_\mu\partial_\nu\phi_a(\mathrm{x}))} + \cdots + \text{B.T.} \tag{B.80c}$$

B.2.5 Exercises for Section B.2

✎ **B.2.1** *Using the results (9.4) and (3.11c), prove that the line element (B.27) is invariant under general coordinate transformations.*

✎ **B.2.2** *Using the standard rules of calculus (B.23)–(B.24), prove that (B.28) is invariant under general coordinate transformations.*

✎ **B.2.3** *Using the standard rules of calculus (B.23)–(B.24), prove that the Levi-Civita symbol (B.29)–(B.30) is invariant under general coordinate transformations.*

✎ **B.2.4** *Prove the relationship (B.31).*

B.3 A telegraphic introduction to Gödelian incompleteness

This sketchy and perforce incomplete account of Kurt Gödel's incompleteness theorem and its corollary, as well as their implications for all sufficiently complex theoretical systems, is meant to alleviate the fact that most physics students are not familiar with it. For a more complete and precise introduction, see Refs. [211, 376].

With excellent prospects of hilariously oversimplifying the historical background and significance of Gödel's theorem, let me just mention as a backdrop the incredible *Principia Mathematica* by A. N. Whitehead and B. Russell: This three-tome opus [571, 568, 569, ∼2,000 pages in total], justifying even a 500-page abridged version of Vol. 1 [570], collects the best efforts to cast the complete and rigorous foundation of all mathematics in Peano's formal symbolic logic and Frege's set theory. The first edition of the *Principia Mathematica* was published in 1910, and was then improved for the second edition in 1927.

The ultimate hope was that all of mathematics could be shown to be deducible from an effectively generable collection of axioms,[4] and by means of perfectly rigorous logic. Whitehead and Russell's opus not only set formidable standards for the rigor of proof (hereafter to be pursued in mathematics), but provided an indelible influence on a century of development in (mathematical) logic and set theory, and *metamathematics* – the mathematics of how mathematics is to be practiced and understood.

In 1931, Kurt Gödel published an announcement of his incompleteness theorem, its corollary (often referred to as the second incompleteness theorem) and an elaborate sketch of proof, deferring the complete proof (to the level of rigor as set by the *Principia Mathematica*). His results were, however, accepted at once and Gödel never did get around to publishing the completely detailed proof [211, 376].

Gödel's incompleteness theorem and its corollary pertain to axiomatic systems that are sufficiently complex to contain the axiomatic system of standard arithmetic. Recall that an axiomatic system is a logical system that has, roughly:

1. a fixed list of symbols,
2. a fixed list of "syntactic/grammatical" rules specifying which strings of symbols represent "well-formed" (meaningful) expressions and statements,
3. a fixed list of adopted logical rules of manipulating and combining statements, and
4. a fixed list of "axioms" (postulates) – statements that are adopted as the "primary statements (truths)" of the given system.

[4] A collection of objects is effectively generable if there exists an algorithm that will enumerate all the objects in the collection without ever enumerating anything else.

Every such axiomatic system then has statements that are spelled out with its symbols (1) and which are well-formed (2); those that can be derived using the rules (3) from the axioms (4) are called "theorems."

Within this framework, it was the hope of the research culminating with Whitehead and Russell's *Principia Mathematica* that a suitable system of axioms could be found for all of mathematics, such that every well-formed mathematical statement could either be proven (by deriving it from the axioms) or disproven (by deriving its logical negation instead).

Gödel's **incompleteness theorem** states that no axiomatic system that is sufficiently complex to contain arithmetics can both be complete and not be self-contradictory. That is, to avoid being self-contradictory, every such axiomatic system must contain statements that can be neither proven nor disproven within the axiomatic system as given. Gödel also drew an immediate **corollary** (oft-cited as his **second incompleteness theorem** owing to its importance), which states that no such axiomatic system can prove/demonstrate its own consistency [211].

Even more remarkably, Gödel's proof is constructive! Within any such axiomatic system, Gödel's proof explicitly shows how to construct a very specific statement, which can neither be proven nor disproven within the given axiomatic system. Although Gödel constructed this particular undecidable statement in his proof, and expressly for the purpose of proving the theorem, it does follow that there exist infinitely many such undecidable statements – and some of those, within physics as a formal axiomatic system, are bound to be of interest. Such undecidable statements are then often called Gödelian, although strictly speaking this name should be reserved for the specific statement constructed in Gödel's proof for the given axiomatic system.

Conclusion B.1 *To any axiomatic system (sufficiently complex so as to contain arithmetic), either a Gödelian undecidable statement or its logical negation may be added as a new axiom – and this extension may be repeated recursively forever [211, 376].*

It is worth noticing that the Popperian notion of falsifiability (at least in an admittedly naive understanding [☞ Digression 1.1 on p. 9]) presupposes all statements that one may spell out within some theory (or theoretical system) necessarily to be either falsified or confirmed – so that there *must* exist provable/derivable statements within that theory, which then Nature (experiment) could falsify. In turn, a theory may well be *undecided* about any particular and otherwise perfectly self-consistent statement being tested. Nature then *may* choose one of the options, so effectively decide the statement – and extend the theory.

Example B.1 Within the standard theoretical system of Newtonian classical mechanics, Bertrand's theorem [☞ textbooks of classical mechanics such as [213]] guarantees that stable circular orbits in $(3+1)$-dimensional spacetime are ensured only by two central potentials:

1. the Kepler/Newton potential, $-\frac{\varkappa}{r}$,
2. the radial harmonic potential, $\frac{1}{2}kr^2$.

However, there is nothing within this theoretical system that could decide which one is *the* one that keeps the planets in stable and nearly circular orbits around the Sun. It is the correlation between the orbital linear velocities of the planets and their distance from the Sun – observed in Nature to be $v \propto r^{-1/2}$[✎ *derive*]– that clearly picks the Kepler/Newton potential over the $v \propto r^{3/2}$ [✎ *derive*] of the harmonic potential.

Example B.2 Within the standard theoretical system of Newtonian classical mechanics, there exists no reason to impose Bohr's ad hoc quantization of the angular momentum for the electron orbiting the proton and so forming a hydrogen atom. However, neither does there exist a reason *against* such a quantization: strictly speaking, the assumed continuous variability of the magnitude of angular momentum in various physical systems is merely an implicit assumption, bolstered by no noticed exemption in the macroscopic world; see, however, the discussion in Section 8.3.1 and Footnote *11* on p. 310 in particular.

Therefore, from within the formal theoretical system of classical mechanics, whether or not the angular momentum of an electron orbiting a proton is to be quantized and in what units is in fact an undecidable statement in Gödel's sense. Nature quite clearly resolves the issue: Experiments show that the angular momentum of any physical system can only change in integral multiples of $\hbar$, and so must be either an integral or a half-integral multiple of this unit. The quantum extension of classical mechanics is in this sense precisely a Gödelian extension of the axiomatic theoretical system of classical physics to the axiomatic theoretical system of quantum physics.

Example B.3 As discussed in Section 9.1.1 and Digression 8.1 on p. 295, attempting to fuse Newtonian mechanics and Maxwell's electrodynamics requires one to either modify electrodynamics so as to become Galilean-symmetric, or mechanics so as to become Lorentz-symmetric. Since the Galilean group is the $c \to \infty$ limit of the Lorentz group, the former of these options is achievable only if we take the $c \to \infty$ limit of the Maxwell equations. Both resulting systems are consistent, so that a choice between them is not decidable from within the theory alone. It is indeed Nature's "choice" that light does propagate at a finite speed, which then implies the latter option for the electrodynamics of moving electric charges.

While one may wish for such a "resolution by Nature," as described in the Examples B.1, B.2 and B.3 above, there is in fact no guarantee that *all* "theoretical" dichotomies in our attempts to describe Nature will be similarly resolvable by observation. Indeed, the discovery of the ever-increasing list of ever more various dualities [☞ Section 11.4, to begin with] seems to indicate that this "plurality" of description is an innate characteristic of our understanding Nature.

Finally, the prospect of perpetual Gödelian extensions – in as much as it seems applicable to physics – seems to agree with some of the historical lessons, seen with the benefit of hindsight. Within the theoretical system of classical vector fields, the model described by the Maxwell equations is "well-formed," but undecidable. There is nothing in classical field theory formalism that could prove or disprove the Maxwell equations from any system of axioms, which does not in fact include either the electrodynamics laws that those differential equations represent or the gauge principle as introduced in Chapter 5.

In this sense then, the gauge principle (or the electrodynamics laws represented by the Maxwell equations) is a Gödelian undecidable statement within the theoretical system of classical fields. By including the gauge principle, we obtain the particular theoretical system of classical fields that is called electrodynamics, in which the vector fields $\vec{E}$, $\vec{B}$, $\vec{A}$ and the scalar field Φ acquire a specific meaning and application. The theoretical system at hand has at once both become more specific and acquired a richer structure (less arbitrariness).

Gödel's incompleteness theorem then implies that the axiomatic system of theoretical physics definitely *can* be extended indefinitely, and in infinitely many ways. Which of those extensions will turn out to be useful towards the intended purpose of theoretical physics, of course, remains an open question – and may well remain so indefinitely.

C

A few more details

C.1 Nobel Prizes

Success in science is, strictly speaking, measured only in ells of time: Democritus' and Leucippus' idea of elementary particles, even after two and a half millennia, serves successfully as a guiding thought and Leitmotif, and Newton's and Leibniz's calculus still forms the basis of the mathematical formulation of the laws of Nature. The fact that more than a third of twentieth century Nobel Prizes were awarded to discoveries relating to the physics of elementary particles and fundamental physics is probably foreordained by the selection effect: in a field where one knows less, the probability of discovering something fundamentally new is higher. Nevertheless, I hope that this, perhaps even pompous, review of major successes in the past century will serve as a convenient reminder.

Table C.1 Nobel Prizes awarded for discoveries and contributions in fundamental physics

Year	Awardee	Award for [*paraphrase*; T.H.]
1901	Wilhelm C. Röntgen	discovery of the remarkable rays subsequently named after him, also known as *X*-rays
1903	A. Henri Becquerel ($\frac{1}{2}$)	discovery of spontaneous radioactivity
	Pierre Curie, Marie Curie, née Sklodowska	their joint researches on radiation phenomena
1906	Joseph J. Thomson	investigations on the conduction of electricity by gases [*i.e., discovery of the electron*; T.H.]
1918	Max K. E. L. Planck	advancement of physics by his discovery of energy quanta [*quantization of electromagnetic radiation **emission***; T.H.]
1921	Albert Einstein	discovery of the law of the photoelectric effect [*not the discovery that electromagnetic radiation **exists** in quanta – photons*; T.H.]
1922	Niels H. D. Bohr	investigation of the structure of atoms and of the radiation emanating from them

Year	Awardee	Award for [*paraphrase*; T.H.]
1923	Robert A. Millikan	work on the elementary charge of electricity and on the photoelectric effect
1925	James Franck, Gustav L. Hertz	discovery of the laws governing the impact of an electron upon an atom [*confirming the quantization of atomic states*; T.H.]
1927	Arthur H. Compton	discovery of the effect named after him
	Charles T. R. Wilson	method of making the paths of electrically charged particles visible by condensation of vapor [*invention of the cloud chamber*; T.H.]
1929	Prince Louis-Victor P. R. de Broglie	discovery of the wave nature of electrons [*and not the universal wave–particle duality*; T.H.]
1932	Werner K. Heisenberg	creation of quantum mechanics
1933	Erwin Schrödinger, Paul A. M. Dirac	discovery of new productive forms of atomic theory
1935	James Chadwick	discovery of the neutron
1936	Victor F. Hess	discovery of cosmic radiation
	Carl D. Anderson	discovery of the positron
1939	Ernest O. Lawrence	invention and development of the cyclotron
1945	Wolfgang Pauli	discovery of the exclusion principle
1949	Hideki Yukawa	prediction of the existence of mesons
1950	Cecil F. Powell	development of the photographic method of studying nuclear processes and his discoveries regarding mesons made with this method
1954	Max Born	statistical interpretation of the wave-function
	Walther W. G Bothe	the coincidence method
1955	Willis E. Lamb	discoveries concerning the fine structure of the hydrogen spectrum
	Polykarp Kusch	precision determination of the magnetic moment of the electron
1957	Chen-Ning Yang, Tsung-Dao Lee	penetrating investigation of the so-called parity laws [*i.e., of C-, P- and CP-violation*; T.H.]
1958	Pavel A. Cherenkov, Il'ja M. Frank, Igor Ye. Tamm	discovery and the interpretation of the Cherenkov effect
1959	Emilio G. Segrè, Owen Chamberlain	discovery of the antiproton
1960	Donald A. Glaser	invention of the bubble chamber
1963	Eugene P. Wigner ($\frac{1}{2}$)	discovery and application of fundamental symmetry principles [$\frac{1}{2}$: *Maria Goeppert-Meyer and J. Hans D. Jensen, nuclear shell structure*; T.H.]
1965	Shin-Ichiro Tomonaga, Julian Schwinger, Richard P. Feynman	fundamental work in quantum electrodynamics, with deep-ploughing consequences for the physics of elementary particles [*renormalization in QED; Freeman Dyson showed the equivalence of the methods of Tomonaga, Schwinger and Feynman*; T.H.]
1968	Luis W. Alvarez	discovery of a large number of resonance states (hadrons)

Year	Awardee	Award for [*paraphrase*; T.H.]
1969	Murray Gell-Mann	classification of elementary particles and their interactions
1976	Burton Richter, Samuel Chao-Chung Ting	discovery of a heavy elementary particle of a new kind
1979	Sheldon L. Glashow, Abdus Salam, Steven Weinberg	theory of the unified weak and electromagnetic interaction between elementary particles, including, inter alia, the prediction of the weak neutral current
1980	James W. Cronin, Val L. Fitch	discovery of violations of fundamental symmetry principles in the decay of neutral K-mesons [*CP-violation*; T.H.]
1982	Kenneth Wilson	theory for critical phenomena in connection with phase transitions [*this theory contains the approach to renormalization that is built into the foundations of contemporary field theory*; T.H.]
1984	Carlo Rubia, Simon van der Meer	decisive contributions to the large project that led to the discovery of the field particles W and Z, communicators of the weak interaction
1988	Leon M. Lederman, Melvin Schwartz, Jack Steinberger	neutrino beam method and the demonstration of $\nu_e \neq \nu_\mu$
1990	Jerome I. Friedman, Henry W. Kendall, Richard E. Taylor	pioneering investigations concerning deep inelastic scattering of electrons on protons and bound neutrons, of essential importance for the development of the quark model
1992	Georges Charpak	invention and development of particle detectors, in particular the multiwire proportional chamber
1995	Martin L. Perl	discovery of the tau lepton
	Frederick Reines	detection of the neutrino [*already in 1956 – 39 years earlier! C. Cowan died in 1974, and was not awarded*; T.H.]
1999	Gerardus 't Hooft, Martinus Veltman	elucidating the quantum structure of electroweak interactions in physics [*renormalization in models with Higgs fields*; T.H.]
2002	Raymond Davis Jr., Masatoshi Koshiba; Riccardo Giacconi	pioneering contributions in astrophysics: detection of cosmic neutrinos and the solar neutrino problem (the Homestake Experiment) pioneering contributions in astrophysics: cosmic X-rays
2004	David J. Gross, H. David Politzer, Frank Wilczek	discovery of asymptotic freedom in the theory of the strong interaction
2006	John C. Mather, George D. Smoot	discovery of the blackbody form and anisotropy of the cosmic microwave background radiation
2008	Yoichiro Nambu ($\frac{1}{2}$)	discovery of the mechanism of spontaneous broken symmetry in subatomic physics
	Makoto Kobayashi, Toshihide Maskawa	discovery of the origin of the broken symmetry that predicts the existence of at least three families of quarks in Nature
2011	Saul Perlmutter ($\frac{1}{2}$), Brian P. Schmidt, Adam G. Riess	discovery of the accelerating expansion of the universe through observations of distant supernovae

It is worth noting that several physicists with very important contributions to fundamental physics were awarded for their contributions in other areas, instead of their main discoveries: For example, Ernest Rutherford was awarded the 1908 prize in chemistry, while his work on classifying radioactivity, identifying α-particles as helium ions, establishment of the exponential decay law and its use as a clock, and – most importantly – the discovery of the atomic nuclei were not so awarded. Similarly, Enrico Fermi was awarded in 1938 for "demonstrations of the existence of new radioactive elements produced by neutron irradiation, and for his related discovery of nuclear reactions brought about by slow neutrons," while his theoretical model of β-decay and his other contributions to fundamental physics remained not so awarded; Vitaly L. Ginzburg was awarded in 2003, together with Alexei A. Abrikosov and Anthony J. Legett, "for pioneering contributions to the theory of superconductors and superfluids," but not for the groundbreaking work with Lev Landau on spontaneous magnetization, which eventually led to the general idea of spontaneous symmetry breaking and the so-called Higgs mechanism [☞ Section 7.1]. Bohr's principle of complementarity, Pauli's prediction of the neutrino, and even Einstein's theory of relativity, among others, remained similarly un-awarded by the Nobel committee. After all, Nobel Prizes are also a testament to the socio-political milieu. Finally, it is important to keep in mind the defined limitations: "In no case may a [Nobel] prize amount be divided between more than three persons." Also, "a [Nobel] Prize cannot be awarded posthumously, unless death has occurred after the announcement of the Nobel Prize" [517].

C.2 Some numerical values and useful formulae

While following the narrative in this book, numerical values of various constants are mostly unnecessary, but it is useful to have an idea about the relative numerical values of the various results, so that the Reader is expected to work through the derivations and complete the skipped steps, as well as to complete the exercises. Tables C.2, C.3 and C.4 should help in this endeavor.

When including electromagnetic phenomena in a study, note that the electric charge (divided by the natural constant $\sqrt{4\pi\epsilon_0}$) may be measured in purely "mechanical" units, as shown in equations (1.12). However, it is frequently useful to extend the unit system based on the measurement of the physical quantities of mass, length and time (M, L, T) by adding, minimally, the measurement of electric charge, C, and then consistently retaining all factors of $\sqrt{4\pi\epsilon_0}$. Owing to the identity $c^2 = 1/\epsilon_0\mu_0$, the constant μ_0 may always be expressed as $\mu_0 = 1/\epsilon_0 c^2$. However, in order to emphasize the electro-magnetic duality, Table C.4 on p. 527 retains both ϵ_0 and $\mu_0 = 1/\epsilon_0 c^2 = 4\pi \times 10^{-7}\,\mathrm{kg\,m/C^2}$.

Table C.2 Natural constants and some useful characteristic values

$\hbar$	$1.054\,572 \times 10^{-34}$ J s	$6.582\,119 \times 10^{-16}$ eV s	M_P	$2.176\,45 \times 10^{-8}$ kg	$1.220\,90 \times 10^{19}$ GeV/c^2
c	299, 792, 458 m/s		m_e	$9.109\,382 \times 10^{-31}$ kg	0.510 999 MeV/c^2
ϵ_0	$8.854\,187\,817 \times 10^{-12}\ \frac{\mathrm{C^2\,s^2}}{\mathrm{kg\,m^2}}$		m_μ	$1.883\,531 \times 10^{-28}$ kg	105.658 MeV/c^2
e	$1.602\,176 \times 10^{-19}$ C		m_τ	$3.167\,772 \times 10^{-27}$ kg	1.776 99 GeV/c^2
G_N	$6.674\,2 \times 10^{-11}\ \frac{\mathrm{m^3}}{\mathrm{kg\,s^2}}$	$6.708\,7 \times 10^{-39}\ \frac{\hbar c^5}{\mathrm{GeV^2}}$	m_p	$1.672\,621 \times 10^{-27}$ kg	938.272 MeV/c^2
N_A	$6.022\,141\,5 \times 10^{23}$ /mol		m_n	$1.674\,927 \times 10^{-27}$ kg	939.566 MeV/c^2
k_B	$1.380\,650\,5 \times 10^{-23}$ J/K	$8.617\,343 \times 10^{-5}$ eV/K	m_W	$1.433\,3 \times 10^{-25}$ kg	80.403 GeV/c^2
θ_w	$(28.74 \pm 0.01)^\circ$	("weak" mixing angle, θ_w)	m_Z	$1.625\,57 \times 10^{-25}$ kg	91.187 6 GeV/c^2
δ_{13}	$(1.20 \pm 0.08)^\circ$	(the CKM matrix phase, δ_{13})	m_H	2.244×10^{-25} kg	125.9 GeV/c^2

Table C.3 Some useful abbreviations and numerical values

α_e	$\frac{e^2}{4\pi\epsilon_0\hbar c} = \frac{g_e^2}{4\pi}$	$\frac{1}{137.035\,999}$	fine structure constant
r_e	$\frac{e^2}{4\pi\epsilon_0 m_e c^2}$	$2.817\,940\,325 \times 10^{-15}$ m	classical electron radius
Ry	$\frac{m_e e^4}{2(4\pi\epsilon_0)^2\hbar^2} = \frac{\alpha_e}{2} m_e c^2$	$13.605\,692\,2$ eV	Rydberg, H-atom ion. energy
$\lambdabar_e$	$\frac{\hbar}{m_e c} = \frac{r_e}{\alpha_e}$	$3.861\,592\,678 \times 10^{-13}$ m	Compton electron wavelength
μ_B	$\frac{e\hbar}{2m_e}$	$5.788\,381\,804 \times 10^{-11}$ MeV/T	Bohr magneton
a_0	$\frac{4\pi\epsilon_0\hbar^2}{m_e e^2} = \frac{\hbar}{\alpha_e m_e c} = \frac{r_e}{\alpha_e^2}$	$5.291\,772\,108 \times 10^{-11}$ m	Bohr radius

Other electromagnetic units (farad, tesla, volt, ampere, etc.) are expressed in terms of N, m, s, C. The unit C and the constants ϵ_0 and μ_0 may be eliminated by using the relation $c = 1/\sqrt{\epsilon_0\mu_0}$, and by redefining the electric charge $q \to q/\sqrt{4\pi\epsilon_0}$, which then is expressed in purely "mechanical" units. In general, note that precisely three base units are required in any system of units, and it is merely a tradition to choose units of mass, length and time.

Alternatively, as practiced in fundamental physics, one chooses a unit of speed (c), a unit of the Hamilton action or angular momentum ($\hbar$) and a unit of the gravitational force per product of the gravitating masses times the square of the distance between them (G_N). In addition to adopting this choice, the first two of these units are not even written in high energy particle physics practice, which is often phrased by stating (somewhat confusingly) that "$\hbar = 1 = c$." Every physical quantity is now expressible in terms (and units) of, say, energy – which is convenient in particle physics, since energy is in most cases the measured and controlled quantity [☞ Table 1.2 on p. 25]; Table C.5 could be helpful in this.

This practice is in fact no different than if one chose to adhere to a limited version of the SI system of units where (1) all distances are expressed in meters and all masses in kilograms, (2) no derivative units are ever used, and (3) one agrees to not even write the powers of 'm' and 'kg.' Every physical quantity would then be expressed in terms of time, and measured in units of suitable powers of seconds. In this system, length, mass and volume-specific mass (density) would have no written dimensions, speed and linear momentum would be measured in s^{-1} alike, while s^{-2} would be the appropriate (written) unit for acceleration, force and energy.

The ultimately natural (and parsimonious) unit system is then the one attributed to Planck, in which the natural constants c, $\hbar$ and G_N are implied but never written. This results, for example,

Table C.4 Comparative listing of primary (mechanical) SI units, minimally extended by the unit of electric charge, coulomb (C), and the dimensions of some oft-used electromagnetic quantities

	ϵ_0	$\vec{E}, F_{\mu\nu}$	Φ, A_μ	ρ_e	$\vec{\jmath}_e$	μ_0	$\vec{B}$	$\vec{A}$	ρ_m	$\vec{\jmath}_m$
Primary SI units	$\frac{\mathrm{s^2\,C^2}}{\mathrm{kg\,m^3}}$	$\frac{\mathrm{kg\,m}}{\mathrm{s^2\,C}}$	$\frac{\mathrm{kg\,m^2}}{\mathrm{s^2\,C}}$	$\frac{\mathrm{C}}{\mathrm{m^3}}$	$\frac{\mathrm{C}}{\mathrm{s\,m^2}}$	$\frac{\mathrm{kg\,m}}{\mathrm{C^2}}$	$\frac{\mathrm{kg}}{\mathrm{s\,C}}$	$\frac{\mathrm{kg\,m}}{\mathrm{s\,C}}$	$\frac{\mathrm{C}}{\mathrm{s^2\,m}}$	$\frac{\mathrm{C}}{\mathrm{s^3}}$
SI units (kg→N s^2/m)	$\frac{\mathrm{C^2}}{\mathrm{N\,m^2}}$	$\frac{\mathrm{N}}{\mathrm{C}}$	$\frac{\mathrm{N\,m}}{\mathrm{C}}$	$\frac{\mathrm{C}}{\mathrm{m^3}}$	$\frac{\mathrm{C}}{\mathrm{s\,m^2}}$	$\frac{\mathrm{N\,s^2}}{\mathrm{C^2}}$	$\frac{\mathrm{N\,s}}{\mathrm{m\,C}}$	$\frac{\mathrm{N\,s}}{\mathrm{C}}$	$\frac{\mathrm{C}}{\mathrm{s^2\,m}}$	$\frac{\mathrm{C}}{\mathrm{s^3}}$
Dimensions	$\frac{T^2C^2}{ML^3}$	$\frac{ML}{T^2C}$	$\frac{ML^2}{T^2C}$	$\frac{C}{L^3}$	$\frac{C}{TL^2}$	$\frac{ML}{C^2}$	$\frac{M}{TC}$	$\frac{ML}{TC}$	$\frac{C}{T^2L}$	$\frac{C}{T^3}$

Table C.5 Dimensions of some oft-used physical quantities, in the general $M^x L^y T^z$ format (first row), and the power-of-energy (particle physics) convention where $\hbar$ and c are implied and unwritten units (second row); e.g., $[\mathscr{L}] = 4$ means $[\mathscr{L}] = \text{MeV}^4$ up to powers of $\hbar$ and c

Basic units			In Lagrangian densities						Feynman calculus			
c	$\hbar$	G_N	$\mathscr{L}^a$	ϕ	$\mathbb{A}_\mu$	$\mathbb{F}_{\mu\nu}$	$\mathbb{J}_\mu$	$(\overline{\Psi}\Psi)$	$(\overline{u}\,u)$	$\mathfrak{M}$	Γ	σ
$\frac{L}{T}$	$\frac{ML^2}{T}$	$\frac{L^3}{MT^2}$	$\frac{M}{L^2T}$	$\frac{M^{1/2}}{T^{1/2}}$	$\frac{ML^2}{T^2}$	$\frac{ML}{T^2}$	$\frac{M}{T^2}$	$\frac{T}{L^4}$	$\frac{ML}{T}$	—	$\frac{1}{T}$	L^2
0	0	2	4	1	1	2	3	3	1	0	1	−2

[a] Relativistic Lagrangian densities $\mathscr{L}$ are normalized so that $[\int \mathrm{d}^4\mathrm{x}\,\mathscr{L}] = [\hbar]$, with $x^0 = ct$ and $[\mathrm{d}^4\mathrm{x}] = [L^4]$. Similarly, $[\int \mathrm{d}^4\mathrm{x}\,\overline{\Psi} mc^2 \Psi] = [\hbar]$, and Feynman calculus uses $u \propto \sqrt{\hbar c^3} \int \mathrm{d}t\, e^{-i\omega t}\Psi(\mathrm{x})$; see also equation (5.53).

Table C.6 Natural (Planck) units and their SI equivalent value

Name	Expression	SI equivalent	Practical equivalent
Length	$\ell_P = \sqrt{\frac{\hbar G_N}{c^3}}$	$1.616\,25 \times 10^{-35}$ m	
Mass	$M_P = \sqrt{\frac{\hbar c}{G_N}}$	$2.176\,44 \times 10^{-8}$ kg	$1.220\,86 \times 10^{19}$ GeV/c^2
Time	$t_P = \sqrt{\frac{\hbar G_N}{c^5}}$	$5.391\,24 \times 10^{-44}$ s	
Charge[a]	$q_P = \sqrt{4\pi\epsilon_0\,\hbar c}$	$1.875\,55 \times 10^{-18}$ C	$e\sqrt{\alpha_e} \approx 11.706\,2\,e$
Temperature	$T_P = \frac{1}{k_B} M_P c^2$	$1.416\,79 \times 10^{32}$ K	

[a] $\alpha_e \approx 1/137.035\,999\,679$ in low-energy scattering experiments, but grows to about 1/127 near $\sim$200 GeV energies [☞ Section 5.3.3].

in the units for physical quantities that are listed in Table C.6 on p. 528, and the Reader is invited to compute many more along the lines of the computations practiced in Section 1.2. Notice, however, that once all physical quantities are expressed in units of $\hbar, c, G_N$ – which are not written explicitly – all physical quantities appear to have no (written) dimensions/units! Note that the Boltzmann constant $k_B = 1.38 \times 10^{-23}$ J/K is clearly simply a unit conversion factor, from temperature to energy, and need be written only if one wishes to emphasize the statistical nature of a certain quoted energy (temperature).

Table C.7 lists a few symbols used in this book, many of which are fairly standard in formal logic and set theory, but are not as frequently used in the physics literature. The symbols: $\propto$ ("proportional"), $\cong$ ("isomorphic"), $\simeq$ ("equivalent"), $\approx$ ("approximate," but "homomorphic" for groups and algebras), $\sim$ ("asymptotic" for functions, but "of the order of" for numbers), $\times$ (Cartesian or direct product, but "vector product" for 3-vectors and the usual product of a decimal number and a power of ten), $\otimes$ (Kronecker, i.e., tensor product), $\ltimes$ (semidirect product), $\hookrightarrow$ (injection), $\twoheadrightarrow$ (surjection) and $\mapsto$ ("maps/assigns to") are probably more familiar, but are listed here for completeness; see also the lexicon of jargon in Section B.1.

Finally, Table C.8 lists symbols that have been constructed for their specific indicated purpose in this book, and which to the best of my knowledge do not appear elsewhere in the literature.

Table C.7 Symbols borrowed from formal logic and set theory

Symbol	Meaning of the symbol as used in this book
$\subset$	"subset"; e.g., "$A \subset B$" means "A is a subset[a] of B"
$\subsetneq$	"proper subset"; e.g., "$A \subsetneq B$" means "A is a subset[a] of B and $A \neq B$"
$\cup$	"union"; an element belongs to $A \cup B$ if it belongs to A or B (inclusively)
$\cap$	"intersection"; an element belongs to $A \cap B$ if it belongs to *both* A and B
$\smallsetminus$	"minus"; an element belongs to $A \smallsetminus B$ if it belongs to A but not to B
$\in$	"in" or "is an element of"; e.g., "$x \in X$" means "x is an element of X"
$\varnothing$	"empty set", i.e., the formal set that has no element at all
$\forall$	"for all"; e.g., "$\forall x$" means "for every x"
$\exists$	"exists"; e.g., "$\exists x$" means "there exists an x"
$\Rightarrow$	"implies"; e.g., "$x \Rightarrow y$" means "x implies y" (said of claims x, y)
$\Leftrightarrow$	"is equivalent"; e.g., "$x \Leftrightarrow y$" means "x is equivalent to y" (said of claims x, y)

[a] If B has a structure (of an algebra, a group, ...), A inherits this structure from B – unless noted otherwise.

Table C.8 The definition of some less frequently used or here constructed mathematical symbols

Symbol	Meaning of the symbol as used in this book
$:=$	the left-hand symbol is defined to equal the right-hand expression
$=:$	the previously undefined right-hand symbol is defined so as to make the equality hold for all values of the remaining symbols
$:\simeq$	the left-hand symbol is defined to be equivalent (by an implicit equivalence, such as integration by parts) to the right-hand expression
$\not\mathrel{\mathop{=}}$	need not be equal – in distinction to the "(certainly) not equal" symbol, $\neq$
$\stackrel{!}{=}$	required to be equal
$\stackrel{\text{"...''}}{=}$	equals, owing to (by use of) the relation/property "..."
$\ni\!\!+$	semidirect sum of two algebras $\mathfrak{a} \ni\!\!+ \mathfrak{b}$, the first summand maps $\mathfrak{a} : \mathfrak{b} \to \mathfrak{b}$; e.g., for Lie algebras, $[a, b] \in \mathfrak{b}$, for $a \in \mathfrak{a}$ and $b \in \mathfrak{b}$.
$\wedge$	antisymmetric product of two forms [☞ Digression 5.8 on p. 184]

C.3 Answers to some exercises

A successful solving of the end-of-section exercises should confirm the understanding of the material of that section. For assistance and orientation, some partial and final results to these exercises are listed here.

Ex. 1.2.1 and 1.2.3 Admittedly, these are trick exercises. Let a standing person's *horizontal* linear dimensions be scaled down by a factor of λ_h while the vertical measurements scale by λ_v, and let W denote the person's weight, A the cross-section area of the bones in the legs (femur, tibia, fibula, etc.) and $P = \frac{W}{A}$ the pressure of the person's own weight on these bones. Then,

$$W \propto \lambda_v \cdot \lambda_h^2, \quad A \propto \lambda_h^2, \quad P \propto \lambda_v, \tag{C.1}$$

so that the vertical pressure in the bones is, in this rough estimate, independent of the horizontal scaling factor and only depends on the vertical scaling factor. Therefore, in part 1 of this exercise, for this pressure to be about the same as in ordinary humans, $\lambda_v \sim 1$ and not $\lambda_v = 40$ as stated.

This then implies that, in Lilliputians and small animals, the structure and even chemical composition of bones may be proportionally weaker than in ordinary humans. In turn, in animals larger than humans, bones must support greater pressures than in ordinary humans. Since the structure and chemical composition of bones cannot vary too much, this provides a strong limitation on the height of land-dwelling animals. Sorry: there can exist no 25-foot, 20-ton gorillas.

Ex. 1.2.6 The principal quantum number n becomes continuous.

Ex. 2.4.1 $\triangle y(\ell) = \frac{1}{2}\frac{q}{m}\,\ell^2\,\frac{B_0^2}{E_0}$. $\triangle z = 0$.

Ex. 3.2.4 $T_2 - T_1 = (m_1 - m_2)(1 - \frac{m_1+m_2}{M})c^2$, so that $T_2 - T_1 = \frac{m_1}{M}(M - m_1)c^2$ when $m_2 = 0$.

Ex. 4.2.1 With only the orthonormal states $|a\rangle$ and $|b\rangle$ given, eigenstates must be of the form $\alpha|a\rangle + \beta|b\rangle$. Then $\mathsf{P}[\alpha|a\rangle + \beta|b\rangle] = \pi_\mathsf{P}[\alpha|a\rangle + \beta|b\rangle]$, where π_P is the eigenvalue, so that

$$\pi_\mathsf{P}\big[\alpha|a\rangle + \beta|b\rangle\big] = \mathsf{P}\big[\alpha|a\rangle + \beta|b\rangle\big] = \big[\alpha|b\rangle + \beta|a\rangle\big]. \tag{C.2}$$

Projecting with $\langle a|$ and $\langle b|$ yields

$$\pi_\mathsf{P}\,\alpha = \beta, \quad \pi_\mathsf{P}\,\beta = \alpha, \qquad \Rightarrow \quad \pi_\mathsf{P}^2 = 1, \;\; \pi_\mathsf{P} = \pm 1. \tag{C.3}$$

From that,

$$\pi_\mathsf{P} = +1, \qquad |+\rangle := \tfrac{1}{\sqrt{2}}\big(|a\rangle + |b\rangle\big), \qquad \mathsf{P}|+\rangle = (+1)|+\rangle; \tag{C.4}$$

$$\pi_\mathsf{P} = -1, \qquad |-\rangle := \tfrac{1}{\sqrt{2}}\big(|a\rangle - |b\rangle\big), \qquad \mathsf{P}|-\rangle = (-1)|-\rangle. \tag{C.5}$$

Ex. 5.3.2 Using the relations from Digression 5.9 on p. 191, we have

$$\begin{aligned}
\partial_\alpha \frac{\partial \mathscr{L}_{QED}}{\partial(\partial_\alpha A_\beta)} &= \partial_\alpha \frac{\partial}{\partial(\partial_\alpha A_\beta)}\Big[-\tfrac{4\pi\epsilon_0}{4}(\partial_\mu A_\nu - \partial_\nu A_\mu)\eta^{\mu\rho}\eta^{\nu\sigma}(\partial_\rho A_\sigma - \partial_\sigma A_\rho)\Big] \\
&= -\tfrac{4\pi\epsilon_0}{4}\partial_\alpha\Big[(\delta^{\alpha\beta}_{\mu\nu} - \delta^{\alpha\beta}_{\nu\mu})\eta^{\mu\rho}\eta^{\nu\sigma}(\partial_\rho A_\sigma - \partial_\sigma A_\rho) \\
&\qquad + (\partial_\mu A_\nu - \partial_\nu A_\mu)\eta^{\mu\rho}\eta^{\nu\sigma}(\delta^{\alpha\beta}_{\rho\sigma} - \delta^{\alpha\beta}_{\sigma\rho})\Big] \\
&= -\tfrac{4\pi\epsilon_0}{4}\partial_\alpha\Big[(\delta^{\alpha\beta}_{\mu\nu} - \delta^{\alpha\beta}_{\nu\mu})(\partial^\mu A^\nu - \partial^\nu A^\mu) + (\partial^\rho A^\sigma - \partial^\sigma A^\rho)(\delta^{\alpha\beta}_{\rho\sigma} - \delta^{\alpha\beta}_{\sigma\rho})\Big] \\
&= -\tfrac{4\pi\epsilon_0}{4}\partial_\alpha\Big[(\delta^{\alpha\beta}_{\mu\nu} - \delta^{\alpha\beta}_{\nu\mu})F^{\mu\nu} + F^{\rho\sigma}(\delta^{\alpha\beta}_{\rho\sigma} - \delta^{\alpha\beta}_{\sigma\rho})\Big] \\
&= -\tfrac{4\pi\epsilon_0}{4}\partial_\alpha\Big[F^{\alpha\beta} - F^{\beta\alpha} + F^{\alpha\beta} - F^{\beta\alpha}\Big] = -4\pi\epsilon_0\partial_\alpha F^{\alpha\beta}.
\end{aligned} \tag{C.6}$$

Similarly,

$$\begin{aligned}
\frac{\partial \mathscr{L}_{QED}}{\partial A_\beta} &= \frac{\partial}{\partial A_\beta}\Big[-\overline{\Psi}(\mathbf{x})\Big[i\boldsymbol{\gamma}^\mu(\hbar c\partial_\mu - iq_\Psi A_\mu) - mc^2\Big]\Psi(\mathbf{x})\Big] \\
&= -\overline{\Psi}(\mathbf{x})\Big[i\boldsymbol{\gamma}^\mu(-iq_\Psi\delta^\beta_\mu)\Big]\Psi(\mathbf{x}) = -q_\Psi\overline{\Psi}(\mathbf{x})\boldsymbol{\gamma}^\beta\Psi(\mathbf{x}).
\end{aligned} \tag{C.7}$$

The relation (5.120f) follows upon equating these two results.

Ex. 5.4.4 Using definition $m_i := z_i M$, the property $\delta(ax) = \delta(x)/a$ and that $x_i = x/z_i$ yields

$$W^i_1 = \frac{Q_i^2}{2(Mz_i)}\,\delta\Big(\frac{x}{z_i} - 1\Big) = \frac{Q_i^2}{2M}\,\delta\Big(z_i\frac{x}{z_i} - z_i\Big) = \frac{Q_i^2}{2M}\,\delta(x - z_i). \tag{C.8}$$

Also, using that $\delta(x-1) = x^2\delta(x-1)$ yields

$$\begin{aligned}
W^i_2 &= -\frac{2m_i c^2 Q_i^2}{\mathsf{q}^2}\,x_i^{\,2}\delta\Big(\frac{x}{z_i} - 1\Big) = -\frac{2(Mz_i)c^2 Q_i^2}{\mathsf{q}^2}\,x_i^{\,2} z_i\,\delta(x - z_i) \\
&= -\frac{2Mc^2 Q_i^2}{\mathsf{q}^2}\,x^2\delta(x - z_i).
\end{aligned} \tag{C.9}$$

Ex. 6.1.3 Write the equation $\partial_\mu F^{a\,\mu\nu} = J^{a\,\nu}_{(c)}$ in matrix notation, $\partial_\mu \mathbb{F}^{\mu\nu} = \mathbb{J}^\nu_{(c)}$, where we also have equation (6.16), $\mathbb{F}'_{\mu\nu} = U_\varphi\,\mathbb{F}_{\mu\nu}\,U_\varphi^{-1}$. It then follows that

$$\begin{aligned}\partial_\mu \mathbb{F}'^{\mu\nu} &= \partial_m (U_\varphi\,\mathbb{F}^{\mu\nu}\,U_\varphi^{-1}) &\text{(C.10)}\\ &= (\partial_\mu U_\varphi)\mathbb{F}^{\mu\nu}\,U_\varphi^{-1} + U_\varphi(\partial_\mu \mathbb{F}^{\mu\nu})U_\varphi^{-1} + U_\varphi\,\mathbb{F}^{\mu\nu}(\partial_\mu U_\varphi^{-1}). &\text{(C.11)}\end{aligned}$$

To simplify this result, use that $\mathbb{1} = U_\varphi U_\varphi^{-1}$, the derivative of which gives

$$0 = (\partial_\mu U_\varphi)U_\varphi^{-1} + U_\varphi(\partial_\mu U_\varphi^{-1}) \qquad \Rightarrow \qquad (\partial_\mu U_\varphi^{-1}) = -U_\varphi^{-1}(\partial_\mu U_\varphi)U_\varphi^{-1}. \tag{C.12}$$

Combining, we have

$$\begin{aligned}\partial_\mu \mathbb{F}'^{\mu\nu} &= (\partial_\mu U_\varphi)\mathbb{F}^{\mu\nu}\,U_\varphi^{-1} + U_\varphi(\partial_\mu \mathbb{F}^{\mu\nu})U_\varphi^{-1} - U_\varphi\,\mathbb{F}^{\mu\nu}U_\varphi^{-1}(\partial_\mu U_\varphi)U_\varphi^{-1}\\ &= (\partial_\mu U_\varphi)U_\varphi^{-1}(U\mathbb{F}^{\mu\nu}\,U_\varphi^{-1}) + (U_\varphi \mathbb{J}^\nu_{(c)}U_\varphi^{-1}) - (U_\varphi\,\mathbb{F}^{\mu\nu}U_\varphi^{-1})(\partial_\mu U_\varphi)U_\varphi^{-1}\\ &= \mathbb{J}'^\nu_{(c)} + (\partial_\mu U_\varphi)U_\varphi^{-1}\,\mathbb{F}'^{\mu\nu} - \mathbb{F}'^{\mu\nu}(\partial_\mu U_\varphi)U_\varphi^{-1}\\ &= \mathbb{J}'^\nu_{(c)} + \big[(\partial_\mu U_\varphi)U_\varphi^{-1}\,,\,\mathbb{F}'^{\mu\nu}\big], &\text{(C.13)}\end{aligned}$$

the form of which could have been guessed from relations (6.39) and (6.6c).

Ex. 7.1.2 Motivated by the form of the result to be proven, use the polar coordinates $\phi_1 = \varrho\cos\theta$, $\phi_2 = \varrho\sin\theta$, where the potential density in the Lagrangian density (7.21) becomes

$$\mathscr{V} = -\tfrac{1}{2}\big(\tfrac{mc}{\hbar}\big)^2\varrho^2 + \tfrac{1}{4}\lambda\varrho^4, \tag{C.14}$$

so that the stationary values of the variable ϱ are given by

$$-\big(\tfrac{mc}{\hbar}\big)^2\varrho + \lambda\varrho^3 = 0 \qquad \Rightarrow \qquad \partial_0 = 0,\;\; \varrho_\pm = \pm\tfrac{mc}{\hbar\sqrt{\lambda}}. \tag{C.15}$$

It is not hard to prove that $\varrho_0 = 0$ is a maximum, and $\varrho_+ = \frac{mc}{\hbar\sqrt{\lambda}}$ a minimum; the third solution, $\varrho_- = -\frac{mc}{\hbar\sqrt{\lambda}}$, is unreasonable as a value for the radial polar coordinate. The desired result follows by transforming back into Cartesian parametrization, (ϕ_1, ϕ_2).

Ex. 9.1.4 In the extended equality (9.14) only the last one is not evident, and follows from the fact that

$$g_{\mu\nu}\,g^{\mu\nu} = 4 \quad \overset{\delta}{\Longrightarrow} \quad \delta(g_{\mu\nu}\,g^{\mu\nu}) = 0 \quad \Rightarrow \quad (\delta g_{\mu\nu})g^{\mu\nu} = -g_{\mu\nu}(\delta g^{\mu\nu}). \tag{C.16}$$

With no extra effort, we also have the general result:

$$g_{\mu\nu}\,g^{\mu\sigma} = \delta^\sigma_\nu \quad \overset{\delta}{\Longrightarrow} \quad \delta(g_{\mu\nu}\,g^{\mu\sigma}) = 0 \quad \Rightarrow \quad (\delta g_{\mu\nu})g^{\mu\sigma} = -g_{\mu\nu}(\delta g^{\mu\sigma}). \tag{C.17}$$

Contracting this last equality with $g^{\rho\nu}$ yields [☞ also Digression 9.3 on p. 329]

$$\delta g^{\rho\sigma} = -g^{\rho\nu}(\delta g_{\mu\nu})g^{\mu\sigma}. \tag{C.18}$$

Ex. 10.3.1 Direct computation yields

$$\begin{aligned}\mathrm{Tr}\big[\{Q_i, Q^{\dagger j}\}\big] &= \tfrac{1}{2}\sum_i\{Q_i, Q^{\dagger i}\} + \tfrac{1}{2}\sum_i\{Q^{\dagger i}, Q_i\}\\ &= \tfrac{1}{2}\sum_i\underbrace{\{Q_i, Q_i\}}_{\equiv 0} + \tfrac{1}{2}\sum_i\{Q_i, Q^{\dagger i}\} + \tfrac{1}{2}\sum_i\{Q^{\dagger i}, Q_i\} + \tfrac{1}{2}\sum_i\underbrace{\{Q^{\dagger i}, Q^{\dagger i}\}}_{\equiv 0}\end{aligned}$$

$$= \tfrac{1}{2}\sum_i \{Q_i + Q^{\dagger i}, Q_i + Q^{\dagger i}\}] \overset{(10.32a)}{=} \tfrac{1}{2}\sum_i \{\mathcal{Q}_i, \mathcal{Q}_i\} = \sum_i \mathcal{Q}_i\mathcal{Q}_i$$
$$= \sum_i |\mathcal{Q}_i|^2 \geqslant 0, \tag{C.19}$$

where $\mathcal{Q}_i\mathcal{Q}_i = |\mathcal{Q}_i|^2$ as the operators $\mathcal{Q}_i$ are Hermitian.

Ex. 11.3.1 The Ricci tensor is

$$[R_{mn}] = \begin{bmatrix} -2e^{-2k|y|}[k\,\text{sig}^2(y) - \delta(y)] & 0 & 0 \\ 0 & 2e^{-2k|y|}[k\,\text{sig}^2(y) - \delta(y)] & 0 \\ 0 & & 2k[\delta(y) - k\,\text{sig}^2(y)] \end{bmatrix}, \tag{C.20}$$

and the scalar curvature is $R = 2k[4\,\delta(y) - 3k\,\text{sig}^2(y)]$.

References

[1] Wikipedia article, *Precision tests of QED*. http://en.wikipedia.org/wiki/Precision_tests_of_QED.
[2] M. Abe, S. Ichinose and N. Nakanishi, *Kerr metric, de Donder condition and gravitational energy density*, *Prog. Theor. Phys.* 78(5) (1987) 1186–1201.
[3] E. S. Abers, *Quantum Mechanics*, Pearson Education, 2004.
[4] M. Abraham, *Dynamik des Electrons*, *Göttinger Nachrichten* (1902) 20–41.
[5] E. M. Abreu, M. A. D. Andrade, L. P. de Assis, J. A. Helayel-Neto, A. Nogueira and R. C. Paschoal, *A supersymmetric model for graphene*, *JEHP* 1105 (2011) 001, arxiv:1002.2660.
[6] D. Adams, *Is there an Artificial God?* (speech at *Digital Biota 2*, Cambridge, UK, September 1998). http://www.biota.org/people/douglasadams/.
[7] H. Adroź and L. Hadasz, *Lectures on Classical and Quantum Theory of Fields*, Springer-Verlag, 2010.
[8] I. J. Aitchison, *Supersymmetry in Particle Physics*, Cambridge University Press, 2007.
[9] K. Akama, *Pregeometry*, in K. Kikkawa, N. Nakanishi and H. Nariai (eds.), *Gauge Theory and Gravitation*, Springer-Verlag, 1982, pp. 267–271.
[10] J. Al-Khalili, *Black Holes, Wormholes & Time Machines*, Institute of Physics Publishing, 1999.
[11] S. Alexandrov and P. Roche, *Critical overview of loops and foams*, *Phys. Rep.* 506 (2011) 41–86.
[12] L. Alvarez-Gaume and E. Witten, *Gravitational anomalies*, *Nucl. Phys.* B234 (1984) 269.
[13] L. Anchordoqui, D. C. Dai, H. Goldberg, G. Landsberg, G. Shaughnessy, D. Stojković and T. J. Weiler, *Searching for the layered structure of space at the LHC*, *Phys. Rev.* D83 (2011) 114046.
[14] L. Anchordoqui, D. C. Dai, M. Fairbairn, G. Landsberg and D. Stojković, *Vanishing dimensions and planar events at the LHC*, *Mod. Phys. Lett.* A27 (2012) 1250021, arXiv:1003.5914.
[15] P. Anderson, *Plasmons, gauge invariance, and mass*, *Phys. Rev.* 130 (1963) 439.
[16] D. Andriot, E. Goi, R. Minasian and M. Petrini, *Supersymmetry breaking branes on solvmanifolds and de Sitter vacua in string theory*, *JHEP* 1105 (2011) 028.
[17] H. Arcos and J. Pereira, *Gen. Rel. Grav.* 36 (2004) 2441.
[18] G. B. Arfken and H. Weber, *Mathematical Methods for Physicists*, 6th edn., Academic Press, 2005.
[19] P. C. Argyres, *An introduction to global supersymmetry*. http://www.physics.uc.edu/~argyres/661/susy 2001.pdf.
[20] T. Asaka, W. Buchmuller and L. Covi, *False vacuum decay after inflation*, *Phys. Lett.* B510 (2001) 271–276.
[21] A. Ashtekar, *New variables for classical and quantum gravity*, *Phys. Rev. Lett.* 57 (18) (1986) 2244–2247.
[22] P. Aspinwall, B. R. Greene and D. R. Morrison, *Space-time topology change and stringy geometry*, *J. Math. Phys.* 35 (1994) 5321–5337.
[23] P. S. Aspinwall, B. R. Greene and D. R. Morrison, *Space-time topology change: The physics of Calabi–Yau moduli space*, in *Berkeley 1993, Proceedings, Strings '93*, 1993, pp. 241–262.
[24] P. S. Aspinwall, B. R. Greene and D. R. Morrison, *Calabi–Yau moduli space, mirror manifolds and space-time topology change in string theory*, *Nucl. Phys.* B416 (1994) 414–480.
[25] ATLAS Collaboration, *Observation of a new particle in the search for the Standard Model Higgs boson with the ATLAS detector at the LHC*, *Phys. Lett.* B716 (2012) 1–29.
[26] A. Babichenko and D. Gepner, *Nonstandard parafermions and string compactification*, *Nucl. Phys.* B854 (2012) 375–392.
[27] J. Baggott, *Higgs: The Invention and Discovery of the God Particle*, Oxford University Press, 2012.
[28] D. Bailin and A. Love, *Cosmology in Gauge Field Theory and String Theory*, IOP Publishing, 2004.

[29] L. E. Ballentine, *Quantum Mechanics*, 2nd edn., World Scientific, 1998.

[30] T. Banks, W. Fischler, S. H. Shenker and L. Susskind, *M theory as a matrix model: A conjecture*, *Phys. Rev.* D55 (1997) 5112–5128.

[31] T. Banks and N. Seiberg, *Symmetries and strings in field theory and gravity*, *Phys. Rev.* D83 (2011) 084019.

[32] W. A. Bardeen, H. Fritzsch and M. Gell-Mann, *Light cone current algebra, π^0 decay, and $e^+ e^-$ annihilation*, in R. Gato (ed.), *Scale and Conformal Symmetry in Hadron Physics*, John Wiley & Sons, 1973, p. 139.

[33] R. Barlow, *Introducing gauge invariance*, *Eur. J. Phys.* 11 (1990) 45–46.

[34] J. D. Barrow, *Theories of Everything: The Quest for Ultimate Explanation*, Fawcett Columbine, 1991.

[35] A. Barut, *Electrodynamics and Classical Field Theory of Fields and Particles*, Dover Publications, 1964.

[36] I. A. Batalin and E. Fradkin, *Operator quantization and abelization of dynamical systems subject to first-class constraints*, *Riv. Nuovo Cimento* 9 (10) (1986) 1–48.

[37] I. A. Batalin, E. Fradkin and T. Fradkina, *Generalized canonical quantization of dynamical systems with constraints and curved phase space*, *Nucl. Phys.* B332 (3) (1990) 723–736.

[38] I. A. Batalin and I. Tyutin, *BRST-invariant algebra of constraints in terms of commutators and quantum antibrackets*, *Theor. Math. Phys.* 138 (1) (2004) 1–17.

[39] I. A. Batalin and G. A. Vilkovisky, *Relativistic s-matrix of dynamical systems with boson and fermion constraints*, *Phys. Lett.* B69 (1977) 309–312.

[40] I. A. Batalin and G. A. Vilkovisky, *Gauge algebra and quantization*, *Phys. Lett.* B102 (1981) 27–31.

[41] I. A. Batalin and G. A. Vilkovisky, *Quantization of gauge theories with linearly dependent generators*, *Phys. Rev.* D28 (1983) 2567–2582.

[42] V. V. Batyrev, *Dual polyhedra and mirror symmetry for Calabi–Yau hypersurfaces in toric varieties*, *J. Algebraic Geom.* 3 (3) (1994) 493–535.

[43] V. V. Batyrev and L. A. Borisov, *Mirror duality and string-theoretic Hodge numbers*, *Invent. Math.* 126 (1) (1996) 183–203.

[44] C. Becchi, A. Rouet and R. Stora, *Renormalization of gauge theories*, *Ann. Physics* 98 (2) (1976) 287–321.

[45] P. Becher, M. Böhm and H. Joos, *Gauge Theories of Strong and Electroweak Interactions*, John Wiley & Sons, 1983.

[46] K. Becker, M. Becker and J. H. Schwarz, *String Theory and M-Theory: A Modern Introduction*, Cambridge University Press, 2007.

[47] S. Bellucci, S. Krivonos, O. Lechtenfeld and A. Shcherbakov, *Superfield formulation of nonlinear N=4 supermultiplets*, *Phys. Rev.* D77 (2008) 045026.

[48] V. Berestetskii, E. Lifshitz and L. Pitaevskii, *Quantum Electrodynamics*, Pergamon Press, 1980.

[49] P. Berglund, P. Candelas, X. de la Ossa, A. Font, T. Hübsch, D. Jancic and F. Quevedo, *Periods for Calabi–Yau and Landau–Ginzburg vacua*, *Nucl. Phys.* B419 (1994) 352–403.

[50] P. Berglund, E. Derrick, T. Hübsch and D. Jancic, *On periods for string compactifications*, *Nucl. Phys.* B420 (1994) 268–288.

[51] P. Berglund and T. Hübsch, *A generalized construction of mirror manifolds*, *Nucl. Phys.* B393 (1–2) (1993) 377–391.

[52] P. Berglund and T. Hübsch, *On a residue representation of deformation, koszul and chiral rings*, *Int. J. Mod. Phys.* A10 (1995) 3381–3430.

[53] P. Berglund, T. Hübsch and D. Minic, *de Sitter spacetimes from warped compactifications of IIB string theory*, *Phys. Lett.* B534 (2002) 147–154.

[54] P. Berglund, T. Hübsch and D. Minic, *Relating the cosmological constant and supersymmetry breaking in warped compactifications of IIB string theory*, *Phys. Rev.* D67 (2003) 041901.

[55] P. G. Bergmann, *Introduction to the Theory of Relativity*, Dover Publications, 1975 (original publ. Prentice Hall, 1942).

[56] R. Berndt, *Representations of Linear Groups*, Vieweg, 2007.

[57] E. Berti and K. D. Kokkotas, *Quasinormal modes of Kerr–Newman black holes: Coupling of electromagnetic and gravitational perturbations*, *Phys. Rev.* D71 (2005) 124008.

[58] E. Bertschinger, *Gravitation in the weak field limit*. http://web.mit.edu/edbert/GR/gr6.pdf.

[59] H. A. Bethe and E. Salpeter, *Quantum Mechanics of One- and Two-Electron Atoms*, Springer-Verlag, 1957.

[60] A. Bilal, *Introduction to supersymmetry*, in *Summer School GIF 2000*, 2001.

[61] S. Bilenkii, *Introduction to Feynman Diagrams*, Elsevier, 1974.

[62] R. L. Bishop and S. I. Goldberg, *Tensor Analysis on Manifolds*, Dover Publications, 1980 (original publ. 1960).

[63] J. Bjorken and S. Drell, *Relativistic Quantum Fields*, McGraw-Hill, 1964.

[64] J. Bjorken and S. Drell, *Relativistic Quantum Mechanics*, McGraw-Hill, 1964.

[65] T. Blacker, R. E. Raspe and W. Rushton, *Baron Münchhausen's Narrative of His Marvelous Travels and Campaigns: Surprising Adventures of Baron Münchhausen*, Knight, 1991 (original publ. 1785).

[66] M. Blagojević, *Gravitation and Gauge Symmetries*, IOP Publishing, 2001.

[67] M. Blau and S. Theisen, *String theory as a theory of quantum gravity: A status report, Gen. Rel. Grav.* 41 (2009) 743–755.

[68] D. L. Block, *Georges Lemaitre and Stiglers law of eponymy*, in R. Holder and S. Mitton (eds.), *Georges Lemaitre: Life, Science and Legacy. Proceedings of the 80th Anniversary Conference held by the Faraday Institute, St Edmund's College, Cambridge*, Royal Astronomical Society–Springer, 2012.

[69] D. Bohm, *The Special Theory of Relativity*, Routledge; New Edn, 2006 (original publ. W.A. Benjamin, 1965).

[70] N. Bohr, H. A. Kramers and J. C. Slater, *The quantum theory of radiation, Zeitschrift für Physik* 24 (1924) 69 (also in *Philos. Mag.* 47 (1924) 785–802).

[71] L. Borisov, *Towards the mirror symmetry for Calabi–Yau complete intersections in Gorenstein toric Fano varieties.* http://arXiv.org/abs/alg-geom/9310001.

[72] L. A. Borisov, *Berglund-Hübsch mirror symmetry via vertex algebras, Comm. Math. Phys.* 320 (1) (2013) 73–99, arXiv1007.2633.

[73] W. Boucher, *Positive energy without supersymmetry, Nucl. Phys.* B242 (1984) 282–296.

[74] R. Bousso and J. Polchinski, *Quantization of four-form fluxes and dynamical neutralization of the cosmological constant, JHEP* 0006 (2000) 006.

[75] M. J. Bowick and S. Rajeev, *Anomalies and curvature in complex geometry, Nucl. Phys.* B296 (1988) 1007–1033.

[76] I. L. Buchbinder and S. M. Kuzenko, *Ideas and Methods of Supersymmetry and Supergravity*, Studies in High Energy Physics Cosmology and Gravitation, IOP Publishing, 1998.

[77] C. Bunster and M. Henneaux, *$Sp(2n,\mathbb{R})$ electric-magnetic duality as* off-shell *symmetry of interacting electromagnetic and scalar fields* in L. Bergstrom (ed.), Hector Rubinstein Memorial Symposium on Quarks, Strings and the Cosmos, Proceedings of Science, vol. HRMS2010, 2011, arXiv:1101.6064.

[78] A. Burinskii, *Complex Kerr geometry, twistors and the Dirac electron, J. Phys.* A41 (2008) 164069.

[79] A. Burinskii, *Gravity vs. quantum theory: Is electron really pointlike?, J. Phys.: Conf. Ser.* 343 (2012) 012019.

[80] R. N. Cahn, *Semi-Simple Lie Algebras and Their Representations*, Benjamin/Cummings, 1984.

[81] E. Calzetta, *Spinodal decomposition in quantum field theory, Ann. Phys.* 190 (1) (1989) 32–58.

[82] P. Candelas, *Yukawa couplings between (2,1) forms, Nucl. Phys.* B298 (1988) 458.

[83] P. Candelas, X. de la Ossa and S. H. Katz, *Mirror symmetry for Calabi–Yau hypersurfaces in weighted $\mathbb{P}^4$ and extensions of Landau–Ginzburg theory, Nucl. Phys.* B450 (1995) 267–292.

[84] P. Candelas, X. C. de la Ossa, P. S. Green and L. Parkes, *An exactly soluble superconformal theory from a mirror pair of Calabi–Yau manifolds, Phys. Lett.* B258 (1991) 118–126.

[85] P. Candelas, X. C. de La Ossa, P. S. Green and L. Parkes, *A pair of Calabi–Yau manifolds as an exactly soluble superconformal theory, Nucl. Phys.* B359 (1991) 21–74.

[86] P. Candelas, P. S. Green and T. Hübsch, *Finite distances between distinct Calabi–Yau manifolds, Phys. Rev. Lett.* 62 (1989) 1956–1959.

[87] P. Candelas, P. S. Green and T. Hübsch, *Rolling among Calabi–Yau vacua, Nucl. Phys.* B330 (1990) 49–102.

[88] P. Candelas, G. T. Horowitz, A. Strominger and E. Witten, *Vacuum configurations for superstrings, Nucl. Phys.* B258 (1985) 46–74.

[89] P. Candelas, T. Hübsch and R. Schimmrigk, *Relation between the Weil–Petersson and Zamolodchikov metrics, Nucl. Phys.* B329 (1990) 583.

[90] P. Candelas and D. J. Raine, *Compactification and supersymmetry in $d = 11$ supergravity*, *Nucl. Phys.* B248 (1984) 415–422.

[91] F. Capra, *The Tao of Physics: An Exploration of the Parallels between Modern Physics and Eastern Mysticism*, 4th edn., Shambhala Publications, 2000.

[92] S. Carneiro and M. C. Nemes, *Spacetime quantization induced by axial currents*, *Chaos Solitons Fractals* 24 (2005) 1183–1187.

[93] S. Carroll, *From Eternity to Here: The Quest for the Ultimate Theory of Time*, Plume, 2010.

[94] S. Carroll, *The Particle at the End of the Universe: How the Hunt for the Higgs Boson Leads Us to the Edge of a New World*, Dutton Adult, 2012.

[95] S. M. Carroll, *Lecture notes on general relativity*. http://arxiv.org/abs/gr-qc/9712019.

[96] S. M. Carroll, *Spacetime and Geometry: An Introduction to General Relativity*, Addison-Wesley, 2004.

[97] P. Carruthers and M. M. Nieto, *Phase and angle variables in quantum mechanics*, *Rev. Math. Phys.* 40 (1968) 411–440.

[98] B. Carter, *Phys. Rev.* 174 (1968) 1559.

[99] R. Casadio, O. Micu and F. Scardigli, *Quantum hoop conjecture: Black hole formation by particle collisions*, *Phys. Lett.* B732 (2014) 105–109.

[100] S. Catto, *Miyzawa supersymmetry*, in *New Facet Of Three Nucleon Force: 50 Years Of Fujita Miyazawa Three Nucleon Force (Fm50)*, 1996, pp. 85–153. http://link.aip.org/link/?APCPCS/1011/253/1.

[101] S. Cecotti, C. Cordova, J. J. Heckman and C. Vafa, *T-branes and monodromy*, *JHEP* 1107 (2011) 030.

[102] M. Chaichian, A. D. Dolgov, V. A. Novikov and A. Tureanu, *CPT violation does not lead to violation of Lorentz invariance and vice versa*. http://arxiv.org/abs/1103.0168.

[103] S. Chandrasekar, *The Mathematical Theory of Black Holes*, Oxford University Press, 1983.

[104] D. Chang, T. Hübsch and R. Mohapatra, *Grand unification of three light generations*, *Phys. Rev. Lett.* 55 (1985) 673.

[105] J.-Q. Chen, *Group Representation Theory for Physicists*, World Scientific, 1989.

[106] T.-P. Cheng and L.-F. Li, *Gauge Theory of Elementary Particle Physics*, Clarendon Press, 1982.

[107] T.-P. Cheng and L.-F. Li, *Gauge Theory of Elementary Particle Physics: Problems and Solutions*, Clarendon Press, 2000.

[108] A. Chiodo and Y. Ruan, *LG/CY correspondence: the state space isomorphism*. http://arxiv.org/abs/0908.0908.

[109] CMS Collaboration, *Observation of a new boson at a mass of 125 GeV with the CMS experiment at the LHC*, *Phys. Lett.* B716 (2012) 30.

[110] C. Cohen-Tannoudji, B. Diu and F. Laloe, *Quantum Mechanics (Vol. I & II)*, Wiley-Interscience, 2006.

[111] S. Coleman and J. Mandula, *All possible symmetries of the S matrix*, *Phys. Rev.* 159 (1967) 1251–1256.

[112] S. Coleman and E. Weinberg, *Radiative corrections as the origin of spontaneous symmetry breaking*, *Phys. Rev.* D7 (1973) 1888–1910.

[113] J. C. Collins, *Renormalization: An Introduction to Renormalization, the Renormalization Group, and the Operator-Product Expansion*, Cambridge University Press, 1984.

[114] J. Conway and V. Pless, *On the enumeration of self-dual codes*, *J. Combinatorial Theory, Series A* 28 (1980) 26–53.

[115] F. Cooper, A. Khare and U. Sukhatme, *Supersymmetry and quantum mechanics*, *Phys. Rep.* 251 (1995) 267–285.

[116] W. Cottingham and D. Greenwood, *An Introduction to the Standard Model of Particle Physics*, 2nd edn., Cambridge University Press, 2007.

[117] D. A. Cox and S. Katz, *Mirror Symmetry and Algebraic Geometry*, Vol. 68 of Mathematical Surveys and Monographs, American Mathematical Society, 1999.

[118] R. Cox, C. McIlwraith and B. Kurrelmeyer, *Apparent evidence of polarization in a beam of β-rays*, *Proc. Natl. Acad. Sci. USA* 14 (1928) 544–547.

[119] R. P. Crease and C. C. Mann, *The Second Creation: Makers of the Revolution in Twentieth-Century Physics*, 2nd edn., Rutgers University Press, 1996.

[120] U. H. Danielsson, P. Koerber and T. V. Riet, *Universal de Sitter solutions at tree-level*, *JHEP* 1005 (2010) 090.

[121] C. Darwin, *On the magnetic moment of the electron*, *Proc. Roy. Soc.* A120 (1928) 621–631.

[122] C. Darwin, *The wave equations of the electron, Proc. Roy. Soc.* A118 (1928) 654.
[123] A. Das, *Field Theory: A Path Integral Approach*, 2nd edn., World Scientific, 2006.
[124] P. Deligne, P. Etingof, D. S. Freed, L. C. Jeffrey, D. K. Kazhdan, J. W. Morgan, D. R. Morrison and E. Witten (eds.), *Quantum Fields and Strings: A Course for Mathematicians (Vol. 1, 2)*, American Mathematical Society, 1999.
[125] P. Deligne and D. S. Freed, *Supersolutions*, in P. Deligne, P. Etingof, D. S. Freed, L. C. Jeffrey, D. K. Kazhdan, J. W. Morgan, D. R. Morrison and E. Witten (eds.), *Quantum Fields and Strings: A Course for Mathematicians (Vol. 1, 2)*, American Mathematical Society, 1999, pp. 227–355.
[126] S. Deser, *No local Maxwell duality invariance, Class. Quant. Grav.* 28 (2011) 085009.
[127] S. Deser and J. Franklin, *Birkhoff for Lovelock redux, Class. Quant. Grav.* 22 (2005) L103–L106.
[128] S. Deser and J. Franklin, *Schwarzschild and Birkhoff à la Weyl, Am. J. Phys.* 73 (2005) 261–264.
[129] B. DeWitt, *Supermanifolds*, Cambridge University Press, 1992.
[130] A. Dimakis and F. Müller-Hoissen, *Discrete differential calculus, graphs, topologies and gauge theory, J. Math. Phys.* 35 (1994) 6703–6735.
[131] R. d'Inverno, *Introducing Einstein's Relativity*, Clarendon Press, 1992.
[132] P. A. M. Dirac, *Generalized Hamiltonian dynamics, Can. J. Math.* 2 (1950) 129–148.
[133] P. A. M. Dirac, *The Hamiltonian form of field dynamics, Can. J. Math.* 3 (1951) 1–23.
[134] P. A. M. Dirac, *Principles of Quantum Mechanics*, Oxford University Press, 1982.
[135] P. A. M. Dirac, *General Theory of Relativity*, Princeton University Press, 1996, (original publ. Wiley, 1975).
[136] J. Distler and S. Kachru, *(0,2) Landau–Ginzburg theory, Nucl. Phys.* B413 (1994) 213–243.
[137] L. Dixon, J. A. Harvey, C. Vafa and E. Witten, *Strings on orbifolds (II), Nucl. Phys.* B274 (1986) 285–314.
[138] L. J. Dixon, J. A. Harvey, C. Vafa and E. Witten, *Strings on orbifolds, Nucl. Phys.* B261 (1985) 678–686.
[139] C. F. Doran, M. G. Faux, S. J. Gates, Jr., T. Hübsch, K. M. Iga and G. D. Landweber, *On graph-theoretic identifications of Adinkras, supersymmetry representations and superfields, Int. J. Mod. Phys.* A22 (2007) 869–930.
[140] C. F. Doran, M. G. Faux, S. J. Gates, Jr., T. Hübsch, K. M. Iga and G. D. Landweber, *Relating doubly-even error-correcting codes, graphs, and irreducible representations of N-extended supersymmetry*, in F. Liu *et al.* (eds.), *Discrete and Computational Mathematics*, Nova Science, 2008.
[141] C. F. Doran, M. G. Faux, S. J. Gates, Jr., T. Hübsch, K. M. Iga and G. D. Landweber, *A superfield for every dash chromotopology, Int. J. Mod. Phys.* A24 (2009) 5681–5695.
[142] C. F. Doran, M. G. Faux, S. J. Gates, Jr., T. Hübsch, K. M. Iga, G. D. Landweber and R. L. Miller, *Codes and supersymmetry in one dimension, Adv. Theor. Math. Phys.* 15 (2011) 1909–1970.
[143] C. F. Doran, T. Hübsch, K. M. Iga and G. D. Landweber, *On general off-shell representations of worldline (1D) supersymmetry, Symmetry* 6 (1) (2014) 67–88.
[144] M. Drees, *An introduction to supersymmetry*, in *Inauguration Conference of the Asia Pacific Center for Theoretical Physics (APCTP), Seoul, Korea, 4–19 June 1996*, 1996.
[145] M. J. Duff, *String and M-theory: Answering the critics, Foundations of Physics* 43 (1) (2011) 182–200.
[146] M. Dütsch, *Connection between the renormalization groups of Stückelberg-Petermann and Wilson, Confl. Math.* 4 (1) (2012) 12400014.
[147] G. Dvali, D. Flassig, C. Gomez, A. Pritzel and N. Wintergerst, *Scrambling in the black hole portrait, Phys. Rev.* D88 (2013) 124041.
[148] G. Dvali and C. Gomez, *Black holes as critical point of quantum phase transition. Eur. Phys. J.* C74 (2014) 2752, arXiv.1207.4059.
[149] G. Dvali and C. Gomez, *Landau–Ginzburg limit of black hole's quantum portrait: Self similarity and critical exponent, Phys. Lett.* B716 (2012) 240–242.
[150] F. Dyson, *Advanced Quantum Mechanics*, 2nd edn., World Scientific, 2011, (transcribed by David Derbes).
[151] C. Eckart and G. Young, *A principal axis transformation for non-Hermitian matrices, Bull. Amer. Math. Soc.* 45 (1939) 118–121. http://www.ams.org/journals/bull/1939-45-02/S0002-9904-1939-06910-3/home.html.
[152] G. F. Ellis, *On horizons and the cosmic landscape, Gen. Rel. Grav.* 38 (2006) 1209–1213.

[153] M. Ezawa, *Supersymmetry and unconventional quantum Hall effect in graphene*, *Phys. Lett.* A372 (2008) 924–929.

[154] L. Faddeev and A. Slavnov, *Gauge Fields: An Introduction to Quantum Theory*, Addison-Wesley, 1991.

[155] B. Famaey and S. McGaugh, *Modified Newtonian dynamics: A review*, *Living Rev. Rel.* 15 (2012) 10.

[156] B. Famaey and S. McGaugh, *Modified Newtonian dynamics (MOND): Observational phenomenology and relativistic extensions*, *Living Rev. Rel.* 15 (2012) 10.

[157] M. G. Faux, K. M. Iga and G. D. Landweber, *Dimensional enhancement via supersymmetry*, *Adv. Math. Phys.* 2011 (2011) 259089.

[158] M. G. Faux and G. D. Landweber, *Spin holography via dimensional enhancement*, *Phys. Lett.* B681 (2009) 161–165.

[159] Fayyazuddin and Riazuddin, *A Modern Introduction to Particle Physics (High Energy Physics)*, World Scientific, 2000.

[160] J. L. Feng, J.-F. Grivaz and J. Nachtman, *Searches for supersymmetry at high-energy colliders*, *Rev. Mod. Phys.* 82 (2010) 699–727.

[161] K. Ferguson and S. W. Hawking, *Quest for a Theory of Everything*, Bantam Books, 1991.

[162] R. P. Feynman, *The Character of Physical Law*, The Massachusetts Institute of Technology Press, 1965.

[163] R. P. Feynman, *The Pleasure of Finding Things Out*, Perseus Publishing, 1999.

[164] R. P. Feynman, *Feynman Lectures on Gravitation*, Westview Press, 2002 (original publ. Addison-Wesley, 1995).

[165] R. P. Feynman, A. R. Hibbs and D. F. Styer, *Quantum Mechanics and Path Integrals*, emended edn., Dover Publications, 2005.

[166] R. P. Feynman and S. Weinberg, *Elementary Particles and the Laws of Physics*, Cambridge University Press, 1987.

[167] B. Figueiredo, I. D. Soares and J. Tiomno, *Gravitational coupling of Klein–Gordon and Dirac particles to matter vorticity and space-time torsion*, *Class. Quant. Grav.* 9 (1992) 1593–1617.

[168] M. Fischler, *Young-tableau methods for Kronecker products of representations of classical groups*, *J. Math. Phys.* 22 (4) (1981) 637–648.

[169] R. Foot and A. Kobakhidze, *Alternative implementation of the Higgs boson*, *Mod. Phys. Lett.* A26 (2011) 461–467.

[170] S. Forste, *Strings, branes and extra dimensions*, *Fortsch. Phys.* 50 (2002) 221–403.

[171] R. L. Forward, *Indistinguishable from Magic*, Baen Books, 1995.

[172] E. Fradkin and T. Fradkina, *Quantization of relativistic systems with boson and fermion first- and second-class constraints*, *Phys. Lett.* B72 (1978) 343–348.

[173] E. Fradkin and V. Linetsky, *BFV approach to geometric quantization*, *Nucl. Phys.* 431 (3) (1994) 569–621.

[174] E. Fradkin and G. A. Vilkovisky, *Quantization of relativistic systems with constraints*, *Phys. Lett.* B55 (1975) 224–226.

[175] P. H. Frampton, *Dual Resonance Models and Superstrings*, World Scientific, 1986.

[176] T. Frankel, *The Geometry of Physics: An Introduction*, Cambridge University Press, 1997.

[177] P. Fré, *Introduction to harmonic expansions on coset manifolds and in particular on coset manifolds with Killing spinors*, in *Supersymmetry and Supergravity '84 (Trieste, 1984)*, World Scientific, 1984, pp. 324–367.

[178] D. S. Freed, *Five Lectures on Supersymmetry*, American Mathematical Society, 1999.

[179] A. French, *Special Relativity*, W.W. Norton & Company, 1968.

[180] C. S. Frenk and S. D. White, *Dark matter and cosmic structure*, *Ann. Phys.* 524 (2012) 507.

[181] C. Fronsdal, *Completion and embedding of the Schwarzschild solution*, *Phys. Rev.* 116 (3) (1959) 778–781.

[182] B. Fuks, *Supersymmetry: When theory inspires experimental searches*. http://arxiv.org/abs/1401.6277.

[183] G. Gamov, *Mr. Tompkins in Wonderland* with *Mr. Tompkins Explores the Atom*, Cambridge University Press, 1965 (original publ. 1940 and 1945, resp.).

[184] G. Gamov, *One Two Three... Infinity*, Bantam Books, 1971 (original publ. 1947).

[185] H. Ganev and S. Brant, *Structure of the doublet bands in doubly odd nuclei: The case of* ^{128}Cs, *Phys. Rev.* C82 (2010) 034328.

[186] A. Gangopadhyaya, J. V. Mallow and C. Rasinariu, *Supersymmetric Quantum Mechanics*, World Scientific, 2011.

[187] M. Gardner, *Relativity Simply Explained*, Dover Publications, 1997 (original publ. *Relativity for the Million*, MacMillan, 1962; also *The Relativity Explosion*, Vintage Books, 1976).

[188] S. J. Gates, Jr., *Vector multiplets and the phases of N=2 theories in 2-D: Through the looking glass, Physics Lett.* B352 (1995) 43–49.

[189] S. J. Gates, Jr., M. Grisaru, M. Roček and W. Siegel, *Superspace*, Benjamin Cummings, 1983.

[190] S. J. Gates, Jr., J. Hallett, T. Hübsch and K. Stiffler, *The real anatomy of complex linear superfields, Int. J. Mod. Phys.* A27 (2012) 1250143.

[191] S. J. Gates, Jr. and T. Hübsch, *On dimensional extension of supersymmetry: From worldlines to worldsheets, Adv. Theor. Math. Phys.* 16 (6) (2012) 1619–1667.

[192] S. J. Gates, Jr. and S. Ketov, *2D(4,4) hypermultiplets. II: Field theory origins of dualities, Phys. Lett.* B418 (1998) 119–124.

[193] S. J. Gates, Jr., W. D. Linch, III and J. Phillips, *When superspace is not enough*. http://arxiv.org/abs/hep-th/0211034.

[194] S. J. Gates, Jr., W. D. Linch, III, J. Phillips and L. Rana, *The fundamental supersymmetry challenge remains, Gravit. Cosmol.* 8 (1–2) (2002) 96–100.

[195] S. J. Gates, Jr. and L. Rana, *On extended supersymmetric quantum mechanics*, University of Maryland Report: UMDPP 93-194 (1994) unpublished.

[196] S. J. Gates, Jr. and L. Rana, *A theory of spinning particles for large N-extended supersymmetry, Phys. Lett.* B352 (1–2) (1995) 50–58.

[197] S. J. Gates, Jr. and L. Rana, *Ultramultiplets: A new representation of rigid 2-d, N=8 supersymmetry, Phys. Lett.* B342 (1995) 132–137.

[198] S. J. Gates, Jr. and L. Rana, *A theory of spinning particles for large N-extended supersymmetry. II Phys. Lett.* B369 (3–4) (1996) 262–268.

[199] S. J. Gates, Jr. and L. Rana, *Tuning the RADIO to the off-shell 2D fayet hypermultiplet problem*, http://arXiv.org/abs/hep-th/9602072.

[200] J. Gegenberg, H. Liu, S. S. Seahra and B. K. Tippett, *A Tomimatsu-Sato/CFT correspondence Class. Quant. Grav.* 28 (2011) 085004, arXiv:1010.2803.

[201] H. Georgi, *Lie Algebras in Particle Physics*, Addison-Wesley, 1982.

[202] H. Georgi and S. L. Glashow, *Unity of all elementary-particle forces, Phys. Rev. Lett.* 32 (1974) 438–441.

[203] D. Gepner and Z.-A. Qiu, *Modular invariant partition functions for parafermionic field theories, Nucl. Phys.* B285 (1987) 423.

[204] R. Geroch, *General Relativity*, University of Chicago Press, 1978.

[205] J.-L. Gervais and B. Sakita, *Field theory interpretation of supergauges in dual models, Nucl. Phys.* B34 (1971) 632.

[206] G. Giachetta, L. Mangiarotti and G. Sardanashvily, *New Lagrangian and Hamiltonian Methods in Field Theory*, World Scientific, 1997.

[207] V. Ginzburg and L. D. Landau, *On the theory of superconductivity, Zhurnal Eksperimental'noi i Teoreticheskoi Fiziki* 20 (1950) 1064–1082.

[208] M. Gitterman and V. Halpern, *Qualitative Analysis of Physical Problems*, Academic Press, 1981.

[209] S. L. Glashow, *Towards a unified theory: Threads in a tapestry*. http://nobelprize.org/nobel_prizes/physics/laureates/1979/glashow-lecture.pdf.

[210] M. Göckler and T. Schücker, *Differential Geometry, Gauge Theory and Gravity*, Cambridge Monographs on Mathematical Physics, Cambridge University Press, 1987.

[211] K. Gödel, *On Formally Undecidable Propositions of Principia Mathematica and Related Systems*, Dover Publications, 1992 (original publ. 1962).

[212] N. Goldenfeld, *Lectures on Phase Transitions and the Renormalization Group*, Addison-Wesley, 1992.

[213] H. Goldstein, C. P. Poole and J. L. Safko, *Classical Mechanics*, 3rd edn., Addison Wesley, 2001.

[214] J. Goldstone, A. Salam and S. Weinberg, *Broken symmetries, Phys. Rev.* 127 (1962) 965–970.

[215] Y. Goncharov, *Estimates for parameters and characteristics of the confining SU(3)-gluonic field in the ground state of toponium: Relativistic and nonrelativistic approaches, Nucl. Phys.* A808 (2008) 73–94.

[216] W. Gordon, *Die Energieniveaus des Wasserstoffatoms nach der Diracschen Quantentheorie des Elektrons*, *Z. Phys.* 48 (1928) 11–14.

[217] M. H. Goroff and A. Sagnotti, *Quantum gravity at two loops*, *Phys. Lett.* B160 (1985) 81–86.

[218] K. Gottfried and V. F. Weiskopf, *Concepts of Particle Physics*, Clarendon Press, 1984.

[219] F. Grave, M. Buser, T. Mueller, G. Wunner and W. P. Schleich, *The Godel universe: Exact geometrical optics and analytical investigations on motion*, *Phys. Rev.* D80 (2009) 103002. http://www.vis.uni-stuttgart.de/%7Egrave/paper/GoedelAnalytic.pdf.

[220] H. Greaves and T. Thomas, *The CPT theorem*, *Stud. Hist. Phil. Mod. Phys.* 45 (2014) 44–66.

[221] D. Green, *Lectures In Particle Physics*, Lecture Notes in Physics, World Scientific, 1994.

[222] D. R. Green, A. Lawrence, J. McGreevy, D. R. Morrison and E. Silverstein, *Dimensional duality*, *Phys. Rev.* D76 (2007) 066004.

[223] M. B. Green and J. H. Schwarz, *Anomaly cancellation in supersymmetric D=10 gauge theory and superstring theory*, *Phys. Lett.* B149 (1984) 117–122.

[224] M. B. Green, J. H. Schwarz and E. Witten, *Superstring Theory, Vol. 2: Loop Amplitudes, Anomalies and Phenomenology*, Cambridge Monographs on Mathematical Physics, Cambridge University Press, 1987.

[225] M. B. Green, J. H. Schwarz and E. Witten, *Superstring Theory, Vol. 1: Introduction*, Cambridge Monographs on Mathematical Physics, Cambridge University Press, 1987.

[226] P. S. Green and T. Hübsch, *Connecting moduli spaces of Calabi–Yau threefolds*, *Comm. Math. Phys.* 119 (1988) 431–441.

[227] P. S. Green and T. Hübsch, *Possible phase transitions among Calabi–Yau compactifications*, *Phys. Rev. Lett.* 61 (1988) 1163–1166.

[228] P. S. Green and T. Hübsch, *Spacetime variable string vacua*, *Int. J. Mod. Phys.* A9 (1994) 3203–3228.

[229] O. W. Greenberg, *Spin and unitary spin independence in a paraquark model of baryons and mesons*, *Phys. Rev. Lett.* 13 (1964) 598–602.

[230] O. W. Greenberg, *CPT violation implies violation of Lorentz invariance*, *Phys. Rev. Lett.* 89 (2002) 231602.

[231] B. Greene, *The Elegant Universe*, Vintage Books, 1999.

[232] B. Greene, *The Fabric of the Cosmos: Space, Time, and the Texture of Reality*, Vintage Books, 2005.

[233] B. Greene, *The universe on a string*, *The New York Times* (2006) 3 pp. http://www.nytimes.com/2006/10/20/opinion/20greenehed.html.

[234] B. Greene, *The Hidden Reality: Parallel Universes and the Deep Laws of the Cosmos*, Vintage Books, 2011.

[235] B. Greene and S.-T. Yau (eds.), *Mirror Manifolds II*, International Press, 1996.

[236] B. R. Greene and M. Plesser, *Duality in Calabi–Yau moduli space*, *Nucl. Phys.* B338 (1990) 15–37.

[237] B. R. Greene, A. D. Shapere, C. Vafa and S.-T. Yau, *Stringy cosmic strings and noncompact Calabi–Yau manifolds*, *Nucl. Phys.* B337 (1990) 1.

[238] W. Greiner, *Relativistic Quantum Mechanics*, Springer-Verlag, 1990.

[239] W. Greiner and B. Müller, *Gauge Theory of Weak Interactions*, Springer-Verlag, 1993.

[240] W. Greiner and B. Müller, *Quantum Chromodynamics*, Springer-Verlag, 1994.

[241] W. Greiner and J. Reinhardt, *Quantum Electrodynamics*, Springer-Verlag, 1992.

[242] D. J. Griffiths, *Introduction to Quantum Mechanics*, 2nd edn., Benjamin Cummings, 2004.

[243] D. J. Griffiths, *Introduction to Elementary Particles*, 2nd edn., Wiley-VCH, 2008.

[244] G. Grimvall, *Quantify! A Crash Course in Smart Thinking*, Johns Hopkins University Press, 2011.

[245] L. Grodzins, *The history of double scattering of electrons and evidence for the polarization of beta rays*, *Proc. Natl. Acad. Sci. USA* 45 (3) (1959) 399–405.

[246] D. J. Gross, J. A. Harvey, E. J. Martinec and R. Rohm, *Heterotic string theory. 1. The free heterotic string*, *Nucl. Phys.* B256 (1985) 253.

[247] D. J. Gross, J. A. Harvey, E. J. Martinec and R. Rohm, *The heterotic string*, *Phys. Rev. Lett.* 54 (1985) 502–505.

[248] D. J. Gross, J. A. Harvey, E. J. Martinec and R. Rohm, *Heterotic string theory. 2. The Interacting heterotic string*, *Nucl. Phys.* B267 (1986) 75.

[249] F. Gross, *Relativistic Quantum Mechanics and Field Theory*, Wiley-VCH, 1993.

[250] C. Grupen, *Astroparticle Physics*, Springer-Verlag, 2005.

[251] S. S. Gubser, *The Little Book of String Theory*, Princeton University Press, 2010.

[252] G. S. Guralnik, *The history of the Guralnik, Hagen and Kibble development of the theory of spontaneous symmetry breaking and gauge particles*, *Int. J. Mod. Phys.* 24 (2009) 2601–2627.
[253] R. Gurau, V. Rivasseau and A. Sfondrini, *Renormalization: An advanced overview*, http://arXiv.org/abs/1401.5003[hep-th].
[254] R. Haag, *Local Quantum Physics: Fields, Particles, Algebras*, Springer-Verlag, 1992.
[255] R. Haag, J. Łopuszański and M. Sohnius, *All possible generators of supersymmetries of the S matrix*, *Nucl. Phys.* B88 (1975) 257.
[256] B. C. Hall, *Lie Groups, Lie Algebras, and Representations*, Springer-Verlag, 2003.
[257] F. Halzen and A. D. Martin, *Quarks and Leptons*, Wiley, 1984.
[258] M. Hammermesh, *Group Theory and its Application to Physical Problems*, Addison-Wesley, 1964.
[259] G. H. Hardy, *Divergent Series*, 2nd edn., American Mathematical Society, 1992 (original publ. 1929).
[260] F. R. Harvey, *Spinors and Calibrations*, Vol. 9 of Perspectives in Mathematics, Academic Press, 1990.
[261] J. A. Harvey, *Lectures on anomalies*, in J. Maldacena (ed.), *Progress in String Theory: Tasi 2003 Lecture Notes, Boulder, CO*, World Scientific, 2005.
[262] M. Harwitt, *Astrophysical Concepts*, 4th edn., Springer-Verlag, 2006.
[263] S. W. Hawking, *The Universe in a Nutshell*, Bantam Books, 2001.
[264] S. W. Hawking and G. F. R. Ellis, *The Large Scale Structure of Space-Time*, Cambridge University Press, 1973.
[265] S. W. Hawking, K. S. Thorne, I. Novikov, T. Ferris and A. Lightman, *The Future of Spacetime*, W.W. Norton & Company, 2003.
[266] R. A. Heinlein, *Stranger in a Strange Land*, uncut edn., Ace Books, 1991 (1st, 27.24% cut edn.: Putnam Publishing Group, 1961).
[267] W. Heisenberg, *Physics and Beyond*, Harper and Row, 1972.
[268] M. Henneaux and C. Teitelboim, *Quantization of Gauge Systems*, Princeton University Press, 1994.
[269] J. A. Heras, *Electromagnetism in euclidean four space: A discussion between god and the devil*, *Am. J. Phys.* 62 (10) (1994) 914–916.
[270] W. Hikida and H. Kodama, *An investigation of the Tomimatsu–Sato spacetime*, in *Proceedings of the 12th Workshop on General Relativity and Gravitation*, 2002.
[271] C. M. Hirata, *Ph 236: General Relativity*. http://www.tapir.caltech.edu/~chirata/ph236/2011-12/.
[272] M. Hobson, G. Efstathiou and A. Lasenby, *General Relativity: An Introduction for Physicists*, Cambridge University Press, 2007.
[273] P. Hořava, *Quantum gravity at a Lifshitz point*, *Phys. Rev.* D79 (2009) 084008.
[274] K. Hori, S. Katz, A. Klemm, R. Pandharipande, R. Thomas, C. Vafa, R. Vakil and E. Zaslow, *Mirror Symmetry*, Vol. 1 of Clay Mathematics Monographs, American Mathematical Society, 2003.
[275] A. Horzela, E. Kapuscik and C. A. Uzes, *Comment on the paper 'Introducing gauge invariance' by R. Barlow*, *Eur. J. Phys.* 14 (1993) 190.
[276] J.-P. Hsu, *Yang–Mills gravity in flat space-time, I. classical gravity with translation gauge symmetry*, *Int. J. Mod. Phys.* A21 (2006) 5119–5139.
[277] K. Huang, *Quantum Field Theory: From Operators to Path Integrals*, Wiley-VCH, 2010.
[278] T. Hübsch, *How singular a space can superstrings thread?*, *Mod. Phys. Lett.* A6 (1991) 207–216.
[279] T. Hübsch, *Calabi–Yau manifolds*, 2nd edn., World Scientific, 1994.
[280] T. Hübsch, *A hitchhiker's guide to superstring jump gates and other worlds*, *Nucl. Phys. Proc. Suppl.* 52A (1997) 347–351.
[281] T. Hübsch, *Haploid (2,2)-superfields in 2-dimensional space-time*, *Nucl. Phys.* B555 (3) (1999) 567–628.
[282] T. Hübsch, *Superspace: A comfortably vast algebraic variety*, in L. Ji (ed.), *Geometry and Analysis*, Vol. 2 of Advanced Lectures in Mathematics, International Press, 2010, pp. 39–67 (closing address at the Conference "Geometric Analysis: Present and Future", Harvard University, 2008).
[283] T. Hübsch, *Weaving worldsheet supermultiplets from the worldlines within*, *Adv. Theor. Math. Phys.* 17 (2013) 1–72.
[284] T. Hübsch and G. A. Katona, *On the construction and the structure of off-shell supermultiplet quotients*, *Int. J. Mod. Phys.* A27 (29) (2012) 1250173.
[285] T. Hübsch and P. B. Pal, *Economical unification of three families in SO(18)*, *Phys. Rev.* D34 (1986) 1606.
[286] W. C. Huffman and V. Pless, *Fundamentals of Error-Correcting Codes*, Cambridge University Press, 2003.

[287] J. E. Humphreys, *Introduction to Lie Algebra and Representation Theory*, 3rd edn., Springer-Verlag, 1972.

[288] N. E. Hurt, *Geometric Quantization in Action*, D. Reidel, 1983.

[289] J. Iizuka, *A systematics and phenomenology of meson family*, *Prog. Theor. Phys. Suppl.* 37 & 38 (1966) 21–34.

[290] K. Intriligator, H. Jockers, P. Mayr, D. R. Morrison and M. R. Plesser, *Conifold transitions in M-theory on Calabi–Yau fourfolds with background fluxes*, *Adv. Theor. Math. Phys.* 17 (2013) 601–699.

[291] C. J. Isham, *Prima facie questions in quantum gravity*, in J. Ehlers and H. Friedrich (eds.), *Canonical Gravity: From Classical to Quantum*, Springer-Verlag, 1994 (proceedings of the 117th WE Heraeus Seminar held at Bad Honnef, Germany, 13–17 September 1993).

[292] E. Ivanov, O. Lechtenfeld and A. Sutulin, *Hierarchy of N=8 mechanics models*, *Nucl. Phys.* B790 (2008) 493–523.

[293] K. J. Beringer *et al.* (Particle Data Group), *Review of particle physics*, *Phys. Rev.* D86 (2012) 010001. http://pdg.lbl.gov/.

[294] K. J. Beringer *et al.* (Particle Data Group), *Review of particle physics*, *Phys. Rev.* D86 (2012) 010001. http://pdg.lbl.gov/2012/reviews/rpp2012-rev-clebsch-gordan-coefs.pdf.

[295] R. Jackiw, *Minimum uncertainty product, number phase uncertainty product, and coherent states*, *J. Math. Phys.* 9 (1968) 339.

[296] J. D. Jackson, *Classical Electrodynamics*, 3rd edn., John Wiley & Sons, 1999.

[297] N. V. Johansen and F. Ravndal, *On the discovery of Birkhoff's theorem*, *Gen. Rel. Grav.* 38 (2006) 537–540.

[298] C. Johnson, *D-branes*, Cambridge University Press, 2006.

[299] A. Z. Jones and D. Robbins, *String Theory for Dummies*, Wiley, 2010.

[300] R. Joost, *A remark on the CTP theorem*, *Helv. Phys. Acta* 30 (1957) 409.

[301] P. Jordan, *Beitraege zur Neutrinotheorie des Lichts*, *Z. Phys.* 114 (1937) 229 (and references therein to earlier works).

[302] S. Kachru, R. Kallosh, A. Linde, J. Maldacena, L. McAllister and S. P. Trivedi, *Towards inflation in string theory*, *JCAP* 0310 (2003) 013.

[303] S. Kachru, R. Kallosh, A. Linde and S. P. Trivedi, *de Sitter vacua in string theory*, *Phys. Rev.* D68 (2003) 046005.

[304] S. Kachru and E. Silverstein, *Chirality changing phase transitions in 4-D string vacua*, *Nucl. Phys.* B504 (1997) 272–284.

[305] D. Kaiser, *Drawing Theories Apart: The Dispersion of Feynman Diagrams in Postwar Physics*, University of Chicago Press, 2005.

[306] G. Källén, *Elementary Particle Physics*, Addison-Wesley, 1964.

[307] G. Kane, *Modern Elementary Particle Physics*, Perseus Books, 1993.

[308] G. Kane, *Supersymmetry*, Perseus Books, 2000.

[309] S. G. Karshenboim, *Precision study of positronium: Testing bound state QED theory*, *Int. J. Mod. Phys.* A19 (2004) 3879–3896.

[310] C. Kiefer, *Quantum Gravity*, Clarendon Press, 2004.

[311] W. Kilian, *Electroweak Symmetry Breaking: The Bottom-Up Approach*, No. 198 in Springer Tracts in Modern Physics, Springer-Verlag, 2003.

[312] E. Kiritsis, *String Theory in a Nutshell*, Princeton University Press, 2007.

[313] F. Kirwan and J. Woolf, *An Introduction to Intersection Homology Theory*, Chapman & Hall/CRC, 2006.

[314] S. C. Kleene, *Introduction to Metamathematics*, North-Holland, 1971.

[315] H. Kleinert, *Multivalued Fields in Condensed Matter, Electromagnetism, and Gravitation*, World Scientific, 2008.

[316] H. Kleinert, *Path Integrals in Quantum Mechanics, Statistics, Polymer Physics, and Financial Markets*, 5th edn., World Scientific, 2009.

[317] S. Kobayashi and K. Nomizu, *Foundations of Differential Geometry*, vol. 1, Wiley-Interscience, 1963.

[318] S. Kobayashi and K. Nomizu, *Foundations of Differential Geometry*, vol. 2, Wiley-Interscience, 1963.

[319] E. Konopinski and H. Mahmoud, *The universal Fermi interaction*, *Phys. Rev.* 92 (1953) 1045–1049.

[320] M. Krawitz, *FJRW rings and Landau–Ginzburg mirror symmetry*. http://arxiv.org/abs/0906.0796.

[321] M. Kreuzer, *The making of Calabi–Yau spaces: Beyond toric hypersurfaces, Fortsch. Phys.* 57 (2009) 625–631.
[322] K. Kuchař, *Canonical quantization of gravity*, in W. Israel (ed.), *Relativity, Astrophysics and Cosmology*, D. Reidel, 1973, pp. 237–288.
[323] T. S. Kuhn, *The Structure of Scientific Revolutions*, 3rd edn., University Of Chicago Press, 1996.
[324] K. S. Lam, *Non-Relativstic Quantum Theory*, World Scientific, 2009.
[325] C. Lanczos, *Über eine stationäre Kosmologie im Sinne der Einsteinischen Gravitationstheories, Z. Phys.* 21 (1924) 73.
[326] L. Landau and E. Lifshitz, *The Classical Theory of Fields*, Pergamon Press, 1975.
[327] L. Landau and E. Lifshitz, *Mechanics*, Pergamon Press, 1976.
[328] L. Landau and E. Lifshitz, *Quantum Mechanics*, Pergamon Press, 1977.
[329] L. Landau and Y. Rumer, *What is the Theory of Relativity*, Foreign Languages Publishing House (English translation by A. Zdornykh), 1959 (reprinted by Neo Press).
[330] L. D. Landau, *Zh. Eksp. Teor. Fiz.* 7 (1937) 627.
[331] S. Lang, *Algebra*, 3rd edn., Vol. 211 of Graduate Texts in Mathematics, Springer-Verlag, 2002.
[332] S. Lang, *Undergraduate Algebra*, 3rd edn., Springer-Verlag, 2002.
[333] H. B. Lawson, Jr. and M.-L. Michelsohn, *Spin Geometry*, vol. 38 of Princeton Mathematical Series, Princeton University Press, 1989.
[334] E. Leader, *Spin in Particle Physics*, Cambridge University Press, 2001.
[335] P. Lecheminant and H. Nonne, *Exotic quantum criticality in one-dimensional coupled dipolar bosons tubes, Phys. Rev.* B85 (2012) 195121, arXiv:1202.6541.
[336] T. Ledvinka, M. Zofka and J. Bicak, *Relativistic disks as sources of Kerr-Newman fields*, in T. Piran (ed.), *Proceedings of the Eighth Marcel Grossmann Meeting on Recent Developments in Theoretical and Experimental General Relativity, Gravitation and Relativistic Field Theories*, World Scientific, 1999.
[337] S.-S. Lee, *Emergence of supersymmetry, gauge theory and string in condensed matter systems*, in M. Dine, T. Banks and T. Sachdev (eds.), *String Theory and Its Applications: From meV to the Planck Scale: TASI 2010 Lecture Notes, Boulder CO*, World Scientific, 2010, pp. 667–706.
[338] M. Leigh, J. Darion and L. Flato, *Man of La Mancha: Vocal Score*, Cherry Lane Music, 1986.
[339] R. L. Liboff, *Introductory Quantum Mechanics*, Addison Wesley Longman, 1998.
[340] E. Lifshitz and L. Pitaevskii, *Statistical Physics, Part 2*, Pergamon Press, 1980.
[341] Y.-K. Lim, *Problems and Solutions in Atomic, Nuclear and Particle Physics*, World Scientific, 2000.
[342] J. Lindesay, *Foundations of Quantum Gravity*, Cambridge University Press, 2013.
[343] D. F. Litim, *Renormalisation group and the Planck scale, Phil. Trans. Roy. Soc.* A369 (2011) 2759–2778.
[344] J. Łopuszański, *An Introduction to Symmetry and Supersymmetry in Quantum Field Theory*, World Scientific, 1991.
[345] A. Lopez and A. Rogers, *BRST cohomology of systems with secondary constraints*. http://arxiv.org/abs/1208.0056.
[346] H. A. Lorentz, *The Theory of Electrons and its Applications to the Phenomena of Light and Radiant Heat*, B. G. Teubner (Leipzig), 1916 (a course of lectures delivered in Columbia University, New York, in March and April, 1906).
[347] C.-K. Lu and I. F. Herbut, *Supersymmetric Runge–Lenz–Pauli vector for Dirac vortex in topological insulators and graphene, J. Phys.* A44 (2011) 295003.
[348] Z. Lu and X. Sun, *On the Weil–Petersson volume and the first Chern class of the moduli space of Calabi–Yau manifolds, Commun. Math. Phys.* 261 (2006) 297–322.
[349] B. Lucini and M. Panero, *SU(N) gauge theories at large N, Phys. Rep.* 526 (2013) 93–163.
[350] G. Lüders, *On the equivalence of invariance under time-reversal and under particle–antiparticle conjugation for relativistic field theories, Det. Kong. Danske Videnskabernes Selskab, Mat.-fys. Medd.* 28 (5).
[351] J. D. Lykken, *Introduction to supersymmetry*, in C. Efthimiou and B. Greene (eds.), *Fields, Strings and Duality: TASI 1996 Lecture Notes, Boulder, CO*, World Scientific, 1997, pp. 85–153.
[352] R. Maartens and K. Koyama, *Brane-world gravity, Living Rev. Rel.* 13 (2010) 5.
[353] Z. Maki, M. Nakagawa and S. Sakata, *Remarks on the unified model of elementary particles, Prog. Theor. Phys.* 28 (5) (1962) 870–880.

[354] J. Maldacena, *The large N limit of superconformal field theories and supergravity*, *Adv. Theor. Math. Phys.* 2 (1998) 231–252.
[355] B. Martin and G. Shaw, *Particle Physics*, 3rd edn., Wiley, 2008.
[356] S. P. Martin, *A supersymmetry primer*, in G. L. Kane (ed.), *Perspectives on Supersymmetry*, World Scientific, 1998, pp. 1–98.
[357] R. D. Mattuck, *A Guide to Feynman Diagrams in the Many-Body Problem*, Dover Publications, 1992.
[358] D. McMahon, *String Theory Demystified*, McGraw-Hill, 2008.
[359] N. Mee, *Higgs Force: The Symmetry-Breaking Force that Makes the World an Interesting Place*, Lutterworth Press, 2012.
[360] E. Merzbacher, *Quantum Mechanics*, 3rd edn., John Wiley & Sons, 1998.
[361] D. Meschini, M. Lehto and J. Piilonen, *Geometry, pregeometry and beyond*, *Stud. Hist. Phil. Mod. Phys.* 36 (2005) 435–464. http://dx.doi.org/doi:10.1016/j.shpsb.2005.01.002.
[362] A. Messiah, *Quantum Mechanics*, Vol. 1, John Wiley & Sons, 1958.
[363] A. Messiah, *Quantum Mechanics*, Vol. 2, John Wiley & Sons, 1958.
[364] A. Metz, J. Jolie, G. Graw, R. Hertenberger, J. Groger *et al.*, *Evidence for the existence of supersymmetry in atomic nuclei*, *Phys. Rev. Lett.* 83 (1999) 1542–1545.
[365] M. Mia and F. Chen, *Non extremal geometries and holographic phase transitions*, *JHEP* 1301 (2013) 083.
[366] J. Milutinović, (student of David Caswell), private communication (2007).
[367] C. Misner, K. S. Thorne and J. A. Wheeler, *Gravitation*, 2nd edn., W. H. Freeman, 1973.
[368] P. Mitra, *Symmetries and Symmetry Breaking in Field Theory*, CRC Press, 2014.
[369] R. N. Mohapatra and P. B. Pal, *Massive Neutrinos in Physics and Astrophysics*, Vol. 60 of *World Scientific Lecture Notes in Physics*, 2nd edn., World Scientific, 1998.
[370] R. N. Mohapatra and A. Smirnov, *Neutrino mass and new physics*, *Ann. Rev. Nucl. Part. Sci.* 56 (2006) 569–628.
[371] D. R. Morrison and M. R. Plesser, *Towards mirror symmetry as duality for two dimensional abelian gauge theories*, *Nucl. Phys. Proc. Suppl.* 46 (1996) 177–186.
[372] T. Mueller and F. Grave, *Catalogue of spacetimes*. http://arxiv.org/abs/0904.4184.
[373] S. Mukhi, *String theory: A perspective over the last 25 years*, *Class. Quant. Grav.* 28 (2011) 153001.
[374] J. Mureika and D. Stojković, *Detecting vanishing dimensions via primordial gravitational wave astronomy*, *Phys. Rev. Lett.* 106 (2011) 101101.
[375] G. Musser, *The Complete Idiot's Guide to String Theory*, Penguin, 2008.
[376] E. Nagel and J. R. Neumann, *Gödel's Proof*, 2nd edn., New York University Press, 2001 (edited and with a new foreword by D. R. Hofstadter).
[377] K. Narain, M. Sarmadi and C. Vafa, *Asymmetric orbifolds*, *Nucl. Phys.* B288 (1987) 551.
[378] K. Narain, M. Sarmadi and C. Vafa, *Asymmetric orbifolds: Path integral and operator formulations*, *Nucl. Phys.* B356 (1991) 163–207.
[379] C. Nash and S. Sen, *Topology and Geometry for Physicists*, Academic Press, 1988.
[380] H. Nastase, *Introduction to supergravity* (lectures given at the IFT-UNESP in 2011). http://arxiv.org/abs/1112.3502.
[381] P. Nelson, *Lectures on strings and moduli spaces*, *Phys. Rep.* 149 (6) (1987) 337–375.
[382] R. Netz and W. Noel, *The Archimedes Codex: Revealing the Blueprint for Modern Science*, Phoenix, 2008.
[383] D. E. Neuenschwander, *Emmy Noether's Wonderful Theorem*, Johns Hopkins University Press, 2011.
[384] R. Nevzorov, *Phenomenological aspects of supersymmetry: SUSY models and electroweak symmetry breaking*, in *Proceedings of the Dynasty Foundation Summer School*, 2011, pp. 108–154.
[385] H. Nicolai and K. Peeters, *Loop and spin foam quantum gravity: A brief guide for beginners*, *Lect. Notes Phys.* 172 (2007) 151–184.
[386] H. Nicolai, K. Peeters and M. Zamaklar, *Loop and spin foam quantum gravity: An outside view*, *Class. Quant. Grav.* 22 (2005) R193.
[387] H. P. Nilles, *Supersymmetry, supergravity and particle physics*, *Phys. Rep.* 110 (1–2) (1984) 1–162.
[388] J. Nyíri (ed.), *The Gribov Theory of Quark Confinement*, World Scientific, 2001.
[389] R. Oerter, *The Theory of Almost Everything*, Plume, 2006.
[390] H. Ohanian and R. Ruffini, *Gravitation and Spacetime*, 2nd edn., W. W. Norton & Company, 1974.
[391] H. C. Ohanian, *Principles of Quantum Mechanics*, Prentice Hall, 1990.

[392] S. Okubo, *φ-meson and unitary symmetry model*, *Phys. Lett.* 5 (1963) 165–168.
[393] L. B. Okun, *The concept of mass*, *Physics Today* 42 (1989) 31–36.
[394] L. B. Okun, *Quantum theory*, in *Proceedings of the VIII UNESCO International School of Physics*, 1998, pp. 13–17.
[395] H. Ooguri and C. Vafa, *On the geometry of the string landscape and the swampland*, *Nucl. Phys.* B766 (2007) 21–33.
[396] L. O'Raifeartaigh, *Internal symmetry and Lorentz invariance*, *Phys. Rev. Lett.* 14 (1965) 332–334.
[397] L. O'Raifeartaigh, *Lorentz invariance and internal symmetry*, *Phys. Rev. B* 139 (1965) 1052–1062.
[398] T. Ortín, *Gravity and Strings*, Cambridge University Press, 2004.
[399] V. Paar, S. Brant, D. Vretenar, D. K. Sunko, A. Balantekin and T. Hübsch, *Algebraic and supersymmetric treatment of odd-odd nuclei*, in A. Raduta (ed.), *Symmetries and Semiclassical Features of Nuclear Dynamics*, Springer Verlag, 1987, p. 179.
[400] V. Paar and T. Hübsch, *Extension of boson–fermion dynamical symmetry to hypernuclei*, *Phys. Lett.* B151 (1984) 1.
[401] V. Paar and T. Hübsch, *Three dynamical boson–fermion symmetries for odd–odd nuclei*, *Z. Phys.* A319 (1984) 111.
[402] J. Pachos and M. Stone, *An index theorem for graphene*, *Int. J. Mod. Phys.* B21 (2007) 5113–5120.
[403] D. N. Page, *Particle emission rates from a black hole: Massless particles from an uncharged, nonrotating hole*, *Phys. Rev. D* 13 (2) (1976) 198–206.
[404] H. Pagels, *The Cosmic Code*, Bantam Books, New York, 1983 (original publ. 1982).
[405] H. Pagels, *Perfect Symmetry*, Bantam Books, New York, 1986 (original publ. 1985).
[406] A. Pais, *Subtle is the Lord: The Science and the Life of Albert Einstein*, Oxford University Press, 1982.
[407] D. Park, *Introduction to Quantum Physics*, 3rd edn., McGraw-Hill, 1992.
[408] K.-S. Park, *Topological effects, index theorem and supersymmetry in graphene*, http://arXiv.org/abs/1009.6033v1.
[409] J. Parker, *Diamonds and bowties: Supersymmetric graph conditions* (reported at the Mathematic REU program at Bard College during the summer of 2011).
[410] J. C. Pati and A. Salam, *Is baryon number conserved?*, *Phys. Rev. Lett.* 31 (1973) 661.
[411] J. C. Pati and A. Salam, *Unified lepton–hadron symmetry and a gauge theory of the basic interactions*, *Phys. Rev.* D8 (1973) 1240.
[412] J. C. Pati and A. Salam, *Lepton number as the fourth color*, *Phys. Rev.* D10 (1974) 275–289 (*Erratum: ibid.*, D11 (1975) 703).
[413] W. Pauli, *Exclusion principle, Lorentz group and reflexion of space-time and charge*, in W. Pauli, L. Rosenfeld and V. Weisskopf (eds.), *Niels Bohr and the Development of Physics: Essays Dedicated to Niels Bohr on the Occasion of His Seventieth Birthday*, Pergamon Press, 1955.
[414] W. Pauli, *Theory of Relativity*, Dover Publications, 1981, (original publ. Pergamon Press, 1958).
[415] J. A. Paulos, *A Mathematician Reads the Newspaper*, Anchor, 1997.
[416] J. A. Paulos, *Innumeracy: Mathematical Illiteracy and Its Consequences*, Hill and Wang, 2001.
[417] F. Peat, *Superstrings and the Search for the Theory of Everything*, Contemporary Books, 1988.
[418] P. Peebles, *Physical Cosmology*, Princeton University Press, 1971.
[419] P. Peebles, *Principles of Physical Cosmology*, Princeton University Press, 1993.
[420] C. Pekeris, *Phys. Rev.* A35 (1987) 14.
[421] C. Pekeris and K. Frankowski, *Phys. Rev.* A39 (1989) 518.
[422] D. H. Perkins, *Introduction to High Energy Physics*, 4th edn., Cambridge University Press, 2000.
[423] D. H. Perkins, *Particle Astrophysics*, Oxford University Press, 2003.
[424] M. E. Peskin, *Introduction to String and Superstring Theory. 2.*, 1987 (lectures presented at the 1986 Theoretical Advances Study Institute in Particle Physics, UC, Santa Cruz).
[425] M. E. Peskin and D. V. Schroeder, *An Introduction to Quantum Field Theory*, Addison-Wesley, 1995.
[426] D. Phong, L. Vinet and S.-T. Yau (eds.), *Mirror Manifolds III*, American Mathematical Society, 1999.
[427] A. Pich, *The Standard Model of Electroweak Interactions*(6th CERN-Fermilab Hadron Collider Physics Summer School, Geneva, 2011).
[428] M. Planck, *Wissenschaftliche Selbstbiographie. Mit einem Bildnis und der von Max von Laue gehaltenen Traueransprache*, Johann Ambrosius Barth Verlag, Leipzig, 1948, translated by F. Gaynor, in *Scientific Autobiography and Other Papers*, Philosophical Library, 1949.

[429] J. Polchinski, *Guest post: Joe Polchinski on science or sociology?* http://www.kitp.ucsb.edu/%7Ejoep/A%20dialog.html.

[430] J. Polchinski, *Joe's little book of string*. http://www.kitp.ucsb.edu/members/PM/joep/JLBS.pdf.

[431] J. Polchinski, *Renormalization and effective lagrangians*, *Nucl. Phys.* B231 (1984) 269–295.

[432] J. Polchinski, *What is string theory?* http://arxiv.org/abs/hep-th/9411028.

[433] J. Polchinski, *Lectures on D-branes*. http://arxiv.org/abs/hep-th/9611050.

[434] J. Polchinski, *String Theory*, Cambridge Monographs on Mathematical Physics, Cambridge University Press, 1998.

[435] J. Polchinski, *All strung out?(American Scientist*, Jan./Feb. 2007 book review of Lee Smolin, *The Trouble with Physics: The Rise of String Theory, the Fall of a Science, and What Comes Next* and Peter Woit, *Not Even Wrong: The Failure of String Theory and the Search for Unity in Physical Law*; reprinted (with many links to other blogs) in the blog http://blogs.gazine.com/cosmicvariance/2006/12/07/guest-blogger-joe-polchinski-on-the-string-debates/).

[436] J. Polchinski, *All strung out?*, *American Scientist* (2007) (book review: Lee Smolin, *The Trouble with Physics: The Rise of String Theory, the Fall of a Science, and What Comes Next* and Peter Woit, *Not Even Wrong: The Failure of String Theory and the Search for Unity in Physical Law*). http://www.americanscientist.org/bookshelf/pub/all-strung-out.

[437] J. Polchinski, *Introduction to gauge/gravity duality*, in M. Dine, T. Banks and T. Sachdev (eds.), *String Theory and Its Applications: From meV to the Planck Scale: TASI 2010 Lecture Notes, Boulder, CO*, World Scientific, 2012, pp. 3–46.

[438] J. Polchinski, S. Chaudhuri and C. V. Johnson, *Notes on D-branes*. http://arxiv.org/abs/hep-th/9602052.

[439] A. M. Polyakov, *Gauge Fields and Strings*, Gordon and Breach, 1987.

[440] A. M. Polyakov, *String theory and quark confinement*, *Nucl. Phys. Proc. Suppl.* 68 (1998) 1–8.

[441] V. N. Popov, *Functional Integrals in Quantum Field Theory and Statistical Physics*, D. Reidel, 1983 (translated by J. Niederle and L. Hlavatý).

[442] K. Popper, *All Life is Problem Solving*, Routledge, 2001 (a collection of writings 1958–1994).

[443] K. R. Popper, *The Logic of Scientific Discovery*, Basic Books, 1959.

[444] K. R. Popper, *Conjectures and Refutations: The Growth of Scientific Knowledge*, Routledge and Kegan Paul, 1963.

[445] C. Quigg, *Gauge Theories of Strong, Weak and Electromagnetic Interactions*, Addison-Wesley, 1983.

[446] V. Radovanović, *Problem Book in Quantum Field Theory*, 2nd edn., Springer-Verlag, 2008.

[447] P. Ramond, *Group Theory: A Physicist's Survey*, Cambridge University Press, 2010.

[448] P. Ramond, *The five instructions*, in T. M. Tait and K. T. Matchev (eds.), *The Dark Secrets of the Terascale: TASI 2011 Lecture Notes, Boulder CO*, World Scientific, 2012, pp. 1–37.

[449] L. Randall, *Warped Passages: Unraveling the Mysteries of the Universe's Hidden Dimensions*, Harper Perennial, 2005.

[450] L. Randall and R. Sundrum, *An alternative to compactification*, *Phys. Rev. Lett.* 83 (1999) 4690–4693.

[451] A. Randono, *Gauge gravity: A forward-looking introduction*. http://arxiv.org/abs/1010.5822.

[452] M. Reece and L.-T. Wang, *Randall–Sundrum and strings*, *JHEP* 1007 (2010) 040.

[453] M. Reid, *The moduli space of 3-folds with $K = 0$ may nevertheless be irreducible*, *Math. Ann.* 278 (1–4) (1987) 329–334.

[454] M. Reid, *Young person's guide to canonical singularities*, *Proc. Symp. Pure Math.* 46 (1987) 345–416.

[455] D. Rickles, *Mirror symmetry and other miracles in superstring theory*, *Found. Phys.* 43 (2013) 54–80.

[456] J. S. Rigden (ed.), *Building Blocks Of Matter: A Supplement to the Macmillan Encyclopedia of Physics*, Macmillan Reference USA, 2003.

[457] K. Riley, M. Hobson and S. Bence, *Mathematical Methods for Physicists and Engineering*, 3rd edn., Cambridge University Press, 2006.

[458] M. Riordan, *The Hunting of the Quark*, Simon & Shuster, 1987.

[459] R. Rivers, *Path Integral Methods in Quantum Field Theory*, Cambridge University Press, 1988.

[460] H. P. Robertson, *The uncertainty principle*, *Phys. Rev.* 34 (1929) 163–164.

[461] A. Rogers, *Supermanifolds: Theory and Applications*, World Scientific, 2007.

[462] P. G. Roll, R. Krotkov and R. H. Dicke, *Ann. Phys.* 26 (1964) 442.

[463] P. Roman and C. Koh, *A sharpening of O'Raifeartaigh's theorem, Il Nuovo Cimento* 39 (3) (1965) 1015–1016.
[464] K. Rosquist, *Gravitationally induced electromagnetism at the Compton scale, Class. Quant. Grav.* 23 (2006) 3111–3122.
[465] C. Rovelli, *A critical look at strings*. http://arxiv.org/abs/1108.0868.
[466] C. Rovelli, *Loop quantum gravity: The first 25 years, Class. Quant. Grav.* 28 (2011) 153002.
[467] C. Rovelli, *A new look at loop quantum gravity, Class. Quant. Grav.* 28 (2011) 114005.
[468] C. Rovelli, *Zakopane lectures on loop quantum gravity*, in J. Barrett, K. Giesel, F. Hellmann, L. Jonke, T. Krajewski, J. Lewandowski, C. Rovelli, H. Sahlmann and H. Steinacker (eds.), *3rd Quantum Geometry and Quantum Gravity School*, 2011, p. 003.
[469] B. Russell, *The ABC of Relativity*, Signet, 1969 (original publ. 1962).
[470] A. Saa, *Local electromagnetic duality and gauge invariance, Class. Quant. Grav.* 28 (2011) 127002.
[471] J. J. Sakurai, *Advanced Quantum Mechanics*, Addison-Wesley, 1967.
[472] J. J. Sakurai, *Modern Quantum Mechanics*, Addison-Wesley, 1994.
[473] A. Salam, *Gauge unification of fundamental forces*. http://nobelprize.org/nobel_prizes/physics/laureates/1979/salam-lecture.pdf.
[474] M. D. Scadron, *Advanced Quantum Theory and Its Applications Through Feynman Diagrams*, Springer-Verlag, 1991.
[475] W.-D. Schlatter and P. Zerwas, *Searching for Higgs: From LEP towards LHC*. http://arxiv.org/abs/1112.5127.
[476] E. Schrödinger, *Zum Heisenbergschen unschärfeprinzip, Proc. Preuss. Acad. Sci.* 14 (1930) 296–303.
[477] E. Schrödinger, *What is Life?* together with *Mind and Matter*, Cambridge University Press, 1992 (original publ. 1944 and 1958, resp.).
[478] B. Schroer, *Unexplored regions in QFT and the conceptual foundations of the Standard Model*, http://arXiv.org/abs/1006.3543.
[479] B. Schroer, *An alternative to the gauge theoretic setting, Found. Phys.* 41 (2011) 1543–1568.
[480] R. Shankar, *Principles of Quantum Mechanics*, 2nd edn., Kluwer Academic/Plenum, 1980.
[481] H. L. Shipman, *Black Holes, Quasars and the Universe*, Houghton Mifflin, 1980.
[482] W. Siegel, *Strings with dimension-dependent intercept, Nucl. Phys.* B109 (1976) 244.
[483] W. Siegel, *Introduction to String Field Theory*, World Scientific, 1988.
[484] W. Siegel, *Fields*. http://arxiv.org/abs/hep-th/9912205.
[485] E. Silverstein, *Simple de Sitter solutions, Phys. Rev.* D77 (2008) 106006.
[486] L. Sindoni, *Emergent models for gravity: An overview of microscopic models, SIGMA* 8 (2012) 027.
[487] R. Siu, *The Tao of Science*, The Massachusetts Institute of Technology Press, 1957.
[488] R. Slansky, *Group theory for unified model building, Phys. Rep.* 79 (1) (1981) 1–128.
[489] L. Smolin, *Three Roads to Quantum Gravity*, Basic Books, 2001.
[490] L. Smolin, *Trouble with Physics*, Houghton Mifflin, 2006.
[491] L. Smolin, *Response to review of The Trouble with Physics by Joe Polchinski*, http://www.kitp.ucsb.edu/%7Ejoep/Response%20to%20Polchinski.html. http://www.thetroublewithphysics.com/Response%20to%20Polchinski.html.
[492] J. Song, *On a conjecture of Candelas and de la Ossa*. http://arxiv.org/abs/1201.4358.
[493] A. Sopczak, *Status of Higgs boson searches at the Tevatron*, at *XII International Workshop on Nuclear Physics (WONP'09), Havana, Cuba*, 2009.
[494] A. Sopczak, *Highlights of current Higgs boson searches*, at *The NExT Phase of Particle Physics (iNExT'10), Sussex, UK*, 2010.
[495] R. D. Sorkin, *Forks in the road, on the way to quantum gravity, Int. J. Theor. Phys.* 36 (1997) 2759–2781.
[496] M. Srednicki, *Quantum Field Theory*, Cambridge University Press, 2007.
[497] H. Stephani, D. Kramer, M. MacCallum, C. Hoenselaers and E. Herlt, *Exact Solutions to Einstein's Field Equations*, 2nd edn., Cambridge University Press, 2003.
[498] L. E. Strigari, *Galactic searches for dark matter, Phys. Rep.* 531 (2013) 1–88.
[499] F. Strocchi, *Symmetry Breaking*, No. 643 in Lecture Notes in Physics, Springer-Verlag, 2005.
[500] A. Strominger and J. Polchinski, *New vacua for type II string theory, Phys. Lett.* B388 (1996) 736–742.
[501] A. Strominger, S.-T. Yau and E. Zaslow, *Mirror symmetry is T-duality, Nucl. Phys.* B479 (1996) 243–259.

[502] E. Stückelberg and A. Petermann, *La normalization des constantes dans la theorie des quanta*, *Helvetica Physica Acta* 26 (1953) 499.

[503] Y. Sugawara, *Comments on non-holomorphic modular forms and non-compact superconformal field theories*, *JHEP* 1201 (2012) 098.

[504] M. Sundaresan, *Handbook of Particle Physics*, CRC Press, 2001.

[505] L. Susskind, *The Cosmic Landscape: String Theory and the Illusion of Intelligent Design*, Back Bay Books, 2006.

[506] L. Susskind and J. Lindesay, *An Introduction to Black Holes, Information and the String Theory Revolution: The Holographic Universe*, World Scientific, 2004.

[507] L. Susskind, The world as a hologram, *J. Math. Phys.* 36 (1995) 6377–6396.

[508] J. Synge and A. Schild, *Tensor Calculus*, Dover Publications, 1949, 1978.

[509] R. J. Szabó, *Equivartiant Cohomology and Localization of Path Integrals*, Springer-Verlag, 2000.

[510] R. J. Szabó, *An Introduction to String Theory and D-brane Dynamics: With Problems and Solutions*, 2nd edn., Imperial College Press, 2011.

[511] G. 't Hooft, *A planar diagram theory for strong interactions*, *Nucl. Phys.* B72 (1974) 461.

[512] E. F. Taylor and J. A. Wheeler, *Spacetime Physics*, W. H. Freeman and Company, 1992.

[513] M. Tegmark, *Is "the theory of everything" merely the ultimate ensemble theory?*, *Ann. Phys.* 270 (1998) 1–51.

[514] M. Tegmark, *Parallel universes*, in J. Barrow, P. Davies and C. Harper (eds.), *Science and Ultimate Reality: From Quantum to Cosmos*, Cambridge University Press, 2003.

[515] M. Tegmark, *The multiverse hierarchy*, in B. Carr (ed.), *Universe or Multiverse?*, Cambridge University Press, 2007.

[516] J. Terning, *Modern Supersymmetry: Dynamics and Duality*, Clarendon Press, 2006.

[517] The Nobel Prize, *The official website*. http://www.nobelprize.org/nobel_prizes/nobelprize_facts.html.

[518] K. Thorne, *Nonspherical gravitational collapse: A short review*, in J. Klauder (ed.), *Magic Without Magic*, W. H. Freeman, 1972, p. 231.

[519] K. S. Thorne, *Black Holes and Time Warps: Einstein's Outrageous Legacy*, W. W. Norton & Company, 1995.

[520] G. Tian and S.-T. Yau, *Complete Kähler manifolds with zero Ricci curvature. I.*, *Amer. Math. Soc.* 3 (3) (1990) 579–609.

[521] G. Tian and S.-T. Yau, *Complete Kähler manifolds with zero Ricci curvature. II.*, *Invent. Math.* 106 (1) (1991) 27–60.

[522] A. Todorov, *Weil–Petersson volumes of the moduli spaces of CY manifolds*, *Commun. Anal. Geom.* 15 (2) (2007) 407–434.

[523] A. Tomimatsu and H. Sato, *New exact solution for the gravitational field of a spinning mass*, *Phys. Rev.* 29 (1972) 1344–1345.

[524] A. Tomimatsu and H. Sato, *New series of exact solutions for gravitational fields of spinning masses*, *Prog. Theor. Phys.* 50 (1) (1973) 95–110.

[525] S. Troitsky, *Unsolved problems in particle physics*, *Phys. Usp.* 55 (2012) 72–95.

[526] I. V. Tyutin, *Gauge invariance in field theory and statistical physics in operator formalism*, P. N. Lebedev Physical Institute preprint No. 39, 1975.

[527] L. Tzu, *Lao Tzu: Tao Te Ching*, Shambhala, 1997 (translated by Ursula K. Le Guin, with the collaboration of Jerome P. Seaton).

[528] G. E. Uhlenbeck and S. A. Goudsmit, *Ersetzung der Hypothese vom unmechanischen Zwang durch eine Forderung bezüglich des inneren Verhaltens jedes einzelnen Elektrons*, *Die Naturwissenschaften* 13 (47) (1925) 953–954.

[529] T. Vachaspati, *Cosmic problems for condensed matter experiment*, *J. Low Temp. Phys.* 136 (2004) 361–377.

[530] C. Vafa, *Evidence for F-theory*, *Nucl. Phys.* B469 (3) (1996) 403–415.

[531] C. Vafa, *The string landscape and the swampland*. http://arxiv.org/abs/hep-th/0509212.

[532] S. van den Bergh, *The curious case of Lemaitre's equation no. 24*, *J. Roy. Astron. Soc. Can.* 105 (4) (2011) 151.

[533] B. C. van Fraassen, *Laws and Symmetry*, Clarendon Press, 1990.

[534] W. J. van Stockum, *The gravitational field of a distribution of particles rotating around an axis of symmetry*, *Proc. Roy. Soc. Edinburgh* A57 (1937) 135.

[535] V. S. Varadarajan, *Supersymmetry for Mathematicians: An Introduction*, American Mathematical Society, 2004.

[536] M. T. Vaughn, *Introduction to Mathematical Physics*, Wiley-VCH, 2007.

[537] M. Veltman, *Diagrammatica: The Path to Feynman Diagrams*, Cambridge University Press, 1994.

[538] M. Veltman, *Facts and Mysteries in Elementary Particle Physics*, World Scientific, 2003.

[539] S. K. Vempati, *Introduction to the MSSM*, Lectures presented at SERC school, held at IIT-Bombay, Mumbai, http://arxiv.org/abs/1201.0334.

[540] M. Visser, *The Kerr spacetime: A brief introduction*. http://arxiv.org/abs/0706.0622.

[541] M. Visser, *Traversable wormholes: Some simple examples*, *Phys. Rev.* D39 (10) (1989) 3182–3184.

[542] M. Visser, *From wormhole to time machine: Comments on Hawking's chronology protection conjecture*, *Phys. Rev.* D47 (1993) 554–565.

[543] M. Visser, *Lorentzian Wormholes: From Einstein to Hawking*, AIP Series in Computational and Applied Mathematical Physics, American Institute of Physics, 1996.

[544] M. Visser, *The quantum physics of chronology protection*, in *Cambridge 2002, The Future of Theoretical Physics and Cosmology*, 2002, pp. 161–176.

[545] M. Visser, *Status of Hořava gravity: A personal perspective*, *J. Phys. Conf. Ser.* 314 (2011) 012002, arXiv:1103.5587.

[546] M. Visser and C. Barceló, *Energy conditions and their cosmological implications*, in *COSMO 99: 3rd International Conference on Particle Physics and the Early Universe*, 1999.

[547] W. von Heisenberg, *The development of quantum mechanics*. http://nobelprize.org/nobel_prizes/physics/laureates/1932/heisenberg-lecture.html.

[548] R. Wald, *General Relativity*, University of Chicago Press, 1984.

[549] D. Wasserman, M. Leigh and J. Darion, *Man of La Mancha*, Random House, 1966.

[550] M. J. Way and H. Nussbaumer, *Lemaitre's Hubble relationship*, *Phys. Today* 64N8 (2011) 8.

[551] S. Weinberg, *The First Three Minutes*, Basic Books, 1977.

[552] S. Weinberg, *Conceptual foundations of the unified theory of weak and electromagnetic interactions*. http://nobelprize.org/nobel_prizes/physics/laureates/1979/weinberg-lecture.pdf.

[553] S. Weinberg, *Dreams of a Final Theory*, Random House, 1993.

[554] S. Weinberg, *The Quantum Theory of Fields, Vol. 1: Foundations*, Cambridge University Press, 1995.

[555] S. Weinberg, *The Quantum Theory of Fields, Vol. 2: Modern Applications*, Cambridge University Press, 1996.

[556] S. Weinberg, *The Quantum Theory of Fields, Vol. 3: Supersymmetry*, Cambridge University Press, 2005.

[557] S. Weinberg, *Living in the multiverse*, in B. Carr (ed.), *Universe or Multiverse?*, Cambridge University Press, 2007.

[558] S. Weinberg, *Cosmology*, Cambridge University Press, 2008.

[559] S. Weinberg and E. Witten, *Limits on massless particles*, *Phys. Lett.* B96 (1980) 59–62.

[560] J. Wess and J. Bagger, *Supersymmetry and Supergravity*, Princeton Series in Physics, 2nd edn., Princeton University Press, 1992.

[561] G. B. West, J. H. Brown and B. J. Enquist, *A general model for the origin of allometric scaling laws in biology*, *Science* 276 (5309) (1997) 122–126.

[562] P. West, *Introduction to Supersymmetry and Supergravity*, World Scientific, 1990.

[563] C. v. Westenhoz, *Differential Forms in Mathematical Physics*, North-Holland, 1978.

[564] H. Weyl, *Eine neue Erweiterung der Relativitätstheorie*, *Ann. Phys.* 59 (1919) 101.

[565] H. Weyl, *The Theory of Groups and Quantum Mechanics*, Kessinger Publishing, 2008.

[566] H. Weyl, *Space-Time-Matter*, Nabu Press, 2010 (original publ. 1918).

[567] T. H. White, *The Once and Future King*, 2nd edn., Ace Books, 1987.

[568] A. N. Whitehead and B. Russell, *Principia Mathematica, Vol. 2*, Merchant Books, 2009 (reprint of the 1st edition by Cambridge University Press, 1910).

[569] A. N. Whitehead and B. Russell, *Principia Mathematica, Vol. 3*, Merchant Books, 2009 (reprint of the 1st edition by Cambridge University Press, 1910).

[570] A. N. Whitehead and B. Russell, *Principia Mathematica*, Nabu Press, 2010 (abridged text of Volume I; reprint of the 2nd edition by Cambridge University Press, 1927; first published in 1910).

[571] A. N. Whitehead and B. Russell, *Principia Mathematica, Vol. 1*, Rough Draft Printing, 2011 (reprint of the 1st edition by Cambridge University Press, 1910).
[572] A. S. Wightman and R. F. Streater, *PCT, Spin and Statistics and All That*, Landmarks in Mathematics and Physics, Princeton University Press, 2000.
[573] E. Witten, *Constraints on supersymmetry breaking*, *Nucl. Phys.* B202 (1982) 253.
[574] E. Witten, *Phases of $N = 2$ theories in two-dimensions*, *Nucl. Phys.* B403 (1993) 159–222.
[575] E. Witten, *String theory dynamics in various dimensions*, *Nucl. Phys.* B443 (1995) 85–126.
[576] P. Woit, *Not Even Wrong ('blog)*. http://www.math.columbia.edu/~woit/wordpress/.
[577] P. Woit, *Not Even Wrong*, Basic Books, 2006.
[578] D. Wolff, *Chinese for Beginners*, Barnes and Noble Books, 1985 (original publ. 1974).
[579] N. Woodhouse, *Geometric Quantization*, 2nd edn., Oxford University Press, 1997.
[580] T.-Y. Wu and W.-Y. P. Hwang, *Relativistic Quantum Mechanics and Quantum Fields*, World Scientific, 1991.
[581] B. G. Wybourne, *Classical Groups for Physicists*, John Wiley & Sons, 1974.
[582] J. Xu, *Conifold transitions for complete intersection Calabi–Yau 3-folds in products of projective spaces*, arXiv:1202.1110.
[583] S.-T. Yau (ed.), *Mirror Manifolds*, International Press, 1990.
[584] S.-T. Yau and S. Nadis, *The Shape of Inner Space*, Basic Books, 2010.
[585] A. Zee, *Fearful Symmetry*, Macmillan, 1986.
[586] A. Zee, *Quantum Field Theory in a Nutshell*, 2nd edn., Princeton University Press, 2010.
[587] R. Zegers, *Birkhoff's theorem in Lovelock gravity*, *J. Math. Phys.* 46 (2005) 072502.
[588] E. Zeidler, *Quantum Field Theory I: Basics in Mathematics and Physics*, Springer-Verlag, 2006.
[589] E. Zeidler, *Quantum Field Theory II: Quantum Electrodynamics*, Springer-Verlag, 2009.
[590] E. Zeidler, *Quantum Field Theory III: Gauge Theory*, Springer-Verlag, 2011.
[591] G. Zukav, *Dancing Wu Li Masters: An Overview of the New Physics*, HarperOne, 2001 (first edition by William Morrow and Co., 1979).
[592] G. Zweig, *An SU_3 model for strong interaction symmetry and its breaking*, Technical Report 8419/TH.401, CERN-TH (January 17 1964).
[593] G. Zweig, *An SU_3 model for strong interaction symmetry and its breaking; Part II*, Technical Report TH-412, CERN-TH (presented at *Developments in the Quark Theory of Hadrons*, pp. 22–101) (February 21 1964).
[594] B. Zwiebach, *A First Course in String Theory*, Cambridge University Press, 2004.

Index

Boldface numbers indicate locations of definitions, italic numbers indicate locations of lexicon entries

To be continued...

Printed in the United States
by Baker & Taylor Publisher Services